VERTEBRATE Biology

VERTEBRATE *Biology*

Donald Linzey

Wytheville Community College

McGraw Hill

Boston Burr Ridge, IL Dubuque, IA Madison, WI New York San Francisco St. Louis
Bangkok Bogotá Caracas Lisbon London Madrid
Mexico City Milan New Delhi Seoul Singapore Sydney Taipei Toronto

McGraw-Hill Higher Education

A Division of The **McGraw-Hill** Companies

VERTEBRATE BIOLOGY

This book is printed on recycled, acid-free paper containing 10% postconsumer waste.

1 2 3 4 5 6 7 8 9 0 KGP/KGP 0 9 8 7 6 5 4 3 2 1 0

ISBN 0-697-36387-2

Cover image: The snowy owl (*Nyctea scandiaca*) is a diurnal arctic owl that winters as far south as the northern United States but which may occasionally winter as far south as South Carolina, Georgia, Louisiana, Texas, and California. Most adult birds are almost pure white. The large size (length to 27 inches; wingspan to 66 inches), pale plumage, and lack of ear tufts are diagnostic. Snowy owls prefer rolling to flat tundra and spend much of their time on such lookouts as banks, boulders, or knolls. When they visit the United States in winter, they are found in marshes, in dune areas, and in open farmland, where they perch on haystacks and buildings, seldom on trees. They feed on lemmings and other rodents and rabbits. Populations are cyclic, with population peaks occurring about every 4 years.

The snowy owl breeds from the Bering Sea to Greenland as well as in northern Russia and Siberia. The nest is a depression in the soil or is located on a rocky shelf and is thinly lined with moss and feathers. It is located on the higher, drier spots in rolling tundra. A normal clutch consists of 5 to 8 white eggs.

Vice president and editor-in-chief: *Kevin T. Kane*
Publisher: *Michael D. Lange*
Senior sponsoring editor: *Margaret J. Kemp*
Developmental editor: *Donna Nemmers*
Editorial assistant: *Dianne Berning*
Marketing managers: *Michelle Watnick/Heather K. Wagner*
Senior project manager: *Kay J. Brimeyer*
Senior production supervisor: *Mary E. Haas*
Coordinator of freelance design: *David W. Hash*
Interior designer: *Rokusek Design*
Cover image: *©PhotoDisc*
Photo research coordinator: *John C. Leland*
Photo research: *Mary Reeg Photo Research*
Compositor: *Electronic Publishing Services Inc., NYC*
Typeface: *10/12 A. Caslon Regular*
Printer: *Quebecor Printing Book Group/Kingsport*

The credits section for this book begins on page 504 and is considered an extension of the copyright page.

Library of Congress Cataloging-in-Publication Data

Linzey, Donald W.
Vertebrate biology / Donald Linzey. — 1st ed.
p. cm.
ISBN 0-697-36387-2
1. Vertebrates. I. Title.

QL605 .L48 2001
596—dc21 00-023088
CIP

www.mhhe.com

To my wife, Nita, whose love and enduring patience have made this work possible.

Brief Table of Contents

Preface to the Instructor x
Preface to the Student xiii
About the Author xv

CHAPTER 1 **The Vertebrate Story: An Overview** 1

CHAPTER 2 **Systematics and Vertebrate Evolution** 23

CHAPTER 3 **Vertebrate Zoogeography** 45

CHAPTER 4 **Early Chordates and Jawless Fishes** 74

CHAPTER 5 **Gnathostome Fishes** 90

CHAPTER 6 **Amphibians** 129

CHAPTER 7 **Evolution of Reptiles** 169

CHAPTER 8 **Reptiles: Morphology, Reproduction, and Development** 198

CHAPTER 9 **Mammals** 264

CHAPTER 10 **Population Dynamics** 324

CHAPTER 11 **Movements** 337

CHAPTER 12 **Intraspecific Behavior and Ecology** 357

CHAPTER 13 **Interspecific Interactions** 375

CHAPTER 14 **Techniques for Ecological and Behavioral Studies** 390

CHAPTER 15 **Extinction and Extirpation** 402

CHAPTER 16 **Conservation and Management** 421

APPENDICES 443

GLOSSARY 459

BIBLIOGRAPHY 471

CREDITS 504

INDEX 509

Contents

Preface to the Instructor x
Preface to the Student xiii
About the Author xv

CHAPTER 1
The Vertebrate Story: An Overview **1**
Introduction 1
Vertebrate Features 4
Role of Vertebrates 18

CHAPTER 2
Systematics and Vertebrate Evolution **23**
Introduction 23
Binomial Nomenclature 23
Classification 25
Methods of Classification 26
Evolution 32

CHAPTER 3
Vertebrate Zoogeography **45**
Introduction 45
Distribution 47
Marine 55
Fresh Water 58
Terrestrial 60

CHAPTER 4
Early Chordates and Jawless Fishes **74**
Introduction 74
Calcichordates 76
Conodonts 76
Early Cambrian Fish-Like Fossils 77
Evolution 78
Morphology of Jawless Fishes 82
Reproduction 88
Growth and Development 88

CHAPTER 5
Gnathostome Fishes **90**
Introduction 90
Evolution 90
Morphology 95
Reproduction 121
Growth and Development 125

CHAPTER 6
Amphibians **129**
Introduction 129
Evolution 129
Morphology 136
Reproduction 158
Growth and Development 161

CHAPTER 7
Evolution of Reptiles **169**
Introduction 169
Evolution 169
Ancestral Reptiles *169*
Ancient and Living Reptiles *173*
Ancestral Birds *191*
Modern Birds *195*
Origin of Flight *195*

CHAPTER 8
Reptiles: Morphology, Reproduction, and Development **198**
Introduction 198
Turtles, Tuataras, Lizards, and Snakes (Testudomorpha and Lepidosauromorpha) 198
Morphology 198
Reproduction 218
Growth and Development 222
Crocodilians and Birds (Archosauromorpha) 226
Morphology 226
Reproduction 252
Growth and Development 255

CHAPTER 9
Mammals **264**
Introduction 264
Evolution 264
Morphology 271
Reproduction 314
Growth and Development 317

CHAPTER 10
Population Dynamics **324**
Introduction 324
Population Density 324
Reproductive (Biotic) Potential 324
Environmental Resistance 326
Cycles 332
Irruptions 335

CHAPTER 11
Movements **337**
Introduction 337
Home Range 337
Dispersal/Invasion 341
Migration 343
Homing 352
Emigration 355

CHAPTER 12
Intraspecific Behavior and Ecology **357**
Introduction 357
Social Interactions 357
Sensory Reception and Communication 360
Feeding Behavior 367
Torpor (Dormancy) 368

CHAPTER 13
Interspecific Interactions **375**
Introduction 375
Competition 375
Symbiosis 377
Predation 382
Human Interactions 387

CHAPTER 14
Techniques for Ecological and Behavioral Studies **390**
Introduction 390
Capture Techniques 391
Identification Techniques 391
Mapping Techniques 397
Censusing Techniques 397
Aging Techniques 398

CHAPTER 15
Extinction and Extirpation **402**
Introduction 402
Natural Extinction 402
Human Impacts on Extirpation and Extinction 409

CHAPTER 16
Conservation and Management **421**
Introduction 421
Regulatory Legislation Affecting Vertebrates in the United States 421
Endangered Species in the United States 423
Sanctuaries and Refuges 424
Value of Museum Collections 426
Wildlife Conservation in a Modern World 427

Appendices 443

Glossary 459

Bibliography 471

Credits 504

Index 509

Preface to the Instructor

Vertebrate Biology is designed to provide a firm foundation for students interested in the natural history of vertebrates. While this may be the only course that some students will take dealing with the biology of vertebrates, othes will subsequently enroll in more specialized courses such as ichthyology, herpetology, ornithology, and mammology.

In writing *Vertebrate Biology,* I have tried to keep the needs and wishes of both instructors and students in mind. As an instructor for more than 35 years, I know how challenging it is to teach a dynamic, balanced, up-to-date course that is both relevant and interesting to students. I'm sure you feel the same way. A book that draws the reader's attention facilitates learning and an interested student is an avid learner. As instructors, our job is to guide students in their learning process.

As this text was taking shape, I pondered its length. If I wrote a short textbook, many topics would have to be omitted, and it would not meet the needs of many instructors. If I wrote a larger book, I could include more topics, discuss certain ones in more detail, and even include some supplemental information that would be interesting but not absolutely essential. I chose the latter course.

In order to keep the book to a reasonable length, most concepts are illustrated by one or two good examples for each group to which the concept applies rather than a catalog of many possible ones. It is my hope that you will utilize these examples to stimulate discussion of these concepts with your students. The structure of the text allows for flexibility in presenting the material and permits it to be adapted to a wide range of teaching styles. Chapters need not be taught in order but can be adapted to any sequence that best meets your curriculum.

Vertebrate biology is a broad field: there is no reason to ignore parts that may be currently unpopular or that do not engage us personally. Therefore, I have tried to achieve balance in terms of the whole field of vertebrate biology: systematics, paleontology, physiology, ecology, etc. I have also tried to achieve balance both taxonomically and geographically.

Following each chapter are several questions designed to encourage your students to apply what they have learned. You can use these in various ways: for discussion, for out-of-class assignment, or as potential short essay questions on tests.

Also at the end of each chapter is a list of supplemental readings and a listing of World Wide Web sites relevant to the material in that particular chapter that you (as well as your students) might find of interest. These can be used as a guide to the literature for student papers or used as a source of additional information on a topic of particular interest. All web addresses listed are up-to-date as of press time, and are, for the most part, maintained by universities, governmental agencies, institutions, museums, etc. Thus, the information they contain should be accurate and the web site should be more-or-less permanent. These addresses can be found as "hot links" by visiting the publisher's web page at www.mhhe.com/zoology.

I have tried to keep the vocabulary as plain as possible for the sake of clarity. I see no reason to use extensive technical terminology when simpler words will suffice. The use of technical terms is unavoidable in some cases, but when overdone, it makes the learning process more difficult for the student. Lucidity has been one of the two major goals of my writing. Interest for the reader has been the second goal.

Almost any textbook is certain to have some typographical and other errors. Although all errors are my sole responsibility, I would like to request your help in improving future editions of this book. Please send any errors you find, and your comments and suggestions for improvements, to Marge Kemp, Sponsoring Editor, Life Sciences, c/o McGraw-Hill Publishers, 2460 Kerper Blvd., Dubuque, Iowa 52001. She will send them to me. Please be assured that all comments and suggestions will be given serious consideration. As time allows, I will respond personally.

Thomas Jefferson once said that, "...difference of opinion leads to inquiry, and inquiry to truth." It is my hope that *Vertebrate Biology* will stimulate inquiry among our students and will enable them to become better informed future biologists.

Donald W. Linzey

ACKNOWLEDGEMENTS

We gratefully recognize the academicians who offered their time and beneficial comments in reviewing the many manuscript pages that make up this text. The author appreciates your contributions and has incorporated them into this work.

Kevin Alexander, *Wayne State College*
Jane Aloi, *Saddleback College*
William H. Barnard, *Norwich University*
Michael Barton, *Centre College*
Paul L. Belanger, *Lake Erie College*
Barbara N. Benson, *Cedar Crest College*
Ronald Beumer, *Armstrong Atlantic State University*

Del Blackburn, *Clark College*
Fred J. Brenner, *Grove City College*
Thomas M. Buchanan, *Westark College*
Brian R. Chapman, *University of Georgia*
Phillip D. Clem, *University of Charleston*
George R. Cline, *Jacksonville State University*
Sneed B. Collard, *The University of West Florida*
Clara Cotton, *Indiana University–Bloomington*
Doyle Crosswhite, *St. Gregory's University*
James W. Demastes, *University of Northern Iowa*
James des Lauriers, *Chaffey College*
Elizabeth A. Desy, *Southwest State University*
Donald Dorfman, *Monmouth University*
Peter K. Ducey, *State University of New York at Cortland*
Ruth E. Ebeling, *Biola University*
Dennis Englin, *The Master's College*
Stephen Ervin, *California State University*
Dale Fishbeck, *Youngstown State University*
M. J. Fouquette, Jr., *Arizona State University*
Margaret Fusari, *University of California–Santa Cruz*
James C. Gibson, *Chadron State College*
Steven Goldmsith, *Austin College*
Brent M. Graves, *Northern Michigan University*
Stephen B. Hager, *Augustana College*
Gregory D. Hartman, *McNeese State University*
Virginia Hayssen, *Smith College*
Herbert T. Henderickson, *University of North Carolina–Greensboro*
Frank Heppner, *University of Rhode Island*
Fritz Hertel, *University of California–Los Angeles*
Michael Hosking, *Davidson College*
Robert J. Howe, *Suffolk University*
Christine Janis, *Brown University*
Edmund D. Keiser, *The University of Mississippi*
Kenneth P. Kinsey, *Rhode Island College*
Mark Konikoff, *University of Southwestern Louisiana*
Jeffrey G. Kopachena, *Texas A & M University–Commerce*
John L. Loprowski, *Willamette University*
Julia E. Krebs, *Francis Marion University*
Allen Kurta, *Eastern Michigan University*
Allan J. Landwer, *Hardin-Simmons University*
Larry N. Latson, *Lipscomb University*
Richard Lewis, *Grant MacEwen Community College*
Susan E. Lewis, *Carroll College*
Michael P. Lombardo, *Grand Valley State University*
David Lonzarich, *University of Wisconsin-Eau Claire*
Bonnie McCormick, *University of the Incarnate Word*
Michael L. McMahan, *Union University*
Michael T. Mengak, *Ferrum College*
Gary D. Miller, *University of New Mexico*
Sue Ann Miller, *Hamilton College*
Margaret A. O'Connell, *Eastern Washington University*
Kathleen O'Reilly, *University of Portland*
Charles M. Page, *El Camino College*
Michael S. Parker, *Southern Oregon University*
Robert Powell, *Avila College*
David Quadagno, *Florida State University*
David O. Ribble, *Trinity University*
S. Laurie Sanderson, *College of William and Mary*
David A. Shealer, *Loras College*
Mitchell Smith, *Willamette University*
L. D. Spears, *John A. Logan Community College*
C. David Vanicek, *California State University–Sacramento*
James F. Waters, *Humboldt State University*
Gregory J. Watkins-Colwell, *Sacred Heart University*
Louise Weber, *Warren Wilson College*
Wm. David Webster, *University of North Carolina at Wilmington*
John O. Whitaker, Jr., *Indiana State University*
Clayton M. White, *Brigham Young University*
Kenneth T. Wilkins, *Baylor University*
Norman R. Williams, *Central Texas College*
Kathy Winnett-Murray, *Hope College*
Neil D. Woffinden, *University of Pittsburgh–Johnstown*
Catherine J. Woodworth, *Eastern Illinois University*
Kelly R. Zamudio, *Cornell University*

The resources of the Wytheville Community College library were used extensively. My thanks are extended to Anna Ray Roberts, George Mattes, Jr., and Brenda King, without whose help the acquisition of vital data would have been considerably more difficult.

Finally, I thank my wife, Nita, for putting up with the piles of books, correspondence, and manuscript sections that cluttered several rooms of our home for varying periods of time. Her understanding and patience made possible the completion of this book. Her greatest joy came when the manuscript was completed and she was able to once again see the surface of our dining room table and to resume using our dining room for its intended purpose!

SUPPLEMENTS

For the Instructor

Please contact your McGraw-Hill sales representative to learn more about these helpful supplementary products for your course:

Life Science Animations Visual Resource Library (VRL) CD-ROM contains 125 animations of important biological concepts and processes. This CD-ROM is perfect for use to support your lectures.

Zoology Visual Resource Library (VRL) CD-ROM contains 1000 photos and figures to support your zoology course. These can be inserted into your own lecture presentations, printed, and can be easily found through a key word search function.

For the Student

A variety of print and media products are available from McGraw-Hill to supplement student learning, in addition to the titles below. Visit www.mhhe.com or contact your bookstore to order these products:

Kardong/Zalisko: *Comparative Anatomy Laboratory Dissection Guide* (0-697-37879-9) is a laboratory manual that weaves functional and evolutionary concepts into the morphological details of laboratory exercises.

Chiasson/Radke: *Laboratory Anatomy of the Vertebrates* (0-697-10160-6) uses a systemic approach to the study of vertebrate morphology and discusses the shark, *Necturus*, and cat in each system chapter.

Marchuk: *Life Sciences Living Lexicon CD ROM* (0-697-37993-0) is an interactive CD-ROM that includes complete glossaries for all life science disciplines, a section describing the classification system, an overview of word construction, and more than 500 vivid illustrations and animations of key processes.

Waters/Janssen: *Web-Based Cat Dissection Review for Human Anatomy and Physiology* (0-07-232157-1) is a new web-based dissection program that identifies approximately 600 different anatomical structures, with sound profiles to provide pronunciations for each of these 600 terms. Also featured on the site are 100 high-resolution photographs. By clicking anywhere on the structures of the cat, an information frame will provide the name of the structure, as well as a smaller image that highlights and puts the structure in its human anatomical context. In addition, approximately 30 QuickTime movies demonstrate joint movements.

TITLES OF RELATED INTEREST

Comparative Anatomy Laboratory Dissection Guide, Second Edition ©1998, by Kardong/Zalisko (ISBN: 0-697-37879-9) is a complete dissection guide that is organized systematically to present basic animal architecture and anatomy. The exercises can be customized to suit most vertebrate course needs.

Mammalogy: Adaptation, Diversity and Ecology, ©1999, by Feldhamer, Drickamer, Vessey, and Merritt (ISBN: 0-697-16733-X) is an introductory mammalogy text for use in upper level undergraduate or graduate level courses, and assumes students have a basic background in zoology of the vertebrates.

Zoology, Fourth Edition ©1999, by Miller/Harley (ISBN: 0-697-34555-6) is a comprehensive, principles-oriented full-color text with an unmatched pedagogical system, Internet references, and a comprehensive web page.

Integrated Principles in Zoology, Eleventh Edition ©2001, by Hickman/Roberts/Larson (ISBN: 0-072-90961-7) remains the leading text in zoology. With outstanding artwork and a continued comprehensive and straightforward approach, the new edition includes current issues such as biotechnology, immunity, and conservation, and provides chapter web links for further study. This text is also supported by a comprehensive web page.

Biology of Animals, Seventh Edition ©1998, by Hickman/Roberts/Larson (ISBN: 0-697-28933-8) is an introductory zoology text which features a variety of pedagogical aids and a supporting web page.

Animal Diversity, Second Edition ©2000, by Hickman/Roberts/Larson (ISBN: 0-070-12200-8) covers the animal kingdom in sixteen survey chapters and is prefaced by evolution, animal architecture, and classification. Internet references are included in the text for further research, and the text is supported by a web site that includes additional student aids.

Laboratory Studies in Integrated Principles of Zoology, Tenth Edition ©2001, by Hickman/Kats (ISBN: 0-072-90966-8) uses a comprehensive, phylogenetic approach in emphasizing basic biological principles, animal form and function, and evolutionary concepts. This introductory lab manual is ideal for a one- or two-semester course. The new edition incorporates more interactivity for students throughout the laboratory exercises, and also has new molecular exercises. This text may be customized to include only the exercises that best suit your course needs.

General Zoology Laboratory Guide, Thirteenth Edition ©2000, by Lytle (ISBN: 0-07-012220-2) emphasizes the dissection and microscopic study of live and preserved specimens. This laboratory guide is also customizable to contain only the exercises that best fit your course.

Understanding Evolution, Sixth Edition ©2000, by Volpe/Rosenbaum (ISBN: 0-697-05137-4) is an introduction to principles of evolution text that is ideally suited as a main text for general evolution or as a supplement for General Biology, Genetics, Zoology, Botany, Anthropology, or any life science course that utilizes evolution as the underlying theme of all life.

Foundations of Parasitology, Sixth Edition ©2000, by Roberts/Janovy (ISBN: 0-697-42430-8) is written for biological/zoological students at the undergraduate level and emphasizes principles with related information on the biology, physiology, morphology, and ecology of the major parasites of humans and domestic animals.

Biology of the Invertebrates, Fourth Edition ©2000, by Pechenik (ISBN: 0-070-12204-0) is the most concise and readable invertebrates book in terms of detail and pedagogy. All phyla of invertebrates covered (comprehensive) with an emphasis on unifying characteristics of each group.

Animal Behavior: Mechanisms, Ecology, Evolution, Fifth Edition ©2001, by Drickamer/Vessey/Jakob (ISBN: 0-070-12199-0) is a unique balance of the necessary elements of mechanisms, ecology, and evolution that support a comprehensive study of behavior. This one-semester text is suitable for upper-level courses.

Marine Biology, Third Edition ©2000, by Castro/Huber (ISBN: 0-07-012197-4) is appropriate for introductory courses and features a global approach, using examples from numerous regions and ecosystems. This edition also includes Internet references for further study, and is supported by a complete web site with additional student study aids.

Preface to the Student

The field of vertebrate biology is changing constantly. As discoveries are made, old concepts fade away and new ones emerge. We are now at a point in time where acceptable solutions must be found to major environmental problems such as global warming, fragmentation of habitats, isolation of populations, and the decline in biodiversity—all of which play a role in the distribution and abundance of vertebrate populations. Trying to discover the knowledge that is really worth knowing is an exciting and never-ending process.

This is why vertebrate biology is so important: it is not just another college course to be passed for credit. It is an opportunity for you to learn about the other organisms most closely related to us with whom we share this ever-more-crowded planet. During this course we will use information from physical sciences such as chemistry and geology as well as from a wide variety of biological specializations such as botany, genetics, cell biology, physiology, anatomy, and ecology.

While many biologists become more specialized and restrict their reading and research to a narrow area of interest, a general textbook author is forced to be a generalist, who understands (in my case) the whole field of vertebrate biology: to this end the author must illustrate the interrelationships and synthesize available information into some kind of coherent whole. My objective has been to write a general text that reflects the broad diversity of subjects that make vertebrate biology such an exciting field. The need for a readable, user-friendly book covering the biology of vertebrates has been recognized for many years. *Vertebrate Biology* is designed to provide a broad and basic background of vertebrate biology: to explain how vertebrates function, evolve, and interact with each other and their nonliving surroundings.

A second objective of writing *Vertebrate Biology* is to provide a balanced overview of the various areas of vertebrate biology—origin, phylogeny, fossil history, adaptations, distribution, and population dynamics; to explore these areas in some detail; to provide a basic reference source; and to produce a readable, user-friendly text. All sources are documented in order to assist you in your search for more detailed information than this text provides.

I could have written a short textbook that omitted many topics, but that strategy would not meet the needs of many instructors. On the other hand, I could have written a larger book and designed it to be flexible enough to be used in many different ways. I chose the latter. This text has been designed to be used in courses with different lengths and emphases. As a result, your instructor has great flexibility in designing the course you are taking according to the time available.

Numerous features enable this textbook to stand out above all others. Among these are review questions at the end of each chapter, supplemental sources of information including relevant journals, a glossary of well over 300 terms, an extensive bibliography, an appendix of endangered species, and a directory of World Wide Web sites dealing with various aspects of vertebrate biology. This latter feature is unique because if you become interested in a particular topic, you can access relevant web sites throughout the world. The World Wide Web is an excellent resource for vertebrate biology and contains a wide array of information. The web addresses listed are up-to-date as of press time, and are, for the most part, maintained by universities, governmental agencies, institutions, museums, etc. The information they contain should be factual and these sites will be kept up-to-date by the publisher. You can access these weblinks at www.mhhe.com/zoology.

Vertebrate Biology begins with an introductory chapter relating vertebrates to other chordates. This chapter also gives a broad overview and comparison of major vertebrate characters, introduces relevant terminology, discusses the significance of vertebrates, and highlights future research topics. Chapter 2 traces the development of systematics as it relates to vertebrates and discusses evolutionary concepts including natural selection, speciation, and geographic variation. Zoogeography is covered in Chapter 3. Chapters 4 through 9 discuss the evolution and biology of each vertebrate group—jawless fishes, fishes, amphibians, reptiles, and mammals. The remaining chapters draw out common themes across all of the vertebrates: population dynamics (10); movements (11); intraspecific behavior (12); interspecific interactions (13); techniques for ecological and behavioral studies (14); extinction and extirpation (15); and conservation and management (16). Subsection headings are included where appropriate to make it easier for you to retrieve specific pieces of information. Thus, I have attempted to blend the two approaches (taxonomic and concept). At the end of each chapter is a set of discussion questions designed to help you develop your critical thinking skills and apply what you have learned.

My goal is to communicate with you, not confuse you. I have tried to write in a clear, interesting, and informal style. I keep sentences and paragraphs fairly short. I try not to use long words and excessive technical terms when short words and nontechnical terms can express an idea just as clearly. Information is presented in a very straightforward, readable manner. It is not weighed down with excessive terminology.

Writing and publishing a textbook is an extremely complex process. Even though the manuscript has been reviewed, copy edited, checked, and rechecked by a large number of teachers and experts, almost any textbook is certain to have some errors. Although all errors are my sole responsibility, I need your help in improving future editions of this book. Please send any errors you find and your comments and suggestions for improvement to Marge Kemp, Sponsoring Editor, Life Sciences, c/o McGraw-Hill Publishers, 2460 Kerper Blvd., Dubuque, Iowa 52001. She will send them to me. Please be assured that all comments and suggestions will be seriously evaluated and considered. As time allows, I will respond personally to your suggestions.

It is my hope that this text will give you a deeper understanding of the biology of a unique group of animals—including ourselves—and an appreciation of the complexity of issues such as the problems of speciation and phylogeny, the thermal physiology of dinosaurs, and the origin of flight in birds, to name but a few.

Relax and enjoy yourself as you learn more about one of the exciting groups of animals with whom we share this planet.

Donald W. Linzey

SUPPLEMENTARY MATERIALS

For the Student

A variety of print and media products are available from McGraw-Hill to supplement your learning, in addition to the titles listed below. Visit www.mhhe.com or contact your bookstore to order these products:

Kardong/Zalisko: *Comparative Anatomy Laboratory Dissection Guide* (0-697-37879-9) is a laboratory manual that weaves functional and evolutionary concepts into the morphological details of laboratory exercises.

Chiasson/Radke: *Laboratory Anatomy of the Vertebrates* (0-697-10160-6) uses a systemic approach to the study of vertebrate morphology and discusses the shark, *Necturus*, and cat in each chapter.

Marchuk: *Life Sciences Living Lexicon CD ROM* (0-697-37993-0) is an interactive CD-ROM that includes complete glossaries for all life science disciplines, a section describing the classification system, an overview of word construction, and more than 500 vivid illustrations and animations of key processes.

Waters/Janssen: *Web-Based Cat Dissection Review for Human Anatomy and Physiology* (0-07-232157-1) is a new web-based dissection program that identifies approximately 600 different anatomical structures, with sound profiles to provide pronunciations for each of these 600 terms. Also featured on the site are 100 high-resolution photographs. By clicking anywhere on the structures of the cat, an information frame will provide the name of the structure, as well as a smaller image that highlights and puts the structure in its human anatomical context. In addition, approximately 30 QuickTime movies demonstrate joint movements.

About the Author

Dr. Donald W. Linzey, a native of Baltimore, Maryland, is a biologist living and working in southwestern Virginia. He has received degrees from Western Maryland College in Westminster, Maryland (B.A., 1961), and Cornell University in Ithaca, New York (M.S., 1963, PhD., 1966). Since the early 1960s, he has been studying mammals and other vertebrates in the eastern United States. In the early 1960s, he worked as a park ranger and naturalist in the Great Smoky Mountains National Park, and he has taught biological and environmental sciences at Cornell University, The University of South Alabama, Virginia Tech, and, since 1986, Wytheville Community College in Wytheville, Virginia. In addition to authoring nine texts and over 50 articles for scientific journals, Dr. Linzey is presently coordinating a long-term multidisciplinary study of the decline of amphibians in Bermuda and coordinating all mammal investigations for the All Taxa Biological Inventory in the Great Smoky Mountains National Park.

For the past 10 years, Dr. Linzey has served as Director of the Blue Ridge Highlands Regional Science Fair and has recently completed a three-year term on the Science Service Advisory Board, which oversees the annual Intel International Science and Engineering Fair. Dr. Linzey's recent awards include Outstanding Faculty Award from the Virginia State Council of Higher Education (1996), Distinguished Service Award for Wytheville Community College (1998), the Chancellor's Professorship Award from the Virginia Community College System (1998), and the 1999 Virginia Professor of the Year Award from the Carnegie Foundation for the Advancement of Teaching.

Dr. Linzey resides in Blacksburg, Virginia with his wife, Juanita. They have four sons.

CHAPTER 1

The Vertebrate Story: An Overview

INTRODUCTION

Life on Earth began some 3.5 billion years ago when a series of reactions culminated in a molecule that could reproduce itself. Although life forms may exist elsewhere in our universe or even beyond, life as we know it occurs only on the planet Earth. From this beginning have arisen all of the vast variety of living organisms—viruses, bacteria, fungi, protozoans, plants, and multicellular animals—that inhabit all parts of our planet. The diversity of life and the ability of life forms to adapt to seemingly harsh environments is astounding. Bacteria live in the hot thermal springs in Yellowstone National Park and in the deepest parts of the Pacific Ocean. Plants inhabit the oceans to the lower limit of light penetration and also cover land areas from the tropics to the icepacks in both the Northern and Southern Hemispheres. Unicellular and multicellular animals are found worldwide. Life on Earth is truly amazing!

Our knowledge of the processes that create and sustain life has grown over the years and continues to grow steadily as new discoveries are announced by scientists. But much remains to be discovered—new species, new drugs, improved understanding of basic processes, and much more.

All forms of life are classified into five major groups known as **kingdoms**. The generally recognized kingdoms are Monera (bacteria), Fungi (fungi), Protista (single-celled organisms), Plant (plants), and Animal (multicellular animals). Within each kingdom, each group of organisms with similar characteristics is classified into a category known as a **phylum**.

Whereas many members of the Animal kingdom possess skeletal, muscular, digestive, respiratory, nervous, and reproductive systems, there is only one group of multicellular animals that possess the following combination of structures: (1) a dorsal, hollow nerve cord; (2) a flexible supportive rod (notochord) running longitudinally through the dorsum just ventral to the nerve cord; (3) pharyngeal slits or pharyngeal pouches; and (4) a postanal tail. These morphological characteristics may be transitory and may be present only during a particular stage of development, or they may be present throughout the animal's life. This group of animals forms the phylum **Chordata**. This phylum is divided into three **subphyla**: Urochordata, Cephalochordata, and Vertebrata. The Urochordata and Cephalochordata consist of small, nonvertebrate marine animals and are often referred to collectively as protochordates. To clearly understand and compare their evolutionary significance in relation to the vertebrates, it is necessary to briefly discuss their characteristics.

Subphylum **Urochordata** (tunicates): Adult tunicates, also known as sea squirts, are mostly sessile, filter-feeding marine animals whose gill slits function in both gas exchange and feeding (Fig. 1.1). Water is taken in through

FIGURE 1.1

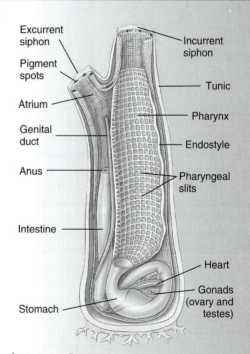

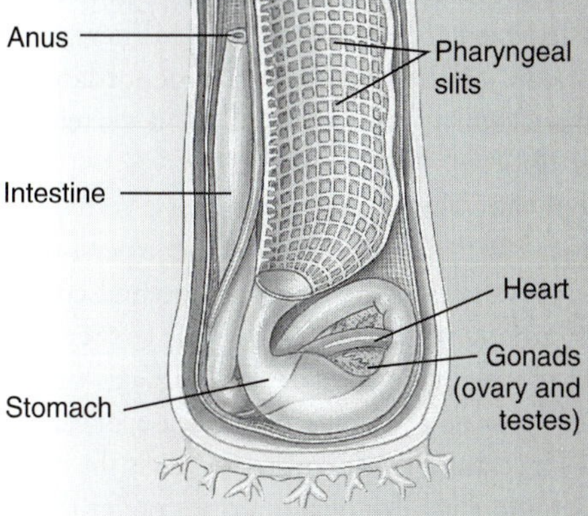

Structure of a tunicate, *Ciona* sp.

an incurrent siphon, goes into a chamber known as the pharynx, and then filters through slits into the surrounding atrium. Larval tunicates, which are free-swimming, possess a muscular larval tail that is used for propulsion. This tail contains a well-developed notochord and a dorsal hollow nerve cord. The name Urochordate is derived from the Greek *oura,* meaning tail, and the Latin *chorda,* meaning cord; thus, the "tail-chordates." When the larva transforms or metamorphoses into an adult, the tail, along with its accompanying notochord and most of the nerve cord, is reabsorbed (Fig. 1.2).

Subphylum **Cephalochordata** (lancelet; amphioxus): Cephalochordates are small (usually less than 5 cm long), fusiform (torpedo-shaped) marine organisms that spend most of their time buried in sand in shallow water. Their bodies are oriented vertically with the tail in the sand and the anterior end exposed. A well-developed notochord and long dorsal hollow nerve cord extend from the head (*cephalo* means head) to the tail and are retained throughout life. The numerous pharyngeal gill slits are used for both respiration and filter-feeding (Fig. 1.3). Cephalochordates have a superficial resemblance to the larvae of lampreys (ammocoete), which are true vertebrates (Fig. 1.3).

Serially arranged blocks of muscle known as **myomeres** occur along both sides of the body of the lancelet. Because the notochord is flexible, alternate contraction and relaxation of the myomeres bend the body and propel it. Other similarities to vertebrates include a closed cardiovascular system with a two-chambered heart, similar muscle proteins, and the organization of cranial and spinal nerves. No other group of living animals shows closer structural and developmental affinities with vertebrates. However, even though cephalochordates now are believed to be the closest living relatives of vertebrates, there are some fundamental differences. For example, the functioning units of the excretory system in cephalochordates are known as protonephridia. They represent a primitive type of kidney design that removes wastes from the coelom. In contrast, the functional units of vertebrate kidneys, which are known as nephrons, are designed to remove wastes by filtering the blood. What long had been thought to be ventral roots of spinal nerves in cephalochordates have now been shown to be muscle fibers (Flood, 1966). Spinal nerves alternate on the two sides of the body in cephalochordates rather than lying in successive pairs as they do in vertebrates (Hildebrand, 1995).

Subphylum **Vertebrata** (vertebrates): Vertebrates (Fig. 1.4) are chordates with a "backbone"—either a persistent notochord as in lampreys and hagfishes, or a vertebral column of cartilaginous or bony vertebrae that more or less replaces the notochord as the main support of the long axis of the body. All vertebrates possess a cranium, or braincase, of cartilage or bone, or both. The cranium supports and protects the brain and major special sense organs. Many authorities prefer the term Craniata instead of Vertebrata, because it recognizes that hagfish and lampreys have a cranium but no vertebrae. In addition, all vertebrate embryos pass through a stage when pharyngeal pouches

FIGURE 1.2

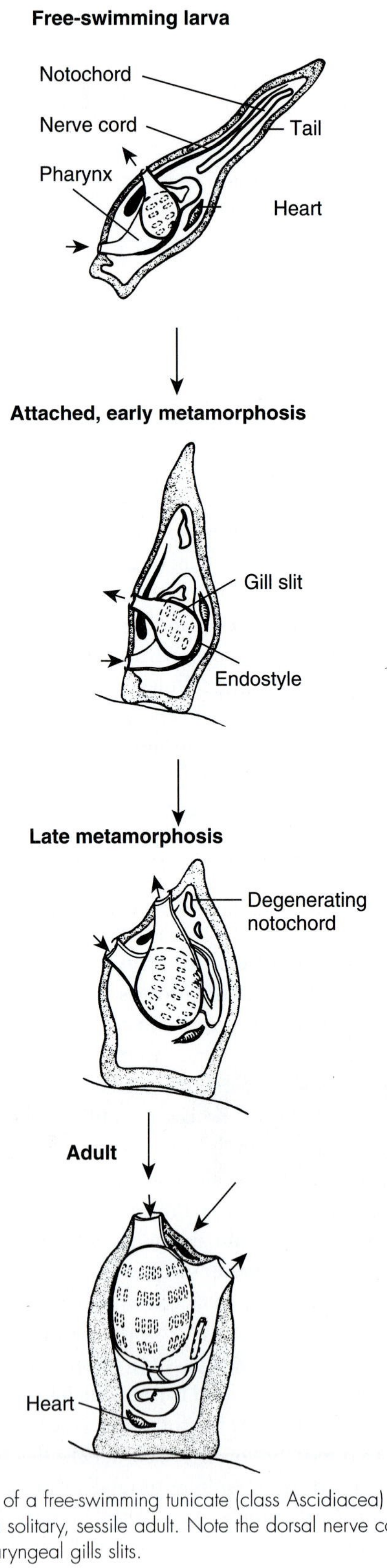

Metamorphosis of a free-swimming tunicate (class Ascidiacea) tadpole-like larva into a solitary, sessile adult. Note the dorsal nerve cord, notochord, and pharyngeal gills slits.

FIGURE 1.3

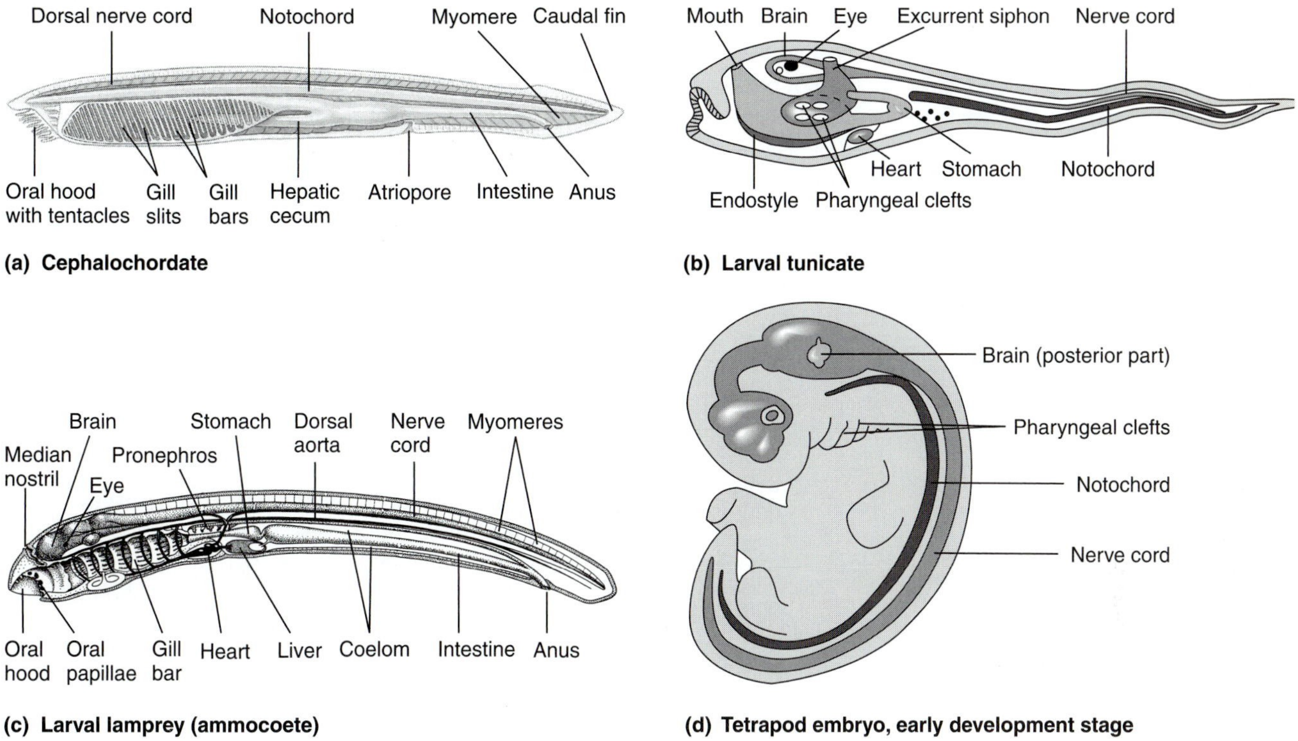

Three chordate characters (dorsal tubular nerve cord, notochord, and pharyngeal clefts) as seen in (*a*) a cephalochordate (amphioxus), (*b*) a larval tunicate, (*c*) a larval lamprey, and (*d*) a tetrapod embryo.

FIGURE 1.4

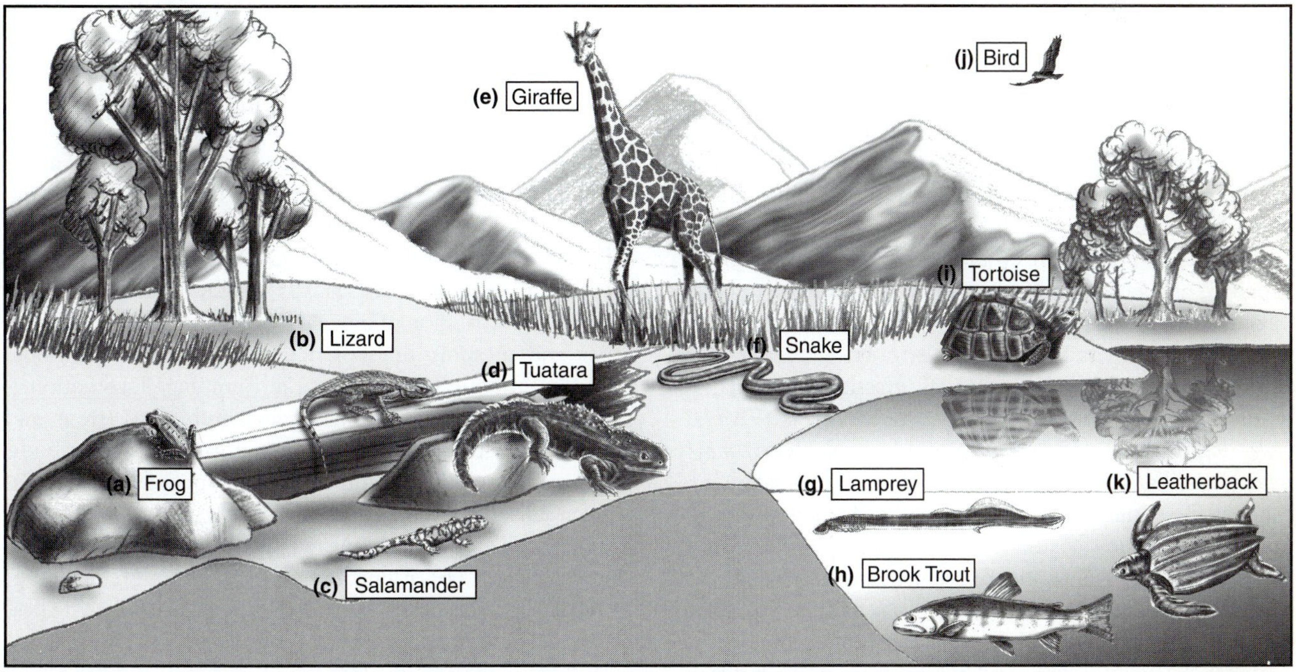

Representative vertebrates: (*a*) wood frog, class Amphibia; (*b*) fence lizard, class Reptilia; (*c*) spotted salamander, class Amphibia; (*d*) tuatara, class Reptilia; (*e*) giraffe, class Mammalia; (*f*) garter snake, class Reptilia; (*g*) lamprey, class Cephalaspidomorphi; (*h*) brook trout, class Osteichthyes; (*i*) gopher tortoise, class Reptilia; (*j*) red-tailed hawk, class Aves; and (*k*) leatherback sea turtle, class Reptilia.

are present (Fig. 1.3). Most living forms of vertebrates also possess paired appendages and limb girdles.

Vertebrate classification is ever-changing as relationships among organisms are continually being clarified. For example, hagfish and lampreys, which were formerly classified together, each have numerous unique characters that are not present in the other. They have probably been evolving independently for many millions of years. Reptiles are no longer a valid taxonomic category, because they have not all arisen from a common ancestor (monophyletic lineage). Although differences of opinion still exist, most vertebrate biologists now divide the more than 53,000 living vertebrates into the following major groups:

Group	Approx. # of Species
Hagfish (Myxinoidea)	43
Lampreys (Petromyzontoidea)	41
Sharks, skates/rays, and ratfish (Chondrichthyes)	850
Ray-finned fish (Actinopterygii)	25,000+
Lobe-finned fish (Sarcopterygii)	4
Salamanders, caecilians (Microsauria)	552
Frogs (Temnospondyli)	3,800
Turtles (Anapsida or Testudomorpha)	230
Diapsids (Diapsida)	
Tuatara, lizards, snakes (Lepidosauromorpha)	8,702
Crocodiles, birds (Archosauromorpha)	9,624
Mammals (Synapsida)	4,629
Total	53,475

Adult vertebrates range in size from the tiny Brazilian brachycephalid frog (*Psyllophryne didactyla*) and the Cuban leptodactylid frog (*Eleutherodactylus iberia*), with total lengths of only 9.8 mm, to the blue whale (*Balaenoptera musculus*), which can attain a length of 30 m and a mass of 123,000 kg (Vergano, 1996; Estrada and Hedges, 1996).

Wide-ranging and diverse, vertebrates successfully inhabit areas from the Arctic (e.g., polar bears) to the Antarctic (e.g., penguins). During the course of vertebrate evolution, which dates back some 500 million years, species within each vertebrate group have evolved unique anatomical, physiological, and behavioral characteristics that have enabled them to successfully inhabit a wide variety of habitats. Many vertebrates are aquatic (living in salt water or fresh water); others are terrestrial (living in forests, grasslands, deserts, or tundra). Some forms, such as blind salamanders (*Typhlomolge, Typhlotriton, Haideotriton*), mole salamanders (*Ambystoma*), caecilians (Gymnophiona), and moles (Talpidae) live beneath the surface of the Earth and spend most or all of their lives in burrows or caves.

Most fishes, salamanders, caecilians, frogs, turtles, and snakes are unable to maintain a constant body temperature independent of their surrounding environmental temperature. Thus, they have a variable body temperature, a condition known as **poikilothermy**, derived from heat acquired from the environment, a situation called **ectothermy**. Although lizards are poikilothermic, many species are very good thermoregulators. Birds and mammals, on the other hand, are able to maintain relatively high and relatively constant body temperatures, a condition known as **homeothermy**, using heat derived from their own oxidative metabolism, a situation called **endothermy**. During periods of inactivity during the summer (torpor) or winter (hibernation), some birds and mammals often become poikilothermic. Under certain conditions, some poikilotherms, such as pythons (*Python*), are able to increase their body's temperature above that of the environmental temperature when incubating their clutch of eggs (see discussion of egg incubation in Chapter 8).

VERTEBRATE FEATURES

Although vertebrates have many characteristics in common, they are very diverse in body form, structure, and the manner in which they survive and reproduce. A brief overview and comparison of these aspects of vertebrate biology at this point, as well as the introduction of terminology that applies to all classes, will provide a firm foundation for more substantive discussions throughout the remainder of the text. Specific adaptations of each class are discussed in Chapters 4–6, 8, and 9.

Body Form. Most fish are fusiform (Fig. 1.5a), which permits the body to pass through the dense medium of water with minimal resistance. The tapered head grades into the trunk with no constriction or neck, and the trunk narrows gradually into the caudal (tail) region. The greatest diameter is near the middle of the body. Various modifications on this plan include the dorsoventrally flattened bodies of skates and rays; the laterally compressed bodies of angelfish; and the greatly elongated (anguilliform) bodies of eels (Fig. 1.5g). Many larval amphibians also possess a fusiform body; however, adult salamanders may be fusiform or anguilliform. Aquatic mammals, such as whales, whose ancestral forms reinvaded water, also tend to be fusiform.

As vertebrates evolved, changes to terrestrial and aerial locomotion brought major changes in body form. The head became readily movable on the constricted and more or less elongated neck. The caudal region became progressively constricted in diameter, but usually remained as a balancing organ. The evolution of bipedal locomotion in ancestral reptiles and in some lines of mammals brought additional changes in body form. Saltatorial (jumping) locomotion is well developed in modern anuran amphibians (frogs and toads), and it brought additional shortening of the body, increased development of the posterior appendages, and loss of the tail (Fig. 1.6a). In saltatorial mammals such as kangaroos and kangaroo rats, the tail has been retained to provide balance (Fig. 1.6b). Elongation of the body and reduction or loss of limbs occurred in some lineages (caecilians, legless lizards, snakes) as adaptations for burrowing.

Aerial locomotion occurred in flying reptiles (pterosaurs), and it is currently a method of locomotion in birds and some mammals. Although pterosaurs became extinct, flying has

FIGURE 1.5

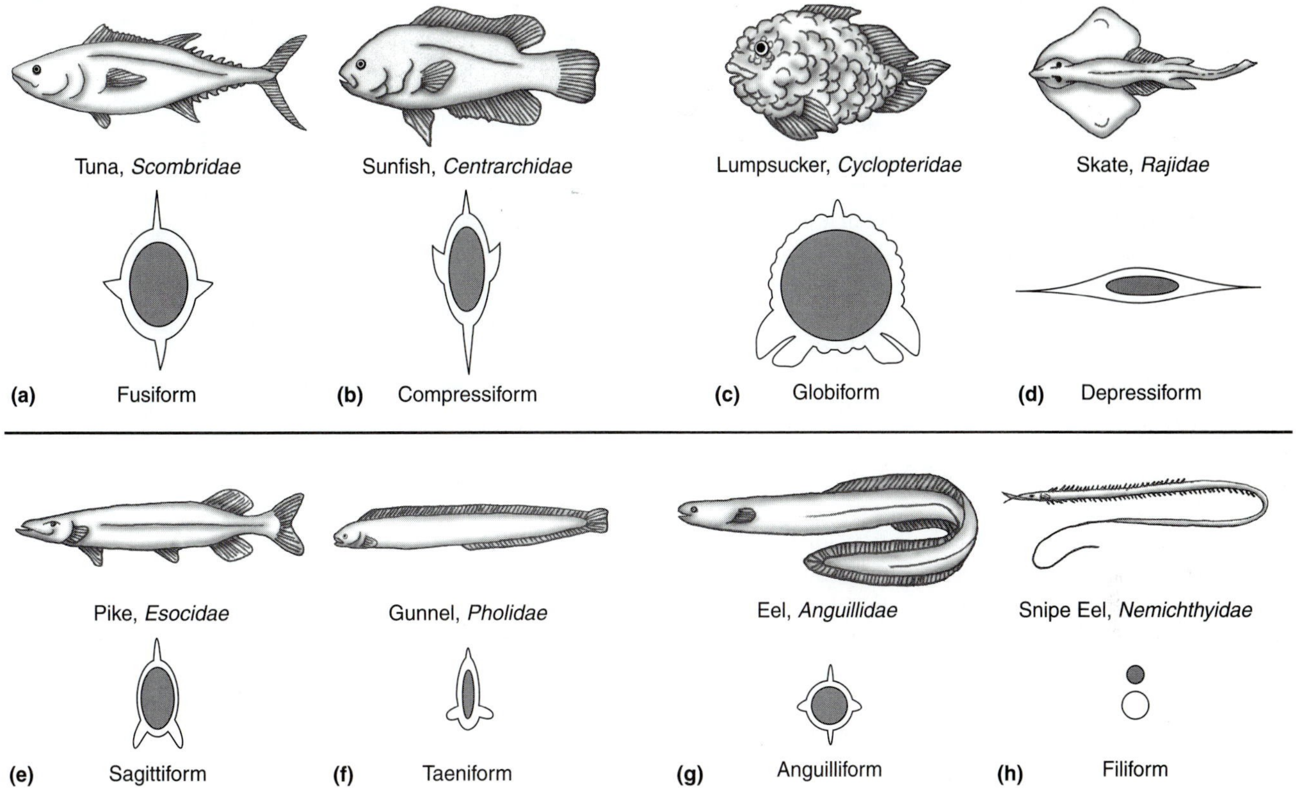

Representative body shapes and typical cross sections of fishes: (*a*) fusiform (tuna, Scombridae); (*b*) compressiform (sunfish, Centrarchidae); (*c*) globiform (lumpsucker, Cyclopteridae); (*d*) depressiform (skate, Rajidae), dorsal view; (*e*) sagittiform (pike, Esocidae); (*f*) taeniform (gunnel, Pholidae); (*g*) anguilliform (eel, Anguillidae); (*h*) filiform (snipe eel, Nemichthyidae).

FIGURE 1.6

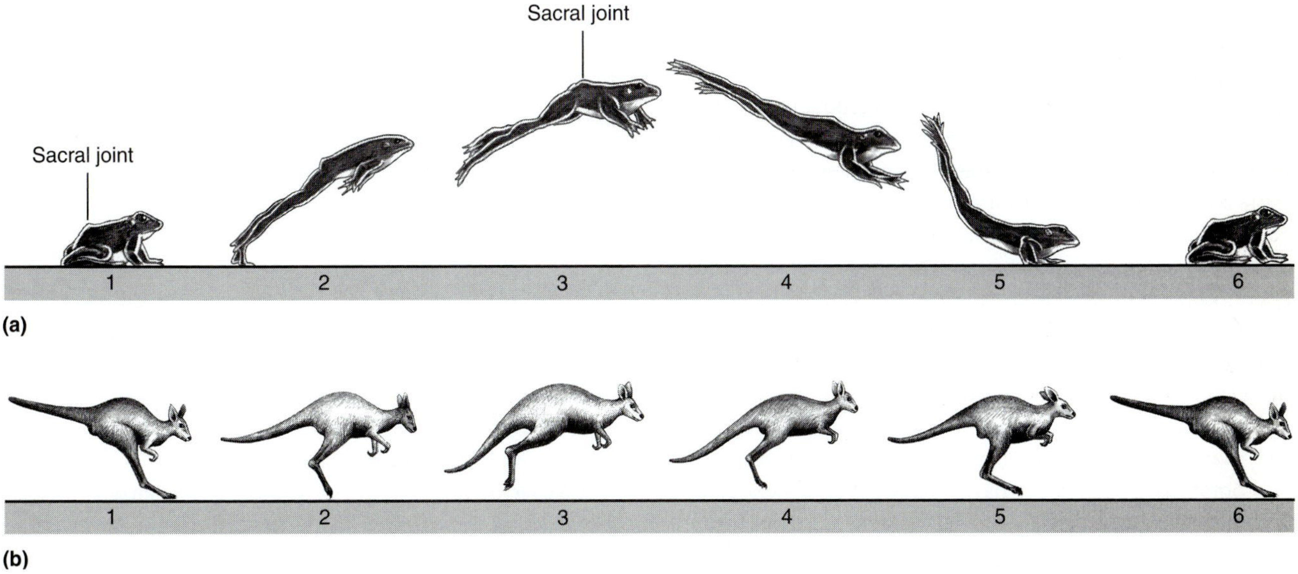

Saltatorial locomotion in (*a*) a frog and (*b*) a kangaroo. Saltatorial locomotion provides a rapid means of travel, but requires enormous development of hind limb muscles. The large muscular tail of the kangaroo is used for balance.

become the principal method of locomotion in birds and bats. The bodies of gliding and flying vertebrates tend to be shortened and relatively rigid, although the neck is quite long in many birds (see Fig. 8.63).

Integument. The skin of vertebrates is composed of an outer layer known as **epidermis** and an inner layer known as **dermis** and serves as the boundary between the animal and its environment. Among vertebrates, skin collectively functions in protection, temperature regulation, storage of calcium, synthesis of vitamin D, maintenance of a suitable water and electrolyte balance, excretion, gas exchange, defense against invasion by microorganisms, reception of sensory stimuli, and production of pheromones (chemical substances released by one organism that influence the behavior or physiological processes of another organism). The condition of an animal's skin often reflects its general health and well-being. Significant changes, particularly in the epidermis, occurred as vertebrates adapted to life in water and later to the new life on land.

The entire epidermis of fishes consists of living cells. Numerous epidermal glands secrete a mucus coating that retards the passage of water through the skin, resists the entrance of foreign organisms and compounds, and reduces friction as the fish moves through water. The protective function of the skin is augmented by dermal scales in most fishes.

The move to land brought a subdivision of the epidermis into an inner layer of living cells, called the **stratum germinativum**, and an outer layer of dead cornified cells, called the **stratum corneum**. In some vertebrates, an additional two to three layers may be present between the stratum germinativum and stratum corneum. The stratum corneum is thin in amphibians, but relatively thick in the more terrestrial lizards, snakes, crocodilians, birds, and mammals, where it serves to retard water loss through the skin. Terrestrial vertebrates developed various accessory structures to their integument such as scales, feathers, and hair as adaptations to life on land. Many ancient amphibians were well covered with scales, but dermal scales occur in modern amphibians only in the tropical, legless, burrowing caecilians, in which they are rudimentary or degenerate (vestigial) and embedded in the dermis. The epidermal scales of turtles, lizards, snakes, and crocodilians serve in part to reduce water loss through the skin, serve as protection from aggressors, and in some cases (snakes), aid in locomotion. The evolution of endothermy in birds and mammals is associated with epidermal insulation that arose with the development of feathers and hair, respectively. Feathers are modified reptilian scales that provide an insulative and contouring cover for the body; they also form the flight surfaces of the wings and tail. Unlike feathers, mammalian hair is an evolutionarily unique epidermal structure that serves primarily for protection and insulation.

Some land vertebrates have epidermal scales underlain by bony plates to form a body armor. For example, turtles have been especially successful with this type of integumental structure. Among mammals, armadillos (*Dasypus*) and pangolins (*Manis*) have similar body armor (Fig. 1.7).

FIGURE 1.7

The interlocking plates of the nine-banded armadillo (*Dasypus novemcinctus*) provide protection for the back and soft undersides. Armadillos, which can run rapidly and burrow into loose soil with lightninglike speed, are also good swimmers.

FIGURE 1.8

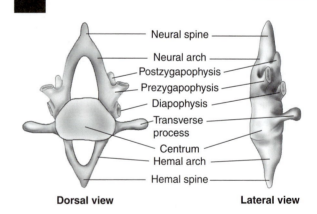

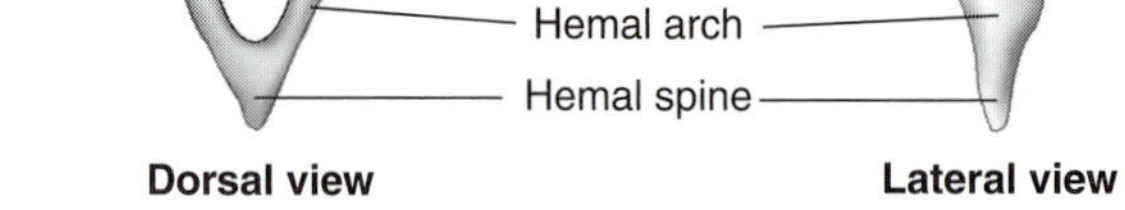

A composite vertebra. The neural arch is dorsal to the centrum and encloses the spinal cord. The hemal arch, when present, is ventral to the centrum and encloses blood vessels.

Cornified (keratinized) epidermal tissue has been modified into various adaptive structures in terrestrial vertebrates, including scales, feathers, and hair. The tips of the digits are protected by this material in the form of claws, nails, or hooves. The horny beaks of various extinct diapsids, living turtles, and birds have the same origin.

Skeleton. The central element of the skeleton is the vertebral column, which is made up of individual vertebrae. There is no typical vertebra; a composite is shown in Fig. 1.8. Each vertebra consists of a main element, the centrum, and various processes.

The vertebral column of fish consists of trunk and caudal vertebrae, whereas in tetrapods (four-legged vertebrates), the vertebral column is differentiated into a neck (cervical) region, trunk region, sacral region, and tail (caudal) region. In some lizards and in birds and mammals, the trunk is divided into a rib-bearing thoracic region and a ribless lumbar region. Two or more sacral vertebrae often are fused in tetrapods for better support of body weight through the attached pelvic girdle; this is carried to an extreme in birds with the fusion of lumbars, sacrals, and some caudals. Neural arches project dorsally to enclose and protect the nerve cord, and in fishes, hemal arches project ventrally to enclose the caudal artery and vein.

The skull supports and protects the brain and the major special sense organs. In hagfish, lampreys, and cartilaginous fish, the skull is cartilaginous and is known as the **chondrocranium**, but in other vertebrates, bones of dermal origin invade the chondrocranium and tend to progressively obscure it. It is believed that primitive vertebrates had seven gill arches and that elements of the most anterior gill arch evolved into the vertebrate jaw, which was braced by elements of the second gill arch (see discussion in Chapter 5). As vertebrates continued to evolve, dermal plates enclosed the old cartilaginous jaw and eventually replaced it.

Teeth are associated with the skull, although they are derived embryologically from the integument and, functionally, are a part of the digestive system. The original function of teeth was probably simple grasping and holding of food organisms. These teeth were simple, conical, and usually numerous. All were similar in shape, a condition called **homodont dentition**. In fish, teeth may be located on various bones of the palate and even on the tongue and in the pharynx, in addition to those along the margin of the jaw. Teeth adapted for different functions, a situation called **heterodont dentition**, have developed in most vertebrate lines from cartilaginous fish to mammals (Fig. 1.9). The teeth of modern amphibians, lizards, snakes, and crocodilians are of the conical type. The teeth of mammals are restricted to the margins of the jaw and are typically (but not always) differentiated into **incisors** (chisel-shaped for biting), **canines** (conical for tearing flesh), **premolars** (flattened for grinding), and **molars** (flattened for grinding). Many modifications occur, such as the tusks of elephants (modified incisors) and the tusks of walruses (modified canines). Teeth have been lost completely by representatives of some vertebrate lines, such as turtles and birds, where the teeth have been replaced by a horny beak.

Appendages. All available evidence (Rosen et al., 1981; Forey, 1986, 1991; Panchen and Smithson, 1987; Edwards, 1989; Gorr et al., 1991; Meyer and Wilson, 1991; Ahlberg, 1995; and many others) suggests that tetrapods evolved from lobe-finned fishes; therefore, tetrapod limbs most likely evolved from the paired lobe fins. Fins of fishes typically are thin webs of membranous tissue, with an inner support of hardened tissue, that propel and stabilize the fish in its aquatic environment. With the move to land, the unpaired fins (dorsal, anal) were lost, and the paired fins became modified into limbs for support and movement. Lobed-finned fishes of today still possess muscular tissue that extends into the base of each fin, and a fin skeleton that in ancestral forms could have been modified into that found in the limbs of tetrapods by losing some of its elements (Fig. 1.10). The earliest known amphibians had a limb skeletal structure intermediate between a lobe-finned fish and the limb skeleton of a terrestrial tetrapod.

FIGURE 1.9

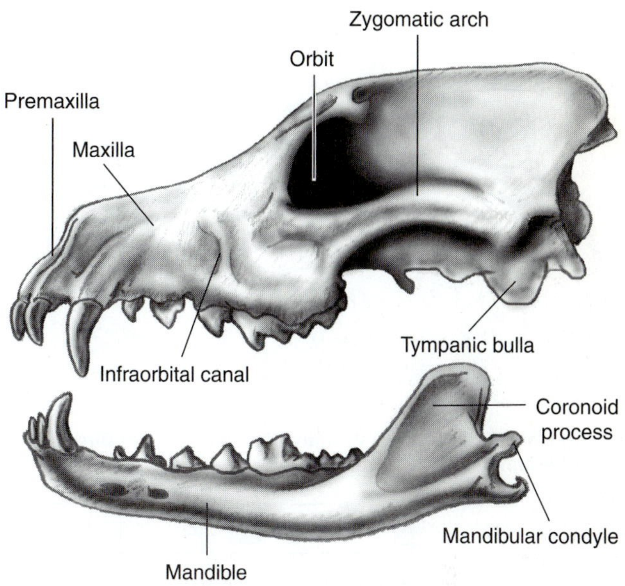

Heterodont dentition of a wolf (*Canis lupus*).

Tetrapod limbs differ from fish fins in that the former are segmented into proximal, intermediate, and terminal parts, often with highly developed joints between the segments. Limbs of tetrapods generally contain large amounts of muscular tissue, because their principal function is to support and move the body. Posterior limbs are usually larger than the anterior pair, because they provide for rapid acceleration and often support a greater part of the body weight. Enormous modifications occurred in the types of locomotion used by tetrapods as they exploited the many ecological niches available on land; this is especially evident in mammals (Fig. 1.11). Mammals may be graviportal (adapted for supporting great body weight; e.g., elephants), cursorial (running; e.g., deer), volant (gliding; e.g., flying squirrels), aerial (flying; e.g., bats), saltatorial (jumping; e.g., kangaroos), aquatic (swimming; e.g., whales), fossorial (adapted for digging; e.g., moles), scansorial (climbing; e.g., gray squirrels), or arboreal (adapted for life in trees; e.g., monkeys). A drastic reduction in the number of functional digits tends to be associated with the development of running types of locomotion, as in various ancient diapsids, in ostriches among living birds, and in horses, deer, and their relatives among living mammals.

A similar structure found in two or more organisms may have formed either from the same embryonic tissues in

FIGURE 1.10

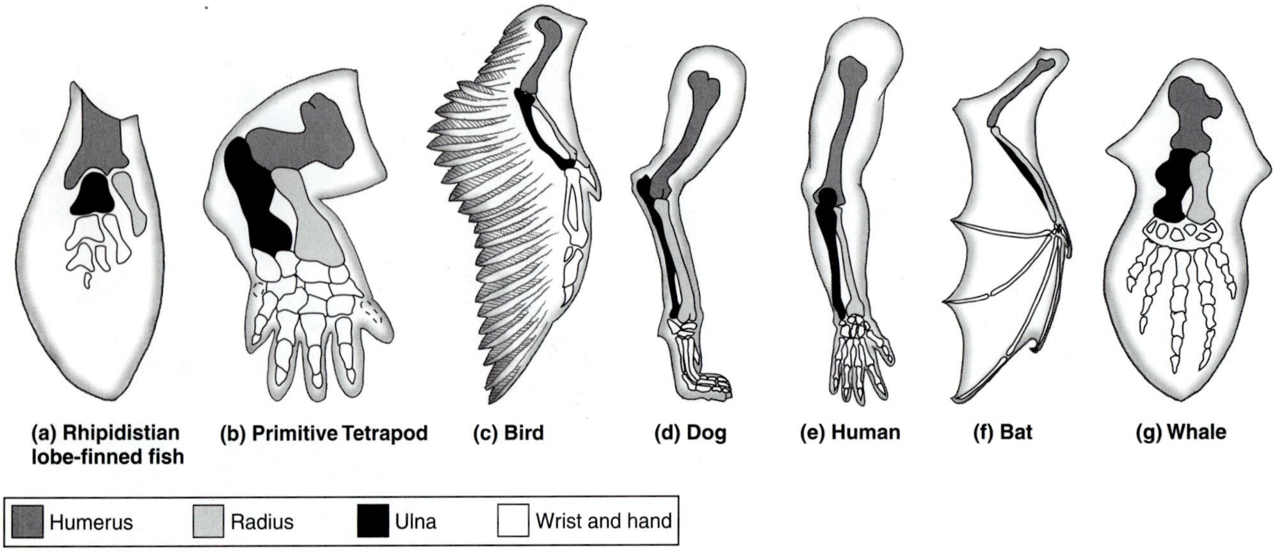

Homologous bones in the front limbs of various vertebrates: (*a*) rhipidistian lobe-finned fish (*Eusthenopteron*); (*b*) primitive tetrapod (*Eryops*); (*c*) bird; (*d*) dog; (*e*) human; (*f*) bat; (*g*) whale. (Key: dark shading: humerus; light shading: radius; black: ulna; white: wrist and hand.)

each organism or from different embryonic tissues. A structure that arises from the same embryonic tissues in two or more organisms sharing a common ancestor is said to be **homologous**. Even though the limb bones may differ in size, and some may be reduced or fused, these bones of the forelimb and hindlimb of amphibians, diapsids, and mammals are homologous to their counterparts (Fig. 1.12a). The wings of insects and bats, however, are said to be **analogous** to one another (Fig. 1.12b). Although they resemble each other superficially and are used for the same purpose (flying), the flight surfaces and internal anatomy have different embryological origins.

The return of various lines of tetrapods to an aquatic environment resulted in modification of the tetrapod limbs into finlike structures, but without the loss of the internal tetrapod structure. This is seen in various lines of extinct plesiosaurs, in sea turtles, in birds such as penguins, and in mammals such as whales, seals, and manatees. All are considered to be homologous structures, because they arise from modifications of tetrapod limb-buds during embryogenesis.

The forelimbs of sharks, penguins, and porpoises provide examples of convergent evolution. When organisms that are not closely related become more similar in one or more characters because of independent adaptation to similar environmental situations, they are said to have undergone **convergent** evolution, and the phenomenon is called **convergence**. Sharks use their fins as body stabilizers; penguins use their "wings" as fins; porpoises, which are mammals, use their "front legs" as fins. All three types of fins have become similar in proportion, position, and function. The overall shape of penguins and porpoises also converged toward that of the shark. All three vertebrates have a streamlined shape that reduces drag during rapid swimming.

Musculature. The greatest bulk of the musculature of fishes is made up of chevron-shaped (V-shaped) masses of muscles (myomeres) arranged segmentally (metamerically) along the long axis of the body and separated by thin sheets of connective tissue known as **myosepta** (Fig. 1.13). A horizontal septum divides the myomeres into dorsal, or **epaxial**, and ventral, or **hypaxial**, muscles. Coordinated contractions of the body (axial) wall musculature provide the main means of locomotion in fish. In the change to terrestrial life, the axial musculature decreased in bulk as the locomotory function was taken over by appendages and their musculature. The original segmentation became obscured as the musculature of the limbs and limb girdles (pectoral and pelvic) spread out over the axial muscles. In fishes, the muscles that move the fins are essentially within the body and are, therefore, **extrinsic** (originating outside the part on which it acts) to the appendages. As vertebrates evolved the abilities to walk, hop, or climb, many other muscles developed, some of which are located entirely within the limb itself and are referred to as **intrinsic** muscles. In flying vertebrates such as birds and bats, the appendicular musculature reaches enormous development, and the axial musculature is proportionately reduced.

Respiration. Gas exchange involves the diffusion of oxygen from either water or air into the bloodstream and carbon dioxide from the bloodstream into the external medium. Fish acquire dissolved oxygen from the water that bathes the

FIGURE 1.11

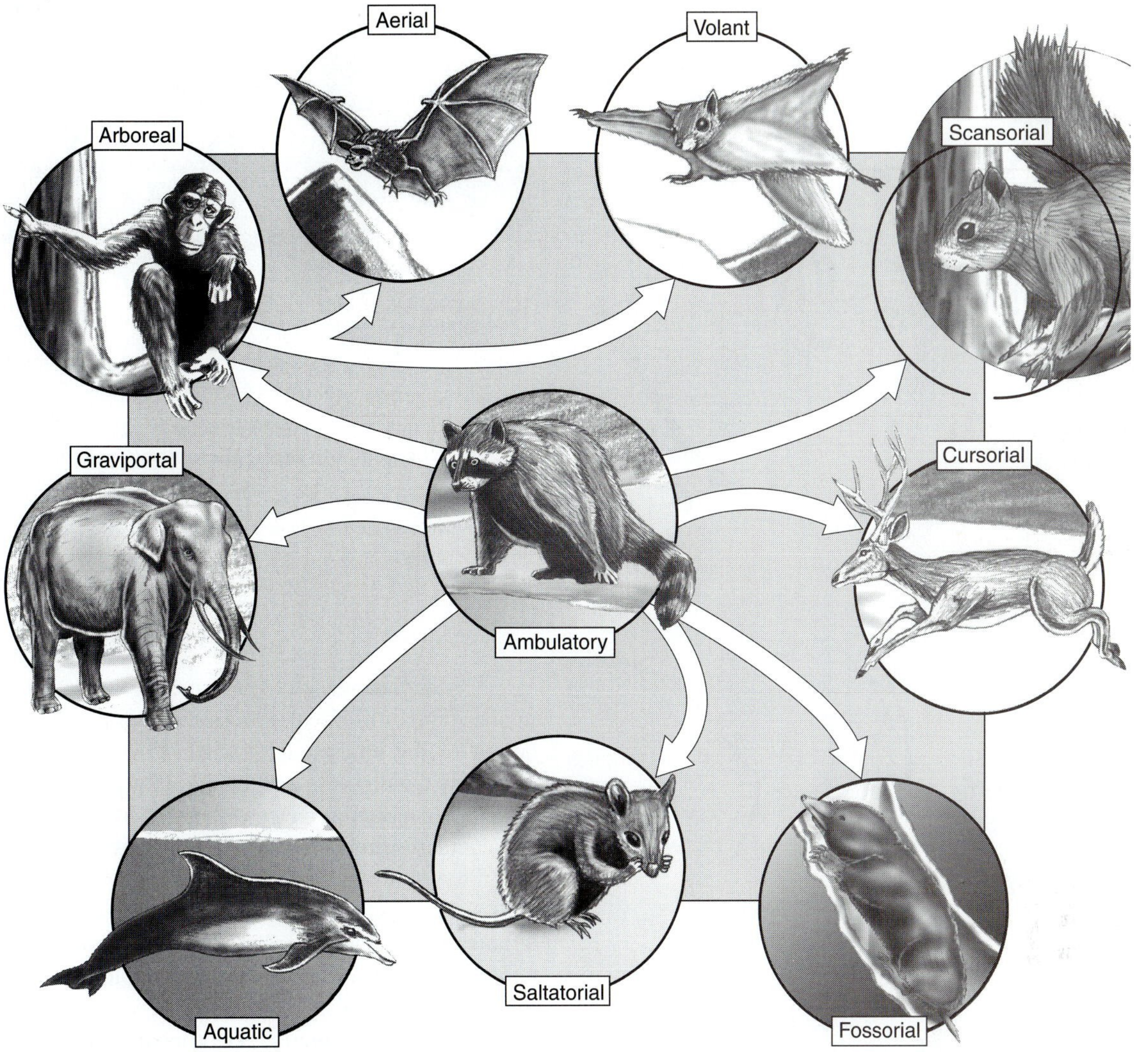

Types of locomotion in mammals. The specialized types of locomotion probably resulted from modifications of the primitive ambulatory (walking) method of locomotion.

gills located in the pharyngeal region. Gas exchange is accomplished by diffusion through the highly vascularized gills, which are arranged as lamellar (platelike) structures in the pharynx (Fig. 1.14). An efficient oxygen uptake mechanism is vital, because the average dissolved oxygen concentration of water is only 1/30 that of the atmosphere.

In most air-breathing vertebrates, oxygen from a mixture of gases diffuses through moist, respiratory membranes of the lungs that are located deep within the body. Filling of the lungs can take place either by forcing air into the lungs as in amphibians or by lowering the pressure in and around the lungs below the atmospheric pressure, thus allowing air to be pulled into the lungs as is the case with turtles, lizards, snakes, and crocodilians as well as with all birds and mammals. The moist skin of amphibians permits a considerable amount of integumental gas exchange with land-living members of one large family of lungless salamanders (Plethodontidae) using no other method of respiration as adults. Structures known as swim bladders that are homologous to the lungs of land vertebrates first appeared in bony fish; some living groups of fish (lungfishes, crossopterygians, garfishes, bowfins) use swim bladders as a supplement to gill breathing. In most living bony fish, however, these structures either serve as hydrostatic (gas-regulating) buoyancy organs, or they are lost.

Circulation. Vertebrate cardiovascular systems consist of a heart, arteries, veins, and blood. The blood, which consists of cells (erythrocytes or red blood cells, leucocytes or white blood cells, thrombocytes or platelets) and a liquid (plasma), is designed to transport substances (e.g., oxygen, waste products of metabolism, nutrients, hormones, and

FIGURE 1.12

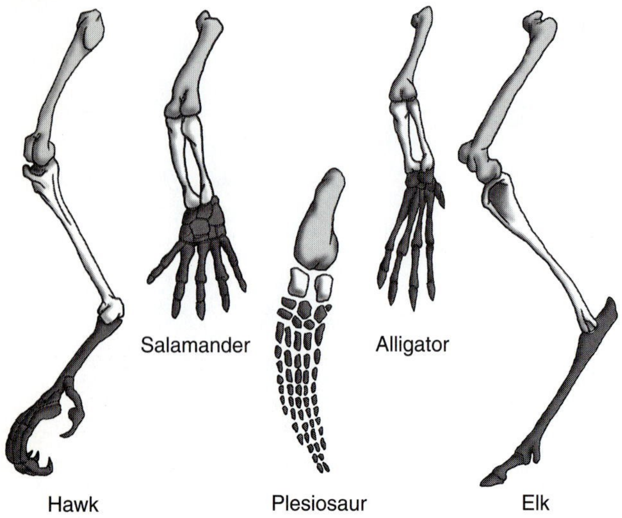

(a) Homology

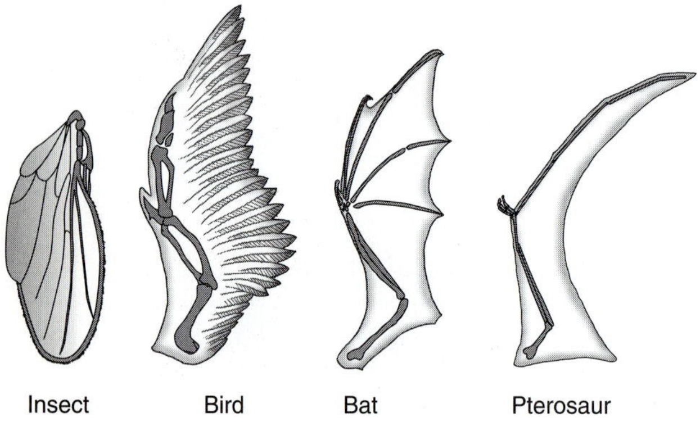

(b) Analogy

(*a*) Homology: hindlimbs of a hawk, a salamander, a plesiosaur, an alligator, and an elk. Bones with the same intensity of shading are homologous, although they are modified in size and in details of shape by reduction or, even, fusion of bones (as in the elk and the hawk). Identical structures have been modified by natural selection to serve the needs of quite different animals. (*b*) Analogy: wings of an insect, a bird, a bat, and a pterosaur. In each, the flight surfaces and internal anatomy have different embryological origins; thus, the resemblances are only superficial and are not based on common ancestry or embryonic origin.

antibodies) rapidly to and from all cells in the body. In homeotherms, cardiovascular systems also regulate and equalize internal temperatures by conducting heat to and from the body surface. In fish, a two-chambered (atrium and ventricle) tubular heart pumps blood anteriorly, where it passes through aortic arches and capillaries of the gill tissues before being distributed throughout the body (Fig. 1.15a). The blood is oxygenated once before each systemic circuit through the body.

FIGURE 1.13

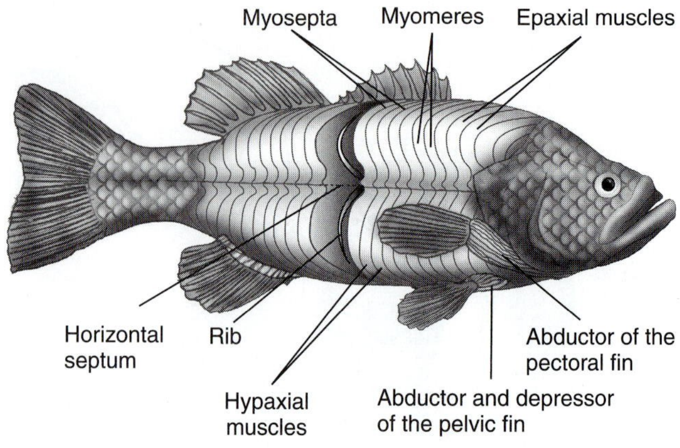

Largemouth Bass, *Micropterus salmoides*

Musculature of a teleost with two myomeres removed to show the shape of the myosepta. Abductor muscles move a fin away from the midline of the body; depressors lower the fin. The horizontal septum divides the myomeres into dorsal (epaxial) and ventral (hypaxial) muscles.

The evolutionary change to lung breathing involved major changes in circulation, mainly to provide a separate circuit to the lungs (Fig. 1.15b). The heart became progressively divided into a right side that pumps blood to the lungs after receiving oxygen-depleted blood from the general circulation and a left side that pumps oxygen-rich blood into the systemic circulation after receiving it from the lungs. This separation of the heart into four chambers (right and left atria, right and left ventricles) first arose in some of the bony fish (lungfishes) and became complete in crocodilians, birds, and mammals.

Digestion. Vertebrates, like other animals, obtain most of their food by eating parts of plants or by eating other animals that previously consumed plants. Fish may ingest food along with some of the water that they use for respiration. In terrestrial vertebrates, mucous glands are either present in the mouth or empty into the mouth to lubricate the recently ingested food.

The digestive tube is modified variously in vertebrates, mostly in relation to the kinds of foods consumed and to the problems of food absorption. The short esophagus of fish became elongated as terrestrial vertebrates developed a neck, and as digestive organs moved posteriorly with the development of lungs. In most vertebrate groups, the stomach has been a relatively unspecialized structure; however, it has become highly specialized in many birds, where it serves to both grind and process food, and in ruminant mammals, where a portion of the stomach has been modified into a fermentation chamber. The intestine, which generally is longer in herbivorous vertebrates than in carnivorous vertebrates as an adaptation for digesting vegetation, is modified

FIGURE 1.14

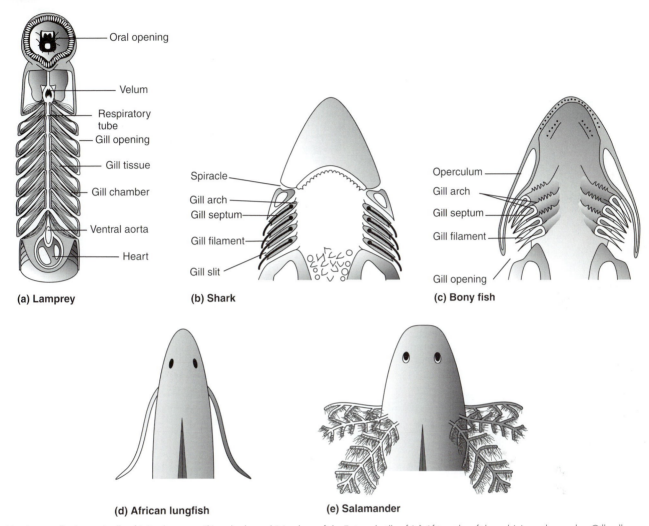

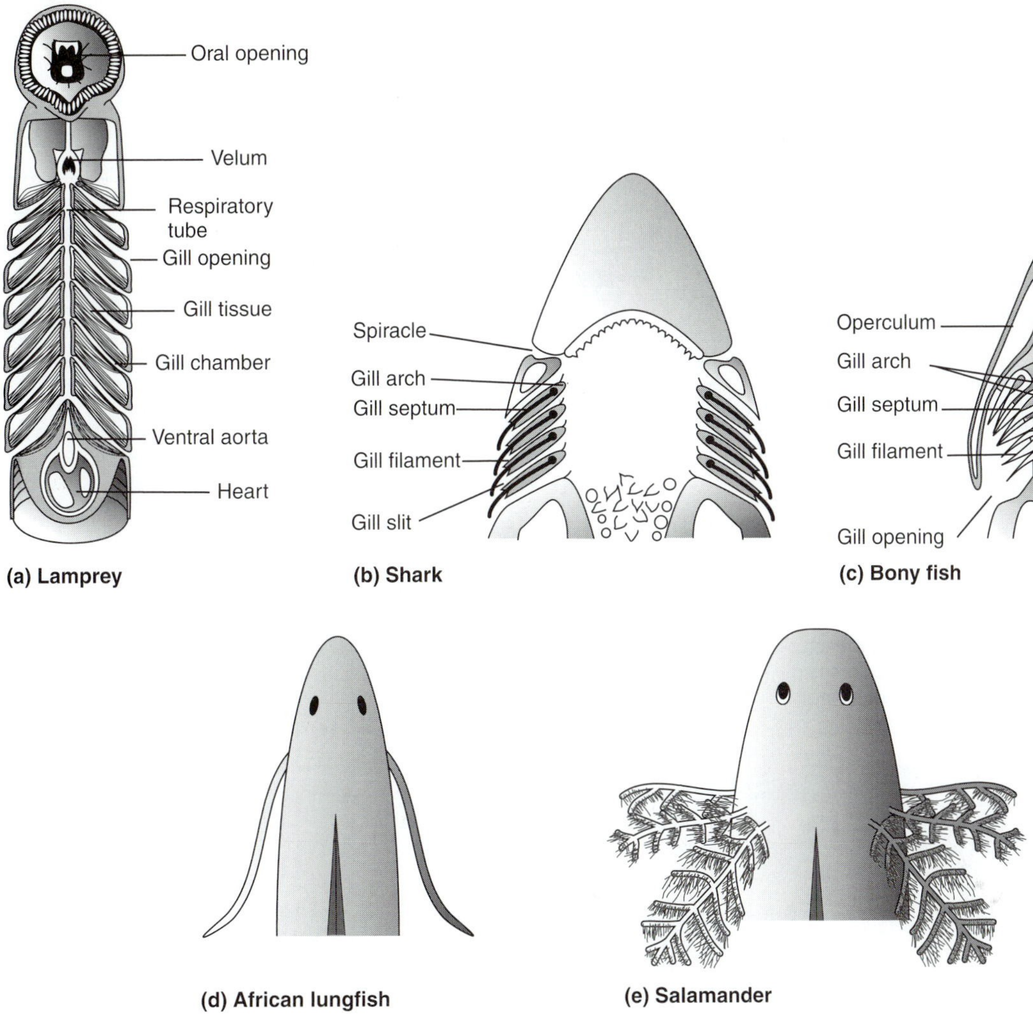

Vertebrate gills: Internal gills of (*a*) a lamprey, (*b*) a shark, and (*c*) a bony fish. External gills of (*d*) African lungfish and (*e*) a salamander. Gills allow aquatic vertebrates to acquire oxygen from water by diffusion across the gill lamellae.

variously internally to slow the passage of food materials and to increase the area available for absorption. A spiral valve that also increases the absorptive area of the intestine is present in cartilaginous and some bony fishes and in some lizards. Pyloric caecae (blind-ended passages at the junction between the stomach and first part of the intestine) serve the same function in most bony fishes (teleosts are one type of bony fish) and also may be present in some diapsids and mammals. In teleosts, caecae number from several to nearly 200 and serve as areas for digestion and absorption of food. The mammalian small intestine is lined with tiny fingerlike projections known as villi that serve to increase the absorptive surface area.

Control and Coordination. The nervous and endocrine systems control and coordinate the activities of the vertebrate body. The brain, as the most important center of nervous coordination, has undergone great changes in the course of vertebrate evolution. In addition, various sense organs have developed to assist in coordinating the activities of the vertebrate with its external environment.

The relative development of the different regions of the brain in vertebrates is related largely to which sense organs are primarily used in obtaining food and mates. The forebrain (telencephalon and diencephalon) consists of the olfactory bulb, cerebrum, optic lobe, parietal eye, pineal body, thalamus, hypothalamus, and hypophysis (pituitary). In hagfish, lampreys, and cartilaginous fishes, the forebrain is highly developed because these vertebrates locate food mainly through olfactory stimuli (Fig. 1.16). The cerebral hemispheres of the forebrain (formerly olfactory in function only)

FIGURE 1.15

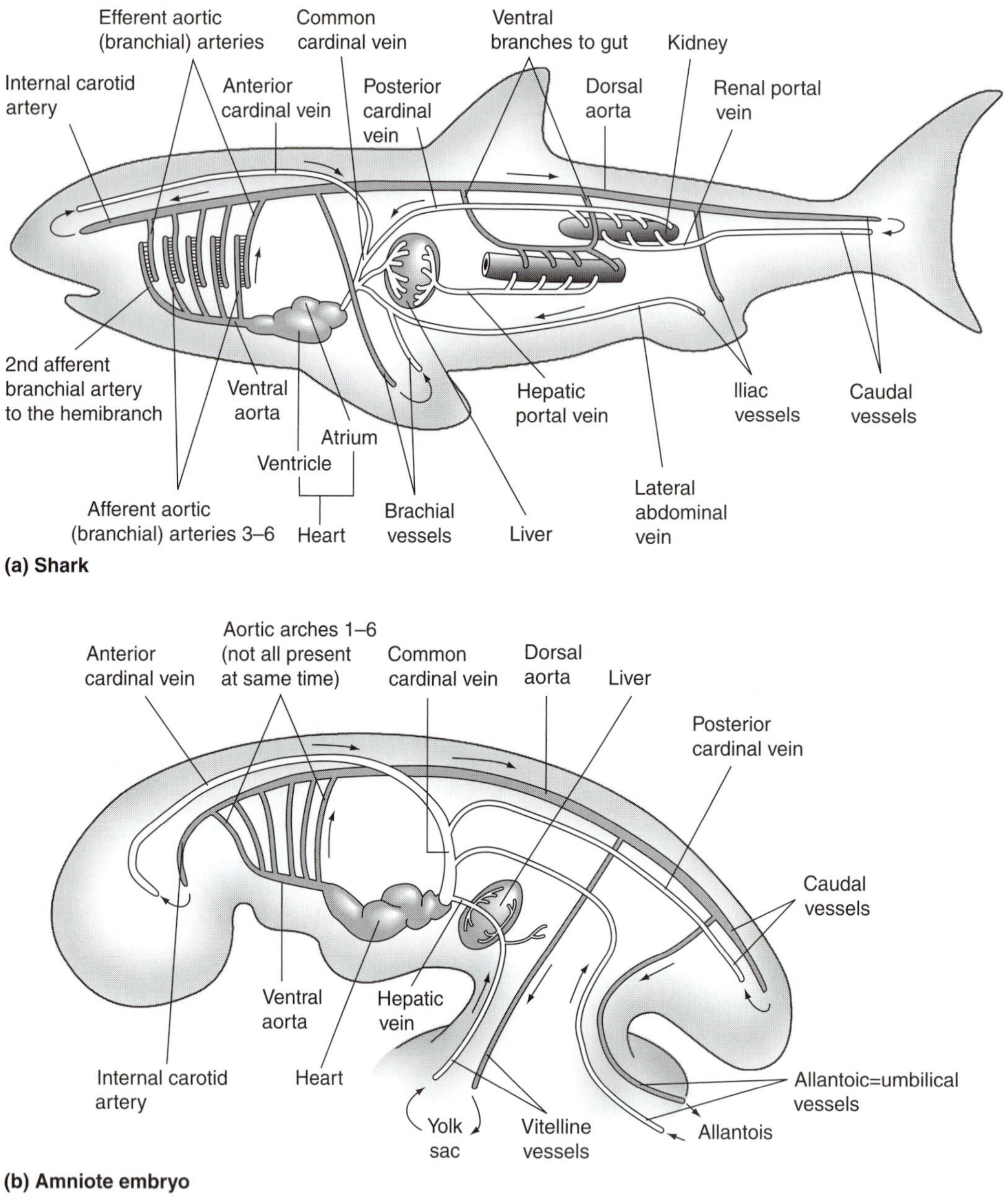

Basic pattern of the vertebrate circulatory system as seen in (*a*) a shark and (*b*) an amniote embryo. All vessels are paired except the dorsal and ventral aortas, the caudal vein, and the vessels of the gut.

From Hildebrand, Analysis of Vertebrate Structure, *4th edition. Copyright © 1995 John Wiley & Sons, Inc. Reprinted by permission of John Wiley & Sons, Inc.*

become increasingly important association centers of the brain. The midbrain (mesencephalon) is most highly developed in many bony fish and in birds because of the importance of vision in obtaining food and for flight. The hindbrain (rhombencephalon) consists of the cerebellum, medulla oblongata, and pons. The cerebellum is responsible for muscular control and coordination; the medulla and pons serve as relay centers and also contain control centers that regulate such functions as respiration and blood pressure. Both the brain and spinal cord are enclosed in protective membranes known as **meninges**.

Olfaction

All vertebrates possess a sense of smell (olfaction). In hagfish, in lampreys, and in all fish except the sarcopterygians (lobe-finned fish), the olfactory receptors are recessed in paired, blind-ended pits known as nasal sacs. In all other vertebrates, the olfactory region is connected to the oral cavity.

FIGURE 1.16

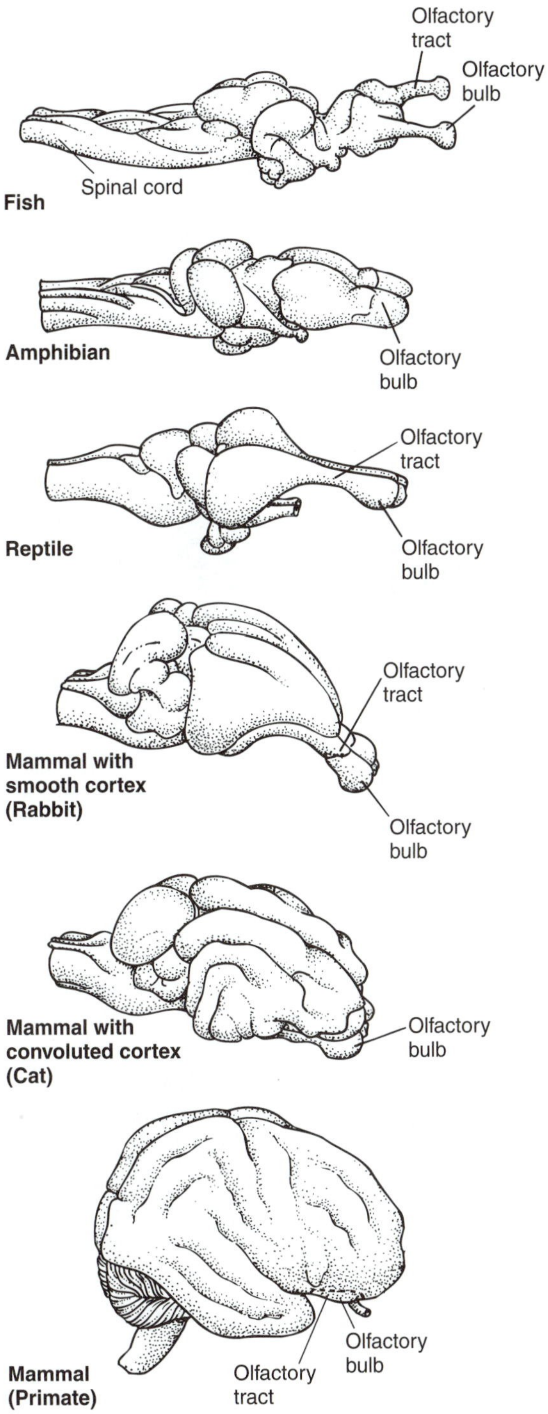

Comparison of the olfactory tract portion of the brain in representative vertebrates. The end of the tract is usually expanded into an olfactory bulb, which receives the olfactory nerves leading from the olfactory epithelium in the nasal region.

Vision

The paired eyes of vertebrates are remarkably constant structures throughout the vertebrates (Fig. 1.17). They tend to be reduced or lost, however, in vertebrates that have adapted to cave or subterranean life where light is dim or absent. An additional medial, unpaired eye is present in hagfish, lampreys, and some diapsids. Among diapsids, this well-developed parietal eye functions as a light-sensitive organ in the tuatara (*Sphenodon*), and a vestigial parietal eye may be seen as a light-colored spot beneath a medial head scale in many kinds of lizards.

Hearing and Vibration Receptors

The ability to detect sound is essential to most vertebrates. The receptors for sound waves, as well as the receptors for equilibrium, are located within the labyrinth in the inner ear (Fig. 1.18). Sound may be used as a warning, for attracting mates, for aggression, for locating food, and for maintaining contact between members of a group. Some vertebrates can detect sound below and above the range of human hearing, called **infrasound** and **ultrasound**, respectively. Some aquatic vertebrates have systems of **neuromasts**, hair cells imbedded in a gelatinous matrix widely distributed over the body surface. Neuromasts open to the outside and are responsive to vibrations in the water; they have been lost in terrestrial vertebrates.

Endocrine System

Chemical control of coordination is accomplished by means of hormones secreted by endocrine glands. In most cases, endocrine organs of different groups of vertebrates are homologous, and similar endocrine controls operate throughout all vertebrates. However, similarities among hormones of different vertebrate groups do not necessarily imply similar function. Prolactin, for example, regulates such activities as nest building, incubation of eggs, and protection of young in many vertebrates. In female mammals, however, prolactin stimulates milk production by the mammary glands.

Kidney Excretion. The vertebrate kidney has evolved through several stages: pronephros, opisthonephros, and metanephros. The pronephros develops from the anterior portion of tissue (nephrogenic mesoderm) that gives rise to the kidney and forms as a developmental stage in all vertebrates. It is functional, however, only in larval fish and amphibians, and it remains throughout life only in lampreys, hagfishes, and a few teleosts. Even then it functions as an adult kidney only in hagfishes; in all other vertebrates, it ceases to function as a kidney and becomes a mass of lymphoid tissue. An opisthonephros serves as the functional kidney of adult lampreys, as well as fishes and amphibians. The kidneys of birds and mammals (metanephros) develop from the posterior portion of the nephrogenic mesoderm.

Nitrogenous wastes from metabolism and excess salts mostly are removed through the kidney by functional units called nephrons (Fig. 1.19). Excretion maintains proper concentrations of salts and other dissolved materials in body fluids. Freshwater fish live in water that has lower salt concentrations than their own body fluids; they have large nephrons and use water freely to dilute metabolic wastes

FIGURE 1.17

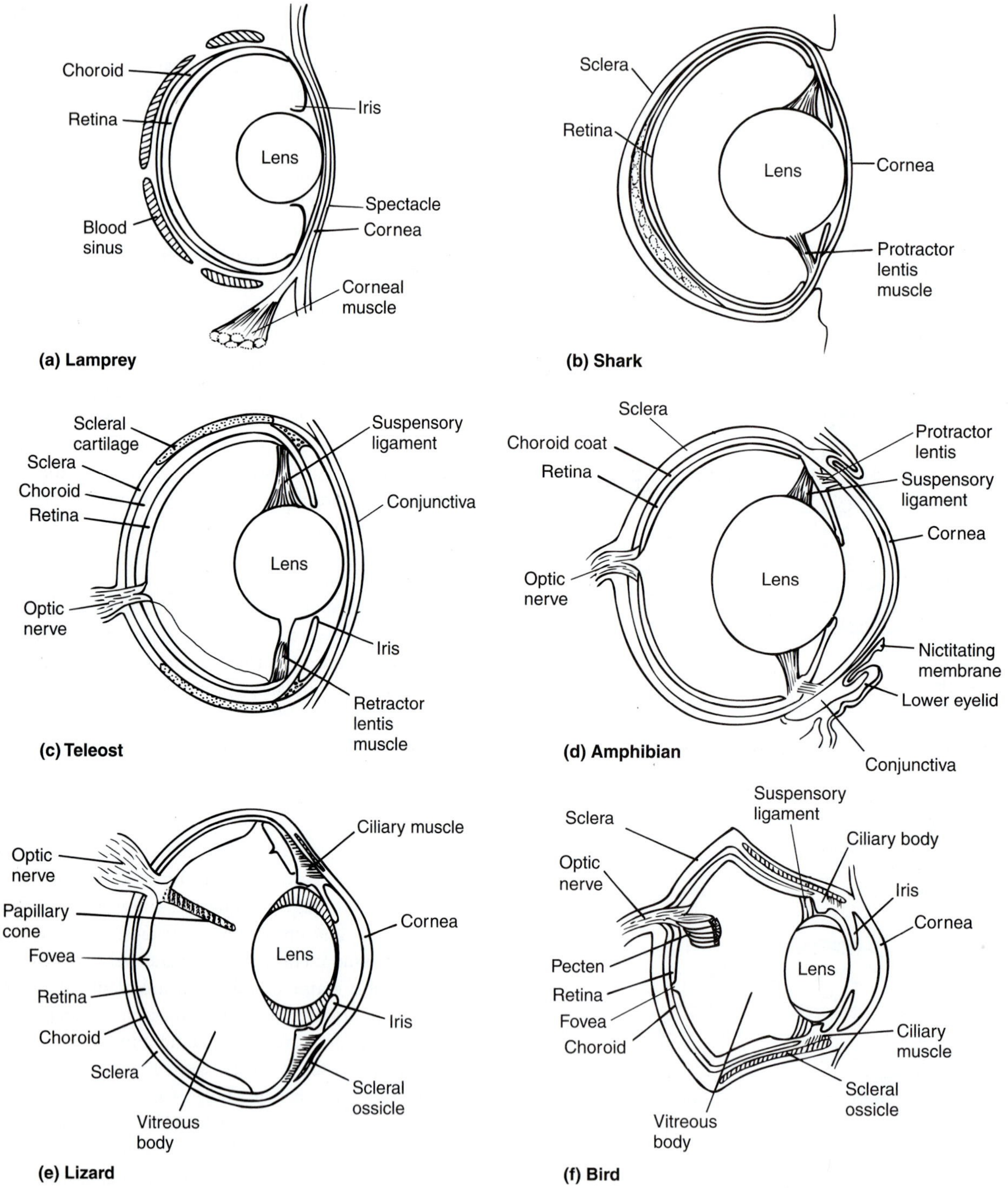

Comparison of the eye in representative vertebrates: (*a*) lamprey; (*b*) shark; (*c*) teleost; (*d*) amphibian; (*e*) lizard; (*f*) bird.

during excretion. Marine fish, on the other hand, live in water in which the salt concentrations are higher than in their own body fluids, and as a result, they are in danger of losing water to their environment. Bony marine fish solve this problem by reducing the size of their nephrons and by excreting salt through their gills. Cartilaginous marine fish solve the problem by retaining nitrogenous wastes in the body fluids in the form of urea, thereby raising the total osmotic pressure of their internal fluids without increasing the salt concentration.

FIGURE 1.18

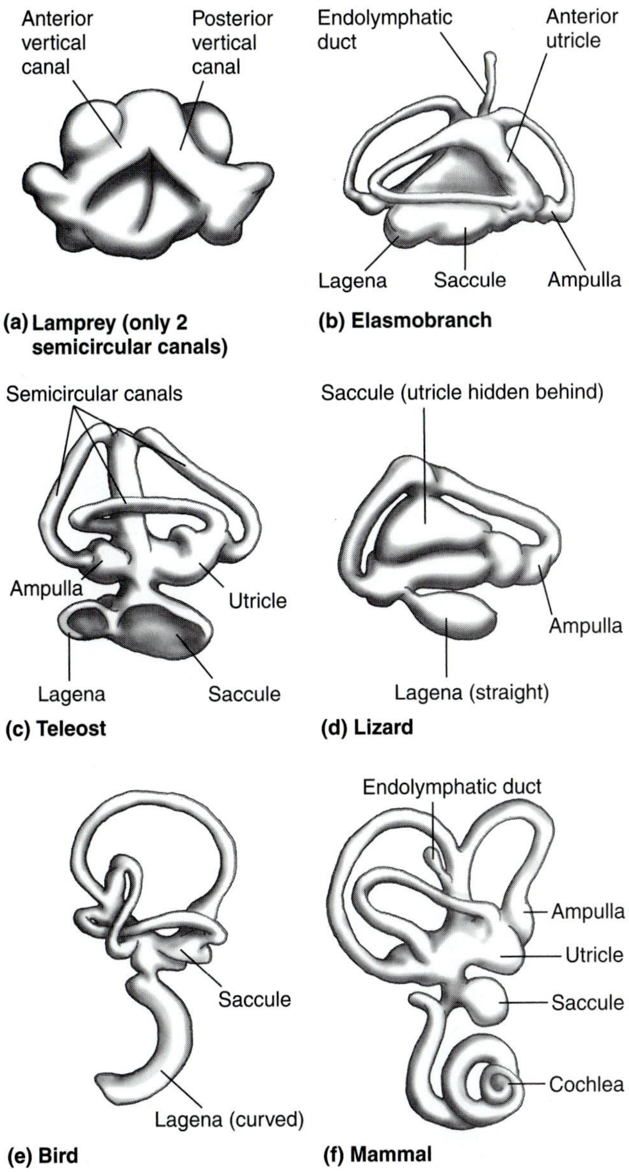

Comparison of the structure of the labyrinth of the inner ear in representative vertebrates (lateral view of the right organ): (*a*) lamprey; (*b*) elasmobranch; (*c*) teleost; (*d*) lizard; (*e*) bird; (*f*) mammal.

From Hildebrand, Analysis of Vertebrate Structure, *4th edition. Copyright © 1995 John Wiley & Sons, Inc. Reprinted by permission of John Wiley & Sons, Inc.*

Terrestrial vertebrates also face the problem of water conservation when excreting metabolic wastes. The filtering portion of the nephron (renal corpuscle) is relatively small, and much water is reabsorbed in the tubule portion of the nephron. Many turtles, lizards, snakes, crocodilians, and most birds excrete crystalline uric acid; mammals, however, excrete a solution of urea, although the solution may be very concentrated in desert inhabitants such as kangaroo rats (*Dipodomys*).

Reproduction. Reproductive output among vertebrates is influenced by sexual state, method of fertilization, and environmental factors such as temperature, photoperiod, and availability of water. Much of this variation in reproductive output arises because some vertebrates are ectothermic (animals whose body temperature is variable and fluctuates with that of the thermal environment), whereas others are endothermic (animals that use heat derived from their own oxidative metabolism to elevate their internal body temperature independently of the thermal environment). Hormones and environmental factors such as temperature, rainfall, and sunlight control the periodicity of breeding and exert a much greater influence on ectothermic species. The age at which an organism reaches sexual maturity and can breed is a major factor in determining growth and size of its population, whereas factors such as floods, droughts, extreme temperatures, parasites, predators, and availability of food can significantly affect the number of individuals reaching sexual maturity.

Vertebrate young typically are born or hatched during the period of the year when environmental conditions are most favorable for their survival. In tropical and subtropical areas, many species are able to breed throughout the year, so that distinct periods of breeding (breeding seasons) are not as pronounced as they are in areas of greater latitude or altitude. In nonequatorial regions, breeding is controlled by cyclic environmental factors such as photoperiod, temperature, and the availability of water. Other factors, such as availability of food and the production of hormones, are influenced by these cyclic environmental factors. Thus, breeding in many vertebrates has a periodicity correlated with the environmental conditions in their region of the world.

Some species have high reproductive rates and produce a large number of offspring with rapid developmental rates but low survival. Such species, which are generally small and provide minimal parental care, are known as *r*-strategists (Table 1.1) They are opportunistic and often inhabit unstable or unpredictable environments where mortality is environmentally caused and is relatively independent of population density (Smith, 1990). Species that are *r*-strategists allocate more energy to reproductive activities and less to growth and maintenance. They are good colonizers and are usually characteristic of early successional stages.

Other species have low reproductive rates and produce relatively few young that mature slowly but are long-lived. These species, known as *K*-strategists, are relatively large and provide care for their young. They are competitive species whose stable populations are limited by available resources. Mortality is generally caused by density-dependent factors rather than by unpredictable environmental factors. *K*-strategists allocate more energy to nonreproductive activities, are poor colonizers, and are characteristic of later stages of succession.

The ability of the sexes of a given species to recognize one another is of utmost importance for generating offspring. This is accomplished in a variety of ways that involve one or more of the sense organs of smell, sight, touch, and hearing. Colorful features, pheromones, vocalizations, and courtship

FIGURE 1.19

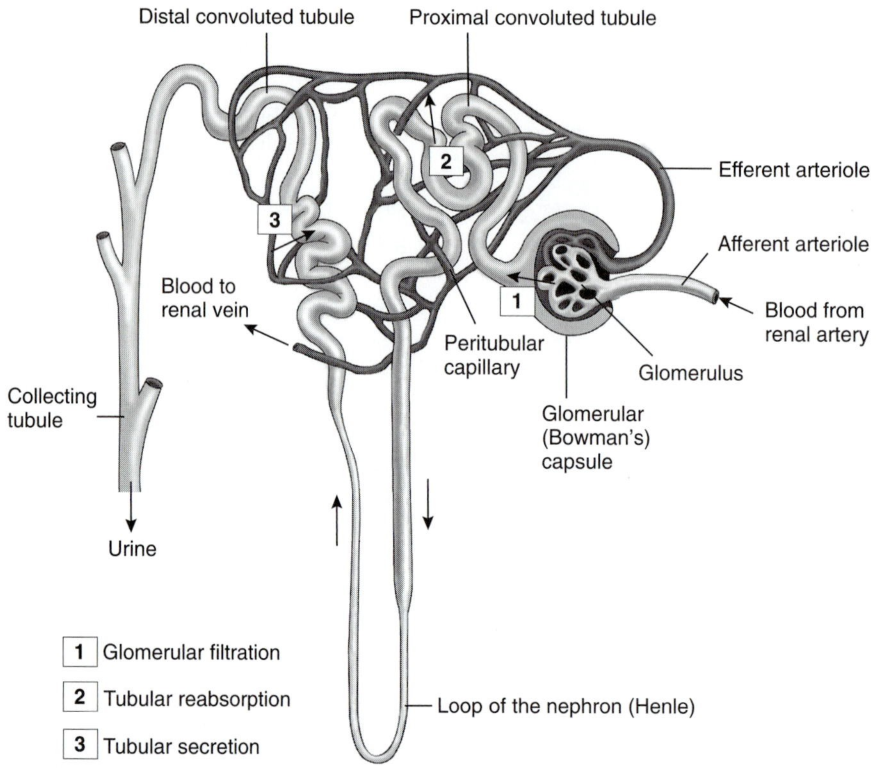

A mammalian nephron. Wastes are filtered out of the glomerulus and travel through the nephron to the collecting tubule.

From Tortora, et al., Principles of Anatomy and Physiology, *6th edition. Copyright © 1990 John Wiley & Sons, Inc. Reprinted by permission of John Wiley & Sons, Inc.*

behavior may all be employed by one sex in their search for a suitable mate.

The complexity of courtship ranges from being almost nonexistent to very elaborate and extensive, such as that found in humans. In each vertebrate group members of some species come together solely to breed, whereas members of other species mate for life. A great deal of variation between these two extremes also occurs.

Most vertebrates are **dioecious**, meaning that male and female reproductive organs are in separate individuals. A few hagfish and lampreys, as well as some fish, are **hermaphroditic** (both male and female reproductive organs develop in the same individual, but normally do not function simultaneously). A few genera of bony fish and lizards have **parthenogenetic** species in which females produce young without being fertilized by males.

Modes of reproduction vary among vertebrates. They include **oviparous** development (egg laying) and **viviparous** development (giving birth to nonshelled young). Oviparity is probably the ancestral mode of reproduction, whereas viviparity represents an evolutionary advance, because a smaller number of larger offspring that have a better chance of survival are produced. According to Blackburn (1992), viviparity originated on at least 132 independent occasions among vertebrates, with 98 of these having occurred in reptiles.

Ova are fertilized in a variety of ways. Fertilization occurs outside the body of the female, called **external fertilization**, in some species; in others, it occurs within the female's body, called **internal fertilization**. In some species, sperm is stored within the body of the female for extended periods of time. Howarth (1974) reports that the extended storage of sperm and the resultant **delayed fertilization** is represented in every vertebrate class with the exception of jawless fishes (classes Myxini and Cephalaspidomorphi). Female diamondback terrapins (*Malaclemys terrapin*), for example, have been reported to lay fertile eggs 4 years following mating (Hildebrand, 1929). Other examples include most temperate species of bats, which mate in the fall just prior to entering hibernation, with viable sperm remaining in the female's reproductive tract until her emergence from hibernation in the spring.

Most ray-finned fish and many amphibians reproduce by external fertilization. Eggs are discharged into the water, and sperm are released in the general vicinity of the eggs. Many eggs and sperm must be produced to ensure that enough of the eggs are fertilized; even so, fertilized eggs (zygotes) may be exposed immediately to the uncertainties of independent

TABLE 1.1

Demographic and Life-History Attributes Associated with *r*- and *K*-Type Populations of Amphibians and Reptiles

Attributes	*r*-type	*K*-type
Population size (density)	Seasonally variable; highest after breeding season, lowest at beginning of breeding season	High to low, but relatively stable from year to year
Age structure	Seasonally and annually variable; most numerous in younger classes, least in adults	Adult age classes relatively stable; most numerous in adult classes
Sex ratio	Variable, often balanced	Variable, often balanced
Population turnover	Usually annual, rarely beyond 2 years	Variable, often >1.5 times age of sexual maturity; to decades
Age at sexual maturity	Usually ≤2 years	Usually ≥4 years
Longevity	Rarely ≥4 years	Commonly >8 years
Body size	Small, relative to taxonomic group	Small to large
Clutch size	Moderate to large	Small to large
Clutch frequency	Usually single breeding season, often multiple times within season	Multiple breeding seasons, usually once each season
Annual reproductive effort	High	Low to moderate

From G.R. Zug, Herpetology, *1993. Copyright © Academic Press, New York. Reprinted by permission.*

existence. Internal fertilization, on the other hand, increases the chances of fertilization and consequently reduces the number of eggs and sperm that must be produced. Internal fertilization has appeared in various groups of ray-finned fish, some amphibians, and universally in cartilaginous fish, turtles, lizards, snakes, crocodilians, birds, and mammals. Retention of developing zygotes within the reproductive tract of the mother (viviparous development) provides a more stable environment for development and has the advantage of protecting the developing young at a stage when they cannot escape predators or unfavorable environmental conditions.

Most fish that use internal fertilization are viviparous. Zygotes are retained within the mother's body until they are ready to emerge as free-swimming juveniles. Among terrestrial vertebrates, turtles, lizards, and crocodilians lay eggs (oviparous). Snakes may be oviparous or viviparous. All birds are oviparous. Two mammals, the duck-billed platypus and the spiny anteater are oviparous; all other mammals are viviparous. Mammalian zygotes retained by the mother must be attached to the wall of the reproductive tract by a highly efficient connection (placenta) so that they can receive nourishment and oxygen from the mother and have their wastes removed.

Embryos of reptiles, birds, and mammals are enclosed in a protective membrane known as an **amnion**. The amnion, which forms a fluid-filled sac in which the embryo floats during its development, is one of four extraembryonic membranes that are present in these groups of vertebrates. Therefore, reptiles, birds, and mammals are referred to as **amniotes**; fishes and amphibians, which lack an amnion, are known as **anamniotes**.

Some kinds of fish, such as bluegills, protect their nest (redd) until the young have hatched, and some even carry the zygotes in their mouth (some catfishes, mouthbreeders in the family Cichlidae) or in a pouch (sea horses) until they hatch (see Fig. 4.53). Many salamanders (Fig. 1.20), some anurans, some lizards, some snakes, all crocodilians, most birds, and all egg-laying mammals guard and protect their eggs during incubation. Some birds and mammals are well developed at birth, have their eyes and ears open, are covered with feathers or hair, and can walk or swim shortly after birth

FIGURE 1.20

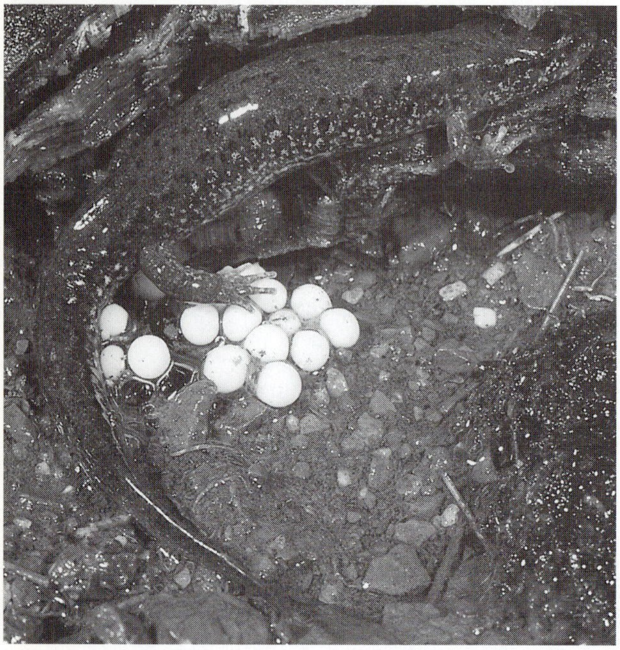

Female dusky salamander (*Desmognathus*) guarding her eggs. Many salamanders and some frogs remain with their eggs to prevent predation by arthropods (such as ants, beetles, and millipedes) and by other salamanders and frogs. In some cases, the male parent guards the eggs.

FIGURE 1.21

Comparison of 1-day-old altricial and precocial young. The altricial meadlowlark (left) is born nearly naked, blind, and helpless. The precocial ruffed grouse (right), however, is born covered with down, is alert, and is able to walk and feed itself.

(Fig. 1.21). Ducks, geese, jackrabbits, and deer are examples of such **precocial** young. Other birds and mammals are born naked and with their eyes and ears sealed, called **altricial** young. Parents of altricial young show more highly developed parental care than parents of precocial young, feeding and caring for the young during their early helpless stages of development. In birds, extensive parental care seems related to the fact that young are mostly helpless until they have learned to fly. In mammals, nourishment is provided by the mammary glands of the mother.

Growth and Development. Prenatal (embryonic) development in all vertebrates follows the same basic pattern: general characters develop first, then the more specific characters. For example, the dorsal nerve tube, notochord, and pharynx are among the first structures to develop. These are followed by gill pouches, aortic arches, and pronephric kidneys in all vertebrates. Then, in tetrapods, pentadactyl limbs, mesonephric and metanephric kidneys, and specific amphibian, reptilian, avian, and mammalian characters appear.

Parental care among vertebrates ranges from being nonexistent in many species to lasting many years in some higher primates. The young of many species are born as miniature adults and do not pass through a larval stage of development. Others, such as many salamanders and frogs, pass through a larval stage of development before transforming or metamorphosing into the adult form. The time required to reach sexual maturity ranges from several weeks in some fishes to several years in some birds and mammals.

The length of time an animal survives depends on its species as well as on factors such as food availability, shelter, and competition. Few animals die of old age in the wild. They may be eaten, killed by hunters, succumb to parasites and/or disease, suffer from climatic events such as drought or flooding, or die because their habitat has become polluted, reduced, or destroyed.

ROLE OF VERTEBRATES

Vertebrates play major roles in the ecosystems of the Earth. They form an essential link in the ecological processes of life and often have close-knit interactions with plants and invertebrates. For example, hummingbirds and some bats (Fig. 1.22) pollinate plants, whereas other birds and mammals assist in transporting seeds (Chapters 9 and 10). Seeds may pass through the digestive tract and are often dispersed long distances from their place of origin, or they may be transported by attachment to the fur of mammals. Some species, such as gopher tortoises (*Gopherus*) and woodchucks and marmots (*Marmota*), excavate burrows that may be used by a wide array of invertebrates as well as by other vertebrate species (Chapter 13). Many feed on invertebrates, including insects. Conversely, many vertebrates serve as food for other species (Chapter 13).

Humans are playing an ever-increasing role in the distribution and abundance of vertebrates. The contamination of natural resources (soil, water, air) has a negative impact on most forms of life. The release of human-made chemicals, the destruction of ozone molecules, global warming, the buildup of estrogen compounds—all have had detrimental effects on other species (Chapter 16). Thus, a critical role for

FIGURE 1.22

Although insects and hummingbirds are widely known as pollinators, some bats, such as this lesser long-nosed bat (*Leptonycteris curasoae*) also facilitate the transfer of pollen. Note the Saguaro cactus pollen on the bat's face.

humanity is to develop a sustainable, nondestructive lifestyle in order to live in harmony with all other vertebrates.

Humans long have hunted other vertebrates for their meat, fur, skin, feathers, ivory, and oil (Chapters 13 and 15). The commercial fishing industry forms a major component of the economy of many nations, and whaling was formerly a significant activity in many countries. Sport fishing and hunting are of major importance in some regions. In recent years, many countries have commercialized wildlife species in order to attract tourists for tours and photographic safaris, a practice called **ecotourism** (Chapter 16). Such activities offer an excellent means of using renewable resources in a less destructive way, while providing educational and economic opportunities.

The domestication of many species has provided humans with food, clothing, work animals, and companionship. Many new laws and regulations have affected the collection of native vertebrates, as well as the importation of certain foreign species that are considered endangered (Chapters 13 and 15, Appendix II). Zoos, theme parks, and aquariums typically feature vertebrate species. In the past, zoos have been composed largely of caged animals whose natural instincts and behavior generally deteriorated the longer they were in captivity. The emphasis now is on providing natural habitats for as many species as possible, supplying educational information about each species, and creating suitable conditions for selected species to breed and produce offspring. Captive breeding programs, which have been established at various zoos around the world, exchange zoo-reared offspring as a means of helping maintain genetic diversity within the species.

FUTURE RESEARCH

The literature on vertebrates is voluminous. In 1991 alone, *Wildlife Review* (formerly an abstracting service of the U.S. Fish and Wildlife Service) listed 13,632 citations just for articles on amphibians, reptiles, birds, and mammals. By 1995, the number of citations had risen to 15,586. A great deal of time and effort is required to keep current with research and developments involving even one group of vertebrates. For this reason, most vertebrate biologists concentrate their attention on only one group, on one aspect of vertebrate life such as reproductive physiology, or on a particular aspect of comparison among two or more groups, such as their systematic relationships.

Much important and significant information has yet to be discovered concerning the biology, ecology, genetics, evolution, and behavior of vertebrate species. For example, what mechanisms are used for communication? How do many species communicate with one another? Is infrasound important in more than just a few species? Which species possess color vision and/or vision outside of the visible spectrum, and exactly what can they see? Why can some vertebrates regenerate limbs and other portions of their bodies, whereas others cannot? Can domestic animals produce beneficial substances such as hemoglobin and hormones in significant quantities for human use? Can squalamine, a compound discovered in the stomachs and livers of dogfish sharks, cure cancer in humans? Squalamine inhibits the growth of tumor-induced new blood vessels in animal systems and also reduces the spread of tumor metastases. Can the toxic alkaloid components of dart poison frogs be beneficial as a drug for humans (Fig. 1.23)? Research currently under way with poison dart frogs indicates that at least one epidermal opioid compound—epibatidine—may act as a potent painkiller (Bradley, 1993).

Efforts to control wild populations of certain abundant mammals such as white-tailed deer (*Odocoileus*), wild horses

FIGURE 1.23

Blue dart poison frogs (*Dendrobates azureus*) produce toxic alkaloids in their skin as a chemical defense against predation.

(*Equus*), raccoons (*Procyon*), skunks (*Mephitis*, *Spilogale*), and woodland voles (*Microtus*) continue. Fertility-inhibiting implants and contraceptive vaccines are the latest techniques being tested for birth control purposes and are discussed in Chapter 14.

The mysteries of migration have yet to be fully understood. How do some young birds, migrating alone for the first time and with no previous knowledge of the terrain, successfully migrate to their overwintering grounds?

New undescribed species continue to be discovered, particularly in tropical areas (Fig. 1.24). New techniques such as DNA sequencing and hybridization will continue to provide data concerning the relationships of living populations and also of some forms now extinct. DNA dated at least 47,000 years B.P. (before the present) has been recovered from Siberian woolly mammoths (*Mammuthus primagenius*) (Hagelberg et al., 1994). These are the oldest dated vertebrate remains from which intact DNA has been amplified. Paleontological discoveries will continue to add to our knowledge of vertebrate species that previously inhabited the earth.

A great deal of future research will be directed toward saving endangered species—both wild populations and those in captive breeding programs. New techniques and procedures will need to be developed to enhance the success of these programs. The reintroduction of such species as the red wolf and timber wolf into suitable areas must be based on sound biological data and not political rhetoric. Public education will be critical to the success of every one of these programs.

Many exciting and challenging research activities await interested researchers. You may be one of those researchers who one day will add to our knowledge of this large and fascinating group of animals.

FIGURE 1.24

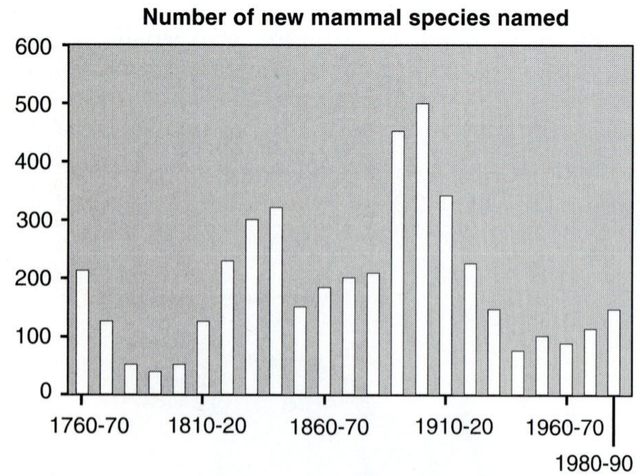

The number of new mammal species discovered (some resulting from taxonomic revisions) from 1760 to 1990. While the biggest burst of discovery is over, the number of new mammals is rising again, with additions from mice to monkeys.

Source: Data from V. Morrell, "New Mammals Discovered by Biology's New Explorers," in Science, *273:1491, September 13, 1996.*

FIGURE 1.25

(a)

(b)

(*a*) The black-headed sagui dwarf marmoset (*Callithrix humulis*) is the seventh new monkey discovered in Brazil since 1990. (*b*) The tree kangaroo (*Dendrolagus mbaiso*), another newly discovered mammal, inhabits an area of Indonesia so remote that the kangaroo had never before been seen by scientists.

BIO-NOTE 1.1

Discovering New Vertebrate Species

The discovery of a new species is an exciting part of scientific research. Although the majority of newly described animals are invertebrates, many vertebrates are still "unknown" to science. In 1993, the smallest known tetrapod, a leptodactylid frog (*Eleutherodactylus iberia*) was discovered in Cuba (Estrada and Hedges, 1996). In November 1994, a tiny acrobatic bird (*Acrobatornis fonsecai*) that spends most of its time upside down, running back and forth along the undersides of branches, was discovered in Brazil (Pacheco, 1996). Between 1937 and 1993 at least 16 "new" large mammal species were discovered, three of which also represented undescribed genera. They included two porpoises (*Lagenodelphis hosei* , *Phocoena sinus*), four beaked whales (*Tasmacetus shepherdi, Mesoplodon ginkgodens, M. carlhubbsi, M. peruvianus*), a wild pig (*Sus heureni*), the Chacoan peccary (*Catagonus wagneri*), four deer (*Mazama chunyi, Moschus fuscus, Muntiacus atherodes, Muntiacus gongshanensis*), the kouprey (*Bos sauveli*), a gazelle (*Gazella bilkis*), a wild sheep (*Pseudois schaeferi*), and a bovid (*Pseudoryx nghetinhensis*).

Smaller mammals also are being discovered. In just the last decade, 11 new species of primates, several bats, and several genera and species of rodents have been identified. In 1988, a new species of tarsier (*Tarsius dianae*) was recorded from Indonesia. In 1990, Bernhard Meier captured the world's second smallest lemur, the hairy-eared dwarf lemur (*Allocebus trichotis*), in Madagascar. This was the first time that scientists had ever seen this animal alive.

Seven new species of Brazilian primates have been found since 1990. The black-faced lion tamarin (*Callithrix caissara*) was found in 1990. The Rio Maues marmoset (*Callithrix mauesi*) was discovered in 1992, in an undisturbed area near the Maues River, a tributary of the Amazon 800 miles upriver from the Amazon delta. The Satere marmoset (*Callithrix saterei*) was found in the rain forests of Amazonia in 1996. Other recently discovered species of primates from Brazil include Kaapor's capuchin (*Cebus kaapori*), the black-headed marmoset (*Callithrix nigriceps*), and Marca's marmoset (*Callithrix marcai*). The most recent new species from Brazil is the black-headed sagui dwarf marmoset (*Callithrix humulis*) (Fig. 1.25a); its discovery was announced in August 1997 (Pennisi, 1997c), and a full scientific description was published in the Brazilian journal *Goeldiana* (Roosmalen et al., 1998). It is the second smallest monkey species, with an average adult measuring 9–10 cm and weighing between 170 and 190 g. This newly discovered monkey may also have the world's smallest distribution for a primate: It is found only between the Amazon tributaries Rio Madeira and Rio Aripuana, in an area 250,000 to 300,000 hectare in size, an area smaller than the state of Rhode Island. This is by far the smallest distribution of any primate in the Amazon.

During a two-week period in July 1996, evolutionary biologist James L. Patton discovered four new species of mice, a shrew, and a marsupial in Colombia's central Andes. In 1991, Patton discovered a new species of spiny mouse (*Scolomys juaraense*) in Brazil whose nearest relatives had been known only from the Andean foothills in Ecuador, 1,500 km away. Lawrence Heaney of Chicago's Field Museum recently discovered 11 new mammals in the Philippine Islands. Between 1991 and 1996, Philip Hershkovitz, also of the Field Museum, discovered two new genera and 16 new species of field mice in Brazil's Cerrado grasslands. In late 1994, a new species of tree-dwelling kangaroo (*Dendrolagus mbaiso*) was discovered in Indonesia (Fig. 1.25b). In late 1997, a plump, 9-in.-long, almost tailless bird known as an antpitta was discovered for the first time in the Ecuadorian Andes (Milius, 1998b). In August 1999, the Ammonite rabbit, a previously unknown species, was discovered in the remote, forested mountains between Laos and Vietnam (in press).

When all of the new genera are officially named and described, researchers estimate that the number of known mammals alone will jump by at least 15%. Russell Mittermeir, a primatologist and president of Conservation International in Washington, D.C., estimates that ten more species of primates will be found in the next decade.

Niemitz et al., 1991
Wilson and Reeder, 1993
Chan, 1994
Pine, 1994
Flannery et al., 1995
Morell, 1996
Anonymous, 1997a

Review Questions

1. Why are tunicates and cephalochordates classified in the phylum Chordata? What do they have in common with vertebrates?
2. Differentiate between poikilothermy and homeothermy. Give several examples of vertebrates exhibitng each type of thermoregulation.
3. Compare the adaptive advantages of hair, feathers, and reptilian scales.
4. List the four types of teeth that may be found in mammals. Give the function of each type.
5. Define the terms *homologous* and *analogous*. Give two examples for each.
6. Distinguish among the following types of locomotion in mammals: cursorial, volant, arboreal, aerial, saltatorial, and fossorial. Give an example for each.
7. What are the two main control systems in the body of a vertebrate?
8. Discuss the adaptations that freshwater bony fish, marine bony fish, and cartilaginous fish have evolved to maintain the proper concentrations of salts and other dissolved materials in their body fluids.
9. Differentiate between viviparous and oviparous. Give examples of each.
10. List several characteristics that distinguish altricial from precocial species. Give several examples.

Supplemental Reading

Bell, G. H., and D. B. Rhodes. 1994. *A Guide to the Zoological Literature*. Englewood, Colorado: Libraries Unlimited, Inc.

Crispins, C. G. 1978. *The Vertebrates: Their Forms and Functions*. Springfield, Illinois: Charles C. Thomas.

Hildebrand, M., D. M. Bramble, K. F. Liem, and D. B. Wake (eds.). 1985. *Functional Vertebrate Morphology*. Cambridge, Massachusetts: Harvard University Press.

Kardong, K. V. 1998. *Vertebrates: Comparative Anatomy, Function, Evolution*. Dubuque, Iowa: W. C. Brown/McGraw-Hill.

Radinsky, L. B. 1987. *The Evolution of Vertebrate Design*. Chicago: University of Chicago Press.

Rogers, E. 1986. *Looking at Vertebrates*. Essex, England: Longman Group Limited.

Vertebrate Internet Sites

Visit the zoology website at http://www.mhhe.com/zoology to find live Internet links for each of the references listed below.

1. **Introduction to the Urochordata.**
   Provides an introduction to the urochordates.
2. **Introduction to the Cephalochordata.**
   Provides an introduction to the cephalochordates.
3. **Morphology of the Chordata.**
   Discusses morphological characteristics of the chordates and serves as a link to additional information.
4. **The Tree of Life.**
   Contains information about phylogenetic relationships, characteristics of vertebrates, their origin and evolution, and a bibliography.
5. **Paleontology Without Walls.**
   Provides links to phylogeny, geologic time, morphology, systematics, and evolutionary thought.
6. **Careers in Biology: Emporia State University.**
   Serves as a link to sites listing job opportunities in the biological sciences.
7. **CalPhotos: Animals.**
   An immense database, which has information and photos of nearly any animal you could imagine. A good resource for photos to include in research papers.
8. **Links to Many Specific Career Descriptions.**
   At least 200 links to websites can be found through this site, which is updated frequently. An alphabetical listing of occupations in biology allows the user to see web sites under many of the listings that include detailed descriptions of careers.
9. **Iternet Resource Guide for Zoology.**
   From Biosis, a searchable index.

CHAPTER 2

Systematics and Vertebrate Evolution

■ INTRODUCTION

Biologists attempt to classify living things according to their evolutionary relationships. Because these relationships probably can never be known exactly, several systematic schools of thought have developed, each of which has developed its own classification system.

The first step in classification is the grouping together of related forms; the second is the application of names to the groups. Some refer to the first step as systematics and the second as taxonomy; others use the two terms interchangeably to describe the entire process of classification.

Systematics comes from the Latinized Greek word *systema,* which was applied to early systems of classification. It is the development of classification schemes in which related kinds of animals are grouped together and separated from less-related kinds. Simpson (1961) defined systematics as, "the scientific study of the kinds and diversity of organisms and of any and all relationships among them." Systematics, which endeavors to order the rich diversity of the animal world and to develop methods and principles to make this task possible, is built on the basic fields of morphology, embryology, physiology, ecology, and genetics.

Taxonomy is derived from two Greek words: *taxis,* meaning "arrangement," and *homos,* meaning "law." It is the branch of biology concerned with applying names to each of the different kinds of organisms. Taxonomy can be regarded as that part of systematics dealing with the theory and practice of describing diversity and erecting classifications. Thus, systematics is the scientific study of classification, whereas taxonomy is the business and laws of classifying organisms.

Frequently, the two disciplines overlap. Taxonomists may attempt to indicate the relationships of the organism they are describing; systematists often have to name a new form before discussing its relationships with other forms. In both disciplines, distinction must be made among various levels of differences. Individual differences must be eliminated from consideration, and features characteristic of the populations of different species must be used as the basis for forming groups. A **population** is a group of organisms of the same species sharing a particular space, the size and boundaries of which are highly variable. Similar and related populations are grouped into **species**, and species are then described. Thus, the species, not individuals, are the fundamental units of systematics and are the basis of classification.

If the fossil record was complete and all of the ancestors of living animals were known, it would be straightforward to arrange them according to their actual relationship. Unfortunately, the fossil record is not complete. Many gaps exist. As a result, the classification of organisms is based primarily on the presence of similarities and differences among groups of *living* organisms. These similarities and differences reflect genetic similarities and differences, and in turn genetic similarities and differences reflect evolutionary origins. Fossilized remains are used whenever possible to extend lineages back into geologic time and to clarify the evolution of groups. For example, paleontological discoveries have clarified our understanding of the development of the tetrapod limb as well as the groups from which birds and mammals arose. Many controversies currently exist due to differences in interpreting the paleontological evidence (Gould, 1989). As techniques improve and more fossils are discovered, the gaps in the fossil record will become fewer, and our understanding of vertebrate evolution and the relationships among the different taxa will increase.

■ BINOMIAL NOMENCLATURE

The current system of naming organisms is based on a method gradually developed over several centuries. It was not finally formalized, however, until the mid-18th century.

In 1753, the Swedish naturalist Carl von Linne, better known as Carolus Linnaeus (1707–1778), published a book, *Species Plantarum,* in which he attempted to list all known kinds of plants. In 1758, he published the tenth edition of a similar book on animals entitled *Systema Naturae.* In that edition, the binomial system of nomenclature (two names) was applied consistently for the first time. The scientific name (binomen) of every species consisted of two Latin or Latinized words: The first was the name of the **genus** to which the organism was assigned, and the second was the **trivial** name. In addition, this work was characterized by clear-cut species descriptions and by the adoption of a hierarchy of higher groupings, or taxa, including family, order, and class.

Linnaeus's methods by no means were entirely original. Even before Linnaeus, there was recognition of the categories "genus" and "species," which in part goes back to the nomenclature of primitive peoples (Bartlett, 1940). Plato definitely recognized two categories, the genus and the species, and so did his pupil Aristotle. But Linnaeus's system was quickly adopted by zoologists and expanded because of his personal prestige and influence. Thus, this was the beginning of the binomial system of nomenclature and of the modern method of classifying organisms. Any zoological binomial published in the year 1758 or later can be considered a valid scientific name; those published prior to 1758 are not. For this reason, Linnaeus is often called the father of taxonomy.

In his tenth edition of *Systema Naturae*, Linnaeus listed 4,387 species of animals. This was a substantial increase over the 549 species mentioned in the first edition in 1735. Since these represented a large variety of different forms, shapes, and sizes of organisms, Linnaeus adopted a system of grouping similar genera together as orders, and groups of similar orders as classes. He grouped all the classes of animals together as members of the animal kingdom, as distinct from the plant kingdom.

The classes established by Linnaeus were as follows:

I.	Quadrupeds	Hairy body; four feet; females viviparous, milk-producing
II.	Birds	Feathered body; two wings; two feet; bony beak; females oviparous
III.	Amphibia	Body naked or scaly; no molar teeth; other teeth always present; no feathers
IV.	Fishes	Body footless; possessing real fins; naked or scaly
V.	Insects	Body covered with bony shell instead of skin; head equipped with antennae
VI.	Worms	Body muscles attached at a single point to a quasi-solid base

Classes I, II, and IV correspond to the traditional evolutionary taxonomic classes (mammals, birds, and fishes) used today. Class III, however, included both amphibians and reptiles.

Common names create difficulties because they often vary with locality, country, or other geographic subdivision. For example, the term *salamander* may mean an aquatic amphibian, or (to many persons in the southeastern United States) it may refer to a mammal, the pocket gopher (*Geomys*). In the latter instance, it is probably a contraction of "sandy-mounder," which refers to the characteristic mounds constructed by the pocket gopher. The word *lizard* is used by many persons to refer to a salamander. The word *gopher* may be used to refer to a ground squirrel, to a pocket gopher, to a mole, and in the southeastern United States, to a turtle, the gopher tortoise (*Gopherus polyphemus*).

Scientific names are recognized internationally and allow for more precise and uniform communication. Because Latin is not a language in current use, it does not change and is intelligible to scientific workers of all nationalities. An important asset of the scientific name is its relative stability. Once an animal is named, the name remains, or if it is changed, the change is made according to established zoological rules. The scientific name is the same throughout the world.

The mammal that once had the largest range of any mammal in the Western Hemisphere is known variously as puma, mountain lion, catamount, deer tiger, Mexican lion, panther, painter, chim blea, Leon, and leopardo in various parts of its range in Canada, the United States, and Central and South America (Fig. 2.1). It is known to biologists in all

FIGURE 2.1

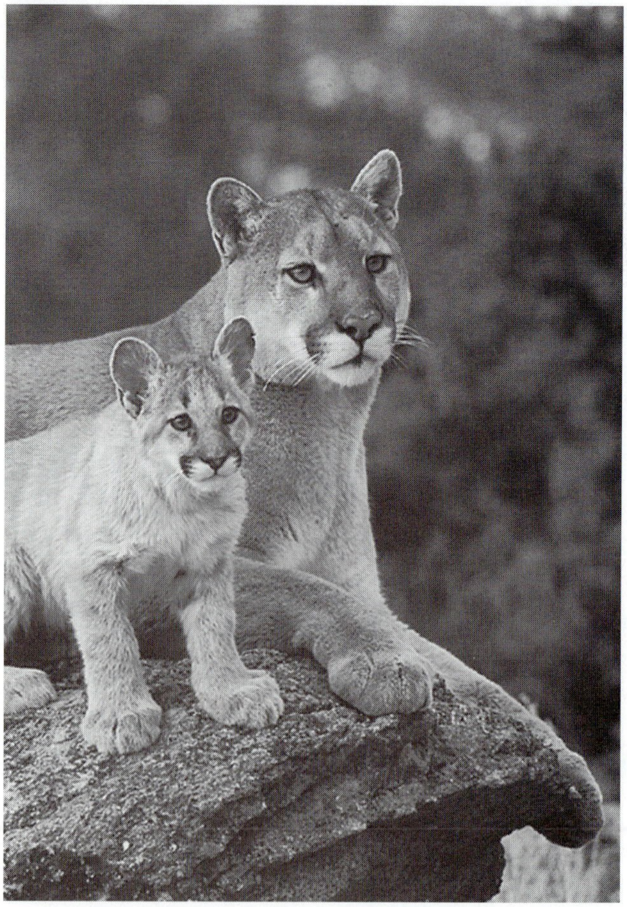

The mountain lion (*Puma concolor*) once had the largest range of any mammal in the Western Hemisphere. Other common names for this species include puma, panther, painter, catamount, and deer tiger.

of these countries, however, as *Puma concolor*. Other members of the cat family (Felidae) are placed in different genera such as the domestic cat (*Felis*), the ocelot and margay (*Leopardus*), the jaguarundi (*Herpailurus*), the Canada lynx and the bobcat (*Lynx*), and the jaguar (*Panthera*). In its complete, official format, the name of the author who described a species may follow the name of the species. For example, the mink is designated as *Mustela vison* Schreber. If a species was described in a given genus and later transferred to another genus, the name of the author of the original species, if cited, is enclosed in parentheses. *Puma concolor* (Linnaeus) indicates that Linnaeus originally classified and named this species. He classified it in the genus *Felis*, but it was later reclassified in the genus *Puma*.

■ CLASSIFICATION

The basic unit of classification, and the most important taxonomic category, is the species. Species are the "types" of organisms. Each type is different from all others, yet the species concept probably has been discussed and debated more than any other concept in biology (Rennie, 1991; Gibbons, 1996a). An understanding of the concept of species is indispensable for taxonomic work.

Through the early part of this century, a morphological species concept was used. Populations were grouped together as species based on how much alike they looked. In the 1930s and 1940s, a more meaningful biological definition of a species emerged. The **biological species concept** was first enunciated by Mayr (1942), as follows: "Species are groups of actually or potentially interbreeding natural populations, which are reproductively isolated from other such groups." Later, Mayr (1969) reformulated his definition: "Species are groups of interbreeding natural populations that are reproductively isolated from other such groups." Thus, a species is a group of organisms that has reached the stage of evolutionary divergence where the members ordinarily do not interbreed with other such groups even when there is opportunity to do so, or if they do, then the resulting progeny are selected against.

Classification involves the recognition of species and the placing of species in a system of higher categories (taxa) that reflect phylogenetic relationships. Mayr (1969) referred to classification as "a communication system, and the best one is that which combines greatest information content with greatest ease of information retrieval." Related species are grouped together in a genus. A **genus**, therefore, is a group of closely related species or a group of species that have descended from a common ancestral group (or species). Because morphological and physiological features are, in part, the result of gene action, more identical genes should be shared by members of a given genus than by members of different genera. In general, members of the various species of a given genus have more morphological and functional features in common than they have in common with species of a related genus. For example, the domestic dog together with wolves and jackals make up the genus *Canis*. When referring to the dog, the trivial name is added—*Canis familiaris*; the wolf, a close relative, is *Canis lupus*. The name of a species is always a binomen and consists of the genus and the trivial name. This system is not unlike our usage of given names and surnames, except that the order is reversed.

In a similar way, a **family** is a group of related genera; an **order** is a group of related families; a **class** is a group of related orders; and a **phylum** is a group of related classes. Related phyla are grouped as a **kingdom**.

These various taxonomic categories traditionally have been arranged in a branching hierarchical order that expresses the various levels of genetic kinship. The sequence from top to bottom indicates decreasing scope or inclusiveness of the various levels. For example:

Kingdom — Animalia
  Phylum — Chordata
    Class — Mammalia
      Order — Carnivora
        Family — Felidae
          Genus — *Puma*
            species — *Puma concolor*

Our present classification scheme has been devised by using the genus and trivial name as a base and then grouping them in a hierarchical system. For example, dogs (*Canis familiaris*) are related in a single genus, and these in turn are related to foxes (*Vulpes, Urocyon*); and all of these are united in one family, Canidae. This group is somewhat more distantly related to the cats, bears, and other flesh-eaters; and all these forms are united in an order, the Carnivora. This order shares many features such as mammary glands and hair, with forms as diverse as bats and whales, and all are grouped in one class, the Mammalia. In turn, mammals have numerous characteristics such as an internal skeleton and a dorsal hollow nerve cord that are also present in fishes, amphibians, and reptiles; thus, all are grouped in one of the major subdivisions of the animal kingdom, the phylum Chordata.

These seven categories are considered essential to defining the relationships of a given organism. Often, however, taxonomists find it necessary because of great variation and large numbers of species to recognize intermediate, or extra, levels between these seven categories of the taxonomic hierarchy by adding the prefixes "super-," "infra-," and "sub-" to the names of the seven major categories just listed (see classifications in Appendix I).

The delineation of taxa higher than the species level is rather arbitrary: A taxonomist may divide a group of species into two genera if he or she is impressed by differences, or combine them into one genus if the similarities are emphasized. For example, some authorities have included the tiger and other large cats in the genus *Felis* with the small cats, whereas other authorities have segregated them as the separate genus *Panthera*.

With many different organisms being named by many different taxonomists throughout the world, biologists recognized the need for a set of rules governing scientific nomenclature. In 1895, the Third International Zoological Congress appointed a committee that drew up the Règles Internationales de la Nomenclature Zoologique (International Rules of Zoological Nomenclature) (Mayr et al., 1953). The Rules, which were adopted by the Fifth International Zoological Congress in 1901, became the universal Code of Zoological Nomenclature. The adoption of the Rules (Code) has helped to produce stability in nomenclature, and it has also helped to standardize certain taxonomic procedures. The Code established a permanent International Commission of Zoological Nomenclature that serves in a judiciary capacity to render decisions concerning difficult cases—"special cases" when the rules do not clearly solve a particular situation. It is vested with the power to interpret, amend, or suspend provisions of the Code. Some of the Code's basic rules include:

1. The generic or specific name applied to a given taxon is the one first published in a generally acceptable book or periodical and in which the name is associated with a recognizable description of the animal.
2. No two genera of animals can have the same name, and within a genus no two species can have the same name.
3. The species name of an animal consists of the generic name plus the trivial name.
4. Names must be either Latin or Latinized and are italicized.
5. The name of a genus must be a single word and must begin with a capital letter, while the specific, or trivial, name must be a single or compound word beginning with a lower case letter.
6. The name of a higher category (family, order, class, etc.) begins with a capital letter, but is not italicized.
7. No names for animals are recognized that were published prior to 1758, the year of publication of the *Systema Naturae*, tenth edition.
8. The name of a family is formed by adding -idae to the stem of the name of one of the genera in the group. This genus is considered the *type genus* of the family.

A complete revision of the Rules was authorized at the International Zoological Congress held in Paris in 1948. All interpretations of the Rules made since 1901 were incorporated into the Revised Rules. The code was rewritten in 1958, as the International Code of Zoological Nomenclature. The fourth and latest edition was published in 1999 (Pennisi, 2000).

METHODS OF CLASSIFICATION

Several methods of grouping organisms together in a hierarchical system of classification have been used during the past 2,300 years. These include Aristotelean essentialism, as well as evolutionary, phenetic, and phylogenetic (cladistic) methods of classification. The latter two methods "can be viewed as late-coming developments that at least partly represent reactions against evolutionary systematics" (Eldredge and Cracraft, 1980).

A **taxon** is a taxonomic group of any rank that is sufficiently distinct to be worthy of being assigned to a definite category. Taxa are often subject to the judgment of the taxonomist. The relationship of taxa may be expressed in one of the following forms: monophyly, paraphyly, or polyphyly. A taxon is **monophyletic** (Fig. 2.2a) if it contains the most recent common ancestor of the group and all of its descendants. It is **paraphyletic** (Fig. 2.2b) if it contains the most recent common ancestor of all members of the group but excludes some descendants of that ancestor. A taxon is **polyphyletic** (Fig. 2.2c) if it does not contain the most recent common ancestor of all members of the group, implying that it has multiple evolutionary origins. Both evolutionary and cladistic taxonomy accept monophyletic groups and reject polyphyletic groups in their classifications. They differ on the acceptance of paraphyletic groups, a difference that has important evolutionary implications.

Aristotelean Essentialism

Pre-Darwinian systems of classification were arbitrarily based on only one or a few convenient (i.e., essential) morphological characters. Aristotle (384–322 B.C.) did not propose a formal classification of animals, but he provided the basis for such a classification by stating that "animals may be characterized according to their way of living, their actions, their habits, and their bodily parts." In other words, animals could be characterized based on the degree of similarity of shared "essential" traits (e.g., birds have feathers, mammals have hair) of those animals. "According to Aristotle, all nature can be subdivided into natural kinds that are, with appropriate

FIGURE 2.2

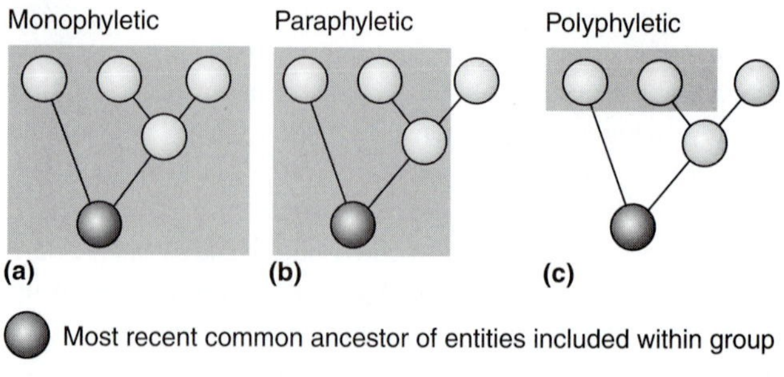

Biological taxonomists currently distinguish among three classes of taxa: monophyletic, paraphyletic, and polyphyletic. (*a*) A monophyletic group includes a common ancestor and all of its descendants. (*b*) A paraphyletic group includes a common ancestor and some but not all of its descendants. (*c*) A polyphyletic group is a group in which the most recent common ancestor of the entities included within the group is not itself included within the group.
Source: deQueiroz, "Systematics and the Darwinian Revolution" in Philosophy of Science, *Vol. 55, 1988.*

provisions, eternal, immutable, and discrete. For example, living organisms are of two sorts—plants and animals.... He subdivided animals into those that have red blood and give birth to their young alive and those that do not. He further subdivided each of these groups until finally he reached the lowest level of the hierarchy—the species" (Hull, 1988).

Aristotle's "classification" is known as the "A and not-A" system of classification:

- Animals with blood (the "A" group)
  - Viviparous quadrupeds
  - Oviparous quadrupeds
  - Fishes
  - Birds
- Animals without blood (the "not-A" group)
  - Mollusks
  - Crustaceans
  - Testaceans
  - Insects

Since Aristotle, philosophers have divided organisms into animals (sensible, motile) and plants (insensible, nonmotile)—a perfect example of "A and not-A" groups. Pliny (A.D. 23–79) divided animals into Aquatilia, Terrestria, and Volatilia based on their habitat. The classification of Linnaeus was similar. The class Worms (Vermes) of Linnaeus was reserved for those animals lacking both skeletons and articulated legs.

Evolutionary (Classical or Traditional) Classification

In the years following Darwin's *Origin of Species* (1859), the theory of evolution replaced the concept of special creation in the scientific community. It was found that living species are not fixed and unchanging, but had evolved from preexisting species during geological time. In other words, organisms in a "natural" systematic category shared characteristics because they were descendants of a common ancestor. The more recent the divergence from a common ancestor, the more characteristics two groups would normally share. It is now considered that, in general, similarities in structure are evidence of evolutionary relationships. This is because similarities in structure are caused by similar genetic material. Organisms that share the greatest number of similar characteristics are assumed to be most closely related to one another and are grouped together. A certain degree of subjectivity is present in this system; therefore, experience and judgment on the part of the taxonomist is important.

Phenetic (Numerical) Classification

Phenetics strives to reduce the degree of subjectivity used in the development of the classification. Phenetic systematists argue that organisms should be classified according to their overall similarity (phenotypic characters). In the 1950s and 1960s, pheneticists (see Sneath and Sokal, 1973) argued that a classification scheme would be most informative if it were based on the overall similarity among species, measured by as many characteristics as possible, even if such a classification did not exactly reflect common ancestry. Their main concern was "a desire to reformulate the process of delineating life's orderliness in a more standardized, repeatable, rigorous, and objective fashion" (Sokal and Sneath, 1963).

As many anatomical and physiological characteristics as possible are examined, with each character being given equal weight. Each character in each species is assigned a number, and all the numbers are entered into a computer. The computer then groups the organisms into clusters based on similarity. There is no attempt to infer phylogeny from the result. It is believed that basing classification on similarity results in a stable and convenient classification. Those organisms that share the greatest number of similar characteristics are assumed to be most closely related to one another. The same characters are compared among taxa, which then are clustered in a hierarchical arrangement on the basis of percentage of shared similarities. Because evolution produces both adaptive radiation and convergent evolution, it is often difficult to distinguish closely related organisms from those that are not closely related but look alike because they have adapted to similar niches. Therefore, classifications that rely exclusively on structural similarities do not always reflect evolutionary history. This system of classification has useful applications at lower taxonomic levels, but it is not as reliable in classifications above the level of species or genus. For instance, pheneticists have developed elaborate numerical methods for grouping species on the bases of overall similarity and portraying this similarity as a **phenogram** (dendrogram), which is almost always generated by computer. A **phenon** is thus a taxonomic unit of a phenogram; "species" do not exist.

Numerical taxonomy expanded rapidly in the early 1960s, but its influence in biological classification then waned (Eldredge and Cracraft, 1980). However, with the development of "molecular taxonomy" and the molecular sequencing of genes, phenetic techniques have been revived. Each amino acid in a protein or each nucleotide in a gene is treated as a "trait," with the potential number of traits within one gene running into the millions (see discussion, pp. 38–39).

Cladistic (Phylogenetic) Classification

In 1950, the German entomologist Willi Hennig proposed a systematic approach emphasizing common descent based on the cladogram of the group being classified. This approach, cladistic analysis, is a systematic method that focuses on shared, derived characters. Derived traits are new characteristics that appear as a new species arises from its ancestor, and hence they represent recent rather than ancient adaptations. Cladistics holds that a classification should express the branching (cladistic) relationships among species, regardless of their degree of morphological similarity or difference.

Cladistics aims specifically to create taxonomic groupings that more accurately reflect organisms' evolutionary histories (de Queiroz, 1988; de Queiroz and Gauthier, 1992). It recognizes only monophyletic taxa (all taxa evolved from a single parent stock) that include all the descendants from a single ancestral group. Cladists feel that their methods allow for better analyses and testing than those of earlier systematists. Shared characteristics are separated into three clearly defined groups: those shared by living organisms because they have evolved from recent common ancestors, called shared derived characters or **synapomorphies** (Gr. *synapsis*, joining together + *apo*, away + *morphe*, form); primitive traits inherited from an ancestor, called **plesiomorphies** (Gr. *plesi*, near); and primitive traits shared by larger groups of organisms because they have been inherited from an ancient common ancestor that had them, which are known as **symplesiomorphies** (Gr. *synapsis*, joining together + *plesio*, near + *morphe*, form).

A character state present in all members of a group is **ancestral** for the group as a whole. Those characters that have newly evolved from the ancestral state, are shared by a more limited set of taxa, and therefore define related subsets of the total set are known as **derived characters.** The organisms or species that share derived character states, called **clades** (Gr. *klados*, branch), form subsets within the overall group. Relationships among species are portrayed in a **cladogram** (Figs. 2.3 and 2.4, and Bio-Note 2.1). A cladogram is an evolutionary diagram that depicts a sequence in the origin of uniquely derived characteristics: traits that are found in *all* of the members of the clade and *not* in any others. It therefore represents the sequence of origin of new groups of organisms. Although its branching pattern is somewhat similar to that of a phylogenetic tree, a cladogram is different because it does not incorporate information on the time of origin of new groups nor how different closely related groups are. A cladogram is not based on overall similarity of species, and so it may differ substantially from a phenogram.

A cladogram uses a method known as outgroup comparison to examine a variable character. A group of organisms that is phylogenetically close but not within the group being studied is included in the cladogram and is known as the **outgroup**. Any character state found both within the outgroup and in the group being studied is considered to be ancestral for the study group. For example, if the study group consisted of four vertebrates (frog, snake, fox, and antelope), *Amphioxus* could serve as the outgroup. In this example, characters such as vertebrae and jaws are common only to the study group and are not found in the outgroup.

Species within a single genus resemble each other because they share a recent common ancestor. Similarly, members of a family represent a larger evolutionary lineage descended from common stock in the more remote past. Because cladistic classifications are based on shared derived character states, they may radically regroup some well-recognized taxa. Furthermore, because a cladogram is based on monophyletic taxa, each group that arises from a particular branch point along a cladogram is related through the characters that define that branch point. A group of organisms most closely related to the study taxon is known as a **sister group**. Traditional evolutionary taxonomy using such characteristics as scales, feathers, and hair is compared with a cladistic classification linking the same organisms through shared characteristics in Fig. 2.4.

Phenetic approaches focus on degrees of difference, whereas cladists concentrate on specific differences or character states (derived traits). Each synapomorphic trait is given equal weight, with the number of trait differences between each pair of organisms being used to create the simplest branching diagram.

To represent the phylogeny of vertebrates in a cladistic classification, animals are arranged on the basis of their historical divergences from a common ancestral species. Animals with similar derived characters are considered more closely related than animals that do not share the characters. The results of such an analysis should produce a cladogram that approximates the phylogeny of the animals considered. Unfortunately, problems arise in actual practice. Evolution may not always occur by what appears to be the simplest route. As in all forms of systematics, similarities and differences such as convergent evolution (the evolution of similar adaptations in unrelated organisms to similar environmental challenges), loss or reversal of characters, and parallelism (evolution of similar structures in related [derived] organisms) can be misinterpreted easily. The greatest problem in creating groupings is the difficulty of determining which character states are primitive and which are derived.

A major difference between evolutionary and phylogenetic systematics is seen, for example, in the classification of reptiles and birds (Fig. 2.5). The tuatara, lizards, snakes,

FIGURE 2.3

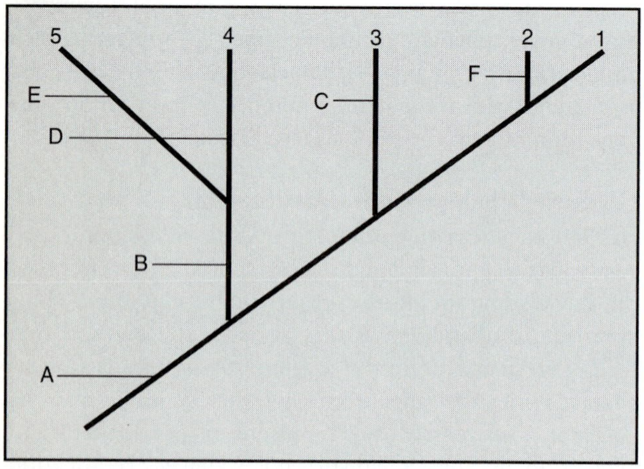

This hypothetical cladogram shows five taxa (1–5) and the characters (A–F) used in deriving the taxonomic relationships. Character A is a symplesiomorphy (shared, ancestral characteristic) because it is shared by members of all five taxa. Because it is present in all taxa, character A cannot be used to distinguish members of this monophyletic lineage from each other. Character B is a synapomorphy (derived, ancestral character) because it is present in taxa 4 and 5 and can be used to distinguish these taxa from 1–3. Character B, however, is on the common branch giving rise to taxa 4 and 5. Character B is, therefore, symplesiomorphic for those two taxa. Characters D and E are derived traits and can be used to distinguish members of taxa 4 and 5.

FIGURE 2.4

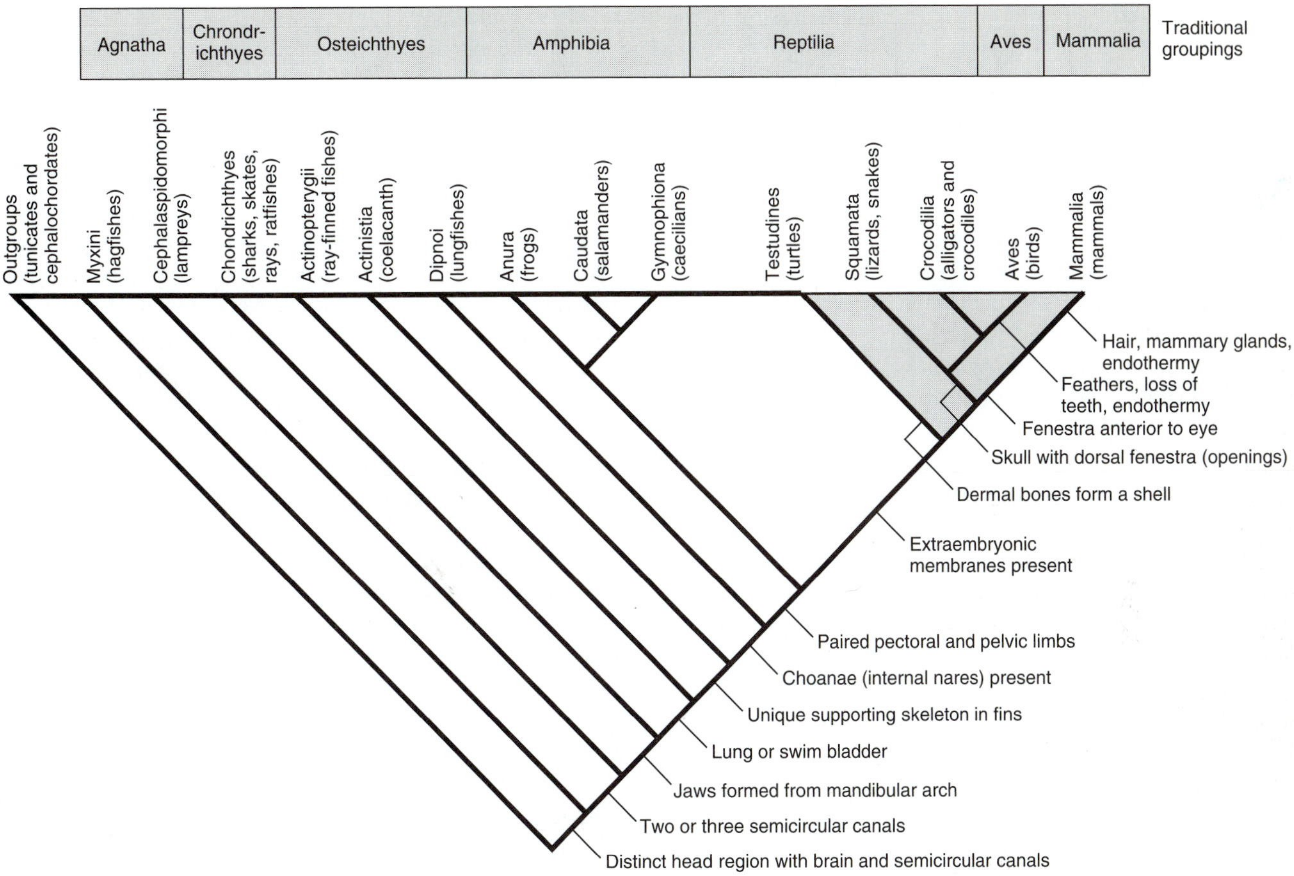

A cladogram is constructed by identifying the point or node at which two groups diverged. Animals that share a branching point are included in the same taxon. Time scales are not given or implied, and the relative abundances of taxa are not shown. This diagram of extant (living) vertebrates shows birds and crocodilians sharing a common branch, indicating that these two groups share many common characters and are more closely related to each other than either is to any other group of extant animals.

FIGURE 2.5

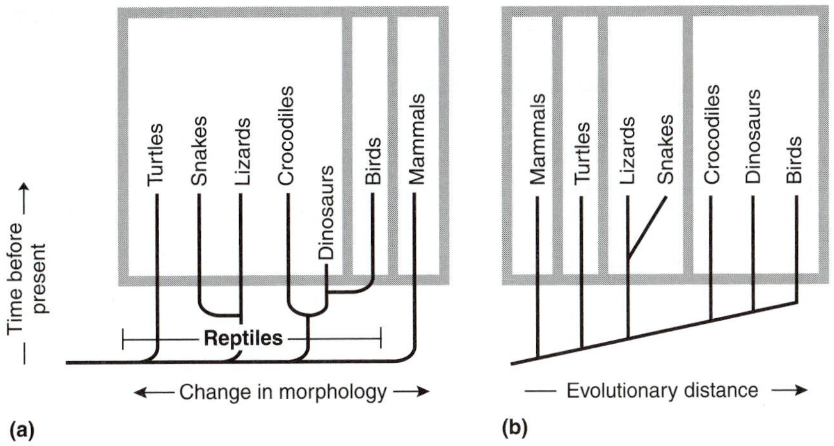

Comparison of evolutionary and cladistic systematics among the amniotes. (*a*) In evolutionary taxonomy, traditional key characteristics such as scales for reptiles, feathers for birds, or fur for mammals are used to differentiate the groups. (*b*) A cladistic classification links organisms with uniquely derived characters and shared ancestries.

crocodilians, and birds all possess a skull with two pairs of depressions in the temporal region (diapsid condition). Phylogenetic systematists (cladists) place all of these forms in one monophyletic group (Diapsida). When this group is subdivided, the birds and crocodilians (Archosauromorpha) and the tuatara, snakes, and lizards (Lepidosauromorpha) are placed in a separate taxonomic rank. Evolutionary systematists, on the other hand, place crocodilians, tuataras, lizards, snakes, and turtles (which are anapsids) in the class Reptilia and birds in a separate class (Aves). Evolutionary systematists attribute great significance to such "key characteristics" in birds as the presence of feathers and endothermy, and they group the diapsid crocodilians and squamates with the turtles, which are morphologically distinct, because they share many primitive characters. Cladists, however, make the point that the use of "key characteristics" involves value judgments by systematists that cannot be tested scientifically.

"Traditional evolutionary" systematists are attempting to achieve the same goal as "phylogenetic" systematists: the accurate interpretation of the pattern of evolutionary descent of specific groups of organisms, such as vertebrates. Thus, both current approaches are phylogenetic and evolutionary. While the goal is the same, the methods differ. Each method has its proponents and its critics. Some have even attempted to combine the best features of both evolutionary and cladistic methods. Wiley (1981) summarized the principles of cladistics, and Cracraft (1983) described the use of cladistic classifications in studying evolution. Additional information concerning phylogenetic systematics can be found in Eldredge and Cracraft (1980), Nelson and Platnick (1981), Halstead (1982), Nelson (1984), Ghiselin (1984), Abbott et al. (1985), and Hull (1988).

To the extent possible, classifications in this text will use monophyletic taxa that are consistent with the criteria of both evolutionary and cladistic taxonomy. Complete revision of vertebrate taxonomy utilizing cladistic criteria would result in vast changes, including the probable abandonment of Linnaean ranks. In many cases, classifications based strictly on cladistics would require numerous taxonomic levels and be too complex for convenience (Fig. 2.6). A separate category must be created for every branch derived from every node in the tree. Not only must many new taxonomic categories be employed, but older ones must be used in unfamiliar ways. For example, in cladistic usage, "reptiles" include birds with traditional reptiles (turtles, lizards, snakes, crocodilians) but exclude some fossil forms, such as the mammal-like reptiles, that have traditionally been classified in the Reptilia.

Some cladistic classifications require compromises. For example, a cladogram showing the evolutionary history of the tuna, lungfish, and pig requires that the lungfish and pig be placed in a group separate from the tuna (Fig. 2.7). The lungfish is obviously a fish, but the pig and all mammals (including humans) have shared a common ancestor with it more recently than its common ancestor with the tuna.

Cladograms for each class of vertebrates are given in Chapters 4–6, 7, and 9.

FIGURE 2.6

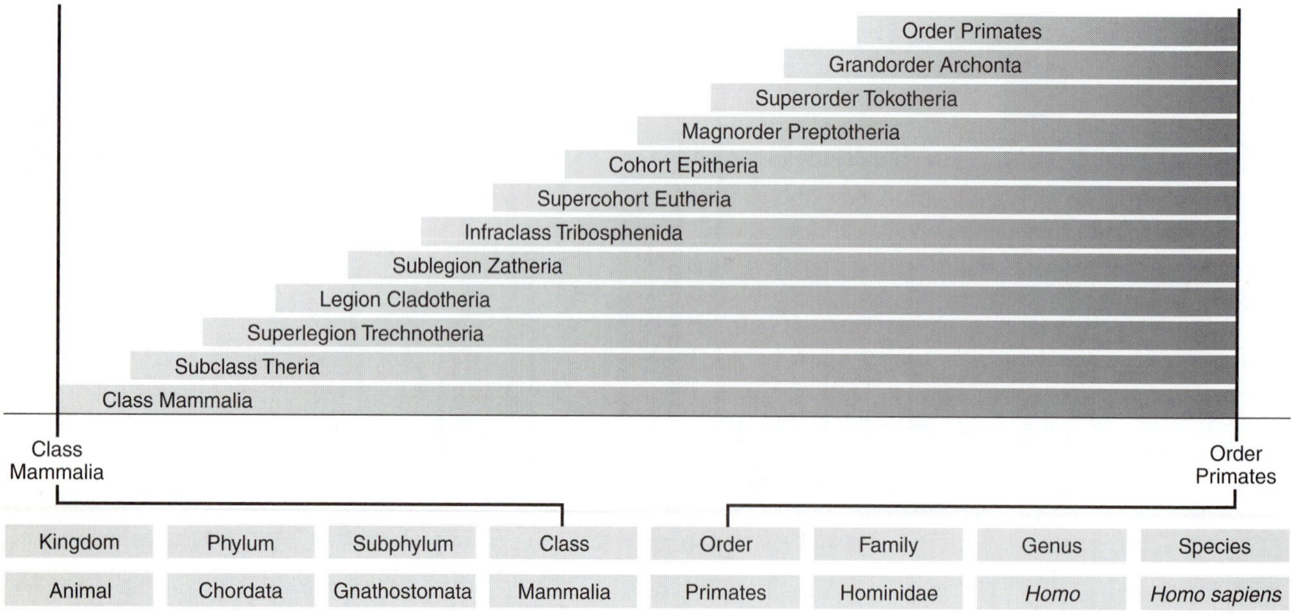

A classification of the primates based on cladistics. Bottom: The major taxonomic categories as they are used in the classification of humans without regard to cladistics.

FIGURE 2.7

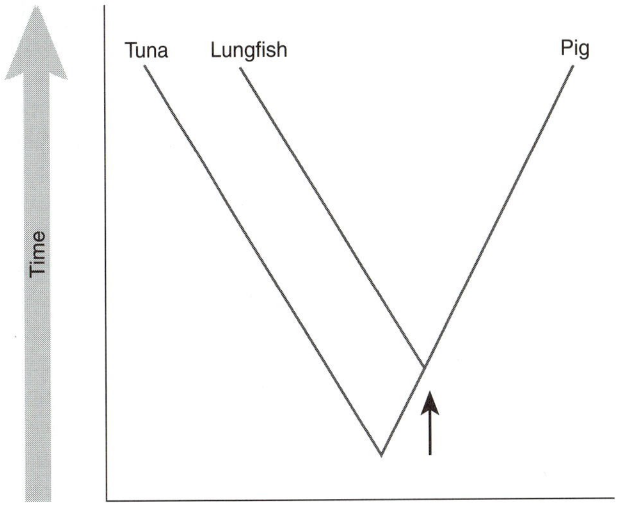

Cladogram showing the evolutionary relationship between the tuna, lungfish, and pig. It is traditional to classify the tuna and lungfish together in the class Osteichthyes (bony fishes) and to classify the pig in the class Mammalia (mammals). However, this violates the basic rule of cladistics: all members of a taxonomic group must have shared a common ancestor with each other more recently than they have with members of any other group. The lungfish, which possesses internal nostrils and an epiglottis, is descended from an ancestor (*arrow*) that is also the ancestor of all land-living vertebrates (including humans).
Source: John Kimball, Biology, *6th edition, 1994, McGraw-Hill Company, Inc.*

BIO-NOTE 2.1

Constructing a Cladogram

The first step in constructing a cladogram is to summarize the characters of the taxa being compared. Knowledge of the organisms is essential for choosing the characters for analysis, since a cladogram is constructed on the basis of unique derived characters. In the following example (Fig. 2.8), the study group consists of four vertebrates: brook trout, tiger salamander, giraffe, and gray squirrel. The lancelet is included as the **outgroup**, a taxon outside of the study group but consisting of one or more of the study group's closest and more primitive relatives. Any character found in both the outgroup and the study group is considered to be primitive, or **plesiomorphic** (ancestral), for the study group. Traits that are common to some, but not all, of the species in the study group are used to construct the simplest and most direct (parsimonious) branching diagram.

This cladogram consists of three clades, with each clade consisting of all the species descended from a common ancestor. Clades differ in size because the first clade (vertebrae and jaws) includes the other two, and the second clade (four legs, lungs) includes the third clade, which contains the giraffe and squirrel.

All of the study groups belong to the first clade, because they all possess vertebrae and jaws. The tiger salamander, giraffe, and gray squirrel are in the clade that

Continued on page 32

FIGURE 2.8

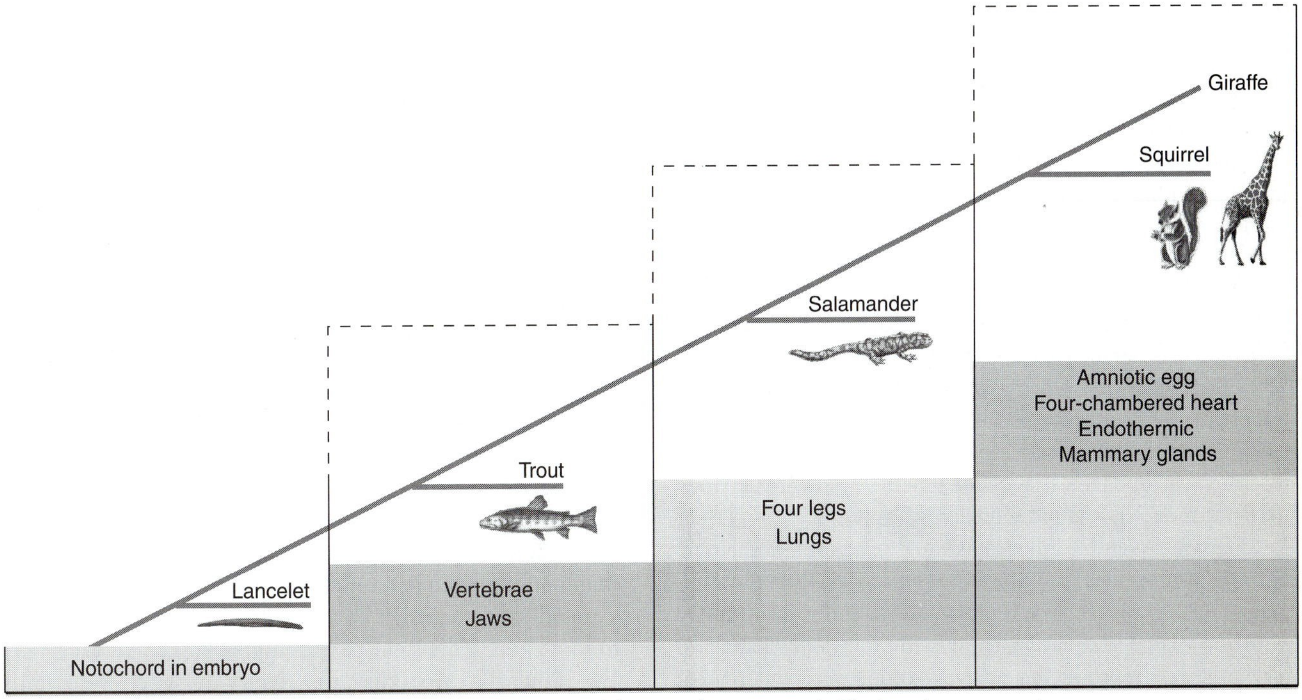

Construction of a cladogram involving four vertebrates: a fish, an amphibian, and two mammals. The lancelet serves as the outgroup.

Continued from page 31

has lungs and four legs. Only the giraffe and gray squirrel (of the animals considered here) have a four-chambered heart, are endothermic, and have an embryo surrounded by an amnion.

EVOLUTION

Evolution is the underlying principle of biology. The modern theory of evolution includes two basic concepts: first, that the characteristics of living things change with time; second, that the change is directed by natural selection. **Natural selection** is the nonrandom reproduction of organisms in a population that results in the survival of those best adapted to their environment and elimination of those less well adapted. If the variation is heritable, natural selection leads to evolutionary change. The change referred to here is not change in an individual during its lifetime, but change in the characteristics of populations over the course of many generations. An individual cannot evolve, but a population can. The genetic makeup of an individual is set from the moment of conception; in populations, though, both the genetic makeup and the expression of the developmental potential can change. Natural selection is thoroughly opportunistic. A population responds to a new environmental challenge by appropriate adaptations or becomes extinct. The fossil record bears witness that a majority of the species living in the past eventually became extinct. The organisms likely to leave more descendants are those whose variations are most advantageous as adaptations to the environment. Natural selection occurs in reference to the environment where the population presently lives; evolutionary adaptations are not anticipatory of the future. The change in the genetic makeup of a population over successive generations is **evolution**.

A population is made up of a large number of individuals that share some important features but differ from one another in numerous ways, some rather obvious, some very subtle. In human beings, for example, we are well aware of the uniqueness of the individual, for we are accustomed to recognizing different individuals on sight, and we know from experience that each person has distinctive anatomical and physiological characteristics as well as distinctive abilities and behavioral traits. It follows that if there is selection against certain variants within a population and selection for other variants within it, the overall makeup of that population may change with time, since its characteristics at any given time are determined by the individuals within it.

Darwin recognized that in nature the majority of the offspring of any species die before they reproduce. If survival of the young organisms were totally random and if every individual in a large population had exactly the same chance of surviving and reproducing as every other individual, then there would probably be no significant evolutionary change in the population. But survival and reproduction are never totally random. Some individuals are born with such gross defects that they stand almost no chance of surviving to reproduce. In addition, differences in the ability to escape predators to obtain nutrients, to withstand the rigors of the climate, to find a mate, and so forth ensure that survival will not be totally random. The individuals with characteristics that weaken their capacity to escape predators, to obtain nutrients, to withstand the rigors of the climate, and the like will have a poorer chance of surviving and reproducing than individuals with characteristics enhancing these capabilities. In each generation, therefore, a slightly higher percentage of the well-adapted individuals will leave progeny. If the characteristics are inherited, those favorable to survival will slowly become more common as the generations pass, and those unfavorable to survival will become less common. Given enough time, these slow shifts can produce major evolutionary changes.

Thus, Darwin's explanation of evolutionary change in terms of natural selection depends on five basic assumptions:

1. Many more individuals are born in each generation than will survive and reproduce.
2. There is variation among individuals; they are not identical in all their characteristics.
3. Individuals with certain characteristics have a better chance of surviving and reproducing than individuals with other characteristics.
4. At least some of the characteristics resulting in differential reproduction are heritable.
5. Enormous spans of time are available for slow, gradual change.

Natural selection is a creative process that generates novel features from the small individual variations that occur among organisms within a population. It is the process whereby organisms adapt to the demands of their environment. Over many generations, favorable new traits will spread through the population. Accumulation of such changes leads, over long periods of time, to the production of new organismal features and new species.

Species and Speciation

Speciation, the process by which new species of organisms evolve in nature from an ancestral species, is generally considered to be a population phenomenon. A small local population, such as all the perch in a given pond or all the deer mice in a certain woodlot, is known as a **deme**. Although no two individuals in a deme are exactly alike, the members of a deme do usually resemble one another more closely than they resemble the members of other demes for two reasons: (1) they are more closely related genetically because pairings occur more frequently between members of the same deme than between members of different demes; and (2) they are exposed to more similar environmental influences and hence to more nearly the same selection pressures.

It must be emphasized that demes are not clear-cut units of population. Although the deer mice in one woodlot are

BIO-NOTE 2.2

High-Speed Evolution

In certain situations, evolution may proceed at a rapid rate. For example, in Trinidad's Aripo River, a species of cichlid fish feeds primarily on relatively large sexually mature guppies (*Poecilia reticulata*); in nearby tributaries, killifish prefer tender young fish. In response to these different pressures, the guppies have evolved two different life-history strategies. Those in the Aripo River reach sexual maturity at an early age and bear many young, while the guppies in the tributaries bear fewer young as well as delayed sexual maturity. By transplanting guppies from the Aripo River to a tributary that happened to be empty of guppies and where killifish were the only predators, researchers were able to prove that predation caused this pattern. Within 4 years, transplanted male guppies were already detectably larger and older at maturity when compared with the control population; they had switched strategies, delaying their sexual maturity and living longer. Seven years later, females were also noticeably larger and older. Some of these adaptations occurred in just 4 years—a rate of evolutionary change some 10,000 to 10 million times faster than the average rates determined from the fossil record.

In another study, small populations of the brown anole (*Anolis sagrei*) were transplanted from Staniel Cay in the Bahamas to several nearby islands in 1977. Staniel Cay has scrubby to moderately tall forests, whereas the experimental islands have few trees and are mostly covered by vegetation with narrow stems. Within a 10- to 14-year period, the displaced lizards were found to have shorter rear legs than their ancestors, an apparent adaptation to the bushy vegetation that dominated their new island. Whereas species living on tree trunks have longer legs for increased speed, shorter legs provide increased agility for species living on bushy vegetation. The more different the recipient island's vegetation from that of Staniel Cay, the greater the magnitude of adaptation. Such changes could in time turn each island's population into a separate species.

The house sparrow (*Passer domesticus*) was introduced into North America from western Europe during the period 1852–1860. Studies of color and of 16 skeletal characters from 1,752 specimens from 33 localities taken between 1962 and 1967 throughout North America revealed color and size differentiation in all 16 characters. This adaptive radiation occurred in just 50 to 115 generations.

Geographic variation in the house sparrows was most pronounced in color. In many cases, the color differences were both marked and consistent, permitting specimens from several localities to be consistently identified solely on the basis of color. One measurable component, gross size, showed strong inverse relationships with measures of winter temperature. This adaptation (larger body size in colder regions) is consistent with the ecogeographic rule of Bergmann. The adaptive variation found in limb size (shorter limb size in colder regions) was consistent with Allen's Rule. These latter adaptations are designed to conserve heat in colder climates and radiate heat in warmer regions. Since sparrows did not reach Mexico City until 1933, Death Valley before 1914, or Vancouver before 1900, the data suggest that racial differentiation in house sparrow populations may require no more than 50 years.

Reznick et al., 1997
Losos et al., 1997
Morell, 1997b
Case, 1997
Johnston and Selander, 1964, 1971

more likely to mate among themselves than with deer mice in the next woodlot down the road, there will almost certainly be occasional matings between mice from different woodlots. And the woodlots themselves are not permanent ecological features. They have only a transient existence as separate and distinct ecological units: Neighboring woodlots may fuse after a few years, or a single large woodlot may become divided into two or more separate smaller ones. Such changes in ecological features will produce corresponding changes in the demes of deer mice. Demes, then, are usually temporary units of population that intergrade with other similar units.

The deme is the ultimate systematic unit of species in nature. In some cases, a deme may correspond to a subspecies, but it is almost always a decidedly smaller group. Demes do not enter into classification, because they do not have long-continuing evolutionary roles and because adjacent demes often have no observable differentiation.

Demes often differ from one another in a geographic series of gradual changes. A gradual geographic shift in any one genetically controlled trait is known as a character **cline** (Fig. 2.9). A series of samples from along a cline reveals a gradual shift in a particular character such as body size, tail length, number of scales, or even intensity of coloration. Because such situations add to the difficulty of deciding the true phylogenetic relationships of populations, the experience and judgment of the systematist play an important role.

Intergradation occurs between "similar" demes. Some interbreeding can be expected between deer mice from adjacent demes, but we do not expect interbreeding between deer mice and house mice or between deer mice and gray squirrels. We recognize the existence of units of population larger than demes that are more distinct from each other and longer lasting than demes. One such unit of population is that containing all the demes of deer mice. We call these larger units species. A species is a genetically distinctive group of natural populations (demes) that share a common gene pool and that are reproductively isolated from all such groups. In other words, a species is the largest unit of population within which

FIGURE 2.9

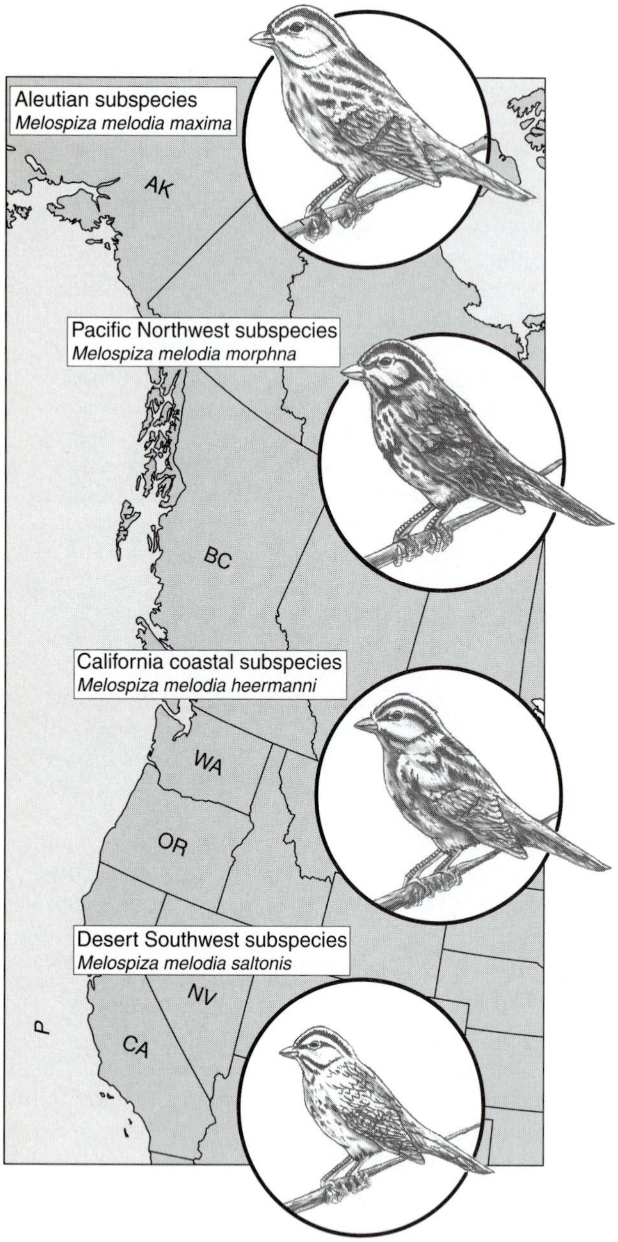

On the West coast of North America, subspecies of the song sparrow form a cline in body size, plumage coloration, and song characteristics. There is a dramatic difference in appearance between the small, pale *Melospiza melodia saltonis* subspecies of the southwestern desert region and the large, dark Aleutian subspecies, *M. m. maxima*. It seems unlikely that the desert-dwelling subspecies *saltonis* would easily breed with the large Alaskan subspecies *maxima*, even if the ranges of the two subspecies were to overlap in the future. If some catastrophe completely eliminated the central West coast populations of the song sparrow, the northern and southern ends of the cline would likely become two distinct species of sparrow.

effective **gene flow** (exchange of genetic material) occurs or can occur.

According to Mayr's biological species concept, a species consists of a collection of interbreeding individuals. The biological species concept is based on reproductive isolation—related species may occasionally hybridize but cannot and do not reproduce regularly.

Sometimes, there may be a rather abrupt shift in some character in a particular part of the species' range. When such an abrupt shift in a genetically determined character occurs in a geographically variable species, some biologists designate the populations on the two sides of the step as **subspecies** or **races**.

Subspecies (and species) were formerly described solely on the basis of their morphological characteristics—primarily differences in external and skeletal form and appearance. Whitaker (1970) proposed a biological subspecies concept to replace the morphological subspecies concept. Whitaker proposed that subspecies should be restricted to situations wherein populations or groups of populations appear to be on their way toward becoming new species.

The initial step in speciation is the introduction of a **primary isolating barrier** that tends to prevent or reduce the opportunity for interbreeding between closely related species. Physical barriers such as mountain ranges, deserts, bodies of water, canyons, and differences in physical size would be examples of primary isolating barriers.

Allopatric populations, those that live in different regions, may have no common borders between their distributional areas; such isolated populations have no natural means of gene exchange. This **geographic isolation** serves as a primary isolating barrier. Elevation often serves as a geographic barrier, particularly in the tropics. Many of the isolated mountains in the southwestern United States exist as cool, moist forests in a sea of hot, dry desert. The small mammals, reptiles, and amphibians currently inhabiting these forests traveled there when the mountains were connected by corridors of suitable forested habitat during the Pleistocene. Today, these animals exist as isolated populations.

Islands also provide geographic isolation for many species. The California Channel Island fox (*Urocyon littoralis*) is a dwarf island species found only on six of the Channel Islands off the coast of California (Gilbert et al., 1990). It is thought that the foxes probably dispersed to the northern islands by swimming through the water from the mainland more than 16,500 years ago, when the distance would have been approximately 6.5 km. Populations probably were transported to the two southern islands by native Americans not more than 10,000 years ago. These are now genetically isolated populations that are evolving independently.

When a primary isolating barrier is in place, the development of **secondary isolating barriers** can occur. These are mechanisms that, over time, prohibit interbreeding even if the primary isolating barrier is removed (Fig. 2.10). The advertisement calls of most male anurans are excellent exam-

FIGURE 2.10

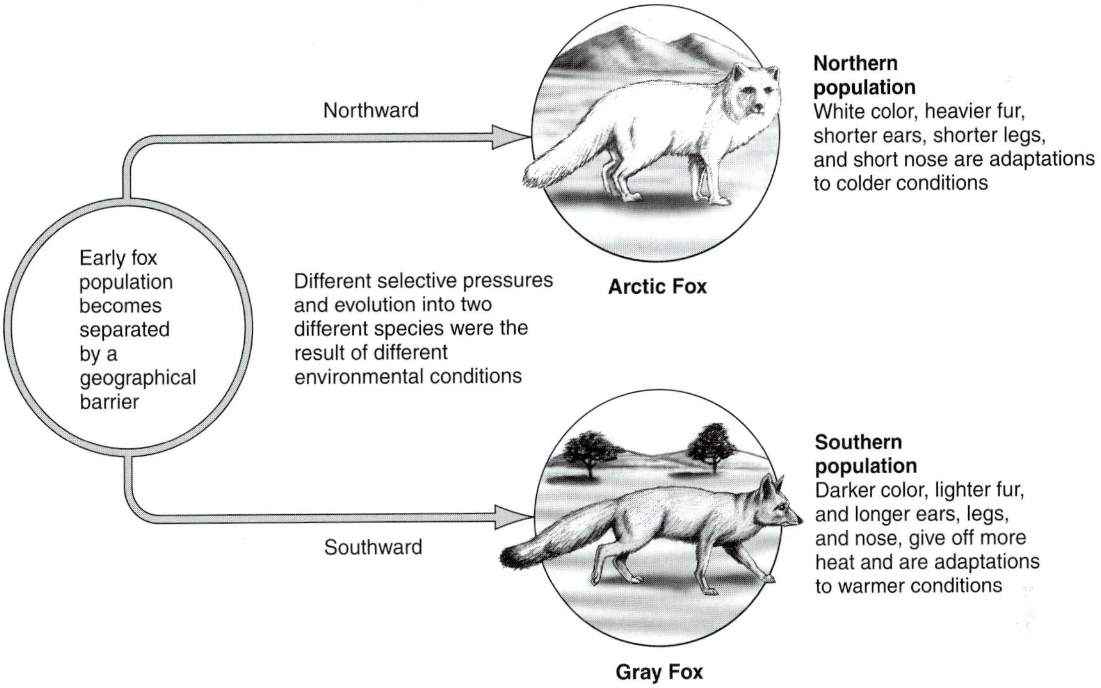
Northward
Northern population
White color, heavier fur, shorter ears, shorter legs, and short nose are adaptations to colder conditions
Early fox population becomes separated by a geographical barrier
Different selective pressures and evolution into two different species were the result of different environmental conditions
Arctic Fox
Southern population
Darker color, lighter fur, and longer ears, legs, and nose, give off more heat and are adaptations to warmer conditions
Southward
Gray Fox

Speciation of an ancestral species of fox into two different species, resulting from migration of portions of the original fox population into geographically isolated areas with different climates.

ples of secondary reproductive isolating barriers. If the secondary isolating barriers have developed to the point that the two populations would be unable to breed even if the primary isolating barrier were removed and they were in contact with one another, then speciation has occurred.

Pre-mating or **prezygotic** barriers prevent successful fertilization. They may be **ecological** (differences in minor habitat requirements), **ethological** (differences in mating behavior), **morphological** (differences in structure that prevent mating), or **physiological** (different mating seasons, gamete incompatibility). **Post-mating** or **postzygotic** barriers operate after fertilization. They may be **physiological** (hybrid inviability, hybrid sterility, etc.) or **cytological** (differences in chromosome structure may prevent development of the fertilized egg). Even if successful mating occurs, offspring may be weak or abnormally developed, they may fail to reach sexual maturity, or they may be sterile. Any of these conditions would disrupt the continuity of genetic exchange between populations.

Thus, Whitaker's (1970) criteria for the recognition of subspecies are: (1) a primary isolating barrier is in place but secondary isolating barriers are not; and (2) the results of evolution can be observed as morphological (or other) variation between the mutually isolated populations.

BIO-NOTE 2.3

Songbird Divergence

Songbird species that have distinct eastern and western North American populations have long been thought to have diverged during the Late Pleistocene glaciations, approximately 100,000 years ago. However, analyses of mitochondrial DNA (mtDNA) of 35 pairs of songbirds indicate that while some diverged relatively recently (e.g., the timberline and Brewer's sparrows about 35,000 years ago), many diverged much earlier (Steller's jay and blue jay as long as 5 million years ago). The average mitochondrial DNA difference was 5.1 percent for all pairs of eastern and western songbirds, suggesting that they have been evolving separately for 2.5 million years.

Klicka and Zink, 1997

When segments of a population become isolated geographically, the two isolated segments of the population might well accumulate enough genetic differences (secondary isolating barrier) to prevent interbreeding and the exchange of genetic information even if the original barrier were removed. This is known as **allopatric speciation** and is the most common type of speciation among animals (Fig. 2.11). Allopatric speciation may occur when climatic or geological changes produce impenetrable barriers that separate a species into different populations. It also may occur when a small number of individuals either disperse or are transported to a new and distant environment. Known as the **founder effect,** speciation can proceed rapidly since only a portion of the original gene pool is normally present in the small, newly relocated population, and natural selection can work more quickly on smaller gene pools. The founder effect may allow the new population to establish itself in a much shorter period of time than might be expected; however, if the gene pool is so limited that the individuals cannot adapt to the new environment, the entire population may be lost through the process of natural selection.

Two major types of allopatry are recognized. **Contiguous allopatry** (Fig. 2.11a) occurs when the ranges of two populations meet and interdigitate (but do not overlap) without interbreeding between the two groups. The usual evidence of contiguous allopatry is the absence of morphologically distinct intermediate forms. Different ecological requirements for sandy and clay soils, for example, may account for interdigitation of the two populations/species where the respective environments meet. As long as interbreeding does not occur, each population is defined as a separate species.

Disjunct allopatry (Fig. 2.11b) occurs when two more or less morphologically differentiated populations are separated by a wide geographic gap where neither population occurs. Such instances are numerous, because geographic separation of populations is one important method by which speciation is initiated. Although such populations are prevented from interbreeding by one or more geographic barriers, the natural situation gives no clue as to whether they have developed reproductive isolating mechanisms. A fairly reliable decision could be reached if individuals of both populations are brought together under controlled conditions and given the opportunity to breed. For example, orangutangs (*Pongo pygmaeus*) from the island of Sumatra and the mainland appear to be distinct species, yet in captivity they interbreed and have healthy hybrid offspring. Zoos are now keeping the two groups separate and destroying any hybrids. Some species, however, refuse to mate in captivity. If successful mating results in inviable offspring because of genetic, ethological, physiological, or other differences, it demonstrates that the parents belong to different species. Little information of this sort is available for disjunct allopatric populations of vertebrates.

Once isolation occurs, each population (initially a subspecies) follows its own evolutionary course. If two populations are **sympatric**—that is, they exist in the same region, with either a broad or narrow zone of overlap—and do not interbreed, this demonstrates that they have evolved into two distinct species (Fig. 2.11c). By their coexistence, such populations show that they have developed effective secondary isolating barriers.

In some cases, speciation can proceed at rapid rates. Members of the freshwater family Cichlidae exhibit exceptional diversity wherever they occur in Africa, Madagascar, southern India, Sri Lanka, South and Central America, and the southwestern United States. By far, however, the most abundant diversity occurs in the East African lakes of Victoria, Malawi, and Tanganyika (Stiassny and Meyer, 1999). Lake Victoria formed between 250,000 and 750,000 years ago; it contains approximately 400 species of cichlids. Lake Malawi is about 4 million years old and contains 300 to 500 cichlid species. Lake Tanganyika is 9 to 12 million years old and has approximately 200 species.

Several factors have allowed cichlids to diversify and exploit a variety of habitats. Cichlids are the only freshwater fishes to possess a modified second set of jaws (remodeled gill arches) (Fig. 2.12a). The jaws in the mouth are used to suck, scrape, or bite off bits of food; the jaws in the throat are used to crush and macerate food. Both the jaws and the teeth (sharp pointed piercers and flat, molarlike crushers [Fig. 2.12b]) can change shape within the lifetime of a single animal. The unique jaws and teeth allow each species to specialize and occupy its own specific ecological niche (see discussion of scale-eating cichlids in Bio-Note 5.8) All cichlids care for their broods long after hatching; many are mouthbrooders and hold fertilized eggs or young in their mouths. Studies of mitochondrial DNA show that the cichlids in Lake Victoria are genetically very close to one another—far closer than to morphologically similar cichlids in the other two lakes (Fig. 2.13). Thus, almost identical evolutionary adaptations can and did evolve many times independently of one another (Stiassny and Meyer, 1999).

FIGURE 2.11

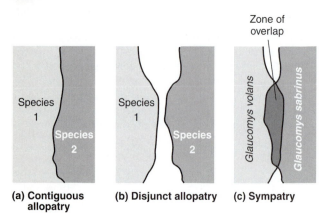

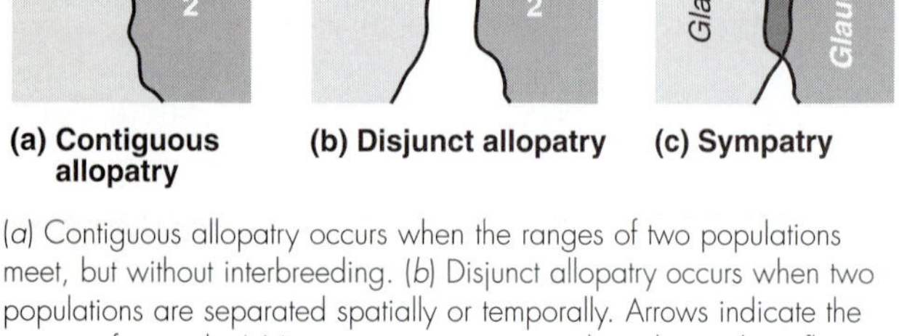

(*a*) Contiguous allopatry occurs when the ranges of two populations meet, but without interbreeding. (*b*) Disjunct allopatry occurs when two populations are separated spatially or temporally. Arrows indicate the ranges of animals. (*c*) In sympatric species, such as the southern flying squirrel (*Glaucomys volans*) and the northern flying squirrel (*Glaucomys sabrinus*), no interbreeding occurs even though ranges overlap.

FIGURE 2.12

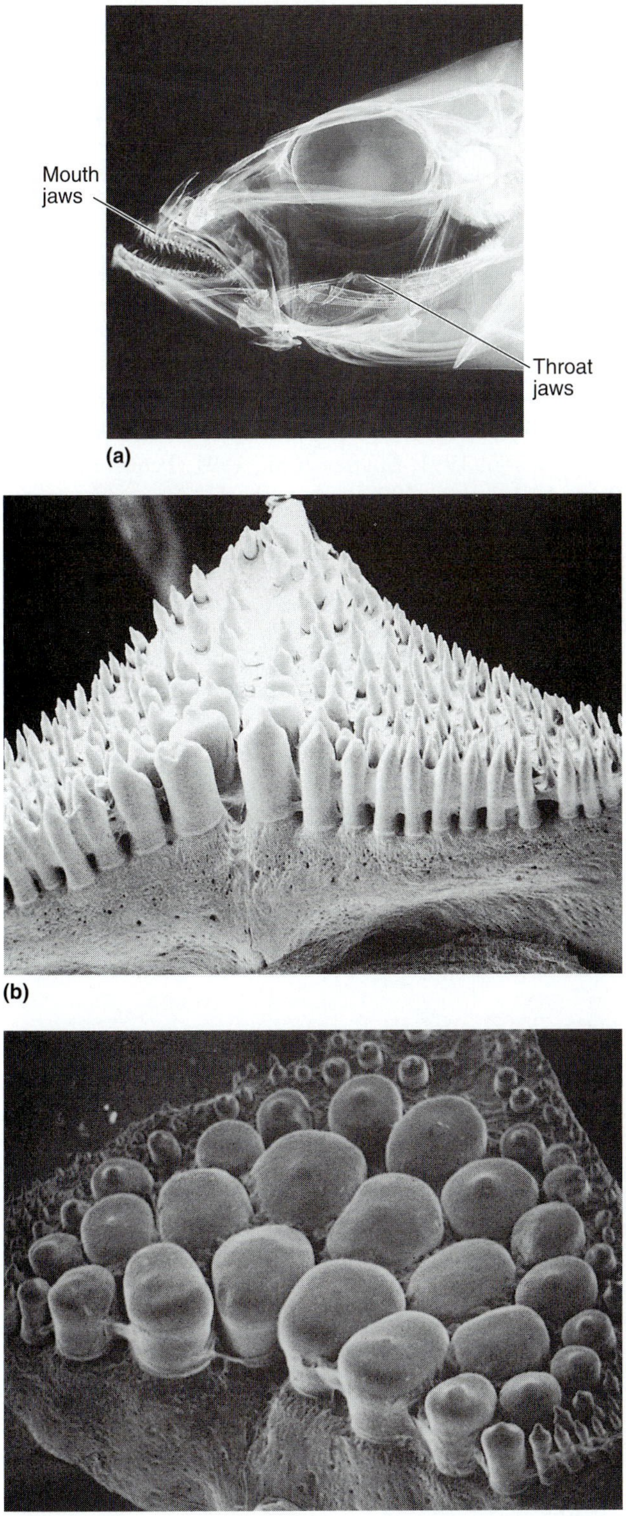

(*a*) Radiograph of the cichlid (*Cichlasoma citrinellum*) showing the two sets of jaws, one in the mouth, the other in the throat. The teeth of *Cichlasoma* may be in the form of sharp pointed piercers (*b*) or flat, molarlike crushers. (*c*) The teeth can change shape within the lifetime of a single animal.

FIGURE 2.13

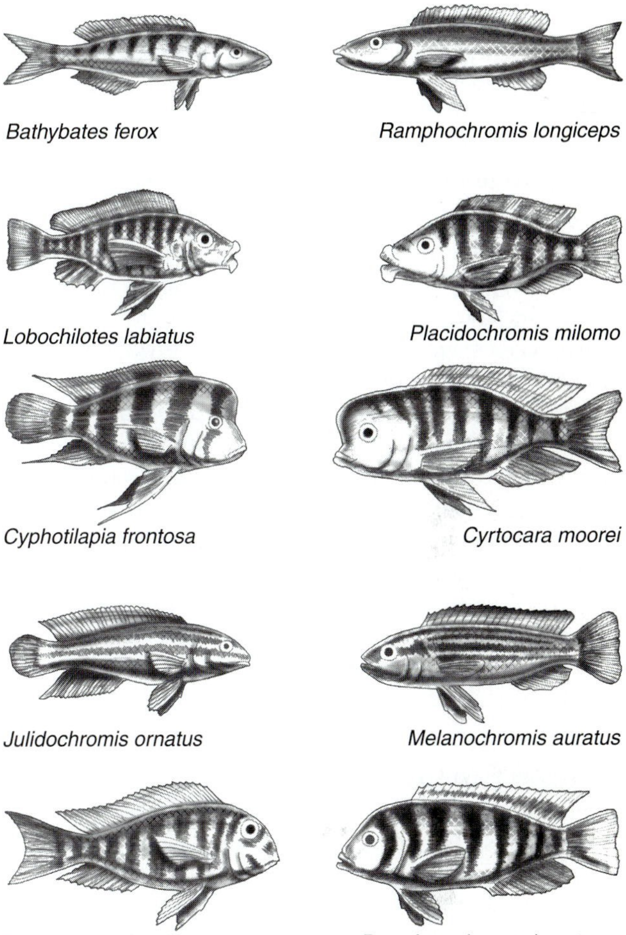

Distantly related cichlids from Lakes Tanganyika and Malawi have evolved to become morphologically similar by occupying similar ecological niches. They demonstrate that there may be little correlation between morphological resemblance and genetic closeness or evolutionary lineage. All of the cichlids of Lake Malawi are more closely related to one another than to any cichlids in Lake Tanganyika.

Molecular clocks that are roughly calibrated on the rate of mutations in mitochondrial DNA suggest that the 400+ species in Lake Victoria arose within the past 200,000 years—an amazingly short period of time.

Geographic Variation

Many species of birds and mammals vary in a somewhat similar manner under similar environmental conditions. These differences have resulted in several "rules" of variation.

1. **Bergmann's Rule.** Many endotherms tend to be larger in colder climates than their relatives in warmer regions. This tendency toward less body surface area in pro-

portion to volume in northern areas is thought to be a means of conserving body heat. One of the best examples of Bergmann's Rule is the song sparrow (*Melospiza melodia*) (see Fig. 2.9). Specimens from the northern part of their range in North America are much larger than their paler relatives from the southwestern United States.

2. **Allen's Rule.** The extremities of many endotherms show a tendency to vary inversely with body size (Fig. 2.14). The ears, feet, and tail of many northern species are proportionately smaller in order to conserve body heat, whereas these parts of the body are larger in members of the same species living in warmer environments in order to lose additional heat by convection and radiation.

3. **Gloger's Rule.** Some endotherms living in arid regions tend to be lighter in color than their relatives living in more humid regions (Fig. 2.14). Fox sparrows (*Passerella iliaca*) illustrate this principle with eastern forms being reddish and their western relatives that breed on arid mountains being pale grayish. The black-tailed jackrabbit (*Lepus californicus*), which ranges from the humid Pacific coast to the arid southwestern deserts, is another species in which the individuals exhibit marked contrast in coloration. Lower humidity does not always result in paler coloration. Benson (1933) noted that in certain desert areas where extensive black lava beds exist, some species of small mammals tend to be very dark, perhaps to blend with the color of the habitat and to enhance the absorption of radiant energy for thermoregulation.

Molecular Evolution

In many cases, the degree of variation in morphological characters (e.g., scale counts, skull measurements) has been the most frequently used taxonomic character, because this type of variation is most easily seen, measured, and compared in preserved material. However, all evolutionary change results from alterations in the sequence of bases of the DNA in the nucleus of reproductive cells and, therefore, the subsequent sequence of amino acids in the proteins of offspring. Studies of molecular evolution compare gene and protein sequences in different species, subspecies, or populations and are providing systematists with much new evolutionary information. Macromolecules—mitochondrial (mt)DNA, RNA, enzymes (isozymes), and the amino acid sequences of proteins such as hemoglobin and cytochrome c—are being analyzed in order to determine how distantly or closely related the organisms in the study groups are to one another (Highton and Webster, 1976; Highton and Larson, 1979; Larson et al., 1981; Sibley and Ahlquist, 1984; Maxson and Heyer, 1988; Meyer et al., 1990; Highton, 1991; McKnight et al., 1991; Bowen et al., 1991; Larson, 1991; Shaffer et al., 1991; Peterson, 1992; Kumar and Hedges, 1998; and many others).

Because DNA and proteins are composed of many bits of information (nucleotides and amino acids, respectively), molecular similarities between two or more species likely reflect a shared, common ancestor in the same way that morphological or other similarities do. Thus, the greater the molecular similarities between two species, the closer their likely evolutionary ancestry. Because differences in nucleotide sequences arise through mutations, and mutations accumulate with time, if two organisms once shared a common ancestor, the time that has passed since they diverged into two distinct species can be estimated by analyzing the mutations that have accumulated in their genes (Fig. 2.15). Evolutionary trees based on biochemical data

FIGURE 2.14

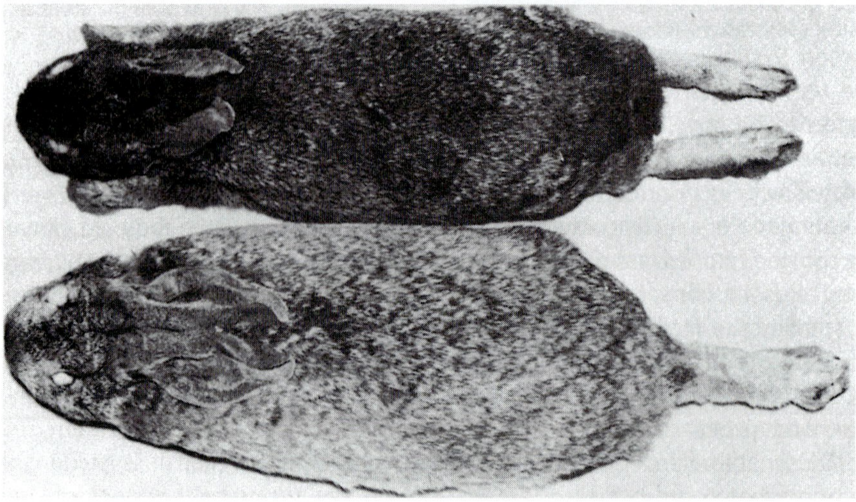

The brush rabbit (*Sylvilagus bachmani*) illustrates both Allen's Rule and Gloger's Rule. The dark, small-eared specimen on top is from the cool, humid coast of northern California; the pale, large-eared specimen is from the hot, arid interior of Baja California.

FIGURE 2.15

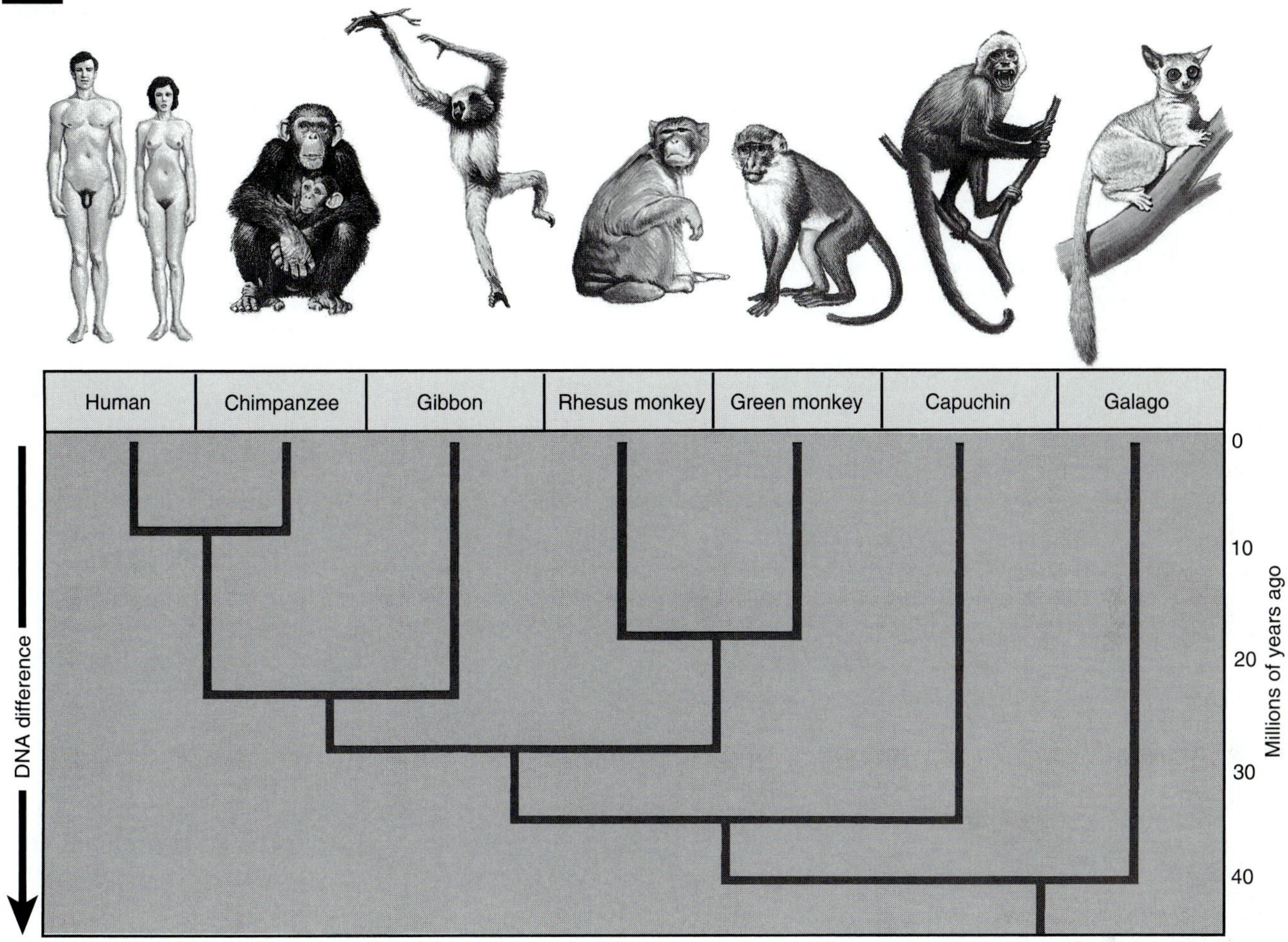

The relationship of certain primate species, based on a molecular comparison of their genomes. The lengths of the branches indicate the relative number of nucleotide pair differences that were found among groups. By using the fossil record as a comparative tool, it is possible to approximate the date at which each group diverged from the other.

are, in most cases, quite similar in appearance to those based solely on anatomical data.

For example, scientists have long thought that the forest elephant of Africa is a separate subspecies from the more familiar savanna elephant. New analyses of mitochondrial DNA now suggest that the two may be entirely different species (Tangley, 1997). The forest elephant has rounder ears and thinner, straighter tusks than its larger relative. Unlike savanna elephants that feed on grass, forest elephants feed on the leaves, twigs, bark, and fruit of rain forest trees.

On the other hand, extensive genetic analysis has revealed just 6, rather than 32, subspecies of puma (*Felis concolor*) in the Western Hemisphere (Culver, 1999). Blood and tissue samples were secured from 209 pumas in zoos, museums, and the wild across North and Central America, and from 106 of the animals in South America. The researchers found no differences in the mitochondrial DNA from North American pumas, suggesting that only one kind of puma inhabits North America, rather than the 15 subspecies previously identified on the basis of where they live and differences in appearance. The DNA analyses also showed that only one subspecies lives in Central America and that just four others roam South America. The team found the most genetically diverse pumas in Paraguay and Brazil south of the Amazon River. This indicates that these populations are the oldest, dating back some 250,000 years, and that northward migrations gave rise to the others over time.

Biochemists are now able to extract proteins and DNA from some extinct organisms, such as frozen mammoths (*Mammuthus*), the dung of a 20,000-year-old extinct ground sloth, and even fossils in order to establish their relationship to living organisms. These studies are possible because all organisms, living or extinct, share some of the same types of molecules. Comparative biochemistry, immunology, protein sequencing, and DNA studies have corroborated, for the most part, earlier evolutionary findings and, at the same time, have provided new understanding of molecular processes in evolution.

BIO-NOTE 2.4

Classifying the Quagga

The quagga (*Equus quagga*) (Fig. 2.16) is an extinct southern African mammal that resembled a zebra.The last known individuals died in captivity in 1875 (in Berlin) and 1883 (in Amsterdam). Some observers considered it to be most closely related to the horse based on analyses of mainly cranial characters. Others thought it was a distinct species of zebra related to the three living species. Still others felt it was merely the southern end of a cline and a subspecies of the plains zebra. Both DNA and protein analyses of samples from quagga skin confirmed that it was, indeed, related to the plains zebra (*E. burchelli*). A breeding program is now under way in an attempt to "re-create" the quagga by repeated inbreeding of the most quaggalike plains zebras. Molecular studies may thus prove responsible, if only indirectly, for the return of one extinct subspecies to its native habitat (at least one that superficially resembles the quagga, since we can never be sure of the evolutionary pathways that created it).

Lowenstein, 1991
Burroughs, 1999

FIGURE 2.16

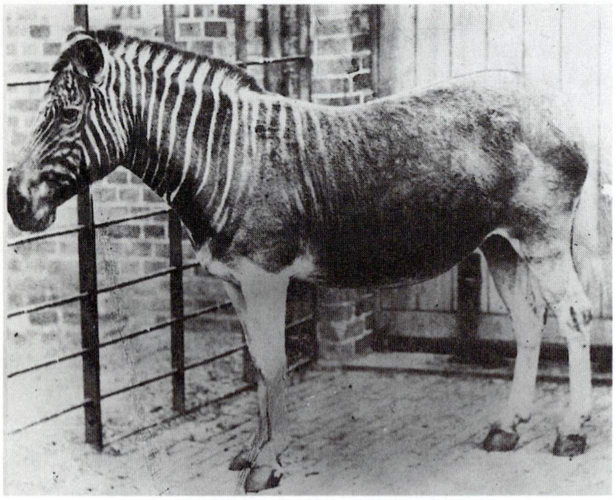

The quagga (*Equus quagga*), an extinct southern African mammal resembling a zebra. The last known individual died in 1883.

DNA fingerprinting (profiling) has proved valuable in studying paternity and genetic variability in whooping cranes. This information helps prevent inbreeding in captive breeding programs (Longmire et al., 1992). DNA analysis of the 52 remaining California condors in 1992 (50 in captivity, 2 released in January 1992) revealed that they were divided into three distinct ancestral groups (Hedrick, 1992). DNA probes have revealed population differentiation in California Channel Island foxes (*Urocyon littoralis*) inhabiting six of the Channel Islands off the coast of southern California (Gilbert et al., 1990). DNA analysis has also revealed the general rule for kinship in lion (*Panthera leo*) prides: Female companions are always closely related, male companions are either closely related or unrelated, and mating partners are usually unrelated (Packer et al., 1991). By applying a genetic probe that selectively binds to gender-specific DNA fragments, Demas et al. (1990) could distinguish sexes in the hatchlings of endangered sea turtles.

Species relatedness also can be estimated by comparing the number of chromosomes, chromosome band patterns, and the order of genes on stained chromosomes. **Karyotypes**, or chromosome pictures, are used to depict gross relationships between species. When chromosome banding patterns between humans and six other species of mammals are compared, regions of apparently conserved chromosomal banding are revealed (Sawyer, 1991). Among primates, the banding pattern of humans most closely matches those of chimpanzees, then gorillas, then orangutans.

Comparison of amino acid sequences in different species often supports fossil or anatomical evidence, but it also may provide evidence refuting long-held traditional schemes of classification. For example, the nucleotide and amino acid sequences of six different genes show that the elephant shrew, the elephant, the aardvark, and the hyrax are all closely related (Balter, 1997). Researchers used genes and proteins with widely different structures and functions to construct their proposed phylogenies. These included a protein that aids water transport across cell membranes, a component of the lens of the eye, and a blood-clotting protein. Rabbits and rodents also appear to be close cousins. Many proteins in humans and chimpanzees are 99% similar in their amino acid sequencing, and several sequences are identical. Cytochrome c, an ancient and well-studied protein, is identical in humans and chimpanzees, as are the alpha and beta chains of hemoglobin from both species. Human and chimpanzee cytochrome c differs from horse cytochrome c by 12 amino acids and from kangaroo cytochrome c by only 8 amino acids (Fig. 2.17a, b). The amino acid sequence between humans and rhesus monkeys varies by only one amino acid.

For centuries, the guinea pig (*Cavia*) had been classified in the suborder Hystricomorpha in the monophyletic order Rodentia. In 1990, 15 proteins, including insulin, were analyzed from a variety of mammals (Graur et al., 1991). Of the 51 amino acids that make up insulin, humans and mice had all but 4 in precisely the same sequence. Guinea pigs, however, had insulin that differed from mice and humans by 18 amino acids. In addition, guinea pigs differed from cows by 19 amino acids and from the opossum—a marsupial—by 20 amino acids. This pattern was repeated in a number of other proteins as well. The protein analysis revealed that the distance between the two "rodents"—guinea pigs and mice—was significantly longer than the distance between other animals that are not even in the same order. Thus, Graur et al. (1991) proposed that the guinea pig be placed in its own order (Caviomorpha). This suggestion, which is still highly controversial, radically contradicts the traditional view of

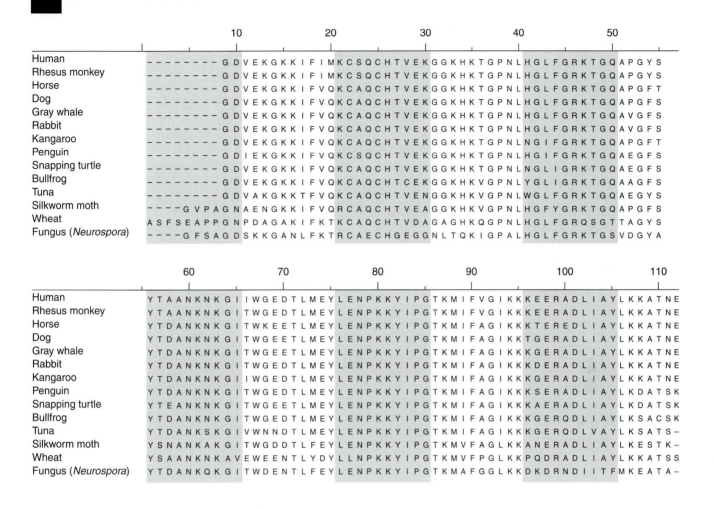

	1–55 (positions 10, 20, 30, 40, 50)
Human	--------GDVEKGKKIFIMKCSQCHTVEKGGKHKTGPNLHGLFGRKTGQAPGYS
Rhesus monkey	--------GDVEKGKKIFIMKCSQCHTVEKGGKHKTGPNLHGLFGRKTGQAPGYS
Horse	--------GDVEKGKKIFVQKCAQCHTVEKGGKHKTGPNLHGLFGRKTGQAPGFT
Dog	--------GDVEKGKKIFVQKCAQCHTVEKGGKHKTGPNLHGLFGRKTGQAPGFS
Gray whale	--------GDVEKGKKIFVQKCAQCHTVEKGGKHKTGPNLHGLFGRKTGQAVGFS
Rabbit	--------GDVEKGKKIFVQKCAQCHTVEKGGKHKTGPNLHGLFGRKTGQAVGFS
Kangaroo	--------GDVEKGKKIFVQKCAQCHTVEKGGKHKTGPNLNGIFGRKTGQAPGFT
Penguin	--------GDIEKGKKIFVQKCSQCHTVEKGGKHKTGPNLHGIFGRKTGQAEGFS
Snapping turtle	--------GDVEKGKKIFVQKCAQCHTVEKGGKHKTGPNLNGLIGRKTGQAEGFS
Bullfrog	--------GDVEKGKKIFVQKCAQCHTCEKGGKHKVGPNLYGLIGRKTGQAAGFS
Tuna	--------GDVAKGKKTFVQKCAQCHTVENGGKHKVGPNLWGLFGRKTGQAEGYS
Silkworm moth	----GVPAGNAENGKKIFVQRCAQCHTVEAGGKHKVGPNLHGFYGRKTGQAPGFS
Wheat	ASFSEAPPGNPDAGAKIFKTKCAQCHTVDAGAGHKQGPNLHGLFGRQSGTTAGYS
Fungus (*Neurospora*)	----GFSAGDSKKGANLFKTRCAECHGEGGNLTQKIGPALHGLFGRKTGSVDGYA

	56–110 (positions 60, 70, 80, 90, 100, 110)
Human	YTAANKNKGIIWGEDTLMEYLENPKKYIPGTKMIFVGIKKKEERADLIAYLKKATNE
Rhesus monkey	YTAANKNKGITWGEDTLMEYLENPKKYIPGTKMIFVGIKKKEERADLIAYLKKATNE
Horse	YTDANKNKGITWKEETLMEYLENPKKYIPGTKMIFAGIKKKTEREDLIAYLKKATNE
Dog	YTDANKNKGITWGEETLMEYLENPKKYIPGTKMIFAGIKKTGERADLIAYLKKATNE
Gray whale	YTDANKNKGITWGEETLMEYLENPKKYIPGTKMIFAGIKKKGERADLIAYLKKATNE
Rabbit	YTDANKNKGITWGEDTLMEYLENPKKYIPGTKMIFAGIKKKDERADLIAYLKKATNE
Kangaroo	YTDANKNKGIIWGEDTLMEYLENPKKYIPGTKMIFAGIKKKGERADLIAYLKKATNE
Penguin	YTDANKNKGITWGEDTLMEYLENPKKYIPGTKMIFAGIKKKSERADLIAYLKDATSK
Snapping turtle	YTEANKNKGITWGEETLMEYLENPKKYIPGTKMIFAGIKKKAERADLIAYLKDATSK
Bullfrog	YTDANKNKGITWGEDTLMEYLENPKKYIPGTKMIFAGIKKKGERQDLIAYLKSACSK
Tuna	YTDANKSKGIVWNNDTLMEYLENPKKYIPGTKMIFAGIKKKGERQDLVAYLKSATS-
Silkworm moth	YSNANKAKGITWGDDTLFEYLENPKKYIPGTKMVFAGLKKANERADLIAYLKESTK-
Wheat	YSAANKNKAVEWEENTLYDYLLNPKKYIPGTKMVFPGLKKPQDRADLIAYLKKATSS
Fungus (*Neurospora*)	YTDANKQKGITWDENTLFEYLENPKKYIPGTKMAFGGLKKDKDRNDIITFMKEATA-

A	Alanine	F	Phenylalanine	K	Lysine	P	Proline	T	Threonine
C	Cysteine	G	Glycine	L	Leucine	Q	Glutamine	V	Valine
D	Aspartic acid	H	Histidine	M	Methionine	R	Arginine	W	Tryptophan
E	Glutamic acid	I	Isoleucine	N	Asparagine	S	Serine	Y	Tyrosine

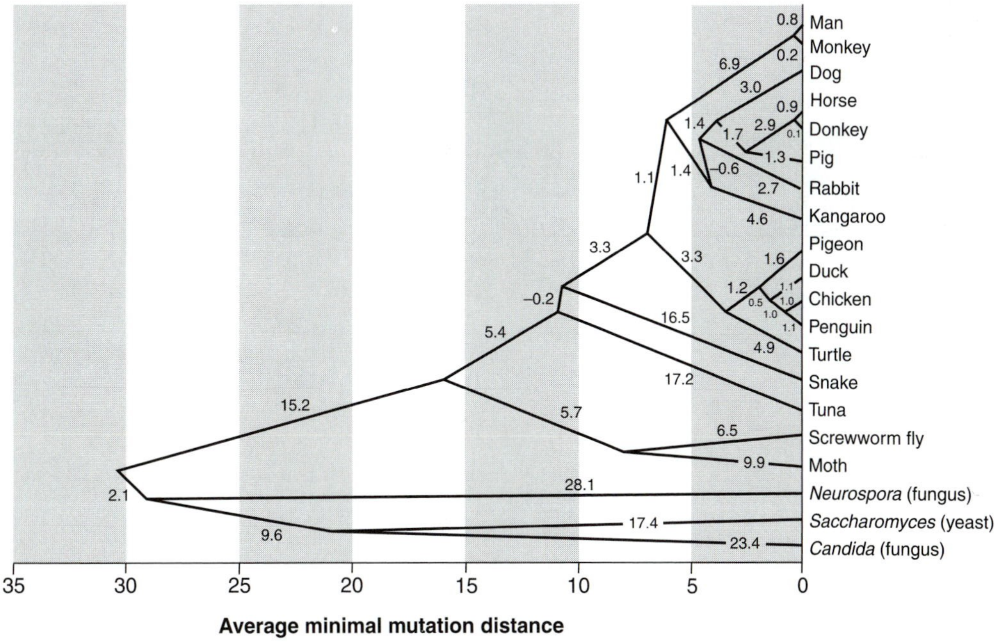

The amino acid sequences of cytochrome c (top) show a striking similarity among living organisms. Each amino acid is represented by a single letter. A phylogenetic table (bottom) depicting evolutionary relationships derived by comparing the differences among amino acid sequences of cytochrome c in different species. The numbers on the lines show the number of nucleotide substitutions that have taken place.

rodent monophyly, which until recently has been based primarily on comparative morphology.

Evidence from comparative DNA sequencing suggests that monotremes (duck-billed platypus and echidna), long thought to represent one of the earliest mammalian lineages, should be placed with the other mammals, specifically on the lineage leading to marsupials (Janke et al., 1996) (see Chapter 9). The link with marsupials is strongly questioned by most nonmolecular mammalogists. Also through DNA analyses, Dragoo and Honeycutt (1997) presented data showing that skunks should be removed from the family Mustelidae and placed in their own family (Mephitidae).

A technique called **DNA hybridization** uses the degree of complementary base pairing to determine the genetic similarity between two species. DNA–DNA comparisons contain evidence of phylogeny, because genetic information is encoded in base sequences and genetic evolution is recorded in the form of changes in base sequences over time. DNA hybridization measures degrees of genealogical relationship among species by comparing the similarity between the nucleotide sequences of their DNA. In this technique, "hybrid" DNA molecules are formed from single strands of DNA taken from two different species in order to determine the degree of similarity in nucleotide sequences between them. When the hybrid DNA is heated, the weaker mismatched areas more readily dissociate than the matched regions (Fig. 2.18). This provides a measure of the changes in the two genomes that have occurred since the two lines diverged from their most recent common ancestor (Fig. 2.19). For example, human DNA differs in 1.8% of its base pairs from chimpanzee DNA; in 2.3% from gorilla DNA;

FIGURE 2.19

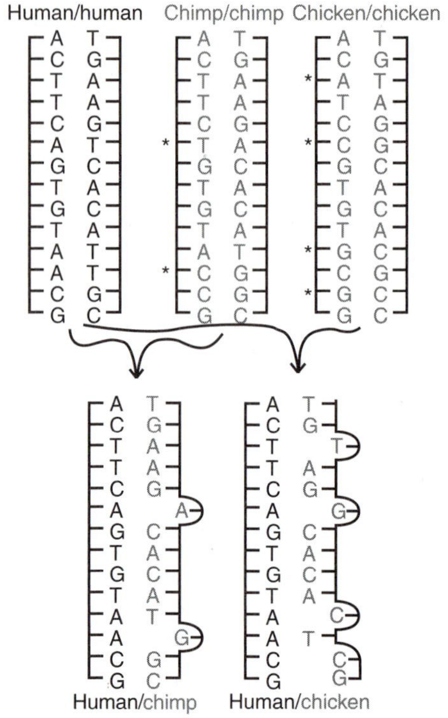

The rate of DNA hybridization reflects the degree of evolutionary relatedness. This schematic diagram shows that DNA from a human hybridizes more rapidly and more closely with chimpanzee DNA than it does with chicken DNA. Mammals and birds diverged from a common ancestor longer ago than humans and chimps, thus accounting for the greater number of changes in their genomes.

FIGURE 2.18

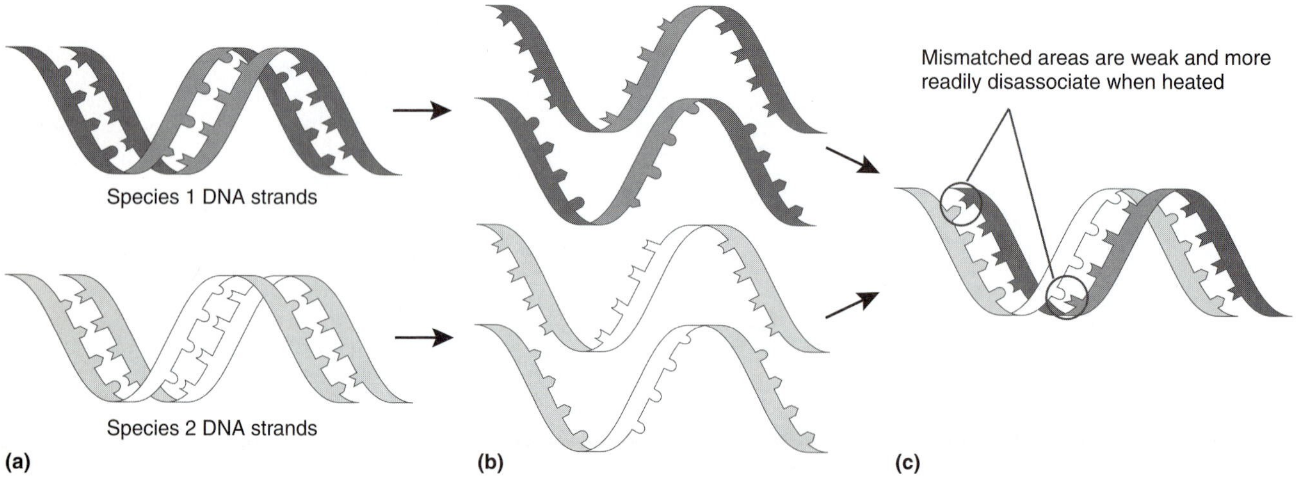

DNA hybridization technique. (*a*) DNA samples from two species are collected and purified. (*b*) The DNA is heated to dissociate the molecules into individual strands. The samples are then combined to form a hybrid DNA. (*c*) The hybrid DNA is then reheated, and the "melting temperature" is observed. The weaker, more mismatched areas dissociate more readily than the more closely matched regions and therefore exhibit a lower melting temperature. This provides a measure of the change in the two genomes that has occurred since the two species diverged from their most recent common ancestor. (Different shapes represent different nucleotides.)

and in 3.7% from orangutan DNA (Sibley and Ahlquist, 1984, 1987; Diamond, 1988; Caccone and Powell, 1989; Marks, 1992). DNA hybridization studies also provide evidence for the relative rate of evolution (Catzeflis et al., 1987). One of the systems of classification for birds used in this book (Chapter 9) is based on DNA studies (Sibley and Ahlquist, 1990).

The analysis of molecular data is not as straightforward as many would have wished. Frequent discrepancies exist between, and even among, morphological and molecular data sets. Some researchers have shown that sequence data can sometimes mislead or even give an entirely wrong answer (Balter, 1997). The significance and implications of DNA studies are highly controversial. Do they really shed light on the evolution of groups? Could DNA sequencing be looking at a piece of genetic material that is too small to be evolutionarily significant? Some researchers prefer DNA–DNA hybridization of the whole genome, as opposed to sequencing of genome fragments, as a phylogenetic method. Further research will gradually shed light on these intriguing questions.

Decisions made by systematists are not final. When new data show that forms originally described as separate species are really geographic races of the same species or that a subspecies should be elevated to the rank of a full species, appropriate changes are made in the nomenclature. Similarly, reclassification is also often necessary at higher taxonomic levels. Vertebrate classification is in a dynamic state, and all groups are subject to change. Because particular classifications eventually become obsolete, they should be regarded as frameworks that will provide a basis for building as advances are made (Nelson, 1984).

The classification of vertebrates is an extraordinarily difficult undertaking because of their diversity (53,475 living species, and perhaps many times that number of species now extinct) and also because of convergent evolution. Species are not static and immutable; thus, simply classifying species is no longer an adequate means of comprehending vertebrate diversity. It is now necessary to include genealogical information in the system of classification in order to express the evolutionary relationships among species.

Review Questions

1. What is meant by the binomial system of nomenclature?
2. Why do biologists use scientific names for species?
3. Describe the hierarchical system used to classify organisms. List in order, from most inclusive to least inclusive, the principal taxa.
4. How does a phylogenetic tree differ from a cladogram?
5. Define the term *species*, using the various schools of systematic thought.
6. In your opinion, should birds be classified in a separate class (Aves) or should they be classified with the reptiles in one monophyletic group (Diapsida)? Why?
7. How do new species evolve? List several types of isolating barriers.
8. Give two examples of reproductive isolating barriers, and discuss what they accomplish.
9. Distinguish between allopatry and sympatry.
10. Differentiate among a deme, a cline, and a species.
11. List several molecular techniques that are providing systematists with new evolutionary information.
12. What data are evolutionary biologists gathering when they heat hybrid molecules of DNA from two species? What do these data indicate?

Supplemental Reading

Allman, J. M. 1999. *Evolving Brains.* New York: Scientific American Library.

Blackwelder, R. E. 1972. *Guide to the Taxonomic Literature of Vertebrates.* Ames: Iowa State University Press.

Carroll, R. L. 1997. *Patterns and Processes of Vertebrate Evolution.* Cambridge Paleobiology Series 2. New York: Cambridge University Press.

Charig, A. 1982. Systematics in biology: A fundamental comparison of some major schools of thought. Pages 363–440. In: Joysey, K. A., and A. E. Friday (eds.). *Problems of Phylogenetic Reconstruction.* New York: Academic Press.

Eldredge, N. 1999. *The Pattern of Evolution.* San Francisco: W. H. Freeman and Company.

Frost, D. R., and D. M. Hillis. 1990. Species in concept and practice: Herpetological applications. *Herpetologica* 46:8–104.

Futuyma, D. J. 1998. *Evolutionary Biology.* Third edition. Sunderland, Massachusetts: Sinauer Associates.

Hennig, W. 1966. *Phylogenetic Systematics.* Urbana: University of Illinois Press.

Hillis, D. M., C. Moritz, and B. K. Mable (eds.) 1996. *Molecular Systematics.* Second edition. Sunderland, Massachusetts: Sinauer Associates.

Hull, D. L. 1988. *Science as a Process.* Chicago: University of Chicago Press.

Jones, S. 1999. *Almost Like a Whale: The Origin of Species, Updated.* New York: Doubleday.

Kocher, T. D., and C. A. Stepien (eds.). 1997. *Molecular Systematics of Fishes.* San Diego: Academic Press.

Mayr, E., and P. D. Ashlock. 1991. *Principles of Systematic Zoology.* New York: McGraw-Hill Publishing Co.

O'Brien, S. J., M. Menotti-Raymond, W. J. Murphy, W. G. Nash, J. Wienberg, R. Stanyon, N. G. Copeland, N. A. Jenkins, J. E. Womack, and J. A. M. Graves. 1999. The promise of comparative genomics in mammals. *Science* 286:458–462, 479–481.

Panchen, A. L. 1992. *Classification, Evolution, and the Nature of Biology.* New York: Cambridge University Press.

Whitaker, J. O., Jr. 1970. The biological subspecies: An adjunct to the biological species. *The Biologist* 52(1):12–15.

Wiley, E. O. 1978. The evolutionary species concept reconsidered. *Systematic Zoology* 27:17–26.

Vertebrate Internet Sites

Visit the zoology website at http://www.mhhe.com/zoology to find live Internet links for each of the references listed below.

1. **Chordate Systematics.**
   Information on chordate classification.
2. **Taxonomy Resources.**
   This site, maintained by the National Center of Biotechnology Information, provides information on systematics and molecular genetics.
3. **Animal Diversity Web, University of Michigan.**
   Kingdom Animalia. More links than you could ever check out!
4. **Taxonomic Classification, from the University of Minnesota.**
5. **Sea World, Busch Gardens; Diversity of Life.**
   An introduction to the animal phyla, including photos and characteristics of the phyla.
6. **Taxonomic Resources and Expertise Directory (TRED).**
   Find a new species in your back yard? This is the place to find information regarding how to classify it.
7. **Journey into the World of Cladistics.**
   This site discusses the introduction, methodology, implication, and the need for cladistics.
8. **Species 2000.**
   The goal of this organization is to provide a uniform and validated index of names of all known species of organisms.
9. **NCBI Taxonomy Homepage.**
   Taxonomic database maintained by the National Center of Biotechnology Information and GenBank.
10. **Phylum/Major Group Index to Zoological Record Taxonomic Hierarchy.**
    Information on taxonomic groups from Biosis.

CHAPTER 3

Vertebrate Zoogeography

■ INTRODUCTION

The study of the geographic distribution of animals and the mutual influence of the environment and animals on each other is known as **zoogeography**. Because animals and plants of a community are interdependent ecologically, zoogeographic studies usually must further include a consideration of plants. Zoogeography also attempts to explain how species have come to be distributed as they are, which requires a knowledge of historical changes in climates, geography, and the distributions of species. Thus, zoogeography is related intimately to both ecology and geology. Four major branches of zoogeographic research are recognized: faunal, comparative, historical, and ecological.

Faunal zoogeography includes the preparation of faunal lists of animal populations for specific areas and forms the basis on which all other zoogeographic research relies. **Comparative zoogeography** attempts to classify the distribution of animals according to their external features. When fauna from different areas are compared, their distribution may not be consistent with the present geologic and geographic divisions of the Earth. For example, amphibians, reptiles, and birds of North Africa are much more closely related to forms in southern Europe than to those in Africa south of the Sahara. The fauna of southern Asia is more closely related to that of trans-Saharan Africa than it is to the fauna of Asia north of the Himalayas. Many groups of North American birds and mammals differ more widely from their corresponding groups in Central and South America than from those in Europe and northern Asia. Homologies among such comparable faunas are based on genetic relationships and common evolutionary origins.

A few species of vertebrates have natural ranges that are virtually cosmopolitan in distribution: mallards (*Anas platyrhynchos*), ospreys (*Pandion haliaetus*), common terns (*Sterna hirundo*), bank swallows (*Riparia riparia*), sperm whales (*Physeter catodon*), and blue whales (*Balaenoptera musculus*). The ranges of others such as starlings (*Sturnus vulgaris*), rock doves (*Columba livia*), house sparrows (*Passer domesticus*), house mice (*Mus musculus*), and Norway rats (*Rattus norvegicus*), which now occur in all but the coldest parts of all of the continents, are the result of human-induced dispersal.

The geographic range of most species, however, is limited in varying degrees to a particular geographic region. In addition, species are restricted to specific communities or groups of communities, with some being very restricted, a situation called being **narrowly endemic**. For example, the giant panda (*Ailuropoda melanoleuca*) inhabits only the few remaining bamboo forests of China; tuataras (*Sphenodon*) are found only on islands along the coast of New Zealand; the koala (*Phascolarctos cinereus*) is restricted to small areas of eastern Australia; and the entire family Todidae (small birds related to kingfishers) is restricted to Caribbean islands.

The distribution of a vertebrate species can be expressed in terms of its current geographic range, its geologic range, and its ecological distribution. The **geographic range** designates the specific land or water area where the vertebrate currently occurs. Taxa that occur in widely separated localities are said to have a disjunct distribution. The family Camellidae, for example, has representatives in Asia (camels) and in South America (llamas and their relatives). The **geologic range** refers to a taxon's past and present distribution in time. A description of the **ecological distribution** provides information on the major biotic communities of which the species is a member.

Closely related species of animals generally will have adjacent ranges; it is assumed that the area in which they evolved from their common ancestor was the common origin of their current distribution. Two major factors affect the subsequent distribution of a new species: the means of dispersal available to an animal, and the existing physical and biotic barriers to such dispersal. External barriers affect the different groups of vertebrates in diverse ways. Aquatic animals are limited in their dispersal by land, and terrestrial animals often are limited by water. Many land animals also are unable to cross mountain ranges, whereas flying species are the least affected by barriers of any kind. Barriers are always relative, and a barrier for one species may well be a main dispersal route for another. Water might be a barrier for a terrestrial species, but normally not for a fish. Topographic barriers, such as mountains, may form more effective barriers in the tropics than in temperate regions, because tropical species are not as well adapted for the cooler temperatures present at higher elevations (Janzen, 1967). Climate, the lack of suitable food, the presence of more successful competitors, or the presence of enemies may present barriers to the dispersal of any group.

Barriers may be of three types: physical, climatic, or biological. Land, water, elevation, soil types, and topography are examples of **physical barriers**. Certain types of physical barriers may serve as "psychological" barriers to some species. For example, some small mammals will not cross roads (even dirt roads), some Neotropical birds will not cross open spaces, and some birds will not cross relatively narrow bodies of water. **Climatic barriers** include temperature, humidity, rainfall, and sunlight, whereas such things as lack of food and the presence of either predators or effective competitors represent **biological barriers**.

The means of dispersal for a species remain relatively unchanged through long periods of time; however, the position of barriers to dispersal become altered with geologic changes in the Earth's surface. For example, many river courses have changed over time; new mountain ranges have arisen, while others have eroded; and formerly well-watered areas now may be desert.

Land connections (bridges) either currently exist or once existed between such areas as North America and Eurasia, North America and South America, and North Africa and southern Europe. Some evidence exists for the presence of a land bridge across the Mozambique Channel between Africa and Madagascar (Anonymous, 1997f). Such a connection would help answer the question of how the ancestors of such unique mammals as lemurs, tenrecs, and fossas got to Madagascar. The appearance and disappearance of connections such as these mean that formerly continuous ranges of related animals may be separated, or alternatively, regions may be united whose faunas were only distantly allied. The older the division between groups of vertebrates, the more times such changes of barriers probably will have occurred during their history.

The marsupials of Australia are thought to have migrated there some 70 million years ago by way of land bridges from South America to Antarctica and across that continent to Australia (Fig. 3.1a). The fossilized remains of a marsupial belonging to the extinct family Polydolopidae were found in late Eocene deposits on Seymour Island, off the northern Antarc-

FIGURE 3.1

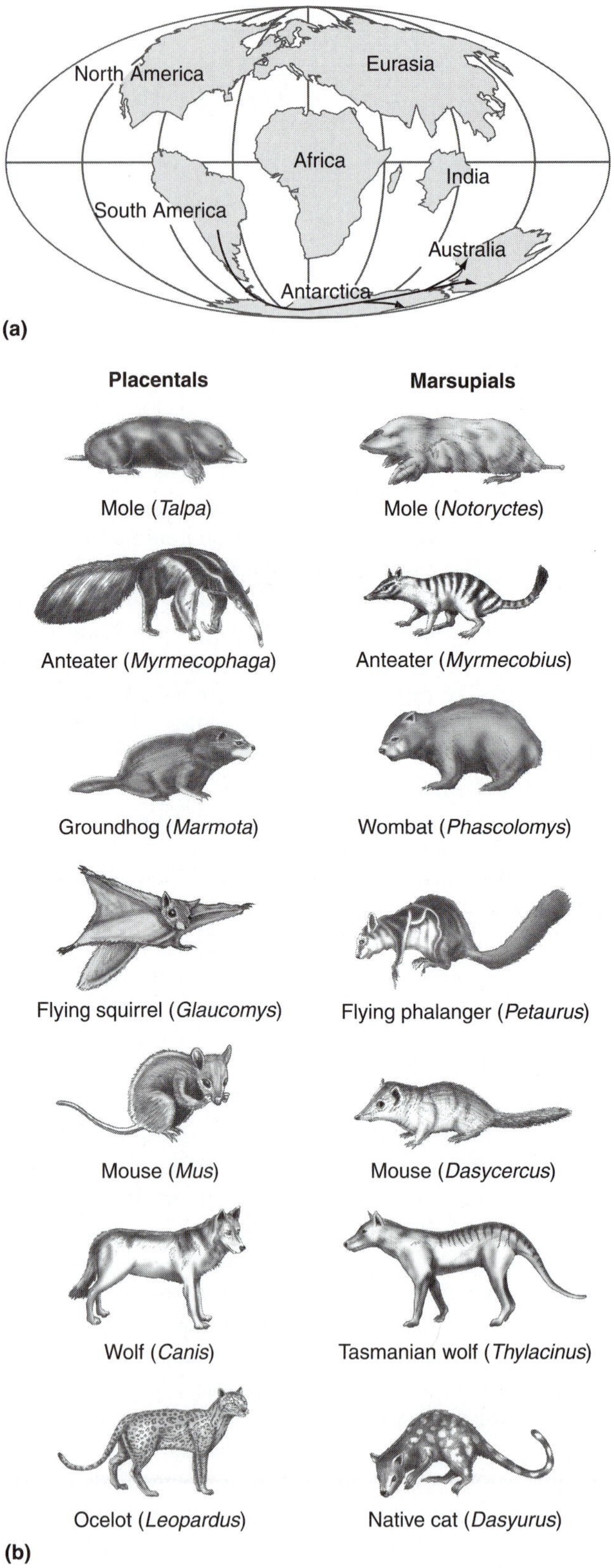

Figure 3.1 ⟶
(*a*) Marsupials are thought to have emigrated from South America across Antarctica (arrow) by way of land bridges some 70 million years ago. After they reached Australia, the continent became isolated from all other continents. Because they encountered no competition from placental mammals, marsupials underwent extensive adaptive radiation. (*b*) Australian marsupial mammals evolved in parallel with placental mammals on other continents.

tic Peninsula (Woodburne and Zinsmeister, 1982, 1984). This discovery, the first land mammal in Antarctica, supports the hypothesis of an early dispersal of marsupials across the southern continents. Because the Australian continent was isolated from all others, and because no placental mammals were present, marsupials underwent extensive adaptive radiation without competition. Because they were filling niches similar to those filled in the rest of the world by placentals and were subject to similar selection pressures, they evolved striking convergent similarities to the placentals (Fig. 3.1b).

Historical zoogeography attempts to work out the development in geologic time of present-day distributions by studying the historical similarities of animal distribution. Studies of spatial and temporal distributions of organisms attempt to provide explanations for these distributions based on past events.

In contrast to historical zoogeography, **ecological zoogeography** investigates the analogies between animal communities in similar habitats. It looks at how animals adapt to the conditions of their native regions. It deals with present conditions, which are more easily subjected to analyses and testing.

■ DISTRIBUTION

Distribution in time depends on continued reproduction, whereas distribution in space depends on the active or passive transport of animals to areas with suitable environments. Vertebrates have the ability to move from one place to another on their own. Some, such as salamanders and frogs, usually move only short distances; others, such as geese, ducks, plovers (Fig. 3.2), warblers, hummingbirds, bats, and whales, may move thousands of miles twice a year in order to reach suitable living or breeding conditions. Some species may be transported passively by storms or by rafting on floating islands of vegetation (Fig. 3.3). Most accounts of rafting have recorded instances of single vertebrates (summarized by King, 1962; Hardy, 1982). However, Boyd (1962) reported multiple individuals of toads (*Bufo woodhousii fowleri*) in Mississippi. Censky et al. (1998) documented the overwater dispersal of a group of large vertebrates (green iguanas, *Iguana iguana*) on islands of the Lesser Antilles in the Caribbean. The distribution of vertebrates has been, and currently is, influenced by geography, geology, climate, ecology, and human activity.

Geographic Distribution

Although a few species are almost cosmopolitan in distribution, the ranges of most species are restricted to a particular geographic region. A. R. Wallace (1876) and other early biogeographers recognized that many taxa have more or less congruent distributions. In an attempt to divide the land masses into a classification reflecting the affinities of the terrestrial flora and fauna, Wallace recognized six major **biogeographic regions**, each of which possessed a characteristic fauna (Fig. 3.4). These regions were named the Palearctic, Nearctic, Oriental, Neotropical, Ethiopian, and Australian.

FIGURE 3.2

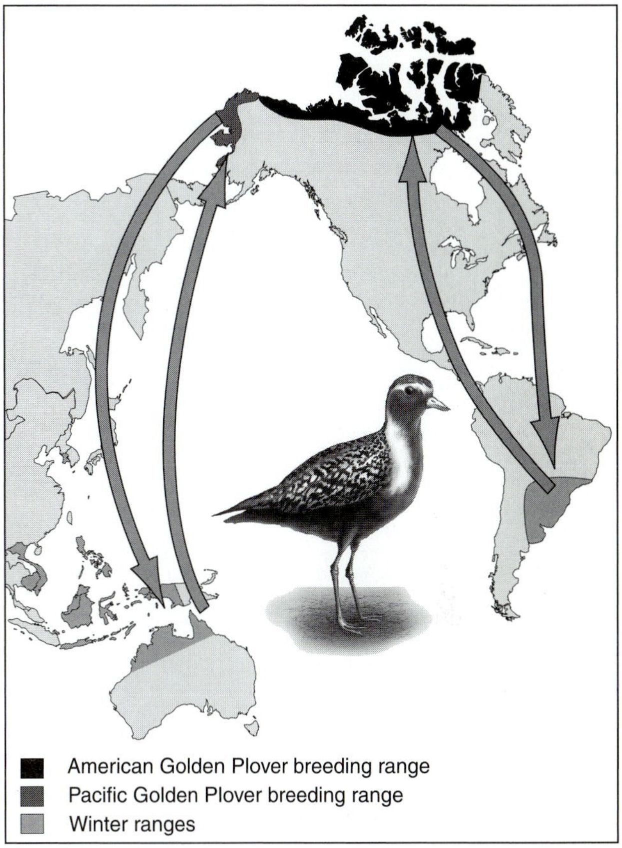

Distribution and migration of the American golden plover (*Pluvialis dominica*). Adult American golden plovers migrate across northeastern Canada and then by a nonstop flight reach South America. In the spring, they return by way of the Mississippi Valley. Their entire route forms a great ellipse. The Pacific golden plover (*Pluvialis fulva*) breeds in Alaska and makes a nonstop flight across the Pacific Ocean to southeastern Asia, Australia, and various Pacific islands. It returns in the spring along the same route.
Source: Data from Orr, Vertebrate Biology, *4th edition, W.B. Saunders & Co.*

This system still is widely accepted today. Wallace's system of classification was similar to the classification proposed for birds in 1856, by Sclater.

Faunal regions do not always correspond to continental boundaries. They sometimes meet along a major barrier to dispersal, such as the Sahara Desert and the Himalayan Mountains.

Holarctic Region (*Palearctic Region, Nearctic Region*)

This region encompasses most of the Northern Hemisphere in both the Old World and the New World. Although the two continents have been connected many times and have many species in common, they usually are separated by vertebrate zoologists (but not by many others, e.g., botanists) into two subregions. The Palearctic Region comprises all of Eurasia (Europe and Asia) not included in the Oriental Region and northern Africa, whereas the Nearctic Region

FIGURE 3.3

Rafting is thought to be a means of dispersal for some vertebrate species. Entire trees, portions of trees, or clumps of vegetation may serve as rafts. Numerous islands of vegetation can be seen floating down the Guayas River toward the Gulf of Guayaquil in Ecuador after heavy rains in the Andes. These islands may harbor many forms of life. The native dugout canoe gives some idea of the size of these floating masses.

FIGURE 3.4

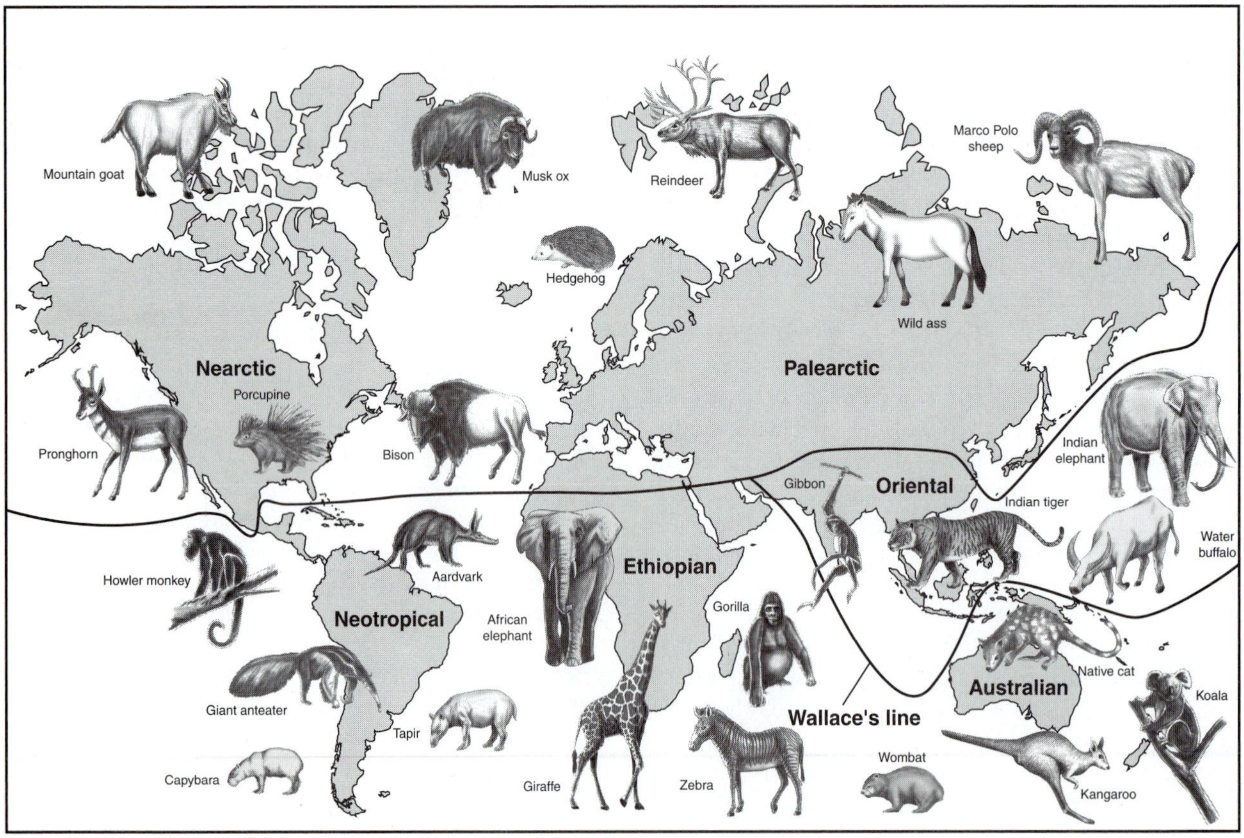

Present and past climates and dispersal pathways have largely been responsible for determining the six biogeographic regions of the world. Different native types of mammals inhabit those regions that were separated from each other during the Mesozoic era, a time when early mammals were diversifying. Although Australia and southeast Asia lie in close proximity to one another, their faunas are quite different because these regions have always been separated. This separation has been recognized by a boundary known as Wallace's Line, which crosses through the island of Sulawezi.

consists of North America down to the southern edge of the Mexican plateau (central Mexico). Because a number of kinds of plants and animals occur in both the Nearctic and Palearctic, the term *Holarctic* is frequently used. This term implies circumpolar distribution. Species that occur in both subregions include snowy owls (*Nyctea scandiaca*), northern harriers (*Circus cyaneus*), lemmings (*Lemmus* sp.), Arctic foxes (*Alopex lagopus*), polar bears (*Ursus maritimus*), and caribou (*Rangifer tarandus*).

Oriental Region

The portion of Asia south of the Himalayan Mountains comprises the Oriental Region. This includes the Indian subcontinent, southeast Asia, most of Indonesia, and the Philippines. Unique species include reticulated and Indian pythons (*Python reticulatus* and *P. molurus*), king cobras (*Ophiophagus hannah*), big-headed turtles (*Platysternon megacephalum*), peafowl (*Pavo*), tarsiers (*Tarsier*), gibbons (*Hylobates*), orangutans (*Pongo pygmaeus*), Indian elephants (*Elephas maximus*), Javan and Indian rhinoceroses (*Rhinoceros sondaicus* and *R. unicornis*), and water buffalo (*Bubalus*).

Neotropical Region

This region has two distinct zoogeographic subregions: mainland South America and Central America. The Central American subregion has been alternately submerged and exposed and has served as a major corridor between the Neotropical and Nearctic regions. The Neotropical Region contains such unique vertebrates as the South American lungfish (Lepidosirenidae), aquatic caecilians (Typhlonectidae), poison-dart frogs (*Dendrobates* and *Phyllobates*), anacondas (*Eunectes murinus*), rheas (*Rhea*), toucans (*Ramphastidae*), llamas and alpacas (*Lama*), giant anteaters (*Myrmecophago, Tamandua,* and *Cyclopes*), howler monkeys (*Alouatta*), capybaras (*Hydrochaeris*), and sloths (*Bradypus* and *Choloepus*).

Ethiopian Region

This is the region of Africa south of the Sahara Desert. Ostriches (*Struthio camelus*), guinea fowl (Numididae), zebras (*Equus*), African elephants (*Loxodonta africana*), gorillas (*Gorilla gorilla*), aardvarks (*Orycteropus afer*), giraffes (*Giraffa camelopardalis*), and hippopotamuses (*Hippopotamus amphibius*) are characteristic species. Fish zoogeographers consider the entire African continent to make up the African Region. This has been a rather stable tropical region with intermittent connections to other continents. It contains many ancient fishes, including the lungfishes (Protopteridae), the bichirs (Polypteridae), and the osteoglossiforms (Moyle and Cech, 1996).

Australian Region

The most distinct of the faunal regions is that of Australia, New Zealand, New Guinea, Tasmania, and surrounding islands. This region is separated from the Oriental Region by Wallace's Line, an invisible line running through the ocean (Fig. 3.4). The native mammalian fauna of the Australian Region consists of marsupials of diverse morphologies and ecological preferences. In addition, emus (*Dromiceius*), cassowaries (*Casuariusi*), lyrebirds (*Menura*), and the egg-laying monotremes (duck-billed platypus [*Ornithorhynchus anatinus*], short-nosed echidna [*Tachyglossus aculeatus*], and long-nosed echidna [*Zaglossus bruijni*]) inhabit the region. No native placental species occur naturally, although many have been introduced. Only two fully freshwater species, both of ancient ancestry, occur here: the Australian lungfish (*Neoceratodus forsteri*) and the barramundi (*Scleropages leichardti*), an osteoglossid. The remainder of the native fish fauna is made up of diadromous or marine families (Moyle and Cech, 1996).

Geologic Distribution

It is estimated that the Earth formed some 4.6 billion years ago (Futuyma, 1986). The oldest rocks discovered on Earth are dated at 3.85 billion years (Woese, 1981; Hayes, 1996), with the earliest fossilized living organisms being marine microbes that were found in rock dated radioactively at 3.3 to 3.5 billion years ago from western Australia. These microfossils are the remains of giant colonies of presumably photosynthetic bacteria. A search for geochemical evidence of past biotic activity (preserved within minerals that are resistant to metamorphism) used ion-microprobe measurements of the carbon-isotope composition of carbonaceous inclusions in rock dating to 3.80–3.85 billion years ago (Mojzsis et al., 1996). These studies yielded isotopically light carbonaceous inclusions, suggesting that life was present on Earth at least 3.8 billion years ago (Mojzsis et al., 1996).

Since its inception, Earth has been undergoing continuous geological changes. Some of these processes, such as volcanoes and earthquakes, are evident and easily observed. In addition, the Earth's crust consists of rigid, slablike plates about 100 km thick that float on the underlying mantle. These plates are constantly in motion due to a process known as seafloor spreading, in which material from the mantle arises along oceanic ridges and pushes the plates apart (Fig. 3.5). Where they converge, one plate may plunge beneath another. Plates also may move laterally past one another along a fault. Because most plates move only a few centimeters a year, their movement is not easily observable and must be measured with sophisticated devices such as lasers. The arrangement of these plates and their movements is known as **plate tectonics**. Although F. B. Taylor and Alfred Wegener first scientifically formulated the theory of continental drift in 1910 and 1912, the basic elements of the concept were suggested by Abraham Ortelius in 1596 (Romm, 1994). The movement of these plates and the continents has significantly affected climates, sea levels, and the geographic distribution of vertebrates throughout time.

From Cambrian through Silurian times (420–600 million years ago), most paleogeologists agree, six ancient continents probably existed. These primitive blocks of land were known as **Laurentia** (most of modern North America, Greenland, Scotland, and part of northwestern Asia); **Baltica** (central Europe and Scandinavia); **Kazakhstania** (central southern Asia); **Siberia** (northeastern Asia); **China** (China, Mongolia, and Indochina); and **Gondwana** (southeastern

FIGURE 3.5

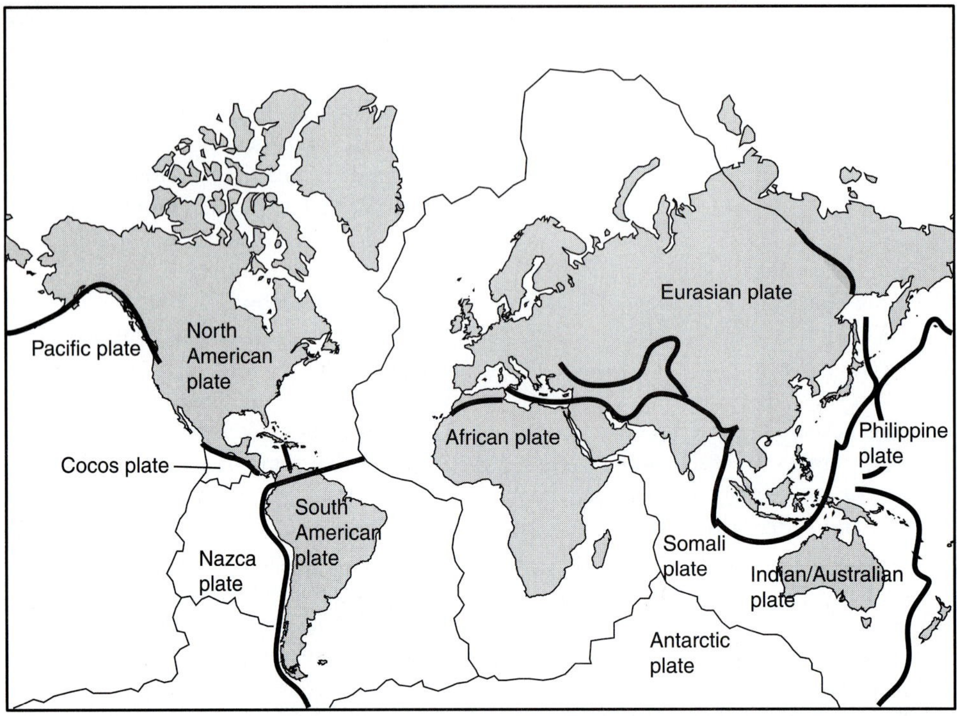

Major plates of the Earth's crust. As the plates push against continental margins, they are often thrust beneath them. This thrusting causes the crumpling and upheavals that have created most of the Earth's major mountain ranges. On a regional scale, many major plates are fractured or composed of numerous smaller plates. As these smaller plates shift, they create Earth tremors and earthquakes. The boundary between two plates is known as a fault. One well-known fault is the San Andreas Fault in California.

United States, South America, Africa, Saudi Arabia, Turkey, southern Europe, Iran, Tibet, India, Australia, and Antarctica) (Murphy and Nance, 1992).

Due to the continuing movement of these plates, the land masses collided to form supercontinents and then split apart to allow new oceans to form. Between 380 and 420 million years ago, the continental land mass known as Laurentia collided with Baltica, forming a supercontinent known as **Laurasia**. Between 360 and 270 million years ago, Laurasia collided with Gondwana, thereby forming the world continent **Pangaea** (Murphy and Nance, 1992). Siberia and China remained as subcontinents until the late Triassic, when they also became part of Pangaea.

Pangaea, the result of multiple collisions that took place over many millions of years, consisted of a single large land mass extending northward along one face of the Earth from near the South Pole to the Arctic Circle (Fig. 3.6). A world ocean stretched from pole to pole and was twice as wide at the Equator as the Pacific Ocean is at present (Futuyma, 1986). Pangaea was not static; it slowly drifted northward from Carboniferous through Triassic times, causing climatic changes in various areas.

During the Permian and Triassic, the eastern part of what is now North America was in contact with Europe and Africa, and South America was joined to Africa. During much of this period, the higher latitudes were relatively warm and moist, while the low and middle latitudes were probably much drier. Triassic deposits from Antarctica and Greenland, for example, have yielded specimens of large amphibians. Regional differences in rainfall and temperature, as well as the formation of the Appalachian Mountains, led to the development of specific floras and faunas.

During the late Triassic, Pangaea began splitting apart into separate continents. This marked the beginning of the independent development of regional biotas. Biota is defined as the combined plants (flora) and animals (fauna) of a region. Animals depend on the plants growing in a particular region; thus a biota is a group of complex interrelationships. Asia and Africa began separating from each other. This was followed in the early Jurassic by the beginning of a westward movement of North America away from Africa and South America, although North America still was connected to Europe in the north. This separation, which began in the Jurassic, continues today. By the late Jurassic, a narrow seaway had formed between North America and Eurasia, connecting the Arctic Ocean to the Tethys Sea, a broad seaway that separated Eurasia from Gondwana. During the Jurassic and Cretaceous, sea levels began to rise, resulting in the flooding of low-lying areas and the formation of shallow inland seas across much of western North America, Canada, and central Eurasia. The Sierra Nevada,

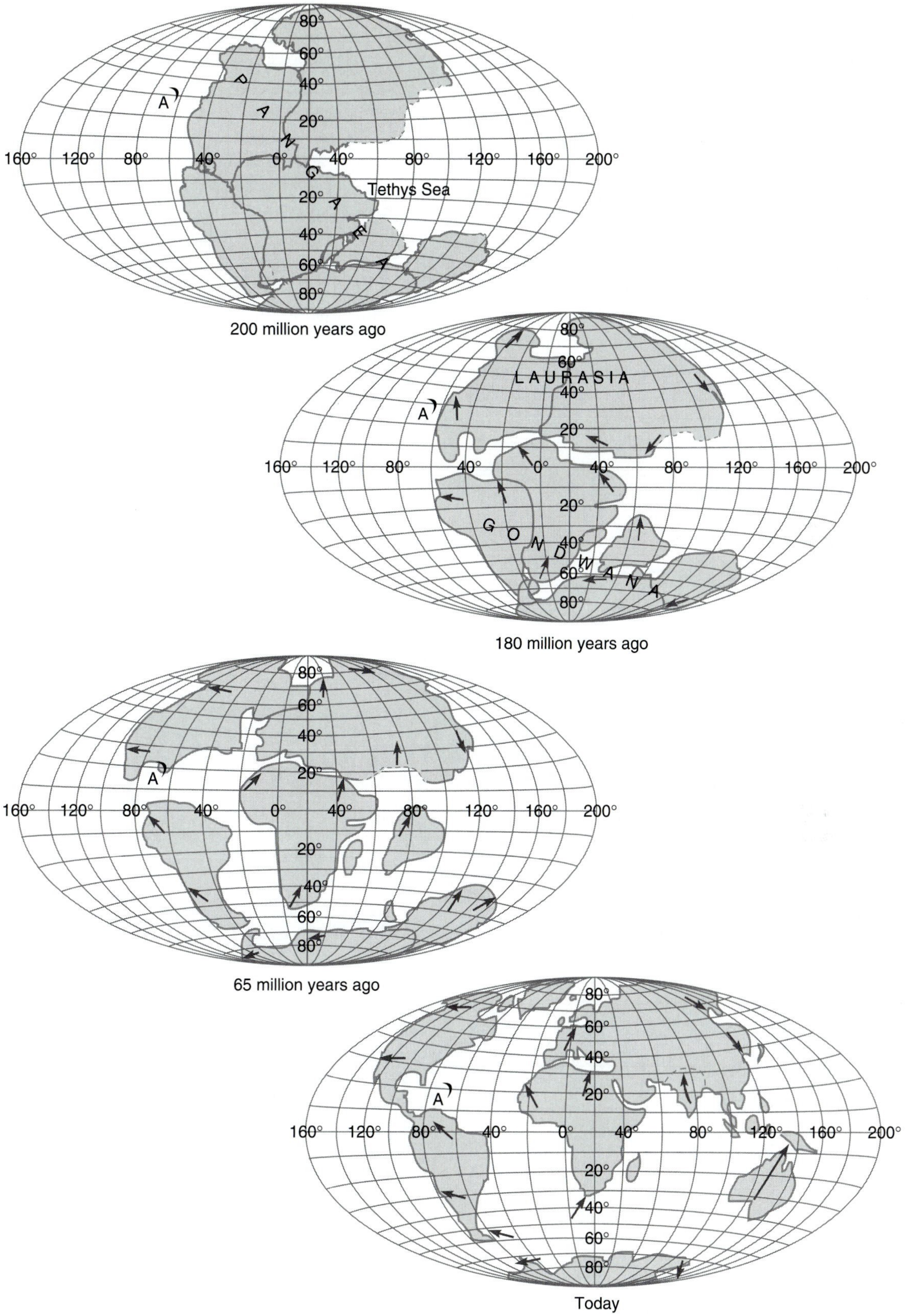

The current position of the continents is not where they have always been. The continents have drifted over the past 200 million years from an original single land mass (Pangaea) to their present positions. Pangaea separated into two supercontinents known as Laurasia and Gondwana, which later broke up into smaller continents. The arrows indicate vector movements of the continents. The black crescent labeled "A" is a modern geographical reference point representing the Antilles arc in the West Indies.

Andes, Himalayas, and Rocky Mountains were formed during this period. Since then, the paths of drifting continents are fairly well understood from evidence in the sea floor. Because no part of the ocean floor is older than 200 million years, the earlier paths must be deduced from other evidence. This process of drifting and colliding has resulted in the current location of our continents and is continuing at the present time.

Gondwana, which began breaking up by mid-Cretaceous, split into Africa, South America (which remained in contact with Africa in the north), and a land mass consisting of Australia, Antarctica, and India. By the late Cretaceous, India had broken free and was moving northward, eventually to collide with Asia, resulting in the formation of the Himalayan Mountains; South America had separated completely from Africa but was still narrowly joined in the south to the Antarctica–Australia land mass; and North America had moved so far to the west that it was separated fully from western Europe, but it had made contact with northeastern Asia to form the Bering land bridge in the region of Alaska and Siberia. (Bridge is somewhat a misnomer, for the land mass ranged up to 1,600 km in width.) A massive regression of **epicontinental seas** (seas covering portions of continents) in the late Cretaceous exposed a great deal of land.

The Atlantic Ocean, which was much narrower during the early Tertiary than at present, continues to expand due to the westward movement of the Americas away from the mid-oceanic ridge. During the late Pliocene, the Panamerican isthmus (Central America) arose, connecting North and South America. The isthmus provided a corridor for the migration of animals between North and South America, forever changing the fauna of both continents (Fig. 3.7). It also blocked a current that once flowed west from Africa to Asia, thus dividing the marine biota of the eastern Pacific from that of the Caribbean.

Steven Stanley, a paleobiologist at Johns Hopkins University, has hypothesized that when the isthmus interrupted the flow of water between the Atlantic and Pacific oceans, it triggered an Ice Age that had a crucial impact on the evolution of hominids in Africa (Svitil, 1996). The dry trade winds that blow west off the Sahara Desert cause water to evaporate from

FIGURE 3.7

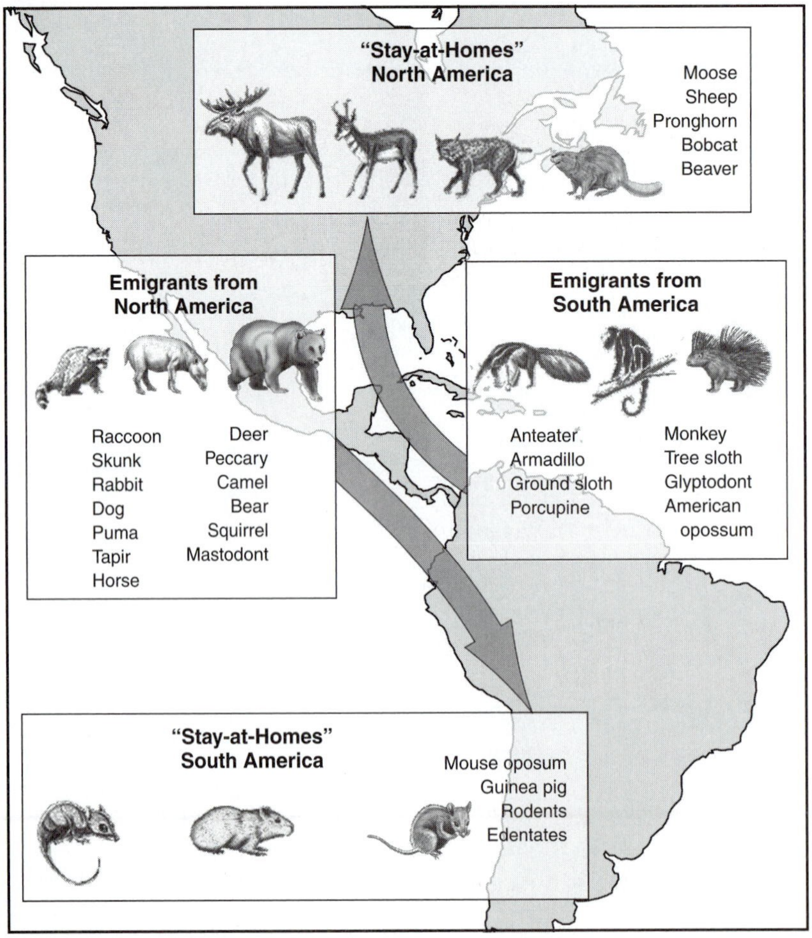

The Great American Interchange. The Plio-Pleistocene land bridge allowed the exchange of fauna between North America and South America. This bridge ended millions of years of isolation on the southern continent and changed forever the fauna of both continents.

the Atlantic, making it saltier as the water evaporates. Prior to the emergence of the isthmus of Panama, saltier Atlantic waters could mix with Pacific waters. After the emergence, such mixing could not occur, so that at present the salinity is low on the Pacific side but very high on the Atlantic side. The salty Atlantic water moves northward to the vicinity of Iceland, where it sinks to the ocean floor and moves southward to Antarctica. Formerly, the Atlantic water was not salty enough to sink, and it continued northward to the Arctic. This flow kept the Arctic relatively warm. After the isthmus formed and the saltier Atlantic water began to sink near Iceland, the Arctic region began to cool. Pack ice, which reflected the sun's rays, formed, cooling the region even further. As the cold spread south, glaciers formed and the Ice Age began. The Ice Age—a long period of waxing and waning of ice sheets—caused Africa to become colder, windier, and drier. This caused rain forests to shrink, and desert and grassland regions to expand (see Fig. 3.25). If *Australopithecus* was semiarboreal, these changing conditions would have made it necessary to develop mechanisms for survival on the ground. Approximately 3 million years ago, *Homo* appeared as a branch of *Australopithecus*. The climatic changes proposed by Stanley are supported by marine sediment cores drilled off the African coast.

Continental drift may account for the distribution of certain vertebrates such as the ratite birds, which include the kiwis (*Apteryx australis*), African ostriches (Struthionidae), the South American rheas (Rheidae) and tinamous (Tinamidae), and emus (Dromiceiidae) and cassowaries (Casuariidae) of the Australian region (Fig. 3.8). Some ornithologists argued against this, claiming that these birds had converged from different ancestors, and so had evolved independently, *in situ*. Evidence from morphology and from DNA hybridization has indicated that the ratites are likely monophyletic (Cracraft, 1974; Sibley and Ahlquist, 1991), and continental drift is now favored as the most plausible explanation for their current zoogeographic distribution.

FIGURE 3.8

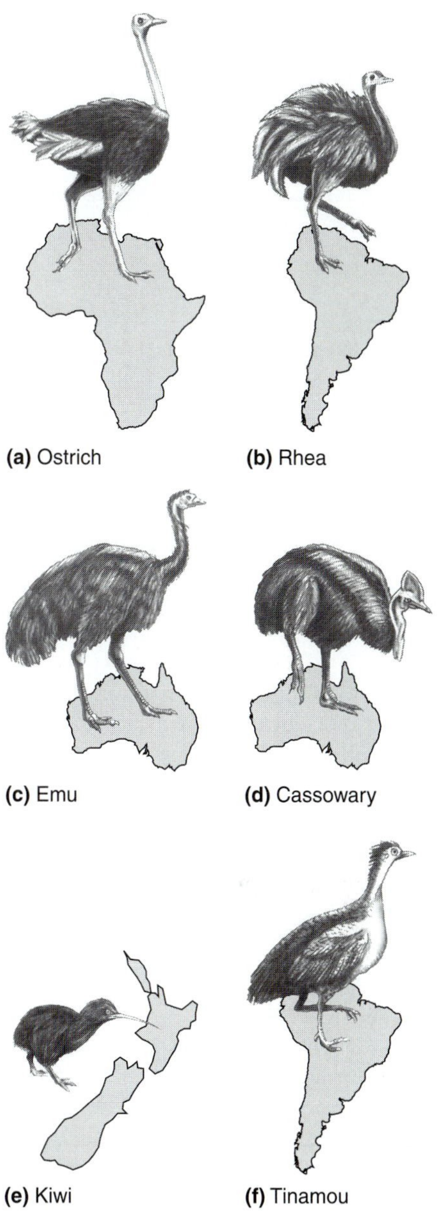

The living families of ratite birds: (*a*) Struthionidae (ostrich), Africa; (*b*) Rheidae (rhea), South America; (*c*) Dromiceiidae (emu), Australia; (*d*) Casuariidae (cassowary), Australia and New Guinea; (*e*) Apterygidae (kiwi), New Zealand; (*f*) Tinamidae (tinamou), tropical America. Despite their disjunct distribution, these birds (which, except for the tinamou, are flightless) are a monophyletic group. Continental drift is now favored as the most plausible explanation for their current zoogeographic distribution.

Climatic Changes

The Earth has undergone a series of climatic shifts throughout its evolution. Periods of rain have alternated with periods of drought, and periods of cooler temperatures have alternated with periods of warmer temperatures. Changes in global temperature have caused sea levels to rise and fall and glaciations to occur. Falling ocean levels have permitted land (filter) bridges to appear in various parts of the world (Fig. 3.9). Land bridges, which permit an exchange of some species from one continent or area to another (hence the term "filter"), have served as major passageways for the dispersal of many vertebrates (see discussion in Chapter 14).

Glaciations have occurred several times during the history of the Earth, including the late Precambrian, Carboniferous, Permian, and Pleistocene. A cooling trend began in the early Tertiary. By Oligocene time, northern latitudes were considerably cooler and drier than they had been, and grasslands and deciduous forests had become widespread. In the Pleistocene, beginning about 1.7 million years ago, the climate changed drastically. The mild climate of the polar and subpolar regions became much colder, polar ice caps formed, and glaciers spread and withdrew repeatedly over the northern portions of the northern continents and parts of western South America. Numerous minor glacial episodes and at least four major ones are known to have occurred. The most recent, called the **Wisconsin glaciation**, withdrew only a little more than 10,000 years ago (Fig. 3.10). During the glacial episodes, sea level dropped throughout the world by as much

FIGURE 3.9

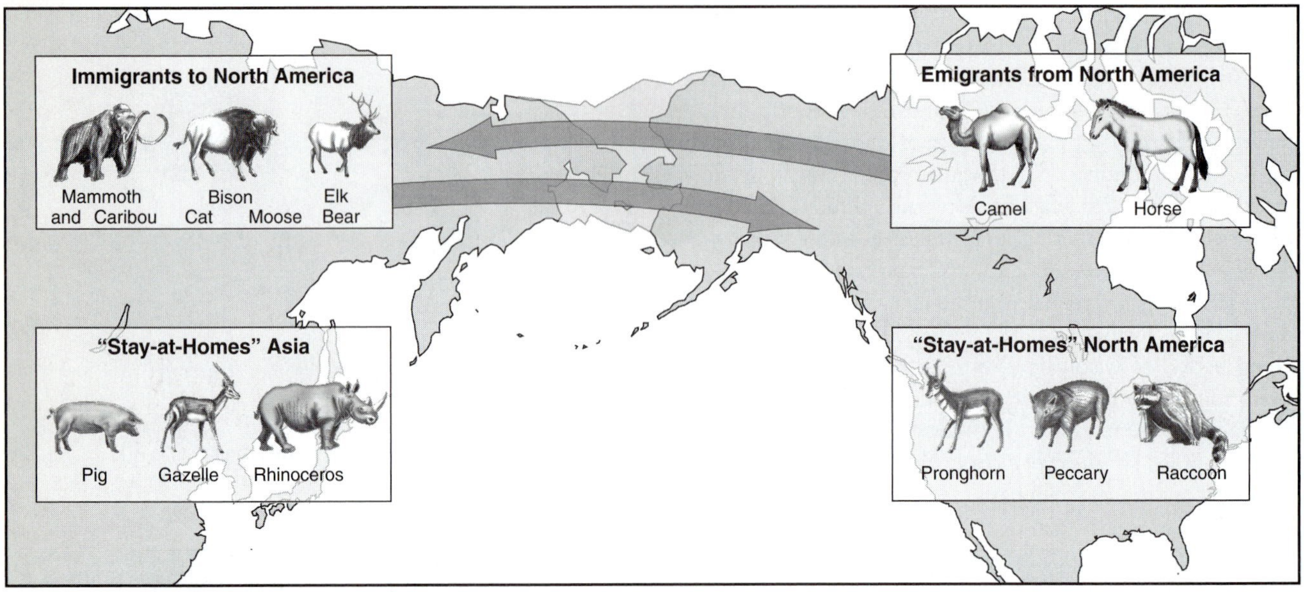

The Pleistocene filter bridge across the Bering Strait permitted the exchange of species between Asia and North America.

as 100 m as water became locked up in ice caps, and the climate in tropical and subtropical areas became drier. During interglacial episodes, climates became warmer and wetter, and sea level rose (Futuyma, 1986).

Many vertebrates ranged far north of their present distribution during the warm interglacial periods. For example, elephants, hippopotamuses, and lions once occurred in England (Futuyma, 1986). Fossils of large (2.4 m long) champosaurs, which are extinct crocodilelike reptiles, have been found at a site just 960 km from the North Pole (Tarduno et al., 1998). Temperatures at the fossil site now routinely drop to minus 30°C in the winter, but when the champosaur lived there 86 to 92 million years ago, temperatures rarely reached freezing and summertime readings of 26°C were common. During glacial episodes, the climate throughout much of the world became cooler and drier, and parts of the Earth were covered by sheets of ice 2 to 3.5 km thick (Kerr, 1994; Peltier, 1994). During these periods, the distributions of animal and plant species became more restricted and shifted toward the tropics.

During glacial maxima, biomes were shifted from 10° to as much as 20° latitude south from their present locations, but they still occupied the same relative positions north-to-south as they do today. North–south mountain ranges in North and South America permitted arctic and boreal taxa to extend their ranges far southward in cool situations (Brown and Gibson, 1983), whereas Europe's east–west mountain ranges served as barriers to many species, and prevented their southward dispersal.

Many species could not adapt and became extinct. Those that could adapt shifted their ranges southward ahead of the advancing ice sheets, with many northern species finding suitable living conditions in the Appalachian and Rocky Mountain ranges. After glaciers receded, many of these populations reinvaded their original northern ranges. Some, however, remained at the higher elevations in the mountains far south of their ancestral ranges and were restricted to "refuges"—local pockets of favorable habitat. Today, these species are known as **disjunct** forms and are usually uncommon in the southern parts of their ranges. Four-toed salamanders (*Hemidactylium scutatum*), wood frogs (*Rana sylvaticai*), northern water shrews (*Sorex palustris*) and northern flying squirrels (*Glaucomys sabrinus*) (Fig. 3.11) are examples of species with disjunct distributions.

For many years, biologists thought that tropical climates remained virtually unchanged while the great ice sheets of North America and Europe waxed and waned through the Pleistocene. Rain forests were once thought to have continuously clothed South America's equatorial lowlands since its Cretaceous separation from Africa about 80 million years ago. Current evidence, however, suggests that the Amazon may have been quite a bit drier and that the rain forest may, at times, have been segmented and was therefore much smaller (Schneider, 1996). Snow lines, even at equatorial latitudes, were substantially lower during ice ages. During nearly 2 million years of the Quaternary period, as glacial climates became more zonal and more extreme, savannas expanded, forcing rain forests into refugia (isolated pockets of suitable habitat). Savanna versus rain forest dominance oscillated as semiarid glacial epochs alternated with humid interglacials at least three times during the last 100,000 years. As forested areas were reduced to isolated patches, they became refuges for populations of tropical plants and animals. Following the next wetter episode, forest patches expanded and coalesced to cover large areas of tropical forest. Formerly isolated animal populations, which had reached various degrees of genetic divergence in their refugia, followed the spread of the forest. Some

FIGURE 3.10

Distribution of glaciers during the last Ice Age

During the last Ice Age, some 15,000 to 20,000 years ago, extensive ice sheets formed over eastern Canada and moved south and west into northern portions of the United States. Other glaciers formed over Scandinavia and moved southward over northwestern Europe and over the arctic regions of Eurasia and North America. The ice sheets covering Greenland and Antarctica also grew larger. Glaciers locked up so much water in the form of ice that sea levels worldwide were lowered by approximately 100 m. We are now in a period of global warming. Temperatures are increasing, remaining glaciers are receding, and large pieces of ice are breaking free from Antarctica. If this warming trend continues, it will undoubtedly affect the worldwide distribution of plants and animals.

FIGURE 3.11

Northern Flying Squirrel, *Glaucomys sabrinus*

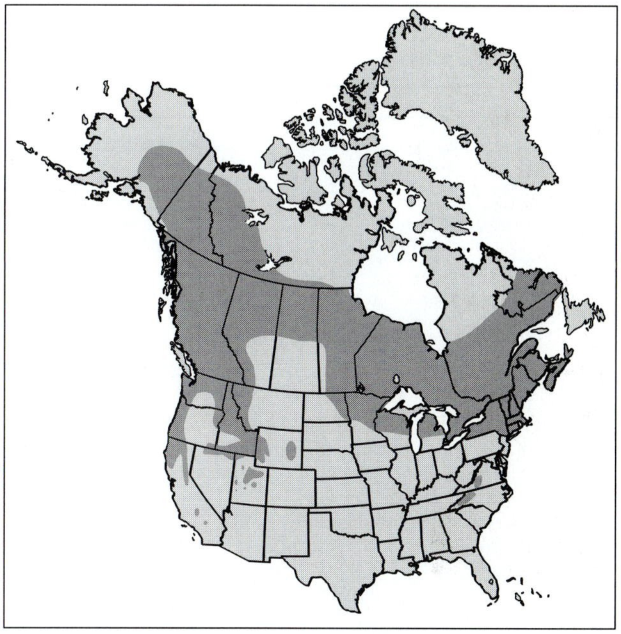

Present distribution of the northern flying squirrel (*Glaucomys sabrinus*). The southern extension of the range of this species along the Appalachian Mountains in the east and the Rocky Mountains in the west consists of disjunct populations.
Source: Data from Burt and Grossenheider, Field Guide to the Mammals, *1964, Houghton Mifflin & Co.*

had become full species, whereas others had reached the level of subspecies. Thus, geologists and biologists now have documented a dynamic evolutionary history to explain the rich Neotropical fauna and flora (Vanzolini, 1995).

Ecological Distribution

Basic environmental factors affecting a species' existence include water, salinity, humidity, temperature, light, oxygen, pressure, and food. Each of these factors has upper and lower limiting values for each species of vertebrate. If the tolerance ranges (within which the species exists) are extensive, the species can live in a variety of habitats. If the limits within which the species can exist are narrow, it will be limited to one or a few types of environments.

Three major ecological environments are recognized. These are marine, fresh water, and terrestrial.

MARINE

The marine environment is the largest. It consists of the oceans, seas, and bays, and it covers 70 percent of the Earth's surface. Oceans form a single vast interconnected water mass only partially separated into divisions by the continents. The unbroken connection of the oceans and the continuous diffusion of sea water by means of currents, tides, and storms result in a *general* equality in the composition and amount of the substances dissolved in sea water. Although temperature varies, this widespread uniformity of conditions is accompanied by an extremely wide distribution of many marine species of animals.

Ocean depth varies from intertidal zones, which are covered by water only part of the time, to depths of 10,800 m (10.8 km). Water temperature varies from 32°C in the tropics to −2.2°C in the Arctic. Even though this worldwide range is great, water temperature in any given area rarely fluctuates more than 5°C. Dissolved salts are relatively constant in concentration (approximately 35 parts per thousand), whereas dissolved gases such as oxygen, carbon dioxide, and nitrogen vary with temperature and depth.

Based on the penetration of sunlight, the sea can be divided into two vertical zones: **photic** and **aphotic** (Fig. 3.12). The depth of the photic zone increases from coastal waters, where light rarely penetrates more than 30 m because of organisms and inanimate particles suspended in the water, to the open ocean, where it may extend to a depth of 100 m or more (Brown and Gibson, 1983). Light penetration also is affected by surface motion. Most producer organisms inhabit the upper 150 m of water, but traces of light can be detected

FIGURE 3.12

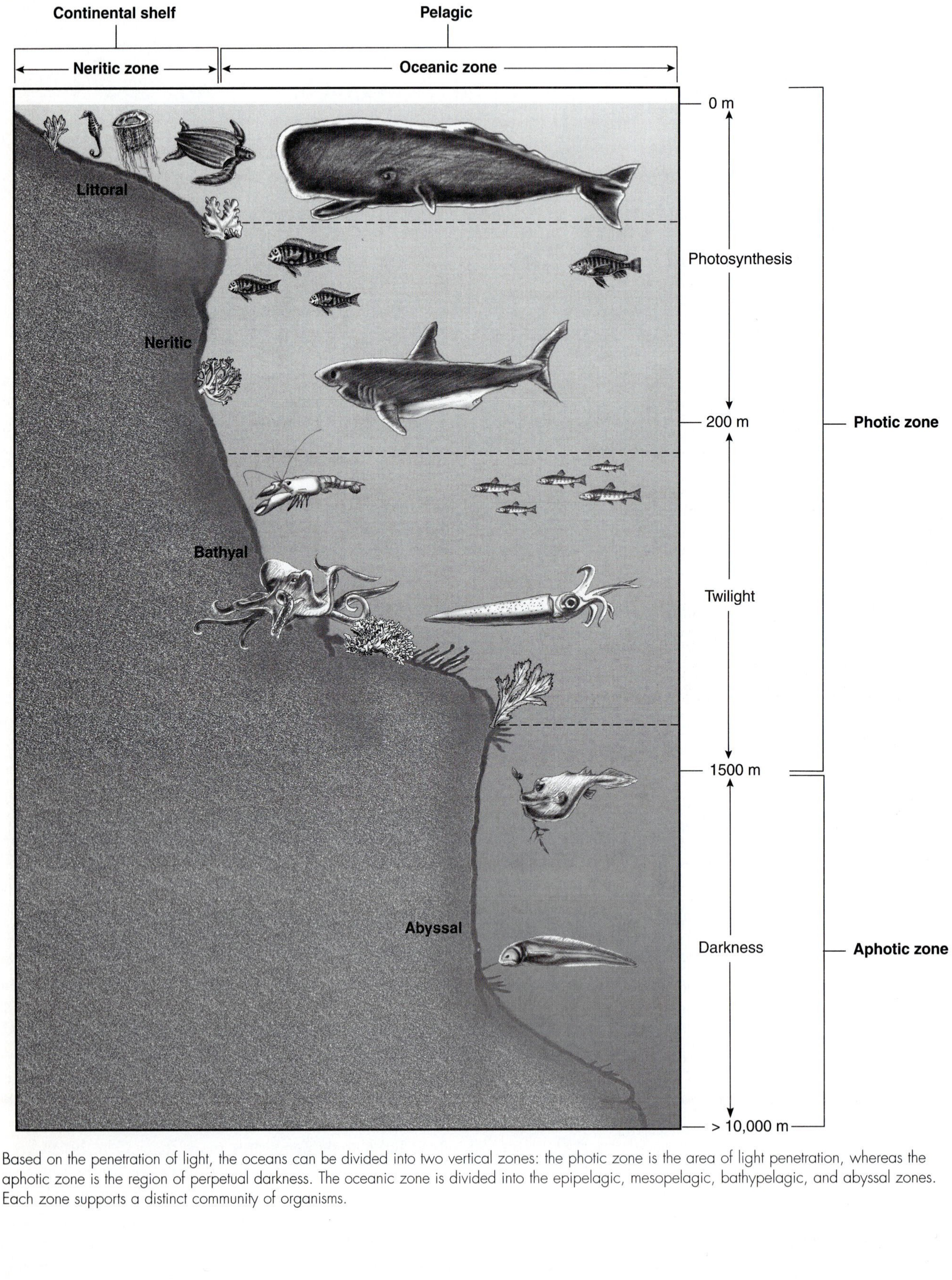

Based on the penetration of light, the oceans can be divided into two vertical zones: the photic zone is the area of light penetration, whereas the aphotic zone is the region of perpetual darkness. The oceanic zone is divided into the epipelagic, mesopelagic, bathypelagic, and abyssal zones. Each zone supports a distinct community of organisms.

at a depth of 1,000 m in the open ocean in the subtropical zone by means of photographic plates (Allee and Schmidt, 1951). The wavelengths of ordinary light are differentially absorbed by water, but below 1,000 m there is permanent darkness. Pressure increases at the rate of 1 atmosphere (14.7 pounds per square inch) for every 10 m depth (Riedman, 1990), so that organisms living in deeper parts of the ocean are exposed to great pressures. At a depth of 600 m, for example, the underwater pressure is equivalent to that of about 60 atmospheres, or 60 times the pressure at the surface.

The marine environment is divided into **pelagic** (open waters) and **benthic** (ocean floor) regions (Fig. 3.12). Two groups of pelagic organisms are recognized: plankton and nekton. **Plankton** are the mostly passive floating plants, called **phytoplankton**, and animals, called **zooplankton**, that move by means of wind, waves, and currents. Plankton serve as food for a variety of vertebrates of all sizes including the blue whale (*Balaenoptera musculus*), the largest living animal. All free-swimming marine organisms are classed as **nekton**. Nekton include fishes, turtles, and marine mammals such as sea otters, porpoises, manatees, and whales.

The pelagic region includes the **neritic zone** (above the continental shelf) and the **oceanic zone**, which extends outward from the continental shelf. The neritic zone has a much greater density of organisms, because nutrients are more abundant and sunlight penetrates the shallower waters. Both producers and consumers are abundant.

The oceanic zone is further divided into the epipelagic, mesopelagic, bathypelagic, and abyssal zones (Fig. 3.12). The **epipelagic zone** receives abundant sunlight. Phytoplankton and zooplankton are abundant and serve as the bases of food webs. The semidark **mesopelagic zone** (200–1,000 m) is known as the twilight zone. Fishes are the primary vertebrate inhabitants there; cetaceans visit, but cannot live there.

The **bathypelagic zone** (1,000–4,000 m) is inhabited by a lesser number of fishes—some of the latter have developed bioluminescent organs. This is an area of cold, quiet water, permanent darkness, and increased pressure. The **abyssopelagic zone** (3,000–6,000+ m) comprises a region with an almost constant physical environment. It is continually dark, cold (4°C), and virtually unchanging in chemical composition. Rattails (Macrouridae), brotulas (Ophidiidae), lumpfishes (Cyclopteridae), and batfishes (Ogcocephalidae) are representative of the benthic fish fauna inhabiting the abyssopelagic zone. Many are eel-like, blind, and have well-developed cephalic lateral-line canals and senses of smell and touch. The area beyond 6,000 m is known as the **hadopelagic zone**. The deepest living fish known (Ophidiidae, *Abyssobrotula galatheae)* is a blind, elongate fish that feeds on benthic invertebrates and lives at depths ranging from 3,100 to 8,370 m (Nielson, 1977; Bond, 1996) (Fig. 3.13).

The benthic division includes organisms that live on the floor of the continental shelf, of the continental slope, and of the abyssal plain. This includes the **intertidal** (littoral) **zones** that alternately are covered by water and exposed to air twice a day. The majority of benthic forms are invertebrates, and many serve as food for species of marine vertebrates. A few vertebrates, such as flounders and sole, spend much of their time on the ocean floor.

FIGURE 3.13

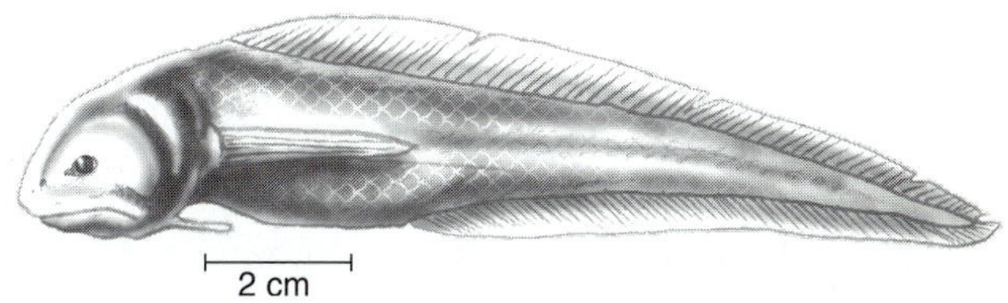

The deepest-living fish known, *Abyssobrotula galatheae*, is a blind, elongate fish that feeds on benthic invertebrates and lives at depths ranging from 3,100 to 8,370 m.

Estuaries, salt marshes, rocky shores, and coral reefs are unique shallow-water marine communities. Estuaries (Fig. 3.14) contain brackish water, a mixture of fresh water and salt water, because they are located where rivers flow into the oceans. Estuaries trap the nutrients brought in from the sea by the tides and prevent the escape of nutrients carried by the rivers. These areas, which are particularly rich in larval shrimps, mollusks, and fishes, often are referred to as the "nurseries of the sea" because well over

FIGURE 3.14

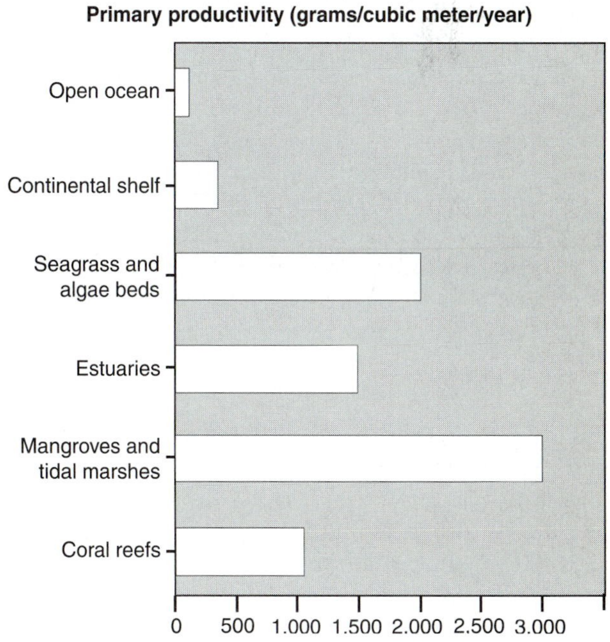

The biological productivity of mangroves, tidal marshes, seagrass and algae beds, estuaries, and coral reefs far surpasses that of all other marine environments. These areas not only provide protection for many marine vertebrates but are also rich in nutrients.

Source: Data from A.P. McGinn, "Safeguarding the Health of Oceans" in World Watch Paper 145, *fig 1, p. 14. World Watch Institute, Washington, D.C.*

FIGURE 3.15

New England salt marsh. An estuary is a partially enclosed coastal region where sea water mixes with nutrient-rich fresh water from rivers, streams, and runoff from the land. Estuaries and coastal marshes are vital feeding grounds and nurseries for many invertebrate and vertebrate species. The major producer in coastal salt marshes is cordgrass or marsh grass (*Spartina*). Broad, shallow estuaries include Chesapeake Bay, Mobile Bay, and San Francisco Bay. Estuaries in Alaska, British Columbia, and Norway are narrow and deep; in Norway, they are known as fjords. From Maryland to Texas, many coastal marshes lie behind narrow strips of sand known as barrier islands.

half of all marine fishes develop in the protective environment of estuaries (Fig. 3.14; Moyle and Cech, 1996).

Salt marshes, which are an important habitat for many vertebrates, often are adjacent to estuaries and contribute nutrients to them (Fig. 3.15). Rocky shores and sandy beaches serve as habitats for many forms of invertebrates that provide food for a wide variety of shorebirds. Coral reefs, which form in warm, shallow seas, provide a haven for many invertebrates as well as many fishes.

FRESH WATER

This is the smallest of the three major environmental types (Fig. 3.16a, b). It includes rain puddles, swamps, ponds, hot springs, lakes, streams, and rivers. Freshwater habitats may consist of flowing water, called a **lotic system**, or still water, called a **lentic system**, and range in temperature from below freezing to boiling. They may be clear or turbid, and the amount of dissolved oxygen may vary widely. Even the hardest fresh waters have salinities of less than 0.5 parts per thousand. Maximum depth may range to 1,700 m or more.

The degree of acidity or alkalinity, or pH, of the water reflects the carbon dioxide content as well as the presence of organic acids and pollution. The higher the pH of stream water, the richer natural waters generally are in carbonates, bicarbonates, and associated salts. Such streams support more abundant aquatic life and larger fish populations than streams with acid waters, which are generally low in nutrients.

The variable temperature of a stream is ecologically important because temperature affects the stream community, influencing the presence or absence of cool-water and warm-water organisms. Streams shaded by trees, shrubs, and high banks will be cooler than those with large areas exposed to sunlight. Small, shallow streams tend to follow, but lag behind, air temperatures, warming and cooling with the seasons but rarely falling below freezing in winter. The constant churning and swirling of stream water over riffles and falls give greater contact with the atmosphere; thus, the oxygen content of the water is high, often near the saturation point for existing temperatures.

In slow-flowing streams where current is at a minimum, streamlined forms of fish give way to species such as smallmouth bass (*Micropterus dolomieu*), shiners (*Notropis, Cyprinella*), and darters (*Ethiostoma, Percina, Ammocrypta*). They trade strong lateral muscles needed in fast current for compressed bodies that enable them to move through beds of aquatic vegetation. Bottom-feeding fish, such as catfish, feed on life in the silty bottom.

BIO-NOTE 3.1

Lake Formation

Lakes are formed by several processes. Those formed in basins created by movements of the Earth's crust are known as **tectonic lakes**. Reelfoot Lake in Tennessee, which was formed as a result of an earthquake in 1811, is an example of a tectonic lake. As the Earth's crust shifts, fissures may form or a narrow strip of land may sink, a process known as faulting. Such fissures or sinks may fill with water and form long, narrow lakes known as **rift lakes**. The deepest lakes known—Lake Baikal in Russia (1,741 m) and Lake Tanganyika in Africa (1,435 m)—are of this type.

Volcanic lakes form when a volcano becomes extinct and its hollow interior fills with water. Also known as crater lakes, they are usually circular in outline and also may be deep. Crater Lake in Oregon, for example, is 608 m deep.

Glacial action has formed many of the world's lakes. As glaciers move over the land, they gouge out previously existing valleys. When the glacier melts, its waters often collect in these valleys to form deep **glacial lakes** with steep sides. Minnesota has about 11,000 lakes formed by the action of glaciers. Many of the smaller lakes (e.g., pot-hole lakes) in northeastern North America were formed in this manner, including the Finger Lakes of central New York. The Great Lakes were formed largely by glacial action.

Lakes contain four life zones (Fig. 3.17). The **littoral zone** is closest to shore and supports rooted plant growth. The **limnetic zone,** which is the main body of a lake, is the

FIGURE 3.16

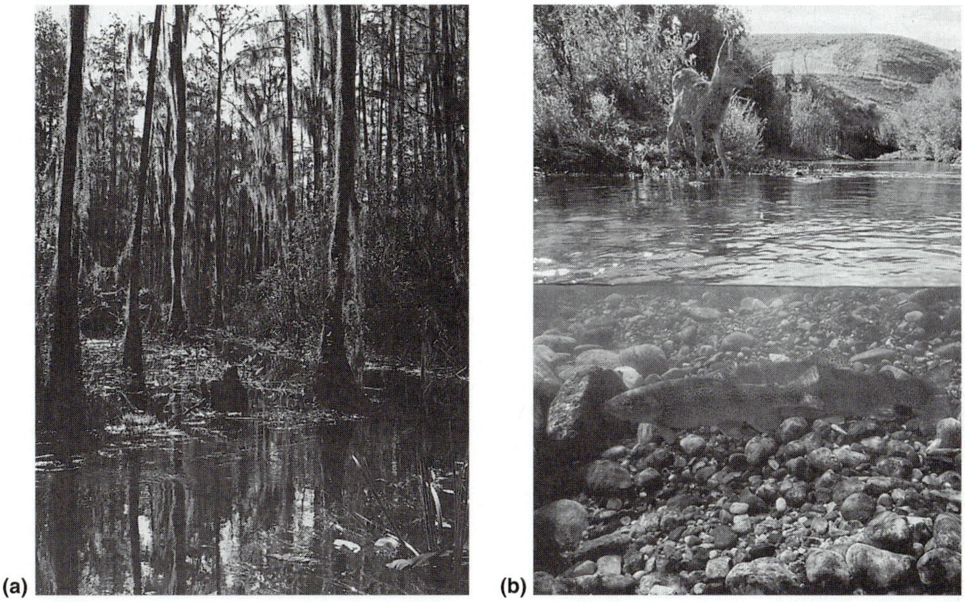

(*a*) Woodland swamp. (*b*) Two worlds of land and water meet in a Montana stream as a fawn encounters trout—a striking example of the barriers that separate two animals so near in space yet so isolated in their ways of life.

FIGURE 3.17

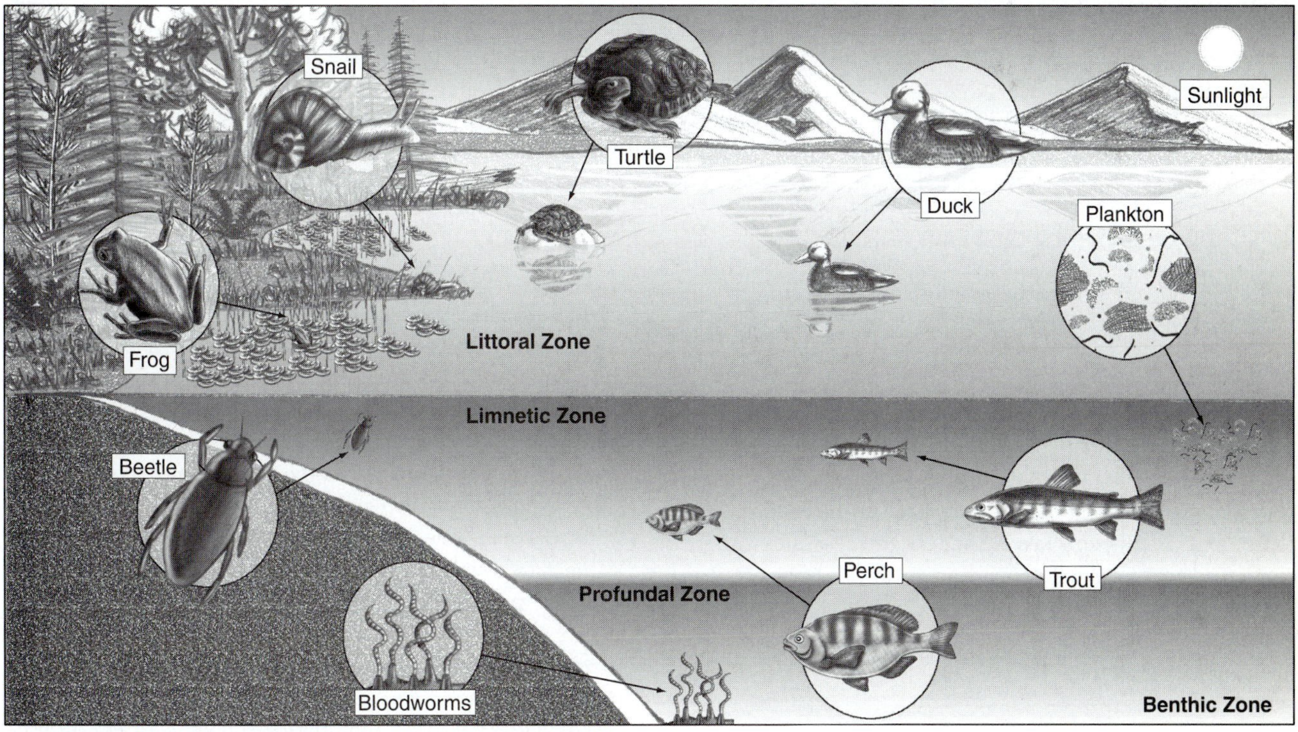

Four major zones of life in a lake. The *littoral* zone includes the shore and the shallow, nutrient-rich waters near the shore where sunlight penetrates to the lake bottom. It contains free-floating producers, rooted aquatic plants, and other aquatic life such as snails, frogs, and turtles. The *limnetic* zone is the open water that receives enough sunlight for photosynthesis to take place. It contains floating phytoplankton, plant-eating zooplankton, and fish. The *profundal* zone is the deep, open water beyond the limit of light penetration. It is too dark for photosynthesis to take place and is inhabited by fish adapted to its cooler, darker water. The *benthic* zone at the bottom of the lake is inhabited by large numbers of decomposers and detritus-feeding clams and wormlike insect larvae. These detritivores feed on dead plant debris, animal remains, and animal wastes that descend from above.

water too deep to support rooted plants but that receives sufficient light to support photosynthesis. Both of these zones contain phytoplankton, zooplankton, and nekton. In deep lakes, the region of open water below the depth of light penetration is known as the **profundal zone**. It is inhabited primarily by invertebrates and by fish adapted to cooler, darker water. The **benthic zone** at the bottom of a lake is inhabited primarily by decomposers that feed on animal wastes as well as dead plants and animals.

Although the freshwater environment is the smallest of the three types, it is vital to most classes of vertebrates. Some fishes and some salamanders live their entire lives in bodies of fresh water. Some fishes, such as eels, are **catadromous**—they spend most of their adult lives in freshwater lakes and streams, but return to the ocean to spawn; others, such as salmon, are **anadromous**—they spend most of their adult lives in the ocean, but return to the rivers where they were born in order to spawn. Many salamanders and anurans live in or near streams, ponds, marshes, and swamps their entire lives. These areas assist in keeping the skin of these vertebrates moist, and they also provide sites for egg deposition, hibernation, and a means of escape from predators.

BIO-NOTE 3.2

A Cold-Water Frog

The tailed frog (*Ascaphus truei*) (see Fig. 6.37) lives in and along cold, well-oxygenated mountains streams in northwestern North America. It has evolved numerous adaptations for life in rapid-flowing streams including strongly webbed hind feet, a copulatory organ for internal fertilization, the lack of vocal sacs and tympana, and tadpoles with oral disks modified as ventral suckers. The large ova have extremely low rates of oxygen consumption due to the low temperatures at which they develop. Furthermore, *Ascaphus* requires a 3-year period to complete its larval development in the cold mountain waters before it metamorphoses to reproduce in its fourth year.

Noble and Putnam, 1931
Brown, 1977

Alligators, many snakes, and many turtles are aquatic or semiaquatic. Many turtles hibernate in the mud on the bottoms of ponds and lakes. Ducks, geese, swans, shorebirds, and many others depend on freshwater habitats for protection, for nesting, and for food resources. Muskrats, beaver, mink, otters, water shrews, and many other mammals are semiaquatic.

TERRESTRIAL

This is the most variable of the three major environments. Elevations in the terrestrial environment range from below sea level to over 8,500 m in the Himalayas. Air temperature varies from −60°C to 60°C (140°F) and decreases approximately 1.2°C for every 305 m increase in elevation above sea level. Precipitation varies from only a fraction of a centimeter over several years to over 1,250 cm per year in some tropical areas (Fig. 3.18).

Various systems have been devised to classify terrestrial environments. Among these are the biome, life zone, and biotic province systems.

Biomes

The biome system of classification is based on natural climax communities and originally was proposed by Clements and Shelford in 1939 (Fig. 3.19). Because these communities are influenced primarily by temperature, rainfall, and soil conditions, they correspond closely to the distribution of climatic zones and soil types. For these reasons, they vary with latitude as well as with altitude, with the number of animals decreasing steadily with increasing latitude and altitude (Fig. 3.20). The size of species that range from the lowlands to the tops of mountains also often decreases with altitude (Allee and Schmidt, 1951).

The transition area between two adjacent communities or between two biomes is known as an **ecotone**. Species diversity of such areas is usually high because they contain resources from both biomes. Where forest grades into grassland, for example, the ecotone will have some characteristic forest species, some grassland species, and some additional species that require resources from both kinds of vegetation.

Tundra

This is a treeless biome that encircles the Earth just south of the ice-covered polar seas in the Northern Hemisphere (Fig. 3.21). For the most part, the tundra exists within the Arctic Circle, but it extends southward in Canada and along the western Alaskan shore. It also occurs at high elevations much farther south in the western United States; it is the area above treeline. It extends as far as 1,900 km south of the Arctic Circle along the shores of Hudson Bay. This cold biome, which receives approximately 20 cm of precipitation a year, has extreme seasonal fluctuations in photoperiod ranging from a 24-hour period of sunlight at the summer solstice (June 21–22) to one 24-hour period of nighttime at the winter solstice (December 21–22). Only during a short summer growing season does the temperature rise above the freezing point of water, and even the warmest summer months have a mean temperature of no more than 10°C. Table 3.1 shows typical air and soil temperatures for winter and summer. Even during the summer, only the top few inches or so of soil thaws; the remainder is permanently frozen and is known as **permafrost**. This makes tree growth impossible. The frozen subsoil prevents the escape of water by seepage, so that the melting ice and snow of winter produce innumerable lakes and ponds. Sedges, shortgrasses, lichens, and mosses—all plants that do not require a deep root system—cover the ground and must complete their flowering and fruiting cycles quickly. Lemmings (*Lemmus, Dicrostonyx*), ptarmigan (*Lagopus*), musk-ox (*Ovibos moscha-*

FIGURE 3.18

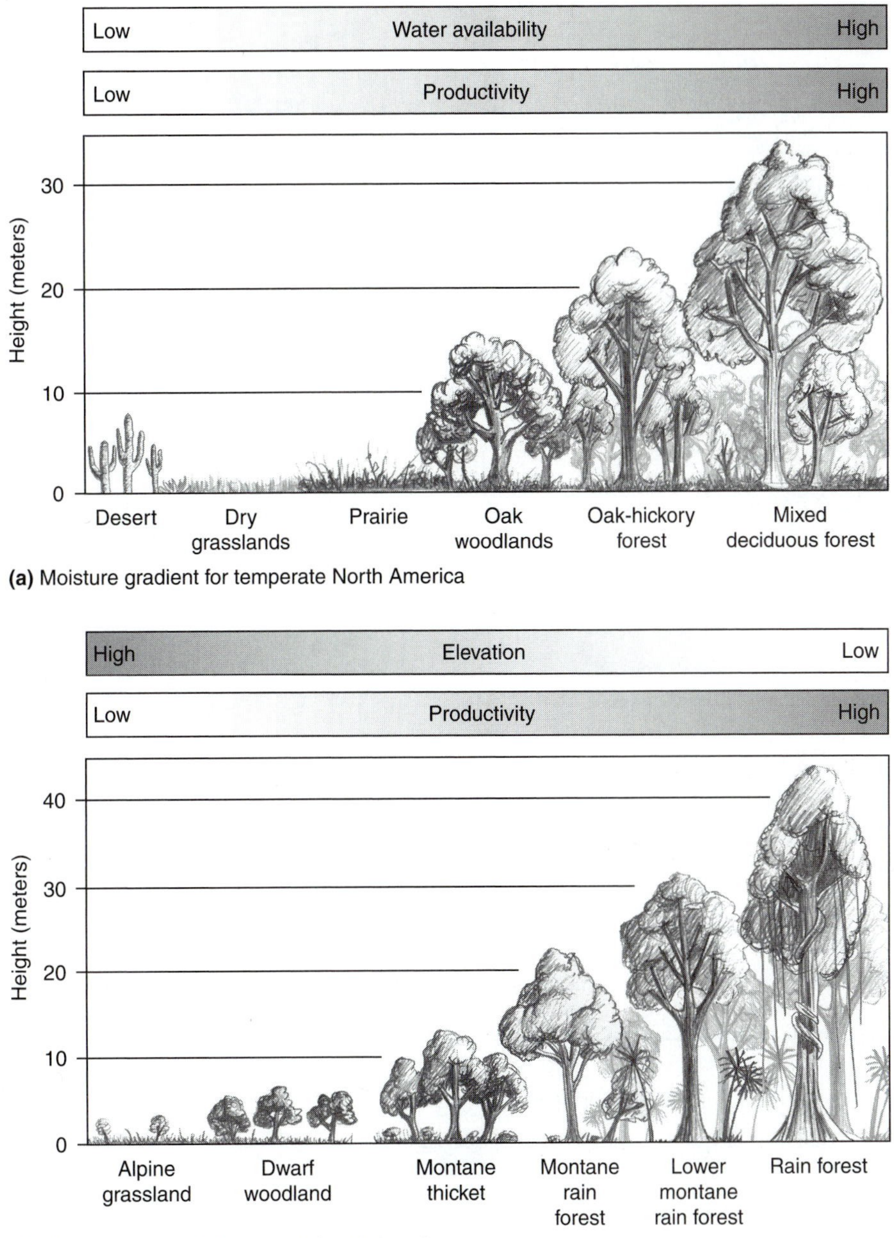

Plant species diversity and distribution, as well as plant forms, are influenced by water availability, elevation, and other factors such as temperature and soil drainage characteristics. *(a)* Moisture gradient for temperate North America. *(b)* Elevation gradient for tropical South America.

tus), polar bears (*Ursus maritimus*), and caribou (*Rangifer tarandus*) are permanent residents. Many shorebirds and waterfowl are summer residents.

Taiga

This is a coniferous forest biome that extends in a broad belt across northern Eurasia and North America. It also occurs at high elevations in more southern areas such as Clingman's Dome in the Great Smoky Mountains National Park in Tennessee, Mount Mitchell along the Blue Ridge Parkway in North Carolina, and in numerous areas in the western states. Spruce (*Picea*), fir (*Abies*), hemlock (*Tsuga*), and pine (*Pinus*) are the dominant tree species in this region of long, cold winters and short, cool summers. These evergreen trees shed one-fourth to one-third of their leaves (needles) annually and reproduce by forming cones. Little light reaches the ground, which is covered with ground-level vegetation such as lichens, mosses, and ferns. Soils are thin and

FIGURE 3.19

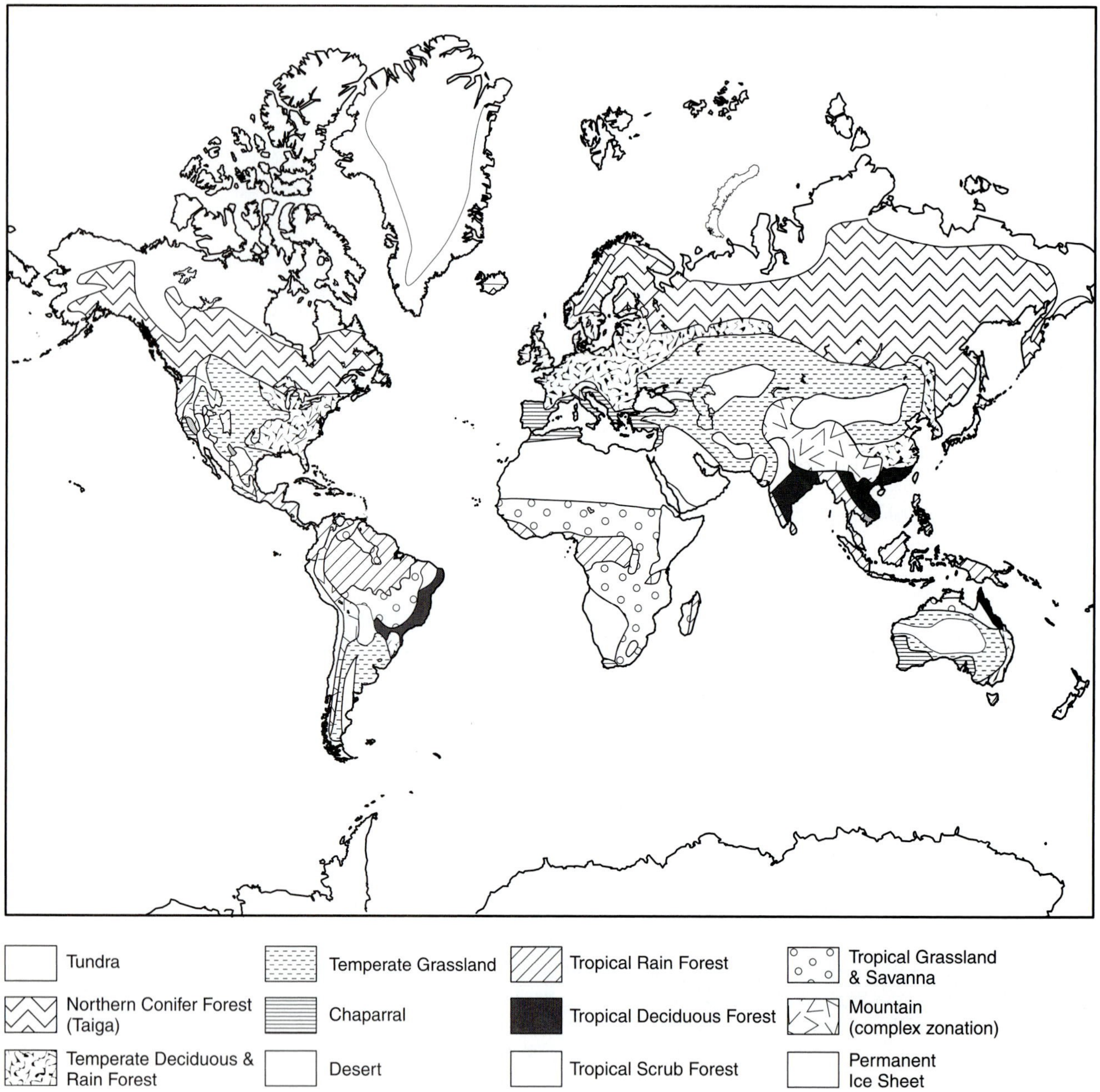

The biomes of the world.

nutrient-poor. Many lakes and swamps are present. Rainfall ranges between 40 and 100 cm per year, much of it in the form of heavy snow. The growing season is about 130 days. Snowy owls (*Nyctea scandiaca*), crossbills (*Loxia*), evening grosbeaks (*Hesperiphona vespertina*), lynx (*Lynx canadensis*), wolves (*Canis lupus*), moose (*Alces alces*), and bears (*Ursus*) are typical resident species.

Temperate Deciduous Forests

South of the taiga in eastern North America, eastern Asia, and much of Europe is a biome dominated by deciduous trees such as beech (*Fagus*), oak (*Quercus*), and maple (*Acer*). These areas have well-defined seasons, and most trees lose their leaves during the colder winter months. The growing season ranges between 140 and 300 days, with annual pre-

FIGURE 3.20

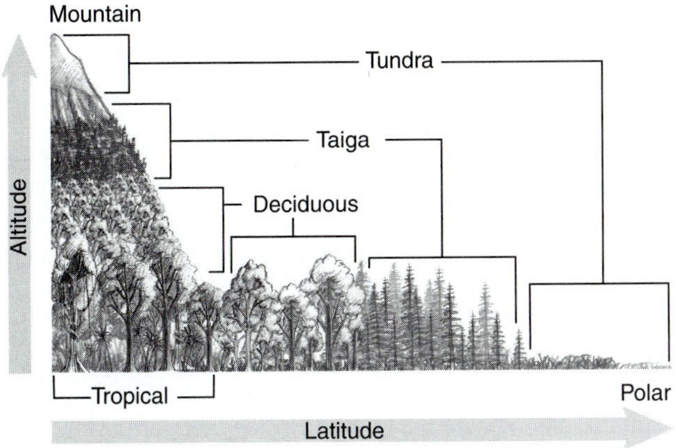

Biomes change not only with latitude (bottom), but also with altitude (left). Numbers of vertebrate species decrease steadily with increasing latitude and altitude.

FIGURE 3.21

Aerial view of the arctic tundra near Point Barrow, Alaska. Numerous small lakes are characteristic of this region.

cipitation being relatively evenly distributed throughout the year and ranging between 75 and 150 cm. Enough light penetrates the canopy so that shrubs, herbaceous plants, mosses, and ferns make up the understory. Typical vertebrate species include lungless salamanders (Plethodontidae), skinks (*Eumeces*), garter snakes (*Thamnophis*), turkeys (*Meleagris gallopavo*), pileated woodpeckers (*Hylatomus pileatus*), black-capped chickadees (*Parus atricapillus*), squirrels (*Sciurus*), bobcats (*Lynx rufus*), foxes (*Vulpes* and *Urocyon*), and white-tailed deer (*Odocoileus virginianus*).

TABLE 3.1

Typical Temperature Gradients in Winter and Summer in Tundra[a]

Height	Winter (Feb.)	Summer (July)
8 m (air)	–25°C	**8°C**
1 m (air)	–31°C	**9°C**
Snow surface (about 0.3 m)	–35°C	—
Soil surface	–22°C	**11°C**
–0.25 m (soil)	–19°C	**1°C**
–1 m (soil)	–16°C	–3°
–4 m (soil)	–10°C	–8°

[a]Temperatures above freezing are in bold type.

Source: Data from R. Brewer, The Science of Ecology, *1988, W.B. Saunders, Orlando, FL.*

Grasslands

Grasses are the climax species in areas that receive more than 20 cm of rainfall annually but not enough precipitation to support tree growth. Extensive root systems absorb water efficiently and allow rapid recovery from droughts and fires. The buildup of organic matter forms rich soil.

The central portion of North America has vast grasslands known as **prairies** (Fig. 3.22). The easternmost grasslands are tallgrass prairies and receive greater annual rainfall than the more western shortgrass prairies. Plains spadefoot toads (*Scaphiopus bombifrons*), Great Plains narrow-mouthed

FIGURE 3.22

Rolling shortgrass prairie to the east of the Rocky Mountains. Grasses are the climax species in areas that receive more than 20 cm of rainfall annually but not enough precipitation to support tree growth.

toads (*Gastrophryne olivacea*), prairie skinks (*Eumeces septentrionalis*), Great Plains skinks (*Eumeces obsoletus*), prairie racerunners (*Cnemidophorus sexlineatus*), prairie lizards (*Sceloporus undulatus*), prairie rattlesnakes (*Crotalus viridis*), Plains garter snakes (*Thamnophis radix*), Graham's water snakes (*Natrix grahami*), burrowing owls (*Speotyto cunicularia*), short-eared owls (*Asio flammeus*), dickcissels (*Spiza americana*), horned larks (*Eremophila alpestris*), prairie dogs (*Cynomys*), black-footed ferrets (*Mustela nigripes*), badgers (*Taxidea taxus*), coyotes (*Canis latrans*), and large grazing animals such as bison (*Bison bison*) and pronghorn antelope (*Antilocapra americanai*) are typical inhabitants.

Tropical grasslands are known as **savannas** (Fig. 3.23). Trees are distributed sparsely because of a severe summer dry season. The African savanna supports the greatest variety and number of large herbivores of all the biomes. These include elephants (*Loxodonta africana*), giraffes (*Giraffa camelopardalis*), many kinds of gazelles and antelopes, zebras (*Equus* spp.), wildebeests (*Connochaetes* spp.), African buffalo (*Syncerus caffer*), and rhinoceroses (*Diceros* and *Ceratotherium*). Typical predators include lions (*Panthera leo*), cheetahs (*Acinonyx jubatusi*), and leopards (*Panthera pardus*), and scavengers such as hyenas (*Hyaena brunnea*).

In some regions of the world, most of the rain falls in the cool winter, and the summers are hot and dry. This results in dense, shrubby, fire-resistant scrubland known as **chaparral.** These areas occur in parts of southern Africa, western Australia, central Chile, around the Mediterranean Sea, and in California. The shrubs have small, thick evergreen leaves and thick underground stems.

FIGURE 3.23

African savanna—a region of warm grasslands with isolated stands of shrubs and trees. More large ungulates such as giraffes, Cape buffalo, zebras, impalas, gnus, and gazelles live here than anywhere else. Predators such as lions and cheetahs and scavengers play important roles in the food web. Herds undertake annual migrations that coincide with rainy and dry seasons. When the rains come, herds return to the plains for the lush new grasses that are the preferred forage. Births are timed to coincide with the re-emergence of the grass, a fact well known to the predators.

Deserts

Large areas of desert occur in Africa, Asia, and Australia, with smaller areas in North and South America (Fig. 3.24a). These habitats experience hot days and cold nights and receive less than 25 cm of rainfall annually. When it does rain, much of it runs off or evaporates. Vegetation is sparse with cacti, Joshua trees (*Yucca*), sagebrush (*Artemis*), creosote (*Larrea*), and mesquite trees (*Prosopis*) among the typical plants in North American deserts. Following a rain, many desert plants burst into bloom and complete their life cycle in a short time. Horned lizards (*Phrynosoma* spp.), earless lizards (*Holbrookia* spp.), Gila monsters (*Heloderma suspectum*), western diamondback rattlesnakes (*Crotalus atrox*), Mojave rattlesnakes (*Crotalus scutulatus*), cactus wrens (*Campylorhynchus brunneicapillus*), Chichuahuan ravens (*Corvus cryptoleucus*), roadrunners (*Geococcyx californianus*), and kangaroo rats (*Dipodomys* spp.) are among the species that have adapted to the desert environment. Many are burrowers or live in burrows of other species. Adaptations include the reduction of cutaneous glands, concentrated urine, and dry feces. Many species can survive for months without drinking. They obtain limited water from their plant and animal food resources together with water produced by the processes of metabolism. Estivation (a period of inactivity during the summer) occurs in some species.

Some geographically separated, seed-eating desert rodents have undergone convergent evolution and developed similar morphological, physiological, ecological, and behavioral adaptations to their xeric habitats (Fig. 3.24b). These bipedal saltatorial forms have independently evolved (convergent evolution) on four continents (North America, Asia, Africa, and Australia).

Tropical Forests

Two major types of tropical forests occur. Tropical deciduous forests consist of broad-leaved trees that lose their leaves because of a dry season. Tropical rain forests, which occur in areas of abundant rainfall (over 250 cm per year) and continual warmth (20 to 25°C), are the richest of all of the biomes (Fig. 3.25). The canopy consists of several layers, with epiphytes common and lianas, or woody vines, encircling the trees. In open areas or clearings below the canopy, vegetation is dense and forms a thick jungle. Most animals spend a substantial amount of time in the trees. Treefrogs, iguanas (*Iguana*), anacondas (*Eunectes*), caimans (*Caiman*), parakeets, parrots, toucans, monkeys, and jaguars (*Panthera onca*) are representative of the vertebrate life.

Morphological convergence is strikingly evident in some African and South American rain forest mammals that occupy similar ecological niches (Fig. 3.26). Many of the species are in different families and even different orders.

FIGURE 3.24

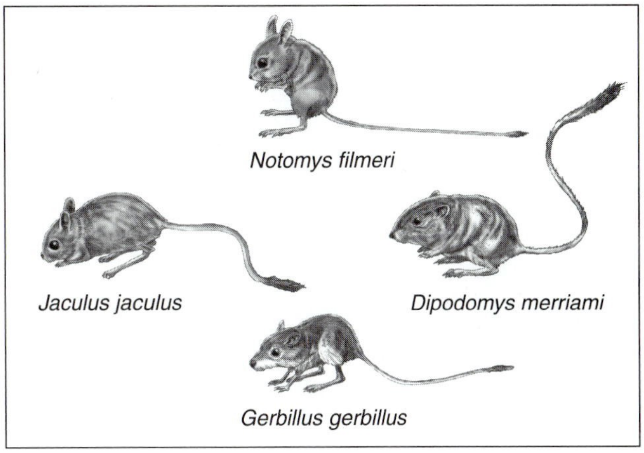

(b)

(a)

(*a*) Warm desert near Tucson, Arizona. Vegetation includes columnlike saguaro cacti, prickly pear cacti, creosote bushes, and tall, multistemmed ocotillo. Deserts experience hot days, cold nights, and receive less than 25 cm of rainfall annually. (*b*) Morphological convergence of seed-eating desert rodents on four continents. Illustrated forms include *Dipodomys* (Heteromyidae) from North America, *Jaculus* (Dipodidae) from Asia, *Gerbillus* (Muridae) from Africa, and *Notomys* (Muridae) from Australia. Each of these groups has independently evolved bipedal, hopping rodents that are similar in morphology.

FIGURE 3.25

Tropical rain forests are the richest of all of the biomes. They receive abundant rainfall (over 250 cm per year) and are continually warm (20 to 25°C). Epiphytes and woody vines are characteristic of the rain forest biome.

In 1990, Friends of the Earth estimated the annual loss of tropical forest at 142,000 km^2 globally, an area the size of Florida or Wisconsin. Also in 1990, the World Resources Institute put the annual loss at 160,000 to 200,000 km^2, roughly equivalent to the state of Washington. If the rate of loss continues to increase as it did between 1980 and 1990, the world's remaining tropical forests will be depleted by the year 2030.

Life Zones

In an attempt to combine the regional distributions of North American plants and animals into one ecological classification scheme, C. Hart Merriam (1890, 1898) proposed the concept of **life zones** (Fig. 3.27). Dr. Merriam was the founder and chief of the U.S. Bureau of Biological Survey, which later became the U.S. Fish and Wildlife Service. Merriam developed his life zone concept around the following two principles: (1) animals and plants are restricted in northward distribution by the total quantity of heat during the season of growth and reproduction; and (2) animals and plants are restricted in southward distribution by the mean temperature of a brief period covering the hottest part of the year. Life zones are broad belts that run longitudinally across the continent with southern horizontal extensions along the mountain chains or around individual mountains. These zones were biotic in concept, but their boundaries (isotherms) were determined by a given mean temperature. Each zone has a theoretical temperature range of approximately 4°C.

FIGURE 3.26

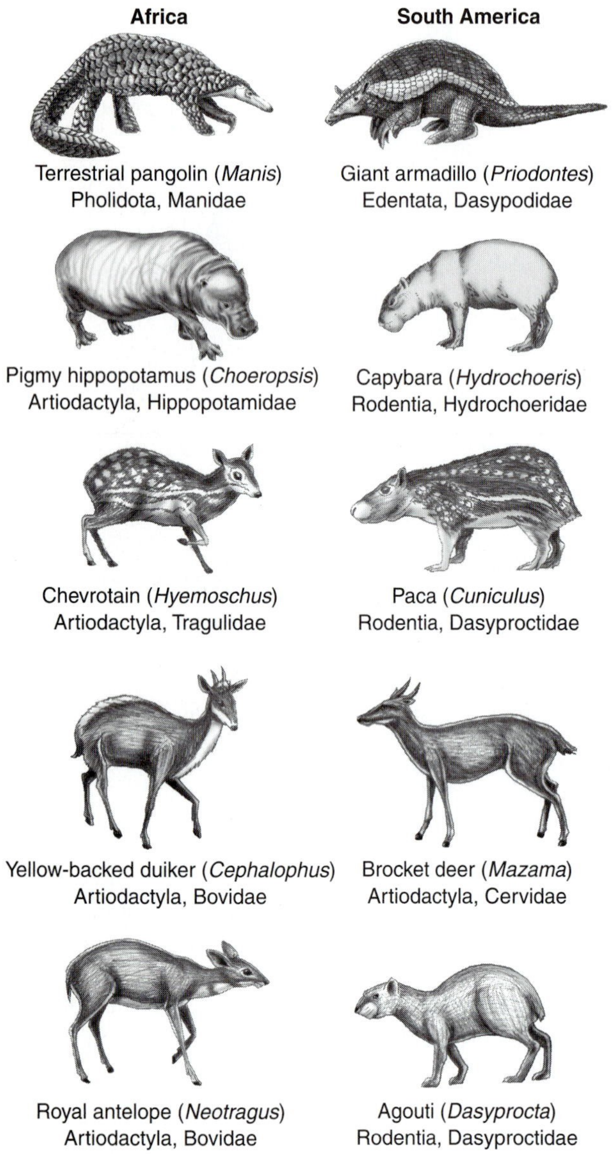

Morphological convergence among African and South American rain forest mammals that occupy similar ecological niches. Many of the species are classified in different families and even different orders.
Source: Gibson and Brown, Biogeography, *1983, Mosby Co.*

FIGURE 3.27

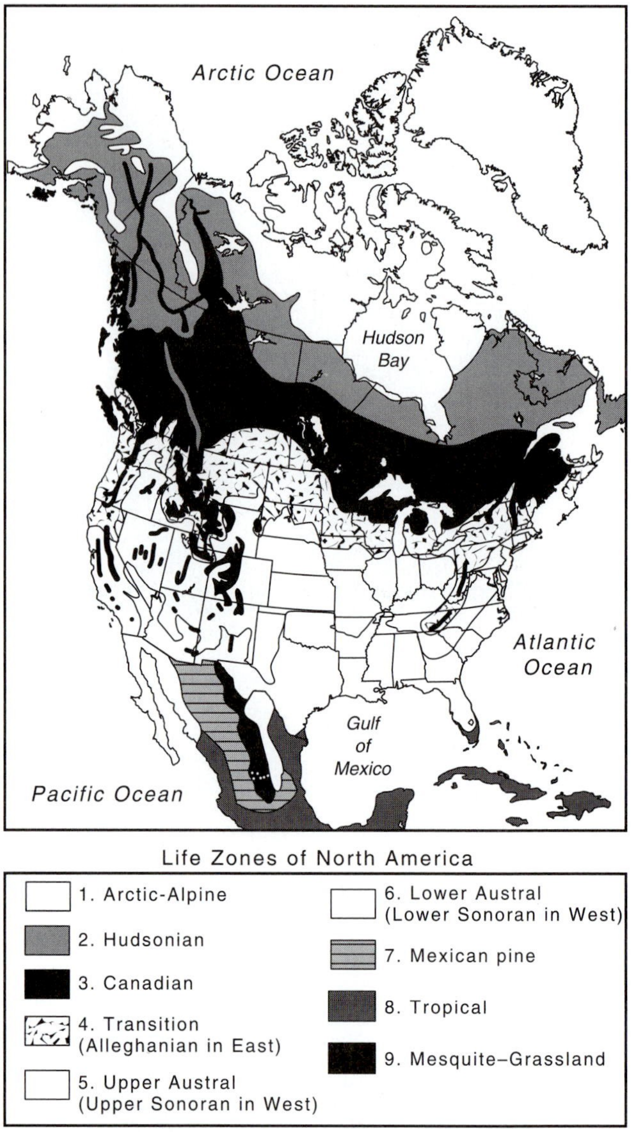

Life zones of North America.
Source: Data from C.F.W. Muesebeck and Arthur D. Cushman, U.S. Bureau of Entomology and Plant Quarantine.

Merriam initially divided the North American continent into three primary transcontinental regions: Boreal, Austral, and Tropical. The Boreal region extends from the northern polar area to southern Canada, with southward extensions along the Appalachians, the Rocky Mountains, and the Cascade–Sierra Nevada Range. The Austral region comprises most of the United States and a large portion of Mexico, and the Tropical region includes southern Florida, extreme southern Texas, some of the lowlands of Mexico, and most of Central America. Each of these regions is further subdivided into life zones.

Boreal Region

Arctic–Alpine Zone. This zone encompasses the area north of the limit of tree growth, including the tundra (Fig. 3.28), and mean temperature ranges from 6°C to 10°C. Lichens, grasses, Arctic willow, and Arctic poppy are typical plants, while Arctic foxes (*Alopex lagopus*), Arctic hares (*Lepus arcticus*), and lemmings (*Lemmus, Dicrostonyx*) are permanent residents. Snowy owls (*Nyctea scandiaca*) are also generally considered permanent residents. However, about every 4 years, large flocks of owls migrate southward in the fall to the northern United States and adjacent Canadian provinces. These migrations coincide with the low point in the population cycles of lemmings, their major prey (see Chapter 10 for further discussion of cycles).

FIGURE 3.28

Arctic–Alpine Life Zone in the Sierra Nevada of California at approximately 3,300 m. Alpine willow (*Salix anglorum*) inhabits this area but rarely attains a height of more than 15 cm.

FIGURE 3.29

Hudsonian Life Zone in the Sierra Nevada of California. Lodgepole pine (*Pinus contorta*) and mountain hemlock (*Tsuga mertenstana*) form the major vegetation near lake level with prostrate forms of the whitebark pine (*Pinus albicaulis*) covering the peaks in the distance.

Hudsonian Zone. This is the region of northern coniferous forests with southern extensions along the higher mountain ranges (Fig. 3.29). Mean temperatures range from 10°C to 14°C. Spruce and fir are the dominant trees, with typical vertebrates including great gray owls (*Strix nebulosa*), pine grosbeaks (*Pinicola enucleator*), olive-sided flycatchers (*Nuttallornis borealis*), moose (*Alces alces*), wolverines (*Gulo gulo*), and timber wolves (*Canis lupus*).

Canadian Zone. This zone includes the southern part of the boreal forest and the coniferous forests that cover the higher mountain ranges extending southward (Fig. 3.30). Mean temperatures range from 14°C to 18°C. Characteristic vertebrates include Canada jays (*Perisoreus canadensis*), spruce grouse (*Canachites canadensis*), varying hares (*Lepus americanus*), and lynx (*Lynx canadensis*).

FIGURE 3.30

A Canadian Life Zone forest consisting of red fir (*Abies magnifica*) in the Sierra Nevada of California. This zone includes the southern part of the boreal forests and the coniferous forests on large mountain ranges extending much farther south.

Austral Region

Transition Zone

This zone extends across the northern United States and runs south on the major mountain ranges (Fig. 3.31). As its name implies, this is a zone where the coniferous forest and deciduous forest intermingle. Mean temperatures range from 18°C to 22°C. In the east, this zone is known as the Alleghenian Zone. Typical Transition Zone plants include hemlock (*Tsuga*), walnut (*Juglans*), beech (*Fagus*), birch (*Betula*), oak (*Quercus*), ponderosa pine (*Pinus ponderosa*), and sagebrush (*Artemisia tridentata*). The American chestnut (*Castanea dentata*) was one of the most important eastern hardwoods until the early 1900s, when a fungus reached the

FIGURE 3.31

Ponderosa pine (*Pinus ponderosa*) is the dominant conifer in the Transition Life Zone over much of western North America.

United States from eastern Asia. The fungus proved to be lethal, and most chestnuts succumbed. Vertebrates in this zone include sage grouse (*Centrocercus urophasianus*), sharp-tailed grouse (*Pedioecetes phasianellusi*), Columbian ground squirrels (*Spermophilus columbianus*), northern flying squirrels (*Glaucomys sabrinus*), and black bears (*Ursus americanus*).

Upper Austral Zone

This broad, irregular transcontinental belt is subdivided into the Carolinian area (east of the Great Plains) and the Upper Sonoran in the semiarid west (Fig. 3.32a, b). Mean temperatures range from 22°C to 26°C. The Carolinian area is humid in summer and is characterized by such species as oaks (*Quercus*), hickory (*Carya*), sweet gum (*Liquedambar*), sycamore (*Platanus*), eastern wood frogs (*Rana sylvatica*), Carolina wrens (*Thryothorus ludovicianus*), cardinals (*Richmondena cardinalis*), eastern wood rats (*Neotoma*), and fox squirrels (*Sciurus niger*). Typical species of the Upper Sonoran include pinyon pine (*Pinus cembroides*), buckeye (*Aesculus glabra*), sagebrush (*Artemisia tridentata*), junipers (*Juniperus* spp.), scrub jays (*Aphelocoma coerulescens*), Bewick's wren (*Thryomanes bewickii*), ring-tailed cats (*Bassariscus astutus*), and northern grasshopper mice (*Onychomys*).

Lower Austral Zone

This region encompasses the southern United States from the Carolinas and the Gulf states to California (Fig. 3.33a, b). Mean temperatures range from 26°C to approximately 30°C. The more humid eastern area (Austroriparian) is characterized by such plants as bald cypress (*Taxodium distichum*),

FIGURE 3.32

(a) (b)

(*a*) Carolinian Life Zone in Ohio. Dominant trees are white and red oaks (*Quercus* sp.) with an understory of hickory (*Carya* sp.). (*b*) Upper Sonoran Life Zone in western Nevada. Sagebrush (*Artemisia tridentata*) and Utah juniper (*Juniperus osteosperma*) are common in this region.

FIGURE 3.33

(a)

(b)

(*a*) Austroriparian Life Zone in South Carolina. View of a bald cypress (Taxodium distichum) swamp forest. (*b*) Lower Sonoran Life Zone of a desert in Western North America. Creosote bush (Larrea divaricata) is the dominant perennial plant. It produces chemicals that prevent the growth of other plants.

magnolia (*Magnolia* spp.), and long-leaf pine (*Pinus palustris*), and by such vertebrates as pig frogs (*Rana grylio*), river frogs (*Rana heckscheri*), alligator snapping turtles (*Macroclemys temminki*), green anoles (*Anolis carolinensis*), brown pelicans (*Pelicanus occidentalis*), boat-tailed grackles (*Cassidix mexicanus*), chuck-will's widow (*Caprimulgus carolinensis*), rice rats (*Oryzomys palustrisi*), and cotton rats (*Sigmodon hispidus*). Indicator species of the arid western area (Lower Sonoran) include many types of cacti, ocotillo, mesquite, creosote bush, patch-nosed snakes (*Salvadora*), Gambel's quail (*Lophortyx gambelii*), cactus wrens (*Campylorhynchus brunneicapillus*), kit foxes (*Vulpes macrotis*), kangaroo rats (*Dipodomys*), and some species of pocket mice (*Perognathus*).

Tropical

This region is often divided into the Arid Tropical and Humid Tropical life zones. This, however, represents an oversimplification of a complex biotic region that more accurately is subdivided into seven major tropical zones: arid tropical scrub, thorn forest, tropical deciduous forest, savanna, tropical evergreen forest, rain forest, and cloud forest.

Biotic Provinces

A third method of classifying North American plant and animal communities and their distribution is by **biotic provinces**. A biotic province "covers a considerable and continuous geographic area and is characterized by the occurrence of one or more important ecologic associations that differ, at least in proportional area covered, from the associations of adjacent provinces" (Dice, 1943). The classification of biotic provinces is based largely on vegetation, because plants often indicate the characters of climate and soil on which animals also are dependent. Animals also directly or indirectly depend on plants for food and often for shelter and breeding places.

Each province is characterized by vegetation type, ecological climax, fauna, climate, physiography, and soil. Each includes both the climax communities and all the successional stages within its geographic area, including the freshwater communities. One or more climax associations may be present. Each biotic province is subdivided further into ecologically unique subunits known as **biotic districts** (subdivisions covering a definite and continuous part of the geographic area of a biotic province) and **life belts** (vertical subdivisions). Life belts, such as a grassland belt, a forest belt, or an alpine belt, are based primarily on altitude. They frequently are not continuous, but may occur under proper conditions of altitude and slope exposure on widely separated mountains within one biotic province. Boundaries of the provinces, districts, and life belts often coincide with physiographic barriers rather than with vegetation types, and they are often difficult to locate precisely because the different areas merge gradually into one another. The area covered by a particular biotic province varies from time to time, because of the production of new habitats through ecologic succession, and also because of slow, but more or less permanent modifications of climate (Dice, 1943).

Changes Caused by Human Activity

Humans have had a major impact on the distribution of vertebrates through domestication of some species, extinction of others, alteration of habitat, and both intentional and unintentional transport and introduction of great numbers of animals to regions where they were originally absent.

Dogs, cats, cattle, goats, horses, pigs, rabbits, and others have been domesticated from their wild ancestors. Domesticated vertebrates have served as invaluable human companions as well as important agricultural resources. Some of these species were taken on early sailing ships to provide food. When released on foreign shores and islands, they frequently survived, multiplied, and became feral (wild). Horses and cattle were introduced into various parts of North and South America in this manner. Cattle were brought to Australia, and pigs and goats established feral populations on many islands including the Galapagos Islands and New Zealand. European rabbits (*Oryctolagus cuniculus*) were released in Australia with devastating effects on both the flora and fauna (Fig. 3.34) (see discussion in Chapter 10).

The Norway rat (*Rattus norvegicaus*), black rat (*Rattus rattus*), and house mouse (*Mus musculus*) are native to Europe. By stowing away aboard early sailing ships, they were transported around the world. They are highly adaptable and flourish in most areas today.

The starling (*Sturnus vulgaris*), native to western Europe, was intentionally released into New York in 1896 (Fig. 3.35). These birds are largely commensal with humans, using our structures for nesting sites and agricultural habitats for food sources. They are aggressive, bullying birds that often drive native hole-nesting birds such as flickers, purple martins, and bluebirds away from suitable nesting sites.

FIGURE 3.34

European rabbits (*Oryctolagus cuniculus*) around a water hole in Australia. In 1859 a farmer in southern Australia imported a dozen pairs of wild European hares with the intention of introducing them as game animals. Within 6 years the 24 hares had mushroomed to 22 million, and by 1907 they had reached every corner of the continent. By the 1930s their population had reached an estimated 750 million.

FIGURE 3.35

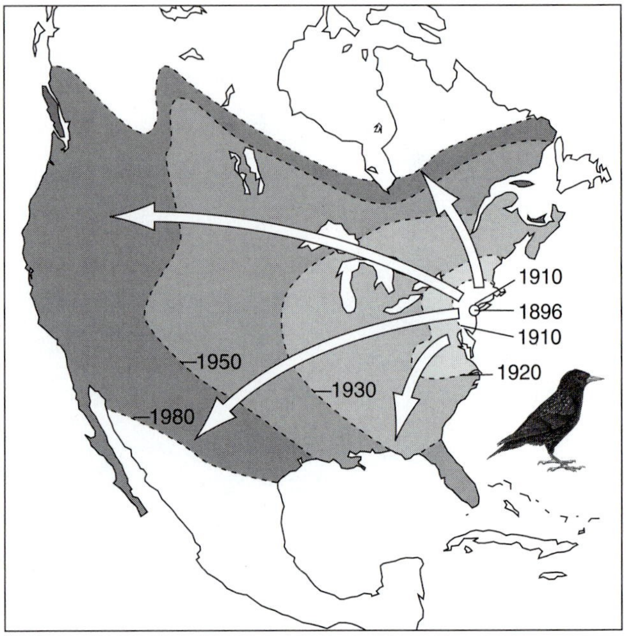

The European starling (*Sturnus vulgaris*) (inset) which was intentionally introduced into New York in 1896, has rapidly expanded its range in North America. These birds are extremely aggressive and often drive native hole-nesting birds, such as bluebirds (*Sialia sialis*), away from suitable nesting sites

Source: Gibson and Brown, Biogeography, *1983, Mosby Co.*

The English (house) sparrow (*Passer domesticus*) was introduced into New York in 1852. It reached the Pacific Coast in the early 1900s and is presently one of the most successful North American birds, ranging from coast to coast and from Guatemala to central Canada. Like the starling, it is usually abundant only around areas of human habitation. This species is undergoing extemely rapid evolution at the racial level. Local populations already have evolved distinctive combinations of traits. For example, birds in the northeastern United States are large and dark-colored, whereas those inhabiting the southwestern deserts are small and pale. These differences appear to represent adaptations to local climates and other factors.

Mongooses (*Herpestes* spp.) have been widely introduced to kill rats and snakes (Fig. 3.36a). Nowak (1991) noted that populations of mongooses (*H. auropunctatus*, native to the region from Iraq to the Malay Peninsula) had become established on most Caribbean islands, the northeastern coast of South America, and the Hawaiian and Fiji islands. He noted that several individuals also had been taken on the mainland of North America. In addition to killing rats, mongooses destroy harmless reptiles, birds, and mammals. A great deal of circumstantial evidence points to the mongoose being responsible for the extinction of many native West Indian birds and reptiles, but definitive proof is lack-

FIGURE 3.36

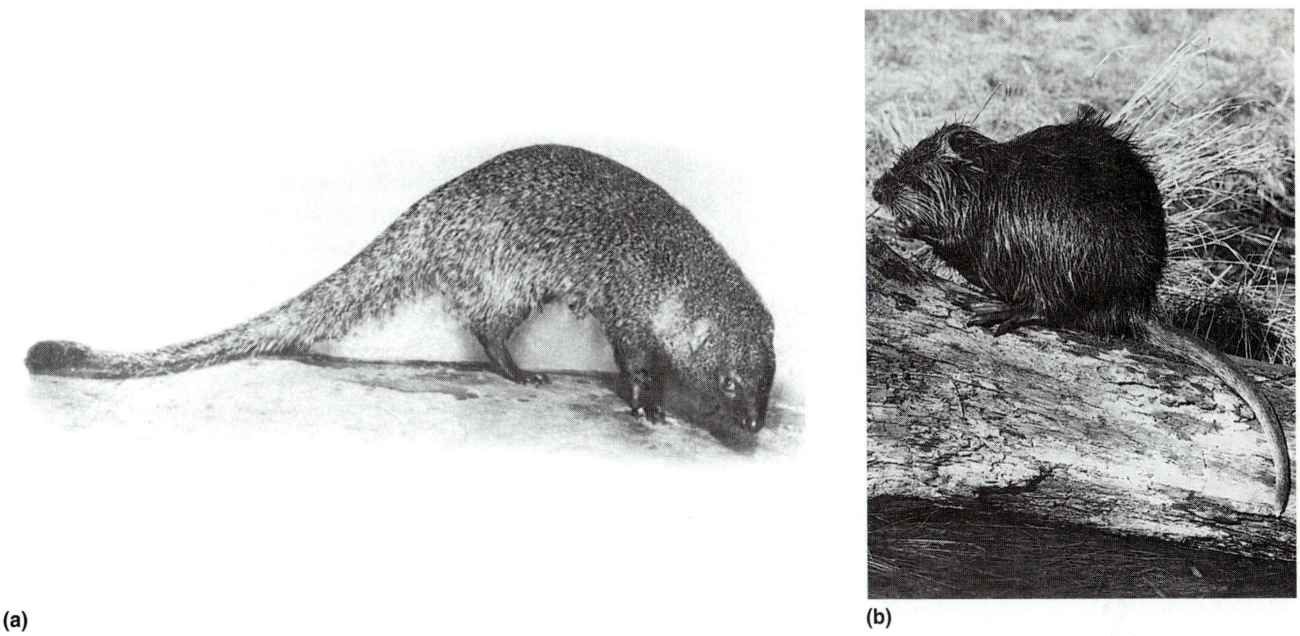

(*a*) The African mongoose (*Herpestes ichneumon*) was introduced into Portugal and Spain to kill rats and snakes. (*b*) The nutria (*Myocastor coypu*) is a South American rodent that was brought to North America as a potential commercial fur-bearer.

ing. They also have become pests by preying on poultry. The importation or possession of mongooses now is forbidden by law in some countries. In many areas, rats now live in trees, where they are safe from mongooses.

The nutria (*Myocastor coypu*), a South American rodent, was brought to North America as a potential fur-bearer (Fig. 3.36b). Most of the animals in the United States are thought to have originated from 20 animals brought to Louisiana in 1938 (Nowak, 1991). These mammals have established themselves in marshes in more than 20 states, with the most extensive populations being in the south-central part of the country. They are direct competitors of the native muskrat, as well as with waterfowl.

Many of the small native fishes in the southwestern United States, such as pupfishes, have suffered great reductions in their geographic ranges and complete extinction of local populations as a result of introductions of large predatory game fish, especially largemouth black bass (*Micropterus salmoides*), into their habitats. The native fishes are not adapted to large specialized predators because they have evolved in isolated lakes, streams, and springs for thousands of years during post glacial periods.

The construction of reservoirs has a drastic effect on riverine fauna. Some species can adapt to live in the warmer, more stable hydrological regime; others cannot. Most reservoir fishes stay close to shore, the mouth of tributaries, and in shallows. Deeper waters are often poorly utilized.

Lake trout (*Salvelinus namaycush*) were distributed widely over northern North America, but they were absent from bodies of water occupied by lampreys. Until the construction of the Welland Canal in 1829, the upper Great Lakes were protected by Niagara Falls, which formed a natural barrier impenetrable by lampreys. Following the opening of the canal, however, sea lampreys had access to the Great Lakes, and populations of trout were decimated. The first lamprey was reported in Lake Erie in 1921 (Dymond,1922). By the late 1940s, the sea lamprey was established in all of the Great Lakes and was causing severe damage to the native fishes, particularly lake trout, which experienced major declines.

Numerous tropical and semitropical vertebrates from fishes to mammals have been released or have escaped in Florida and have become established. The walking catfish (*Clarias batrachus*), marine toad (*Bufo marinus*), brown anole (*Anolis sagrei*), and several species of Neotropical monkeys are just a few examples.

Although most vertebrate introductions have been detrimental to the habitat and/or to other species, some have proved successful. Many of the trouts, including rainbow, brown, and brook trout, have been transported all over the world. The ringnecked pheasant (*Phasianus colchicus*) of Asia and eastern Europe has established itself in some regions of North America. The eastern gray partridge (*Perdix perdix*) became an important game bird in the wheat country of western Canada.

Human activities have been responsible for increasing the rates of extirpation and extinction of species throughout the world. These topics will be discussed in Chapter 14.

FIGURE 3.37

The coyote (*Canis latrans*) has extended its range eastward to the Atlantic Ocean and now occurs in all 48 contiguous states. It has benefited from timber cutting, which has opened vast new areas that were formerly forested.

Cutting of forested areas for lumber, for farmland, and for villages, towns, and cities has destroyed the habitat of many forest-dwelling species. It has, however, allowed a few species to become more widely distributed. The prairie deermouse (*Peromyscus maniculatus bairdii*) originally inhabited the prairie grasslands of central North America, but as eastern forests were cleared, it gradually extended its range eastward and colonized agricultural cropland and old fields (Baker, 1968). White-tailed deer in North America are more abundant now than they have ever been. Part of the reason for their abundance has been the elimination of predators such as cougars (*Felis concolor*) and wolves (*Canis lupus* and *C. rufus*). Another factor, however, has been the cutting of the forests, which opened vast new areas of fields and second-growth timber that are ideal deer habitat. This same activity has allowed the coyote (*Canis latrans*) to spread eastward to the Atlantic Ocean and to occur in all 48 contiguous states (Fig. 3.37).

Review Questions

1. Discuss the three types of barriers to vertebrate distribution. Give several specific examples of each.
2. Distinguish between biogeographic realms and biomes. In which biome do you live?
3. List the six major biogeographic regions as proposed by Wallace. Give several species of vertebrates that inhabit each region.
4. Define plate tectonics. How has this process affected climate and the distribution of vertebrates?
5. Land bridges have played a major role in the distribution of vertebrates. Name two land bridges and discuss the significance of each.
6. Contrast the effect of mountain ranges in Europe and North America on the distribution of vertebrates during glacial episodes.
7. List several survival adaptations developed by desert species.
8. Differentiate between plankton and nekton.
9. Explain the importance of estuaries to marine communities.
10. What is the difference between the littoral and profundal zones of a lake ecosystem?
11. Which life zone do you live in?
12. Define ecotones and describe their significance.
13. Differentiate between tundra and taiga.
14. Why do you think that amphibians and reptiles are unable to inhabit the arctic tundra?

Supplemental Reading

Brown, D. E., F. Reichenbacher, and S. E. Franson. 1998. *A Classification of North American Biotic Communities.* Salt Lake City: University of Utah Press.

Brown, J. H., and M. V. Lomolino. 1998. *Biogeography.* Second edition. Sunderland, Massachusetts: Sinauer Associates.

Elias, S. A. 1995. *The Ice-Age History of Alaskan National Parks.* Washington, D. C.: Smithsonian Institution Press.

Elias, S. A. 1995. *The Ice-Age History of National Parks in the Rocky Mountains.* Washington, D.C.: Smithsonian Institution Press.

Hocutt, C. H., and E. O. Wiley (eds.). 1986. *The Zoogeography of North American Freshwater Fishes.* New York: John Wiley and Sons.

Leigh, E. C., Jr., A. S. Rand, and D. M. Windsor (eds.). 1996. *The Ecology of a Tropical Forest.* Washington, D.C.: Smithsonian Institution Press.

Matthews, W. J., D. J. Hough, and H. W. Robison. 1992. Similarities in fish distribution and water quality patterns in streams of Arkansas: congruence of multivariate analysis. *Copeia* 1992: 296–305.

McGinn, A.P. 1999. *Safeguarding the Health of Oceans.* Worldwatch Paper 145. Washington D.C.: Worldwatch Institute.

McHugh, J. L. 1967. Estuarine nekton. Pages 581–620. In: Lauff, G. H. (ed.). *Estuaries.* Washington, D.C.: American Association for the Advancement of Science Publication 83.

Naiman, R. J., and D. L. Soltz. 1981. *Fishes in North American Deserts*. New York: John Wiley and Sons.

Newman, A. 1990. *Tropical Rainforest*. New York: Facts On File.

Rogers, R. A., L. A. Rogers, R. S. Hoffman, and L. D. Martin. 1991. North American biological diversity and the biogeographic influence of Ice Age refugia. *Journal of Biogeography* 18:623–630.

Tarduno, J. A., D. B. Brinkman, P. R. Renne, R. D. Cottrell, H. Scher, and P. Castillo. 1998. Evidence for extreme climatic warmth from Late Cretaceous arctic vertebrates. *Science* 282:2241–2243.

Wagner, W. L. and V. A. Funk (eds.). 1995. *Hawaiian Biogeography*. Washington, D.C.: Smithsonian Institution Press.

JOURNAL

Mangroves and Saltmarshes. SPB Academic Publishing. Papers addressing both pure and applied research topics. Four issues per year with six research articles per issue.

Vertebrate Internet Sites

Visit the zoology web site at http://www.mhhe.com/zoology to find live Internet links for each of the references listed below.

1. **Biomes.**
   This site, from the University of California at Berkeley's Museum of Paleontology, has much information on the 5 primary biomes.
2. **Biomes.**
   This web site is from NASA's Classroom of the Future.
3. **Clickable Map: U.S. Fish and Wildlife Service.**
   Links to sites with more information related to the 7 regions of the United States.
4. **Habitat Conservation Plans.**
   This page, from the National Wildlife Federation, contains fact sheets, articles, and links to more information on saving species by protecting habitats.
5. **Ecology of the Intertidal and Subtidal Areas.**
   Information on both the intertidal and subtidal estuarine subsystems. From the Coastal Services Center of NOAA.
6. **Desert Animal Survival.**
   Information on adaptations of vertebrates to life in the desert. Excellent images and links to organizations.
7. **SeaWeb.**
   Information designed to raise awareness of the world's oceans and the life within it.
8. **This Dynamic Earth.**
   Information on plate tectonics from the U.S. Geological Survey.
9. **Plate Tectonics.**
   Information from the University of California Museum of Paleontology, including a number of animations.
10. **Alfred Wegener's Theory of Continental Drift.**
    The page on German geophysicist Alfred Wegener's studies. Wegener put forth his theory of continental drift that was met with considerable hostility from his colleagues and not fully accepted until the late 1950s.
11. **Modern Tectonic Plate Boundaries.**
    Nice GIF image showing the various plates.
12. **Rainforest Action Network.**
    The state of the world's rainforests and temperate moist forests, and what you can do! For example, where should you purchase lumber in order to be environmentally conscious? See this site for the answer!
13. **Rainforest Preservation Foundation.**
    Information, downloads, and how you can help to preserve this imperiled ecosystem.
14. **Home Page of the Nature Conservancy—Help Protect Endangered Species.**
    Go to the site index to find all of the topics covered.
15. **Historical and Cultural Significance of the Bering Land Bridge National Preserve.**
    Excellent source of information on the region, climate, continental glaciation, native peoples.

CHAPTER 4

Early Chordates and Jawless Fishes

INTRODUCTION

There are many hypotheses concerning the evolution of vertebrates. These hypotheses are continually being changed and refined as new studies uncover additional evidence of evolutionary relationships and force reassessments of some earlier ideas about vertebrate evolution (Fig. 4.1). New fossil evidence, morphological studies, and comparative studies of DNA and RNA are gradually filling gaps in our knowledge and providing a more complete understanding of the relationships among vertebrates.

Evolution takes place on many scales of time. Gingerich (1993) noted that field and laboratory experiments usually are designed to study morphological and ecological changes on short time scales; in contrast, fossils provide the most direct and best information about evolution on long time scales. The principal problem with the fossil record is that the time scales involved, typically millions of years, are so long that they are difficult to relate to the time scales of our lifetimes and those of other organisms. Many biologists have difficulty understanding evolution on a geological scale of time, and many paleontologists have difficulty understanding evolution on a biological scale of time. One reason for this is that we have almost no record of changes on intermediate scales of time—scales of hundreds or thousands of years—that would permit evolution on a laboratory scale of time to be related to evolution on a geological scale.

No living protochordate (tunicate and lancelet) is regarded as being ancestral to the vertebrates, but their common ancestry is evident. In 1928, Garstang proposed a hypothesis by which larval tunicates could have given rise to cephalochordates and vertebrates (Fig. 4.2). Garstang suggested that the sessile adult tunicate was the ancestral stock and that the tadpolelike larvae evolved as an adaptation for spreading to new habitats. Furthermore, Garstang suggested that larval tunicates failed to metamorphose into adults but developed functional gonads and reproduced while still in the larval stage. As larval evolution continued, the sessile adult stage was lost, and a new group of free-swimming animals appeared. This hypothesis, known as **paedomorphosis** (the presence of evolutionary juvenile or larval traits in the adult body), allowed traits of larval tunicates to be passed on to succeeding generations of adult animals.

The first vertebrate is thought to have used internal gills for respiration and feeding while swimming through shallow water. It was probably similar in appearance and mode of living to the lancelet or amphioxus, *Branchiostoma*, which currently lives in shallow coastal waters. Cephalochordates possess symplesiomorphic (Ch.2, p.28) features that ancestral vertebrates are presumed to have inherited, such as a notochord, a dorsal hollow nerve cord, and pharyngeal gill slits, and they occurred earlier in geological time than the first known fossil vertebrates. Even though the lancelet is primitive, its asymmetry and unusual pattern of nerves appear to make it too specialized to be considered a truly ancestral type.

Feduccia and McCrary (1991), however, believed that cephalochordates were the probable vertebrate ancestors. As evidence, they cited the discovery of the mid-Cambrian 520-million-year-old *Pikaia gracilens*, a cephalochordate fossil found in the Burgess Shale formation in British Columbia, Canada. *Pikaia* possessed a notochord and segmented muscles and, in 1991, was the earliest known chordate (Fig. 4.3). Since that time, an even earlier possible chordate, *Yunnanozoon lividum*, from the Early Cambrian (525 million years ago), has been reported from the Chengjiang fauna in China (Chen et al., 1995). It possessed a spinelike rod believed to be a notochord, metameric (segmental) branchial arches that possibly supported gills, segmented musculature, and a row of gonads on each side of the body. Not everyone is convinced that *Yunnanozoon* is a chordate. In fact, another Chinese researcher (Shu et al., 1996a) has classified it in another closely related phylum—the phylum Hemichordata (acorn worms).

In 1996, researchers discovered a 530-million-year-old fossil from the same Chengjiang fossil site and proclaimed it to be the oldest chordate fossil (Monastersky, 1996c; Shu et al., 1996b). *Cathaymyrus diadexus* (Fig. 4.4a) is 2.2 cm long, has V-shaped segments that closely resemble the stacked

FIGURE 4.1

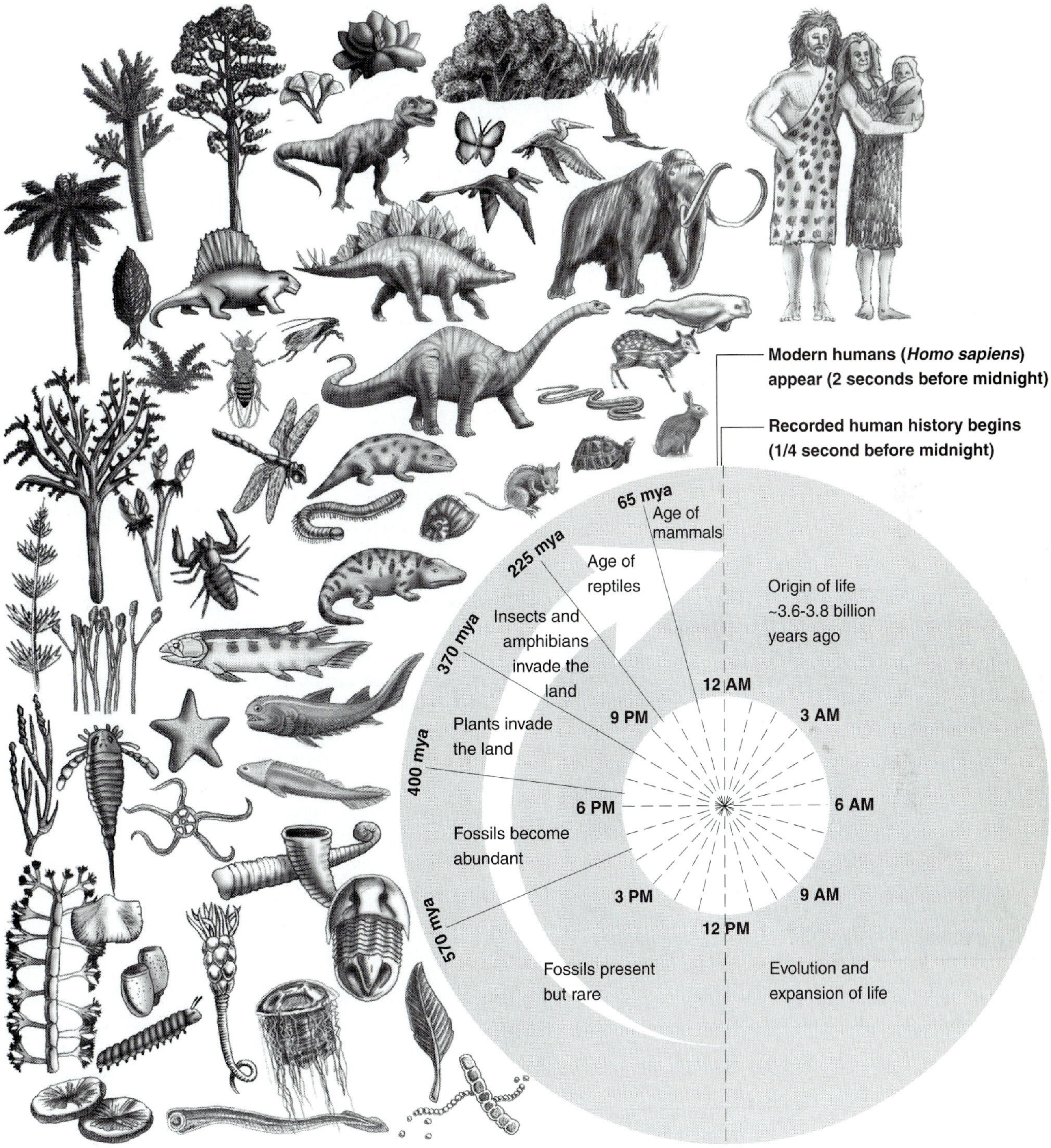

Greatly simplified timeline showing the history of the evolution of different forms of life on Earth compared to a 24-hour time scale. The human species evolved only about 2 seconds before the end of this 24-hour period.

muscle blocks in primitive living chordates such as amphioxus, and a creaselike impression running partway down the back of the body that scientists interpret as the imprint left by the animal's notochord.

More than 300 fossil specimens of another craniate-like chordate, *Haikouella lanceolata*, were recovered from Lower Cambrian (530 million year old) shale in central Yunnan in southern China (Chen et al., 1999). The 3-centimeter

FIGURE 4.2

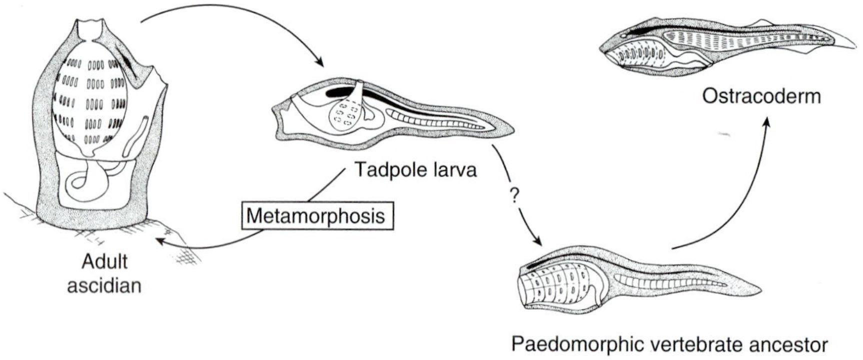

Garstang's hypothesis of larval evolution from paedomorphic urochordate larvae. Adult tunicates live on the sea floor but reproduce through a free-swimming "tadpole" stage. More than 500 million years ago, some larvae began to reproduce in the swimming stage. These are believed to have evolved into the ostracoderms, the first known vertebrates.

FIGURE 4.3

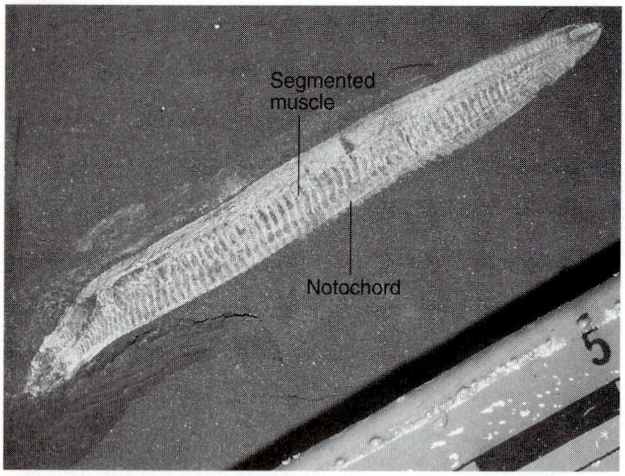

Pikaia gracilens, an early chordate, from the Burgess Shale of British Columbia, Canada.

Haikouella fossils are similar to *Yunnanozoon*, but they have several additional features: a heart, ventral and dorsal aortae, gill filaments, a caudal projection, a neural cord with a relatively large brain, a head with possible lateral eyes, and a ventrally situated buccal cavity with short tentacles.

Researchers continue to search for the earliest vertebrate (Janvier, 1999). Several groups of organisms have been proposed as "possible" chordates and vertebrates. Their inclusion in the vertebrate group is still uncertain, and their significance to the vertebrate story remains unclear.

CALCICHORDATES

One of these groups, the **calcichordates**, comprise marine organisms, usually classified as echinoderms, known only from fossils dated from 600–400 million years ago (Jefferies, 1986) (Fig. 4.4b). Calcichordates were covered by small plates of calcium carbonate, possibly representing incipient bone. Although they possessed indentations on their sides and an expanded anterior chamber, there is no evidence that these structures formed a pharyngeal gill apparatus. Other vertebrate-like characteristics pointed out by proponents include an expanded anterior nervous system (brain?) and a whiplike stalk (postanal tail?). However, there is no evidence of a notochord, nerve cord, or segmented musculature.

CONODONTS

The second group recently proposed as possible vertebrates are the **conodonts** (Fig. 4.5). These were small (4 cm) wormlike marine organisms, known only from some fossils with small teeth containing calcium phosphate. Some segmented muscle was present in a bilaterally symmetrical body. They appeared in the Cambrian (510 million years ago) approximately 40 million years before the earliest vertebrate fossils and lasted until the Triassic (200 million years ago). Recent evidence of large eyes with their associated muscles; fossilized muscle fibers strikingly similar to fibers in fossil fishes; a mineralized exoskeleton; the presence of dentine; and the presence of bone cells make it a likely candidate as a neargnathostome (jawed) vertebrate. The absence of a gill apparatus, however, is still puzzling (Sansom et al., 1994; Gabbott et al., 1995; Janvier, 1995). The discovery of microscopic

FIGURE 4.4

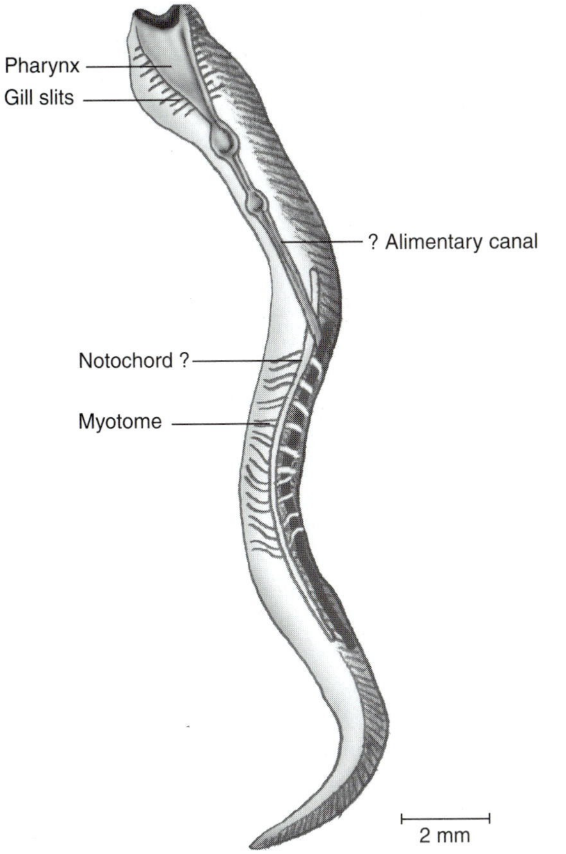

(a)

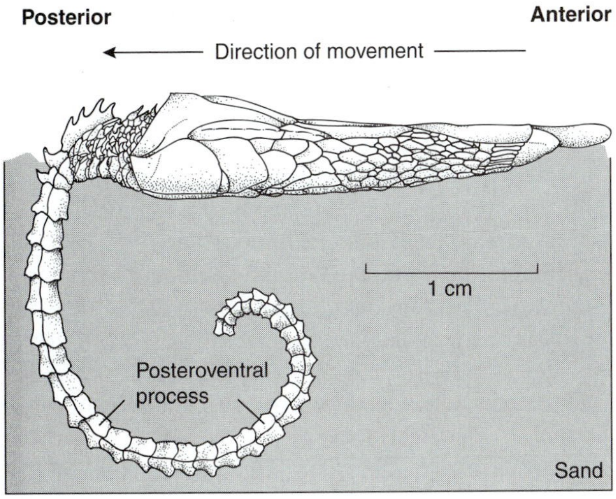

(b)

(*a*) Camera lucida drawing of *Cathaymyrus diadexus*, a new species. (*b*) Lateral view of a calcichordate, showing small overlapping plates of calcium carbonate covering the surface of the animal's body.

FIGURE 4.5

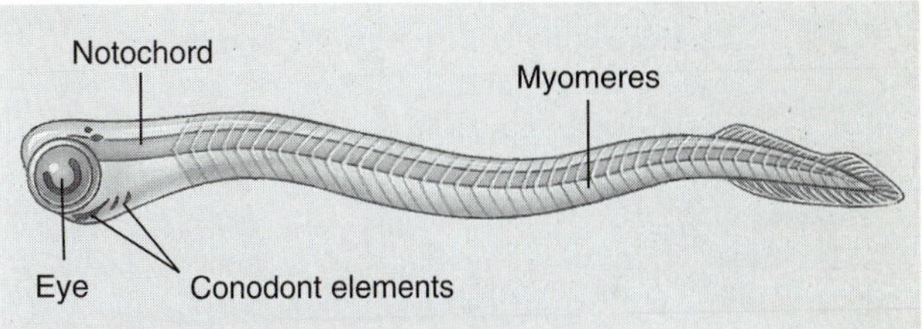

(a)

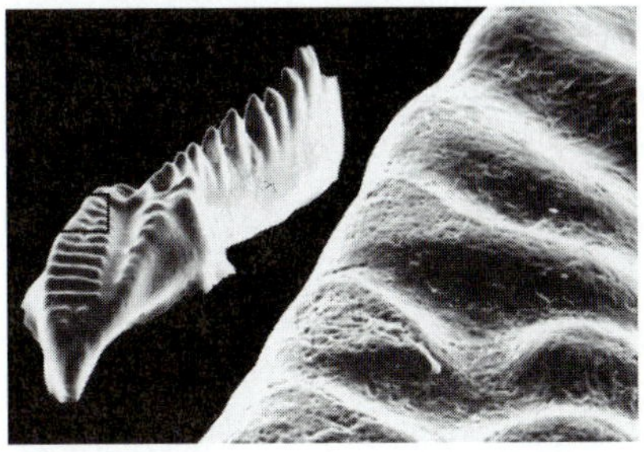

(b)

(*a*) Restoration of a living conodont. Although superficially resembling an amphioxus, the conodont possessed a much greater degree of encephalization (large, paired eyes; possible auditory capsules) and bonelike mineralized elements—all indicating that the conodont was a vertebrate. The conodont elements are believed to be gill-supporting structures or part of a suspension-feeding apparatus. (*b*) Micrograph shows single conodont tooth with closeup of ridges worn down by crushing food.
(a) *From Cleveland P. Hickman, Jr., et al.,* Integrated Principles of Zoology, *10th edition. Copyright © 1997 McGraw-Hill Company, Inc. All Rights Reserved. Reprinted by permission.*

wear patterns on the teeth, perhaps produced as food was sheared and crushed, supports the hypothesis that these early forms were predators (Purnell, 1995).

EARLY CAMBRIAN FISH-LIKE FOSSILS

Shu et al. (1999) described two distinct types of agnathan from the mid-Lower Cambrian (530 million years ago) Chenjiang fossil site. One form, *Haikouichthys ercaicunensis*, has structures resembling a branchial basket and a dorsal fin with prominent fin-radials, and is lamprey-like. The second fossil, *Myllokunmingia fengjiaoa*, has well developed gill pouches with probable hemibranchs and is closer to the hagfish. Shared features include complex myomeres and a notochord, as well as probable paired ventral finfolds and a pericard. The zigzag arrangement of segmented muscles is the same type pattern seen in fish today. The arrangement of the gills is more complex than the simple slits used by amphioxus. These agnathan vertebrates predate previous records by at least 20 and possibly as many as 50 million years (Shu et al., 1999).

Although both *Haikouichthys* and *Myllokunmingia* lack the bony skeleton and teeth seen in most, but not all, members of the subphylum Vertebrata, they appeared to have had skulls and other skeletal structures made of cartilage. Shu et al. (1999) proposed that vertebrates evolved during the explosive period of animal evolution at the start of the Cambrian and only some 30 million years later developed the ability to accumulate minerals in their bodies to form bones, teeth, and scales.

BIO-NOTE 4.1

Homeobox Genes

Whereas the preceding discussion focused on recent (direct) ancestors of vertebrates, some researchers believe that *all* animals are descended from a common ancestor and share a special family of genes (the homeobox, or *Hox*, genes) that are important for determining overall body pattern. The protein product of *Hox* genes controls the activation of other genes, ensuring that various body parts develop in the appropriate places. *Hox* genes are "organizer" genes; they switch other genes "on" and "off." Garcia-Fernandez and Holland (1994) have described a single cluster of *Hox* genes from an amphioxus, *Branchiostoma floridae,* that matches the 38 *Hox* genes in four clusters on different chromosomes known from mammals. Each amphioxus *Hox* gene can be assigned to one of the four clusters, and they are even arranged in the same order along the main axis of each chromosome. These genes are involved in embryonic patterning and development and serve as blueprint genes. Patterns of *Hox* gene expression are established that give cells a positional address, and then the interpretation of this positional information leads to the appropriate development of particular bones, appendages, and other structures. Most vertebrates, including mammals, have four *Hox* clusters, suggesting that two genome duplications occurred since these lineages split from the invertebrates, which typically have only one *Hox* cluster.

A change in *Hox* gene number has been hypothesized as a significant factor in the evolution of vertebrate structures. For example, at the 1998 meeting of the Canadian Institute for Advanced Research Programs in Evolutionary Biology, John Postlethwait and his colleagues at the University of Oregon announced that they had found that zebra fish have seven *Hox* clusters on seven different chromosomes. They hypothesize that the doubling might have occurred very early in the ray-finned fish (Actinopterygii) lineage and might explain how the 25,000+ species came to evolve such diverse forms. Although their respective evolutionary histories are unique, vertebrate, insect, and other animal appendages are organized via a similar genetic regulatory system that may have been established in a common ancestor.

Garcia-Fernandez and Holland, 1994
Gee, 1994
Shubin et al., 1997
Vogel, 1998

EVOLUTION

The evolution of the major groups of hagfishes, lampreys, and fishes and their relationships to each other, to the amphibians, and to amniotes is shown in Fig. 4.6. A cladogram showing probable relationships among the major groups of fishes is shown in Fig. 4.7. Because taxonomy is constantly undergoing refinement and change, the relationships depicted in this cladogram, along with others used in this text, are subject to considerable controversy and differences of opinion among researchers (see Supplemental Readings at end of chapter).

The earliest vertebrate remains were thought to consist of fossil remnants of bony armor of an ostracoderm (*Anatolepis*) recovered from marine deposits in Upper Cambrian rocks dating from approximately 510 million years ago (Repetski, 1978). Recent studies, however, have identified these remains of "bone" as the hardened external cuticles of early fossil arthropods (Long, 1995). Since bone is found only in vertebrates, the presence of bone in a fossil is highly significant. Young et al. (1996) and Janvier (1996) reported fragments of bony armor from a possible Late Cambrian (510 million years ago) early armored fish from Australia. The fragments bear rounded projections, or tubercles, that bear a striking resemblance to those of arandaspids, a group of jawless vertebrates from the Ordovician period. The Australian fragments, unlike arandaspid armor, which is composed of bone, are made up of enamel-like material. Both arandaspids and the Australian fragments also lack dentin (a substance softer than enamel but harder than bone). Dentin is deposited by specialized cells derived from ectomesoderm, thus providing indirect evidence of the presence of a neural crest, a unique vertebrate tissue found nowhere else in the Animal kingdom (Kardong, 1998).

At present, the *oldest identifiable vertebrate fossils with real bone* are fragmentary ostracoderm fossils (*Arandaspis*) that have been found in sedimentary rocks formed in fresh water near Alice Springs in central Australia during the Ordovician period, approximately 470 million years ago (Long, 1995) (Fig. 4.8a). The bony shields were not preserved as bone but as impressions in the ancient sandstones. The *first complete* Ordovician ostracoderm fossils (*Sacabambaspis*) were discovered in central Bolivia in the mid-1980s by Pierre Yves-Gagnier (Long, 1995) (Fig. 4.8b). They have been dated at about 450 million years ago and, thus, are slightly younger than the Australian fossils, but they are much more completely preserved.

Although ostracoderms presumably possessed a cartilaginous endoskeleton, the head and front part of the body of many forms were encased in a shieldlike, bony, external cover (Fig. 4.9). Bony armor, together with a lack of jaws and paired fins, characterized these early vertebrates (heterostracans), which presumably moved along the bottom sucking up organic material containing food. Their tails consisted of two lobes, with the distal end of the notochord extending into the larger lobe. If the larger lobe was dorsal, the tail was known as an epicercal tail; if ventral, it was known as a hypocercal tail. Later ostracoderms (cephalaspidiforms) developed paired "stabilizers" behind their gill openings that probably improved maneuverability. Most of these stabilizers were extensions of the head shield rather than true fins, although some contained muscle and a shoulder joint homologous with that of gnathostomes.

FIGURE 4.6

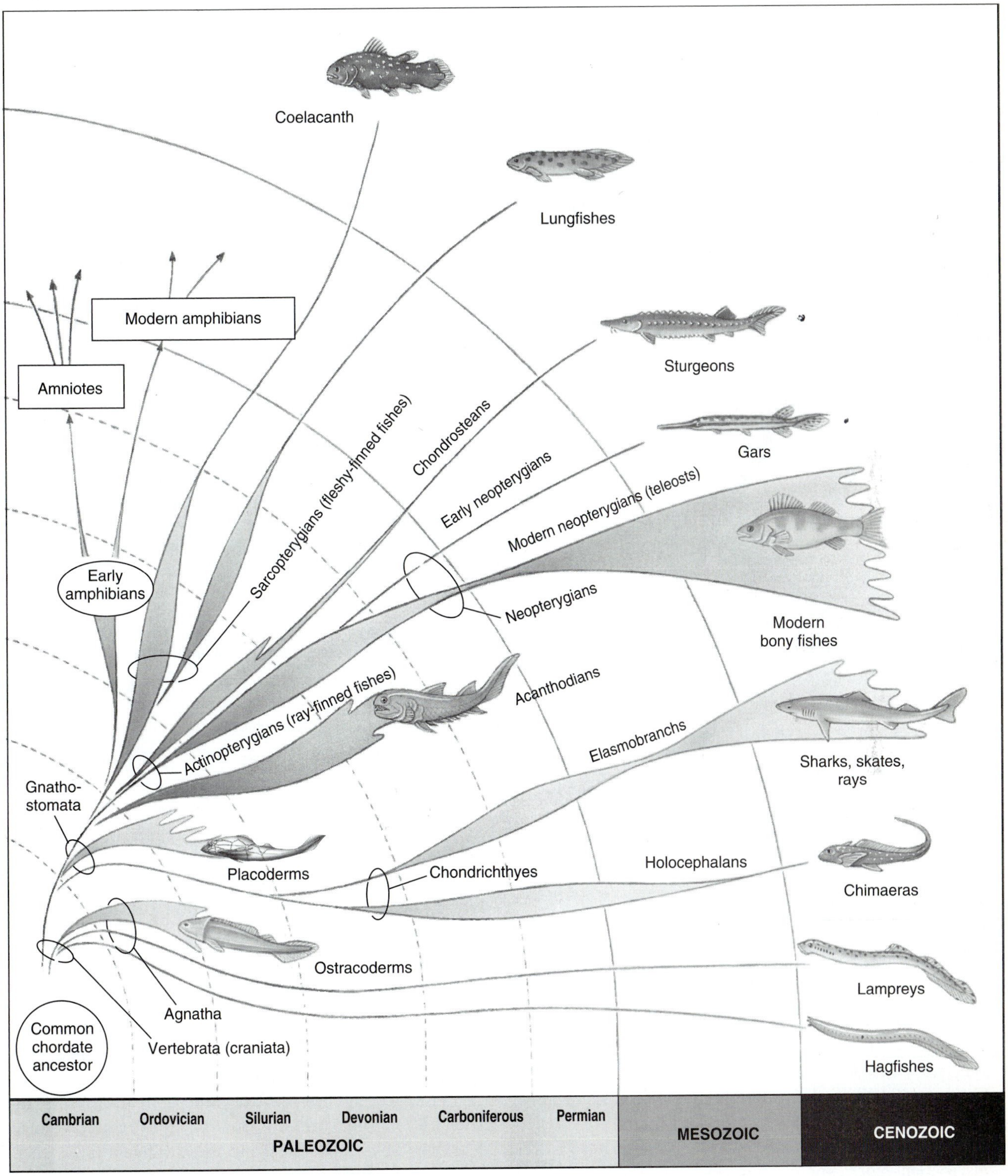

Graphic representation of the family tree of fishes, showing the evolution of major groups through geological time. Many lineages of extinct fishes are not shown. Widths of lines of descent indicate relative numbers of species. Widened regions of the lines indicate periods of adaptive radiation. The fleshy-finned fishes (sarcopterygians), for example, flourished in the Devonian period, but declined and are today represented by only four surviving genera (lungfishes and the coelacanth). Homologies shared by the sarcopterygians and tetrapods suggest that they are sister groups. The sharks and rays, which radiated during the Carboniferous period, apparently came close to extinction during the Permian period but recovered in the Mesozoic era. The diverse modern fishes, or teleosts, currently make up most of the living fishes.

FIGURE 4.7

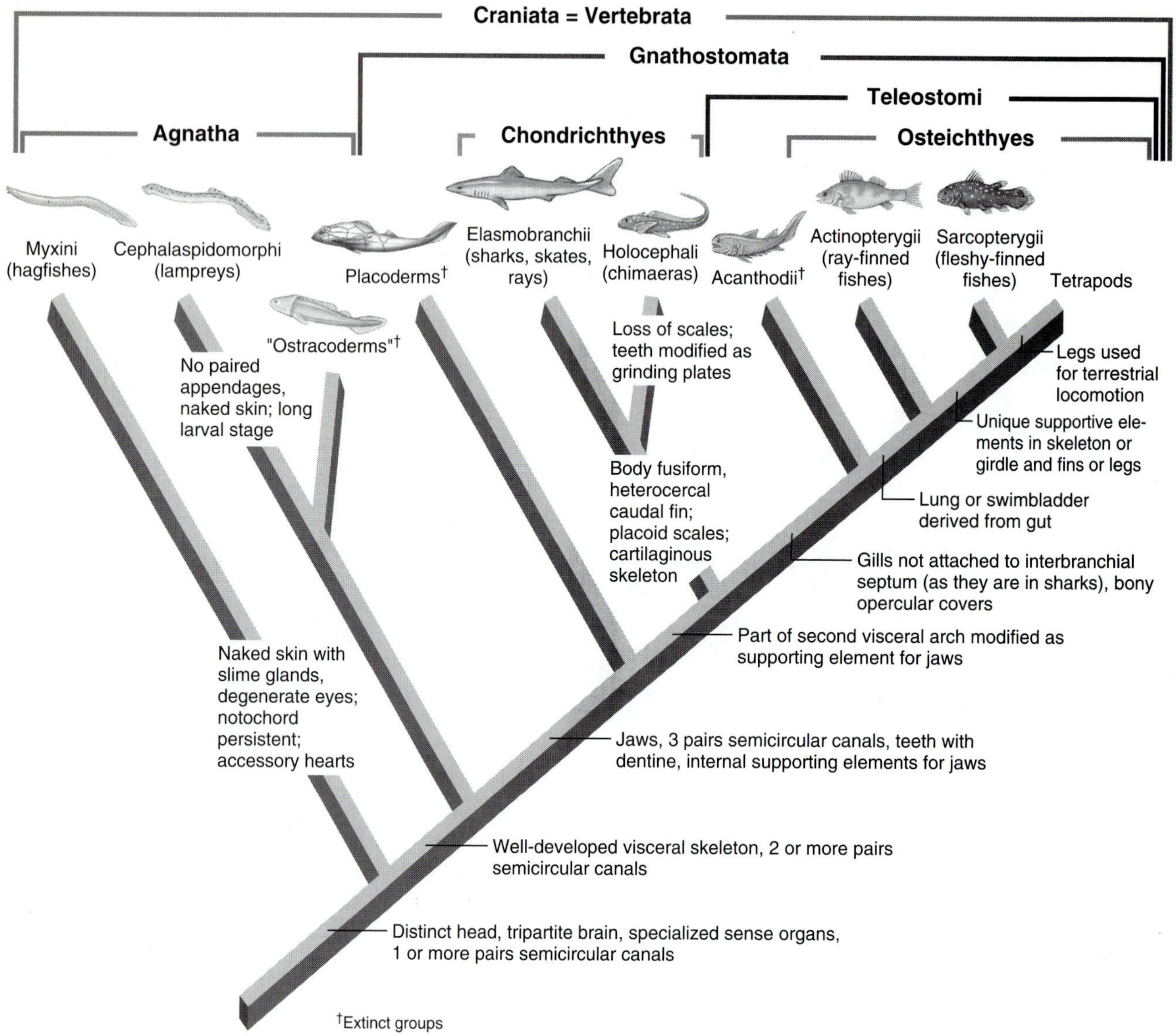

Cladogram of the fishes, showing the probable relationships of major monophyletic fish taxa. Several alternative relationships have been proposed. Extinct groups are designated by a dagger (†). Some of the shared derived characters that mark the branchings are shown to the right of the branch points.

Ostracoderms, which are considered to be a sister group to the lampreys (Cephalaspidomorphi), survived some 100 million years before becoming extinct at the end of the Devonian period. Two relatives of this group—hagfishes and lampreys—exist today.

The earliest hagfish (class Myxini) fossil comes from the Pennsylvanian epoch, approximately 330 million years ago (Bardack, 1991). Whereas lampreys occur in both freshwater and marine habitats, hagfishes are strictly marine and live in burrows on the ocean bottom in waters cooler than 22°C (Marini, 1998). They occur worldwide, except in the Arctic and Antarctic oceans, and serve as prey for many marine animals including codfish, dogfish sharks, octopuses, cormorants, harbor porpoises, harbor seals, elephant seals, and some species of dolphins (Marini, 1998).

Hagfishes have been evolving independently for such an extremely long time (probably more than 530 million years, according to Martini [1998]) and are so different from other vertebrates that many researchers question their relationship to vertebrates. They appear to have changed little over the past 330 million years. Some researchers, such as Janvier (1981), do not classify hagfishes as vertebrates because there is no evidence of vertebrae either during their embryonic development or as adults. However, because they have a cra-

FIGURE 4.8

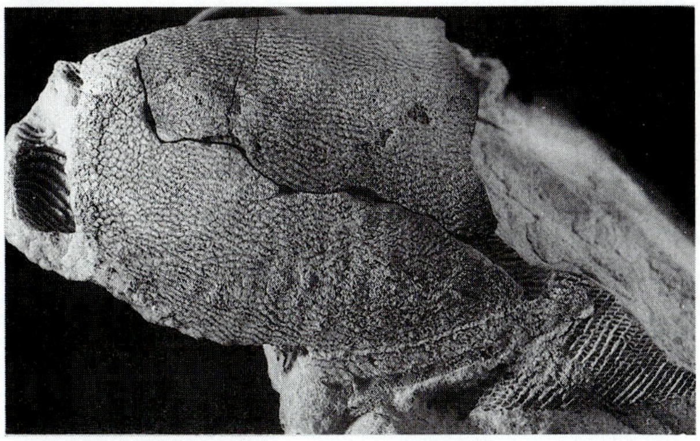

(a)

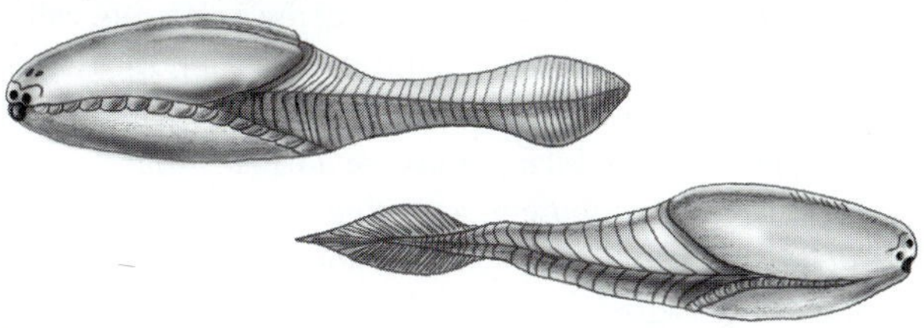

(b)

(*a*) *Arandaspis*, a 470-million-year-old jawless fish found near Alice Springs in central Australia. The fossilized impression of the bony plates was preserved in sandstone. The impression of the ribbed clam shell is approximately where the mouth of the fish would have been. The length of this specimen is approximately 20 cm. (*b*) Reconstructions of the primitive Ordovician fishes *Arandaspis* (above) and *Sacabambaspis* (below).
(b) Source: Long, The Rise of Fishes, *Johns Hopkins University Press, 1995.*

nium, they are included in the "Craniata" by phylogenetic systematists; they are considered the most primitive living craniates. The Craniata includes all members of the subphylum Vertebrata in the traditional method of classification.

The earliest fossils of lampreys (class Cephalaspidomorphi) also come from the Pennsylvanian epoch, approximately 300 million years ago (Bardack and Zangerl, 1968). Cephalaspids possess a distinctive dorsally placed nasohypophyseal opening. The single nasal opening merges with a single opening of the hypophysis to form a common keyhole-shaped opening. This is a synapomorphy of the group. In addition, the brain and cranial nerves are strikingly similar. Fossils differ little from modern forms and share characteristics and presumably ancestry with two groups of ostracoderms (anaspids and cephalaspids).

As is the case with many issues discussed in this text, there is considerable controversy concerning the evolutionary history of these groups. Both lampreys and hagfishes possess many primitive features. Besides the absence of jaws and

FIGURE 4.9

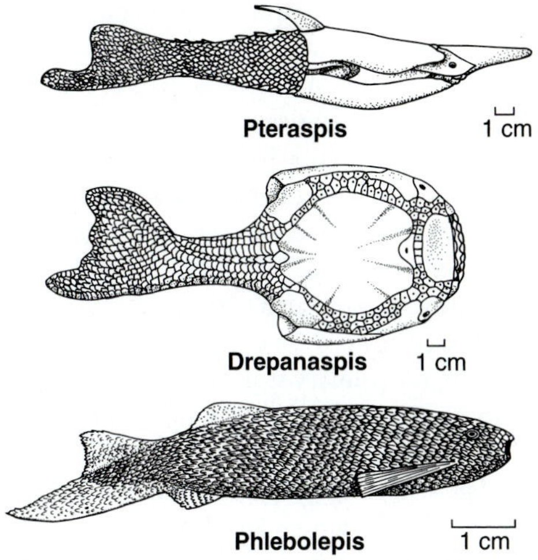

(a)

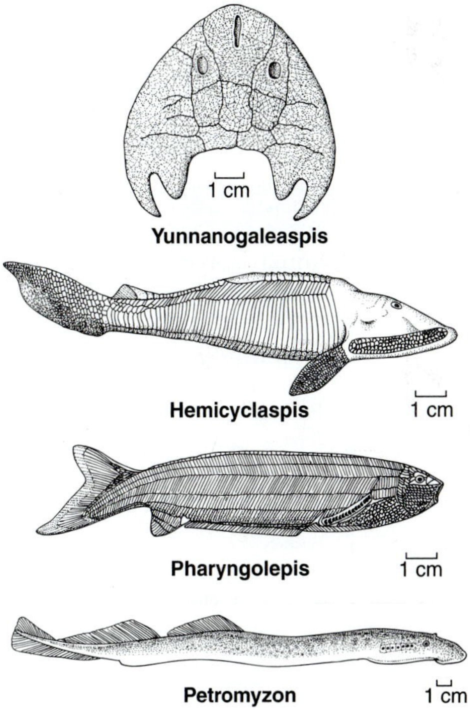

(b)

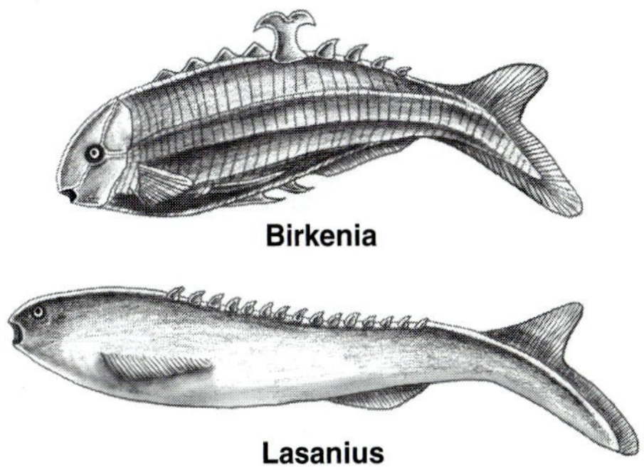

(c)

Figure 4.9 ⟶
Representative ostracoderms. (*a*) Pteraspidomorphs from the early Paleozoic, with plates of bony armor that developed in the head. All are extinct. (*b*) Representative cephalaspidomorphs. All are extinct except the lamprey. (*c*) Representative anaspidomorphs. All are extinct.

paired fins, both groups lack ribs, vertebrae, a thymus, lymphatic vessels, and genital ducts. Both possess cartilaginous skeletons. Based on these shared primitive characteristics, many researchers and taxonomists feel that lampreys and hagfishes form a monophyletic group—the agnathans. Recent phylogenetic comparisons of ribosomal RNA sequences from hagfishes, lampreys, a tunicate, a lancelet, and several gnathostomes provide additional evidence to support the proposed monophyly of the agnathans (Stock and Whitt, 1992).

Hagfishes, however, lack some structures found in lampreys, such as well-developed eyes, extrinsic eyeball muscles, and the radial muscles associated with the median fins (Stock and Whitt, 1992). They possess only a rudimentary braincase, or cranium. Also, the primary structure of insulin, a hormone secreted by the pancreas, has been found to differ in the two groups, leading researchers to note that the most likely conclusion would be that lampreys and hagfishes descended from different ancestors (Mommsen and Plisetskaya, 1991). Differences between adult lampreys and hagfishes are presented in Table 4.1. Based on such morphological analyses, other researchers believe that agnathans are paraphyletic, with lampreys being more closely related to gnathostomes than either group is to hagfishes (Janvier, 1981; Hardisty, 1982; Forey, 1984; Maisey, 1986). Additional studies, including analyses of sequences from other genes, are needed to clarify the phylogenetic relationships of the agnathans.

■ MORPHOLOGY OF JAWLESS FISHES

Integumentary System

The outer surface of the body of extant jawless fishes is smooth and scaleless (Figs. 4.10 and 4.11). The skin consists of a thin epidermis composed of living cells and a thicker, more complex dermis consisting of multiple, dense layers of collagen fibers. The skin of hagfishes is attached to underlying muscles only along the dorsal midline and along the ventral surface at the level of the slime glands (Marini, 1998). Tanned hagfish skin is sold as "eel-skin" and is used to produce designer handbags, shoes, wallets, purses, and briefcases (Marini, 1998). A nonliving secretion of the epidermis, called cuticle, covers the epidermis in lampreys. Within the dermis are sensory receptors, blood vessels, and chromatophores. Several types of unicellular glands are normally found in the epidermis; they contribute to a coating of mucus that covers the outside of the body. A series of pores along the sides of the body of a hagfish connect to approximately 200 slime glands that produce the defensive slime (mucus) that can coat

TABLE 4.1

Comparison of Anatomical and Physiological Characteristics Between Adult Lampreys and Hagfishes

Characteristics	Lampreys	Hagfishes
Dorsal fin	One or two	None
Pre-anal fin	Absent	Present
Eyes	Well developed	Rudimentary
Extrinsic eye muscles	Present	Absent
Lateral-line system	Well developed	Degenerate
Semicircular canals	Two on each side of head	One on each side of head
Barbels	Absent	Three pairs
Intestine	Ciliated	Unciliated
Spiral valve intestine	Present	Absent
Buccal funnel	Present	Absent
Buccal glands	Present	Absent
Nostril location	Top of head	Front of head
Nasohypophyseal sac	Does not open into pharynx	Opens into pharynx
External gill openings	7	1 to 14
Internal gill openings	United into single tube connecting to oral cavity	Each enters directly into pharynx
Cranium	Cartilaginous	Poorly developed
Branchial skeleton	Well developed	Rudimentary
Vertebrae (cartilaginous)	Neural cartilages	Neural cartilages only in tail
Pairs of spinal nerves per body segment	Two	One
Kidney	Mesonephros	Pronephros anterior, mesonephros posterior
Osmoregulation	Hyper- or hypoosmotic	Isosmotic
Eggs	Small, without hooks	Very large, with hooks
Cleavage of embryos	Holoblastic	Meroblastic

From Moyle/Cech, Fishes: An Introduction to Ichthyology, *3/e, Copyright ©1996. Adapted by permission of Prentice-Hall, Inc., Upper Saddle River, NJ.*

FIGURE 4.10

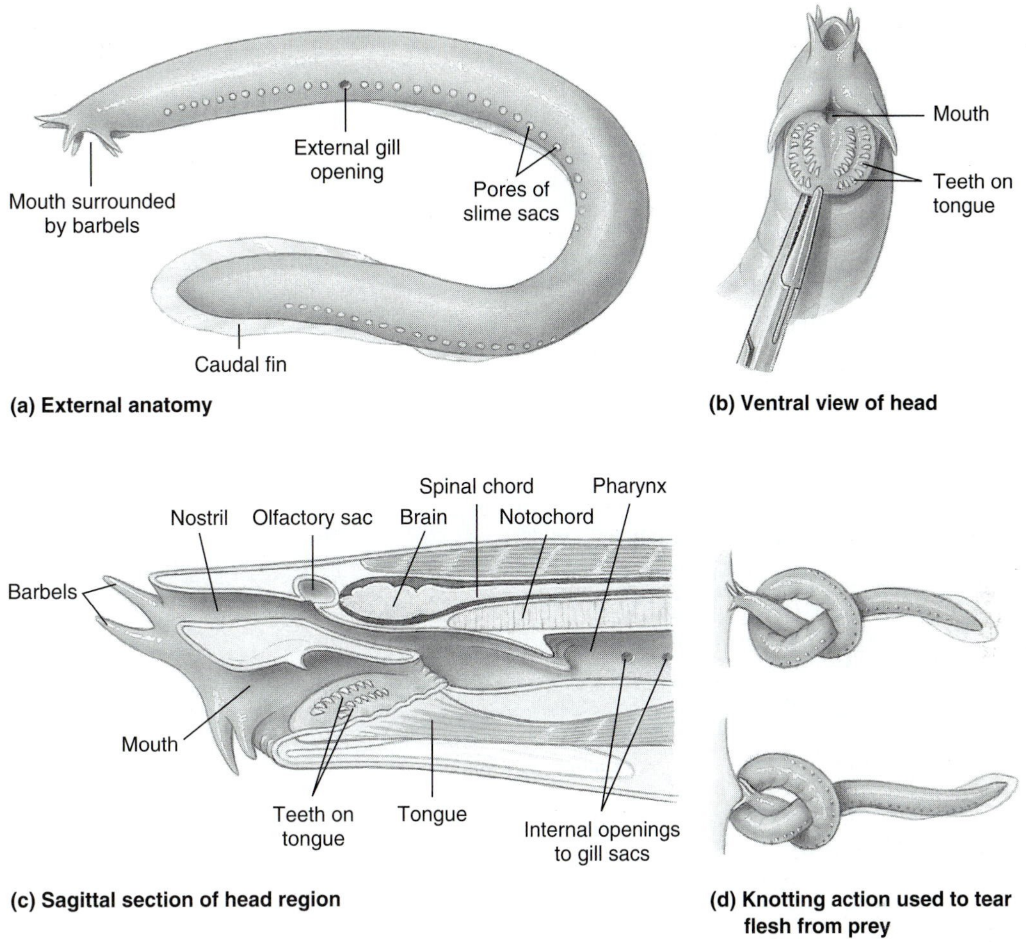

The Atlantic hagfish *Myxine glutinosa*: (*a*) external anatomy; (*b*) ventral view of head with mouth held open, showing horny plates used to grasp food during feeding; (*c*) sagittal section of head region; (*d*) knotting action, illustrating how the hagfish obtains leverage to tear flesh from its prey.

From Cleveland P. Hickman, Jr., et al., Integrated Principles of Zoology, *10th edition. Copyright © 1997 McGraw-Hill Company, Inc. All Rights Reserved. Reprinted by permission.*

the gills of predatory fish and either suffocate them or cause them to leave the hagfish alone (Fig. 4.10a). To clean the mucus off their own bodies, hagfishes have developed the remarkable ability to tie themselves in a knot, which passes down the body, pushing the mucus away (Fig. 4.10d). The knotting behavior is also useful in giving hagfishes extra leverage when feeding on large fish (Moyle and Cech, 1996).

Skeletal System

Cartilages supporting the mouth parts and the gills are suspended from the skull, which is little more than a troughlike plate of cartilage on which the brain rests. The rest of the branchial (gill) skeleton consists of a fenestrated basketlike framework under the skin surrounding the gill slits (Fig. 4.11). This branchial basket supports the gill region.

Although a true vertebral column is lacking in jawless fishes, paired lateral neural cartilages are located on top of the notochord lateral to the spinal cord in lampreys. These cartilaginous segments are the first evolutionary rudiments of a backbone, or vertebral column. In hagfishes, however, lateral neural cartilages are found only in the tail. While reminiscent of neural arches, it is unclear whether they represent primitive vertebrae, vestigial vertebrae, or entirely different structures. Anteriorly, only an incomplete cartilaginous sheath covers the notochord in hagfishes.

All jawless fishes lack paired appendages, although all possess a caudal fin. In addition, one or two dorsal fins are present in lampreys. Hagfishes lack dorsal fins but have a preanal fin.

Muscular System

Body muscles are segmentally arranged in a series of **myomeres**, each of which consists of bundles of longitudinal muscle fibers that attach to thin sheets of connective tissue, called **myosepta**, between the myomeres (Fig. 4.11). There is no further division of body wall musculature in these primitive

FIGURE 4.11

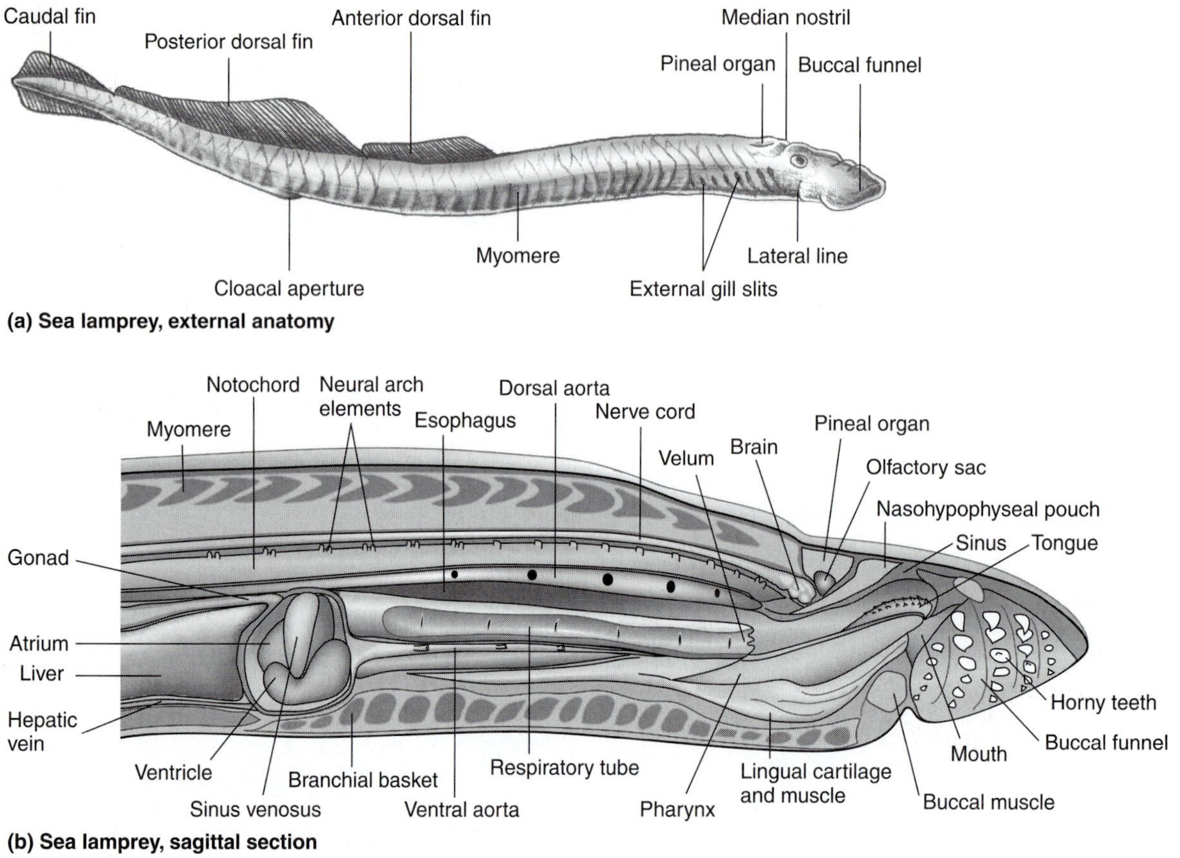

(*a*) Lateral view of the sea lamprey, *Petromyzon marinus*. The dorsal fins, eyes, and lateral-line system are well developed. These structures are either absent or rudimentary in hagfishes. (*b*) Sagittal section through the anterior portion of a lamprey. Note the prominent buccal funnel.

vertebrates. Waves of contraction passing alternately down the two sides of the body cause the lateral undulation of the trunk and tail. Jets of water expelled from the gill slits may also aid in locomotion. Buccal and lingual muscles are situated in the buccal funnel and pharyngeal regions (Fig. 4.11).

Cardiovascular System

The heart in hagfishes and lampreys is located in the **pericardial cavity** ventral to the pharynx (Fig. 4.11). It consists of four parts: From posterior to anterior, these are the **sinus venosus, atrium, ventricle,** and **conus arteriosus**, ,through which blood flows in that sequence. Hagfishes have three additional sets of accessory hearts along their venous system: the **portal heart**, which receives venous blood from the cardinal vein and from the intestine and pumps this blood to the liver; **cardinal hearts**, which are located in the cardinal veins and help to propel the blood; and **caudal hearts**, which are paired expansions of the caudal veins (Johansen, 1960). These accessory hearts have their own intrinsic pumping rhythm and are not coordinated by the central nervous system. Despite the presence of these accessory hearts, blood pressure in hagfishes is very low (Randall, 1970). The pooling of blood in large blood sinuses also contributes to the low blood pressure in hagfishes.

Respiratory System

Living jawless fishes have between 5 and 15 pairs of external gill openings (Figs. 4.10 and 4.11). In hagfishes, water enters the nostril and passes through the nasopharyngeal duct to the pharynx. Afferent branchial ducts lead from the pharynx to the gill pouches, while water is carried from the pouches to the outside through efferent branchial ducts. In parasitic lampreys, however, the direction of water flow has been modified. Water enters the external gill slits and is ejected by the same route. This modification is essential so that the lamprey can carry on respiration when it is attached by its buccal funnel to a host fish. The nasal duct ends in a **nasohypophyseal sac** and does not lead to the pharynx. At metamorphosis in the lamprey, the pharynx differentiates into an esophagus dorsally and pharynx ventrally, so that, in adults, the pharynx terminates blindly.

Digestive System

Lampreys have a round, suctorial mouth (oral disk) located inside a **buccal funnel** (Fig. 4.11b). Within the buccal funnel is a thick, fleshy rasping "tongue" armed with horny epidermal "teeth" for scraping flesh. Many lampreys are parasitic, attaching to a host with their oral disks and using their rasping teeth to bore into the host's body wall (Fig. 4.12). Blood and other body fluids are sucked from the host while the lamprey secretes an anticoagulant. Parasitic lampreys generally do not kill their hosts, but detach, leaving a weakened animal with an open wound. For this reason, some biologists prefer to regard parasitic lampreys as predators, rather than parasites. During its lifetime, each sea lamprey can kill 40 or more pounds of fish.

Food passes through the pharyngeal region and into the esophagus (Fig. 4.11b). Because lampreys have no true stomach, the esophagus leads directly into a straight intestine that contains a region of longitudinal folds known as a typhlosole, whose function is to increase the absorptive area. The intestine opens into the cloaca. Swim bladders, which are derivatives of the digestive tract in many fishes, are not present in agnathans.

Hagfishes are usually scavengers and lack oral disks (Fig. 4.10c). Their feeding apparatus consists of two dental plates, hinged along the midline, each with two curved rows of sharp, horny cusps (Fig. 4.10b). They feed primarily on small, soft-bodied living invertebrates and also on dead and dying fish, making them important scavengers on the ocean floor. In the Gulf of Maine, Martini (1998) recorded an average of 59,700 *Myxine glutinosa* per square kilometer of sea floor. Although individual hagfishes have extremely low metabolic rates, Marini (1998) calculated that 59,700 animals needed to consume the caloric equivalent of 18.25 metric tons of shrimp, 11.7 metric tons of sea worms, or 9.9 metric tons of fish every year to maintain themselves at rest. When actively swimming or burrowing, their energy demands increase four- to fivefold. Like lampreys, hagfishes also lack a stomach. Food passes through the pharynx and esophagus into a straight intestine, which opens into the cloaca.

FIGURE 4.12

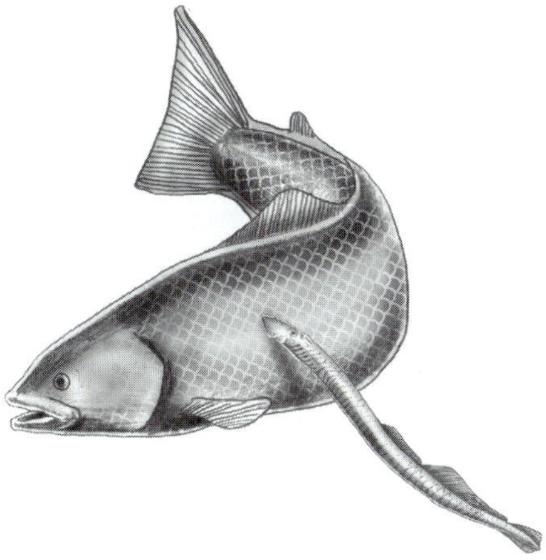

Sea lampreys are parasites that attach themselves to the body of fish such as lake trout and salmon. Using their horny teeth, they scrape a hole in the skin of their host and suck out its blood and other body fluids. An adult sea lamprey can kill approximately 40 pounds of fish annually.

BIO-NOTE 4.2

Lampreys in the Great Lakes

The only natural connection between Lake Ontario and Lake Erie is the Niagara River. Falls and rapids, including Niagara Falls, made the river useless as a commercial waterway and had always served as a barrier to the upstream movement of aquatic organisms. The completion of the Welland Canal by Canada in 1829, however, created a navigable waterway 43 km long between the two lakes. The canal also provided a means by which the sea lamprey (*Petromyzon marinus*) was able to invade the remaining four Great Lakes. The lamprey was first recorded in Lake Erie about 1921, in Lake Michigan in 1936, in Lake Huron in 1937, and in Lake Superior in 1938. As the sea lamprey invaded each lake, populations of larger fish species such as lake whitefish, lake trout, and burbot collapsed, as did the commercial fisheries that depended on them. Each year, the Great Lakes Fishery Commission traps 50,000 to 100,000 lampreys—sterilizing and releasing the males and killing the females. Intensive research has resulted in the development of a lampricide (poison) that is known as TFM (3-trifluoromethyl-4-nitrophenol) and is specific for the ammocoete larvae. As a result of these and other control measures, lamprey populations have been reduced and most fish populations are recovering. Eventually, the species may coadapt to one another and reach an equlibrium so that they can coexist without the need for any control programs.

Lampreys spawn in streams draining into the Great Lakes. Electrical barriers are effective in preventing adult lampreys from entering tributary streams to spawn (Fig. 4.13). Researchers have found that migrating adult lampreys are attracted to particular spawning streams by the smell of two bile acids secreted by larval lampreys in those streams. By using this knowledge, lamprey migration patterns can be altered, and they can be more efficiently lured into traps.

Nicholson, 1996

FIGURE 4.13

Electrical barriers spanning tributary streams of the Great Lakes have proven effective in preventing adult lampreys from entering the streams to spawn.

Nervous System

Even though the brains of vertebrates have undergone great changes in the course of vertebrate evolution, they are all similar in basic design. The anterior portion of the lamprey brain consists of inconspicuous cerebral hemispheres and rather prominent olfactory bulbs (Fig. 4.14). The pineal organ (epiphysis), a photoreceptor, is absent in hagfishes but is present in lampreys (Wurtman et al., 1968), where it lies just beneath the connective tissue covering most of the brain; it usually is connected to the posterior roof of the brain's diencephalon by a stalk.

A prominent parietal (parapineal) organ is found only in lampreys, where it serves as a photoreceptor. Even though the pineal and parapineal organs are light-sensitive, they do not form images like the lateral eyes. Rather, they monitor the intensity and duration of light. Feduccia and McCrady (1991) suggested that the single median "eye" present in many vertebrates is not homologous, because in some the eye corresponds to the parietal body, and in others, to the pineal body. The cerebellum, which aids in maintaining balance and posture, is not well developed in lampreys and hagfishes. A discussion of the cranial nerves is included in Chapter 5.

The dorsoventrally flattened spinal cord lies dorsal to the notochord and is surrounded by a fibrous neural sheath. Paired spinal nerves arise from the spinal cord. In lampreys, spinal nerves are completely separated into dorsal and ventral portions, a characteristic not found in any other living vertebrate. Grillner (1996) found that the neural controls for the lateral undulatory mode of locomotion utilized by the lamprey were distributed throughout the spinal cord. Axons extend from cells in the brain stem and lead to the specialized motor neurons involved in locomotion. In response to signals from the brain, local networks of cells generate bursts of neural activity. These networks act as specialized circuits, stimulating the neurons on one side of a segment of the lamprey's body while suppressing the neurons on the opposite side. The resultant bursts of muscle activity occur in smooth waves that alternately bend segments of the body from one side to the other.

FIGURE 4.14

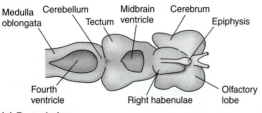

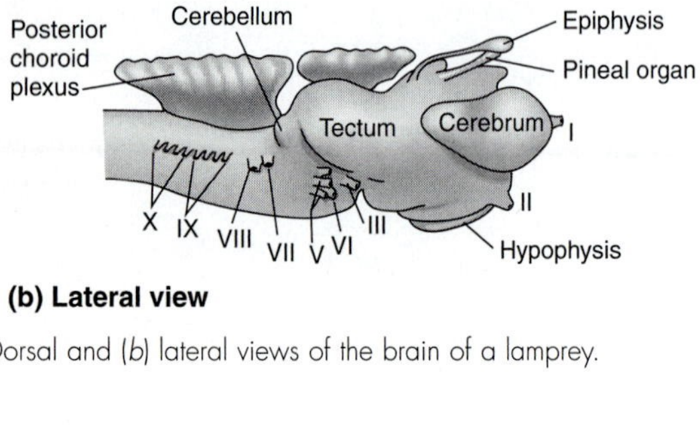

(*a*) Dorsal and (*b*) lateral views of the brain of a lamprey.

Sense Organs

The lamprey is the only living jawless fish that has an evident lateral-line system (Fig. 4.11), consisting of superficial sensory organs called **neuromasts**, arranged in several noncontinuous lines on the head and around the branchial chambers. Indistinct dorsal, lateral, and ventral portions of the lateral-line system are present on the body behind and above the branchial openings.

Lampreys, like most vertebrates, possess two lateral eyes that serve as their primary receptors of light. In hagfishes, however, the eyes, which are rudimentary, are light-sensitive. Small light-colored areas often mark the site for the eyes, although no eyeball ever forms.

Olfactory organs are well developed in lampreys and hagfishes. The single, median, dorsal nostril leads into a short nasal tube that ends in a nasohypophyseal sac at the level of the second internal gill slit. In lampreys the pouch ends blindly, but in hagfishes the pouch connects with the gut. Three pairs of sensory barbels are located around the mouth and nostril of hagfishes.

In most vertebrates, three semicircular canals located in the inner ear provide the organism with its sense of dynamic equilibrium. One end of each canal is enlarged into a swelling, called an **ampulla**, that contains a patch of sensory cells known as **cristae**. Lampreys, however, have only the anterior and posterior semicircular canals, whereas in hagfishes the two semicircular canals are connected in such a manner that they appear as one (Berg, 1947).

Endocrine System

Endocrine organs and their functions in jawless fishes and gnathostomes are discussed in Chapter 5.

Urogenital System

With the exception of hagfishes, whose body fluids have salt concentrations similar to that in sea water (= isotonic), all other marine vertebrates maintain salt concentrations in their body fluids at a fraction of the level in the water (= hypotonic) (Schmidt-Nielsen, 1990).

Lampreys can live in both sea water and fresh water. In the ocean, they prevent osmotic water loss by having a tough skin, by using salt-excreting cells in their gills to rid themselves of salt absorbed in their gut, and by reabsorbing water in their kidneys. In fresh water, the kidneys excrete large amounts of excess water while retaining essential proteins and salts.

Hagfish embryos possess a primitive kidney known as an archinephros (Fig. 4.15a). It is replaced by a pronephric kidney (Fig. 4.15b), which forms from the anterior portion of

FIGURE 4.15

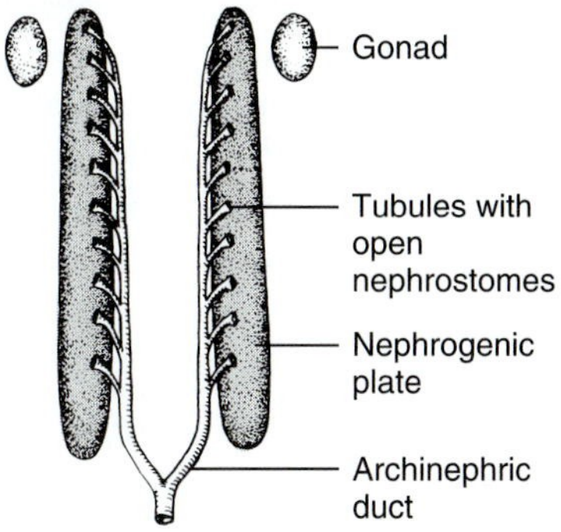

(a) Archinephros: Kidney found in embryo of hagfish; this is the inferred ancestral condition of the vertebrate kidney.

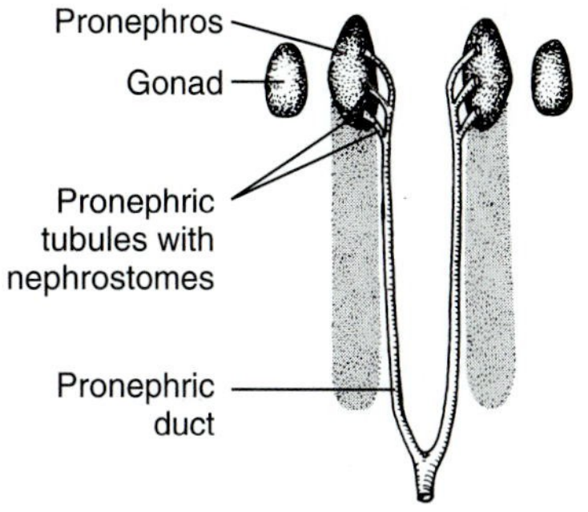

(b) Pronephros: Functional kidney in adult hagfish and embryonic fishes and amphibians; fleeting existence in embryonic reptiles, birds, and mammals

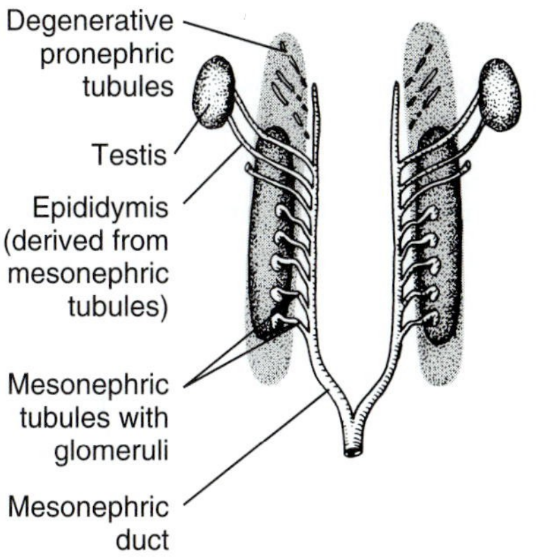

(c) Mesonephros (Opisthonephros): Functional kidney of adult lampreys, fishes, and amphibians; transient function in embryonic reptiles, birds, and mammals

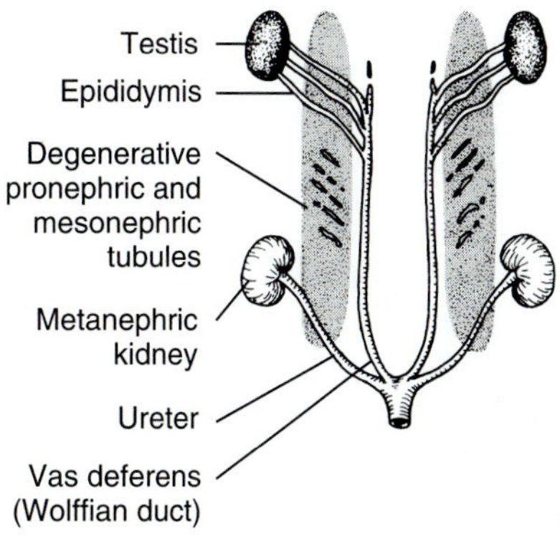

(d) Metanephros: Functional kidney of adult reptiles, birds, and mammals

(*a*) Archinephric kidney found in hagfish embryos. (*b*) The pronephros is the functional kidney found in adult hagfishes and embryonic fishes and amphibians. It is present only for a short time in embryonic reptiles, birds, and mammals. (*c*) The functional kidney of adult lampreys, fishes, and amphibians is the mesonephros (opisthonephros). It functions for only a short time in embryonic reptiles, birds, and mammals. (*d*) Metanephric kidney of adult reptiles, birds, and mammals.

the nephrogenic mesoderm and functions as the adult kidney in hagfishes. In lampreys, the anterior portion of the nephrogenic mesoderm forms a pronephric kidney and the posterior portion forms an opisthonephric kidney (Fig. 4.15c). The pronephric kidney consists of segmentally arranged pronephric tubules with ciliated funnels that extend into the peritoneal, or body, cavity and receive materials directly from this cavity. The opisthonephric kidney consists of longer tubules (nephrons) that usually lack peritoneal funnels. Each tubule partially encloses a specialized cluster of capillaries known as a glomerulus, thereby increasing the efficiency of the filtering process. A pronephros forms as a developmental stage in all vertebrates. It is functional, however, only in larval fish and amphibians and remains throughout life only in lampreys, hagfishes, and a few teleosts. Even then, it functions as an adult kidney only in hagfishes; in all other vertebrates, it ceases to function as a kidney and becomes a mass of lymphoid tissue. An opisthonephros serves as the functional kidney of adult lampreys, as well as fishes and amphibians.

Hagfishes and lampreys possess only a single gonad (ovary or testis). Occasionally, individuals with both an ovary and a testis are found, but only one gonad is functional. There are no oviducts or sperm ducts; thus, eggs and sperm are released directly into the abdominal cavity and pass into the cloaca through an abdominal pore.

REPRODUCTION

Little is known about hagfish reproduction. There is no information about when breeding occurs or how the eggs are fertilized. Few fertilized eggs have ever been found (Martini, 1998). It is known that female hagfishes produce between 20 and 30 large, yolky eggs, which possess hooked filaments by which they can be attached to the sea bottom and to each other (Fig. 4.16).

Adult lampreys may live in either salt water or fresh water, but all species spawn only in fresh water, indicating that this group evolved in fresh water and secondarily invaded marine environments. Both eggs and sperm are shed directly into the body cavity and exit through one or more abdominal pores. Female parasitic sea lampreys (*Petromyzon marinus*) may deposit as many as 260,000 eggs, whereas most nonpredatory female lampreys produce only 1,000 to 2,000 eggs (Moyle and Cech, 1996). Most lampreys spawn in shallow, gravel-bottomed streams in late winter or early spring. Following their release from the female, eggs require approximately 2 weeks to hatch.

Cleavage in the large, yolky egg of hagfishes is termed **meroblastic** because only a portion of the cytoplasm is cleaved (divided). Cleavage is **holoblastic** in the sparsely yolked egg of lampreys, in which the mitotic cleavage furrows pass through the entire egg.

FIGURE 4.16

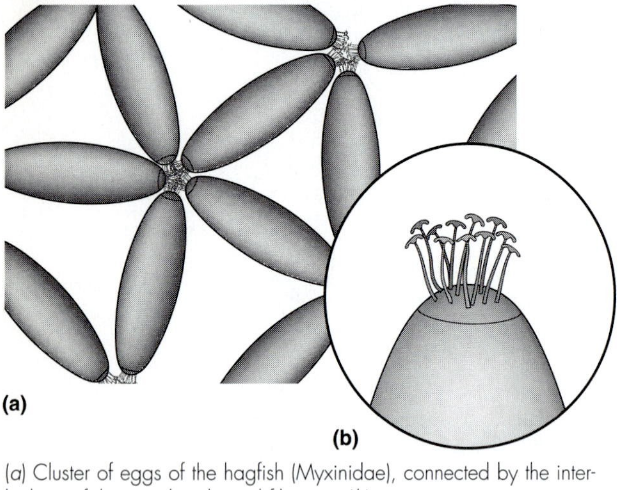

(*a*) Cluster of eggs of the hagfish (Myxinidae), connected by the interlocking of their anchor-shaped filaments (*b*).
Source: Bridge and Boulenger, Macmillan and Co., England, 1904.

GROWTH AND DEVELOPMENT

Young hagfishes hatch as small, fully formed hagfishes. Since no larval hagfishes have ever been found, it is presumed that development is direct and without metamorphosis. Details of their life history (e.g., life span, age at reproduction, breeding sites, juvenile habitat) are unknown.

After hatching, lampreys undergo an extended larval period, during which they are known as ammocoetes. The blind, toothless ammocoete larvae (see Fig. 1.3c) drift with the current until they come to an area of quiet water with a muddy bottom. They burrow into the mud until only the oral hood is left exposed and feed by straining microorganisms, organic detritus from the water. Larval development may last from 3 to 7 years depending on the species (Hardisty and Potter, 1971). Those species that are parasitic generally have a shorter larval life and longer adult life than nonparasitic species. Following metamorphosis, some will migrate to the ocean; others will migrate to large bodies of fresh water. The adult stage generally lasts from 5 to 6 months in nonparasitic species and up to 2 years in parasitic species. Lampreys die after breeding once. Some nonparasitic lampreys live as larvae for 7 years and then metamorphose into nonfeeding adults. The adults live for a few weeks, reproduce, and die.

Adult hagfishes are generally less than 1 m in total length. Parasitic marine lampreys attain a larger adult size (30–80 cm) than parasitic species living in fresh water (usually less than 30 cm). Adults of both fresh water and marine parasitic forms are larger than adults of nonparasitic forms (usually less than 20 cm).

Review Questions

1. Do you feel that calcichordates and conodonts should be classified as vertebrates? Explain.
2. What characteristics do lampreys and hagfishes have in common?
3. Compare and contrast the digestive systems of hagfishes and lampreys.
4. Explain (or attempt to explain) why lampreys spawn in fresh water environments but may live their adult lives in marine environments.

Supplemental Reading

Brodal, A., and R. Fange (eds). 1963. *The Biology of* Myxine. Oslo: Universitetsforlaget.

Forey, P. L., and P. Janvier. 1993. Agnathans and the origin of jawed vertebrates. *Nature* 361:129–134.

Forey, P. L., and P. Janvier. 1994. Evolution of the early vertebrates. *American Scientist* 82:554–565.

Gee, H. 1996. *Before the Backbone: Views on the Origin of the Vertebrates.* New York: Chapman and Hall.

Gorbman, A., H. Kobayashi, Y. Honma, and M. Matsuyama. 1990. The hagfishery of Japan. *Fisheries* 15(4):12–18.

Janvier, P. 1981. The phylogeny of the Craniata, with particular reference to the significance of fossil "agnathans." *Journal of Vertebrate Paleontology* 1(2):121–159.

Jefferies, R. P. S. 1986. *The Ancestry of the Vertebrates.* London: British Museum of Natural History.

Jensen, D. 1966. The hagfish. *Scientific American* 214(2):82–90.

Jorgensen, J. M., J. P. Lomholt, R. E. Weber, and H. Malte (eds.) 1998. *The Biology of Hagfishes.* New York: Chapman and Hall.

Martini, F. H. 1998. Secrets of the slime hag. *Scientific American* 279(4):70–75.

Matthews, W. J., and D. C. Heins (eds.) 1987. *Community and Evolutionary Ecology of North American Stream Fishes.* Norman: University of Oklahoma.

Vertebrate Internet Sites

Visit the zoology web site at http://www.mhhe.com/zoology to find live Internet links for each of the references listed below.

1. **Hyperotreti (Hagfishes).**
   Arizona's Tree of Life Web Page. An introduction, pictures, characteristics, discussion of the skull, phylogenetic relationships, and references on hagfishes.
2. **Hyperoartia (Lampreys).**
   Arizona's Tree of Life Web Page. An introduction, pictures, characteristics, discussion of the skull, phylogenetic relationships, and references on lampreys.
3. **Introduction to the Myxini.**
   University of California at Berkeley, Museum of Paleontology. Images, photos, systematics, more information, and links.
4. **Hagfish.**
   Information from the Sea Grant on the hagfish fishery.
5. **The Lowly Hag.**
   Information on hagfish.

CHAPTER 5

Gnathostome Fishes

■ INTRODUCTION

The two groups of living gnathostome (jawed) fishes are the Chondrichthyes or cartilaginous fishes (sharks, skates, rays, and ratfishes), and the Osteichthyes or bony fishes (Fig. 5.1). Both groups may have evolved in separate but parallel fashion from placoderm ancestors and are the survivors of hundreds of millions of years of evolution from more ancient forms. Fishes are the most diverse group of vertebrates, with approximately 26,000 species of bony and cartilaginous fishes extant in the world today (Bond, 1996).

■ EVOLUTION

The evolution of the major groups of hagfishes, lampreys, and gnathostome fishes and their relationships to each other, to the amphibians, and to amniotes are shown in Fig. 4.6. In Fig. 4.7 is presented a cladogram showing probable relationships among the major groups of fishes. Because taxonomy is constantly undergoing refinement and change, the relationships depicted in this cladogram, along with others used in this text, are subject to considerable controversy and differences of opinion among researchers (see Supplemental Readings at end of chapter).

Evolution of Jaws

The development of hinged jaws from the most anterior pair of primitive pharyngeal arches (see discussion on page 99 of this chapter) was one of the most important events in vertebrate evolution. Jaws permitted the capture and ingestion of a much wider array of food than was available to the jawless ostracoderms, and they also permitted the development of predatory lifestyles. Fish with jaws could selectively capture more food and occupy more niches than ostracoderms and, thus, were more likely to survive and leave offspring. They

FIGURE 5.1

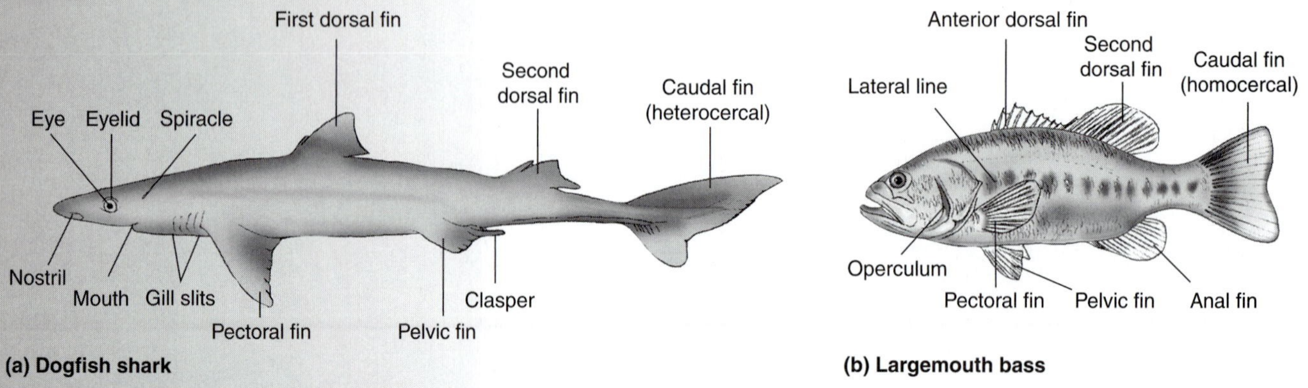

External anatomy of (*a*) dogfish shark (Chondrichthyes) and (*b*) largemouth bass (Osteichthyes).

could venture into new habitats in search of food, breeding sites, and retreats. Jaws, which also could be used for defensive purposes, could have aided these primitive fish in both intraspecific and interspecific combat. Thus, hinged jaws made possible a revolution in the method of feeding and hence in the entire mode of life of early fishes. The term **gnathostome** includes all of the jawed fishes and the tetrapods.

Mallatt (1996) reassessed homologies between the oropharyngeal regions of jawless fishes and Chondrichthyes and proposed that jaws originally evolved and enlarged for a ventilatory function—namely, closing the jaws prevented reflux of water through the mouth during forceful expiration. As the jaws enlarged further to participate in feeding, they nearly obliterated the ancestral mouth in front of them, leading to the formation of a new pharyngeal mouth behind the jaws. The secondary function of jaws was to grasp prey in feeding. Thus, Mallatt (1996) proposed the following stages in the evolution of gnathostomes: (1) ancestral vertebrate (with unjointed branchial arches); (2) early pre-gnathostome (jointed internal arches and stronger ventilation); (3) late pre-gnathostome (with mouth-closing, ventilatory "jaws"); and (4) early gnathostome (feeding jaws).

Evolution of Paired Fins

A second major development in the evolution of vertebrates was the evolution of paired appendages. As early fishes became more active, they would have experienced instability while in motion. Presumably, just such conditions favored any body projection that resisted roll (rotation around the body axis), pitch (tilting up or down), or yaw (swinging from side to side) and led to the evolution of the first paired fins (pectoral and pelvic). Force applied by a fin in one direction against the water is opposed by an equal force in the opposite direction. Thus, fins can resist roll if pressed on the water in the direction of the roll; fins projecting horizontally near the anterior end of the body similarly counteract pitch. (Yaw is controlled by vertical fins along the mid-dorsal and mid-ventral lines.) Thus, fins bring stability to a streamlined body. Pectoral fins, which project laterally from the sides of the body, are used for balancing and turning, whereas pelvic fins serve as stabilizers. The associated girdles stabilized the fins, served as sites for muscle attachment, and transmitted propulsive forces to the body.

The origin of paired fins has long been debated and even today remains unresolved. The **Gill Arch Theory** of Gegenbaur (1872, 1876) proposed that posterior gill arches became modified to form pectoral and pelvic girdles and that modified gill rays formed the skeletons of the fins. Pectoral girdles superficially resemble gill arches and are located behind the last gill in some fish, which provided early support for this theory. However, a rearward migration of branchial parts would have been necessary to form the pelvic girdle. There is no embryological or morphological evidence to support this theory.

A second theory, the **Fin Fold Theory**, was originally proposed independently in 1876 by J. K. Thacher and F. M. Balfour. It has been further developed and modified by later investigators including Goodrich (1930) and Ekman (1941), who provided evidence that the paired fins of sharks develop from a continuous thickening of the ectoderm. This theory suggests that paired fins arose within a paired but continuous set of ventrolateral folds in the body wall. This continuous fold became interrupted at intervals, forming a series of paired appendages. Intermediate ones were lost, and the remaining portions supposedly evolved into pectoral and pelvic fins. Some primitive ostracoderms had such folds, although they were higher on the sides of the body. The primitive shark *Cladoselache* (class Chondrichthyes), whose paired fins are hardly more than lateral folds of the body wall, is cited often as possible evidence of this theory. However, there is no supporting fossil evidence.

The most recent hypothesis is the **Fin Spine Theory**. Spiny sharks (acanthodians) possessed as many as seven pairs of spiny appendages along their trunks (Fig. 5.2). These appendages are thought to have served as stabilizers. In some forms, a fleshy weblike membrane was attached to each spine (Romer, 1966). All of the spines may have been lost except for two pairs—an anterior pair that would develop into pectoral fins and a posterior pair that would become pelvic fins.

Although paired fins are the phylogenetic source of tetrapod limbs, a definitive explanation for their origin is lacking, and the fossil record provides no clear answer. The possibility exists that paired fins may have originated independently more than once (convergent evolution); if so, more than one of these theories could be accurate.

FIGURE 5.2

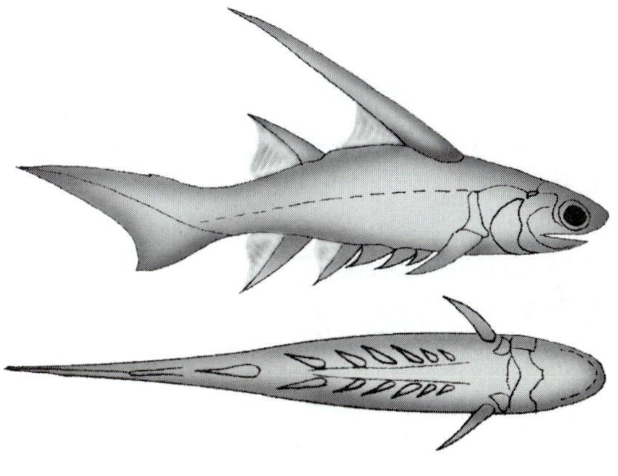

Parexus

Parexus, a typical acanthodian genus whose members often had a series of spiny appendages along the trunk. A fleshy weblike membrane was attached to some of the spines.

BIO-NOTE 5.1

Counting Genes in Vertebrates and Invertebrates

Molecular zoologists have found that all vertebrates have roughly the same number of genes and that all invertebrates have roughly the same number of genes. However, there was a distinct jump in the *total number* of genes from invertebrates to vertebrates. Peter Holland has suggested that a mutation in an animal similar to a lancelet resulted in a doubling of chromosomes and a second copy of all genes. This initial gene doubling occurred more than 500 million years ago, just before vertebrates originated. It is hypothesized that the additional genes enabled the hypothetical vertebrate ancestor to evolve entirely new body structures—in particular, a more complex head and brain. There is some evidence that a second genome duplication occurred later and resulted in the appearance of jaws.

Holland, 1992

Acanthodians and Placoderms

Long before ostracoderms became extinct, the jawed vertebrates (gnathostomes) appeared. The earliest known jawed vertebrates were the spiny sharks or acanthodians (class Acanthodii), which appeared approximately 440 million years ago in the Silurian period (Fig. 5.3). These were mostly small fishes, with the majority of individuals less than 20 cm in length. They had large eyes, small nostrils, an internal skeleton composed partly of bone, and a well-developed lateral-line system. Their bodies were covered with a series of small, flat, bony, diamond-shaped ganoid scales, so called because overlying the basal plate of each scale were layers of a shiny enamel-like substance known as ganoin. The gill region typically was covered by a flap (operculum), presumably composed of folds of skin reinforced by small dermal scales. A row of ventral paired fins was present along each side of the body of some individuals. All fins, both paired and unpaired (except the caudal fin), had strong and apparently immovable dermal spines at their front edges that are believed to have been highly developed scales. These active swimming fish, which were adapted to open water, have sometimes been included with the placoderms in the class Placodermi. Romer (1966) considered acanthodians as an early branch from the unknown ancestral stock from which the Osteichthyes (bony fishes) arose. Moyle and Cech (1988) noted that acanthodians may represent an independent evolutionary line intermediate between Osteichthyes and Chondrichthyes (cartilaginous fishes). Most researchers now regard them either as a separate class of early vertebrates or as a subclass of the class Osteichthyes (Feduccia and McCrady, 1991). Although acanthodians survived into the Lower Permian period, they were never a dominant group and were overshadowed by the placoderms.

FIGURE 5.3

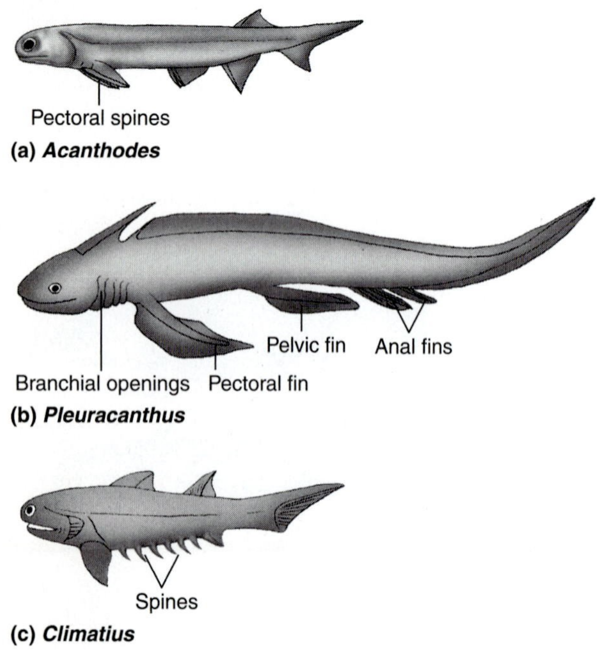

Representative acanthodians or spiny sharks, the earliest-known jawed vertebrates.

Placoderms (Fig. 5.4), which also possessed jaws and whose bodies were covered with dermal bony plates, became the dominant fishes during most of the Devonian period. In addition, they possessed an internal skeleton of bone and cartilage and sharp dermal armor on the margins of their jaws, which functioned like teeth for seizing, tearing, and crushing a wide variety of food. Fundamental differences in jaw structure and musculature together with the absence of true teeth are often thought to indicate that placoderms are the most primitive of the gnathostomes. The dorsoventrally flattened body in many forms suggests they were primarily bottom-dwellers.

The largest group of placoderms and the most common Devonian vertebrates were the jointed-necked, armored fishes (arthrodires), which ranged in length from 0.3 to 9.0 m. Their bony armor was arranged in two rigid parts: one covering the head and gill region, and the second enclosing much of the trunk. The latter segment articulated with the anterior shield by ball-and-socket joints. Thus, the head was for the first time freely movable up and down on the trunk, allowing for a wider field of vision, a wider gape, and increased efficiency in securing food.

Placoderms were too specialized to be directly intermediate between ostracoderms and modern groups of fishes. Although they dominated the Devonian seas, they were rather abruptly replaced in the early Carboniferous by the cartilaginous fishes (Chondrichthyes) and the bony fishes (Osteichthyes). Placoderms became extinct in the Mississippian period (approximately 345 million years ago) and left no modern living descendants.

FIGURE 5.4

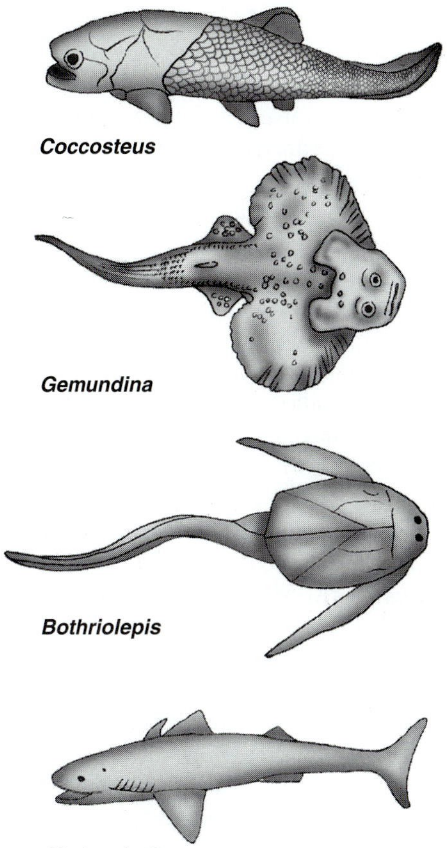

Representative placoderms with jaws and paired appendages. Most possessed a dermal armor composed of bony plates that were broken up into small scales on the midbody and tail. Most placoderms were active predators.

BIO-NOTE 5.2

Color in Ancient Fishes

Red and silver pigment cells have been found in a 370-million-year-old placoderm found in the Antarctic. Previously, the oldest known animal pigment cells were from a 50-million-year-old frog found in Mesel, Germany. When transparently thin sections of fragments of the fish were prepared, silver iridescence-producing cells were found on the fish's belly and red pigment cells were found on its back. By mapping the cells' distribution, a partial color model of the ancient fish was prepared. The finding of color cells on the fossil fish provides evidence that Devonian animals or their predators may have had color vision.

Parker, In preparation.

Chondrichthyes

The class Chondrichthyes consists of sharks, skates, rays, and chimaeras (Fig. 5.5). These fish are distinguished by their predominantly cartilaginous skeletons and placoid scales with a posteriorly projecting spine of dentin (see Fig. 5.9). The near absence of bone in the skeleton, except for traces of bone in the placoid scales and teeth, apparently represents a secondary loss, because bone was more extensive in the ostracoderms (largely in the dermis).

Cartilaginous fishes are thought to have arisen from placoderm ancestors. Recent fossil finds from China indicate the existence of several different jawed fishes in the Silurian, which began approximately 438 million years ago (Monastersky, 1996a). These discoveries imply that the first jaws appeared well before that time. The presence of sharks, possible acanthodians, conodonts, and heterostracan-like fish presumably indicates that the major period of diversification within these vertebrates was well under way during the Ordovician period.

In spite of a rather good fossil record, the taxonomic relationships of cartilaginous fish remain unclear. By the Cenozoic era, however, they were present in large numbers and had diversified greatly (Fig. 5.5). Approximately 850 species, mostly marine, are living today. They comprise two subclasses: Elasmobranchii (sharks, skates, rays) and Holocephali (chimaeras or ratfishes). Male chondrichthyans possess claspers on their pelvic fins, which are specializations associated with the practice of internal fertilization.

Skates and rays (superorder Batoidea) are primarily adapted for bottom-living. Rays make up over half of all elasmobranchs and include skates, electric rays, sawfishes, stingrays, manta rays, and eagle rays. Skates differ from rays in that skates have a more muscular tail, usually have two dorsal fins and sometimes a caudal fin, and lay eggs rather than giving birth to living young. Skates and rays differ from sharks in having enlarged pectoral fins that attach to the side of the head, no anal fin, horizontal gill openings, and eyes and spiracles located on the top of the head; in sharks, the eyes and spiracles are situated laterally. With the exception of whales, sharks include the largest living marine vertebrates. The whale shark (*Rhinocodon typus*), which may attain a length of up to 15 m, is the world's largest fish. Manta rays (*Manta* sp.) and devil rays (*Mobula* sp.) may measure up to 7 m in width from fin tip to fin tip.

The Holocephali contains the chimaeras (ratfishes), which have a long evolutionary history independent of that of the elasmobranchs. They have large heads, long, slender tails, and a gill flap over the gill slits similar to the operculum in bony fish. In addition to pelvic claspers, males possess a single clasper on their head, which is thought to clench the female during mating.

Osteichthyes

Bony fish, the largest group of living fishes, have been the dominant form of aquatic vertebrate life for the last 180 million years. Comprising approximately 97 percent of all known

FIGURE 5.5

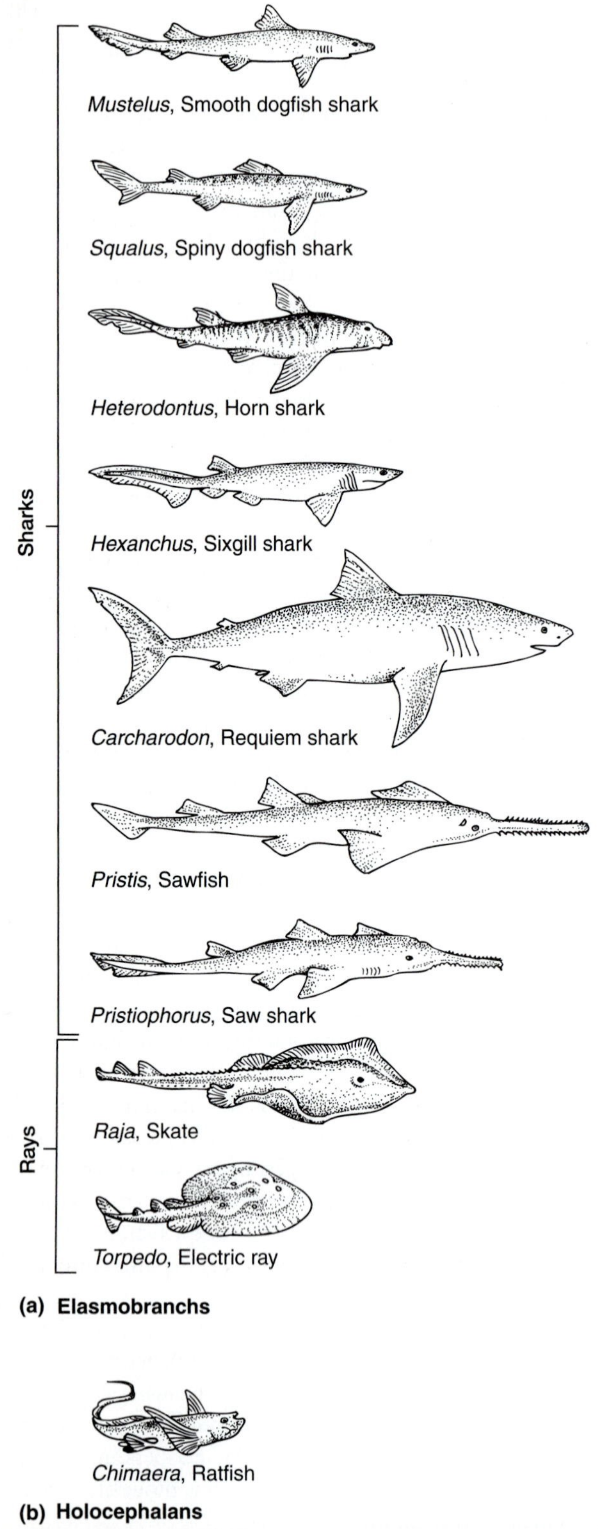

Representative chondrichthyans: (*a*) elasmobranchs, including sharks, skates, and rays; (*b*) holocephalans.

species of fishes, they first appear in the fossil record in the Late Silurian period, and they are very closely related to acanthodians. Because most early fossils are from freshwater deposits, it is thought that bony fishes, which possess well-ossified internal skeletons, probably evolved in fresh water.

It is unclear how the common osteichthyan ancestor of actinopterygians and sarcopterygians arose from non-osteichthyan gnathostome ancestors. Zhu et al. (1999) reported a 400-million-year-old sarcopterygian-like fish (*Psarolepis*) from China with an unusual combination of osteichthyan and non-osteichthyan features. Zhu and colleagues feel that this early bony fish provides a morphological link between osteichthyans and non-osteichthyan groups. Whether *Psarolepis* turns out to be a stem-group osteichthyan or a stem-group sarcopterygian, its combination of unique characters will probably have a marked impact on studies of osteichthyan evolution.

Two major groups currently are recognized: lobe-finned fishes (subclass Sarcopterygii) and ray-finned fishes (subclass Actinopterygii) (Figs. 4.6, 4.7, and 5.6). The subclass Sarcopterygii contains the lungfishes and the coelacanths. These fish possess muscular, lobed, paired fins supported by internal skeletal elements. Moyle and Cech (1996) treat each group separately because recent studies indicate that each is derived from a long independent evolutionary line. The evolutionary histories of lungfishes and coelacanths are of great interest because one or the other is considered by different investigators to be a sister group of all land vertebrates (tetrapods). In addition, the only living coelacanth, *Latimeria*, is the only living animal with a functional intracranial joint (a complete division running through the braincase and separating the nasal organs and eye from the ear and brain)

FIGURE 5.6

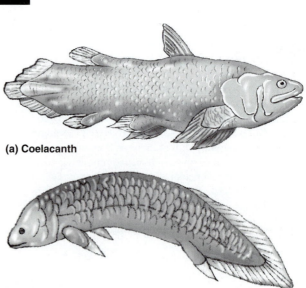

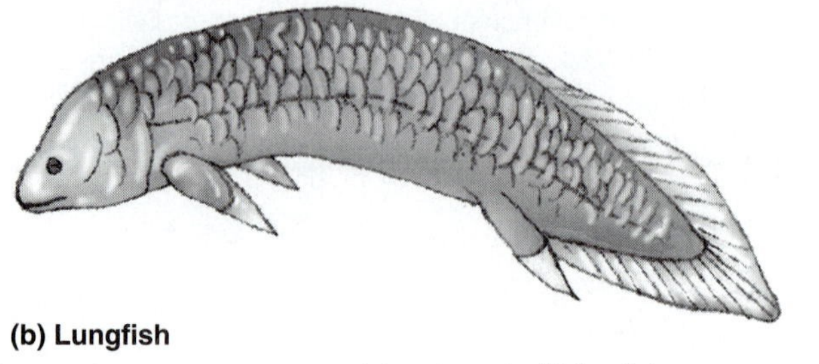

Representative sarcopterygians: (*a*) coelacanth; (*b*) lungfish.

BIO-NOTE 5.3

Coelacanths

The first living coelacanth (*Latimeria chalumnae*) (Fig. 5.6a) was discovered in 1938, when natives caught one while fishing in deep water off the coast of South Africa in the Indian Ocean. Prior to this time, coelacanths were known only from Mesozoic fossils and were thought to have become extinct some 75 million years ago.

The 1938 specimen was taken in a trawling net in water approximately 73 m deep near the mouth of the Chalumna River (Fig. 5.7). It initially was examined by Ms. M. Courtenay-Latimer, the curator of the museum in nearby East London, South Africa. Although she could not make a positive identification, she notified J. L. B. Smith, an ichthyologist, who identified the fish as a coelacanth and named it in honor of the curator and the river.

Since 1938, numerous coelacanths have been taken in deep waters (73 to 146 m) around the Comoro Islands off the coast of Madagascar. Known coelacanth populations have been monitored for a number of years and show an alarming decline in numbers. A study of underwater caves along 8 km of coastline off Grande Comore revealed a decline from an average of 20.5 individuals in all underwater caves in 1991 to an average of 6.5 in 1994. A total of 59 coelacanths were counted in 1991, but only 40 in 1994. The total estimated population near Grande Comore is about 200 individuals. The decline is thought to be due to overfishing by native Comorans, who often get paid by scientists eager to obtain a specimen.

The coelacanth is listed as an endangered species by the International Union for the Conservation of Nature. To coordinate and promote research and conservation efforts, a Coelacanth Conservation Council has been formed. The address for the Council is J. L. B. Smith Institute of Ichthyology, Private Bag 1015, Grahamstown 6140, South Africa.

A previously unrecorded population of coelacanths was discovered off the Indonesian island of Manado Tua, some 10,000 km east of Africa's Comora Archipelago, by Mark Erdmann in 1997. The Indonesian coelacanth has been described as a new species, *Latimeria menadoensis*.

Fricke et al., 1995
Erdmann et al., 1998
Pouyard et al., 1999

FIGURE 5.7

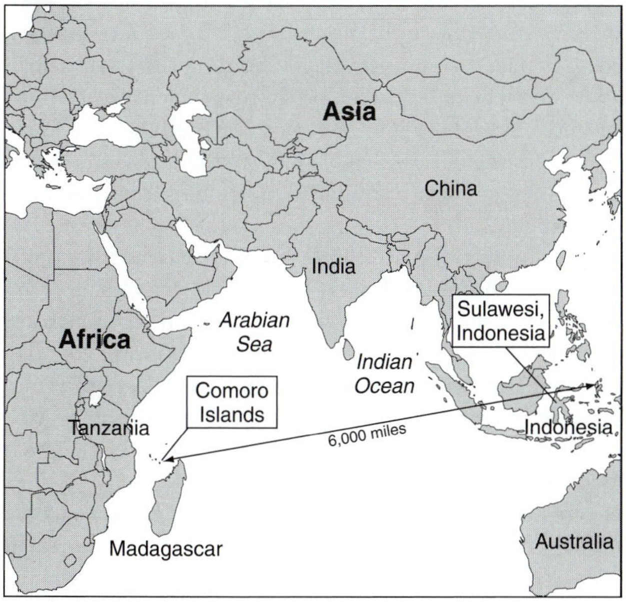

The first living coelacanth was taken near the mouth of the Chalumna River, southeast of East London in South Africa's Cape Province. A second population was discovered in Indonesia 10,000 km east of the Comoro Islands by Mark Erdmann in 1997.

and paired fins that are coordinated, not like most fishes, but in a fashion identical to human limbs.

The Actinopterygii formerly were classified into three groups: Chondrostei (primitive ray-finned fishes), Holostei (intermediate ray-finned fishes), and Teleostei (advanced ray-finned fishes). Currently, two major divisions of Actinopterygii are recognized: Chondrostei (primitive ray-finned fishes) and Neopterygii (advanced ray-finned fishes).

MORPHOLOGY

Integumentary System

Unlike most other vertebrates, most fishes have an epidermis that consists entirely of living cells. Multicellular glands that produce mucus, various toxic secretions, and other substances are present in most species and are particularly abundant in those fish that lack scales. These glands may be confined to the epidermis, or they may grow into the dermis.

The dermis in most fishes is characterized by the presence of scales composed of bony and fibrous material (Fig. 5.8). Broad plates of dermal bone were present in the earliest known vertebrates, the ostracoderms or armored fishes, and they were well developed in the extinct placoderms. These large bony plates have gradually been reduced to smaller bony plates or scales in modern fishes. **Cosmoid scales** are small, thick scales consisting of a dentinelike material, known as cosmine, overlaid by a thin layer of enamel. Although many extinct lobe-finned fish possessed cosmoid scales, the only living fish having this type of scale is the lobe-finned coelacanth (*Latimeria*).

Placoid scales (Fig. 5.8a) are characteristic of elasmobranchs and consist of a basal plate embedded in the

FIGURE 5.8

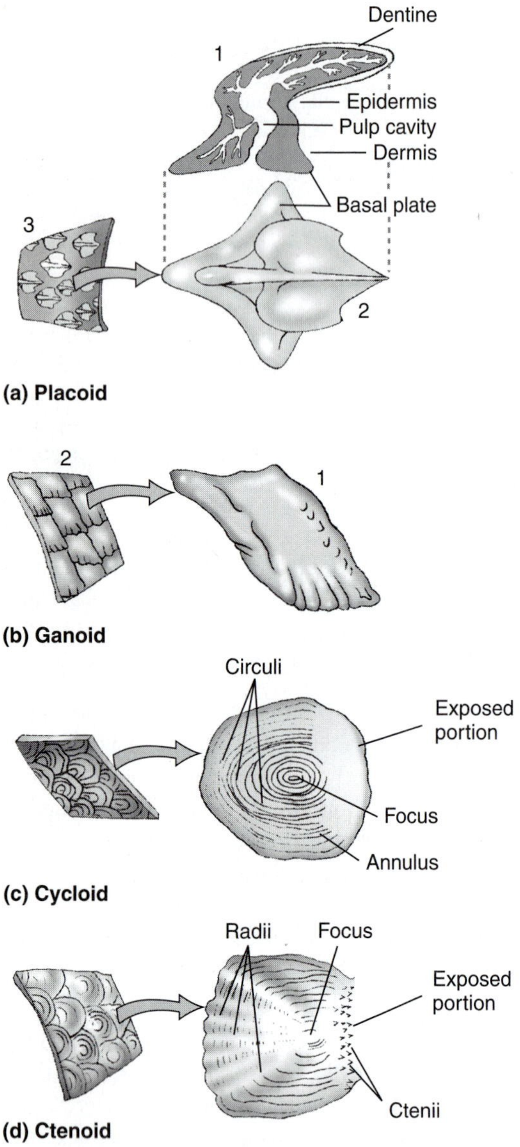

Scale types. (*a*) Placoid: 1, sagittal section; 2, dorsal view; 3, normal arrangement on skin, i.e., not overlapping. (*b*) Ganoid: 1, single scale; 2, normal arrangement on skin, i.e., slightly overlapping. (*c*) Cycloid. (*d*) Ctenoid. Cycloid and ctenoid scales overlap extensively.

dermis with a caudally directed spine projecting through the epidermis. Both the plate and spine are composed of dentine, a hard, bonelike substance. Each spine is covered by enamel and contains a central pulp cavity of blood vessels, nerve endings, and lymph channels from the dermis. Modified placoid scales form a variety of structures including shark teeth, dorsal fin spines, barbs, sawteeth, and some gill rakers.

Ganoid scales (Fig. 5.8b) are rhomboidal in shape and composed of bone. On the surface of the bone is a hard, shiny, inorganic substance known as ganoin. Today, these scales are found only on bichirs and reedfish (*Polypterus* and *Erpetoichthys*), sturgeons (*Acipenser*), paddlefishes (*Polyodon* and *Psephurus*), and gars (*Lepisosteus*). In gars, these scales fit against each other like bricks on a wall, whereas in sturgeons five rows of scales form ridges of armor along portions of their sides and back.

Cycloid and **ctenoid scales** (Fig. 5.8c, d) closely resemble one another, and both may occur on the same fish. They consist of an outer layer of bone and a thin inner layer of connective tissue. The bony layer is usually characterized by concentric ridges that represent growth increments during the life of the fish. Ctenoid scales possess comblike or serrated edges along their rear margins, whereas cycloid scales have smooth rear margins. They both are thin and flexible, have their anterior portions embedded in the dermis, and overlap each other like shingles on a roof. Cycloid and ctenoid scales are characteristic of teleost fishes. Together with reduction in heaviness and complexity, these scales allow increased flexibility of the body.

Considerable variation exists in both the abundance and size of fish scales. Most species of North American catfishes (Ictaluridae) are "naked," or smooth-skinned, whereas the scales of eels are widely separated and buried deep in the skin. Paddlefishes and sculpins have only a few scales. The scales of trout are tiny (more than 110 in the lateral line), and those of mackerels are even smaller.

Fishes that either lack scales entirely or have a reduced number of scales are typically bottom-dwellers in moving water (such as sculpin); fishes that frequently hide in caves, crevices, and other tight places (such as many catfishes and eels); or fast-swimming pelagic fishes (such as swordfish and some mackerels). The loss of scales increases flexibility and decreases friction. Many ecologically similar fishes that appear to be scaleless, such as most tunas and anguillid eels, in fact have a complete covering of deeply embedded scales.

Coloration is produced by pigment-bearing cells known as **chromatophores.** Many kinds of pigments are found in fishes, but the most common are melanins, carotenoids, and purines. Those chromatophores containing the pigment melanin are known as **melanophores** and produce brown, gray, or black colors. **Lipophores** are the pigment-bearing cells that contain the carotenoids, which are responsible for yellow, orange, and red colors. **Purines** are crystalline substances that reflect light. The most common purine in fishes is guanine, which is contained in special chromatophores known as **iridophores** or guanophores. Iridophores reflect and disperse light and are responsible for iridescence.

Color change is controlled by the nervous and endocrine systems. It involves reflex activities brought about by visual stimulation of the eyes and/or the pineal body, through hormones such as adrenalin and acetylcholine, and through the stimulus of light on the skin and/or chromatophores. Color change may be brought about either by a change in the shape of the chromatophores or by a redistribution of the pigment within the chromatophores.

BIO-NOTE 5.4

Color Change in Flounders

Flounders (order Pleuronectiformes) are famous for their ability to match their background either to avoid predators or to enhance their ability to capture prey. The initiation of a color change usually comes from visual cues. A flounder with its head on one background and its body on another will have a body color matching that of the background around its head.

In the laboratory, tropical flounders (*Bothus ocellatus*) can transform their markings in less than 8 seconds to match even unusual patterns put on the floor of their laboratory tanks. They changed their markings even faster—in as little as 2 seconds—when exposed to the same pattern for the second or third time. When swimming over sand, flounders look like sand. Above a pattern of polka dots, the fishes develop a pattern of dots. They can even match a checkerboard fairly well when placed on one in the laboratory.

Bothus ocellatus possesses at least six types of skin markings, including H-shaped blotches, small dark rings, and small spots. The darkness of these figures is adjusted to blend into the different backgrounds. The neural mechanisms that enable a flounder to alter its spots are still not known, but it is thought that cells in its visual system may respond specifically to shapes in its environment.

Ramachandran et al., 1996

Multicellular epidermal glands of at least 42 families of fishes are modified to function as light-emitting organs known as **photophores** (Fig. 5.9). Most of these families are teleosts (bony fishes); only two families of elasmobranchs (sharks, skates, and rays) are known to be luminous. Most live at depths of 300 to 1,000 m, although many move vertically into surface waters on nightly feeding migrations.

Light in some luminous fishes is produced chemically by the interaction of an enzyme (luciferase) with a phenol (luciferin) (Bond, 1996). In others, including many marine species that live in deeper waters (orders Stomiiformes, Myctophiformes, Batrachoidiformes, Lophiiformes, and others), bioluminescent bacteria reside in specialized glandlike organs (Foran, 1991). Because these bacteria glow continuously, fishes have evolved methods of covering and uncovering the pouches to produce light signals for intraspecific communication, camouflage, and attracting food. Some have evolved a pigmented irislike shutter to conceal the light; others rotate the light organ into a black-pigmented pocket (Fig. 5.9).

Lanternfishes (Myctophidae) are small, blunt-headed fishes with large eyes and rows of photophores on the body and head. Photophore patterns are different for each species, and also different for the sexes of each species. This sexual dimorphism led some early investigators to describe males and females of the same species as separate species.

FIGURE 5.9

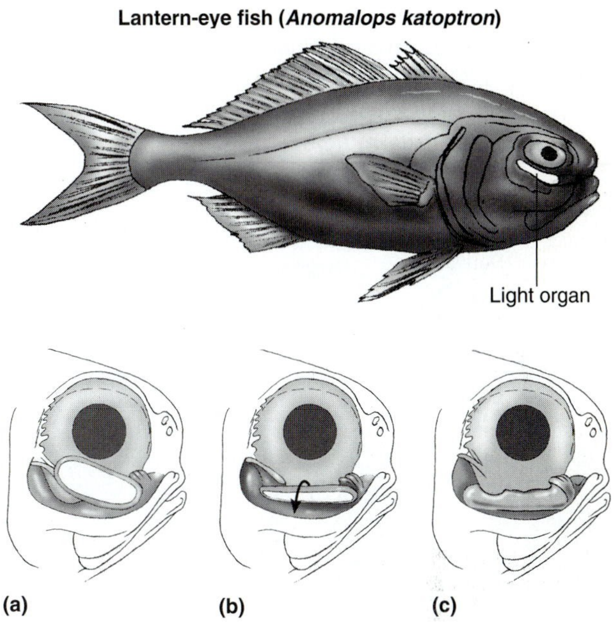

The bioluminescent light organ of the lantern-eye fish (*Anomalops katoptron*) is hinged at the front by a muscle (*a*). This muscle is used by the fish to rotate the organ downward into a pouch (*b* and *c*). These fish blink several times per minute.

BIO-NOTE 5.5

Light Organs in Predatory Fishes

Anglerfish have a long "fishing rod" attached to the skull, with a luminous bulbous light lure at the tip that can be wiggled about. Viperfish, on the other hand, have light organs directly inside their mouths to lure prey into a waiting stomach. The most specialized light source, however, may belong to a small predatory fish in the genus *Pachystomias*, which emits a red beam from an organ directly under its eye. Because most fishes cannot see red, this fish can use its beam like a sniperscope, sighting and then moving in on its target without detection.

Skeletal System

A fish's skeleton is composed of cartilage and/or bone. It provides a foundation for the body and fins, encases and protects the brain and spinal cord, and serves as an attachment site for muscles. The axial skeleton of a fish consists of the skull and vertebral column; the appendicular skeleton consists of the fin skeleton.

Skull

The skull consists of the chondrocranium, splanchnocranium, and dermatocranium. The **chondrocranium** (neurocranium) surrounds the brain and the special sense organs. It develops from paired cartilages, most of which eventually fuse with one another. The **splanchnocranium** arises from arches of cartilage that develop in association with the pharynx. It develops into the branchial (visceral, pharyngeal) arches that support the gills and make up the skeleton of the jaws and gills in fishes and amphibians that breathe by means of gills. The splanchnocranium may remain cartilaginous or become ensheathed by dermal bones. The **dermatocranium** (Fig. 5.10), which develops in the dermis, is formed of dermal bones that overlie the chondrocranium and splanchnocranium and completes the protective cover of the brain and jaws.

FIGURE 5.10

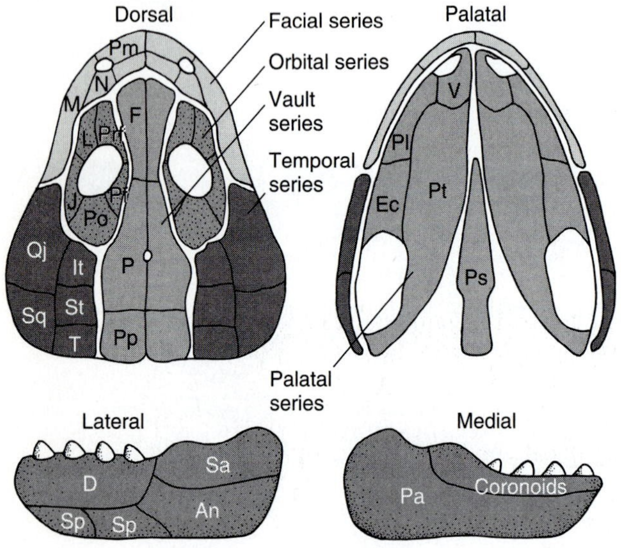

Major bones of the dermatocranium. Meckel's cartilage (not shown) is encased by the bones forming the mandible. Key: An, angular; D, dentary; Ec, ectopterygoid; F, frontal; It, intertemporal; J, jugal; L, lacrimal; M, maxilla; N, nasal; P, parietal; Pa, prearticular, Pl, palatine; Pm, premaxilla; Po, postorbital; Pp, postparietal; Prf, prefrontal; Ps, parasphenoid; Pt, pterygoid; Qj, quadratojugal; Sa, surangular; Sp, splenial; Sq, squamosal; St, supratemporal; T, tabular; and V, vomer.

In the Chondrichthyes, the skull consists of a cartilaginous chondrocranium and splanchnocranium (Fig. 5.11). The splanchnocranium in Chondrichthyes includes seven pairs of branchial cartilages and a series of median cartilages in the pharyngeal floor. The first pair of branchial cartilages, called the **mandibular arch**, consists of a dorsal **palatoquadrate (pterygoquadrate) cartilage** and a ventral **Meckel's cartilage** on each side. The upper jaw is formed by the palatoquadrates, and the lower jaw is formed on each side by Meckel's cartilages. The second pair of visceral cartilages, called the **hyoid arch**, consists of several elements, with the most dorsal being known as **hyomandibular** cartilages. Ligaments hold the jaws together and bind them to the hyomandibular cartilages, which suspend the entire splanchnocranium from the skull. The last five pairs of visceral cartilages consist of four segments each (pharyngobranchial, epibranchial, ceratobranchial, and basibranchial) and are similar to one another. Embryological evidence and comparative anatomy studies indicate that jaws evolved from the first gill arch (Feduccia and McCrady, 1991).

The skulls of bony fish are compressed laterally. They are cartilaginous initially, but are partly or wholly replaced by bone as development progresses. The only portions of the embryonic palatoquadrate cartilages that contribute to the upper jaws in bony fish are the caudal ends, which become **quadrate bones** (Fig. 5.12); the remainder of the palatoquadrate cartilages are replaced by several bones, including the **premaxillae** and **maxillae**. Teeth are usually present on the premaxillae and maxillae (as well as on many bones forming the palate), but in teleosts, maxillae may be toothless, reduced, or even lost from the upper jaw margin. The posterior tip of Meckel's cartilage ossifies and becomes the **articular bone**; the remainder of

FIGURE 5.11

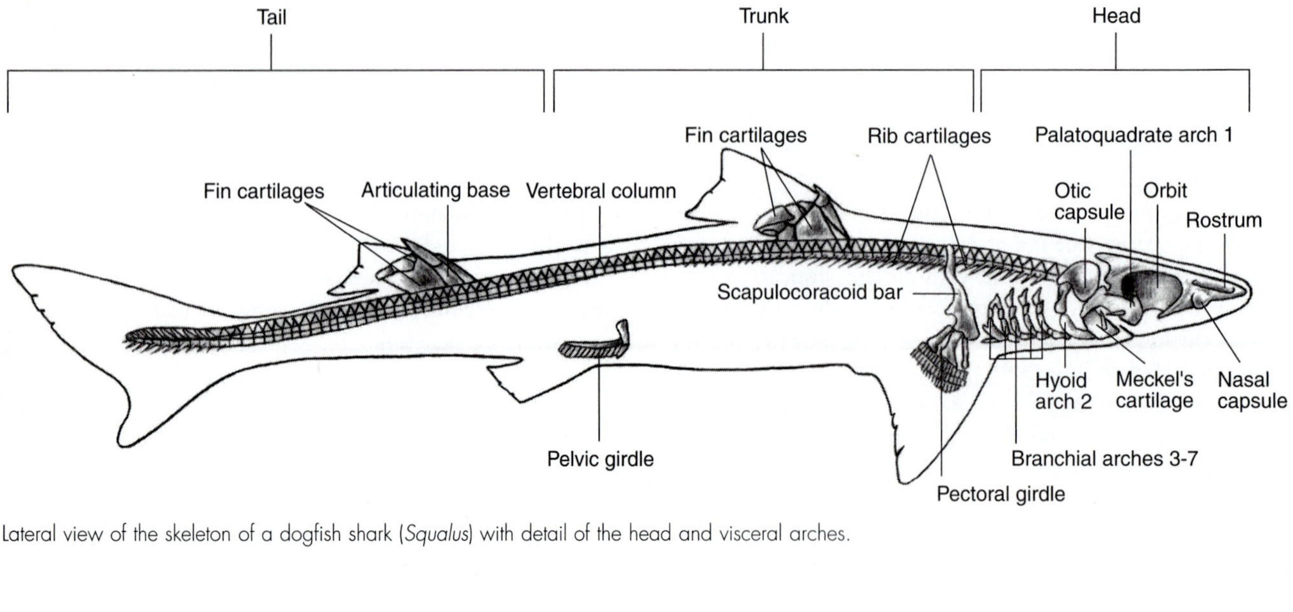

Lateral view of the skeleton of a dogfish shark (*Squalus*) with detail of the head and visceral arches.

Meckel's cartilage becomes ensheathed by dermal bones such as the **dentaries** and **angulars** (Fig. 5.12).

The hyoid skeleton of bony fish undergoes extensive ossification and performs key roles in the specialized movements of ingestion and respiration. The **operculum**, which is of dermal origin, extends backward over the gill slits and regulates the flow of water across the gills. Movements of the operculum and hyoid, therefore, must be well coordinated. An operculum is absent in most cartilaginous fishes.

Jaw suspension in fishes is accomplished in three ways (Fig. 5.13). In some sharks, the jaws and hyoid arch are braced directly against the braincase, an arrangement called **amphistylic** suspension. In lungfish and chimaeras, the hyomandibular cartilage is not involved in bracing the jaws. This "self-bracing" condition, known as **autostylic** suspension, is also utilized by all of the tetrapods. In most of the Chondrichthyes and in some of the bony fishes, the hyomandibular cartilage is braced against the chondrocranium,

FIGURE 5.12

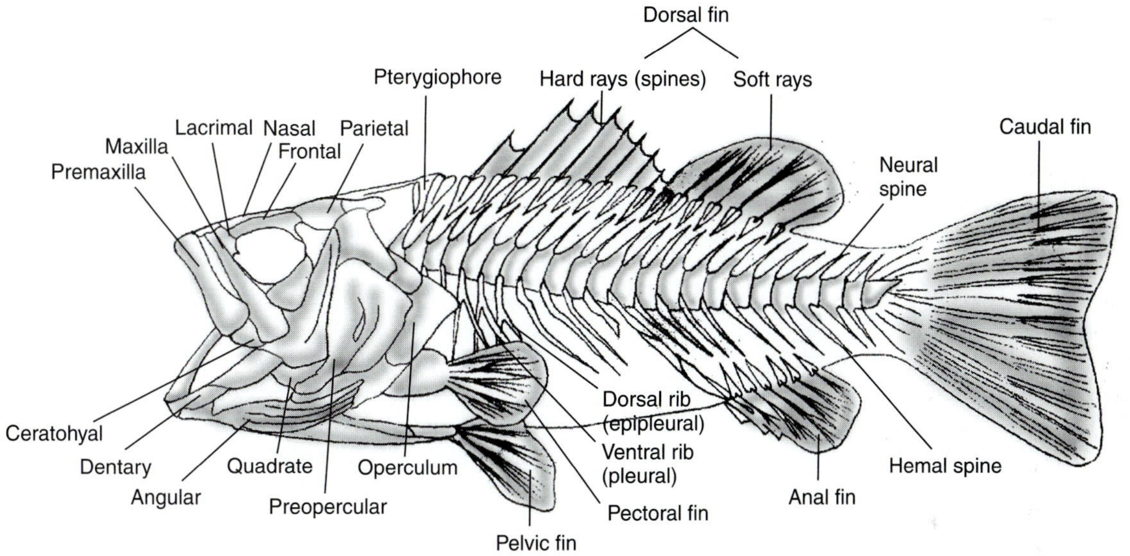

Lateral view of the skeleton of a bony fish (Teleostei). Note the position of the paired and unpaired fins and the hyostylic method of jaw suspension.

FIGURE 5.13

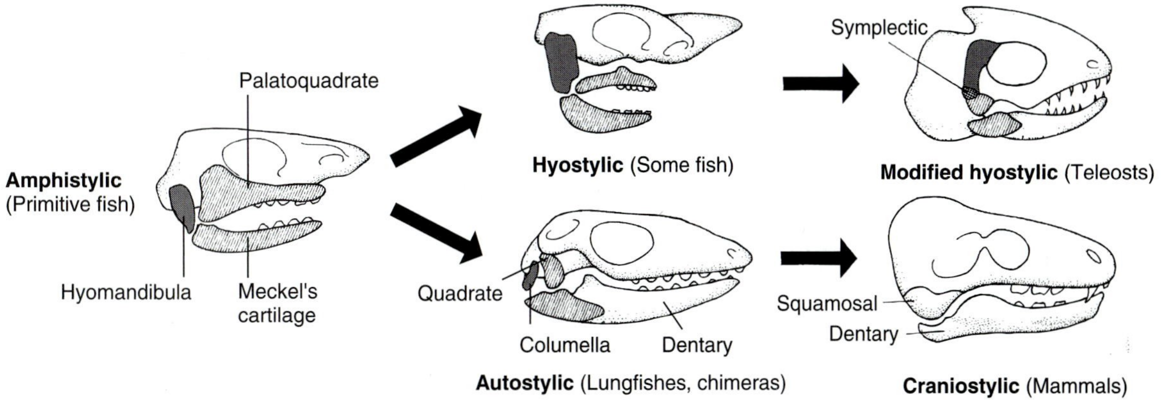

Evolution of jaws and jaw suspension. The types of jaw suspension are defined by the points at which the jaws attach to the rest of the skull. Note the mandibular arches (crosshatched areas) and hyoid arches (shaded areas). The dermal bone (white areas) of the lower jaw is the dentary.

and the jaws are braced against the hyomandibular cartilage, a condition known as **hyostylic** jaw suspension.

Vertebral Column

The vertebral column in fishes ranges from a column having cartilaginous vertebrae with centra in elasmobranchs (Fig. 5.11) to one having vertebrae of solid bone in teleosts (Fig. 5.12). Extending from the skull to the tip of the tail, fish vertebrae are differentiated into **trunk vertebrae** and **caudal** (tail) **vertebrae**. Both ends of the centra (body) of a vertebra in most fishes are concave, a condition known as **amphicoelous** (Fig. 5.14). A greatly constricted notochord runs through the center of each centrum and also fills the spaces between adjacent vertebrae.

Fin Skeleton

The pectoral and pelvic girdles together with the skeleton of the paired fins make up the appendicular skeleton of fishes (Figs. 5.11 and 5.12). Most gnathostome fishes have both pectoral and pelvic fins, although pelvic fins are lost in elongate fishes, such as eels, that wriggle along the bottom.

The **pectoral girdle** braces the anterior pair of appendages (pectoral fins) of fishes. Pectoral fins may be located high on the sides of the body, more toward the midline, or below the midline (Fig. 5.15). They may be long and pointed or broader and more rounded. In most fish, these fins operate not only as stabilizers, but also as "diving planes." They are set at an angle to generate lift for the anterior part of the body and, in some species, are important in thrust generation. In threadfins (Polynemidae) (Fig. 5.15d), the pectoral fins are divided into two parts with the lower portion consisting of several long filaments that are thought to function as tactile organs. The pectoral fins of batfish (Ogcocephalidae) (Fig. 5.15f) are located posterior to the pelvic fins. They are used for "walking" over the bottom.

The pelvic girdle braces the posterior pair of appendages (pelvic fins). In sharks and in the more ancestral bony fishes, such as salmon, shad, and carp, pelvic fins are located ventrally, toward the rear of the fish; this is called the **abdominal position** (Fig. 5.16a). In more recently evolved teleosts, many of which are deep-bodied, the pelvic fins are more anterior and are located either slightly behind the pectoral fins, in the **subabdominal position** (Fig. 5.16b); below the pectoral fins, in the **thoracic position** (Fig. 5.16c); or even in front of the pectoral fins, in the **jugular position** (Fig. 5.16d). In some, such as eels and eel-like fishes, the pectoral and pelvic fins are frequently absent or greatly reduced in size, whereas in bottom-dwelling fishes, pelvic fins are frequently modified into organs for holding onto the substrate (Figs. 5.16e–g).

Most fishes also have unpaired **median fins** that assist in stabilizing their bodies during swimming (Figs. 5.11 and 5.12). These include one or two **dorsal fins**, a ventral **anal fin** behind the anus or vent, and a **caudal fin**. Some primitive bony fishes including salmon, trout, and smelts (Salmoniformes), as well as catfishes (Siluriformes) and characins (Characiformes), possess an **adipose fin**, a median, fleshy dorsal fin that lies near the caudal fin and has no internal stiffening rays or bony elements. It probably plays a minor role in propulsion. Eel-like fishes have long dorsal and anal fins that frequently run most of the length of the body.

Caudal fins are modified in three major ways. They are unlobed, or **diphycercal**, in lungfishes and bichirs (Fig. 5.17a). Sharks, in contrast, possess **heterocercal** fins (Fig. 5.17b), in which one lobe is larger than the other. If the vertebral column extends into the dorsal lobe, the caudal fin is **epicercal**; if the vertebral column extends into the ventral lobe, it is **hypocercal**. Such fins, which provide lift for the posterior part of the body, counter the shark's tendency to sink and also assist in lifting the body off the substrate following periods of rest. In most bony fishes, the upper and lower lobes of the caudal fin are about the same size, or

FIGURE 5.14

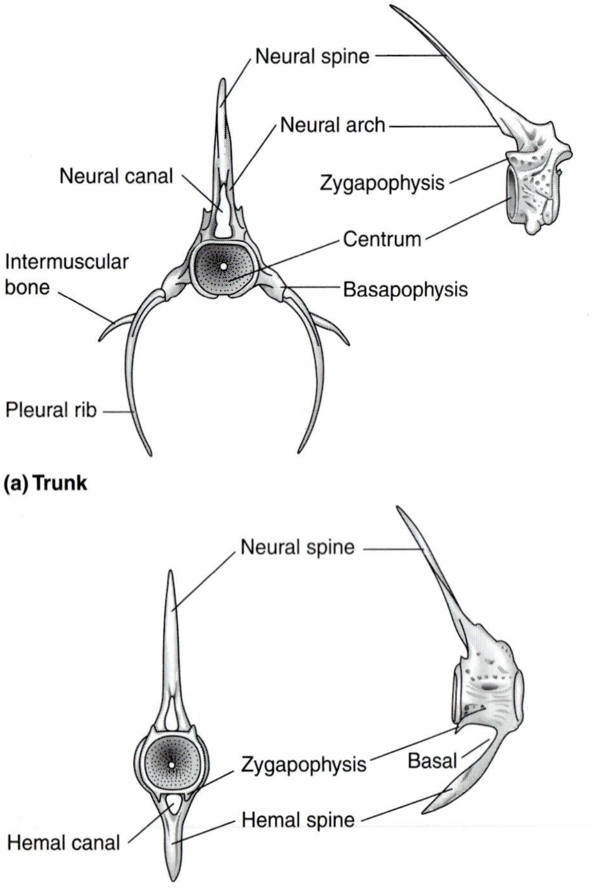

Structure of the two kinds of vertebrae found in teleosts: (*a*) trunk; (*b*) caudal.

FIGURE 5.15

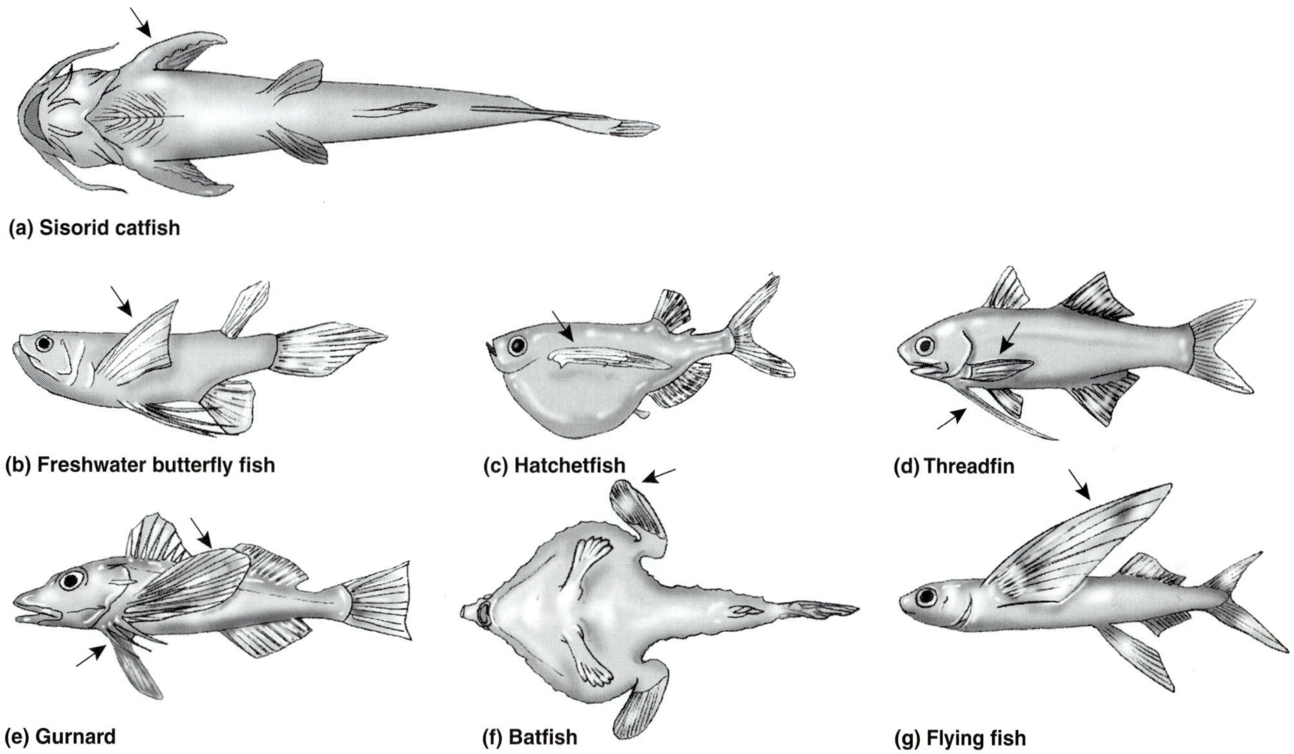

Modifications of pectoral fins (indicated by arrows) in several fish genera: (*a*) ventral view of sisorid catfish (*Glyptothorax*); (*b*) freshwater butterfly fish (*Pantodon*); (*c*) hatchetfish (*Gastropelecus*); (*d*) threadfin (Polynemidae); (*e*) gurnard (Triglidae); (*f*) ventral view of batfish (Ogcocephalidae) with armlike pectoral fins well behind pelvic fins; (*g*) flying fish (Exocoetidae).

homocercal (Fig. 5.17c). The vertebral column does not extend into either lobe.

Pelvic fins in male chimaeras, skates, and oviparous sharks that utilize internal fertilization have been modified by the addition of skeletal elements to form intromittent organs known as **claspers** (see Fig. 5.35). The anal fin is modified into an intromittent organ known as a **gonopodium** in some male teleosts, such as guppies and mollies (*Poecilia*), swordtails (*Xiphophorus*), and mosquitofish (*Gambusia*). These organs, which have evolved to improve fertilization of the eggs, are inserted into the genital openings of females and guide sperm into the female reproductive tract.

Fishes are propelled through the water by fins, body movement, or both. In most fishes, both paired and unpaired fins serve primarily for steering and stabilizing rather than for propulsion. In general, the main moving force is created by the caudal fin and the area immediately adjacent to it, known as the **caudal peduncle**. It long had been hypothesized that the anterior musculature generated most of the power and that the posterior musculature transmitted the force to the tail (Lighthill, 1971; Wainwright, 1983). By analogy, the anterior muscle was thought to act as the "motor," the tail as the "propeller," and the posterior muscle as the "drive shaft." However, through a combination of filming, electrical impulse recordings, and mathematical modeling of red muscle bundles in the scup (*Stenotomus chrysops*), Rome et al. (1993) showed that most of the power for normal swimming came from muscle in the posterior region of this fish, and relatively little came from the anterior musculature. Eels rely on extreme, serpentlike body undulations to swim, with fin movement assisting to a minor extent. Fishes with a fairly rigid body such as the filefish, trunkfish, triggerfish, manta, and skate, however, depend mostly on fin action for propulsion.

Muscular System

The metamerically arranged body wall muscles are composed of a series of zig-zag-shaped **myomeres** (Fig. 5.18a, b), with each myomere constituting one muscle segment. Coordinated contractions (contraction on one side accompanied by relaxation on the opposite side) of posterior myomeres produce waves of contraction that provide the main locomotor mechanism of most fishes. As this propulsive wave moves posteriorly, the water adjacent to the fish is accelerated backward until it passes over the posterior margin of the caudal fin, producing thrust (Lighthill, 1969).

In most fishes, white muscles predominate and may comprise up to 90 percent or more of the entire body weight (Bond, 1995). White muscle has relatively thick

FIGURE 5.16

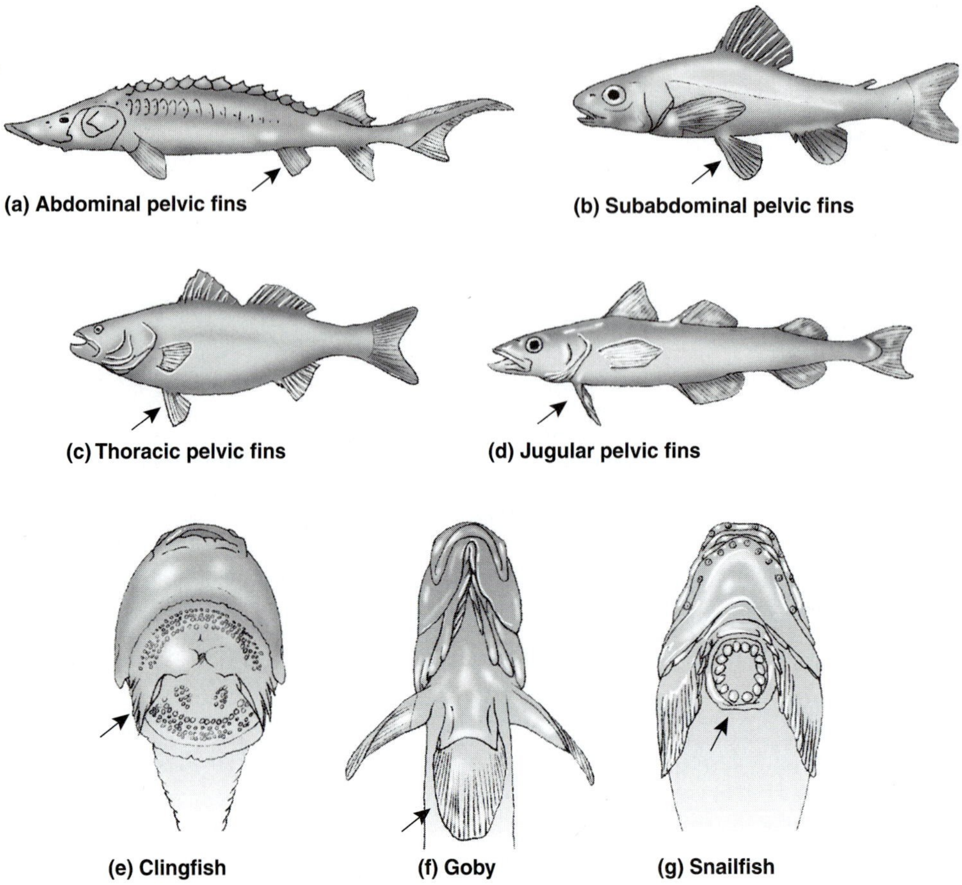

Modifications of pelvic fins and their positions (pelvic fins indicated by arrows): (*a*) abdominal (sturgeon, Acipenseridae); (*b*) subabdominal (sand roller, Percopsidae); (*c*) thoracic (bass, Moronidae); (*d*) jugular (pollock, Gadidae). Some pelvic fins have been modified for holding onto the substrate: (*e*) clingfish (Gobiesocidae); (*f*) goby (Gobiidae); (*g*) snailfish (Liparidae).

FIGURE 5.17

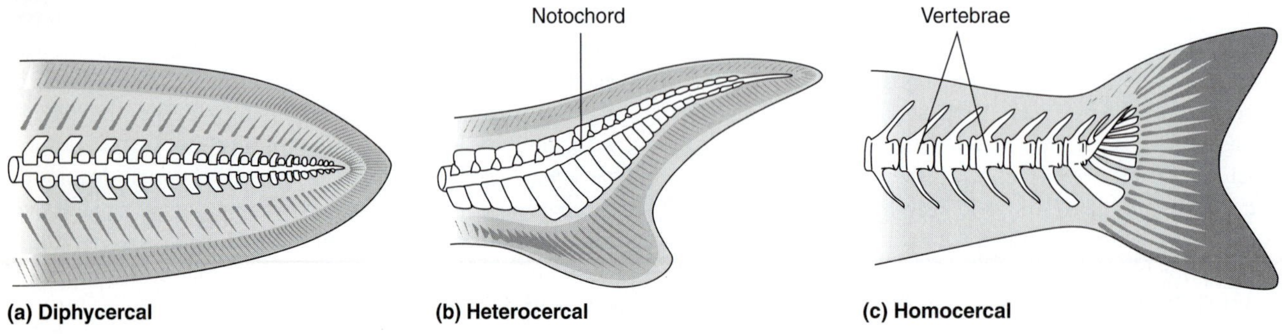

Major caudal fin (tail) modifications in fishes: (*a*) diphycercal (lungfishes and bichirs); (*b*) heterocercal (sharks); (*c*) homocercal (most bony fishes). Heterocercal tails may be further subdivided: if the dorsal lobe is larger than the ventral lobe, it is designated as an epicercal tail; if the ventral lobe is larger, it is known as a hypocercal tail.

FIGURE 5.18

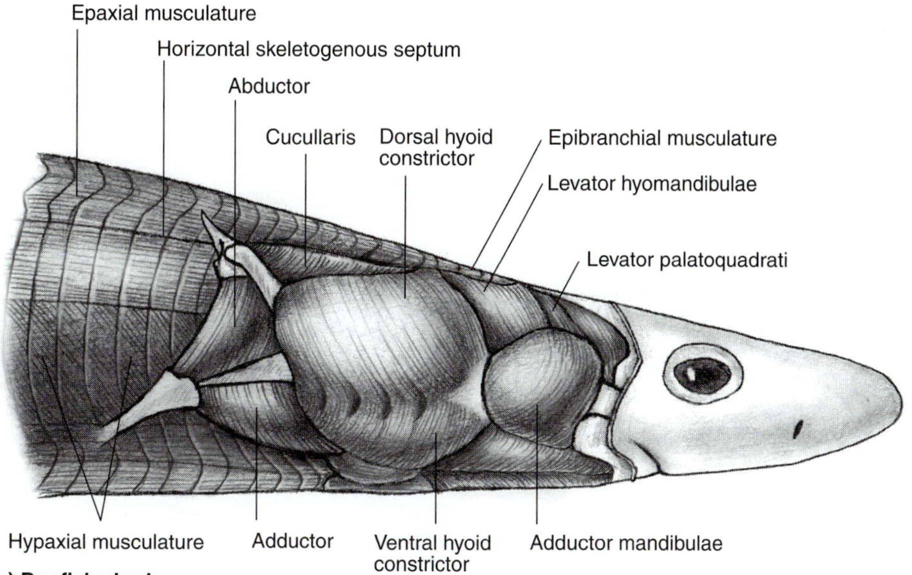

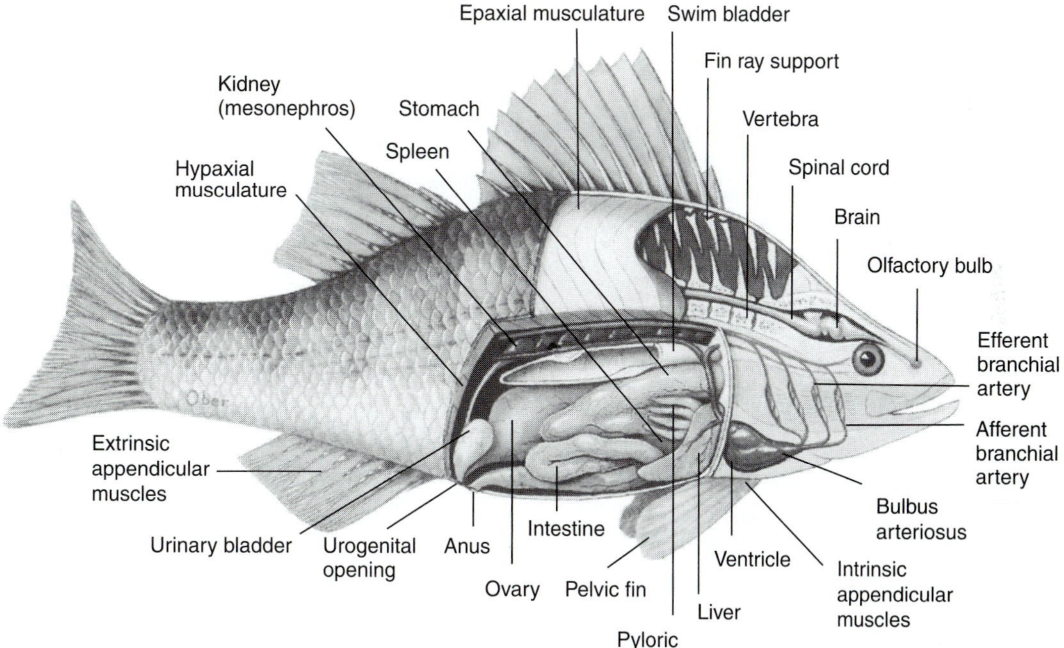

(*a*) Lateroventral view of the head of a dogfish shark (*Squalus*) showing epibranchial (above the gills), hypobranchial (below the gills), and branchiomeric musculature. (*b*) Muscles and internal anatomy of the yellow perch (*Perca flavescens*), a freshwater teleost fish. (*c*) Cross section of body musculature in a chinook salmon (*Oncorhynchus tshawytscha*). (*d*) Diagram showing approximate extent of red muscle (stippled) in skipjack tuna (*Katsuwonus pelamis*).

Continued on page 104

fibers, no fat or myoglobin (a protein that bonds with oxygen), and primarily utilizes anaerobic metabolism. In those fishes that swim most of the time and have an adequate oxygen supply, however, such as tuna, bonito, and marlin, red muscles make up a greater portion of the body muscle mass (Cailliet et al., 1986) (Fig. 5.18d). Red muscle consists of thin-diameter fibers, contains fat and myoglobin, and utilizes aerobic respiration.

Six groups of fishes (Rajidae, Torpedinidae, Mormyriformes, Gymnotiformes, Melapteruridae, and Uranoscopidae)

FIGURE 5.18 *Continued from page 103*

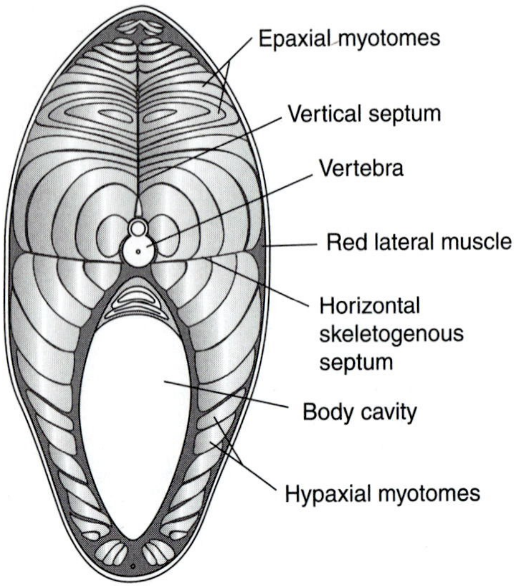

(c) Chinook salmon

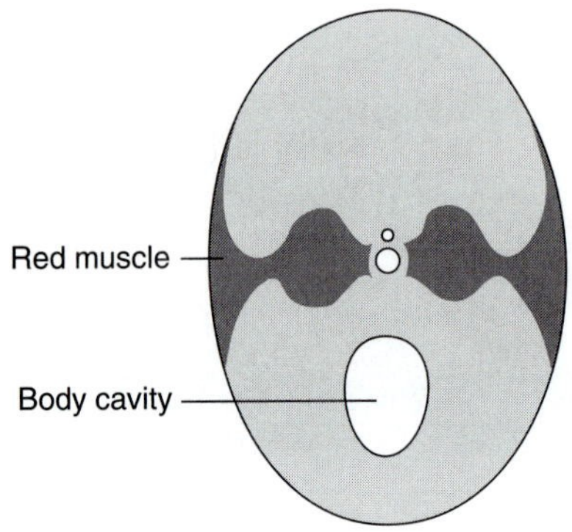

(d) Skipjack tuna

BIO-NOTE 5.6

Endothermy

Some degree of endothermy is present in sharks of the family Lamnidae and Alopiidae and in certain oceanic teleost fishes such as mackerels, tunas, and billfishes. The development of endothermy requires the elevation of aerobic capacity and the reduction of heat loss. Tunas have exceptionally high metabolic rates and are able to reduce their overall heat loss. They retain metabolic heat by way of vascular countercurrent heat exchangers located in the brain, muscle, and viscera. Red aerobic muscle contributes the majority of metabolically derived heat and is centrally located near the vertebral column rather than laterally as in most teleosts. Thus, these fishes warm their brain, muscle, and viscera. Billfishes, however, use cranial endothermy and warm only the brain and eyes by passing blood through the superior rectus eyeball muscle. A countercurrent heat exchanger retains the heat beneath the brain, and a distinct arterial supply directs warm blood to the retina. The butterfly mackerel also uses cranial endothermy, but the thermogenic tissue is derived from the lateral rectus eyeball muscle. The development of endothermy may have permitted range expansion into cooler waters.

Block et al., 1993

are known to possess electric organs derived from muscle fibers. Certain muscle masses in these fishes are highly modified to produce, store, and discharge electricity. Specialized cells known as electrocytes are stimulated by signals from spinal nerves to generate small voltage gradients. Because electrocytes are arranged in columns surrounded by insulating tissues, voltages are linearly increased, similar to a series of small batteries (Heiligenberg, 1977). Because these groups of fishes are only remotely related, electric organs appear to have evolved several times independently in Africa and South America after the two continents separated. Although this capability evolved in early vertebrates, only some of the primitive fishes living today have retained this ability.

Some marine and freshwater fishes can produce charges up to approximately 500 volts, although most species are limited to weak electric discharges in the range of millivolts to volts. Most are nocturnal, have poorly developed eyes, and live in dark, murky water where visibility is poor.

The electric ray (*Torpedo*) has two dorsal electric organs in the pectoral fins, which are apparently used to immobilize prey. In another ray (*Raja*) and the electric eel (*Electrophorus*), electric organs lie in the tail and are modifications of the hypaxial musculature. Electric organs can be used for defense or to stun prey, to scan the environment and locate enemies or prey, and for social communication. (See the section on Sense Organs, page 114, for additional information on electroreceptors.)

Cardiovascular System

In fishes, the **sinus venosus** is a thin-walled sac that serves chiefly as a collecting chamber for venous blood it receives from all parts of the body (Fig. 5.19a, b). Blood flows from the sinus venosus into a large, thin-walled muscular sac, the **atrium** (auricle). From the atrium, blood enters the **ventricle** through an atrioventricular aperture guarded by valves. The ventricle, a relatively large chamber with heavy

FIGURE 5.19

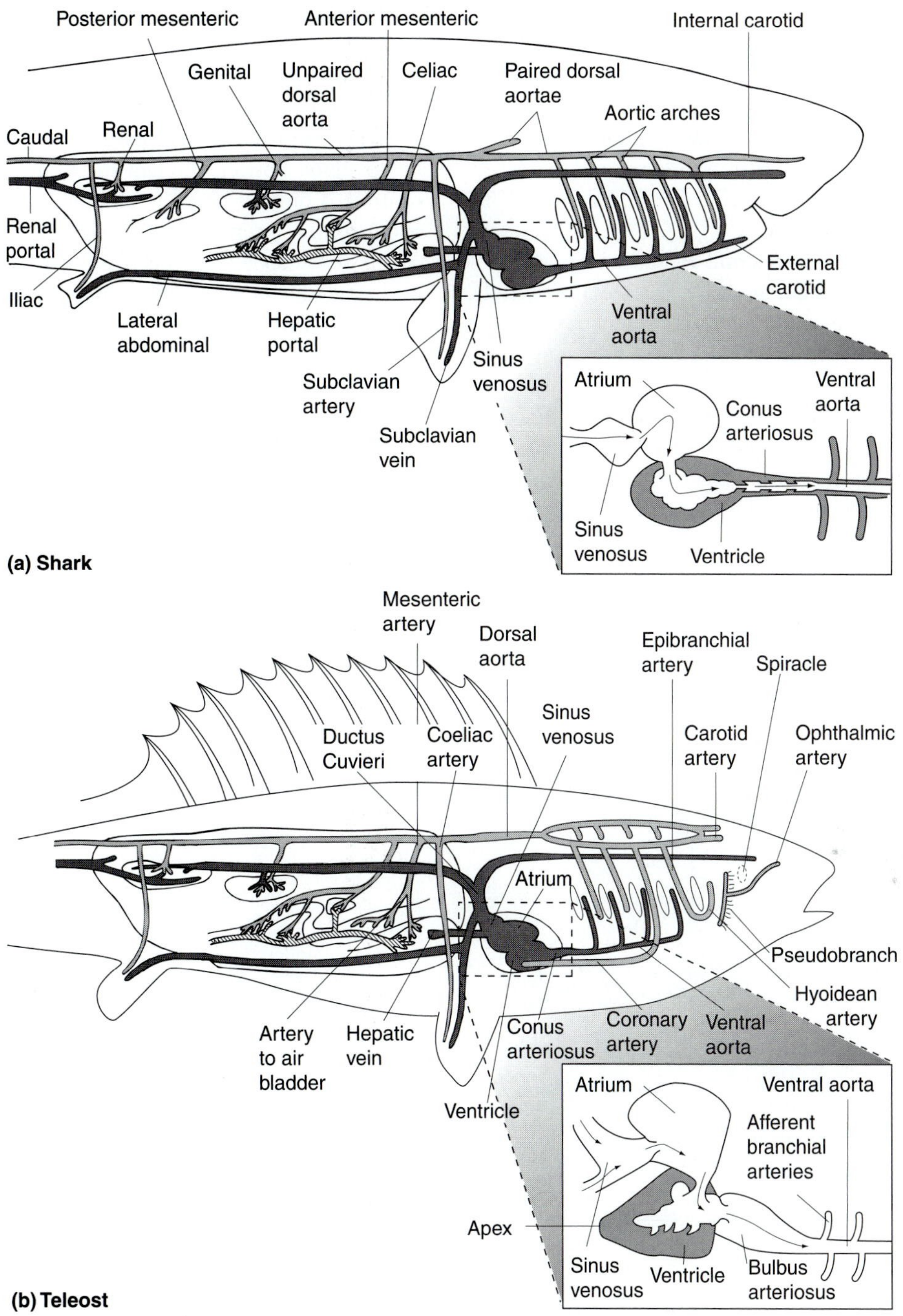

(*a*) Basic vertebrate circulatory pattern in a shark. Blood is pumped by the heart to the ventral aorta. It flows through the gill region via the branchial arteries and the paired aortic arches, which lead to the dorsal aorta. The dorsal aorta carries blood anteriorly to the head and posteriorly to the remainder of the body. The aorta gives off major branches to the viscera and somatic tissues. (*b*) Diagram of the branchial circulation of a teleost fish. Blood is pumped anteriorly by the heart into the ventral aorta. After being aerated by passing through the gills, the blood flows into the dorsal aorta.

walls of cardiac muscle, functions as the primary pump distributing blood anteriorly. The anterior end of the ventricle becomes a muscular tube, the **conus arteriosus**, which connects with the **ventral aorta** and serves to moderate blood pressure. In teleosts, the conus is short, and its function is assumed by the **bulbus arteriosus**, an expansion of the ventral aorta. A series of semilunar valves in the conus arteriosus prevent the backflow of blood. Cameron (1975) found that the teleost heart in three species of freshwater fish requires up to 4.4 percent of the total energy of the fish.

The ventral aorta carries blood forward beneath the pharynx, where six pairs of **aortic arches** connect the ventral aorta with the **dorsal aorta**. The dorsal aorta carries blood above the digestive tract toward the tail. It continues into the tail as the **caudal artery**.

In most Chondrichthyes, branchial arteries form in the aortic arches. Blood entering an aortic arch from the ventral aorta must pass through gill capillaries before continuing to the dorsal aorta. This arrangement allows aortic arches to serve a gas exchange (respiratory) function. In most teleosts, the first and second aortic arches disappear during development. The remaining pairs are converted to branchial arteries.

In fishes, all blood must traverse through at least two capillary systems (gills and tissue) before returning to the heart (Fig. 5.20a). A drop of blood passes through the heart of a fish only one time during each circuit of the body. After passing through the heart, the blood is carried to the gills for aeration and then is distributed to all parts of the body. In most fish, all of the blood flowing through the heart is venous blood. In contrast, a drop of blood in amphibians, reptiles, birds, and mammals must pass through the heart twice during any single circuit of the body (Fig. 5.20b). As a result of this difference in circulation pattern, the pressure of blood supplying the tissues is lower in fishes than in reptiles, birds, and mammals.

The blood of fishes contains nucleated erythrocytes, leucocytes, and thrombocytes. Seasonal changes in red blood cell production have been reported in some fish (Hevesy et al., 1964). For example, when oxygen demands of tissues are relatively low, as when water temperatures are low and the fish is not very active, large numbers of erythrocytes are not required and the number tends to drop.

FIGURE 5.20

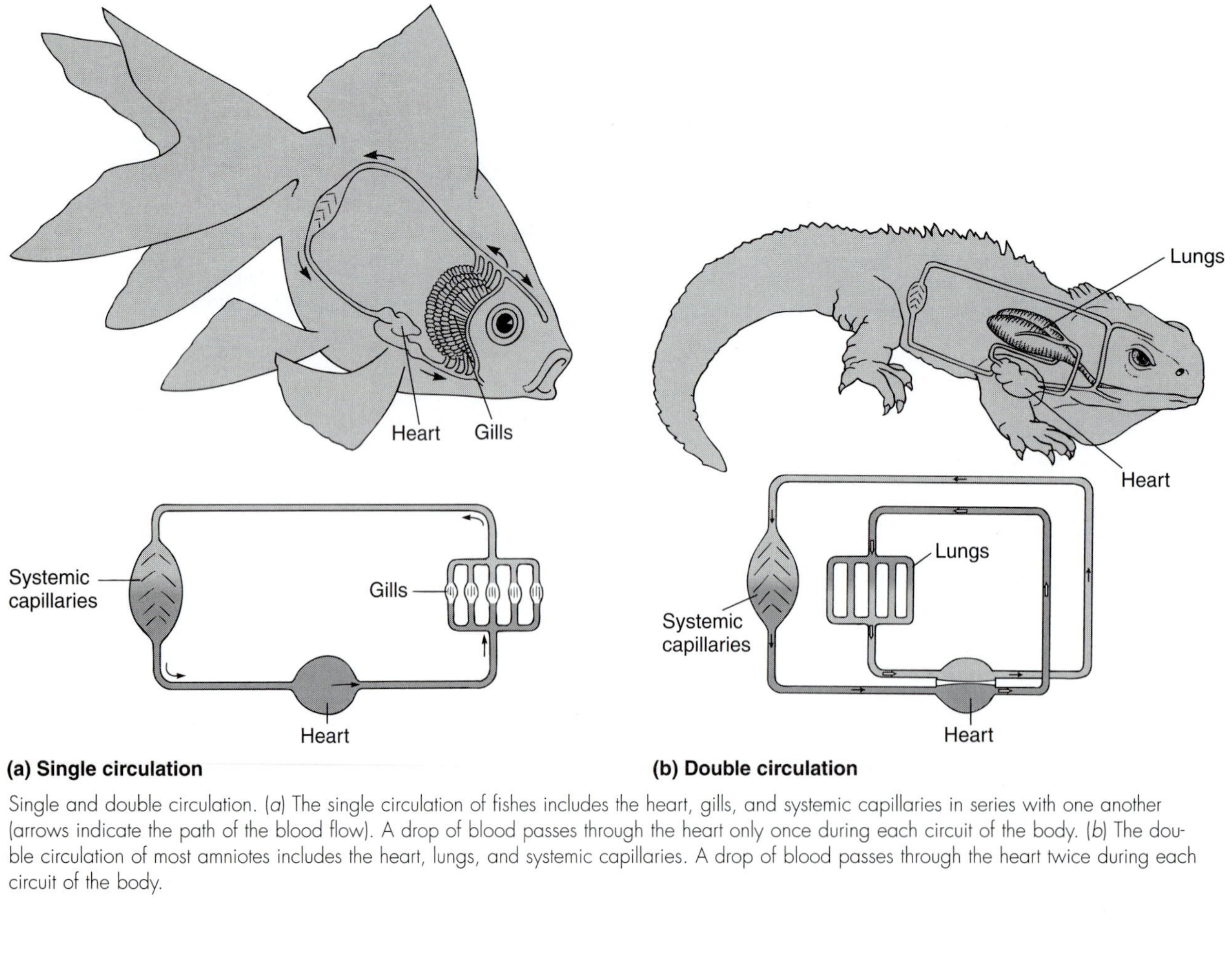

Single and double circulation. (*a*) The single circulation of fishes includes the heart, gills, and systemic capillaries in series with one another (arrows indicate the path of the blood flow). A drop of blood passes through the heart only once during each circuit of the body. (*b*) The double circulation of most amniotes includes the heart, lungs, and systemic capillaries. A drop of blood passes through the heart twice during each circuit of the body.

BIO-NOTE 5.7

Icefishes

Antarctica's marine fish fauna (there are no freshwater species because there is no permanent liquid water on the continent) comprises approximately 275 species, 95 of which belong to the perciform suborder Notothenioidei. This group contains many species with unusual adaptations. Members of one family—Channichthyidae—are known as icefishes and are unique among vertebrates in that they totally lack the respiratory pigment hemoglobin (although some nonpigmented erythrocytes are present), and their muscles contain only minute traces of myoglobin. These fishes, also commonly known as "white-blooded fish," possess creamy-white gills, yellowish-tinted blood, and yellow muscles. Oxygen is carried throughout the body in simple dissolved solution, a process that reduces the oxygen-carrying capacity of the blood to only about 10 percent of that of red-blooded fishes. Although dissolved oxygen is high in the consistently cold Antarctic waters, these sluggish fishes have low metabolic oxygen requirements. Physiological compensation is achieved through adaptations such as large ventricles, low arterial pressure, large-diameter vessels, and low erythrocyte densities that serve to increase blood volume and flow rate.

Douglas et al., 1985
Harrison et al., 1991
Eastman, 1993

Respiratory System

In most fishes, **external nares** lead to blind olfactory sacs that contain the olfactory epithelium. Water usually enters the external nares through an incurrent aperture, flows over the olfactory epithelium, and exits through an excurrent aperture. In many lobe-finned fishes, nasal canals lead from the olfactory sacs and open into the oral cavity via **internal nares (choanae)**. However, these are not used in aquatic respiration.

Internal nares are thought to have first served to increase the effectiveness of olfaction by making possible more efficient sampling of the environment. Their first respiratory function was probably to help prevent desiccation of the gills and lungs by serving as devices for aerial respiration. Thus, internal nares in crossopterygians may have been preadapted for use in aerial respiration.

In fishes, gills function primarily for respiration. They are ventilated by a unidirectional flow of water, created either actively by branchial pumping or passively by simply opening the mouth and operculum while swimming forward. The gill system consists of several major gill arches on each side of the head (Fig. 5.21) with two rows of gill filaments extending from each gill arch. Each filament consists of rows of densely packed flat **lamellae** (primary and secondary). **Gill rakers**, which project from gill arch cartilages into the pharynx, serve to protect the gills and to direct food in the water toward the esophagus. Tips of filaments from adjoining arches meet, forcing water to flow between the filaments.

FIGURE 5.21

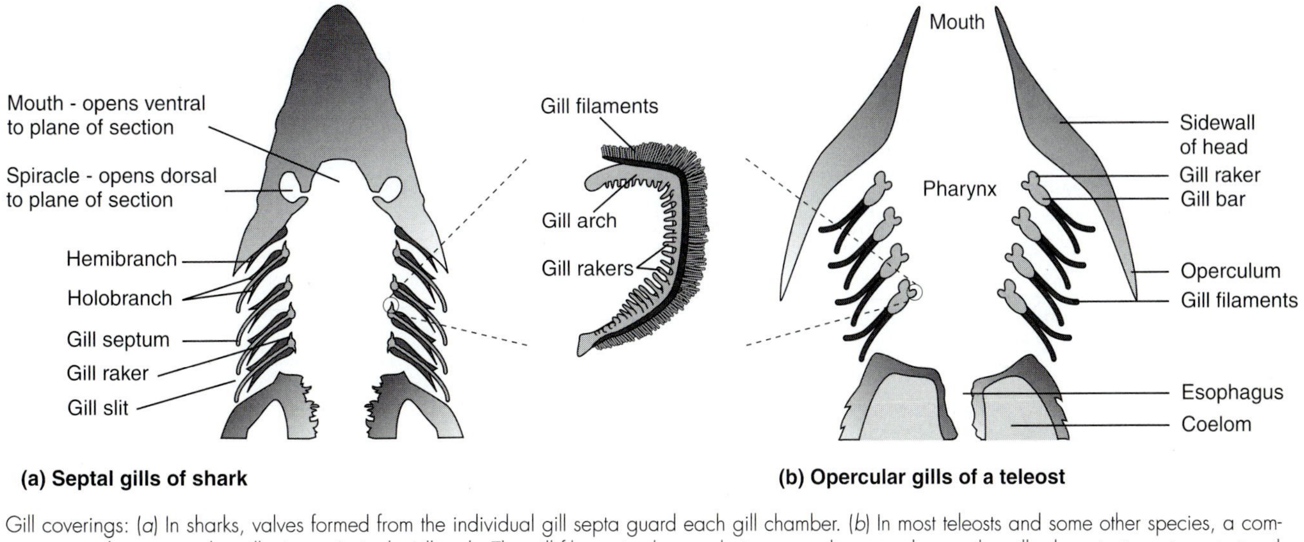

Gill coverings: (*a*) In sharks, valves formed from the individual gill septa guard each gill chamber. (*b*) In most teleosts and some other species, a common operculum covers the gills. Inset: A single gill arch. The gill filaments play a role in gas exchange, whereas the gill rakers strain water entering the gill chamber from the pharynx.

From Hildebrand, Analysis of Vertebrate Structure. *Copyright © 1986 John Wiley & Sons, Inc. Reprinted by permission of John Wiley & Sons, Inc.*

Most elasmobranchs possess five exposed (naked) gill slits that are visible on the surface of the pharyngeal region (Fig. 5.21a). They are exposed because no operculum is present. Each gill slit opens into a gill chamber whose anterior and posterior walls possess gills that are supported by gill arches. These are the sites of gas exchange. Water enters the pharynx through the mouth or spiracle and passes into the gill chambers, where it bathes the gill surfaces (Fig. 5.22). As water flows from front to rear through the slits between the lamellae, gas exchange takes place in the lamellae. At the same time, blood flows through capillaries in the opposite direction (rear to front). This **countercurrent flow** greatly increases the efficiency of gills as gas exchangers by allowing better exploitation of the low oxygen content of the water. Water is forced from the gill chambers by contraction of branchiomeric muscles. Water may leave the gills of bony fishes with a loss of as much as 80 percent or 90 percent of its initial oxygen content (Hazelhoff and Evenhuis, 1952). In contrast, mammalian lungs remove only about 25 percent of the oxygen present in inhaled air. (While gill respiration is more efficient than mammalian respiration in terms of percent saturation, it must be remembered that the amount of oxygen available in air is approximately 20 times that in an equal volume of water.)

Gills have become modified in some species such as the "walking catfish" (*Clarias batrachus*) of southeastern Asia and now introduced to southern Florida. In these species, the second and fourth gill arches possess modified gill filaments that do not collapse when exposed to the air. Ordinarily, gills tend to adhere to one another and lose much of their effective surface area when removed from water. Walking catfish usually leave the water during periods of rain so that the gills can be kept moist while moving about on land (Jordan, 1976).

Digestive System

Fishes may consume a variety of foods: filter-feeders feed on plankton, herbivores feed on plant material, detritivores consume partly decomposed organic matter, carnivores feed on animal material, and omnivores consume a variety of plant and animal material.

In most fishes, the mouth is terminal in position (Figs. 5.1b and 5.12a), although in some, especially sharks and rays, it is located ventrally and often well back from the tip of the head (subterminal) (Figs. 5.1a and 5.11). Still others, like barracudas, have projecting lower jaws, and some, like the swordfish, have elongated upper jaws.

Most fishes possess a flat, rigid, cartilaginous tongue that arises from the floor of the oral cavity. It is not always sharply demarcated and is not freely movable.

The roof of the oral cavity is formed by the primary palate. If internal nares are present, they open into the anterior portion of the oral cavity. Oral glands are sparse and consist primarily of mucus-secreting cells.

Teeth are numerous and may occur on the jaws, palate, and pharyngeal bones. Teeth composed of epidermal cells are present in lampreys, hagfishes, and adult sturgeons. Teeth may be attached to the outer surface, or summit, of the jawbone, a situation called **acrodont dentition,** or rooted in individual bony sockets, a situation called **thecodont dentition**. Most fishes have **polyphyodont dentition**—that is, they can replace damaged or injured teeth.

FIGURE 5.22

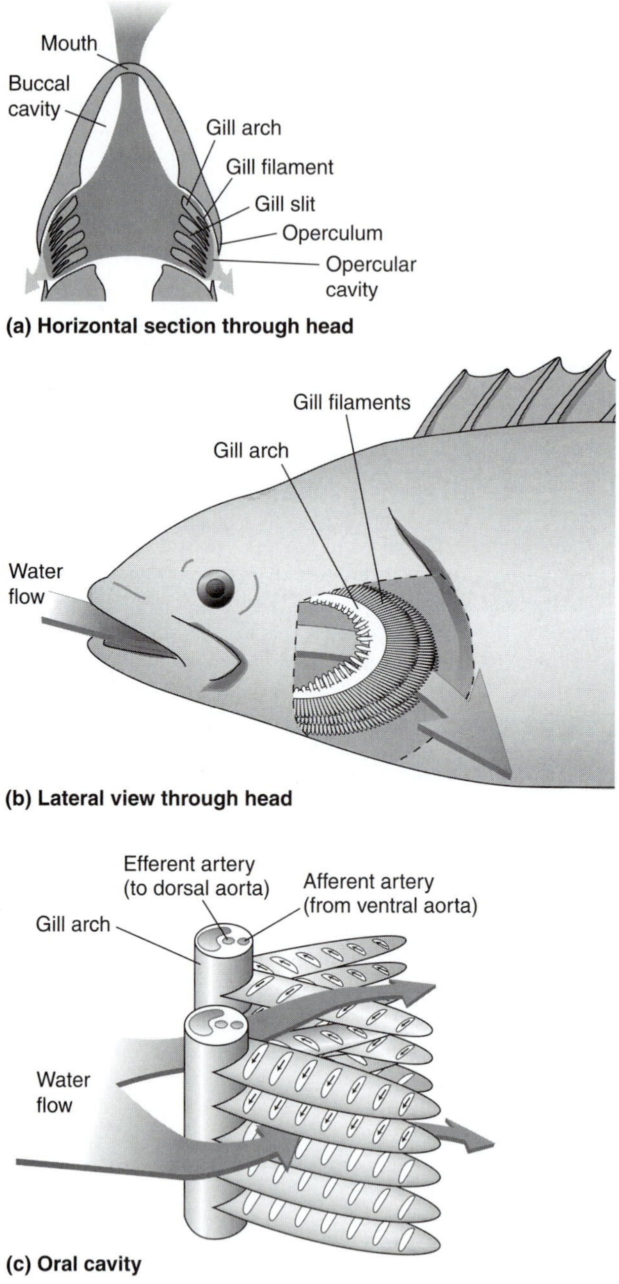

Morphology of teleost gills: (*a*) position of the gills in the head and the general flow of water; (*b,c*) water flow (shaded arrow) and blood flow (solid arrows) patterns through the gills.

Source: After Hughes, Comparative Physiology of Vertebrate Respiration, *1963, Harvard University Press, Cambridge, MA.*

BIO-NOTE 5.8

Left- and Right-Mouthed Scale-Eaters

Seven of the thousands of species of cichlid fishes in the lakes of Africa's Rift Valley are scale-eaters—species that exhibit a peculiar feeding habit involving eating the scales from other fish. The prey's behavior (see below) serves to maintain a nearly perfect 1:1 ratio between two genetically determined forms of predatory cichlids. These forms are distinguished by mouths that twist either to the left or to the right. Predators often have been shown to exert such frequency-dependent selection on their prey populations, but this is the first indirect evidence of frequency-dependent selection by prey on predators.

The mouth of a scale-eating cichlid opens either rightward or leftward as a result of an asymmetrical joint of the jaw to the suspensory apparatus. As a result of this lateral asymmetry, the mouth is "frozen open" on one side, which makes the cichlid's attacks more efficient. However, this specialization, which is considered an adaptation for efficiently tearing off a prey's scales, restricts each fish to attacking from just one side. Left-mouthed scale-eaters only attack the right side of their prey, and vice versa. Stomach analyses from left-mouthed fish were found to contain mostly scales from their prey's right sides.

This restriction may explain why left-mouthed and right-mouthed populations stay in balance. If, for example, right-mouthed fish became more common, their prey would change their behavior, becoming more alert to activity on their left sides and thus fending off more attacks by right-mouthed fishes. Meanwhile, left-mouthed fish would gain the advantage, and the balance would be restored.

Hori, 1993

Food and water, which are taken into the oral cavity together, are separated in the pharynx, with gill rakers preventing the food from passing out through the gill slits. The pharyngeal region is thus a vital portion of both the digestive and respiratory systems.

The stomach, a muscular organ at the end of the esophagus, secretes digestive enzymes and serves as a temporary storage site (Fig. 5.18b). The distinction between the esophagus and stomach is poorly delineated in fishes. Blind pouches known as **pyloric caeca** are found near the junction of the stomach and intestine in ray-finned fishes, especially teleosts (Fig. 5.18b). Their primary function appears to involve fat absorption, although they secrete a variety of digestive enzymes. No other vertebrate group possesses pyloric caeca.

The digestive tract of most fishes is a relatively straight tube terminating at the cloaca or anus. Predators have relatively shorter intestines than herbivorous species because meaty foods are more easily digested than plant foods. Sharks, chimaeras, lungfish, and some primitive teleosts have intestines that incorporate a **spiral valve** that aids in digestion by increasing the absorptive area (Fig. 5.23b-d).

In most fishes, the colon (large intestine) leads to the rectum, which then opens into the cloaca. The cloaca is the common chamber at the end of the digestive tract in elasmobranchs

FIGURE 5.23

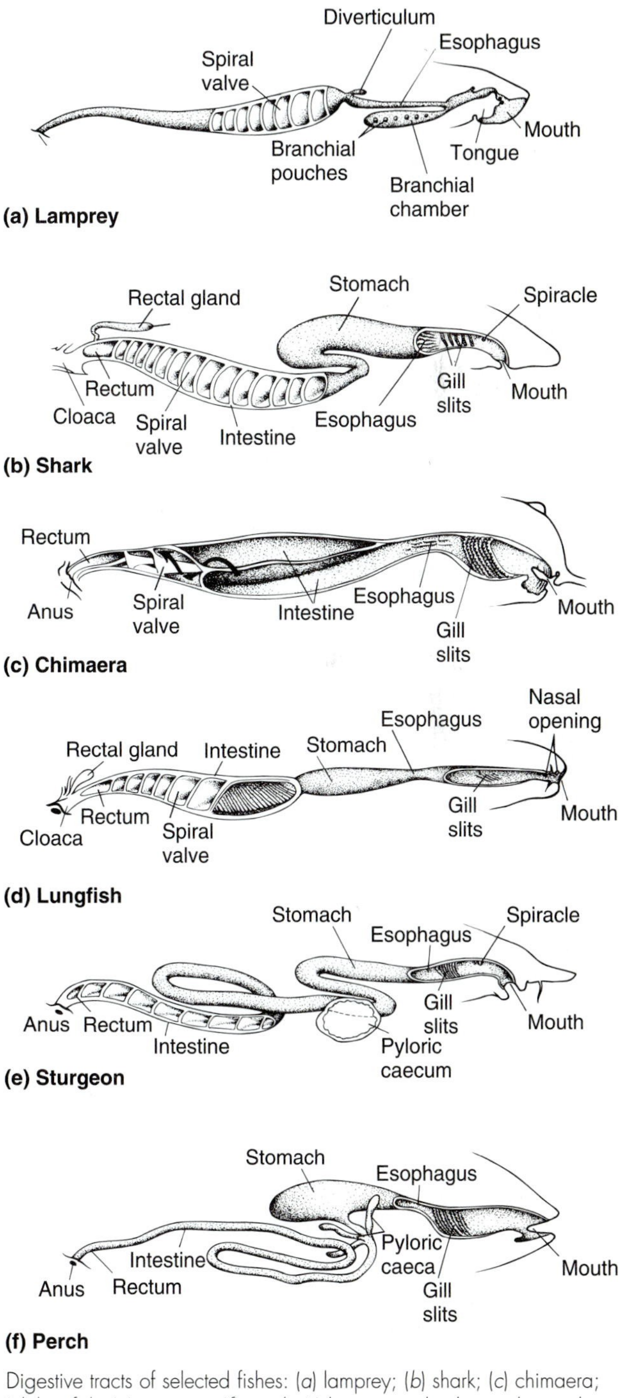

Digestive tracts of selected fishes: (*a*) lamprey; (*b*) shark; (*c*) chimaera; (*d*) lungfish; (*e*) sturgeon; (*f*) perch. When a spiral valve is absent, the intestine is often lengthened, as in the perch. A stomach is absent in lampreys and chimaeras.

and lungfishes. It receives the colon, the urinary tract, and the genital tract and opens to the exterior via the **vent**. In most of the remaining fishes, the digestive, genital, and urinary systems tend to open separately to the exterior; in some teleosts, the urinary and reproductive systems may unite and empty into a common sinus before leaving the body. The alimentary canal and digestion in teleosts has been discussed in detail by Kapoor et al. (1975).

Air (gas) bladders probably originated as modified, paired pharyngeal pouches that were ventral in position (Feduccia and McCrady, 1991). They originally evolved as organs of respiration and, in some species, eventually evolved into true lungs. In many ray-finned fishes (Actinopterygii), they developed into paired swim bladders and then into a single organ on the dorsal side of the gut (Fig. 5.24). The climbing perch (*Erythrinus*) has a lateral attachment of its swim bladder to the gut (Fig. 5.24c) and often is cited as an intermediate stage in the migration of the swim bladder to a dorsal position (Feduccia and McCrady, 1991).

Because the density of most animal tissues is greater than that of water, water-dwelling species tend to sink. This can be prevented either actively by swimming movements or passively by the development of a gas-filled swim bladder. The primary function of the swim bladder, therefore, is to serve as a hydrostatic organ in which buoyancy can be regulated with water depth. Swim bladders allow a relatively precise control of buoyancy because the volume and pressure of gas they contain can be regulated with comparative ease, thereby altering the specific gravity of the fish and increasing or decreasing its buoyancy. By the processes of diffusion and/or active transport, gas enters the swim

FIGURE 5.24

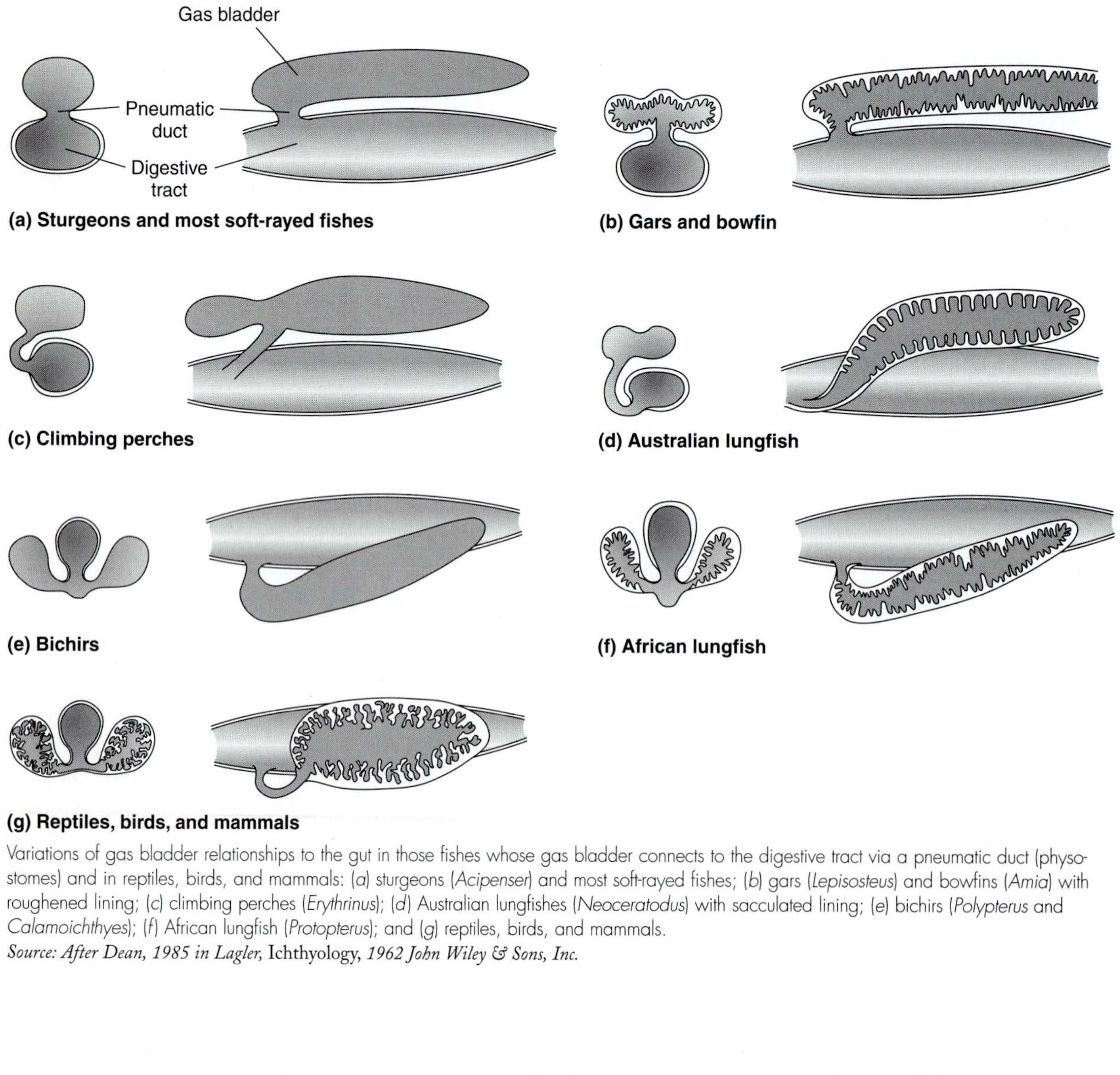

Variations of gas bladder relationships to the gut in those fishes whose gas bladder connects to the digestive tract via a pneumatic duct (physostomes) and in reptiles, birds, and mammals: (a) sturgeons (*Acipenser*) and most soft-rayed fishes; (b) gars (*Lepisosteus*) and bowfins (*Amia*) with roughened lining; (c) climbing perches (*Erythrinus*); (d) Australian lungfishes (*Neoceratodus*) with sacculated lining; (e) bichirs (*Polypterus* and *Calamoichthyes*); (f) African lungfish (*Protopterus*); and (g) reptiles, birds, and mammals.
Source: After Dean, 1985 in Lagler, Ichthyology, *1962 John Wiley & Sons, Inc.*

bladder. The gases used differ among fishes. Oxygen, carbon dioxide, and nitrogen are found in swim bladders in higher concentrations than would be expected on the basis of their partial pressure in water.

Swim bladders are not found in lampreys, hagfishes, and cartilaginous fishes, but they are present in about half of the teleost fishes. Many bottom-dwelling species have lost them as an adaptation to this lifestyle. Some teleosts that lack these organs as adults have them during embryonic development. To retain buoyancy, cartilaginous fishes such as sharks never stop swimming. They use their pectoral fins as hydrofoils and their asymmetrically shaped tails to generate lift. To increase their buoyancy, they also accumulate low-density oils and the hydrocarbon squalene in their tissues, particularly the liver.

Swim bladders are more or less oval, soft-walled sacs in the abdominal cavity just below the spinal column (Fig. 5.25). They may be paired or unpaired and may be partitioned into chambers. They lie retroperitoneally (behind the peritoneum) near the kidneys and bulge into the coelomic cavity. A pneumatic duct, when present in adults, usually connects the swim bladder to the esophagus; it may, however, connect the swim bladder to the pharynx or even to the stomach. Fishes in which the swim bladder connects to the digestive tract via a pneumatic duct are termed **physostomes** and include many of the more ancestral soft-rayed teleosts such as herrings, salmon, catfishes, eels, pike, and cyprinids. Physostomous species gulp air at the water's surface and push it through the pneumatic duct into the swim bladder using a force supplied by the buccal cavity (Fange, 1976). Gas is removed from the swim bladder by bubbling it through the pneumatic duct and mouth.

Bony fish possessing a closed swim bladder that lacks a pneumatic duct are known as **physoclists**. These fishes have special structures associated with the circulatory system for inflating or deflating the swim bladder. The source of gas is normally from the blood contained in a network of capillaries (*rete mirabile* or red gland) found in the lining of the swim bladder (Fig. 5.25). Gas from the blood is moved into the gas gland by the *rete mirabile*. The capillaries of the red gland are usually supplied with arterial blood through branches of the celiac artery and are drained by veins that empty into the hepatic portal vein. Over two-thirds of all teleosts with swim bladders are physoclists.

Because the density of the gases is much less than that of the body tissues, slight changes in pressure will have a pronounced effect on a fish's buoyancy. Pressure on the body of a fish increases significantly after a descent of only a fraction of a meter. The pressure compresses the swim bladder, making the fish denser and less buoyant, thereby increasing the rate of its descent. To counteract this descent, the fish must expend energy either by secreting more gas into its swim bladder or by actively swimming. Thus, gas must be secreted into the swim bladder against a pressure gradient; the deeper the fish swims, the greater the gradient.

FIGURE 5.25

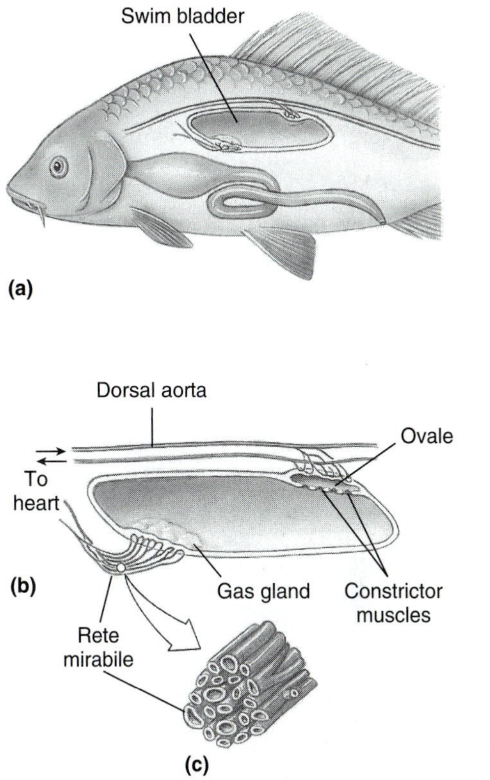

(*a*) The swim bladder in a teleost fish lies just beneath the vertebral column. (*b*) Gas is secreted into the swim bladder by the gas gland. Gas from the blood is moved into the gas gland by the *rete mirabile*, an array of tightly packed capillaries that acts as a countercurrent multiplier to build up the gas concentration in the blood vessels. The arrangement of venous and arterial capillaries in the *rete mirabile* is shown in (*c*). Gas is released during ascent by opening a muscular valve; this allows the gas to enter an absorptive area (ovale), from which the gas is removed by blood circulation.

BIO-NOTE 5.9

The Root Effect

The hemoglobins of certain fishes have structural characteristics that make them efficient oxygen pumps, able to drive gases into the swim bladder. A network of capillaries (*rete mirabile*) secretes lactic acid, which forces nitrogen out of solution and also causes a marked reduction in the oxygen affinity of blood hemoglobin—a phenomenon known as the Root effect. It is this pH-dependent release of oxygen from Root-effect hemoglobin that largely accounts for the ability of fish to "compress" oxygen and force it into the swim bladder when extra buoyancy is needed.

Mylvaganam et al., 1996

Besides their role in regulating buoyancy, swim bladders are modified to perform other functions in some fishes. In minnows, carp, catfish, and other members of the order Cypriniformes, the swim bladder is involved in the sense of hearing (Fig. 5.26b). A chain of small bones known as **Weberian ossicles** joins together the anterior end of the swim bladder and a Y-shaped lymph sinus known as the **sinus impar** (Fig. 5.26b). The connecting bones are modified portions of the first four vertebrae. The sinus impar lies adjacent to a lymph-filled canal that joins the sacculi of the right and left ears. Movements in the water cause vibrations of the gas within the swim bladder. These vibrations are transmitted by the Weberian ossicles to the inner ear. Some squirrelfishes (Holocentridae), the tarpon (Elopidae), featherbacks (Notopteridae), deepsea cods (Moridae), and sea breams (Sparidae) have forked, forward extensions of the swim bladder that end near the ear and serve to amplify sound waves (Moyle and Cech, 1996) (Fig. 5.26a). Herrings (Clupeidae)

BIO-NOTE 5.10

Courtship and the Swim Bladder

The "fastest" muscle among vertebrates is used neither for fleeing predators nor for capturing food, but for sex—the courtship part. The swim bladder muscle of the toadfish (*Opsanus tau*) contracts and relaxes 200 times per second (i.e., at 200 Hz) as it makes its boat-whistle-like mating call—compared with 0.5 to 5.0 Hz for the fish's locomotory muscles. Calcium, which triggers muscle contraction, cycles through the swim bladder muscle 50 times as fast as in locomotory muscle. The muscle also may have an unusually fast version of a protein—troponin—that binds and releases calcium; and the rate at which myosin cross-bridges detach from actin filaments in the muscle is 100 times as fast as in muscles used for locomotion.

Rome et al., 1996

FIGURE 5.26

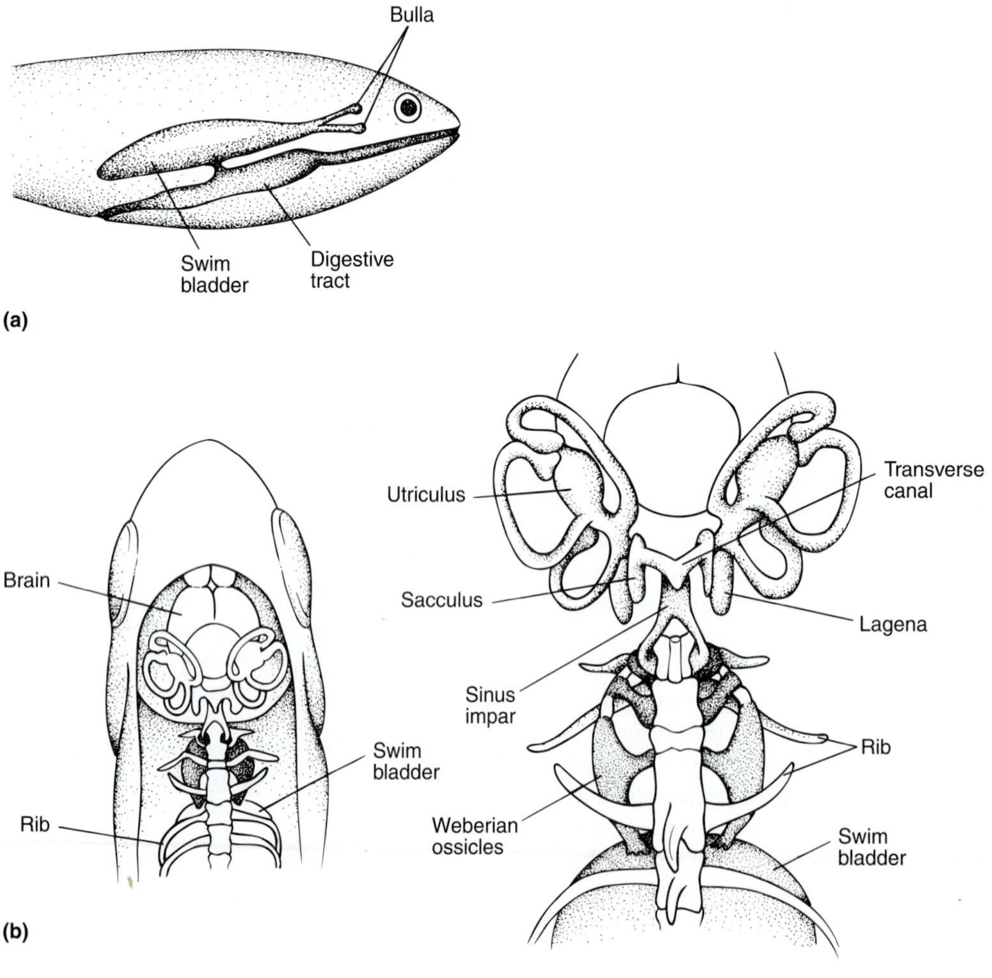

Route of sound transfer in fishes. (*a*) In some fishes, the swim bladder includes anterior extensions that contact the inner ear. (*b*) In other fishes, the Weberian ossicles, a tiny series of bones, connect the swim bladder to the inner ear.

and elephantfishes (Mormyridae) have similar extensions that enter the auditory capsule and are in contact with the inner ear. Their function also is to amplify sound.

Many fishes make rasping, squeaking, grunting, or squealing noises. In some, such as croakers and grunts, the sounds are the result of muscles attached to the swim bladder. These contractions may cause the swim bladder to emit thumping sounds, or they may cause air to be forced back and forth between the chambers within the swim bladder.

Nervous System

Brain

In fishes, the anterior portion of the brain (telencephalon) (Fig. 5.27) consists of rather prominent olfactory bulbs and inconspicuous cerebral hemispheres. A pineal organ (epiphysis), which serves as a photoreceptor, is present in most fishes.

BIO-NOTE 5.11

Imprinting

Imprinting is a process of rapid, irreversible learning of a particular visual, auditory, or olfactory stimulus that occurs in the cerebrum at a "critical" or "sensitive" period during development and that influences the future behavior of the animal. Imprinting is known to occur in fishes. Juvenile salmon, for example, imprint on the distinctive odor of their natal stream before migrating to sea. Several years later, as adults, this olfactory imprinting enables them to return to their home tributary.

Hasler and Wisby, 1951
Hasler et al., 1978

FIGURE 5.27

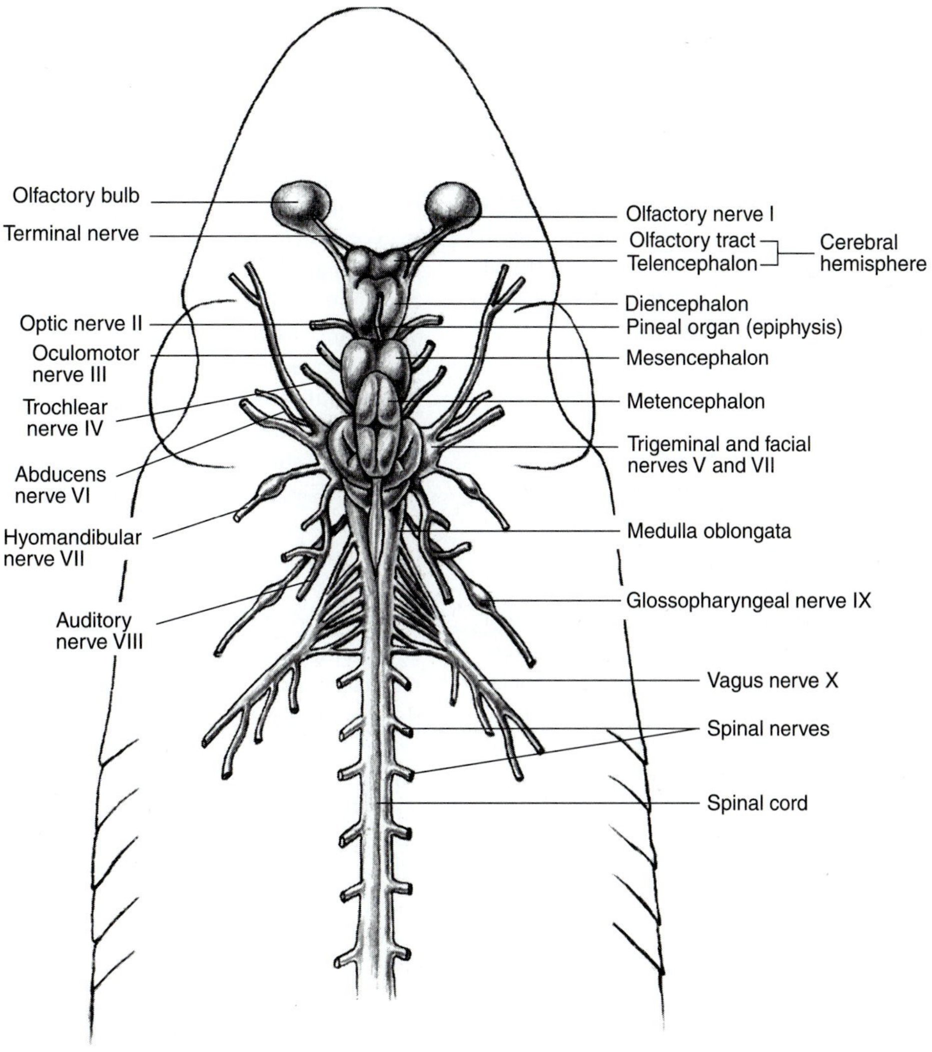

Brain, cranial nerves, and anterior portion of the spinal cord of a shark.

Remnants of the parapineal organ (parietal eye) are found in some bony fishes. The cerebellum is particularly large and assists in the maintenance of posture and balance. In some species of fish, it overlies the medulla and the midbrain.

Cranial Nerves

The conventional view that vertebrates possess 12 cranial nerves is based on mammalian anatomy, and its application to the vertebrates as a group is inaccurate. In fishes and amphibians (anamniotes), the 11th nerve is an integral part of the 10th and whether or not a 12th nerve should be recognized is controversial. Furthermore, following establishment of the nomenclature for the basic 12, a 13th anteriormost nerve was identified, and to avoid disrupting the time-honored numbering system, it was given the number 0. So anamniotes actually possess 11 cranial nerves, and amniotes have 13.

The first 10 (or 11) cranial nerves of all vertebrates (Fig. 5.27) may be grouped into three categories: predominately sensory nerves (0: terminal; I: olfactory; II: optic; and VIII: vestibulocochlear); mixed nerves containing both sensory and motor fibers and serving the eyeball muscles (III: oculomotor; IV: trochlear; and VI: abducens); and mixed branchiomeric nerves (V: trigeminal; VII: facial; IX: glossopharyngeal; and X: vagus) supplying the head, neck, stomach, heart, and lungs.

The spinal cord, a portion of the central nervous system, technically begins at the **foramen magnum**, where it merges imperceptibly with the brain. It is protected by the centra and neural arches of the vertebrae as well as by fat and one or more connective tissue membranes (meninges). Paired spinal nerves emerge at regular intervals along the length of the spinal cord and innervate each body segment. In all vertebrates except lampreys, each nerve arises by a dorsal and ventral root, which then join to form the spinal nerve. In lampreys, dorsal and ventral roots alternate.

In most vertebrates, the spinal cord occupies the bony vertebral canal and extends for most of the length of the vertebral column. In fishes, however, the length of the spinal cord varies considerably. In fishes with abundant tail musculature, the cord tapers gradually and extends to the end of the tail. In some teleosts, however, the spinal cord is shortened and terminates at various distances from the end of the vertebral column. In a few species, the spinal cord is actually shorter than the length of the brain. For example, the marine sunfish *Orthagoriscus* spp. may reach a length of 2.5 m and weigh 900 kg or more. Its spinal cord, however, is less than 2 cm in length (Weichert, 1965).

Sense Organs

Sensory receptors monitor the external and internal environments. Some receptors are widely distributed over the body; others are localized. Most receptors transmit nerve impulses to the brain, where they are interpreted as sensations; some impulses are transmitted to the spinal cord and result in reflex actions.

Neuromast System

Fish possess a **neuromast system** consisting of fluid-filled **pits**, **ampullae**, and **canals**. Some neuromasts open onto the surface of the skin via pores, whereas others consist of sunken canals beneath the skin. Examples of neuromast organs include lateral-line canals, cephalic canals, ampullae of Lorenzini, and pit organs. The neuromast system, which responds primarily to mechanical stimulation such as water movement and sound waves, enables fish to avoid potential enemies, to orient themselves, to locate food, and to participate in schooling.

Portions of the neuromast system in some fishes, such as the ampullae of Lorenzini in elasmobranchs, are sensitive to minute electrical currents in the water and are known to be **electroreceptors**. Electric currents are generated by such processes as the exchange of ions across the gills and by the contraction of respiratory muscles (see discussion on pages 103-104). Whenever an electric pulse is emitted, the fish is surrounded by an electric field. When the pulse stops, the field disappears. Any object near the fish will distort the electric field and change the way the current flows over the fish's body (Cooper, 1996). This creates an electric "shadow" on the fish's body surface, which is covered with electroreceptors. From this information, the fish will sense an image and discern its size, shape, movement, and electrical properties. The weakly electric fish (*Gnathonemus petersii*) can measure the distance of stationary objects, independently of the objects' size, shape, or constituent material, by instantaneously analyzing the electric image of objects with a single array of electroreceptors embedded in its skin in the tail region (von der Emde et al., 1998). Because most electric fish are nocturnal, the evolution of an electric sense was a valuable adaptation. The ability to perceive such electric fields helps these fishes orient themselves, communicate with one another, and locate prey, and it assists males in locating females during the mating season (Anonymous, 1992).

In a school of fish, movements of individuals are extremely well coordinated and almost perfectly synchronized. Fish in a school often swim at a constant pace and maintain characteristic individual distances. They are able to monitor the swimming speed and direction of travel of neighboring fish. Thus, the school as a whole can execute complicated maneuvers that require individuals to respond quickly to changes in the velocity and direction of their neighbors.

Ear

All fishes have a series of closed tubes and sacs that compose the membranous labyrinths (inner ears) embedded in the auditory region of the skull (see Fig. 1.16). The fluid within the labyrinth is known as **endolymph**. Each labyrinth consists of three semicircular canals and two sacs (utriculus and sacculus), with each of the sacs containing particles of calcium carbonate known as **otoliths**. The utriculus and sacculus are involved primarily in maintaining **static equilibrium**, or posture: the position of the body (mainly the head) relative to the earth's surface (i.e., with respect to gravity). The semicircular canals are involved primarily with **dynamic equilibrium**: the maintenance of the body's position in response to sudden movements such as rotation, acceleration, and deceleration.

Although the membranous labyrinth serves primarily as an organ of equilibrium in fishes, in a few species it has developed the accessory function of sound transmission. A series of modified transverse processes of the first four trunk vertebrae in some fishes serve as auditory structures (Fig. 5.26). Portions of the first four vertebrae are actually separated from the vertebrae themselves and form a chain of tiny bones that connect the swim bladder to the inner ear. This arrangement provides acute hearing. The tiny bones are known as Weberian ossicles; their action in conjunction with the swim bladder was discussed on page 112.

Sound consists of regular compression waves that can be transmitted in air, water, or solids; thus, "hearing" cannot be restricted to the perception of compression in air only. Because of its greater density, water is a much more effective conductor of sound than is air. Underwater, sounds will carry much farther and travel on average 4.8 times as fast (Moyle and Cech, 1996).

It has long been thought that most teleost fishes cannot detect sounds higher than 2 or 3 kHz (Fay, 1988). However, Mann et al. (1997) report that the American shad (*Alosa sapidissima*) can detect ultrasounds up to 180 kHz. These researchers speculate that the American shad as well as other clupeid fishes may be able to detect the ultrasonic clicks of one of their major predators, echolocating cetaceans (whales).

Eye

The structure of the vertebrate eye has remained remarkably constant throughout vertebrate evolution. Most fishes have two lateral eyes that serve as primary receptor sites of light (see Fig. 5.18a, b on page 103). Fishes that live in caves have vestigial eyes, are frequently blind, and in many cases, have eyelids that do not open.

In fishes, adjustments in focus for near and far vision (called optical accommodation) are accomplished by movement of the spherical lens closer to or farther away from the cornea by muscles within the eye. This is accomplished without changing the shape of the eye. The **retina**, located at the rear of the vitreous chamber, contains light-sensitive cells known as **rods** and **cones.** Colorless oil droplets are found in the cone photoreceptors of some fish, an indication that they lack color vision (Bowmaker, 1986). Robinson (1994), however, reported colored oil droplets, and possibly tetrachromatic (color) vision in the lungfish (*Neoceratodus forsteri*).

Morphological adaptations to amphibious vision are found in several groups of teleost fishes. *Anableps*, the "four-eyed fish," uses an aspherical lens and two pupils for amphibious vision (Fig. 5.28). The division of the eye into two functionally separated areas allows the fish to have optical accommodation both in water and in air. Other fishes, such as the amphibious clinids *Dialommus* and *Mnierpes* and the blenniid *Coryphoblennius*, have flattened corneas that reduce refraction in air. The Atlantic flying fish *Cypselurus* uses a prismlike cornea (Jermann and Senn, 1992).

The only fishes in which adults are not bilaterally symmetrical are the soles and flounders (Pleuronectiformes). Larval flatfish are bilaterally symmetrical and possess one eye on each side of their head. As they mature and adapt to a bottom-dwelling existence, however, they become asymmetrical: one eye moves to the opposite side of the body. This movement causes modifications of muscles, skull bones, blood vessels, and nerves. Those species of flatfishes that lie on the right side of their body have both eyes on their left side; those that lie on their left side have both eyes on their right side. Although the movement of the eye is generally species-specific, there are some species that have both left- and right-eyed forms.

Pineal and Parapineal Organs

Light-sensitive organs develop as a pair of sacs in the roof of the diencephalon of most fish (see Fig. 5.27). The larger and more dorsal sac develops on the right side and is known as the **pineal organ (epiphysis)**. A **parapineal organ** develops on the left side. In lampreys and some larval fishes, the pineal and parapineal organs are both light-sensitive, but unlike lateral eyes, they do not form images. Instead, they monitor the duration of the photoperiod and intensity of solar radiation. Pineal organs are also sensitive to light in most adult fish and appear to have importance in the control of daily and seasonal circadian rhythms (Kavaliers, 1979). These organs appear to be ultrasensitive light sensors closely connected to the brain. Gruber et al. (1975) showed that the chondrocrania of three species of sharks

FIGURE 5.28

The "four-eyed" fish has the upper half of its eye adapted for vision in air above the water line while the lower half is adjusted to seeing in the water. A dumbbell-shaped pupil makes it possible to use both parts of the eye simultaneously. Most of the time, this fish is found at or close to the surface.

were modified for light transmission, and that seven times more light reached the pineal receptors than surrounding areas of the brain.

Olfaction (Smell)

Olfactory organs are well developed in fish, and chemoreception in general plays an important and indispensable role in their behavior. Chemoreception is involved in the procurement of food, the recognition of sex, the discrimination between individuals of the same or different species, defense against predators, parental behavior, and migration and orientation, as well as many other activities.

The olfactory region is a blind sac in all fishes except lobe-finned fish; in all other vertebrates, the olfactory region is connected to the oral, or pharyngeal, cavity. In fishes, each naris is divided into incurrent and excurrent apertures, so that the forward motion of the fish propels a stream of water into one aperture and out the other. Water is induced to flow through the olfactory sacs by the action of cilia within the sac, by the muscular movement of the branchial pump, by swimming, or by a combination of these actions. Olfactory cells in the olfactory epithelium monitoring the water stream can detect extremely small concentrations of certain chemicals dissolved in water. Olfactory cues have been shown to be extremely important to salmon in locating their natal stream once they have reached the vicinity of the river mouth (Hasler and Wisby, 1951; Hasler and Scholz, 1983; Brannon and Quinn, 1990; Barinaga, 1999). Sharks also possess an extremely acute sense of smell.

Magnetoreceptor cells have been located in the nose of the rainbow trout (*Oncorhynchus mykiss*) (Walker et al., 1997). Behavioral and electrophysiological responses to magnetic fields have been recorded.

Taste

Gustatory cells (taste buds) are located in the roof, side walls, and floor of the oral cavity and pharynx, where they monitor incoming water, as well as on the fins and body of some species of fishes. The number of taste buds may vary greatly from one region of the body to another, but the greatest number are found in regions most closely associated with food contact. Bottom-feeders or scavengers such as catfish, carp, and suckers have sensitive chemoreceptors distributed over the entire surface of the body. Gustatory cells are also abundant on the barbels surrounding the mouth of a catfish. Whereas taste is primarily a close-range sense, it is used in food item discrimination after other senses such as smell, sight, or hearing have recognized its presence. Besides its function in identifying nutrients and verifying their palatability, taste receptors also permit fish to avoid noxious substances. The gustatory system in fishes has been reviewed by Kapoor et al. (1975).

Endocrine System

Since many of the endocrine glands and hormones are similar in the different groups of vertebrates, they will be discussed here. In the other vertebrate groups, discussion will focus only on instances where a hormone has a different function from that discussed here.

The anterior lobe (adenohypophysis) of the **pituitary gland** secretes a number of hormones, some of which control the activities of other endocrine glands. **Growth hormone (GH)** stimulates body cells to grow and to maintain their size once growth is attained. **Adrenocorticotropic hormone (ACTH)** stimulates the adrenal cortex to secrete its hormones. **Thyroid-stimulating hormone (TSH)** stimulates the thyroid gland to synthesize and release thyroid hormones. Two **gonadotropic hormones** are released cyclically in response to exterior stimuli. **Follicle-stimulating hormone (FSH)** stimulates the development of ovarian follicles in the female and the development of spermatozoa in the seminiferous tubules in the male. In the female, **luteinizing hormone (LH)** induces development of a new endocrine gland, the corpus luteum, following ovulation. In males, luteinizing hormone is better known as **interstitial cell-stimulating hormone** and induces interstitial cells of the testes to produce testosterone. **Melanocyte-stimulating hormone (MSH)** controls skin pigmentation by regulating production of the pigment melanin. **Prolactin (PRL)** regulates a wide range of processes in the different groups of vertebrates. Certain parental behavior patterns are brought about by the effects of prolactin. These include activities such as nest building, the incubation of eggs, and the protection of young.

The posterior lobe (neurohypophysis) of the pituitary serves solely for the storage and release of neurosecretions synthesized in the hypothalamus. These include **argenine vasotocin** (oxytocin), which is a smooth muscle contractor, and **vasopressin** (antidiuretic hormone [**ADH**]), which regulates water loss.

The **thyroid gland** may be paired or unpaired. It secretes **thyroxin** and **triiodothyronine**, which control the processes of metabolism, metamorphosis, and maturation. Production of thyroxin is regulated by thyroid-stimulating hormone secreted by the anterior lobe of the pituitary gland. An increase in TSH causes an increase in thyroid output, which in turn depresses the secretion of TSH. A sufficient quantity of iodine is also necessary for thyroid hormones to be produced. A third hormone, **calcitonin**, permits calcium in the circulating blood to be used for metabolic functions such as bone formation, muscle contraction, and nerve transmission. Calcitonin reduces calcium levels in the blood and prevents bone resorption. Fishes lack parathyroid glands (Duellman and Trueb, 1986).

Adrenal glands are located on or near the kidneys. The inner portion (adrenal medulla), or its equivalent, produces **epinephrine (adrenaline)** and **norepinephrine (noradrenaline)** in all vertebrates. Epinephrine is a vasodilator, and norepinephrine a vasoconstrictor, of the circulatory system. They increase the amount of blood sugar in times of sudden metabolic need and stimulate increased production of adrenocortical hormones in times of prolonged stress. These hormones cause acceleration of the heartbeat and increased blood pressure, with increased blood flow to the heart muscle, skeletal muscle, and lungs, and decreased blood flow to smooth muscle of the digestive tract and skin. They are the primary hormones involved in the "fight-or-flight" response to fear, pain, and aggression. They help mobilize the physical resources of the body in response to emergency situations.

The outer portion of the adrenal gland (adrenal cortex) is essential to life. It produces **corticoids** such as **aldosterone**, which regulates sodium levels by acting on the gills and/or kidneys. Secretion of aldosterone promotes the retention of sodium and the excretion of potassium by the kidney. Other corticoids secreted include the glucocorticoids **cortisone, cortisol**, and **corticosterone**. Together with aldosterone, glucocorticoids regulate the metabolism of carbohydrates, proteins, and fats, as well as the use of electrolytes and water.

The pancreas is a unique organ in that some cells, the **acinar** cells, function as an **exocrine gland** to aid in digestion, while other cells are part of the endocrine system and secrete hormones. Hormone-secreting cells are known as **pancreatic islets** (islets of Langerhans), and in fishes, like most vertebrates, the islets are dispersed throughout the secretory tissue of the pancreas. Islet cells in fishes may also be found in scattered locations around the gallbladder and between the pyloric caeca. Although the location of islet cells may vary somewhat in different vertebrates, the function of the pancreas is similar in all groups.

The two major pancreatic hormones are **insulin** and **glucagon**, which function antagonistically to regulate blood glucose levels. Insulin, secreted by beta cells of the islets, facilitates the assimilation of sugar by tissues and stimulates the formation of glycogen in muscle tissue and in the liver from blood glucose. Glucagon, a product of alpha cells, has the opposite action and increases blood glucose concentration.

Gonads act as endocrine organs. The ovaries and testes of most vertebrates produce three types of steroid hormones: estrogens, androgens, and progesterone. Secretion of these hormones is controlled by pituitary gonadotropins. **Estrogen** is produced by ovarian follicles. It causes formation of primary follicles and also the development of the accessory sex organs, including the reproductive tract. Maturation of follicles depends on follicle-stimulating hormones from the pituitary gland. Secretion of estrogen is partly responsible for differentiation of the Müllerian ducts into uteri and oviducts in female embryos. Estrogen also regulates reproductive behaviors. In males, luteinizing hormone (LH) (also known as interstitial-cell stimulating hormone, or ICSH) from the anterior pituitary gland causes the interstitial cells of the testes to produce the male androgen, **testosterone**. Increased levels of testosterone cause the development of secondary sex characteristics and initiate the process of spermatogenesis. The failure of Müllerian ducts to develop in males is caused by increased levels of androgens. Larger muscles and bones in males are also a result of androgens. In addition, androgens regulate reproductive behavior.

Ultimobranchial bodies develop from the last pair of pharyngeal (branchial) pouches in all vertebrates. They may be known as postbranchial bodies or, in bony fish, as suprapericardial bodies. In some fishes, amphibians, reptiles, and birds, they produce calcitonin, which removes calcium and phosphates from the circulating blood so these ions can be utilized for essential functions such as bone formation, muscle contraction, and nerve transmission.

A **thymus gland** is found in all vertebrates except lampreys and hagfishes. The thymus secretes the hormone **thymosin**, which stimulates lymph glands to produce lymphocytes. It is also involved in producing antibodies and in the maturation of **T lymphocytes**, or T cells, which can distinguish foreign cells and/or substances in the body (e.g., viruses, bacteria) and destroy them; thus, the thymus provides a major defense mechanism to keep bodies free of foreign substances. It increases in size until the animal reaches sexual maturity, and then it steadily shrinks.

The pineal region in some lower vertebrates, such as lampreys, contains highly specialized cells that function as photoreceptors. They function in a similar manner to that of the mammalian retina and generate impulses that are carried to the brain. As vertebrates evolved and the cerebrum grew back over the dorsal portion of the brain, the pineal organ lost much of its photoreceptivity and became primarily an endocrine organ.

The pineal gland secretes the hormone **melatonin**, which causes melanophores to aggregate in lower vertebrates. The pineal is negatively light-sensitive and secretes melatonin during darkness. Thus, pineal and blood melatonin levels exhibit daily (diurnal) rhythms. The weight of the pineal gland varies within a 24-hour cycle: Its weight is lowest at the end of the daily light period (Wurtman et al., 1968; Falcon et al., 1987; Kezuka et al., 1988), and continuous light reduces the weight of the pineal gland. Considering its 24-hour cycle, the pineal may play a role in the synchronization of other 24-hour **circadian cycles** such as sleeping, eating, and adrenocortical function. Although diurnal rhythms of the melatonin level in the retina have been reported in amphibians, birds, and mammals, the lateral eyes of trout apparently have no significant endocrine function as far as regulating melatonin levels in the blood (Zachmann et al., 1992).

Urogenital System

Freshwater fishes live in water that has a lower salt concentration (hypotonic) than that of their own body fluids (Fig. 5.29). They have large renal corpuscles and use water freely in excreting nitrogenous wastes. Marine fishes, on the other hand, live in water in which the salt concentration is normally higher (hypertonic) than in their own body fluids. They are in danger of losing water to their environment.

With the exception of hagfish, whose body fluids have salt concentrations similar to that of sea water, all other marine vertebrates maintain salt concentrations in their body fluids at a fraction of the level in the water (Schmidt-Nielsen, 1990). Bony marine fishes maintain osmotic concentrations of about

FIGURE 5.29

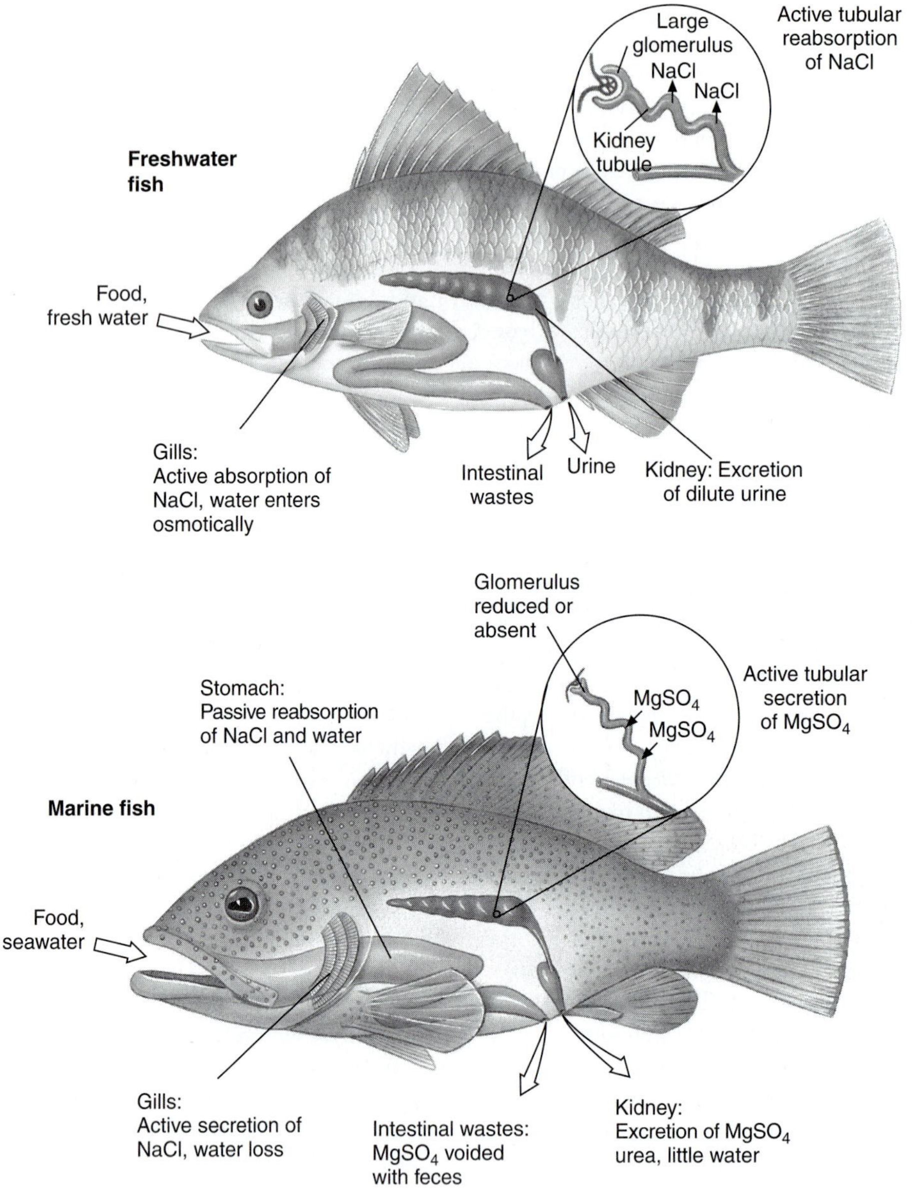

Osmotic regulation in freshwater and marine bony fishes. Freshwater fishes maintain osmotic and ionic balance in their dilute environment by actively absorbing sodium chloride across their gills (some salt is also gained with food). To flush out excess water that constantly enters the body, a glomerular kidney produces a dilute urine by reabsorbing sodium chloride. Marine fishes drink sea water to replace water lost osmotically to their salty environment. Sodium chloride and water are absorbed from the stomach. Excess sodium chloride is actively transported outward by the gills. Other sea salts, mostly magnesium sulfate, are eliminated with the feces and secreted by the tubular kidney.

one-fourth to one-third the level in sea water. They have reduced the size of their renal corpuscles, and they excrete excess salt through the gills. Cartilaginous fishes maintain salt concentrations in their body fluids at roughly one-third the level in sea water. They maintain osmotic equilibrium by retaining large amounts of organic compounds, primarily urea, in their body fluids. This allows these fishes to raise the total osmotic concentration of their blood so that it is equal to, or slightly exceeds, that of sea water without increasing the overall salt concentration of their body (Schmidt-Nielsen, 1990).

Many modern fishes (e.g., salmon, eels, shad, smelt) migrate during their lives between oceanic and fresh waters.

Animals that do this require adaptations to control the osmotic pressure of their internal fluids. Osmoregulation is thought to be controlled by hormones: thyroxin from the thyroid gland, and corticosteroids from the adrenal glands. Fishes that are **anadromous**—that is, spend most of their life in the sea and migrate to fresh water to breed, such as Chinook salmon (*Oncorhynchus tshawytscha*)—must be able to prevent dilution of their fluids by the inward diffusion of water from the less concentrated external medium. In contrast, freshwater fishes passing into the sea to spawn—called **catadromous**—must prevent concentration of their fluids by loss of water to the more concentrated external medium. The most common catadromous fish in North America is the American eel (*Anguilla rostrata*). Fishes whose migration from fresh water to the sea, or vice versa, is not for the purpose of breeding but occurs regularly at some other stage of the life cycle are referred to as **diadromous**. For example, bull sharks (*Carcharhinus*) and sawfishes (*Pristis*) move in and out of Lake Nicaragua in Central America.

A pronephric kidney forms as a developmental stage in all vertebrates but is functional only in larval fish and amphibians. It remains throughout life only in lampreys, hagfishes, and a few teleosts as a mass of lymphoid tissue.

As differentiation of the embryo proceeds, more and more tubules form behind the pronephric tubules (see Fig. 4.15). The anteriormost tubules usually are little different from the pronephric tubules (segmentally arranged with nephrostomes), but soon the segmented arrangement (metamerism) is lost and internal glomeruli and tubules develop that lack nephrostomes. With the disappearance of pronephric tubules, the pronephric duct becomes the **opisthonephric** (mesonephric) **duct** and the newer kidney is known as the **opisthonephros** (mesonephros). The latter is the functional kidney of adult fish and amphibians (opisthonephros) and functions as an embryonic kidney in reptiles, birds, and mammals (mesonephros). The term *mesonephros* is used in reference to the kidney that serves as an intermediate stage in the development of a vertebrate, whereas the term *opisthonephros* is used in reference to the adult kidney. They both develop from the same portion of the nephrogenic mesoderm.

The opisthonephros may extend for most of the length of the coelom as in sharks, or it may be confined to more caudal regions. In males, the anteriormost tubules of the opisthonephros have no association with glomeruli and are used solely to conduct sperm to the opisthonephric duct. This part of the kidney becomes known as the epididymal kidney, and the highly coiled part of the opisthonephric duct that drains it is the **epididymis**. The corresponding part of the opisthonephros in females may or may not degenerate.

Although most teleosts excrete ammonia, African lungfishes (*Protopterus*) are capable of excreting either ammonia, while living in an aquatic environment, or urea, while in **estivation** (i.e., in a dormant condition during droughts). Elasmobranchs, as well as the coelacanth (*Latimeria*), filter urea from the blood plasma at the glomerulus and excrete it as the primary nitrogenous end product. However, much of the urea is reabsorbed from the filtrate by the kidney tubules, thus preventing major losses of urea in the urine (Schmidt-Nielsen, 1990). The urea is retained and gives their body fluids a nearly isosmotic relationship with their environment.

Most fishes, with the exception of lampreys, hagfishes, and elasmobranchs, have a **urinary bladder**. Bladders of most fishes are terminal enlargements or evaginations of the mesonephric ducts known as **tubal bladders**.

BIO-NOTE 5.12

Natural "Antifreeze" in Antarctic Fishes

Small ice crystals, inadvertently swallowed when drinking seawater, presumably exist in the bodies of notothenioid Antarctic fishes (suborder Notothenioidei). A major physiological–biochemical adaptation of these fishes is the presence of macromolecular "antifreeze" substances (sugars and amino acids) in their body fluids. Glycopeptides (glycoproteins) inhibit ice growth through a process known as adsorption–inhibition. By adsorption onto existing ice crystals, the glycopeptides produce a barrier between the ice surface and water molecules, thereby preventing further growth of the ice crystals, which could cause tissue damage. Antifreeze glycopeptides are synthesized in the liver and are found in most of the body fluids. Peak effectiveness is reached when antifreeze levels reach about 4 percent by mass.

The same molecules that keep fish from freezing in frigid polar waters may one day enable blood banks to refrigerate platelets for weeks at a time. Human platelets are disk-shaped cell fragments that help clot blood after injury. They are often given to surgery patients to limit bleeding. But unlike red blood cells, they cannot be frozen and must be used within 5 days. However, platelets kept in a solution containing glycoproteins derived from the blood of Antarctic or Arctic fish last up to 21 days at 5°C. This is a completely new and innovative approach to the long-term storage problem of blood products.

Eastman, 1993
Tablin et al., 1996

The caudal ends of the opisthonephric ducts may enlarge to form urinary bladders or seminal vesicles for the temporary storage of sperm. In some fishes, such as the lamprey and dogfish shark, the two opisthonephric ducts empty caudally into a urinary, or urogenital, papilla. When accessory urinary ducts are numerous in males, the opisthonephric ducts may be used chiefly, or entirely, for sperm transport.

Both testes and ovaries are usually paired and are suspended by mesenteries from the wall of the body cavity near the kidneys. During the spawning season, the testes are smooth, white structures that rarely account for more than 12 percent of the weight of the fish, whereas the ovaries are large, yellowish structures, granular in appearance, that may make up 30 to 70 percent of a fish's weight (Moyle and Cech,

1996). The ovaries of most teleosts are hollow and saccular, whereas the ovaries of most other fishes are solid. A few hermaphroditic species have ovotestes, which are part ovary and part testis, and these species are capable of self-fertilization. Synchronous (or simultaneous) hermaphrodites have ripe ovaries and testes at the same time but usually spawn with one or more other individuals, alternately taking the role of male and female. Some, such as the belted sandfish (*Serranus subligarius*), a small bass found only along the Florida coast, have fertilized their own eggs in captivity, but in the wild, self-fertilization is unlikely to happen because other spawning individuals are usually in close proximity. The neotropical cyprinodont *Rivulus marmoratus* can fertilize its eggs internally prior to oviposition (Wheeler, 1985; Bond, 1979). Synchronous hermaphrodites are known from the following families: Chlorophthalmidae, Bathypteroidae, Alepisauridae, Paralepididae, Ipnopidae, Evermannellidae, Cyprinodontidae, Serranidae, Maenidae, and Labridae.

Female elasmobranchs have a pair of ovaries, but the left one may be greatly reduced in size in some genera (*Scyliorhinus, Pristiophorus, Carcharhinus, Galeus, Mustelus,* and *Sphyrna*). In rays, the left ovary is functional in *Urolophus*; the right ovary is completely absent in *Dasyatis* (Kardong, 1995).

A pair of oviducts and a pair of uteri arise from the Müllerian ducts (originally called pronephric ducts) in female elasmobranchs (Fig. 5.30). Oviducts possess **oviducal** (nidamental or shell) **glands** that secrete albumen (protein) and shells around the eggs. The paired ostia (anterior openings of the oviducts) unite to form a single ostium in the falciform ligament.

Most female teleosts have paired ovaries. Fusion of the ovaries occurs in some genera, such as the perches (*Perca*), so that there is only one functional ovary. In the Japanese ricefish (*Oryzias latipes*) and the guppy (*Poecilia reticulata*), only one gonad develops.

In contrast with most vertebrates, most teleosts have oviducts that are continuous with the covering of the ovaries, so that the ova are not shed into the body cavity. Ovaries in this saccular ovary–oviduct system are known as **cystovarian ovaries** (Hoar, 1969). Most nonteleosts and a few teleosts have **gymnovarian ovaries,** in which the ovaries open into the body cavity and the ova are conveyed through an open funnel to the oviduct. These include the loach (*Misgurnus*), as well as members of the Anguillidae, Salmonidae, and Galaxiidae (Bond, 1996).

In some male fishes, such as the bowfin (*Amia*), sturgeon (*Acipenser*), and gar (*Lepisosteus*), the opisthonephric duct conducts sperm as well as urine (Fig. 5.31). Connections between the opisthonephros and testes are established early in embryonic life when some of the anterior mesonephric tubules connect with a network of channels within the testes. These modified mesonephric tubules become structures known as **vasa efferentia** and carry sperm from the testes to the opisthonephric duct.

FIGURE 5.30

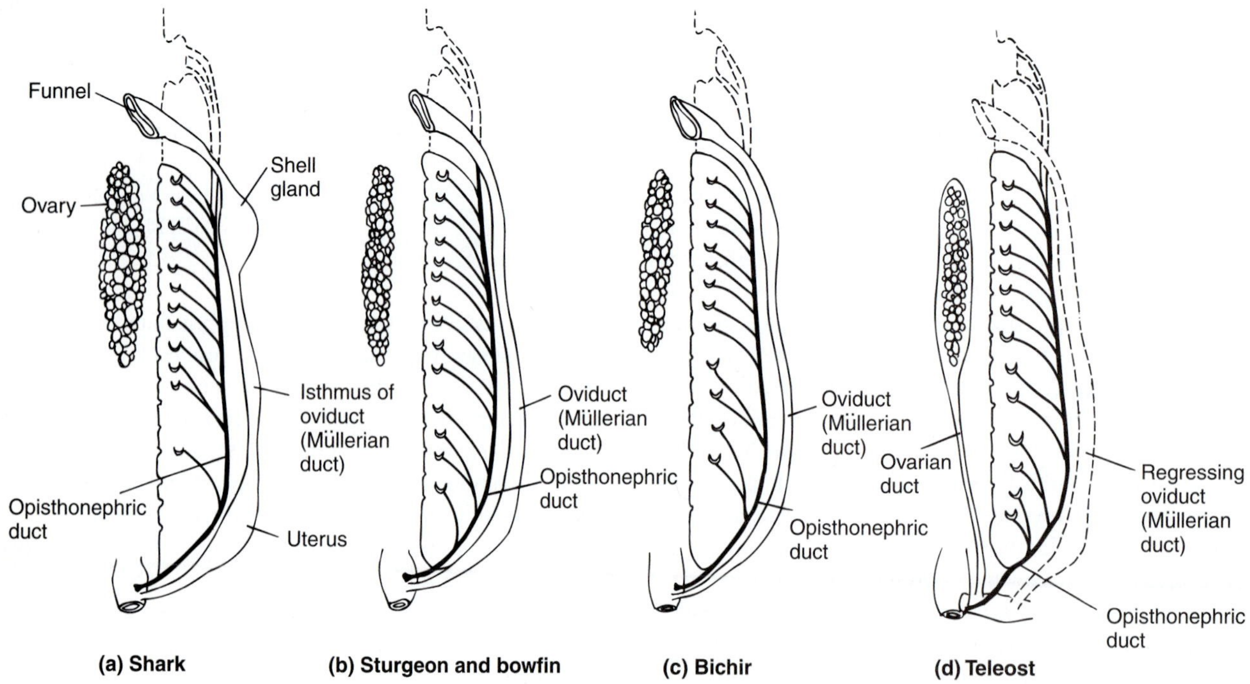

Oviducts of female fishes: (*a*) shark; (*b*) sturgeon (*Acipenser*) and bowfin (*Amia*); (*c*) bichir (*Polypterus*); (*d*) teleost. The oviduct (Müllerian duct) arises adjacent to and parallel with the opisthonephric duct in most fishes. In teleosts, the oviduct is usually replaced by an ovarian duct that is derived separately.

FIGURE 5.31

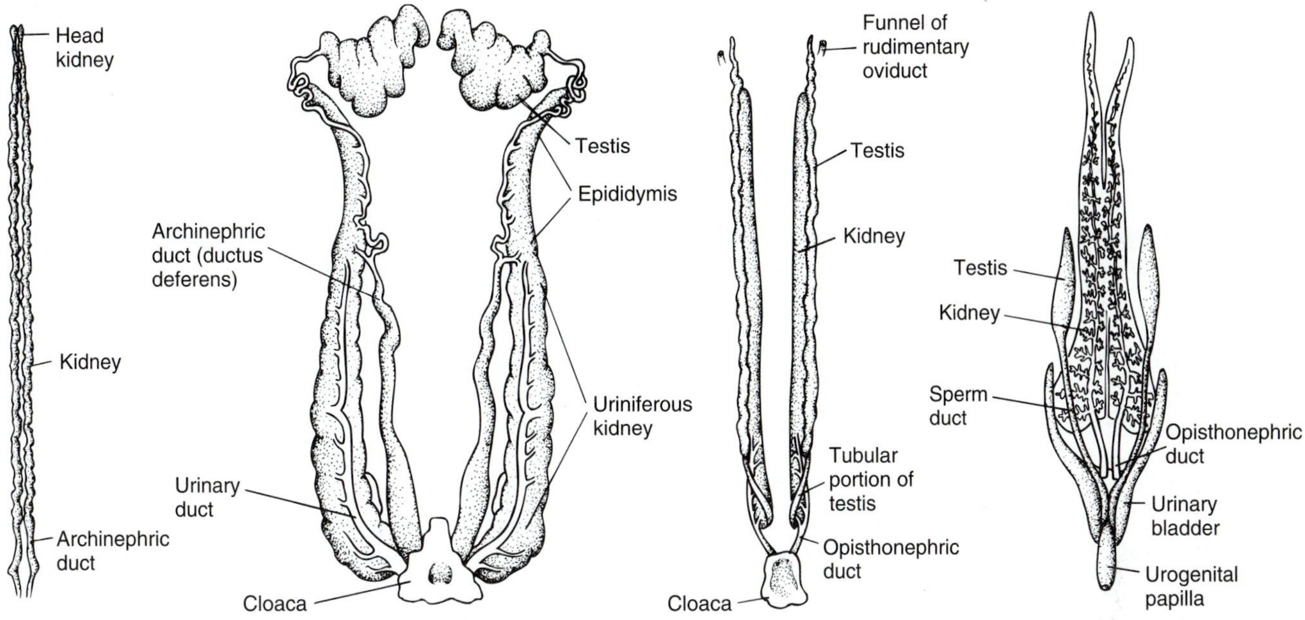

Urogenital systems of male fishes: (*a*) hagfish (*Bdellostoma*)—the single testis of the hagfish hangs in the dorsal body wall between the kidneys; (*b*) elasmobranch (*Torpedo*); (*c*) lungfish (*Protopterus*); (*d*) teleost, sea horse (*Hippocampus*).

In the Salmonidae, sperm are shed directly into the body cavity and exit through an abdominal pore or pores. In Chondrichthyes, sperm pass through a duct shared with the kidney (opisthonephric duct) and may be stored in a seminal vesicle for a short time before being expelled. It is the same situation in most nonteleost bony fishes, except that a seminal vesicle is lacking. Most teleosts possess separate sperm ducts (Moyle and Cech, 1996).

Some male fishes, including sharks, skates, rays, sawfish, and guitarfish, possess intromittent organs. Elasmobranchs possess claspers—grooved digitiform appendages of the pelvic fins (see Fig. 5.35). Associated with each clasper is a muscular saclike siphon that lies just beneath the skin and fills with water just prior to mating. During copulation, the male inserts one of his claspers into the female's reproductive tract. Sperm are washed along the groove in the clasper and into the female by sea water discharged from one of the siphon sacs. The siphon sac also contributes large quantities of an energy-rich mucopolysaccharide to the seminal fluid. In many teleosts, the anal fin is modified for sperm transport and is known as a gonopodium.

■ REPRODUCTION

Reproductive cycles of temperate fishes are closely tied to seasonal changes in temperature, day length, and in some cases, acidity of the water (Lam, 1983; Weiner et al., 1986). Under the influence of increasing light and temperature, the hypothalamus and pituitary glands are stimulated to produce hormones that control the development of the gonads and initiate reproductive behavioral activities and the development of secondary sexual characteristics. Temperatures above a certain level will inhibit spawning (Moyle and Cech, 1996), whereas low pH caused by acid rain has been shown to delay maturation of gonads in rainbow trout (*Salmo gairdneri*) (Weiner et al., 1986). Some fishes spawn only once as an adult and then die (e.g., Chinook salmon [*Onchorynchus tshawytscha*]); others reproduce for several years (e.g., bass [*Micropterus*]). In other species, sperm can be stored in females and remain viable for many months and possibly for several years.

The onset of reproductive activity (generation time) in fishes varies with sex. In many species, especially those with sexual differences in size (e.g., striped bass [*Morone saxatilis*] and sturgeon [*Acipenser*]) and live-bearers (e.g., guppies [*Poecilia*] and many species of sharks), males mature at a smaller size and younger age than do females. Reproductive activity also varies with the nature of the population, because a larger size may be attained at a younger age in expanding populations inhabiting favorable environments. Where the environment is favorable for growth and favors high adult survival, fishes tend to delay maturation, whereas if conditions are unfavorable, so that growth and adult survival are low, reproduction tends to take place at a younger age. Females that belong to an expanding population tend to reproduce at an earlier age than those in more stable populations (Moyle and Cech, 1996).

Various means of sexual recognition have evolved among fishes. Some, such as female stingrays (*Dasyatis*), passively emit electric fields that attract males (Anonymous, 1992). A difference in form or color between the sexes of a given species is known as **sexual dimorphism** (Fig. 5.32). In many fishes, adult females are larger than adult males. This sexual dimorphism is particularly evident in those fish that bear live young, such as members of the family Poeciliidae (guppies, mollies, swordtails, and mosquitofish). In others, especially species in which males compete for females, the males are the larger sex (e.g., many coral-reef fish).

Sexual dimorphism also may involve the development of tubercles on the heads and/or scales of males during the breeding season (Fig. 5.32c). These structures often develop on body parts that come into contact with females and may stimulate the female to lay eggs and assist males in maintaining contact during spawning (Moyle and Cech, 1996). Males of some species, such as three-spined sticklebacks (*Gasterosteus aculeatus*) and guppies (*Poecilia reticularia*), become brightly colored during the breeding season. Male sticklebacks develop a red belly or throat and a whitish back. Female sticklebacks exhibit a significant preference for males having the most intense coloration (Milinski and Bakker, 1990). Parasitization of male sticklebacks by the ciliate that causes "white spot" significantly decreases the intensity of the male's coloration and reduces the male's acceptance by females. Because tropical fishes breed all year, males typically retain their bright colors throughout the year.

Some fishes engage in elaborate courtship behavior (Fig. 5.33). The male three-spine stickleback, for example, constructs a tubelike nest and then, using a complex courtship ritual, entices the female to enter the nest. Her occupation of the nest, in turn, stimulates the male to thrust his snout against her rump in a series of quick rhythmic trembling movements. This induces the female to spawn. Without the stimulus of the male's tremble-thrusts, the female is incapable of spawning. Male gouramis (Helostomatidae) and Siamese fighting fish (Belontiidae) build bubble nests as part of their courtship behavior (Fig. 5.34b). As a female expels eggs from her body, the male blows them upward and into the floating nest, which he then maintains until the eggs hatch.

Most fishes do not establish long-lasting pair bonds. Males and females come together to breed and then go their separate ways. Some female mosquitofish (*Gambusia*) will mate with dozens of males during one breeding cycle. Offspring of these multiple unions are genetically more diverse than offspring of females who mate with just one male (Brown, 1991). In addition, larger females breed many times and produce larger broods, both of which have definite survival value. Female Mediterranean blenniid fish (*Aidablennius sphynx*) may use "test" eggs to assess paternal quality (Kraak and Van Den Berghe, 1992). Females initially lay a small quantity of eggs at a male-guarded site. If the male guards the eggs successfully, she will continue laying at his site.

Most female bony fishes release eggs into the water, where sperm are deposited by males (Fig. 5.34a). Although all sperm consist of a head containing the nucleus, a midpiece containing mitochondria that serve as an energy source, and a tail (flagellum) for locomotion, they exhibit much variation in shape. These species of fishes usually spawn in groups and do not have elaborate courtship displays or specialized reproductive structures. They depend on water currents for egg dispersal. Other fishes hide their eggs among gravel or rocks, or build nests in which eggs are buried (Fig. 5.34b). Female salmon and trout (Salmonidae), for example, excavate redds (nests) in gravel substrates by digging with their tails. Redds are defended by both sexes. Some fishes protect the eggs while spawning and also guard the embryos until they hatch (e.g., many North American minnows [Cyprinidae]); in some cases, guarding extends through larval development (e.g., sunfish and bass [Centrarchidae]) (Fig. 5.34b). In addition

FIGURE 5.32

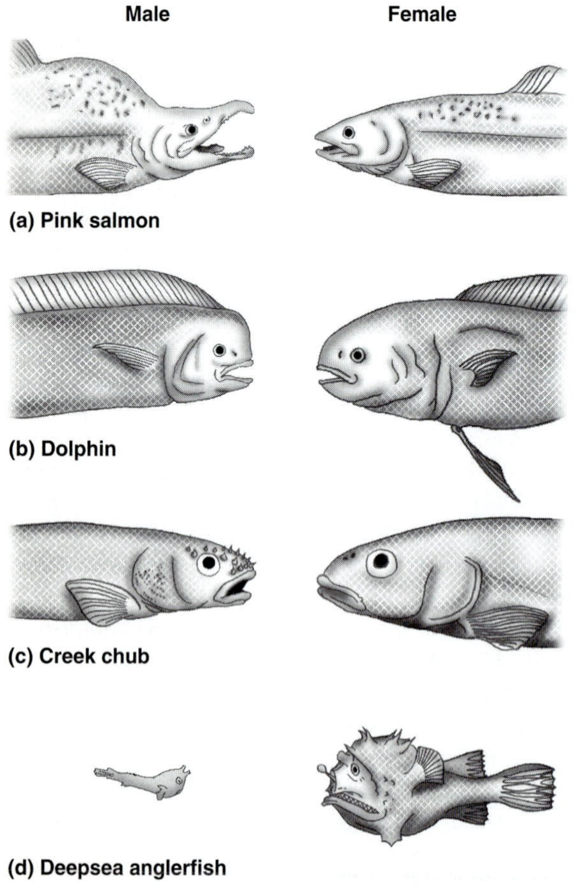

Sexual dimorphism in fishes: (*a*) humped back and hooked jaws of the male pink salmon (*Oncorhynchus gorbuscha*); (*b*) domed forehead and anterior position of the dorsal fin in a dolphin (*Coryphaena hippurus*); (*c*) nuptial tubercles on the snout and forehead of a male creek chub (*Semotilus atromaculatus*); (*d*) parasitic male of the deepsea anglerfish (*Photocorynus spiniceps*).

FIGURE 5.33

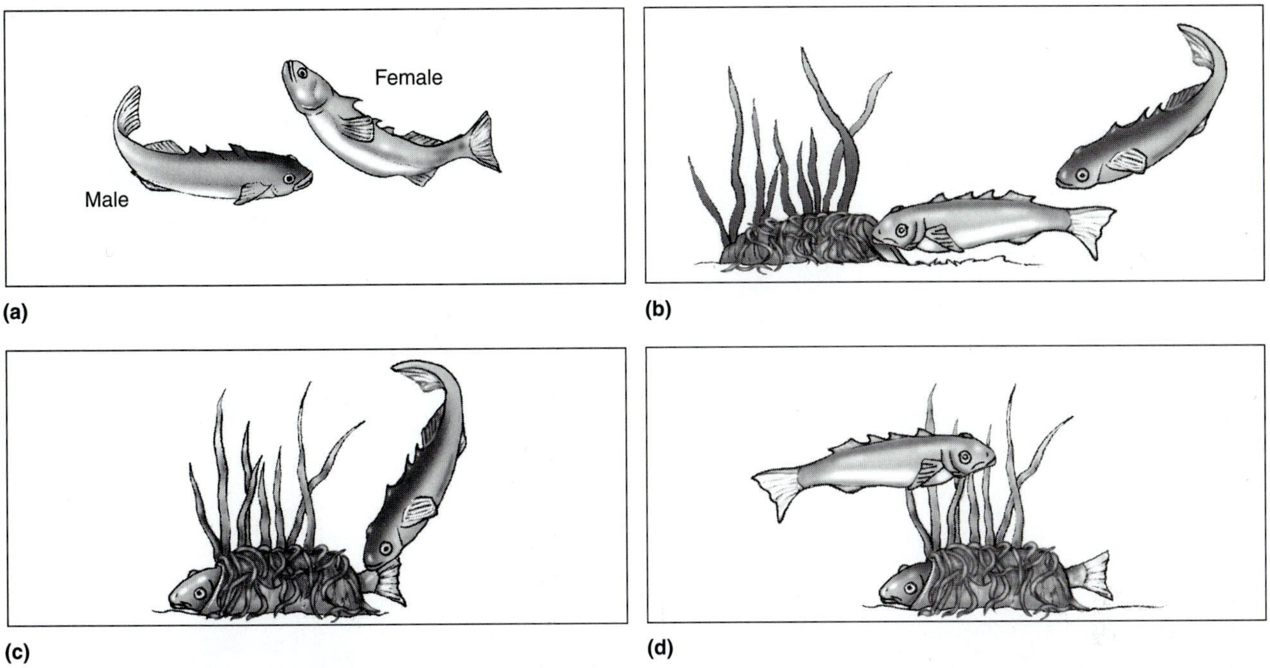

Mating behavior in the three-spined stickleback (*Gasterosteus aculeatus*). The movements that make up the courtship behavior of this species can be traced down to the activity of single muscles or groups of muscles such as those that move a single fin. It has repeatedly been shown that in this species one particular movement in the behavior train acts as a stimulus for the next. The instinctive movements imply a course of patterned rather than haphazard activity. In early spring in shallow ponds and lakes, the male builds a nest of vegetation, which is held together with a sticky secretion from the male's kidneys. After excavating a hole through the mass of vegetation, the male changes color, with his chin becoming red and his back a bluish white. (*a*) Upon finding a female, the male swims to his nest, thrusts his red snout into it, and then he turns on his side and presents the fins across his back to the female. (*b*) The female then swims into the burrow in the nest. (*c*) The male follows and touches the base of her tail, an action that stimulates the female to deposit 50 to 100 eggs. (*d*) The male then enters the nest, fertilizes the eggs, and by using his fins and tail, moves water over them to bring in additional oxygen. With fertilization accomplished, the male leaves the nest.

to providing extensive protection for their young, these fishes usually exhibit elaborate courtship behavior.

Fertilization in fishes is usually external. As females release eggs, sperm are released by one or more males in the same general area. In some teleosts, such as *Fundulus*, the male and female interlock their anal and pelvic fins and then execute external fertilization.

Internal fertilization occurs in most cartilaginous fishes and in some teleost fishes. Males of some teleost species possess an intromittent organ known as a gonopodium that may be formed from a thickened anal spine (e.g., Embiotocidae) or a modified anal ray (e.g., Poeciliidae). Gonopodia have grooved passageways that guide sperm into the female during copulation. The shape and structure of the gonopodium in poeciliids is a key feature in the classification of this family. Sharks and rays have modified pelvic fins known as claspers (Fig. 5.35). A clasper is inserted into the female, and sperm are transported along a groove into her cloaca. Many fishes with internal fertilization, such as guppies (*Poecilia reticulata*), dogfish sharks (*Squalus*), and coelacanths (*Latimeria*), carry the embryos until the time of birth.

Some fishes protect their eggs and/or offspring by carrying them in their mouths or in special pouches (Fig. 5.34c). Maternal mouthbrooding is found in sea catfishes (Ariidae), cichlids (Cichlidae), cardinal fishes (Apogonidae), and bonytongues (Osteoglossidae). Females take eggs into their mouths during spawning and inhale sperm to ensure fertilization in the mouth. The brood is carried in the mouth until the young are able to swim and feed. Male sea horses (Syngnathidae), as well as some pipefishes (Syngnathidae), have a brood pouch (marsupium) on the belly into which females deposit eggs (Fig. 5.34c and 5.37). Following incubation, the young are carried in the pouch until they can swim actively on their own.

Some fishes are **hermaphroditic**, a condition in which one individual is both male and female. Some hermaphrodites possess both male and female sex organs at the same time and are called **synchronous hermaphrodites**; others change sex as they grow and are called **sequential hermaphrodites**. Sea basses (Serranidae), for example, are synchronous hermaphrodites. Fertilization is external, and each member of the pair assumes a specific sex role associated with the release of eggs or sperm. Individuals release only one type of gamete during a spawning episode. Individuals of other species, such as black hamlets (*Hypoplecturus nigricans*), take turns releasing eggs for their partners to fertilize, a mating system referred to as "serial monogamy." Serial monogamy is

FIGURE 5.34

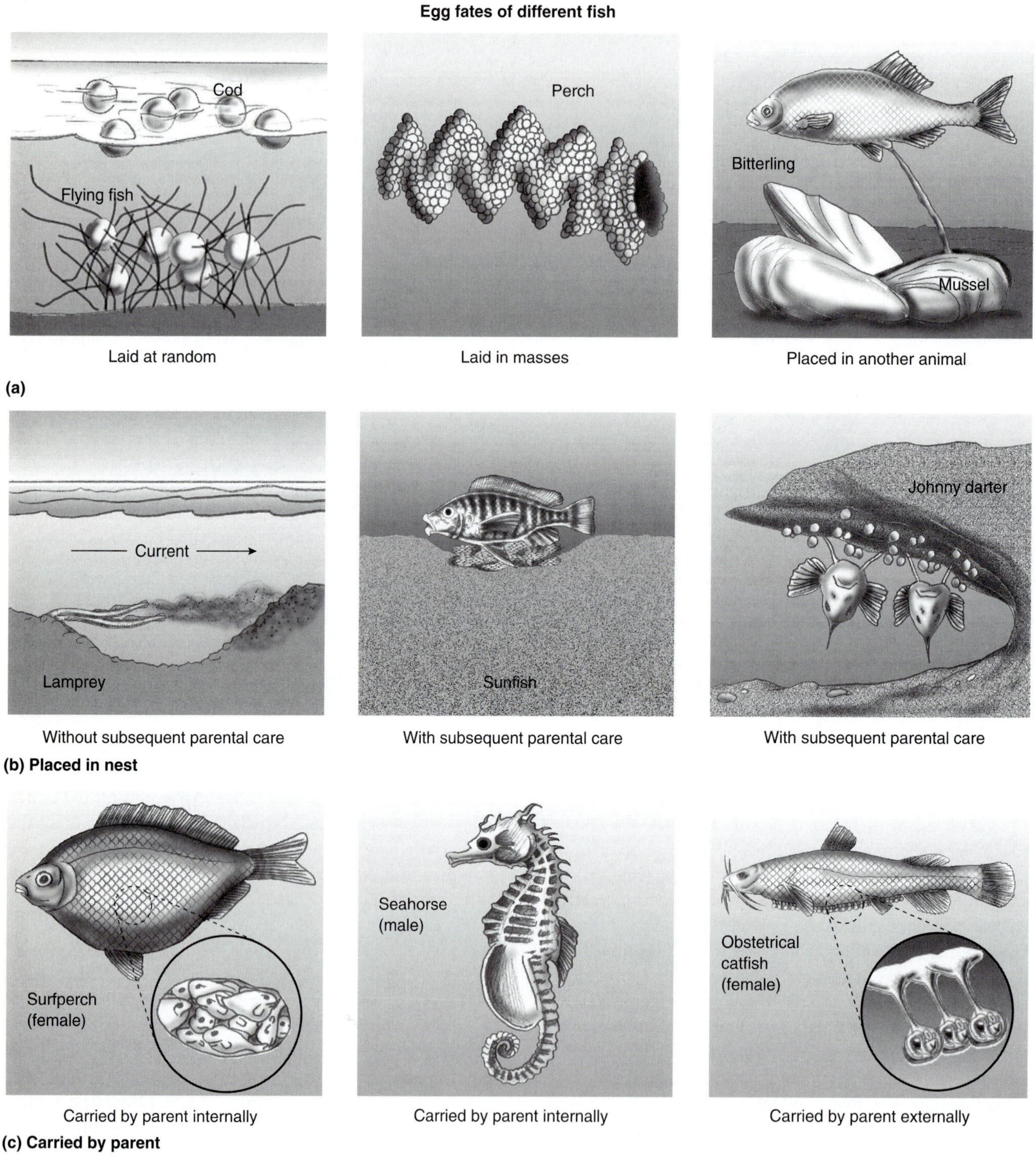

Methods of egg deposition in fishes are diverse. They range from (*a*) simply releasing the eggs into the water to (*b*) depositing the eggs in a carefully constructed nest. (*c*) In some species, the eggs are carried on or within the body of one of the parents.

FIGURE 5.35

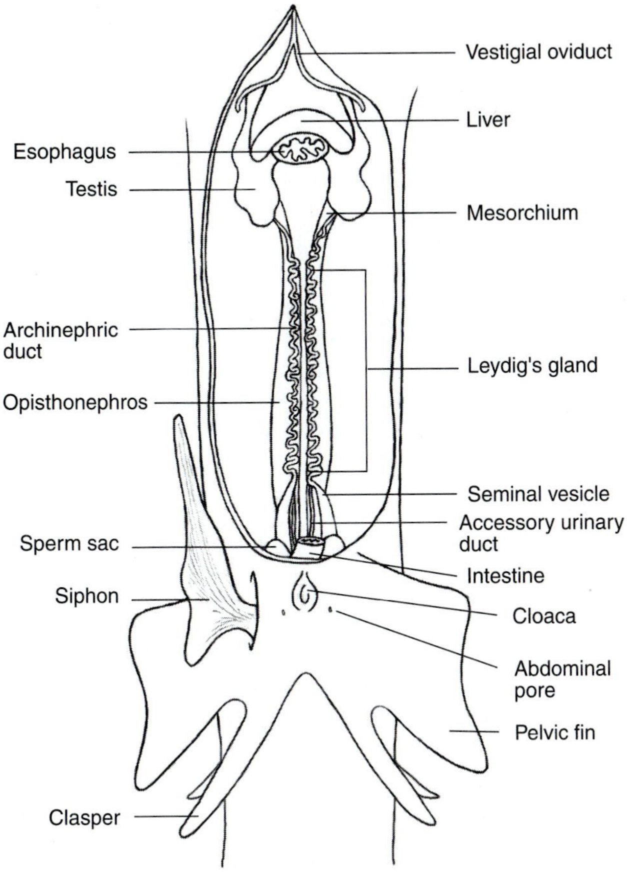

Male intromittent organs of the horn shark (*Heterodontus francisci*). Each pelvic fin is modified into a clasper (myxopterygium).

FIGURE 5.36

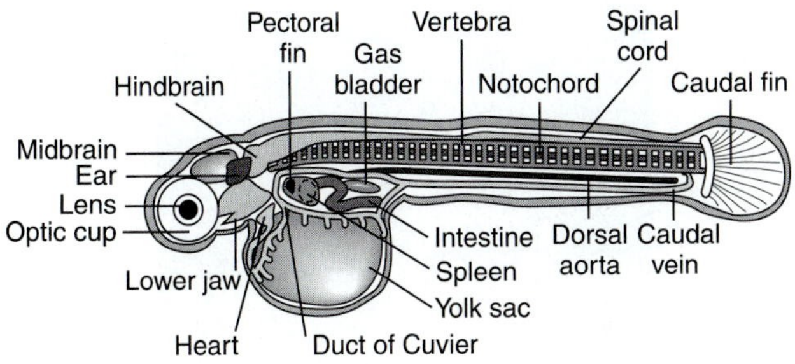

Embryo of the mummichog (*Fundulus heteroclitus*) at hatching. The embryo is nourished by the yolk contained within the egg. At hatching, a small amount of yolk remains within the yolk sac.
Source: Lagler, et al., Ichthyology, *1962, John Wiley & Sons, Inc.*

common among parrotfishes (Scaridae), wrasses (Labridae), and groupers (Serranidae). The most common form of sequential hermaphroditism, called **protogyny**, occurs when females turn into males. A less common form is **protandry**, a change from male to female. Protandrous species are found among the moray eels (Muraenidae), damselfishes (Pomacentridae), and anemonefishes and porgies (Sparidae). These processes allow eggs and/or sperm to be produced when the numbers of one sex in the population are low.

Most marine teleosts release large numbers of small, buoyant eggs that contain relatively small amounts of yolk. Some sturgeons (*Acipenser, Huso*), for example, may deposit as many as 5 million eggs (Herald, 1967) and the mackerel (*Scomberomicrus cavalla*) lays more than 6 million eggs (Elgar, 1990) during each annual breeding cycle. The eggs are fertilized externally and develop and hatch while drifting in the water. This method of reproduction, which provides for wide dispersal of the young, may have evolved as a means of decreasing predation and the possibly adverse effects of local environmental changes (Moyle and Cech, 1996). Marine larvae, which usually are very different in appearance from their parents, are specialized to feed and grow for weeks or months while adrift at sea.

Freshwater fishes, on the other hand, deposit fewer eggs than marine species; these eggs are significantly larger than those of marine species because of a greater amount of yolk. Eggs hatch into juveniles with adultlike body forms. In addition, nest building and parental care are more common in freshwater species than in marine species. Clutch size in both freshwater and marine fishes is correlated positively with fish length (Elgar, 1990): larger fishes tend to lay a greater number of relatively small eggs.

GROWTH AND DEVELOPMENT

Prenatal Development

Oviparous Species

All skates, some sharks, and many bony fishes are oviparous. Some female sharks that can store sperm release fertilized eggs over an extended period of time. These shark eggs are large and covered with tough, leathery cases, many of which have tendrils at the corners, apparently to attach the egg cases to seaweeds. Other species of oviparous fishes tend simultaneously to release vast quantities of eggs and sperm into the water. Fertilized eggs may drift with the current, or they may be deposited within a nest constructed by one or both parents. The developing embryo is nourished by yolk contained within the egg.

Viviparous Species

Viviparity is defined as the retention of fertilized eggs within the female reproductive tract. Embryos may be enclosed in a thin membrane or envelope along with a yolk supply, or they may be nourished by direct transfer of nutrients from the mother. Young are born alive. Most sharks, all rays, sawfish, guitarfish, and some bony fishes, including the coelacanth (*Latimeria chalumnae*), are viviparous. The eggs of the coelacanth are reported to be as "big as a grapefruit, the largest in any animal" by David Noakes of the University of Guelph in Ontario (Milius, 1998). The 10-m-long whale shark (*Rhincodon typus*), the largest fish in the oceans, is viviparous.

In some sharks, such as the hammerhead sharks (*Sphyrna* spp.) and requiem sharks (*Carcharinus* spp.), embryos initially are nourished by yolk, but the yolk supply is used up early in development. Remnants of the yolk sac form a placenta-like connection to the wall of the uterus, which supplies the developing young with nourishment during the remainder of the 9- to 12-month gestation period.

In the lamnoid sharks (order Lamniformes), which includes the makos (*Isurus* spp.), threshers (*Alopias* spp.), crocodile sharks (*Pseudocarcharias* spp.), great white (*Carcharodon carcharias*), and sand tiger (*Carcharias taurus*), the embryos use up their yolk supply and "hatch" in the uterus after the first several months of their 10- to 12-month gestation period. The female continues to ovulate throughout the gestation period, and when the yolk supply runs out, the young ingest the nutritious eggs (Luer and Gilbert, 1991). In some species of rays that use up their supply of yolk before development is complete, young are nourished by a fluid known as **uterine milk** secreted by stringlike "trophonemata" that line the inner wall of the uterus (Luer and Gilbert, 1991).

Several methods of nutrient exchange exist in viviparous bony fishes. In some, placenta-like growths form in the posterior portion of the maternal reproductive tract. Surfperches (Embiotocidae) are a small group of fishes found only along the northern Pacific coasts of North America and Asia. Females carry young in uterus-like sacs in the ovary. Body fluids of the mother that contain nutrients diffuse through greatly enlarged and highly vascularized dorsal, pelvic, and anal fins of the embryos, which are in close contact with the mother's tissues (Moyle and Cech, 1996).

Duration of Embryonic Development

Live-bearing sharks and rays have gestation periods of 6 to 22 months (Moyle and Cech, 1996). Hatching in oviparous fishes, however, may be as short as 1–2 days in gouramis and coral fishes or as long as 16 months in some sharks and rays (Moyle and Cech, 1996). The rate of development generally is temperature-dependent (Bond, 1996). For example, at 10°C, trout and salmon eggs hatch in about 50 days, but at 2°C, incubation requires about 6 months. Incubation temperature also influences the development of certain segmental features of the body such as vertebrae, scale rows, and fin rays. Typically, individuals from a selected batch of eggs incubated at low temperatures will have more vertebrae, scale rows, and fin rays than their siblings incubated at higher temperatures (Bond, 1996). The opposite is true in a few species.

Hatching and Birth

Most fish embryos produce hatching enzymes that assist in digesting the membrane enclosing them at the time of hatching. In many bony fishes, secretions come from special glands on the head or inside the mouth (Bond, 1996).

The sex of some fishes, such as the Atlantic silverside (*Menidia menidia*), is determined in part by the temperature of the water when the young are born (Conover and Kynard, 1981). Low temperatures early in the season produce females; higher temperatures later in the year yield males.

Parental Care

Parental care is lacking in oviparous elasmobranchs and in most other fishes. However, some species spend a great deal of time preparing a nest, oxygenating the eggs, and guarding the eggs and young. All three genera of lungfishes lay large, yolky eggs in nests formed by scooping out depressions in the pond bottom where the water is shallow. In the African lungfish *Protopterus*, males guard their nests and may beat the water with their tails to scare away predators and, perhaps, to aerate the nest (Thomson, 1991).

Salmon and trout excavate a nest (redd) and bury the eggs for protection. Several species, such as gouramis, build bubble nests for their floating eggs. Eggs of some species adhere to the bodies of adult fish and are transported with them. Some fishes carry eggs and/or young with them in brood pouches (sea horses, pipefishes, some catfishes) (Fig. 5.37). Mouthbrooders, such as sea catfishes (Aridae), cichlids (Cichlidae), cardinal fishes (Apogonidae), and bonytongues (Osteoglossidae), carry the eggs, embryos, and sometimes even larval fishes in their mouth cavities.

Growth and Metamorphosis

The life of an oviparous fish is divided into four stages: embryo, larva, juvenile, and adult. Most of the **embryonic** stage is passed within the egg, although it may continue for a short time after hatching while the embryo's nutrition is still derived from yolk (Fig. 5.36). The **larval** period begins when the fish is able to catch food organisms. During this time, the skeleton and other organ systems become fully formed,

FIGURE 5.37

Female sea horses deposit their eggs in a brood pouch on the abdomen of the male. The eggs are incubated and hatched inside of the sealed brood pouch. The young are expelled once they are capable of swimming on their own.

and the fins develop. In most species, there is no dramatic development that signals the end of the larval stage and the beginning of the juvenile stage; it is a gradual, progressive transition. In the **juvenile** stage, the fish and its organ systems grow considerably in size; thus, this is usually a period of rapid growth. During this stage, which lasts until the gonads mature, the fishes may be distinctively colored. When gonads mature, the fish is able to spawn and is considered to be an **adult**.

Those species of fishes that have undergone their development within the female's reproductive tract are well developed at birth and are born as miniature adults. Thus, there is no larval life stage.

Attainment of Sexual Maturity

The time required to reach sexual maturity in most fishes ranges from several weeks to several years. Some, such as members of the viviparous perch family (Embiotocidae), become sexually mature a few weeks after birth (Moyle and Cech, 1996). Other species, such as salmon (*Oncorhynchus*), usually spawn when 4 or 5 years of age.

Longevity

Maximum longevity in fishes is probably about 50 years. Some sharks along with such species as sturgeons and paddlefishes are probably among the longest-lived.

Review Questions

1. What advantages are conferred on an organism by having vertebrae that are differentiated into various types?
2. Distinguish between ostracoderms and placoderms. What important evolutionary advances did each contribute to vertebrate evolution?
3. Discuss the theories concerning the origin of paired fins.
4. Why was the discovery of the first living coelacanth (*Latimeria*) in 1938 of such great importance to scientists?
5. Compare the functions of paired fins and unpaired median fins such as the dorsal and caudal fins.
6. What is the function of the bulbous arteriosus in fish? Why is it not present in birds and mammals?
7. Why is an operculum important to bony fish and not to most cartilaginous fish? What functions does it perform?
8. Explain the mechanism by which bony fish control their buoyancy.
9. Trace the pathway that food would follow from the time it enters the mouth of a largemouth bass until it is absorbed into the bloodstream.
10. How are spiral valves and villi similar?
11. Why wouldn't you expect birds to have taste buds distributed over their entire body surface, as does a catfish? What adaptive advantage does this arrangement afford the catfish?
12. When (or under what conditions) are freshwater fishes in danger of excessive water intake ("drowning") and marine fishes in danger of dehydration? What mechanisms has each group evolved to counteract these problems?
13. Which hormones influence male reproductive function?
14. Differentiate between oviparous and viviparous. Give an example for each.
15. Compare and contrast the advantages and disadvantages to the female of external versus internal fertilization.

Supplemental Reading

Bruton, M. N. (ed.). 1990. *Alternative Life-History Styles of Fishes.* Dordrecht, Netherlands: Kluwer Academic Publishers.

Cushing, D. 1995. *Population Production and Regulation in the Sea: A Fisheries Perspective.* Cambridge: Cambridge University Press.

Donovan, S. K., and C. R. C. Paul (eds.). 1998. *The Adequacy of the Fossil Record.* New York: John Wiley.

Graham, J. B. 1997. *Air-Breathing Fishes: Evolution, Diversity and Adaptation.* San Diego: Academic Press.

Groot, C., L. Margolis, and W. C. Clarke. 1995. *Physiological Ecology of Pacific Salmon.* Vancouver: UBC Press.

Hara, T. J. 1992. *Fish Chemoreception.* London: Chapman and Hall.

Helfman, G. S., B. B. Collette, and D. E. Facey. 1997. *The Biodiversity of Fishes.* Malden, Massachusetts: Blackwell Science.

Hoar, W. S., A. P. Farrell, and D. J. Randall (eds.). 1969–1994. *Fish Physiology.* 14 volumes. New York: Academic Press.

Jamieson, B. G. M. 1991. *Fish Evolution and Systematics: Evidence from Spermatoozoa.* Cambridge, England: Cambridge University Press.

Jobling, M. 1994. *Fish Bioenergetics.* London: Chapman and Hall.

Lee, D. S., C. R. Gilbert, C. H. Hocutt, R. E. Jenkins, D. E. McAllister, and J. R. Stauffer, Jr. 1981. *Atlas of North American Freshwater Fishes.* Raleigh: North Carolina State Museum of Natural History.

Mead, G. W., E. Bertelsen, and D. M. Cohen. 1964. Reproduction among deep-sea fishes. *Deep-Sea Research* 11(4):569–596.

Murdy, E. O., R. S. Birdsong, and J. A. Musick. 1996. *Fishes of Chesapeake Bay.* Washington, D.C.: Smithsonian Institution Press.

Musick, J. A., M. N. Bruton, and E. K. Balon. 1991. *The Biology of* Latimeria chalumnae *and Evolution of Coelacanths.* Boston: Kluwer Academic Publishers.

Norris, D. O., and R. E. Jones (eds.). 1987. *Hormones and Reproduction in Fishes, Amphibians, and Reptiles.* New York: Plenum Press.

Paxton, J. R., and W. N. Eschmeyer (eds.). 1995. *Encyclopedia of Fishes.* New York: Academic Press.

Radinsky, L. B. 1987. *The Evolution of Vertebrate Design.* Chicago: The University of Chicago Press.

Rankin, J. C., and F. B. Jensen (eds.). 1993. *Fish Ecophysiology.* London: Chapman and Hall.

Stahl, B. J. 1974. *Vertebrate History: Problems in Evolution.* New York: McGraw-Hill Book Company.

Thomson, K. S. 1991. *Living Fossil. The Story of the Coelacanth.* New York: W. W. Norton and Company.

Weinberg, S. 1999. *A Fish Caught in Time: The Search for the Coelacanth.* London, England: Fourth Estate.

Wootton, R. J. 1990. *Ecology of Teleost Fishes.* London: Chapman and Hall.

Young, J. Z. 1981. *The Life of Vertebrates.* Third edition. Oxford, England: Clarendon Press.

SELECTED JOURNALS

Copeia. A broadly based scientific journal of the American Society of Ichthyologists and Herpetologists. It publishes results of original research performed by members in which fish, amphibians, or reptiles are utilized as study organisms.

Environmental Biology of Fishes. Published by Kluwer Academic Publishers, Dordrecht, Netherlands. Publishes original studies of the ecology, life history, epigenetics, behavior, physiology, morphology, systematics, and evolution of marine and freshwater fishes.

Fish Physiology and Biochemistry. An international journal. Amsterdam: Kugler Publications.

Journal of Fish Biology. Published by Fisheries Society of the British Isles. London: Academic Press. Publishes articles with new biological insight into any aspect of fish biology.

Journal of Paleontology. Published by the Paleontological Society. Research in all areas of paleontology.

Vertebrate Internet Sites

Visit the zoology web site at http://www.mhhe.com/zoology to find live Internet links for each of the references listed below.

1. **Introduction to the Chondrichthyes.**
   University of California at Berkeley, Museum of Paleontology. Images, photos, systematics.
2. **The Great White Shark.**
   Images and information. University of California at Berkeley, Museum of Paleontology. Images, photos, systematics, more information, and links.
3. **Sharks and Their Relatives.**
   Sea World Education Department information on sharks.
4. **FAO Fisheries Department Homepage.**
   Many links and resources on fisheries.
5. **Reel Time: Saltwater Fish.**
   Information on common fishes of the North Atlantic.
6. **Class Actinopterygii.**
   University of Michigan site on actinopterygiian fish. Pictures, much information on the morphology, distribution, and ecology of a large number of fish. Each fish is linked to additional web pages.
7. **Introduction to the Actinopterygii.**
   University of California at Berkeley, Museum of Paleontology. Images, photos, systematics, more information, and links.
8. **Neopterygii.**
   University of California at Berkeley, Museum of Paleontology. Images, photos, systematics, more information, and links.
9. **Teleostei.**
   University of California at Berkeley, Museum of Paleontology. Images, photos, systematics, more information, and links.
10. **Teleostei.**
    Arizona's Tree of Life Web Page. Pictures, characteristics, phylogenetic relationships, references on teleost fishes. A cladogram and references on teleosts. Some links to various groups of teleosts. http://phylogeny.arizona.edu/tree/eukaryotes/animals/chordata/actinopterygii/teleostei.html
11. **Marine Fishes of Hawaii.**
    A pictorial guide to the families of marine fishes found in the waters surrounding the Hawaiian Islands.
12. **Dissection of the Perch.**
    Several pictures of internal organs.
13. **Subphylum Vertebrata, Class Osteichthyes, from the Universityof Minnesota.**
14. **American Society of Ichthyologists and Herpetologists.**
    This organization publishes *Copeia*, and includes related societies, links and publications. Many links to other sites focusing on ichthyology and herpetology.
15. **Vertebrate Systematics.**
    University of California at Berkeley, Museum of Paleontology. Information on the systematics and natural history of each of the major groups of fishes and links to other fish sites. Click on photographs for more information about each group.
16. **Great Lakes Fishery Commission.**
    Information on fishery management in the Great Lakes. This site contains recent information on the lamprey problem in the Great Lakes, as well as information on sport and commercial fishing. It also includes a newsletter and information of fisheries research.
17. **Fins.**
    This site, provided by the Fish Information Service, is an archive of information about aquariums.
18. **The Audubon Guide to Seafood.**
    This interesting chart tells of recent trends in populations of seafood that humans commonly consume, and gives recommendations of what to eat and what to avoid to be ecologically responsible.

CHAPTER 6

Amphibians

INTRODUCTION

Amphibians are the first quadrupedal vertebrates that can support themselves and move about on land. They have a strong, mostly bony, skeleton and usually four limbs (tetrapod), although some are legless. Webbed feet are often present, and no claws or true nails are present. The glandular skin is smooth and moist. Scales are absent, except in some caecilians that possess concealed dermal scales. Gas exchange is accomplished either through lungs (absent in some salamanders), gills, or directly through the skin. Amphibians have a double circulation consisting of separate pulmonary and systemic circuits, with blood being pumped through the body by a three-chambered heart (two atria, one ventricle). They are able to pick up airborne sounds because of their tympanum and columella and to detect odors because of their well-developed olfactory epithelium.

The emergence of a vertebrate form onto land was a dramatic development in the evolution of vertebrates. Some ancestral vertebrate evolved a radically different type of limb skeleton with a strong central axis perpendicular to the body and numerous lateral branches radiating from this common focus. This transition had its beginnings during the early to middle Devonian period and took place over many millions of years (Fig. 6.1). It involved significant morphological, physiological, and behavioral modifications. A cladogram showing presumed relationships of early amphibians with their aquatic ancestors as well as with those amphibians that arose later is shown in Fig. 6.2. Phylogenetic relationships depicted in such diagrams are controversial and subject to a wide range of interpretations.

EVOLUTION

Controversy surrounds the ancestor of the amphibians. Was it a lungfish, a lobe-finned rhipidistian, or a lobe-finned coelacanth? Rhipidistians, which are now extinct, were dominant freshwater predators among bony fishes. Did amphibians arise from more than one ancestor and have a polyphyletic origin, or did they all arise from a common ancestor, illustrating a monophyletic origin? Are salamanders and caecilians more closely related to each other than either group is to the anurans?

Great gaps in the fossil record make it difficult to connect major extinct groups and to link extinct groups to modern amphibians. These so-called "missing links" are a natural result of the conditions under which divergence takes place. Evolution at that point is likely to have been rapid. Any significant step in evolution probably would take place in a relatively small population isolated from the rest of the species. Under such conditions, new species can evolve without being swamped by interbreeding with the ancestral species, and the new species and new habits of life have more chance of survival. The chances of finding fossils from such populations, however, are minute. In addition, as amphibians became smaller, their skeletons became less robust and more delicate due to an evolutionary trend toward reduced ossification. These factors increased the likelihood of the skeletons being crushed before they could fossilize intact.

The extinct lobe-finned rhipidistian fishes, which were abundant and widely distributed in the Devonian period some 400 million years ago, have been regarded by some investigators as the closest relatives of the tetrapods (Panchen and Smithson, 1987). One group of rhipidistians, the osteolepiforms (named in reference to the earliest described genus *Osteolepis*, from the Devonian rocks of Scotland), had several unique anatomical characters. One of the best known osteolepiforms was *Eusthenopteron foordi* (Fig. 6.3). These fishes possessed a combination of unique characteristics in common with the earliest amphibians (labyrinthodonts) (Figs. 6.4 and 6.5). Along with most of the bony fishes (Osteichthyes), rhipidistians both had gills and had air passageways leading from their external nares to their lungs, so that they presumably (there is no concrete evidence, because no fossils of lungs exist) could breathe atmospheric air. If the

FIGURE 6.1

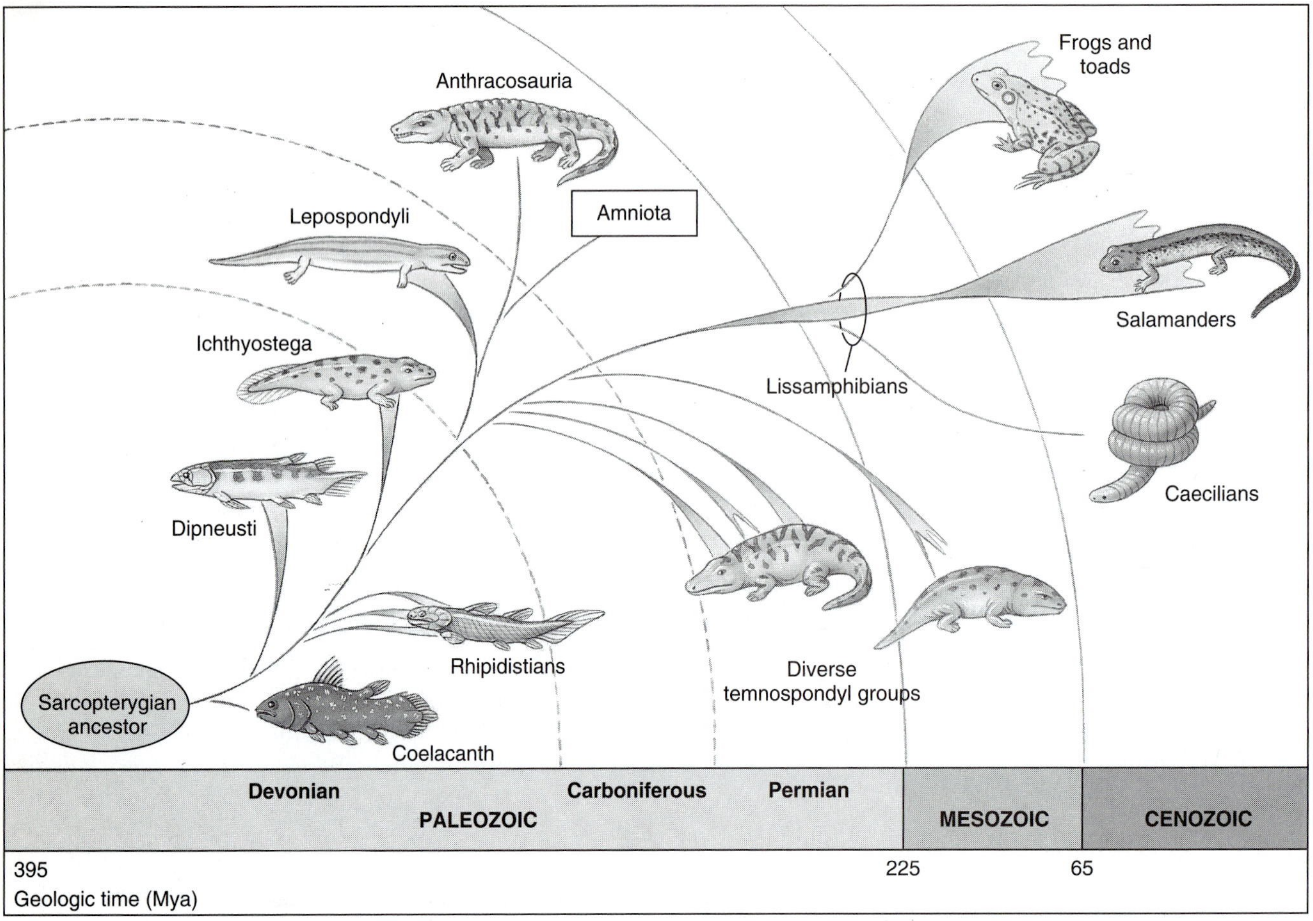

Early tetrapod evolution and the rise of amphibians. The tetrapods share their most recent common ancestry with the rhipidistians of the Devonian. Amphibians share their most recent common ancestry with the temnospondyls of the Carboniferous and Permian periods of the Paleozoic and the Triassic period of the Mesozoic.

oxygen content of the stagnant water decreased, respiration could be supplemented by using the lungs to breathe air. The skeletons of rhipidistians were well ossified, and their muscular, lobed fins contained a skeletal structure amazingly comparable to the bones of the tetrapod limb (Fig. 6.5). Such fins may have given these fish an adaptive advantage by facilitating mobility on the bottoms of warm, shallow ponds or swamps with abundant vegetation (Edwards, 1989), to move short distances over land to new bodies of water, and/or to escape aquatic predators. Palatal and jaw structures, as well as the structure of the vertebrae, were identical to early amphibians. The teeth have the complex foldings of the enamel—visible as grooves on the outside of each tooth—that are also found in the earliest labyrinthodont ("labyrinth tooth") amphibians (Fig. 6.4).

The skull and jaw bones of *Elginerpeton pancheni* from the Upper Devonian (approximately 368 million years ago) in Scotland exhibit a mosaic of fish and amphibian features, making it the oldest known stem tetrapod (Ahlberg, 1995). Appendicular bones (amphibian-like tibia, robust ilium, incomplete pectoral girdles) exhibit some tetrapod features, but whether this genus had feet like later amphibians or fish-like fins has not been established. The genera *Elginerpeton* and *Obruchevichthys* from Latvia and Russia possess several unique derived cranial characters, and so they cannot be closely related to any of the Upper Devonian or Carboniferous amphibians. Instead, they form a clade that is the sister group of all other Tetrapoda.

Some researchers feel that the sole surviving crossopterygian, the coelacanth (see Fig. 5.6), is the closest extant relative of tetrapods. Evidence supporting this hypothesis has been presented by Gorr et al. (1991), who analyzed the sequence of amino acids in hemoglobin, the protein that carries oxygen through the bloodstream. This study concluded that coelacanth hemoglobin matched larval amphibian hemoglobin more closely than it matched the hemoglobin of any other vertebrate tested (several cartilaginous and bony fishes, larval and adult amphibians). As might be expected,

FIGURE 6.2

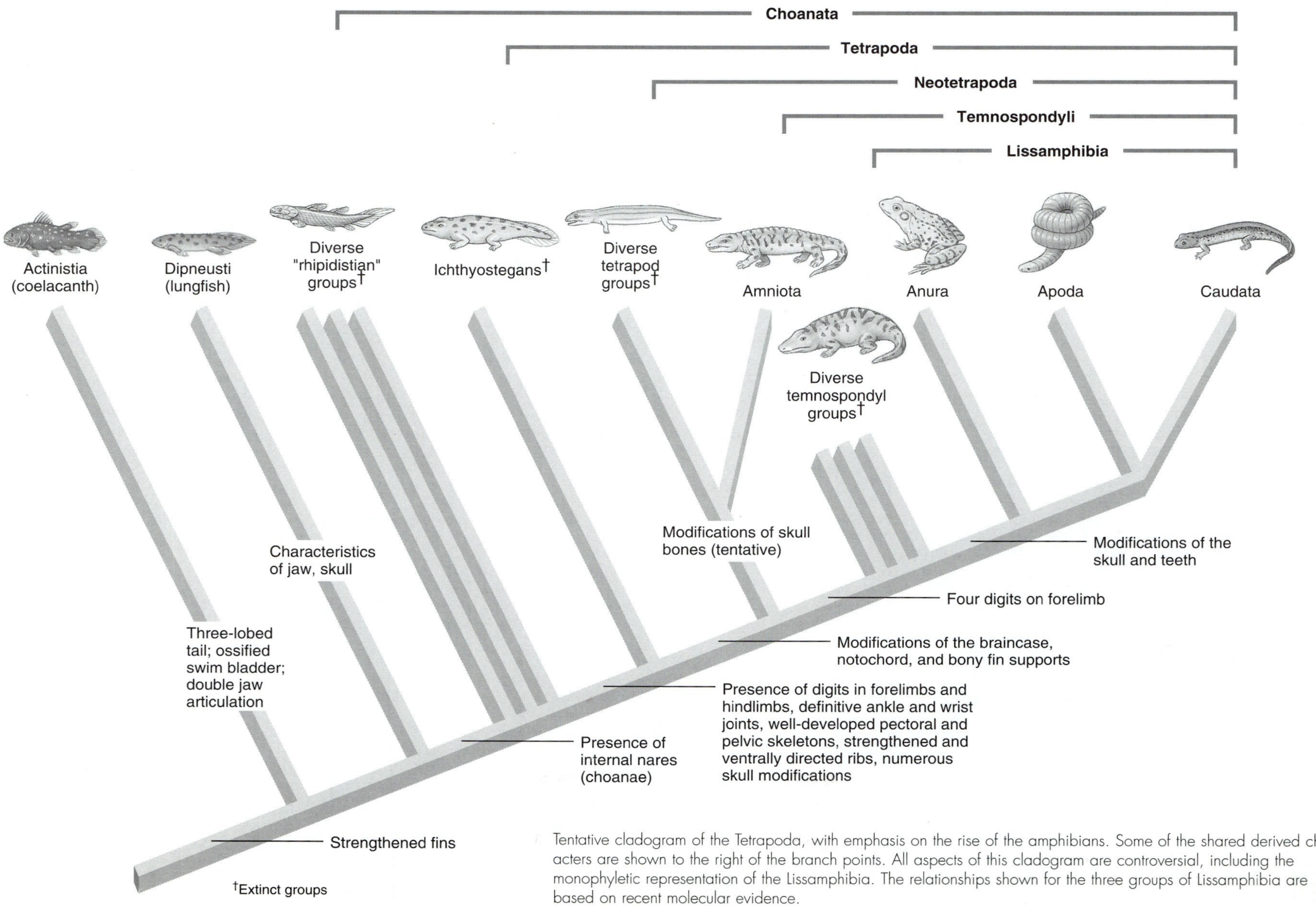

Tentative cladogram of the Tetrapoda, with emphasis on the rise of the amphibians. Some of the shared derived characters are shown to the right of the branch points. All aspects of this cladogram are controversial, including the monophyletic representation of the Lissamphibia. The relationships shown for the three groups of Lissamphibia are based on recent molecular evidence.

FIGURE 6.3

Eusthenopteron, a lobe-finned rhipidistian that is a possible early ancestor of the tetrapods.

FIGURE 6.4

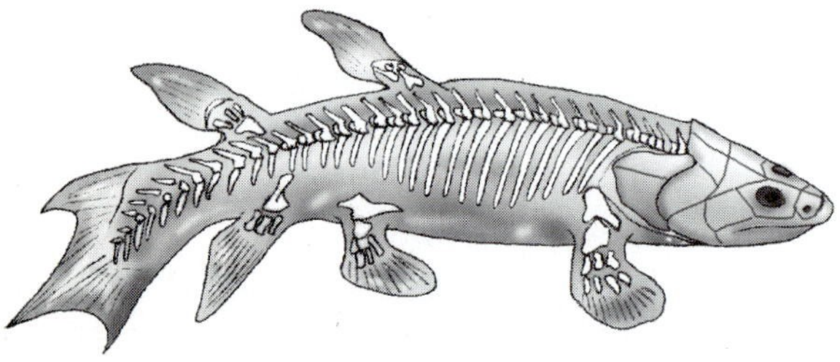

(a) Upper Devonian lobe-finned fish

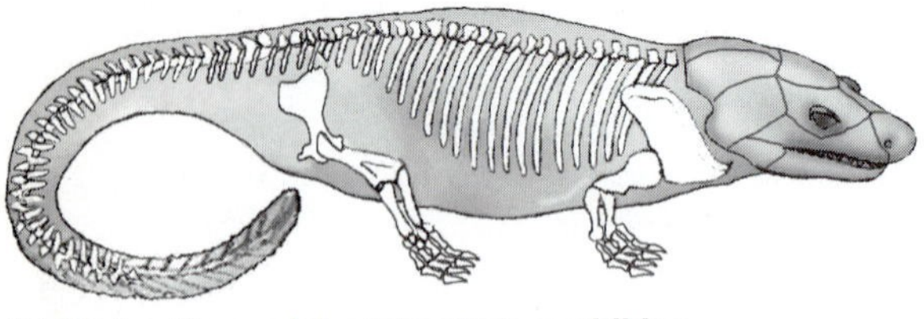

(b) Carboniferous labyrinthodont amphibian

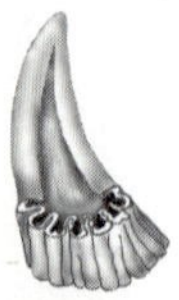

(c) Labyrinthodont tooth

(*a*) An Upper Devonian lobe-finned fish (*Eusthenopteron*) and (*b*) a Carboniferous labyrinthodont amphibian (*Diplovertebron*). Note in the amphibian the loss of median fins, the transformation of paired paddles into limbs, the development of strong ribs, and the spread of the dorsal blade of the pelvic girdle. (*c*) Labyrinthodont tooth characteristic of crossopterygians and labyrinthodont amphibians.

considerable controversy has been generated by these findings, since extinct forms such as rhipidistians could not be analyzed for comparison.

Based on the most extensive character set ever used to analyze osteolepiform relationships, Ahlberg and Johanson (1998) presented evidence showing that osteolepiforms were paraphyletic, not monophyletic, to tetrapods. Their analyses revealed that tetrapod-like character complexes (reduced median fins, elaborate anterior dentition, morphology of a large predator) evolved three times in parallel within closely related groups of fishes (rhizodonts, tristicopterids, and elpistostegids). Thus, Ahlberg and Johanson concluded that tetrapods are believed to have arisen from one of several similar evolutionary "experiments" with a large aquatic predator.

Still other researchers (Rosen et al., 1981; Forey, 1986, 1991; Meyer and Wilson, 1991) have presented convincing anatomical and molecular evidence favoring lungfishes as the ancestor. Forey (1986) concluded that, "among Recent taxa, lungfishes and tetrapods are sister-groups, with coelacanths as the plesiomorphic sister-group to that combined group." Meyer and Wilson (1991) found lungfish mitochondrial DNA (mtDNA) was more closely related to that of the frog than is the mtDNA of the coelacanth. Zardoya and Meyer (1997a) reported that a statistical comparison using the complete coelacanth mtDNA sequence did not point unambiguously to either lungfish or coelacanths as the tetrapods' closest sister group. However, when Zardoya and Meyer (1997b) reanalyzed their data, they concluded that they could "clearly reject" the possibility that coelacanths are the closest sister group to tetrapods. (The possibility that coelacanths and lungfish are equally close relations of tetrapods, although unlikely, could not be formally ruled out.) At present, most paleontologists and ichthyologists reject the lungfish hypothesis.

Some researchers consider tetrapods to have arisen from two ancestral groups. Holmgren (1933, 1939, 1949, 1952) considered tetrapods to be diphyletic, with the majority being derived from one group of fossil fish, the Rhipidistia, and the rest (the salamanders) being derived from lungfishes (Dipneusti). As recently as 1986, Jarvik (1980, 1986) continued to argue that tetrapods were diphyletic with salamanders, being separately derived from a different group of rhipidistians, the Porolepiformes, than were other tetrapods, whose ancestry is traced to the rhipidistian Osteolepiformes. Benton (1990) considered the class Amphibia to be "clearly a paraphyletic group if it is assumed to include the ancestor of the reptiles, birds, and mammals (the Amniota)."

The Devonian period saw great climatic fluctuations, with wet periods followed by severe droughts. As bodies of water became smaller, they probably became stagnant and more eutrophic as dissolved oxygen dropped dramatically. They also probably became overcrowded with competing fishes. With their lobed fins and their ability to breathe air, ancestors to the tetrapods could have moved themselves about in the shallow waters and onto the muddy shores (see Fig. 6.3). Lobed fins with their bony skeletal elements, along

FIGURE 6.5

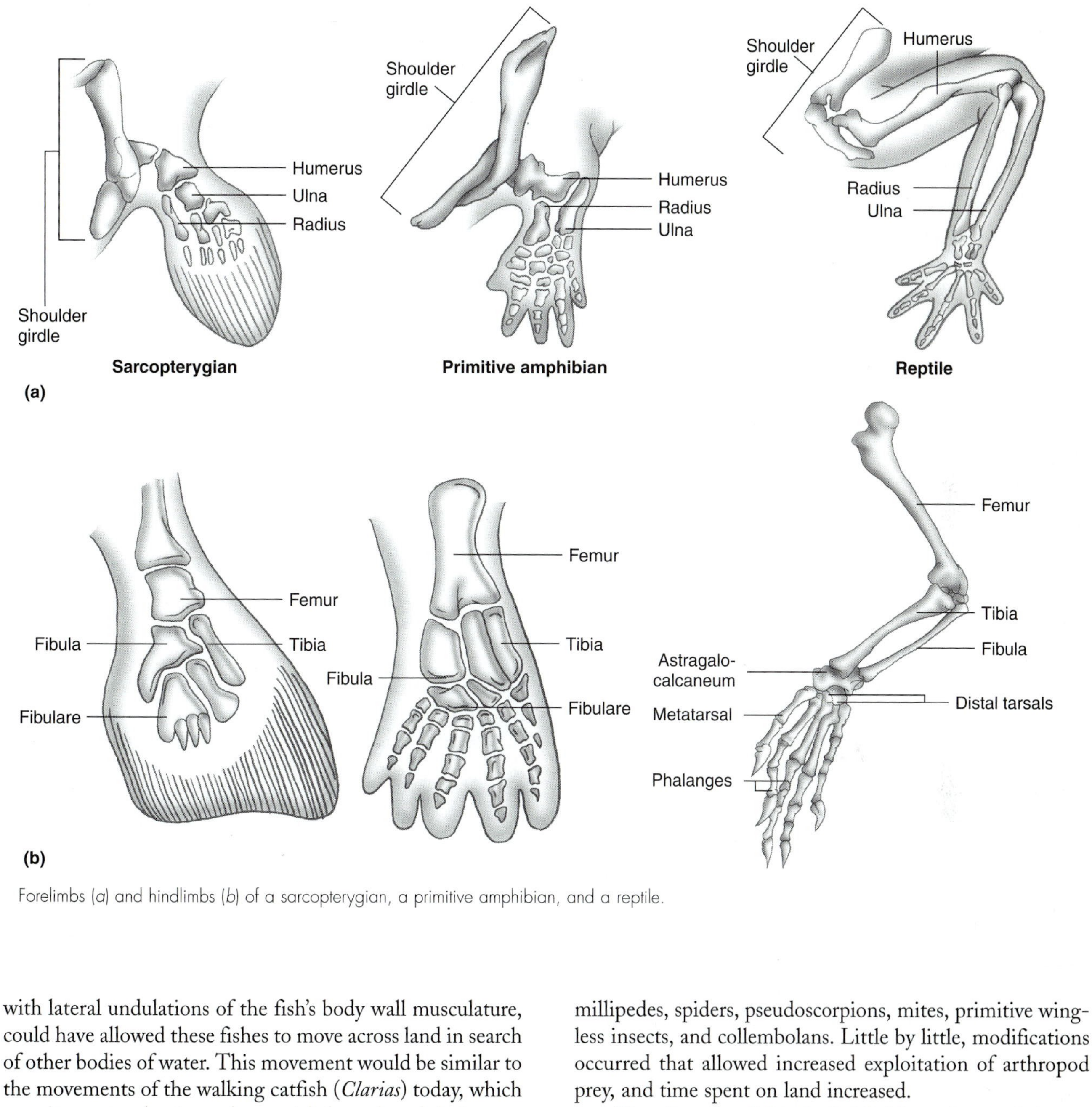

Forelimbs (*a*) and hindlimbs (*b*) of a sarcopterygian, a primitive amphibian, and a reptile.

with lateral undulations of the fish's body wall musculature, could have allowed these fishes to move across land in search of other bodies of water. This movement would be similar to the movements of the walking catfish (*Clarias*) today, which uses its pectoral spines along with lateral undulations to "walk" on land, or mudskippers (*Perioptbalmus*), which climb out of the water and "walk" on mudflats and along mangrove roots on their pectoral fins. Thus, lobed fins and the ability to breathe air may have allowed increased survival as an aquatic animal, and then later allowed movement overland. These ancestral semiamphibious groups may have been moving temporarily onto land to avoid predators or to seek arthropod prey. Early Devonian arthropod faunas are known from North America, Germany, and the United Kingdom and may well have been an abundant food source (Kenrick and Crane, 1997). These arthropods included centipedes, millipedes, spiders, pseudoscorpions, mites, primitive wingless insects, and collembolans. Little by little, modifications occurred that allowed increased exploitation of arthropod prey, and time spent on land increased.

The class Amphibia is divided into three subclasses: Labyrinthodontia, Lepospondyli, and the subclass containing all living amphibians, Lissamphibia.

Labyrinthodontia

The earliest known amphibians are the labyrinthodonts (order Ichthyostegalia) (Fig. 6.6), and the earliest known labyrinthodont fossils are from Upper Devonian freshwater deposits in Greenland. Labyrinthodonts appear to have been the most abundant and diverse amphibians of the Carboniferous, Permian, and Triassic periods. At the present time, two families and three genera are recognized, with the best known

FIGURE 6.6

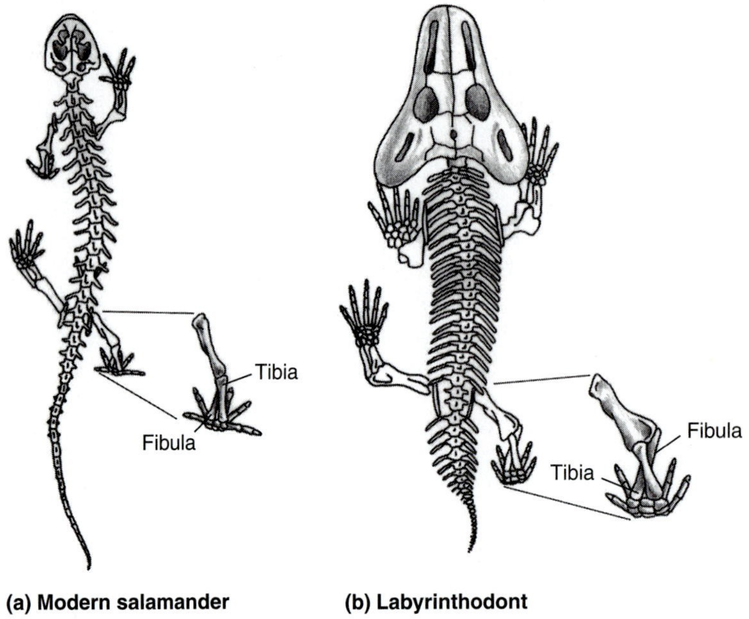

Modern salamander (*a*) and ancient labyrinthodont (*b*). Lateral undulations of the body are used to extend the stride of the limbs. The forward planting of the feet requires the crossing of the tibia by the fibula and thus places twisting stress on the tarsus.

genera being *Ichthyostega* and *Acanthostega*. The name *Ichthyostega* means "fish with a roof," referring to its primitive fishlike structure and the thick roof of its skull. The first *Ichthyostega* fossils were discovered in 1932.

Ichthyostega was a fairly large animal (approximately 65 to 70 cm) that exhibited characters intermediate between crossopterygians and later tetrapods (see Figs. 6.1, and 6.2). It had short, stocky limbs instead of fins. Jarvik (1996) provided evidence of pentadactyl hind feet (five digits) and refuted the statements of Coates and Clack (1990) that each hind foot contained seven digits. The pentadactyl limb is an ancestral vertebrate characteristic. The skull was broad, heavily roofed, and flattened, and it possessed only a single occipital condyle (rounded process on the base of the skull that articulates with the first vertebra). Ichthyostegids possessed rhachitomous "arch vertebrae" similar to those of some crossopterygians. The snout was short and rounded, and an opercular fold was present on each side of the head. The tail was fishlike and had a small dorsomedial tail fin partially supported by dermal rays. *Ichthyostega* probably was primarily aquatic, as evidenced by the presence of lateral line canals, but it likely could move about on land using its short, but effective, limbs.

The branchial (gill) skeleton of *Acanthostega gunnari* from the Upper Devonian (about 363 million years ago) has revealed structural details similar to those of modern fishes (Coates and Clack, 1991; Coates, 1996). These features indicate that *Acanthostega* "retained fish-like internal gills and an open opercular chamber for use in aquatic respiration, implying that the earliest tetrapods were not fully terrestrial" (Coates and Clack, 1991). Fish differ from tetrapods in that their pectoral girdles are firmly attached to the back of the skull by a series of dermal bones; these bones are reduced or lost in tetrapods. *Acanthostega* retains a fishlike shoulder girdle, similar to that in the lungfish, *Neoceratodus*. Both forelimbs and hindlimbs are thought to have been flipperlike, and the forelimb contained eight fingers (Coates and Clack, 1990, 1991). Limbs with digits probably evolved initially in aquatic ancestors rather than terrestrial ones. They could have provided increased maneuverability among aquatic plants and fallen debris in shallow waters near the edges of ponds and streams.

The discovery in Upper Devonian deposits in Scotland of the tibia of *Elginerpeton* bearing articular facets for ankle bones (and thus feet) is strongly suggestive of tetrapod affinity and represents the earliest known tetrapod-type limb (Ahlberg, 1991). This find pushed back the origin of tetrapods by about 10 million years. Because tetrapod or near-tetrapod fossils have been described from the Upper Devonian (about 370 million years ago) of Pennsylvania in the United States, Greenland, Scotland, Latvia, Russia, and Australia (Ahlberg, 1991; Daeschler et al., 1994), a virtually global equatorial distribution of these early forms was established by the end of the Devonian.

Two other groups of labyrinthodonts evolved: the temnospondyls and the anthracosaurs. Members of the order Temnospondyli had two occipital condyles and a tendency toward a flattened skull. They were more successful as amphibians than the order Anthracosauria, which was a short-lived group (but which were ancestral to the turtles and diapsids). The ancestor of turtles and diapsids is thought to have diverged from the main anthracosaur line during the Late Mississippian period (approximately 370 million years ago). The temnospondyls, which may have given rise to the living amphibians, died out by the end of the Triassic (245 million years ago).

Numerous problems had to be overcome in order to survive on land. Some have been solved by the amphibians; others were not overcome until reptiles evolved. One major problem was locomotion. The weight of the body in a terrestrial vertebrate is passed to the legs through the pectoral and pelvic girdles. The general consensus is that the primitive bony elements of the ancestral fish fin gradually differentiated into the bones of the tetrapod forelimb (humerus, radius, ulna, carpals, metacarpals, and phalanges) and hindlimb (femur, tibia, fibula, tarsals, metatarsals, and phalanges). The girdles and their musculature were modified and strengthened. Even today, however, most salamanders cannot fully support the weight of their bodies with their limbs. They still primarily used a lateral undulatory method of locomotion, with their ventral surfaces dragging on the ground. Salamander appendages project nearly at right angles to the body, thus making the limbs inefficient structures for support or rapid locomotion. Not until reptiles evolved did the limbs rotate to a position more beneath the body.

Although the earliest amphibians probably were covered by scales, the evolution of the integument and the subsequent loss of scales in most forms made dessiccation a significant threat to survival. The problem of dessiccation was solved partly by the development of a **stratum corneum** (outermost layer of the epidermis) and by the presence of mucous glands in the epidermis. The entire epidermis of fishes consists of living cells, whereas the stratum corneum in amphibians is a single layer of dead keratinized cells. The keratinized layer is thin and does not prevent the skin from being permeable. These developments were especially vital in preventing dessiccation in derived groups that used cutaneous gas exchange to supplement oxygen obtained through their lungs. In forms that lost their lungs completely and now rely solely on cutaneous gas exchange (family Plethodontidae), these changes became absolutely critical.

Most fishes deposit eggs and sperm in water, and fertilization is external. One problem that most amphibians did not solve was the ability to reproduce away from water. Desiccation risk to eggs greatly limits the distribution of amphibians and the habitats that can be exploited. Fertilization of eggs is external in some salamanders and most anurans. In most salamanders, however, fertilization occurs internally but without copulation. In these forms, males deposit spermatophores (see Fig. 6.33) whose caps are full of sperm. The caps are removed by the female's cloaca (the posterior chamber of the digestive tract, which receives feces and urogenital products), and sperm are stored in a chamber of the cloaca known as the spermatheca. As eggs pass through the cloaca, they are fertilized and must be deposited in a moist site. Many amphibians undergo larval development within the egg, called **direct development**, and hatch as immature versions of the adult form. Others hatch into aquatic larvae and undergo metamorphosis into terrestrial adults. Some, however, remain completely aquatic as adults. A few species are viviparous, a method of reproduction in which fertilized eggs develop within the mother's body and hatch within the parent or immediately after laying.

Lepospondyli

Lepospondyls were small, salamander-like amphibians that appear in the fossil record during the Carboniferous and Permian periods. They are distinguished from the labyrinthodonts primarily on the basis of their vertebral construction. The vertebral centra were formed by the direct deposition of bone around the notochord; their formation was not preceded by cartilaginous elements as in the temnospondyls and anthracosaurs. Little is known regarding their relationships to each other or to other groups of amphibians.

Lissamphibia

Lissamphibia include the salamanders, frogs, toads, and caecilians. Fossil salamanders are represented reasonably well in the fossil record beginning in the Upper Jurassic of North America and Eurasia (approximately 145 million years ago) (Estes, 1981). Blair (1976) noted that all fossil salamanders were from land masses of the Northern Hemisphere. Currently, the oldest known fossils of the most successful family in North America, the Plethodontidae, date back only to the Lower Miocene of North America (Duellman and Trueb, 1986).

Salamander-like fossil amphibians, the albanerpetontids, are known from the mid-Jurassic to mid-Tertiary (Miocene epoch) across North America, Europe, and Central Asia (McGowan and Evans, 1995). Some investigators place this group within the salamanders, whereas others consider them to be a separate amphibian group. Although they resemble salamanders by having an unspecialized tailed body form, cladistic analysis using a data matrix of 30 skeletal characters suggests that they represent a distinct lissamphibian lineage (McGowan and Evans, 1995).

Caecilians were unknown as fossils until Estes and Wake (1972) described a single vertebra from Brazil. It was recovered from Paleocene deposits approximately 55 million years old. Since then, additional fossils have been recovered from Jurassic deposits, pushing the age of caecilians back to approximately 195 million years ago (Benton, 1990; Monastersky, 1990c). Jurassic specimens apparently had well-developed eyes, sensory tentacles, small functional limbs, and were about 4 cm long. Because of the diminished role of the limbs for terrestrial locomotion, most researchers presume that these ancient caecilians also burrowed underground.

The nature and origin of caecilians continues to be open to debate. We still do not know whether caecilians evolved from a group of early lepospondyl amphibians known as microsaurs and developed separately from salamanders and anurans, or whether the three groups of amphibians are more closely related (Feduccia and McCrady, 1991).

The oldest known froglike vertebrate was taken from a Triassic deposit (200 million years ago) in Madagascar (Estes and Reig, 1973). Its relationship to modern frogs is still unclear; therefore, it is placed in a separate order, the Proanura. The 190-million-year-old *Prosalirus bitis*, the oldest true frog yet discovered, comes from the Jurassic period in Arizona (Shubin and Jenkins, 1995). The fossil includes hind legs, which were long enough to give it a powerful forward spring, and a well-preserved pelvis.

In the end, the primitive paired fins of an ancestral fish, used originally for steering and maneuverability, evolved into appendages able to support the weight of an animal and provide locomotion on land. Additional limb modifications have evolved in the turtles, diapsids, and mammals.

MORPHOLOGY

Integumentary System

An amphibian's skin is permeable to water and gases and also provides protection against injury and abrasion. Many species of salamanders and anurans absorb moisture from the soil or other substrates via their skins (Packer, 1963; Dole, 1967; Ruibal et al., 1969; Spotila, 1972; Marshall and Hughes, 1980; Shoemaker et al., 1992). Water uptake in anurans occurs primarily through the pelvic region of the ventral skin, a region that is heavily vascularized and typically thinner than the dorsal skin. Called the "seat patch" or "pelvic patch," it accounts for only 10 percent of the surface area but 70 percent of the water uptake in dehydrated red-spotted toads (*Bufo punctatus*) (McClanahan and Baldwin, 1969). In dehydrated giant toads (*Bufo marinus*), the hydraulic conductance of pelvic skin is six times that of pectoral skin (Parsons and Mobin, 1989). In addition, some minerals, such as sodium, are absorbed from the aqueous environment through the skin. Rates of absorption depend on soil moisture and the animal's internal osmotic concentration. Thus, in addition to protection, amphibian skin is important in respiration, osmoregulation, and to some extent, thermoregulation.

The skin consists of an outer, thin epidermis and an inner, thicker dermis (Fig. 6.7a). The epidermis is composed of an outermost single layer of keratinized cells that form a distinct stratum corneum, a middle transitional layer (stratum spinosum and stratum granulosum), and an innermost germinative layer (stratum germinativum or stratum basale), which is the region that gives rise to all epidermal cells. Mucous and granular (poison) glands may also be present. Aquatic amphibians have many mucus-secreting glands and usually few keratinized cells in their epidermis. Terrestrial forms, however, have fewer mucus-secreting glands and a single layer of keratinized cells. The keratinized layer is thin and does not prevent the skin from being permeable. As in fishes, the epidermis of most amphibians lacks blood vessels and nerves.

Molting or shedding of outer keratinized epidermal tissue occurs in both aquatic and terrestrial salamanders and anurans. It involves the separation of the upper keratinized layer (stratum corneum) from the underlying transitional layer. Prior to shedding, mucus is secreted beneath the layer of stratum corneum about to be shed in order to serve as a lubricant. The separated stratum corneum is shed either in bits and pieces or in its entirety, and it is consumed by most species immediately after sloughing. The period between molts is known as the **intermolt**, and its duration is species-specific. Both the shedding of the stratum corneum and the intermolt frequency are under endocrine control, with molting being less frequent in adult amphibians than in juveniles (Jorgensen and Larsen, 1961). In the laboratory, molt frequency has been shown to increase with temperature (Stefano and Donoso, 1964). Photoperiod is less important (Taylor and Ewer, 1956), whereas the relationship of food intake to molting is variable and unclear.

Multicellular mucous and granular glands are numerous and well developed (Fig. 6.7b). These glands originate in the epidermis and are embedded in the dermis. Mucous glands, which continuously secrete mucopolysaccharides to keep the skin moist in air and allow it to continue serving as a respiratory surface, are especially advantageous to aquatic species that spend some time out of water. Excessive secretion of mucus when an animal is captured can serve as a protective mechanism by making the animal slimy, slippery, and difficult to restrain.

Granular glands produce noxious or even toxic secretions. Such secretions benefit their possessors by making them unpalatable to some predators. These glands often occur in masses and give a roughened texture to the skin. The warts and parotoid glands of toads (Fig. 6.7c) and the dorsolateral ridges of ranid frogs (Fig. 6.7d) are examples. Secretions of these integumentary glands consist of amines such as histamine and norepinephrine, peptides, and steroidal alkaloids. In some groups of frogs, such as the poison-dart frogs of Central and South America, phylogenetic relationships have been based on the biochemical differences of integumentary gland secretions.

Toxin-secreting granular glands are most abundant in anurans, but also occur in some caecilians and salamanders. Members of the family Salamandridae and the genera *Pseudotriton* and *Bolitoglossa* (Plethodontidae) are known to secrete toxins (Brodie et al., 1974; Brandon and Huheey, 1981). Toxins, which can be vasoconstrictors, hemolytic agents, hallucinogens, or neurotoxins, may cause muscle convulsions, hypothermia, or just local irritation in a potential predator. For example, *Salamandra* secretes a toxin that causes muscle convulsions, whereas the newts *Notophthalmus* and *Taricha* possess a neurotoxic tetrodotoxin. Sufficient toxin is present in one adult *Taricha granulosa* to kill approximately 25,000 white mice (Brodie et al., 1974). Skin secretions of *Bolitoglossa* cause snakes of the genus *Thamnophis* to pause during ingestion, paralyzes their mouth, and may render them incapable of moving or responding to external stimuli.

FIGURE 6.7

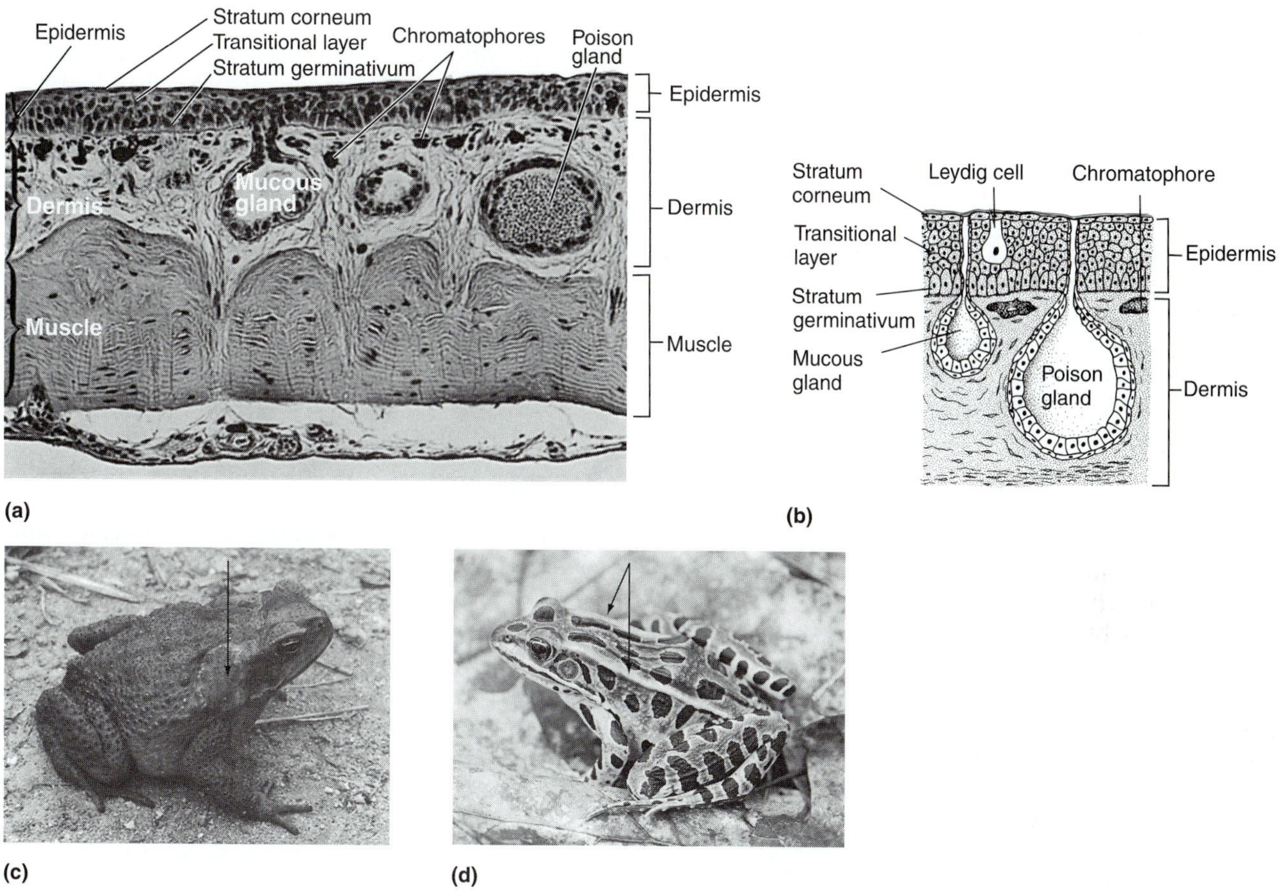

Amphibian skin. (*a*) Section through the skin of an adult frog. The epidermis consists of a basal stratum germinativum (stratum basale), a transitional layer consisting of a stratum spinosum and a stratum granulosum, and a thin, superficial stratum corneum. (*b*) Diagrammatic view of amphibian skin showing the mucous and poison glands that empty their secretions through short ducts onto the surface of the epidermis. (*c*) Warts and parotoid glands (arrow) of the giant toad (*Bufo marinus*). (*d*) Dorsolateral ridges (arrows) of the leopard frog (*Rana pipiens*).

Snakes often die after attempting to eat *Bolitoglossa rostrata* (Brodie et al., 1991).

Bacteria-killing antibiotic peptides—small strings of amino acids, which are the building blocks of all proteins—were originally discovered in the skin of African clawed frogs (*Xenopus laevis*) (Glausiusz, 1998). The peptide was named *magainin* by its discoverer, Michael Zasloff. Magainin filters urea from the blood plasma at the glomerulus; it is discharged onto the frog's skin in response to adrenaline, which is released when pain receptors in the skin send the brain a message that an injury has occurred. Magainins have now been found in many species, ranging from plants and insects to fish, birds, and humans. These peptides are being turned into antibiotic drugs in hopes of providing an alternative to currently available antibiotics. They can kill a wide range of microorganisms, including Gram-positive and Gram-negative bacteria, fungi, parasites, and enveloped viruses, without harming mammalian cells. In addition, some can selectively destroy tumor cells. Their mechanism of action is completely different from that of most conventional antibiotics. Instead of disabling a vital bacterial enzyme, as penicillin does, antimicrobial peptides appear to selectively disrupt bacterial membranes by punching holes in them, making them porous and leaky. Efforts are currently under way to chemically synthesize the peptides and make them available for clinical trials.

Although a wide variety of toxic secretions have been identified in many species of anurans, several genera of tropical frogs—*Dendrobates*, *Phyllobates*, and *Epipedobates*—possess extremely toxic steroidal alkaloids in their skin, apparently as a chemical defense against predation (Daly et al., 1978; Myers and Daly, 1983). Some 300 alkaloid compounds affecting the nervous and muscular systems have been identified. The alkaloids, which render neurons incapable of transmitting nerve impulses and induce muscle cells to remain in a contracted state, may cause cardiac failure and death. Other alkaloids block acetylcholine receptors in muscles, block potassium channels in cell membranes, or affect calcium transport in the body. Although these frogs rarely exceed 5 cm in length, the combination of toxic alkaloids in the body of a single frog is sufficient to kill several humans (Kluger, 1991). Members of the same species, however, are immune to each other's toxins.

BIO-NOTE 6.1

Drugs from Tropical Frogs

Alkaloid substances from tropical frogs may be a source of new drugs for humans. In 1992, J. W. Daly and the U.S. National Institutes of Health patented an opioid compound from a poison-arrow frog (*Epipedobates tricolor*). The compound, epibatidine, acts as a painkiller that is 200 times more powerful than morphine. Development of epibatidine as an analgesic agent has been precluded, however, because its use is accompanied by adverse effects such as hypertension, neuromuscular paralysis, and seizures. By using nuclear magnetic resonance spectroscopy to determine epibatidine's structure, researchers have been able to produce a potential new painkiller, ABT-594, that lacks some of the opioid's drawbacks. It apparently acts not through opioid receptors but through a receptor for the neurotransmitter acetylcholine, blocking both acute and chronic pain in rats. Safety trials to determine whether the drug is safe and effective in humans have already begun.

Research in natural products chemistry involving dendrobatid frogs has become more difficult because these frogs, native to several South American countries, have become rare and have been accorded protection as threatened species under the Convention on International Trade in Endangered Species of Flora and Fauna.

Bradley, 1993
Myers and Daly, 1993
Bannon et al., 1998
Strauss, 1998

Several western Colombian Indian tribes utilize the deadly toxic secretions of three species of *Phyllobates* for lacing blowgun darts with poison (Myers and Daly, 1993). Frogs are impaled on sticks and held over open fires. The heat causes the glands to secrete their toxin, which is collected and allowed to ferment. Darts are dipped into the solution and allowed to dry. The small amount of poison on the tip of a dart is sufficient to instantly paralyze small birds and mammals that are sought for food.

In the wild, about half of the 135 species in the family Dendrobatidae produce poisons. These alkaloids persist for years in frogs kept in captivity but are not present in captive-raised frogs. The alkaloids vanish in the first generation raised outside their natural habitat. Studies at the National Institutes of Health, the National Aquarium, and elsewhere are attempting to find the cause of this intriguing situation. One hypothesis is that the wild diet may include some "cofactor," an organism such as an ant or another substance that is not an alkaloid itself but is needed to produce the frogs' alkaloids (Daly et al., 1992; 1994a, b). For example, offspring of wild-caught parents of *Dendrobates auratus* from Hawaii, Panama, or Costa Rica raised in indoor terrariums on a diet of crickets and fruit flies do not contain detectable amounts of skin alkaloids. Offspring raised in large outside terrariums and fed mainly wild-caught termites and fruit flies do contain the same alkaloids as their wild-caught parents, but at reduced levels. Another hypothesis suggests the frogs need some kind of unknown environmental factor to trigger the production of the toxins, such as a combination of sunlight and variable temperatures, or the stress of hunting for food.

Most species that possess noxious or toxic secretions are predominately or uniformly red, orange, or yellow. Such bright **aposematic** (warning) **coloration** is thought to provide visual warning to a predator. Supposedly, predators learn to associate the foul taste with the warning color and thereafter avoid the distasteful species. In some species, these colors are present along with a contrasting background color such as black.

Because their skin has little resistance to evaporation, amphibians experience high rates of water loss when exposed to dessiccating conditions. Heat is lost as water evaporates, resulting in decreased skin temperatures (Wygoda and Williams, 1991). Most amphibians are unable to control the physiological processes that result in heat gain and/or loss; thus, thermoregulation is accomplished through changes in their position or location. Some arboreal anurans, such as the green tree frog (*Hyla cinerea*), have been shown to have reduced rates of evaporative water loss through the skin, and their body temperatures may be as much as 9°C higher than typical terrestrial species (Wygoda and Williams, 1991). The adaptive significance of lower rates of evaporative water loss may be to allow these frogs to remain away from water for longer periods, thus making them less susceptible to predators.

The skin of many amphibians is modified and serves a variety of functions. These modifications include the highly vascularized skin folds of some aquatic amphibians, the annuli or dermal folds of caecilians, and the costal grooves in many salamanders, all of which serve to increase the surface area available for gas exchange. The male hairy frog (*Astylosternus robustus*) of Africa possesses glandular filaments resembling hairs on its sides and hind legs (Fig. 6.8). These cutaneous vascular papillae develop only during the breeding season and are thought to be accessory respiratory structures that are used when increased activity triggers an increased demand for oxygen. Other integumentary structures, such as superciliary processes, cranial crests, and flaps on the heels of some frogs (calcars), are thought to aid in concealment. Metatarsal tubercles that occur on some fossorial forms aid in digging, and toe pads assist in locomotion. Brood pouches occur in South American hylid "marsupial" frogs (*Gastrotheca*) and in the Australian myobatrachine (*Assa*).

During the breeding season, some male salamanders (ambystomatids, plethodontids, and some salamandrids) develop glands on various parts of their bodies. Such glands may be on the head, neck, chin (mental), or tail. During

FIGURE 6.8

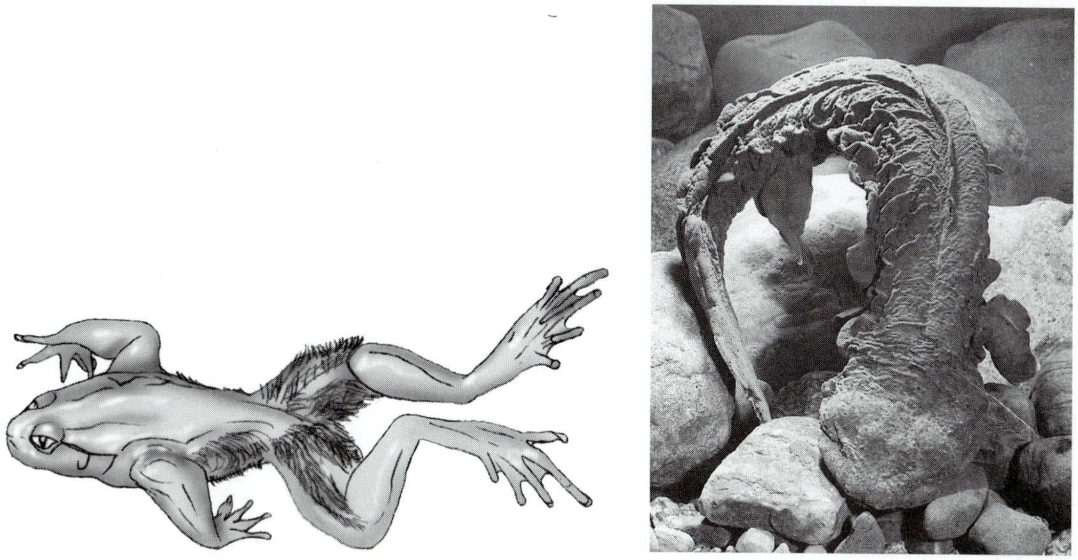

(a) (b)

(a) The "hairy frog" (*Astylosternus robustus*) receives its name from the thick growth of vascular filaments resembling hair that develops in the male during the breeding season. These are respiratory organs that compensate for the reduced lungs of this species at the time of the year when the metabolism increases. (b) A hellbender (*Cryptobranchus alleganiensis*), an aquatic salamander, with highly vascularized skin folds.

courtship, these glands come in contact with the female's body. Their secretions, known as pheromones, presumably aid in stimulating the female. The biochemical identification of one such pheromone from the mental gland of a salamander (*Plethodon jordani*) was reported by Rollman et al. (1999). Similar glands are present on various parts of the bodies of male anurans. In addition, the thumb pads of many breeding male anurans consist of clusters of keratinized mucous glands that help them clasp females.

Webbing between the fingers and toes of anurans is part of the integument. It is most extensively developed on the rear feet of the more aquatic species and provides a broader surface to the foot when swimming. In some species, such as the Malaysian flying frog (*Rhacophorus reinwardtii*), both hands and feet are fully webbed and are used in a parachute fashion for controlled jumping from a higher perch to a lower one. The tips of the digits of some salamanders and anurans are modified with thickened, keratinized epidermis.

Many tree frogs possess expanded adhesive **toe pads** with glandular disks at the tips of their toes, which aid in grasping and climbing (Fig. 6.9). Toe pads consist of columnar epithelium whose cells feature stout, hexagonal, flat-topped apices that are separated from each other by deep crypts (Fig. 6.10) (Ernst, 1973; Green, 1979). Studies by Emerson and Diehl (1980), Green (1981), and Green and Carson (1988) show that surface tension created by mucus secretions is the primary factor in allowing anurans with toe pads to cling to smooth surfaces. The strength of the adhesive bond, produced by the surface tension of the fluid that lies between the toe pad and the substrate, is a function of the area of contact with the substrate. An **intercalary bone** allows the adhesive toe disk to be offset from the end of the digit so that the entire surface of the toe pad can be in contact with the substrate (Fig. 6.11b). Arboreal salamanders lack toe pads, but may have recurved, spatulate terminal phalanges to assist in grasping (Fig. 6.11c).

The dermis of amphibians contains a rich network of capillaries that supply nutrients to the epidermis. Dermal scales,

FIGURE 6.9

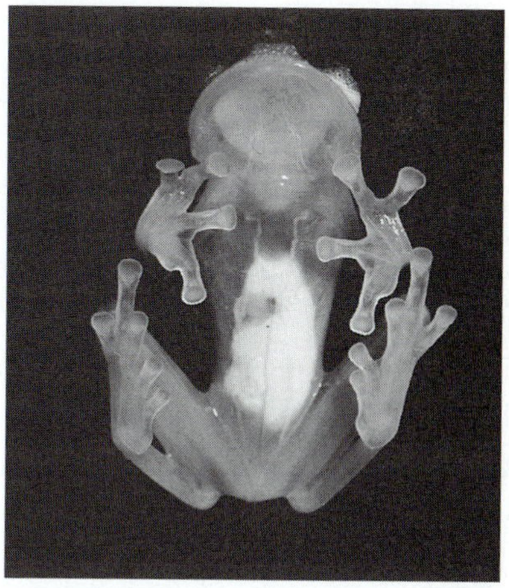

Glass frog (*Centrolenella*) with the heart visible through the skin. Adhesive toe pads aid in climbing.

FIGURE 6.10

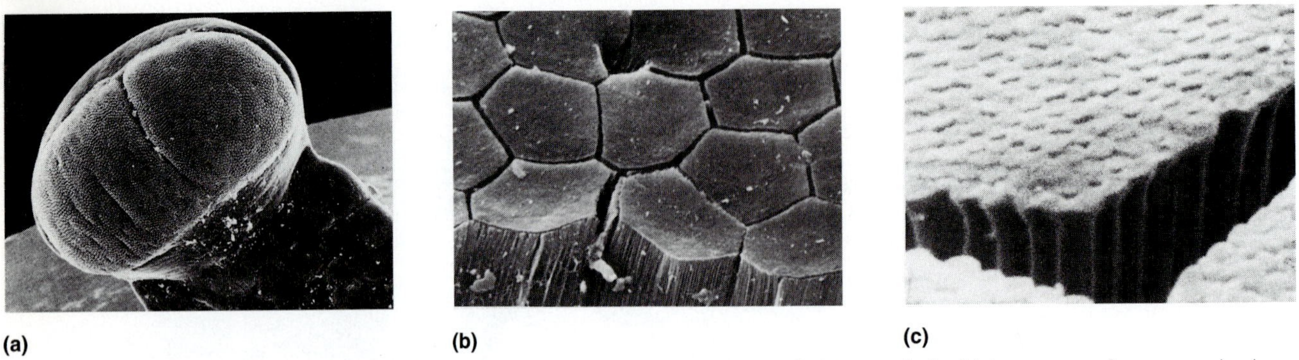

Scanning electron micrographs of the toe pad of a frog: (*a*) ventral view of the entire toe pad of *Litoria rubella*; (*b*) the opening of a mucous gland on the epidermal surface of the toe pad in *Eleutherodactylus coqui*; (*c*) fibrous epithelium of individual toe pad cells in *Hyla picta*.

or ossicles, are present in several kinds of anurans (*Brachycephalus*, *Ceratophrys*, *Gastrotheca*, *Phyllomedusa*, and others) and in caecilians. The ability to change skin color is advantageous to amphibians, both in providing protective coloration and in temperature regulation. Three types of chromatophores—melanophores, iridophores, and xanthophores (erythrophores)—are present in the epidermis and/or in the dermis. Color change may be effected by the amoeboid movement of the chromatophores or by a shifting of pigment granules within the cell. Color change in adult amphibians appears to be controlled primarily by melanocyte-stimulating hormone (MSH) secreted by the anterior lobe (adenohypophysis) of the pituitary gland (Duellman and Trueb, 1986). Coloration may be the result of the dispersion or concentration of pigments, or a combination of pigments and dermal structures. For example, lightening of the integument is due to secretion of melatonin, a hormone found in the pineal gland, brain, and retina that aggregates melanin granules in dermal melanophores, thus causing the skin to appear lighter in color (Baker et al., 1965; Pang et al., 1985). Melatonin also appears to be responsible for color change in amphibian larvae (Bagnara, 1960).

BIO-NOTE 6.2

Why Frogs Are Green

Why do many frogs appear green? Because the epithelium is transparent, a portion of skin appears green from the outside when light of long wavelength passes through the iridophores and is then absorbed by melanophores, whereas light of short wavelength is diffracted and refracted back by the iridophores. Only the green component of this refracted light escapes absorption in the yellow color screen of the lipophores. Other colors such as blue, yellow, and black are seen either where the pigment layers are not continuous, or where they are irregularly arranged.

Lindemann and Voute, 1976

FIGURE 6.11

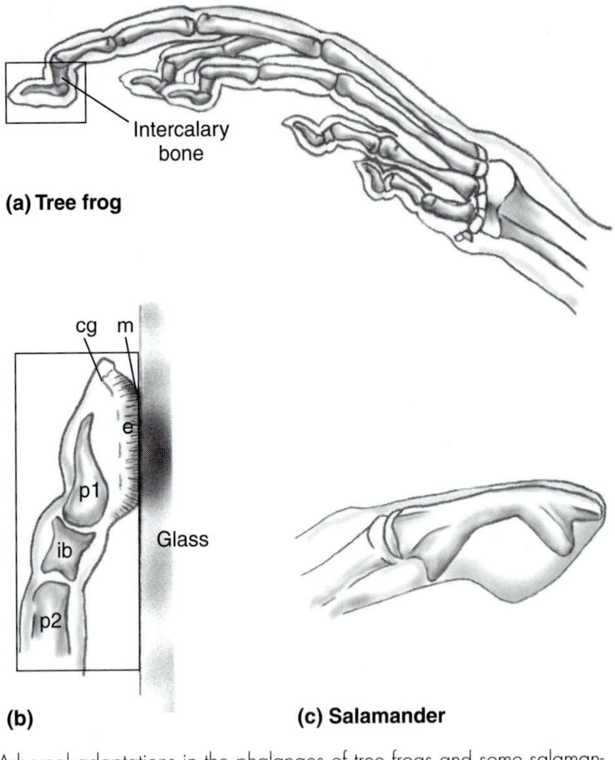

Arboreal adaptations in the phalanges of tree frogs and some salamanders. (*a*) Tree frogs have terminal phalanges that rotate on the intercalary bones. (*b*) A diagrammatic cross section of a tree frog's toe pad in contact with a smooth glass surface illustrating the mechanism of adhesion by surface tension. *Key:* e, adhesive epidermis; cg, circumferal groove of the toe pad; m, meniscus; p1, first phalange; ib, intercalary bone or cartilage; p2, second phalange. (*c*) Arboreal salamanders such as *Aneides lugubris* may have recurved, spatulated terminal phalanges.

Skeletal System

Compared with that of fishes, the amphibian skeleton exhibits increased ossification, loss and fusion of elements, and extensive modification of the appendicular skeleton for terrestrial locomotion (Fig. 6.12).

FIGURE 6.12

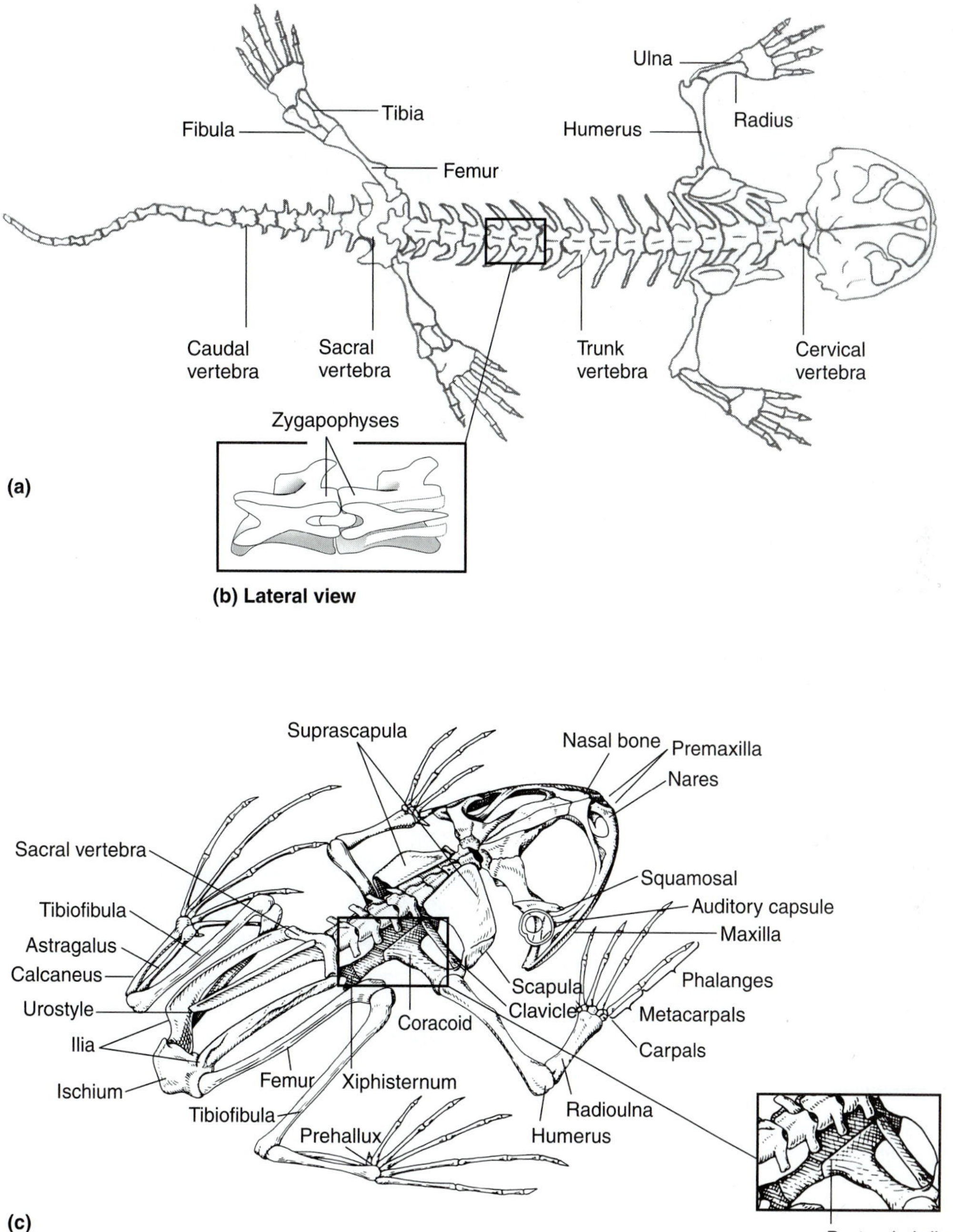

(*a*) Dorsal view of a salamander skeleton. (*b*) Lateral view of salamander trunk vertebrae. (*c*) Skeleton of a bullfrog (*Rana catesbeiana*).

Terrestrial salamanders have a somewhat arched and narrow skull, whereas in aquatic forms the skull is flatter. Salamander skulls, which may be partly or wholly ossified, contain fewer bones than skulls of teleost fishes. Through loss and fusion, skulls of caecilians and anurans contain even fewer bones than those of salamanders (Fig. 6.13a, b). The broad, flat head of anurans is almost as wide as the body.

The upper jaw of anurans is composed of a pair of premaxillae and a pair of maxillae. Meckel's cartilage in the lower jaw is ensheathed primarily by the dentary and angular bones, with the latter articulating with the quadrate of the skull.

The posterior ends of the embryonic palatoquadrate cartilages serve as the posterior tips of the upper jaws. They may remain as quadrate cartilages, or they may ossify to

FIGURE 6.13

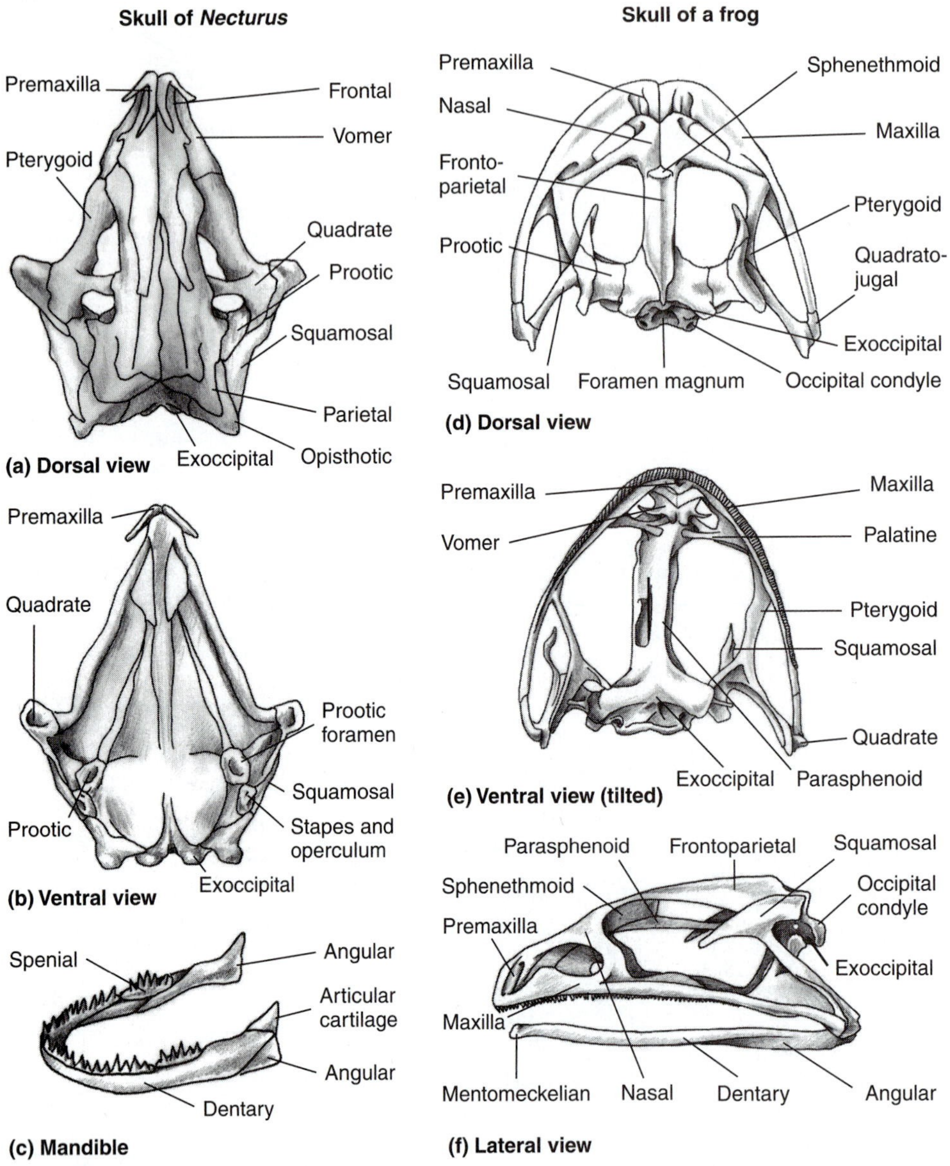

Left—Skull of *Necturus*: (*a*) dorsal view; (*b*) ventral view; (*c*) mandible. Right—Skull of a frog: (*d*) dorsal view; (*e*) ventral view, tilted laterally to the left side; (*f*) lateral view.
(a–c) *Source: Warren F. Walker, Jr.,* Vertebrate Dissection, *5th edition, 1975, Saunders College Publishing;*
(d–f) *Source: Wingert,* Frog Dissection Manual, *Johns Hopkins University Press.*

become quadrate bones. The more anterior part of the palatoquadrate cartilages become ensheathed by dermal bones such as the premaxilla and maxilla. The upper jaw is connected directly to the skull in amphibians, a method of jaw suspension known as **autostylic**. The dentary forms the major portion of the mandible (lower jaw).

The hyomandibular cartilage, which in sharks is located between the quadrate region of the upper jaw and the otic capsule, ossifies in tetrapods and becomes the **columella** of the middle ear (see Fig. 6.23). It transmits sound waves from the quadrate bone to the inner ear. The columella serves as an evolutionary stage in conducting airborne sounds in terrestrial vertebrates, a process culminating in the presence of three ear ossicles in mammals.

Larval gill-bearing amphibians have visceral arches that support gills. During metamorphosis, changes occur that result in a pharyngeal skeleton (that initially was adapted for branchial respiration) being converted in the span of a few days to one characteristic of animals that live on land and breathe air. Those amphibians (salamanders) that remain aquatic as adults retain an essentially fishlike branchial skeleton throughout life, except that the number of gill-bearing arches is fewer than in fishes.

As vertebrates became increasingly specialized for life on land, the ancestral branchial skeleton underwent substantial adaptive modifications. Some previously functional parts were deleted, and those that persisted perform new and sometimes surprising functions. For example, the hyobranchial apparatus supports gills in larval salamanders and the complex, projectile tongue in metamorphosed adults. In anurans, however, vocalization is possible because of modifications of the hyobranchial apparatus to form laryngeal cartilages.

The vertebral column in amphibians varies considerably in length. Some salamanders have as many as 100 vertebrae, and caecilians may have up to 285 (Wake, 1980a). Anurans usually have 8 (excluding the urostyle), though the number may range from 6 to 10.

With the evolution of tetrapods and life on land, the vertebral column has become more specialized. It serves to support the head and viscera and acts as a brace for the suspension of the appendicular skeleton. Four (sometimes five) types of vertebrae are present in most salamanders (Fig. 6.12a), whereas the anuran vertebral column normally is divided into four regions (Fig. 6.12c). In salamanders, the first **trunk** vertebra became a **cervical** vertebra, which now provides for an increasingly flexible neck. This single cervical vertebra, the **atlas**, has two concave facets for articulation with the two occipital condyles of the skull. Trunk vertebrae vary in number from approximately 10 to 60 depending on the species.

Because of the force generated against the vertebral column by the tetrapod hindlimbs and pelvic girdle, the terminal trunk vertebra has become enlarged and modified as a **sacral** vertebra. Salamanders have a sacrum consisting of one sacral vertebra, which serves to brace the pelvic girdle and hindlimbs against the vertebral column. This arrangement does not provide very strong support for the hindlimbs; therefore, most salamanders have difficulty completely raising their bodies off the ground when walking. Their sprawl-legged stance and sinusoidal method of locomotion also contribute to their inability to keep their bellies off the substrate. Most salamanders "wriggle." A **caudal–sacral** region consisting of 2 to 4 vertebrae immediately posterior to the sacrum is recognized by some authors. The **caudal**, or tail, vertebrae may range up to 20 or more in salamanders. Some salamanders have weak articulations between their caudal vertebrae that allow them to shed their tails (caudal autotomy) when attacked by predators (Wake and Dresner, 1967).

Caecilians have one cervical vertebra (atlas) and a variable number of trunk vertebrae. They lack a sacrum, and most species lack a tail. With the exception of the atlas, all vertebrae of caecilians are nearly identical in shape.

The anuran vertebral column consists of cervical, trunk, sacral, and postsacral regions (Fig. 6.12c). The presacral region consists of 5 to 8 vertebrae, with the first being modified as a cervical vertebra, the atlas. A single vertebra, the sacrum, is modified for articulation with the pelvic girdle. Postsacral vertebrae are fused into a **urostyle**, an unsegmented part of the vertebral column that is homologous to the separate postsacral vertebrae of early amphibians.

Amphicoelous vertebrae in which both anterior and posterior faces of the centra are concave are found in caecilians, a few primitive anurans, and some salamanders. Most salamanders and a few anurans possess **opisthocoelous** vertebrae, in which the centrum is concave on its posterior face and convex on its anterior face. Most anurans possess **procoelous** vertebrae, in which the concave surface faces anteriorly and the posterior face is convex. Intervertebral joints of amphibians are reinforced by two pairs of processes (zygapophyses) arising from the neural arch (Fig. 6.12b).

The earliest amphibians had well-developed ribs on both trunk and tail vertebrae (Fig. 6.4b). In modern amphibians, however, ribs are always absent on the atlas and are either reduced or absent on the other vertebrae. When present, they are usually shortened structures that are fused with transverse processes. They are longest in caecilians, shorter in salamanders, and vestigial or absent in most anurans.

A true **sternum**, characteristic of higher tetrapods, appears for the first time in amphibians. It is absent in caecilians and in some salamanders. In other salamanders, it is poorly developed and exists as a simple, medial triangular plate that articulates with the pectoral girdle. It is poorly developed in primitive frogs, but in more advanced frogs, it may exist as a rod-shaped structure consisting of four elements or as an ossified plate. Although ribs do not attach to it, the amphibian sternum functions as a site for muscle attachment.

The evolutionary origin of the sternum is unclear. One hypothesis is that it results from the fusion of the ventral ends of the thoracic ribs. A second hypothesis proposes that the sternum developed independently of the ribs, a view that is supported by the embryonic origin of the sternum in reptiles and mammals. Feduccia and McCrady (1991) noted that it "may even be possible that amphibian and amniote sterna have evolved independently and are not homologous structures."

Early amphibians, which were not truly terrestrial and spent much of their time in water, possessed two pairs of limbs. The pectoral girdle of early tetrapods closely resembled the basic pattern of their crossopterygian ancestors; it did not articulate with the vertebral column, and the coracoid braced the girdle against the newly acquired sternum (Fig. 6.4).

In modern salamanders, the pectoral girdle is mostly cartilaginous, with one-half of the girdle overlapping the other and moving independently. A small ventral, cartilaginous sternum lies posterior to the pectoral girdle in some salamanders.

In most anurans, the scapula and other elements may be ossified or cartilaginous; the girdle is suspended from both the skull and the vertebral column and is designed to absorb the shock of landing on the forelimbs.

The structure of the pectoral girdle of anurans has been used as an important taxonomic tool. Those families in which the two halves of the pectoral girdle overlap and that possess posteriorly directed epicoracoid horns (Bufonidae, Discoglossidae, Hylidae, Pelobatidae, Pipidae, and Leptodactylidae) have an **arciferous** type pectoral girdle (Fig. 6.14a). Here, the epicoracoids articulate with the sternum by means of grooves, pouches, or fossae in the dorsal surface of the sternum. Those families in which the sternum is fused

FIGURE 6.14

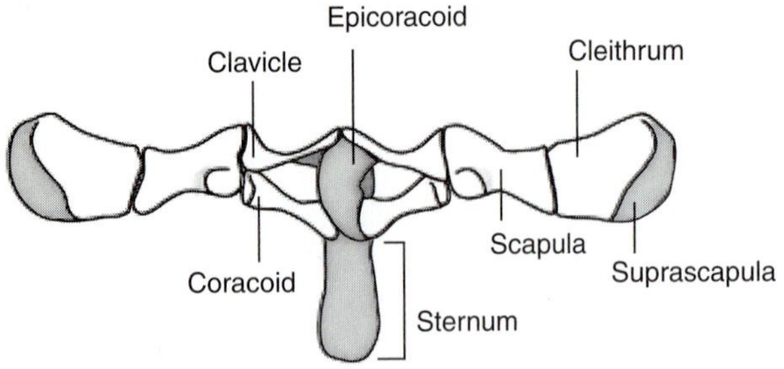

(a) Arciferous girdle

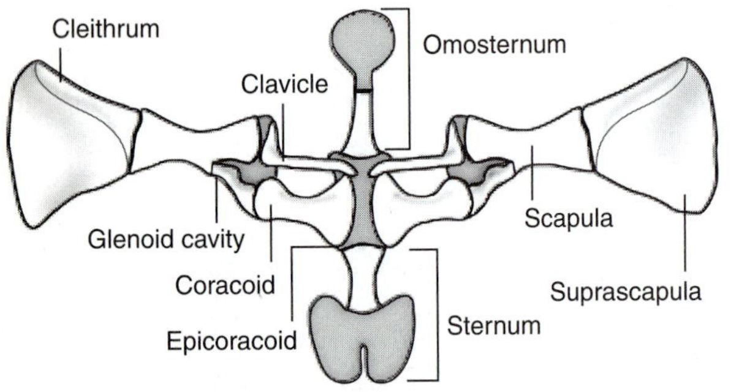

(b) Firmisternal girdle

Anuran pectoral girdles in ventral view. Stippled areas are cartilaginous. (*a*) Arciferous girdle with overlapping halves (*Bufo coccifer*). (*b*) Firmisternal girdle with two halves of girdle fusing in midline (*Rana esculenta*).

to the pectoral arch and the epicoracoid cartilages of each half of the pectoral girdle are fused to one another (Ranidae, Rhacophoridae, and Microhylidae) have a **firmisternal** type of girdle (Fig. 6.14b).

Considerable diversity in limbs exists among modern amphibians as a result of their locomotion (hopping) and their various adaptations to aquatic, burrowing, and arboreal habits. Limbs of modern salamanders are short, stout, and directed outward at right angles to the body. Anterior limbs consist of a single upper bone, the **humerus**, two lower forearm bones, the **radius** and **ulna**, as well as **carpals**, **metacarpals**, and **phalanges** (Fig. 6.12a). The primary function of the forelimbs in salamanders is to raise the body and assist the hind limbs in moving the body forward. In anurans, the forelimb is considerably shorter than the hindlimb (Fig. 6.12c). Instead of having two foreleg bones (radius, ulna), the ossification of the ligament between the radius and ulna creates a single bone: the **radio-ulna**. Carpals, metacarpals, and phalanges complete the skeleton of the forelimb.

Modifications to the front limb in amphibians involve a reduction of bones by loss or fusion. Most modern amphibians have reduced or lost at least one digit and one metacarpal, so that four functional digits are present on each front foot. Others, such as members of the genus *Amphiuma*, have girdles, but both forelimbs and hindlimbs are vestigial. Both girdles and limbs are absent in caecilians.

The pelvic girdle of salamanders may be partially ossified and consists of a ventral **puboischiac plate** and a dorsal pair of **ilia** on each side. A median Y-shaped **ypsiloid (prepubic) cartilage** develops just anterior to the pubic area in most salamanders. The ypsiloid cartilage is associated with the hydrostatic function of the lungs. By elevating the cartilage, the salamander is thought to be able to compress the posterior end of its body cavity and force air in its lungs forward, thereby causing its head to rise in the water. When the ypsiloid cartilage is depressed, air is thought to move posteriorly in the lungs, thereby reducing the buoyancy of the head so that it tends to sink in the water (Duellman and Trueb, 1986).

In anurans, each half of the pelvic girdle consists of an ilium, ischium, and pubis (Fig. 6.12c). Ilia are greatly elongated and articulate with the sacrum. They extend to the end of the urostyle, where they meet the ischia and pubis. Ilia are thus adapted to absorb the shock of impact when frogs land after a jump.

Hind limbs in salamanders consist of a single upper bone, the **femur**, two lower leg bones, the **tibia** and **fibula**, as well as **tarsals**, **metatarsals**, and **phalanges** (Fig. 6.12a). Sirens (family Sirenidae) have a pectoral girdle and small forelimbs, but lack pelvic girdles and hindlimbs.

The well-developed hindlimbs of anurans are specialized for jumping and swimming (Fig. 6.12c). The head of the upper leg bone (femur) articulates with the acetabulum (socket) of the pelvic girdle. Distally, the femur articulates with the **tibiofibula**, representing the fusion of the separate tibia and fibula and forming a stronger and more efficient structure for leaping. As in salamanders, the knee joint is directed anteriorly to provide better support and power for forward propulsion. A series of tarsal bones constitutes the ankle. Four or five metatarsals form the foot, and phalanges form the toes. A small additional bone, the **prehallux**, frequently occurs on the inner side of the foot. It commonly supports a sharp-edged tubercle used for digging by burrowing species like spadefoot toads (*Scaphiopus*). Most amphibians have five digits on each of the rear feet. The primary function of the hindlimbs is to provide the power for locomotion.

All anurans, whether primarily walkers, hoppers, or swimmers, use some form of jumping or leaping (saltatorial) locomotion. For this, forelimbs must be positioned differently than those of salamanders and fulfill a different role in locomotion. Duellman and Trueb (1986) describe the mechanism of a frog's leap in the following manner:

> *At rest, the shoulder joint tends to be extended with the upper arm lying against the flank rather than held out at a right angle to the body as in salamanders. The elbow joint is flexed and the forearm directed in an anteromedial direction rather than directly forward. Thus, the entire lower arm and hand are rotated inward toward the center of the body. As the animal thrusts*

itself forward in a leap, it probably rolls off the palmar surface of the hand while straightening the elbow and wrist joints. Thus, the forelimb lies parallel to the body for maximum streamlining. After full thrust has been developed from the hindlimbs, the forelimb is flexed at the elbow, and the upper arm is pulled as far forward as possible. Subsequent flexion of the wrist allows the animal to land on its hands, the force of landing presumably being absorbed by the pectoral girdle.

Muscular System

The body musculature of amphibians varies widely; that of aquatic salamanders is similar to the pattern in fishes, whereas the body musculature of terrestrial species, especially anurans, is markedly different. Metamerism is clearly evident in salamanders, caecilians, and in larval anurans. Epaxial myomeres have begun to form elongated bundles of muscle that extend through many body segments. These muscles, which are partially buried under the expanding appendicular muscles, extend along the vertebral column from the base of the skull to the tip of the tail. In salamanders, these muscles are known as the **dorsalis trunci** and allow for side-to-side movement of the vertebral column, the same locomotor pattern as in fishes.

Those amphibians that utilize lateral undulations of their hypaxial muscles for swimming, such as most larval forms and adult aquatic salamanders, retain a more fishlike, segmented hypaxial musculature. Even terrestrial salamanders utilize lateral undulations to a great extent. In other amphibians, hypaxial muscle masses begin to lose their segmental pattern and form sheets of muscle (external oblique, internal oblique, transversus), especially in the abdominal region.

As vertebrates evolved into more efficient land-dwelling forms, the axial musculature decreased in bulk as the locomotor function was taken over by the appendages and their musculature. The original segmentation becomes obscured as the musculature of the limbs and limb girdles spreads out over the axial muscles.

The appendicular muscles of most amphibians are far more complicated than those of fishes due to the greater leverage required on land. In amphibians, the limbs (for the first time in the evolution of the vertebrates) must support the entire weight of the body. Due to the difference in locomotion between salamanders and anurans, considerable variation exists in the musculature of the girdles and limbs between these two groups. Even so, many salamanders still drag their bellies over the substrate when they walk. Lateral undulatory movements of the body wall assist the appendicular muscles in this movement.

Hindlimb muscles of frogs that jump must generate maximum mechanical power during jumping. Maximum power is generated by the rapid release of calcium from sarcoplasmic reticula in muscle fibers, which initiates cross-bridge formation between actin and myosin filaments in the sarcomeres, and by having the maximum number of muscle fibers contracting (Lutz and Rome, 1994).

BIO-NOTE 6.3

Forward Motion in Caecilians

Caecilians are legless, wormlike, burrowing tropical amphibians. Unlike other vertebrates, caecilians have muscles that ring the body wall, running from the belly to the back (the muscles in most vertebrates tend to run lengthwise, from head to tail). By contracting these muscles, caecilians pressurize the fluid in their body cavity, creating a hydrostatic force that goes in the direction of the head, driving the animal forward and causing it to become longer and thinner. This remarkably efficient technique permits the caecilian to generate about twice the force of a similar-size burrowing snake, which uses the muscles that run along the vertebral column to twist and arch itself through the soil. By using its entire body as a single-chambered hydrostatic organ, a caecilian applies nearly 100 percent of its muscular energy toward forward motion.

O'Reilly et al., 1997

In amphibians, muscles of the first visceral arch continue to operate the jaws. Some of the muscles of the second arch retain their association with the lower jaw, whereas muscles of the third and successive arches operate gill cartilages in those amphibians with gills. In amphibians without gills, these muscles are reduced. They assume new functions such as assisting in swallowing and opening and closing of the pharynx and larynx.

Cardiovascular System

The evolution of lungs was a significant development in the evolution of vertebrates. Those mechanisms must have evolved to enable the best use of the oxygenated blood returning from the lungs via pulmonary veins. Development of an interatrial septum in the heart of most amphibians was essential in helping keep oxygenated blood separated from deoxygenated blood.

Instead of the simple two-chambered heart (atrium, ventricle) characteristic of most fishes, many amphibians have a heart with two atria and a single ventricle (Fig. 6.15). Although the interatrial septum is incomplete (fenestrated) in most salamanders and caecilians and is lacking completely in lungless salamanders, it is complete in anurans (Fig. 6.16). The right atrium receives deoxygenated blood from the sinus venosus; the left atrium receives the pulmonary veins (absent in lungless forms) and oxygenated blood. Some blood travels from the heart via pulmonary arteries to cutaneous arteries in the skin in order for cutaneous respiration to occur. Once aerated, the blood returns to the heart via cutaneous and pulmonary veins. **Ventricular trabeculae** (ridges in the ventricular wall) are common in many amphibians and help to keep oxygenated and deoxygenated blood separated in the ventricle. A few salamanders have partial interventricular septa, but no living amphibian is known to have a complete interventricular septum.

FIGURE 6.15

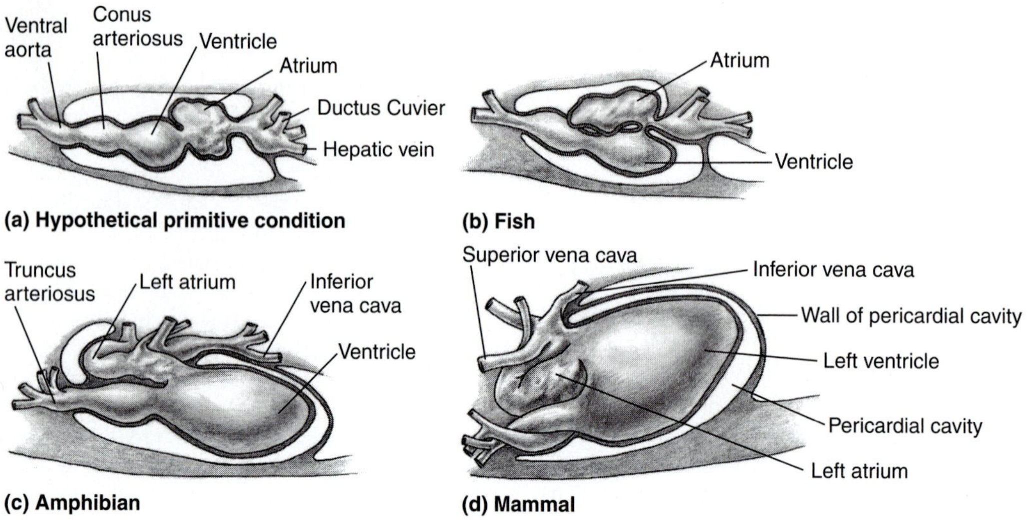

Stages in the evolution of the vertebrate heart: (*a*) hypothetical primitive condition; (*b*) fish; (*c*) amphibian; (*d*) mammal. The atrium, which was posterior to the ventricle, moves anteriorly. The original atrium and ventricle become partitioned into right and left chambers.

FIGURE 6.16

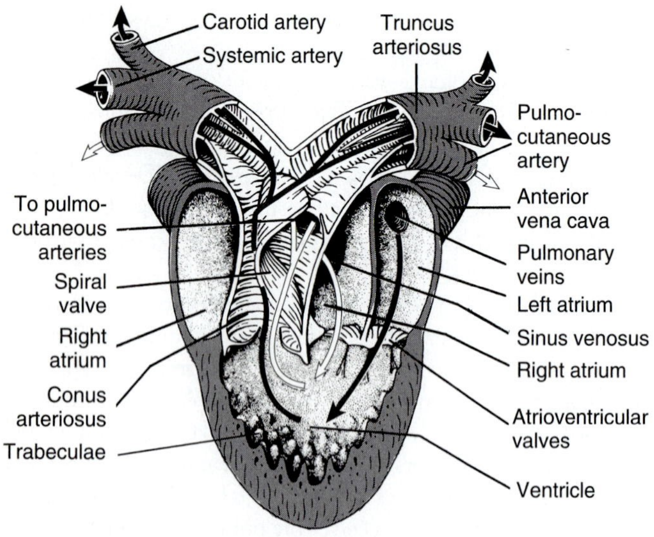

Structure of the frog heart. Oxygenated blood is indicated by dark arrows, deoxygenated blood by white arrows.

In most fishes, six aortic arches appear between the developing gill slits in embryos (Fig. 6.17). The most anterior aortic arch disappears during embryonic development, so that adult elasmobranchs are left with five arches. Adult teleosts have four aortic arches, the second usually disappearing as well during development. Lungfishes have the same four arches, and the lungs are supplied from the most posterior of these. This is equivalent to the sixth of the original embryonic series. The lungs of all land vertebrates are supplied with blood from this source, indicating common ancestry and homology.

During development, most larval salamanders and all tadpoles pass through a stage in which the arches form gill capillaries and also may supply the external gills. Later, the gill circulations are lost and the adult pattern develops. Aortic arches 3 (carotid), 4 (systemic), and 6 (pulmonary) always are retained, and arch 5 (systemic) is present in some salamanders. All anurans and some salamanders have a spiral valve in the conus arteriosus that shunts oxygenated blood to arches 3 and 4 (to the head and dorsal aorta) and deoxygenated blood to arch 6.

All amphibians utilize cutaneous gas exchange to some degree. The moist skin may play only a minor role in oxygen uptake in some species, whereas in others, such as plethodontid (lungless) salamanders, it plays a major role. Branches of the pulmonary artery transport blood to the skin, so that many amphibians lose most of their carbon dioxide through their skin. Blood returning from the skin through the cutaneous vein and into the right atrium is oxygenated just as that returning from the lungs into the left atrium is oxygenated. Depending on the extent to which cutaneous respiration is being utilized, keeping the two bloodstreams separate may or may not be an advantage.

The blood of many amphibians consists of plasma, erythrocytes, leucocytes, and thrombocytes. Frogs, however, lack thrombocytes. Normal erythrocytes are elliptical, nucleated

FIGURE 6.17

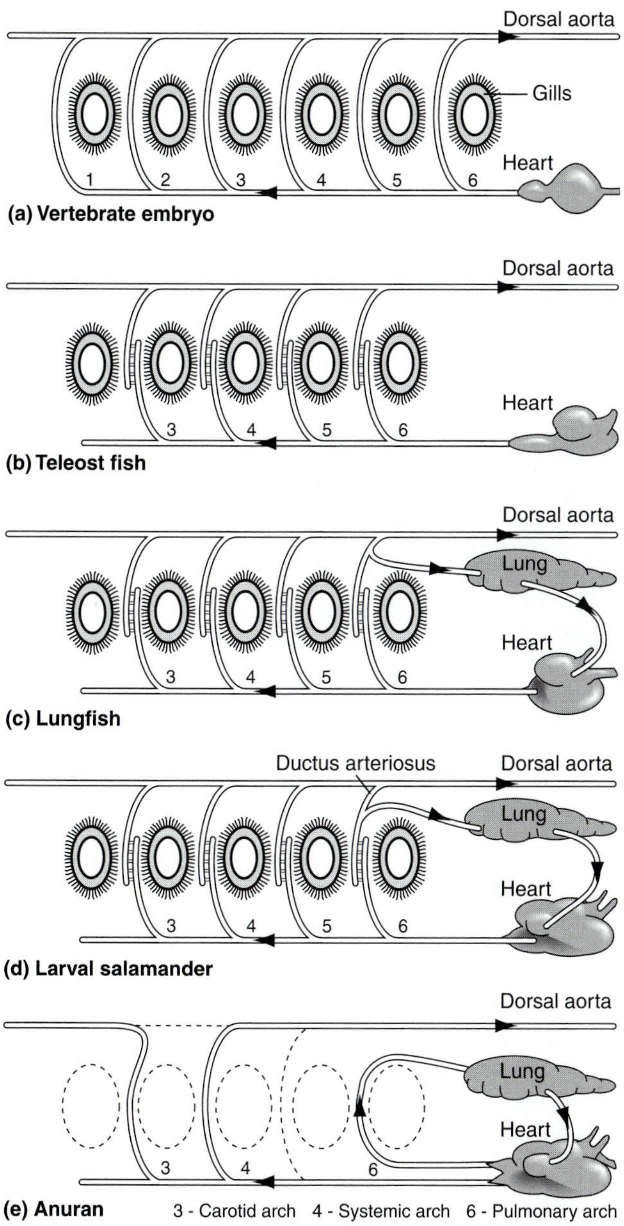

Arrangement of the aortic arches in (*a*) vertebrate embryo; (*b*) teleost fish; (*c*) lungfish; (*d*) larval salamander; and (*e*) anuran.

disks varying in size from less than 10 Mm in diameter in some species to over 70 Mm (in *Amphiuma*), the largest known erythrocyte of any vertebrate.

Hematopoiesis (production of all formed elements in the blood—i.e., all red and white blood cells) in salamanders takes place primarily in the spleen, whereas in anurans it occurs in the spleen and in the marrow of the long bones at metamorphosis and upon emerging from hibernation (Duellman and Trueb, 1986). Leucocytes may be formed in the liver, in the submucosa of the intestines, and in the bone marrow.

Respiratory System

Body size and temperature influence gas exchange in amphibians. In general, as mass increases, oxygen consumption and carbon dioxide production increase, although the consumption rate declines with increasing mass. Thus, respiratory surfaces may be unable to meet the metabolic needs without modification. Modifications include increasing the surface by additional folds of skin or partitioning of the lungs; increasing vascularization of the skin and/or having blood vessels closer to the surface; increasing the gas transport capacity of the blood and increasing flow rate; and/or similar respiration-enhancing devices.

External nares (nostrils) lead via nasal passages to internal nares (choanae) (Fig. 6.18a). Because amphibians lack a secondary palate, the internal nares usually open far forward in the roof of the mouth just inside the upper jaw. From the pharynx, air passes through the **glottis** into a short **trachea.**

Amphibians are the most primitive vertebrates to have the anterior end of the trachea modified to form a voice box or **larynx**. Voice is well developed in most male frogs and toads which have two muscular bands stretching across the laryngeal chamber; these form vocal cords that vibrate when air passes over them. Tightening or relaxing these vocal cords causes variations in pitch. Many male anurans have paired or median **vocal sacs**, or resonating chambers (Fig. 6.18). The size, shape, and position of vocal sacs is species-specific.

Calls have long been thought to radiate from the vocal sac. However, Alejandro Purgue of the University of California at Los Angeles discovered that the ears account for up to 90 percent of the sound output in the American bullfrog (*Rana catesbeiana*) (Purgue, 1997; Pennisi, 1997a). The ears act as loudspeakers amplifying the sound of the frog's vocal cords. The vocal sac serves primarily to store the air used by the vocal cords. Six additional, closely related frog species have loudspeaker ears, whereas western chorus frogs and California tree frogs use other body parts as resonators.

Although a larynx is present in the mudpuppy (*Necturus*) and a few other salamanders, most lack vocal cords and are

BIO-NOTE 6.4

Sounds Without Vocal Cords

The totally aquatic pipid anuran *Xenopus borealis* lacks vocal cords yet produces long series of clicklike sounds underwater at night. Although it retains an essentially terrestrial respiratory tract, the larynx is highly modified. Unlike all other anurans, sound production does not involve a moving air column. Rather, calcified rods with disklike enlargements in the larynx are held tightly together. When muscle tension is developed and exceeds the adhesive force, the disks rapidly separate, leaving a vacuum. A click is produced by air rushing at high speed into the space between the disks.

Yager, 1992a, b

FIGURE 6.18

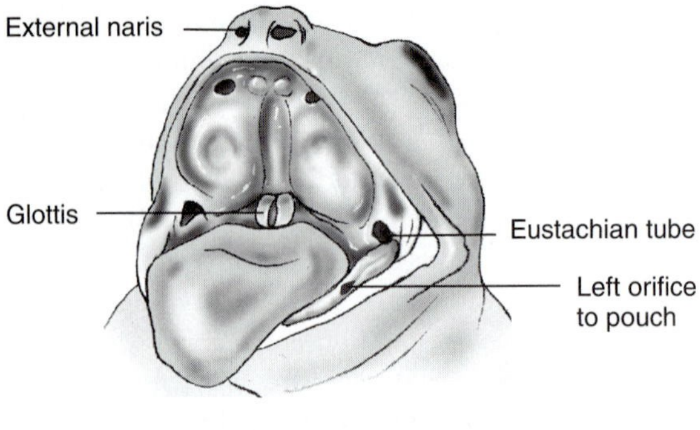

(a) The oral cavity of *Scaphiopus holbrookii*

(b)

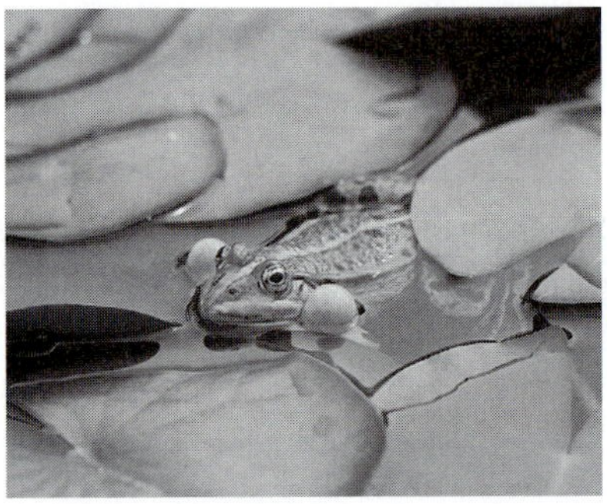

(c)

(*a*) Oral cavity of toad (*Scaphiopus holbrookii*) showing location of certain respiratory structures. (*b*) Distended median vocal sac of the spring peeper (*Pseudacris crucifer*). (*c*) Distended paired vocal sacs of the edible frog (*Rana esculenta*).

voiceless. Sounds reported from salamanders are probably produced by the inspiration and expiration of air. A few, such as the Pacific giant salamander (*Dicamptodon ensatus*), have a large larynx and bands, known as **plicae vocales**, that resemble anuran vocal cords. Air from the lungs passes over the plicae, causing them to vibrate. Lungless salamanders (plethodontids) lack both a trachea and a larynx.

A force-pump mechanism (Fig. 6.19) is used by amphibians to get air into their lungs. Air enters the oral cavity through the internal nares. When the nostrils close and the floor of the oral cavity is raised, air is forced through the glottis into the lungs and is retained by closure of the glottis sphincter. While air is in the lungs and the glottis is closed, "throat flutters" can provide additional aeration of oral surfaces (Fig. 6.19c). By taking repeated volumes of air into its lungs several times in succession without letting air out, a frog or toad can blow itself up to a considerable size as a defensive maneuver when confronted by a potential predator.

Amphibians utilize several different methods of gas exchange: cutaneous, buccopharyngeal, branchial, and pulmonary. Some salamanders and one caecilian (*Atretochoana*) (Anonymous, 1996a) are the only tetrapods in which the evolutionary loss of lungs has occurred. Land-living members of one large family of salamanders (Plethodontidae), which constitute about 70 percent of existing salamander species, utilize *only* cutaneous and buccopharyngeal gas exchange. They depend entirely on gas exchange through the moist, well-vascularized skin (cutaneous gas exchange) and through the lining of the mouth and pharynx (buccopharyngeal gas exchange). Lunglessness, which reduces buoyancy, has been proposed to be adaptive, particularly for larval survival, in flowing, well-oxygenated streams (Wilder and Dunn, 1920; Beachy and Bruce, 1992). Ruben and Boucot (1989), however, suggested terrestrial or semiterrestrial ancestors for plethodontids, which would mean that lungs were lost for reasons other than ballast.

Larval amphibians breathe by means of external gills (branchial gas exchange). In anuran tadpoles, gills are enclosed in an atrial chamber, which may be either ventral or lateral and which opens via a **spiracle**. The position of the spiracle is a generic characteristic. In tadpoles, water enters the atrial chamber via the mouth, flows over the gills, and passes to the outside through the spiracle. Gills of tadpoles are usually smaller and simpler than those of salamander larvae. During metamorphosis, gills of anurans are reabsorbed, the gill slits close, and gas exchange using lungs takes over.

In larval salamanders and caecilians, gills are exposed on each side behind the head. No atrial chamber develops. As they mature, aquatic amphiumas (*Amphiuma* spp.) and hellbenders (*Cryptobranchus alleganiensis*) develop lungs and lose their gills, but retain the openings of one pair of gill slits.

FIGURE 6.19

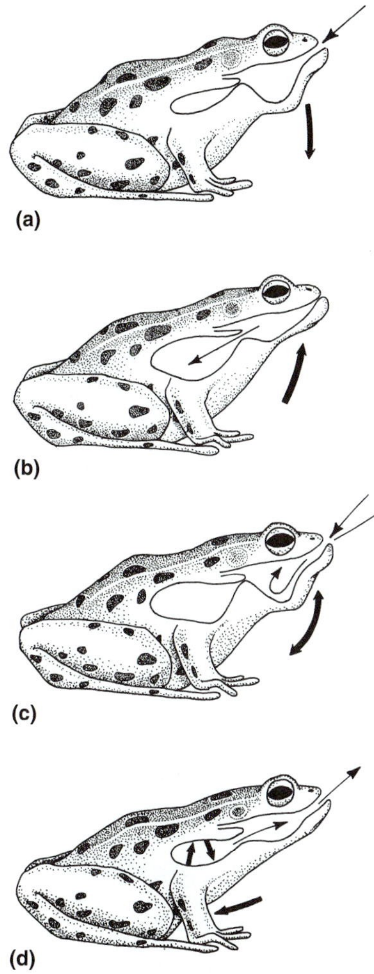

Breathing in the frog. Frogs use positive pressure to force air into their lungs. In the sequence shown, air is drawn in through the nostrils by lowering the floor of the mouth (*a*). By closing the nostrils, opening the glottis, and elevating the floor of the mouth, the frog forces the air into its lungs (*b*). The mouth cavity is ventilated rhythmically for a period (*c*), after which the air is forced out of the lungs by contraction of the body wall musculature and by the elastic recoil of the lungs (*d*).

FIGURE 6.20

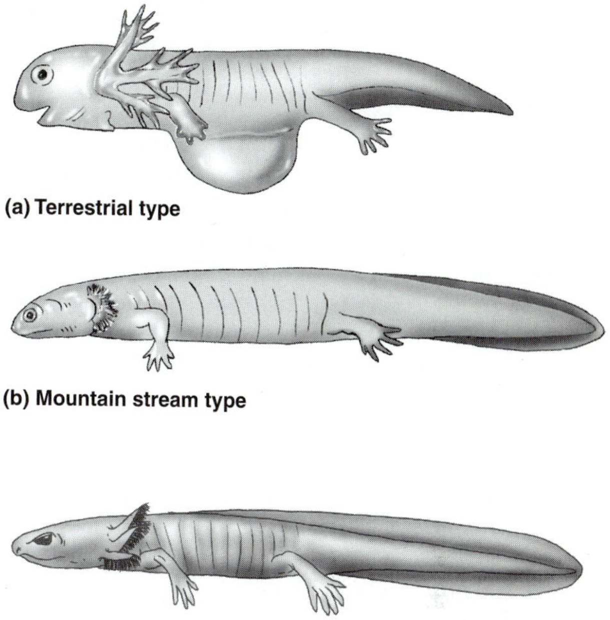

Gills in larval salamanders. (*a*) terrestrial type (*Plethodon vandykei*); (*b*) mountain stream type (*Dicamptodon ensatus*); (*c*) pond type (*Ambystoma gracile*).

Some species, however, retain gills throughout their lives (perennibranchiates). The retention of larval or embryonic characters is known as **neoteny**. Adult *Necturus*, for example, possess both gills and lungs, and two gill slits remain open. Adult sirens (Sirenidae) also have lungs, gills, and gill slits. *Necturus* and *Cryptobranchus* are water-breathing aquatic salamanders that utilize aerial gas exchange primarily under stress conditions such as environmental hypoxia and possibly during recovery from strenuous activity. Sirens and amphiumas, both of which have highly vascularized lungs, are aquatic salamanders that are primarily air breathers and are known to enter drought-induced estivation, during which time they breathe atmospheric air exclusively. Their aquatic gas exchange mechanism is primarily limited to the integument.

Larval gills in salamanders vary as a result of adaptation to the larval habitat (Fig. 6.20). Terrestrial forms within the family Plethodontidae lay their eggs on land, and the young undergo larval development in the egg. They have staghorn-shaped gills. Stream salamanders have reduced gills with short, broad gill filaments. Pond salamanders have larger, featherlike gills for life in quiet water with reduced oxygen.

With the development of lungs, oxygen from a mixture of gases (air) passes through moist, gas exchange membranes deep within the body (pulmonary gas exchange), and gas exchange takes place with minimal loss of water through evaporation. Internal lungs must be ventilated by a tidal movement of air to replenish the oxygen supply at the gas exchange surfaces.

The paired lungs of amphibians develop within the pleuroperitoneal (coelomic) cavity before metamorphosis. Although some salamanders (plethodontids) lack lungs and the lungs of some mountain stream salamanders are extremely small, lungs are present in all other adult amphibians. In caecilians, the right lung is functional, and the left lung is rudimentary—presumably an adaptation associated with the elongate body form of caecilians. This adaptation is similar to that found in snakes (Chapter 8).

The internal lining of amphibian lungs may be either smooth, or it may be pocketed to increase the surface area available for gas exchange. Lung linings are more complex in anurans, where the lungs may be made up of many folds lined with **alveoli** (respiratory pockets) that are supplied by dense capillary networks. Pulmonary oxygen uptake (lung and buccopharyngeal surfaces) accounts for only 26 to 50 percent of the total gas exchange in mole salamanders (Ambystomatidae)

(Whitford and Hutchison, 1966); however, approximately 80 percent of the carbon dioxide release is through the skin.

Some neotenic salamanders, some adult newts, and pipid frogs apparently utilize the lungs more as hydrostatic organs than as organs for gas exchange. In *Necturus*, for example, only about 2 percent of the oxygen is obtained via lungs when water is well oxygenated. Some of these forms take air in through their mouth, which along with the pharynx, is lined with highly vascularized epithelium, called the **buccopharyngeal mucosa**. In winter, when the oxygen uptake is quite low, the skin takes up more oxygen than the lungs. In summer, when oxygen consumption is high, uptake through the lungs increases severalfold and far exceeds cutaneous uptake (Schmidt-Nielsen, 1990). Oxygen uptake through the skin remains nearly constant throughout the year (Dolk and Postma, 1927).

Rates of oxygen consumption by larval and adult amphibians at rest and during locomotion have been presented by Gatten et al. (1992). Absolute levels of oxygen uptake during rest and exercise and the difference between these two measures were found to be consistently lower in salamanders than in anurans.

Digestive System

Most species of amphibians possess a tongue in their oral cavity. It may be attached by its anterior end or its posterior end, or it may be mushroom-shaped (boletoid) and consist of a pedestal with a free upper edge. These variations permit the tongue to be used in taxonomic classifications. The tongue is poorly developed in aquatic forms and is absent in pipid frogs (Pipidae).

In those salamanders with protrusible tongues, the tongue is mounted on the hyoid, and hyoid movement serves to evert the tongue beyond the mouth. Tongues of some plethodontid salamanders can be extended several times the length of the head. Anurans lack such an intrinsic lingual skeleton.

BIO-NOTE 6.5

A Projectile Tongue

Salamanders of the genus *Hydromantes* possess tongues that can shoot out about 6 cm, or 80 percent of the salamander's body length. The tongue is fired from the mouth by a ballistic mechanism and is retracted by muscles that originate at the pelvis. When the tongue is extended, the entire tongue skeleton leaves the mouth completely. *Hydromantes* is the only vertebrate known to shoot part of its visceral skeleton completely out of its body as a projectile.

Deban, Wake, and Roth, 1997

The anuran tongue is a well-developed, sticky prehensile organ that is important in gathering food, particularly insects. Numerous glands and secretory cells maintain a layer of sticky mucus that coats the tongue and assists in capturing prey. Tongues of most anurans are attached anteriorly, are highly flexible, and are used for securing food. Because considerable diversity exists in tongue structure, the mechanism of protraction varies (Gans and Gorniak, 1982; Nishikawa and Roth, 1991; Nishikawa and Cannatella, 1991; Deban and Nishikawa, 1992) (Fig. 6.21). Protrusion involves muscular action with the tongues of some, such as *Rana* and *Bufo*, being highly protrusible, whereas those of *Ascaphus*, *Discoglossus*, and most hylids are weakly protrusible (Deban and Nishikawa, 1992). Food capture involves a lingual flip in which the posterodorsal surface of the retracted tongue becomes the anteroventral surface of the fully extended tongue. The tongue of caecilians is rudimentary, cannot be protruded from the oral cavity, and is capable of only limited movement.

Most amphibians have small teeth (Figs. 6.13a–c) that are shaped alike, a situation called **homodont dentition**, and are found on the palate as well as on the jaws. Teeth are attached to the inner side of the jawbone, which is called **pleurodont dentition**, and are replaced an indefinite number of times if lost or injured, which is known as **polyphyodont dentition**. Because amphibians do not chew their food,

FIGURE 6.21

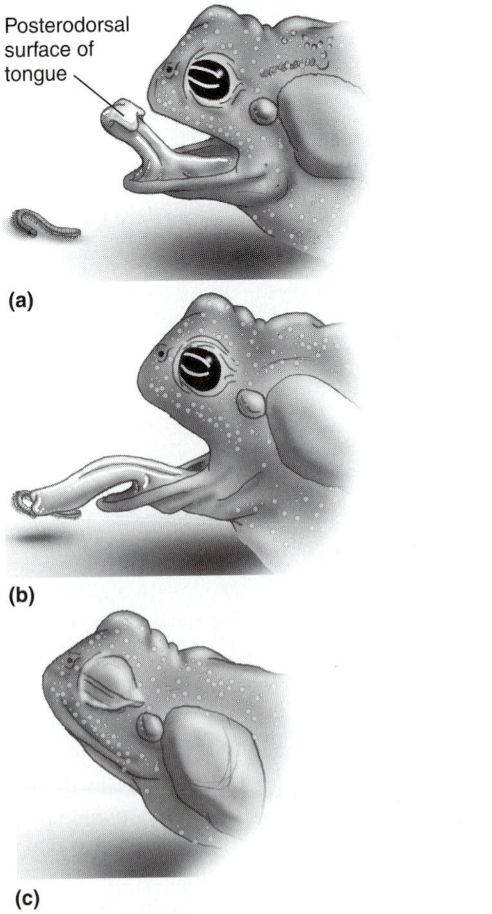

Lingual flipping feeding mechanism in the toad *Bufo marinus*. Note the depressed anterior part of the jaw.

the function of teeth is to grasp and hold food until it is swallowed. Most frogs lack teeth in the lower jaw.

The boundary between the esophagus and stomach is indistinct. The stomach is generally unspecialized and retains food items for 8 to 24 hours, during which the food mixes with gastric secretions and digestion begins (Larsen, 1992).

Feeding habits and digestive systems change drastically with metamorphosis. Larval forms with herbivorous diets have longer intestines than those with carnivorous diets in order to more efficiently break down the cellulose cell walls of plant cells.

Anuran larvae have much longer mid- and hindguts than do larval salamanders. Anuran digestive tracts are coiled within the abdominal cavity, and their total length is several times greater than the length of the animal. The maximum length of the gut is reached when the hind legs are well developed. Reduction in length of larval intestines comes from contraction of the circular and longitudinal muscles at both ends. Shortening and reorganization of the gut in *Bufo* requires 24 hours and occurs within 10 days after the front legs break through (Bowers, 1909). Larval salamanders tend to be carnivorous and feed on larger prey than do anuran tadpoles.

All amphibians have a cloaca that receives the contents of the digestive, urinary, and reproductive systems. A urinary bladder is connected to the ventral side of the cloaca. The comparative anatomy and phylogeny of the cloacae of salamanders has been discussed by Sever (1991a, b, 1992). Kikuyama et al. (1995) isolated sodefrin from the abdominal gland of the cloaca of the male red-bellied newt (*Cynops pyrrhogaster*). Sodefrin is a species-specific, female-attracting pheromone (a secretion that elicits a behavioral response in another member of the same species), the first ever identified in an amphibian. It is also the first peptide pheromone identified in a vertebrate.

BIO-NOTE 6.6

Right Forelimb Dominance in Toads

Behavioral asymmetry in forelimb usage has been demonstrated in European toads (*Bufo bufo*), which showed a bias for right forepaw use. Toads (and frogs) that ingest undesirable objects such as ants and wasps (whose bodies may contain toxins) empty their stomachs by regurgitating (everting) their entire stomach. The stomach hangs out of the side of their mouth, and they use their hand to wipe away remaining vomitus from the surface of the prolapsed stomach before reswallowing it. The right hand is always used for this "gastric" grooming. Why? Because the anuran stomach, like ours, lies somewhat left of center and is held in place by membranes. Because the membrane attached to the right side of the stomach is shorter, it pulls the stomach to the right as it is everted. Toads and frogs cannot reach over to the right corner of their mouth with their left hand because their arms are too short, and so they use the right hand.

Bisazza et al., 1996
Naitoh and Wassersug, 1996

Nervous System

The anterior portion of the brain consists of a pair of olfactory lobes and a pair of cerebral hemispheres (Fig. 6.22). A pineal organ is present and may serve as a photoreceptor, but only remnants of the parapineal organ are found in amphibians. Optic lobes are present; however, the cerebellum is relatively inconspicuous—a condition presumably correlated with the comparatively simple locomotor activities of many amphibians. Impulses from the lateral-line system are directed to the cerebellum, which coordinates and controls voluntary muscular activity. The cerebellum is very poorly developed in those amphibians with a reduced lateral-line system.

Cranial nerves in anamniotes were discussed in Chapter 5. Amphibians have the same 10 basic cranial nerves as fishes, and the same terminalis nerve. However, some authorities recognize 2 additional nerves: the **accessory nerve** (XI), which supplies the cucullaris muscle in amphibians; and the **hypoglossal nerve** (XII), which innervates muscles of the tongue and supplies hypobranchial muscles in the neck. Primitive fossil amphibians apparently had 12 cranial nerves emerging from their skull. Due to a shortening of the cranium, the 12th cranial nerve is now associated with the first two spinal nerves (Duellman and Trueb, 1986).

Two meninges—an outer dura mater and an inner vascular pia-arachnoid membrane—surround the spinal cord. In tailed amphibians, the spinal cord extends to the caudal end of the vertebral column, whereas in most frogs it consists of just 11 segments and ends anterior to the urostyle. Cervical and lumbar enlargements occur for the first time, because these are the first forms to have appendages modified into true limbs. Eleven pairs of spinal nerves emerge from the spinal cord of anurans by means of ventral and dorsal roots. An autonomic nervous system, which controls activities of smooth muscles, glands, and viscera, is well developed.

FIGURE 6.22

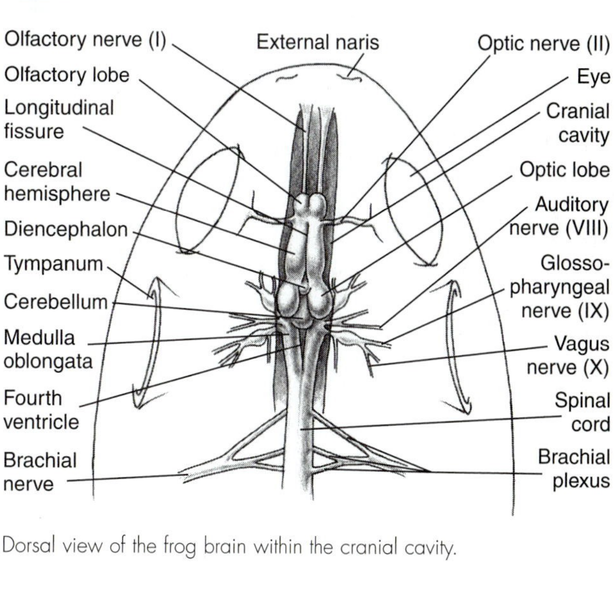

Dorsal view of the frog brain within the cranial cavity.

Sense Organs

Neuromast Organs

Larval and adult aquatic amphibians possess **neuromast organs** in the form of lateral-line canals and cephalic canals. Receptors are distributed either singly or in small groups over the dorsal and lateral surfaces of the body and head. Receptors are especially abundant on the head, where they form distinct patterns.

Each neuromast consists of a pear- or onion-shaped group of hair cells embedded in the epidermis. They perceive low-frequency vibrations of the water and water currents and may also be sensitive to pressure (Russell, 1976). Collectively, they enable the animal to maintain equilibrium and posture. In some salamanders, such as the axolotl (neotenic *Ambystoma* spp.), the lateral-line system also provides electroreception. In these salamanders, two types of sensory units are present: electrosensitive and mechanosensitive. Electrosensitive units react to minute voltage gradients, whereas mechanosensitive units are extremely sensitive to movements in the water (Munz et al., 1984).

The lateral-line organs of some salamanders regress and regenerate in an annual cycle. Regeneration is associated with the return of the amphibians to an aquatic existence during the breeding season. It is the only special sensory system in vertebrates that alternately regresses and regenerates during the life of the animal (Russell, 1976).

Ear

The amphibian ear shows several advances over the ear of fishes. Amphibians possess an auditory system with three main divisions: an outer ear, a middle ear, and an inner ear (Fig. 6.23). The system is sensitive to both ground vibrations and airborne sound waves, with the ears of most anurans being more highly developed than those of salamanders and caecilians.

The outer ear of most anurans consists of a **tympanic membrane**, or **tympanum**, which initially receives airborne vibrations; it is absent in larval and adult salamanders and caecilians (Jaslow et al., 1988). In some species, such as *Rana*, this membrane may be much larger in males than in females, even though the two sexes may be of approximately equal body size (Fig. 6.24). Although Capranica (1976) noted that the functional significance of the size of the eardrum was not clear, new studies (e.g., Purgue, 1997) suggest that its larger size in some male anurans is due to male's using their tympanum as an amplification device, as discussed on pages 147-148 in this chapter.

In anuran tadpoles, the developing lungs serve as eardrums. A columella connects the round window membrane of the inner ear with the bronchus and lung sac on the same side of the body. Changes in lung volume result in displacement of the bronchial membranes (Caprancia, 1976).

Amphibians are the first group of vertebrates in which the first pair of pharyngeal pouches becomes involved in forming the middle ear. The distal end of each pouch expands to form the tympanic cavity in anurans, while the Eustachian

FIGURE 6.23

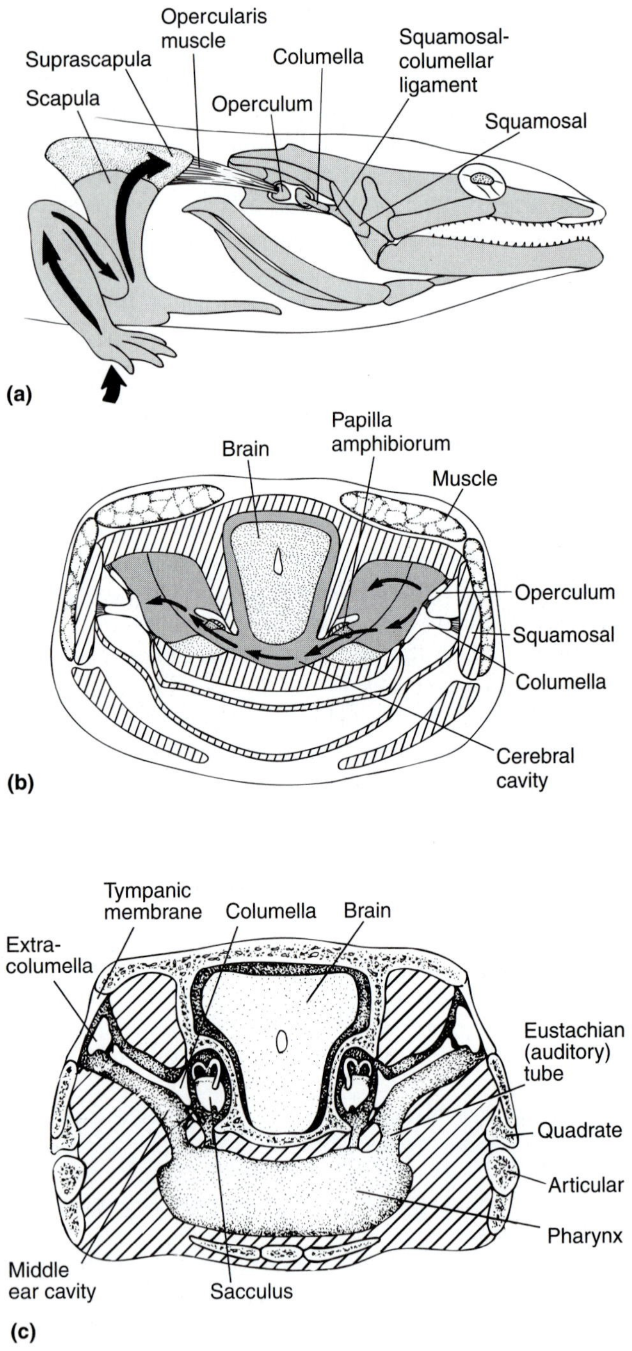

(*a*) The inner ear of many salamanders receives sounds via a squamosal–columella route and/or via the opecularis muscle from the scapula. (*b*) As in frogs, the two inner ears on the opposite sides of the head are connected by means of a fluid-filled channel that passes through the cerebral cavity. This arrangement may allow sound vibrations to spread from one ear to the other (black arrows). (*c*) In a frog, both ears are connected via the Eustachian tubes and pharnyx; thus, any sound that sets one tympanum in motion also affects the ear on the opposite side. This unique structure is thought to allow frogs to localize the source of sounds.

FIGURE 6.24

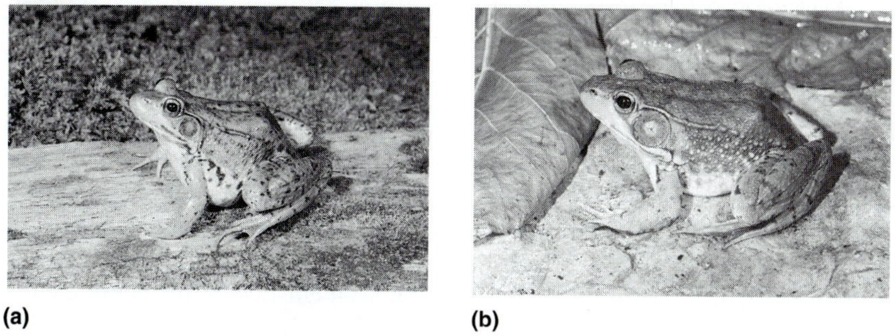

Sexual dimorphism. The tympanum is markedly larger in male green frogs (*Rana clamitans*) (*b*) than in the females (*a*).

(auditory) tubes form a passageway from the middle ear to the pharynx (Feduccia and McCrady, 1991).

The middle ear, or **tympanic cavity**, is an air-filled chamber that contains the small, rod-shaped columella (Fig. 6.23) and another small movable bone, the **operculum**. A small opercularis muscle joins the operculum to the pectoral girdle. The primary function of the columella, which is homologous to the dorsal segment of the hyoid arch (hyomandibula) in fishes and transmits vibrations from the tympanum to the oval window, is to convey sound from the external environment to the fluid-filled inner ear. In anurans lacking a tympanum, the columella may be reduced or absent. All salamanders lack tympanic cavities (Fig. 6.23). The columella, often degenerate, is joined to the squamosal bone by a short squamosal–columellar ligament (Fig. 6.23), so that sounds may reach the inner ear via a squamosal–columella route.

In addition, sound waves may travel from the ground to the inner ear via the scapula–opercularis muscle–operculum route. In anurans, a **Eustachian (auditory) tube** leads from the middle ear to the pharynx and serves to equalize pressure on both sides of the tympanum. Because salamanders and caecilians lack a middle ear, they also lack a Eustachian tube.

The inner ear consists of a utriculus, a sacculus, a lagena (slight bulge in the ventral wall of the saccule), and three semicircular canals each lying in a different plane. Two fluids—endolymph and perilymph—are present in the inner ear and function in both hearing and the maintenance of equilibrium. Endolymph is enclosed within the inner ear membranes, whereas perilymph is external to the membranes. Movement of the endolymph stimulates sensory hair receptors and allows vibrations to be transmitted to the brain. The receptor cells, located in ampullae at the base of each canal (at the point where each canal enters the utriculus), are known as **cristae**. By having each canal oriented in a different plane, the endolymph in one or more of the canals will shift with even the slightest movement. Patches of sensory epithelia known as **maculae** are present within the utriculus and sacculus.

Eyes

Eyes of terrestrial amphibians are large and well developed, and they show a number of advances over those of fishes (see Fig. 1.18d, page 15). Salamanders have good color vision; anurans probably have some color vision (Porter, 1972). Colorless oil droplets are found between the inner and outer segments of cone photoreceptor cells of some species (Bowmaker, 1986). They probably filter out damaging ultraviolet radiation, but they do not appear to contribute to acuity of vision (Hailman, 1976). Their function may be chiefly chemical storage, perhaps in relation to the visual pigment cycle, or they may make wavelength perception more "even" by spreading out the photons.

At times, the eyes may be partially retracted into the orbit, which facilitates the swallowing of large objects. Because the eyeballs protrude into the oral cavity, they assist in forcing food into the esophagus.

Movable eyelids and **orbital glands** (Harderian and lacrimal) are present to afford protection for the eyes in most terrestrial forms. The eyelids and glands develop at metamorphosis in most salamanders and anurans. The lower eyelid has a much greater range of motion than the upper and is better developed in anurans than in salamanders. Eyelids are absent in purely aquatic salamanders and in all amphibian larvae. Harderian glands, which secrete an oily substance, and lacrimal glands, which secrete a watery fluid (tears), are present evolutionarily for the first time in the vertebrates. They serve to lubricate and cleanse the outer surfaces of the eyes.

In many frogs, the lower eyelid has become modified into a translucent or transparent fold of skin called the **nictitating membrane** (Fig. 6.25). This membrane can be drawn up over the retracted eye by tendons encircling most of the eyeball and gives the frog a certain amount of vision even when it appears to be sleeping with partly closed eyes. This membrane is often marbled with a pattern of colored lines or spots in designs characteristic of the species. In water, the nictitating membrane is drawn over the eye to protect it while allowing the frog some degree of vision.

FIGURE 6.25

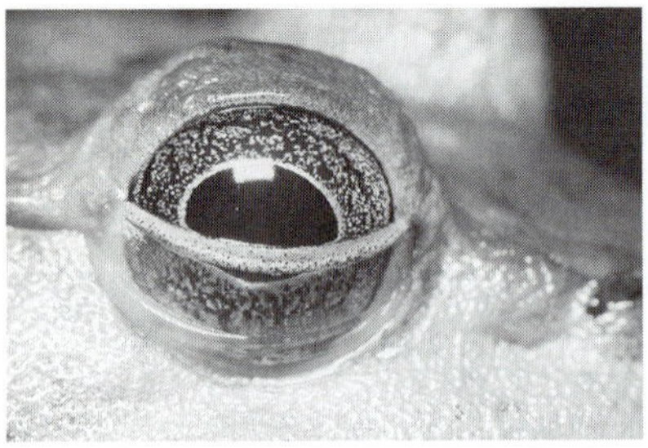

Bullfrog (*Rana catesbeiana*) showing the eye partly covered by the nictitating membrane.

The amphibian lens, hard and almost spherical, cannot change shape. Accommodation for near vision is accomplished by contraction of the protractor lentis muscle, which moves the lens closer to the cornea. Relaxation of the protractor muscle allows the eye to focus on distant objects. Further discussion of the amphibian visual system can be found in Fite (1976).

Some cave-dwelling and subterranean salamanders and caecilians may be blind, or the eyes may be vestigial or poorly developed (Fig. 6.26). Cave-dwelling genera include *Typhlomolge, Typhlotriton,* and *Haideotriton*—all members of the family Plethodontidae.

Pineal Organ

A pineal organ is present in all amphibians; the parietal organ, however, is present only in anurans. Both organs function as photoreceptors and are located in the top of the head between the eyes. They both appear to be important in the establishment of endogenous rhythms, in thermoregulation, in gonadal development, and in compass orientation (Adler, 1970). The parietal organ is known to be sensitive both to light intensity and to different wavelengths of light (Dodt and Heerd, 1962; Dodt and Jacobson, 1963).

FIGURE 6.26

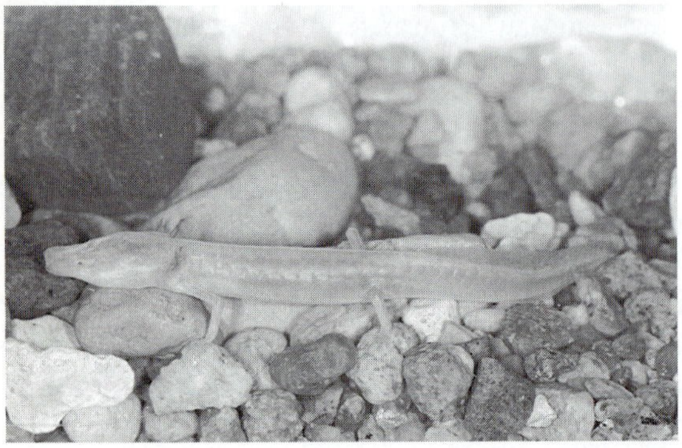

The Texas blind salamander (*Typhlomolge rathbuni*), a plethodontid. Note the poorly-developed eyes, external gills, and lack of pigment in the skin.

Nose

Amphibians have a double olfactory system. Air enters the nasal chamber via external nares and exits via internal nares into the oral cavity. Air entering the external nares must flow past the olfactory epithelium on its way to the lungs, thus allowing the olfactory epithelium to sample chemicals in the airflow. The olfactory epithelium, therefore, monitors an airstream instead of a water stream. In addition, rudimentary vomeronasal, or Jacobson's, organs (first described in 1811, by L. Jacobson, a Danish physician), are present for the first time (Fig. 6.27). Each vomeronasal organ consists of a ventral segment of olfactory epithelium that has become isolated from the nasal passageway and usually is located dorsal

FIGURE 6.27

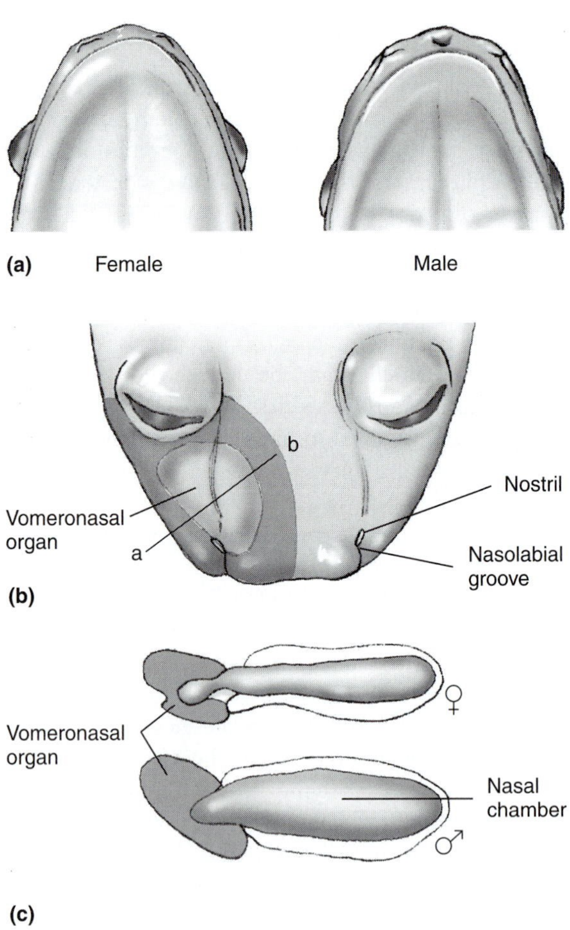

Sexual dimorphism in the nasolabial grooves and vomeronasal organs of plethodontid salamanders. (*a*) Note the difference in forking of the grooves at the edge of the upper lip in ensatina (*Ensatina eschscholtzii*): (left) female; (right) male. (*b*) Nasal passage of the red-backed salamander (*Plethodon cinereus*) exposed to show the position of the vomeronasal organ. (*c*) Cross section through female (top) and male (bottom) nasal passages at line a–b in (*b*), showing differences in the size of the vomeronasal organ.
Source: Stebbins and Cohen, A Natural History of Amphibians, *1995, Princeton University Press.*

to the vomer bone in the oral cavity. The organs, consisting of a pair of deep grooves in salamanders and a pair of sacs in anurans, are innervated by a branch of the olfactory nerve. They probably play a role in the recognition of food and are used to test substances held in the mouth. The presence of a vomeronasal system is often an indication of the use of olfactory signals, or **pheromones**, for intraspecific communication. Indeed, chemical signals associated with courtship or territoriality have been demonstrated repeatedly in adult salamanders (Dawley, 1984; Horne and Jaeger, 1988), but not in adult caecilians. An aquatic, female-attracting pheromone named *splendipherin* is the first pheromone to be identified from an adult anuran (Wabnitz et al., 1999). It is secreted from the parotoid and rostral glands of the male magnificent tree frog (*Litoria splendida*).

The sense of smell has been shown to be particularly well developed among plethodontid salamanders (see Jaeger and Gergits, 1979, for review). The presence of **nasolabial grooves** in these salamanders is thought to aid in their olfactory reception and is unique among terrestrial vertebrates (Fig. 6.27). A small furrow extends from the edge of each upper lip to the nostril. The lower ends of the grooves may divide into long branches, break up into capillary networks, or end in nipplelike palps projecting downward from the upper lip. When the snout comes into contact with moist surfaces, fluid rapidly passes up the grooves by capillary action to the external nares, where ciliary action draws it into the nose and over the chemoreceptors of the vomeronasal organs (Brown, 1968; Dawley, 1987; Dawley and Bass, 1989). Nasolabial grooves are found in both sexes, but they are larger, better developed, and more elaborate in males (Dawley, 1992). They are thought to play a significant role in reproductive behavior by allowing males to track and identify females. Plethodontids frequently engage in nose-tapping, a behavior in which the snout is used to tap the substrate or other individuals. Plethodontid salamanders are the only nonmammalian vertebrates in which sexual dimorphism of the vomeronasal system has been documented.

Caecilians have an opening on each side of the head between the eye and nostril, through which a tentacle with both chemoreceptive and tactile functions can be extended. The tentacle, which is unique among vertebrates, contains a duct that opens to the exterior. Its primary function is to convey airborne chemoreceptive information to the vomeronasal organs. Development of the tentacle is thought to be an adaptation to caecilians' subterranean, burrowing existence, and the tentacle allows chemoreception during burrowing and swimming when the nostrils are closed. Tentacles appear to have evolved through modification of muscles and other structures associated with the eyes, which are degenerate (Billo and Wake, 1987; Schmidt and Wake, 1990).

In many caecilians, the paired tentacles can be protracted a considerable distance out of the head. In the East African caecilian *Scolecomorphus kirkii*, the eye is attached laterally to the base of the tentacle. As the tentacle protracts and retracts, the eye moves with it and can actually be protruded beyond the skull (O'Reilly et al., 1996). This is the only known vertebrate with highly mobile, protrusible eyes.

Taste

Taste buds primarily are found scattered over the dorsal surface of the tongue, the floor of the mouth, the jaws, and the palate. Those on the dorsal surface of the tongue in frogs are in the form of epithelial disks (Sato, 1976; Jaeger and Hillman, 1976). Each taste bud consists of a cluster of gustatory (taste) cells opening via a taste pore.

Endocrine System

Because, in most cases, the endocrine organs are homologous in the vertebrates, the discussion in Chapter 5 included hormones and their respective actions for all groups. Only specific examples of hormonal action will be covered here and in the chapters on reptiles and mammals.

The morphological and physiological changes that occur during metamorphosis are the result of hormones secreted by a finely tuned integration of endocrine glands. The hypothalamus in the brain controls the release of pituitary hormones such as thyroid-stimulating hormone, prolactin, and possibly a growth hormone—all secreted by the adenohypophysis (anterior pituitary gland)—that appear to control growth and development by regulating the activity of the thyroid gland. The thyroid gland, which secretes thyroxin and triiodothyronine, is considered to be the keystone of amphibian metamorphosis. Calcitonin, which lowers blood calcium levels, is produced in the ultimobranchial bodies, which are located near the larynx in amphibians. The ultimobranchial bodies conserve calcium to ensure an adequate supply for calcification during metamorphosis.

Changes during metamorphosis include regression of the gills, degeneration of the tail and tail muscles in anurans, development of limbs, formation of dermal glands, and reorganization of the intestinal tract (Duellman and Trueb, 1986). If the thyroid gland is removed from a tadpole, the tadpole will grow into an abnormally large, fat tadpole with lungs and reproductive organs, but it will never metamorphose into an adult. If thyroid extract is administered following the thyroidectomy, however, metamorphosis will take place. In adults, thyroid hormones help control the rate of metabolism, heart rate, and shedding of the skin. Molting and intermolt frequency are also under the control of adrenocorticotropic hormone (ACTH) and corticosterone produced by the adrenal glands.

Amphibians present the first evolutionary appearance of parathyroid glands, which develop from pharyngeal pouches in anurans and in many salamanders. Parathyroid hormone (parathormone) raises the calcium and phosphate levels in circulating blood by withdrawing these minerals from storage sites such as bone. Seasonal differences in the parathyroid gland have been found in leopard frogs (*Rana pipiens*), in which the glands degenerate during the winter (Cortelyou et al., 1960; Cortelyou and McWhinnie, 1967). Some paedomorphic salamanders lack parathyroid glands (Duellman and Trueb, 1986). Adrenal glands, which are diffuse in salamanders, appear as strips of golden yellow tissue partially embedded in the ventral surfaces of each kidney in anurans.

The hypothalamus and pituitary gland regulate reproductive behavior in both male and female amphibians. Estrogens and progesterone control the development of eggs and breeding behavior in females. Androgens, such as testosterone, at least partially control the swelling of the thumb pads, the enlargement of the mental (chin) gland, changes in color pattern, the development of the dorsal crest in some salamanders, and the enlargement of the cloacal glands in males (Fig. 6.28). Stimuli from external social or environmental cues are transformed along a multistep pathway known as the hypothalamic–pituitary–gonadal axis into neural and endocrine information, ultimately affecting androgen production by the testes (Houck and Woodley, 1995). Typically, external cues provide stimulation that is converted into neural signals, which are integrated in several regions of the brain, including the hypothalamus. The hypothalamus produces gonadotropin-releasing hormone (GnRH), which in turn, causes the pituitary to release two peptide hormones—follicle-stimulating hormone (FSH) and luteinizing hormone (LH)—which are carried by the circulatory system to the gonads. FSH prepares the Sertoli cells in the seminiferous tubules for spermatogenesis in the presence of androgens (male sex steroids), whose production by Leydig cells within the testis is stimulated by LH. Androgens most typically found in amphibians include testosterone (T) and dihydrotestosterone (DHT). These androgens diffuse within the testes and are also transported by proteins in the circulatory system to steroid-sensitive targets, including the brain. Circulating androgens may influence male reproductive behavior and also affect certain peripheral tissues, thus resulting in the development of secondary sexual characters.

FIGURE 6.28

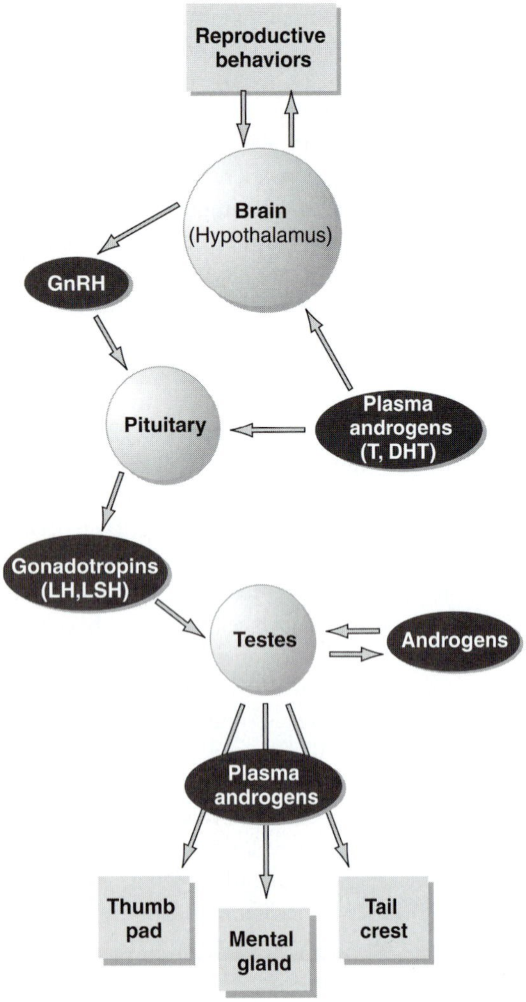

The hypothalamic–pituitary–gonad axis influences androgen production in male amphibians. Hormonal influences on androgen production by the testes are illustrated, as well as targets potentially affected by changes in androgen levels.

Diurnal rhythms of the melatonin level in the retina have been reported in amphibians, birds, and mammals (reviewed by Zachmann et al., 1992). Melatonin causes melanophores to aggregate.

Urogenital System

The pronephric kidney is functional during larval development, after which it is replaced by the opisthonephros, which serves as the functional kidney of most adult amphibians. In salamanders and anurans, the pronephric kidney consists of 2 to 4 tubules with nephrostomes that drain the coelomic cavities into a pronephric duct. In most other amphibians, the kidney is considerably shorter. Anuran kidneys are less than half the length of the coelom and are located posteriorly. In males, the anterior portion of the mesonephric duct becomes the epididymis and drains the

BIO-NOTE 6.7

Using Bladders to Store Water

Large bladders in many amphibians allow them to store water in their bodies. The African bullfrog (*Rana adspersus*), which weighs up to 2 pounds, inhabits the Kalahari Desert. When pools dry up, the bullfrog digs into the earth and survives in a protective cocoon made from layers of shed skin. A large bladder allows for storage of water in amounts up to one-third of its body weight, and it can tolerate a loss of 45 percent of that original body weight. As water is lost, the viscosity of its blood can thicken to four times its normal concentration. The bullfrog can wait in semidormancy for up to 2 years until rains begin filling the fleeting pools.

Crowe, 1995

testes through efferent tubules. Caecilians have a pronephric kidney consisting of 10 to 13 tubules and an opisthonephric kidney in which both anterior and posterior nephrostomes persist. The opisthonephros may extend most of the length of the coelom in these forms.

The bladder in all amphibians is a thin-walled evagination of the ventral wall of the cloaca. It receives and stores urine and participates in water and ion exchange processes. Glucose reabsorption from the urinary bladder in the freeze-tolerant wood frog (*Rana sylvatica*) permits the recovery of sugar destined for excretion (Costanzo et al., 1997). Urine flows down the mesonephric duct (original pronephric duct) into the cloaca and then backs up into the bladder for storage. The bladder, along with the alimentary canal, mesonephric ducts, and oviducts, empties into the cloaca.

Female Reproductive System

During embryonic development in female salamanders, each pronephric duct divides longitudinally into a duct, which becomes an oviduct, and an opisthonephric duct, which drains the kidney (Fig. 6.29c).

The ovary, a hollow sac with an enclosed lymphatic cavity, ranges in shape from short and compact in anurans to elongate in caecilians. Its surface contains germinal epithelium that gives rise to eggs, which are shed into the coelom. Oviducts are long and straight in caecilians, slightly convoluted in salamanders, and greatly convoluted in anurans (Fig. 6.29). The opening of the oviduct (ostium) is ciliated, and its walls contain smooth muscle. The lumen of the oviduct is lined with ciliated, glandular epithelium. The glandular lining of the oviducts secretes several jelly envelopes around each egg as peristaltic contractions of the smooth muscle carry the eggs through the oviducts. Caudal portions of the oviducts may enlarge to form **ovisacs** where eggs are stored temporarily prior to oviposition. In those salamanders in which fertilization is internal and occurs as the eggs pass down the oviducts, the roof of the cloaca in females is modified into a **spermatheca**. This functions as a storage receptacle for spermatozoa prior to ovulation.

Male Reproductive System

In some male salamanders, such as *Necturus*, the mesonephric duct carries both urine and sperm. In many salamanders,

FIGURE 6.29

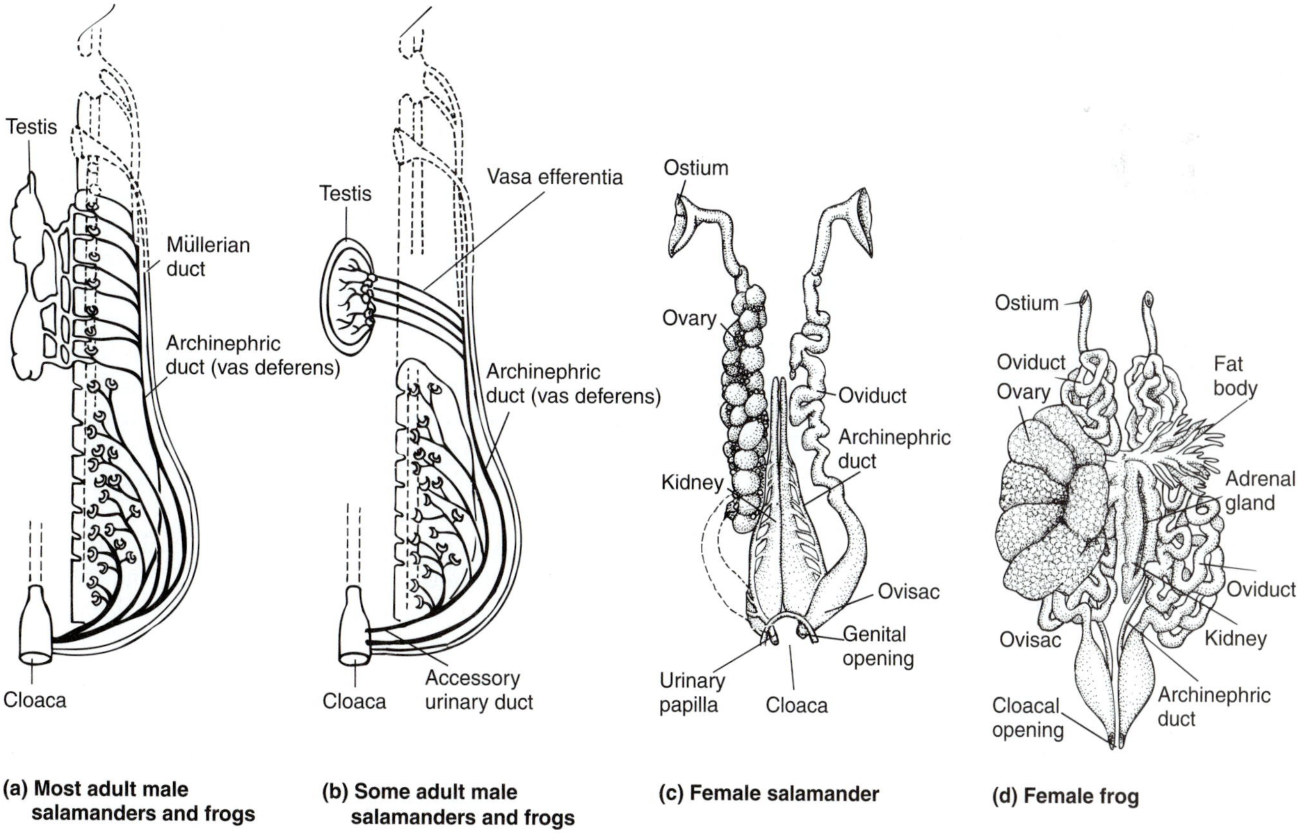

Urogenital ducts of male and female amphibians: (*a*) most adult male salamanders and frogs; (*b*) some adult male salamanders and frogs; (*c*) a female salamander; (*d*) a female frog. Note the relatively straight oviducts in the female salamander as opposed to the highly convoluted oviducts in the frog.

however, including plethodontids, the mesonephric duct has lost its excretory function and carries only sperm (6.29b). A duct carrying only sperm is known as a **ductus deferens** or **vas deferens**.

In male anurans, a series of modified kidney tubules called **vasa efferentia** grow from the anterior tissue of the kidney to the testes (Fig. 6.29b). The anterior part of the mesonephric duct, the vas deferens, functions primarily to transport spermatozoa, whereas the posterior part serves also for the elimination of urinary wastes. In some anurans, the lower portion of the mesonephric duct is modified into a seminal vesicle.

Testes, which serve as endocrine organs and produce the male sex hormone testosterone, are composed of vast numbers of seminiferous tubules. During the breeding season, when spermatogenesis activity is at its peak, the testes of all amphibians increase in size.

Fat bodies usually are associated with the gonads in both males and females, and both have a common embryological origin. According to Noble (1931), fat bodies are a source of nutrients for the gonads. They are largest just before hibernation, begin to shrink as ova and testes mature and enlarge, and are smallest just after the breeding season.

An intromittent organ that facilitates transfer of sperm from the male into the female's reproductive tract is present in caecilians and in one anuran (*Ascaphus*) (Fig. 6.30). In *Ascaphus*, the intromittent organ is a permanent tubular extension of the cloaca and resembles a tail, whereas male caecilians evert the cloaca to produce a temporary intromittent organ.

Male toads in the family Bufonidae possess rudimentary ovaries known as Bidder's organs in addition to functional testes. If testes are removed, Bidder's organ develops into a functional ovary, so that the result is complete sex reversal.

FIGURE 6.30

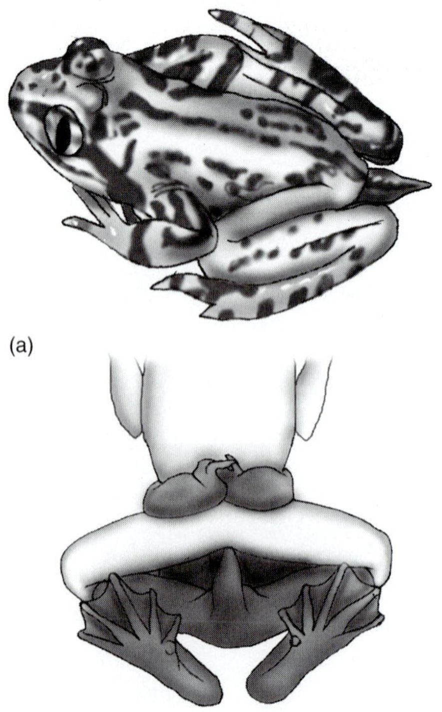

Male *Ascaphus truei*, the only anuran with an intromittent organ.

REPRODUCTION

Amphibian reproduction can be divided into three broad categories (Crump, 1974):

1. *Aquatic development:* Eggs deposited in water; larvae develop in water.
2. *Semiterrestrial development:* Eggs deposited out of water; larvae develop in water.
3. *Terrestrial development:* Eggs and young completely independent of standing water.

Precipitation and temperature are major climatic factors that affect breeding in amphibians. Photoperiod has little effect on regulation of sexual cycles in amphibians that are nocturnal or those that remain underground when inactive (Duellman and Trueb, 1986). Genetically controlled innate sexual rhythms, which would not be under the direct influence of environmental factors, may exist in some species.

Most North American amphibians have an annual cycle of breeding and become active with the increase in temperature in late winter and spring. Rainfall stimulates early breeders such as spring peepers (*Pseudacris [Hyla] crucifer*), wood frogs (*Rana sylvatica*), and spotted salamanders (*Ambystoma maculatum*). Although most amphibians breed in late winter or spring, some, like the bullfrog (*Rana catesbeiana*) may wait until June to deposit their 10,000 to 20,000 eggs. Others, such as marbled salamanders (*Ambystoma opacum*), breed in late summer or early autumn at dry pond sites. After breeding, females deposit their eggs in nest cavities under vegetation and logs and in crayfish holes. Female marbled salamanders stay with the eggs for as long as a month. Once the ponds start to fill and the nests are flooded, the eggs hatch. Even though most anurans have a primary breeding period, some may be stimulated to call, and even deposit additional eggs, later in the summer during periods of heavy rain.

Many amphibians may grow continuously throughout their lifetime (called indeterminate growth), and many can breed at 1 to 2 years of age (Duellman and Trueb, 1986). Ranid frogs, however, may remain as tadpoles for 2 or 3 years before metamorphosing. Newts may spend from 4 to 8 years in a terrestrial juvenile stage known as an eft before returning to water and maturing. Some salamanders become sexually mature but retain some larval characters, a condition known as **paedomorphosis**. Paedomorphic forms are found in all families of salamanders and may require longer than a year before they are reproductively mature. Some hellbenders

(*Cryptobranchus*) and amphiumas (*Amphiuma*) may require as long as 5 to 6 years to mature.

Salamanders recognize other members of their species by using visual and olfactory cues. Extensive courtship rituals involve nuzzling and rubbing the body of a potential mate. Sexual dimorphism is pronounced in some species such as newts (*Triturus*), in which males develop large dorsal and caudal fins during the breeding season (Fig. 6.31).

A variety of courtship activities occur in anurans. Some of the most complex and elaborate mating behaviors have been recorded for many species of poison-dart frogs (*Dendrobates*), including tactile interactions, specific postural displays, and species-specific calls (Silverstone, 1973; Wells, 1977). Some females guard their mates by remaining in or near their mates' territories, courting them frequently, and attacking any other females they encounter courting their mates (Summers, 1989). The spotted poison-dart frog (*D. vanzolinii*) forms pair bonds, and both parents provide extended care of the young under natural conditions (Caldwell, 1997).

After touching a female, most male frogs and toads will tightly grasp her behind her forelimbs, a copulatory embrace known as **amplexus** (Fig. 6.32). Swollen glandular thumb pads of males (see Fig. 6.9) aid in this process, and the pressure applied to the body of the female assists in expelling the eggs. Size and behavior are important sex recognition factors: Females usually are larger than males of the same species and generally will be receptive to the male's clasping efforts. If a male frog attempts to clasp another male, it will meet with resistance and an entirely different pattern of behavior.

Since most anurans breed at night, auditory rather than visual cues are most advantageous for attracting members of the opposite sex; thus, most male anurans have developed species-specific calls. Air passes through either a single median vocal pouch under the throat (Fig. 6.18a, b) or paired pouches on either side of the head (Fig. 6.18c). Ryan (1991) provided evidence that females show a preference for conspecific over heterospecific calls. The female auditory system decodes species-specific information contained in the male's advertisement calls (Capranica, 1976). For example, female coqui frogs hear the "co" in the "co-qui" call; males hear the "qui." Males of some species can alter the frequency of their calls as well as determine the size of neighboring frogs by assessing the tones of their voices (Wagner, 1989a, 1989b). Because calling consumes energy that might otherwise be used for growth, some species reduce calling when food levels decrease (Ryan, 1991). In addition, calling is potentially dangerous because it marks the location of the calling male to potential predators.

FIGURE 6.31

(a) Female

(b) Male in nuptial dress

Sexual dimorphism in the newt *Triturus cristatus*. During the breeding season, males develop large dorsal and caudal fins.
Source: Gadow Amphibians and Reptiles *1901, Macmillan and Co., England*

FIGURE 6.32

(a) **(b)**

(*a*) Male wood frog (*Rana sylvatica*) clasping the female in amplexus, which aids external fertilization. As the female releases eggs into the water, the male releases sperm over them. Note the eggs in a globular cluster. (*b*) Toads (*Bufo*) in amplexus. Note eggs in "string-of-pearls" formation.

BIO-NOTE 6.8

Underwater Mating Calls

The South American frog *Leptodactylus ocellatus* may have evolved an underwater mating call in response to airwave competition from another frog that uses the same frequency above water. Alejandro Purgue, a herpetologist at the University of Utah, documented the underwater calls using U.S. Navy hydrophones and computer analysis. The underwater calls are thought to minimize competition with *Physalaemus cuvieri*, whose calls are in a similar range (250–500 Hz) but above water.

Rand subaquavocalis, a distinctive leopard frog species of the *Rana pipiens* complex known only from the Huachuca Mountains of southeastern Arizona, can be distinguished from other leopard frog species on the basis of morphology and mating call characteristics. Males offer the mating call entirely underwater from a depth of more than a meter, making it completely inaudible in the air.

Weiss, 1990
Platz, 1993

FIGURE 6.33

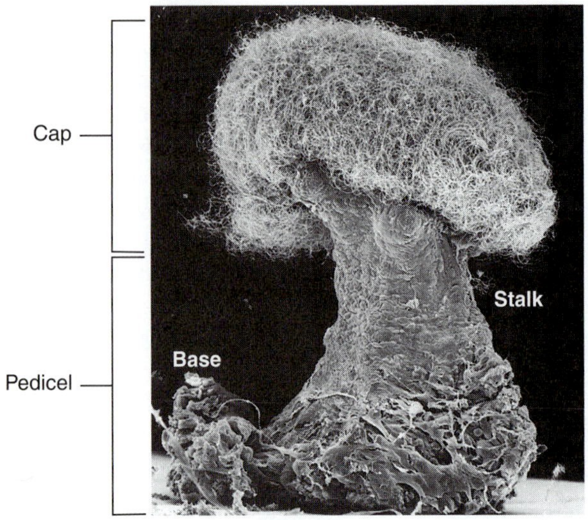

Salamander spermatophore. Whole spermatophore deposited by male *Ambystoma macrodactylum*. Sperm heads generally point outward, tails inward.

Red-backed salamanders (*Plethodon cinereus*) are completely terrestrial, and both mating and egg laying occur in moist microhabitats. Males with high-quality territories achieve greater access to females than males with low-quality territories (Walls et al., 1989). In addition, feces of males with high-quality diets are attractive to females, and gravid females prefer to remain in areas near fecal pellets containing the remains of highly nutritious prey (Jaeger and Wise, 1991). In addition, males found near females are significantly larger than adult males found alone and are more aggressive when paired with smaller males. Large body size in males may positively affect both intra- and intersexual interactions and, ultimately, the mating success of male *P. cinereus* (Mathis, 1991b).

Approximately 90 percent of all salamanders use internal fertilization (Duellman and Trueb, 1986). A single male salamander may deposit packets of sperm known as **spermatophores** on submerged leaves and twigs (Fig. 6.33). The number of spermatophores produced varies widely, with one male spotted salamander (*Ambystoma maculatum*) depositing up to 81 spermatophores in a single evening with one female (Arnold, 1976). Newts of the genus *Triturus* deposit a few spermatophores each day for a few weeks, whereas members of the genus *Plethodon* are much more conservative and deposit one spermatophore each week for several weeks (Arnold, 1977). Because the breeding season for plethodontids may range from several months in temperate zone species to the majority of the year in tropical species, and because plethodontids expend a great deal of energy per spermatophore in courtship activities and in defending a territory, it is evolutionarily advantageous for males not to expend all of their spermatophores on the first few females encountered.

Females grasp spermatophores with the lips of their cloacas and take the sperm packages inside their cloacas so that eggs can be fertilized as they pass out of the female's body. Some female salamanders may store sperm in specialized cloacal sacs (spermathecae) for months or years. Baylis (1939) reported fertilization of ova in *Salamandra salamandra* (Salamandridae) at least 2 years after the last possible mating.

In most anurans, males in amplexus deposit sperm on the eggs as the eggs are released by the female (Fig. 6.32). However, internal fertilization occurs in a few genera (*Ascaphus, Nectophrynoides, Mertensophryne,* and *Eleutherodactylus*) (Townsend et al., 1981; Duellman and Trueb, 1986). The "tail" of the tailed frog, *Ascaphus*, is a posterior extension of the cloaca partially supported by paired rods. The "tail" serves as an intromittent organ for transferring sperm into the female (Fig. 6.30). *Ascaphus* is the only

BIO-NOTE 6.9

Changing Sex

The ability to change sex exists in some fish and amphibians. A few fish species switch sex apparently to maximize breeding in areas where breeding individuals of the opposite sex are not abundant. The only amphibian known to naturally change its sex and successfully breed is the African reed frog (*Hyperolius viridiflavus ommatostictus*). Male toads have a vestige of ovarian tissue (Bidder's organ), which will develop into functional ovaries if testicular tissue is surgically removed. Under experimental conditions, tadpoles of many species consistently develop into males after experimental exposure to the male hormone testosterone.

Grafe and Linsenhair, 1989

known anuran to engage in copulation, internal fertilization that apparently ensures conception in the cold, fast-moving, turbulent streams of the Pacific Northwest where this frog resides.

All caecilians have internal fertilization (Duellman and Trueb, 1986). Some are oviparous with aquatic larvae; some are oviparous with embryos undergoing direct development into terrestrial young; still others are viviparous. The developing young of viviparous species have specialized teeth that are used to scrape the epithelial lining of the oviduct and obtain nutrients secreted by the oviducal cells (Wake, 1977). These teeth are shed after birth.

■ GROWTH AND DEVELOPMENT

Oviparous

Most amphibians deposit their eggs in water. Salamander eggs may be attached individually or in small clusters to underwater vegetation (Fig. 6.34a, b) or to the undersides of partially or completely submerged rocks or boulders; they may be part of a globular gelatinous mass; or they may be part of a floating surface mass. Some terrestrial salamanders, such as ensatina (*Ensatina*), the arboreal salamander (*Aneides*), and the woodland salamanders (*Plethodon* spp.), deposit their eggs in moist sites on land. Some caecilians (Rhinatrematidae, Ichthyophiidae) are oviparous and deposit their eggs in mud near water. The caecilian eggs hatch into free-swimming larvae.

Anurans exhibit a wider variety of larval development than any other group of vertebrates. Many anurans lay their eggs in water, either singly, in globular masses (Fig. 6.34c), or in strings (Fig. 6.34d). There are, however, many exceptions (Table 6.1). Some tropical forms deposit their eggs on leaves overhanging water, so that as the eggs hatch, the larvae drop into the water. Some utilize the pools of water contained in bromeliads—epiphytic plants growing on trees in tropical forests. Some anurans encase their eggs in a frothy, protective foam mass consisting of air, sperm, eggs, cloacal secretions, and sometimes water. The outer surface of the mass dries and turns tacky (sometimes hard), protecting the eggs from physical harm, while the egg-filled interior remains liquid for as long as 10 days to enable the eggs to survive periods of drought. Other anurans possess dorsal pouches on their backs in which eggs are incubated

FIGURE 6.34

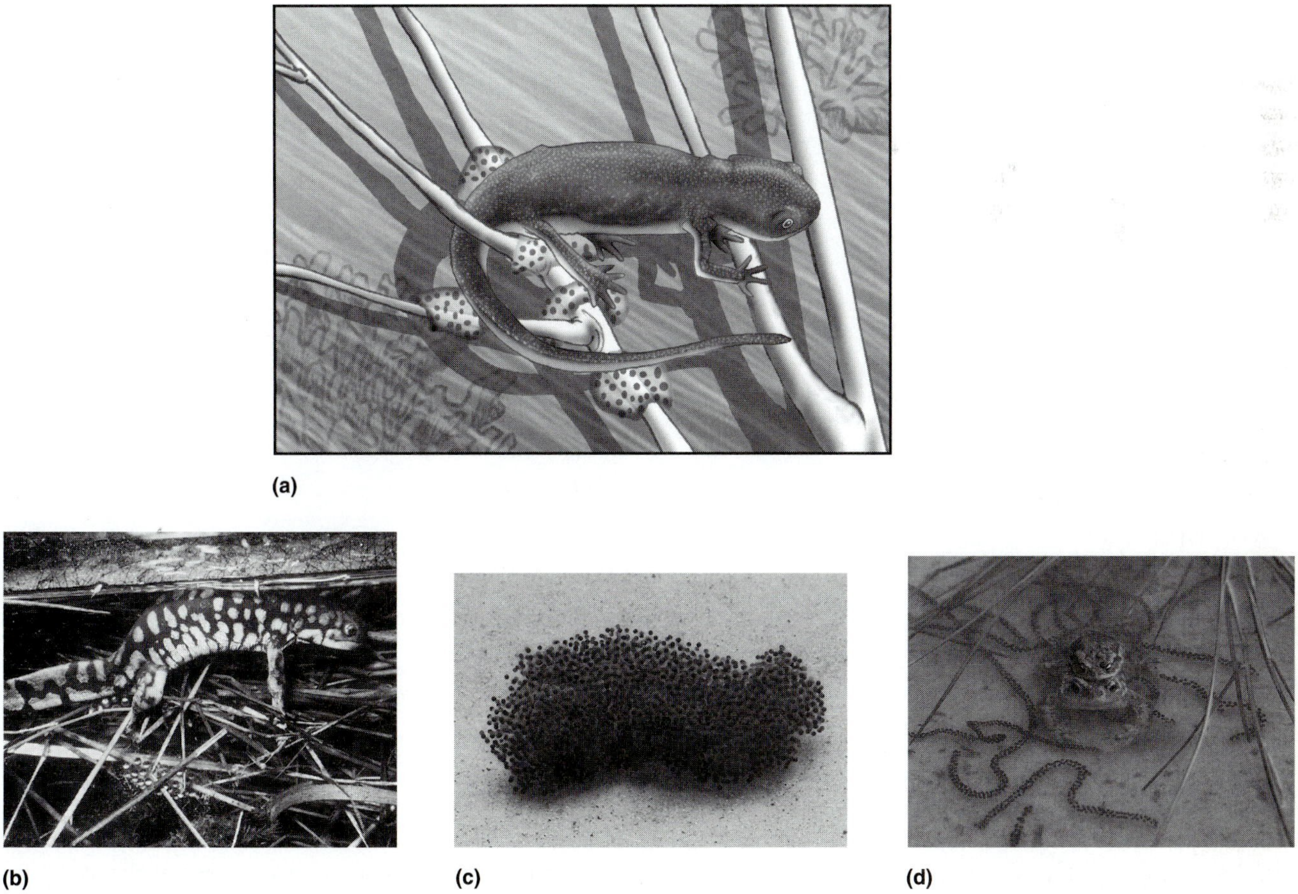

Egg deposition by amphibians. (*a*) An aquatic salamander, the California newt (*Taricha torosa*) laying her eggs in water. The clusters of eggs are usually attached to roots and stems. (*b*) A female eastern tiger salamander (*Ambystoma tigrinum*) attaching eggs to submerged vegetation. (*c*) Globular egg clusters of the leopard frog (*Rana pipiens*) in shallow water. (*d*) Egg strings of the common European toad (*Bufo bufo*) in shallow water.

TABLE 6.1

Distribution of Major Reproductive Modes Among Family Groups of Anurans

Family Group	Aquatic Eggs and Larvae	Nonaquatic Eggs, Aquatic Larvae	Foam Nest	Terrestrial Eggs, Direct Develop.	Eggs Carried by Adult
Allophrynidae	–	–	–	–	–
Arthroleptinae	–	–	–	+	–
Ascaphidae	+	–	–	–	–
Asterophrynidae	–	–	–	+	–
Astylosternidae	+	–	–	–	–
Bombinatoridae	+	–	–	–	–
Brachycephalidae	–	–	–	+	–
Brevicipitinae	–	–	–	+	–
Bufonidae	+	+	–	+[a]	–
Centrolenidae	–	+	–	–	–
Ceratophrynidae	–	–	–	–	–
Cophylinae	–	+	–	–	–
Dendrobatidae	–	–	–	–	+
Discoglossidae	+	–	–	–	–
Dyscophinae	+	–	–	–	–
Geneophryninae	–	–	–	+	–
Heleophrynidae	+	–	–	–	–
Hemiphractinae	–	+	–	–	+[b]
Hemisotidae	–	+	–	–	+
Hylinae	+	+	–	–	–
Hylodinae	–	+	–	–	–
Hyperoliidae	+	+	+	–	–
Kassininae	+	+	–	–	–
Leiopelmatidae	–	–	–	+	–
Leptodactylinae	+	–	+	+	–
Leptopelinae	–	+	–	–	–
Limnodynastinae	+	+	+	–	–
Mantellidae	+	+	–	+	–
Megophrynidae	+	–	–	–	–
Melanobatrachinae	+	+	–	–	–
Microhylinae	–	–	–	+	–
Myobatrachinae	+	+	–	+	–
Pelobatidae	+	–	–	–	–
Pelodryadinae	+	–	–	–	–
Pelodytidae	+	–	–	–	–
Petropedetinae	–	–	–	–	–
Philautinae	–	–	–	–	–
Phrynomerinae	–	–	–	–	–
Phyllomedusinae	–	–	–	–	–
Pipidae	–	–	–	–	–[b]
Pseudidae	–	–	–	–	–
Raninae	–	–	–	–	–
Rhacophoridae	+	–	–	–	–
Rhinodermatidae	–	–	–	–	–
Rhinophrynidae	+	–	–	–	–
Scaphiophryninae	+	–	–	–	–
Sooglossidae	–	–	–	+	+
Telmatobiinae	+	+	–	+[a]	–

From W.E. Duellman, "Alternative Life-history Styles in Anuran Amphibians" in M.N. Bruton, editor, Alternative Life-history Styles of Animals, *1989. With kind permission of Kluwer Academic Publishers, The Netherlands.*

[a] Also some species ovoviviparous or viviparous.
[b] Direct development (no free-living larvae) in some species.

(del Pino et al., 1975), or eggs may be attached in some other manner to the back of a parent, as in the Surinam toad (*Pipa pipa*), where they remain until hatching.

Nearly 800 species of anurans throughout the world (20 percent of the known species) have eliminated a free-living, feeding, tadpole stage. These frogs, which lay eggs that hatch into four-legged froglets, reproduce by **direct development** (Duellman, 1992). Eggs are deposited in moist sites, and in some cases such as certain species of whistling frogs (*Eleutherodactylus* sp.), they are guarded by the male parent. Whereas this method of reproduction provides the developing frogs with a food supply and protects them from aquatic predators, they are still susceptible to predation by ants and other invertebrates. The male Chilean rhinodermatid frog (*Rhinoderma darwinii*) incubates the eggs inside his vocal pouch until the young complete their metamorphosis and emerge as miniature adults. *Rheobatrachus silus* of Australia is a "gastric brooder" (Fig. 6.35). Adult females of this stream-dwelling species swallow their eggs, hatching and protecting the development of the young in their stomachs. Soon after the offspring lose their tadpolelike tails, the mothers eject the baby frogs out their mouth.

Some caecilians in the families Caeciliidae and Uraeotyphlidae are thought to be oviparous with direct development into terrestrial young (Duellman and Trueb, 1986).

Viviparous

Four African toads (*Nectophrynoides liberiensis, N. occidentalis, N. tornieri, and N. viviparus*) and one frog (*Eleutherodactylus jasperi*) are the only known anurans to give birth to live young. Female toads supplement yolk with secretions from their oviduct and give birth to newly metamorphosed toadlets (Wake, 1980b; Duellman, 1992). Some species of caecilians in the families Caeciliidae and Typhlonectidae are viviparous (Duellman and Trueb, 1986). In these species, fetuses have specialized teeth that are used for scraping the inner lining of the oviducts in order to release nutrients for ingestion. Among salamanders, this mode of reproduction occurs only in *Salamandra atra* and *Mertensiella luschani antelyana.* It may occur in montane populations of *Salamandra* and in females of *M. caucasica* subjected to prolonged drought (Duellman and Trueb, 1986).

Duration of Embryonic Development

Embryonic development in amphibians ranges from 1-2 days to approximately 9 months and is temperature-dependent (Fig. 6.36). The time from fertilization to hatching in salamanders ranges from about 13 days in some mole salamanders (*Ambystoma* spp.) to 275 days in the Pacific giant salamander (*Dicamptodon*) (Duellman and Trueb, 1986) (Table 6.2). The young of species that lay eggs early in the season, when environmental temperatures are cold, or species that live near the extremes of their ranges in both the Northern and Southern Hemispheres develop more rapidly than those that spawn later under warmer conditions. Most

FIGURE 6.35

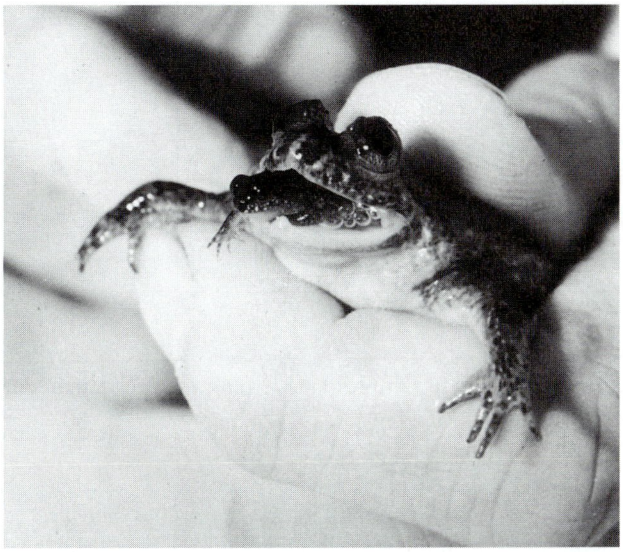

The gastric brooding frog *Rheobatrachus silus* of Australia. A froglet emerges from the mouth of its mother after having developed in her stomach for 37 days. During this time the mother does not eat, and the secretion of acid and digestive enzymes from the stomach is suppressed. This species has not been seen for several years and may be extinct.

FIGURE 6.36

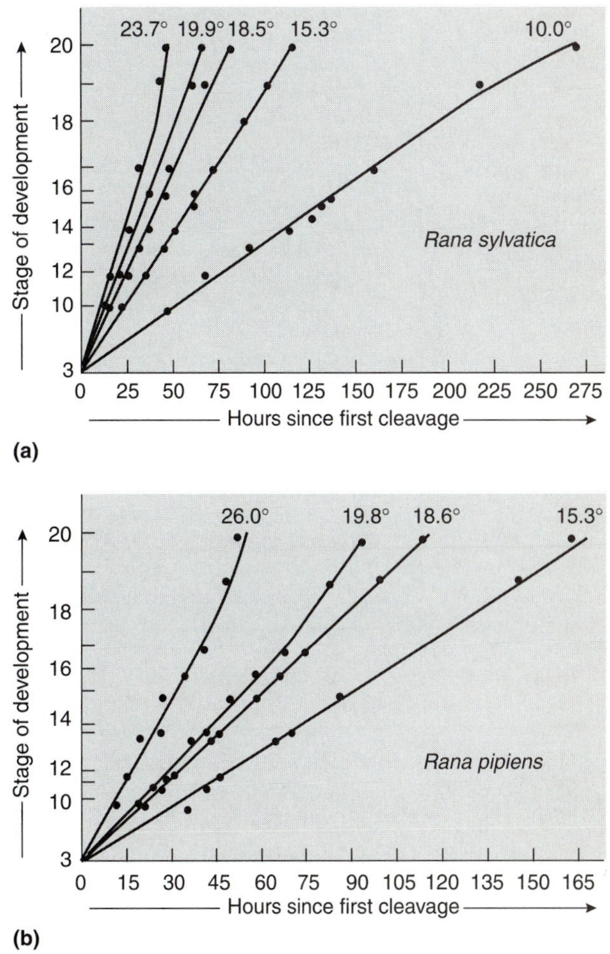

Rates of development in (a) the wood frog (*Rana sylvatica*) and (b) the leopard frog (*Rana pipiens*). The rate of development may show considerable variation among closely related species developing at the same temperature.

TABLE 6.2

Sample Salamander Reproduction Data

Family and Species	Number of Eggs	Incubation Period	Guarded by Parent
Cryptobranchidae			
Hellbender (*Cryptobranchus a. allegeniensis*)	300–450	68–84 days	Yes, male
Proteidae			
Mudpuppy (*Necturus m. maculosus*)	18–180	38–63 days	Yes, female
Sirenidae			
Dwarf siren (*Siren intermedia*)	Up to 555	?	?
Ambystomidae			
Mountain salamander (*Rhyacotriton olympicus*)	3–15	?	No
Northwestern salamander (*Ambystoma gracile*)	30–270	14–35 days	No
Marbled salamander (*Ambystoma opacum*)	50–232	15–207 days	Yes, female
Jefferson's salamander (*Ambystoma jeffersonianum*)	107–286	13–45 days	No
Salamandridae			
Red-spotted newt (*Notophthalmus v. viridescens*)	200–375	20–35 days	No
California newt (*Taricha torosa*)	7–29	18–52 days	No
Amphiumidae			
Three-toed amphiuma (*Amphiuma means tridactylum*)	42–150	30–70 days	Yes, female
Plethodontidae			
Red-backed salamander (*Plethodon c. cinereus*)	3–13	30–60 days	Yes, female
Eastern four-toed salamander (*Hemidactylium scutatum*)	30	38–60 days	Yes, female
Northern purple salamander (*Gyrinophilus p. porphyriticus*)	44–132	?	No
Arboreal salamander (*Aneides l. lugubris*)	12–19	30–60 days	Yes, female

Source: Data from J.A. Oliver The Natural History of North American Amphibians *1955, Kluwer Academic Publishers, Boston, MA.*

anurans, on the other hand, develop faster than salamanders (Table 6.3). Some, such as members of the genera *Bufo, Scaphiopus,* and *Hyla,* which have aquatic eggs, may hatch in 1 to 2 days, whereas some ranids may require 40 or more days. Rates of development at the same temperature may show considerable variation among closely related species (Fig. 6.36). Most amphibians that undergo direct development in terrestrial eggs require between 15 and about 50 days to hatch, although an average of 107 days are required for *Pipa pipa* eggs (carried on the female's back) to hatch into froglets (Duellman and Trueb, 1986). The rate of development in all amphibians is temperature-dependent (Fig. 6.36).

Hatching and Birth

Amphibian embryos produce hatching enzymes that assist in digesting the membrane enclosing them at the time of hatching. These enzymes are produced by frontal glands located primarily on the snout of the embryo. Only frogs in the genus *Eleutherodactylus* are known to possess an egg tooth with which they can cut through the egg capsule. The egg tooth is reabsorbed shortly after hatching.

TABLE 6.3

Sample Frog Reproduction Data

Family and Species	Number of Eggs	Incubation Period
Ascaphiidae		
Tailed frog (*Ascaphus truei*)	28–50	30 days
Pelobatidae		
Eastern spadefoot toad (*Scaphiopus h. holbrooki*)	1,000–2,500	5–15 days
Western spadefoot toad (*Scaphipous h. hammondi*)	1,000–2,000	2–7 days
Leptodactylidae		
Mexican white-lipped frog (*Leptodactylus labialis*)	86	40 hours
Greenhouse frog (*Eleutherodactylus ricordii planirostris*)	19–25	10–11 days
Bufonidae		
American toad (*Bufo terrestris americanus*)	4,000–8,000	3–12 days
Red-spotted toad (*Bufo punctatus*)	?	1.5–3 days
Oak toad (*Bufo quercicus*)	610–766	?
Woodhouse's toad (*Bufo w. woodhousei*)	Up to 25,650	2–4 days
Hylidae		
Northern spring peeper (*Pseudacris [Hyla] c. crucifer*)	800–1,000	?
Pacific tree frog (*Hyla regilla*)	500–1,500	7–14 days
Gray tree frog (*Hyla v. versicolor*)	Up to 1,800	4–5 days
Microhylidae		
Narrow-mouthed toad (*Microhyla c. carolinensis*)	Up to 869	?
Ranidae		
California yellow-legged frog (*Rana b. boylii*)	900–1,050	?
Bullfrog (*Rana catesbeiana*)	10,000–20,000	5–20 days
Pickerel frog (*Rana palustris*)	2,000–3,000	?
Eastern wood frog (*Rana s. sylvatica*)	2,000–3,000	10–30 days

Source: Data from J.A. Oliver The Natural History of North American Amphibians *1955, Kluwer Academic Publishers, Boston, MA.*

Parental Care

For many years, only a few amphibians were thought to exhibit parental care toward their eggs and/or offspring. In reality, however, parental care is exhibited by many species, especially those with terrestrial reproductive strategies. Extended parental care is more common in salamanders than in anurans and caecilians (Fig. 6.37). It may be provided by either sex and is shown in a variety of ways: guarding eggs against predators, moistening and/or aerating eggs, and transporting eggs and larvae. All instances involving the transport of eggs and/or larvae involve frogs. The duration of parental care for a given clutch of eggs in salamanders may extend from 5 to 6 weeks to as long as 275 days or more under certain conditions in *Dicamptodon* (Nussbaum, 1969). In anurans, parental care ranges from several days to as long as 4 months in *Gastrotheca riobambae,* which inhabits cool habitats in the Andes Mountains of South America (del Pino et al., 1975). Some female caecilians also are known to guard their eggs.

FIGURE 6.37

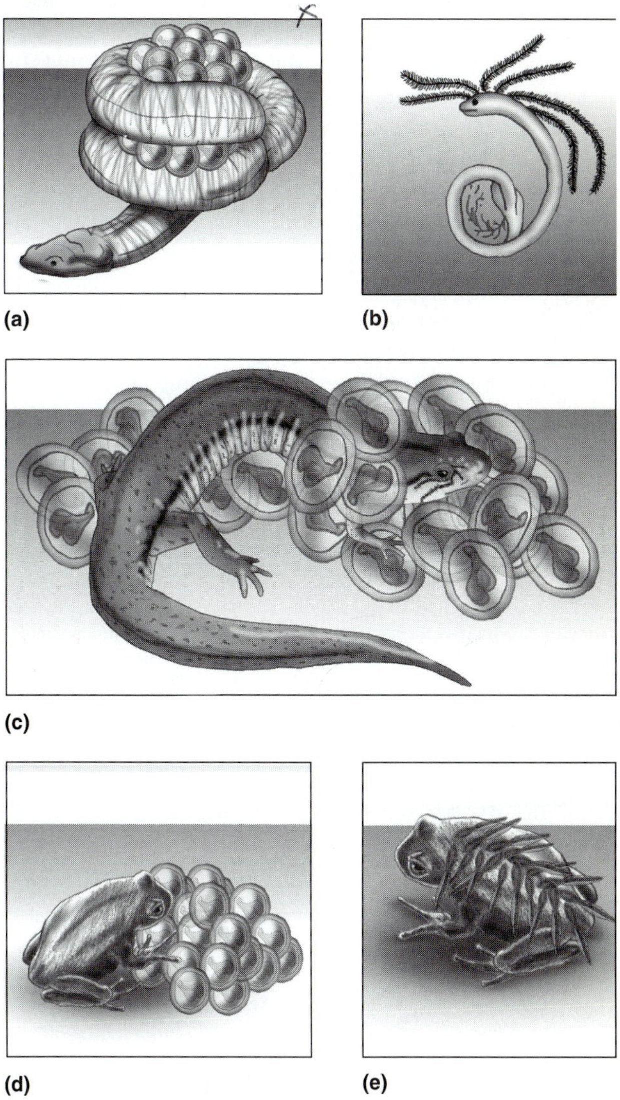

Parental care in amphibians. (*a*) Female *Ichthyophis glutinosa,* a caecilian, coiled up in a hole underground guarding her eggs. (*b*) An embryo of *Ichthyophis* nearly ready to hatch. Note gills, tail-fin, and considerable amount of yolk. (*c*) Female *Desmognathus fuscus* brooding her eggs. (*d*) The dendrobatid frog *Colostethus subpunctatus* guarding its eggs and (*e*) transporting its tadpoles to water. The tadpoles attach to the back of the frog by a sticky mucus secreted by the skin of the female.

Growth and Metamorphosis

Amphibian eggs that are deposited in water will hatch into gill-bearing larvae (salamanders) or tadpoles (anurans). In most terrestrial habitats, those eggs undergoing direct development will hatch into miniature adults. Morphological variation in salamander larvae and in tadpoles is related to their habitats (see Fig. 6.20) and/or to their methods of feeding (see Fig. 13.1).

The larval stage in anurans may be as short as 7 to 10 days in spadefoot toads (*Scaphiopus*) or as long as 2 to 3 years in bullfrogs (*Rana catesbeiana*). At the end of this time, most amphibian larvae undergo a process called **metamorphosis**—a dramatic change in their body shape as well as in their critical life support systems (Fig. 6.38). The role of hormones in controlling metamorphosis was discussed earlier in this chapter (see the section Endocrine System). During metamorphosis, lungs form so that atmospheric air can be used for respiration, and gills gradually are reabsorbed. The digestive system becomes adapted for a carnivorous diet. In anurans, the tail is reabsorbed and the limbs form. Fins that may be present on the tails of larval salamanders usually are lost. Shortly before the onset of metamorphic climax, when forelimbs and other adult features emerge, aquatic ranid tadpoles experience a deaf period of 2 to 4 days during which no auditory activity can be detected (Boatright-Horowitz and Simmons, 1998). Research suggests that a growing bit of cartilage important to adult hearing disables the tadpole's hearing before the system matures.

Some amphibians do not undergo complete metamorphosis, but retain some of their larval characteristics as adults. The attainment of sexual maturity with the retention of at least some larval morphology is known as **neoteny**. Reproduction by larval individuals is known as **paedogenesis**. Neotenic species include sirens (*Siren* spp.) and mudpuppies (*Necturus* spp.), both of which retain their external gills and never leave the water.

The eastern red-spotted newt (*Notophthalmus viridescens*) may metamorphose into an adult by passing through two "larval" stages. Eggs hatch into aquatic larvae. After several months, the aquatic larva metamorphoses into a terrestrial "larval," or juvenile, stage known as a red eft. This carnivorous stage develops lungs, leaves the pond, and lives a terrestrial existence for several years. After that time, it returns to water, develops fins on the dorsal and ventral surfaces of its tail, and lives the remainder of its adult life as a lung-breathing aquatic salamander.

Attainment of Sexual Maturity

Most amphibians can breed the year following their birth. There can be, however, significant differences in the age of sexual maturation within the same species at different geographic sites (Table 6.4). Jorgensen (1992) noted: "The difference in age at sexual maturation among toads from various geographical locations may not arise entirely from different climatic environments, but may also be genetically fixed, determined by relationships between growth, body size, and sexual maturation that are established independently of particular climatic conditions."

The bullfrog (*Rana catesbeiana*), which may spend 1 year as a tadpole in the southern portion of its range, may remain

FIGURE 6.38

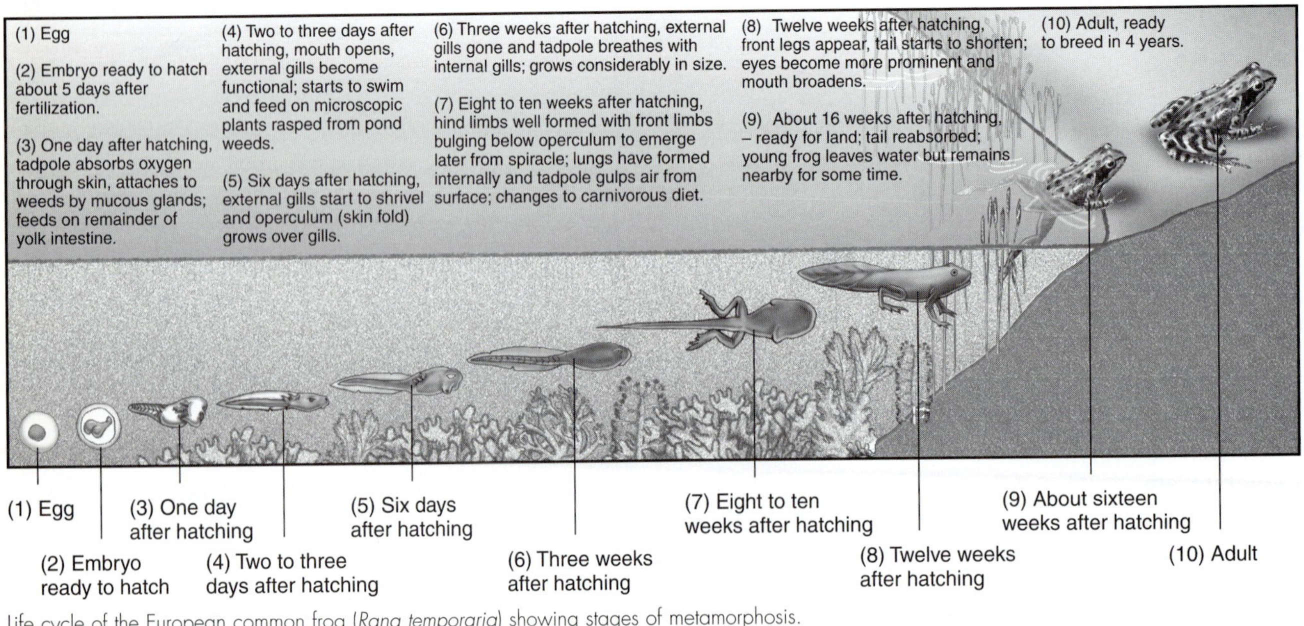

Life cycle of the European common frog (*Rana temporaria*) showing stages of metamorphosis.
Source: Halliday and Adler (eds.), The Encyclopedia of Reptiles and Amphibians, *1986, Facts on File.*

TABLE 6.4

Age at Sexual Maturation in Populations of the Toad *Bufo bufo* from Different Sites in Europe

Locality	Age at Sexual Maturation (years) Males	Females
Norway	4–7	6–9
Netherlands	2–3	3–4
Germany	3–4	4–5
Switzerland	6–9	8–10
France	3–5	4–7

From M.E. Feder and W.W Burggren, editors, Environmental Physiology of the Amphibians, *1992. Copyright © 1992 University of Chicago Press. Reprinted by permission.*

a tadpole for up to 3 years in the northern part. Thus, it might require 4 or 5 years for this species to become sexually mature. The European common frog (*Rana temporaria*) normally metamorphoses in about 16 weeks but does not become sexually mature for 4 years (Fig. 6.38).

Review Questions

1. What adaptive advantage is gained by having two bones fuse into a single structure such as the tibiofibula in anurans?
2. What adaptive advantages are there in having the limbs positioned beneath the body rather than out to the sides as in salamanders?
3. Why has the metameric arrangement of muscle, as seen in fishes, gradually been lost in higher vertebrates?
4. Compare and contrast the flow of blood through the heart of a shark with the flow through the heart of an adult anuran during a single circuit through the body.
5. Describe several different methods of gas exchange used by amphibians.
6. Describe the force-pump mechanism used by amphibians.
7. How is the length of the intestine correlated with herbivorous and carnivorous species? Why is this so?
8. What is the significance of having three semicircular canals in the inner ear, each lying in a different plane?
9. What is meant by sexual dimorphism? Give several examples.
10. Explain how environmental factors control the periodicity of breeding in ectotherms.
11. Define paedomorphosis.
12. Discuss several methods of egg deposition used by oviparous amphibians.
13. Give several advantages and several disadvantages of direct development in amphibians.
14. How do environmental factors such as rainfall and temperature affect egg and larval development in amphibians?
15. Give some examples of evidence supporting either the lungfish, rhipidistian, or coelacanth as the ancestor of the amphibians.
16. Discuss the problems faced by a vertebrate emerging from water to live on land. Have amphibians successfully solved all of the problems? If so, how?

Supplemental Reading

Bogert, C. M. 1998. *Sounds of North American Frogs: The Biological Significance of Voice in Frogs.* Washington, D.C.: Smithsonian Folkways Recordings.

Cochran, D. M. 1961. *Living Amphibians of the World.* New York: Doubleday and Company.

Duellman, W. E. 1989. Alternative life-history styles in anuran amphibians: Evolutionary and ecological implications. In: Bruton, M. N. (ed.). *Alternative Life History Styles of Animals.* New York: Kluwer Academic Publishers.

Duellman, W. E. 1992. Reproductive strategies of frogs. *Scientific American* 267(1):80–87.

Duellman, W. E. (ed.) 1999. *Patterns of Distribution of Amphibians.* Baltimore: The Johns Hopkins University Press.

Duellman, W. E. and L. Trueb. 1986. *Biology of Amphibians.* New York: McGraw-Hill Publishing Company.

Feder, M. E., and W. W. Burggren. 1992. *Environmental Physiology of the Amphibians.* Chicago: University of Chicago Press.

Fritzsch, B., M. J. Ryan, W. Wilczynski, T. E. Hetherington, and W. Walkowiak. 1988. *The Evolution of the Amphibian Auditory System.* New York: John Wiley and Sons.

Halliday, T. R., and K. Adler. 1986. *The Encyclopedia of Reptiles and Amphibians.* New York: Facts on File, Inc.

McDiarmid, R.W. and R. Altig. 1999 *Tadpoles: The Biology of Anuran Larvae.* Chicago: University of Chicago Press.

Marks, S. B., and A. Collazo. 1998. Direct development in *Desmognathus aeneus. Copeia* 1998 (3):637–648.

Moffett, M. W. 1995. Poison-dart frogs. *National Geographic* 187(5):98–111.

Phillips, K. 1994. *Tracking the Vanishing Frog: An Ecological Mystery.* New York: St. Martin's Press.

Pough, F. H., R. M. Andrews, J. E. Cadle, M. L. Crump, A. H. Savitzky, and K. D. Wells. 1998. *Herpetology.* Upper Saddle River, New Jersey: Prentice Hall.

Schwartz, A., and R. W. Henderson. 1991. *Amphibians and Reptiles of the West Indies: Descriptions, Distributions, and Natural History.* Gainesville, Florida: University of Florida Press.

Stebbins, R. C., and N. W. Cohen. 1995. *A Natural History of Amphibians.* Princeton: Princeton University Press.

Taylor, E. H. 1968. *The Caecilians of the World: A Taxonomic Review.* Lawrence: University of Kansas Press.

Tyler, M. J. (ed.). 1983. *The Gastric Brooding Frog.* Croom Helm Ltd., distributed by Routledge, New York.

Zimmer, C. 1998. *At The Water's Edge: Macroevolution and theTransformation of Life.* New York: Free Press.

Zug, G. R. 1993. *Herpetology. An Introductory Biology of Amphibians and Reptiles.* New York: Academic Press.

SELECTED JOURNALS

Amphibia-Reptilia. Published by E. J. Brill, Leiden, New York. Multidisciplinary journal devoted to all aspects of herpetology. Seeks to further the interchange of knowledge and ideas between specialists on different subjects, ranging from the use of biochemical techniques in herpetological research to electron microscopy, from amphibian and reptilian genetics to the fossil history of the modern amphibians and reptiles, including systematics and taxonomy as well as behavior, physiology, and ecology.

Copeia. Published by the American Society of Ichthyologists and Herpetologists. Publishes results of original research performed by members in which fish, amphibians, or reptiles are utilized as study organisms.

Herpetologica. Published by the Herpetologists' League. Dedicated to furthering knowledge of the biology of amphibians and reptiles.

Herpetological Natural History. Published by La Sierra University, La Sierra, California. Peer-reviewed journal publishing manuscripts containing new results or observations concerning the natural history of amphibians and reptiles (e.g., ecology, behavior, evolution, life history, biogeography, paleontology, conservation biology, surveys).

Journal of Herpetology. Published by the Society for the Study of Amphibians and Reptiles. Aims to increase knowledge of amphibians and reptiles and provide effective communication among herpetologists and other biologists interested in amphibians and reptiles.

Vertebrate Internet Sites

Visit the zoology website at http://www.mhhe.com to find live Internet links for each of the references listed below.

1. **Amphibia: Fossil Record.**
   Information on extinct amphibians.
2. **Introduction to the Tetrapoda.**
   Information on tetrapod fossils, natural history, ecology, systematics, and morphology.
3. **International Reptiles and Amphibian Websites Directory.**
   "Herp Hot Links."
4. **Amphibia.**
   From Biosis, many links with information on amphibians.
5. **Animal Diversity Web, University of Michigan.**
   Class Lissamphibia. Links containing information and pictures of caecilians, frogs and toads, and salamanders are included.
6. **Subphylum Vertebrata, Class Amphibia, from the University of Minnesota.**
7. **Society for the Study of Amphibians and Reptiles.**
   Resources, links, information on conservation, and more.
8. **American Society of Ichthyologists and Herpetologists.**
   Links, also information on their publication, *Copeia.*
9. **Savannah River Ecology Laboratory's Herpetology Lab Home Page.**
   The University of Georgia's SREL has been involved in research on reptiles and amphibians since the 1960s. This site has pictures, research summaries, and links to more information on "herps."
10. **North American Amphibian Monitoring Program.**
    Information and links to information on the study and conservation efforts for amphibians. Supported by the USGS and the Patuxent Wildlife Research Center.
11. **Great Lakes Declining Amphibians Working Group.**
    A plethora of links available from this site.
12. **The Virtual Frog Project.**
    This was designed at the Lawrence Berkeley National Laboratory using X-ray CT imaging, MRI imaging, and other visualization techniques to make clickable 3-D images of the frog and its internal organs.

CHAPTER 7

Evolution of Reptiles

■ INTRODUCTION

The class Reptilia is no longer recognized by phylogenetic systematists, because it is not a monophyletic group. Traditionally, the class Reptilia included the turtles, tuatara, lizards, snakes, and crocodilians. Birds, which descend from the most recent common ancestor of reptiles, have traditionally been classified by themselves in the class Aves. Reptiles, therefore, are a paraphyletic group unless birds are included. Furthermore, based on shared derived characteristics, crocodilians and birds are more recently descended from a common ancestor than either is from any living reptilian lineage; thus, they are sister groups.

In phylogenetic systematics (cladistics), turtles, tuataras, lizards, snakes, crocodilians, and birds are placed in the monophyletic group Sauropsida. The Sauropsida include three groups: turtles (Testudomorpha); tuataras, lizards, and snakes (Lepidosauromorpha); and the crocodilians and birds (Archosauromorpha). In this method of classification, turtles are placed at the base of the tree. New evidence from 2 nuclear genes and analyses of mitochondrial DNA and 22 additional nuclear genes join crocodilians with turtles and place squamates at the base of the tree (Hedges and Poling, 1999; Rieppel, 1999). Morphological and paleontological evidence for this phylogeny are unclear at the present time.

Considerable disagreement continues between proponents of evolutionary (traditional) taxonomy and cladistics. The classification used in this text, for the most part, will follow the cladistic method. Comparisons between the two classification methods will be presented at appropriate points. For ease of discussion, we will divide the reptiles (sauropsids) into two chapters: Evolution (this chapter) and Morphology, Reproduction, and Growth and Development (Chapter 8).

■ EVOLUTION

The fossil record for reptiles is much more complete than the one for amphibians. Based on current evidence, all lineages of modern reptiles can be traced back to the Triassic period (Fig. 7.1). Disagreement, however, exists concerning origins and relationships prior to the Triassic and whether reptiles had a monophyletic, diphyletic, or even a polyphyletic origin. Molecular investigations, including comparative protein sequence studies of amniote (sauropsids and mammals) myoglobins and hemoglobins (Bishop and Friday, 1988), are shedding new light on reptilian relationships. A cladogram giving one interpretation of the relationships among the amniotes is presented in Fig. 7.2.

Molecular geneticists are attempting to extract intact DNA from dinosaur bones and from vertebrate blood in the gut of amber-preserved biting insects whose last meal might have been taken from a dinosaur (Morrell, 1993a). Although a report exists of DNA being extracted from 80-million-year-old dinosaur bones (Woodward, 1994), most molecular evolutionists feel that the DNA came instead from human genes that contaminated the sample (Stewart and Collura, 1995; Zischler, et al., 1995).

Ancestral Reptiles

The earliest amniote skeleton comes from the Lower Carboniferous of Scotland, approximately 338 million years ago (Smithson, 1989). More recently, the same site yielded another Lower Carboniferous tetrapod, *Eucritta melanolimnetes*, which exhibits characters from three different types of primitive tetrapods: temnospondyls (relatives of living amphibians), anthracosaurs (amniotes and their close relatives), and baphetids (crocodile-like body with a unique keyhole-shaped orbit) (Clack, 1998). Since temnospondyls and anthracosaurs have previously been found at this site between Glasgow and Edinburgh, it has been hypothesized that at least three different lineages of early tetrapod may have independently evolved into medium-sized fish-eating animals. This is but one of numerous examples of parallel evolution in vertebrates.

Most recently, the smallest of all known Lower Carboniferous tetrapods, *Casineria kiddi* with an estimated snout-vent length of 85 mm, was reported from East Lothian, Scotland (Paton et al., 1999). *Casineria* shows a variety of

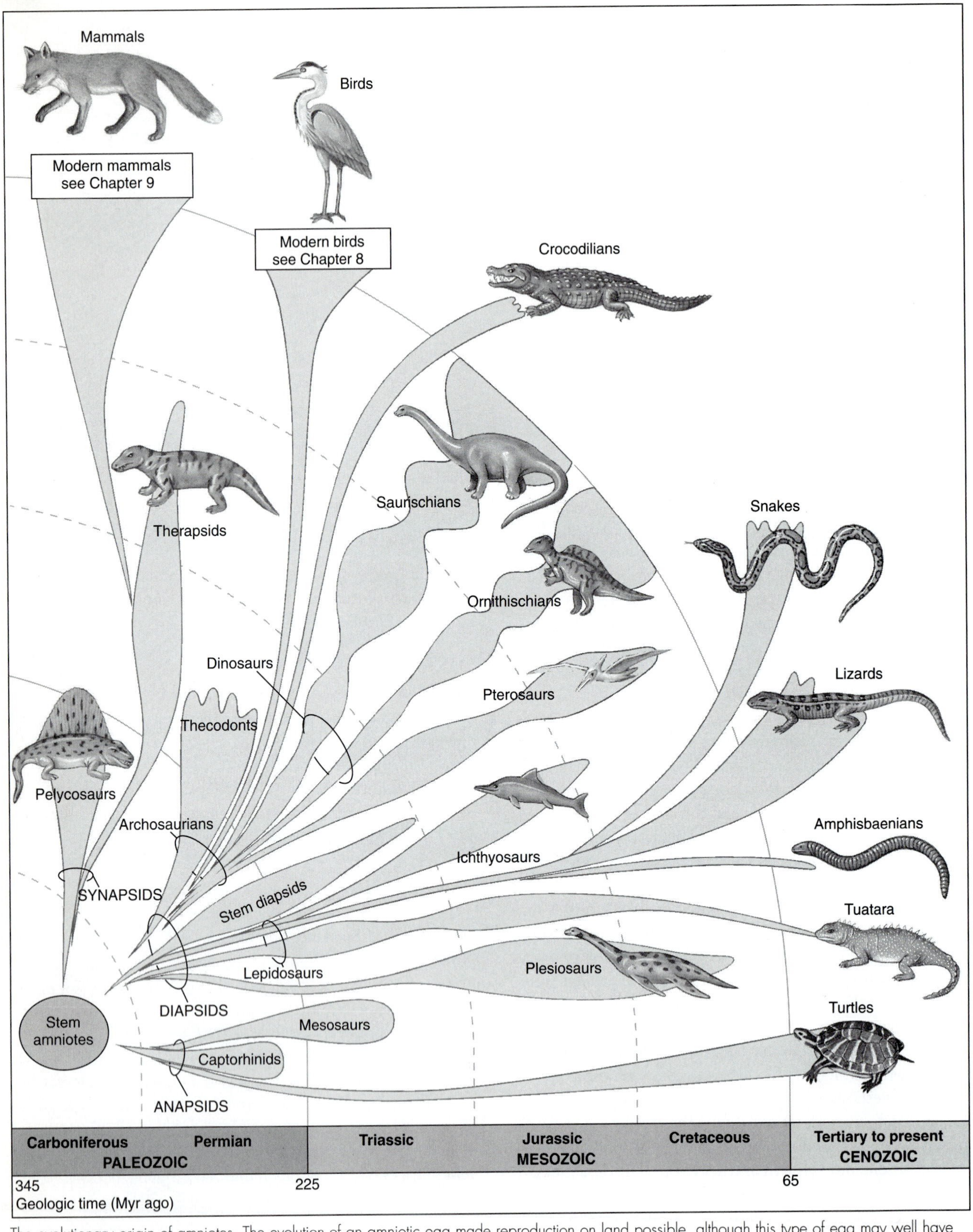

The evolutionary origin of amniotes. The evolution of an amniotic egg made reproduction on land possible, although this type of egg may well have developed before the earliest amniotes had ventured far onto land. The amniotes (reptiles, birds, and mammals) evolved from small lizardlike forms known as captorhinids that retained the skull pattern of the early tetrapods. The mammal-like reptiles, which were the first to diverge from the primitive stock, possessed synapsid skulls. All other amniotes, except turtles, have a diapsid skull. Turtle skulls are of the anapsid type. The great Mesozoic radiation of reptiles may have been caused partly by the increased variety of ecological habitats available for the amniotes.

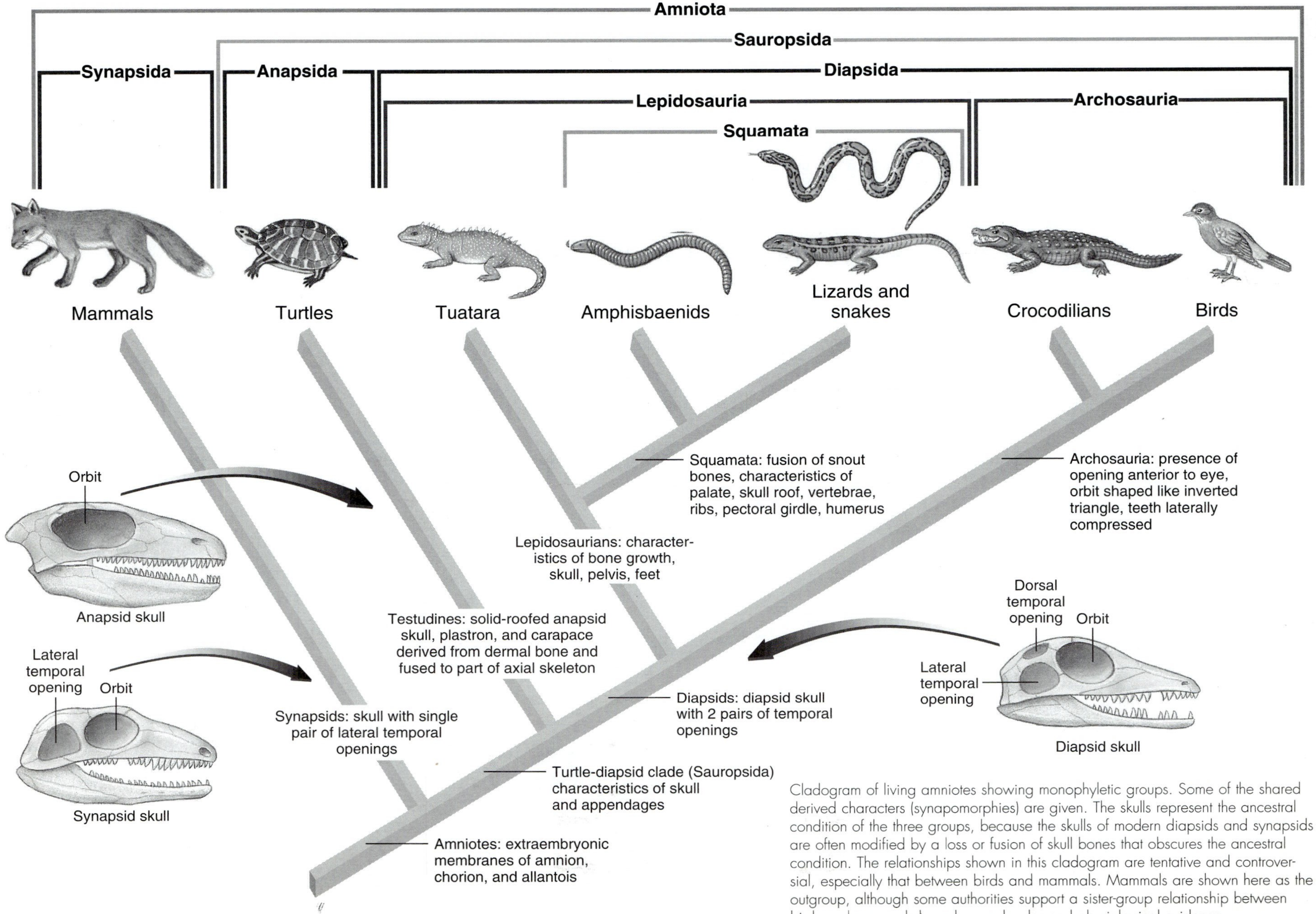

Cladogram of living amniotes showing monophyletic groups. Some of the shared derived characters (synapomorphies) are given. The skulls represent the ancestral condition of the three groups, because the skulls of modern diapsids and synapsids are often modified by a loss or fusion of skull bones that obscures the ancestral condition. The relationships shown in this cladogram are tentative and controversial, especially that between birds and mammals. Mammals are shown here as the outgroup, although some authorities support a sister-group relationship between birds and mammals based on molecular and physiological evidence.

adaptations to terrestrial life. For example, vertebrae are connected to each other to form a relatively stiff backbone, which would have served as a suspension bridge to hold up the animal's body. *Casineria* also possessed the earliest pentadactyl limb, which is clearly terrestrially adapted. The humerus had a constricted shaft and exhibited torsion between proximal and distal articulations, features associated with the maintenance of postural support and strong evidence of locomotion on land. All limbs described from earlier Late Devonian animals, such as *Ichthyostega* and *Acanthostega*, possessed more than five digits and belonged to arguably aquatic forms (Paton et al., 1999). The authors note that the degree of terrestriality exhibited by *Casineria* indicates that the transition to land-dwelling may have taken place within a period of about 20 million years.

By the end of the Carboniferous (about 286 million years ago), at least two phylogenetic lines of reptiles existed: the pelycosaurs (order Pelycosauria) and the more primitive captorhinids (suborder Captorhinomorpha of the order Cotylosauria). Both of these forms have been found together in deposits approximately 300 million years old in Nova Scotia. Because of their similarity, some investigators believe that they probably evolved from a common ancestor in the Early Carboniferous (Carroll, 1988). Romer's (1966) observation, that the development of the amniote egg was so complex and so uniform among reptiles that it is not likely it could have evolved independently in two or more different groups of amphibians, lends additional weight to the belief that the origin of reptiles was monophyletic. Carroll (1988) noted that by the Upper Carboniferous, amniotes had diverged into three major lineages: synapsids gave rise to mammals, anapsids to turtles, and diapsids to all of the other reptilian groups including birds.

Members of the order Anthracosauria (subclass Labyrinthodontia) most closely resemble the primitive captorhinomorphs. One group of these amphibians, the seymouriamorphs (suborder Seymouriamorpha), possessed a combination of amphibian and reptilian characteristics. The best known genus of this group is *Seymouria*, discovered in lower Permian deposits near Seymour, Texas (Fig. 7.3). Although *Seymouria* lived too recently to have been ancestral to the reptiles, it is thought to be an advanced member of a more primitive group of amphibians that did give rise to the original reptiles. *Seymouria* had a relatively short vertebral column, an amphibian-like skull, and well-developed limbs and girdles (Fig. 7.3). The neural arches, however, were similar to those found in reptiles, and the dentition had a distinctly reptilian aspect with teeth set in shallow pits. *Seymouria* had a single occipital condyle, as did primitive amphibians and reptiles.

Seymouria appears to have been clearly capable of living on land and probably of supporting its body above the ground. *Seymouria* probably lived part of the time on land and part in pools and swamps, where it fed on small fish as well as on aquatic and terrestrial invertebrates. Carroll (1969) believed that, although adults appeared to be adapted for life on dry land, they were phylogenetically, morphologically, and physiologically amphibian.

A fundamental difference between amphibians and reptiles involves the type of egg produced and the method of development of the young. Amphibians have an anamniotic embryo (one without an amnion) that must always be deposited in water or in a moist habitat. In most species of amphibians, fertilized eggs will develop into aquatic larvae. Numerous labyrinthodont amphibians are known to have

FIGURE 7.3

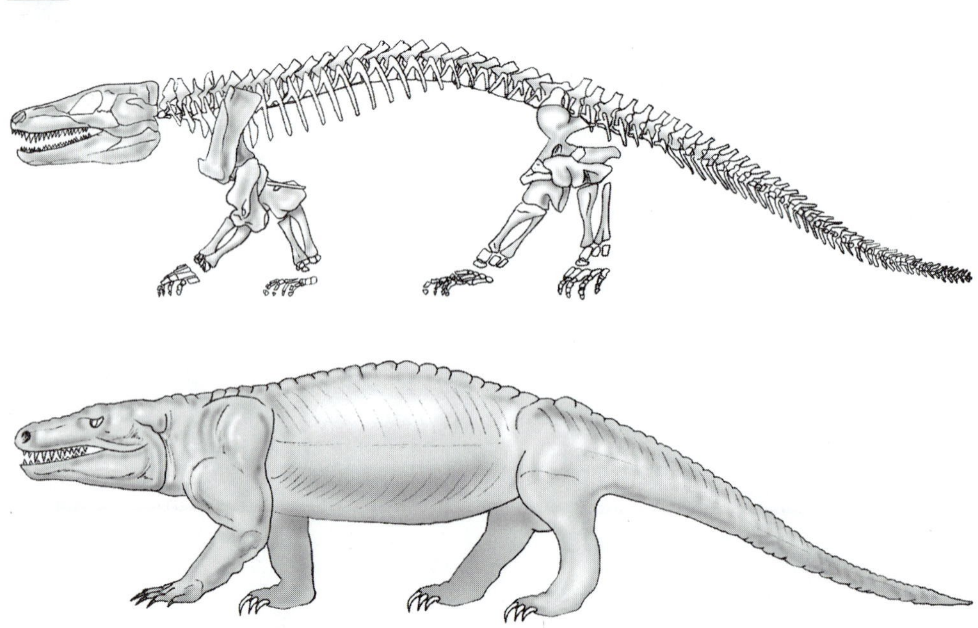

Seymouria, a primitive genus of reptile with well-developed limbs positioned beneath the body, providing better support. Estimated total length of the skeleton is approximately 0.8 m.

had larval stages with external gills, as do many living amphibians (Carroll, 1969). Most reptiles, on the other hand, produce an egg sealed in a leathery shell that is much more resistant to dessiccation (Fig. 7.4). Four extraembryonic membranes are present inside the leathery shell: a **chorion** (outer membrane surrounding the embryo that assists in gas exchange and in forming blood vessels); an **amnion** (inner membrane surrounding the embryo forming the amniotic cavity and containing amniotic fluid); a **yolk sac** (enclosing the yolk); and an **allantois** (forming a respiratory structure and storing nitrogenous waste). Reptiles lack a larval stage and, following hatching, develop directly into the adult form.

Unfortunately, little fossil evidence is available concerning eggs and early developmental stages of primitive reptiles, because eggs do not generally fossilize well. The oldest fossil amniote egg was found in Early Permian deposits in Texas (Romer and Price, 1939). It was 59 mm in length and was probably laid by a pelycosaur, the most common member of the fauna (Romer and Price, 1940).

How long young dinosaurs remained in their nest has been debated for many years. Some scientists have argued that the thigh bones of newly hatched dinosaurs were not formed well enough to support their weight. Geist and Jones (1996), however, examined the pelvic girdles of some living relatives of dinosaurs—crocodiles and birds. The pelvis starts out as soft cartilage, and later it becomes hard due to the deposit of minerals. Geist and Jones found that in animals that can walk immediately after birth—such as crocodiles, emus, and ducks—the pelvis is bony by hatching time. But in animals that cannot walk immediately, the pelvis is not fully hardened at birth. Of the five dinosaur species for which embryos have been found, all had bony pelvises while they were still in the egg, implying that they could stand upright at birth.

FIGURE 7.4

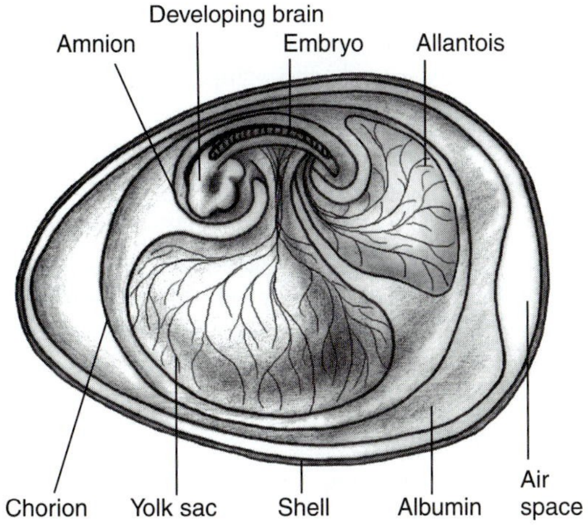

Generalized structure of the amniotic egg. Its membranes—chorion, amnion, yolk sac, and allantois—protect the embryo and provide it with metabolic support.

Romer (1957) expressed the belief that the earliest reptiles were amphibious or semiaquatic, as were their immediate amphibian ancestors. The amniotic egg was developed by such semiaquatic animals, not by a group of animals in which the adults had already become terrestrial. Romer stated, "although the terrestrial egg-laying habit evolved at the beginning of reptilian evolution, adult reptiles at that stage were still essentially aquatic forms, and many remained aquatic or amphibious long after the amniote egg opened up to them the full potentialities of terrestrial existence. It was the egg which came ashore first; the adult followed."

Tihen (1960) agreed with Romer regarding the origin of the amniote egg. He pointed out that the terrestrial egg probably developed in order to avoid "the necessity for an aquatic existence during the particularly vulnerable immature stages of the life history." In addition, Tihen suggested that the development of the terrestrial egg occurred under "very humid, probably swampy and tropical, climatic conditions," rather than during a period of drought. A generalization such as "drought" during a portion of a geological period does not accurately indicate conditions on a regional and/or local level. Areas in close proximity to one another can have vastly different environmental conditions. In support of his theory, Tihen cited examples of modern amphibians living in areas where the water supply is intermittent and undependable. Rather than deposit their eggs on the fringes of the water, they deposit them "more positively within" the available bodies of water. Because most amphibians that deposit terrestrial eggs live in humid habitats, Tihen believed terrestrial eggs evolved as a device for escaping predation, *not* for avoiding dessiccation. Furthermore, he noted that in the early stages of its evolution, the amniote egg must have been quite susceptible to dessiccation and that only after the specializations that now protect it (extraembryonic membranes) had been developed could it have been deposited in even moderately dry surroundings.

Eggs and young of *Seymouria* are unknown. However, gilled larvae of a closely related seymouriamorph (*Discosauriscus*) have been discovered (Porter, 1972). The presence of gilled larvae indicates that these were definitely amphibians even though they were quite close to the reptilian phylogenetic line of development.

Were the earliest reptiles aquatic, coming onto land only to deposit their amniotic eggs as turtles do today, or were they primarily terrestrial animals? Did the amniotic egg evolve in response to drought conditions, or did it evolve as a means to protect the young from the dangers of aquatic predation? These questions continue to be the subject of much debate.

Ancient and Living Reptiles

Reptiles were the dominant terrestrial vertebrates during most of the Mesozoic era. There were terrestrial, aquatic, and aerial groups. Quadrupedal and bipedal groups existed, as did carnivorous and herbivorous groups. One group gave rise to the mammals in the late Triassic. As many as 22 orders

BIO-NOTE 7.1

Dinosaur Nests and Eggs

Although the first publicized dinosaur nests and eggs were discovered in Mongolia in 1923 (Andrews, 1932; Brown and Schlaikjer, 1940; Norman, 1991), Carpenter et al. (1994) noted that dinosaur eggs have been known for thousands of years and that the first dinosaur egg shell in historical times can be traced back to 1859, in southern France (Buffetaut and LeLoewff, 1989). The Mongolian eggs were originally identified as being from *Protoceratops*, a small ceratopsian dinosaur, but later were reidentified as being from a theropod dinosaur in the family Oviraptoridae (Norrell et al., 1994). The first nest containing the remains of a baby dinosaur (*Mussaurus*) was reported in 1974 from Argentina (Bonaparte and Vince, 1974).

The best known dinosaur nest (containing crushed egg shells as well as the skeletons of baby hadrosaurs) was discovered in 1978, in Montana (Horner, 1984; Horner and Gorman, 1988). The nest was approximately 1.8 m in diameter and 0.9 m deep and contained the fossilized remains of 15 one-meter-long duckbill dinosaurs (*Maiasaura*, meaning "good mother"). It provided evidence that, unlike most reptiles, these young had stayed in the nest while they were growing and that one or both parents had cared for them. The teeth were well worn, indicating that the young had been in the nest and had been eating there for some time. Analysis of the hatchlings' bones revealed bone tissue that grows rapidly, the same way the bones of modern birds and mammals grow. The implications are that the young must have been developing rapidly and that they were probably homeothermic (Horner and Gorman, 1988).

Clusters of nests that were found indicate that female *Maiasaura* and *Orodromeus* laid their eggs and raised their young in colonies, as do some species of birds. The discovery of large fossil beds containing individuals of all ages led Bakker (1986), Horner and Gorman (1988), and Horner (1998, 1999) to conclude that some dinosaurs, including *Apatosaurus* (*Brontosaurus*) and *Maiasaura*, lived in large herds. Many of the bones of these dinosaurs were either unbroken or showed clean breaks indicating they had been broken *after* fossilization. In 1979, a clutch of 19 eggs containing embryonic skeletons of *Troodon* (originally misidentified as *Orodromeus*; Moffat, 1997) was found in Montana. One was fully articulated and was the first such embryonic dinosaur skeleton ever unearthed (Horner and Gorman, 1988). Carpenter and Alf (1994) surveyed the global distribution of dinosaur eggs, nests, and young. More recently, numerous nests and eggs containing embryos have been recovered from exceptionally rich fossil sources in China (O'Brien, 1995), along the seashore in Spain (Sanz et al., 1995), and in Mongolia (Dashzeveg et al., 1995). The oldest dinosaur embryo, probably a theropod, was reported from 140-million-year-old Jurassic sediments from Lourinha, Portugal (Holden, 1997).

In 1994, researchers from the American Museum of Natural History and the Mongolian Academy of Sciences announced the discovery of the fossilized remains of a 3-m carnivorous dinosaur (*Oviraptor*) nesting on its eggs like a brooding bird (Gibbons, 1994; Norell et al., 1994). This nest and its brood of unhatched young were discovered in the Gobi Desert of Mongolia and represent the first concrete proof that dinosaurs actively protected and cared for their young.

Thousands of sauropod dinosaur eggs were discovered at Auca Mahuevo in Patagonia, Argentina (Chiappe et al., 1998). The proportion of eggs containing embryonic remains is high at this Upper Cretaceous site—more than a dozen *in situ* eggs and nearly 40 egg fragments encasing embryonic remains. In addition, many specimens contained large patches of fossil skin casts, the first portions of integument ever reported for a nonavian dinosaur embryo.

of reptiles have, at one time or another, inhabited the Earth, but their numbers have decreased until living representatives of only 4 orders remain. Living reptiles (and mammals) are thus the descendents of the great Mesozoic differentiation of the ancestral reptiles.

The traditional classification of reptiles is based on a single key character: the presence and position of **temporal fenestrae**, which are openings in the temporal region of the skull that accommodate the jaw musculature (Fig. 7.5). These criteria, using only Paleozoic taxa, yield three groups:

Anapsida: turtles, captorhinomorphs, procolophonids, and pareiasaurs
Diapsida: dinosaurs, tuataras, lizards, snakes, crocodiles, and birds
Synapsida: mammal-like reptiles

Rieppel and deBraga (1996), however, adopted a more inclusive perspective by adding Mesozoic and extant taxa to the analysis. Their studies support diapsid affinities for turtles and require the reassessment of categorizing turtles as "primitive" reptiles in phylogenetic reconstructions. Platz and Conlon (1997) also concluded that turtles should be considered diapsids, by determining the amino acid sequence of pancreatic polypeptide for a turtle and comparing it with published sequences for 14 additional tetrapod taxa. Other researchers (Wilkinson et al., 1997; Lee, 1997), however, question the analysis of the data presented by Rieppel and deBraga.

In the phylogenetic (cladistic) classification, anapsid turtles are placed in the Testudomorpha, whereas all of the diapsid forms (tuataras, lizards, and snakes) make up the Lepidosauromorpha (lepidosaurs), and crocodilians and birds compose the Archosauromorpha (archosaurs).

Turtles (Testudomorpha)

Turtles (see Figs. 1.4, page 3, and 7.2) are anapsid reptiles that lack fenestrae (openings) in the temporal regions of their skulls. Cotylosaurs, or stem reptiles (order Cotylosauria), first appeared in the early Carboniferous and had anapsid skulls. One of the oldest known cotylosaur reptiles, *Hylonomus* is a captorhinomorph—a group frequently cited

FIGURE 7.5

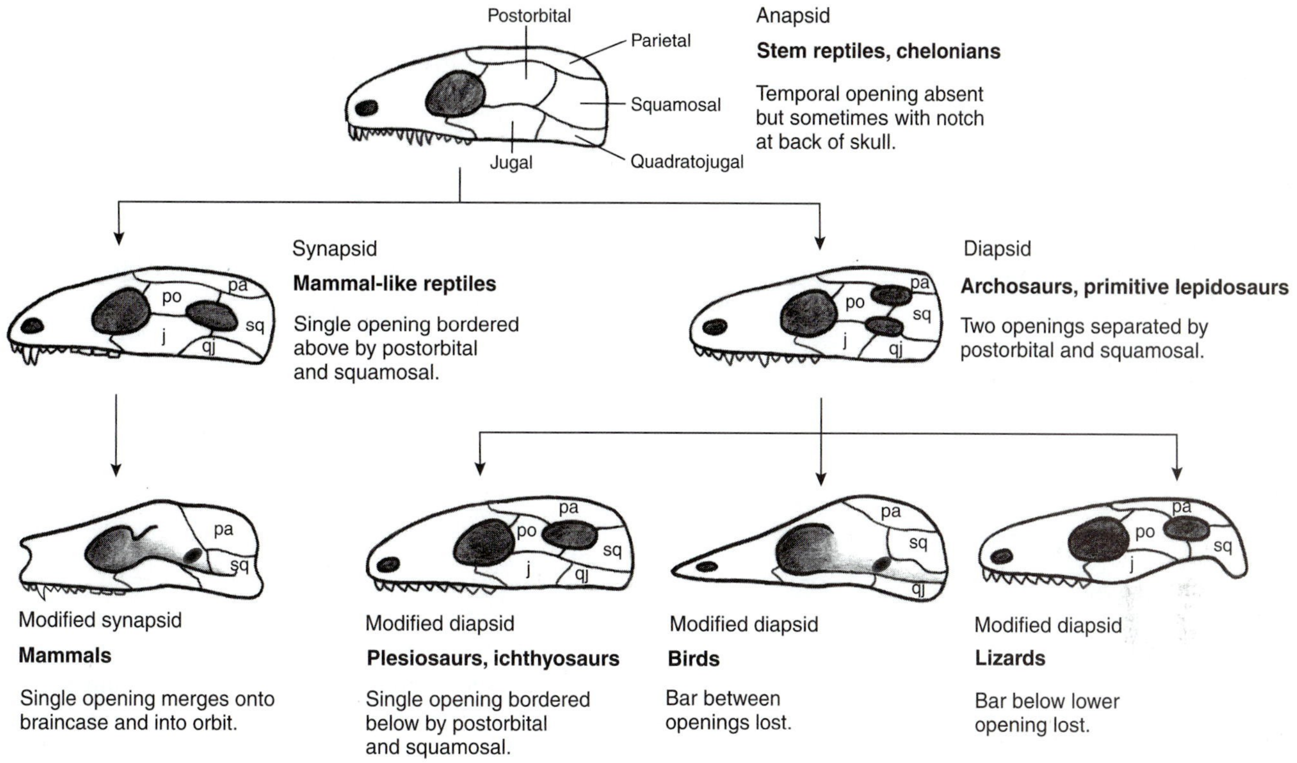

Phylogeny constructed by comparing temporal fenestrae of reptiles and their descendants.
From Hildebrand, Analysis of Vertebrate Structure, *4th edition. Copyright © 1995 John Wiley & Sons, Inc. Reprinted by permission of John Wiley & Sons, Inc.*

as the possible primitive relatives of turtles. Reisz and Laurin (1991), however, present new evidence showing that a group of primitive amniotes, the procolophonids (Fig. 7.6), were the closest sister group of turtles. If true, the origin of turtles may be as late as the Late Permian. Lee (1993), however, considered the evidence uniting captorhinid and procolophonoids with turtles to be weak and instead proposed the pareiasaurs as the nearest relatives of turtles. Pareiasaurs were large anapsid reptiles that flourished briefly during the Late Permian. They were ponderous, heavily armored herbivores. Cladistic analyses reveal that pareiasaurs shared 16 derived features with turtles.

The only living reptiles with anapsid skulls are the turtles (Testudomorpha), which first appeared in Triassic deposits (Fig. 7.1). Prior to 1995, the oldest turtle fossils, about 210 million years old, came from Thailand, Greenland, and Germany—all of which at that time (210 million years ago) were part of the northern half of the supercontinent Pangaea. In 1995, turtle fossils were described from Argentina that were also 210 million years old, indicating that turtles had already spread over the planet by that time (Rougier, 1995). The Argentinian turtles were different from their northern contemporaries in that their shell extended over the neck (early turtles could not retract their necks), whereas other turtles had evolved external spines to protect their necks. The oldest known chelonioid sea turtle is from the Early Cretaceous period of eastern Brazil (Hirayama, 1998). The turtle is primitive in the sense that the bones in its wrists, ankles, and digits have not become consolidated into rigid paddles. However, it possessed enormous salt glands around the eyes. The fossilized remains of the largest turtle ever recorded (*Archelon*) were found along the south fork of the Cheyenne River in South Dakota (Fig. 7.7c). It was approximately 3.3 m long and 3.6 m across at the flippers.

Ichthyosaurs, Plesiosaurs, Tuatara, Lizards, and Snakes (Lepidosauromorpha)

The lepidosauromorpha include those reptiles having two pairs of temporal fenestrae (diapsid) separated by the postorbital and squamosal bones. Some species, however, have lost one or both temporal arches, so that the skull has a dorsal temporal opening but lacks a lower temporal fenestra (Fig. 7.5). The earliest known diapsid fossil is a member of the genus *Petrolacosaurus* from the Upper Pennsylvanian of Kansas (Reisz, 1981). The lepidosaurs include two major extinct groups (ichthyosaurs and plesiosaurs) and one group (Squamata) containing three subgroups that survive today: Sphenodontia (tuataras); Lacertilia (lizards); and Serpentes (snakes).

Ichthyosauria. One extinct group, the Ichthyosauria (Fig. 7.8), comprised highly specialized marine lepidosauromorphs that probably occupied the niche in nature now taken by dolphins and porpoises. Limbs were modified into paddlelike appendages, and a sharklike dorsal fin was present. Specimens of *Utatsusaurus hataii* from the Lower

FIGURE 7.6

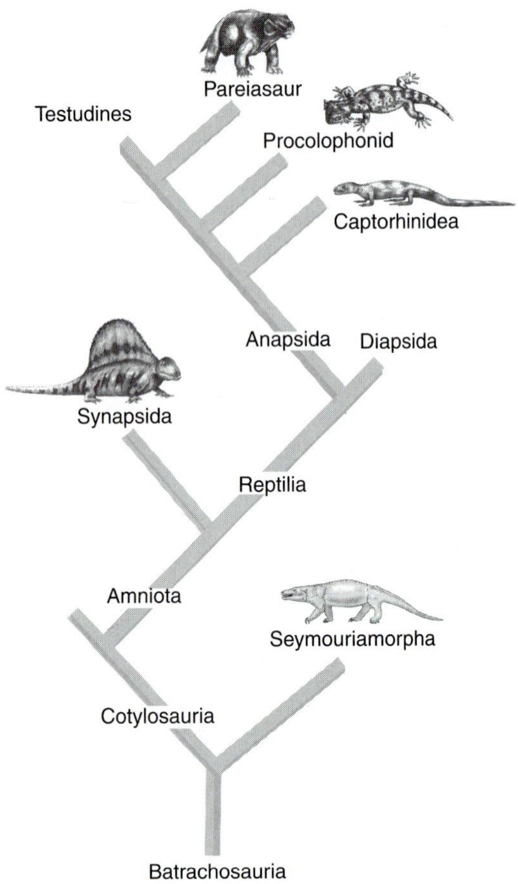

Reisz and Laurin (1991) proposed the procolophonids as the closest sister group to turtles. Lee (1993), however, proposed the pareiasaurs as the nearest relatives.

Triassic of Japan show that this species retained features of terrestrial amniotes in both the skull and the postcranial skeleton, such as the connection between the vertebral column and the pelvic girdle (Motani et al., 1998). Appendages were used primarily for steering, because an ichthyosaur swam by undulations of its body and tail. These "fish lepidosauromorphs" became extinct near the end of the Cretaceous.

Plesiosauria. Plesiosaurs (Fig. 7.9) formed a second extinct group of diapsids. They were marine lepidosauromorphs that had broad, flattened forelimbs and hindlimbs which served as oars to row the body through the water. The trunk was dorsoventrally compressed, and the tail served as a rudder. Some had long necks and small heads, whereas others had short necks and long skulls. Nostrils were located high on the head, and the paddlelike limbs had additional phalanges. Like the Ichthyosauria, plesiosaurs became extinct near the end of the Cretaceous.

Sphenodontidae. Tuataras (*Sphenodon* spp.) (Fig. 7.10) are relics from the Triassic that survive today on about 20 small islands in the Bay of Plenty and in Cook Strait north of Auckland, New Zealand. The two living species (*Sphenodon punctatus* and *S. guntheri*) have been called "living fossils" and are considered the most primitive of living reptiles. Fossil remains have been dated as far back as the Triassic (Carroll, 1988).

The tuatara's teeth are attached to the summit of the jaws (dentition) and are not replaced during the animal's lifetime. The palate contains an additional row of teeth running parallel to the teeth on the maxilla. When the mouth is closed, teeth in the lower jaw fit between the two rows of teeth in the upper jaw. A parietal foramen for the pineal, or third eye, is present.

By day, the tuatara lives in a burrow, venturing forth after sunset to feed on snails, crickets, and even small vertebrates. Up to 14 eggs are deposited in the earth, where they remain for almost a year. Newly hatched tuataras are about 11 cm long, and several years are required to reach the maximum length of slightly over 0.6 m. Tuataras have been known to survive over 20 years. The long gestation and longevity are probably the result of the cold climate in this region of the world.

Squamata. Lizards and snakes (see Fig. 1.4, page 3, and 7.2) are thought to have evolved from an eosuchian (order Eosuchia) ancestor, probably during the Triassic. Eosuchians were primitive lepidosaurs with a diapsid skull and slender limbs. Some taxonomists place a group of tropical and subtropical (mostly legless) reptiles known as amphisbaenans with the lizards; others classify them as a distinct group. Snakes, which arose from lizards before the end of the Jurassic (Carroll, 1988), represent a group of highly modified legless lizards. Although all known snakes lack well-developed legs, the Cretaceous marine squamate *Pachyrhachis problematicus* possessed a well-developed pelvis and hindlimbs and is considered to be a primitive snake (Caldwell and Lee, 1997). The body was slender and elongated, and the head exhibited most of the derived features of modern snakes. Snakes are considered to be the most recently evolved group of reptiles (Romer, 1966; Carroll, 1988).

Thecodonts, Nonavian Dinosaurs, Pterosaurs, Crocodilians, and Birds (Archosauromorpha)

The diapsid archosaurs possess two fenestrae, each with an arch in the temporal region of their skull. The archosaurs include several extinct groups (thecodonts, most of the familiar dinosaurs, and the pterosaurs) and two living groups (crocodilians and birds). In discussing the evolution of dinosaurs, Sereno (1999) noted that the ascendancy of dinosaurs near the close of the Triassic appears to have been as accidental and opportunistic as their demise and replacement by therian mammals at the end of the Cretaceous.

Thecodontia (= Proterosuchia). One of the extinct groups of archosaurians, the Thecodontia, is considered to be ancestral to the dinosaurs, pterosaurs, and birds (Fig. 7.11). Thecodonts ranged in size from around 20 kg to as much as 80,000 kg. In many groups, limbs were positioned directly beneath the body—similar to the limb position in birds and mammals. In some groups, hindlimbs were much larger than forelimbs. Some bipedal species have left track pathways (Fig. 7.12) from which their running speed has been computed (up to 64 km per hour; Bakker, 1986).

Dinosaurs have traditionally been divided into the Saurischia and Ornithischia (Fig. 7.13 and 7.15). Half of the

FIGURE 7.7

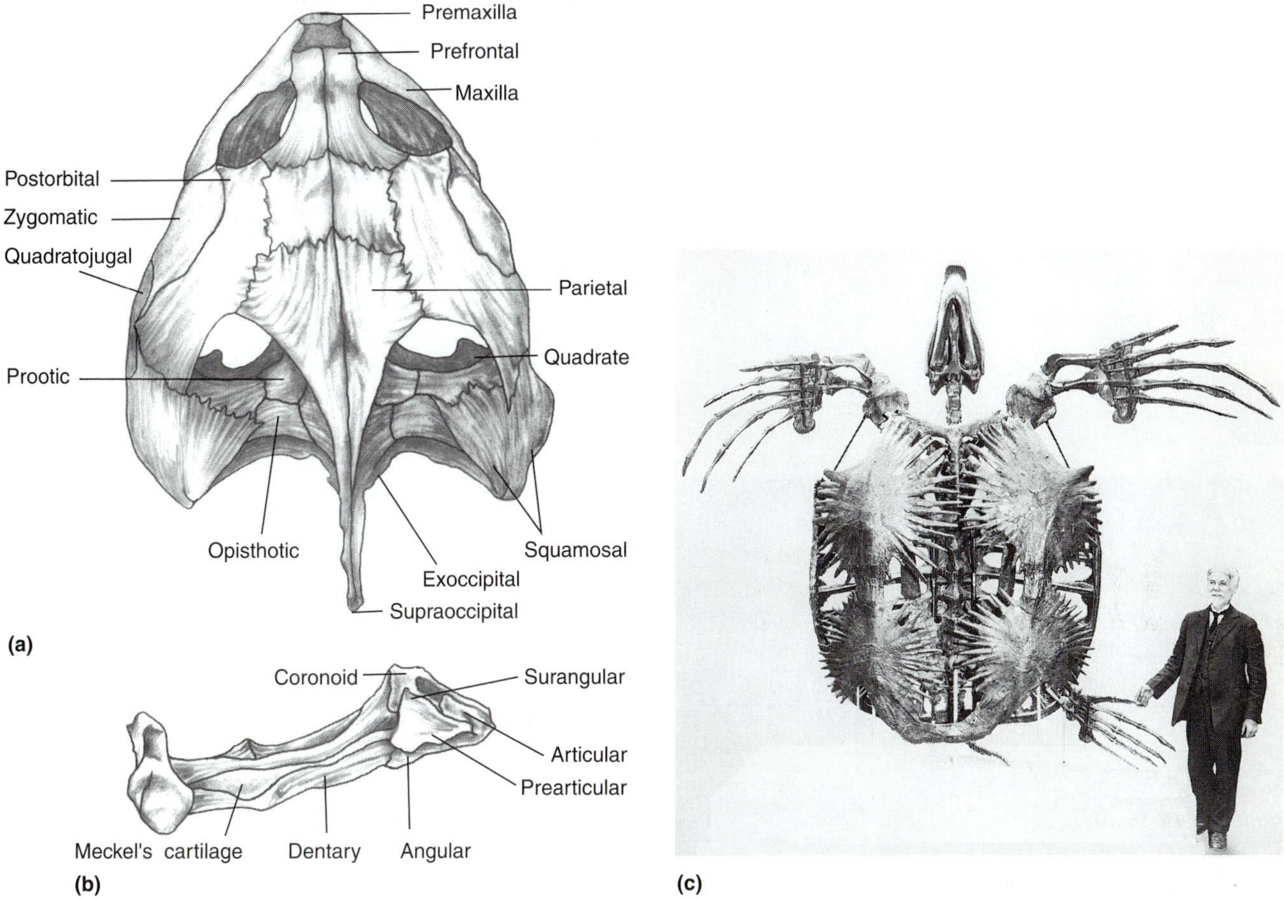

Snapping turtle (*Chelydra*) skull: (*a*) dorsal view of skull and (*b*) posteromedial view of lower jaw; (*c*) *Archelon*, the largest turtle ever found. From the Pierre shale on the south fork of the Cheyenne River approximately 35 miles southeast of the Black Hills of South Dakota. It was approximately 3.3 m long and 3.6 m across at the flippers.

FIGURE 7.8

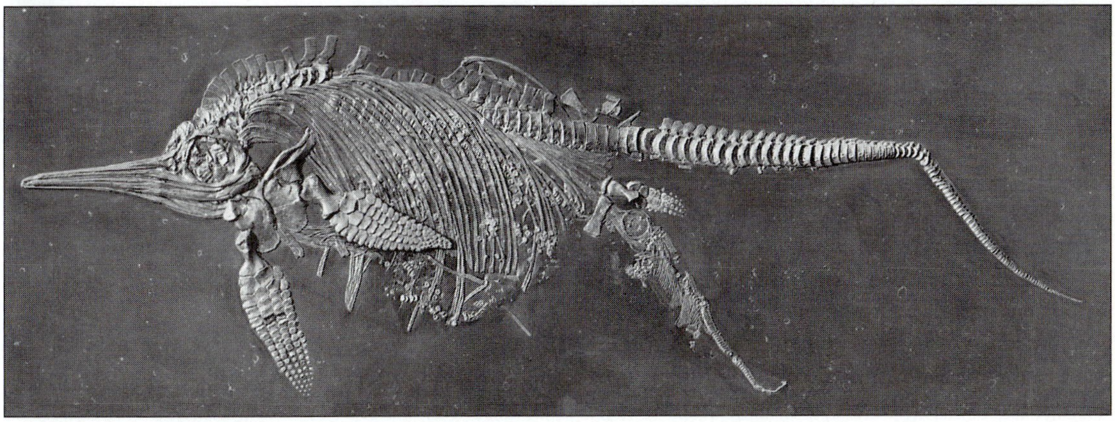

Complete fossil of a female ichthyosaur, about 200 million years old, that died while giving birth.

FIGURE 7.9

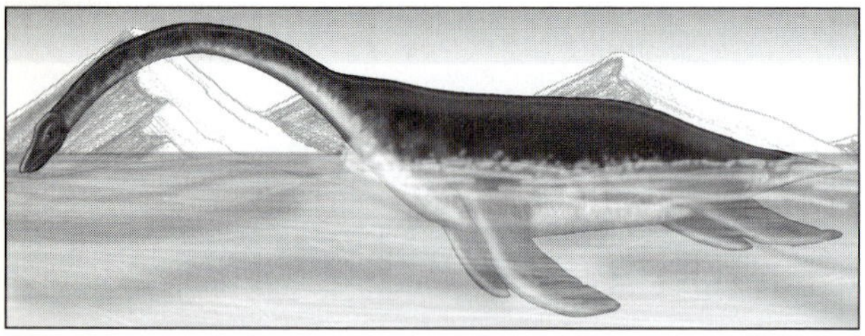

Plesiosaurs were marine diapsids that had flattened forelimbs and hindlimbs that served as "oars." They became extinct near the end of the Cretaceous.

FIGURE 7.10

Tuatara (*Sphenodon punctatum*).

350 species of known dinosaurs have been identified in the past 25 years. Recent discoveries have unearthed genera such as *Herrerasaurus* (Fig. 7.14) and *Eoraptor* in Argentina (Sereno and Novas, 1992; Sereno et al., 1993) that cannot currently be classified as belonging to either of these groups. The skulls have a unique heterodont dentition and do not exhibit any of the specializations of the Saurischia or Ornithischia. They are tentatively classed as "protodinosaurs." Two prosauropod dinosaurs, primitive plant-eaters with long necks, from the Middle to Late Triassic (225 to 230 million years old) fauna of Madagascar (Flynn et al., 1999), may possibly represent the most primitive dinosaurs ever found.

Saurischia. **Saurischians** (L. *saur*, lizard, + *ischia*, hip) were one of the two main groups of dinosaurs that evolved during the Triassic from the Thecodontia. The members of these groups included both quadrupedal and bipedal herbivores and carnivores. They all possessed a triradiate ("lizard-hipped") pelvic girdle (Fig. 7.15), with the ilium connected to the vertebral column by strong ribs. The pubis was located beneath the ilium and extended downward and forward. The ischium, also below the ilium, extended backward. The hip socket was formed at the junction of the three bones. Two types of dinosaurs—theropods and sauropodomorphs—had this type of hip structure. Norman (1991) noted that it seemed highly likely that modern birds were derived from one group of theropod dinosaurs. Even though the avian hip has a backwardly turned pubis, it is derived from the saurischian condition.

Theropods included birds and all of the carnivorous dinosaur genera such as *Ornitholestes, Megalosaurus, Tyrannosaurus, Allosaurus, Ceratosaurus, Deinonychus, Struthiomimus, Utahraptor,* and *Afrovenator* (Sereno et al., 1994) (Fig. 7.16). Theropods are characterized by a sharply curved and very flexible neck; slender or lightly built arms; a rather short and compact chest; long, powerful hind limbs ending in sharply clawed birdlike feet; a body balanced at the hip by a long, muscular tail; and a head equipped with large eyes and long jaws. Most were equipped with numerous serrated teeth (Abler, 1999), although some genera such as *Oviraptor, Struthiomimus,* and *Ornithomimus* were toothless.

The Saurischia included the largest terrestrial carnivores that have ever lived, such as *Giganotosaurus carolinii* from Argentina whose estimated length was between 13.7 and 14.3 m and may have weighed as much as 9,000 kg (Coria and Salgado, 1995; Monastersky, 1997c), and *Tyrannosaurus*, with a length up to 16 m, a height of approximately 5.8 m, and a weight of 6,500 to 9,000 kg (Romer, 1966) (Fig. 7.16). Coria and Salgado (1995) noted that these two enormous dinosaurs evolved independently—*Tyrannosaurus* in the Northern Hemisphere, *Giganotosaurus* in the Southern Hemisphere; consequently, gigantism may have been linked to common environmental conditions of their ecosystems.

FIGURE 7.11

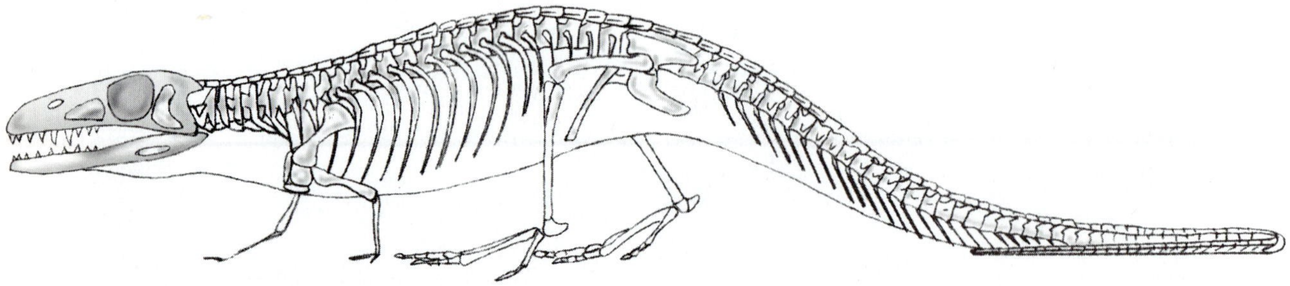

Saltoposuchus, a genus of primitive thecodont from Connecticut.

FIGURE 7.12

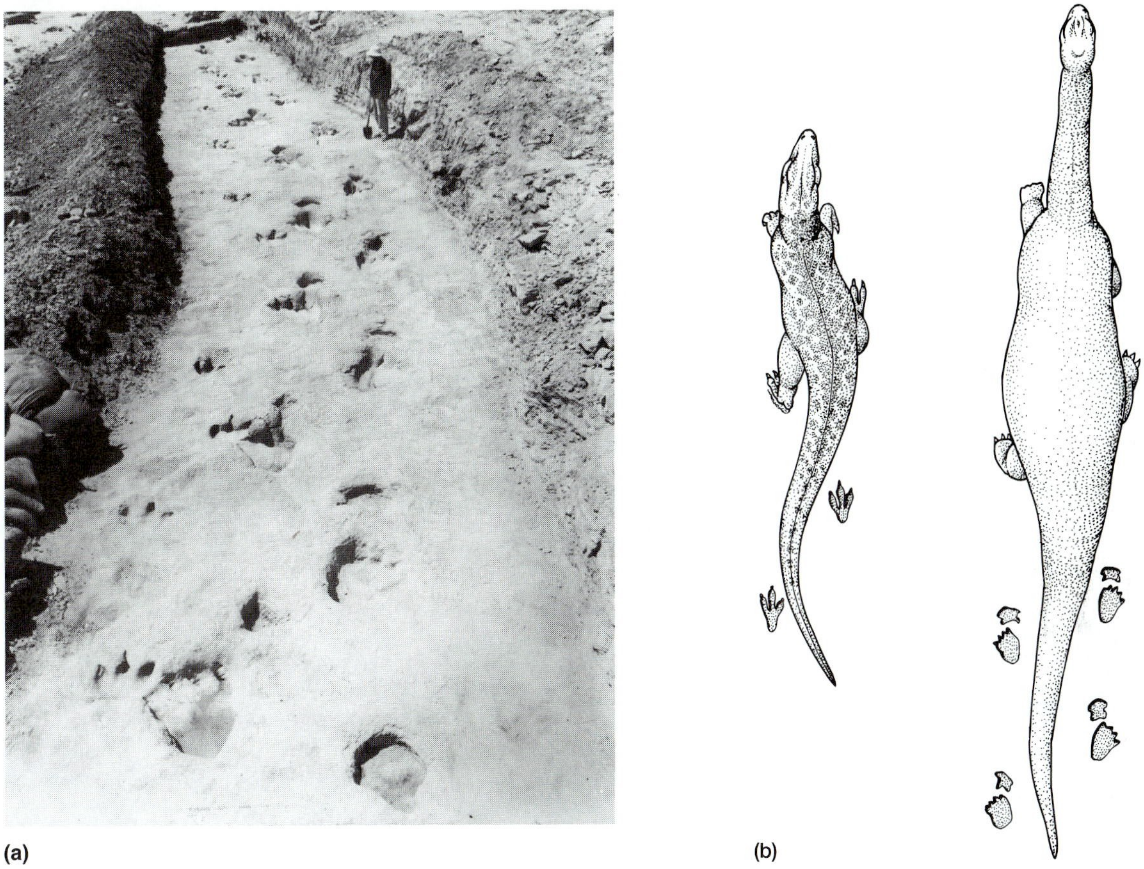

(a) (b)

Dinosaur tracks. (*a*) Tracks from the late Jurassic that were originally made in soft sand which later hardened to form rock. (*b*) The large tracks are those of a sauropod; the three-toed tracks are those of a smaller carnosaur, a bipedal carnivorous dinosaur.

BIO-NOTE 7.2

An Extraordinary Fossil

The first theropod dinosaur ever to be found in Italy was a 24-cm theropod identified as *Scipionyx samniticus.* It represents a young dinosaur just hatched from its egg before it died. Fossilization normally preserves only hard body parts, such as bones and teeth. However, this specimen is so well preserved that it displays the intestine, muscle fibers, and the cartilage that once housed its windpipe—details of soft anatomy never seen previously in any dinosaur. The exceptional quality of the preservation of the soft parts makes this one of the most important fossil vertebrates ever discovered.

Dal Sasso and Signore, 1998

BIO-NOTE 7.3

A Deadly Dinosaur

Utahraptor roamed the Colorado Plateau approximately 130 million years ago. It stood approximately 2.5 m tall, reached a length of about 6 m, and weighed about 900 kg. It has been nicknamed "super slasher"—the deadliest land creature the Earth has seen. *Utahraptor* was a swift runner, and it was armed with a 38-mm slashing claw that stood upright and apart from the other claws on each hind foot. The animal's forelegs were tipped with powerful claws suitable for grasping prey, while the dinosaur kicked its victim with its sickle-clawed hind feet. *Utahraptor* was described by its finders as a "Ginsu-knife-pawed kick-boxer" that could disembowel a much larger dinosaur with a single kick.

Browne, 1993

FIGURE 7.13

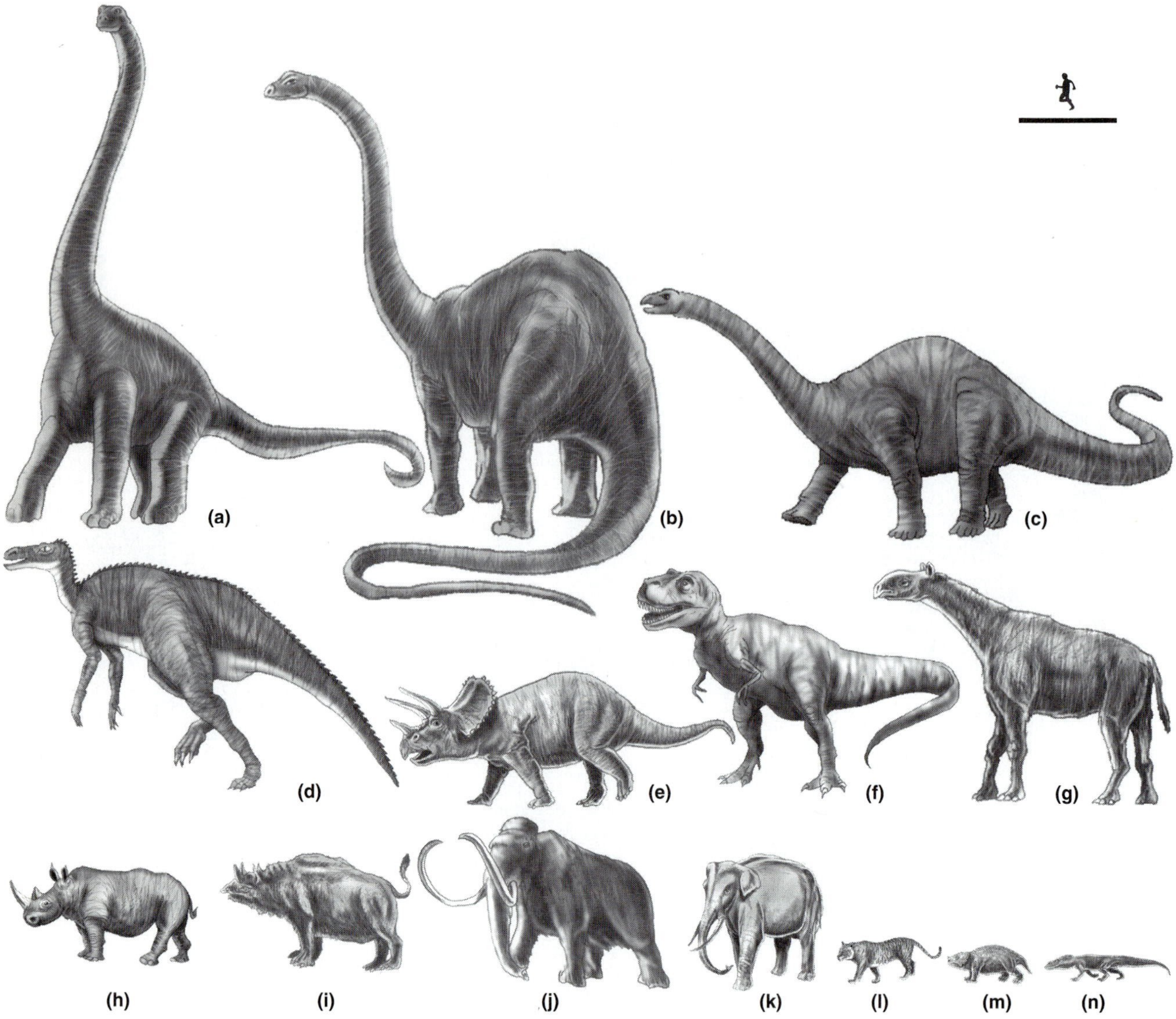

Size comparison of dinosaurs, mammals, and reptiles drawn to the same scale. Comparison of extinct taxa are based on the largest known specimens and masses from volumetric models. Comparison of extant and recent taxa are based on the sizes of large adult males. (*a*) 60- to 80-ton titanosaur; (*b*) 55-ton *Supersaurus*; (*c*) 45-ton *Brachiosaurus* (= *Ultrasaurus*); (*d*) 13-ton *Shantungosaurus*; (*e*) 6-ton *Triceratops*; (*f*) 7-ton *Tyrannosaurus*; (*g*) 16-ton *Indricotherium*; (*h*) 2-ton *Rhinoceros*; (*i*) 5-ton *Megacerops*; (*j*) 10-ton *Mammuthus*; (*k*) 6-ton *Loxodonta*; (*l*) 0.3-ton *Panthera*; (*m*) 1-ton *Scutosaurus*; (*n*) 1-ton *Megalania*. Human figure 1.62 m tall. Scale bar = 4 m

Source: Carpenter, et al., Dinosaur Eggs and Babies, *Cambridge University Press.*

BIO-NOTE 7.4

Coprolites

Paleontologists have previously found numerous coprolites (fossil feces) from herbivorous dinosaurs. Assigning coprolites to theropods has been difficult, because sites with dinosaur fossils often also contain skeletons of other carnivorous animals that could have produced bone-filled feces.

The first example of fossilized feces that clearly came from a carnivorous dinosaur was found in Saskatchewan, Canada. The whitish-green rock is so massive—44 cm long—that it must have come from a large theropod. The only large theropod known from these Saskatchewan deposits is *Tyrannosaurus rex*. The coprolite contains fragments of bone from a juvenile ornithischian dinosaur. It indicates that *T. rex*'s teeth were strong enough to crunch through bone, a topic of much debate in the past. The bone fragments indicate that tyrannosaurs repeatedly crushed mouthfuls of food before swallowing, unlike living reptiles that often swallow large pieces of prey.

Chin et al., 1998

FIGURE 7.14

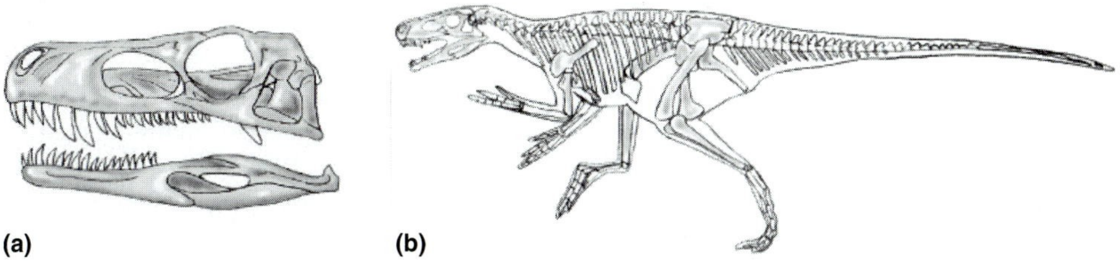

(a) (b)

(*a*) Reconstruction of the skull of *Herrerasaurus ischiqualastensis* from Argentina. (*b*) Skeletal reconstruction of *Herrerasaurus*.
Source: Sereno and Novas in Science, *258:1138, November 13, 1992.*

FIGURE 7.15

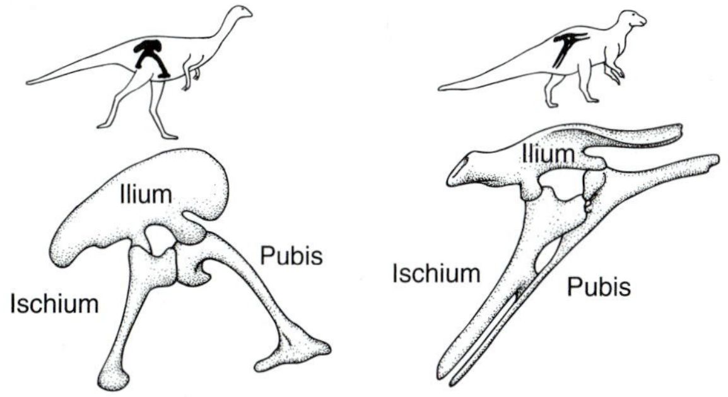

(a) Saurischian hip **(b) Ornithischian hip**

Dinosaur hips. (*a*) Saurischians possessed a pelvic girdle with three radiating bones. (*b*) Ornithischians had a hip with pubis and ischium bones lying parallel and next to each other.

Another huge dinosaur, *Carcharodontosaurus* (shark-toothed reptile), was discovered by Sereno in Morocco (Sinha, 1996). Its head was 1.6 m long, just slightly larger than that of *T. rex*. The Moroccan bones represent the first major dinosaur fossils to be unearthed in Africa and are being used by paleogeographers and biogeographers in their quest to understand exactly when the continents split apart during the Jurassic (see Chapter 3).

Some interesting revelations concerning dinosaurs have been discovered by using sophisticated equipment. For example, computed tomography (CT) scanning utilizes an x-ray source moving in an arc around the body. X-rays are converted to electronic signals to produce a cross-sectional picture, called a CT scan. Formerly known as computerized axial tomography (CAT) scanning, this technique shows that both *Tyrannosaurus* and the smaller *Nanotyrannus* shared a trait still found in such diverse modern animals as crocodiles, elephants, and birds: a sophisticated system of air canals ramifying through their skulls. These large air pockets and

FIGURE 7.16

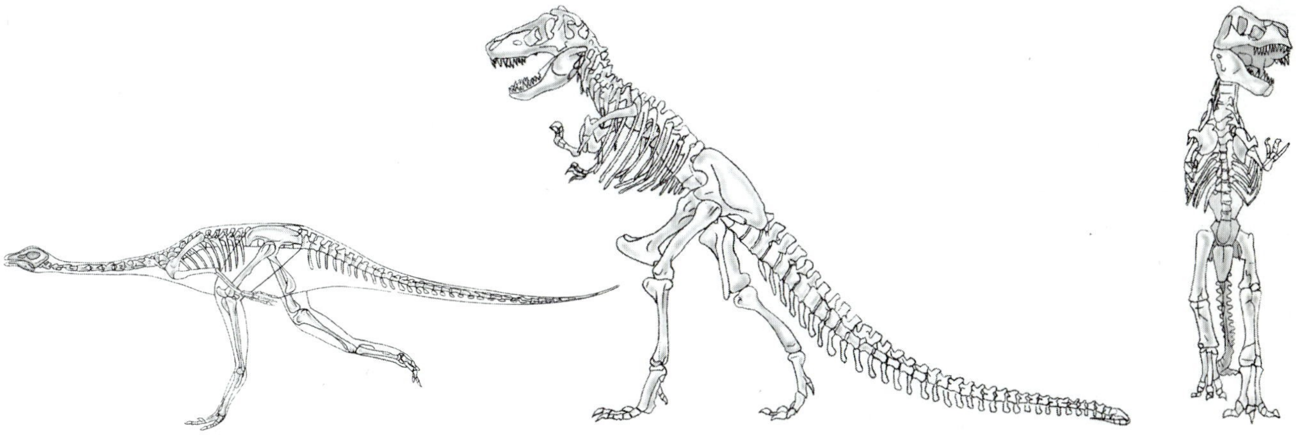

(a) (b) (c)

(*a*) A theropod: *Struthiomimus*. Theropods had flexible necks, slender arms, long, powerful hindlimbs, sharply curved birdlike feet, and a body balanced at the hip by a long muscular tail. Most had serrated teeth, but some were toothless. (*b*) Side view of *Tyrannosaurus*—members of this genus are among the largest dinosaurs that ever lived. (*c*) Front view showing orientation of pelvic girdle and hindlimbs.
Source: W. C. Gregory, Evolution Emerging, *1974, Ayer Company.*

tubes allowed dinosaurs to move air between their lungs and brain, presumably to help regulate the temperature of the brain. Such a need for temperature regulation has been cited as evidence by some researchers that these animals may have been homeothermic.

However, Hillenius (1994) used the absence of scroll-like turbinate bones in the nose as evidence that at least some of the dinosaurs were poikilothermic. CT scans of several theropod dinosaurs showed no evidence of respiratory turbinates in these active predators. Turbinate bones slow down the passage of incoming air so that it can be warmed and moistened. When the animal exhales, the turbinates recapture heat and moisture before it leaves the body. Over 99 percent of living mammals and birds have turbinate bones, but they are completely absent in living sauropsids. By using turbinate bones, Hillenius was able to trace endothermy back about 250 million years in the mammal lineage and 70 million years in birds. Although the absence of respiratory turbinates does not negate the possibility of other thermoregulatory strategies, these bones may represent an important anatomical clue to endothermy (Fischman, 1995a).

Reptilian bones (and the bones of some Mesozoic birds; Chinsamy et al., 1994) generally grow in spurts, thus producing annual growth rings. In contrast, avian and mammalian bones form rapidly and produce fibrolamellar bone tissue in which the collagen (protein) fibers are haphazardly arranged and form a fibrous, or woven, bony matrix and no annual rings. Chinsamy (1995) conducted histological studies on the bones of a prosauropod and a theropod dinosaur. He found distinct reptilian-like growth rings, but also a type of fibrolamellar bone (Fig. 7.17). Thus, the bones showed both reptilian and mammalian characteristics. Studies of growth rings also indicate that some dinosaurs continued growing throughout their lives, whereas others stopped growing when they reached maturity, as is the case with mammals and birds.

FIGURE 7.17

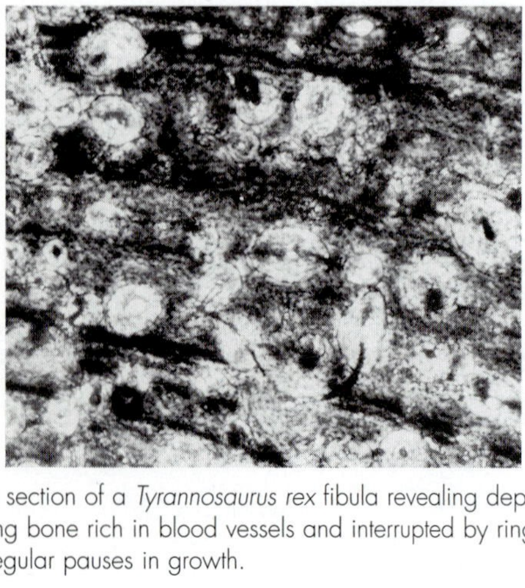

Transverse section of a *Tyrannosaurus rex* fibula revealing deposits of fast-growing bone rich in blood vessels and interrupted by rings which indicate regular pauses in growth.

The growth rate of *Apatosaurus*, a sauropod that reached its full growth in 8 to 11 years, implies that sauropods deposited about 10.1 μm of bone tissue per day—about the same rate as living ducks, which deposit an average of 10.0 μm of bone per day (Stokstad, 1998). Ducks, however, reach their adult size in about 22 weeks, whereas dinosaurs maintained this growth rate for many years.

Ruben et al. (1997, 1999) examined the fossilized soft tissue of the Chinese theropod *Sinosauropteryx* and the Italian theropod *Scipionyx samniticus*. By using ultraviolet (UV) light, the researchers were able to distinguish the outlines of the intestines, liver, trachea, and muscles; they discovered that these two theropods had the same kind of compartmentalization of lungs, liver, and intestines as the crocodile—not a bird.

Theropods had two major cavities—the thoracic cavity containing the lungs and heart, and the abdominal cavity containing the liver, intestines, and other organs. These were completely separated from each other by a hepatic-piston diaphragm, as is the case in crocodiles. Most reptiles maintain a low resting metabolic rate and breathe by expanding their rib cages; they lack the power of a hepatic-piston diaphragm. Mammals and birds use both rib-based and diaphragm-driven respiration. The diaphragm system provides extra oxygen for sustained, intense activity.

The liver in *Scipionyx* extended from the top to the bottom of the abdominal cavity. A muscle located next to the pubic bone appeared similar to those in some modern reptiles that run from the pubis to the liver. It helps move the liver back and forth like a piston, causing the lungs to expand and contract. In *Scipionyx* the diaphragm formed an airtight layer separating the liver and lungs.

Ruben et al. (1999) concluded that although these theropods were basically poikilothermic, diaphragm-assisted lung ventilation was present, and their lungs might have been able to power periods of high metabolism and intense activity. This dual-metabolism hypothesis, which remains controversial, would have allowed highly active theropods to have had an economical resting metabolism with a capacity for bursts of activity.

Chemical analyses of the bones of a 70-million-year-old *Tyrannosaurus rex* by a research team from North Carolina State University revealed bone growth by an animal with a very narrow range of internal temperatures (Barrick and Showers, 1994). The researchers measured the ratio of two naturally occurring isotopes of oxygen that are part of the phosphate compounds normally found in bone. This ratio in bone varies with the temperature at which the bone formed. Bone from deep inside a homeothermic animal will have formed at nearly the same temperature as bone near its surface—the result of a metabolic process that keeps the entire body in a temperature range within which muscles can work at peak activity. Barrick and Showers interpreted their evidence as indicating that *T. rex*'s bones all formed at nearly the same temperature. The core body temperature and the temperature in the extremities varied by only 4°C or less. Such a homeothermic animal could have been active at night when the temperature was cool and could have been active at high altitudes. Hence, they suggested that it was homeothermic. Critics point out, however, that in the 70 million years that the bones lay in the ground, their oxygen isotope ratios could have been altered by groundwater and other substances; that bone tissue must be tested individually and not in groups; and that the animal's bulk alone could have meant that it retained more body heat than any of today's reptiles, all of which are smaller (Millard, 1995).

All **sauropodomorphs** were herbivorous (Fig. 7.18a) and included the largest quadrupeds that have ever existed—*Diplodocus*, *Apatosaurus* (*Brontosaurus*), *Brachiosaurus*, *Seismosaurus*, *Ultrasaurus*, and *Argentinosaurus*—with some forms reaching lengths of nearly 40 m and estimated weights as great as 80,000 kg (Colbert, 1962; Carroll,1988; Norman, 1991; Appenzeller, 1994). The tallest of all dinosaurs, *Sauroposeidon*, was over 18 m tall, 30 m long, and weighed approximately 54,000 kg. (Journal of Vertebrate Paleontology, in press, March 2000). Limb bones of sauropods were thick, solid, and nearly vertical, and little bending occurred at the elbow and knee joints. Some, such as *Supersaurus*, may even have had hollow bones (Monastersky, 1989a), an adaptation to reduce weight, yet maybe being stronger than solid bone. Paleontologists and computer scientists have recently joined forces in a new field of research called cyberpaleontology that uses computer-generated images to better understand the biomechanical movements of sauropods (Zimmer, 1997). By the end of the Cretaceous, all theropods and sauropods had become extinct.

BIO-NOTE 7.5

Dinosaurs in Antarctica

Early Jurassic tetrapods have been collected near the Beardmore Glacier in the Transantarctic Mountains in Antarctica, approximately 650 km from the geographic South Pole. These fossils, which are similar to Early Jurassic fossils from other continents, indicate that no geographic or climatic barriers prevented dinosaurs from populating high southern latitudes during the Jurassic. The fossils included two dinosaurs (a large crested theropod, *Cryolophosaurus ellioti*, and a large prosauropod), a pterosaur, and a large tritylodont (synapsid). Antarctica's location and climate have not always been as they are today. The changing positions of the continents (continental drift) and the resulting effects on vertebrate distribution were discussed in Chapter 3.

Hammer and Hickerson, 1994

FIGURE 7.18

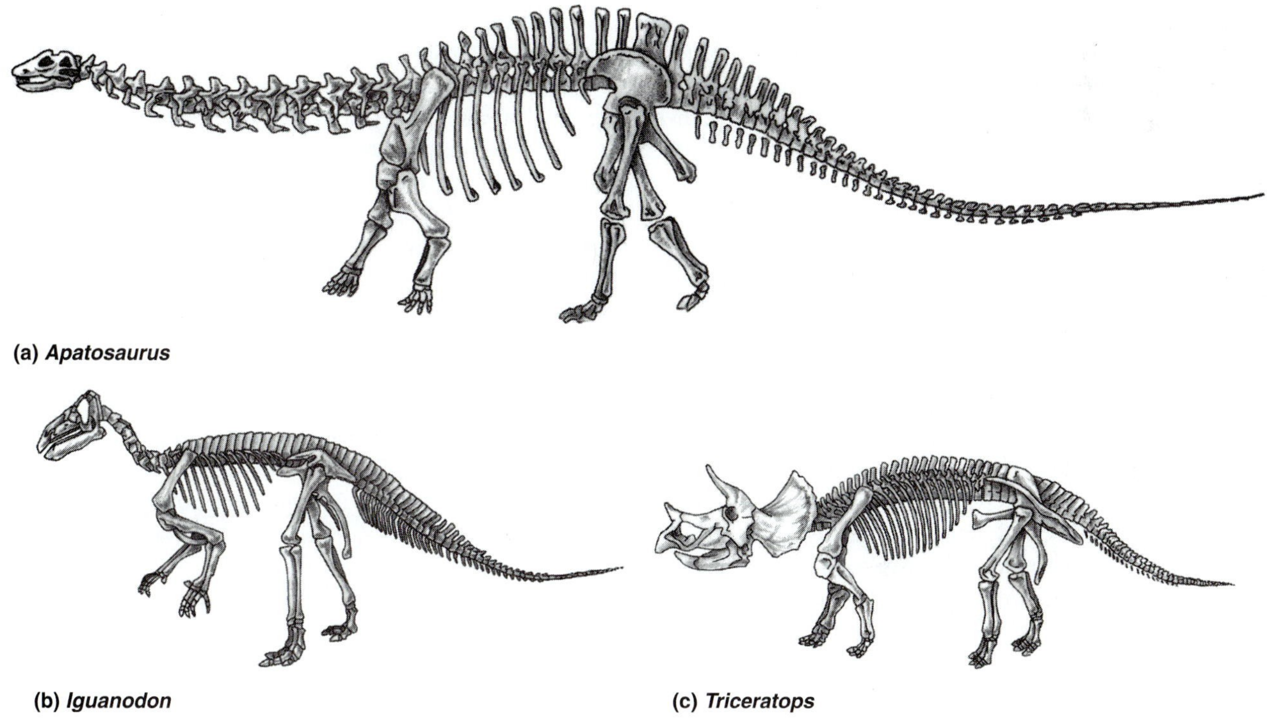

(*a*) One of the largest sauropods: *Apatosaurus* (formerly known as *Brontosaurus*). All sauropods were herbivorous. (*b*) *Iguanodon*, a genus of ornithopod. Ornithopods were mostly small- to medium-sized reptiles that walked on their hind legs most of the time. Some may have lived in large herds, × 1/80. (*c*) *Triceratops*, a genus of ceratopsian. The frill may have served as an anchor site for powerful lower jaw muscles. It may also have played a role in agonistic and sexual behavior, × 1/70.

Ornithischia. Dinosaurs in the order Ornithischia (L. *ornithos*, bird, + *ischia*, hip) tended to have thin, pencil-shaped teeth, long, slender bodies, and whiplike tails. Ornithischians had a birdlike pelvis in which the pubis, instead of extending downward and forward, extended posteriorly alongside the ischium (Fig. 7.15b). The pubis of some forms also developed an anterior projection. This arrangement is similar (convergent) to that of living birds, although no evidence exists that birds evolved from this group.

Ornithischians were either bipedal or quadrupedal herbivores. The lower jaw of all forms consisted of a small, horn-covered beak. Unusual features found in specific groups also included ducklike bills (hadrosaurs); overlapping plates of bony armor (ankylosaurs); rows of protective plates and spines down their backs and tails (stegosaurs); and parrotlike beaks along with bony frills (neck shields) and horns on their heads (ceratopsians). Although some ornithischians were larger than elephants (*Stegosaurus*, for example, was 6.5 m in length and weighed at least 9,000 kg; Feduccia and McCrady, 1991), they had relatively small brains for their size. By the end of the Cretaceous period, all ornithischians, like saurischians, had become extinct.

Barreto et al. (1993) have shown that the cells within the growth plates (disks of cartilage near the ends of the bones) of *Maiasaurus*, an ornithischian, bear a striking resemblance to the cells of chicken growth plates and look very different from the growth plates of living reptiles and mammals. The plate zone boundary is very irregular, the cells (chondrocytes) are shorter and ovoid in shape, and all cell membranes are calcified. The researchers concluded that the similarity of the growth plates points to a common ancestor for dinosaurs and birds, because it is too complex a morphological character to have evolved twice. In addition, this synapomorphy (shared derived anatomical character) supports the inclusion of birds along with reptiles in a group known as Dinosauria. The Dinosauria was first proposed in 1841, by Richard Owen, the first head of the British Museum of Natural History. Although it fell out of favor in the late 19th century, it was resurrected in the 1970s by Bakker and Galton, who argued that it should include not only the ornithischians and saurischians, but birds as well. However, not all paleontologists agree (Fischman, 1993).

Five groups of ornithischians—ornithopods, ceratopsians, pachycephalosaurs, stegosaurs, and ankylosaurs—have been defined. Ornithopods were mostly small-to medium-sized genera such as *Camptosaurus* and *Iguanodon* (Fig. 7.18b), although hadrosaurs, or duck-billed dinosaurs, reached lengths of 13 m. Ornithopods walked on their hind legs most of the time. Some, especially the hadrosaurs, may have lived in large herds. In Massachusetts, John Ostrom found tracks of significant numbers of individuals moving in the same direction at the same time (Ostrom, 1972; Norman, 1991). These findings provided evidence for herding and possible migratory movements as socially integrated groups.

Ceratopsians were distinctive because of their parrotlike beaks and their horns and frills. The frills are thought to have served as anchor sites for powerful muscles that attached to the lower jaw and also were of great significance in agonistic (aggressive) and sexual behavior (Farlow, 1975). Because the frills contained networks of blood vessels, they may also have served to help regulate body temperature by cooling the blood before it returned to the interior of the body (Monastersky, 1989b). Ceratopsians included genera such as *Protoceratops*, *Triceratops* (Fig. 7.18c), and *Centrosaurus*. They also are thought to have lived in large herds.

Pachycephalosaurs are poorly known (Fig. 7.19a). They had "curiously domed and massively reinforced heads," with the bulge of the head being filled with solid bone. The head is thought to have been used as a battering device (Norman, 1991).

Stegosaurs were the plated dinosaurs (Fig. 7.19b). The large plates and spines of such animals as *Stegosaurus* may have acted as panels to gain heat from the absorption of solar radiation and to lose heat by convection to wind currents, thereby regulating body temperature (Stuart, 1992). They were light honeycomb structures that seemed to be designed to allow large quantities of blood to pour through the plates and out onto the surface of the plates beneath the skin. These structures figure prominently in the debate over whether some dinosaurs were homeothermic or poikilothermic.

One *Stegosaurus* skeleton was so well preserved that researchers were able to confirm that dorsal plates were arranged in an alternating pattern rather than in matched pairs and that the animal had even more body armor than had been previously thought, including a disk-shaped plate near its hip and a web of ossicles—small coin-sized bony plates—in its throat region. The size of the dorsal plates may indicate gender.

Investigations of dinosaur spinal canals show how dinosaurs may have stood and moved (Giffin, 1990, 1991). The varying thickness of the spinal cord (spinal quotient) is reflected in the varying width of the spinal canal, and the presence and relative size of neural bundles along the spinal cord provide information concerning the posture of a given species. Some dinosaurs carried themselves with their legs straight up and down—in a so-called improved posture—whereas others moved in a more lizardlike sprawl. The ratio of neural development between the limb and the torso region can show how an animal held its body. For example, stegosaurs possessed a smaller than expected spinal cord serving the front legs, an indication that the animal had a somewhat bowlegged, rather than an upright, posture.

The fifth group of ornithischians, the ankylosaurs, were heavily armored to provide protection from the larger carnivorous dinosaurs (Fig. 7.19c). Some also had large rounded clubs at the ends of their tails.

Bakker (1986) believed that all plant-eating dinosaurs constituted a single natural group—Phytosauria ("plant dinosaurs")—that branched out from a single ancestor. In addition, Bakker believed that dinosaurs developed in a similar fashion to mammals—growing quickly and breeding early. The legs and muscles of many species were built for speed (with deep shoulder and hip sockets; the crests of the knee joints were massively developed to support the extensive muscles of the knee), so that they needed powerful hearts and lungs of high capacity. They had a mammal-like bone texture. The presence of densely packed Haversian systems in bone is only found in dinosaurs and mammals. On the basis of these characters, Bakker (1986) concluded that dinosaurs must have been homeothermic. As might be expected, considerable discussion and controversy have been generated by Bakker's hypothesis. Studies of oxygen isotopes and infrared spectroscopy currently are being employed in an attempt to provide additional evidence concerning the possibility of endothermy in the dinosaurs.

Pterosauria. Another extinct order of archosaurians—Pterosauria—included the first flying vertebrates (Fig. 7.20). Many of the bones of pterosaurs were hollow and air-filled; their skull bones were thin and fused; their jaws were elongated and contained teeth; a large sternum was present; and their anterior appendages were modified into wings. It is now generally accepted that pterosaurs (pterodactyls) were fliers, but whether they had broad, batlike wings connected to both forelimbs and hindlimbs or narrow, stiff wings free of the legs has long been a subject of debate (Peters, 1995).

The discovery of well-preserved wing membranes on a long-tailed pterosaur (*Sordes pilosus*) from Khazakhstan shows that the hind limbs were intimately involved in the flight apparatus (Unwin and Bakhurina, 1994) (Fig. 7.21). The hindlimbs connected externally to the wing membrane and internally were connected by a uropatagium controlled by the fifth toe. Furthermore, the flight surface was nonhomogeneous with a stiffened outer half and a softer, more extensible inner portion.

The earliest known flying vertebrate, *Coelurosauravus jaekeli*, glided on a unique set of wings unlike any other known in living or extinct animals (Frey et al., 1997) (Fig. 7.22). The long, hollow bones that strengthened its wings formed directly in the skin itself, unlike the wing bones of birds and bats, which are converted front limbs.

The hip socket of pterosaurs was unlike that of birds in that it was shallow and had no central hole for a ligament (Unwin, 1987; Boxer, 1987). The femur extended outward and slightly upward from the pelvis, so that the animal presumably had a sprawling gait. The entire foot, rather than just the toes, contacted the ground during terrestrial locomotion (Clark et al., 1998).

FIGURE 7.19

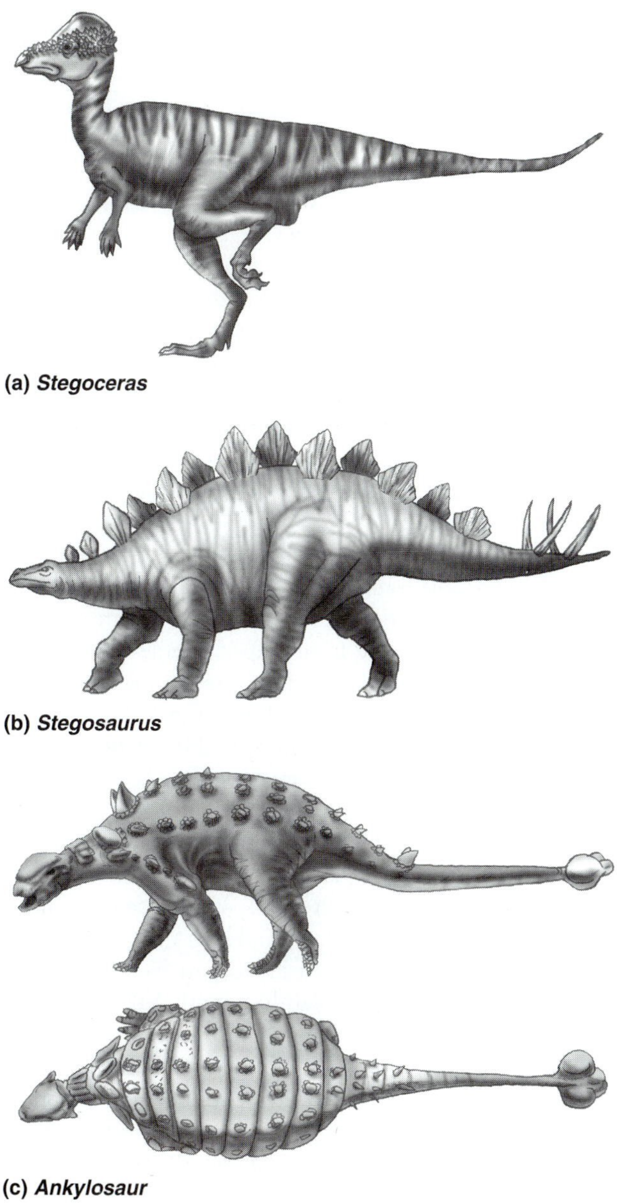

(a) *Stegoceras*, a pachycephalosaurid genus. These dinosaurs looked somewhat similar to the ornithopods except for their domed heads. (b) *Stegosaurus*. The large plates may have acted as solar panels to help control body temperature by collecting solar radiation for heat and also acting as radiators for cooling. (c) An ankylosaur: top, lateral view; bottom, dorsal view. The heavy armor provided protection from larger carnivorous dinosaurs.

FIGURE 7.20

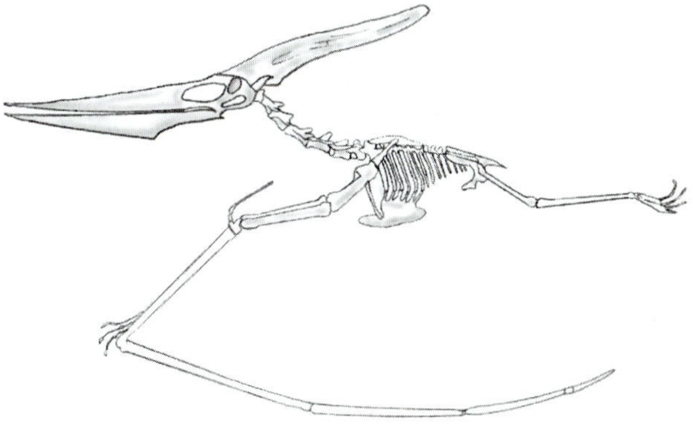

Pteranodon, a giant pterosaur from the Upper Cretaceous of Kansas. The wingspread was up to 6.7 m. The head, which was 3 2/3 times the length of the body, was exceedingly light and strong.

FIGURE 7.21

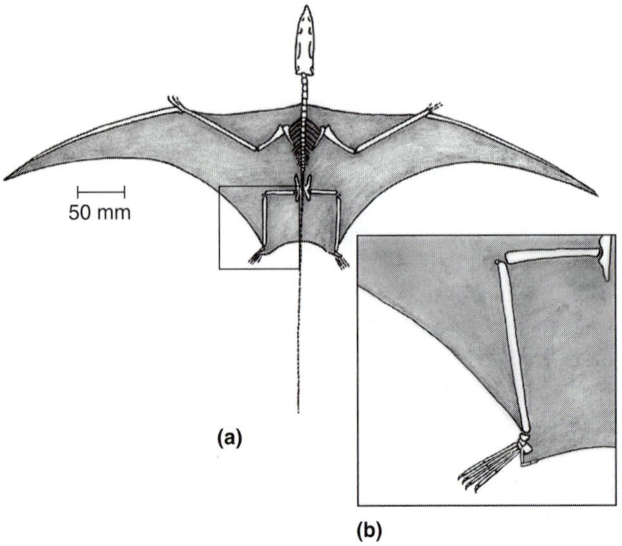

(a) Restoration of *Sordes pilosus*, a pterosaur, in dorsal view showing the relationship of the skeleton to the flight membranes. Key: pr, propatagium; ch, cheiropatagium; u, uropatagium. Scale bar = 50 mm. (b) The hindlimb of *Sordes pilosus* in "flight" position with the fifth metatarsal located dorsomedial to the foot, the first phalange of the fifth toe directed laterally, and the second phalanx reflected medially to insert into the rear edge of the uropatagium.
Source: Unwin and Bakhurina, "Sordes Pilosus" in Nature, *371, September 1, 1994.*

FIGURE 7.22

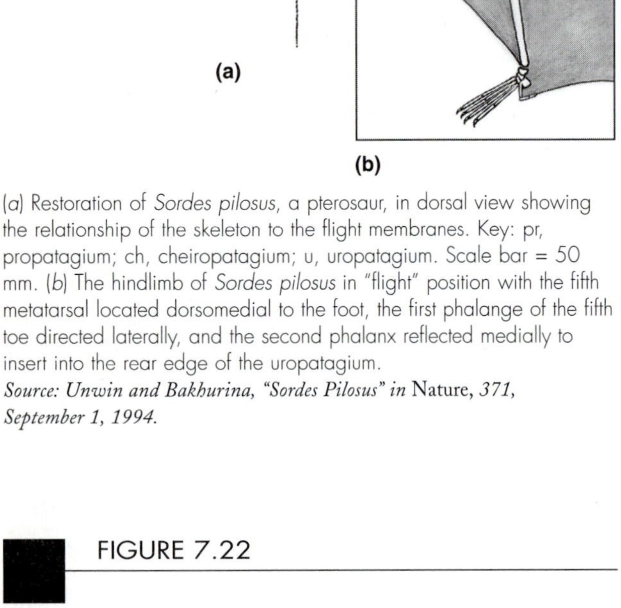

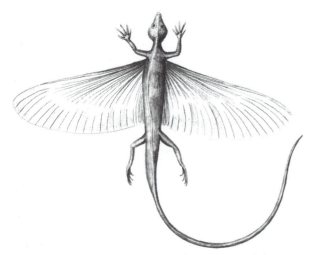

The earliest known flying vertebrate, *Coelurosauravus jaekeli*. Reconstruction in dorsal view. Note the numerous long rods for support of the lateral gliding membrane and the very long tail. Distal portions of the larger rods may have been curved backward as a result of tension produced by the intervening gliding membrane when the wing was spread. Reconstruction is based on a fossil with a snout-vent length of 18 cm.

Bakker (1986) presented evidence that if pterodactyls actively flapped their wings during flight, heat generated by their muscles would have warmed their body cores to temperatures higher than that of the air. In addition, the bodies of some pterodactyls were covered with a dense coat of long, hairlike scales, which presumably could have served to insulate the body.

Competition with birds is thought to be a primary reason for the extinction of the pterosaurs prior to the end of the Cretaceous period. They did not give rise to any other group of vertebrates.

Crocodilia. The order Crocodilia, which includes the alligators, crocodiles, caimans, gavials, and their relatives, is thought to have arisen from thecodont stock (Figs. 7.1 and 7.2). In 1986, the skull and jawbone of an extremely large fossil crocodilian (*Purussaurus*) was discovered in the Amazon region along the border between Peru and Brazil (Campbell and Frailey, 1991). This giant crocodilian had an estimated length of 12 m and stood 2.5 m tall. It is estimated to have weighed 10,000 to 12,000 kg, which would have made it even more massive than *Tyrannosaurus rex*, the largest known terrestrial carnivore. *Deinosuchus rugosus*, a 9 m crocodile weighing 2,700 kg, inhabited the southeastern coastal swamps of North America during the Cretaceous period (Anonymous, 1997e). A possible plant-eating crocodiliform archosaur from the Cretaceous of China (*Chimaerasuchus paradoxus*) was reported by Wu (1995). The presence of teeth possessing three longitudinal rows of cusps (multicuspid molariform) may make it the first known herbivorous member of the Crocodiliformes.

Aves. As early as 1868, Thomas Huxley and others had discussed a possible connection between dinosaurs and birds. Much of the current evidence indicates that birds are a monophyletic group that arose from diapsid reptiles (theropods) during the Jurassic period. Birds still retain many traces of their reptilian ancestry (Norman, 1991) (Fig. 7.23). A cladogram of the Archosauria showing possible relationships of several archosaurian groups to modern birds is presented in Fig. 7.24.

Today, the origin of birds remains ornithology's longest-running debate. Some researchers, including Philip Currie, the dinosaur curator of the Royal Tyrrell Museum of Paleontology in Alberta; Mark Norell and Luis Chiappe from the American Museum of Natural History in New York City; John Ostrom and Jacques Gauthier, both Yale University paleontologists; and Paul Sereno, a University of Chicago paleontologist, are proponents of a dinosaur–bird link with the ancestral dinosaur being a theropod. Sereno has stated: "Everywhere we look, from their skeletal features to their behaviors to even the microstructure of their eggs, we see evidence that birds are descended from dinosaurs" (Morell, 1997e). In fact, paleontologists have identified some 200 anatomical features shared by birds and dinosaurs—a far greater number than those linking birds to any other type of reptile, ancient or living (Monastersky, 1997b). Even the furcula ("wishbone"), whose absence in dinosaurs was considered

FIGURE 7.23

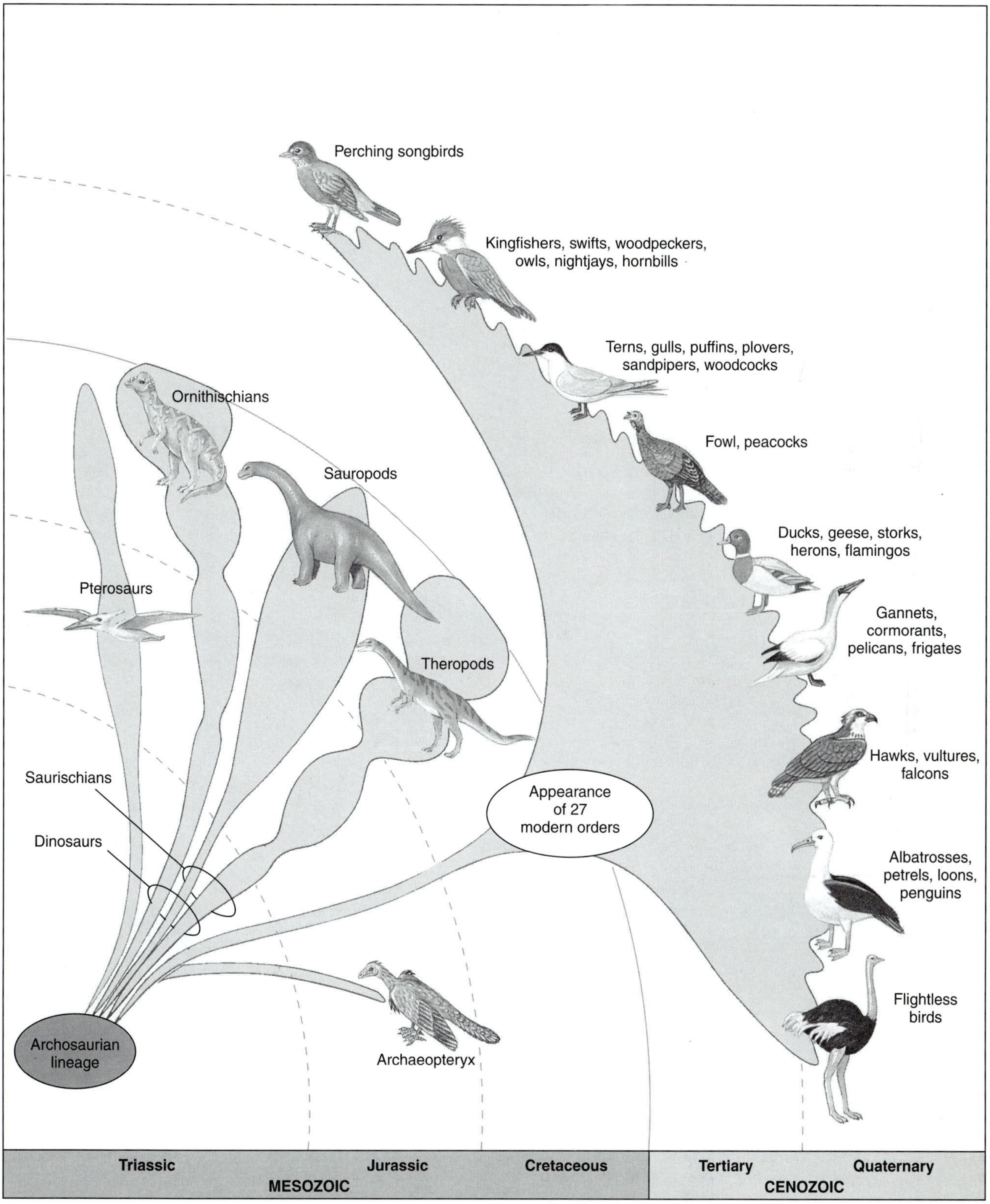

Evolution of modern birds. Nine of the largest of the 27 living orders of birds are shown. The earliest known bird, *Archeopteryx lithographica*, lived in the Upper Jurassic, about 147 million years ago. *Archeopteryx* shares many specialized aspects of its skeleton with the smaller theropod dinosaurs and is considered by many researchers to have evolved within the theropod lineage. Evolution of modern bird orders occurred rapidly during the Cretaceous and early Tertiary periods.

FIGURE 7.24

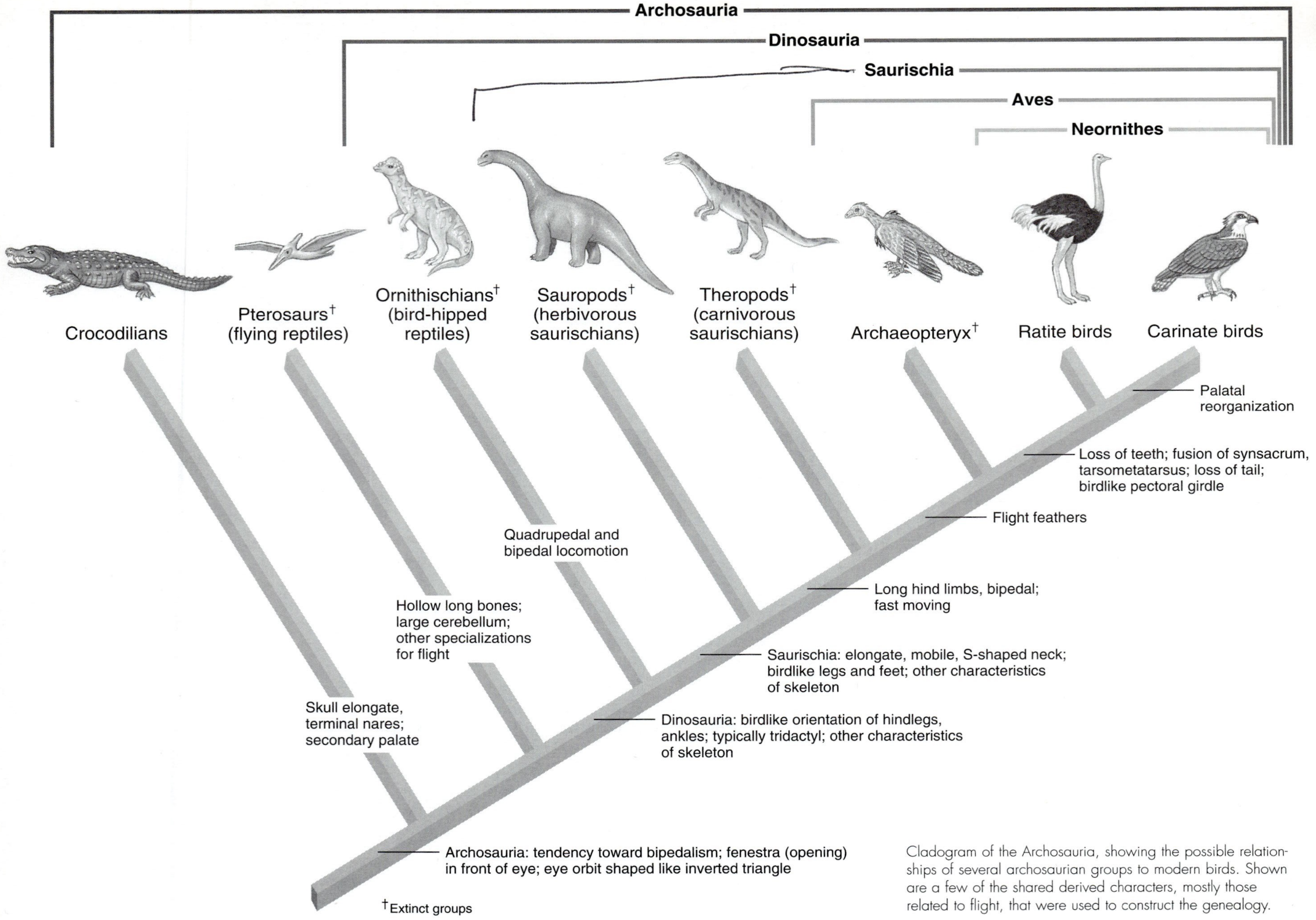

Cladogram of the Archosauria, showing the possible relationships of several archosaurian groups to modern birds. Shown are a few of the shared derived characters, mostly those related to flight, that were used to construct the genealogy.

powerful evidence barring them from bird ancestry, has now been found in several theropod dinosaurs (Norell et al., 1997).

Other researchers, such as Alan Feduccia of the University of North Carolina, and Larry Martin of the University of Kansas, however, believe that dinosaurs and birds shared a common thecodont ancestor. Feduccia postulates that some as-yet-undiscovered, lightly built, tree-living reptile produced the avian line (Feduccia, 1980; 1996). Feduccia and other evolutionary biologists argue that dinosaurs and birds had a similar way of life that could account for a coincidental similarity of appearance—a process known as convergent evolution. Major elements of disagreement involve lung structure and ventilation (Ruben et al., 1997; Gibbons, 1997d), whether some theropods and early birds were ectothermic or endothermic (Ruben et al., 1997; Gibbons, 1997d), and the developmental patterns and homologies in the avian wing (hand). Theropod hands retain only digits 1–2–3, whereas some researchers claim that birds supposedly have a 2–3–4 digital formula (Burke and Feduccia, 1997).

Bird skulls are essentially reptilian with a single occipital condyle, only one auditory ossicle or middle ear bone (columella), and a lower jaw (mandible) composed of several bones. The lower jaw is hinged on a movable quadrate bone, as in snakes and some extinct reptiles. Most birds have flat processes on their ribs (uncinate processes), presumably to strengthen the thorax and prevent it from collapsing because of the force exerted by the powerful flight muscles as they contract with every beat of the wing during flight. The only other animals to possess uncinate processes are tuataras and crocodilians. In these lepidosaurs, uncinate processes provide support for muscle attachment and serve to strengthen the wall of the thoracic cavity. The ankle joint of birds is between two rows of tarsal bones (intratarsal), instead of being between the tibia and the tarsal bones as in the reptiles, and the foot retains the primitive phalangeal formula of 2–3–4–5 phalanges for the first four digits, similar to sauropsids. Scales are present on the legs and feet of birds. Only crocodilians, birds, and mammals have a four-chambered heart. Both reptiles and birds have nucleated erythrocytes, an egg tooth on the upper jaw at hatching, and the same general type of shelled telolecithal egg (having the yolk concentrated at one pole of the egg) with four extraembryonic membranes. Embryological development is also basically similar in both groups. Molecular evidence, including DNA sequences from four genes, provides strong statistical support for a bird–crocodilian relationship (Hedges, 1994).

The most obvious features that distinguish birds from modern reptiles are that birds are endothermic and possess feathers. Recent evidence indicates, however, that some, if not all, dinosaurs may have had high metabolic rates and also may have been endothermic (Bakker, 1986). Developmentally, feathers and reptilian scales are homologous structures. Feathers are produced by papillae in the skin, and their early development is quite similar to that of a scale. Romer (1966) noted that birds are structurally similar to reptiles and "are so close to the archosaurians that we are tempted to include them in that group." Some taxonomists (Gauthier et al., 1988a, b; Kemp, 1988; Benton, 1990) place reptiles and birds in a single class, the Sauropsida. Others (Gardiner, 1982; Lovtrop, 1985) have argued that Aves is the sister group (most closely related) of Mammalia, forming a larger clade, the Haemothermia.

Bakker (1986), another proponent of the dinosaur–bird link, claims that the traditional grouping that places birds in a class of their own and dinosaurs together with reptiles is "neither fair nor accurate." He stated:

> *The small advanced predators like* Deinonychus *were so close to* Archeopteryx *in nearly every detail that* Archeopteryx *might be called a flying* Deinonychus, *and* Deinonychus *a flightless* Archeopteryx. *There simply was no great anatomical gulf separating birds from dinosaurs. And that implies dinosaurs are not extinct. One great, advanced clan of them still survives in today's ecosystem and the more than eight thousand species of modern bird are an eloquent testimony to the success in aerial form of the dinosaurs' heritage.*

Bakker further proposed to resurrect the name Dinosauria for a class that would include the dinosaurs, the therapsids, and the birds. He stated: "And let us squarely face the dinosaurness of birds and the birdness of the Dinosauria. When the Canada geese honk their way northward, we can say: 'The dinosaurs are migrating, it must be spring!' "

Although reptiles have a common ancestry, the class Reptilia is not monophyletic. Based solely on shared derived characters, crocodilians and birds are living sister groups; they are descended from a common ancestor more recently than either is from any other living reptilian lineage (see Figs. 7.1 and 7.2). This is the reason cladists believe birds should be classified as reptiles, and crocodilians and birds should be placed in a separate clade, the Archosauria, which also includes the extinct dinosaurs. However, traditional evolutionary taxonomists point out that birds possess many unique morphological characteristics, whereas crocodilians have more features in common with reptiles. In this view, the morphological and ecological uniqueness of birds should be recognized by maintaining the traditional classification that places crocodilians in the Reptilia and birds in the Aves. This does not represent true evolutionary affinities and is, therefore, artificial systematics. However, the standard taxonomic practice in most texts is to classify crocodilians as reptiles. Keep in mind, however, that systematics is based on new techniques and discoveries and is continually being refined.

In October 1996, Pei-Ji Chen of the Nanjing Paleontology Institute in China showed photographs of two recently discovered "feathered" dinosaurs at the annual meeting of the Society of Vertebrate Paleontology (Gibbons, 1996c, 1997b; Monastersky, 1996b; Morell, 1997e). The 121-million-year-old fossils, known as *Sinosauropteryx prima*, could be the most graphic evidence yet that birds are descended from dinosaurs (Chen et al., 1998). Arrayed down the chicken-sized dinosaur's back, from the nape of its neck to the tip of its tail, is what appears to be an almost manelike row of feathers that have left their impression in the rock. The hollow filaments, up to 40 mm long, resemble extremely simple feathers, called plumules, found on some modern birds. The authors suggest that the fibers could either represent protofeathers that helped trap body heat, or they could have served as a colorful display for attracting mates. Luis Chiappe of the American Museum of Natural History speculated that the "feathered" dinosaur may have developed feathers because it was on the road to warm-bloodedness, but had not gotten far.

In life, the "feathered" dinosaur was about 90 cm long and had numerous serrated teeth in its mouth. It ran on its hindlimbs, holding its forelimbs in front of it. It appears closely related to a species of *Compsognathus*, a small dinosaur that ate insects and other small animals. One specimen has two oval shapes inside its abdomen—the first clear case of eggs found inside a dinosaur.

In April 1997, several researchers, including John Ostrom, Larry Martin, ornithologist Alan Brush, paleontologist Peter Wellnhofer, and photographer David Bubier, traveled to China to examine the fossil. Their conclusion was that the "feathers" were fibers either within the skin or above the skin (Monastersky, 1997a). If the fibers are within the skin, they could have been part of a ridge similar to the frill of an iguana. If the fibers are above the skin, they could be bristles or protofeathers—structures that preceded the evolution of true feathers. To demonstrate conclusively whether the impressions represent feathers, featherlike scales, or hairlike structures, the Chinese researchers will have to examine the fossil in much greater detail, perhaps by performing biochemical analyses.

Three theropod dinosaurs with feathers—*Protarchaeopteryx robusta, Caudipteryx zoui,* and *Sinornithosaurus millenii*—have been described from the Upper Jurassic/Lower Cretaceous formations of Liaoning, China (Qiang et al., 1998; Padian, 1998; Swisher et al., 1999; and Xu et al., 1999). These turkey-sized animals had strong legs, stubby arms, and down-covered bodies. Although feathers covered the body of *Protarchaeopteryx*, no preserved evidence of wing feathers exists. *Caudipteryx*, or "tail feather," has more plumage, including a generous tail fan. The postcranial skeleton of the dromaeosaurid *Sinornithosaurus* is remarkably similar to early birds. The structure of the shoulder girdle shows that terrestrial dromaeosaurid dinosaurs had attained the prerequisites for powered, flapping flight. The body was apparently covered by a layer of integumentary filaments generally reaching 40 mm in length that differ little from the external filaments of other theropod dinosaurs or even the plumule-like feathers of *Confuciusornis* from the same locality. These three theropod dinosaurs are thought to have been capable of running swiftly, flapping feathered wings, and fanning out impressive tail feathers, but were unable to actually fly. Phylogenetic analysis indicates that all are more primitive than the earliest known bird, *Archaeopteryx*. These fossils are thought to represent stages in the evolution of birds from feathered, ground-living, bipedal dinosaurs.

The announcement of a turkey-sized animal that may have been the first flying dinosaur created much excitement in 1999 (Sloan, 1999). Fossils of the animal, named *Archeoraptor liaoningensis*, came from the Liaoning, China site. The shoulder girdle and breastbone resembled those of modern birds. In addition, it had hollow bones and a long, stiff tail. In January 2000, however, CT scans confirmed "anomalies" in the reconstruction of the fossil; the "feathered dinosaur" combined the tail of a dinosaur with the body of a bird (Monastersky, 2000).

Although birds were once believed to have descended from the birdlike dinosaurs (Ornithischia), they are now thought to have branched from theropod ancestors (Ostrom, 1985, 1994; reviewed by Norman, 1991; Padian and Chiappe, 1998). A unique theropod skull discovered in Mongolia in 1965 from the Late Cretaceous shows a combination of theropod and primitive avian characters (Elzanowski and Wellnhofer, 1992). It has been named *Archeornithoides deinosauriscus* and probably belongs to the closest of the known nonavian relatives of *Archeopteryx* and other birds.

BIO-NOTE 7.6

Mammal-Eating Dinosaur

While studying a specimen of *Compsognathus* from northeast China, researchers found the jawbone of a tiny mammal in the digestive tract of the dinosaur. This is the first evidence of a dinosaur preying on a mammal. The stomach of the larger specimen of *Sinosauropteryx* contained a semiarticulated skeleton of a lizard, complete with skull.

Monastersky, 1997a
Chen et al., 1998

Through the use of CT scans, Bakker (1992) has found a number of similarities between carnivorous dinosaurs and birds. For example, members of the genus *Nanotyrannus*, a group of smaller dinosaurs related to those giant dinosaurs in the genus *Tyrannosaurus*, possessed cranial air canals that looked remarkably like those of *Troodon*, a small carnivorous dinosaur whose canals resembled those of modern birds. *Troodon*'s canals, along with its birdlike wrists and inner ear structure, have led some scientists to consider it to be the nearest known relative of modern birds. In addition, egg clutches and nests of *Troodon* indicate that two eggs were produced simultaneously at daily or longer intervals and that eggs were incubated using a combination of soil and direct body contact (Varruchio et al., 1997). *Troodon* egg shape, size, and microstructure suggest a more avian than crocodilian reproductive tract.

Bakker (1992) speculated that approximately 140 to 160 million years ago, birdlike innovations appeared and expressed themselves in both large and small dinosaurs. Suddenly a whole range of animals exhibited avian features. He stated:

> *Some of those creatures, perhaps including* Archeopteryx, *actually* were *birds, while others stayed on the ground, keeping their birdness in terrestrial mode. The descendants of these bird-dinos*—Nanotyrannus,

Troodon, *and others—would die out at the end of the Cretaceous, while the true birds would fly on into the evolutionary future.*

The relationship between birds and mammals also has generated considerable discussion. Comparative protein sequence studies of amniote myoglobin and hemoglobin by Bishop and Friday (1988) show that bird and mammal globins "frequently resemble one another biochemically more than would be expected on the majority view of their separate evolutionary histories." The reasons for these similarities are not clear. Gardner (1982) and Lovtrup (1985) presented cladistic arguments that mammals and birds are the nearest sister groups among living tetrapods. They noted about 20 characters shared by birds and mammals, but their arguments have been criticized by Kemp (1988) and others.

The fossil record has yielded few complete skeletons of birds. Because bird bones are fragile and many are hollow, they are easily broken and fragmented. As a result, much of the paleontological research on birds has been accomplished by studying fragments of bones, many of which may have been fragmented by carnivorous animals.

A remarkably well-preserved nestling bird dating from about 135 million years ago was discovered in the Pyrenees of northern Spain (Sanz et al., 1997; Morell, 1997c). It is the earliest hatchling bird yet discovered and comes just 10 million years after *Archaeopteryx*, the first undisputed bird. The toothed skull of this nestling looks dinosaurian, but other features resemble those of modern birds. Postorbital bones found in small theropod dinosaurs but not in modern birds are still present in the nestling, but they show signs of breaking down.

A 90-million-year-old theropod dinosaur that folded up its front limbs as if they were wings was reported from Patagonia (Novas and Puerta, 1997; Morell, 1997c). *Unenlagia comahuensis* could have stretched its forelimbs out as if taking flight, but probably extended them for balance instead. Changes in arm and shoulder anatomy coupled with its very birdlike pelvic girdle suggest the kind of changes that dinosaurs would have undergone during their transition to birds.

Ancestral Birds

In 1861, the impression of a single feather was discovered in a German limestone quarry. This was the first evidence that birds existed in the Jurassic period, approximately 150 million years ago. The next year, a fossilized pigeon-sized skeleton with imprints of feathers was discovered in the same quarry. A second skeleton was found in 1876, about 16 km away (Fig. 7.25). At the present time, a total of seven specimens have been recovered, the most recent in 1993 (Wellnhofer, 1990; Fraser, 1996). Although the first two skeletons originally were assigned to different genera, they later were classified in a single genus and species: *Archaeopteryx lithographica. Archaeopteryx* means "ancient wing," and *lithographica* refers to the limestone that was used for lithographic plates during the 19th century.

In almost all of its structure, *Archaeopteryx* is intermediate between modern birds and thecodonts. The modified diapsid skull and hindlimbs are reptilelike; well-developed wing claws and a long lizardlike tail are present, as are teeth set within sockets in both jaws. If clear imprints of feathers had not been present, the fossils very easily might have been classified as reptiles. In fact, the fifth specimen was classified as a small dinosaur for 20 years after its discovery in 1951 (Wellnhofer, 1990).

FIGURE 7.25

A restoration of the second specimen of *Archeopteryx lithographica* to be discovered. Positions of the body parts correspond to the positions of the fossilized bones.

As late as the mid-1940s, some investigators still denied that *Archaeopteryx* was a bird. Lowe (1944), for example, suggested that *Archaeopteryx* was an arboreal, climbing dinosaur that should "take its place not at the bottom of the avian phylum [class] but at the top of the reptilian."

The vertebral column of *Archaeopteryx* consists of cervical, thoracic, lumbar, sacral, and caudal vertebrae. The centra (body) of each vertebra is amphicoelous (biconcave). The presence of pneumatic foramina in the cervical and anterior thoracic vertebrae confirm the phylogenetic continuity between the pneumatic systems of some theropods such as *Compsognathus, Allosaurus, and Ornithomimus* and living birds (Britt et al., 1998). Five of the sacral vertebrae are fused into a primitive synsacrum but no pygostyle (fused caudal vertebrae) is present. Although the structure of the pelvis is similar to that of ornithischian dinosaurs, the resemblances are thought to be due to parallel evolution (Carter, 1967). The bones of *Archaeopteryx* are solid, not pneumatic. Metacarpals are not fused, but metatarsals are partially fused; thus, no carpometacarpus (fused wrist and hand bones so characteristic of all birds except *Archaeopteryx*) exists (Vazquez, 1992). The two clavicles have fused to form a furcula, but no sternum has been found in any of the six specimens. Modern flying birds possess a broad-keeled sternum for the attachment of enlarged flight (pectoral) muscles. The lack of indications of well-developed pectoral muscles in *Archaeopteryx* suggests that its ability to fly would have been limited.

Recent studies of feather asymmetry in *Archaeopteryx* and extant birds have not resolved whether or not *Archaeopteryx* would have been capable of sustained flapping flight (Speakman and Thomson, 1994; Norberg, 1995). Gastral (abdominal) ribs, similar to those of its thecodont ancestors, were present. The feet were adapted for running and showed features intermediate between reptiles and birds: a reduced digit 1 (hallux), which was diverted to the rear; a fused metatarsus; a mesotarsal joint; and a claw curvature typical of perching and trunk-climbing birds (Feduccia, 1993). Contour feathers were well developed, and tail feathers arose from the lateral surfaces of the caudal vertebrae. Because the body appears to have been covered with feathers, Carter (1967) suggested that feathers may have evolved first as an insulating cover for the body and not for flight. If feathers had evolved for flight, they would be expected to be primarily on the wings, not covering the entire body in a primitive bird.

The first Jurassic-Cretaceous birds from outside Germany were reported from northeastern China (Hou et al., 1995; Swisher et al., 1999). *Confuciusornis sanctus* has a primitive wing skeleton similar to that found in *Archaeopteryx*, including unfused carpal elements and long digits (Fig. 7.26). The pelvis and the climbing adaptations in the hands indicate a vertical, *Archaeopteryx*-like posture (Hou et al., 1996). The foot is similar to *Archaeopteryx*, with a reflexed hallux and large recurved claws supposedly reflecting an arboreal lifestyle. The wings were well developed and a long, feathered tail was present. Teeth were absent. Contour feathers are thought to have covered the entire body. A second species, *C. dui*, is based on a remarkably well-preserved skeleton with feathers and for the first time in the Mesozoic record, direct evidence of the shape of a horny beak (Hou et al., 1999).

A partial skeleton of a primitive bird, *Rahona ostromi*, was recovered from the Late Cretaceous of Madagascar (Forster et al., 1998). The skeleton exhibits a mosaic of theropod and derived avian features. For example, it possesses avian features, such as an avian antebrachium (forearm), feathered wings, a reversed hallux, and an avian-like synsacrum, but it retains characteristics that indicate a theropod ancestry, such as a sicklelike claw on the second digit, a unique characteristic of certain theropod groups. Although it lived 80 million years after the first known bird, *Archaeopteryx*, phylogenetic analysis places *Rahona* with *Archaeopteryx*, making *Rahona* one of the most primitive birds yet discovered.

Fossils discovered in 1984, 1990, and 1994 in Spain (*Iberomesornis, Concornis, Noguerornis,* and *Eoalulavis*) and northeast China (*Sinornis, Cathayornis,* and *Boluochia*) date from the Early Cretaceous, between 130 and 120 million years ago (Shipman, 1989; Monastersky, 1990a; Wellnhofer, 1990; Sereno, 1991; Zhou et al., 1992; Fischman, 1993a; Hou et al., 1996; Sanz et al., 1996). These fossils reveal unexpected diversity in early birds starting at about 135 million years ago.

FIGURE 7.26

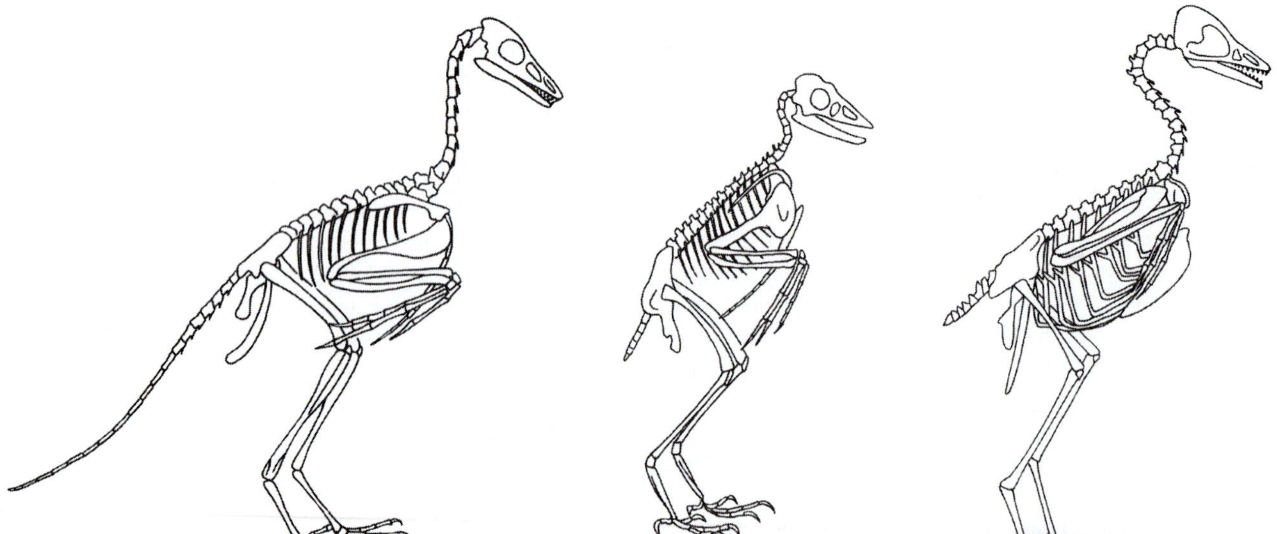

Restorations of Late Jurassic primitve birds: (a) *Archeopteryx lithographica*; (b) *Confuciusornis sanctus*; (c) reconstruction of the *Chaoyangia* skeleton.
Source: Hou, et al., in Science, *274:1165–1166, 1996.*

They are intermediate both chronologically and anatomically between *Archaeopteryx* and modern birds. They represent the earliest known examples of birds with a toothless beak and modernized flying ability. The fossils resemble *Archaeopteryx* in having a primitive avian wing with an unfused carpometacarpus; a robust ischium with an anterodorsal ischial process; a posteriorly projecting pubis; a long, bowed femur; partially fused metatarsals; and a reflexed hallux. The third metatarsal is the longest and the fifth metatarsal is present. Gastral ribs are present, and claws on the feet are long and curved in the Chinese specimens. However, they also show adaptations for flight, such as a pygostyle to strengthen the tail, a collarbone strongly connected to the sternum, and a reduced first digit and enlarged second digit on the hand. These birds were known as enantiornithes, or "opposite" birds; they were the dominant group of land birds during the Mesozoic. They are so named because three bones of their feet are partially fused from the top down, rather than from the bottom up as in modern birds (Fig. 7.27b, c).

This new clade of birds, the enantiornithine (opposite) birds, was proposed by Walker (1981). Opposite birds closely resemble *Archaeopteryx* with their primitive pelvic region and toothed skull. However, instead of a long reptilian tail, caudal vertebrae were fused into a long pygostyle, and these birds were quite capable of flying. These were the dominant birds of the Mesozoic and included such genera as *Sinornis* and *Ichthyonis* (Fig. 7.27a).

Until recently, many paleontologists thought that *Archaeopteryx* itself gave rise to opposite birds, which in turn evolved into modern birds. That view has faded, but Chiappe and others still hold that opposite and modern birds are closely related sister taxa, with a recent common ancestor that lived at about the time of *Archaeopteryx* or a bit earlier (Fig. 7.28).

Hou et al. (1996) now challenge that view with fossils of sparrow-sized birds called *Liaoningornis* from northeastern China's Liaoning Province. These specimens, taken from volcanic rock dated between 121 and 142 million years ago, possess foot bones and a keeled sternum that resemble those of modern birds. The presence of a keeled sternum is the earliest evidence for this distinctly avian structure. The bones represent what may be the oldest modern-looking bird, or ornithurine. If the dating is confirmed, it provides evidence for a pre-*Archaeopteryx* or a rapid post-*Archaeopteryx* evolution in birds. It could cause *Archaeopteryx* and the enantiornithines to be moved off the evolutionary branch that leads to modern birds. It could even imply an earlier origin for all birds. Feduccia stated: "It shows that there was a dichotomy, and that *Archaeopteryx* and most of the other early birds were a side line of avian evolution" (Gibbons, 1996d). Feduccia and Larry Martin believe birds had already diverged into two lineages by the time of *Archaeopteryx*, but that the fossil record is still missing. Feduccia and Martin noted that one lineage led to modern birds; the other led to *Archaeopteryx* and the opposite birds, which they view as sister taxa. They believe that both bird lineages must have descended from a much earlier ancestral bird.

The specimens of *Liaoningornis* come from the same fossil beds that yielded the magpie-sized primitive bird (*Confuciusornis*) and the controversial "feathered" dinosaur *Compsognathus prima*. These fossil beds are approximately 124 million years old, placing them within middle Early Cretaceous time (Swisher et al., 1999). The next oldest ornithurine bird is *Chaoyangia* from the Early Cretaceous of China

FIGURE 7.27

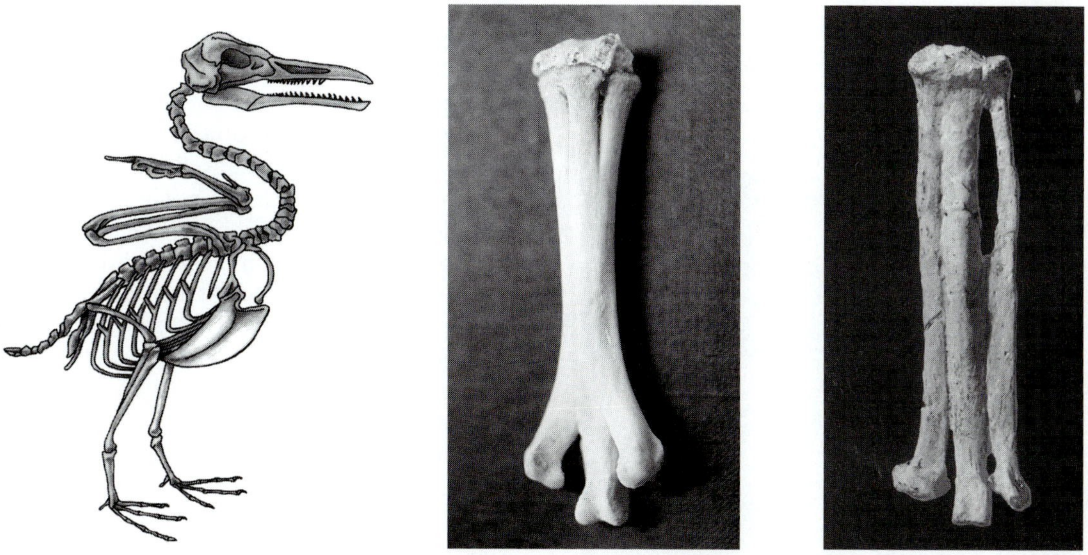

(a) (b) (c)

(a) Skeleton of the Upper Cretaceous bird *Ichthyornis*. Note the teeth and the well-developed keel on the sternum. (b) Modern birds have the foot bones fused from the bottom up (c); in opposite birds, the fusion is top down.
Source: Carroll, Vertebrate Paleontology and Evolution, *1998, W. H. Freeman and Co.; after Marsh, 1880.*

FIGURE 7.28

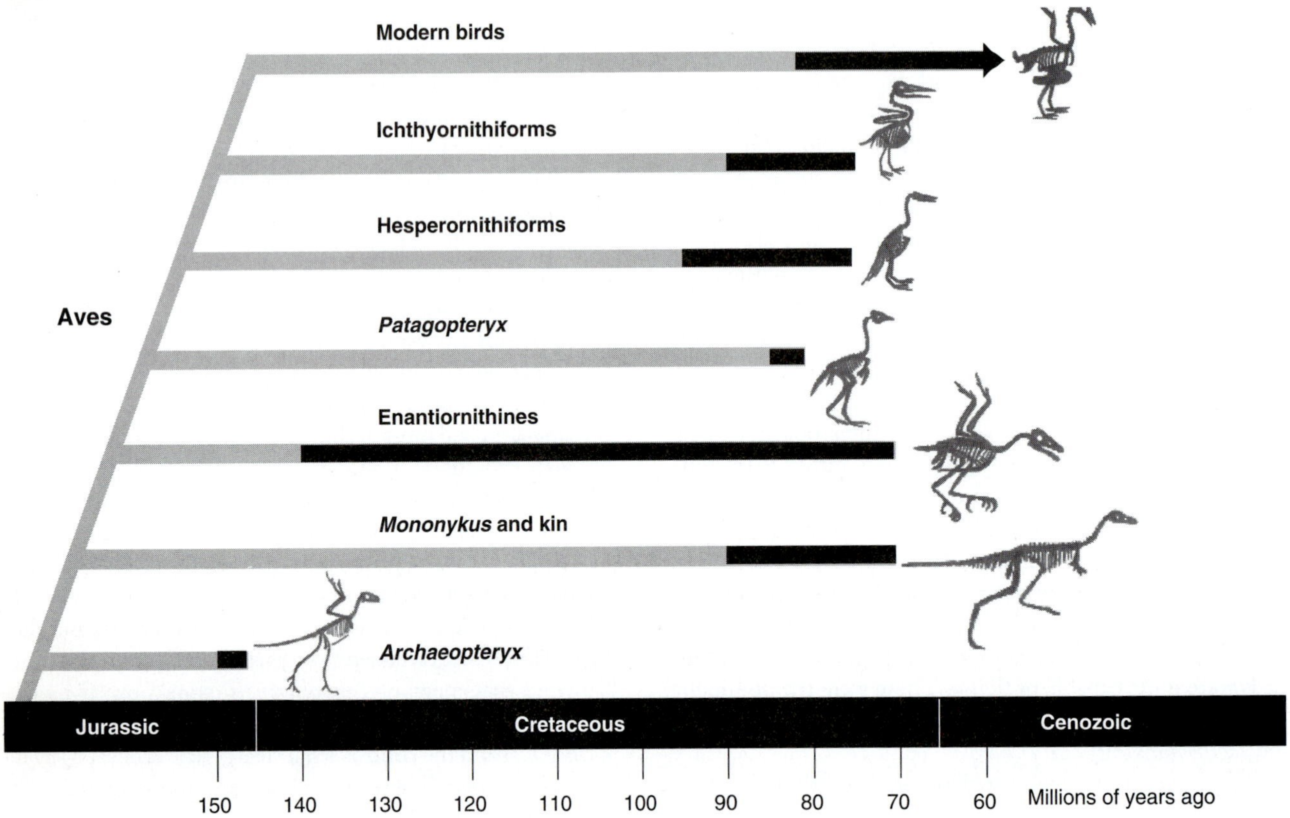

The fossil record reveals that living birds are just one branch on the avian family tree.

FIGURE 7.29

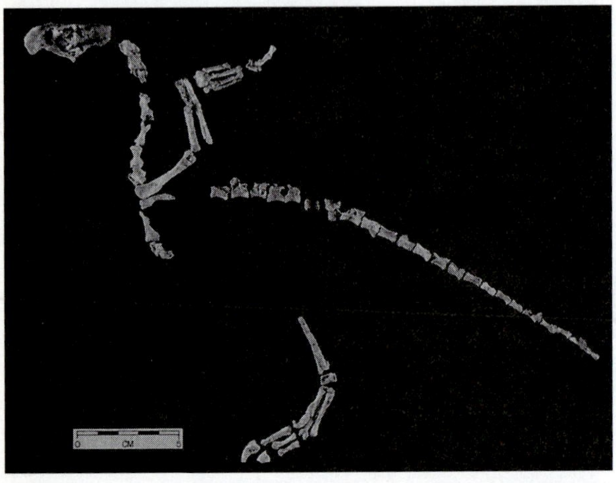

Protoavis texensis from the Triassic in Texas.

(Fig. 7.26c). It possesses a keeled sternum as well as premaxillary teeth.

In 1986, the fossil remains of a crowlike bird were discovered in Triassic deposits in Texas (Chatterjee, 1991). It is 75 million years older than *Archaeopteryx* and tentatively has been placed in the genus *Protoavis* (Fig. 7.29). Chatterjee's discovery remains highly controversial due primarily to similarities he sees between the skull and neck bones of *Protoavis* and the bones of modern birds. These structural similarities include a lightly built and pneumaticized skull; evidence of movable upper jaws as in modern birds; a temporal region similar to that in modern birds; heterocoelous (saddle-shaped) neck vertebrae; and a relatively large braincase. No feather impressions have been found, nor has a carpometacarpus been found. Circumstances surrounding this specimen and its scrutiny by qualified scientists have resulted in considerable controversy (Ostrom, 1991; Chatterjee, 1991). Critics speculate that Chatterjee's poorly preserved specimen was completely disarticulated when it was found and that it is likely to be a collection of bones from several unrelated species of small dinosaurs rather than the bones of a single individual. Many of the crucial skull bones are mere fragments, providing ambiguous evidence for the supposedly birdlike features of the skull. However, some prominent avian paleontologists see *Protoavis* as important new evidence that birds arose as a distinct group before the theropod dinosaurs arose, and they strongly support Chatterjee's position that *Protoavis* has too many birdlike features to be written off as a misinterpretation of a poor specimen (Anderson, 1991; Zimmer, 1992). Whether *Protoavis* was a bird ancestor or a genuine "early bird" remains to be answered.

BIO-NOTE 7.7

A Giant Bird

The 450-kg elephant bird (*Aepyornis*) formerly inhabited Madagascar. When humans arrived on the island (probably less than 2,000 years ago), *Aepyornis* was subjected to intense ecological pressures (forest destruction and hunting of the birds and their eggs) that led to their final extinction around the year 1000. However, their gigantic eggs (Fig. 7.30), weighing more than 9 kg and equal in size and weight to eight ostrich eggs, are still found.

Wetmore, 1967
Page and Morton, 1989

In 1993, a turkey-sized predator (*Mononykus*, meaning "one claw") with a mouthful of sharp teeth and a long tail was reported from the Upper Cretaceous of the Gobi Desert of Mongolia (Perle et al., 1993; Dashzeveg et al., 1995). It looked like a modern flightless bird but had bone structure characteristic of both birds and theropod dinosaurs. However, in *Archaeopteryx*, the fibula touches the ankle; in modern birds and *Mononykus*, it does not. Both modern birds and *Mononykus* have a keeled sternum. Some of *Mononykus*'s wristbones were fused together, providing another adaptation for flight. Some researchers are skeptical: should *Mononykus* be considered a primitive flightless bird? Zhou (1995) refutes the avian status of *Mononykus* by concluding that many of its resemblances to extant birds, such as the large, ossified, keeled sternum, and reduced fibula, are digging adaptations or adaptations for bipedalism. Chiappe et al. (1997), however, claim that Zhou's methodology and anatomical comparisons fail to support his claim. The controversy continues!

FIGURE 7.30

An elephant bird (*Aepyornis*) egg from Madagascar.

Modern Birds

Few modern orders of birds existed in the Late Cretaceous. The modern orders that did exist all belonged to what are termed "transitional shorebirds." Bird fossils from the Cretaceous period (approximately 125 million years ago) have yielded toothed birds with widely divergent skeletal structures that represent intermediate stages between *Archaeopteryx* and modern birds. These differences are indicative of the great radiation that took place in birds between the Jurassic and Cretaceous periods. The greatest radiation, however, occurred in the Cenozoic era. Most modern orders were fully differentiated by the Eocene epoch, but some made their first appearance in the fossil record of the Oligocene. The avian fauna during the Pleistocene was even richer than it is today. Feduccia (1995) hypothesized that birds endured massive late Mesozoic extinctions, underwent a bottleneck at the Cretaceous–Tertiary boundary, and then experienced an explosive evolution in the Early Tertiary. He suggested that the explosive evolution that took place in the Early Tertiary produced all of the modern lineages within about 10 million years. If this is true, then former hypotheses attributing bird biogeography to drifting continents will need to be reexamined (Cracraft, 1971, 1973, 1986; Sibley and Ahlquist, 1990; and others).

Origin of Flight

Theories concerning the origin of flight have been discussed and disputed since 1879–80, when two hypotheses were proposed. One hypothesis supposes that the theropod ancestor was a cursorial, ground-dwelling form that held its arms out horizontally as it ran to capture insects and other food (Ostrom, 1974, 1985, 1994; Padian and Chiappe, 1998). The upward pressure of the air on the arms (held at a specific angle so that lift exceeded drag) is thought to have taken some of the weight off the feet. Some researchers conjecture that this would help the animal run faster, which would lead later to more continuous gliding flight, and finally to active flapping flight (Burgers and Chiappe, 1999). Others argue that it would slow the speed of running by removing the "traction" necessary for high speed. Frayed scales are thought to have gradually evolved into feathers for insulation. Those on the arms would increase the surface area and allow the theropods to take even more pressure off their feet. Albatrosses, geese, and swans today use a similar running–flapping method in order to get airborne but are forced to use longer takeoffs than other birds due to their relatively slow ground speed.

A second hypothesis presupposes that the ancestors to birds were arboreal and that the arms and frayed scales served as parachute-like structures to aid the animal in jumping from branch to branch and to help break its falls. As feathered wings evolved, gliding from tree to tree or from tree to ground became possible. Only later would these ancestors evolve the ability to flap their wings and fly actively. Parachute-type structures and body modifications can be seen today in certain frogs, lizards, flying squirrels, and phalangers. Carter (1967), Bakker (1986), Norman (1991), and others have supported the climbing-and-gliding flier theory. Feduccia (1993) used claw geometry to conclude that *Archeopteryx* appeared to have been a perching bird rather than a cursorial predator. The claws of both the foot and hand exhibit degrees of curvature typical of perching and trunk-climbing birds. Not all paleontologists agree (Morrell, 1993). Wellnhofer (1990) combined features of both the arboreal and cursorial theories into an arbocursorial, or climber–runner, theory.

Review Questions

1. What is the significance of the extraembryonic membranes in reptiles and birds?
2. What adaptations did reptiles evolve that permitted them to live in xeric habitats?
3. What are the advantages and disadvantages of dinosaurs depositing their eggs in nests and caring for their young?
4. What advantages were gained by some dinosaurs growing to enormous size?
5. Why do reptilian bones generally grow in spurts and produce annual rings of growth, whereas bird and mammal bones do not form annual rings?
6. Which group is considered the most recently evolved group of reptiles—lizards, snakes, or crocodilians? Why?
7. Which characteristics are shared by reptiles and birds?
8. (a) List some of the reptilelike characteristics of *Archaeopteryx*. (b) List some of the birdlike characteristics of *Archaeopteryx*.
9. Discuss the two major theories concerning the origin of flight in birds.
10. Do you think that the fossil record will ever be complete? Why or why not?

Supplemental Readings

Abler, W. L. 1992. The serrated teeth of tyrannosaurid dinosaurs, and biting structures in other animals. *Paleobiology* 18(2):161–183.

Ackerman, J. 1998. Dinosaurs take wing. *National Geographic* 194(1):74–99.

Callaway, J. M., and E. L. Nicholls (eds.). 1997. *Ancient Marine Reptiles.* New York: Academic Press.

Carpenter, K., and J. Currie (eds.). 1992. *Dinosaur Systematics. Approaches and Perspectives.* New York: Cambridge University Press.

Chatterjee, S. 1998. *The Rise of Birds: 225 Million Years of Evolution.* Baltimore: The Johns Hopkins University Press.

Currie, P. J. 1996. The great dinosaur egg hunt. *National Geographic* 189(5):96–111.

Currie, P. J., and K. Padian (eds). 1997. *Encyclopedia of Dinosaurs.* San Diego: Academic Press.

Dingus, L., and T. Rowe. 1997. *The Mistaken Extinction: Dinosaur Evolution and the Origin of Birds.* San Francisco: W. H. Freeman.

Dodson, P. 1996. *The Horned Dinosaurs: A Natural History.* Princeton, New Jersey: Princeton University Press.

Elias, S. A. 1995. *The Ice-Age History of National Parks in the Rocky Mountains.* Washington, D.C.: Smithsonian Institution Press.

Elias, S. A. 1995. *The Ice-Age History of Alaskan National Parks.* Washington, D.C.: Smithsonian Institution Press.

Erickson, G. M. 1999. Breathing life into *Tyrannosaurus rex*. *Scientific American* 281(3):42–49.

Farlow, J. O., and M. K. Brett-Surman (eds.). 1997. *The Complete Dinosaur.* Bloomington: Indiana University Press.

Fastovsky, D. E. and D. B. Weishampel. 1996. *The Evolution and Extinction of Dinosaurs.* New York: Cambridge University Press.

Fortey, R. 1998. Life: *A Natural History of the First Four Billion Years of Life on Earth.* New York: Alfred A. Knopf.

Fuller, E. 1987. *Extinct Birds.* New York: Facts on File, Inc..

Halliday, T. R. and K. Adler. 1986. *The Encyclopedia of Reptiles and Amphibians.* New York: Facts on File, Inc.

Natural History. June 1995. (Entire issue is devoted to dinosaurs and birds.) 104(6):1–88.

Novacek, M. 1996. *Dinosaurs of the Flaming Cliffs.* New York: Doubleday.

Sereno, P. C. 1999. The evolution of dinosaurs. *Science* 284:2137–2147.
Shipman, P. 1998. *Taking Wing:* Archeopteryx *and the Evolution of Bird Flight.* New York: Simon and Schuster.
Stokstad, E. 1998. Young dinos grew up fast. *Science* 282:603–604.
Sumida, S. S., and K. L. M. Martin (eds.). 1996. *Amniote Origins.* San Diego: Academic Press.
Weishampel, D. B., and L. Young. 1996. *Dinosaurs of the East Coast.* Baltimore: The Johns Hopkins University Press.

Vertebrate Internet Sites

Visit the zoology website at http://www.mhhe.com to find live Internet links for each of the references listed below.

1. **The Dinosauria.**
   University of California at Berkeley, Museum of Paleontology site contains information on morphology, systematics, and links to more information about subjects such as hadrosaurs and dinosaur locomotion.

2. **Dinobuzz.**
   Current topics concerning dinosaurs from the University of California at Berkeley Museum of Paleontology.

3. **Dinosaur Physiology—the Crazy Debate.**
   Hot-Blooded or Cold-Blooded?

4. **Dinosaur-Bird Relationships.**
   Are Birds Really Dinosaurs?

5. **Dino Russ's Lair.**
   Much information on dinosaurs from Russ Jacobson at the Illinois State Geological Survey.

6. **ZoomDinosaurs.com.**
   This Disney-sponsored dino-dictionary contains write-ups on all sorts of obscure facts pertaining to dinosaurs.

7. **Dinosauria-On-Line.**
   Lots of links and information.

8. **Paurasaurolophus Page at NMMNH (the New Mexico Museum of Natural History and Science).**
   Hear the supposed sounds from this dinosaur with a nasal resonator.

9. **Dinosaurs in Cyberspace: Dinolinks.**
   University of California at Berkeley Museum of Paleontology. A list of dinosaur-oriented web sites, scientific and otherwise. This site has pages and pages of links!

10. **National Museum of Natural History/Smithsonian Institution.**
    Interesting misconceptions about dinosaurs.

CHAPTER 8

Reptiles: Morphology, Reproduction, and Development

■ INTRODUCTION

Reptiles, comprising over 16,150 of the 53,000+ species of vertebrates, include turtles (230 species), tuataras (2 species), lizards (3,900 species), snakes (2,400 species), crocodilians (21 species), and birds (9,600 species). With the evolution of internal fertilization and the amniote egg, reptiles became the first fully terrestrial vertebrates.

TURTLES, TUATARAS, LIZARDS, AND SNAKES (TESTUDOMORPHA AND LEPIDOSAUROMORPHA)

■ MORPHOLOGY

Integumentary System

Turtles and lepidosaurs possess scales that, unlike those of fishes, are formed mainly from epidermal layers. The dry, scaly epidermis, which may be six or more layers in thickness, serves primarily for protection and to reduce water loss. Some snakes utilize the broad, flat scales, known as scutes, on the undersides of their bellies to aid in locomotion. In turtles, the shell of dermal plates (Fig. 8.1a, b) is covered by horny, keratinized scales (also known as shields or scutes) (Fig. 8.1c, d). The embryonic origin and morphology of the integument in reptiles is discussed by Maderson (1985).

In squamates (snakes and lizards), the epidermis consists of a stratum corneum (outer tissue layer) and a stratum intermedium (middle tissue layer) above the stratum basale (basal cell layer) (Fig. 8.2a, b). In most lizards and snakes a continual body covering of scales develops from the stratum corneum, with each scale projecting backward to overlap part of the one behind. In turtles, however, each epidermal scale develops separately, so that the scales do not form a solid sheet. The number and arrangement of epidermal scales on the body is usually species-specific and is used extensively in classification.

The stratum corneum is sloughed and replaced either a few cells at a time or in patches, or it is shed at intervals in

FIGURE 8.1

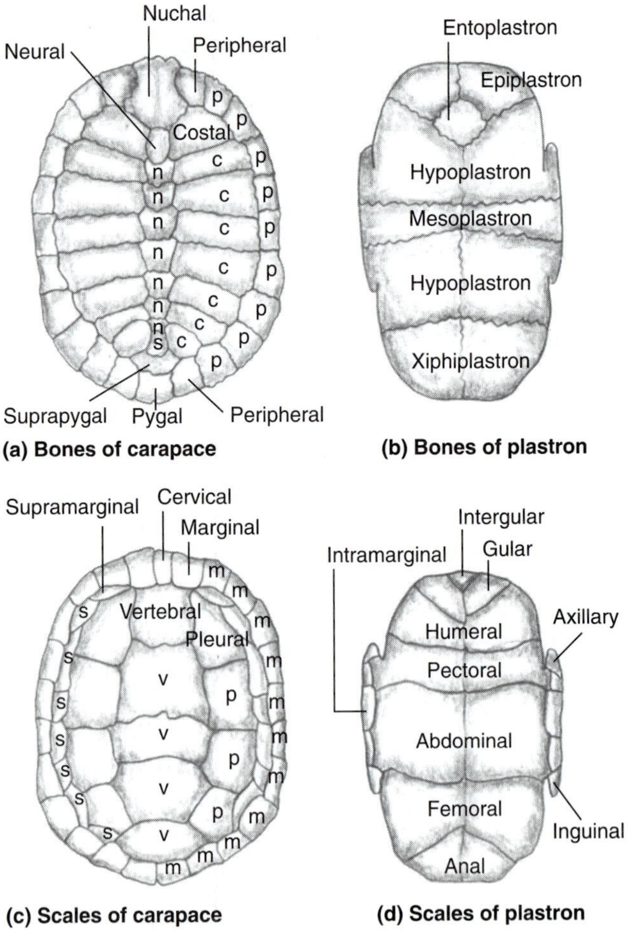

Upper: Dermal bones forming the carapace (*a*) and plastron (*b*) of a turtle. Lower: Epidermal scales covering the carapace (*c*) and plastron (*d*) of a turtle.

FIGURE 8.2

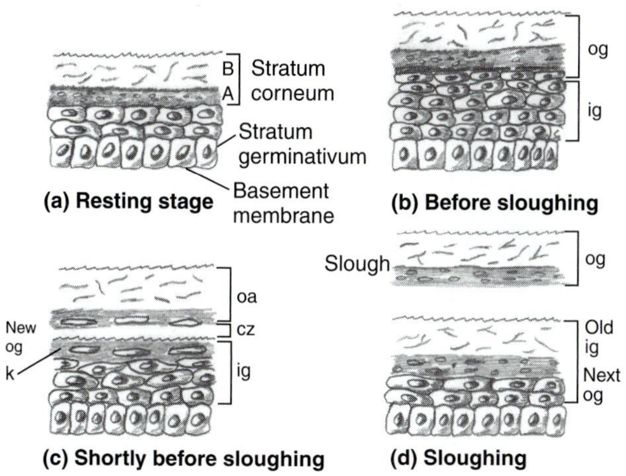

Simplified diagrams showing changes in the epidermis during the sloughing cycle of a snake. (*a*) Resting stage. (*b*) Before sloughing. The basal cells have divided to form a new, inner epidermal generation (ig). The snake's color is dulled. (*c*) Shortly before sloughing. A cleavage zone appears between the two generations; the superficial part of the inner generation is becoming keratinized (*k*), and a new serrated outer epidermal generation (og) is being formed. This stage probably coincides with the clearing of the skin. (*d*) Sloughing. The original outer generation is shed, and the old inner generation becomes the next outer generation.

FIGURE 8.3

Ecdysis in a lizard. Most lizards shed their skin in pieces, whereas healthy snakes usually shed their skins in one piece.

one piece, a process known as **ecdysis**. Healthy snakes usually shed their skins in one piece, whereas most lizards normally shed their skin in a number of pieces (Fig. 8.3). Old scales on some turtles peel off; in other species of turtles, the scales remain and give the shell a roughened texture.

When ecdysis begins, the two epidermal layers separate simultaneously over the entire body, and the outer layer is removed (Fig. 8.2). The stratum corneum is loosened primarily through the diffusion of lymph and white blood cells between the old and new layers. Shortly before shedding begins, the separation of the scales covering the eyes of snakes causes the eyes to become cloudy. As the outer layer of epidermis is removed, the inner layer becomes the new outer layer. Ecdysis normally begins in the head region and is initiated by an increase in blood pressure that causes the head to enlarge. This swelling causes the outermost layer of cells to loosen and rupture. Then, by means of the animal's rubbing and crawling movements, the remainder of the outer layer is removed. After a week or so, the stratum germinativum will have produced enough new cells through mitosis to form a new inner tissue layer. Frequency of shedding varies with species, age, and the health of the animal.

Few integumentary glands are present in reptiles. Those that are present (musk, femoral, pre-anal, cloacal, and nuchodorsal) either secrete strong-smelling substances that may be obnoxious to potential predators or serve for species and sex recognition (pheromones) during breeding. No sweat glands are present.

Scales, claws, rattles, horny protuberances, and spines are all keratinized modifications of the epidermis. **Claws**, which first appeared in turtles and have persisted in birds and most mammals, are shed periodically in turtles and lepidosaurs; however, in birds and mammals, claws, nails, and hooves are worn down by abrasion.

Many lizards can climb vertical surfaces readily by using their sharp claws. Some climbing lizards, such as geckos and anoles, utilize "dry" adhesion systems on their toes as additional aids when climbing on steep smooth surfaces and overhangs (Cartmill, 1985) (Fig. 8.4a, b). On the underside of each toe are approximately 20 broad overlapping scales (lamellae) consisting of numerous minute setae composed of keratin (Ruibal and Ernst, 1965; Ernst and Ruibal, 1966). Up to 150,000 hairlike setae are located on the exposed surface of each lamella. The setae are so small that their intermolecular interactions with the surface enable the gecko to adhere to the surface. A gecko's ability to walk on a vertical surface is directly correlated with that surface's molecular polarity and not with the degree of microscopic abrasions on that surface (Hiller, 1975). The many close contacts assure that the animal adheres as if it had climbing shoes with thousands of minute cleats imbedded in the substrate.

Rattlesnake **rattles** are modified portions of the stratum corneum that remain attached to the tip of the tail following each ecdysis (Fig. 8.4c). The rattles interlock in such a way that they fit together more tightly dorsally than ventrally. The rapid oscillation of the tail and its horny appendage produces the characteristic buzzing sound of an agitated rattlesnake. Because lobes of the rattle often are broken off accidentally, and because snakes may shed their skins several times a year, the age of a rattlesnake cannot be determined accurately by simply counting the number of segments in the rattle.

The **horns** of the horned "toad" (a lizard) (*Phrynosoma* sp.) are bony projections of the occipital bone of the skull covered with scaly integument, whereas the hornlike processes on the head of the sidewinder rattlesnake (*Crotalus cerastes*) are exclusively integumental in origin. A horny sheath of

FIGURE 8.4

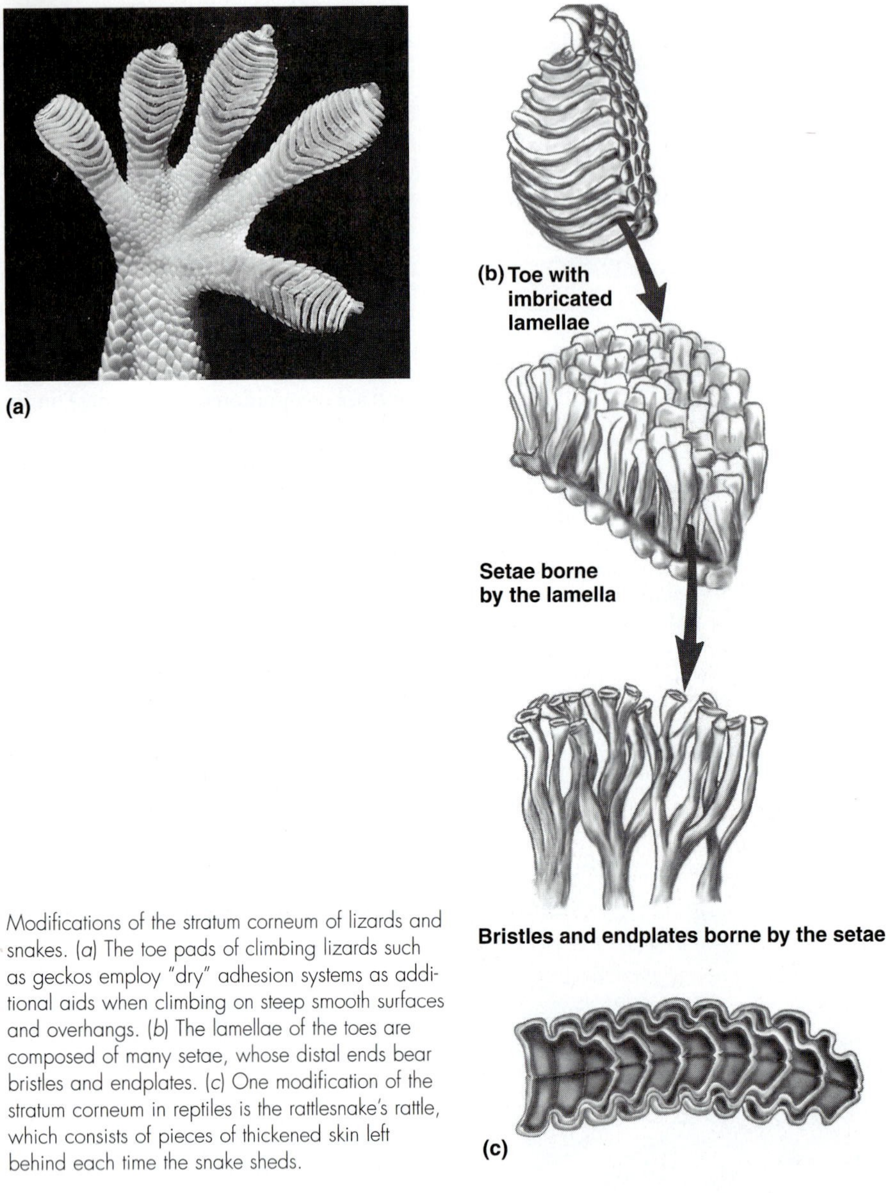

Modifications of the stratum corneum of lizards and snakes. (*a*) The toe pads of climbing lizards such as geckos employ "dry" adhesion systems as additional aids when climbing on steep smooth surfaces and overhangs. (*b*) The lamellae of the toes are composed of many setae, whose distal ends bear bristles and endplates. (*c*) One modification of the stratum corneum in reptiles is the rattlesnake's rattle, which consists of pieces of thickened skin left behind each time the snake sheds.

stratum corneum covered the beaks of some extinct reptiles and continues to cover the beaks of living turtles.

The dermis consists of a relatively loosely packed superficial layer and a much more densely packed deeper layer. As in other vertebrates, it consists of fat cells, nerve fibers, blood vessels, and chromatophores.

The skin of cotylosaurs (stem reptiles) was heavily armored with large bony dermal scales. Today, turtles are the most armored members of this group, with a shell consisting of large, bony, dermal plates (Fig. 8.1a, b). The dorsal arched portion of the shell is the **carapace**; the ventral, flattened portion is the **plastron**. These are united by bony **lateral bridges**. Nuchal and costal plates of the carapace are fused with vertebrae, and each costal plate is united with a rib. Soft-shelled and leatherback sea turtles have leathery shells, because the dense collagenous connective tissue of the dermis does not become ossified. A few lizards, such as skinks and glass lizards, have similar, but smaller, bony dermal scales (osteoderms) that serve to reinforce their scales.

The dermis has an abundance of **chromatophores**, which are responsible for elaborate color patterns. Although several speculative mechanisms for pigment pattern formation have been proposed, the actual mechanism of color pattern formation is not known. Some lizards, such as chameleons, change their color in response to environmental stimuli by concentrating and dispersing pigment granules. Coloration may be protective (camouflage and warning patterns), may reflect dominant social status, may serve for sex recognition, or may be important in thermoregulation.

Color patterns of many snakes involve spots or stripes. Spotted patterns are usually thought to serve as camouflage and are most common in species that feign immobility when approached. Stripes, which make it hard to judge the speed of a moving object, are characteristic of species that flee when threatened. Brodie (1989) demonstrated a genetic link between behavior and pattern in newborn garter snakes (*Thamnophis ordinoides*) in which striped individuals tended to flee, whereas spotted individuals tended to remain motionless.

Skeletal System

Skeletal modifications for terrestrial life that originated in amphibians are developed further in reptiles. The skeleton shows numerous modifications for muscle attachment, varied dietary habits, and terrestrial locomotion. Reptiles can better support their body weight and, in many cases, can move with great speed. Papers discussing various aspects of the skeleton are included in Gans et al. (1969).

Adult reptilian skulls differ from those of amphibians in many ways (see Fig. 6.13). A single occipital condyle is present, the skull has a higher and narrower shape, a greater degree of ossification is present, and a reduction of bones through loss and/or fusion has occurred. A partial secondary palate is present in many turtles.

Quadrate and articular bones form in the skull from gill arch supports. The hyomandibular becomes the columella (stapes) in the middle ear, and the roof of the skull, palate, and lower jaw become ensheathed in dermal bones. The symphysis of the anterior ends of the two dentaries, the major tooth-bearing bones in the lower jaw, is a rigid suture in some turtles and some lizards, but in snakes and many lizards it is connected by ligaments. Independent movement of the upper jaw on the braincase is well developed in some snakes and allows for great distensibility of the mouth, an adaptation necessary for swallowing large prey. Some palatine bones as well as the jaws are connected so loosely to the skull that each half of the upper and lower jaws can move independently of each other (see discussion in Digestive System, page 208).

Amphibians and many reptiles swallow their food whole; thus, having internal nares in the anterior portion of the oral cavity presents no problem. Others (especially nonmammalian synapsids) tore, crushed, and chewed their food before swallowing. To avoid interrupting their breathing, it became necessary to get air into the pharynx posterior to the chewing mechanism.

Turtles developed a secondary palate below the primary palate. The secondary palate, which separates the nasal passages from the oral cavity, serves to increase the length of the nasal passages and permits an animal to breathe while processing food in its mouth. This latter adaptation (ability to breathe and chew food simultaneously) is seen as a link to the development of higher metabolic rates leading to the origin of endothermy in Therapsida. Lepidosaurians (lizards, tuataras, and snakes) lack a secondary palate.

A feature seen for the first time in the evolution of vertebrates is the development of either one or two pairs of **fenestrae** (fossae) in the temporal region of the skull (see Figs. 7.2 and 7.5). These fenestrae provide additional surface area for stout muscles that originate from the temporal region of the skull and insert on the lower jaw, enabling it to close with increased pressure. Temporal fenestrae are major characters used in reptilian classification systems.

Stem reptiles (order Cotylosauria) lacked fossae, a condition referred to as **anapsid**. Today, the only living reptiles with anapsid skulls are turtles (order Chelonia). Extinct mammal-like reptiles (order Therapsida) developed a single pair of temporal fossae low on each side of the skull. This **synapsid** type of skull is found in extant mammals. **Diapsid** skulls have two pairs of fossae (superior and inferior) on each side of the skull. Diapsid skulls are characteristic of extinct archosaurs, surviving archosaurs (crocodilians), and the tuatara (*Sphenodon*). Lizards, snakes, and birds have modified diapsid skulls.

The reptilian vertebral column has undergone additional modifications and exhibits a wide range of types and arrangements of vertebrae. Snakes have the longest vertebral columns, with up to 500 vertebrae (Fig. 8.5). Most turtles and lepidosaurs have a variable number of cervical vertebrae. In addition, for the first time in the evolution of vertebrates, the first two cervical vertebrae have become modified to permit movements of the head in several directions. The first vertebra (atlas) is ringlike because most of its centrum has been detached. It articulates with the single occipital condyle on the skull. The second cervical vertebra (axis) has an anterior projection, known as the **dens** or **odontoid process**, that rests on the floor of the atlas when the two vertebrae are

FIGURE 8.5

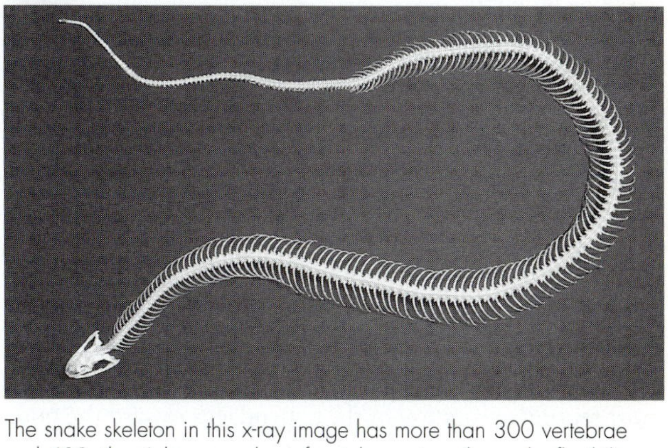

The snake skeleton in this x-ray image has more than 300 vertebrae and 400 ribs. A large number of vertebrae gives the snake flexibility and allows it to bend and twist more than most vertebrates.

articulated. It represents the detached centrum of the atlas. Cervical vertebrae of turtles are loosely articulated in order to allow the neck to be pulled back into the shell.

In most lizards, trunk vertebrae have differentiated into **thoracic** vertebrae that bear ribs, and **lumbar** vertebrae with ribs greatly reduced or absent. The vertebral column of snakes and legless lizards is divided into two regions: a **precaudal** (anterior to the vent) series of vertebrae, which bear free ribs; and a **postcaudal** (posterior to the vent) series with ribs either fused to the vertebrae or absent altogether. **Zygapophyses** (processes by which adjacent vertebrae articulate to one another) strengthen intervertebral joints. Most living reptiles possess two sacral vertebrae, which usually are fused to form a single bony complex, the **sacrum**, to support the pelvic girdle. The stronger sacrum provides the support necessary for raising the body off the ground as reptiles walk; some lizards and dinosaurs even adopted a bipedal mode of locomotion.

All reptiles have a distinct tail composed of many caudal vertebrae. A unique characteristic possessed by many lizards and a few snakes is the ability to break off their tails in order to avoid capture, known as **caudal autotomy**. This ability is possible because the centrum and part of the neural arch of each caudal vertebra are divided in half by an area of soft tissue. The plane of fracture occurs at this point, and most of the lost portion of the tail will regenerate. Caudal autotomy as a defense mechanism has been discussed by Dial and Fitzpatrick (1983) and Arnold (1988).

All vertebrae from cervicals to caudals may have ribs. The ribs of turtles, as well as the neural arches of the dorsal, sacral, and first caudal vertebrae, are fused with the carapace. Posterior cervical and anterior dorsal ribs of tuataras each bear a curved cartilaginous **uncinate process**, which projects posteriorly to overlap the rib behind, presumably giving strength to the thoracic body wall. In "flying dragons" (family Agamidae), five to seven posterior trunk ribs are greatly elongated to support large, thin membranes, which allow these lizards to glide distances up to 60 m. A mostly cartilaginous sternum is present in some lizards, although it is absent in snakes, legless lizards, and turtles. When present, the sternum consists of a plate to which thoracic ribs attach.

The limb structure shows considerable variation related to the burrowing, terrestrial, arboreal, or aquatic habits of these reptiles. Methods of locomotion include lateral undulation on both land and in water (swimming), bipedal and quadrupedal gaits, and flight.

Most snakes and some lizards lack an appendicular skeleton. Clavicles are reduced or absent from the pectoral girdle in legless lizards, and the entire pectoral girdle is missing in snakes. In turtles, the clavicles are fused with the carapace.

Limbs are typically pentadactyl, with five digits normally present on both the front and rear feet. Elements of the anterior limb are similar to those in salamanders (Fig. 8.6a). In most reptiles, a rotation of the appendages toward the body causes the long axis of the humerus and femur to lie more nearly parallel the body. A moderate bend at the elbow and knee allows the front and hindlimbs to be directed somewhat vertically, and the elbow is directed caudad (toward the tail). Limbs oriented in this fashion can better support the weight of the body and serve as more efficient shock absorbers. In addition, by having the body moderately elevated above the ground, greater speed and agility is possible. Such reorientation was an essential step toward bipedalism in reptiles. Even with this reorientation, the position of the limbs among modern quadrupedal reptiles remains a sprawling one. Complete and more efficient elevation comes from the limb–girdle arrangement found in birds and mammals (Fig. 8.6b).

Modification of the front limb occurred in aquatic reptiles such as ichthyosaurs, plesiosaurs, and sea turtles (Fig. 8.7). The short, stout appendages tend to become flattened and paddle-shaped, and in some, the number of phalanges is greatly increased. Aquatic and semiaquatic reptiles often have webbing between the toes.

FIGURE 8.6

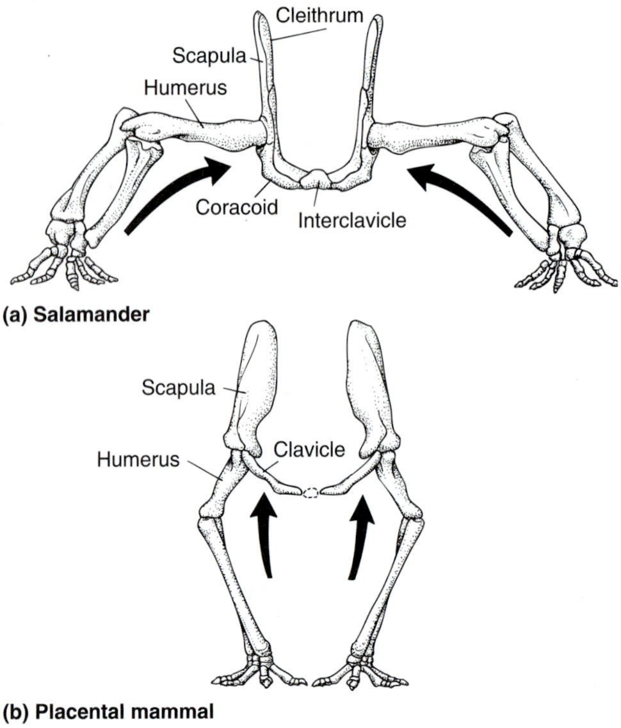

Evolution of limb structure. (*a*) Salamander. The sprawled posture of the salamander was typical of fossil amphibians as well as of most reptiles. (*b*) Placental mammal. This posture began to change in synapsids, so that in late therapsid reptiles the limbs were thought to be carried more beneath the body, resulting in better support and more rapid locomotion. As a result of the change in limb posture, the shoulder girdle also became modified. A sprawled posture brings a medially directed force toward the shoulder girdle, with medial elements playing a major role in resisting these forces. As limbs were brought under the body, these forces were directed less toward the midline and more in a vertical direction. This position of the limbs might account for the loss of some elements of the pectoral girdle in those phylogenetic lines in which the limb posture shifted.

FIGURE 8.7

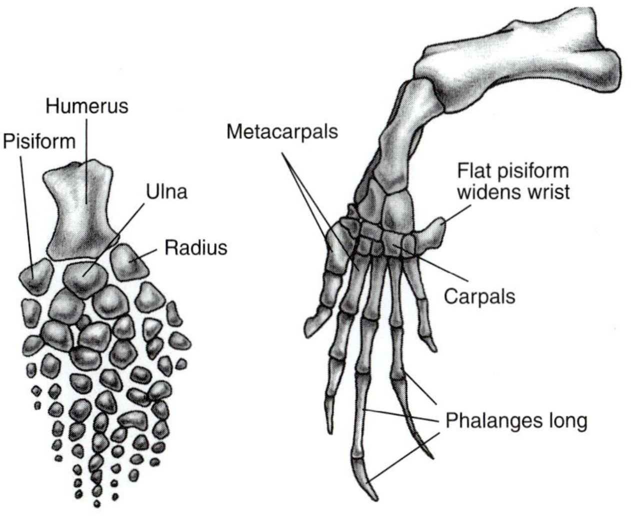

(a) Ichthyosaur **(b) Sea turtle**

Front limb of (*a*) an ichthyosaur and (*b*) a sea turtle (*Chelonia*). The limbs are short, stout, and often have an increased number of phalanges.

FIGURE 8.8

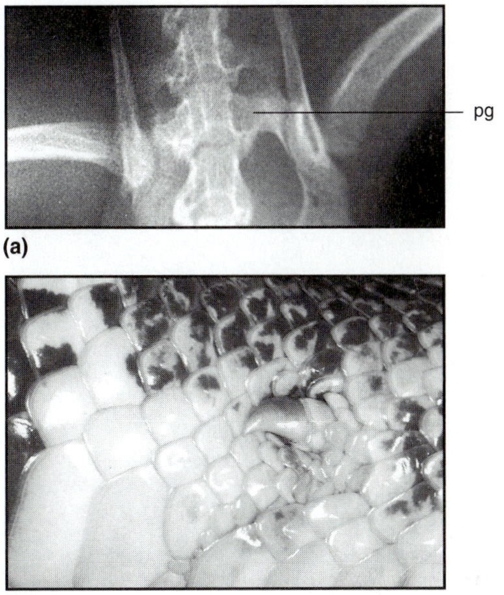

(*a*) The pelvic region of a boa constrictor, showing its vestigial pelvic girdle; pg, vestigial pelvic girdle. (*b*) The presence of a vestigial pelvic limb in boas, pythons, and a few other groups is associated with small claws or spurs located on either side of the vent. Male boas and pythons use their spurs, which are often longer than those of females, to stimulate females during courtship.

Each half of the pelvic girdle consists of a pubis, ischium, and ilium. The ilium is braced against two sacral vertebrae and tends to become broader for the attachment of larger hindlimb muscles, particularly in dinosaurs and lizards that carry the trunk elevated well above the ground. In reptiles that have lost the hindlimbs, the pelvic girdle has been reduced (legless lizards) or lost (most snakes).

Bony elements in the posterior limbs are similar to those in amphibians. The knee is directed anteriorly, and the ankle joint is located between two rows of tarsal bones rather than between the tarsals and the tibia and fibula. A **patella**, or kneecap, is present for the first time and occurs in certain lizards.

The order Squamata (lizards and snakes) is the only group of vertebrates in which there has been evolutionary losses of limbs and redevelopment of undulatory body movements. Snakes are thought to have evolved from lizards and to have lost both pairs of limbs as well as both girdles. Some primitive families such as the Boidae, however, still possess vestigial pelvic girdles and/or rear legs (Fig. 8.8a, b). Different stages of reduction and loss of limbs are found in lizards. Some possess only vestiges of forelimbs, some have only hindlimbs, and others, such as glass lizards (Anguidae), have lost all of their limbs.

Most species that lack limbs (snakes, legless lizards) move by **horizontal undulations** (Fig. 8.9a). This method, in which all parts of the body move along the same wavy track, is efficient on the ground as well as in trees. Its effectiveness is diminished, however, when the substrate lacks fixed surfaces or when the fixed surfaces are too widely spaced. Under these latter conditions, most snakes use **concertina**-like movements in which the stationary portion of the body is bent into a series of S-shaped coils from which the moving anterior portion straightens and then bends again (Fig. 8.9b). The posterior coils press downward and backward against the substrate, relying on friction to prevent slipping. This method is used extensively by climbing and burrowing species such as boids and rat snakes (*Elaphe*).

During **rectilinear** locomotion, the skin is drawn forward, the belly scales (scutes) make contact with the surface and provide stationary points, and then the body is pulled forward by the scutes (Fig. 8.9c). Muscles that slant backward and downward from the ribs to the scutes cause the ventral skin, which fits loosely and is very distensible, to bunch at several regions so that the scutes overlap. Between these regions, the skin is stretched. Where scutes are bunched, they rest on the ground to support the animal's weight; where stretched, they are lifted off the ground. One by one, additional scutes are drawn into each bunched region from behind as others are stretched away anteriorly. Friction of the bunched scutes against the substrate prevents slipping. Muscles that slant backward and upward from the scutes to the ribs pull the body along within the skin. The body is held in a straight line and moves directly ahead. Rectilinear movement is slow and may be used by snakes and worm lizards (amphisbaenians) when stalking prey or moving in narrow tunnels.

Sidewinding is unique to snakes, especially desert species (Fig. 8.9d). It provides for rapid travel over a smooth,

FIGURE 8.9

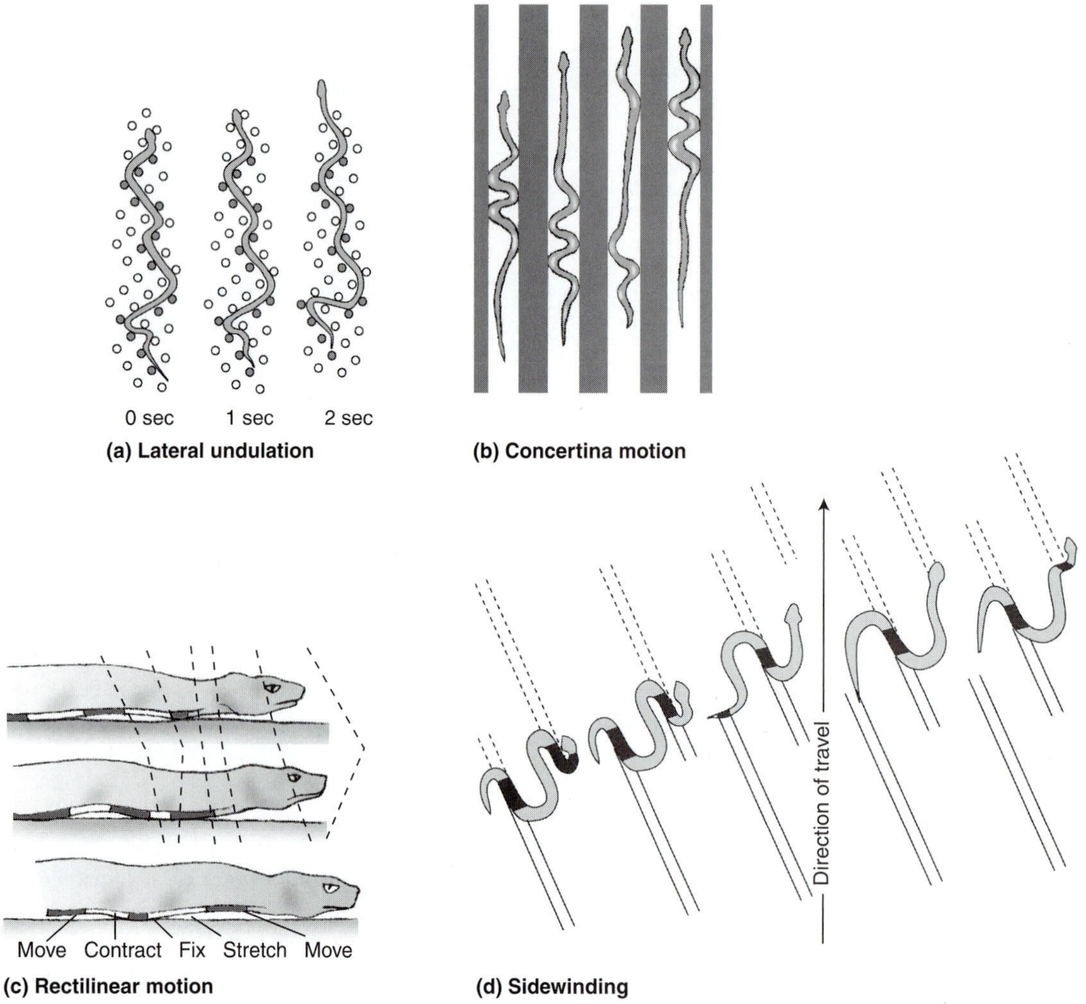

Snake locomotion: (*a*) lateral undulation; (*b*) concertina locomotion; (*c*) rectilinear locomotion; (*d*) sidewinding. Refer to the text for a discussion of each type.

unstable, and often hot substrate. The snake makes a series of tracks that are more or less straight lines, parallel to one another and angled to the direction of travel. Movement begins by the snake swinging its head, neck, and anterior body through an angle of 90° to 120° and placing its head on the ground in a new position. The remainder of the body then is lifted rapidly, section by section, from the old line of rest to the new position. The snake's body is in contact with two or three tracks at any given time, which are constantly changing position. Parts of its body are within the tracks; other parts are arching between tracks and are held above the substrate. As each new track is made by anterior portions of the body, posterior segments are released from the previous track. The head normally starts swinging to a third track ahead of the snake's position before the tail comes to rest on the second. The body moves forward at an angle of about 60° to the direction of travel. Some snakes, such as the African desert viper (*Bitis caudalis*), have modified the basic sidewinding locomotion and are able to jump short distances in order to reduce their contact with the intense heat of the desert floor.

Muscular System

The muscular system of reptiles has become more differentiated and better adapted to terrestrial life than that of amphibians. Muscles have become modified, not only to support the viscera and the weight of the body, but also to allow for various methods of locomotion. In addition, respiratory muscles have further differentiated and become better developed.

Epaxial muscles show less modification than hypaxial muscles, with some epaxial muscles losing their metamerism and differentiating into bundles. Besides their function of allowing side-to-side movement of the vertebral column, epaxial muscles take on new functions including support and vertical bending, or arching, of the back. In turtles, epaxial muscles of the rigid trunk region are poorly developed, but those of the neck and tail are well developed.

The presence of ribs on most trunk vertebrae causes increased modification of the hypaxial muscles. Ribs form in the myosepta of the body wall muscles along most of the length of the vertebral column in snakes and legless lizards. The dissections of Mosauer (1935) and Gasc (1967) indicate the presence of as many as 20 discrete muscles on each side of a single snake vertebra. These muscles connect vertebra to vertebra, vertebra to rib, rib to rib, and both rib and vertebra to skin, as well as attach to longitudinal tendons that help form and control the curvatures of the body. In other reptiles, myosepta and ribs have become confined to the anterior portion of the trunk, now known as the **thorax**. Abdominal wall muscles lack segmentation, and they have differentiated into three layers: external oblique, internal oblique, and transversus abdominis. Hypaxial muscles of the thoracic body wall, known as **intercostal muscles**, assist in respiration by raising and lowering the rib cage.

The body of reptiles is suspended from the scapulae by muscles that show much more differentiation than those of amphibians. Muscles of the limbs and girdles consist of dorsal extensor and ventral flexor muscles. Increased specialization of the intrinsic muscles allows for more precise and powerful movement of the limbs as well as greater support for the body. In those forms utilizing quadrupedal locomotion, muscles attached to the humerus and femur must rotate these bones forward and backward, as well as hold the bones steady in a horizontal position at the appropriate angle to the horizontal so that the body can be held above the substrate.

Muscles of the first pharyngeal arch continue to operate the jaws, and muscles of the second arch are attached to the hyoid skeleton. Muscles of the remaining arches continue to be associated primarily with the pharynx and larynx.

Extrinsic integumentary muscles insert on the underside of the dermis and allow independent movement of the skin. This is the first group of vertebrates to have integumentary muscles capable of moving the skin.

Cardiovascular System

Because reptiles are the first truly terrestrial vertebrates, many differences between the reptilian and amphibian cardiovascular systems are associated with the loss of functional gills and the need for efficient pulmonary circulation to bring blood to and from the lungs. Reptiles exhibit three different modes of circulation (Fig. 8.10). The ventricle of reptiles other than crocodilians is incompletely divided into dorsal and ventral chambers by a horizontal septum. A smaller vertical septum divides the ventricles into right and left chambers. The pulmonary trunk leaves the right ventricle. Both systemic trunks exit from the left ventricle in the Squamata (snakes and lizards); in turtles, however, one systemic trunk leaves the left ventricle and the other leaves the right ventricle. Because the interventricular septum is not complete, and because both atria open into the left ventricle, blood can flow from the left ventricle into the right ventricle.

The atrioventricular valve consists of two flaps that partially subdivide the left ventricle into a cavum arteriosum on the left and a cavum venosum on the right. When the atria contract, the cavum venosum becomes filled with deoxygenated blood from the right atrium and the cavum arteriosum becomes filled with oxygenated blood from the left atrium. Most of the deoxygenated blood in the cavum venosum flows into the right atrium. When the right ventricle contracts, the blood flows out through the pulmonary trunk, and in turtles through the right systemic trunk also. The oxygenated blood, together with some deoxygenated blood, is pumped out from the cavum arteriosum through the left systemic trunk (turtles) or both systemic trunks (Squamata).

FIGURE 8.10

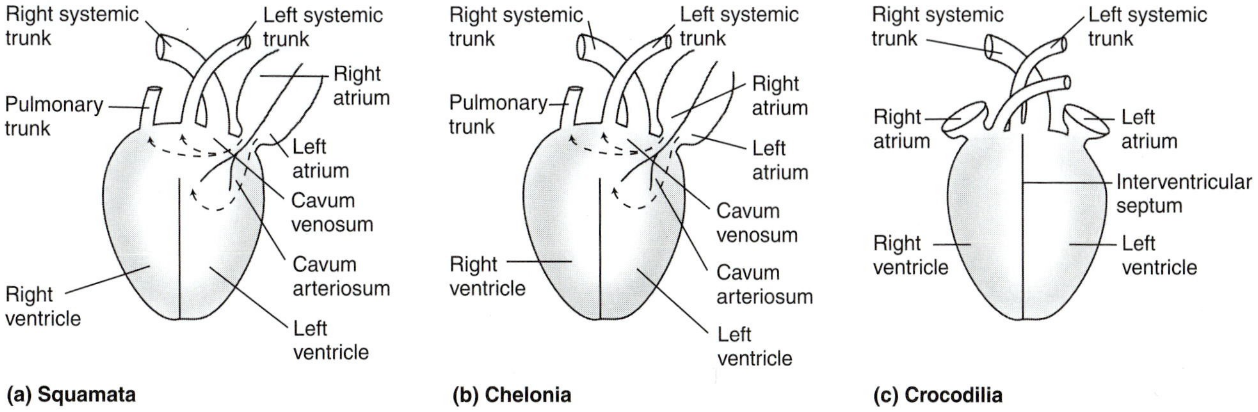

Diagrammatic representation of cardiac circulation in two lepidosaurs (*a*, *b*) and an archosaur (*c*). Note the complete separation of the ventricles in crocodilians.

From Feduccia and McGrady, Torrey's Morphegenesis of the Vertebrates. *Copyright © 1991 John Wiley & Sons, Inc. Reprinted by permission of John Wiley & Sons, Inc.*

Aortic arches III, IV, and the ventral part of VI remain in most adult turtles, lizards, and snakes, but their connections have been modified considerably (Fig. 8.11). The primitive ventral aorta splits into three channels: a pulmonary trunk and two aortic trunks. The pulmonary trunk leaves the right ventricle and carries blood to the lungs. One aortic trunk emerges from the left side of the ventricle and leads to the third aortic arch and to the fourth aortic arch on the right side of the body. A second aortic trunk emerges from the right side of the ventricle and leads to the fourth aortic arch on the left side of the body.

Because only one lung is functional in most snakes and legless lizards, the embryonic sixth aortic arch on the lungless side is lost. The third aortic arch on that side also disappears in most snakes. Most adult snakes, therefore, possess only a right pulmonary artery and a right common carotid artery.

The venous system shows little change from that of amphibians, although some modifications have occurred due to changes in the heart and kidneys and due to the elimination of cutaneous respiration. Due to the latter modification, reptiles have larger pulmonary veins and smaller cutaneous veins than amphibians.

Reptilian erythrocytes are oval and nucleated. They are smaller and more numerous than those of amphibians, ranging in length from about 15 to 23 µm. Leucocytes and thrombocytes make up the remaining cellular components of reptilian blood.

Respiratory System

With the formation of a secondary palate in some reptiles, the anterior part of the respiratory tract begins to be separated from the anterior part of the digestive tract. The internal nares in these forms are located farther caudad and nearer the midline than in amphibians, and nasal passages are lengthened.

To keep the air passageway open while large prey is being slowly swallowed, snakes have a glottis that can be protruded. In most reptiles, the trachea is about as long as the neck; it is shortest in lizards. In some turtles, however, it is longer than the neck and convoluted.

Although most reptiles are voiceless, some lizards and turtles possess vocal cords, but they produce few sounds discernible to the human ear. A few turtles make grunting noises, geckos "bark," and many species of anoles emit distinctive squeaks, especially when being captured.

The lungs of most reptiles are located in the **pleuroperitoneal cavity** and are, in most cases, better developed than amphibian lungs. In turtles and most lizards, the lungs consist of numerous large chambers, each composed of many individual subchambers, called **faveoli** (Fig. 8.12a, b). Internal partitioning is best developed in legless lizards, with pockets of trapped air causing the lungs to be spongy. Lungs of snakes are elongate and may be paired or unpaired. In

FIGURE 8.11

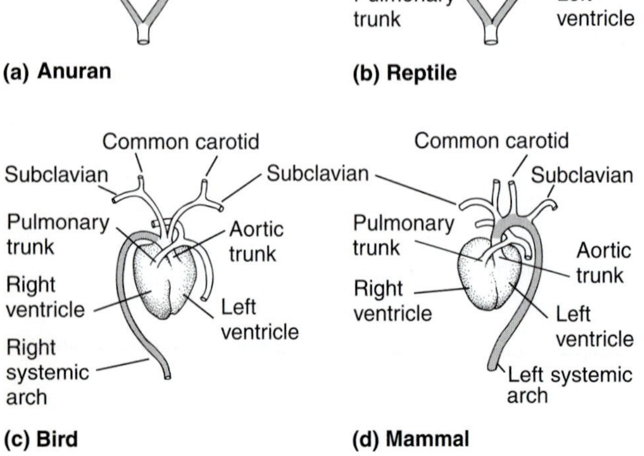

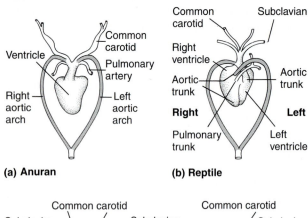

Fate of the systemic arches in tetrapods. The systemic arches of both sides persist in adult anurans (*a*) and reptiles (*b*). Only the right systemic arch persists in birds (*c*), whereas only the left remains in mammals (*d*).

FIGURE 8.12

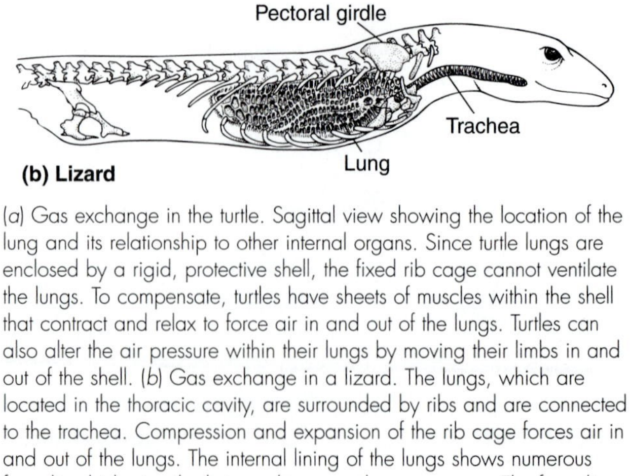

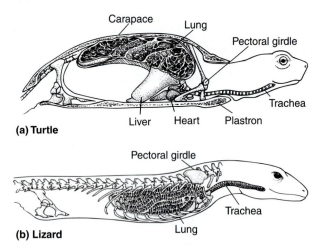

(*a*) Gas exchange in the turtle. Sagittal view showing the location of the lung and its relationship to other internal organs. Since turtle lungs are enclosed by a rigid, protective shell, the fixed rib cage cannot ventilate the lungs. To compensate, turtles have sheets of muscles within the shell that contract and relax to force air in and out of the lungs. Turtles can also alter the air pressure within their lungs by moving their limbs in and out of the shell. (*b*) Gas exchange in a lizard. The lungs, which are located in the thoracic cavity, are surrounded by ribs and are connected to the trachea. Compression and expansion of the rib cage forces air in and out of the lungs. The internal lining of the lungs shows numerous faveoli, which give the lining a honeycomb appearance. The faveoli increase the respiratory surface area of the lung and function in gas exchange along with the capillaries that line their walls.

tuataras (*Sphenodon*), the lungs continue to exist as simple sacs similar to those in amphibians.

Reptiles may use a force-pump system similar to that of amphibians in order to get air into their lungs, or they may inspire air by active suction and expire by pressure. Most reptiles respire by expanding and compressing the pleuroperitoneal cavity through backward and forward movements of the ribs produced through contraction of intercostal muscles.

In a few snakes and lizards, the lungs are in separate **pleural cavities**. A tendinous transverse partition, the **oblique septum**, separates the pleural cavities from the coelom. Contraction of muscles tightens the septum, which lowers pressure within the lungs below that of the external atmosphere. Because turtles are encased in a rigid shell and the ribs are fused to the carapace, the ribs cannot function in gas exchange. Instead, specialized sheets of muscle contract and relax to move air in and out of the lungs (Gans and Hughes, 1967).

Reptiles depend almost entirely on lungs to aerate their blood. Some reptiles supplement gas exchange by utilizing gas exchange membranes in other areas of the body, including the pharynx (pharyngeal gas exchange), cloaca (cloacal gas exchange), and skin (cutaneous gas exchange).

Pharyngeal (buccopharyngeal) gas exchange is particularly well developed in soft-shelled turtles (Trionychidae) (Girgis, 1961; Dunson and Weymouth, 1965). It permits them to obtain dissolved oxygen from water and, hence, to stay underwater for long periods. Pharyngeal gas exchange also is known to occur in the Australian skink (Drummond, 1946).

Cloacal gas exchange may occur in many turtles in the families Chelydridae, Testudinidae, and Pelomedusidae. The accessory cloacal bladders have been proposed as auxiliary gas exchange structures, as water is pumped in and out of the vent when the turtles are submerged (Zug, 1993). Oxygen supposedly diffuses through the smooth and lightly vascularized walls of the bladders, and carbon dioxide passes out.

Cutaneous gas exchange has been reported in several turtles, including the soft-shelled turtle (*Apalone* sp.), musk turtles (*Sternotherus odoratus* and *S. minor*), mud turtles (*Kinosternon subrubrum*), snapping turtles (*Chelydra serpentina*), and pond sliders (*Trachemys scripta*) (Stone et al., 1992). Girgis (1961) reported that soft-shelled turtles obtain up to 70 percent of their oxygen by diffusion through the leathery skin covering their carapace and plastron.

Shallow-diving sea turtles are thought to make aerobic dives relying on the lungs' oxygen store. The leatherback is the largest sea turtle and the deepest diver, able to dive beyond 1,000 m (Eckert et al., 1986). Increased hydrostatic pressure probably collapses their lungs during deeper dives (Berkson, 1967), so that they may have to rely on blood and tissue stores of oxygen. The oxygen-carrying capacity of their blood is twice that of shallow-diving sea turtles (Lutcavage, 1990). Their blood volume is slightly higher, but their lung volume is considerably smaller than in other sea turtles (Lutcavage et al., 1992).

Although one lung is rudimentary or absent in legless lizards and most snakes, both lungs are functional in primitive snakes such as boas and pythons (family Boidae), where the left lung is about 30 to 80 percent as large as the right lung. In some snakes, such as cobras, hognose snakes, and others that inflate their neck region as a defensive maneuver, a large saclike diverticulum of the left lung extends into the neck. Inflation of the sac causes the neck region to greatly expand.

Although the lung is the primary organ of gas exchange in the sea snake (*Pelamis platurus*), this species can take up oxygen through the skin at rates up to 33 percent of its total standard oxygen uptake and excrete carbon dioxide at rates up to 94 percent of its total rate (Graham, 1974). Specially adapted lungs that extend back to the cloaca and tightly sealing, valvular nostrils allow sea snakes to remain submerged for 8 hours or more before surfacing (Cooke, 1991).

Some lizards, such as the desert iguana (*Dipsosaurus dorsalis*) conserve a significant amount of water by exhaling air that is cooler than the body temperature (Murrish and Schmidt-Nielsen, 1970). The distal portion of the nasal passageways forms a slight depression in which fluid secreted from the nasal **salt gland** accumulates, which assists in humidifying incoming air. As the fluid becomes more concentrated, salts crystalize near the opening of the nares.

The chuckwalla (*Sauromalus*), a desert lizard from southwestern North America, often seeks refuge in rocky crevices. By inflating its lungs, it wedges its body in place and defies efforts to remove it from its safe haven. Recent studies have shown that *Sauromalus* "uses the primitive, anamniote, buccal pumping respiration for defense, and the derived, amniote, aspiration mechanism for respiration" (Deban and Theimer, 1991). Buccal pumping, unlike aspiration, can create greater-than-ambient pressures in the abdominal cavity.

Air sacs, which are diverticula of the lungs, extend among the viscera in some chameleons, often as far caudad as the pelvis (Orr, 1976). They serve to increase the air capacity and efficiency of the lungs in somewhat the same manner as the air sacs in birds.

Although five pharyngeal pouches develop in reptilian embryos, gills never develop, and adults never have open gill slits. The first pharyngeal pouch persists in adults as the Eustachian (auditory) tube and middle ear, and in reptiles and birds this temporarily opens to the outside during embryonic development. The other pouches become modified into other structures (thymus, parathyroid glands, and ultimobranchial body) or disappear during embryonic development.

Digestive System

The reptilian digestive tract exhibits numerous modifications as compared with amphibians. The jaws of most reptiles are covered by nonmuscular, immovable, thickened lips. The jaw margins of turtles, however, are covered with a shell of keratin and, together with the jaws, form a beak.

Oral Cavity

With the exception of turtles, some of which have serrations that simulate teeth on their horny beaks, all other groups of reptiles possess teeth. Although an evolutionary trend toward a reduction in the number of teeth and a restriction of the distribution of teeth is seen in the vertebrates, many reptiles still possess teeth on the palate as well as on the jaws.

Most reptiles have homodont dentition, although partial heterodonty is found in snakes and some lizards. Those with heterodont dentition (monitor lizards, *Varanus*; caiman lizards, *Dracaena*) have incisors, caninelike teeth, and molars. Tooth attachment is variable. In most lizards, the marginal teeth are attached to the biting edges of the jaws, an arrangement known as **acrodont dentition** (Fig. 8.13a); in other lizards, however, the jaw teeth are attached to the inner sides of the jawbone, which is called **pleurodont dentition** (Fig. 8.13b). Most snake teeth are recurved and either acrodont or pleurodont. The teeth of crocodilians are rooted in sockets, a condition known as *thecodont dentition.*

The right and left sides of many snakes' mandibles (as well as those of some lizards) are joined only by an elastic ligament so that the symphysis can stretch to accommodate large prey (Fig. 8.14a, b). Further expansion is possible because both halves of the lower jaw are suspended from the skull by a long, folding strut (the quadrate), which is divided into two hinged sections similar to a folding carpenter's rule. As large prey is engulfed, these joints swing outward on their flexible struts, greatly increasing the diameter of the throat region. Recurved teeth are present on the independently moving bones of the palate and jaws. By alternately engaging and disengaging the teeth on each side of the mouth, prey is gradually drawn into the oral cavity and passed to the pharynx and esophagus.

Threadsnakes (family Leptotyphlopidae) grow 6 to 8 inches long, are a bit thicker than a strand of spaghetti, and weigh about one gram. They feed upon insect larvae, pupae, and adults. These snakes have a unique feeding mechanism known as mandibular raking in which the anterior, tooth-bearing halves of the lower jaw rotate rapidly in and out of the mouth like a pair of swinging doors, dragging prey into the esophagus (Kley and Brainerd, 1999). This mechanism is made possible by the triple-jointed, extremely mobile lower jaw of threadsnakes that allows the transversely oriented mandibular tooth rows (the only teeth in the skull) to be rotated backwards. The high-speed gulping is thought to minimize the time spent in insect nests. It is the only vertebrate feeding mechanism known in which prey is transported exclusively by movements of the lower jaw.

FIGURE 8.13

Three methods of tooth attachment in reptiles: (*a*) acrodont—attached to the rim of the jawbone; (*b*) pleurodont—attached to the side of the jawbone; (*c*) thecodont—rooted in sockets in the jawbone.

FIGURE 8.14

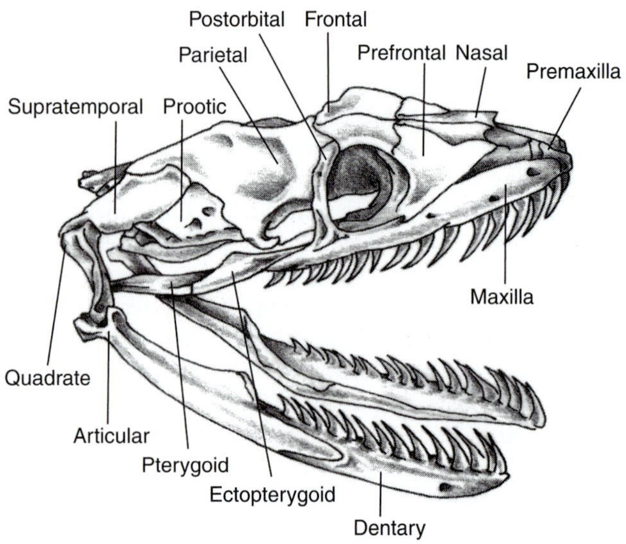

(a)

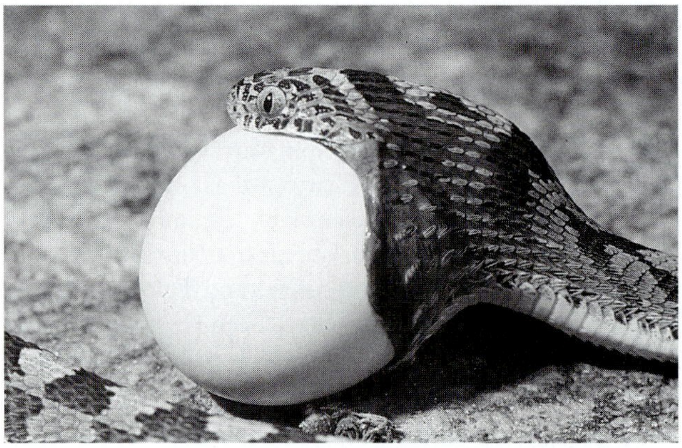

(b)

(*a*) Lateral view of python skull showing teeth and skull bones. The mouth of a snake is modified in order to accommodate large prey. The right and left sides of the mandible (lower jaw) are joined only by an elastic ligament so that the symphysis can stretch. In addition, both halves of the lower jaw are suspended from the skull by a long, folding strut, which is divided into two hinged sections. As large prey is engulfed, these joints swing outward on their flexible struts, thus increasing the diameter of the throat region. Recurved teeth securely hold the prey as upper and lower bones on one side of the jaw slide forward and backward alternately with bones of the other side. (*b*) Yellow rat snake (*Elaphe obsoleta quadrivittata*) swallowing an egg.

BIO-NOTE 8.1

Living on Large Prey

Boa constrictors and pythons are sit-and-wait predators that eat infrequently and undergo long and unpredictable fasts up to several months. When they do ingest prey, it may be extremely large (Fig. 8.15). Pythons, for example, may consume between 50 percent and 160 percent of their own body weight at one time and digest and assimilate such an enormous meal in just a few days. The biggest prey on record is a 58.5-kg antelope that was swallowed by an African rock python (Diamond, 1994). An Asian reticulated python consumed a 12.6-kg goat and a 17.5-kg goat at one sitting, and then downed a 32-kg ibex a few days later.

Oxygen consumption in pythons increases up to 17-fold only 24 hours after feeding and is maintained for several days while the snake remains virtually motionless. Approximately 32 percent of the energy in the food consumed is used during digestion, as opposed to 10 to 23 percent in animals that feed regularly. During periods of fasting, the mucosal lining of the small intestine atrophies, resulting in significant energy savings.

Cossins and Roberts, 1996
Secor and Diamond, 1995

Tuataras (*Sphenodon* spp.) have only one set of teeth during their lifetime, a situation called **monophyodont dentition**. Most reptiles, however, have a constant and steady replacement of their teeth, which is called **polyphyodont dentition**. This occurs in a regular alternating pattern so that adjacent teeth are at different stages in their development (Edmund, 1969).

The only two venomous lizards in the world, the Gila monster (*Heloderma suspectum*) and the Mexican beaded lizard (*Heloderma horridum*) possess grooves on the anteromedial edges of their teeth. Venom from modified salivary glands flows upward along grooves in the teeth of the lower jaw and into the prey as the lizard "chews" its prey. The venom is primarily neurotoxic, affecting the nervous and respiratory systems of the prey.

FIGURE 8.15

The African rock python (*Python sebae*) may feed on mammals as large as this Thomson's gazelle.

Many transitional forms exist between the complete absence of venom and the development of a venom-conducting apparatus. Three modifications occur in the teeth of venomous snakes. Less-specialized venomous snakes lack hollow fangs but have one to three enlarged, grooved teeth at the backs of their upper jaws (maxillae). The normally weak venom from modified salivary glands flows into the prey as it is "chewed." These snakes, known as **opisthoglyphs**, occur in the family Colubridae. Rear-fanged snakes primarily inhabit tropical regions and include the mangrove snake (*Boiga*), long-nosed tree snakes (e.g., *Dryophis oxybelis*), African vine snakes (*Thelotornis* spp.), African sand snakes (*Psammophis* spp.), and tree snakes (*Chrysopelea* spp.) of Malaysia and the East Indies. Lyre snakes (*Trimorphodon* sp.) of the southwestern United States and Mexico have a pair of enlarged, grooved rear fangs. Several other North American genera such as black-headed snakes (*Tantilla* sp.), ground snakes (*Sonora* sp.), annulated snakes (*Leptodeira* sp.), night snakes (*Hypsiglena* sp.), shovel-nosed ground snakes (*Chionactis* sp.), and banded burrowing snakes (*Chilomeniscus* sp.) possess slightly to moderately enlarged posterior maxillary teeth that may or may not be grooved.

Some rear-fanged snakes are potentially dangerous to humans. The boomslang (*Dispholidus typus*) of Africa, a tree snake that can grow to a length of 1.5 m, causes multiple internal hemorrhages. It has been responsible for several human fatalities, including that of Dr. Karl Schmidt in 1957. Dr. Schmidt was Curator Emeritus of the Field Museum of Natural History in Chicago and an internationally known authority on reptiles. He died 24 hours after being bitten by an 83-cm specimen he was handling.

Proteroglyphs and **solenoglyphs** possess a pair of grooved or hollow elongated fangs attached to the maxillae in the front of their upper jaw. The fangs are each connected by a venom duct to a venom gland (Fig. 8.16). Both types of snakes use muscular contraction of the venom gland to inject the venom into the prey. Proteroglyphs, which have rigid fangs that fit into pockets in the outer gum of the lower jaw, include the cobras, coral snakes, and sea snakes (Elapidae). Solenoglyphs include the vipers and pit vipers (Viperidae). These snakes possess the most highly specialized fang mechanism. The maxilla rotates like a hinge on the anterior end of the prefrontal bone. The fangs are only connected to the maxilla. When the mouth is closed, the fangs fold back and lie along the upper jaw. When the mouth opens, the fangs swing down and forward.

Venom glands of snakes are modified labial salivary glands and are homologous with the parotid glands of mammals. Venom of elapids (Elapidae) is primarily neurotoxic, attacking nerve centers and causing paralysis of many parts of the body, including the respiratory center. Venom of the vipers and pit vipers (Viperidae and Crotalidae) are primarily **hemolytic poisons**, which break up formed elements (red blood cells, white blood cells, and platelets) in the blood and attack the lining of blood vessels.

FIGURE 8.16

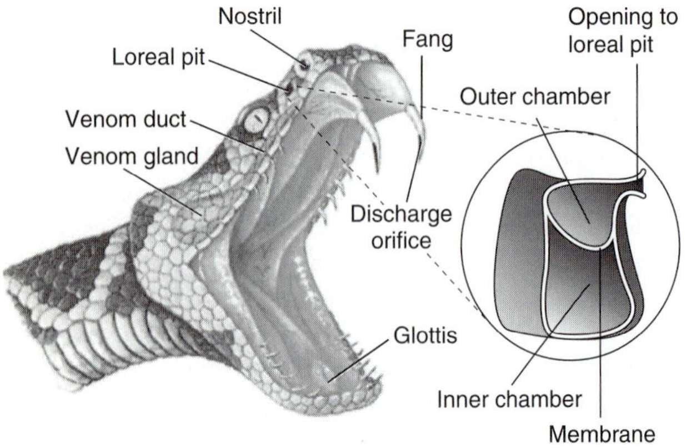

Head of rattlesnake, a pit viper, showing the venom apparatus. The venom gland, a modified salivary gland, is connected by a duct to the hollow fang. Note the loreal pit containing the heat-sensing pit organ between the eye and nostril. The cutaway view of the pit organ shows the location of a deep membrane that divides the pit into inner and outer chambers. Heat-sensitive nerve endings are concentrated in the membrane.

FIGURE 8.17

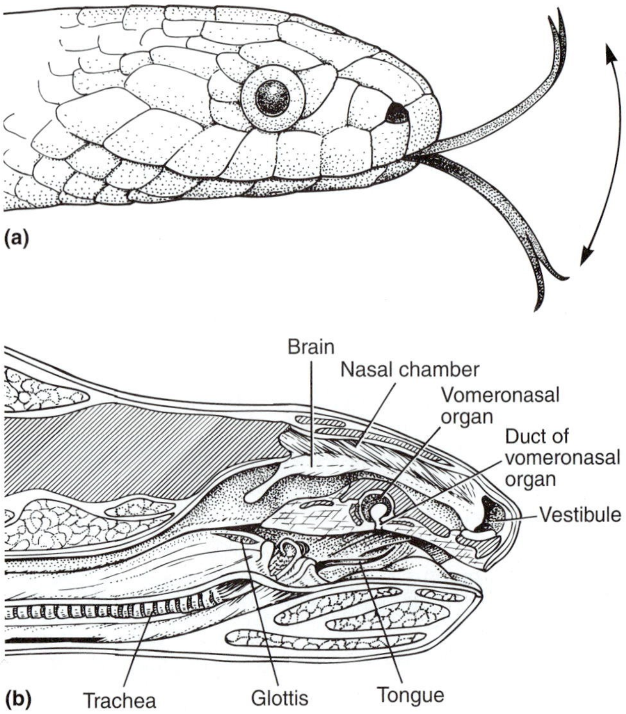

Tongue flicking in snakes. (*a*) Snakes and lizards extend their tongues to sweep air in front of them. The tongue collects and transports air-borne chemical particles into the mouth, where it deposits the particles onto the vomeronasal organ in the roof of the mouth. (*b*) Sagittal section of the head of a boa constrictor. The vomeronasal organ is a blind pocket with a lumen that opens directly into the mouth via a duct. The tip of the retracted tongue can be seen projecting from its sheath beneath the trachea.

In North America, approximately 8,000 cases of human **envenomation** (being bitten by a vertebrate or invertebrate capable of injecting venom) occur annually, most of which involve snakes (Brown, 1987). Fatalities are rare, but permanent damage, such as amputations and partial immobility of a joint near the site of the bite, may result.

Venom composition within some snake species has been found to show considerable geographical variation, presumably associated in some manner with local diets (Daltry et al., 1996). Because prey animals vary in their susceptibility to venom, geographical variation in venom composition may reflect natural selection for feeding on local prey.

The reptilian tongue is better developed than that of amphibians and is quite variable in structure. In turtles, the tongue arises from the floor of the oral cavity and is non-protrusible. In snakes and some lizards, however, the long, slender, forked tongue is flexible and moves with great speed (Fig. 8.17). It is best developed in African chameleons, which feed by projecting their highly extendable tongues (farther than their own body lengths!) and capturing insects on their thickened stocky tips (tongue pads) (Fig. 8.18).

FIGURE 8.18

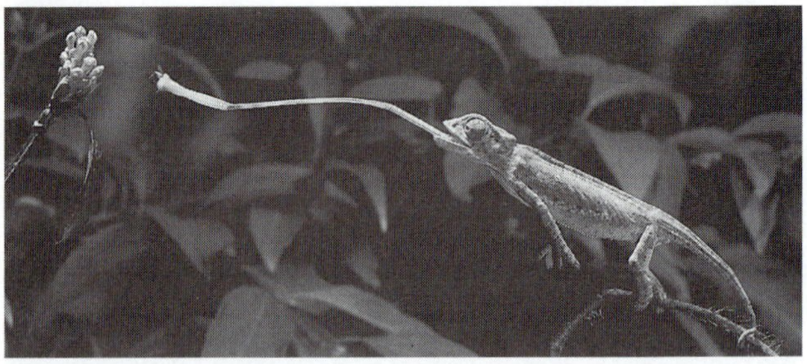

A chameleon (*Chameleo chameleon*) using its tongue to capture prey. Note the prehensile tail.

In most species, the forked tongue samples environmental chemicals by means of **flicking**, a behavior in which the tongue is rapidly protruded, contacts the air, and is then retracted into the mouth (Chiszar et al., 1980; Burghardt, 1970). It does not function in the manipulation of food, but transfers chemicals in the air to chemoreceptors, thereby allowing the snake or lizard to follow pheromone and scent trails of both mates and prey, respectively (Webb and Shine, 1992; Schwenk, 1994, 1995). Odor molecules on the tongue are delivered to the vomeronasal fenestrae in the palate, from where they make their way to the sensory epithelia of the paired vomeronasal (Jacobson's) organ (Burghardt, 1980; Halpern, 1983, 1992) (Fig. 8.17). Vomeronasal organs are small, bulb-shaped structures that contain patches of sensory cells with nerves that connect to the olfactory lobe of the brain. These organs became highly specialized in the lineage leading to lizards and snakes. Vomeronasal organs are greatly reduced or absent in turtles.

The reason for the evolution of a forked tongue remains uncertain, although it may be related to the paired vomeronasal organs. A small notch at the anterior end of the upper jaw allows the tongue to be flicked in and out without the reptile having to open its mouth. The ability to sample simultaneously two points along a chemical gradient provides the basis for instantaneous assessment of a scent trail. Forked tongues have evolved at least twice, and possibly four times, among squamate reptiles. The brightly colored tongues of some African snakes serve as visible threats (Mertens, 1955; Pitman, 1974), and a skink (*Tiliqua scincoides*) threatens other members of its own species (conspecifics) by presenting its cobalt-blue tongue (Carpenter, 1978).

Esophagus

The esophagus of most reptiles is longer than that of amphibians and is clearly differentiated from the stomach. Longitudinal folds permit its expansion during the swallowing of large prey. Some egg-eating snakes have esophageal "teeth"—anterior vertebral processes that project into the esophagus—that form an egg shell-cutting apparatus (Fig. 8.19).

Stomach

The stomach is elongate in lizards and snakes, and a distinct large intestine is usually present (Fig. 8.20a, b). In some herbivorous lizards, a cecum is present near the junction of the small and large intestines. The cecum contains specific bacteria to assist in digesting the cellulose-containing plant material. In many reptiles, the cloaca has become partly subdivided so that the intestine and urogenital ducts open into separate compartments (coprodeum and urodeum, respectively), which then join in a common outlet, the vent.

BIO-NOTE 8.2

Insights from Gastroliths

The discovery of approximately 240 gastroliths—"stomach stones" that certain animals ingest as an aid to grinding food—in and near a fossilized skeleton of a *Seismosaurus* dinosaur in New Mexico has provided possible clues concerning its digestive tract (Fig. 8.21). Stones discovered within the *Seismosaurus*'s rib cage were arranged in two distinct clusters: a larger group near the base of the neck, and a smaller group near the pelvic region. This arrangement suggests that the herbivorous *Seismosaurus* might have had a crop and a gizzard. Researchers hypothesize that the swallowed food passed from the crop, where it was ground by the gastroliths, to a gastrolith-free stomach where it was subjected to enzymatic action. The processed food then passed into the gizzard for additional grinding before finally moving into the intestine. Some living reptiles, particularly crocodilians, use grinding stones.

In 1997, each of twelve skeletons of an ornithomimid dinosaur from China contained a preserved gastrolith mass inside its ribcage. The presence of gastroliths indicates that these non-avian toothless theropods may have had gizzards and been herbivores, like modern herbivorous birds that use grit to grind up plant matter. The gastroliths found in these dinosaurs were mainly composed of grains of silicate, with no bony elements or insect remains as might be expected if the dinosaur were insectivorous or omnivorous.

Monastersky, 1990b
Gillette, 1995
Kobayashi et al., 1999

FIGURE 8.19

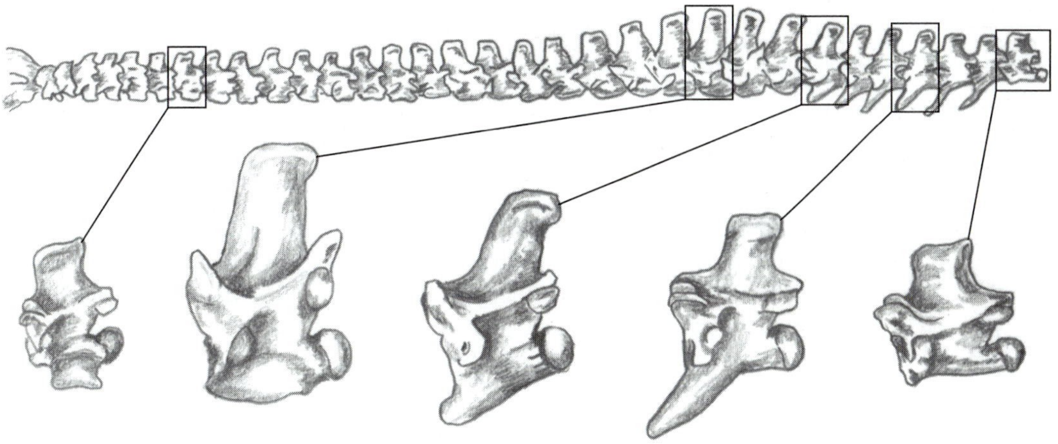

Morphology of vertebrae in the anterior vertebral column of the African egg-eating snake (*Dasypeltus scaber*). Anterior is to the left with the rear of the skull being shown. Note vertebrae with thickened hypapophyses (ventral processes) used for crushing egg shells and those with long, anteriorly directed hypapophyses that slit the egg membranes.
Source: Pough, Herpetology, *1998, Prentice-Hall, Inc.*

FIGURE 8.20

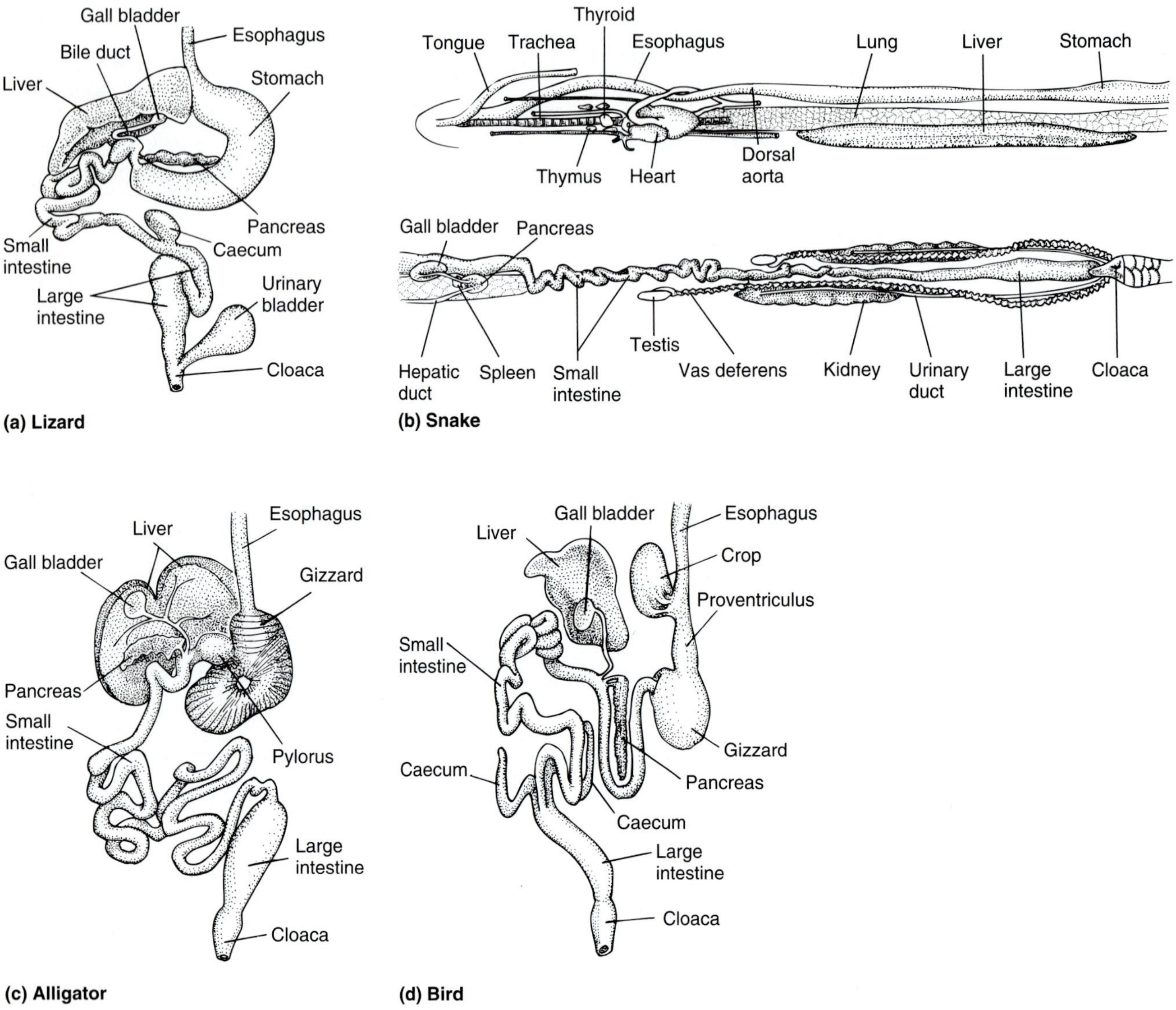

Ventral views of the alimentary canals in reptiles: (*a*) lizard (*Iguana*); (*b*) snake (*Natrix*); (*c*) alligator (*Alligator*); (*d*) bird (*Gallus*).

Nervous System

In all vertebrates, larger animals tend to have smaller brains (in proportion to body size) than small animals. The brain of a shrew, for example, is equal to about 1/23 (4.3%) of the weight of the entire animal, whereas the human brain comprises approximately 1/48 (2.1%) of the total body weight. In a giant sauropod such as *Apatosaurus*, the brain may have weighed only 1/100,000 (0.001%) of a total body weight of approximately 30,450 kg (Bellairs, 1970). Figures for some weights of brain and body in a variety of vertebrates are presented in Table 8.1. Note that the brain weight differs little in three different-sized specimens of the green lizard (*Lacerta viridis*), an indication that the brain stops growing before the rest of the body (Bellairs, 1970).

The brain of reptiles, particularly the cerebral hemispheres and cerebellum, is significantly advanced over those of fishes and amphibians. The cerebral hemispheres are considerably enlarged and bulge laterally, dorsally, and to some extent, posteriorly over the diencephalon. The thalamus and hypothalamus have become more prominent and control the coordination of metabolic activities. The hypothalamus acts as a thermostat to control body temperature, a process known as **basking poikilothermy** (Fig. 8.22). If the hypothalamus becomes cooler than the preferred temperature range, nerve signals will induce the reptile to move to a warmer site; if the hypothalamus becomes warmer, the reptile will move to a cooler site. An increase of 10°C within the biologically meaningful range of 0 to 40°C usually

FIGURE 8.21

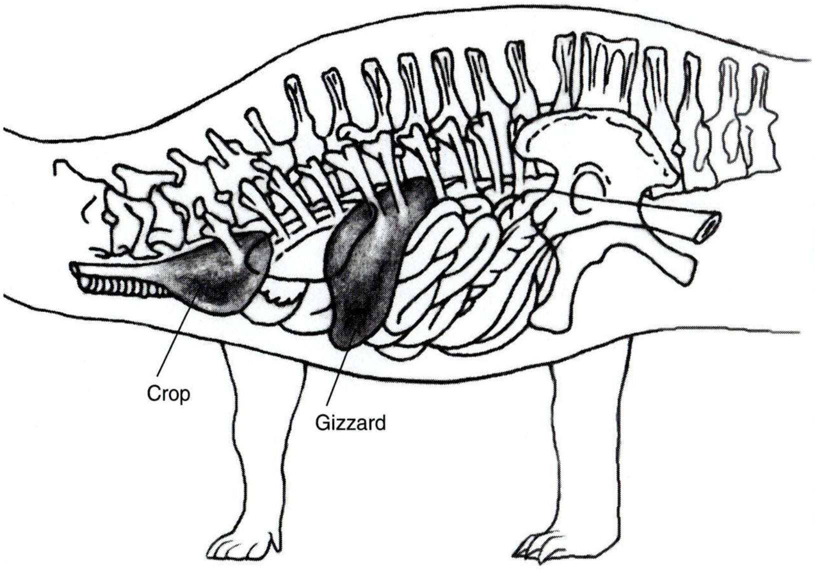

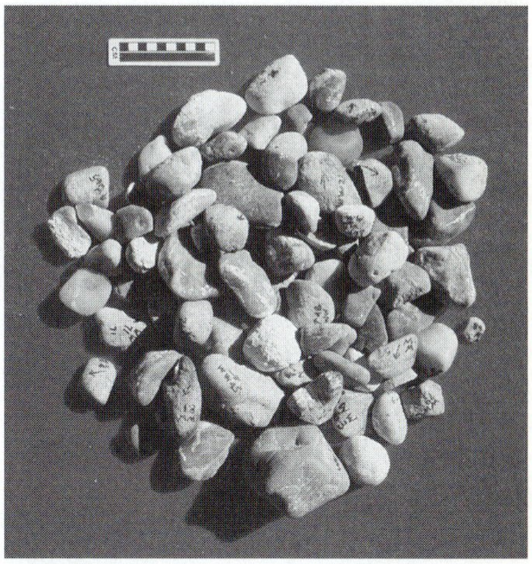

Stomach stones, or gastroliths (inset), were found in two main clusters in association with the *Seismosaurus* skeleton, suggesting that the creature's capacious digestive tract included a crop and a gizzard (shaded areas, left).

TABLE 8.1

Weights of Brain and Body in Representative Adult Vertebrates

Species	Body wt. (g)	Brain wt. (g)	Brain wt. as % body wt.
Esox lucius (pike)	12,700	4.9	0.04
Cyprinus carpio (carp)	1,817	1.3	0.07
Perca fluviatilis (perch)	67	0.16	0.2
Gasterosteus aculeatus (stickleback)	1.4	0.02	1.5
Rana catesbeiana (bullfrog)	244	0.2	0.08
Rana temporaria (frog)	53	0.1	0.2
Bufo bufo (toad)	44	0.1	0.2
Salamandra maculosa (salamander)	25	0.05	0.2
Triturus cristatus (newt)	7	0.02	0.28
Varanus niloticus (monitor)	7,500	2.4	0.03
Testudo hermanni (tortoise)	994	0.4	0.04
Naja melanoleuca (cobra)	1,770	0.65	0.04
Vipera berus (adder)	64	0.1	0.16
Anguis fragilis (slow-worm)	19	0.04	0.2
Lacerta viridis (green lizard)	17	0.093	0.6
Lacerta viridis (green lizard)	24	0.125	0.5
Lacerta viridis (green lizard)	32	0.130	0.4
Lacerta agilis (sand lizard)	12	0.08	0.6
Hemidactylus brooki (gecko)	5	0.043	0.9
Gallus gallus (fowl)	1,665	3.8	0.2
Troglodytes trogodytes (wren)	9	0.48	5.3
Balaenoptera sibbaldi (whale)	100,000,000	12,000	0.01
Homo sapiens (European male)	65,000	1,360	2.1
Felis catus (cat)	3,284	29	0.9
Sorex araneus (shrew)	2.9	0.125	4.3

Source: Data from A. D. Bellairs, The Life of Reptiles, *2 vols., 1970, © Universe Books.*

FIGURE 8.22

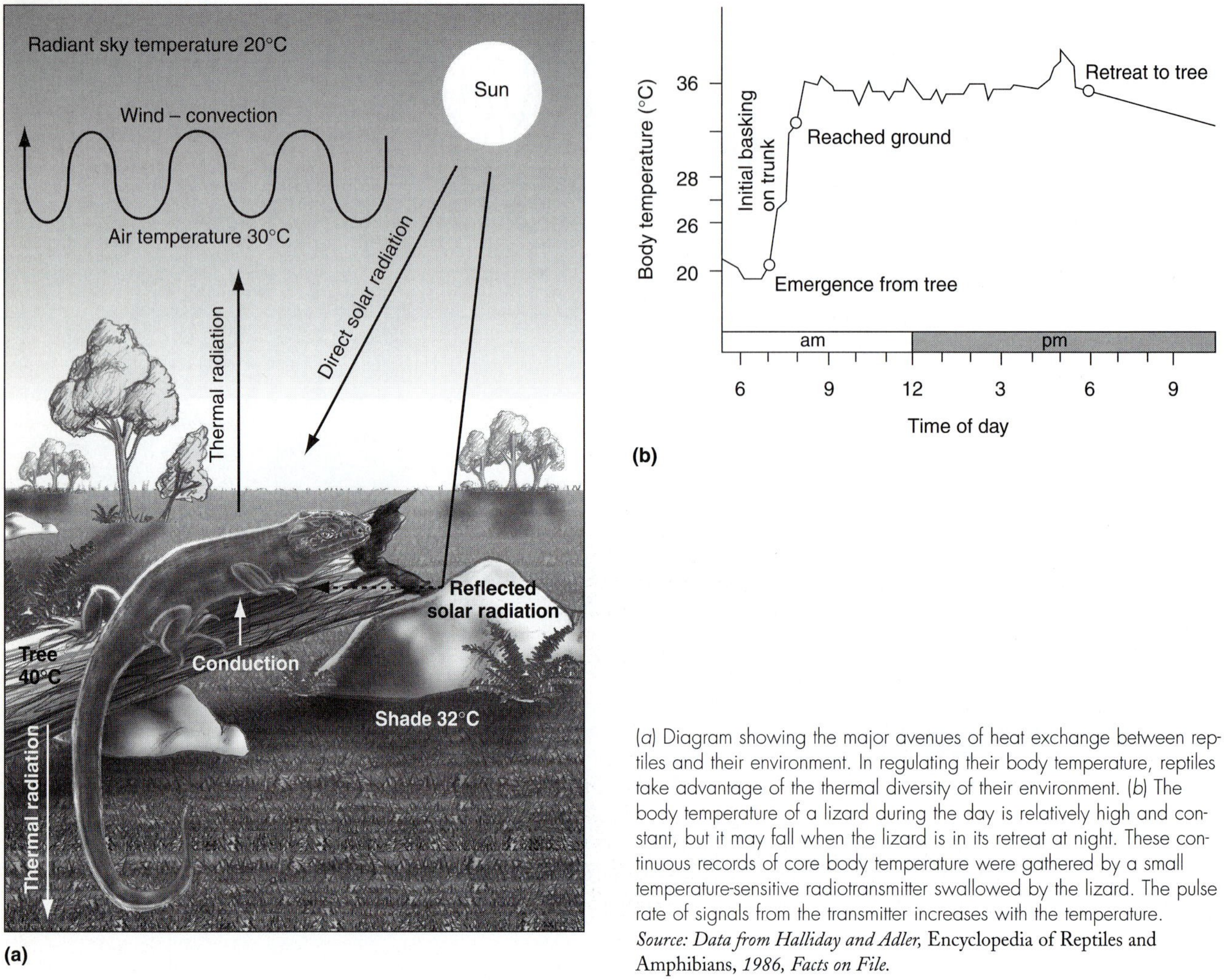

(*a*) Diagram showing the major avenues of heat exchange between reptiles and their environment. In regulating their body temperature, reptiles take advantage of the thermal diversity of their environment. (*b*) The body temperature of a lizard during the day is relatively high and constant, but it may fall when the lizard is in its retreat at night. These continuous records of core body temperature were gathered by a small temperature-sensitive radiotransmitter swallowed by the lizard. The pulse rate of signals from the transmitter increases with the temperature.
Source: Data from Halliday and Adler, Encyclopedia of Reptiles and Amphibians, *1986, Facts on File.*

doubles or triples the rates at which physiological processes occur (Halladay and Adler, 1986).

One or two outgrowths are present from the roof of the diencephalon. The anterior growth is the parapineal organ (parietal eye); the posterior is the pineal organ (epiphysis), which functions chiefly as an endocrine gland. A parapineal organ is prominent in *Sphenodon* and in many lizards, in which it serves as a photoreceptor.

Optic lobes and, for the first time in the vertebrates, distinct auditory lobes are present. The cerebellum, proportionately larger than in amphibians, is slightly larger in lizards, definitely larger in snakes, and best developed and most prominent in crocodilians and turtles. Cranial nerves have been discussed in Chapter 5.

The spinal cord is surrounded by two **meninges** (membranes): the dura mater and the leptomeninx. The spinal cord extends to the caudal end of the vertebral column in reptiles and, in most forms (except snakes and legless lizards), possesses cervical and lumbar enlargements in the two regions where large numbers of nerves that innervate the limbs exit the spinal cord. In turtles, the trunk region of the spinal cord is particularly slender because the thoracic and abdominal musculature is greatly reduced.

Sense Organs

Cutaneous Receptors

Except for some vestiges of lateral-line organs in adult turtles (Feduccia and McCrady, 1991), lateral-line organs do not exist in reptiles. Cutaneous sense organs, however, are especially common in reptiles and occur in a variety of forms. Pain and temperature receptors are widely distributed. Pressure, tension, and stretching of the skin are monitored by intraepidermal and intradermal mechanoreceptors, some of which are present within the hinges that lie between scales of lizards and snakes.

Some snakes have receptors in the form of pits that open onto the general body surface between epidermal scales. A single pair of specialized pits housing infrared receptors are found on the heads of crotalid snakes (family Viperidae) and on boas and pythons (Fig. 8.16). The venomous viperids, which include such North American snakes as rattlesnakes, copperheads, and water moccasins, are often referred to as "pit vipers" because of this feature. These specific receptors, called **loreal pits**, are located between the eye and the nostril and are highly specialized heat-sensing structures. They are deep, directed forward, and covered by a thin transparent membrane. In adult snakes, the pits are about 5 mm deep and several millimeters in diameter; the inner cavity of the pit is larger than the external opening. Loreal pits in boas are slitlike and less obvious. Both types of pits are thermal receptors that respond to radiant heat. Studies of pit vipers have shown the loreal pits can detect temperature changes of 0.003°C (Bullock and Diecke, 1956). Thus, they can detect the presence of objects, including homeothermic animals on which the snakes prey, even if the object is only slightly warmer than the environment. Because of this extreme sensitivity, these snakes, which are at least partly nocturnal, can locate and strike accurately at objects in the dark. They prey on small mammals that are primarily nocturnal and whose temperatures usually differ from those of their surroundings.

Ear

The reptilian ear serves as an organ of both equilibrium and hearing. Some lizards have no eardrum and no middle ear cavity. In snakes, a tympanic membrane, middle ear cavity, and Eustachian tube are missing entirely, although the semicircular canals and simple cochleae are present. A **cochlea**, an organ for detecting sound, has developed from the lagena of more primitive vertebrates. Sound waves are transmitted from the tympanic membrane to the cochlea via the columella in the middle ear cavity.

The distal end of the columella in snakes, which would articulate with the tympanum in other reptiles, articulates with the quadrate bone. Because snakes have no external ear openings, many have thought that snakes could not hear airborne vibrations (sound waves). Wever and Vernon (1960) demonstrated clearly that some colubrid snakes (e.g., *Pituophis melanoleucas, Thamnophis sirtalis,* and *Natrix sipedon*) can hear aerial sounds and are moderately sensitive to frequencies in the range of 100 to 700 Hz. They concluded that the quadrate bone acts as a receiving surface for the aerial sounds and have noted that its removal produces moderate reduction in the inner ear response to sound. Hearing apparently is possible because the muscle and fiber layers over the distal end of the columella continue to transmit sound pressures to it. In working with species of the families Colubridae, Crotalidae, and Boidae, Hartline (1971) reported a more restricted frequency range, from about 150 to 600 Hz. Vibrations also may be conducted from the ground to the membranous labyrinth by way of the bones of the jaws.

In most reptiles other than snakes, the tympanic membrane is visible on the surface of the head. It typically is found at the inner boundary of a cavity (external auditory meatus). In aquatic turtles, the tympanic membrane is thin and transparent, whereas in other turtles it is covered with scales. Studies of turtles (Wever and Vernon, 1956a, b) indicate that their hearing is similar in range and sensitivity to that of snakes. The tympanic membrane of some iguanian lizards (*Holbrookia*, some *Phrynosoma*) is covered with scaly skin. In most burrowing lizards, the middle ear has degenerated and the tympanic membrane is vestigial or absent. In some lizards, the cavity of the middle ear also has been lost. These groups are probably insensitive to airborne sounds, but they probably can detect ground vibrations.

Eye

In snakes, some lizards, and a few turtles, the upper and lower eyelids are permanently united in a "closed" position and are transparent, forming a **spectacle** that resembles a cornea (Fig. 8.23). The lids fuse together during embryonic development. The stratum corneum on the outer surface of the eyelids is shed each time the animal molts. The lower lid is the larger, and its upward movement is primarily what closes the eye. A third eyelid, more or less transparent, is called a **nictitating membrane** and is located medial to the eye in turtles and some lizards; it is absent in snakes. The nictitating membrane serves to cleanse and lubricate the cornea by rapidly sweeping back over its surface.

Lacrimal glands are present in many reptiles (absent in snakes) and secrete a watery fluid (tears) to keep the surface of the eye moist and clean. Although poorly developed in some reptiles, the lacrimal gland is enormous in marine turtles, where it plays an important role in osmotic regulation by controlling the excretion of salt.

Focusing of the eyes in reptiles occurs in two ways. In most snakes, increased pressure in the posterior chamber of the eye pushes the lens forward for focusing. This pressure is the result of the contraction of muscles at the root of the iris, which puts pressure on the vitreous humor. This, in turn, forces the lens forward (Gans and Parsons, 1970). In reptiles other than snakes, lens shape is changed by the contraction of radial muscles in the ciliary body encircling the lens. The lens is flattened for focusing on distant objects, and its curvature is increased for focusing on near objects.

Great variation exists with respect to the presence of photoreceptor cells in the eyes of reptiles. Cones are lost in burrowing snakes such as blind snakes (*Typhlops*), and only tiny rods are present. Many diurnal lizards and snakes in the family Colubridae have lost the rods and are virtually blind at night. Both rods and cones are always present in the retinas of tuataras and turtles.

FIGURE 8.23

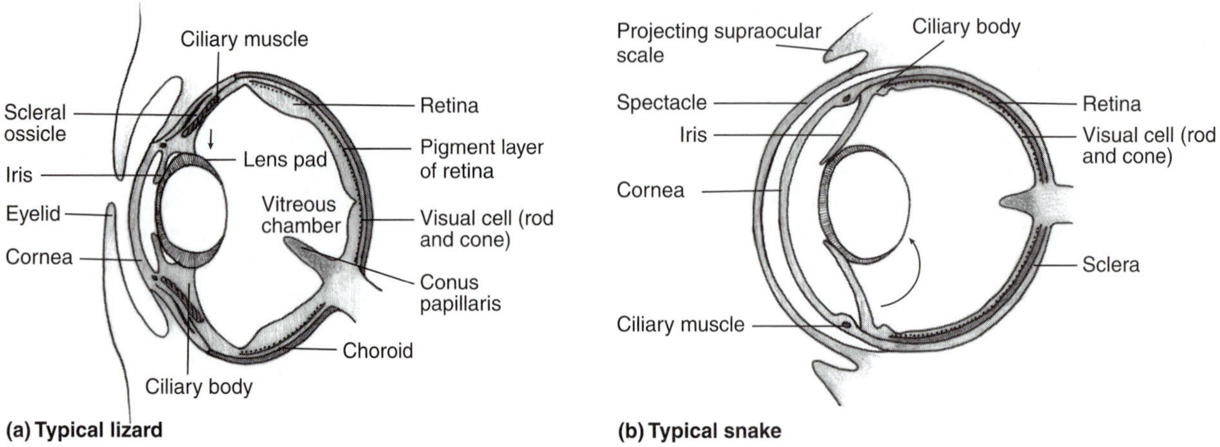

Transverse section of (*a*) a typical lizard eye and (*b*) a typical snake eye. In the lizard, contraction of the ciliary muscles forces the ciliary bodies against the lens pad and squeezes the lens. In the snake, the ciliary muscles have shifted into the iris and those near its roots press the iris against the vitreous body; this raises the intravitreous pressure and pushes the whole lens forward.

Colored oil droplets (yellow, orange, and red) are found between the inner and outer segments of cone photoreceptor cells on the retinas of lizards and turtles (Bowmaker, 1986; Robinson, 1994). The color in the droplets absorbs particular wavelengths of light, thereby narrowing the spectral sensitivity of the four visual pigment proteins while expanding the waveband of color sensitivity. The result is a tetrachromatic (four-color) visual system that extends from near-ultraviolet (350 nm) to infrared (750 nm) in turtles. Thus, turtles appear to have good color vision.

Chameleon eyes move independently until they lock onto prey. In addition, the eyelids make an aperture in front of the cornea that moves with the eye and serves as an iris to control the amount of light entering the eye (Land, 1995). Most remarkable, however, is that the lens, which is slightly convex, has negative refractive power (it causes light rays to diverge rather than converge) (Ott and Schaeffel, 1995). This is the only known example of a lens with negative power occurring in a vertebrate. Such an unusual optical design might have evolved to further extend the range of accommodation or to maximize the relative retinal image size. When retinal image sizes in a variety of vertebrate eyes were compared, chameleons were found to have the largest images.

Many lizards have a vestigial "eye" in the top of their head. The parapineal organ (parietal body) serves as this third, or **parietal**, eye. It lies beneath a single, translucent epidermal scale in the midline of the head and consists of a cornea, a lens, and retina-like receptor cells. In most lizards, it has the structure of a degenerate eye, but it is well developed in tuataras (*Sphenodon*). This medial eye is used to regulate body temperature by regulating the duration of exposure of the body to the Sun. It is also partly responsible for the daily and annual body rhythms necessary for survival.

Nose

Nasal passages of reptiles are longer than those of amphibians. The internal nares normally open into the anterior portion of the oral cavity except in turtles, which have a partial secondary palate, thus extending the nasal passageways farther posteriorly.

In snakes and some lizards, **vomeronasal** (Jacobson's) **organs** are well developed. These tubular organs lose their connection with the nasal canal and open into the anterior roof of the oral cavity (Fig. 8.24). Here, there are moist chemoreceptive depressions that monitor the odoriferous chemicals that accumulate on the tongue from the air. Vomeronasal organs are vestigial or absent in turtles.

Taste

Although the sense of taste in reptiles is poorer than that of fish and amphibians, it still plays a role in the recognition of prey, enemies, or potential mates. Taste buds are found primarily in the lining of the pharynx, with few, if any, located on the tongue. Turtles probably can taste objects held in the mouth.

Endocrine System

The endocrine system has been discussed in Chapter 5. Only those hormones that have specific and unique functions in reptiles are included here. Hormones secreted by the thyroid glands assist in regulating the metabolism of reptiles. They also play an important role in ecdysis (shedding). Parathyroid glands produce parathormone, which helps regulate levels of calcium and phosphate in the blood. Regulation of calcium is important not only for homeostasis, but also for the proper formation of the calcareous shell enclosing some reptilian embryos.

FIGURE 8.24

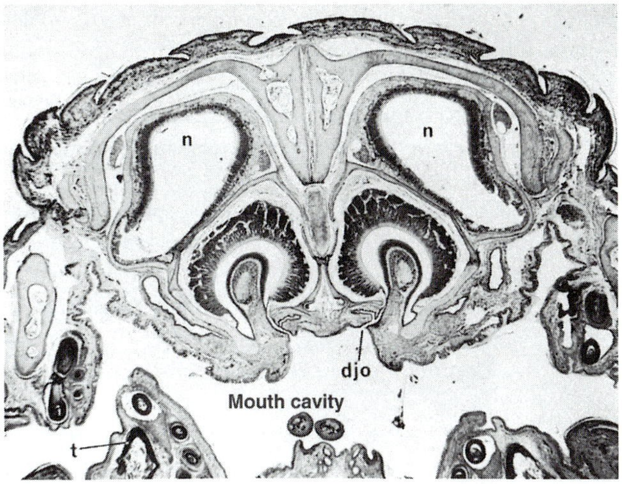

Transverse section through the head of a lizard (*Lacerta*) showing the crescent-shaped vomeronasal (Jacobson's) organs, and their connections to the mouth cavity via ducts (djo). The larger openings (n) above the paired vomeronasal organs are the nasal passages, which are completely independent of the vomeronasal organs.

Testosterone, estrogen, and progesterone regulate reproductive behavior. Studies of the red-sided garter snake (*Thamnophis sirtalis*) show that the testosterone levels in males peak in late summer—more than 8 months before the breeding season, as females are not receptive to mating until spring (Fackelmann, 1991).

Melatonin, secreted by the pineal gland, is responsible for the daily (and possibly seasonal) hypothermia in various lizard species, including the European green lizard (*Lacerta viridis*) (Rismiller and Heldmaier, 1987). These lizards display a pronounced rhythm in body temperature selection with daily voluntary hypothermia in fall, preceding winter dormancy. Because light inhibits the synthesis of melatonin, increasing amounts are produced as photoperiod (daylength) shortens in the fall.

Urogenital System

The reptilian kidney is designed for a terrestrial environment in which the excretory and reproductive tracts become separated. A pronephros develops in the embryo, but it apparently never functions as an excretory organ. The structure of the functional embryonic kidney, the opisthonephros, is essentially the same as the adult kidneys of fishes and amphibians. It continues to function for a short time after hatching—as late as the first hibernation in some lizards and the first molt in some snakes.

During the time that the opisthonephros is functioning, the **metanephros** is in the process of development (see Fig. 4.17d). The metanephros replaces the mesonephros and becomes the adult kidney. Thus, reptiles are the first vertebrates to have a true metanephros as a functioning kidney. As the metanephros begins to function, the mesonephros slowly ceases functioning, so that, during later stages of embryonic development, both kidneys function simultaneously. Soon, only remnants of the mesonephros remain. As long as the mesonephros is functioning, the metanephric duct serves as an accessory urinary duct. When the mesonephros involutes, the metanephric duct becomes the sole urinary duct (ureter).

The more compact metanephric kidney contains renal corpuscles, proximal and distal convoluted tubules, and collecting tubules. Two or more renal arteries bring blood to the kidney to be filtered. No trace of segmentation is evident in the metanephric tubules, and no nephrostomes are present.

Filtration pressures of ectothermic vertebrates, which are much lower than those of endotherms due primarily to lower blood pressures in ectotherms, may decrease even more because of low body temperatures. Hibernation, for example, causes depressed cardiac function and a drop in blood pressure. In some parts of their range, painted turtles (*Chrysemys picta*) may spend as long as 6 to 7 months in ice-covered bodies of water. Herbert and Jackson (1985) reported a reduction in systolic blood pressure in the painted turtle from 28.9 mm Hg at 20°C to 7.7 mm Hg at 3°C during submergence. The lower blood pressure in the renal arterioles during hibernation may limit or eliminate glomerular filtration.

Most reptilian kidneys are lobulate, with each lobe consisting of clusters of many tubules. In snakes and legless lizards, kidneys are elongated to conform to the slender body. In most reptiles, the ureters empty into a cloaca. A urinary bladder is present in tuataras, turtles, and many lizards, but lacking in snakes and crocodilians. When present, the bladder is an evagination of the ventral wall of the cloaca. Urine backs up into the reptilian bladder from the cloaca. In a few male reptiles, however, the ureters continue to empty into the mesonephric ducts, which carry sperm.

Turtles and many lizards have large bladders. Some turtles also have two highly vascularized, thin-walled accessory cloacal bladders that may function as water reservoirs in dry environments (Jorgensen, 1998). This function depends on copious water intake when water becomes available and discontinued voiding of urine in the absence of water. In females, this excess water helps moisten the soil when the female is excavating a nest for the eggs.

The cloaca in reptiles is divided into three chambers: a coprodeum, a urodeum, and a proctodeum. The **coprodeum** receives the terminal portion of the large intestine. The **urodeum** contains the openings of the excretory and genital ducts, whereas the **proctodeum** is a short chamber enclosed by a sphincter muscle at the vent.

Nitrogenous waste excretions of reptiles may be in the form of ammonia, urea, or uric acid. Aquatic and semiaquatic forms are more likely to excrete ammonia, whereas terrestrial forms, which must conserve water, generally excrete uric acid, a concentrated waste product that is inert and insoluble in water. Some reptiles also have extrarenal salt excretion mechanisms that are active only when the osmotic concentration of the plasma becomes high. In the orbit of each eye, marine turtles have a salt-excreting lacrimal gland. A nasal gland performs a similar function in marine iguanas and some desert-dwelling lizards (Templeton, 1966). Sea snakes have a salt-excreting gland in their palate.

In female snakes and lizards, the paired ovaries are saccular with hollow lymph-filled cavities (lacunae) and are basically similar to those found in amphibians (Fig. 8.25a). In turtles and crocodilians, the ovaries are solid structures, similar to the ovaries of birds and mammals. The number of ova maturing at any one time is less than in most amphibians—usually fewer than 100.

Oviducts are similar to those of amphibians (Fig. 8.25a). They form from embryonic Müllerian ducts and are lined with glands that secrete albumen around the ovum. The posterior portions of the oviducts, known as uteri, contain specialized shell glands that secrete the egg shell (which is porous enough to allow for respiratory gas exchange) over the outer layers of the eggs. The two uteri enter the cloaca independently. In some snakes and lizards, special vaginal tubules known as **spermathecae** store sperm over the winter. Some lizards, some snakes, and all crocodilians retain only the right oviduct as a functional oviduct in adults; the left remains rudimentary.

FIGURE 8.25

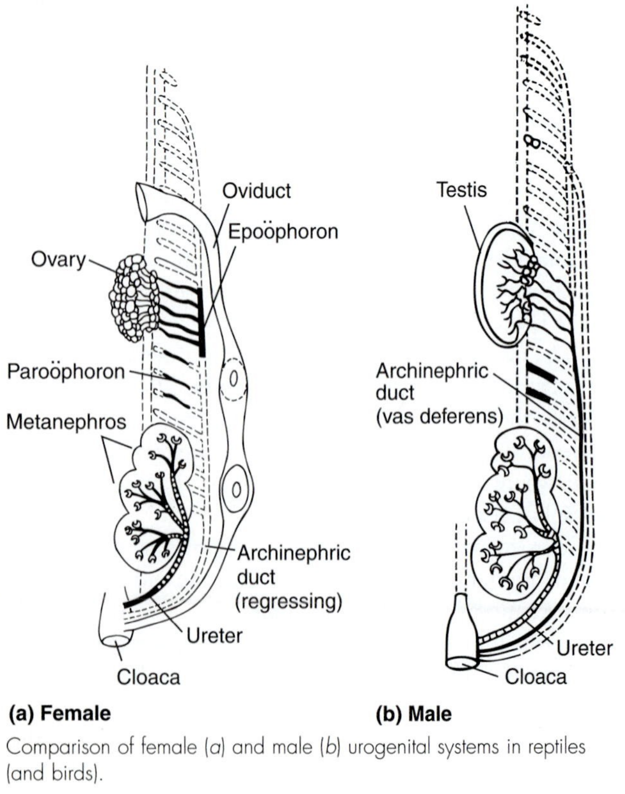

(a) Female **(b) Male**

Comparison of female (*a*) and male (*b*) urogenital systems in reptiles (and birds).

The paired testes of males are suspended in the abdominal cavity (Fig. 8.25b) and undergo marked seasonal fluctuations in size. Although mesonephric kidneys disappear during early development, mesonephric ducts of males remain. The portion closest to the testis forms the highly coiled **epididymis**. The remainder becomes known as the **ductus deferens** (vas deferens) and empties into the cloaca. Thus, a complete separation of excretory and reproductive tracts occurs anterior to the cloaca.

Fertilization, which is internal in all reptiles, occurs in the upper portion of the oviducts. With the development of internal fertilization, intromittent, or copulatory, organs evolved in males to introduce sperm into the female reproductive tract. Although present in a few fishes, amphibians, and birds, intromittent organs are characteristic of mammals and all male reptiles except tuataras (*Sphenodon*).

Two types of intromittent organs are present in reptiles. Paired structures known as **hemipenes** evolved in male squamates (lizards and snakes) and lie concealed in long sacs opening to the outside of the body on each side of the cloacal aperture (Fig. 8.26). Hemipenes lack specific erectile tissue; they are erected when they are turned inside out and everted like the fingers of a glove. This eversion is accomplished by muscle action and by filling of blood sinuses within the hemipenes. When everted, the hemipenes protrude through the vent. The shape of hemipenes is extremely variable and is used in squamate taxonomy. A groove on the surface of each hemipenis (sulcus spermaticus) conducts sperm into the cloaca of the female. Normally, only one hemipenis is inserted. Homologous structures, although much smaller, are present in females.

Male turtles and crocodilians possess an unpaired **penis** (Fig. 8.27). The penis consists of the corpus cavernosum (spongy erectile tissue), which contains blood sinuses. The corpus cavernosum also bears a sulcus spermaticus for the passage of both sperm and urine. A rudimentary penis, the **clitoris**, develops in females.

REPRODUCTION

One of the most important evolutionary developments for long-term adaptation to terrestrial life was the development of the amniotic egg in reptiles (see discussion in Chapter 7; Fig. 7.4). The parchmentlike shell (hard shell in crocodilians) allows for the exchange of gases, whereas the four extraembryonic membranes assist in forming blood vessels (chorion); surround the embryo and enclose the amniotic fluid (amnion); enclose the yolk (yolk sac); and serve as a respiratory organ and a repository for nitrogenous wastes (allantois). The adaptive value of the amniotic egg is to provide an internal aquatic environment for the development of young on land.

Reptilian reproductive behavior is influenced primarily by temperature and photoperiod, with precipitation appearing to play only a minor role in initiating the breeding cycle.

FIGURE 8.26

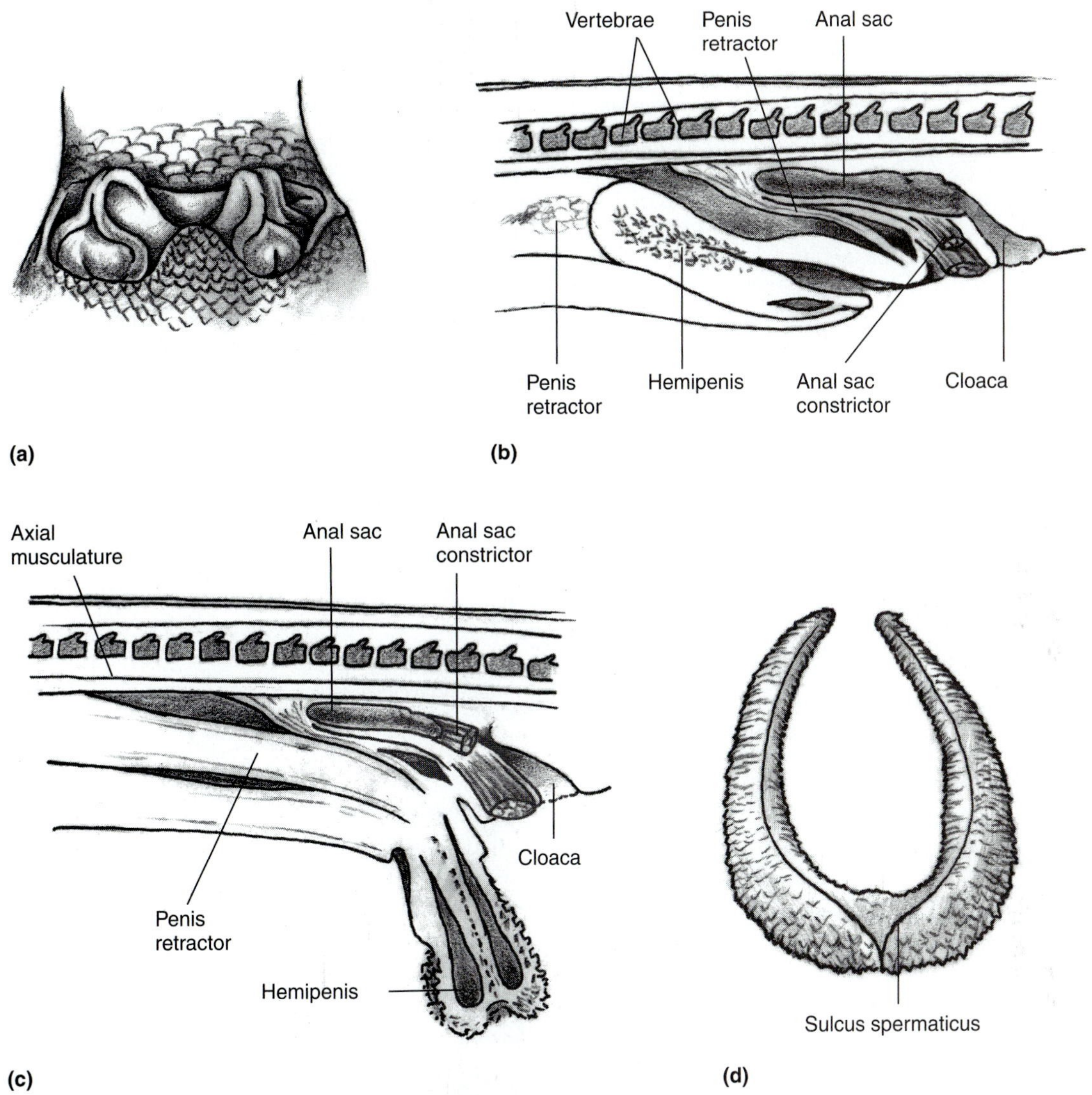

Lizards and snakes have paired hemipenes, but usually only one is used during copulation. (*a*) The hemipenes of a lizard. (*b*) The hemipenis is pulled back into the body by the retractor muscle (sagittal view). (*c*) When erect, the hemipenis's internal sinuses become engorged with blood and it pops through the vent (sagittal view). During copulation, the male inserts its hemipenis into the cloaca of the female. Sperm travel down the sulcus spermaticus into the female. (*d*) One of the two hemipenes from the rattlesnake *Crotalus atrox* is shown everted. This single hemipenis is divided, which gives it a horseshoe shape. Note the divided sulcus spermaticus that runs along each arched branch of the hemipenis.

Some tropical lizards and crocodiles, however, manage to breed, lay eggs, and/or have their young during or following the rainy season—a possible adaptation to seasonally abundant food supplies. In temperate regions, reptiles usually breed within 4 to 8 weeks following their emergence from hibernation and produce a single clutch of eggs annually, whereas in many tropical regions breeding may take place throughout the year.

Vision and olfaction are the most important senses used by most reptiles in seeking a suitable mate. During the breeding season, the musk glands of some turtles (e.g., *Sternotherus*) enlarge and may secrete pheromones. Lizards, which are primarily diurnal and live at fairly high densities, use primarily visual cues for attracting members of the opposite sex. Many lizards and snakes use pheromones for species and sex recognition, identification of eggs, and recognition of individuals. Some, such as male broad-headed skinks (*Eumeces laticeps*), follow female conspecific odor trails (Cooper and Vitt, 1986). Chemical trailing of conspecifics also occurs widely in snakes.

Courtship behavior of most reptiles is poorly known. Although all turtles are oviparous and lay eggs on land, aquatic species mate in the water and appear to rely almost

FIGURE 8.27

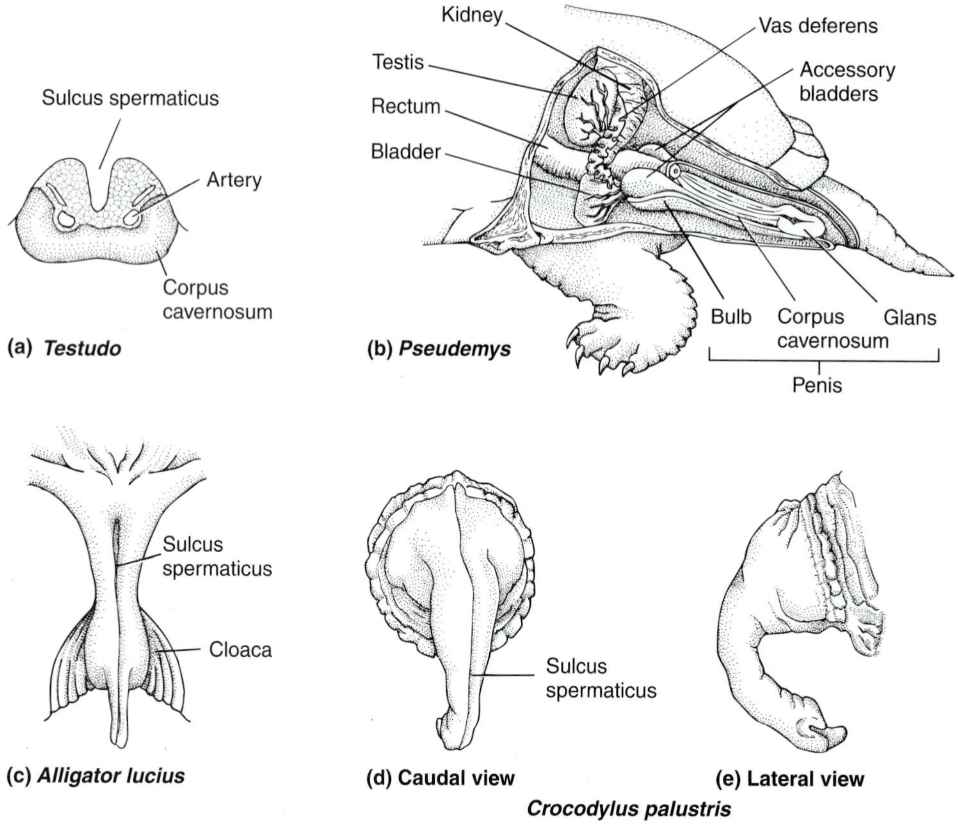

Penes of reptiles. (*a*) Turtle, *Testudo*: cross section of the penis within the cloaca. (*b*) Turtle, *Pseudemys*: sagittal section of the penis. (*c*) Alligator, *Alligator lucius* penis. Caudal (*d*) and lateral (*e*) views of the penis of the crocodile, *Crocodylus palustris*.

entirely on visual means for species and sex recognition. Males swim onto the backs of females and grip their carapaces with their long claws. Terrestrial and semiaquatic turtles mate on land and use visual and olfactory cues for sex recognition. Courtship behavior usually involves head-bobbing by the male, an action that gives an obvious visual signal but that also may be sending scent through the air. Lizards depend primarily on visual clues such as brightly colored patches of skin, conspicuous head crests, and bobbing of the head and body. Both sexes of some anoles such as *Anolis carolinensis* and *A. sagrei* possess an orange-red or pink flap of tissue beneath their chin (Fig. 8.28). This flap, known as a **dewlap**, can be voluntarily extended and retracted. It plays an important role in both courtship and territorial defense (Evans, 1938). Snakes also appear to use mainly visual and olfactory cues for sex and species recognition. Courtship in most species involves tactile stimulation of the female and olfactory stimulation of the male. Many species entwine their bodies around each other prior to copulation.

Lipids extracted from the skin of female red-sided garter snakes (*Thamnophis sirtalis parietalis*) are attractive to sexually active courting males (Mason et al., 1989) (Fig. 8.29). The lipids contain a female sex attractant pheromone, which consists of a series of long-chain methyl ketones (Mason et al., 1989). Females of related groups of snakes have some of the same methyl ketones as well as variations of the com-

FIGURE 8.28

Male anoles (*Anolis carolinensis*) of the southeastern United States possess an orange-red flap of tissue beneath their chins known as a dewlap. The dewlap, which can be voluntarily extended and retracted, plays an important role in courtship and territorial defense.

FIGURE 8.29

Lipids extracted from the skin of female red-sided garter snakes (*Thamnophis sirtalis parietalis*) are attractive to sexually active courting males. The snake on the extreme left is the female.

BIO-NOTE 8.3

A Volcanic Nest

In a study of land iguanas (*Conolophus subcristatus*) on Fernandina Island in the Galapagos, Dagmar Werner found that following mating, 95 percent of the females migrated into the caldera of a volcano to nest, a trip of up to 6 km, involving an initial climb of 1,400 m to the crater rim, followed by a 900-m descent to the caldera floor. Prior to descending, several thousand females circle the crater at the rim searching for the descent access, adding approximately 9 km to their journey. This is the longest distance known to be walked by any species of lizard to its nesting grounds. On average, the time spent between leaving the mating area and nesting was 30 days. The journey to reach the rim required between 4 and 10 days.

Each female lays between 8 and 23 eggs in a fumarole crevice with soft earth at least 40 cm deep. The eggs are heated by hot vapor and remain between 32 and 34°C both day and night. Incubation requires 3 to 4 months, during which the eggs are unattended. After nesting, females slowly return to the proximity of the mating area.

Werner, 1982; 1983

pound forming the pheromone. When extracts of lipids from male skins are added to female extracts, male courtship stops, suggesting that males emit specific chemical cues that identify them as males. One chemical in the male lipid, squalene, caused a significant reduction in courting behavior and is important in the male sex recognition pheromone.

Lipids stored in abdominal fat bodies can play an important role in the reproductive cycle of lizards (Derickson, 1974, 1976). In seasons during which lipid storage is reduced, egg production also is reduced greatly (Ballinger, 1977). Lipids in the fat bodies are depleted immediately upon emerging from hibernation and are not restored until after reproduction is completed. Seasonal changes in lipid composition during the reproductive cycle in fence lizards (*Sceloporus*) were reported by Ballinger et al. (1992). For example, the fatty acid triacylglycerol made up 88 percent of total lipids in fat bodies early in the reproductive season but declined to 66 percent following egg production.

Although most lizards and snakes are oviparous, a few, such as most boas (Boidae), pit vipers (Crotalidae), some skinks (Scincidae), night lizards (Xantusidae), and garter and ribbon snakes (*Thamnophis*), are viviparous. Viviparity provides additional protection for the developing young. In garter snakes (*Thamnophis sirtalis*), the developing young are attached to the reproductive tract of the female. The placenta serves as a respiratory organ and also facilitiates the transfer of amino acids from the mother to the embryo (Clack et al., 1955). Viviparity, however, elicits a substantial cost to the female in reducing mobility, rendering her less able to evade predators and catch food, and energetics.

Multiple copulations are reported in Swedish adders (*Vipera berus*) (Madsen et al., 1992). In this species, females averaged 3.7 copulations (range 1–8) per season, usually with different males. Multiple matings enhanced the genetic diversity of the offspring and significantly reduced the proportions of young that were born dead. Because female snakes can store sperm for months before actual ovulation, collecting sperm from numerous males creates a "competition" within the female's body in which the best-performing and most competitive sperm have the best chance to fertilize eggs. If sperm that are more successful in fertilizing ova are also more effective in producing viable offspring, then the increase in the average viability of offspring would be correlated with multiple matings. However, multiple matings in Swedish adders did not result in increases in the size of offspring, litter size, or total litter mass, nor did females have higher fertility.

Female sand lizards (*Lacerta agilis*) copulate with virtually every male that courts them. Thus, they often mate with close relatives. However, they actively select, via intrauterine sperm competition, the sperm from distantly related males (Olsson et al., 1996). Even if they are unable to avoid mating with close relatives, females "choose" not to use sperm from these matings.

Most reptiles have one breeding cycle per year, which for most temperate species begins upon their emergence from hibernation. Some, however, may not breed for several years at a time. Female tuataras (*Sphenodon*) breed only once every

4 or 5 years and take 2 or 3 years to develop eggs. After mating, eggs are held in the oviduct for another 7 months before they are deposited. Twelve to 16 months later, the eggs finally hatch. The 10-cm-long babies spend a few days in their underground nest before pushing up through the soil.

Some all-female species and some populations of vertebrates reproduce by **parthenogenesis**, a process in which a new individual develops from an unfertilized egg. In some species, parthenogenesis is **facultative**, occurring only when this mode of reproduction is necessary as a last resort for producing offspring. In a few species, however, reproduction is exclusively parthenogenic.

Parthenogenesis may evolve only in populations devoid of the parental bisexual individuals, because such species would prevent newly formed unisexuals from establishing clones, due either to hybridization or competition. Parthenogenesis has its advantages, the most obvious being higher potential population growth, because all individuals (rather than half) can produce eggs. This doubles the intrinsic rate of increase (*r*-strategist) and increases the ability of one individual to establish a new colony. Thus, parthenogenic species hypothetically can invade and occupy habitats faster than bisexuals.

Parthenogenesis, known in fishes, amphibians, and reptiles, occurs most commonly in lizards. Thirteen parthenogenic species occur in the genus *Cnemidophorus* in the United States. The desert grassland whiptail lizard (*C. uniparens*) is one such species. It is a single clone in which the chromosomes of each lizard duplicate just before the oocyte divides, producing the homozygous diploid condition as well as genetically complete and identical eggs. Parthenogenesis has been recorded in snakes on several occasions. A captive 14-year-old timber rattlesnake (*Crotalus horridus*) that had never been with a male gave birth to a litter of young (Schuett et al., 1997). A garter snake (*Thamnophis elegans*) caught as a young adult in 1983 lived only with females for the next year and was kept completely isolated from snakes after that. She produced four litters between 1988 and 1994 (Schuett et al., 1997). Two clutches of baby snakes were born to an aquatic Australian Arafuran filesnake (*Acrochordus arafurae*) in 1988 and 1991 that had been isolated from all males since 1983 (Dubach and Sajewicz, 1997).

BIO-NOTE 8.4

Courtship in a Unisex Lizard

Captive whiptail lizards, although all females, go through the motions of courtship and mounting behavior. One female acts as a receptive (gravid) female; the other takes the male role of mounting the receptive female, slipping her tail under her partner, and assuming the stereotypical male copulatory position. Every 10 to 14 days, the two females switch roles. The lizard that has just ovulated has a surge of progesterone, which causes her to act like a male, and then after a week or so her ovaries become larger and secrete estrogen, and she reassumes the female's part. Individual lizards were more likely to ovulate when they were with a malelike female. Thus, "pseudosexual" behavior apparently serves to maximize the number of eggs produced.

Crews and Young, 1991

GROWTH AND DEVELOPMENT

Prenatal Development

Turtles and most lizards and snakes are oviparous reptiles that deposit their eggs on land (Fig. 8.30). No aquatic larval stage exists.

Reptiles were the first group of vertebrates to have colonized the terrestrial environment successfully. They solved the problem of having to return to water to reproduce by evolving the amniote egg with its leathery shell and four extraembryonic membranes (amnion, chorion, yolk sac, and allantois) (see Fig. 7.5). The **amnion** completely surrounds the embryo and contains the amniotic fluid in which the embryo floats. During later stages of development, it forms part of the chorioamnionic connection where the amnion and chorion remain united (Ewert, 1985). The **chorion** is the outermost membrane and assists primarily in gas exchange

FIGURE 8.30

A female loggerhead sea turtle (*Caretta caretta*) depositing her eggs on a tropical beach.

through the shell. The **yolk sac**, which is attached to the digestive tract, contains the yolk, which serves as the embryo's source of nutrients (a carryover from anamniotes), whereas the **allantois** functions as the primary organ for mid- to advanced embryonic respiratory gas exchange and as a receptacle for the embryo's metabolic wastes. It is the last extraembryonic membrane to form. As the embryo develops, the allantois expands and fuses with the chorion to form the chorioallantoic membrane, which lines the entire shell membrane with blood vessels and provides a rich vascular network.

Amniotic eggs are characteristic of reptiles, birds, monotremes, and in modified form, therian mammals. The development of the amniotic egg was such a remarkable event in the evolution of terrestrial vertebrates that it serves as a major characteristic separating fishes and amphibians—**anamniotes**—from the reptiles, birds, and mammals—**amniotes**.

Most reptilian eggs are enclosed in a leathery shell to prevent dessiccation (Fig. 8.31). They may be buried in sand, mud, or soil, or deposited in decaying vegetation or rotting logs. Most adult female sea turtles breed on 1- to 3-year cycles (reviewed by Ehrhart, 1981) and only return to land for periods of several hours to deposit their eggs in nests excavated on sandy beaches (see Fig. 8.30). The sand heated by the sun incubates the eggs.

Many oviparous lizards and snakes retain eggs in utero until about halfway through their embryonic development (Shine, 1983). One such lizard, *Sphenomorphus fragilis*, exhibits extreme egg retention (Greer and Parker, 1979). The female lays thinly shelled eggs that "hatch" almost immediately after being laid. Greer and Parker (1979) concluded that *S. fragilis* was "effectively a live-bearer but that it has retained the egg shell of its presumed oviparous ancestor, albeit in a greatly thinned condition."

The smooth green snake (*Opheodrys vernalis*) is one of the northernmost oviparous snakes in North America. The time between egg laying and hatching in this species has been shown to be as short as 4 days (Blanchard, 1933).

Viviparity has evolved at least 55 times among lizards and 35 times among snakes (Blackburn, 1982, 1985). Young are retained within the female for their entire gestation period. All North American rattlesnakes are viviparous, as are the well-known garter snake (*Thamnophis sirtalis*) of North America and two species of horned lizards (*Phrynosoma douglassii* and *P. orbiculare*). Viviparity in reptiles serves to protect eggs from predators and to speed development of those species living in cool montane regions, as well as those species living near the extremes of their ranges in both the Northern and Southern Hemispheres.

Charland and Gregory (1990) have shown that the body temperature maintained by gravid prairie rattlesnakes (*Crotalus viridis*) is far more constant than in nongravid females. Regulation of body temperature is accomplished by moving between locations of higher and lower environmental temperature and by influencing rates of heat gain and loss by making subtle changes in basking posture or orientation that alter the surface area exposed to the sun. Adaptive strategies of temperature regulation can result in an increased rate of development and better control of the time of birth.

Two species of lizards—*Sceloporus aeneus* (Iguanidae) and *Lacerta vivipara* (Lacertidae)—have been shown to exhibit both modes (oviparity and viviparity) of reproduction. Viviparity is the usual mode of reproduction for *L. vivipara* in its range from Russia to Scandinavia and northern France (Heulin et al., 1991). However, in the warmer, extreme southwestern part of its range in France and northwest Spain, it is oviparous. Studies of embryonic development and birth dates found one main advantage of viviparity to be that embryos developed more rapidly at high maternal temperatures than at low soil temperatures (Heulin et al., 1991). The difference in birth dates between oviparous and viviparous forms did not exceed 1 week when oviparous eggs were artificially incubated at temperatures close to the body temperatures of pregnant viviparous females. However, oviparous eggs incubated under lower outdoor temperatures hatched approximately 1 month after viviparous females gave birth to their young. The evolutionary origins of viviparity in reptiles have been discussed by Blackburn (1982, 1985), Packard et al. (1991), and Guillette (1991).

FIGURE 8.31

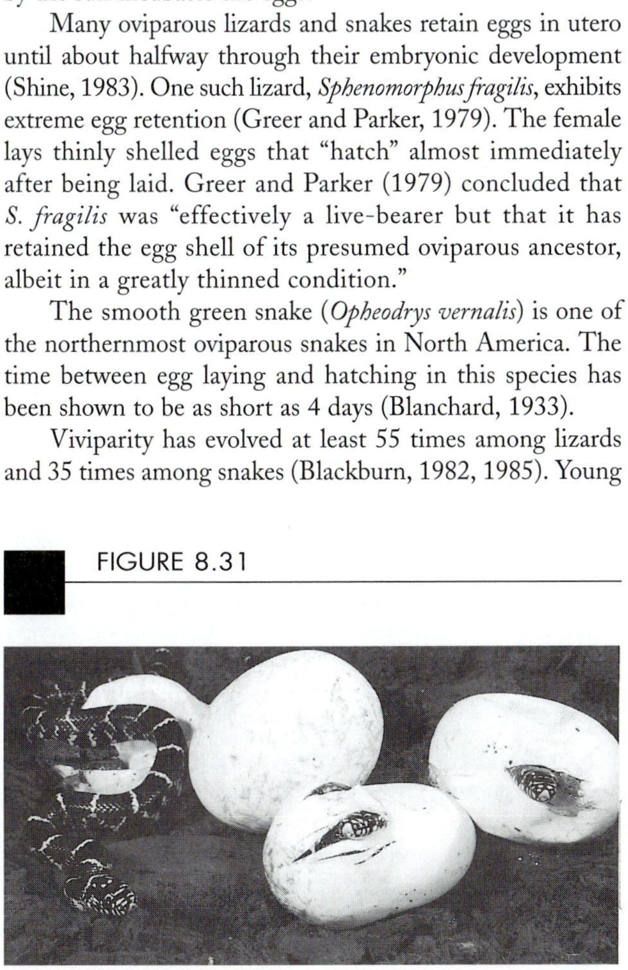

Hatching kingsnakes (*Lampropeltis getulus*). Most snakes are oviparous and lay leathery-shelled eggs.

Duration of Embryonic Development

Reptile eggs require incubation periods ranging from 5 weeks to 16 months. Most turtle eggs hatch between 8 and 16 weeks from the time they are deposited. Lizard eggs require 5 to 12 weeks to hatch, whereas the incubation period for snake eggs ranges from 8 to 12 weeks (Fig. 8.31). Female tuataras retain fertilized eggs within the oviduct for approximately 7 months before laying them, and eggs hatch 12 to 16 months after being deposited in an underground nest.

Hatching and Birth

The young of oviparous reptiles develop a **caruncle** (egg tooth) prior to hatching. In these reptiles, it usually is located on the premaxillary bone. Just prior to hatching, the egg tooth is used to crack or "pip" the shell. The pressure exerted by the young inside the egg, together with the increasing movements of the embryo, causes the weakened shell to crack enough to allow the young reptile to emerge. The egg tooth is reabsorbed shortly after birth.

In many species of reptiles, sex is determined by incubation temperature of the egg during the middle third of incubation (reviewed by Bull, 1983; Janzen and Paukstis, 1991a), which is the time when gonads differentiate (Wibbels et al., 1991). Three patterns of temperature-dependent sex determination (TSD) have been identified: Type A—males produced at high temperatures, females at low temperatures (most crocodilians and some lizards); Type B—females produced at high temperatures, males at low temperatures (many turtles); and Type C—females produced at high and low temperatures with males produced at intermediate temperatures (three crocodiles, one lizard, and three turtle species) (Bull, 1980, 1983; Ewert and Nelson, 1991; Brooks et al., 1991; Etchberger et al., 1992). For example, nest temperatures greater than 33°C have been found to produce a greater percentage of male alligators (Ferguson and Joanen, 1982). The underlying molecular and physiological basis of TSD is unknown (Janzen and Paukstis, 1991a).

The possible adaptive significance of environmental sex determination (ESD) mechanisms in reptiles was investigated by Janzen and Paukstis (1991b). They concluded that comparison of sex-determining mechanisms (hydric, thermal) with sexual dimorphism in adult body size did not support the sexual dimorphism hypothesis that posthatching growth correlates with both incubation temperature and offspring sex. No relationship existed between the type of ESD and patterns of sexual dimorphism in adult body size. They noted, however, that evidence for the possible adaptive significance of ESD in leopard geckos (*Eublepharis macularius*) had previously been reported by Gutzke and Crews (1988). Female geckos incubated at warm temperatures apparently are later functionally sterile, whereas females incubated at cool temperatures are fertile and sexually receptive when courted by males.

Wibbels et al. (1991) reported a sex ratio of 2.1 females to 1.0 males for immature loggerhead sea turtles (*Caretta caretta*) inhabiting the Atlantic coastal waters of Florida. Shepherd (1989) reported that nest temperatures of 29°C and below produced male loggerheads, whereas temperatures above 31°C produced females. Sex-biased ratios may be of importance in managing or manipulating populations of endangered species. For example, researchers working with the endangered Kemp's ridley sea turtle (*Lepidochelys kempi*) along Mexico's Gulf Coast plan to repopulate native waters with egg-laying ridley females.

Until recently, sexing turtles younger than 2 years of age has been difficult. At about 2 years of age, radioimmunoassay for testosterone can reveal the gender with 90 percent accuracy. Most recently, scientists at the University of Tennessee at Memphis have developed a promising alternative based on genetic fingerprinting techniques (Demas et al., 1990). After fragmenting DNA extracted from a small blood sample, a genetic probe is applied that selectively binds to gender-specific DNA fragments. The technique can determine sex even in hatchlings. It has been used successfully in ridley and green sea turtles (*Chelonia mydas*).

The choice of nest sites by female diamondback terrapins (*Malaclemmys terrapin*) alters the incubation temperature and, consequently, the sex of the young that hatch (Roosenburg, 1996). Terrapins have a large variation in egg size between clutches, but little variation within clutches. Egg mass, the primary determinant of hatchling mass, can cause as much as a 3-year difference in reaching the size of first reproduction in females, but may not affect age or size of first reproduction in males. In addition, nesting females apparently can discriminate among nesting sites (sun-open vs. sun-edge habitats). Nest temperature can be adjusted by adjusting the depth of the nest—a difference of 2 to 3 cm can drastically change the temperature at which the eggs incubate. Larger females deposit their eggs in open nest sites that receive lots of sunlight, whereas smaller, younger terrapins lay their eggs in cooler sites. The warmer eggs, which are also larger and have more yolk, hatch into females that can start laying their own eggs earlier.

Incubation temperature may have long-term effects on posthatching survival, growth rates, behavior, and environmental preferences of some reptiles (Lang, 1987; Burger, 1989; Webb and Cooper-Preston, 1989; Van Damme et al., 1992). To determine behavioral differences as a function of incubation temperature, pine snake (*Pituophis melanoleucus*) eggs were incubated at temperatures of 21, 23, 26, 28, 30, and 32°C. Hatchlings from medium-temperature incubation (26 or 28°C) performed all behavioral and physiological tests better than hatchlings from eggs incubated at low temperatures (21 or 23°C). Hatchlings from eggs incubated at high temperatures (30 or 32°C) performed some behavioral tests less well than medium-temperature hatchlings. Physiological and behavioral activities such as shedding time, drinking speeds, and movement were monitored, with some of the behavioral differences persisting for at least 24 weeks following hatching.

Van Damme et al. (1992) incubated eggs of the lacertid lizard, *Podarcis muralis*, at temperatures ranging from 24 to 35°C. Although embryos incubated at 32 and 35°C hatched about 10 days earlier than those incubated at 28°C, and over 5 weeks before those incubated at 24°C, hatching success was highest at 24 and 28°C. Neonates incubated at low temperatures had larger snout-vent lengths and body masses, grew faster, and had higher sprint speeds than hatchlings incubated at higher temperatures.

Oxygen availability to embryos is critical to their development. Low oxygen concentrations extended the incubation period of mud turtles (*Kinosternon subrubrum*), snapping turtles (*Chelydra serpentina*), and painted turtles (*Chrysemys picta*) and killed many embryos prior to hatching (Ewert, 1985). Turtle eggs incubated at higher than atmospheric oxygen (30% oxygen), however, did not show an increase in growth rate.

Parental Care

Numerous lizards and snakes care for their eggs, but parental care of offspring is rare in reptiles (reviewed by Shine, 1988; Branch, 1989). Some female skinks (*Eumeces* spp.) brood their eggs and even retrieve them if necessary, but no apparent association exists between maternal females and their offspring following hatching (Vitt and Cooper, 1989). Egg guarding has been observed in several species of snakes including the king cobra (*Ophiophagus hannah*), some members of the genus *Elaphe*, and the Texas thread snake (*Leptotyphlops dulcis*). All species of pythons brood their eggs, but only female Indian pythons (*Python sebae*) coil around their eggs—they reportedly can increase their body temperature from 5.5 to 7.5°C during incubation by contracting their body muscles (shivering thermogenesis) (Fig. 8.32) (Halliday and Adler, 1986). The number of contractions increases as the ambient temperature decreases so that the temperature within the clutch of eggs is maintained at 32–33°C, which may be as much as 7.5°C above that of the surrounding air. Female mud snakes (*Farancia abacura*) also have been known to coil around their eggs, but it is not known if they can increase their body temperatures.

Growth

Reptiles hatch or are born as miniature adults (see Fig. 8.31); development is completed inside the egg or inside the female's body. At hatching or birth, they are capable of moving about on their own, feeding, and defending themselves.

Attainment of Sexual Maturity

Reptiles continue to grow throughout their lives. As with amphibians, most appear to reach sexual maturity after they have reached a certain size rather than a specific age. Many of the smaller lizards and snakes can breed in the year following their birth. Female utas (*Uta stansburiana*) become sexually mature at 4 months of age. Most snakes reach sexual maturity after 2 to 3 years, and most turtles can breed when 3 to 5 years of age, but female Galapagos tortoises (*Geochelone gigantea*) and female tuataras (*Sphenodon*) require approximately 11 years to reach maturity (Grubb, 1971; Bourne and Coe, 1978; Castanet et al., 1988). The age at which a species reaches maturity is a compromise among many variables, with the "goal" of maximizing an individual's contribution to the next generation. One way of achieving this goal is to mature and reproduce quickly; however, smaller adults are more susceptible to predation and small body size reduces the number and/or size of offspring that can be produced. Maturing later and at a larger body size permits the production of more and/or larger offspring but increases the probability of death prior to reproducing and may yield a smaller total lifetime output of offspring.

FIGURE 8.32

Python (*Python sebae*) coiled around a clutch of eggs and contracting its muscles at a regular rate, thus raising the temperature of its body by as much as 7.5°C and warming the clutch of eggs.

Longevity

Galapagos tortoises have been known to survive for over 150 years, and the tuatara for 77 years (Goin et al., 1978) (Table 8.2). American alligators have reached a maximum

TABLE 8.2

Longevity of Some Species of Reptiles

Species	Maxiumum Age (years)
Testudines	
Testudinidae (*Geochelone gigantea*)	152
Emydidae (*Terrapene Carolina)*	71
Chelydridae (*Macroclemys temmincki*)	59
Pelomedusidae (*Pelusios castaneus*)	41+
Chelidae (*Chelodina longicollis*)	37+
Cheloniidae (*Caretta caretta*)	33
Trionychidae (*Trionyx triunguis*)	25
Sphenodonta	
Sphenodontidae (*Sphenodon punctatus*)	77
Squamata	
Lacertilia	
Anguinidae (*Anguis fragilis*)	54
Scincidae (*Egernia cunninghami*)	20
Helodermatidae (*Heloderma suspectum*)	20
Iguanidae (*Conolophus subcristatus*)	15
Teiidae (*Tupinambis teguixin*)	13
Gekkonidae (*Tarentola mauritanica*)	7
Varanidae (*Varanus varius*)	7
Agamidae (*Physignathus lesueuri*)	6
Cordylidae (*Cordylus giganteus*)	5
Serpentes	
Boidae (*Eunectes murinus*)	29
Elapidae (*Naja melanoleuca*)	29
Boidae (*Boa constrictor*)	25
Colubridae (*Drymarchon corais couperi*)	25
Colubridae (*Elaphe situla*)	23
Viperidae (*Vipera ammodytes*)	22
Viperidae (*Crotalus atrox*)	22
Viperidae (*Agkistrodon piscivorus*)	21

From Introduction to Herpetology, *3E by C. J. Goin, Goin, and Zug © 1978 by W. H. Freeman and Company. Used with permission.*

age of 56 years, and several species of snakes, including boas and cobras, have survived for over 25 years (Goin et al., 1978). At 54 years, the slowworm (*Anguis*) holds the record among lizards (Goin et al., 1978).

Crocodilians and Birds (Archosaurs)

■ MORPHOLOGY

Integumentary System

Crocodilians have heavily armored bodies with long snouts and powerful tails. Each epidermal scale develops separately, so that the scales do not form a solid sheet. The epidermal scales wear away and are gradually replaced. The dorsal armor is formed by heavy plates of bone called osteoderms that lie within the dermis, underneath the epidermal scales. In some species, osteoderms also occur on the ventral surface of the the body. Otherwise, the dermis is thick and relatively soft.

Feathers are modified reptilian scales and distinguish birds from all other animals. They are derived from the keratinized stratum corneum portion of the epidermis (Fig. 8.33). Typical epidermal scales, which occur chiefly on the legs and at the base of the beak (Figs. 8.33 and 8.34), are homologous to those found on other reptiles. The claws and horny covering of the beak are also modifications of the keratinized stratum corneum.

The lightweight feathers help to insulate the body and form the resistant, yet flexible, flight surfaces of the wings and tails of many birds. Feathers are highly diversified, keratinized structures that are specialized in form, color, and arrangement. A large bird such as a whistling swan (*Cygnus columbianus*) may have more than 25,000 feathers on its body, whereas a ruby-throated hummingbird (*Archilochus colubris*) may have less than 1,000, with those from each part of the body being characteristically unique in form. There is no significant sexual variation in the number of feathers. Counts for individuals of the same species often vary by less than 1 percent (Wetmore, 1936; Trainer, 1947; Brodkorb, 1951). Despite their individual lightness, the combined weight of the feathers often exceeds the weight of the entire skeleton.

A typical feather consists of a long, tapering central **shaft** composed of two main parts: the hollow cylindrical basal portion known as the **calamus** (quill) and a solid, squarish portion, the **rachis**. A row of small parallel branches, or **barbs** (which are set at an angle inclined toward the tip of the feather), collectively form the **vane** on each side of the rachis. Barbs possess yet smaller branches known as **barbules**, which are inclined toward the tip of the barb. Barbules have **hamuli (hooklets)**, which interlock with the barbules of adjacent barbs to stiffen the vane.

Five basic types of feathers can be distinguished: contour feathers, semiplumes, down feathers (plumules), bristles, and filoplumes (Fig. 8.34).

FIGURE 8.33

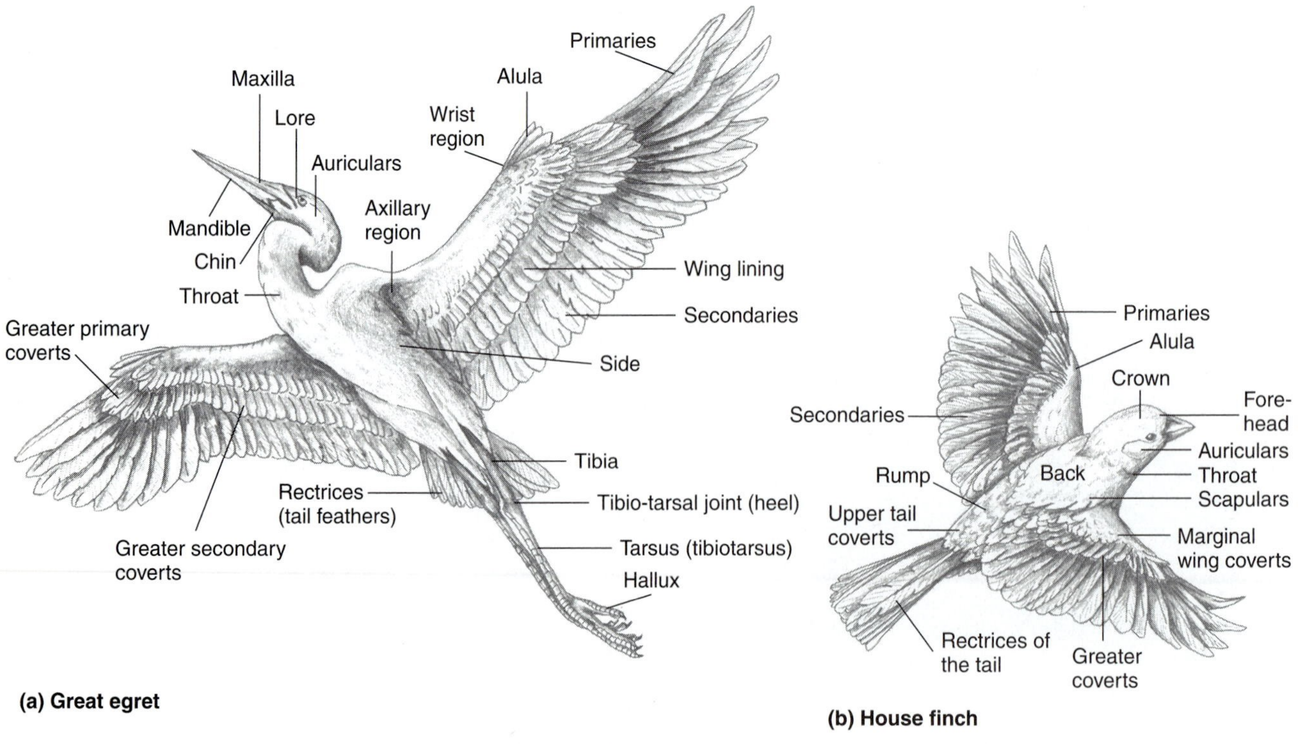

External anatomy of a bird: (*a*) Great egret (*Casmerodus albus*), ventral view; (*b*) House finch (*Carpodacus mexicanus*), dorsal view.

FIGURE 8.34

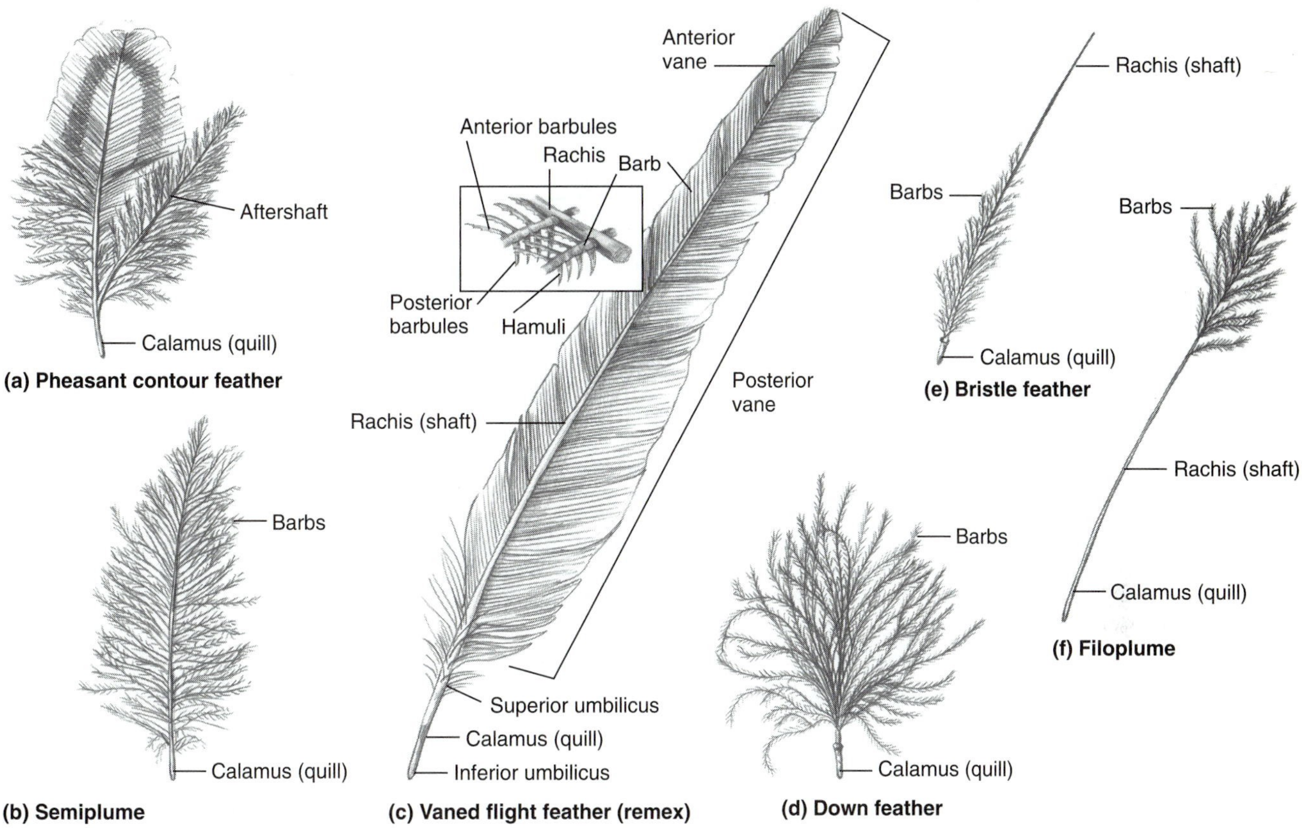

Basic feather types: (*a*) contour feather; (*b*) semiplume; (*c*) flight feather; (*d*) down feather; (*e*) bristle feather; (*f*) filoplume. Inset: Detail of the vane of a flight feather.
Source: Proctor and Lynch, Manual of Ornithology, *1993, Yale University Press.*

Contour feathers (Fig. 8.34a, c) are those that form the contour or outline of the body of the bird. A typical contour feather has a large, firm vane and a downy base. On most birds, these feathers do not arise randomly over the body of the bird. Rather, **feather follicles** arise from definite tracts known as **pterylae** (Fig. 8.35). Bare patches between pterylae are known as **apteria.** A few birds, such as penguins, kiwis, ostriches, and South American screamers, lack feather tracts, and feathers arise randomly over the body. Feather tracts vary in shape and extent in different species and are important in deducing taxonomic relationships. Feathers on the wing and tail have become highly developed for flight and/or display. Feathers borne on the modified hand of the bird and located between the wrist and the tip of the wing are known as **primaries**; those on the forearm between the wrist and the elbow are **secondaries** (see Fig. 8.33). Tail feathers are collectively known as **rectrices.** Small feathers that overlie the bases of the flight feathers (primaries and secondaries) and several rows of small feathers between the bases of the flight feathers and the leading edge of the wing are known as **coverts** (see Fig. 8.33).

Semiplumes (Fig. 8.34b) are loosely webbed contour feathers. Each possesses a definite rachis but no hamuli and, thus, no firm vane. Semiplumes are found most often at the margins of pterylae. They help insulate the body and increase the buoyancy of waterbirds.

Down feathers (Fig. 8.34d) are small, fluffy feathers lying beneath and between the contour feathers. They are fluffy because their barbs do not interlock along any portion of the rachis. They usually are not confined to the pterylae, but are widely distributed over the body. The principal function of down feathers is to provide insulation. Newly hatched birds lack contour feathers and are covered at hatching by a coat of natal (newborn) down.

Powder down feathers are highly modified down feathers that grow continuously from the base and disintegrate at the tip. As the barbs disintegrate, they give off a fine talclike powder composed of minute scalelike particles of keratin. The powder may be used in preening and may serve to protect the feathers from moisture. In some species, the powder may affect the color of the plumage.

Bristles (Fig. 8.34e) are modified, usually vaneless, feathers that consist of only a shaft. They normally are found around the mouth, nostrils, and eyes. Those around the nostrils probably serve to filter incoming air, whereas those around the mouth help swifts and other aerial insect eaters to trap insects. Others may have a sensory function.

FIGURE 8.35

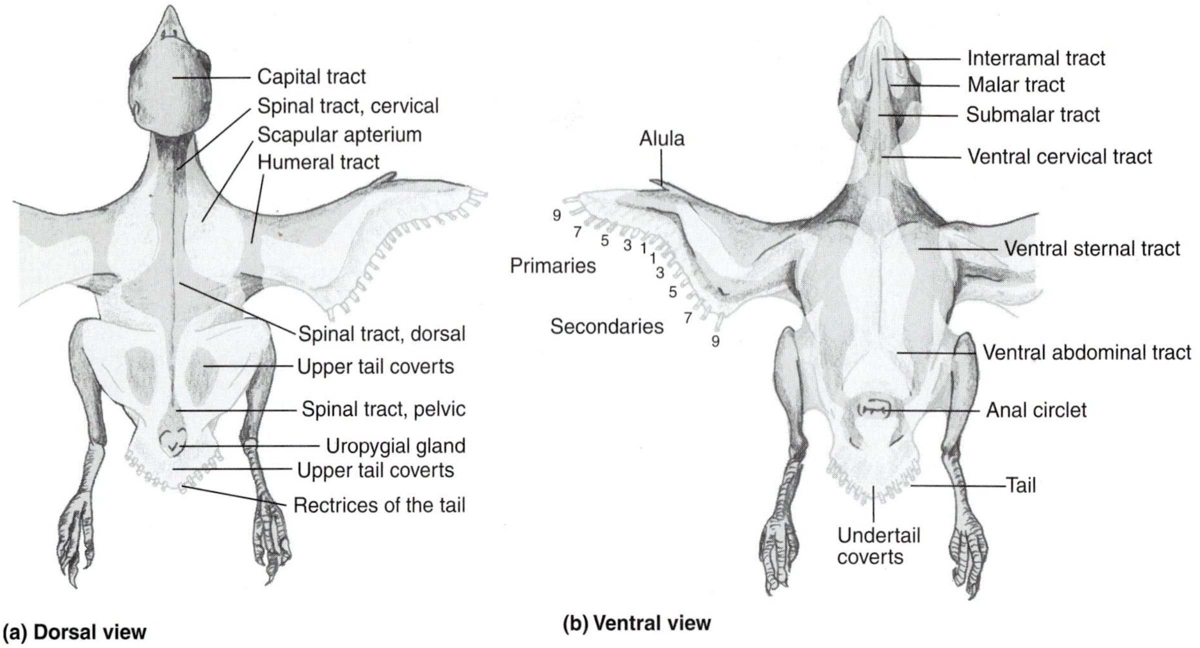

Feather tracts (pterylae) in a typical passerine: (*a*) dorsal view; (*b*) ventral view.

Filoplumes (Fig. 8.34f) are specialized, hairlike feathers. They generally lack vanes and consist mainly of a threadlike shaft. They are usually scattered over the skin between contour feathers and may serve in both decorative and sensory functions. The long colorful feathers of a peacock are examples of filoplumes.

BIO-NOTE 8.5

Poisonous Birds

Three species of pitohuis endemic to New Guinea (the hooded pitohui, *Pitohui dichrous;* the variable pitohui, *P. kirhocephalus;* and the rusty pitohui, *P. ferrugineus*) contain one of nature's most powerful toxins and are the world's only known poisonous birds. Bioassays have shown that the feathers and skin are most toxic, with striated muscle, heart, liver, stomach, intestines, and uropygial glands containing lesser concentrations. These tissues contain the steroidal alkaloid hemobatrachotoxin—the same nerve agent that is secreted by poison-dart frogs (*Phyllobates*) of Central and South America. Batrachotoxins depolarize nerve and muscle cells by activating sodium channels and also irritate sensory neurons in and around the mouth. Extracts from these birds are capable of killing a mouse within just a few minutes. In humans, the alkaloid causes numbness, burning, and sneezing on contact.

Dumbacher et al., 1992

Erector muscles known as **arrectores plumarum** originate in the dermis and insert on the wall of each feather follicle. Along with extrinsic integumentary muscles, the arrectores plumarum muscles enable a bird to fluff its feathers for display and/or to alter their insulating properties by either trapping a layer of air between the feathers and the skin or flattening the feather to reduce insulation when temperatures are high.

Feathers are subjected continually to wear and deterioration. Thus, they must be replaced periodically by a process known as **molting**, which is closely synchronized with reproductive cycles and the seasons of the year. Most adult birds undergo a single annual molt following their breeding and nesting activities called a **postnuptial** molt. The males of some species, however, also undergo a partial molt prior to breeding—a **prenuptial** molt—during which they may acquire their colorful breeding plumage. Most migratory species complete their postnuptial molt prior to the beginning of fall migration. Changing daylength (photoperiod) is primarily responsible for initiating the molting process.

The early development of a feather is similar to that of a scale. Vascularized tissue in the dermis first forms a cone-shaped structure known as a **papilla**. Then the thin, overlying epidermal layer sinks inward and forms a cylindrical pit known as the **feather follicle**, out of which the papilla develops into a feather. During molting, new feathers develop from the same reactivated dermal papilla that gave rise to the previous feathers. Incoming feathers, in many cases, push the old feathers out of the follicles.

Molting is an orderly process in most birds, although the timing of molt varies greatly among species. A complete molt usually begins with the innermost primaries of the wings. Next, the remaining primaries drop out in succession as their predecessors are replaced. At any one time, only a small gap between primaries is present and the bird retains its power of flight. When molting of the primaries is partially completed, molting of secondaries begins. This is followed by the tracts of other feathers on the body. Tail feathers are usually molted a few at a time so that the tail can remain useful as an organ of flight. Some birds (mainly aquatic and semiaquatic species that can elude predators for several weeks without resorting to flight) lose all of their flight feathers at once. These include ducks, geese, and swans (Anatidae), most rails (Rallidae), and many alcids (Alcidae).

Birds are among the most colorful of all animals, with colors being produced by a variety of pigments and structural features of the feathers. Melanins produce black, grays, and browns; carotenoids produce intense reds and yellows; and porphyrins are responsible for a range of reds, browns, and greens. Complex patterns of reflection and refraction in the cell membranes of the barbs and barbules of each feather produce metallic blues, greens, and iridescent colors. Much of the color is modified by abrasion and oxidation. Coloration is important for concealment from predators, for identifying members of the same or related species, and for controlling the absorption of radiant energy (Fig. 8.36). Seasonal changes in color are accomplished through molting or wearing off the tips of feathers.

Owl feathers confer the advantage of silent flight, both at frequencies audible to the human ear and at ultrasonic levels (Thorpe and Griffin, 1962). By reducing their own noise, owls can both make maximal use of their acute hearing, and silently approach their prey. Three structural adaptations that tend to reduce flight noise were identified in barn owls by Graham (1934). First, a very prominent, stiff, comblike fringe is present along the leading edges of the flight feathers, particularly the outer primaries (Fig. 8.37). This fringe reduces turbulence in air flow and hence cuts noise production. Second, the primaries and secondaries along the trailing edge of the wing have a soft hairlike fringe, which probably reduces turbulence where air streams flowing over the top and bottom of the wing meet. Third, the downy upper surfaces of the primaries, secondaries, and coverts must reduce noise that would otherwise be produced when these feathers move over each other during the normal wingbeat. This downy appearance is produced by extremely elongated extensions of the barbules (Fig. 8.37).

Owls and ptarmigans have lower legs and feet that are completely feathered, whereas other species, such as rough-legged hawks, have feathers on their lower legs but not on their feet. Most birds, however, have featherless lower legs and feet that are covered with horny scales (Fig. 8.38). When these scales overlap one another on the anterior surface of the leg (tarsometatarsus), as they do in finches and sparrows, they are said to be *scutellate* (Fig. 8.38a). Individual scales are not present in the smooth appearance of the *booted* tarsometatarsus of thrushes (Fig. 8.38b). In geese and many

FIGURE 8.36

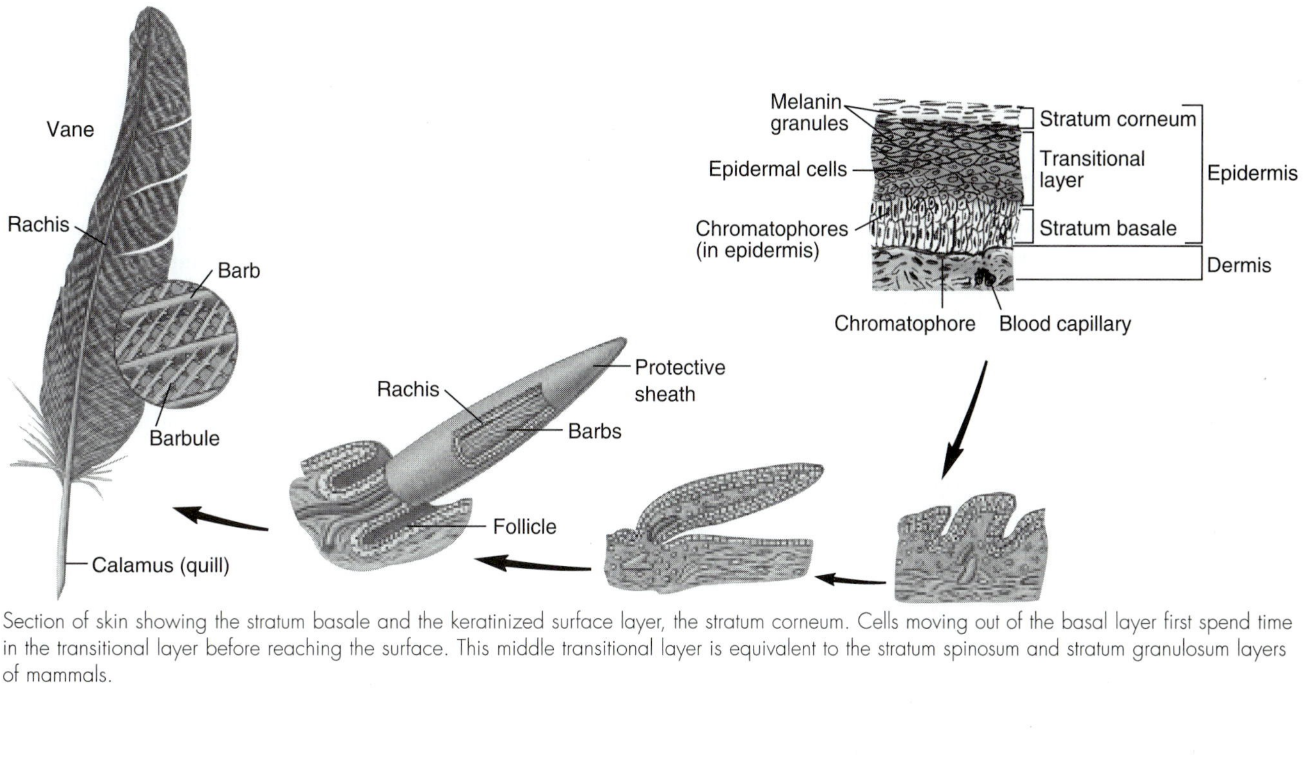

Section of skin showing the stratum basale and the keratinized surface layer, the stratum corneum. Cells moving out of the basal layer first spend time in the transitional layer before reaching the surface. This middle transitional layer is equivalent to the stratum spinosum and stratum granulosum layers of mammals.

FIGURE 8.37

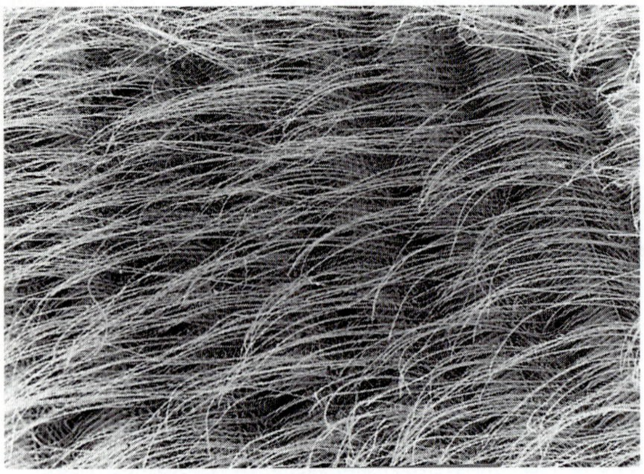

Barn owl feathers. Long hairlike extensions of the barbules seen in this highly magnified scanning electron micrograph enable the barn owl to fly silently. These structures are responsible for the soft downy feel of the barn owl's plumage.

FIGURE 8.38

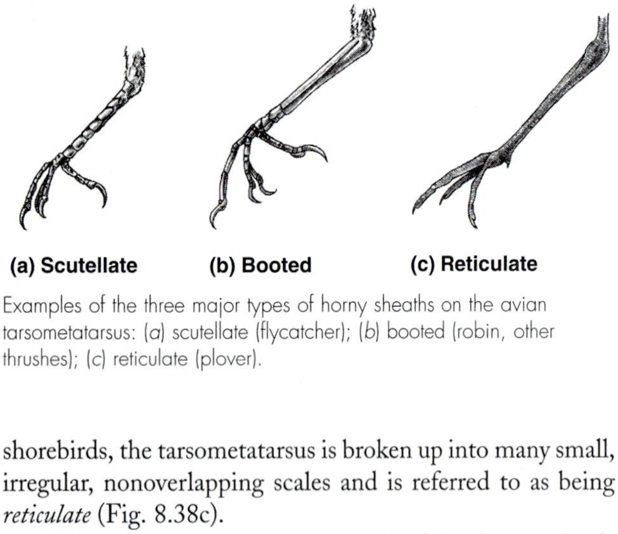

Examples of the three major types of horny sheaths on the avian tarsometatarsus: (*a*) scutellate (flycatcher); (*b*) booted (robin, other thrushes); (*c*) reticulate (plover).

shorebirds, the tarsometatarsus is broken up into many small, irregular, nonoverlapping scales and is referred to as being *reticulate* (Fig. 8.38c).

The stratum corneum at the ends of the digits in birds is modified into claws (Fig. 8.38). All birds possess claws on their feet. Ostriches, geese, some swifts, and others may also possess sharp claws at the ends of one or two digits of the wings. Hoatzins (*Opisthocomus hoatzin*) are chicken-sized birds that live along sluggish tropical rivers in South America. Young hoatzins use two reptilelike claws on each wing to provide increased grasping ability for climbing about in trees (Fig. 8.39). If frightened, young hoatzins may drop out of the nest into the water. They are able to swim underwater before they can fly. They return to the nest by using their claws and crawling up the trunk and limbs in lizardlike fashion. These claws are lost when the birds mature and become capable of flight. Fossilized remains of the earliest known bird (*Archaeopteryx*) have revealed the presence of three claws on each wing (see Fig. 7.27).

FIGURE 8.39

Baby hoatzins of South American rain forests have claws on their wings that enable them to clamber about the limbs of their streamside nesting bushes. At any sign of an aerial predator, a baby hoatzin drops into the water beneath the nest. As soon as the danger has passed, it uses its claws to scramble out of the water and away from other predators lurking there.

Birds have fewer epidermal glands than any other vertebrate group. One of these glands, the **uropygial gland**, is best developed in aquatic birds (see Fig. 8.35). Located dorsally at the base of the tail, this gland exudes an oily secretion of semisolid fatty substances that contain precursors of vitamin D. Birds rub their beaks against these glands or squeeze the nipple of the gland with their beaks and distribute the oil over their feathers as they preen. Some of the oil may be swallowed for its essential vitamin D. The presence or absence of this gland has been used in some taxonomic classifications; it is absent in some woodpeckers, pigeons, parrots, ostriches, and some other birds. Some birds also have small oil glands lining their outer ear canal and the area around the exit from the cloaca. Sweat glands are absent.

As in other vertebrates, the dermis is richly vascularized and supplies oxygen and nutrients to the epidermis. It consists of blood vessels, lymphatic vessels, nerve endings, and sense organs. It is attached to underlying tissues by means of the basement membrane.

Skeletal System

Crocodilians possess a complete secondary palate below the primary palate (Fig. 8.40). The evolution of a secondary palate allows the air passageways to be separated from the oral cavity. Teeth are set into sockets in the jaws, a situation called **thecodont** dentition (see Fig. 8.13c). The vertebral column

FIGURE 8.40

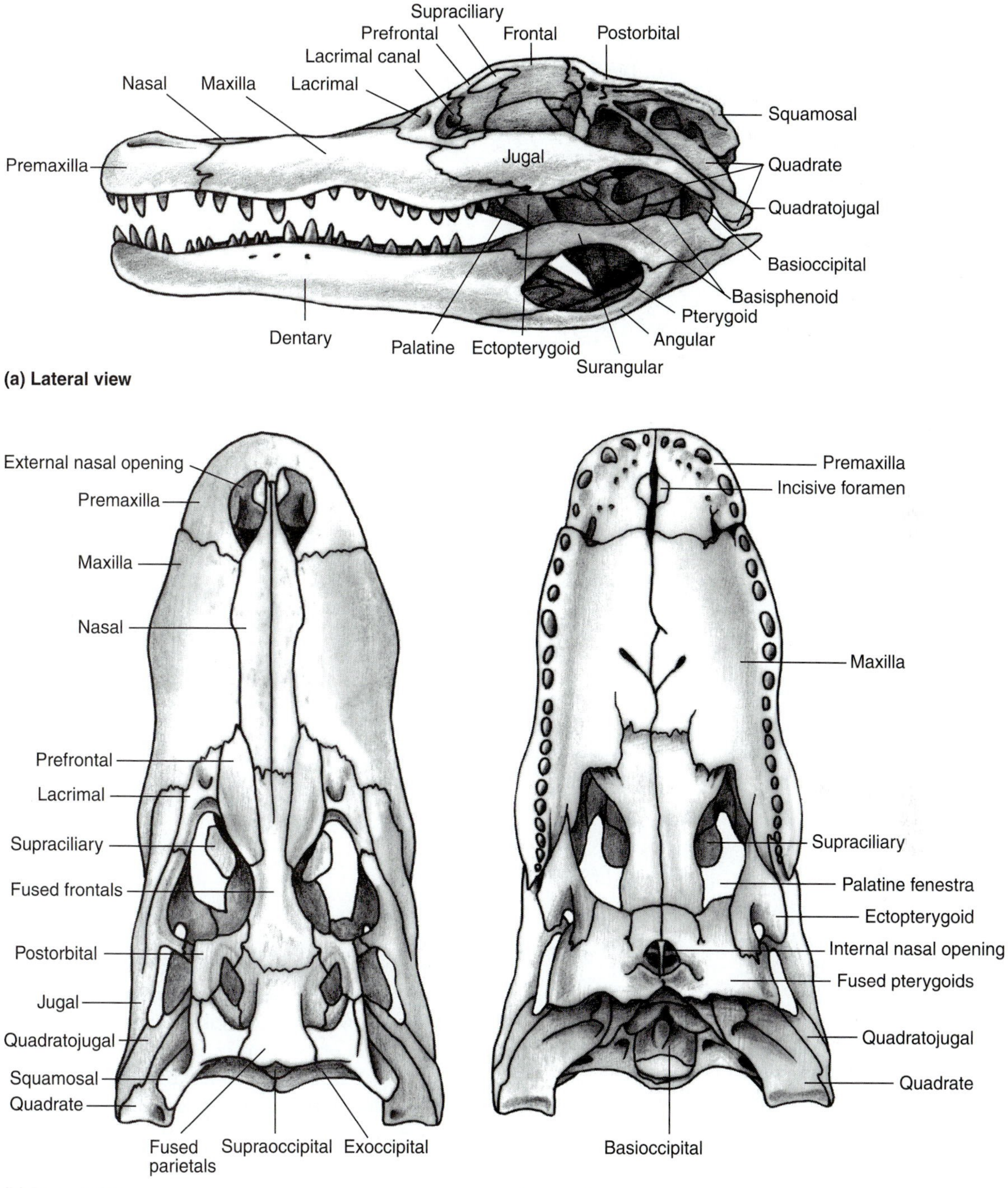

Skull of American alligator (*Alligator mississippiensis*): (*a*) lateral view; (*b*) dorsal view; (*c*) ventral view.
Sources: Jollie, Chordate Morphology, *1962, Wadsworth; and Hildebrand,* Analysis of Vertebrate Structure, *4th ed., 1995, John Wiley & Sons, Inc.*

of crocodilians consists of cervical, thoracic, lumbar, sacral, and caudal vertebrae. Posterior cervical and anterior thoracic ribs each bear a curved cartilaginous **uncinate process**, which projects posteriorly to overlap the rib behind, presumably giving strength to the thoracic body wall. A sternum is present and consists of a plate to which the ribs attach.

The legs of crocodilians are sprawled during normal walking; however, when moving rapidly, the legs hold the body off the ground in a semierect attitude. In water, crocodilians fold the limbs against the body and use lateral undulation of the body and tail to swim. Webbing is present between the toes.

The avian skeleton is specialized for both lightness and strength (Fig. 8.41). Many bones that form the structural framework of the skeleton are paper-thin. In most birds, resorption of the marrow and its replacement by stiffening struts, air spaces, and extensions of the air sac system further strengthen the bone while decreasing its weight (Fig. 8.42). In addition, the fusion of many of the main bones provides strength.

FIGURE 8.41

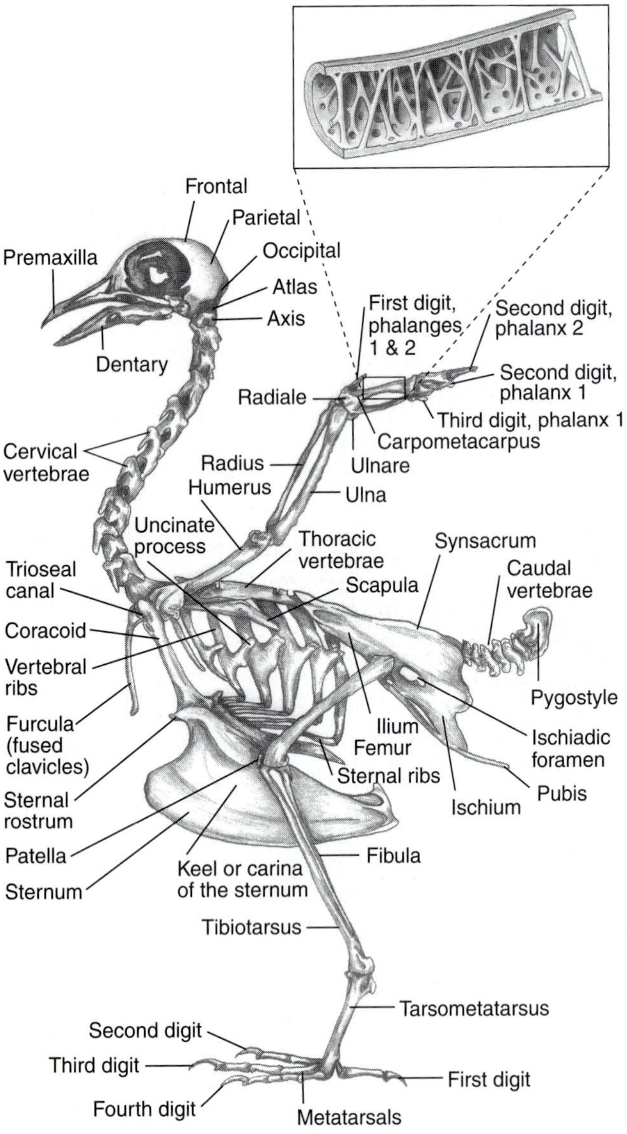

Lateral view of the skeleton of the rock dove (*Columba livia*). Note the many fused bones along the axial skeleton (the skull, spinal column, and pelvis). The fused bones provide a strong, stable central platform for the flight muscles. The bird skeleton is extremely light (see inset). In some birds, the skeleton weighs less than the feathers. Inset: Hollow bone of a bird showing the stiffening struts and air spaces that replace bone marrow. Such "pneumatized" bones are remarkably light and strong.
Source: Proctor and Lynch, Manual of Ornithology, *1993, Yale University Press.*

FIGURE 8.42

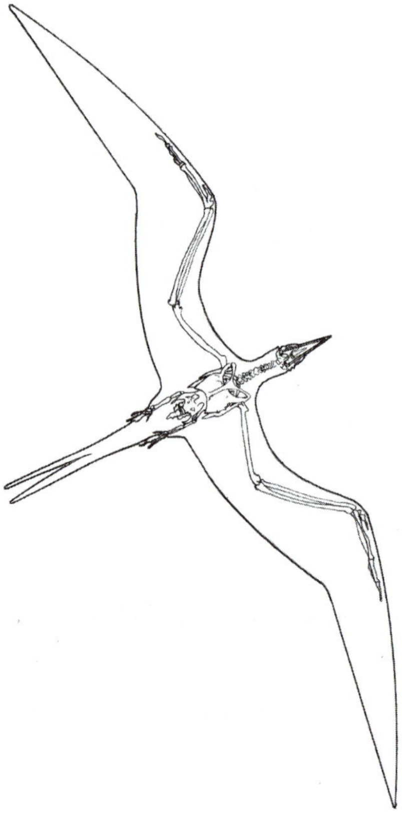

The frigatebird (*Fregata magnificens*) has a 7-foot wing span, but its skeleton weighs only 4 ounces—less than the weight of its feathers. The skeleton is shown against the outline of the bird.
Source: Welty, "Birds as Flying Machines" in Scientific American, *192(3):88–96, 1955.*

The bird skull (see Fig. 8.41) exhibits many features of the diapsid reptilian skull such as a single occipital condyle and an incomplete secondary palate. Unlike the case for reptiles, the major bones making up the braincase in adult birds are very thin and usually are fused completely. Quadrate and articular bones are present in the jaws of birds. The freely movable quadrate bone suspends the mandible from the cranium. In adult birds, few if any remnants of Meckel's cartilage remain within the mandible: it has ossified or become totally ensheathed by membrane bones. The hyomandibular cartilage becomes the columella of the middle ear. This quadrate–articular–columella arrangement is essentially similar to the reptilian condition. However, the brain of a bird is larger than that of a comparable-sized reptile. The orbits are also large. An anterior **sclerotic ring** consisting of 10 to 18 overlapping platelike bones develops in the sclera of the eye (Fig. 8.43). Its function is to reinforce the huge eyeball.

A beak (bill) is formed primarily by the elongation of the premaxillary and dentary bones (Figs. 8.41 and 8.43). The upper beak, supported by the maxilla and other bones of the skull, is composed of a bony framework covered by a tough

FIGURE 8.43

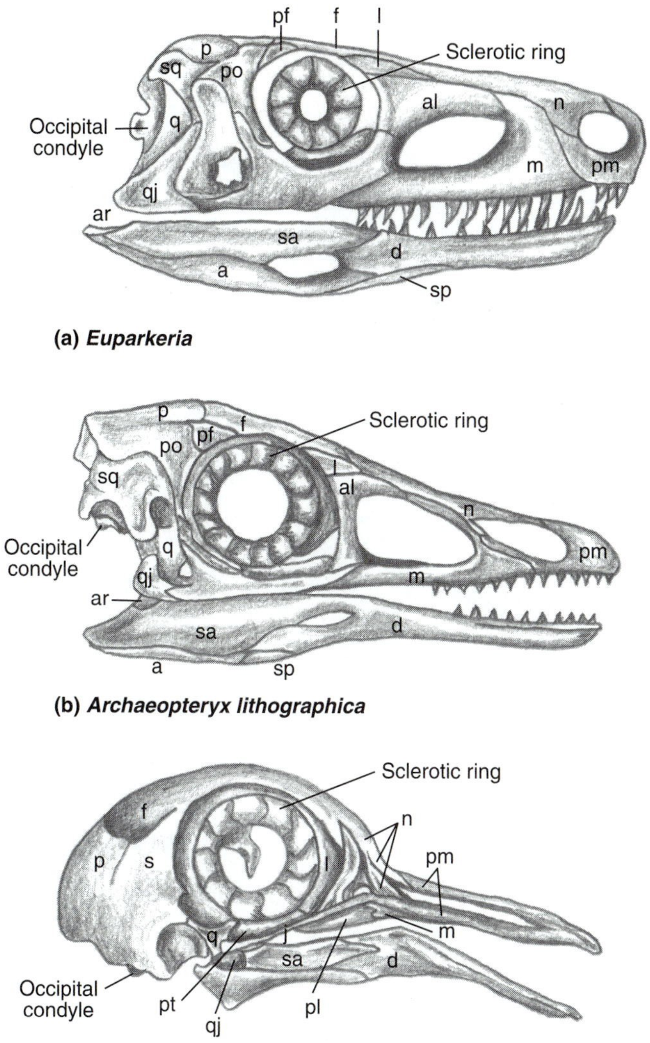

(a) ***Euparkeria***

(b) ***Archaeopteryx lithographica***

(c) ***Columba***

Skulls of (*a*) a pseudosuchian reptile (*Euparkeria*); (*b*) *Archaeopteryx lithographica*; and (*c*) a modern bird (*Columba*) in lateral view. The sclerotic ring within the orbit of the skull is shared by birds and their reptilian ancestors. In birds, the sclerotic ring reinforces the huge eyeball. Key: a, angular; ar, articular; d, dentary; f, frontal; j, jugal; l, lacrimal; m, maxilla; n, nasal; p, parietal; pl, palatine; pm, premaxilla; po, postorbital; q, quadrate; qj, quadratojugal; sq, squamosal.

layer of keratin. Unlike the jaws of mammals, the upper jaw of all birds is slightly mobile, because the pterygoid, quadrate, and zygomatic bones that support the maxilla can slide forward or backward. Birds depend on their beaks not only to obtain food (Fig. 8.44), but also to preen their feathers, build their nests, perform courtship displays, and defend themselves from predators or rivals.

Birds have cervical, thoracic, lumbar, sacral, and caudal vertebrae. Those vertebrae that are connected to the sternum by ribs and that do not fuse with the synsacrum sometimes are known as dorsal vertebrae. The highly flexible neck may consist of as many as 25 vertebrae with saddle-shaped ends, called **heterocoelous** vertebrae. The anterior two cervical vertebrae are modified as the atlas and axis.

Most birds have two sacral vertebrae. These are fused with the last thoracic, the first few caudal, and all of the lumbar vertebrae to form one structure, the **synsacrum** (see Fig. 8.41). The synsacrum, along with the pelvic girdle to which it is more or less fused, forms a rigid framework both for flight and for the bipedal gait of birds. Thoracic vertebrae anterior to the synsacrum also usually fuse together, so that there is little flexibility in the vertebral column behind the neck.

Usually, 10 to 15 caudal vertebrae are present in birds. These vertebrae are the remnants of an ancestral reptilian tail. The first 6 or 7 vertebrae posterior to the sacrum are free, whereas the remaining vertebrae (usually 4 to 7) fuse together to form a **pygostyle**—the skeleton within the tail (see Fig. 8.41).

The ribs of birds are ossified. Thoracic ribs have flat uncinate processes, each of which overlaps the next most posterior rib and provides added support to the rib cage and attachment sites for muscles (see Fig. 8.41). The ossified sternum articulates with the pectoral girdle and the ribs. In **carinate** (flying) birds and penguins, which are powerful swimmers, the sternum has a midventral **keel**, or **carina**, for the attachment of the large flight muscles (see Fig. 8.41). In general, **ratites** (flightless birds) such as ostriches and emus lack a keeled sternum.

Each pectoral girdle is formed by the clavicle, coracoid, and scapula. Clavicles have become prominent in most birds, and they function in bracing the pectoral girdle against the sternum. In most carinate birds, the two clavicles unite in the midline with the interclavicle to form the **furcula**, or wishbone (see Fig. 8.41). In ratites and some carinate birds, including some parrots and pigeons, clavicles are absent or rudimentary.

The skeleton of the anterior appendage consists of the humerus, radius, ulna, carpals, metacarpals, and phalanges; it has been modified for flight through the loss and fusion of bones. Pectoral muscles insert chiefly on the humerus and provide the power for flight, while air moving over the convex surface of the wings provides the necessary lift. Three distal carpal bones usually fuse with three metacarpals to form a rigid **carpometacarpus**. Three digits are usually present, but the number of phalanges in each has been reduced. The anterior appendages of penguins have been further modified as paddles for life in the sea by becoming flattened, shortened, and stout, whereas the appendages of soaring birds such as herring gulls (*Larus argentatus*) have lengthened and become lighter to maximize their efficiency for soaring flight.

The pelvic girdle is formed by the fusion of the pubis, ischium, ilium, and synsacrum. The ilium, which is braced against both lumbar and sacral vertebrae, is broad and expanded for the attachment of the appendicular muscles, which are highly developed for bipedal locomotion. Pubic bones are reduced to long splinters projecting toward the tail.

FIGURE 8.44

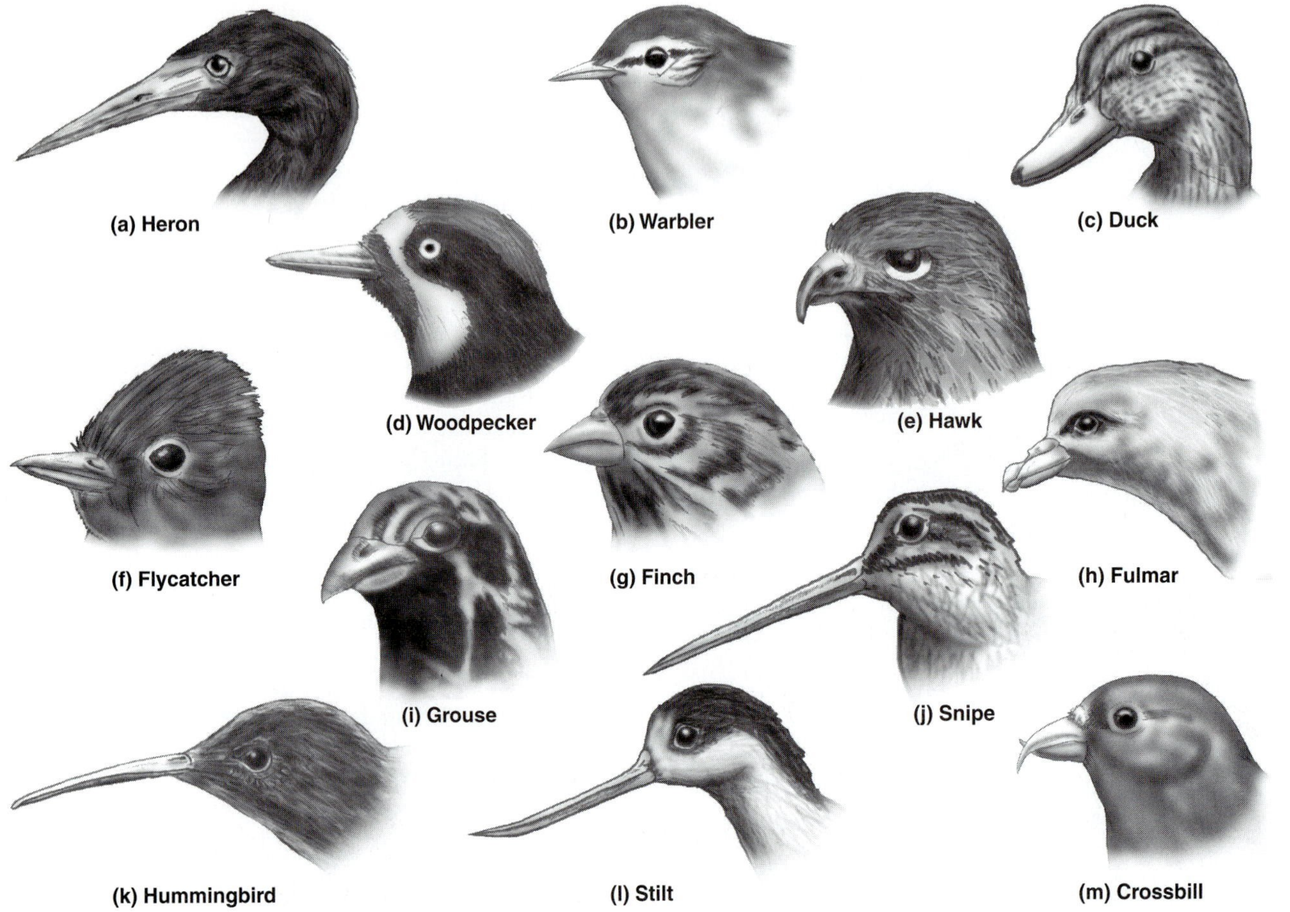

Specialized feeding methods and bill modifications that reflect the adaptations to the ecological niches of various birds: (*a*) heron; (*b*) warbler; (*c*) duck; (*d*) woodpecker; (*e*) hawk; (*f*) flycatcher; (*g*) finch; (*h*) fulmar; (*i*) grouse; (*j*) snipe; (*k*) hummingbird; (*l*) stilt; (*m*) crossbill.

A pubic symphysis is absent in most birds (*Archaeopteryx* and ostriches are exceptions).

The legs of birds contain the same bones as those of reptiles. In birds, however, the tibia fuses with the proximal row of tarsals to form a **tibiotarsus**, and the metatarsals fuse with the distal row of tarsals to form a **tarsometatarsus** (Fig. 8.41). The joints between the tibiotarsus and tarsometatarsus and those between the tarsometatarsus and the toes allow flexion to occur. In some birds, such as parrots, the fibula may be reduced to a splinter. A patella is present and protects the knee joint.

Ostriches are the only living two-toed (didactyl) birds. The feet of all other birds have either three or four functional toes and show a wide variety of modifications (Fig. 8.45). Aquatic birds—ducks, geese, gulls, and others—have feet that are either webbed or have lobed toes to assist in swimming. (The term "semipalmate" is frequently used in describing the degree of webbing on shorebirds and other aquatic birds. It means "half-webbed" and refers to the condition of having the front three toes joined by a web along only the basal half [or less] of their length.) Grebes have flattened claws that are incorporated into a paddlelike foot. Some marsh birds have elongated toes designed to support them as they walk on water plants or mudflats. Gallinaceous birds such as quail, grouse, turkeys, and pheasants have strong feet and legs adapted for running. Their strong claws are used for scratching the ground in search of food. Hawks and owls that seize their prey with powerful feet have long, sharp claws called talons. Perching birds, including most songbirds, have a long, backward-pointing hind toe (hallux) that helps the bird secure a foothold on a branch or other perch. The feet of woodpeckers are adapted for climbing and, except for the three-toed woodpecker (*Picoides tridactylus*) (Fig. 8.45), consist of two toes facing forward and two facing rearward (zygodactylous).

Penguins are the most specialized of all birds for swimming underwater. Their wings are short, narrowed, and flattened—all adaptations for underwater "flying," and their legs are attached more posteriorly on the body. The modified feathers are scalelike, which provides a smooth surface on

FIGURE 8.45

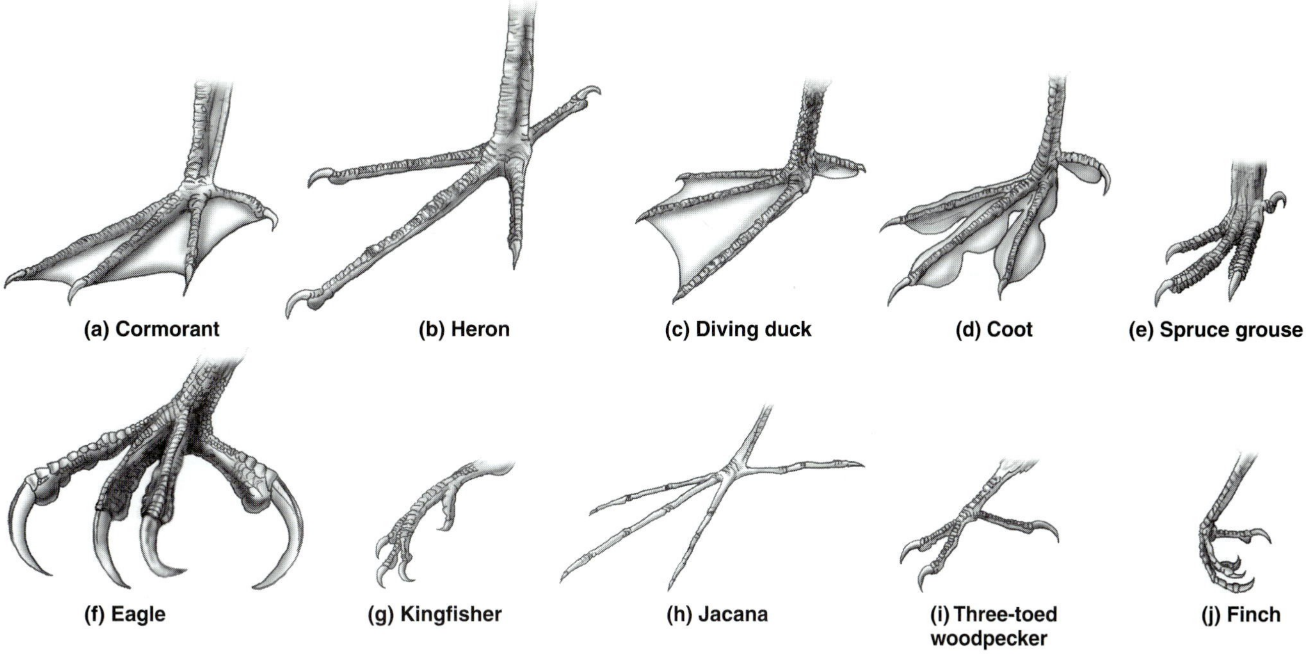

Selected examples of avian feet: (*a*) cormorant; (*b*) heron; (*c*) diving duck; (*d*) coot; (*e*) spruce grouse; (*f*) eagle; (*g*) kingfisher; (*h*) jacana; (*i*) three-toed woodpecker; (*j*) finch. Note the extensive webbing in the aquatic birds (a, c) and the lobing in the coot (*d*), a shorebird.

BIO-NOTE 8.6

The Beak of the Crossbill

Crossbills (*Loxia*) have remarkable beaks (Fig. 8.44m). The lower mandible twists either to the right or left of, and crosses over, the upper one. This unique beak serves as an adaptation for extracting and handling seeds from conifer cones, allowing crossbills access to parts of the cone that other birds cannot reach. Conifer cones vary in structure depending on ripeness. Cone scales are initially closed, but as the cone matures, the scales have wide gaps between them. Cone scales must be separated so that crossbills can extract seeds with their tongue. On closed cones, crossbills create gaps between the scales by biting between the scales with their crossed and pointed mandibles. The lower mandible curves slightly laterally, either to the left or right, whereas the upper mandible is directed more straight ahead. To create a gap between cones scales, the decurved and pointed upper bill slides between the scales in a biting motion. After an initial gap is created, the jaws spread sideways (the lower jaw is abducted laterally in the direction that its tip points). Lateral spread of the jaws widens and deepens the gap between the scales. When the seed is exposed, the tongue is protruded and its spoon-shaped tip carries the loose seed to the bill to be husked. Lateral mobility of the lower mandible probably evolved because it increased the rate with which seeds could be extracted from partly open cones. It provides increased access to seeds in cones whose scales are spread apart. Crossed mandibles further expand the time during which conifer seeds can be used, especially from closed cones.

Crossbills have evolved larger bills, and associated musculature and body mass, to provide the power necessary to separate cone scales. This specialization, however, may cause high mortality during conifer cone failures due to their ineffectiveness in handling nonconifer seeds, because of their narrow mandibles, which reduce the efficiency with which large seeds can be handled, a lowered horny palate, and a large body mass.

Benkman found that small-billed crossbills, such as white-winged crossbills (*Loxia leucoptera*), forage more efficaciously on small conifer cones such as tamarack and spruce with thin and relatively short cone scales, whereas large-billed crossbills such as red crossbills (*Loxia curvirostra*) forage more efficiently on large conifer cones such as pines and black spruce with thicker cone scales. White-winged crossbills rely less on powerful lateral abduction of their mandibles in separating adjacent cone scales to expose seeds.

Based on bill depth and palate structure, Benkman found that each taxon of red crossbill in North America is adapted to a particular species of conifer, a "key conifer." The adaptive value of bill size differences among red crossbill populations shows morphological diversification in bill characteristics in response to adaptive peaks represented by their main food: seeds in conifer cones.

Nestlings have uncrossed bills. Crossing over starts at about the age of 27 days (1 week after fledging) and is complete by about 45 days, when young birds are able to feed effectively on cones.

Robbins, 1932
Griscom, 1937
Benkman, 1987; 1988a, b; 1993

the fusiform body. Their solid bones reduce buoyancy. Diving behavior and physiology have been discussed by Kooyman and Ponganis (1990).

Muscular System

The musculature of crocodilians has become well differentiated into individual powerful muscles. The limb muscles are especially well developed. Although crocodilians usually use a sprawling gait, their limb muscles allow them to raise their body off the ground and travel rapidly for short distances.

Unlike many vertebrates, crocodilians open their mouths by lifting their heads, not by lowering their jaws. The muscles responsible for this action (depressor mandibulae and several dorsal neck muscles) are relatively small and have very little mechanical advantage; thus, it is possible to keep a crocodilian's mouth shut by holding its jaws. Two large pterygoid muscles, however, can generate tremendous power and force when closing the jaws.

The epaxial musculature of birds (Fig. 8.46) is basically similar to that described for lepidosaurs, although many birds have lost their muscular metamerism and developed muscle bundles. Because the vertebral column in the trunk region is more or less rigid, epaxial muscles in this region are poorly developed; those in the neck and tail region, though, are much better developed. Differentiation of the abdominal wall muscles is greatly reduced, and metamerism is no longer evident. External and internal intercostal muscles are present to assist in breathing.

Two "color" types of skeletal muscle tissue occur in birds: red and white. Red muscle fibers are smaller in diameter,

FIGURE 8.46

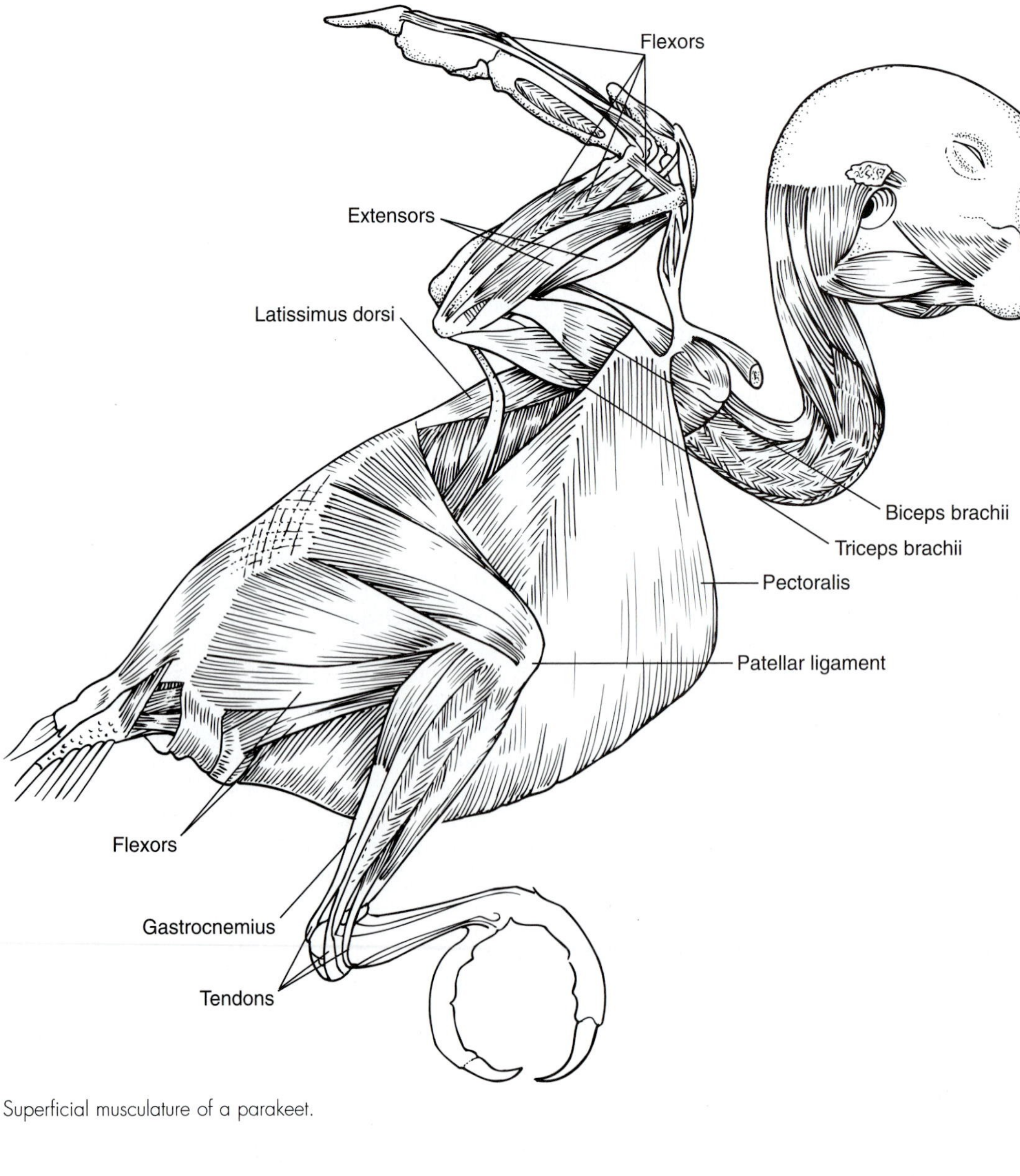

Superficial musculature of a parakeet.

have a richer blood supply, and contain more nuclei and mitochondria than white fibers. In addition to hemoglobin, red muscle also contains the protein myoglobin. Myoglobin, which is found in the flight muscles of many birds, accounts for the designations of "dark" meat and "white" meat. Myoglobin stores oxygen that then can be released to the muscle as required. Because myoglobin has a higher affinity for oxygen than blood hemoglobin, it automatically loads up from this latter source. Some muscles may be composed of red fibers in one species and white fibers in another, depending on their function. White muscle is designed for short bursts of intense activity, such as a turkey taking off and flying a short distance; red muscle is designed for sustained activity. Even in the same species, the type of muscle fiber in a given muscle may vary (e.g., breast muscle in a wild turkey [red] versus a domestic turkey [white]).

Appendicular muscles of birds are highly modified for flight as well as for perching, grasping, walking, hopping, and swimming. The largest and most powerful muscles in carinate birds are the **extrinsic** (originating *outside* the part on which they act) muscles of the wings (Fig. 8.46). These pectoral, or breast, muscles (pectoralis and supracoracoideus) are attached primarily to the keeled sternum as well as to the coracoid and furcula. The avian pectoralis muscle is proportionately the most massive paired muscle found in any tetrapod; it can occupy as much as 25 percent of a bird's total body mass (Greenwalt, 1962). Contraction of the pectoralis muscles pulls the wing downward and forward to provide lift (Fig. 8.47). Elevation of the wings is brought about primarily by contraction of the supracoracoideus, which also originates on the ventral side of the sternum but whose tendon passes through the foramen triosseum (an opening formed by the clavicles, coracoids, and scapulae) to insert on the humerus. This pulleylike action pulls the humerus upward. This unusual arrangement permits the primary abductor and adductor muscles of the wings to originate and insert on the same bones. The **intrinsic** (originating *on* the part on which they act, i.e., the wing) musculature of the wings is reduced. On the other hand, the musculature of the hindlimbs is well developed. Specialized muscles are necessary in the legs and feet of raptors, passerines, and aquatic species to allow for the diverse uses of their legs. Perching birds, for example, have long tendons that extend from muscles on the proximal part of the leg and insert on the digits on the distal end. This arrangement has a double purpose. The bird's energy expenditure is reduced because the muscles do not have to shorten as much as they would if they were extrinsic muscles. In addition, whenever the bird is perching, bending of the leg causes the tendons to tighten, effectively "locking" the toes around the perch.

Among the most important features of the thermal physiology of flight are the greatly increased heat production generated by the flight muscles and the increased oxygen uptake. For example, the metabolic rate of Costa's hummingbird (*Calypte costae*) during hovering flight was 7 times the standard rate (Lasiewski, 1963). The flight metabolism of rock doves (*Columba livia*) exceeded that during rest by a factor of 8.2 (LeFebvre, 1964). Similar increments in oxygen uptake

FIGURE 8.47

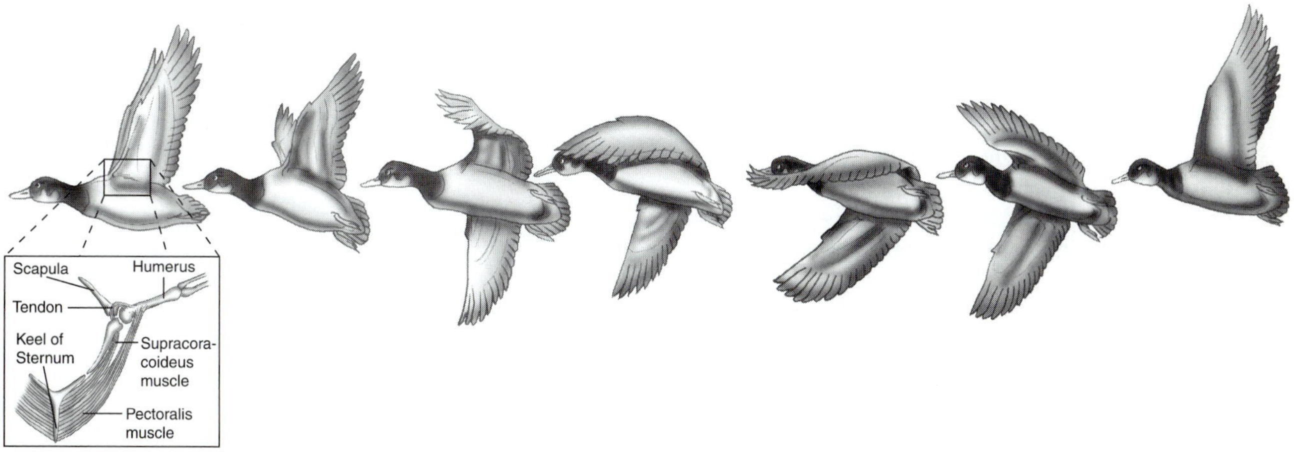

In strong fliers like ducks, normal flapping flight is accomplished by the fully extended wings sweeping downward and forward. The primary feathers at the wing tips provide thrust. The upbeat occurs by bending the wing and bringing it upward and backward. The wing is then extended and ready for the next downbeat. The detail shows the frontal view of the pectoral girdle and attached wing muscles. Flight muscles are arranged to keep the center of gravity low in the body. Thus, both major flight muscles are anchored on the keel of the sternum. Contraction of the pectoralis muscles pulls the wing downward and forward to provide lift. The upstroke is due to the contraction of the supracoracoideus muscle, which is also attached to the sternal keel. The tendon of the supracoracoideus passes through a foramen to insert on the humerus. This pulleylike system pulls the humerus upward. In chickens and most other birds that fly weakly or not at all, the flight muscles are white and produce ATP anaerobically. In good fliers, the muscles, which contain numerous mitochondria, are red and aerobic.

during flight have been recorded in budgerigars (*Melopsittacus undulatus*) and in the laughing gull (*Larus atricilla*) by Tucker (1966, 1972). For very short periods of flight, the oxygen uptake of the evening grosbeak (*Hesperiphora vespertina*), ring-billed gull (*Larus delawarensis*), and black duck (*Anas rubripes*) were 12–13 times the resting values (Berger et al., 1970). Metabolic rates vary with the flight speed. Some of the heat produced during flight is stored in the bird; elevations of body temperature of 1.5–2.6°C have been described after flight (Dawson and Hudson, 1970). Most of the excess heat is lost by evaporative cooling (Tucker, 1968).

BIO-NOTE 8.7

Hummingbird Flight

Hummingbirds typically hover to feed more than a hundred times a day, consuming about 20 percent of their daylight hours with this expensive aerobic activity. Hover-feeding usually lasts less than a minute. Hummingbird heart rates are about 500 beats per minute at rest and may increase to about 1,300 per minute during hovering flight. Wingbeat frequencies up to 80–100 per second have been recorded. Unlike most other birds, lift during hovering is generated in both up- and downstrokes in the wingbeat cycle. The main flight muscles—the pectoralis and the supracoracoideus—are highly developed and constitute about 30 percent of the body mass. Some hummingbirds, such as the ruby-throated hummingbird (*Archilochus colubris*), fly nonstop for several hundred kilometers across the Gulf of Mexico during their annual migration, a trip that may require about 20 hours to complete.

Suarez, 1992

Branchiomeric muscles serve much the same function as they do in reptiles. They operate the jaws, the hyoid skeleton and its derivatives, as well as the pharynx and the larynx.

Both extrinsic and intrinsic integumentary muscles are present in birds. Extrinsic muscles permit independent movement of the skin. Intrinsic muscles are mostly smooth muscles that attach to the feather follicles (arrectores plumarum). These muscles are used to ruffle the feathers for insulation, for display, or as an emotional response to danger.

Cardiovascular System

The crocodilian heart possesses a complete interventricular septum oriented in a vertical position (Fig. 8.10c). Not only are the ventricles totally separated from each other, but the right atrium opens into the right ventricle, and the left atrium opens into the left ventricle. The pulmonary trunk and one systemic trunk leave the right ventricle and one systemic trunk leaves the left ventricle.

Both crocodilian and avian erythrocytes and thrombocytes are nucleated and usually oval in shape. All five types of leucocytes (neutrophils, eosinophils, basophils, monocytes, and lymphocytes) and thrombocytes are also present. Birds have two atria and two ventricles that are completely separated (Fig. 8.48). The sinus venosus is reduced and largely incorporated into the wall of the right atrium during embryonic development. It always is retained in adult birds, however, as a discrete structure delineated externally from the right atrium by a sulcus (groove) and internally by the retention of its valves (Feduccia and McCrady, 1991). Deoxygenated blood, carried by the two precaval veins, postcaval vein, and coronary veins, empties into the right atrium through the sinus venosus. It then flows into the right ventricle and is pumped to the lungs via the pulmonary trunk and the pulmonary arteries. Oxygenated blood returns to the left atrium via the pulmonary veins and then passes to the left ventricle, which pumps the blood through the right aortic arch (the only half of the fourth aortic arch remaining in birds) and then through the body. This makes possible a double circulation, with completely separated pulmonary and systemic streams. The backflow of blood is prevented by four valves: the tricuspid (between right atrium and right ventricle), the pulmonary semilunar (at entrance to pulmonary trunk), the bicuspid (between left atrium and left ventricle), and the aortic semilunar (at entrance to aorta).

The size of the heart (Table 8.3) and the rate of the heartbeat vary with the size and lifestyle of the bird: larger birds generally have relatively smaller and slower-beating hearts than smaller birds. The heart rate of a turkey, for example, may be less than 100 beats per minute, a hen

FIGURE 8.48

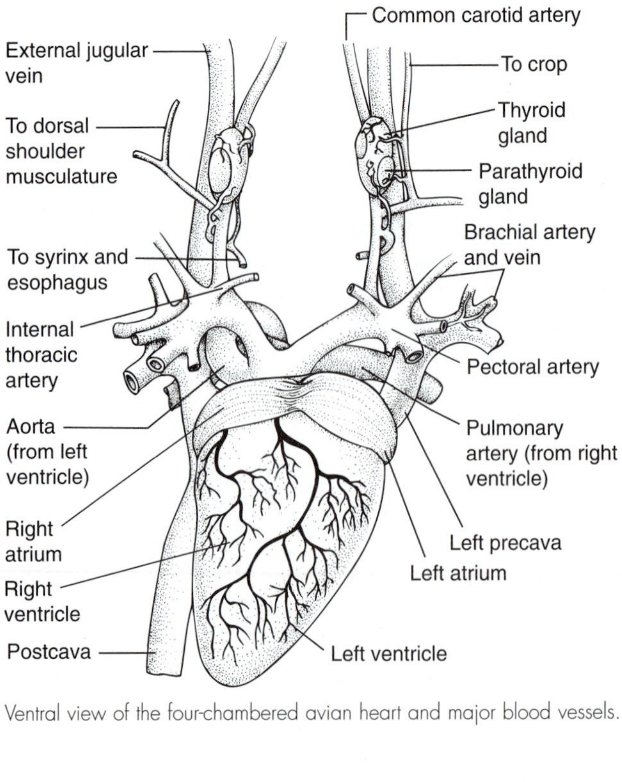

Ventral view of the four-chambered avian heart and major blood vessels.

TABLE 8.3

Heart Size and Body Weight

Species	Body Weight (g)	Heart Weight/ Body Weight (%)
Goose	4,405	0.8
Duck	1,685	0.74
Chicken	3,120	0.44
Ptarmigan	258	1.05
Hummingbird	—	2.4
Coturnix (quail)	119	0.90
Pigeon	458	1.02

From P. D. Sturkie "Heart & Circulation: Anatomy, Hemodynamics, Blood Pressure, Blood Flow & Body Fluids" in Avian Physiology. *Copyright © 1976 Springer-Verlag, NY.*

FIGURE 8.49

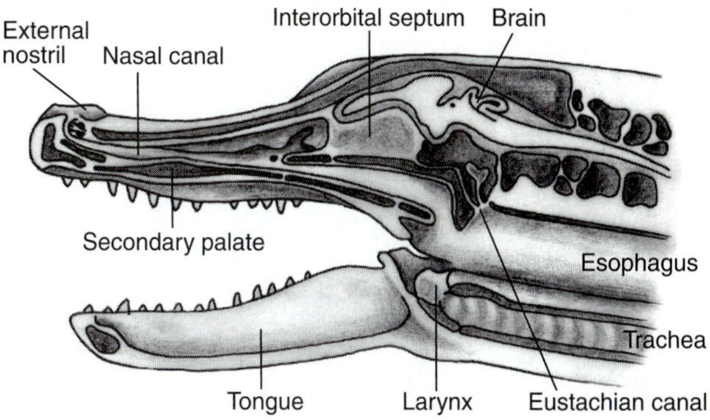

Midsagittal section of the head of the alligator, *Alligator mississippiensis*, showing the right nasal canal running above the bony palate and opening far back into the pharynx opposite the glottis.

chicken about 300 beats per minute, and a sparrow nearly 500 beats per minute (Young, 1962).

A pulmonary trunk emerges from the right ventricle and leads to what in early vertebrates was originally the sixth aortic arch. A single aortic trunk emerges from the left ventricle and leads to derivatives of the third and right fourth aortic arches. The left fourth aortic arch is absent in birds. The unpaired dorsal aorta gives off branches to the body wall, the viscera, and the appendages before continuing into the tail as the caudal artery.

The venous channels of birds are much like those of crocodilians. Internal jugular veins and subclavian veins join to form precaval veins anterior to the heart. All blood from the hindlimbs now bypasses the kidneys and goes directly into the postcaval vein, which is the chief vessel draining the trunk. The renal portal system is reduced to draining the tail. Postcardinal veins are absent.

Heat exchange in the legs is important, especially for birds that stand or swim in cold water. Blood flowing to these thin-skinned peripheral surfaces passes through a **heat exchanger** in order to reduce heat loss. Heat energy from the arterial blood warms the cold venous blood returning to the heart. Arteriovenous ***retia mirabilia*** are located near the junctions of the legs with the body. In these *retia*, small arterioles going to the legs run alongside small veins in which blood is flowing in the opposite direction, and heat energy is exchanged between them. After leaving the *retia*, arterial blood is at approximately the same low temperature as the legs. In this manner, the legs lose much less heat energy than they otherwise would, and heat energy is retained by the body core in the blood of the veins. Through this heat exchanging mechanism, energy loss is minimal. Because blood flows in opposite directions in the two opposing streams, this type of heat exchanger is known as a **countercurrent heat exchanger**.

Respiratory System

The nostrils of crocodilians are located at the tip of the snout and are closed by valves during diving. Due to the presence of a complete secondary palate, the internal nares are located at the rear of the oral cavity (Fig. 8.49). The internal nares can be closed off from the throat by fleshy folds on the back of the tongue and palate so that crocodilians can continue to breathe while holding prey in their mouth. The trachea is longer than the neck and convoluted. The lungs, each of which is located in a separate pleural cavity, consist of numerous large chambers, each composed of many individual subchambers (alveoli). Internal partitioning is well developed in crocodilians, with pockets of trapped air causing the lungs to be spongy.

The back-and-forth movements of the liver in crocodilians act as a piston on the lungs (Fig. 8.50). Inhalation occurs when the rib cage expands and the liver is pulled back; exhalation takes place when the rib cage relaxes and the forward-moving liver compresses the lungs.

Crocodilians are the most vocal of reptiles, using a wide variety of vocalizations for both close-range and long-range communication. For example, adults bellow during the mating season. Young crocodilians emit sounds from within the nest that attract the mother, who assists the newborn in exiting the eggs and the nest. Even after hatching, the young produce "alarm cries" that summon the adults.

In birds, air enters through the external nares, flows through nasal canals above the secondary palate, and exits through internal nares located near the rear of the oral cavity. Air then flows through the glottis and into the trachea. The glottis is surrounded by a series of cartilages that form the larynx. The larynx, however, is not a sound-producing organ; it serves to modulate tones that originate in the **syrinx**, the special voice box unique to birds (see Fig. 8.52). It is an enlargement and modification of the lower end of the trachea. The syrinx contains a pair of semilunar membranes with muscles that alter the pitch of the sound. Contraction of the muscles as air is expelled from the lungs alters the membranes and produces the songs and calls characteristic

FIGURE 8.50

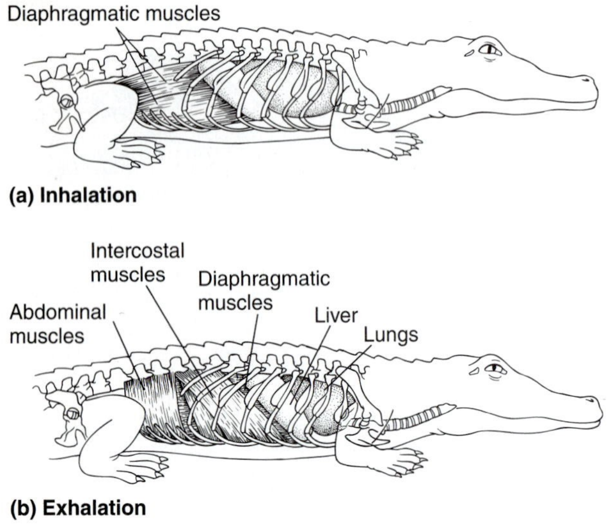

Gas exchange in the crocodile. Crocodiles use their rib cage as well as pistonlike back-and-forth movements of the liver to move air into and out of the lungs. (*a*) During inhalation, the rib cage expands and the liver is pulled back, enabling the crocodile to inhale fresh air into its lungs. (*b*) During exhalation, the lungs are compressed by the rib cage and the forward-moving liver, allowing the crocodile to exhale spent air.

FIGURE 8.51

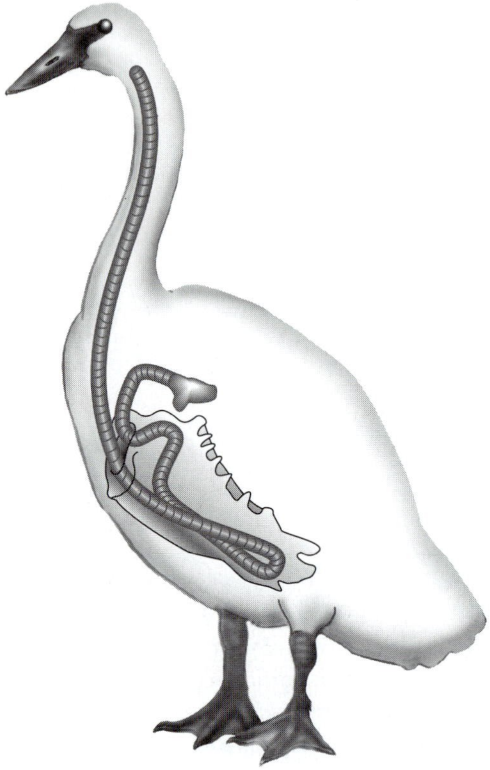

The long, coiled trachea of the trumpeter swan (*Olor buccinator*).

of different species. A series of cartilaginous or bony rings, most of which are complete, support the trachea. The trachea bifurcates (splits) to form two bronchi that then lead to the lungs. The syrinx is located at the bronchial bifurcation.

In some birds, the trachea is essentially straight, but in at least 60 species—including the trumpeter swan (Fig. 8.51)—it consists of coils and loops instead of taking a direct route between the throat and the lungs. This trait is found in six avian orders and has probably evolved several times (Whitfield, 1999). In some species, only males possess elongated tracheae, whereas in others, both sexes do. The elongated trachea serves as a resonator, producing a deeper, more baritone sound.

Birds have no muscular diaphragm as exists in mammals, and the thoracic and abdominal cavities are functionally continuous. The lungs are located in **pleural cavities**. An oblique septum extends caudally from the pericardium to the lateral body wall as a mostly membranous partition that separates the pleural cavities from the rest of the coelom. This membranous "diaphragm" participates in the operation of the suction pump by which the lungs are inflated, though it functions differently from the muscular mammalian diaphragm. Contraction of the external intercostal muscles causes the ribs to move outward and the sternum to move downward. This tightens the septum, which lowers the pressure around and within the lungs to below atmospheric pressure. Air then can be drawn into the air sacs after traversing the lungs on its way. During expiration, air is forced through the lungs on its way back to the trachea.

Bird lungs are highly modified: 7 to 12 diverticula (air sacs) of the lungs invade many parts of the body (Fig. 8.52a, b). These air sacs are thin-walled, distensible diverticula derived from the primary or secondary bronchi. They lie between layers of pectoral muscle, project among the viscera, and even penetrate the marrow cavity of some bones. They may occupy as much as 80 percent of the total body cavity. Anatomically and functionally, air sacs form two groups: a posterior, or caudal, group that includes the large abdominal sacs; and an anterior, or cranial, group that consists of several somewhat smaller sacs. Most air sacs are paired, but some may be united across the midline. Paired air sacs usually occur near the base of the neck (cervical), near the heart (anterior thoracic), within the oblique septum (posterior thoracic), and among the viscera (abdominal). An air sac that may or may not be united across the midline is usually located dorsal to the furcula (interclavicular). Some birds also possess air sacs between the layers of pectoral muscle (axillary). Diverticula of these air sacs may invade almost any part of the skeleton, most frequently the humerus, femur, and vertebrae.

The lung volume of a typical bird is only about half that of a mammal of the same size, but with the air sacs, the total potential volume of the respiratory system is some three times as great as that of a mammal. Air sacs are not designed to function directly in gas exchange. They contain no alveoli or

FIGURE 8.52

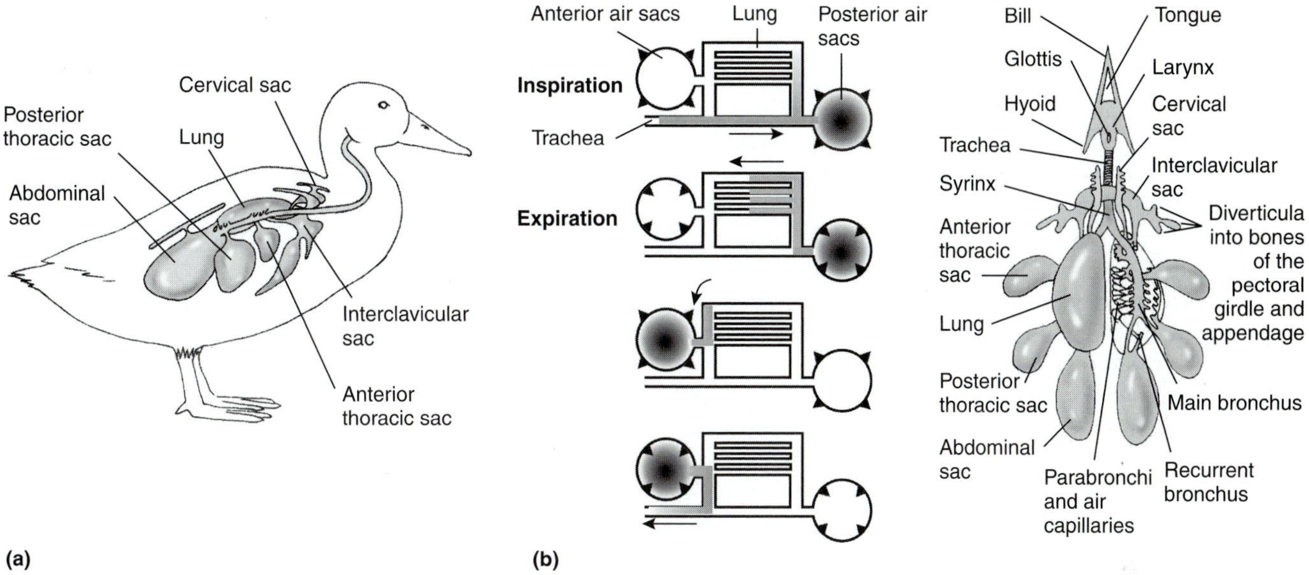

The ventilatory system in a mallard duck. (*a*) The vocal organ is the syrinx, which is located at the point where the two bronchi join to form the trachea. (*b*) Movement of a single volume of air through the bird's respiratory system. One of the most distinctive features of the avian ventilation system is the presence of numerous air sacs that are connected to the bronchial passages and the lungs. Major air sacs are shown in the diagram; additional air sacs that enter into the bones are not shown. Most of the inhaled air goes directly into the posterior sacs. As the ventilatory cycle continues, air from the posterior sacs passes through the lungs and into the anterior sacs. This mechanism provides a continuous flow of air through the lungs, in contrast to all other tetrapods, in which the air is moved into and out of only the lungs, which serve as blind sacs. Two full respiratory cycles are required to move the air through the system.

respiratory epithelia with any network of capillaries necessary for gas exchange. However, they do play an important accessory role in ventilation. During inspiration, incoming air flows through the bronchioles and mainly into the posterior air sacs (Fig. 8.52b). Some air also goes into the lungs. The air that first reaches the posterior sacs is air that was left in the trachea after the previous inhalation, so that it contains more carbon dioxide than does fresh air. On exhalation, the sacs decrease in size, and air from the posterior sacs flows into the lungs, which contain parabronchi (tubes) rather than alveolar sacs. On the next inhalation, the mass of air moves from the lungs into the anterior sacs. On the following exhalation, it is discharged from the air sacs into the trachea and, finally, to the outside. The bellows action of the air sacs provides a continuous, unidirectional flow through the bird's lungs.

The duct system within the bird lung is unique. All of the air ducts are open-ended and interconnecting (parabronchi), and none terminate blindly within the lung. This makes possible a steady flow of oxygen-rich air, permitting cross-current gaseous exchange and increased efficiency. The bellows action of the air sacs causes the air to be constantly and completely replaced. Bird lungs, therefore, contain only fresh air, in contrast to the lungs of other vertebrates, in which there is always residual unexpired air partly depleted of its oxygen content.

Contraction of the intercostal and abdominal muscles maintains an adequate flow of air over the respiratory epithelia when a bird is at rest. During flight, however, alternate contraction and relaxation of pectoral muscles provide additional force for the bellows action. Inspiration is coordinated with the upstroke of the wings and expiration with downstroke. A more rapid respiratory rate is required during flight because of an increased oxygen demand. In addition, in black-billed magpies (*Pica pica*), the respiratory pattern changes as the wingbeat pattern changes from high-amplitude rapid flaps to low-amplitude slower flaps with interspersed brief glides (Boggs et al., 1997). Breath cycle times were longer when associated with low-amplitude, longer duration wingbeats and shorter when associated with high-amplitude, shorter wingbeats. In the absence of sweat glands, an increase in the respiratory rate effectively increases heat loss. Thus, air sacs also aid in thermoregulation by helping dissipate excess internal heat energy produced by the muscular action of flight activity.

Birds, as well as small mammals, normally exhale air that is well below body temperature. Murrish and Schmidt-Nielsen (1970) noted:

> *The cooling of the exhaled air is due to heat exchange in the nasal passageways; it is merely the reverse process of the heating of air that takes place during inhalation. During inhalation the air is heated and humidified and the walls of the passageways are thus cooled by convection and evaporation; on exhalation the returning warm air gives up heat to these cool surfaces and water vapor condenses.*

Dehydration is probably the most important physiological constraint for migrating birds. Studies of blackcaps (*Sylvia atricapilla*) migrating across the Sahara Desert suggest they "go to ground" during the day to minimize evaporative water loss at high altitudes (Izhaki and Maitav, 1998). Although a majority of the birds migrating in autumn had accumulated sufficient fat to enable them to traverse the Sahara in a single flight, they would probably lose at least 12 percent of their initial mass through dehydration by the time they reached their southern destination. Therefore, most birds probably use intermittent migration with stopovers at sites where food and water is available. By resting during the day when evaporative water loss is highest and flying at night, the majority of small migrants can successfully navigate the (at minimum) 1,200-km Sahara crossing.

In the humid tropics, evaporative cooling is relatively ineffective in small birds such as variable seedeaters (*Sporophila aurita*). They employ an elevated body temperature (hyperthermia) to cope with heat (Weathers, 1997). They can survive body temperatures between 46.8 and 47.0°C, among the highest recorded for birds. Tolerance of hyperthermia is advantageous for this species because it allows them to maintain an unusually high body temperature to ambient air temperature gradient in hot environments, and thus they are able to dissipate heat passively. The main response of these birds to heat stress in the field is to avoid it behaviorally rather than to overcome it physiologically.

In a study of oxygen consumption and metabolic rate during locomotion in a large, flightless bird, a rhea (*Rhea americana*) was trained to run on an inclined treadmill. The upper limit to aerobic metabolism was 36 times as great as the minimum resting rate, a factorial increase exceeding that reported for nearly all mammals (Bundle et al., 1999) (Fig. 8.53). The rhea's factorial increase in anaerobic metabolism was 1.7 times as great as the highest previously reported in any bird. Because rheas and other large flightless birds presumably lost the ability to fly because of a lack of predation and because they lead relatively inactive lives, they have little need to generate such great aerobic power. The evolutionary significance of this ability remains unanswered.

FIGURE 8.53

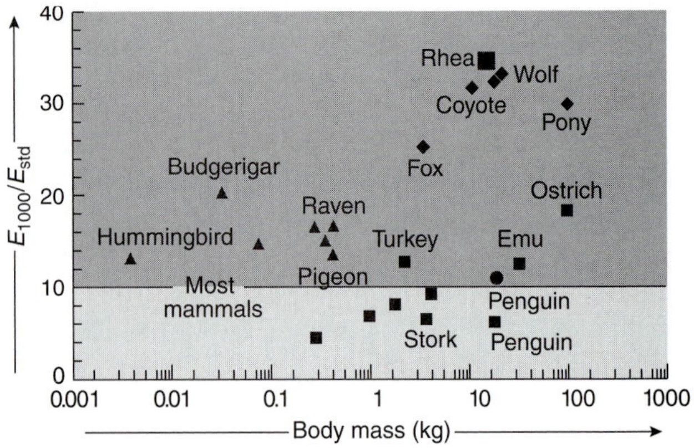

Factorial increases in metabolic rates above resting minimum for rheas (*Rhea americana*) and athletic mammals at the aerobic maximum and for other avian species at the highest rates available. Key: triangles, flying birds; squares, walking and running birds; circles, swimming birds; and diamonds, running mammals.

BIO-NOTE 8.8

The Deep Dives of Penguins

Penguins have solid bones that reduce buoyancy; tightly packed, scalelike feathers; and narrow, flattened wings for underwater "flight." King and emperor penguins are amazing divers—they can dive to depths of 304 and 534 m for as long as 7.5 and 15.8 minutes, respectively. The abdominal temperature of king penguins may fall to as low as 11°C during sustained deep diving. The slower metabolism of cooler tissues resulting from physiological adjustments associated with diving could partially explain why penguins and possibly marine mammals can dive for such long duration. Such dives are longer than their oxygen stores should allow. Research is now in progress to determine how such dives are accomplished.

Ancel et al., 1992
Handrich et al., 1997
Harding, 1993
Kooyman et al., 1992
Kooyman and Kooyman, 1995

Digestive System

Crocodilians possess homodont dentition with the teeth rooted in sockets (thecodont dentition) and confined to the jaws. Teeth are continually replaced during the animal's lifetime (polyphyodont dentition). A nonprotrusible tongue is present that assists in moving food within the oral cavity. The crocodilian stomach (see Fig. 8.20c) is similar to those of grain-eating birds, with the posterior portion being modified into a muscular gizzardlike compartment that grinds food against ingested small stones that have been deliberately swallowed. Gastric juices are added in the thin-walled glandular region of the stomach that lies in front of the gizzard. The small intestine consists of duodenum, jejunum, and ileum regions. A pancreas, located near the junction of the stomach and duodenum, secretes pancreatic juice into the duodenum in order to facilitate digestion and to neutralize the acidic chyme from the stomach. A cloaca leads to the vent. Whereas the vent is a transverse slit in turtles and lepidosaurs, it is oriented longitudinally in crocodilians.

All living birds lack teeth. The horny "egg tooth" developed by young birds (when within the egg) is a temporary, epidermal, toothlike scale located dorsally (externally) on the upper jaw. It functions in helping to crack the egg shell for hatching and is either reabsorbed or shed after hatching.

The tongue of birds is nonmuscular and exhibits great diversity in form and structure. Most bird tongues are short, narrow, and triangular with few taste buds. Woodpeckers use

FIGURE 8.54

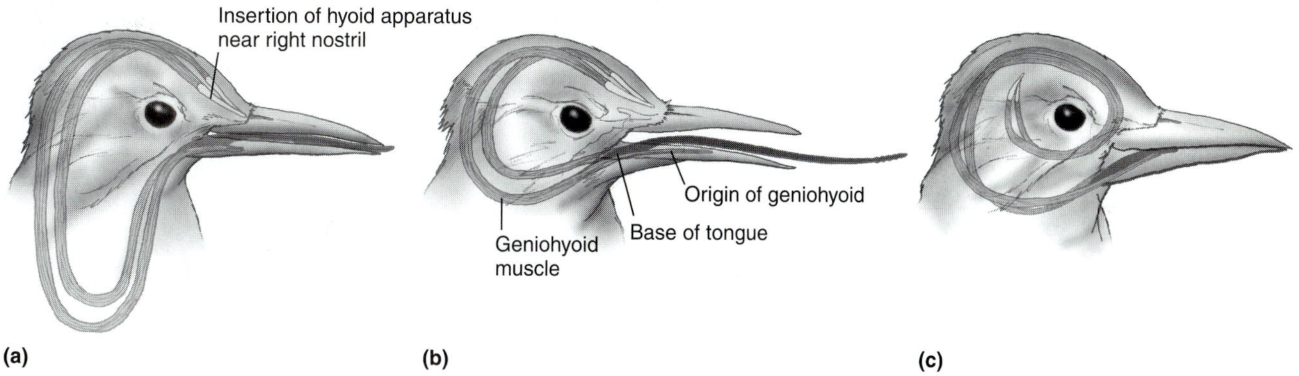

Woodpeckers have an exceptionally large and unusual hyoid apparatus to support their long tongues, which are used to probe into deep crevices of trees in search of insects. To allow the long tongue to be extended and retracted from the mouth, the ceratobranchial and epibranchial horns of the hyoid apparatus wrap completely around the back of the skull and attach near the bones at the base of the upper mandible. (*a*) When the tongue is not extended, retractor muscles (crosshatched) of the hyoid horns insert far forward. (*b*) When the retractor muscle contracts and draws the hyoid horns forward, the loop of the tongue is raised and the tongue is shot far beyond the tip of the bill. (*c*) When fully contracted, the hyoid horns end at or in front of the eyes in the cavity of the upper mandible, or (as seen here) are even wound around the eye.
Source: Proctor and Lynch, Manual of Ornithology *1993, Yale University Press.*

their long specialized tongue (Fig. 8.54) as a probe to obtain insects between cracks in tree bark or in holes they excavate. When retracted, the woodpecker's tongue actually wraps around the back of the head.

A complex system of tactile sensors is present on the tongue, so that birds can make fine distinctions in food items based on the feel of the food to the tongue and hard palate. Well-developed salivary glands are present in most birds, but are reduced or absent in most aquatic birds (ducks are the only prominent exception) (Worden, 1964). Both unicellular and multicellular oral glands are more numerous than in amphibians and reptiles, and these secrete mostly mucus. The salivary secretions of swifts (Apodidae) and crested swifts (Hemiprocnidae) are used in nest construction, and in one species of cave swiftlet (*Collocalia inexpectata*), the entire nest is composed of saliva. These salivary secretions are the source of "bird's nest soup," which is a million dollar a year industry in Indonesia.

The esophagus conducts food from the pharynx to the stomach (see Fig. 8.20d). A portion of the lower part of the esophagus in some birds is modified to form a membranous diverticulum known as a **crop**. A crop occurs primarily in grain-eating birds and is used for the temporary storage of seeds. This permits small birds to feed intensely in short bouts on the ground and then retreat to the safety of a branch for leisurely digestion. In some birds, enzymes may be secreted in the crop for preliminary digestion. In predatory birds, the lower portion of the esophagus may be dilated to form a temporary crop that is used to store partially digested food until it can be regurgitated to nestling young. In some species, the crop is greatly expanded.

The South American hoatzin (*Opisthocomus hoazin*) possesses an unusually large and muscular crop (Grajal et al., 1989) (Fig. 8.55). Leaves constitute over 80 percent of its diet, and it is the only bird known to digest food in the same way as cows, sheep, and other ruminants—by using bacteria to break down the fibrous plant material in a special chamber before the plant material reaches the stomach. Ruminants have a prestomach compartment known as a rumen (see Chapter 9), whereas the crop and lower esophagus of

FIGURE 8.55

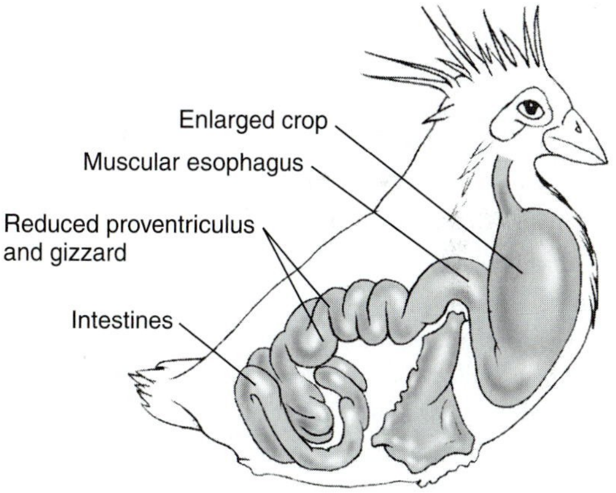

The hoatzin (*Opisthocomus hoazin*) is a true ruminant, feeding almost entirely on leafy foliage and fermenting the ingested leaves with a stomach that functions much like that of a cow. Like cows, hoatzins are foregut fermenters. A series of chambers near the anterior end of the intestinal tract house bacteria that break down the leaves and their indigestible cellulose fibers into simpler digestible carbohydrates. Two such fermentation chambers are located within the hoatzin's large crop. Further fermentation occurs as the fermented mash of leaves passes through an enlarged lower esophagus area. The crop and esophagus are both quite muscular in the hoatzin, contracting to grind the leaf mash much as a cow "chews its cud" to aid fermentation. In the hoatzin, the proventriculus and gizzard areas of the true stomach are relatively small and of secondary importance in digestion.
Source: Proctor and Lynch, Manual of Ornithology *1993, Yale University Press.*

hoatzins, both of which are quite muscular, are used to extract volatile fatty acids from the cellulose of tender young plants and to break down indigestible cellulose fibers into simpler digestible carbohydrates before they reach the small intestine. The crop and esophagus contract to grind the leaf mash much as a cow "chews its cud" to assist in fermentation; therefore, the proventriculus and gizzard are small.

The crop is modified in some birds (pigeons, doves, emperor penguins, and flamingos) to produce "pigeon's milk," the initial food fed to nestlings. In these species, the hormone prolactin stimulates glandular cells lining the crop in both parents to release fatty secretions. This substance then is regurgitated, mixed with partially digested food, and fed to nestlings. The milk is thick and curdlike and consists of 65 to 81% water, 13 to 19% protein, 7 to 13% fat, 1 to 2% mineral matter, and vitamins A, B, and B_2 (Skutch, 1991). The proteins contain a large variety of amino acids, whereas the mineral component is largely sodium, with a small amount of calcium and phosphorus. There are no sugars or other carbohydrates in crop milk. Initially, nestlings receive only crop milk, but later, the milk is mixed with increasing amounts of other foods. The prolactin that stimulates secretion of "pigeon's milk" is the same hormone that causes lactation in mammals.

The stomach is well differentiated from the esophagus. In many birds, especially those that feed on seeds and grain, the stomach is divided into two regions: a glandular **proventriculus** that secretes gastric juice, and a thick, muscular **gizzard** that often contains grit and pebbles (see Fig. 8.20d). The gizzard serves a grinding function in the absence of teeth.

Piersma and Gill (1998) examined bar-tailed godwits (*Limosa lapponica baueri*) that died after colliding with a radar dome in Alaska just after taking off on a trans-Pacific flight of 11,000 km. The birds had relative fat loads (subcutaneous and intraperitoneal) making up approximately 55 percent of their fresh body mass, among the highest ever recorded in birds. These birds also had very small gizzards, livers, kidneys, and guts as compared with northbound godwits from New Zealand. The authors suggested that upon departure, long-distance migrants "dispense" with parts of their body not directly necessary during flight and rebuild these organs after arriving at their destination.

The long, convoluted small intestine consists of a short duodenum and a long, coiled ileum (see Fig. 8.20d). In some birds, the large intestine is divided into a colon and rectum. Most birds have paired ceca at the juncture of the small and large intestines that apparently have a major digestive function. These ceca are large in ostriches, gallinaceous birds, many ducks, and shorebirds and, in species such as the sage grouse (*Centrocercus urophasianus*), may be responsible for one-third to one-half of the absorption of glucose and certain amino acids (Obst, 1991). Ceca are vestigial and probably nonfunctional in herons, gulls, and passerines. They are absent in woodpeckers, hummingbirds, and other species. The absence or presence of ceca and their degree of development is taxonomically useful. The rectum is usually a short, straight tube, but in some species, it may be long and coiled. The rectum and cloaca empty via the vent.

BIO-NOTE 8.9

The Germination of Ingested Seeds

Seed germination may be affected by passage through a bird's digestive system. Studies conducted on white-eyed bulbuls (*Pycnonotus xanthopygos*) and Eurasian blackbirds (*Turdus merula*) revealed that, in most cases, seed ingestion had no consistent influence on germination, although certain seeds ingested by blackbirds showed substantially higher germination percentages than those ingested by bulbuls. Blackbirds retain seeds longer, which may cause greater seed coat abrasion and better germination.

On the Canary Islands, the fruit-eating lizard *Gallotia atlantica* is preyed on by the great gray shrike (*Lanius excubitor*). Seeds of the only fleshy fruited plant on the island (*Lycium intricatum*) (Solanaceae) in the lizard's digestive tract are ingested by the shrike when the lizard is consumed. The seeds in the shrike pellets had a higher germination rate (64%) than those from lizard droppings (50%) or directly from the plant (54%), showing that their experience in passing through two vectors had increased their potential for immediate germination. Where seed transfer involves increased mobility and range, predation can also play a biogeographically significant role in dispersal.

Barnea et al., 1991
Nogales et al., 1998

Nervous System

The cerebral hemispheres of the crocodilian brain are enlarged and partially cover the diencephalon (Fig. 8.56a, b). Although a pineal organ is present in turtles, lizards, and snakes, it is absent in crocodilians (Wurtman, 1968). Both optic lobes and auditory lobes are present. The cerebellum is prominent and well developed.

In birds, the olfactory portion of the brain has become greatly reduced in size, so that most birds, with the exception of some carrion eaters and some seabirds, have a poor sense of smell (Fig. 8.56c, d). Vomeronasal organs are absent. Cerebral hemispheres are better developed than in most reptiles and are much like those of crocodilians.

Optic lobes are greatly enlarged and well developed, and a pineal gland is present. The cerebellum is large and may overlie part of the medulla and the midbrain. It is highly developed in flying birds but is relatively small in some flightless species.

Adult canaries and zebra finches generate a fresh supply of brain cells to replace those lost with age in forebrain regions that control song learning and song production (Alvarez-Buylla, 1990). Newly formed cells are known as projection neurons; these link two related structures in the bird's cerebral song-control center. Thus, adult avian brains possess considerable potential for self-repair as neurons age and gradually deteriorate.

During a short sensitive period just after hatching, young birds of many precocial species form an attachment, called

FIGURE 8.56

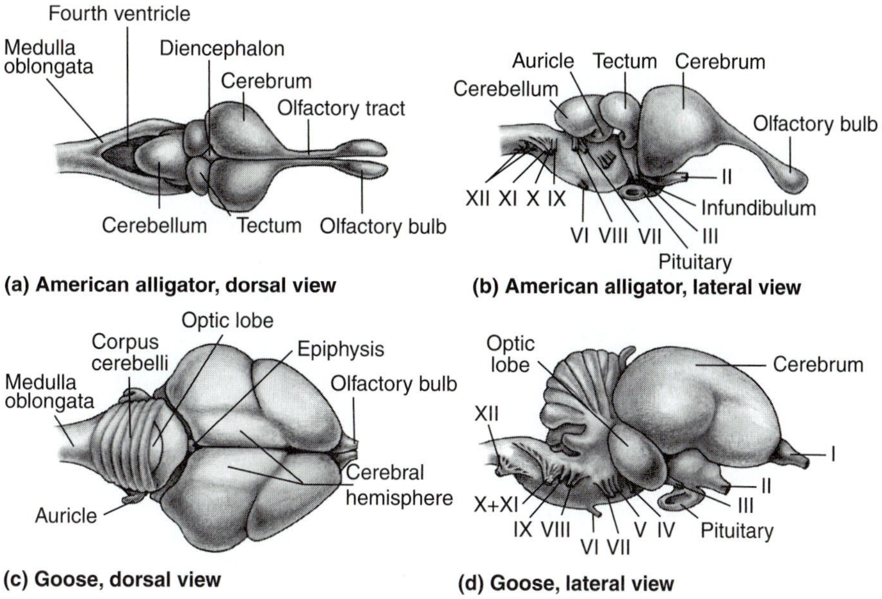

Dorsal (*a*) and lateral (*b*) views of the brain of an American alligator (*Alligator mississippiensis*). Dorsal (*c*) and lateral (*d*) views of the brain of a goose (*Anser*). Compare the size of the cerebrum with Fig. 6.22 (amphibian).

After Romer and Parsons, 1985 in Kenneth V. Kardong, Vertebrates: Comparative Anatomy, Function, Evolution, *2nd edition. Copyright ©1998 McGraw-Hill Company, Inc. All rights reserved. Reprinted by permission.*

imprinting, to the first moving object they see—normally their mother. As early as 1935, Konrad Lorenz found that young goslings separated from their mother shortly after hatching imprinted on other moving objects, including humans (Fig.8.57a). Early social experience plays a major role in survival and subsequent sexual preferences. Lorenz emphasized that, in some birds, the preference of an individual to mate with a member of its own species is subject to its imprinting shortly after hatching. Altricial birds imprint much later.

When young birds of a variety of species are foster reared through the time of fledging, they will subsequently direct their sexual preferences toward the foster species (Fig. 8.57b). Even large amounts of postfledging exposure to their own species may not reverse their preference for their foster species. Sexual imprinting has critical implications for captive breeding programs, including those involving the hand-rearing of young by human caretakers, and those involving foster rearing of chicks by parents of closely related species. For example, extreme care is taken so that young California condors (*Gymnogyps californianus*) being raised in captivity do not imprint on humans. They are provided food by means of a mitt in the shape of the head of an adult California condor. Caretakers of young whooping cranes (*Grus americana*) drape themselves in loose-fitting sheetlike material to prevent young birds from imprinting on them.

Cranial nerves have been discussed in Chapter 5. Some researchers believe that birds possess only 11 cranial nerves and that the "accessory nerve" is only a caudal extension of the vagus (X) nerve.

In some birds, the spinal cord is surrounded by two meninges: a dura mater and leptomeninx. In a few birds, the leptomeninx differentiates into a weblike **arachnoid layer** and an inner **pia mater**. Because the embryonic vertebral column elongates more rapidly than the spinal cord, the cord is somewhat shorter than the backbone. All birds have enlargements of the spinal cord, with the cervical enlargement being most pronounced in carinate birds due to the great development of muscles and nerves needed to operate the wings.

Sense Organs

Eyes

Birds have the best vision and are the most visually dependent of all vertebrates. Many adaptive responses to their environment and their survival depend on precise visual discrimination. Such behaviors as foraging (including the striking of prey with talons), avoiding branches when flying through thickets, landing on branches, defense of a territory and the nest, the selection of mates, orientation, homing, flying in formation in flocks, and navigation depend on a well-developed and highly sensitive visual system.

The eyes of most birds are asymmetrical rather than being spherical as is the mammalian eye (see Fig. 1.17, page 14). The flattened back of a bird's eye forms a much more even "screen" on which the entire image is in focus,

FIGURE 8.57

Imprinting in young birds. During a critical period in a young bird's life, an object becomes fixed (imprinted) in the bird's brain for life. The critical period for imprinting is most likely to occur 13 to 16 hours after birth. (*a*) Imprinted goslings following Konrad Lorenz on land. (*b*) An imprinted rooster wading out to meet a mallard duck to which it was exposed during its critical imprinting period. As the rooster matured and its sexual behavior developed, it sought out ducks and not birds of its own kind.

not blurring toward the periphery. The exact shape of the eye varies with the species and is characterized by a change in shape of the cornea. This resultant loss of strength in the cornea is compensated for by a ring of bony plates, the sclerotic ossicles (see Fig. 8.43).

The eyes of a bird are so large that they may equal or exceed the weight of the brain. Ostrich's eyes are among the largest of all land vertebrates, with an axial length of 50 mm as compared with 24 mm for the human eye. They are the most widely separated eyes of all birds; therefore, they have a surprisingly small binocular field (Martin, 1993; Martin and Katzir, 1995). The binocular retinal visual field spans only 20° as compared with 140° in humans.

The eyes of most birds have very limited mobility. The extrinsic eyeball muscles that rotate the eyeball in its socket are reduced in size and are poorly developed. The lack of eyeball movement in birds is compensated for by an increase in neck mobility, so that a bird must move the entire head to change the direction of its gaze. While foraging for food, birds use visual control of their bill position to make highly accurate pecks. The eyes of owls are immovably fixed in their orbits; however, some owls are able to rotate their heads through an arc of almost 270°.

The position of the eyes in the head differs among species. In some, such as owls and birds of prey, the eyes are positioned toward the front of the head. In others, including many ground-feeding birds such as doves, the eyes are positioned on the side of the head. These two different designs serve different functions. Birds that hunt fast-moving prey must track their prey during the chase. To be successful, they need accurate perceptions of speed and distance. These are achieved through binocular (or stereoscopic) vision—that is, both eyes are locked onto the object simultaneously. Owls have a binocular field of 60 to 70° and hawks up to 50° or more, whereas most granivorous birds have binocular fields of considerably less than 25° (Van Tyne and Berger, 1961; Zeigler and Bischof, 1993).

Birds with laterally positioned eyes have a very large field of view—more than 300° in pigeons—but gain this advantage at the cost of having a much reduced field of binocular vision (Hockey, 1997). A wide field of view is, however, of great value in detecting approaching danger.

In birds, unlike primitive vertebrates, the lens of the eye is elastic and resilient, and its shape is altered by ciliary muscles located at the base of suspensory ligaments. Birds can change their focal distance very rapidly. Accommodation occurs via adjustment of the lens and via changes in the curvature of the cornea. In addition, in order to catch their fish prey, species such as kingfishers, gannets, herons, and ospreys must compensate for the refraction of light at the air–water interface and visual distortion caused by ripples (Fig. 8.58) (Katzir, 1993, 1994). As the bird moves from air to water, refraction increases by some 20 diopters (Hockey, 1997). Humans wear glasses to correct for a deficiency of 1 diopter; glasses needed to correct for 20 diopters would probably require a neck brace. Rapid accommodation can be accomplished because bird lenses are much softer than those of mammals, and their shape can be changed more easily by the ciliary body and zonular fibers attaching to them. In addition, birds can also change the shape of their cornea by draw-

FIGURE 8.58

Pied kingfisher returning from a successful dive with a captured fish (from a 16-mm film taken at 64 frames/sec).

BIO-NOTE 8.10

Using Ultraviolet to Find Prey

Unlike humans, kestrels can see in the ultraviolet. They use this ability to detect voles, their primary prey. Voles communicate with one another by laying highways through grass and vegetation, which they mark with urine and feces. Vole waste products are strong absorbers of ultraviolet light, much stronger than the background absorption of vegetation. To a human's eye, the difference between a vole highway and any other patch of grass is not discernible; however, a kestrel sees every field as streaked with the marks of vole activity. Both laboratory and field experiments have shown that kestrels take a far greater interest in an environment daubed with ultraviolet-absorbing vole excrement than one without. This ability to see and use vole scent marks to assess vole numbers enables kestrels to "screen" large areas in a relatively short time in order to rapidly detect areas of vole abundance.

Viitala et al., 1995

ing the edges backward, thus increasing the curvature in the middle. Sivak (1986) noted that the avian eye generally has the most effective accommodative mechanism among terrestrial vertebrates. Lacrimal glands are present but are usually poorly developed.

Many birds, including hummingbirds, pigeons, zebra finches, starlings, and kestrels, can perceive ultraviolet and polarized light (Delius and Emmerton, 1979; Goldsmith, 1980; Hunt et al., 1997; Hockey, 1997). The pattern of light polarization in the sky changes during the day as the sun's position moves. Especially on overcast days, the ability to detect polarized light may aid migrating birds that use the sun as a compass. Those birds that have had their visual pigments characterized have been found to have four spectrally distinct classes of cones in their retinas and can distinguish colors ranging from ultraviolet (approximately 300 nm) to red (approximately 750 nm) (Bowmaker, 1986; Hunt et al., 1998). For comparison, humans have three distinctive cone types and can distinguish wavelengths between approximately 400 nm and 700 nm. Birds that appear to be monochromatic in the human visible spectrum may be dichromatic in the ultraviolet. This has significant implications for both intra- and interspecific analyses of sexually selected traits (Bennett and Cuthill, 1994; Bennett et al., 1996, 1997). For example, male blue tit (*Parus caeruleus*) crests tend to be brighter than females, particularly in the ultraviolet (Andersson et al., 1997; Hunt et al., 1998). All females tested chose the male with the brightest crest, an indication that ultraviolet plumage reflections are important in intersexual signaling. Electrophysiological studies of bird species from ten different families have revealed that these birds have cones that are maximally sensitive at 370 nm, well within the ultraviolet region (Chen et al., 1984).

The mammalian retina is richly supplied with blood vessels, but the space occupied by these vessels limits the number of rods and cones that can be accommodated. There is no blood supply to the avian retina; rather, nutrition is provided via a small, blood-rich structure known as the pecten. The pecten is anchored to a small area of the retina called the optic disk, the region where the photo-sensitive cells come together and leave the eye via the optic nerve en route to the brain. The optic disk contains no rods or cones, so that bird vision is not impaired. The central fovea acts as a small magnifying glass and projects a slightly enlarged image onto the retina. Some birds that pursue prey at high speed, such as hawks, swallows, and kingfishers, have two fovea on their retinas: a lateral fovea for vision to the side of the head and a temporal fovea that gives these birds better forward binocular vision (Proctor and Lynch, 1993).

The cones of birds contain oil droplets between their inner and outer segments (Robinson, 1994). Six types of avian cone droplets have been identified (Goldsmith et al., 1984). The colored droplets, which are absent from the rods, contain high concentrations of carotenoids dissolved in lipids. The oil may be brightly colored (ranging from deep red through orange to yellow) to transparent. The types of oil droplets vary considerably; some birds (such as pigeons), have up to 50 percent red and orange droplets, whereas other species have only about 10 percent. This may be related to their habits and activity patterns. Nocturnal species have only a few colored droplets, whereas birds active at twilight, such as swallows and swifts, have low percentages of red and yellow droplets. Penguins, which hunt underwater, also have but a few colored droplets (Bowmaker and Martin, 1985), whereas gulls and terns have high percentages of red droplets.

The oil droplets do not produce a color image *per se*, but they filter the light, in the same way as photographic filters do, before it is absorbed and detected by the photosensitive visual pigments. Thus, yellow droplets would remove much of the blue from the background by enhancing contrast between an object and a blue sky. Red droplets would remove much of the green, which would help birds searching for insects in a forest. Color vision can guide behavior in feeding and in a variety of social interactions such as sexual recognition, courtship, and mating. Oil droplets may also act as lenses focusing light onto the outer segment of the photoreceptors, thus enhancing the amount of light striking the visual pigments (Young and Martin, 1984).

Gondo and Ando (1995) examined the retinas of 30 avian species with different ecological habits and found that diurnal birds had red, yellow, orange, and pale green oil droplets, whereas nocturnal birds had only pale green oil droplets. Their findings suggested that each colored oil droplet has specific advantage for visual perception, and that each bird has the retina that best matches its natural environment and feeding behavior in each habitat.

Ear

A tympanic membrane is present in crocodilians, as it is in all reptiles. A special muscle allows crocodilians to close the ear cavity when submerging in water. The middle ear is well developed, with two Eustachian tubes joining just before connecting to the pharynx and opening into it by a single median orifice. The cochlear duct, which is straight in turtles, lizards, and snakes, forms a simple spiral in crocodilians.

Hearing is keen in birds and is next to sight in importance (Fig. 8.59). Pinnae (external ear flaps) are absent. The "ears" of owls are just tufts of feathers. The eardrum (tympanum), which is located at the inner end of the external auditory meatus, transmits vibrations to the columella, which in turn transmits vibrations across the middle ear cavity to the oval window in the lateral wall of the inner ear. An auditory (Eustachian) tube connects the middle ear cavity with the pharynx. The membranous labyrinth of the inner ear consists of a cochlea and a spiral organ, which convert the vibrations of the tympanum and columella into nerve impulses that are sent to the brain via the cochlear nerve, a branch of the auditory cranial nerve (VIII). The inner ear also houses the organs of balance and equilibrium (utriculus, sacculus, and semicircular canals).

Hearing is extremely well developed in owls, which use sounds coming from their prey to determine its location. Barn owls, for example, can locate prey in total darkness, using only their sense of hearing, with an error of less than 1° in both the vertical and the horizontal planes. For such accuracy, owls depend on frequencies of about 5,000 Hz, which provide better directional capabilities than lower frequencies (Payne, 1971). Many owls possess a facial ruff of

FIGURE 8.59

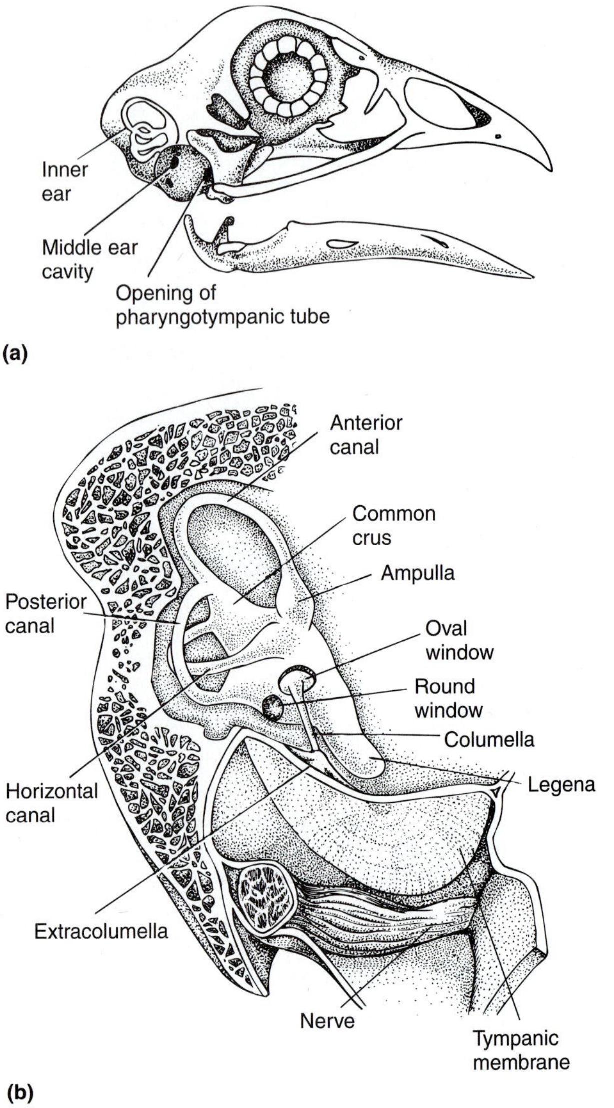

Hearing in birds. (*a*) Lateral view of a bird skull, showing the location of auditory and equilibrium organs. (*b*) The organs of equilibrium (vestibular apparatus) in a chicken.

stiffened feathers that acts as a parabolic sound reflector, focusing sounds on the external auditory meatus and amplifying them by 10 dB (Konishi, 1973) (Fig. 8.60). In addition, the right and left external ear openings as well as the external ear canals are asymmetrical, an adaptation that allows these birds to quickly locate sounds by tilting their head. This difference in size and shape allows the two ears to receive sound stimuli from many directions.

Several birds use **echolocation**, including oilbirds (*Steatornis*) of South America and cave swiftlets (*Collocalia*) of Southeast Asia, both of which live and nest in deep caves. Oilbirds use a sonar system much like that of bats, except their sounds have a frequency of only about 7,000 Hz—well

FIGURE 8.60

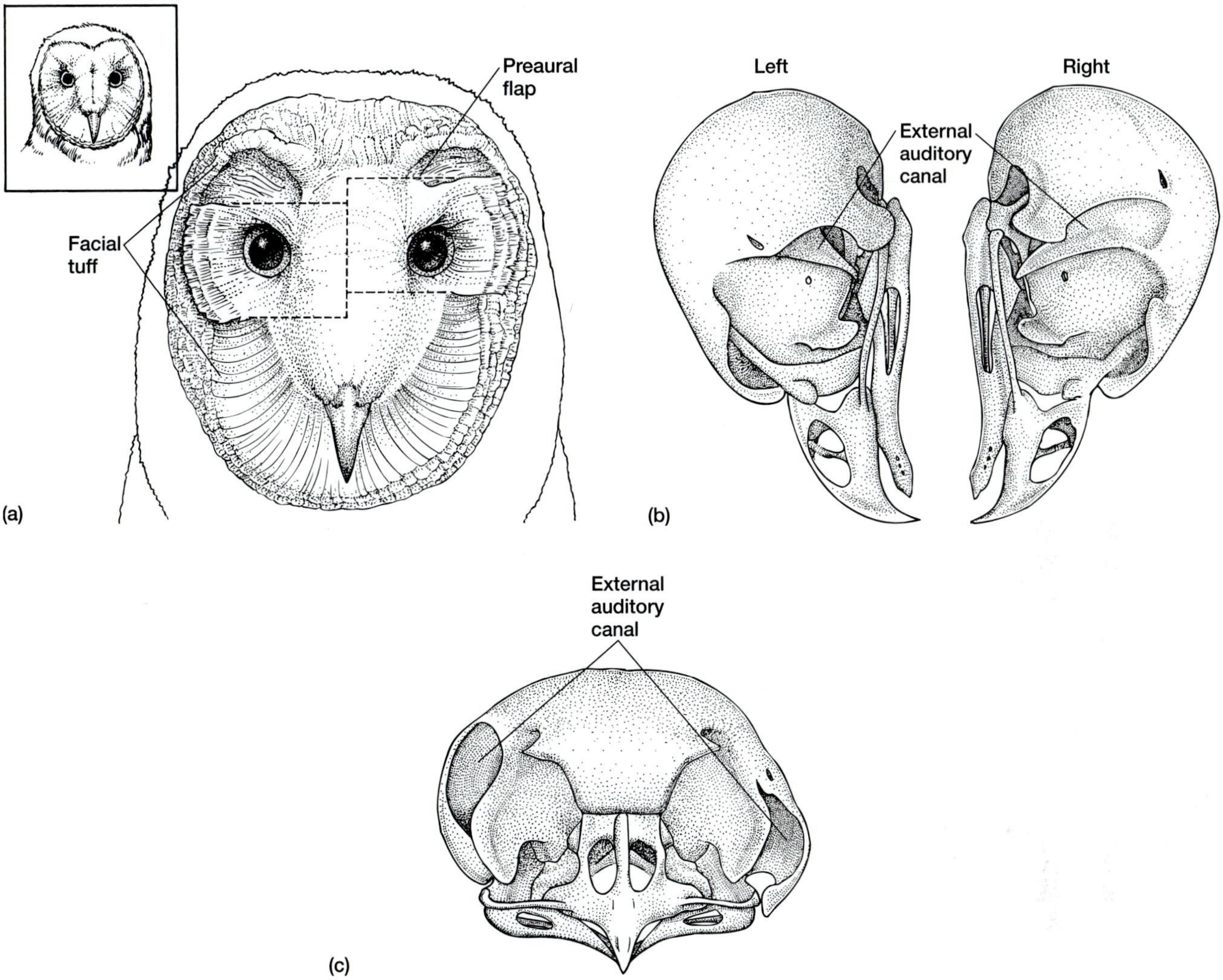

Auditory acuity in owls. (*a*) Facial disk feathers of a barn owl in the normal position (inset) have been removed to reveal the auditory feathers arranged in tightly packed parabolic rims encircling the face and external orifice of the ear. This facial ruff of feathers collects and directs sounds to the external auditory canal of the ear. Note the asymmetrical positioning of the preaural flaps of feathers (dashed lines). (*b*) Left and right sides of the skull of a Tengmalm's owl (*Aegolius funereus*) showing slight differences in the size of the external auditory canal. (*c*) Anterior view of Tengmalm's owl showing the asymmetry of the otic areas.

within the range of human hearing (Griffin, 1953; 1954) (Table 8.4).

Infrasound—sound frequencies below those that humans can hear—may be important in navigating and determining direction in birds. Homing pigeons can detect extremely low frequencies—as low as 0.05 Hz (Kreithen and Quine, 1979). The **lagena**, a specialized portion of the cochlea, may be the area responsive to low frequencies.

Nose

The presence of a secondary palate in crocodilians permits the internal nares to be located at the rear of the oral cavity. Vomeronasal organs are absent.

Olfactory mechanisms are less developed in birds than in other vertebrates; thus, most birds, with the exception of carrion eaters such as vultures and some seabirds such as petrels, are believed to have a poor sense of smell. Vomeronasal organs are absent.

Experimental studies in sub-Antarctic waters have shown that many procellariform seabirds such as white-chinned petrels and prions may use a naturally occurring scented compound, dimethyl sulfide (DMS), as an orientation cue (Nevitt et al., 1995). Dimethyl sulfide is produced by phytoplankton in response to zooplankton grazing. Zooplankton are, in turn, eaten by seabirds. The petrels and prions responding to DMS often forage at night, when olfactory

TABLE 8.4

Comparison of the Hearing Range of Humans and Several Birds

Species	Lower Limit (Hz)	Highest Sensitivity (Hz)	Upper Limit (Hz)
Human (*Homo sapiens*)	16	1,000–3,000	20,000
Budgerigar (*Melopsittacus undulatus*)	40	—	14,000
Starling (*Sturnus vulgaris*)	<100	2,000	15,000
Crossbill (*Loxia curvirostra*)	—	—	20,000
House sparrow (*Passer domesticus*)	—	—	18,000
European robin (*Erithsacus rubecula*)	—	—	21,000
Greenfinch (*Chloris chloris*)	—	—	20,000
Bullfinch (*Pyrrhula pyrrhula*)	<100	3,200	20,000–25,000
Chaffinch (*Fringilla coelebs*)	<200	3,200	29,000
Magpie (*Pica pica*)	<100	800–1,600	21,000
Crow (*Corvus* sp.)	<300	1,000–2,000	< 8,000
Canary (*Serinus canarius*)	250	2,800	10,000
Snowbunting (*Plectophenax nivalis*)	400	—	7,200
Horned lark (*Eremophila alpestris*)	350	—	7,600
Sparrow hawk (*Falco sparverius*)	<300	2,000	<10,000
Mallard (*Anas platyrhyncos*)	<300	2,000–3,000	<8,000
Canvasback (*Aythya valisineria*)	190	—	5,200
Ring-billed gull (*Larus delawarensis*)	100	500–800	3,000
Domestic pigeon (*Columba livia*)	50	1,000–2,400	12,000
Pheasant (*Phasianus colchicus*)	250	—	10,500
Long-eared owl (*Asio otus*)	<100	6,000	18,000
Tawny owl (*Strix aluco*)	<100	3,000–6,000	21,000
Great horned owl (*Bubo virginianus*)	60	1,000	<8,000
Penguin (*Spheniscus dermersus*)	4,000	2,000–4,000	15,000

From M. R. Kare and J. G. Rogers, Jr., "Sense Organs" in P. D. Sturkie (ed.) Avian Philosophy. *Copyright ©1976 Springer-Verlag, NY.*

cues might be especially valuable, whereas albatrosses often forage by locating aggregations of seals, whales, and other visibly conspicuous seabirds. Black-browed albatrosses were not attracted to DMS.

Taste

Taste buds are less abundant in birds than in most other groups of vertebrates (Table 8.5). Only a few taste buds are present; they occur primarily on the palate, pharynx, and epiglottis.

Endocrine System

The endocrine system in vertebrates was discussed in Chapter 5. As noted, certain parental behavior patterns are brought about by the effects of prolactin. These behaviors include such activities as nest building, incubation of eggs, and protection of the young. As discussed previously, prolactin promotes a secretion from the crop of pigeons, emperor penguins, and flamingos that is fed to the young by regurgitation. This nutritive secretion is known as "pigeon's milk."

The size of the thyroid in birds is influenced by age, sex, climatic conditions, diet, activity, and species (Ringer, 1976). Female chickens (*Gallus domesticus*) have a greater thyroid

TABLE 8.5

Comparison of the Number of Taste Buds in Various Animals

Animal	No. of Taste Buds
Chicken	24
Pigeon	37
Bullfinch	46
Starling	200
Duck	200
Parrot	350
Japanese quail	62
Snake	0
Kitten	473
Bat	800
Human	9,000
Pig	15,000
Goat	15,000
Rabbit	17,000
Calf	25,000
Catfish	100,000

From M. R. Kare and J. G. Rogers, Jr., "Sense Organs" in P. D. Sturkie (ed.) Avian Philosophy. *Copyright ©1976 Springer-Verlag, NY.*

weight than males. Early investigators reported that the thyroid weight of chickens and pigeons (*Columba livia*) was greater in the winter than in the summer. However, two studies on ducks in the Mediterranean region (Rosenberg et al., 1967; Astier et al., 1970) have shown greater thyroid weight and activity in the summer (June) than in the cold months. In southern France, for example, Astier et al. (1970) reported a definite annual rhythm of thyroid function in male Pekin ducks, with maximum iodine uptake occurring in June. June is the time of maximum photoperiod, the seasonal decrease of testicular activity, and the onset of the annual molt. Lowest thyroid function was in January and February.

Corticosterone is the primary glucocorticoid and hormone of stress in birds (Wingfield et al., 1983; Wingfield, 1984; 1985a, b; Astheimer et al., 1995). It is released rapidly into the bloodstream in response to a variety of stressors. A rise in corticosterone in the bloodstream promotes glucogenesis and leads to an increase in glucose levels. This additional energy store helps an individual meet increased energy demands that occur during periods of extreme stress such as storms, attacks by predators, or food shortage.

In studies of white-crowned sparrows (*Zonotrichia leucophrys gambelii*), Wingfield et al. (1996) found that long days resulted in the well-known increase in body mass and fat store, indicative of preparations for migration. In females, treatment with low temperatures resulted in a reduction in the premigratory increase in fat and body mass when transferred to long days. This was accompanied by an increase in plasma levels of corticosterone during the early stages of photostimulation. Temperature did not affect photoinduced testicular development, length of the cloacal protuberance (an androgen-dependent copulatory organ) in males, or ovarian follicle development. Temperature regimes had no effect on fattening or body mass in males despite an early increase in plasma corticosterone at low temperature. In addition, temperature treatments—low (5°C), moderate (20°C), and high (30°C)—had no effect on plasma levels of thyroid hormones in males, but low temperatures did inhibit thyroid secretion in females, thus impairing preparations for migration in females but not males. The sexual dimorphism in this response may be related to sexual selection in which males arrive on the breeding grounds ahead of females regardless of local weather conditions.

Increased levels of testosterone initiate the process of spermatogenesis in males and, together with the lack of estrogen, causes the development of secondary sex characteristics such as brightly colored breeding plumage. Estrogens and progesterone regulate female reproductive activity.

The emerging evolutionary perspective of androgen action has been greatly influenced by studies of birds (Wingfield et al., 1990). The role of testosterone in male–male aggression has revealed that, during the mating season, male–male aggression related to territorial defense can result in rapid and profound elevations in plasma levels of testosterone. These surges far surpass any daily variation that might normally occur in plasma testosterone levels, and therefore the response is easily distinguished. This dramatic hormonal response to social interactions has been termed the "challenge response" by Wingfield and colleagues. Wingfield observed that the challenge response is most likely to be observed in monogamous birds in which the male contributes heavily toward parental care. In contrast, species that are polygynous and lack male parental care usually do not exhibit hormonal changes in response to male–male aggression. More broadly, the challenge response also includes other social interactions, including male–female mating behaviors.

Regulation of calcium levels by the parathyroid glands is important for the proper formation of the calcareous shell enclosing bird embryos. If insufficient calcium is present, parathormone will cause it to be released from storage in the bones. Interference with this system by chlorinated hydrocarbon chemicals, such as DDT, caused thinning of egg shells in some species by reducing calcium availability. In some birds, the **ultimobranchial bodies** produce calcitonin, a hormone that lowers blood calcium levels. The pineal gland's role in avian thermoregulation has been discussed by Heldmaier (1991).

Urogenital System

A pronephric developmental stage is present early in embryonic growth of both crocodilians and birds but is supplanted by the mesonephros, which serves as the functional embryonic kidney. The mesonephros reaches its peak of development about halfway through embryonic life. When the metanephros begins to function, the mesonephros gradually ceases functioning. The metanephric kidney consists of many convoluted tubules and collecting tubules, with each kidney being drained by a pair of ureters. Mesonephric ducts remain as sperm ducts in male birds, but they involute in females and remain only as short, blind **Gartner's ducts** embedded in the mesentery of the oviducts.

The avian kidney, like that of reptiles, consists of lobes; each lobe is composed of clusters of many tubules. The kidneys lie in a retroperitoneal position flattened against the sacrum and ilium. In most birds, the ureters empty into the **urodeum**, the middle of the three compartments forming the cloaca. Only ostriches possess a urinary bladder.

Birds produce a semisolid and hypertonic urine, with uric acid as the main nitrogenous excretory product. Urine excreted by the kidneys is concentrated during its passage through the renal tubules, and additional water reabsorption takes place in the **coprodeum**, where the urine is mixed with the feces.

The urogenital anatomy of a female bird is basically similar to that found in most reptiles (Fig. 8.61). Histologically and ultrastructurally, it is comparable to that of a crocodilian (Palmer and Guillette, 1992). Both birds and crocodilians have separate regions, which may be homologous, for formation of the egg shell membrane and the calcareous layer.

The ovaries of birds consist of many fluid-filled cavities. In most, only the left ovary is functional; if present, the right ovary is rudimentary. Most birds also possess only a single

FIGURE 8.61

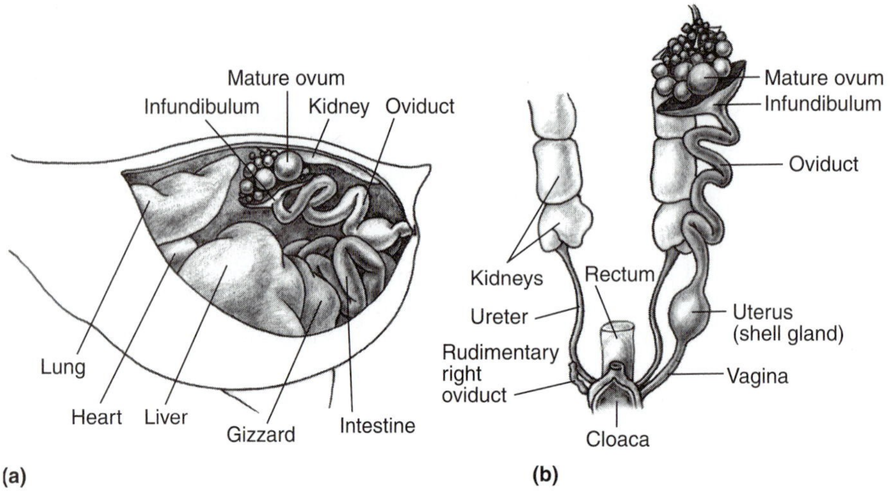

(*a*) Reproductive tract of the chicken in relation to other organs in the body cavity. The single functional ovary and oviduct are on the hen's left side; an undeveloped ovary and an oviduct are sometimes found on the right side, having degenerated in the embryo. (*b*) Both the ovary and oviduct are involved in the formation of the egg. The shell is formed in the uterus, which is also called the shell gland.
Source: Taylor, "How an Eggshell is Made" in Scientific American, *222(3): 88–95, 1970.*

(left) oviduct; the other remains rudimentary. Although the right ovary and oviduct are formed in the embryonic stages, they usually do not persist in adult life. The presence of only one ovary and oviduct is thought to be an evolutionary adaptation to lighten the body for flight.

Fertilization occurs in the upper region of the oviduct. Glands lining the oviduct secrete albumen around the ovum, and the caudal portion of the oviduct becomes modified as a thick-walled shell gland. A short, muscular terminal segment (vagina) empties into the urodeum portion of the cloaca. The vagina secretes mucus that seals the pores of the shell and then expels the egg.

The male reproductive system is also similar to that of reptiles. Two testes are present, in contrast to the single ovary and oviduct in females. Opisthonephric ducts empty into the urodeum. Unpaired copulatory organs (penis in males; clitoris in females) develop from the ventral wall of the proctodeum and are present in swans, ducks, ostriches, and a few other birds.

FIGURE 8.62

Female saltwater crocodile (*Crocodylus acutus*) guarding her nest made of vegetation. The eggs are deposited in the center of the nest and are incubated by heat generated by the decaying vegetation.

REPRODUCTION

Oviparous

Female alligators and crocodiles construct nests of vegetation that may be 1 to 2 m in diameter and 45 to 90 cm high (Fig. 8.62). Oval, hard-shelled eggs are laid in the center and covered by vegetation. Heat generated by the decaying vegetation provides the warmth needed for incubation.

Increasing daylength in late winter appears to be the primary factor initiating gonadal development in temperate zone breeding birds, whose gonads undergo seasonal regression in late summer and fall. Associated hormonal changes cause development of the nuptial plumage, courtship behavior, breeding, and the migration of many species that spend the winter months in warm climates and then fly to nontropical areas to breed and raise young.

Climatic conditions greatly affect breeding efforts and clutch size. Breeding success in some species, especially aquatic feeders, is related to rainfall. Breeding success is high in years with abundant rainfall, but falls during periods of drought (Maddock and Baxter, 1991). Many birds produce only a single brood annually, although some passerines such as bluebirds (*Sialia sialis*), robins (*Turdus migratorius*), starlings (*Sturnus vulgaris*), and sparrows may produce two and even three broods during the summer months under optimal conditions. Clutch size usually numbers between two and five chicks, although albatrosses, shearwaters, petrels, tropicbirds, most penguins, some auks, some nightjars, and others lay only a single egg each year. Loons, gannets, boobies, hummingbirds, whooping cranes, some penguins, most pigeons, and many tropical passerine birds lay just two eggs every year (van Tyne and Berger, 1961). Some birds, however, breed much less frequently. For example, the California condor (*Gymnogyps californianus*) lays only a single egg every other year.

Breeding in most species begins the year following birth, but some take longer to mature. Bald eagles (*Haliaeetus leucocephalus*), for example, require 3 years to mature, and puffins require 4 to 5 years. The California condor does not mature for approximately 9 years, and albatrosses take as long as 10 years to reach adulthood.

Sexual dimorphism among birds involves primarily visual and auditory differences, since birds are primarily diurnal and live at fairly high densities. Some species, such as the American avocet (*Recurvirostra americana*), have permanent morphological differences (Fig. 8.63a): the bills of females curve upward more than those of males. Males of many species, such as wood ducks (*Aix sponsa*) (Fig. 8.63b), develop brightly colored feathers or patches of skin as part of their breeding plumage. Other examples of sexual dimorphism include the red shoulder patches of male red-winged blackbirds (*Agelaius phoeniceus*); the black hood of the male hooded warbler (*Wilsonia citrina*); the orange tail patches of male American redstarts (*Setophaga ruticilla*); and the distinctive black-below, white-above color pattern of male bobolinks (*Dolichonyx oryzivorus*). Female swallows (*Hirundo rustica)* use two criteria in choosing a mate: the size and shape of a potential partner's tail. Female swallows prefer mates with long, perfectly forked tails. Males with elongated symmetrical tails mate earlier and enjoy larger annual reproductive success (Moller, 1992). Female house finches (*Carpodacus mexicanus*) prefer the most colorful males (Hill, 1990).

The ability of birds to see in ultraviolet light may be an important factor in courtship. For example, female starlings (*Sturnus vulgaris*) ranked males in a different and nonrandom order in the absence of ultraviolet light, but plumage reflectance in the human visible spectrum did not predict choice (Bennett et al., 1997). Thus, nocturnal variation in ultraviolet reflectance is important in avian mate assessment; the prevailing light environment may have profound effects on observed mating preferences.

The species-specific songs and calls evolved by many birds identify the male's territory and attract members of the opposite sex for breeding. The same is true of actions such as the "drumming" performed by male grouse. Female great reed warblers (*Acrocephalus arundinaceus*) base their mate choice on the quality of the male's territory and on the range of his song repertoire (Hasselquist et al., 1996). In addition, some male birds such as rollers and falcons put on spectacular aerial displays to impress prospective mates. Others, like weaverbirds (*Ploceidae*), build a basketlike nest to attract a female.

Postural displays, as well as the display of colored, highly specialized breeding plumage are important aspects of courtship in many species (Fig. 8.64). Evidence that vocal stimulation controls endocrine state and follicular growth in ring doves (*Streptopelia risoria*) was presented by Cheng (1992), who showed that female ring doves affect their own endocrine state by cooing. Females prevented from performing the nest-coo

FIGURE 8.63

Sexual dimorphism in birds: (*a*) American avocets (*Recurvirostra americana*). Bill curvature differs between sexes: bills of females (bottom) turn upward more then those of males (top). (*b*) Wood ducks (*Aix sponsa*). Highly colored iridescent male is on the left.

FIGURE 8.64

Courtship behavior of the albatross (*Diomedea*). At a future nesting site in the male's territory, the male initiates the courtship ritual by spreading his wings and pointing his head skyward. As pair bonding proceeds, the birds begin to touch bills. This contact display precedes copulation.

showed little or no growth of ovarian follicles. While male courtship stimulates females to begin nest-cooing, the female's own nest-coos stimulate her hypothalamus to respond. Vocalizations also accelerate ovulation in several species, including white-crowned sparrows (*Zonotrichia leucophrys gambelii*) (Cheng, 1992).

Breeding strategies among birds range from monogamy to polygamy with indiscriminant breeding (promiscuity) by both males and females occurring in some species. The mating of a male with more than one female is thought to enhance the male's genetic fitness by increasing the number of his offspring and their genetic variability, whereas polygamous females are thought to be seeking the best quality genes for their offspring. A male that mates with more than one female is **polygynous (polygyny)**. A female that mates with more than one male is referred to as **polyandrous (polyandry)**.

In many species of birds, males and females jointly build a nest, incubate the eggs, and feed the young. Following the successful fledging of their young, male and female often separate. Some, however, such as swans and Canada geese (*Branta canadensis*), pair for life and will remain together after breeding. Males guard incubating females, assist in caring for the young, and migrate with their mates.

Formerly, researchers believed that approximately 94 percent of all bird species were monogamous (Stone, 1991). The loon (*Gavia immer*), some gulls, and the Florida scrub jay (*Aphelocoma coerulescens*) are among birds known to be monogamous (Quinn, 1999). However, extra-pair copulations (EPCs) have been found to be widespread in so-called monogamous birds and can result in extra-pair fertilizations. In some studies, DNA analyses of a variety of species have shown that "30 percent of baby birds in a nest are not the offspring of the caretaking male" (Stone, 1991). Almost all songbirds have been found to be quite promiscuous. For example, research shows that 50 percent of the eggs in the typical American robin's (*Turdus migratorius*) nest have a different father (Quinn, 1999). The evolutionary significance of EPCs is thought to be an attempt by females to increase genetic diversity for their offspring.

Studies of the indigo bunting (*Passerina cyanea*) revealed that over 30 percent of all offspring were of extra-pair paternity (Westneat, 1987). One study showed that 55 percent of the young from 86 percent of the nests of reed buntings (*Emberiza schoeniclus*) were not the offspring of the territorial male (Dixon et al., 1994). Male reed buntings appeared to adjust their brood care in proportion to their likelihood of paternity and fed illegitimate broods less than broods they fathered. Estimated frequencies of polygynous mating among American redstarts (*Setophaga ruticilla*) in New Hampshire were 16, 5, and 8 percent in 1988, 1989, and 1990, respectively (Secunda and Sherry, 1991).

Extra-pair copulations, therefore, are not unusual among North American passerines as a facultative or opportunistic reproductive strategy. Females may be as promiscuous as males, because they are the ones that often enter an adjoining male's territory for an extra-pair copulation. Males develop counterstrategies to prevent EPCs by females, including frequent copulations and guarding of females in order to ensure paternity.

In some species, such as Smith's longspur (*Calcarius pictus*), a small Canadian subarctic bunting, females averaged 350 copulations per clutch and copulated with two or more males, while males typically mated with one to several females (Briskie, 1992). Multiple matings and the lack of territorial behavior in males are thought to be adaptations to sperm competition.

Most studies of fertilization have looked at the fertilization success of two males after a single copulation from each (Birkhead, 1988). In spotted sandpipers (*Actitis macularia*), however, the mating procedure usually involves repeated copulations by one male, followed by the laying of a clutch of eggs; then repeated copulations by a second male, followed by the laying of a second clutch. Oring et al. (1992) concluded that stored sperm from former mates was still the primary source of fertilizations for these second clutches. Because old males precede younger males to the breeding grounds, early pairing gives older males a reproductive advantage through additional fertilizations from sperm stored by females, thereby increasing their fitness.

A few species, such as the spotted sandpiper (*Actitis macularia*), red phalarope (*Phalaropus fulicarius*), and Wilson's phalarope (*Steganopus tricolor*), reverse sex roles: females are aggressive and play the more active role in courtship, while males provide most or all of the care for the young (Oring, 1995). Some species are polyandrous, a condition in which dominant females mate with multiple males, each of which tend a clutch of eggs. Males have higher prolactin levels (prolactin promotes incubation) than females and low testosterone levels. Female sandpipers are able to store sperm for up to 31 days before using it to fertilize eggs (Oring, 1995).

BIO-NOTE 8.11

Lekking

Males of approximately 200 species of birds assemble in a small area (a lek) during the breeding season for communal courtship display and mating (lekking). Females pass through this group of small territories and choose a male for copulation. The prairie chicken (*Tympanuchus cupido*), sage grouse (*Centrocercus urophasianus*), and sharp-tailed grouse (*Tympanuchus phasianellus*) of the western United States are well-known lekking species. Three hypotheses have been proposed to explain why males aggregate: the hotspot model (clusters form near places females frequently visit), the hotshot model (individuals cluster around attractive males to increase their chances of being noticed), and the female-preference model (males cluster because females like to visit groups, where they can choose a mate quickly and safely). Tests of each hypothesis have found support in some species, but not others.

Long-tailed manakins (*Pipridae*) are lek-mating Central American birds that engage in a unique behavior not typical of other lekking species. Manakins cooperate in multiyear male–male partnerships in which an alpha male is responsible for virtually all mating, and the beta male assists in courtship displays. Alpha and beta males are not usually related. Beta males forgo reproduction for many years in order to maintain this unusual form of male–male cooperation. Their benefits include rare copulations, future ascendancy to alpha status, and female lek fidelity.

Some female birds, including the domestic chicken (*Gallus domesticus*), pheasant (*Phasianus calchicus*), pigeon (*Columba livia*), mallard (*Anas platyrhynchos*), and domestic goose (*Anser anser*), can store sperm for periods ranging up to 52 days (Birkhead, 1988). Sperm storage allows sperm of previous mates to compete with sperm from more recent copulations; however, at least in domestic birds, sperm from the most recent copulation has the greatest probability of fertilizing eggs (Birkhead, 1988). Sperm are stored in sperm storage tubules at the uterovaginal junction of the oviduct (Bakst, 1987). The number of sperm storage tubules ranges from 500 to 20,000, depending on the species (Birkhead and Moller, 1992). Sperm size among species of birds varies considerably, primarily due to variation in the length of the sperm's tail (Briskie and Montgomerie, 1992).

Young Seychelles warblers (*Acrocephalus sechellensis*) often remain in their natal territories and help their parents raise later offspring. Helpers are mostly adult daughters and, on high-quality territories where food is plentiful, they increase their parents' reproductive success. On low-quality territories, however, these helpers decrease their parents' reproductive success through competition for limited food. Biased hatchling sex ratios are caused by biased reproduction (Komdeur et al., 1997). Breeding pairs without helpers on low-quality territories produce 77 percent sons, whereas pairs without helpers on high-quality territories produce 13 percent sons. Breeding pairs that were transferred from low- to high-quality territories switched from the production of male to female eggs. Breeding pairs occupying high-quality territories switched from producing female eggs when less than two helpers were present, to producing male eggs when at least two helpers were present in the territory. These are the largest skews in hatchling sex ratios and adaptive modifications that have ever been recorded in a bird species (Gowaty, 1997).

GROWTH AND DEVELOPMENT

Prenatal Development

Oviparous

Crocodilians lay eggs in which the embryos and the four extraembryonic membranes are enclosed in a leathery shell; all birds lay eggs with hard, calcified shells. Egg size and shape varies greatly among different species of birds. The largest known egg, which measured 34 by 24 cm and had a capacity of 9 liters, was that of the extinct elephant bird (*Aepyornis maxima*) of Madagascar (Faaborg, 1988). The smallest known egg belongs to the Jamaican emerald hummingbird (*Melisuga minima*) whose egg is only 1 by 0.65 cm and whose mass is 0.5 g, or about 1/50,000 of that of an elephant bird egg.

Bird eggs can be classified into three basic shapes: oval, elliptical, and pyriform (Fig. 8.65). Oval eggs, which are round on both ends but widest in the middle, are produced by most birds. Grebes, hummingbirds, and some raptors,

BIO-NOTE 8.12

Finding Calcium for Egg Shells

Female birds need calcium to produce thick, protective egg shells, but those whose diets are restricted to seeds and insects have trouble getting enough calcium from these sources. Therefore, during egg laying, they seek out calcium supplements, foraging for bits of bone, seashells, and other items rich in the mineral.

The first known instance of a bird hoarding a substance for its mineral, rather than caloric, content was reported in the red-cockaded woodpecker (*Picoides borealis*). Several days before and during egg laying, birds would land near a bone they spotted on the ground, consume flakes of it, and then carry larger fragments to a nearby tree and wedge them into the bark. They were observed retrieving the fragments on several occasions during the egg-laying period.

Repasky et al., 1991

FIGURE 8.65

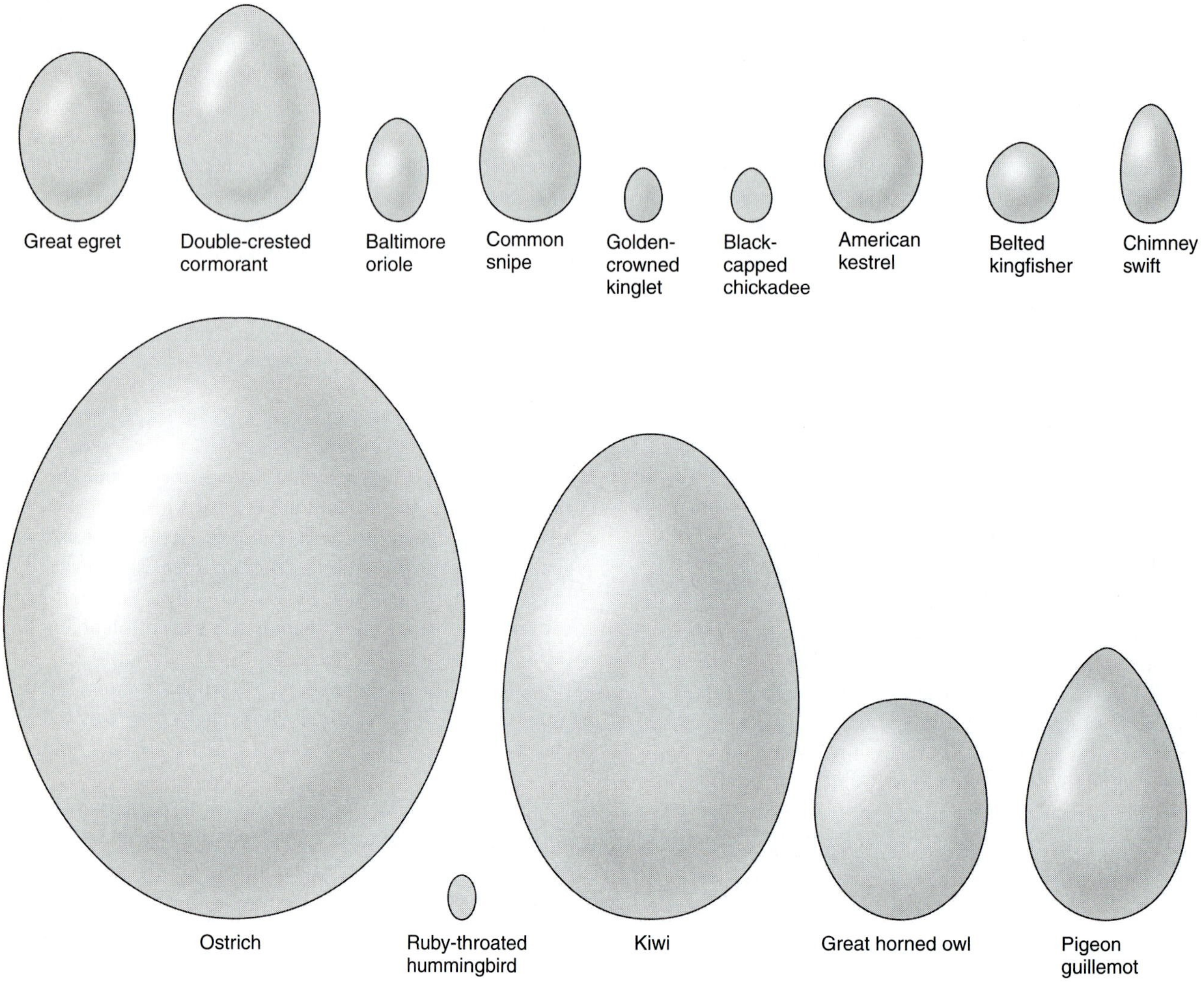

Bird eggs vary greatly in size and shape. The three basic shapes are as follows: oval (round on both ends but widest in the middle—e.g., egret); elliptical (curvature of each end is about the same—e.g., hummingbird); and pyriform (one large end and one smaller end—e.g., snipe).

however, produce eggs in which the curvatures of each end are about the same (elliptical). Pyriform, or pear-shaped eggs, have one large end and one smaller end and are characteristic of most shore birds and many passerines. The pyriform eggs of murres (*Uria lomvia*) roll in tight circles on the rock ledges where the birds nest, thus making it less likely they will fall into the sea.

Although most birds deposit their eggs in nests constructed of plant materials, mud, or even salivary secretions (Fig. 8.66), some lay their eggs directly on the ground with little or no nesting material. Killdeer (*Charadrius vociferus*), for example, lay their well-camouflaged eggs among stones or gravel. Royal terns (*Sterna maxima*) deposit their single egg in a depression in the sand (Fig. 8.67). The fairy, or white, tern (*Gygis alba*) of the Pacific and southern Atlantic oceans lays a single egg on a rock or on a horizontal branch with no nesting material whatever. Auks (*Alcidae*) and sheathbills (*Chionis alba*) lay their eggs on rock ledges and in rock crevices. Some, such as burrowing owls (*Speotyto cunicularia*), fairy penguins (*Eudyptula minor*) (Fig. 8.68), and sooty shearwaters (*Puffinus griseus*) (see Chapter 13 for a discussion of an interesting commensal relationship between shearwaters and tuataras), nest in underground burrows or in holes in banks (e.g., bank swallow, *Riparia riparia;* kingfisher, *Megaceryle alcyon*); some are cavity nesters (purple martins, *Progne subis*; bluebirds, *Sialia* spp.; and woodpeckers, *Picidae*); some nest in chimneys (chimney swift, *Choetura pelagica*); and some have nests that float

FIGURE 8.66

Diversity of nests and nest sites in birds. (*a*) The terrestrial ground nest of the lesser prairie chicken (*Tympanuchus pallidicinctusi*). (*b*) The mud and pebble nests of the house martin (*Delichon orbica*).

FIGURE 8.67

Royal terns (*Sterna maxima*) deposit their single egg in a depression in the sand. Tern eggs are in the foreground.

FIGURE 8.68

Nesting burrow of the fairy penguin (*Eudyptula minor*) taken on Phillip's Island, Australia.

FIGURE 8.69

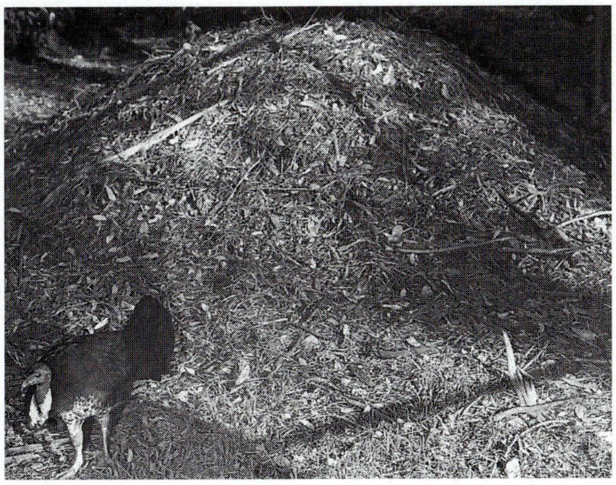

Megapodes (brush-turkeys) deposit their eggs in mounds of soil and debris. The eggs are incubated by the heat generated by the decomposing vegetation.

on the water (grebes, Colymbidae). African social weavers (*Philetarus socius*) construct elaborate nests by weaving grass and plant fibers. Bahama parrots (*Amazonia leucocephala bahamensis*) nest in limestone solution cavities beneath the ground, a habit unique among New World psittacines (Gnam, 1991).

Megapodes, or brush-turkeys, of Australia and the East Indies, are mound builders (Fig. 8.69). They construct mounds of decaying vegetation or volcanic ash somewhat resembling those of the American alligator and deposit their eggs in the center. Eggs are incubated by solar, geothermal, or microbial-generated heat. The Australian brush-turkey (*Alectura lathami*) constructs a mound of decomposing vegetation with an average volume of approximately 12.7 m^3 and a weight of about 6,800 kg (Seymour and Bradford,

1992). The rate of heat production is more than 20 times the heat production of a resting adult.

Some birds, such as the brown-headed cowbird (*Molothrus ater*) and Old World cuckoos (*Cuculus*), are parasites. Females never build a nest; rather, they deposit one or more eggs in the nest of another species when that female is away from her nest (see Chapter 13) .

Prehatching vocalizations by the young beginning just prior to hatching are common in precocial and some altricial birds (Evans, 1988; Bugden and Evans, 1991). Altricial American white pelican (*Pelecanus erythrorhynchos*) embryos that are close to hatching emit harsh squawks from inside the egg if their temperature begins to drop. This is a frequent occurrence with terminal eggs in clutches from altricial species that hatch asynchronously, because the parent is spending an increasing amount of time obtaining food for the first hatchlings and less time incubating the unhatched eggs (Evans, 1990a). Hatching of pelican embryos is significantly retarded by moderate chilling (Evans, 1990b). The squawking gets the attention of the parents and usually results in a parent resettling over the nest and resuming incubation (Evans, 1992). Thus, late-stage embryos influence their own incubation temperature by communicating with their parents (Evans, 1990c, d).

Conditions occurring at the time of breeding (such as changes in food abundance and population density) are known to affect clutch size. In addition, studies of zebra finches (*Taeniopygia guttata*) and great tits (*Parus major*) show that environmental factors affecting the food intake by nestling females (e.g., the ability of parents to locate food, the quality of the territory, the timing of food abundance) may permanently influence the mechanisms controlling the clutch sizes of nestlings when they reach adulthood (Haywood and Perrins, 1992).

Duration of Embryonic Development

American alligator eggs hatch in about 9 or 10 weeks. Temperature-dependent sex determination occurs in all species that have been examined. Incubation of bird eggs ranges from approximately 10 or 11 days in some passerine birds to 11 or 12 weeks in the royal albatross (*Diomedea epomorphora*). Incubation in domestic chickens requires approximately 21 days (Fig. 8.70). In most species, incubation is carried out by one or both parents. In the megapodes, however, decaying vegetation generates the heat necessary for incubation.

Prinzinger and Hinninger (1992) reported a diurnal (circadian) rhythm of energy metabolism present as early as the fourth day of incubation in pigeon (*Columba livia*) embryos. Nighttime values of oxygen consumption were significantly below those for daytime. The mean difference between the daytime value and the value for the following night was 17 percent. Data supporting diurnal rhythms in chicken (*Gallus domesticus*) embryos (Barnwell, 1960) and herring gull (*Larus argentatus*) embryos (Drent, 1970) have also been reported.

FIGURE 8.70

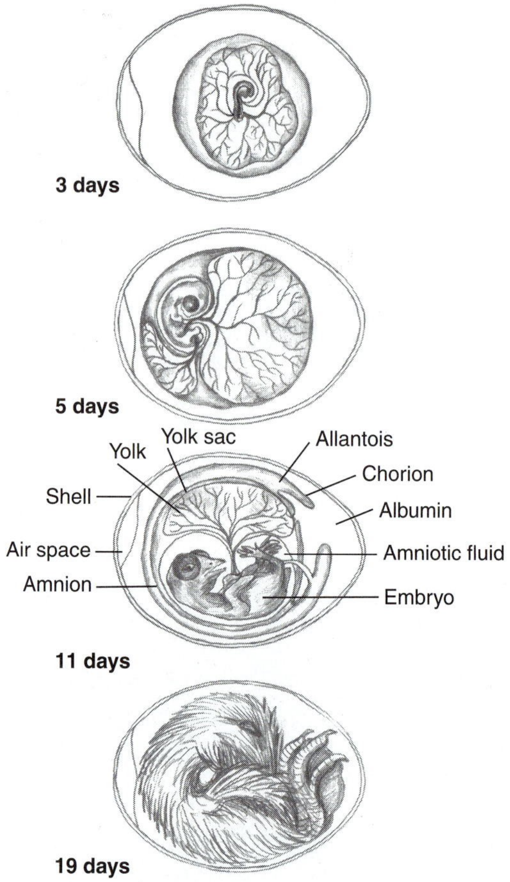

The fetus of a domestic chicken exhibits rapid changes in development. The chorion is the outer boundary of structures derived from the true egg shell. The shell and thick layer of albumen (a protein) surround the developing chick. Shown here are embryos at 3, 5, 11, and 19 days after incubation begins. Hatching requires approximately 21 days in this species.

Hatching

Evaporative water loss from eggs has been found to affect hatching success (Walsberg and Schmidt, 1992). Mourning dove embryos incubated under either arid or humid conditions developed normally up to, and including, pipping; however, only 50 percent of the embryos incubated under humid conditions hatched as opposed to approximately 85 percent incubated under arid conditions and 90 percent at an intermediate relative humidity. Increased mortality of those under humid conditions apparently was caused by the lack of sufficient space inside the egg for the embryo to maneuver in order to break the shell. This lack of space presumably was caused by decreased evaporative water loss.

In birds, pipping of the shell may begin 10 hours or more before hatching due to the random movement of the embryo (Faaborg, 1988). At some point, the movement becomes much more active. Strong thrusts of the beak into the shell are accompanied by movements of the entire body produced by pushing with the feet. During this time, the embryo rotates within the shell, a process that causes a ring of cracks to form near the blunt end of the egg. The full-term embryo possesses a horny knob or egg tooth on its upper mandible and hatching muscles on the back of its head (Fig. 8.71) that are responsible for the vigorous forward thrust of the head during pipping. Both the egg tooth and the hatching muscles are lost or regress shortly after hatching. Continued movement and thrusting by the embryo eventually cause the end of the shell to break free and allow the baby bird's emergence from the egg. The entire hatching process may last from several hours to several days.

The eggs of parasitic birds, such as cowbirds and cuckoos, generally have a shorter gestation period than the eggs of their hosts and usually are the first to hatch. In addition, McMaster and Sealy (1998), who analyzed cowbird eggs laid in nests of yellow warblers, concluded that cowbirds hatch before many hosts by prolonging incubation of smaller eggs and, possibly, hatching early in response to stimuli from host eggs. Because the cowbird egg is roughly twice the size of warbler eggs, it keeps the incubating parent from making optimal contact with those eggs, thus adding about a day and a half to the normal 11-day incubation of yellow warblers. Newly hatched parasitic nestlings often push one or more of the host's eggs out of the nest (Fig. 8.72) and consume much of the food delivered to the nest by the parents. Nestling common cuckoos (*Cuculus canorus*) stimulate their adoptive parents to provide adequate food by possessing a rapid begging call that sounds remarkably like a whole brood of host chicks (Davies et al., 1998).

Parental Care

Female crocodilians often remain in the vicinity of their nests during the incubation period. Hatching young emit grunting noises that attract the female. She assists the young in escaping from the nest by pulling the nest apart with her front feet and mouth. Both male and female crocodilians may pick newly hatched young up in their mouth, one at a time, and gently carry them to a nearby source of water (Fig. 8.73). Females also respond to alarm cries from the young.

FIGURE 8.71

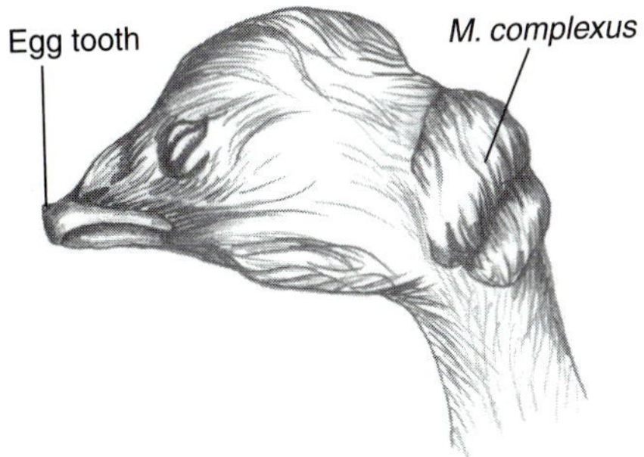

The hatching muscle in a chicken on the day of hatching. This muscle is primarily responsible for the vigorous forward thrusting movements of the head that result in pipping of the shell.
Source: Faaborg, Ornithology, *1988; after Bock and Hikada 1968.*

FIGURE 8.72

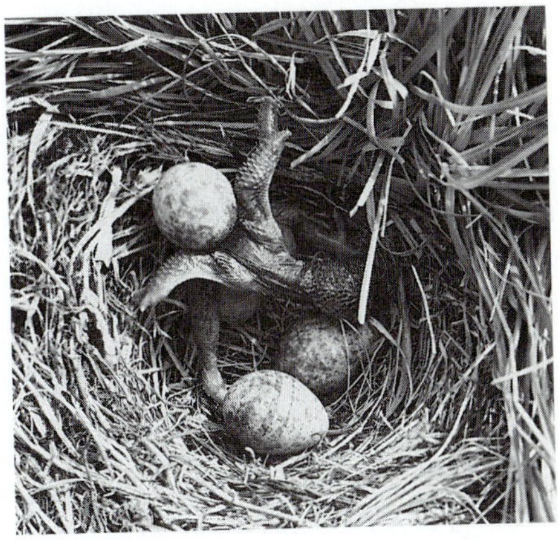

Newly hatched parasitic cuckoo nestling that is in the act of rolling the eggs of its host out of the nest.

Most birds are well known for providing parental care. Depending on the species, either one or both parents incubate the eggs and feed the young. Male barn swallows (*Hirundo rustica*) in North America assist in incubating the eggs, but male incubation has never been recorded in European populations of this species (Smith and Montgomerie, 1992).

Male emperor penguins (*Aptenodytes forsteri*) incubate the single egg for approximately 120 days, often under severe conditions of high winds and temperatures of −30°C. They fast during the entire incubation period (Ancel et al., 1992). Males are relieved by the female near hatching time. Satellite monitoring has revealed males walking up to 296 km to open water to forage, whereas males swimming in light pack ice traveled as far as 895 km from the breeding colony.

In some species, such as bluebirds (*Sialia* spp.) and parakeets (*Melopsittacus* spp.), fathers feed female nestlings more often than male nestlings—in some cases, twice as frequently—whereas mother bluebirds feed sons and daughters equally (Fackelmann, 1992). The reason for this behavior is unknown. There is no difference in metabolic rates between male and female nestlings. Some researchers believe that daughters are fed preferentially in order to instill high standards for selecting a mate. Those females that have been favored as nestlings might look for more generous mates that can provide plenty of food for a hungry brood.

Some birds do not care for their young. Megapodes construct a mound nest in which eggs are incubated, but provide no parental care for their newborn. Only the parasitic species, such as European cuckoos (*Cuculus canorus*) and cowbirds (*Molothrus ater*), neither build a nest nor care for their young.

Incubating birds are known to retrieve eggs displaced from their nests. Waterfowl accomplish this by rolling the egg with their bills. Two incubating mallards are known to have moved their eggs to drier nest sites 55 cm and 30 cm away from their initial nest site (Fleskes, 1991). One new site was 15 cm above the original site, so that the female had to roll the eggs uphill.

FIGURE 8.73

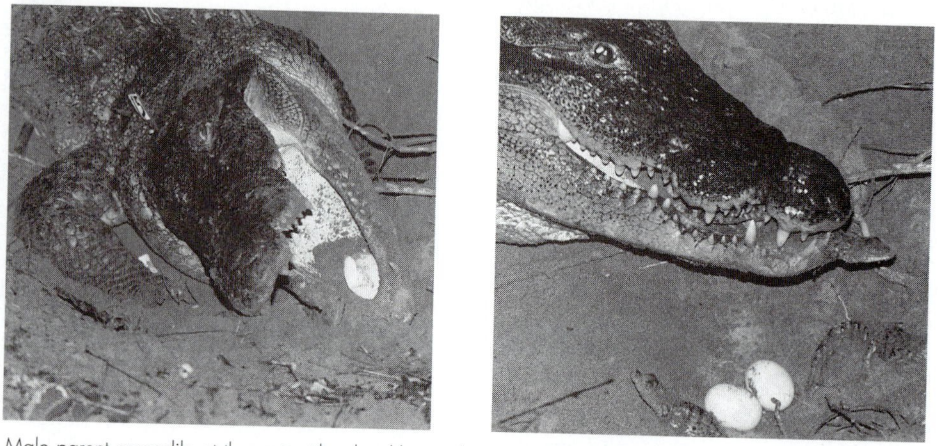

Male parent crocodile at the nest with a hatchling in his jaws. The male was observed transporting the young one at a time in his mouth to the nearby water.

Growth

Neither crocodilians nor birds have a larval stage; they hatch into miniature adult forms. Many birds hatch with their eyes sealed and their bodies naked. These **altricial** species require extensive parental care before they can fend for themselves (see Fig. 1.20). They soon develop a covering of down feathers and, later, contour feathers to aid in thermoregulation. It may take several years for individuals of some species, such as bald eagles (*Haliaeetus leucocephalus*), to acquire their adult plumage. Most species must also learn to fly. The young of other birds, such as ducks, geese, shorebirds, and gamebirds, have their eyes open, are covered with a thick coat of natal down at the time of hatching, and require a lesser degree of parental care than altricial species. The **precocial** young of aquatic birds begin swimming with their parents shortly after birth. Even though the young can thermoregulate well and obtain their own food, the parents still are involved in such activities as brooding and protecting their chicks from predators. The most precocial birds are the megapodes of Australia, whose young require no parental care. Table 8.6 compares characteristics of precocial and altricial species of birds.

Most young birds assume a series of distinct subadult plumages as they mature. The natal down coat that precocial chicks are born with and altricial chicks develop in the first weeks of life is replaced gradually by the **juvenal plumage** through a **postnatal molt**. The juvenal plumage, which includes the flight feathers, is usually retained until late in the first summer. The **postjuvenal molt** is a partial molt in which the juvenal plumage (except for the flight feathers) is replaced by the first winter plumage. The first **nuptial (breeding) plumage** is usually completed by mid-spring of the next year. Birds such as gulls or eagles may require up to 5 years to mature and attain their adult plumage. Thereafter, winter plumages alternate with nuptial plumages.

TABLE 8.6

A Comparison of the Characteristics of Precocial and Altricial Birds

Precocial	*Altricial*
Large eggs	Small eggs
Much yolk	Minimal yolk
Large clutch size	Small clutch size
Delayed incubation	Immediate incubation
Long incubation period	Short incubation period
Short nest period	Long nest period
Large chick	Small chick
Much down	Little or no down
Great attrition of young	Less attrition of young
Short ectothermic period	Long ectothermic period
Partially dependent on adults	Totally dependent on adults
Slow growth rate	Rapid growth rate
Do not remove shells from nest site	Remove shells from nest site

From G.D. Alcorn, Birds and Their Young, *1991. Copyright © Stackpole Books. Reprinted by permission.*

These features serve as a general guide because not all birds of one type will exhibit all of the characteristics listed.

Attainment of Sexual Maturity

Breeding may not occur in crocodilians for 10 or more years, although Bellairs (1970) noted that the American alligator (*Alligator mississippiensis*) is one of the more rapidly growing species and matures when it is about 6 years old. Most passerine birds can breed the spring following their hatching. Larger birds, such as gulls and eagles, may require 3 or more years to mature. The California condor requires at least 5 years to mature, whereas the royal albatross may not begin breeding until it is approximately 8 years of age.

TABLE 8.7

Longevity of the American Alligator and Some Species of Birds

Species	Maximum Age (years)
Alligatoridae	
American alligator (*Alligator mississippiensis*)	56
Fregatidae	
Frigatebird (*Fregata minor*)	33
Diomedeidae	
Black-footed albatross (*Phoebastria nigripes*)	40
Wandering albatross (*Diomedea exulans*)	34
Black-browed albatross (*Thalassarche melanophris*)	32
Phaethontidae	
Red-tailed tropicbird (*Phaethon rubricauda*)	28
Ardeidae	
Great blue heron (*Ardea herodias*)	23
Anatidae	
Canada goose (*Branta canadensis*)	28
Mallard (*Anas platyrhynchos*)	26
Accipitridae	
Golden eagle (*Aquila chrysaetos*)	22
Peregrine falcon (*Falco peregrinus*)	16
Osprey (*Pandion haliaetus*)	26
Vulturidae	
Black vulture (*Coragyps atratus*)	25
Laridae	
Herring gull (*Larus argentatus*)	28
White tern (*Gygis alba*)	35
Sooty tern (*Sterna fuscata*)	35
Arctic tern (*Sterna paradisaea*)	34
Caspian tern (*Sterna caspia*)	29
Columbidae	
Mourning dove (*Zenaidura macroura*)	27
Tytonidae	
Barn owl (*Tyto alba*)	15
Picidae	
Yellow-shafted flicker (*Colaptes auratus*)	12
Hirundinidae	
Barn swallow (*Hirundo rustica*)	16
Corvidae	
Common crow (*Corvus brachyrhynchos*)	14
Blue jay (*Cyanocitta cristata*)	17
Paridae	
Black-capped chickadee (*Parus atricapillus*)	12
Turdidae	
American robin (*Turdus migratorius*)	14
Sturnidae	
Starling (*Sturnus vulgaris*)	16
Icteridae	
Brown-headed cowbird (*Molothrus ater*)	8
Red-winged blackbird (*Agelaius phoeniceus*)	14
Common grackle (*Quiscalus quiscula*)	16
Tanagridae	
Cardinal (*Pyrrhuloxia cardinalis*)	13
Emberizidae	
Song sparrow (*Melospiza melodia*)	8

From J. C. Welty, The Life of Birds, *1963, Alfred Knopf; and G. J. Wallace* An Introduction to Ornithology, *2nd edition, 1963, Macmillan Publishing; A. Bellairs,* The Life of Reptiles, *Universe Books; and Bird Banding Laboratory (www.pwrc.usgs.gov/bbl), 1999.*

Longevity

Among birds, parrots live longer than any other species (Table 8.7). Parrots 40 to 50 years old are not unusual. A Laysan albatross (*Phoebastria immutabilis*) has been recorded with an age of 42 years, 5 months (Bird Banding Laboratory, 1999). A banded Leach's storm-petrel (*Oceanodroma leucorhoa*) with an estimated minimum age of 31 years was recorded by Klimkiewicz and Futcher (1989). Banded great horned owls have lived over 28 years (Nero, 1992). Boobys, frigatebirds, tropicbirds, swans, geese, and some ducks often survive 20 or more years (Klimkiewicz and Futcher, 1989). Birds such as condors and ostriches probably live no more than 15 or 20 years. A banded broad-tailed hummingbird (*Selasphorus platycercus*) lived over 12 years. Banded jays have survived for 17 years.

Review Questions

1. Describe the process of ecdysis in reptiles. Why is it necessary?
2. How does ecdysis differ in turtles, lizards, snakes, and crocodilians?
3. Integumentary glands, which are abundant in amphibians, are sparse in reptiles. What is the adaptive significance of this modification?
4. List several modifications of the stratum corneum in reptiles.
5. Describe the four methods of locomotion used by snakes. In what situations would each be used?
6. How does the crocodilian heart differ from the hearts of other reptiles?
7. List four methods of gas exchange used by reptiles.

8. Discuss the adaptations of a snake's jaws that allow it to engulf large prey.
9. How do opisthoglyphs differ from proteroglyphs? From solenoglyphs? Discuss the poison-conducting apparatus of each.
10. What is the function of the tongue in snakes? Explain how it accomplishes this function.
11. How does body temperature regulate physiological processes in reptiles? How does it affect growth?
12. Differentiate between the two types of intromittent organs found in reptiles. Discuss the structure and function of each.
13. Describe the six major types of feathers and give the function of each.
14. Compare the early development of a feather with that of a reptilian scale.
15. Discuss the modifications of the bird skeleton that allow it to be lightweight yet strong.
16. What is the functional advantage of a synsacrum?
17. Differentiate between red skeletal muscle and white skeletal muscle.
18. How does the heart of a bird differ from the heart of a shark?
19. Explain how the countercurrent heat exchanger works in the leg of a bird.
20. Explain the process of ventilation in a bird. What are the functions of the air sacs?
21. Why do so few birds live as grazers on leaves and grass?
22. Differentiate between determinate and indeterminate growth. Give examples.
23. Why is internal fertilization a necessity for amniotes?
24. What is the evolutionary significance of the amniotic egg?
25. List several advantages of viviparity in reptiles.
26. List several types of nest sites used by birds. Give an example of a bird using each type of site.
27. What is the adaptive significance of prehatching vocalizations in birds?
28. Discuss adaptations used by oviparous vertebrates to escape from the membrane or shell enclosing them at hatching.
29. Discuss environmental sex determination and its possible adaptive significance in managing endangered species populations.
30. Give several examples of parental care in reptiles.
31. Differentiate between an altricial species and a precocial species. Give several examples of each found in bird.

Supplemental Reading

Alcorn, G. D. 1991. *Birds and Their Young*. Harrisburg, Pennsylvania: Stackpole Books.

Beletsky, L. D., and G. H. Orians. 1997. *Red-Winged Blackbirds*. Chicago: The University of Chicago Press.

Bellairs, A. 1970. *The Life of Reptiles*. 2 vols. New York: Universe Books.

Birkhead, T. R., and A. P. Moller (eds.) 1998. *Sperm Competition and Sexual Selection*. New York: Academic Press.

Bjorndal, K. A. 1995. *Biology and Conservation of Sea Turtles*. Washington, D.C.: Smithsonian Institution Press.

Black, J. M. (ed.). 1996. *Partnerships in Birds: The Study of Monogamy*. New York: Oxford University Press.

Bruemmer, F. 1995. La Arribada. *Natural History* 104(8):36–43.

Burley, R. W., and D. V. Vadehra. 1989. *The Avian Egg. Chemistry and Biology*. New York: John Wiley and Sons.

Carr, A. 1980. *The Reptiles*. Life Nature Library. Alexandria, Virginia: Time-Life Books.

Ciofi, C. 1999. The Komodo dragon. *Scientific American* 280(3):84–91.

Cole, C. J. 1984. Unisexual lizards. *Scientific American* 250(1):94–100.

Davis, L. S., and J. T. Darby (eds.). 1990. *Penguin Biology*. San Diego: Academic Press.

Doughty, R. W. 1989. *Return of the Whooping Crane*. Austin: University of Texas Press.

Ernst, C. H. 1992. *Venomous Reptiles of North America*. Washington, D.C.: Smithsonian Institution Press.

Ernst, C. H., and R. W. Barbour. 1984. *Turtles of the World*. Washington, D.C.: Smithsonian Institution Press.

Ernst, C. H., and G. R. Zug. 1997. *Snakes in Question: The Smithsonian Answer Book*. Washington, D.C.: Smithsonian Institution Press.

Ernst, C. H., R. W. Barbour, and J. E. Lovich. 1994. *Turtles of the United States and Canada*. Washington, D.C.: Smithsonian Institution Press.

Estes, R., and G. Pregill. 1988. *Phylogenetic Relationships of the Lizard Families*. Stanford, California: Stanford University Press.

Forsyth, A. 1986. *A Natural History of Sex: The Ecology and Evolution of Sexual Behavior*. New York: Charles Scribner's Sons.

Grenard, S. 1991. *Handbook of Alligators and Crocodiles*. Malabar, Florida: Krieger Publishing Company.

Greene, H. W. 1997. *Snakes: The Evolution of Mystery in Nature*. Berkeley: University of California Press.

Halliday, T. 1982. *Sexual Strategy: Survival in the Wild*. Chicago: University of Chicago Press.

Hamilton, W. J., and G. H. Orians. 1965. The evolution of brood parasitism in altricial birds. *Condor* 67:361–382.

Jellis, R. 1984. *Bird Sounds and Their Meaning*. Ithaca, New York: Cornell University Press.

Johnsgard, P. A. 1990. *Hawks, Eagles, and Falcons of North America*. Washington, D.C.: Smithsonian Institution Press.

Johnsgard, P. A. 1997. *The Hummingbirds of North America*. Second edition. Washington, D.C.: Smithsonian Institution Press.

Johnsgard, P. A. 1997. *North American Owls*. Washington, D.C.: Smithsonian Institution Press.

Kauffman, K. 1996. *Lives of North American Birds*. New York: Houghton Mifflin Company.

King, A. S., and J. McLelland (eds.). 1979–1989. *Form and Function in Birds*. 4 vols. New York: Academic Press.

Klauber, L. M. 1982. *Rattlesnakes*. Berkeley: University of California Press.

Lee, M. S. Y., and R. Shine. 1998. Reptilian viviparity and Dollo's law. *Evolution* 52(5):1441–1450.

Mattison, C. 1996. *Rattler! A Natural History of Rattlesnakes*. London: Blandford Press.

Norris, D. O., and R. E. Jones (eds.). 1987. *Hormones and Reproduction in Fishes, Amphibians, and Reptiles*. New York: Plenum Press.

O'Connor, R. J. 1984. *The Growth and Development of Birds.* New York: John Wiley and Sons.

Padian, K., and L. M. Chiappe. 1998. The origin and early evolution of birds. *Biological Reviews of the Cambridge Philosophical Society* 73(1):1–42.

Pough, F. H., R. M. Andrews, J. E. Cadle, M. L. Crump, A. H. Savitzky, and K. D. Wells. 1998. *Herpetology.* Upper Saddle River, New Jersey: Prentice-Hall.

Rossman, D. A., N. B. Ford, and R. A. Seigel. 1996. *The Garter Snakes: Evolution and Ecology.* Norman: University of Oklahoma Press.

Rubio, M. 1998. *Rattlesnake: Portrait of a Predator.* Washington, D.C.: Smithsonian Institution Press.

Sanders, J. 2000. *Internet Guide to Birds and Birding.* Dubuque, Iowa: McGraw-Hill Publishing Company.

Searcy, W. A., and K. Yasukawa. 1995. *Polygyny and Sexual Selection in Red-Winged Blackbirds.* Princeton, New Jersey: Princeton University Press.

Seigel, R. A., and J. T. Collins. 1993. *Snakes: Ecology and Behavior.* New York: McGraw-Hill.

Shine, R. 1991. *Australian Snakes: A Natural History.* Ithaca, New York: Cornell University Press.

Snyder, N., and H. Snyder. 1991. *Birds of Prey: Natural History of North American Raptors.* Stillwater, Minnesota: Voyageur Press.

Strawn, M. A. 1997. *Alligators: Prehistoric Presence in the American Landscape.* Baltimore, Maryland: The Johns Hopkins Press.

Taylor, I. 1994. *Barn Owls: Predator-Prey Relationships and Conservation.* New York: Cambridge University Press.

Thorpe, R. S., W. Wuster, and A. Malhorta. 1996. *Venomous Snakes: Ecology, Evolution and Snakebites.* New York: Oxford University Press.

Zeigler, H. P., and H.- J. Bischof. 1993. *Vision, Brain, and Behavior in Birds.* Cambridge, Massachusetts: MIT Press.

SELECTED JOURNALS

The Auk. A Quarterly Journal of Ornithology. Published by the American Ornithologists' Union. Publishes original reports on the biology of birds.

The Condor. A Journal of Avian Biology. Published by the Cooper Ornithological Society. Publishes research reports and occasional review articles.

The Wilson Bulletin. Published by the Wilson Ornithological Society. Publishes significant research and review articles in the field of ornithology.

Vertebrate Internet Sites

Visit the zoology website at http://www.mhhe.com to find live Internet links for each of the references listed below.

1. **Class Reptilia.**
   University of Michigan site on amphibians. Pictures, much information on the morphology, distribution, and ecology of a large number of reptiles. Each species is linked to web pages.
2. **Subphylum Vertebrata, Class Reptilia, from the University of Minnesota.**
3. **Reptilia.**
   This site provides links to all kinds of information on reptiles.
4. **Amphibian and Reptile WWW Sites.**
   This site contains many links to herpetological sites.
5. **Reptile Database.**
   This site contains information on living and extinct reptiles.
6. **Snakebite Emergency Information.**
   What to do, and not to do if bitten.
7. **National Wildlife Federation, Endangered Sea Turtles Site.**
   General information, links, conservation issues, and more.
8. **Subphylum Vertebrata, Class Aves, from the University of Minnesota.**
9. **The Birds of North America.**
   The Academy of Natural Sciences supports this site, where you can look up natural history information on almost all breeding birds in North America.
10. **Ornithological Web Library (O.W.L.) Main Page.**
    Links to organizations and other bird sites are found on this huge list of over 1000 URLs.
11. **Birdnet.**
    All about ornithology—links to information and 10 professional organizations on birds.
12. **Peterson Online: Birds.**
    This site presents information that helps the novice or the skilled birder to identify birds. It includes games in bird identification.
13. **Class Aves.**
    University of Michigan site on birds. Pictures, much information on the morphology, distribution and ecology of a large number of birds, with links to nearly all orders. Each taxon is linked to web pages.
14. **Animal Diversity Web, University of Michigan.**
    Passerine birds. Pictures and much information on a variety of passerine (perching) birds.
15. **Bird Banding Laboratory—Patuxent Wildlife Research Center.**
    Information from their bird banding program.
16. **Berkeley Museum Bird Collection.**
    Information on the Berkeley Museum bird collection, including its 14,000 sets of nests and eggs.
17. **Links of Interest in Ornithology.**
    This site provides links to ornithology journals, societies, meetings, and other information.
18. **The North American Breeding Bird Survey.**
    Offers numerous details about the program as well as colorful maps and informative graphs for each species surveyed.

CHAPTER 9

Mammals

INTRODUCTION

Descendants of synapsid reptiles, mammals are vertebrates with hair and mammary glands. Additional characteristics distinguishing mammals from other vertebrates include a lower jaw composed solely of a dentary bone articulating with the squamosal bone; two sets of teeth (deciduous and permanent); three middle ear bones (ossicles); a pinna to funnel sound waves into the ear canal; marrow within the bones; loss of the right fourth aortic arch; nonnucleated red blood cells; and a muscular diaphragm separating the thoracic and abdominal cavities. In addition, most mammals have sweat glands, heterodont dentition, and extensive development of the cerebral cortex. Approximately 4,600 species of mammals currently inhabit the world.

EVOLUTION

Fossil evidence indicates that mammals arose from a synapsid reptilian ancestor (Fig. 9.1). The subclass Synapsida appeared during the Lower Pennsylvanian over 300 million years ago and became extinct about the end of the Triassic period some 190 million years ago. The earliest synapsid, *Archaeothyris,* was a pelycosaur found in Nova Scotia (Reisz, 1972). The climate of Nova Scotia some 300 million years ago was warm and moist, and much of the land was covered by forests dominated by giant lycopods. A cladogram of the synapsids emphasizing mammalian characteristics is presented in Fig. 9.2.

Synapsids were quadrupedal reptiles (Fig. 9.2) that possessed a single temporal fossa whose upper border was formed by the postorbital and squamosal bones (see Fig. 7.5). Some researchers feel that a chain of small bones (articular, quadrate, angular) that formed the hinge attaching jaw and skull in mammal ancestors began moving back along the skull in synapsids. These bones were beginning to do double duty: hinging the jaw and likely picking up higher frequency sounds (perhaps made by insects). They also were destined to join with the columella (stapes) already in the ear to become part of the middle ear in all mammals, a process that would result in a shift in jaw articulation from articular–quadrate to dentary–squamosal. The quadrate became the incus, the articular became the malleus, and the angular became a bony ring, the tympanic, which holds the tympanus (eardrum) (Fig. 9.3). When sound waves strike the tympanum, vibrations are transmitted via the malleus, incus, and stapes to the inner ear.

Brain growth in early mammals could have triggered the migration of these skull bones. Paleontologist Timothy Rowe of the University of Texas at Austin followed brain growth and ossicle position in opossum embryos (Fischman, 1995b). Whereas the ossicles reached their maximum size 3 weeks after conception, the brains continued to enlarge for another 9 weeks, putting pressure on the ear ossicles. The ossicles, whose movement away from the jaw was caused by the expansion of the skull to hold the bigger brain, were pushed backward until they reached the adult position.

As early synapsids increased in size, they adapted by developing proportionately larger heads, longer jaws, and more-advanced jaw muscles. Teeth differentiated into incisors, canines, and grinding cheek teeth (molars). Of all the synapsids, therapsids (order Therapsida) are considered to be the line that branched to the mammals (Fig. 9.2). Therapsids date back to the early Permian, 280 million years ago (Novacek, 1992); fossils of Middle and Late Permian and Triassic age are known from all continents. The temporal fossae of all therapsids were much larger than in pelycosaurs, indicating an increase in size of the jaw-closing musculature (Kemp, 1982). Associated with this was the presence of a single large canine in each jaw, sharply distinct from both incisors and postcanine teeth. The skull is also more robust than in advanced pelycosaurs. Broom's (1910) classic paper demonstrated that the therapsids were closely related to the pelycosaurs, and this affinity has never been questioned.

There were five major groups of mammal-like reptiles: dinocephalians (primitive carnivorous therapsids), gorgonopsians (advanced carnivorous therapsids), anomodonts

FIGURE 9.1

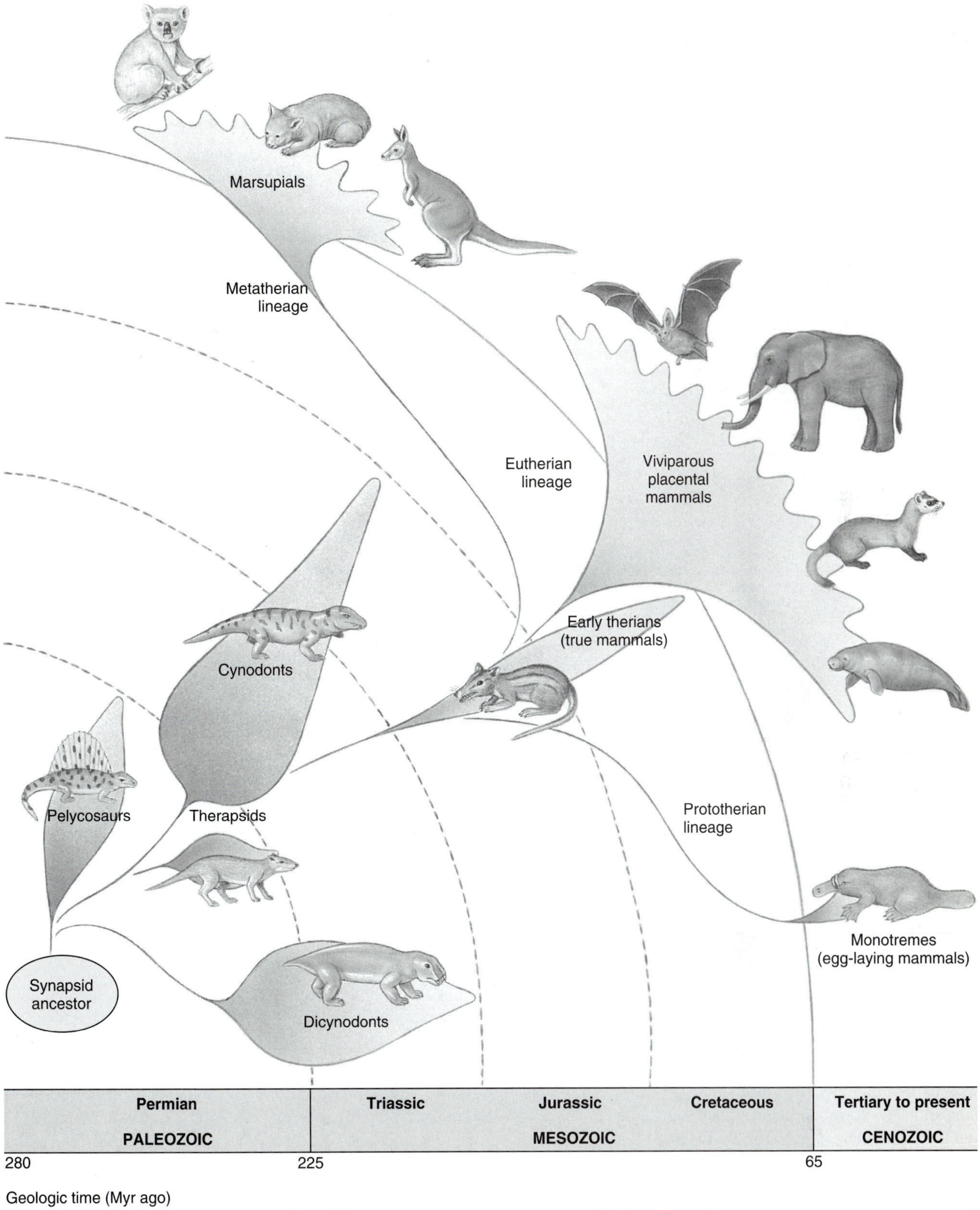

Evolution of major synapsid groups. Synapsids are characterized by a single temporal opening on each side of the skull. Pelycosaurs (early mammal-like amniotes of the Permian) and their successors, the therapsids, gradually evolved changes in their jaws, teeth, and body form that presaged several mammalian characteristics. One group of therapsids, the cynodonts, gave rise to the therians (true mammals) in the Triassic. Current fossil evidence indicates that all three groups of living mammals—monotremes, marsupials, and placentals—are derived from the same lineage. The great radiation of modern placental mammals occurred during the Cretaceous and Tertiary periods.

FIGURE 9.2

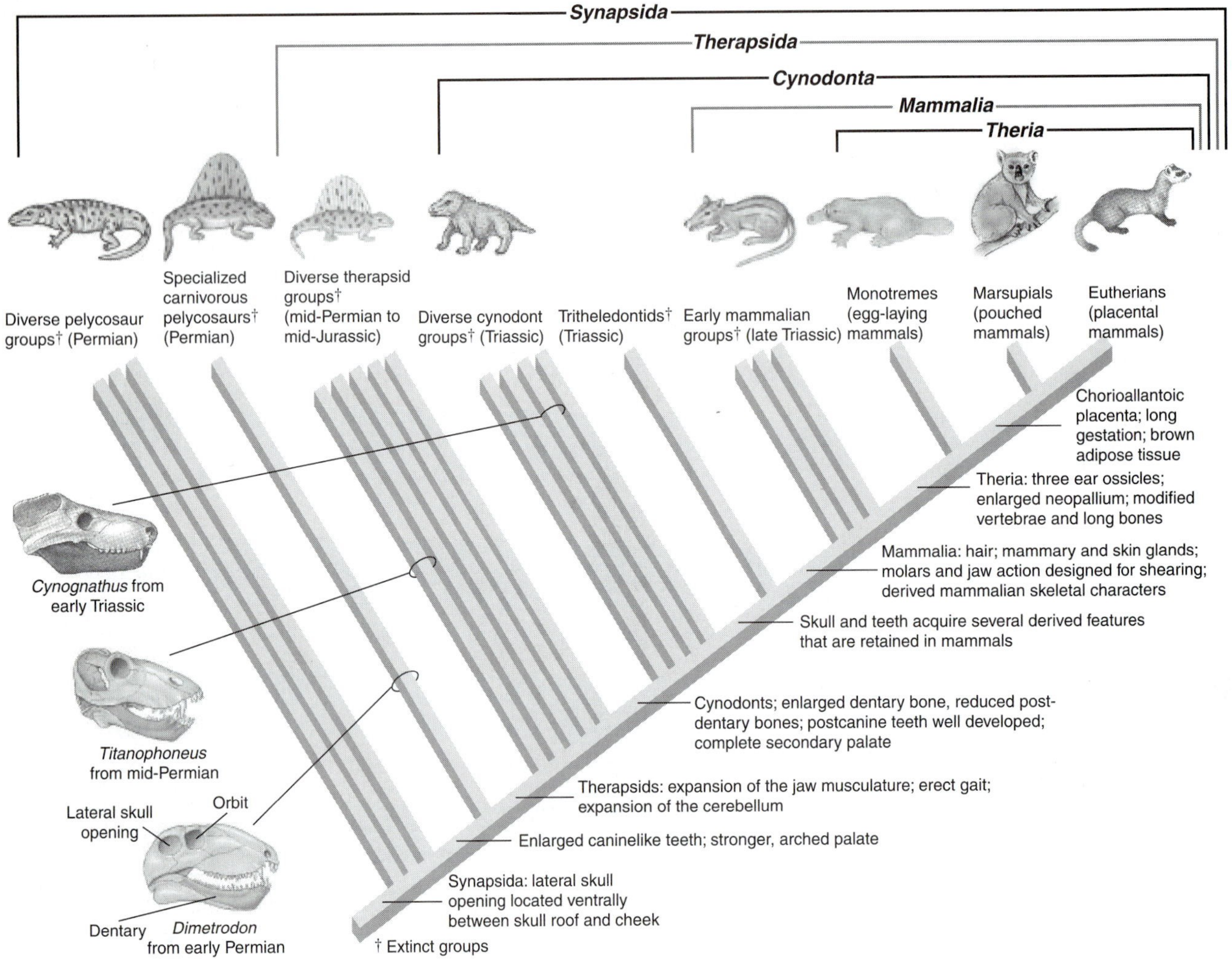

Cladogram of the synapsids emphasizing the origins of important mammalian characteristics, which are shown to the right of the cladogram. Extinct groups are indicated by a dagger †. The skulls show the progressive increase in size of the dentary relative to other bones in the lower jaw.

(herbivorous therapsids), therocephalians (advanced carnivorous therapsids), and cynodonts (advanced carnivorous therapsids). Of these, cynodonts are most closely associated with the lineage that evolved into modern mammals.

Cynognathus was a typical advanced cynodont (Fig. 9.4a). The known members of this genus were the size of a large dog and had powerful jaw muscles (masseter and temporalis). The dentary bone formed most of the lower jaw, in contrast to the typical reptilian mandible, which consisted of several bones. Heterodont dentition was present; instead of swallowing food whole as reptiles do, *Cynognathus* had cheek teeth that were adapted for cutting and crushing food. A well-developed secondary palate and two occipital condyles were present. Although the articulation of the mandible to the skull was still reptilian (articular–quadrate), the articular bone of the lower jaw and the quadrate bone of the skull had decreased in size. Thus, *Cynognathus* had not yet attained the most widely accepted character separating mammals from reptiles—a functional joint between the dentary and squamosal bones.

The limbs of therapsids such as *Cynognathus* had evolved from the primitive sprawling position to a position where the long bones of the limbs were parallel to the body and almost beneath the trunk, thus making support and locomotion easier (Fig. 9.4b, c). The elbow was directed posteriorly and the knee anteriorly. This resulted in changes in bone shape and associated musculature.

Whether *Cynognathus* possessed hair and whether it was warm-blooded are unknown (Romer, 1966). Cynodonts probably did have a high metabolic rate and a more advanced, mammal-like temperature physiology (Kemp, 1982). Regardless of whether or not *Cynognathus* was a direct ancestor of mammals, it definitely appears to have been closely associated with the lineage that was evolving into mammals.

FIGURE 9.3

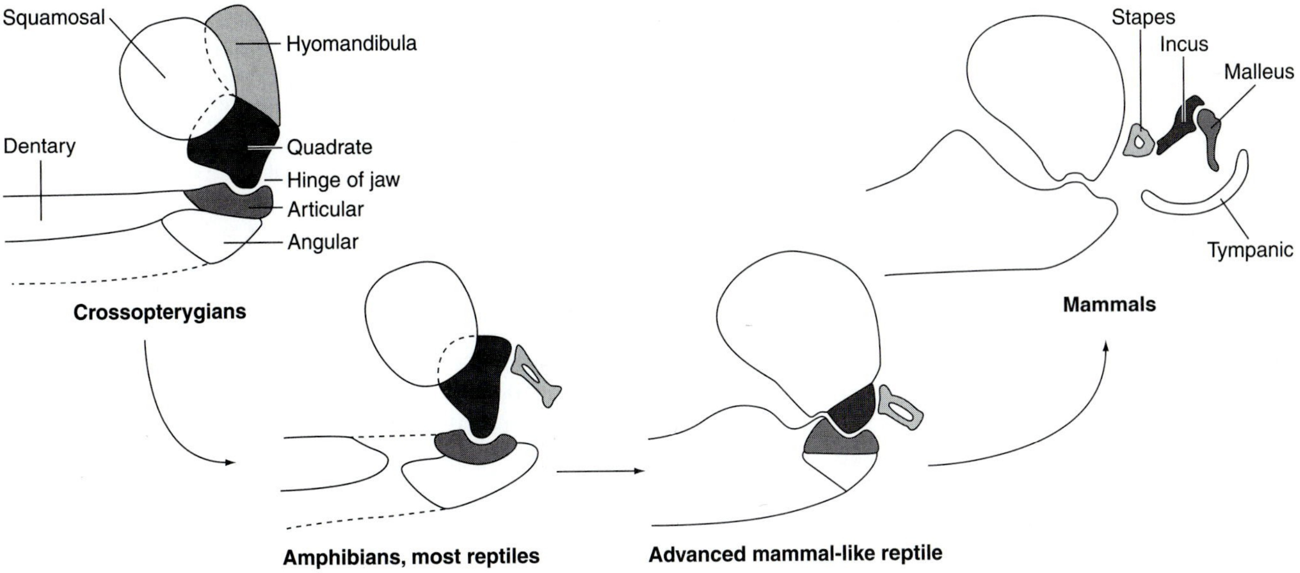

Evolution of jaw articulation and associated structures.
From Hildebrand, Analysis of Vertebrate Structure, *4th edition. Copyright © 1995 John Wiley & Sons, Inc. Reprinted by permission of John Wiley & Sons, Inc.*

FIGURE 9.4

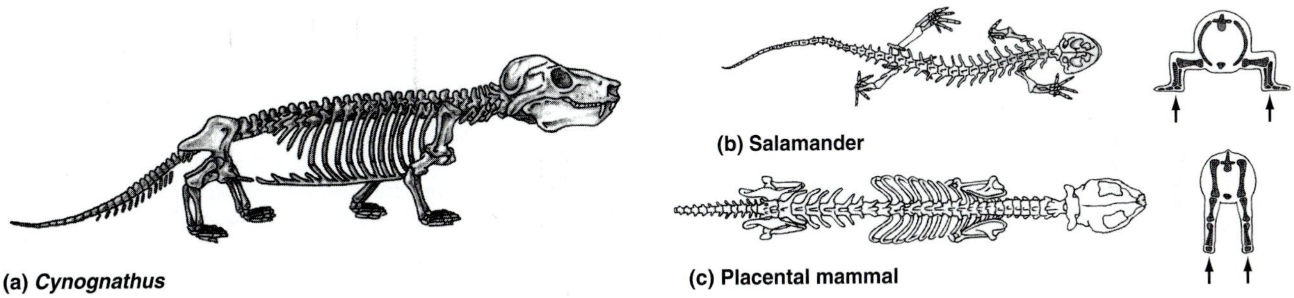

(*a*) *Cynognathus*, an advanced cynodont, was about the size of a large dog. Powerful jaw muscles and heterodont dentition allowed the cutting and crushing of food. A well-developed secondary palate and two occipital condyles were present. Evolution of posture: (*b*) The sprawled posture of the salamander was typical of fossil amphibians as well as of most reptiles. (*c*) Placental mammals. This posture began to change in synapsids, so that in late therapsid reptiles the limbs were thought to be carried more beneath the body, resulting in better support and more rapid locomotion.
From Hildebrand, Analysis of Vertebrate Structure, *4th edition. Copyright © 1995 John Wiley & Sons, Inc. Reprinted by permission of John Wiley & Sons, Inc.*

The discovery of a possible new therapsid, *Chronoperates paradoxus,* from the late Paleocene indicates that "therapsids and mammals were contemporaries for at least the first two-thirds of mammalian history" (Fox et al., 1992). *Chronoperates* is thought to have branched from a primitive cynodont and survived as a relict into the Paleocene. Prior to this discovery, therapsids were thought to have become extinct by the mid-Jurassic. The recent Paleocene fossil extends the existence of therapsids by 100 million years and has generated considerable discussion (Sues, 1992).

Controversy continues as to whether mammals had a monophyletic origin (Moss, 1969; Hopson and Crompton, 1969; Hopson, 1970; Parrington, 1973; Crompton and Jenkins, 1979; Futuyma, 1986) or a polyphyletic origin (Simpson, 1945; Romer, 1966; Kermack, 1967; Marshall, 1979; Kermack et al., 1981). Kemp (1982) noted that all of the various groups of mammals can be traced to a single, hypothetical ancestor that had itself achieved the mammalian organization. He pointed out that mammals share such a range of derived characters with the advanced cynodonts that a relationship between the two seems beyond question.

Carroll (1988) considered it difficult to establish interrelationships among the remaining, nontherian mammals (monotremes, triconodonts, and multituberculates) of the Mesozoic. The entire assemblage, including the monotremes, was placed within the subclass Prototheria. However, as Carroll noted, it is presently impossible to establish that the Prototheria is a natural group. Aside from spiny anteaters (echidnas) and the platypus, all living mammals are included in a single monophyletic assemblage, the Theria (marsupial and placental mammals).

Attempts to determine which of the cynodonts are most closely related to mammals is still open to question. There are three possible candidates: a small, advanced carnivore (i.e., *Probainognathus*); another group of small carnivorous forms, the tritheledontids; or the tritylodontids, which possess the greatest number and range of mammalian characters of any of the cynodonts (Kemp, 1982).

Probainognathus was a small animal with a slender zygomatic arch. The dentary had possibly just made contact with the squamosal, forming the mammalian secondary jaw hinge alongside the reptilian hinge. Canine teeth were present.

Diarthrognathus, the best known of the tritheledontids, possessed the nonmammalian articular–quadrate joint as well as the mammalian dentary–squamosal joint. Postorbital and prefrontal bones were absent; the zygomatic arch was slender; and the teeth were covered with enamel. Dentition indicates that these were highly specialized herbivores, whereas early mammals were carnivorous (Carroll, 1988).

Tritylodontids, which possessed multirooted teeth, also had lost the prefrontal and postorbital bones. They possessed acoelous (flattened centra on both anterior and posterior surfaces) vertebrae. The large acromion process of the scapula permitted the development of a large supraspinatus muscle. The humerus had become slender, and the forelimb now operated in a more erect manner. The pubis had turned posteriorly, and the ischium was reduced and horizontal. The musculature and locomotion were virtually fully mammalian.

Even though the tritylodontids possessed the greatest number and range of mammalian characters, Hopson and Barghusen (1986) and Shubin et al. (1991) presented data supporting the cynodont–tritheledont phylogeny. The decision as to which groups should be included as mammals still is open to conjecture and cannot be answered definitively until additional evidence clarifies the relationships among several key groups. The "answer" ultimately depends on the definition one uses to define a mammal.

Diphyodonty (having two sets of teeth during life) is considered a basic characteristic of the class Mammalia. Parrington (1971) has shown fairly conclusively that diphyodonty occurred in *Eozostrodon* specimens examined from the Triassic. *Eozostrodon,* believed to be an early triconodont (primitive mammal), was similar to a small shrew, with a skull length of 2 to 3 cm and a presacral length of approximately 10 cm. Many features of the skull were mammalian, including tooth structure, the presence of diphyodont dentition, and the form of the lower jaw. However, the articular bone still formed a jaw hinge with the small quadrate bone lying in a pocket in the squamosal. Thus, the postdentary bones had not formed a set of ear ossicles independent of the lower jaw, as occurs in modern mammals. Parrington (1967) and Crompton and Jenkins (1968) independently concluded, from the similarity of the molar teeth, that *Kuehneotherium,* a therian, and *Eozostrodon* were closely related. The molar teeth of *Eozostrodon* appear to be the basic type from which all therian molars have evolved.

During the Triassic, each group of advanced synapsids gave rise to a different group of animals (symmetrodonts, pantotheres, multituberculates, triconodonts) that we can call mammals. The transition from primitive reptile to primitive mammal occurred between the end of the Pennsylvanian period and the close of the Triassic, because the earliest known mammal fossils are from the Late Triassic of Europe.

The Symmetrodonta and Pantotheria were both Jurassic mammals and are probably more closely related to each other than to the other groups. Symmetrodonts had the cusps of their molar teeth arranged in a symmetrical triangle, with the base of the triangle external in the upper jaw and internal in the lower. Pantotheres had molars that were three-cusped, with the cusps arranged in an asymmetrical triangle. The Multituberculata were probably among the earliest herbivorous mammals, although they also probably included insectivorous and omnivorous forms. Ranging in size from a small mouse to as large as a woodchuck, multituberculates appeared in the Upper Triassic and persisted into the Eocene. Some researchers feel they are most closely related to monotremes; others consider them closer to therians (Monastersky, 1996d). Triconodonts were Jurassic mammals that were probably carnivorous; their molars typically had three sharp conical cusps arranged in a row along the long axis of the tooth. The main cusp of the lower molars occluded between the main cusp and the anterior accessory cusp of the corresponding upper molar. This shearing dentition may indicate that they preyed on other vertebrates.

During the Jurassic and Cretaceous, a variety of mammals evolved. Recent analyses indicate that the last common ancestor of living mammals probably lived in the Early or Middle Jurassic (Rowe, 1999). Thus, Mammalia is 20 to 40 million years younger than once believed. Until recently, the earliest details of mammalian history were unknown due to a lack of fossils.

In 1999, however, Ji et al. (1999) described one of the world's oldest complete mammal fossils (*Jeholodens jenkinsi*) (Fig 9.5), dating back at least 20 million years. The fossil is a close relative to the common ancestor of all mammals alive today, from monotremes to opossums to humans. The incredibly complete fossil comes from the same Late Jurassic/Early Cretaceous deposit of Liaoning, China, that recently yielded feathered dinosaurs and one other complete mammal skeleton. Although the teeth identify it as a triconodont, skeletal characteristics largely support the sister-group relationship of multituberculates with therian mammals. The rat-sized animal walked on mammalian front legs and splayed reptilian hind legs (Zimmer, 1999). The elbows point back, whereas the knees point to the side. The limb structure indicates it was probably a ground-dwelling animal, thus indicating that

FIGURE 9.5

The ancestor of all modern mammals may have resembled this reconstruction of *Jeholodens jenkinsi*, a 120-million-year-old mammal from China.

BIO-NOTE 9.1

The Saltville Deposits

Saltville, a small town with extensive saline deposits along the Holston River in southwestern Virginia, has been the site of major paleontological investigations since 1964. Large Pleistocene mammals known from this site include Jefferson's ground sloth (*Megalonyx jeffersonii*), the mastodon (*Mammut americanum*), the woolly mammoth (*Mammuthus primigenius*), the horse (*Equus* sp.), the caribou (*Rangifer tarandus*), the stag-moose (*Cervalces scotti*), and the musk-ox (*Bootherium bombifrons*). Thomas Jefferson mentioned the "salines opened on the North Holston" in his 1787 book *Notes on the State of Virginia,* giving them as the source of a mastodon tooth sent to him. Jefferson's reference makes the Saltville Valley one of the earliest localities on record for fossils of large mammals that lived in North America during the Pleistocene.

mammals arose as terrestrial forms and only later did their therian descendants take to the trees.

Living mammals are classified into 26 orders. The Monotremata contains the only egg-laying mammals (duck-billed platypus and two species of echidnas or spiny anteaters) (Fig. 9.6). All other mammals are viviparous.

The Jurassic and Early Cretaceous were times of "experimentation" for the mammals. Dinosaurs were still abundant. Primitive hooved mammals and even some early primates had evolved. Birds were able to fly. The Late Cretaceous, however, was a time of change. Dinosaurs and most other reptiles became extinct. The extinction of Mesozoic reptiles left empty niches that were exploited by the more efficient mammals. Mammals began to "inherit the Earth." The advantages of homeothermy, viviparity, and the expansion of the brain allowed mammals to spread over most of the land surface of the Earth, to develop flight, and to reinvade the aquatic environment.

Relationships among fossil and extant mammals are being investigated and clarified through new systematic

FIGURE 9.6

(a)

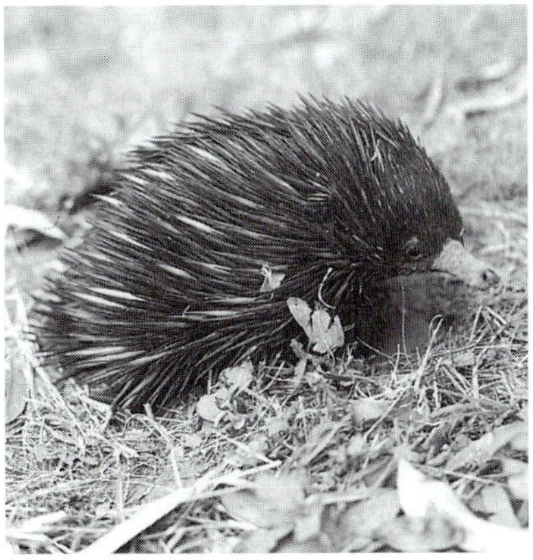

(b)

Monotremes—the only egg-laying mammals: (*a*) duck-billed platypus; (*b*) echidna. The platypus raises its young in a nest; the echidna, or spiny anteater, places them in a pouch on her abdomen.

methods, a growing molecular database, and continuing paleontological discoveries (Novacek, 1992). Cladistics and powerful computer programs have permitted the analyses of diverse anatomical characters and nucleotide sequences, while molecular techniques have produced data through protein sequencing, direct comparisons of DNA sequences from selected genes, and immunological comparisons. For example, Springer et al. (1997) found that the sequence of nucleotides of five genes differed from animal to animal. Of the mammals studied, the most closely related turned out to be elephants, aardvarks, manatees, golden moles, elephant shrews, and hyraxes—small rabbitlike animals of Africa and Asia. Far less similar were the animals that had been considered relatives of golden moles—shrews, common moles, and hedgehogs. The genetic evidence also showed that elephant shrews, thought to be most closely related to rabbits and rodents, are nearer to aardvarks and elephants. Based on the genetic differences observed, Springer et al. estimated that the common ancestor of all these mammals lived about 80 million years ago, probably in Africa, since that is where the earliest fossils of members of these six groups have been found. Embryological studies of the renal, reproductive, and respiratory systems of the elephant confirm that it evolved from an aquatic mammal and that elephants share a common ancestor with sea cows (Sirenia) (Gaeth et al., 1999).

Molecular, paleontological, and morphological studies have suggested that the cetaceans (whales, dolphins, and porpoises) and artiodactyls (even-toed ungulates, including pigs, hippopotamuses, camels, and ruminants) form a clade or monophyletic group; that is, they have a common

BIO-NOTE 9.2

Primate Evolution

The discovery of mouse-sized primates (*Shoshonius cooperi*) in North American deposits 50 million years old (Fig. 9.7) and in 45-million-year-old rocks in eastern China (*Eosimias*) has altered estimates of when early primate groups first evolved. Anatomical features of four nearly complete fossil skulls of *Shoshonius* indicate that it was a primitive form of tarsier—a tree-dwelling primate today found only in the forests of Southwest Asia. The *Eosimias* fossils display several anthropoid characteristics such as small incisors, large canines, and the presence of distinctive premolars and molars. The back corner of its lower jaw was rounded along the bottom, as is the jaw of humans and other higher primates. Scientists had assumed the evolutionary parting of tarsiers and simians had occurred about 40 million years ago. Prior to the discovery of *Shoshonius* in the mid-1980s and *Eosimias* in 1994, the oldest well-documented anthropoids came from 36-million-year-old rocks in Egypt, suggesting that such creatures arose in Africa; however, the *Eosimias* fossils indicate an earlier origin, possibly in Asia. It now appears that *Shoshonius* and modern tarsiers evolved from a common ancestor that split off from the forerunners of simians—monkeys, apes, and humans—sometime before 50 million years ago. The position of *Shoshonius* in the evolution of primates is not yet clear.

Beard et al., 1991
Bower, 1995
Culotta, 1995b
Monastersky, 1995
Simons, 1995

FIGURE 9.7

The mouse-sized primate *Shoshonius cooperi* was discovered in North American deposits aged 50 million years old.

BIO-NOTE 9.3

The Number of Mammalian Genera

John Alroy at the University of Arizona has compiled a massive database showing an "equilibrium" of mammalian diversity in North America. After the Cretaceous–Tertiary extinctions 65 million years ago, the number of mammalian genera shot up to a high of about 130 genera 55 million years ago. Thereafter, the number of genera waxed and waned, sinking to as low as 60 and rising to as high as 120, presumably in response to climate change and immigration. These fluctuations lasted millions of years, but diversity over time always averaged an equilibrium of about 90 genera. This equilibrium may represent the ecological carrying capacity for North America, and resource (e.g., food) availability may be enforcing the limit. When diversity is low, species tend to fare better because they face less competition from other species. As diversity increases, speciation declines and extinction rates go up. The result is a continuous turnover of genera but the maintenance of a relatively stable mammalian diversity over time.

Culotta, 1994

ancestor that is not shared by any other group of mammals. Molecular data presented by Shimamura et al. (1997) and reviewed by Milinkovitch and Thewissen (1997) confirm this close relationship, and also propose that cetaceans, ruminants, and hippopotamuses form a monophyletic group within the artiodactyla.

MORPHOLOGY

Integumentary System

The presence of a lightweight, waterproof epidermal layer has been important in allowing mammals to successfully colonize a variety of terrestrial environments. Some mammals, such as beavers and rats, have epidermal scales on parts of their bodies (feet, tail), but only the armadillo has **dermal scales (plates)** beneath the epidermal scales (see Fig. 1.7). These dermal scales form the protective armor covering of the armadillo.

The epidermis, which is composed of keratinized stratified squamous epithelium, is differentiated into distinct layers (Fig. 9.8). The deepest layer is the **stratum basale (germinativum)** and, as in other vertebrates, is the area of active mitosis. As new cells form, older cells are pushed toward the surface and become successively part of the **stratum spinosum,** the **stratum granulosum,** often the **stratum lucidum,** and finally, the surface **stratum corneum. Keratin,** which is impermeable to water and gases, is produced by **keratinocytes,** the most abundant of the epidermal cells. **Melanocytes** (melanophores) produce the pigment melanin, which is primarily responsible for skin color and for protecting keratinocytes and the underlying dermis from excessive ultraviolet (UV) radiation.

FIGURE 9.8

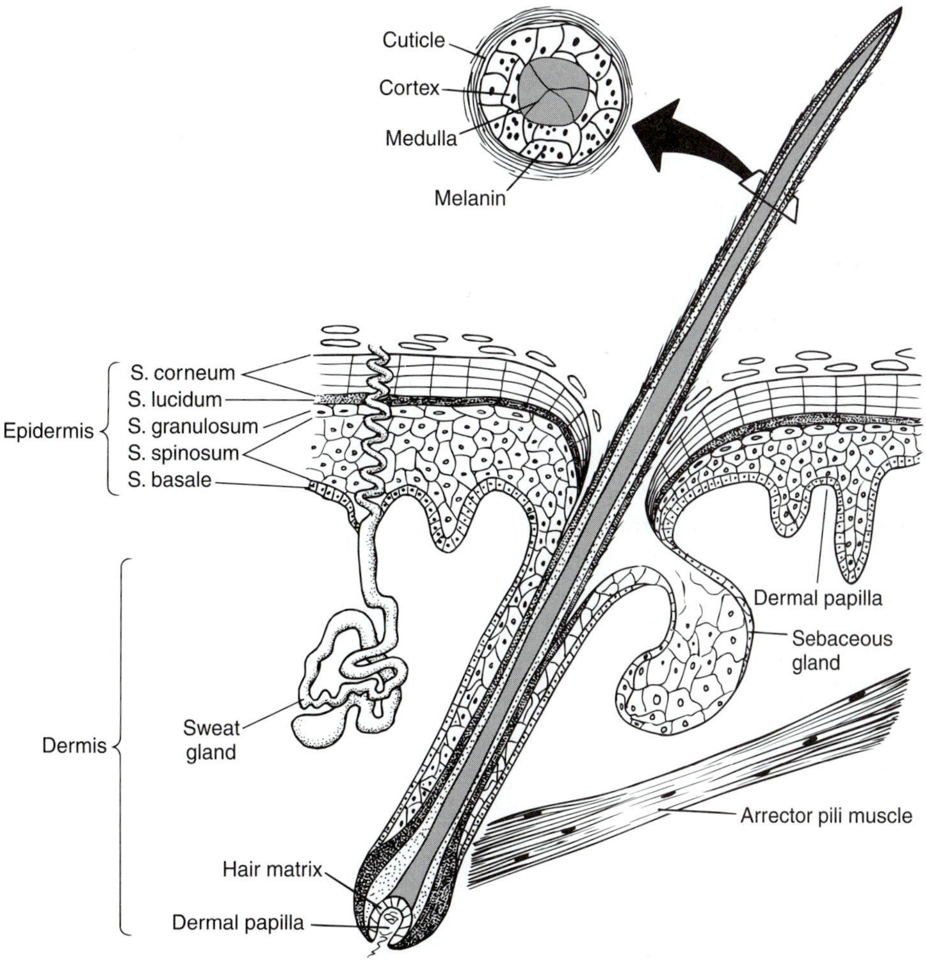

Section of mammalian skin showing the structure of a hair and glands. Sebaceous glands produce sebum, which lubricates the hair and skin. Sweat (sudoriferous) glands secrete either a watery sweat that cools the body as it evaporates or a milky secretion that may play a role in sexual attraction.

The epidermis gives rise to hair, various glands, nails, scales, hooves, baleen, and (together with the dermis) horns. Hair is one of the most characteristic epidermal specializations in mammals. It represents a new development, not a modification of horny scales, as are feathers. Anatomical, developmental, neurological, and paleontological data have been used to support the hypothesis that mammalian hair arose from highly specialized sensory **mechanoreceptors** (receptors that detect mechanical deformation of the receptor itself) found in early synapsids.

Two kinds of hair are present. The outer, coarser, and usually longer hairs that serve primarily a protective function are **guard hairs.** The inner, finer, and usually shorter hairs constitute the **underfur,** which serves to insulate the body. Vibrissae, bristles, and quills are specialized modifications of guard hairs. Specialized hairs around the mouth (mystacial vibrissae) and eyes (superciliary vibrissae) often serve as tactile organs and are sensitive to touch. Each vibrissa is attached beneath the skin to a capsule that loosely surrounds it. In the capsule is a jellylike layer of fatty tissue that can stimulate the capsule membrane, to which up to 150 nerve fibers may be attached.

Hair is present in at least some stage of development in all mammals. Intraspecific variation occurs, especially in the guard hairs and in the scales along the hair shaft. There is little or no sexual or seasonal difference in hair structure.

Hair completely covers the bodies of most species, although it may be restricted to specific areas in others. For example, the naked mole rat (*Heterocephalus glaber*) of Ethiopia, Somalia, and Kenya has only a few pale-colored hairs scattered over the body, vibrissae on the lips, and fringes of hairs on its tail and between the toes of its hind feet (Sherman et al., 1992). In some adult whales, hair may be almost entirely absent, with only a few vibrissae being present around the lips. In some whales, hair may be present only in the young. On the other hand, sea otter fur contains about 100,000 hairs per cm^2—the densest fur of any mammal (Love, 1992; Kruuk, 1995).

An individual hair first appears as a hair primordium. The primordium is a downward-projecting growth from the stratum basale. A dermal papilla forms at the base of the indentation. As epidermal cells continue to proliferate, the hair primordium grows deeper into the dermis and is nourished by blood vessels of the papilla. The hair primordium finally surrounds the dermal papilla as an inflated balloon would surround a finger pushed into it. When the bulb at the base of the primordium is differentiated sufficiently, cells begin to appear, and a hair shaft begins to rise in the follicle. The hair thus is forced out of the skin by growth from below.

A typical hair consists of a **shaft** and a **root** (Fig. 9.8). The shaft lies free within the follicle and projects above the surface of the skin. In general, the shaft points posteriorly in order to minimize friction with the environment. The root is that portion deep within the follicle where the hair has not yet separated from the surrounding epidermal cells of the follicle wall. The swelling at the base of the hair containing the dermal papilla is known as the **bulb.** It is an area of rapid mitosis, which constantly is contributing new cells that make the hair longer.

The shaft of a typical hair consists of an inner **medulla,** which contains most of the pigments that determine the appearance of the hair; a surrounding **cortex** that forms the main bulk of the hair and is usually transparent, but may contain pigments; and an outer **cuticle.** Hair color is determined by the distribution and density of melanin and xanthophyll granules in the keratinized cells and by the number of air vacuoles in the medulla of the hair. Gray and white hairs result from large numbers of air vacuoles and little melanin.

The thin cuticle is devoid of pigment and is composed of **cuticular scales** that completely surround the hair shaft. Scale patterns vary so greatly that their arrangement is often characteristic of a species and has been used to identify loose hairs from animal dens and scats (feces). Large hairs generally contain air spaces that add greatly to the insulating properties of the hair. In cross section, hairs may be round, oval, or flattened. Circular hairs are usually straight or only slightly curved, whereas flattened hairs are curly.

A small smooth muscle, the **arrector pili** (Fig. 9.8), inserts on the wall of the hair follicle in the dermis of many mammals. When the arrector pili contract, the follicles and hair shafts are drawn toward a vertical position. The skin around the base of each hair is pulled into a tiny mound, causing (in humans) "goose bumps" or "goose flesh," a vestigial physiological response no longer capable of serving an insulating function. In most mammals, however, the action of the arrector pili muscles allows a layer of air to be trapped between the skin and the fur to provide increased insulation for both heat gain and heat loss. When frightened or alarmed, some mammals show aggression by erecting their hair.

Hairs, like feathers, are nonliving, keratinized structures that are constantly subjected to wear and must be replaced. In many mammals, hair replacement is seasonal; some molt their fur annually, usually in the fall. Showshoe hares (*Lepus americanus*) and short-tailed weasels (*Mustela erminea*) are examples of mammals that molt twice a year, in the spring and fall; still others molt irregularly. The replacement of specialized hairs such as eyelashes and vibrissae occurs on a continual basis and varies with each individual.

Some species undergo dramatic changes in appearance when they molt (King, 1989). Short-tailed weasels, for example, remain brown throughout the year in the southern part of their range. In the northern part of their range, however, they turn white in winter and are much less conspicuous on snow-covered terrain. This change results from a molt in which the new hairs contain no pigment. Several hares, including the varying hare (Fig. 9.9), weasels, the Arctic fox (*Alopex lapogus*), and collared lemmings (*Dicrostonyx* sp.), exhibit similar seasonal changes.

Photoperiod, in conjunction with melatonin produced by the pineal gland, initiates changes in the central nervous system and in the endocrine glands that induce molting. Molting is a gradual process in which old hairs are not lost

FIGURE 9.9

(a)

(b)

Snowshoe, or varying, hare (*Lepus americanus*) in (*a*) brown summer pelage and (*b*) white winter pelage. In winter, extra hair growth on the hind feet provides the hare with better support in the snow. Snowshoe hares inhabit the northern coniferous forests and serve as important food for lynxes, foxes, and other carnivores.

FIGURE 9.10

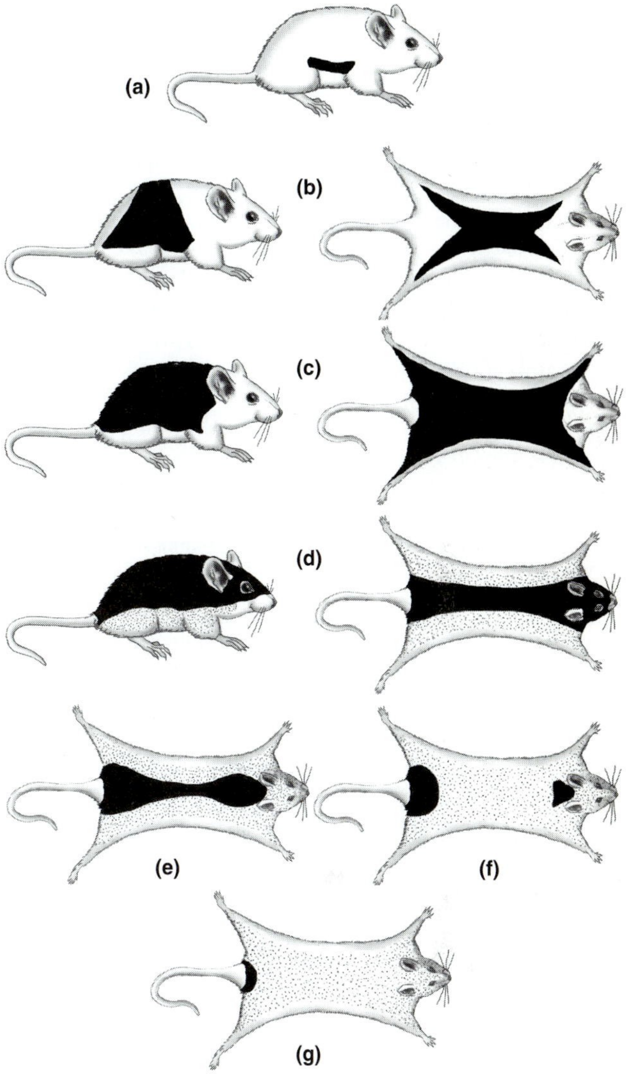

Sequence of postjuvenile molt on the dorsum in the golden mouse (*Ochrotomys nuttalli*). Shaded portions represent areas of active hair replacement. Stippled areas represent adult pelage.
Source: Linzey and Linzey, in Journal of Mammalogy, *Vol. 48(2):236–241, May 1967.*

until new hairs have almost fully formed. The sequence of hair replacement and the molting pattern is species-specific (Fig. 9.10).

The color of a mammal is the result of either the color of the skin due to capillaries and pigments, the color of the fur (pelage), or both. Hair color is determined by the amount and distribution of pigments in addition to the structure of the hair. Two pigments, melanin and xanthophyll, normally are found in mammalian hairs and are deposited while the hair is growing in the follicle. Melanin may be present in the medulla and/or cortex of a hair and produces black or brown hair. Xanthophyll occurs only in the medulla and results in reddish or yellowish colors. Color patterns are caused by genetically controlled variations in the amounts and distribution of pigments present in the hair.

Two color phases may be expressed in different individuals of the same species. This phenomenon is termed **dichromatism.** The occurrence of these color phases (which is genetically controlled) often consists of a black phase (melanistic form) as well as the normal wild type in the same population. Black gray squirrels (*Sciurus carolinensis*) and black fox squirrels (*Sciurus niger*) often have melanistic individuals and normal-colored individuals occurring in the same litter. The darker phase sometimes is more prevalent in the northern part of the range of the species. As many as 12 color phases are known in the fox squirrel (*Sciurus niger*).

Whereas one function of hair is to serve as a tactile organ, other major functions are to protect the body from the elements, to provide insulation, to aid in concealment, to serve as

BIO-NOTE 9.4

Vibrissae in Seals

Extremely well developed vibrissae are found in the ringed seal (*Phoca hispida saimensis*), which lives in perpetually murky water around Finland. The innervations of the vibrissae are more than 10 times as great (1200–1600 fibers) as those normally found in mammals. The seals are thought to maneuver and locate food by echolocation; however, their low-frequency clicks do not reflect well. The sensitive vibrissae may aid echolocation by serving as antennae for monitoring the returning echoes. In addition to spatial orientation, the vibrissae may also provide the seals with information about the diving speed and changes in swimming orientation. Harbor seals (*Phoca vitulina*) also use their vibrissae as a hydrodynamic receptor system to detect minute water movements, allowing them to gain information about aquatic prey, predators, or conspecifics.

Hyvarinen, 1989
Dehnhart et al., 1998

a warning mechanism, to assist with communication, and even to assist in locomotion. Air trapped under the hair also can modify the buoyancy of some aquatic mammals, such as river offers (*Lutra canadensis*); hair on the dorsoventrally flattened tails of flying squirrels (*Glaucomys*) assists in gliding by helping the tail serve as a rudder (Fig. 9.11); water shrews (*Sorex palustris*) and muskrats (*Ondatra zibethicus*) have a fringe of hairs along the outer edge of each foot, which assists their movements in water by providing a greater surface area. Quills of the porcupine (*Erethizon dorsatum*) and scales (modified hairs) of the pangolin (*Manis tricuspis*) serve for protection. In certain situations, hair may be used to show aggression. Some prey species have developed elaborate displays that often may be warning signals directed to other prey. For example, the white rump patch of white-tailed deer (*Odocoileus virginianus*) is normally covered by the tail and is only slightly visible, but when the deer is alarmed, the tail is raised and the exposed white rump patch serves as a warning to other nearby deer (Fig. 9.12). In other species, the prey display may be aimed instead at the predator in an apparent attempt to deter further pursuit (Hasson, 1991).

Claws, nails, and **hooves** are hard, keratinized modifications of the stratum corneum at the ends of digits. Most mammals possess claws, but these structures have evolved into nails in most primates and into hooves in ungulates. None of these structures is shed; they must be worn down by friction.

Several unique epidermal derivatives occur in some mammals. For example, pangolins are covered with epidermal scales that are composed of fused bundles of hair. Hairs also grow between these scales. The scales covering the tails of rodents are also epidermal in origin. Large plates of **baleen** (often known as whalebone), which develop from cornified oral epithelium, are suspended from the palate in toothless whales (Mysticeti). The frayed edges of these sheets serve to strain plankton from water in the oral cavity (Fig. 9.13).

FIGURE 9.11

The nocturnal northern flying squirrel (*Glaucomys sabrinus*) has excellent night vision. When the gliding skin (patagium) is spread, the undersurface is nearly trebled, and glides of 40 to 50 m are possible. Special muscles adjust the position of the patagium during flight, thus providing good maneuverability.

FIGURE 9.12

Alarmed white-tailed deer (*Odocoileus virginianus*) lift their tails high in the air and expose the conspicuous white underside as a means of alerting other members of the herd or perhaps as a signal to a potential predator that it has been seen.

Four types of structures that adorn the heads of some mammals—true horns, rhinoceros horns, antlers, and giraffe horns—play roles in reproductive behavior, defense, and offense. **True horns,** characteristic of most members of the bovine family (cattle, most antelope, sheep, and goats) (Fig. 9.14 a–d), consist of a permanent bony dermal core covered by a permanent horny epidermal sheath. True horns,

FIGURE 9.13

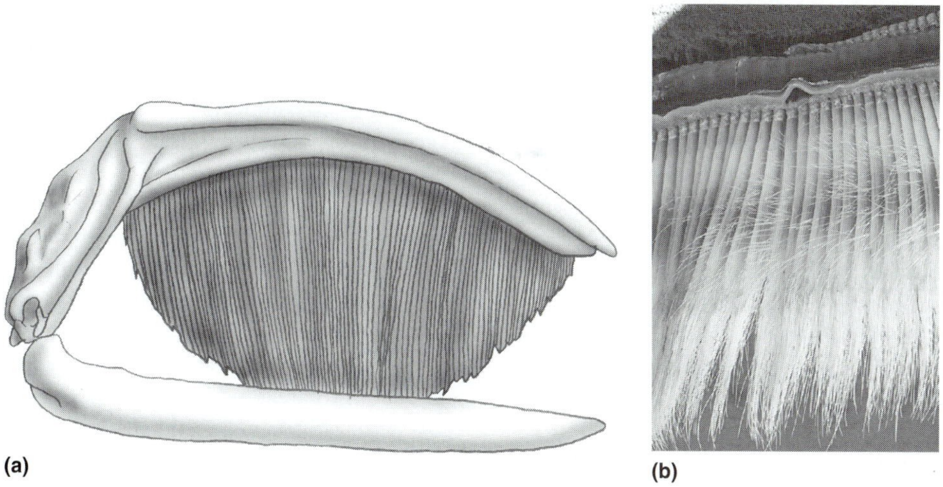

Baleen. The lining of the mouth in some whales includes an epithelium with the ability to form keratinized structures. Groups of outgrowing epithelium become keratinized and frilly to form the baleen. (*a*) Skull of the Atlantic right whale (Eubalaena); length of the skull is approximately 4 m. Note the baleen plates attached to the maxilla. (*b*) Lateral view into the partly opened mouth of a gray whale (*Eschrich gibbosus*) showing the plates of baleen hanging from the palate.

FIGURE 9.14

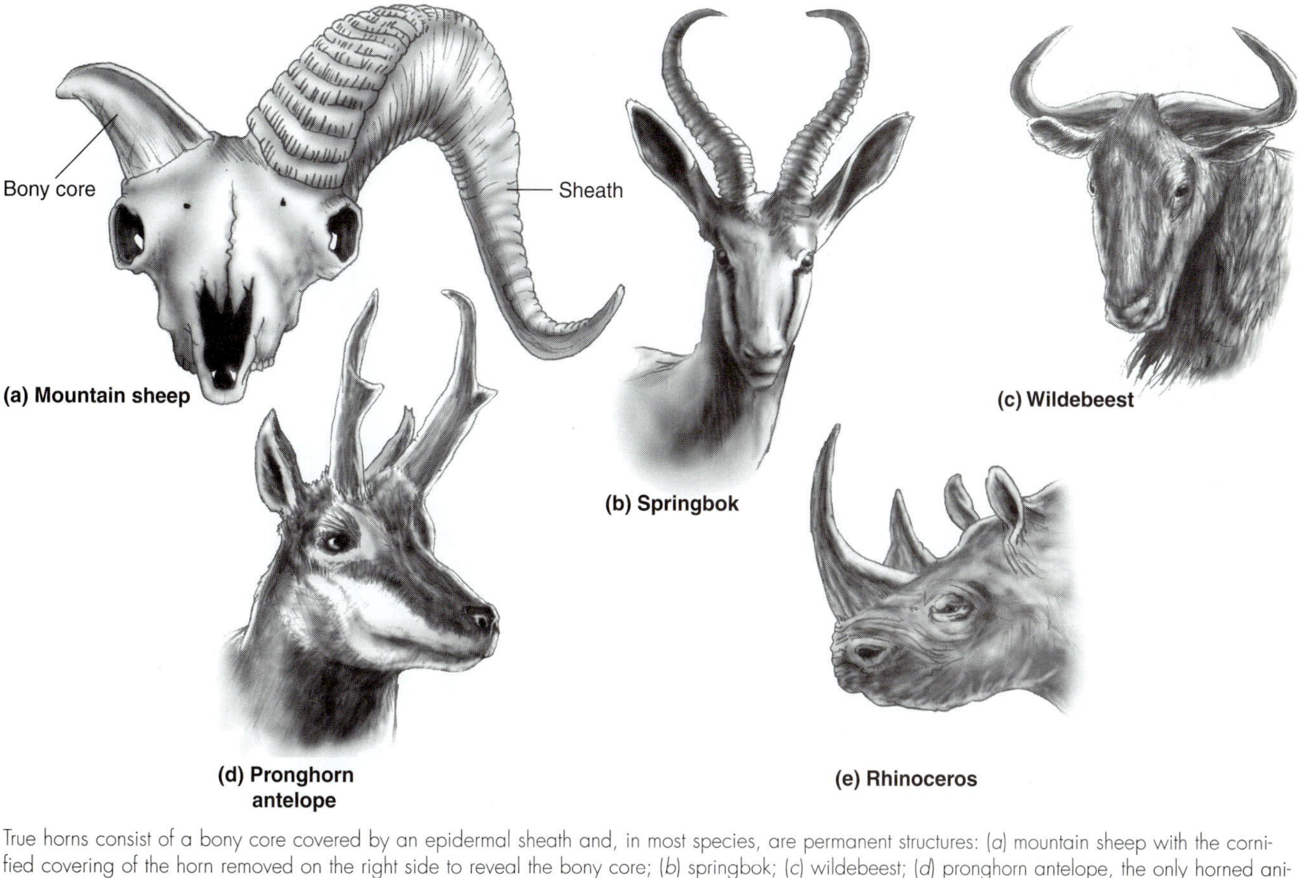

True horns consist of a bony core covered by an epidermal sheath and, in most species, are permanent structures: (*a*) mountain sheep with the cornified covering of the horn removed on the right side to reveal the bony core; (*b*) springbok; (*c*) wildebeest; (*d*) pronghorn antelope, the only horned animal with horns that are deciduous and are periodically shed; (*e*) rhinoceros horns are solid structures composed of compacted keratinized fibers.

which generally are not branched and usually are never shed, occur in both sexes but are typically larger in males. Horns grow from the base, and the dermal core increases in girth as the animal ages (Fig. 9.15a). The growth rings that are formed are somewhat similar to those of a tree and are useful in determining age. Pronghorn antelopes (Fig. 9.14d) possess true horns, but they are forked; the horny covering (but not the bony core) is shed annually.

Rhinoceros horns (Fig. 9.14e), are not true horns; they are composed of tightly packed, keratinized filaments similar to hairs but which differ in that they possess gas spaces and lack a cuticle (Ryder, 1962). Each fiber is separately visible, and there is no bony core. Rhinoceros horns occur on the snouts of both sexes and never are shed. Indian rhinoceroses (*Rhinoceros*) have one horn; African rhinoceroses (*Diceros*) have two horns. These horns grow throughout the animal's life and will regrow if removed.

Antlers are branched structures composed of solid, dead dermal bone and are characteristic of members of the deer family (Cervidae) (Figs. 9.15b and 9.16). Antlers, which are secondary sex characteristics, are affected by annual fluctuations in secretion of sex hormones, primarily testosterone. Photoperiod (daylength) is the primary stimulus for antler replacement (Goss, 1983). Increasing photoperiod stimulates the **adenohypophysis** (anterior pituitary gland), which in turn stimulates the production of testosterone. Antlers normally begin growing in spring, reach their full growth during summer, and are shed in mid-winter (Fig. 9.17a–e). They are renewed annually by apical growth centers; thus, they grow by adding new material at the extremities. While forming, antlers are covered with a layer of skin, or **"velvet."** Blood in the arteries of the velvet supplies the growing antlers, which are innervated by branches of the trigeminal nerve. When the antler reaches its full growth, the velvet is shed or rubbed off, and the bare, dead, branched bone remains. Antlers usually occur only on males and are shed annually. In caribou, or reindeer (*Rangifer tarandus*), antlers occur on both sexes, although those of males are larger and more branched. The only members of the family Cervidae that do not possess antlers are the Chinese water deer (*Hydropotes inermis*) and the musk deer (*Moschus* sp.), both of which have evolved tusks

FIGURE 9.15

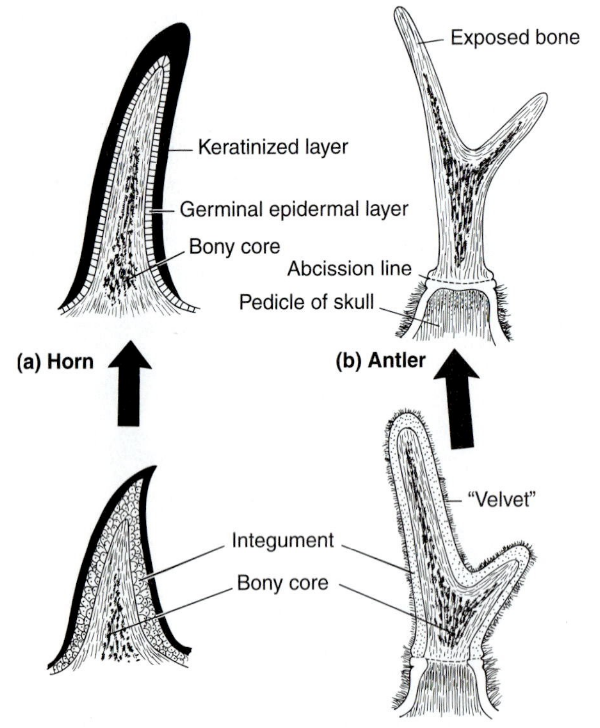

Horns and antlers. (*a*) Horns are bony outgrowths of the skull beneath the integument, which forms a keratinized (cornified) sheath. Horns occur in bovids of both sexes and are usually retained year-round. (*b*) Antlers are also bony outgrowths of the skull beneath the integument. The integument, or skin, which is referred to as "velvet" because of its appearance, gradually dries and falls off, leaving the bone of the antlers exposed. Antlers are found only in members of the deer family (Cervidae) and, except for caribou (reindeer), they are normally present only in males. Antlers are shed and replaced annually.

FIGURE 9.16

The massive antlers of the extinct giant Irish elk weighed more than its entire internal skeleton.

FIGURE 9.17

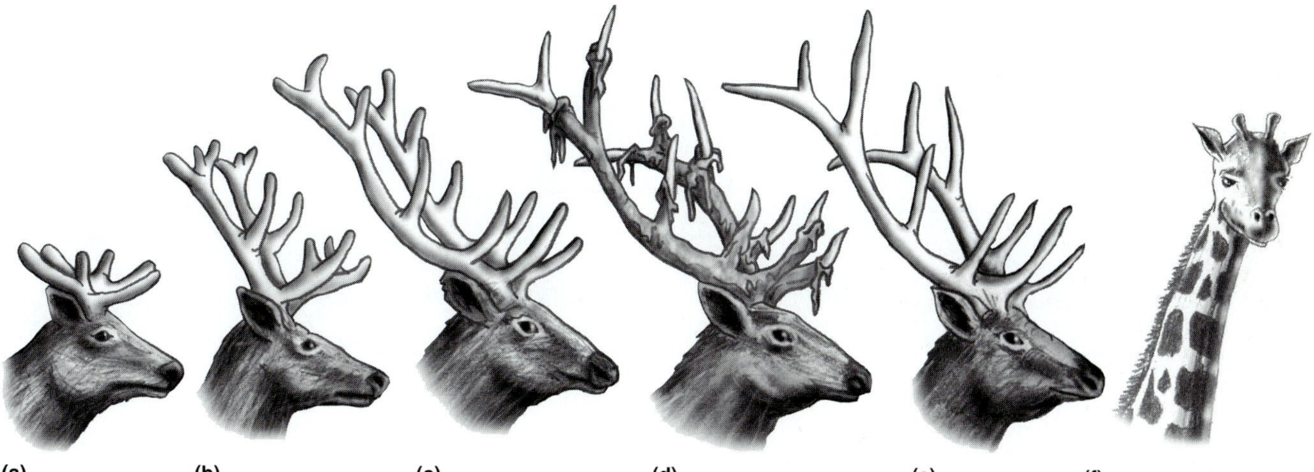

Annual growth of elk antlers: (*a*, *b*) New antlers begin to grow in April. (*c*) By May, antlers are nearly fully formed, even though they are still covered by the living integument (velvet). (*d*) By late summer, the velvet has begun to dry and peel off. (*e*) Fully formed bony antlers are in place. (*f*) Giraffe horns are small ossified knobs covered by the integument.

(Fig. 9.18). A complete discussion of the anatomy, evolution, and function of antlers may be found in Goss (1983).

The fourth type of head structure occurs on giraffes (*Giraffa camelopardalis*) (Fig. 9.17f) and okapi (*Okapi johnstoni*). These stubby "stunted" horns remain covered by living skin (velvet) throughout life and never are shed.

Multicellular skin glands are more abundant and diverse in mammals than in any other vertebrate group. They serve many functions, from sensing changes in the environment to providing milk for the young. Glands such as the mammary and sweat glands are unique to mammals.

FIGURE 9.18

The Chinese water deer (*Hydropotesinermis*) (left) and the musk deer (*Moschus*) (right) are the only members of the family Cervidae that do not possess antlers. These two species, however, have evolved tusks.

Sudoriferous (sweat) glands are long, slender coiled tubes of epidermal cells that extend deep into the dermis and often into the subcutaneous layer (see Fig. 9.8). Their secretion is watery and contains fatty substances, salts, and pigments. Sweat assists in thermoregulation, and also helps eliminate wastes, such as urea and various salts, from the body. Sweat glands, which occur in most mammals, are absent in moles, sloths, scaly anteaters, elephants, and many marine forms. They may be distributed widely over the body or restricted to certain regions such as the soles of the feet (mice, cats), the face (some bats), or the ventral surface of the body (wandering shrew).

Mammary glands (Fig. 9.19), one of the distinguishing characteristics of mammals, are modified sudoriferous (sweat) glands that produce milk for the nourishment of the young. They consist of numerous lobules, each of which is a cluster of secretory alveoli in which milk is produced. The alveoli may empty into a common duct that opens directly to the surface through a raised epidermal papilla, or nipple (Fig. 9.19, Inset). Alveolar ducts also can open into a common chamber, known as a cistern, with a long collar of epidermis, called the teat (Fig. 9.19, Inset). Secretion of milk is due mainly to the hormone prolactin produced by the anterior lobe of the pituitary gland. The distribution and number of mammary glands vary with species, with the number ranging from a single pair to as many as 12 pairs. In general, the number of teats is equal to the maximum litter size or twice the average litter size (Diamond, 1988b). Teats occur in locations appropriate to the habits of the species—thoracic (bats, primates, elephants, manatees), abdominal (many rodents, carnivores), or inguinal (ungulates) (Fig. 9.19a, b). Nutria (*Myocastor coypus*) have nipples high on their sides so that the young can nurse while swimming. Monotremes (duck-billed

FIGURE 9.19

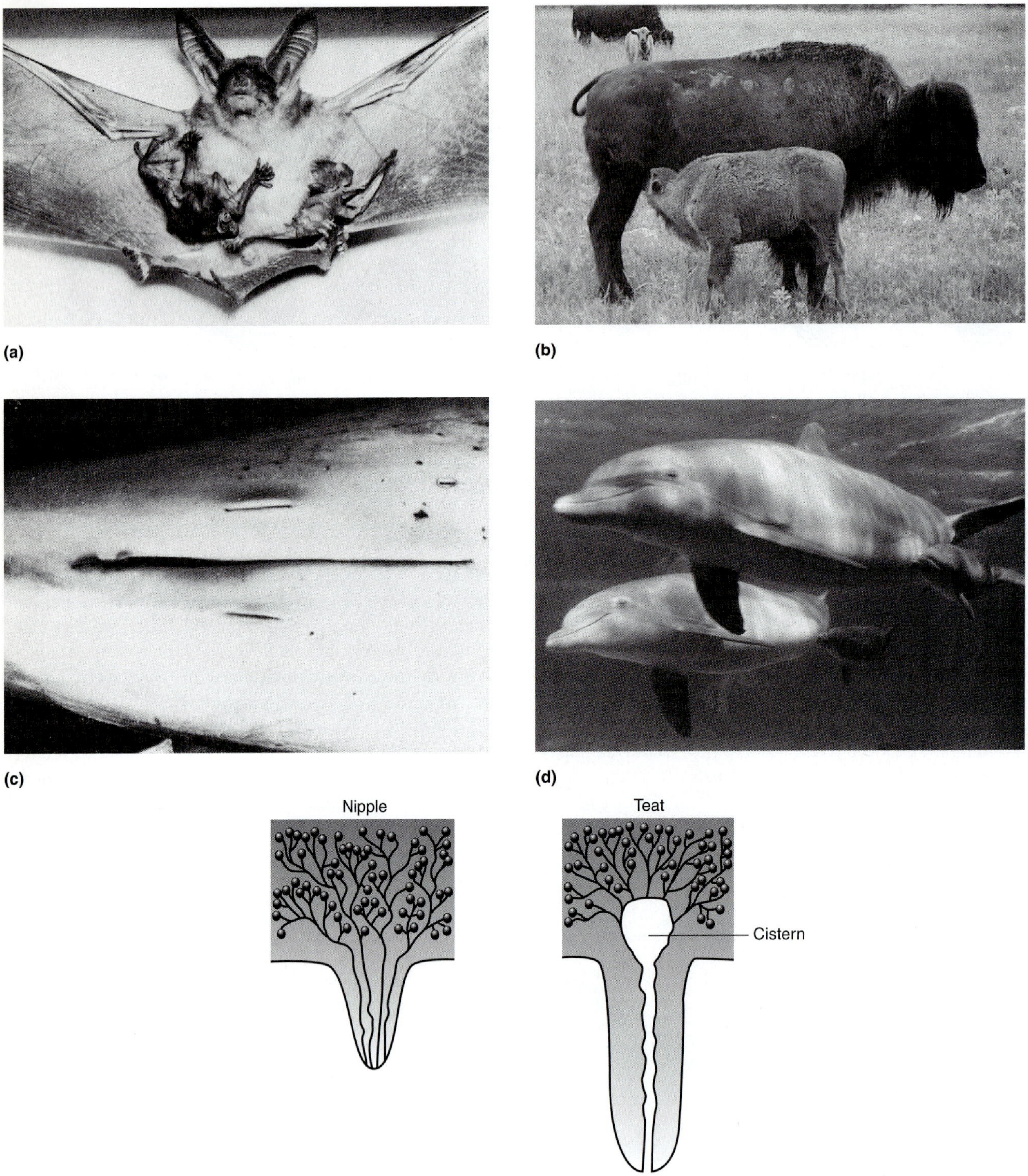

The mammary gland and milk are characters of major importance in the class Mammalia. (*a*) Young pallid bats (*Antrozous pallidus*) are naked and helpless when born. (*b*) American bison (*Bison bison*). (*c*) In cetaceans, the nipples of the female (when not in use) are retracted within slits situated on either side of the vent, as seen on the undersurface of this female harbor porpoise (*Phocoena phocoena*). (*d*) Nursing bottle-nosed dolphins (*Tursiops truncatus*). Inset: Nipple and teat.

platypus and echidna) lack true mammary glands and nipples. Glands resembling modified sweat glands produce a nutritious secretion that is lapped off tufts of hairs. In many cetaceans (whales, dolphins, porpoises) and seals, the nipples may be retracted into slits on either side of the vent when they are not in use (Fig. 9.19c). This improves hydrodynamics during swimming and keeps the nipples warm. The nipples descend when the pup nudges its mother's belly.

Sebaceous, or **oil, glands** normally are associated with hair follicles (see Fig. 9.8). Their secretion, **sebum,** is emptied into the follicle to lubricate the hair and surrounding skin and to act as an antibacterial agent. Many marine mammals possess little hair and lack sebaceous glands.

Scent glands, which may be modified sudoriferous or sebaceous glands, are numerous and widely distributed on the bodies of most mammals. They are most highly developed in those mammals that have the keenest sense of smell and may be used to mark an individual's territory, to attract members of the opposite sex, or for defensive purposes. Glandular secretions that elicit a specific reaction from other individuals of the same species are known as **pheromones** (see Chapter 12).

The function of scent glands varies depending on the sex and physiological state of the species. Perianal glands of skunks secrete a defensive spray consisting of several major components that may cause severe irritation and temporary blindness (Anderson and Bernstein, 1975). Major volatile components differ in the secretions of the striped skunk (*Mephitis mephitis*) and the spotted skunk (*Spilogale putorius*) (Wood, 1990; Wood et al., 1991). The release of scent during times of stress also has been reported in the house shrew, mice, rats, woodchucks, mink, weasels, and the black-tailed deer, among others. Territorial marking is practiced by many species including short-tailed shrews, muskrats, beaver, social rabbits, canids, antelope, and deer. Deer have glands anterior to the eye (preorbital), on the medial side of the tarsal joint (tarsal), on the outside of the metatarsus (metatarsal), and between the main digits (interdigital). Many bats have scent glands in the skin of the face and head.

Because the epidermis lacks blood vessels and nerves, it must be supplied by the highly vascularized dermis. The mammalian dermis contains vast networks of blood vessels, lymphatic vessels, free nerve endings, and encapsulated sense organs sensitive to touch and pressure. Beneath the dermis is a subcutaneous layer that in many mammals has a substantial fat component, which serves as protection, as an insulating layer, and as an emergency energy source. In many marine mammals, the subcutaneous layer serves to minimize heat loss in the water and provide buoyancy; however, it may also serve as an energy reservoir to provide nourishment during long periods of fasting. Some of the larger whales are insulated by a layer of fat that may reach 0.6 m in thickness (Riedman, 1990).

Skeletal System

Axial Skeleton (Skull, Auditory Ossicles, Hyoid, Ribs, Sternum)

The skeleton of mammals supports the body, provides protection for important organs, and serves as a point of attachment for skeletal muscles (Fig. 9.20). The mammalian line of evolution, however, has resulted in significant modifications of skeletal elements. Because of loss and/or fusion, the

FIGURE 9.20

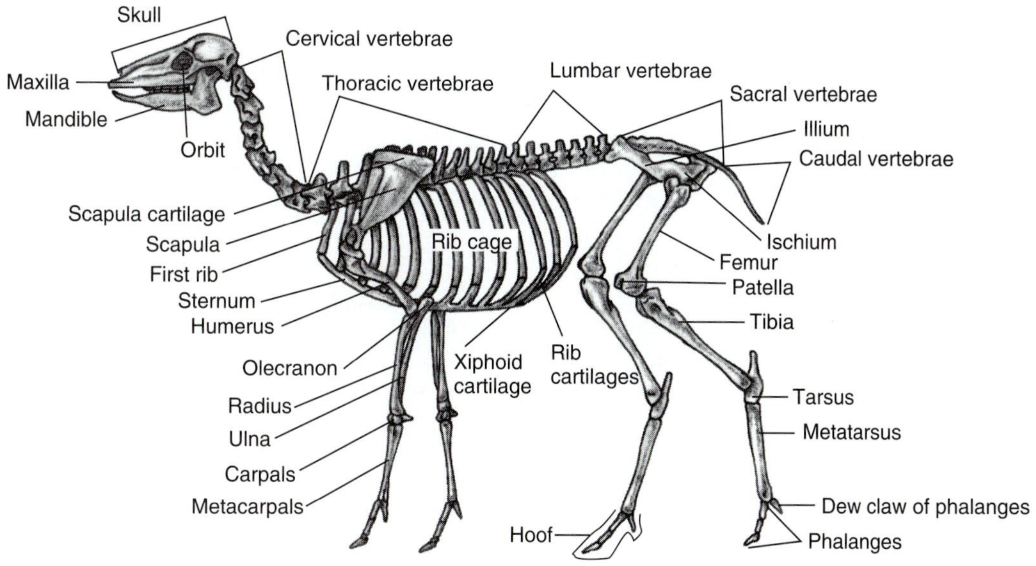

The skeletal structure of a white-tailed deer (*Odocoileus virginianus*). The mammalian skeleton provides support, protection for internal organs, and serves as a point of attachment for muscles.
Source: Halls, White-Tailed Deer Ecology and Management, *1984, Stackpole Books.*

number of bones in the skull and lower jaw of mammals is less than in other vertebrates (Fig. 9.21). The axial skeleton has become stronger and more rigid, most of the skeleton has completely ossified, and skeletal growth generally is restricted to immature mammals.

The mammalian skull, which inherited its basic characteristics from its synapsid reptilian ancestor, includes a single pair of temporal fenestrae bounded ventrally by the zygomatic arches (Fig. 9.22). One of the most remarkable modifications of bones in the history of vertebrate evolution is the transformation and change of articulating elements between the jaw and skull in reptiles to auditory elements in mammals (Rowe, 1996). In mammals, the posterior portion of the palatoquadrate cartilage ossifies as the quadrate bone. It becomes enclosed by the developing middle ear cavity, separates from the remainder of the palatoquadrate cartilage, and becomes the **incus** of the middle ear. Intermediate evolutionary steps can be seen in mammal-like reptiles. Dermal bones ensheath the anterior portion of the palatoquadrate. Meckel's cartilage totally ossifies in adult mammals. The pos-

FIGURE 9.21

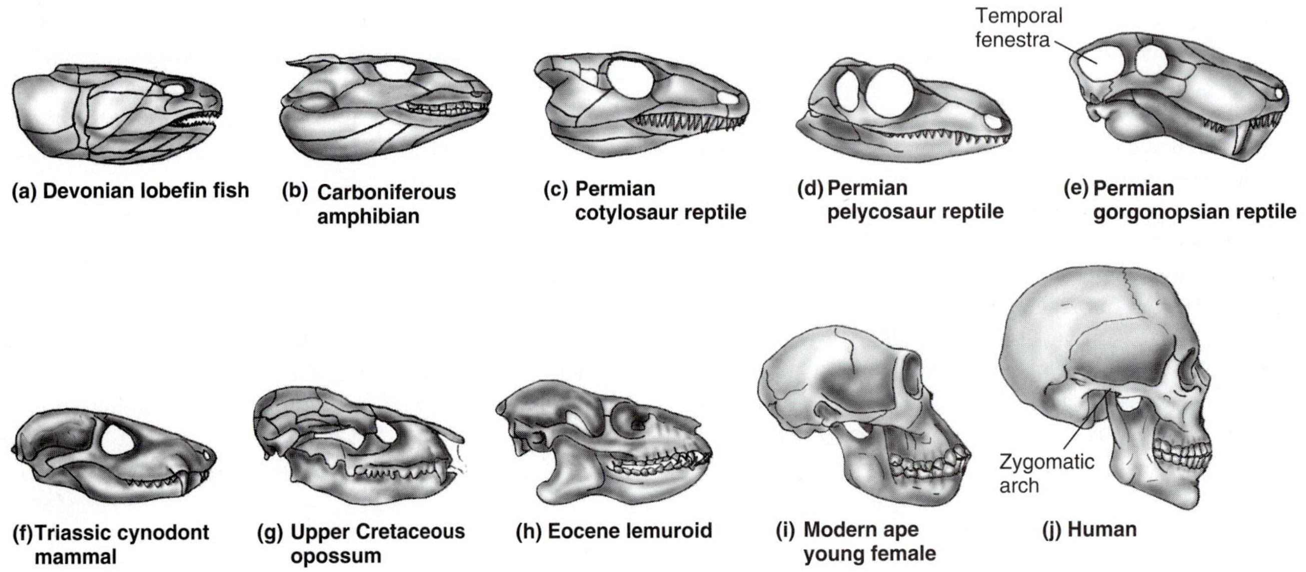

Evolution of the vertebrate skull from fish to man.
Source: W. C. Gregory, Evolution Emerging, *1974, Ayer Company.*

FIGURE 9.22

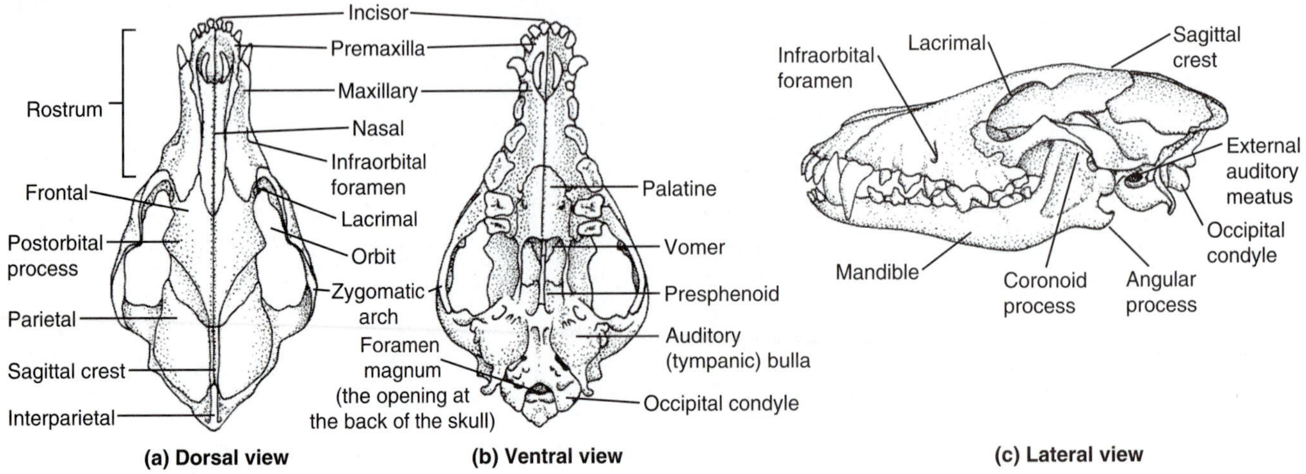

Skull and mandible of the coyote (*Canis latrans*): (*a*) dorsal view; (*b*) ventral view; (*c*) lateral view.

terior tip of Meckel's cartilage—the **articular**—projects into the middle ear cavity, separates from the rest of Meckel's cartilage, and becomes the **malleus.** The malleus and the incus (homologous to the quadrate) still articulate with one another, but the joint is now in the middle ear instead of at the tip of the jaw. The reptilian stapes (columella) was already present in the middle ear. These three ear ossicles conduct vibrations from the eardrum to the inner ear. The dentary, the only remaining dermal bone in the lower jaw, articulates directly with the squamosal and/or temporal bone. Jarvik (1980) cited several problems with the long-held theory of evolution of mammalian ear ossicles, and he proposed a new theory in which the malleus and incus evolve from portions of the hyoid arch. This theory, however, has never been widely accepted.

Most mammals possess a zygomatic arch, which assists in protecting the eye and provides the origin for the masseter muscle (Fig. 9.22). The zygomatic arch is incomplete in some insectivores, anteaters, and some other less derived mammals. It is absent in whales and some insectivores.

Mammals possess a complete secondary palate, which serves to separate the nasal passages from the oral cavity, allowing mammals to continue breathing while chewing food and to strengthen the skull when chewing. The caudal portion fails to ossify and forms a "soft" palate.

The hyobranchial skeleton shows considerable variation, and precise homologies are unclear. Basically, it consists of a **hyoid apparatus** associated with the posterior portion of the tongue and a **larynx** comprising the upper end of the trachea (Feduccia and McCrady, 1991).

The mammalian vertebral column is made up of a series of intersegmentally arranged, acoelous vertebrae: cervical, thoracic, lumbar, sacral, and caudal (see Fig. 9.20). Zygapophyses (articulating processes) overlap adjacent vertebrae to give additional firmness to the backbone and limit the amount of flexion and torsion to which it can be subjected.

Most mammals possess seven cervical vertebrae (Figs. 9.20 and 9.24). The only exceptions are xenarthrans (formerly called edentates)—anteaters, armadillos, and sloths—with six, eight, or nine, and manatees with six. In some mammals, such as cetaceans, some rodents, and armadillos, cervical vertebrae are shortened and more or less fused together. The first two cervical vertebrae in all mammals, the atlas and axis, are specialized for articulation and movement of the skull (Fig. 9.24). The ring-shaped atlas has no centrum and wide transverse processes; anteriorly, it has two concave surfaces that articulate with the occipital condyles of the skull. The axis has a centrum that is elongated anteriorly as the **dens** (odontoid process) and a large neural spine that overlaps the atlas. The dens represents the original centrum of the atlas. When the atlas and axis are articulated, the dens occupies its original position even though it has become a functional part of the axis. The atlas–occipital condyle articulation permits the typical vertical (up-and-down) movement of the head, whereas the atlas–axis articulation allows lateral (side-to-side) movements of the head.

BIO-NOTE 9.5

The Hero Shrew's Vertebral Column

The armored, or hero, shrew (*Scutisorex somereni*) of Congo (formerly Zaire), Rwanda, and Uganda has a remarkably strong vertebral column. Its strength is derived, in part, from a backbone equipped with extra joints for flexibility. Whereas most shrews have 5 lumbar vertebrae, hero shrews have 11, giving them extra bending points (Fig. 9.23). In addition, the shape and size of individual vertebrae, together with an increased number of articular facets, make them unique among mammals. Vertebrae are three times wider than those of other shrews and have interlocking, fingerlike projections that create sturdy links between neighboring vertebrae.

In discussing the extraordinary strength of this shrew, Allen (1917) noted that

> Whenever [the natives] have a chance they take great delight in showing its resistance to weight and pressure. After the usual hubbub of various invocations, a full-grown man weighing some 160 pounds steps barefooted upon the shrew. Steadily trying to balance himself upon one leg, he continues to vociferate several minutes. The poor creature seems certainly to be doomed. But as soon as his tormentor jumps off, the shrew after a few shivering movements tries to escape, none the worse for this mad experience.... The strength of the vertebral column, together with the strong convex curve behind the shoulder... evidently protects the heart and other viscera from being crushed.

Allen, 1917
Pennisi, 1996

FIGURE 9.23

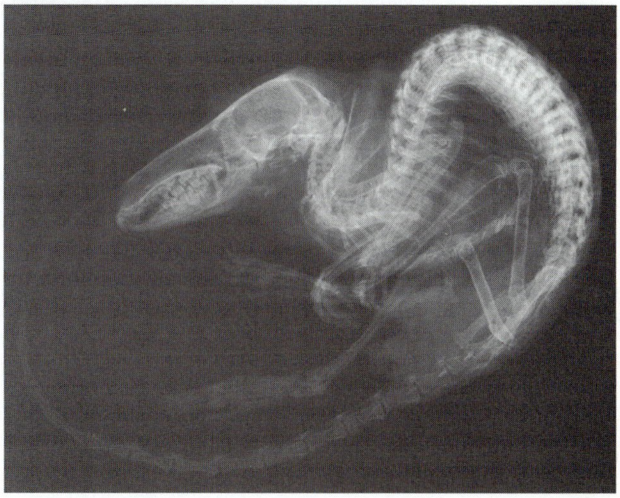

An x-ray image of the body of the armored shrew (*Scutisorex somereni*) of Congo (formerly Zaire), Rwanda, and Uganda. These shrews have 11 (rather than 5) lumbar vertebrae, giving them extra bending points and a remarkably strong vertebral column.

FIGURE 9.24

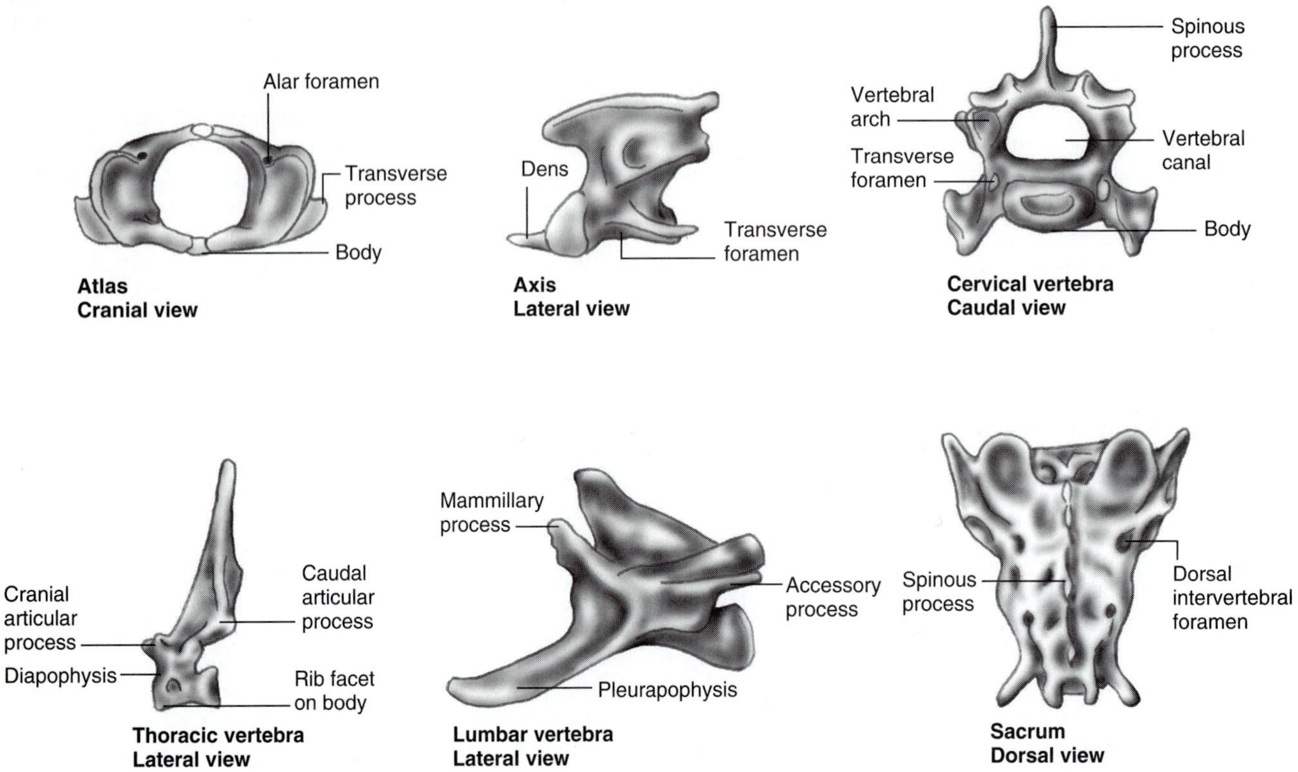

Representative vertebrae from a cat (*Felis*). For those shown in the lateral view, cranial is toward the left.
Figure from Vertebrate Dissection, *7th edition by Warren F. Walker, Jr., Copyright © 1986 by Saunders College Publishing, reproduced by permission of the publisher.*

Thoracic and lumbar vertebrae vary in number and structure. Most thoracic vertebrae have short centra, tall posteriorly directed neural spines, and small zygapophyses. These vertebrae also articulate with the ribs. Lumbar vertebrae vary in number from 2 (in some monotremes) to as many as 21 (in cetaceans). Typical lumbar vertebrae are large and stout with prominent, broad neural spines and long, forward-projecting transverse processes. No ribs articulate with lumbar vertebrae. Whereas the number of thoracic and lumbar vertebrae varies from species to species, and even occasionally within one species, the total number of thoracic and lumbar vertebrae is constant in a given species and even in higher taxonomic groups.

Most mammals have 3 to 5 sacral vertebrae that are fused to form the sacrum, which serves as a point of attachment for the pelvic girdle. There is no sacrum in whales due to the absence of hindlimbs and a pelvic girdle. The number of sacral vertebrae may range up to 13 in some xenarthrans.

Mammals still have remnants of an ancestral reptilian tail and have 3 to 50 caudal (tail) vertebrae (see Fig. 9.20). In apes and humans, the 3 to 5 caudal vertebrae are called coccygeal vertebrae because some or all usually fuse to form a **coccyx.** Caudal autotomy (the ability to break off the tail) is known to occur in a few rodents.

BIO-NOTE 9.6

The Vertebrae of the Queen Naked Mole Rat

In naked mole rat (*Heterocephalus glaber*) colonies, the queen has an elongated body. The elongation is caused by a lengthening of individual vertebrae after a female becomes a breeder—a phenomenon unique among mammals.

Sherman et al., 1992

Most mammals possess 12 to 15 pairs of ribs (see Fig. 9.20). Most are composed of a dorsal (vertebral) portion and a ventral (sternal) portion. The latter remains as a **costal cartilage** in mammals. The costal cartilages connect, either directly or indirectly, with the sternum to form a rib cage that functions in protection as well as in respiration. Floating ribs lack a sternal connection.

The mammalian sternum articulates with the ribs and anteriorly with the pectoral girdle. In all mammals except cetaceans and sirenians, the sternum consists of bony segments known as **sternebrae.** The sternum assists in strengthening the body wall, helps protect the thoracic viscera, serves as a point of attachment for pectoral and limb muscles, and in some amniotes, aids in ventilating the lungs.

Appendicular Skeleton (Pectoral and Pelvic Girdles and Appendages)

The pectoral girdles of most mammals other than monotremes consist of either a pair of clavicles and scapulae or just a pair of scapulae. A clavicle is present in those mammals whose front limbs move in several planes; it is absent in those where the forelimbs move in only one plane, such as deer and horses. When present, the clavicle braces the scapula against the sternum. Some marsupials, insectivores, bats, rodents, and higher primates, including humans, have a clavicle. It is lacking in cetaceans, ungulates, and some carnivores. In other carnivores, such as cats, the clavicle has been reduced to a slender bony splinter that fails to articulate with either the sternum or the scapula.

The efficiency of mammalian limbs has been increased by bringing the limbs to a vertical position and, at the same time, rotating them. Hindlimbs are rotated forward, so that knees and feet point anteriorly; forelimbs are rotated backward so that elbows point to the rear. An additional 180° rotation at the wrist is necessary to allow the front feet to also point forward. The rotation at the wrist resulted in the crossing of the radius and ulna in the forearm.

The anterior appendage contains the same six skeletal elements as in all tetrapods (humerus, radius and ulna, carpals, metacarpals, and phalanges) (see Fig. 9.20). The shape, length, and number of skeletal elements in the appendages have evolved as primitive mammals developed specialized locomotory techniques (Fig. 9.25). Further modifications

FIGURE 9.25

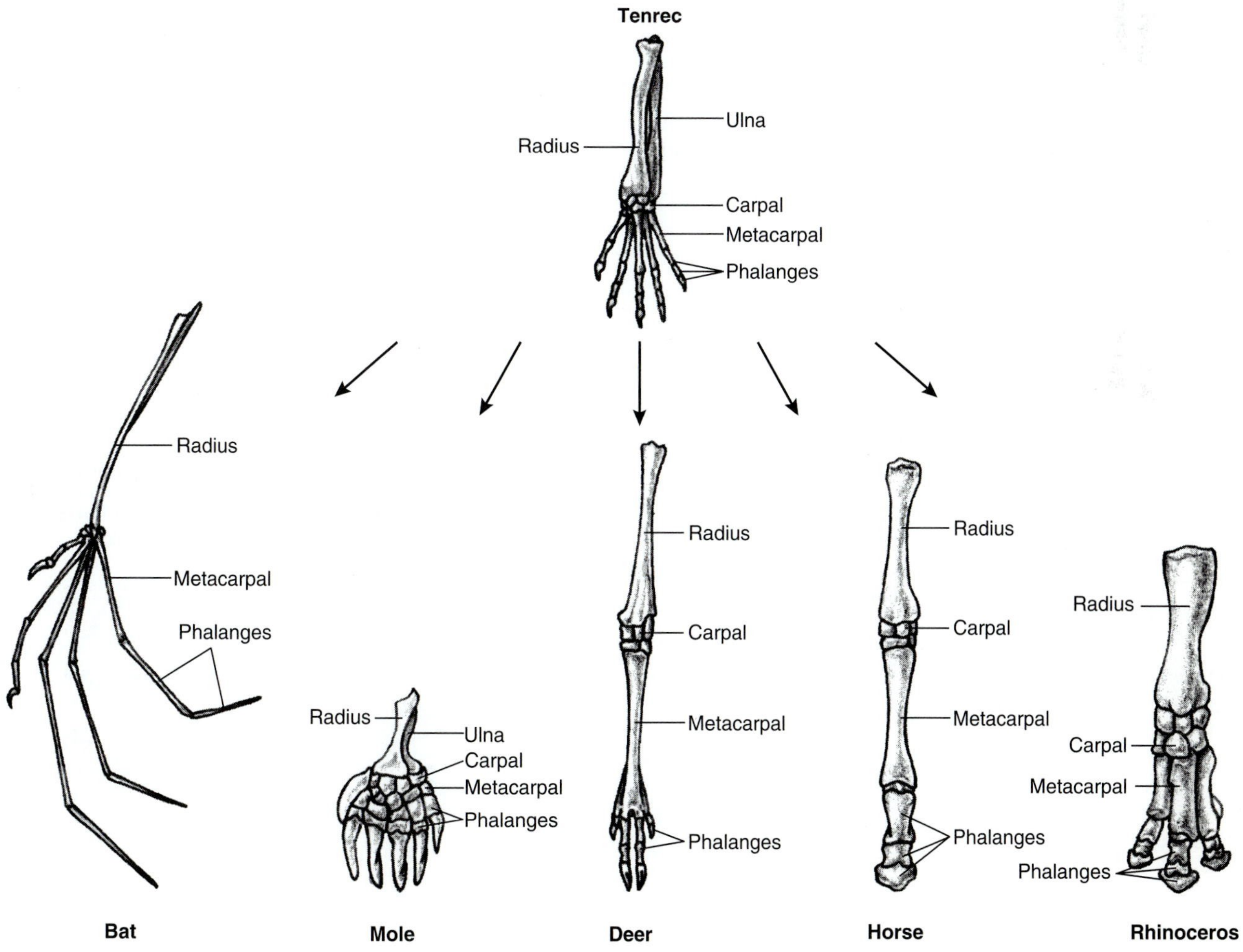

Adaptive radiation of the forelimbs of mammals. The forelimb of a tenrec (top) is used as an example of a primitive form. Although not drawn to scale, portions of the forelimbs of a bat, a mole, a deer, a horse, and a rhinoceros show various modifications by the fusion of parts, the loss of parts, or by changes in the proportion of parts of the limb.

Source: Dodson and Dodson, Evolution: Process and Product, *Prindle, Weber and Schmidt.*

have occurred in association with the occupation of subterranean, arboreal, aquatic, and aerial habitats. For example, the humerus of moles, echidnas, and other burrowing mammals is short and stout and has expanded areas for attachment of large muscles used for digging. In ungulates, the shaft of the ulna may fuse with the radius during embryonic development; in other mammals, such as bats, it may fail to develop fully. The front limb of bats has been modified into a wing for flight (Fig. 9.25). Metacarpals and phalanges of the last four digits have become elongated to support the wing membrane. Although bats typically hang by their feet upside down, the thumb remains free and is sometimes used when crawling.

The front limbs of cetaceans, sirenians, seals, and sea lions are modified for life in the sea and superficially resemble the modified appendages of sea turtles and penguins (convergence) (Fig. 9.26). Appendages become flattened, short, and stout and may have a greatly increased number of phalanges. Insectivores and primates have tended to remain **pentadactyl** (five fingers, five toes) and usually have five carpals and five metacarpals.

Front limbs of many mammals have been modified for grasping, with the thumb developing into an opposable structure capable of touching each of the other digits (Fig. 9.27). The evolution of an opposable thumb was accomplished by the development of a unique joint known as a **saddle joint** at the base of the thumb and by the development of strong skeletal muscles to operate the thumb. In those primates that swing from branch to branch (brachiation), the large clavicle is firmly attached to the sternum (see Fig. 9.32b). In some forms of monkeys that are almost exclusively arboreal, such as spider monkeys (*Ateles*) and woolly spider monkeys (*Brachyteles*), the thumb is rudimentary or has been lost altogether, an evolutionary modification that facilitates movement through the canopy.

Pandas possess a pseudothumb (Fig. 9.28) (Catton, 1990). This is, functionally, a sixth digit formed by a wristbone, the radial sesamoid, and lies beneath a pad on the animal's forepaw. Muscles that normally attach to the thumb attach to the radial sesamoid and enable the panda to grip and manipulate bamboo efficiently.

Adhesive devices have been identified on the thumbs of five genera of vespertilionid bats (*Thyroptera, Glischropus, Tylonycteris, Pipistrellus,* and *Myzopoda*) (Thewissen and Etnier, 1995). These pads may adhere by suction, dry adhesion, or gluing.

Two **innominate** (coxal, hip) **bones** make up the pelvic girdle and articulate with the sacrum dorsally (see Fig. 9.24). Each innominate bone consists of three fused elements: an anterior pubis, a posterior ischium, and a dorsal ilium. In most mammals, the two pubic bones unite to form a pubic symphysis. Because of the pubic symphysis and the uniting of the ilium with the vertebral col-

FIGURE 9.26

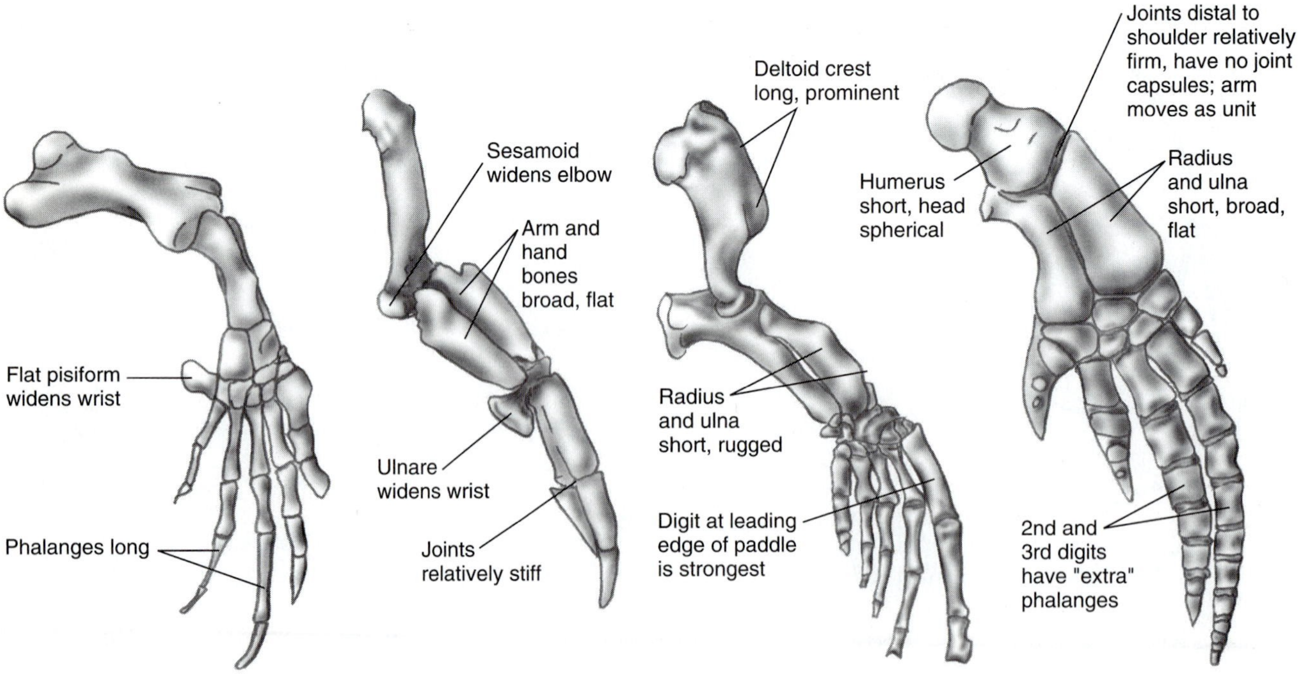

Dorsolateral views of the right forelimb skeletons of some aquatic vertebrates that use the pectoral appendage for propulsion: (*a*) sea turtle (*Chelonia*); (*b*) penguin (*Spheniscus*); (*c*) sea lion (*Zalophus*); (*d*) dolphin (*Lagenorhynchus*).

From Hildebrand, Analysis of Vertebrate Structure, *4th edition. Copyright © 1995 John Wiley & Sons, Inc. Reprinted by permission of John Wiley & Sons. Inc.*

FIGURE 9.27

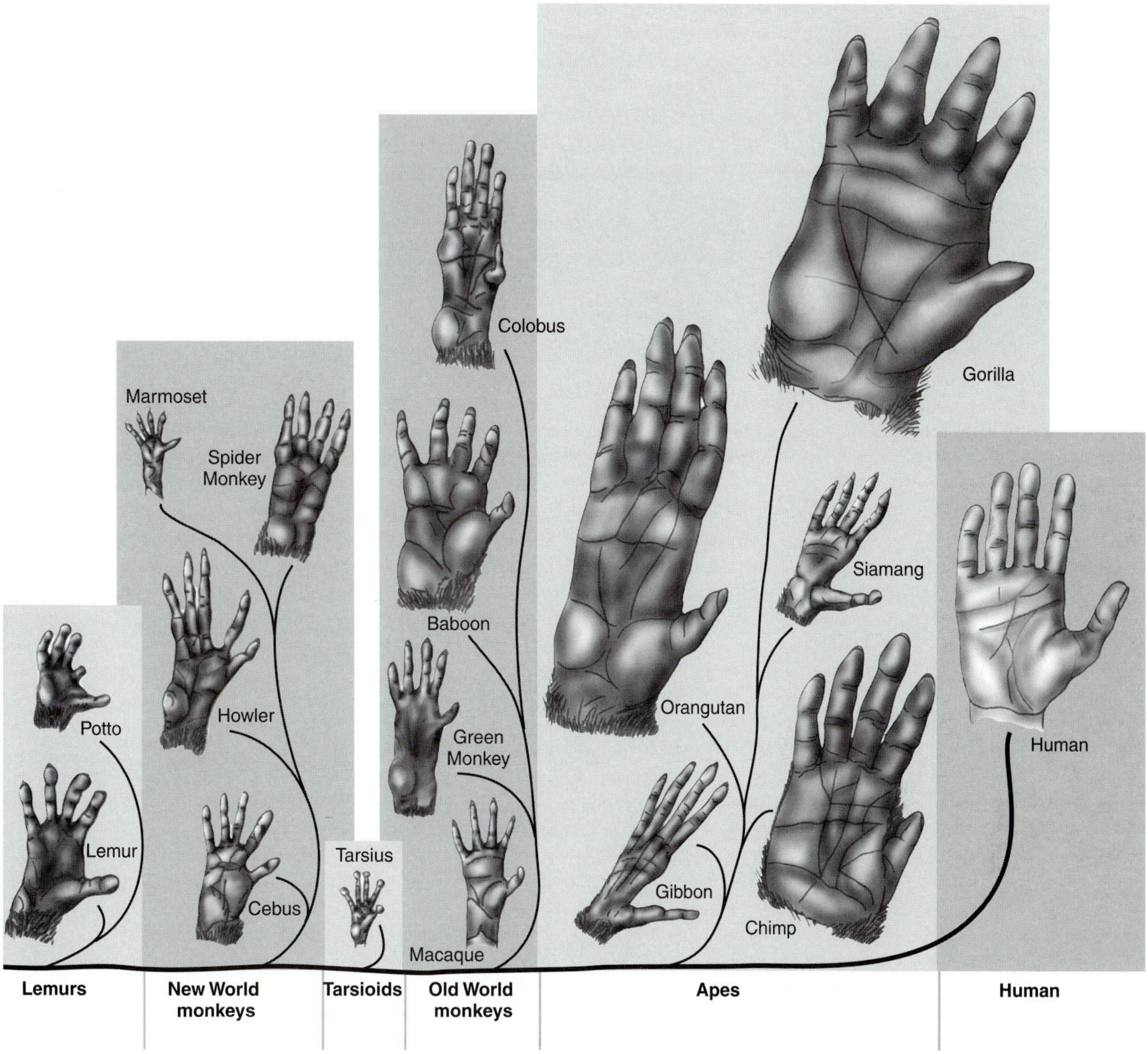

Modifications of the hands of primates for grasping. Note the opposable thumb in most species.

umn, the pelvic girdle forms a ring, the **pelvis,** around the caudal end of the digestive and urogenital systems. The pelvic girdle is vestigial or absent in all living cetaceans (Fig. 9.29) and sirenians. Cetaceans have lost all external manifestations of hindlimbs, but vestiges sometimes remain embedded within the body wall (see discussion of *Basilosaurus* on page 291).

Two small **epipubic bones** articulate with the pubic bones and extend forward in the abdominal wall in marsupials and monotremes (see Fig. 9.31a). They have also been identified in two primitive Cretaceous eutherians (Novacek et al., 1997). Some researchers contend these bones support the abdominal pouch in which the young are transported, but this seems doubtful because the bones are developed in both sexes. These bones do show sexual dimorphism, as they are either longer or broader in females than in males of the same species. Nowak and Paradiso (1983) think it more likely that epipubic bones are a heritage from reptilian ancestors and were associated with the attachment of abdominal muscles that supported large hindquarters. They may also have served to stiffen the median part of the ventral abdominal wall (Presley, 1997) and/or to have aided in locomotion by acting with the hypaxial muscles of the trunk to protract (move forward) the pelvic limbs (White, 1989).

The hindlimbs of mammals are comparable to those of reptiles. They consist of a femur, tibia and fibula, tarsals,

FIGURE 9.28

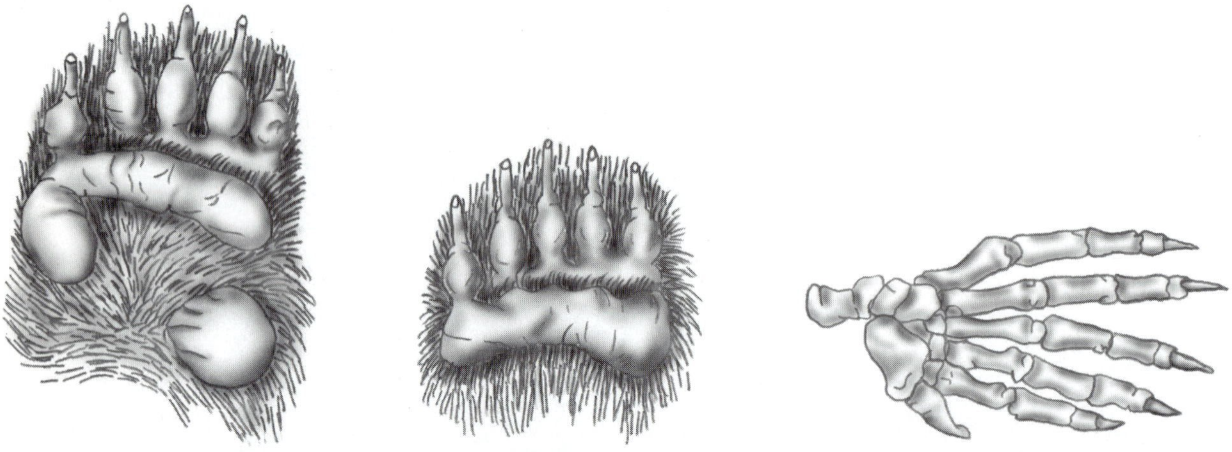

The giant panda's peculiar thumb. The forepaw (top left) shows the pseudothumb used to grasp bamboo stems. The skeleton of the forepaw, far right, shows that the pseudothumb is a modified wristbone, the radial sesamoid. The hindpaw (middle) is included for comparison.
Source: Catton, Pandas, *1990 Facts on File.*

FIGURE 9.29

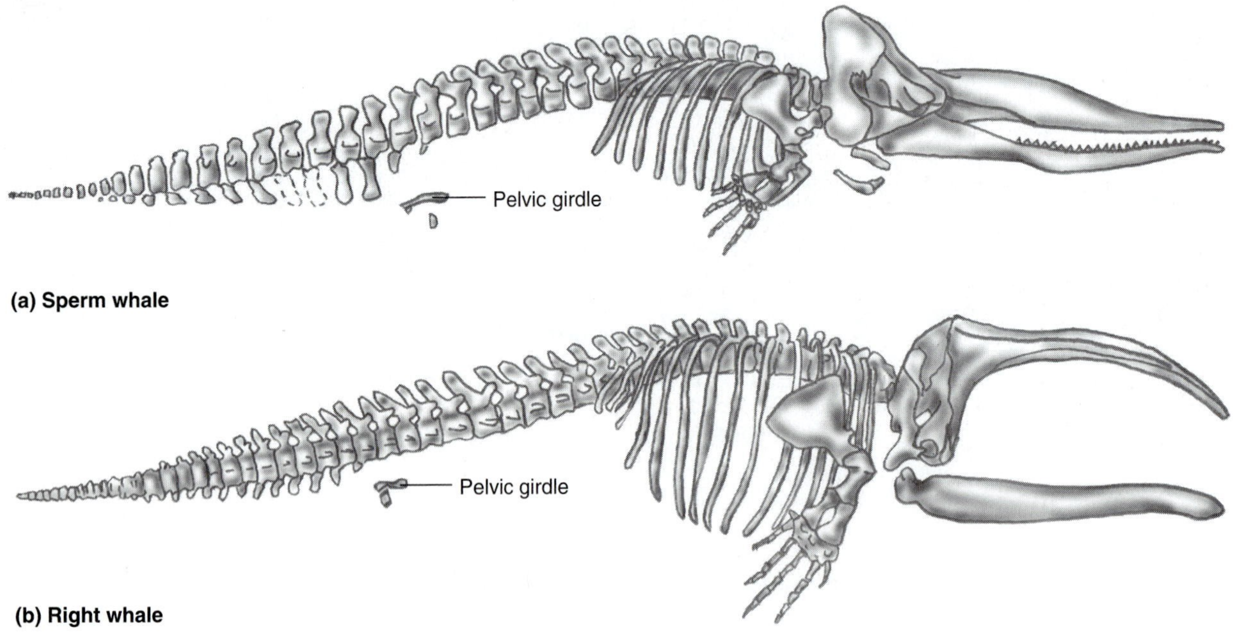

Skeletons of (*a*) sperm whale (*Physeter*) and (*b*) right whale (*Eubalaena*). Note vestiges of the pelvic girdle.
Source: W. C. Gregory, Evolution Emerging, *1974, Ayer Company.*

metatarsals, and phalanges (see Fig. 9.20). In addition, a sesamoid bone (a bone that develops in a tendon), the patella, protects the knee joint. Semiaquatic mammals, such as muskrats, beaver, and nutria, have webbing between their toes. In many primates the big toe or **hallux** is opposable (Fig. 9.30). In opossums (*Didelphis*), the big toe is opposable and assists in climbing (Fig. 9.31).

Mammal limbs are variously modified for different forms of locomotion. Some, such as insectivores, monkeys, apes, humans, and bears walk by placing the entire surface of their foot on the ground with each step (Fig. 9.32a). Such mammals usually possess pentadactyl hands and feet. This ancestral method of locomotion is known as **plantigrade** locomotion.

FIGURE 9.30

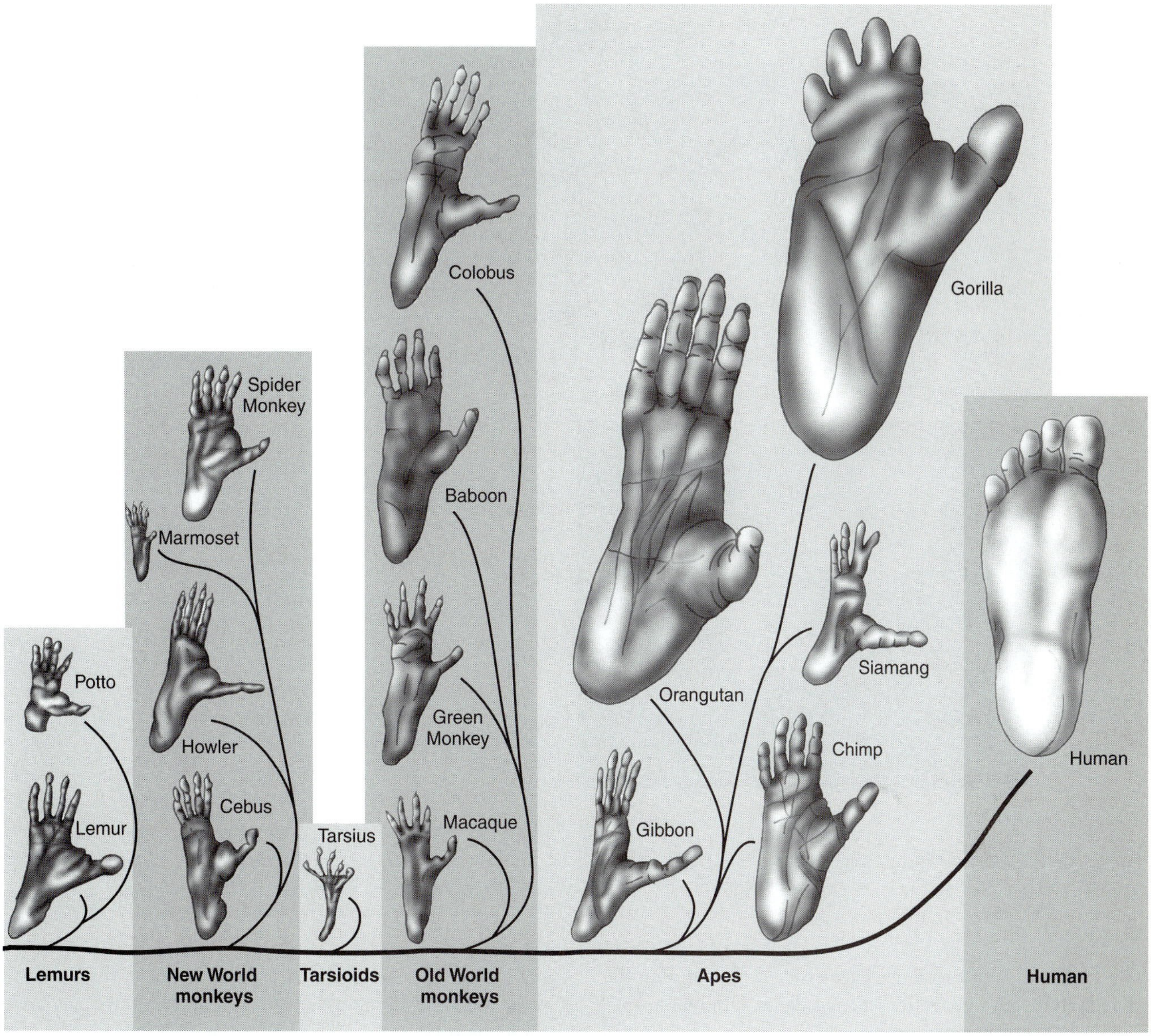

Modifications of the feet of primates. Note the opposable big toe in most species.

Some mammals bear their weight on the ends of their metacarpals and metatarsals (Fig. 9.32b). Their wrists and ankles are elevated, and the thumb has been reduced or lost. They usually can walk and run faster than plantigrade species. They also walk more silently and are more agile. This method of locomotion, called **digitigrade** movement, is common in mammals such as rabbits, rodents, and many carnivores.

Ungulates illustrate the extreme in modification of the distal appendages. The number of digits has been reduced so that ungulates possess either four, three, two, or even one, and the animals walk on the tips of their remaining fingers and toes. This method of locomotion is known as **unguligrade** (Fig. 9.32c). The weight of the body is borne on hooves, which represent modified claws that have become hardened and thickened. The metacarpals corresponding to the missing digits have been either reduced in size or lost, and those that remain are elongated and often united, a modification that greatly strengthens the lower leg and foot. The limbs are capable of only forward and backward movement; no twisting or rotation is possible. Muscles activating the lower portions of the limbs are located close to the body in order to lessen the weight of the limb each time it is raised; the appendicular muscles attach to the limb bones by long, lightweight tendons. Thus, the limbs and feet of hooved mammals, which are long, light, and capable of only fore and aft movements, are highly specialized for running and/or for maneuvering on rocky terrain.

FIGURE 9.31

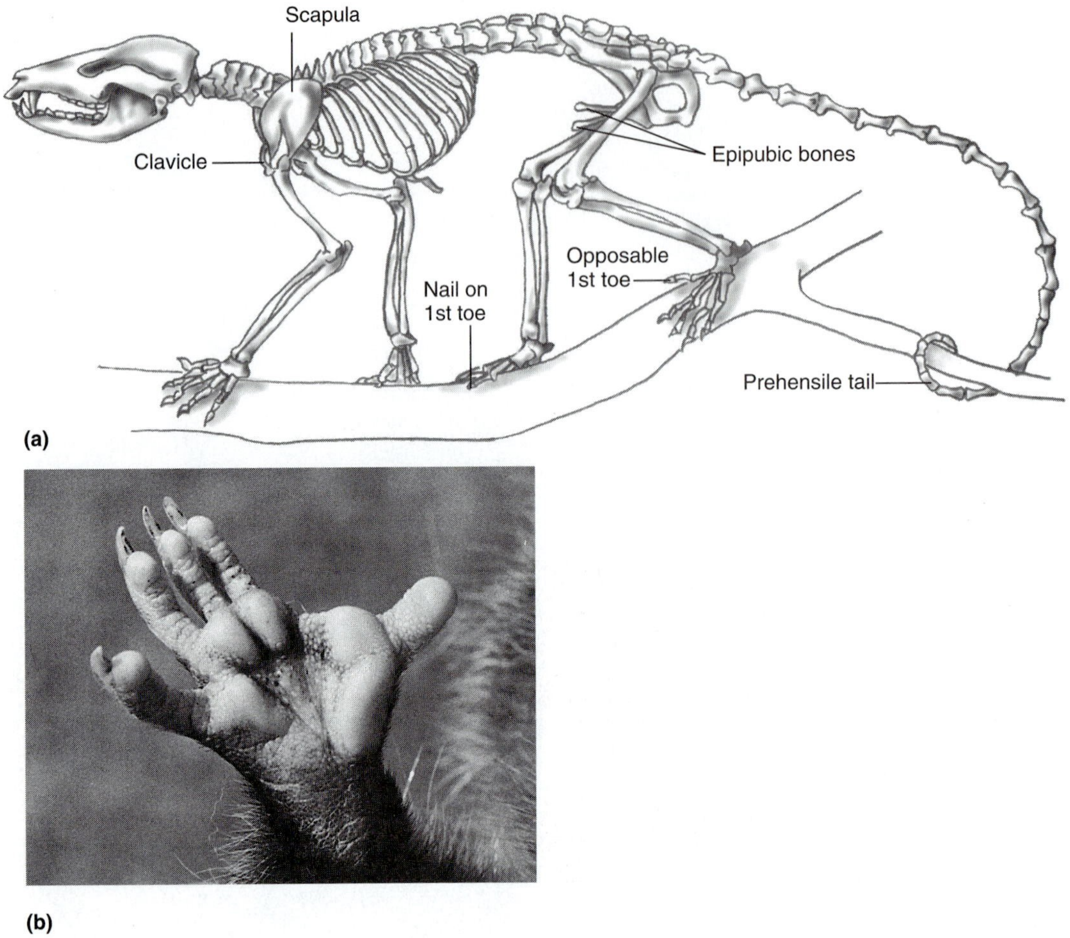

(*a*) Skeleton of the opossum (*Didelphis*) showing the opposable big toe, the epipubic bones, and the prehensile tail. (*b*) Right hind foot of a juvenile opossum with opposable clawless "thumb". It was this handlike foot that prompted the discoverer of the opossum, Pinzon, in 1500 to describe the animal as "part monkey."

Two groups of ungulates have evolved. One group, the **artiodactyls,** have generally retained digits 3 and 4 as functional digits and their weight is equally distributed between them (Fig. 9.33a). Digits 2 and 5 are reduced, and digit 1 is lost. Because the weight of the body is borne on two parallel axes, they are said to have a **paraxonic foot** and a **cloven hoof.** This group includes pigs, peccaries, javelinas, and hippopotamuses, which have four digits; and camels, llamas, chevrotains or mouse deer, deer, elk, caribou, giraffes, okapis, pronghorns, antelopes, bison, buffalo, cattle, gazelles, goats, and sheep which have two digits. In the second group, the **perissodactyls** (Fig. 9.33b), digit 3 has been retained as the primary functional digit in most forms, and it bears all of the body weight; digits 2 and 4 are reduced, and digits 1 and 5 are usually lost. These animals are said to have a **mesaxonic foot.** Thus, perissodactyls have an odd number of digits. They include horses, zebras, asses, tapirs, and rhinoceroses.

In whales, dolphins, and porpoises, the hindlimbs are absent, and forelimbs have been modified into paddles. Fur seals and sea lions have large, naked front flippers and reversible hind flippers that can be brought under the body for locomotion on land (Fig. 9.34a, b, f). Hind flippers also are reversible in walruses. In hair (earless) seals (Fig. 9.34c, d, e), front flippers are haired and are smaller than the hind flippers, which are not reversible.

Muscular System

Epaxial muscles exist in bundles along the vertebral column but have become covered and partially obscured by the greatly expanded extrinsic appendicular muscles (Fig. 9.36). The function of epaxial muscles is the same in mammals as in other vertebrates. They allow side-to-side movement of the vertebral column and provide for the support and arching of the back. Appendicular muscles of mammals are basically similar to those of reptiles. Due to their expansion and differentiation, however, they obscure much of the epaxial musculature. Hypaxial muscles of the abdominal wall are well developed in most mammals and support the abdomen, assist in bending the vertebral column, and serve as the musculature of the tail.

Specialized runners, such as ungulates, have short muscles with long tendons in the lower parts of their legs that are slender in proportion to the forces they have to transmit. An extreme example is the plantaris muscle of the camel, in which the tendon runs almost the entire distance between the femur and the muscle's insertion on the phalanges (Alexander et al., 1982).

FIGURE 9.32

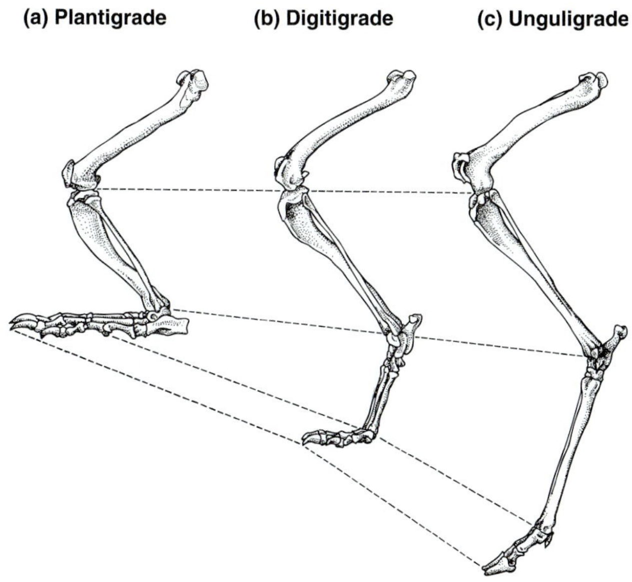

Modifications of mammal limbs for different forms of locomotion: (*a*) plantigrade; (*b*) digitigrade; (*c*) unguligrade. Note how changes in foot posture produce relatively longer limbs.

BIO-NOTE 9.7

The Origins of Whales

It is believed that whales diverged from primitive mammalian stock and that adaptation to a marine life is secondary. Two main hypotheses exist for the relationship of the mammalian order Cetacea (whales, dolphins, and porpoises). The first hypothesis, mainly supported by DNA sequence data, is that one of the groups of artiodactyls (for example, hippopotamuses) is the closest extant relative of whales. The second hypothesis, mainly supported by paleontological data, identified mesonychians as the sister group to whales. The oldest whale (*Pakicetus*) dates from 50 million years ago. Recent evidence from 40-million-year-old Eocene fossils discovered in Egypt seems to confirm a long-suspected connection between cetaceans and early artiodactyl relatives. Specimens of *Basilosaurus isis* include the first functional pelvic limb and foot bones known in the order Cetacea (Fig. 9.35). Distal portions of the hindlimbs show a paraxonic arrangement (the functional axis of the leg passes between the third and fourth digits), which is strikingly similar to that of an extinct group of ungulates, the mesonychid condylarths, as well as to that of modern artiodactyls. The skull and dental structure of mesonychid condylarths are similar to those of primitive whales (archaeocetes). These paleontological data are supported by new protein sequence and cytochrome *b* sequence molecular data. However, new fossils have weakened the links between the whales and the mesonychians and show that whales are probably cousins of the ungulates, if not actual members of that group.

Gingerich et al., 1990
Goodman et al., 1985
Irwin et al., 1991
Normile, 1998
Novacek, 1992
Thewissen et al., 1998

FIGURE 9.33

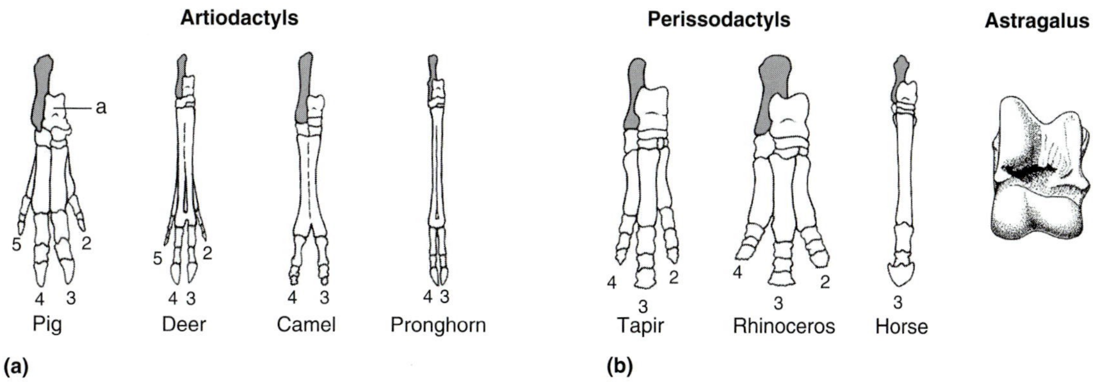

(*a*) Foot structure in artiodactyls. Weight is equally distributed between digits 3 and 4. (*b*) Foot structure of a horse (perissodactyl). Digit 3 is the only functional digit on each limb. In all examples, the heel bone (calcaneum) is shaded and articulates with the astragalus (*a*).

FIGURE 9.34

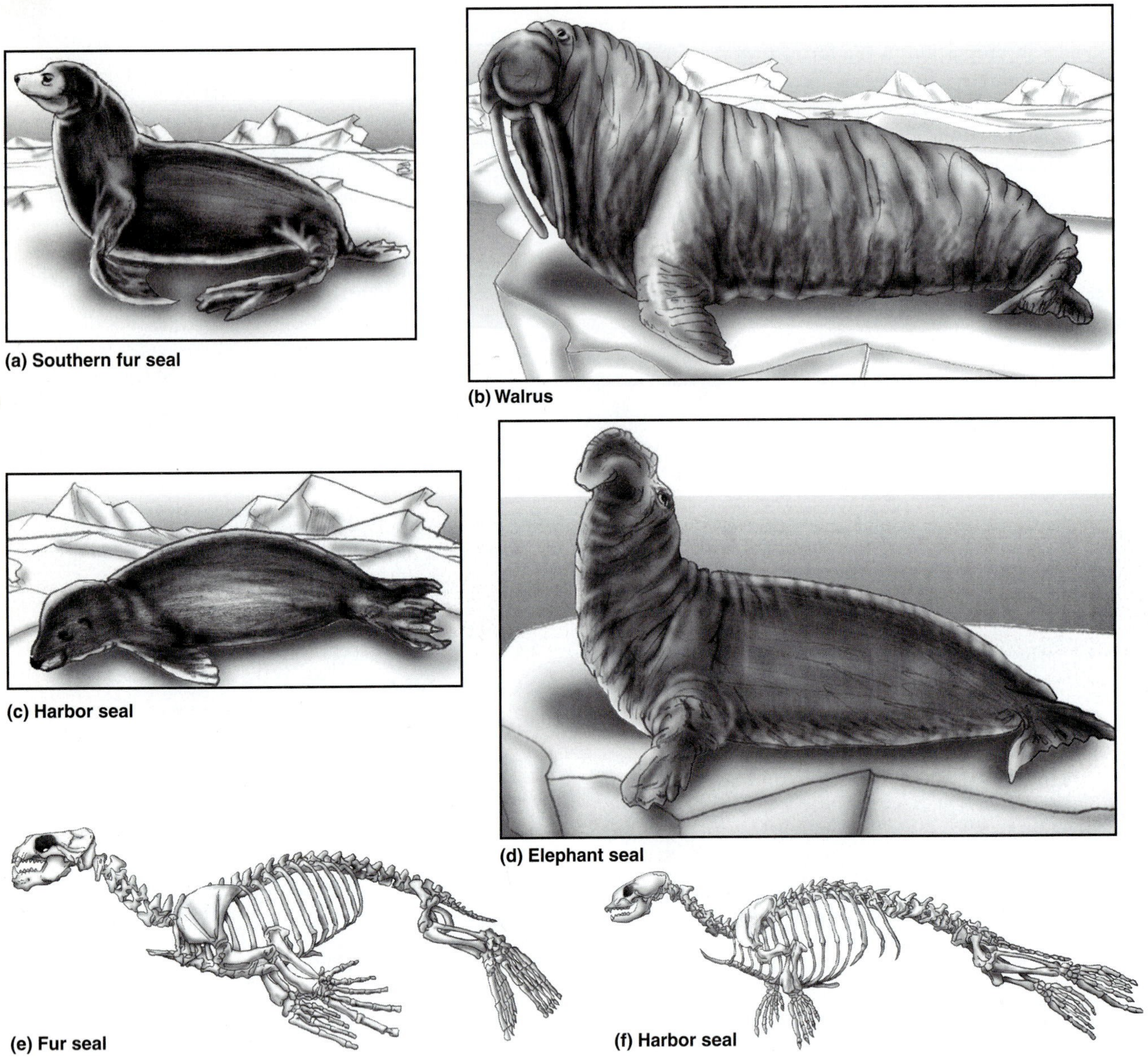

(*a*) Southern fur seal (*Arctocephalus*) and (*b*) walrus (*Odobaenus*) possess hindlimbs that can be brought under the body for locomotion on land. (*c*) Harbor seal (*Phoca*) and (*d*) elephant seal (*Mirounga*) are unable to bring hindlimbs forward. Note the difference in the size of the forelimbs in the skeletons of (*e*) fur seal and (*f*) harbor seal.

Branchiomeric muscles (muscles associated with the sides of the branchial arches) of the first pharyngeal arch continue to operate the jaws. Muscles of the second arch are attached to the hyoid skeleton. Muscles of arches III–VII continue to be associated with the pharynx and larynx.

The branchiomeric muscles in some mammals are extremely well developed. For example, approximately 25 percent of the naked mole rat's muscle mass is concentrated in the jaw region (Sherman et al., 1992). This subterranean burrower uses its incisors and powerful jaws to excavate burrows in the semideserts of Kenya, Somalia, and Ethiopia. In contrast, a human jaw contains less than 1 percent of the body's muscle mass.

Integumentary muscles are best developed in mammals. The arrector pili muscles, which cause the elevation of hairs for insulation or as a response to danger, have already been discussed in this chapter (see pages 271–272). In some mammals, a derivative of the hypaxial musculature, the **panniculus carnosus** (cutaneous maximus), covers much of the trunk. Nipples and teats are surrounded by the compressor mammae muscle, a specialized part of the panniculus carnosus. Armadillos use the panniculus carnosus to roll into a ball

FIGURE 9.35

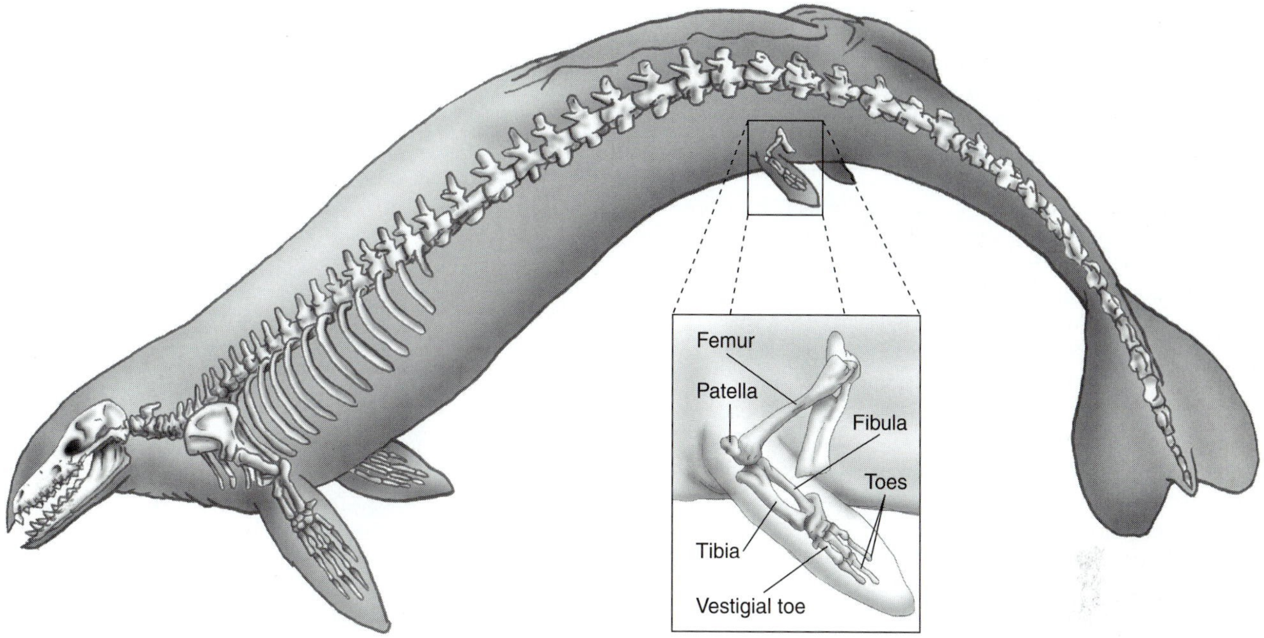

It is believed that whales diverged from primitive mammalian stock and that adaptation to a marine life is secondary. Specimens of *Basilosaurus isis* include the first functional pelvic limb and foot bones known in the order Cetacea. Distal portions of the hindlimbs show a paraxonic arrangement (the functional axis of the leg passes between the third and fourth digits), which is strikingly similar to that of an extinct group of ungulates, the mesonychid condylarths, as well as modern artiodactyls. In addition, the skull and dental structure of mesonychid condylarths are similar to those of primitive whales. These paleontological data are supported by protein sequence and cytochrome b sequence molecular data.
Source: Walter Stuart in Discover Magazine, *May 1991.*

FIGURE 9.36

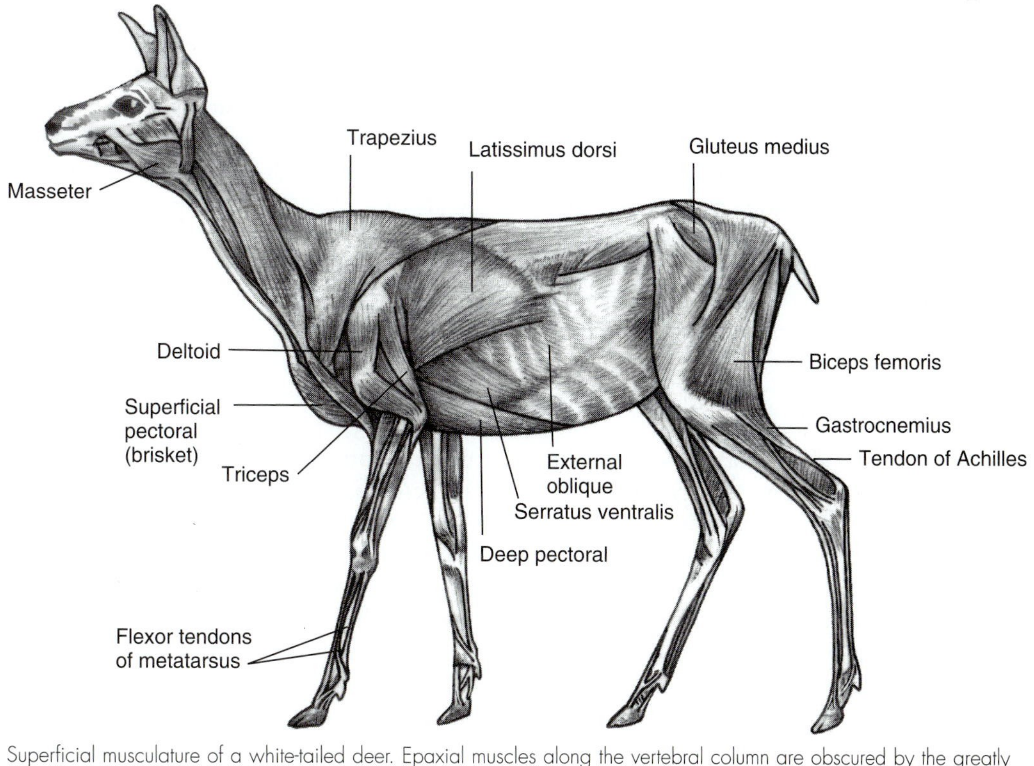

Superficial musculature of a white-tailed deer. Epaxial muscles along the vertebral column are obscured by the greatly expanded, extrinsic appendicular muscles.

BIO-NOTE 9.8

The Aerobic Pronghorn

Cheetahs are fast: over short stretches, they can sprint at 95 km/hr. But for high-speed, long-distance running, nothing beats the pronghorn, or American, antelope. The most reliable estimates show that pronghorns comfortably can cover 11 km in 10 minutes—an average speed of 65 km/hr.

The pronghorn's secret is a series of physiological and structural adaptations that allow it to consume oxygen with more than three times the expected efficiency. For this reason, the pronghorn can process much more oxygen than other mammals its size.

Pronghorns have spectacularly large lungs, which are three times as large as those of comparably sized goats. The heart is unusually large and the blood of the pronghorn is rich in hemoglobin, which means that more oxygen can be delivered to the muscles in less time. The pronghorn's skeletal muscle cells also are densely packed with mitochondria, the intracellular structures involved in aerobic metabolism.

Together, these adaptations provide the pronghorn with speed and endurance, both of which may be essential for escaping from predators in the exposed habitat of the North American prairie.

Lindstedt et al., 1991
Rennie, 1992

when endangered. A portion of this muscle in marsupials forms a sphincter surrounding the entrance to the pouch. Many mammals, including horses and cows, make a portion of this muscle contract and "twitch" the skin in order to shake off flies. It is either poorly developed or absent in primates.

Mimetic muscles, or muscles of facial expression, have evolved from the platysma muscle and have spread onto the faces of mammals. They are best developed in primates, with humans possessing approximately 30 of these muscles, the largest number in any mammal. By contracting one or more of these muscles, actions and expressions can be conveyed: for example, wrinkling the skin of the forehead, raising the eyebrow, closing the eye, depressing the corner of the mouth, elevating the upper lip to expose the canine teeth, and dilating the nostril.

Cardiovascular System

The two atria and two ventricles of the heart are separated by interatrial and interventricular septa, respectively (Fig. 9.37). Atria exhibit unique, earlike lobes—the **auricles.** The sinus venosus is incorporated into the wall of the right atrium.

Normal heart rates can vary from fewer than 25 per minute in the Asiatic elephant (*Elephas maximus*) to more than 1,000 per minute in some shrews. Blood pressure is highest in the aorta and decreases as it flows through the smaller arteries, arterioles, capillaries, venules, and veins.

FIGURE 9.37

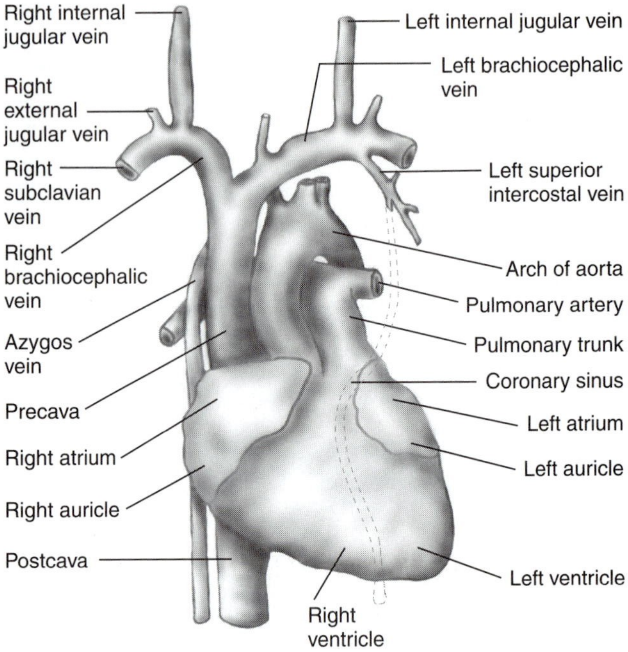

The mammalian heart, ventral view. The four-chambered heart consists of two atria and two ventricles.

All blood being pumped from the left ventricle goes through the former left fourth aortic arch (arch of the aorta) prior to going to the head, to the front limbs, or into the descending aorta (Fig. 9.37). This route is opposite to the condition in birds (i.e., right aortic arch in birds rather than left). The right brachiocephalic artery, when present, is a remnant of the right fourth arch, as is the proximal part of the right subclavian. During embryonic development in mammals, the six homologous aortic arches are represented but are never all present at the same time; some regress before others form.

In mammals that dive to great depths, modifications of the circulatory system are necessary to withstand the increased pressure and to provide oxygen to vital organs while the animal is underwater. Some species may remain underwater as long as 2 hours. Vast networks of arteries (*retia mirabilia*) are located in protected positions along the vertebral column under the transverse processes, within the bony neural canal, and within the thoracic cavity. All of the *retia* are interconnected, are supplied by branches of the aorta, and are drained by efferent arteries. When a whale dives, the abdominal wall is compressed against the vertebral column by external pressure. This increased pressure forces abdominal viscera into the thorax, and air is forced out of the lungs into the trachea. The increased pressure causes the constriction of all arteries except those of the brain, which are protected by the skull. Blood forced out of the organs collects in large quantities in the *retia,* and these pools form protected reservoirs of oxygenated blood that are available for use by the brain.

FIGURE 9.38

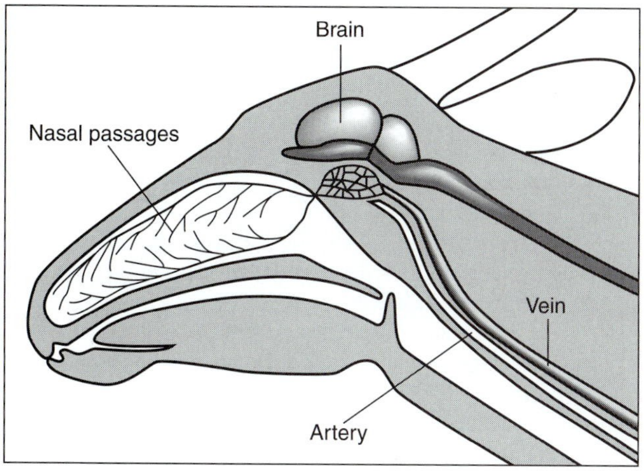

In the oryx (*Oryx* sp.), countercurrent cooling of arterial blood occurs in the cavernous sinus, on its way from the heart to the brain. Within the sinus, the carotid artery ramifies into hundreds of smaller vessels. Also within the sinus, venous blood from the oryx's nasal passages, cooled by respiratory evaporation, lowers the arterial blood temperature. Since oryx inhabit desert areas, having a brain cooler than the body temperature may be vital for its survival.

Retia serving other functions are found in ungulates, xenarthrans, carnivores, and birds and are not uncommon in lower vertebrates, especially on the pathways of arteries leading to the brains of fishes, and in the linings of swim bladders (red glands) (see Chapter 5). *Retia* frequently regulate pressure in arteries distal to themselves. In the desert-dwelling oryx (*Oryx* sp.) and Grant's gazelle (*Gazella granti*), a *rete* system serves to cool arterial blood supplying the brain (Fig. 9.38).

Seals and whales have flippers and flukes that lack blubber and are poorly insulated. These appendages are composed of bone and cartilage and are not well supplied with blood vessels. Nonetheless, these relatively thin structures with their large surfaces can lose substantial amounts of heat and aid in heat dissipation. Excessive heat loss from blood in the flippers is prevented by the arrangement of the arterial and venous vessels, which allows for countercurrent heat exchange. In the *retes,* which are located just inside the contour of the body adjacent to the flippers, each artery is completely surrounded by veins, and as warm arterial blood flows into the flipper, it is cooled by cold venous blood that surrounds it on all sides. The arterial blood, therefore, reaches the flipper precooled and loses little heat to the water. Heat has been transferred to the venous blood, which thus is prewarmed as it enters the body. If the heat exchange is efficient, the venous blood nearly reaches arterial temperatures and thus causes virtually no cooling to the core of the whale's body.

The oral cavity of baleen whales, which is relatively large in order to accommodate the filtering surface composed of baleen, is potentially a major site for heat loss during feeding in colder waters. In the gray whale (*Eschrichtius robustus*), heat loss is substantially reduced by the presence of numerous individual countercurrent heat exchangers found throughout the massive tongue, which may comprise 5 percent of the body's surface area (Heyning and Mead, 1997). Cool venous blood returning from the surface of the tongue flows first ventrally, then posteriorly toward the back of the tongue. Heat exchangers converge at the base of the tongue to form a bilateral pair of large vascular *retia.* Although the tongue is much more vascularized and has much less insulation than any other body surface, temperature measurements indicate that more heat may be lost through the blubber layer over the body than through the tongue. The lingual *retia* of the gray whale form one of the largest countercurrent heat exchangers described in any endotherm.

The blood of mammals contains erythrocytes (Fig. 9.39), leucocytes, and thrombocytes. With only one exception (camels), mature circulating erythrocytes of mammals are nonnucleated, biconcave disks. During the process of erythropoiesis (erythrocyte formation) in red bone marrow, the cells contain a nucleus, ribosomes, and mitochondria, but these gradually disappear before the erythrocytes are released into the circulating blood.

Camels possess nonnucleated *biconvex* erythrocytes that are reinforced by a hoop made of a bundle of microtubules (Weibel, 1984). This is thought to be an adaptation to allow the cells to shrink without being deformed when the animal is subjected to periods of considerable water loss.

Mammals regulate their body temperature by continuously monitoring the outside temperatures on the surface of their skin and at the hypothalamus. Heat-absorbing and heat-transporting capabilities of the blood are vital in maintaining homeostasis. When overheating occurs, some mammals sweat. The liquid drops of perspiration that appear on the surface of the skin cause a cooling effect as

FIGURE 9.39

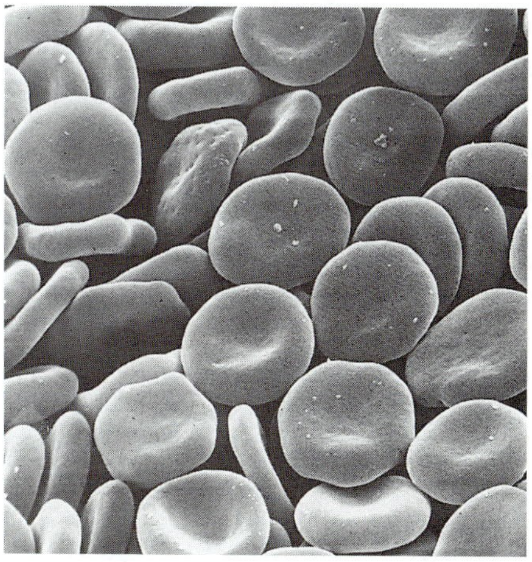

As shown by scanning electron microscopy, mammalian erythrocytes (red blood cells) are biconcave disks.

evaporation occurs. Thermal panting is characteristic of canids (Schmidt-Nielsen et al., 1970). The major portion of inhaled air enters through the nose and exits via the mouth. As exhaled air passes over the moist tongue and mucous membranes, it causes the evaporation of saliva, which in turn cools blood circulating in the capillaries of the tongue and adjacent tissues.

BIO-NOTE 9.9

Heart Size

In mammals, heart size increases with body size, but relative to their body size, small and large mammals have approximately the same heart size. The mass of the mammalian heart, irrespective of the size of the mammal, averages 0.59 percent of the body mass. Heart sizes of birds average 0.82 percent of body size. A reptile's heart size is approximately 0.51 percent of its body weight, whereas that of an amphibian is 0.46 percent. Although the proportions of heart mass to body mass in reptiles and amphibians are only slightly lower than in mammals, metabolic rates of reptiles and amphibians are only approximately one-tenth of mammalian rates. Hearts of fishes average only about 0.2 percent of their body weight.

Schmidt-Nielsen, 1990

Respiratory System

The paired external nares lead into paired nasal passageways (canals) that are partially occluded by several scroll-like bones, called **nasal conchae** or turbinal bones (Figs. 9.40 and 9.41), covered by ciliated mucous epithelium. Conchae slow down air flow in order to allow it to be filtered, warmed, and moistened. The nasal passageways are located dorsal to the hard and soft palates, which effectively separate the respiratory passages from the oral cavity. The olfactory epithelium is located in the posterodorsal portions of each nasal passageway. The nasal passageways open via internal nares into the nasopharynx above the soft palate (Fig. 9.40).

The soft palate separates the dorsal **nasopharynx,** which contains the openings of the auditory (Eustachian) tubes, from the **oropharynx.** Ventral to the oropharynx is the laryngopharynx, which leads to the glottis and esophagus. A cartilaginous flap, the **epiglottis,** covers the glottis when food is being swallowed. The food and air pathways cross in the laryngopharynx. Air must pass through the pharynx and enter the slitlike glottis at the top of the larynx, whereas food must enter the esophagus on its route to the stomach. The larynx, or voice box, which is located near the anterior (superior) end of the trachea, contains vocal cords, which in most mammals may vibrate during exhalation. Cartilaginous rings surround the trachea to prevent it from collapsing.

BIO-NOTE 9.10

Cat Vocalizations

There are 37 species of cats on earth. Most howl or yowl, meow or purr, but only four—the African lion, the tiger, leopard, and jaguar—can roar. The reason: these species have a unique vocal apparatus. Bones supporting the larynx are linked by an elastic ligament rather than more solid bone, and a length of tough cartilage runs up the hyoid bones to the skull. This feature prevents purring, but stretching of this ligament enlarges the air passage, facilitating a full-throated, resonant noise that may be heard up to 8 km away.

The purring of the smaller cats—house cats, bobcats, ocelots, lynxes, and others—occurs when the cat vibrates its larynx, which in turn, allows the delicate hyoid bones to resonate. Muscles create the continuous purring vibration—one twitch per second—during both inhalation and exhalation. It is independent of age, sex, weight, and size and is controlled by the brain. One big cat that purrs but cannot roar is the cheetah.

Sissom et al., 1991
Stewart, 1995

The trachea branches into two bronchi, each of which divide into secondary and tertiary branches. Tertiary branches give rise to the smallest air passageways, the **bronchioles** (see Fig. 9.40). Bronchioles lead into delicate, thin-walled **alveoli,** or respiratory pockets, which vastly increase the surface area of the lung available for gas exchange. Alveoli are surrounded by dense capillary beds. It is here that the exchange of gases between the inspired air and the blood occurs. Large surface areas are essential for the high rates of oxygen uptake required by the high metabolism of homeotherms. An efficient respiratory system enhances the efficiency of maintaining a relatively constant high body temperature.

The lungs of most mammals are divided into lobes, with each lung occupying a separate pleural cavity. In some mammals, including monotremes, only the right lung is lobed. In others, such as whales, sirenians, elephants, and perissodactyls, both lungs lack lobes.

Carbon dioxide normally diffuses across the alveolar–capillary membrane in the lung and is expelled through the process of exhalation. Some bats, however, may lose carbon dioxide through their large, highly vascularized, hairless wing membranes. In the big brown bat (*Eptesicus fuscus*), 0.4 percent of the total carbon dioxide production is lost from the wing skin at 18°C. The amount increases with temperature, and at an air temperature of 27.5°C as much as 11.5 percent of the total carbon dioxide is lost this way (Herreid et al., 1968). Uptake of oxygen through the wing membranes, however, is not sufficiently great to be of any physiological significance.

Most mammals do not carry on gas exchange through their skin because they have high metabolic rates and because

FIGURE 9.40

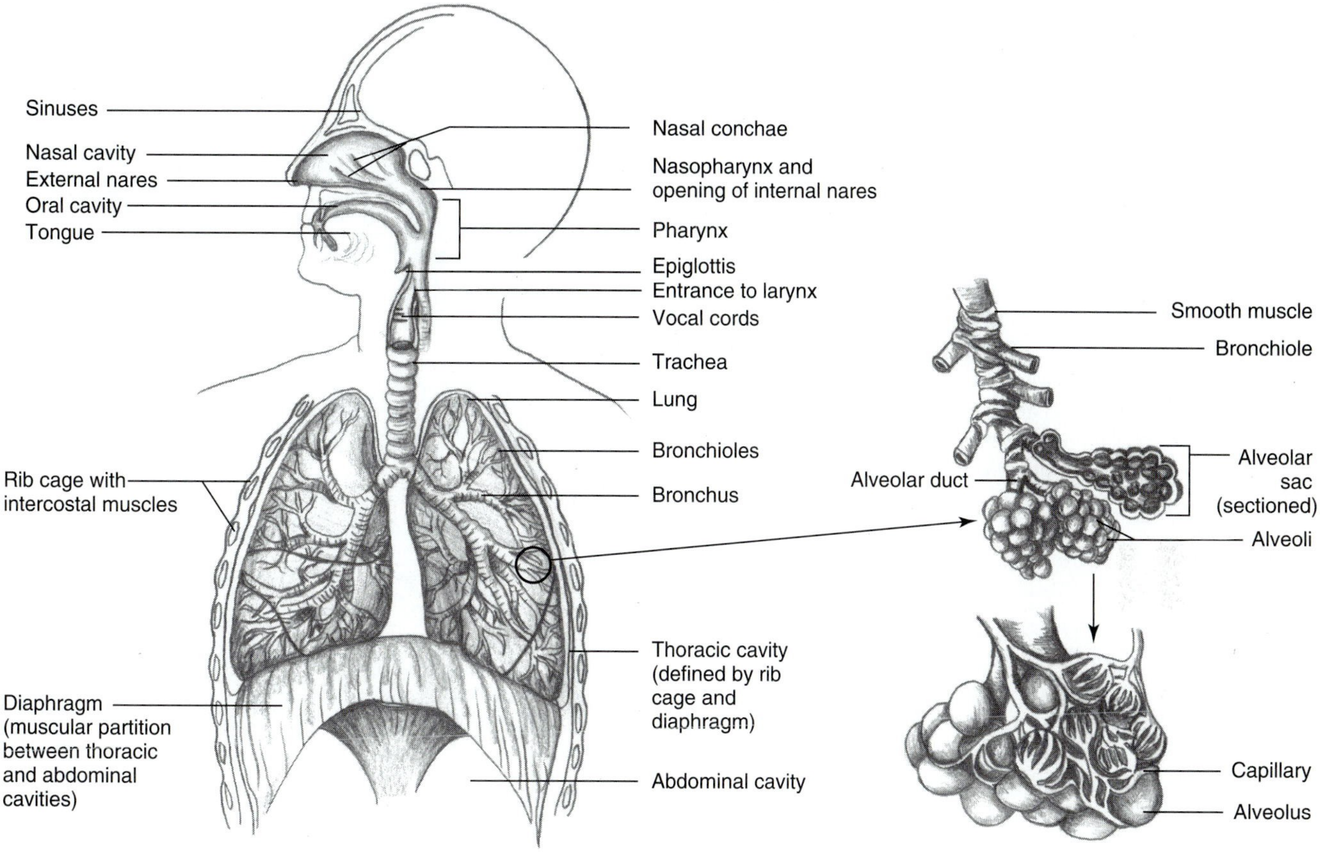

The human respiratory system. Note the muscular diaphragm.
Source: Starr, Biology, *1991, Wadsworth Publishing.*

FIGURE 9.41

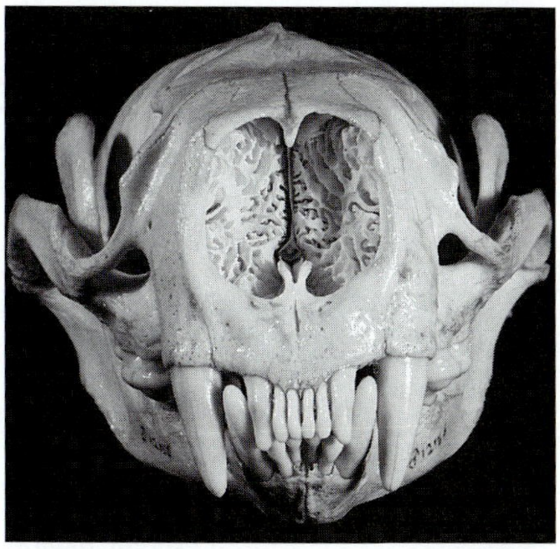

Front view of the skull of a sea otter (*Enhydra lutris*) showing the highly developed, scroll-like structure of the nasal conchae. Nasal conchae slow down the passage of air so that it can be filtered, warmed, and moistened prior to traveling through the respiratory system.

diffusion through the skin is poor. The Julia Creek dunnart (*Sminthopsis douglasi*) is a small marsupial mouse with one of the smallest newborns of any mammal (Mortola et al., 1999). Following a gestation period of approximately 13 days, the newborn is about 4 mm long and weighs about 17 mg. During the first few days following birth, the predominant form of oxygen and carbon dioxide exchange is through the skin. In young animals with body weight below 100 mg, gas exchange through the skin exceeded that through the lungs. In animals 20–21 days of age with an average weight of 290 mg, skin exchange was about one-third that of the lungs. Gas exchange through the skin apparently allows these very small animals to be born before the respiratory apparatus is fully functional.

A muscular, dome-shaped **diaphragm,** an extension of the body wall musculature, separates the pleural cavities from the abdominal cavity. At rest, the diaphragm is dome-shaped (Figs. 9.40 and 9.43). When the diaphragm contracts, it flattens and enlarges the thoracic cavity. In conjunction with the elevation of the ribs by external intercostal muscles,

BIO-NOTE 9.11

Lungs to the Limit

The mammalian respiratory system reaches its ultimate potential in the smallest known mammals such as the Etruscan shrew (*Suncus etruscus*), which weighs about 2 g (Fig. 9.42), and Kitti's hog-nosed bat (otherwise known as the bumble-bee bat) of Thailand (*Craseonycteris thonglongyai*), which weighs approximately 1.7 g. The shrew has the highest metabolic rate of any known mammal. Its heart rate is over 1,000 beats/minute and its respiration rate is over 300 breaths/minute (Weibel, 1984). The heart fills about half of the left chest cavity, leaving little room for the left lung. The lungs, which consist of a maze of tiny alveoli and blood capillaries, have a gas exchange surface whose density is eight times as large as that of a human and three times as large as that of a mouse. The air–blood barrier within the lung consists of a very fine barrier of extremely thin endothelium.

Weibel, 1984

these actions also lower the internal pressure in the pleural cavities below atmospheric pressure and allow air to flow into the lungs. Relaxation of the diaphragm, a passive process, and contraction of the internal intercostal muscles decrease the volume and increase the pressure in the thoracic cavity above atmospheric pressure, thus forcing air out of the lungs.

FIGURE 9.42

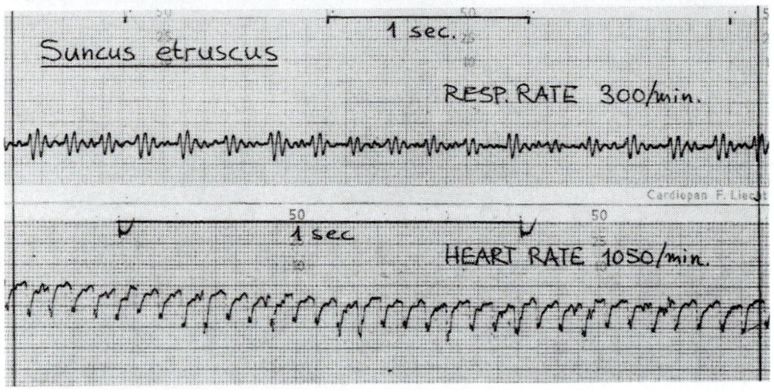

(*a*) One of the smallest mammals, the Etruscan shrew (*Suncus etruscus*), weighing 2.5 g, is shown beside a match. (*b*) Recordings of the respiratory and heart rates of the Etruscan shrew at rest.

FIGURE 9.43

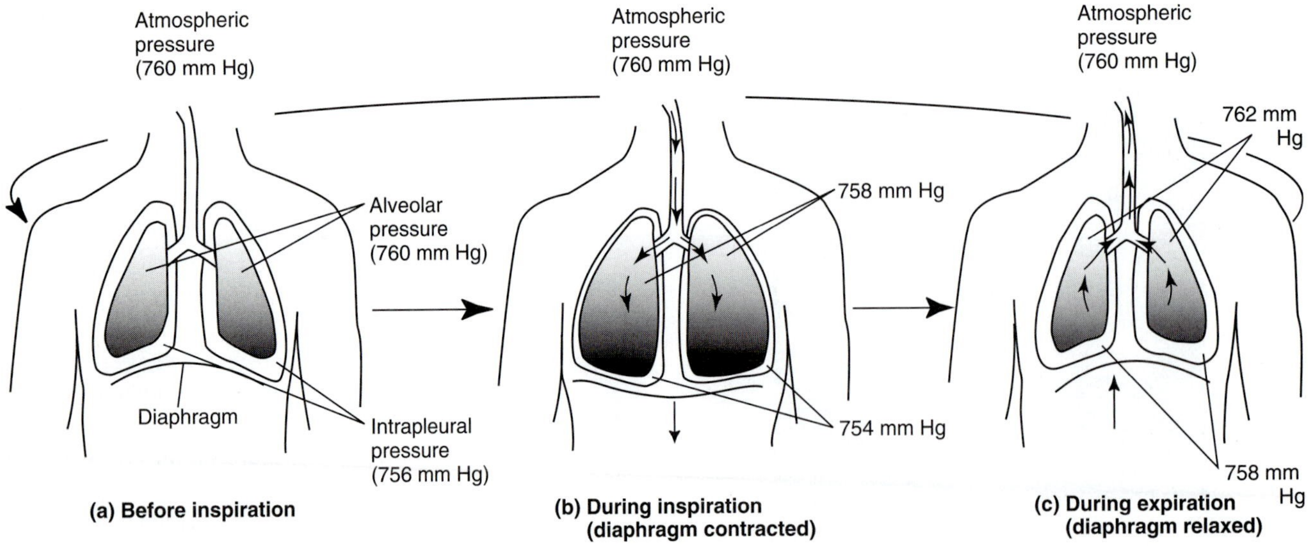

Pulmonary ventilation: pressure changes.(*a*) Prior to inspiration, the diaphragm is dome-shaped and the alveolar and atmospheric pressures are the same. (*b*) At the beginning of inspiration, the diaphragm contracts, the external intercostal muscles elevate the ribs causing the chest to expand, and alveolar pressure decreases. (*c*) As the diaphragm relaxes, internal intercostal muscles lower the ribs, and alveolar pressure rises, forcing air out until the alveolar pressure equals the atmospheric pressure.

BIO-NOTE 9.12

The Spermaceti Organ

One problem that whales have is maintaining neutral buoyancy while they are diving and when they remain at a particular depth. Neutral buoyancy control by sperm whales at various depths and temperatures is maintained by means of the **spermaceti organ** and its oil. The spermaceti organ, which is richly supplied by blood vessels, is a massive heat exchange device that controls the temperature and density of the oil. Spermaceti oil (unlike water, which freezes suddenly at 0°C) becomes increasingly crystalline over a range of several degrees. To swim downward, substantial effort is needed to overcome the initial buoyancy of the lungs. This causes the whale's body temperature to rise significantly. When the time comes for surfacing, blood courses through the spermaceti organ, transferring the surplus heat to the oil and liquefying it. The resulting increase in its buoyancy greatly facilitates the whale's return to the surface. Thus, efficient use is made of the surplus heat generated in diving.

As the whale descends, its lungs continue to collapse until at about 200 m they no longer contribute significantly to the whale's buoyancy. During the descent, cold water circulating through the spermaceti organ continuously cools the oil and reduces its buoyancy (Fig. 9.44). By the time the whale arrives at its hunting depth, it has achieved neutral buoyancy and can, in effect, turn off its engines and conserve energy.

One of the two nasal passages that link the blowhole to the mouth passes through the spermaceti organ. The left passage retains its respiratory function. The right nasal passage, however, has become highly modified and passes controlled amounts of cold sea water through the spermaceti organ, countering the warming effect of the blood. A sperm whale remains submerged for about 50 minutes in a typical dive, although dives of 90 minutes have been observed. Each dive is followed by about 10 minutes of heavy breathing at the surface.

Cooke, 1991

When resting, humans generally renew about 10 percent of the air in the lungs in a single breath. Some mammals, such as the manatee, can renew about 90 percent of the air in their lungs in a single breath while resting (Reynolds and Odell, 1991); this explains why they can remain underwater up to 20 minutes. Weddell seals in Antarctica may plunge hundreds of meters below the ocean's surface to hunt for fish and squid. Although most dives last 5 minutes or less, the longest recorded dive lasted 48 minutes (Harding, 1993). The deepest recorded dive was 600 m. Northern elephant seals have been known to descend 1,530 m and remain submerged for 77 minutes (Harding, 1993). Sperm whales (*Physeter macrocephalus*) often dive to depths of at least 1,000 m and remain submerged up to 90 minutes (Riedman, 1990). An acoustic recording of a sperm whale at a depth of 2,250 m was reported by Norris and Harvey (1972).

When a marine mammal dives, muscles in its nose relax and the nasal openings close. Almost immediately the heart rate drops well over 50 percent, and blood flow is reduced to all but the most vital organs, such as to the heart and brain. For example, in bottlenosed dolphins (*Tursiops truncatus*), the heart rate drops from 110 beats per minute while at the surface to about 50 while submerged. Killer whales (*Orcinus orca*) have a rate of about 60 beats per minute at the surface and 30 per minute while submerged (Rice, 1967). At the same time, extra red blood cells pour into the bloodstream from the spleen and become available to transport increased oxygen.

Deep-diving mammals do not use their lungs as their major oxygen storehouse. Because they have higher concentrations of red blood cells and hemoglobin than other animals, they can store large amounts of oxygen in myoglobin (a form of hemoglobin) in their muscles—as much as 13 times the amount held in the lungs. To cope with extreme pressure,

FIGURE 9.44

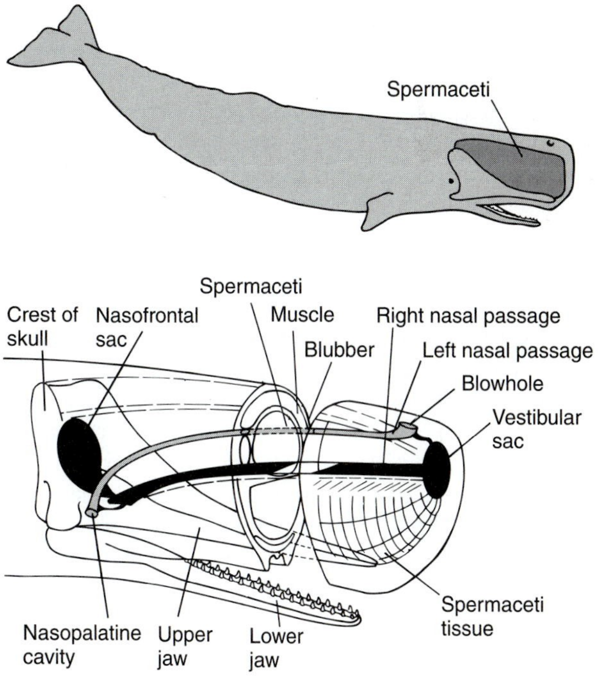

Sperm whales (*Physeter catadon*) regulate their buoyancy by warming or cooling the oil in the spermaceti tissue in their large heads. The two nasal passages are asymmetrical, with the left passage running from the nasopalatine cavity to the blowhole, and the right running through the spermaceti organ. Cool water channeled through the spermaceti organ causes the surrounding oil to crystallize and become more dense, thus enabling the whale to descend. Warming of the organ changes the crystallized oil into a liquid and increases the whale's buoyancy.

most deep-diving mammals have cartilage-reinforced air passageways that allow the lungs to collapse and then reinflate. The lungs contract slowly as pressure increases during the descent, and they gradually open back up during the ascent (William et al, 2000).

Digestive System

The major components of the digestive system are the oral cavity, pharynx, esophagus, stomach, small intestine, large intestine, rectum, and anus. The oral cavity (mouth) is bounded by the lips, cheeks, hard palate, and tongue. The lips of many mammals are highly vascularized and function in determining the texture and temperature of foods. Teeth begin the mechanical breakdown of food; saliva secreted by the salivary glands initiates the chemical breakdown. The tongue, which is roughened by the presence of papillae on its dorsal surface, manipulates the food in the mouth and plays a major role in swallowing by forcing the food back into the pharynx.

Some mammals, such as eastern chipmunks (*Tamias striatus*), have large membranous, internal cheek pouches on the medial surfaces of the cheeks (Fig. 9.45), whereas others, such as pocket gophers (*Geomys* sp.), have external cheek pouches. Cheek pouches are used to transport seeds and other foods to the nest site or storage chambers.

FIGURE 9.45

The eastern chipmunk (*Tamias striatus*) has internal cheek pouches that can be filled with seeds to be carried to a storage cache. Although chipmunks are inactive during severe winter weather, they are active during warmer periods. They attempt to store at least one-half bushel of food to carry them through the winter months.

Most mammals have teeth composed of dentin and enamel. Enamel is the hardest substance in the body, but once damaged or destroyed, it is not replaced. Dentin, which is almost as hard as enamel, surrounds the pulp cavity that contains blood vessels and a nerve.

A few mammals, such as armadillos, sloths, and anteaters, have reduced dentition, or they have lost their teeth entirely (Fig. 9.46e). When present, teeth are confined to the jaws and are much less numerous than in other vertebrate groups. Only in mammals is there a specific number of teeth in a given species.

Some mammals experience a forward progression of their teeth so that new teeth form at the rear of the jaw and gradually move forward to replace worn teeth. This type of tooth replacement is known to occur in one species of manatee, one species of kangaroo, and in elephants (Fig. 9.46b). Elephants have the use of 24 molar (cheek) teeth during their lifetime. Each molar is enormous, measuring 20 cm or more in length and weighing 3.5 to 4.0 kg. Only one, or parts of two, on each side of each jaw are in use, or in existence, at any one time. While the first group of four teeth are being worn down, four new teeth are growing behind them. These gradually move forward, replacing the old teeth, which are eventually shed. The process is then repeated, with each successive tooth being larger than the last. When six teeth have passed through each half of each jaw, no other teeth can be grown, and the animal finds itself unable to chew its food.

Baleen whales are toothless and are specialized for feeding on plankton. They may employ filter-feeding (suspension-feeding) or suction-feeding, or both. A filtering apparatus consisting of 200 to 400 horny plates, called **baleen,** attached to the upper jaws hangs down on both sides (see Fig. 9.13). Depending on the species, each lightweight plate may be from 0.25 to 3 m long. In a form of filter-feeding known as ram-feeding, the whales take water into their mouths as they swim forward. Water that is taken into the oral cavity flows over and between the plates, which serve as massive strainers to remove minute food particles. Plankton caught in the hairlike edges of the plates is directed into the esophagus. Plankton-eating whales include blue whales, the largest living animals, which may attain weights of over 90,000 kg.

Whereas most vertebrates have a succession of teeth throughout their lives, called the **polyphyodont condition,** most mammals develop only two sets of teeth, known as the **diphyodont condition.** The first set is known as the **milk (deciduous) teeth,** which usually erupt following birth but may erupt before birth in some species. These teeth are rooted in individual bony sockets (thecodont dentition) and will gradually be replaced by the **permanent,** or **adult, dentition.** Thereafter, a damaged or injured tooth will not be replaced in most mammals. Incisors, canines, and premolars are preceded by milk teeth; molars are not.

Some mammals have homodont dentition in which teeth are similar in size and appearance (Fig. 9.46). Most mammals, however, have teeth that are modified for different purposes. This **heterodont dentition** often includes incisor, canine, premolar, and molar teeth.

Incisors, the most anterior teeth in the oral cavity, are chisel-like and specialized for cutting and gnawing. In rodents and lagomorphs (rabbits and hares), these teeth continue to grow throughout life (Fig. 9.47b). Enamel is present only on the anterior surface of these enlarged incisors and wears down more slowly than the dentin on the posterior surface; thus, these teeth continually have sharp, chisel-like edges. Incisors may be totally absent, as in sloths, or absent

FIGURE 9.46

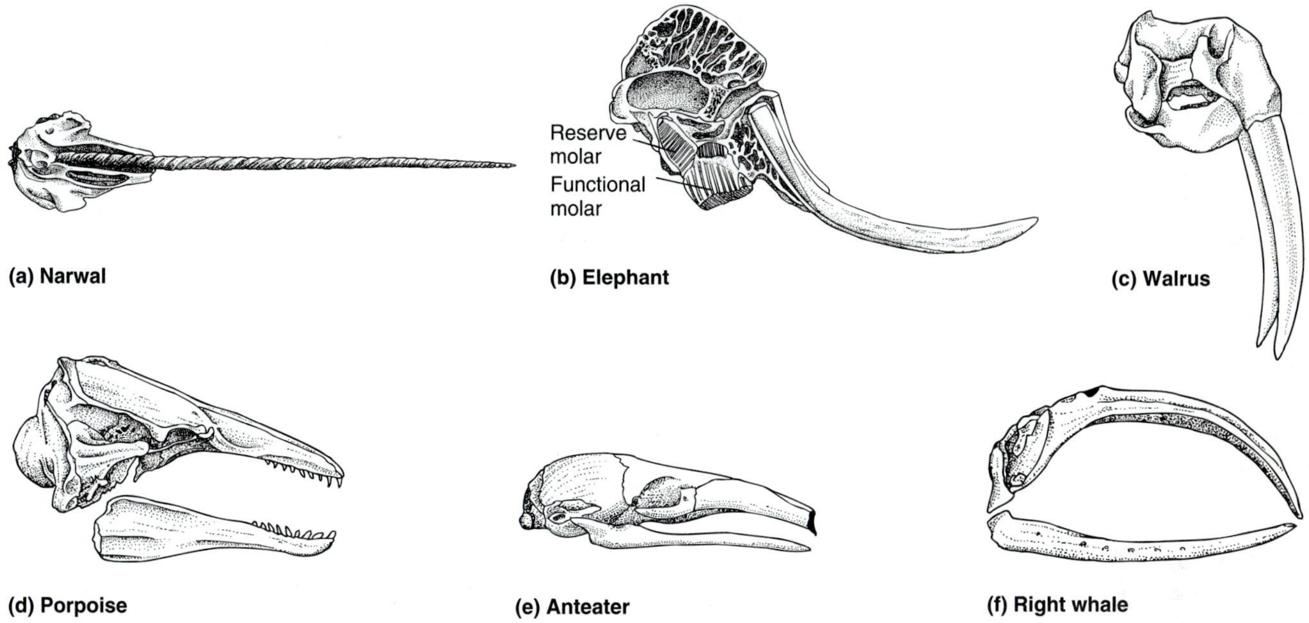

Specialized mammalian teeth. (*a*) The single tusk of the narwhal is a modified upper left incisor. (*b*) The paired tusks of the elephant are modified upper incisors. (*c*) The tusks in the walrus are modified canines. (*d*) Peglike teeth of a porpoise. Teeth are absent in adult anteaters (*e*) and baleen whales (*f*).

just from the upper jaw, as in deer. The single tusk of the narwhal (Fig. 9.46a) arises from the upper left incisor.

The tusks of elephants (Fig. 9.46b) are enlarged upper incisors that continue to grow throughout the life of the animal. In a young elephant, a pair of deciduous tusks grow to a length of approximately 5 cm and remain for about a year before being replaced by the permanent tusks growing behind them (Haynes, 1991). Either a single molar or parts of two molars function at one time, as a rear molar moves *forward* to replace the one being worn away. As a tooth wears out, it is gradually pushed forward and falls out. In its lifetime, an elephant has 24 cheek teeth (molars) in all, the first 12 being milk teeth.

The four lower incisors of flying lemurs (order Dermoptera, family Cynocephalidae) are unique among mammals in that each is pectinate or compressed into a fine-toothed comb. Each lower incisor may have as many as 20 prongs radiating from one root. Comb teeth may serve as food strainers or as scrapers, or they may be used to groom the fur (Aimi and Inagaki, 1988).

Canines are located immediately behind the incisors. In carnivorous mammals, these teeth are used for piercing and tearing flesh. Canine teeth always are absent in rodents and lagomorphs; the resultant gap between the incisors and cheek teeth is called a **diastema.** Walrus tusks are modified canine teeth (Fig. 9.46c).

FIGURE 9.47

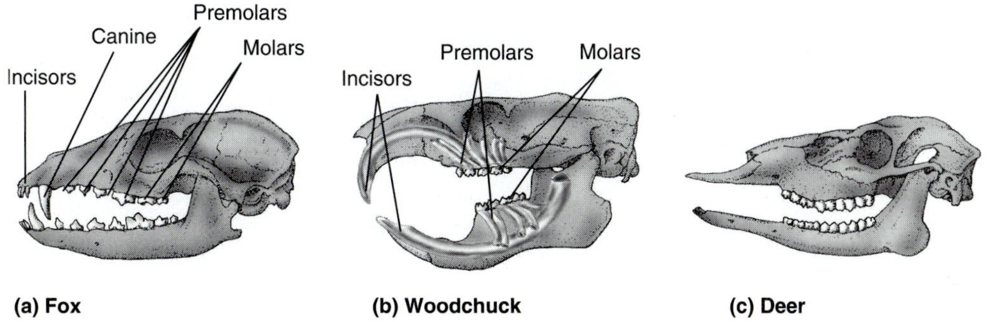

Teeth adapted for processing three principal kinds of food: (*a*) fox; (*b*) woodchuck; (*c*) deer.

Premolars and molars are known as cheek teeth and are modified for crushing and grinding (Fig. 9.47). Cheek teeth in herbivorous mammals possess large, flat grinding surfaces, an evolutionary adaptation for feeding on vegetation. In carnivores, the number of cheek teeth may be reduced, cusps may be present, and at least one pair in each jaw may have very sharp cusps for cracking bones and shearing tendons. These are known as **carnassial teeth** (Fig. 9.47a). Grinding is deemphasized in carnivores, and food is swallowed with little chewing. The cheek teeth of omnivorous mammals are intermediate in form. In some primates, cutting edges form on the upper canines and lower premolars—the **sectorial teeth.** These teeth are used in defense or in fights between individuals.

Because each species of mammal has specific numbers and kinds of teeth, dental formulas are useful for taxonomic purposes. Primitive placental mammals had 3 incisors, 1 canine, 4 premolars, and 3 molars in each half of the upper and lower jaw for a total of 44 teeth; therefore, the dental formula was 3/3, 1/1, 4/4, 3/3. Representative examples of dental formulas for other adult mammals include dogs (*Canis familiaris*), 3/3, 1/1, 4/4, 2/3; rabbits (*Sylvilagus* sp.), 2/1, 0/0, 3/2, 3/3; beavers (*Castor canadensis*), 1/1, 0/0, 1/1, 3/3; horses (*Equus caballus*), 3/3, 0/1, 3/3, 3/3; white-tailed deer (*Odocoileus virginianus*) 0/3, 0/1, 3/3, 3/3; and humans (*Homo sapiens*) 2/2, 1/1, 2/2, 3/3.

Most mammals have a muscular tongue that can be protruded out of the mouth. Some insectivorous mammals, as well as pollen- and nectar-eating bats, have a tongue as long as their entire body. The surface of most mammal tongues usually bears papillae, whose function is to increase friction and assist in the manipulation of food by the tongue. Taste buds are located on or among the papillae. Whales have a tongue, but in most species it cannot be extended.

The giant anteater, *Myrmecophaga tridactyla,* has an elongate, tubular snout, a tiny, toothless mouth, and a long, slender, sticky tongue which can project to a distance greater than the cranial length. A unique hyoid arrangement enables *Myrmecophagus* to project the tongue with great speed and precise positional control so that it can probe deep into the nests of ants and termites. Most mammals open their mouth by contracting the digastric muscle to move the lower jaw. The giant anteater has no digastric muscle, and the lower jaw depresses only slightly during feeding. Instead, it opens its mouth by the medial rotation and depression of the two halves of its elongate lower jaw (the mandibular rami) about their long axes (Naples, 1999). This allows an anteater to open and close its mouth rapidly. The protraction and retraction of the elongated tongue permits feeding in ant and termite nests, where the spaces are too narrow for the mouth to open wide.

Oral glands, which secrete mucus and digestive enzymes, are abundant and serve to begin the chemical digestion of certain foods. Most mammals have three major pairs of salivary glands—the **parotid, mandibular** (also known as submandibular or submaxillary), and **sublingual** (Fig. 9.48). These glands secrete saliva to moisten food and begin the initial

FIGURE 9.48

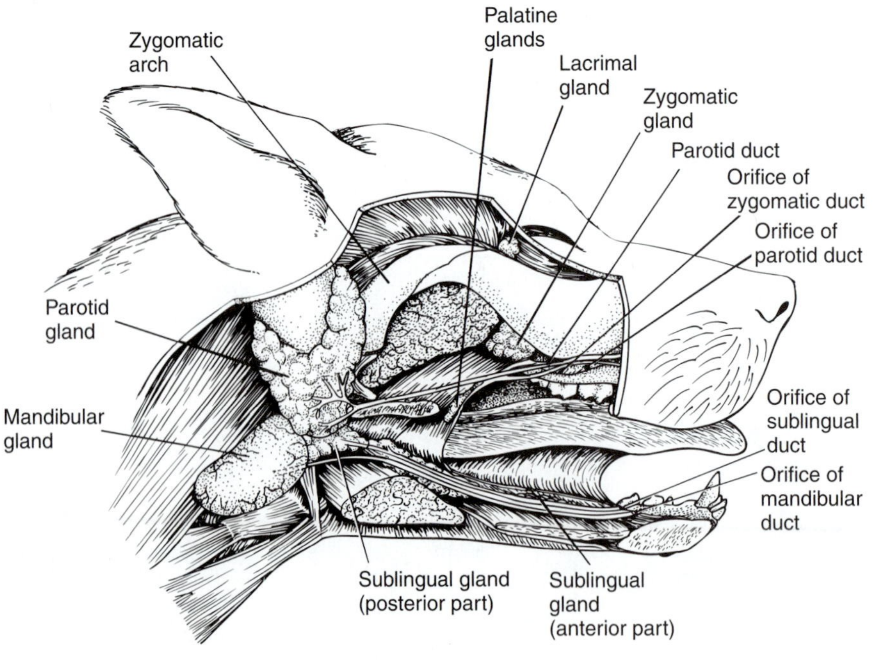

Salivary glands of a dog. All mammals possess three main salivary glands—parotid, mandibular, and sublingual. In addition, some mammals may posses smaller glands, such as palatine glands above the palate. A zygomatic gland is present only in dogs and cats. Each gland has a duct opening into the oral cavity.

digestion of starch. Saliva consists of mucin, water, and the digestive enzyme amylase.

The esophagus shows no special modifications. An esophageal (cardiac) sphincter, whose function is to prevent the backflow of contents from the stomach, is located at the point where the esophagus enters the stomach.

The stomach is a muscular organ that varies from a single saclike compartment to the four-chambered stomach of ruminants, where digestion is aided by microorganisms (Fig. 9.49). Dwarf and pygmy sperm whales (*Kogia* sp.), river dolphins, and beaked whales (Ziphiidae) also have multichambered stomachs. In a typical stomach, a series of longitudinal ridges—the **rugae**—line the inner surface and permit the stomach to distend. Three types of cells are present in the stomach. **Mucous cells** secrete mucus; **chief cells** secrete the principal gastric enzyme precursor, pepsinogen, which is

FIGURE 9.49

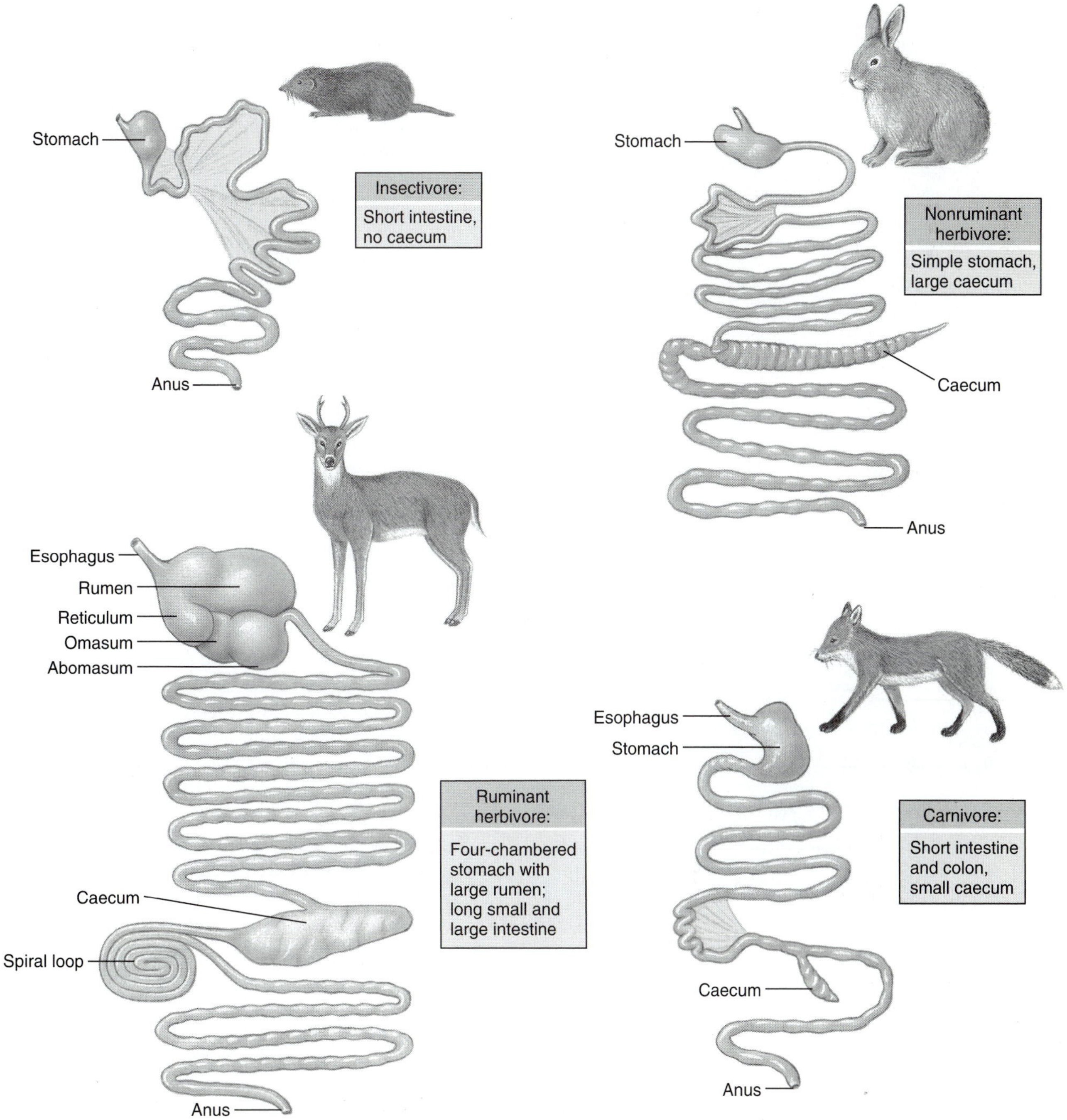

Digestive systems of mammals illustrating different modifications for different diets. The length of the digestive tract of an herbivore is considerably longer than the digestive tract of a similar-size carnivore; this provides additional time for microorganisms to break down the cellulose cell walls of plant cells.

subsequently converted to pepsin; and **parietal cells** secrete hydrochloric acid. The hydrochloric acid provides the acidity necessary for pepsin to be effective. The mucus serves as a protective lining for the stomach wall cells. Together, the secretions of these cells form **gastric juice.** The stomach terminates at the pyloric sphincter, where it joins the small intestine.

BIO-NOTE 9.13

Pika Haypiles

Pikas (*Ochotona* sp.) (Fig. 9.50) are small rodentlike, lagomorph mammals of the mountains of the western United States and Asia. They work prodigiously during the summer gathering enormous amounts of grasses and wildflowers into mounds known as haypiles, which serve as their winter food. Each pika gathers the equivalent of about 29 kg of plants. This is roughly comparable to a 58 kg person collecting 9000 kg of vegetation by making 5,000 shopping trips. As much as 75 percent of some haypiles consists of *Acomastylis rossii,* a plant whose phenolic compounds presumably inhibit bacterial growth and act as a preservative during the long, cold, mountain winter.

Dearing, 1993

Some mammals, such as ruminants (cattle, goats, sheep, bison, camels, deer, giraffes, and others), are foregut fermenters and have a stomach consisting of several chambers (Fig. 9.51a). After being chewed briefly, food is swallowed and passes into the first and largest chamber, the **rumen,** where it is acted on by microorganisms. The rumen serves as a large fermentation vat where food is mixed with large numbers of bacteria and protozoans and undergoes extensive fermentation. Microorganisms break down cellulose, making it available for further digestion. The food then may be regurgitated, formed into a cud, and rechewed. When swallowed, this rechewed food passes first into the **reticulum,** a small accessory chamber with a honeycombed texture, then into the **omasum,** where it is acted on by the salivary enzyme amylase. Finally, it enters the glandular **abomasum,** where gastric glands secrete gastric juice.

FIGURE 9.50

The North American pika (*Ochotona* sp.), also known as little chief, whistling hare, cony, rock rabbit, mouse hare, and haymaker, averages approximately 15 cm in total length. It looks like a miniature rabbit, but its ears are not as long, its hind legs are nearly as short as its front legs, and it lacks a tail. Its voice varies from a sharp, shrill whistle to a bleat like that of a lamb. It dwells among tumbled rock piles near the summits of mountain ridges.

Horses and some other mammals are hindgut fermenters (Fig. 9.51b). There is no four-chambered stomach. Fermentation takes place in the large cecum near the junction of the small and large intestines.

The small intestine in most mammals is long and convoluted (see Fig. 9.49). The absorptive surface, which is the chief site of digestion and absorption, is increased by fingerlike projections known as **villi** and **microvilli.** The small intestine generally is longer in herbivorous mammals than in carnivorous forms in order to provide more opportunity for digesting the cellulose cell walls of plant cells. The small intestine consists of three portions: the **duodenum** (anterior), the **jejunum** (middle), and the **ileum** (posterior). Pancreatic juice and bile are secreted into the duodenum. Pancreatic juice is alkaline and contains enzymes (amylase, lipase, maltase, and sucrase) that continue the digestive processes. Bile, which is also alkaline, emulsifies fats.

The large intestine consists of the colon, cecum, and rectum. Its major functions are the reabsorption of water and of electrolytes, which result in the compaction of wastes that must be eliminated from the body. The **colon** may be divided into **ascending, transverse,** and **descending** portions. The ileum opens into the ascending colon through the ileocecal valve. A blind sac, the **cecum** (see Fig. 9.49) is located at the juncture of the small intestine and colon. In mammals that feed on cellulose, the cecum often contains cellulose-digesting bacteria and ciliate protozoans (Oxford, 1958), which also synthesize vitamin B, amino acids, and proteins. In some mammals, including rabbits, rodents, monkeys, apes, and humans, the distal end of the cecum is mostly vestigial and terminates in a **vermiform appendix.** In humans and possibly some other mammals, the appendix may have a limited lymphatic function. The descending colon terminates in the rectum, which serves as a temporary storage chamber for feces. Feces consist of indigestible materials, water, electrolytes, and some bacteria. Feces are eliminated from the rectum via the anal canal, which opens to the exterior via the anus. Anal control is through internal and external anal sphincter muscles.

Nervous System

The proportion of brain to body size is usually greater in mammals than in other vertebrates. Pound for pound, the record for brain size is probably held by fruit-eating South American squirrel monkeys (genus *Saimiri*), whose brains account for an average of 5 percent of their body weight

BIO-NOTE 9.14

DNA from Coprolites

Researchers studying dung (coprolites) from Gypsum Cave near Las Vegas, Nevada, have been able to unlock DNA trapped in the ancient feces. The mitochondrial DNA, which presumably came from intestinal cells shed into the feces, probably came from an extinct ground sloth (*Nothrotheriops shastensis*) about 20,000 years ago. In addition, researchers were able to extract a wide variety of plant DNA from the coprolite—clues to the vegetarian sloth's diet. Plant DNA sequences were identified from eight plant families and included grasses, yucca, grapes, and mint. The plant assemblage that formed part of the sloth's diet exists today at elevations about 800 m higher than the cave. Such studies may help provide answers as to why these and other large animals vanished from North America about 10,000 years ago.

Poinar et al., 1998

(Nilsson, 1999). For humans, the figure is approximately 2 percent. The big whales have the smallest brains of all the mammals: that of the blue whale (*Balaenoptera musculus*) represents about 0.005 percent of its body mass.

The four major regions of the adult brain include the **brain stem** (medulla oblongata, pons, midbrain), **diencephalon** (epithalamus, thalamus, hypothalamus), **cerebrum,** and **cerebellum** (Fig. 9.52). Four spaces known as **ventricles** are present in

FIGURE 9.51

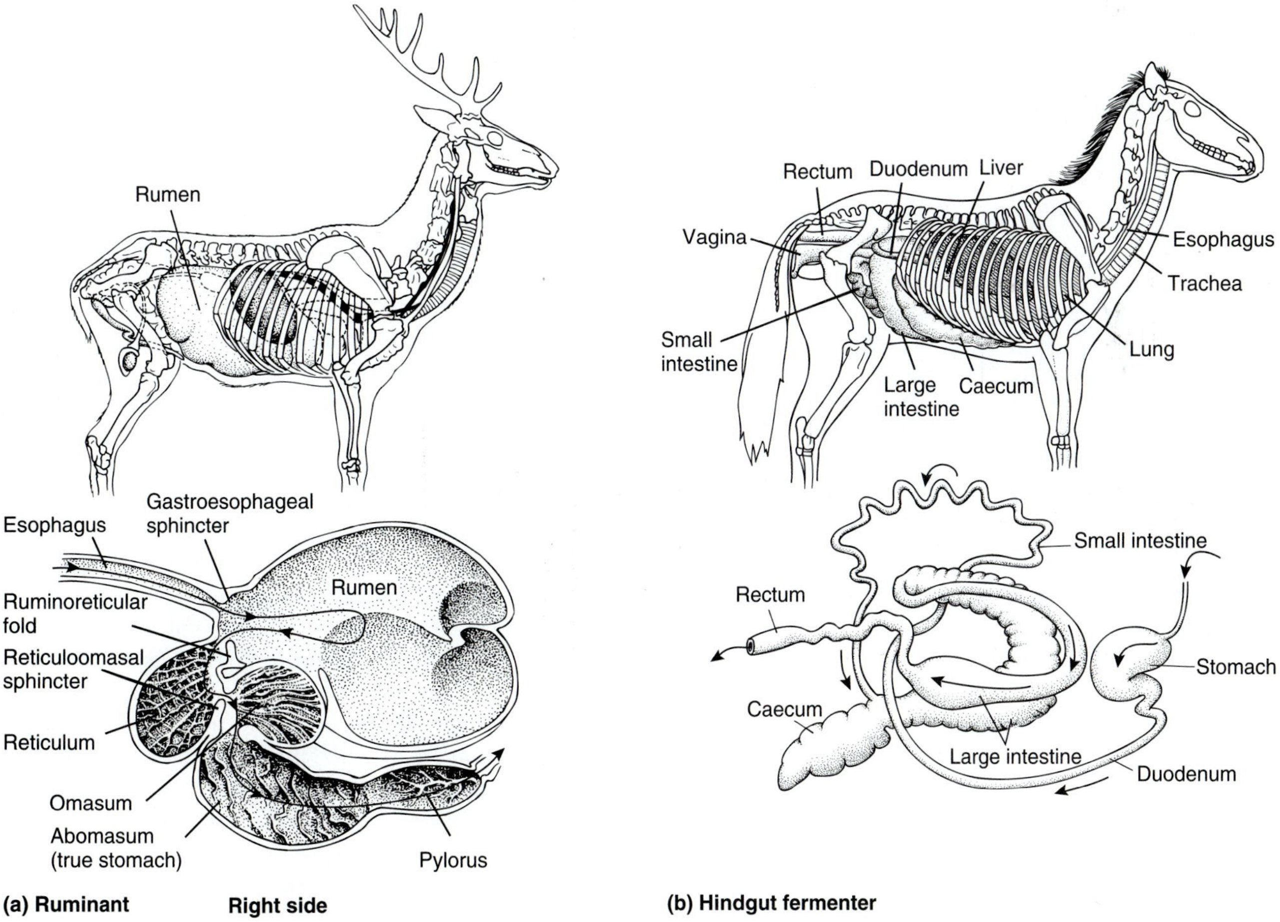

Alimentary canal of foregut and hindgut fermenters. (*a*) Ruminants (deer, cows, and others) ferment food in their foregut. These grazing animals swallow partially chewed plant material, which moves into two stomachlike chambers (rumen and reticulum). Then they regurgitate the material, chew it, and swallow it again. The double chewing (commonly referred to as "chewing a cud") breaks apart the plant tissues and better exposes the cellulose fibers. The cellulose is broken down by digestive enzymes that are produced by symbiotic bacteria living in the gut. The double chewing gives the enzymes more time and more surface area on which to act. The rumen, reticulum, and omasum are derivatives of the esophagus. The fourth compartment, the abomasum, is the true stomach. (*b*) Hindgut fermenters, such as the horse, possess a large cecum near the juncture of the small and large intestines. The cecum is the major site of fermentation. There is no four-chambered stomach in hindgut fermenters.

FIGURE 9.52

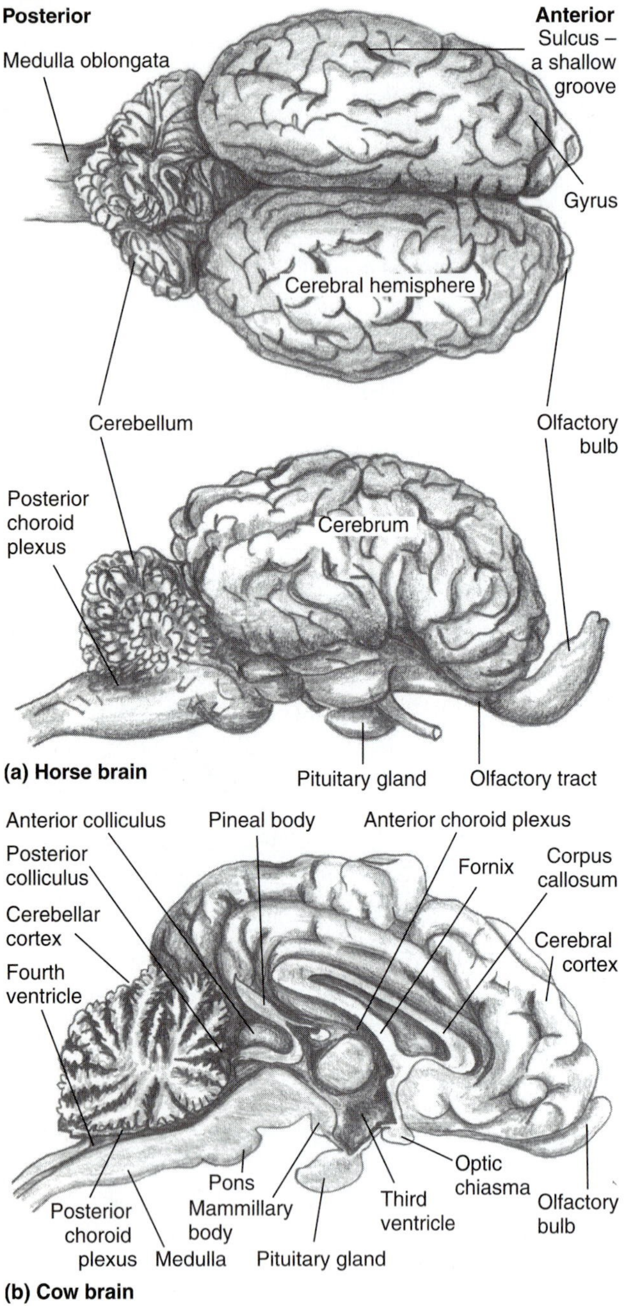

Mammal brains: (*a*) dorsal and lateral views of a horse brain; (*b*) sagittal section of a cow brain.

the mammalian brain. They are filled with **cerebrospinal fluid,** which is secreted by choroid plexuses in the roof of each ventricle. Cerebrospinal fluid serves as a transport medium within and around the brain and spinal cord. The entire brain is enclosed and protected by three meninges (dura mater, arachnoid, and pia mater).

The **medulla oblongata** is continuous with the spinal cord and acts as a major relay center for all ascending and descending tracts. In addition, it contains several specialized areas including the respiratory center and cardiac center. The respiratory center controls the rhythmic movements of the diaphragm and intercostal muscles. It is sensitive to increased carbon dioxide or increased acidity in cerebrospinal fluid caused by an increased concentration of carbon dioxide in the blood (Schmidt-Nielsen, 1990). The pons assists the respiratory center in controlling the rate and depth of breathing; it also serves as a relay center to higher portions of the brain and between the cerebellum and other parts of the nervous system. The mesencephalon in mammals serves in a limited way for visual and auditory reflexes.

The thick walls of the diencephalon make up the **thalamus,** which serves as a relay center to the cerebrum. The thalamus contains clusters of nuclei that are associated with motor fibers leading posteriorly in the brain and with sensory pathways leading anteriorly. Perception of pain and pleasure takes place in the thalamus. The thinner, ventral walls and floor of the diencephalon, the **hypothalamus,** contain clusters of nuclei that control autonomic functions of the body such as water balance, temperature regulation, appetite and digestion, blood pressure, sleep and waking, and sexual behavior. It also serves as an endocrine gland producing antidiuretic hormone, oxytocin, releasing factors, and release-inhibiting factors. The releasing factors and release-inhibiting factors control the release of hormones from the adenohypophysis (anterior pituitary gland). These hormones, in turn, affect other endocrine glands (see Endocrine System, page 309–310).

On the ventral surface of the diencephalon is the **optic chiasma,** the point where the optic nerves converge and cross. In humans, about 50 percent of the nerve fibers cross over; in other mammals, the proportion may be higher (Carter, 1967). Just posterior to the optic chiasma is the **hypophysis** (pituitary gland), which also functions as an endocrine gland. Posterior to the hypophysis are a pair of **mammillary bodies,** which function in olfaction and are a part of the limbic system, a portion of the brain involved in the expression of emotions (which are important to survival) and in short-term memory. Two structures are found on the dorsal portion of the diencephalon in most mammals. The **habenula** contributes to the coordination of olfactory reflexes, whereas the **pineal body (gland),** also known as the epiphysis, functions as an endocrine gland. It secretes the hormone melatonin, which appears to regulate certain cyclical functions of the body. The effects of melatonin are discussed in greater detail later (see Endocrine System). A pineal body is absent in xenarthrans (anteaters, sloths, and armadillos) and, perhaps, dugongs (Wurtman et al., 1968).

The cerebrum consists of a pair of greatly expanded cerebral hemispheres that have grown upward, over, and backward so that the diencephalon and mesencephalon are hidden from dorsal view (Fig. 9.52). In many mammals, the surface of the cerebrum (cerebral cortex) is folded into a series of **gyri** ridges and **sulci** grooves. The **longitudinal fissure** separates the cerebrum into right and left halves, or **hemispheres.** The two cerebral hemispheres are connected internally and are coordinated by a broad transverse sheet of nerve fibers, the **corpus callosum.** The cerebrum contains centers for interpreting sensory impulses arriving from various sense organs as well as centers for initiating voluntary muscular movements; it stores past experiences as memory; it controls certain emotions; and it is a center where data may be correlated, analyzed, and employed in making choices. It is the "thinking" part of the brain. **Olfactory lobes,** responsible for the sense of smell, are located on the anteroventral surface of the cerebrum.

Muscular coordination is regulated by the cerebellum, the second largest region of the brain, which covers the mesencephalon and much of the medulla oblongata.

BIO-NOTE 9.15

Cat Brains, from Past to Present

Dr. Robert Williams of the University of Tennessee Medical School compared the brains of the modern house cat with those of a Spanish wildcat that appears to be almost identical to the species that was domesticated by the ancient Egyptians more than 3,500 years ago. The modern house cat was found to have significantly fewer (50% fewer) brain cells than the wild creature from which it evolved.

Williams et al., 1993

The roof of the telencephalon, known as the **pallium (hippocampus),** is an area critical to processing spatial information and essential in converting short-term memory to long-term memory. Damage to the hippocampus does not destroy memory prior to the injury, but causes loss of recent memory by making it impossible to form long-term memory traces. Jacobs et al. (1990) found that the hippocampus takes up a significantly greater portion of the total brain in the polygamous male meadow vole (*Microtus pennsylvanicus*) than in the monogamous male woodland vole (*Microtus pinetorum*). Females of both species have a hippocampal size closely matching that of the male woodland vole. Breeding male meadow voles range over large areas in search of sexually receptive mates, whereas male woodland voles and females of both species stick close to home. Polygamous males also performed better in laboratory mazes that tested their abilities to discern information about space and their orientation. These voles apparently evolved superior spatial skills—and larger hippocampi to regulate those skills and to remember what they discovered—in order to navigate efficiently throughout their surroundings during the breeding season.

The remainder of the central nervous system, the spinal cord, is protected by vertebrae, fat, and the three meninges. The length of the spinal cord varies among mammals. It may extend as a single structure nearly to the end of the vertebral column, or it may be shorter, as in humans where it extends to about the third lumbar vertebra as a single structure before splaying out and forming the **cauda equina** (horse's tail) portion. Cervical and lumbar enlargements of the spinal cord are present, with bats having an unusually large cervical enlargement associated with their highly developed flight muscles.

Sense Organs

Ear

The membranous labyrinth and the sense of hearing are highly developed in mammals. The human ear, for example, is sensitive to regular compression waves in air in a range of approximately 16 to 20,000 Hz (Kare and Rogers, 1976). A dog, however, can perceive higher frequencies that are completely inaudible to humans, up to 30,000 or 40,000 Hz, and bats can perceive frequencies as high as 100,000 Hz (Schmidt-Nielsen, 1990).

The **pinna** (part of the outer ear) serves to funnel sound waves into the **external auditory meatus** (ear canal), where they are conveyed to the eardrum (tympanum) (Fig. 9.53). Some species, such as galagos and deer, have pinnae that can be directed independently toward the source of the sound. All mammals have a middle ear that contains three bones: the malleus (hammer), **incus** (anvil), and **stapes** (stirrup) (Fig. 9.53). These three auditory ossicles transmit vibrations from the eardrum to the oval window in the cochlea of the inner ear. A **Eustachian** (auditory) **tube** connects the middle ear cavity with the nasopharynx and serves as a means to equalize pressure in the middle ear. The middle ear cavity is enclosed by a bony **auditory bulla.** The **cochlea,** containing the **organ of Corti** (spiral organ), is the portion of the inner ear concerned with hearing. Hair cells in the organ of Corti convert mechanical energy into nerve impulses, which in turn are transmitted to the brain by the auditory nerve (VIII).

The remainder of the inner ear is devoted to balance and equilibrium and consists of the **vestibule** and three **semicircular canals** (Fig. 9.53). Within the vestibule are two regions—the **utricle** and the **saccule**—that contain patches of sensory epithelia known, respectively, as **maculae** and **cristae.** They function in determining static equilibrium (position of the head in relation to the ground surface or gravity). The semicircular canals each lie in different planes and terminate in enlarged **ampullae.** The fluid contained within the semicircular canals (as well as within the vestibule and cochlea) is known as endolymph. Movements of fluid within the semicircular canals is responsible for dynamic equilibrium (movement of the body through space).

Some mammals obtain information from the faint reflections, or echoes, of sound they themselves produce, thus enabling them to avoid obstacles and to locate food. This **echolocation,** or animal sonar, is particularly well developed in bats (Chiroptera) (Fig. 9.54), and exists also in shrews (Insectivora), whales (Cetacea), seals (Pinnipedia), and manatees (Sirenia). The potential to echolocate also may be present in many other kinds of mammals. High-frequency sound (short wavelengths) is particularly useful for mammals that are active in the dark and in deep or murky waters. In those mammals that use echolocation, the auditory bullae have no connection with other bones of the skull; rather, they "float." These loosely connected

FIGURE 9.53

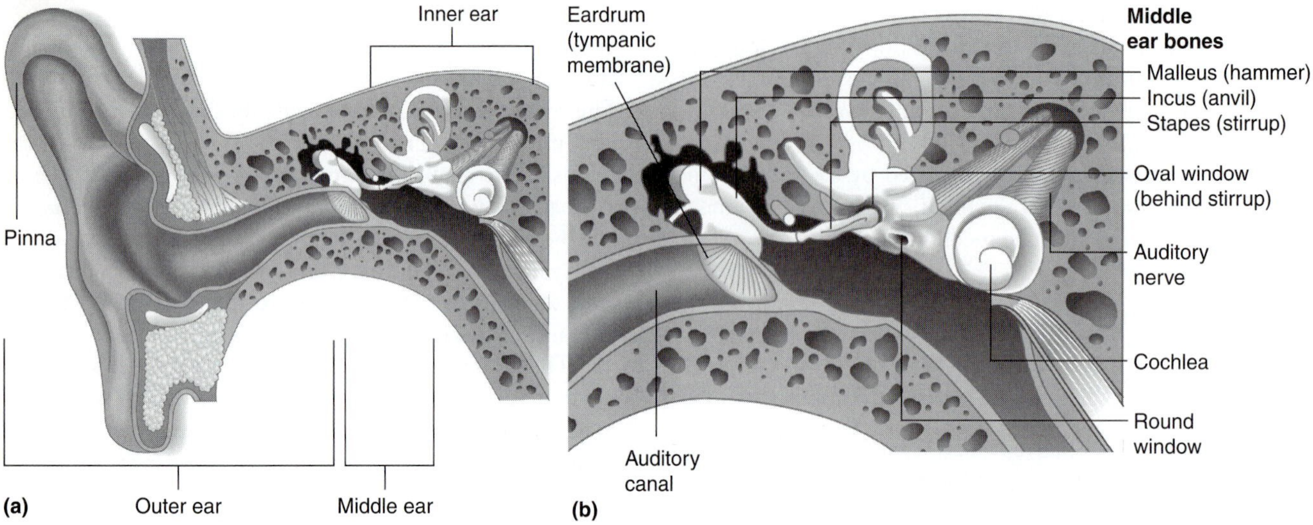

Hearing in humans. (*a, b*) Sound waves are funneled into the external auditory canal by the pinna. Pressure waves strike the eardrum (tympanic membrane), causing it to vibrate at the same frequency as the waves. This, in turn, causes the ear ossicles (malleus, incus, and stapes) to vibrate in the middle ear. The stapes articulates with the oval window, which bows in and out causing fluid pressure waves in the perilymph of the scala vestibuli and scala tympani. These waves cause the round window to bulge under pressure. The sensory receptors for hearing are located in the coiled cochlear duct portion of the inner ear. The organ of Corti, which contains sensory hair cells, sits atop the basilar membrane within the cochlea. Vibrations of the basilar membrane move patches of hair cells against an overhanging flap (the tectorial membrane). Signals from these hair cells initiate action potentials that are carried to the brain by the auditory nerve. The inner ear consists of an outer area known as the bony labyrinth and an inner area, the membranous labyrinth. The semicircular canals are responsible for dynamic equilibrium (movement of the body through space).

FIGURE 9.54

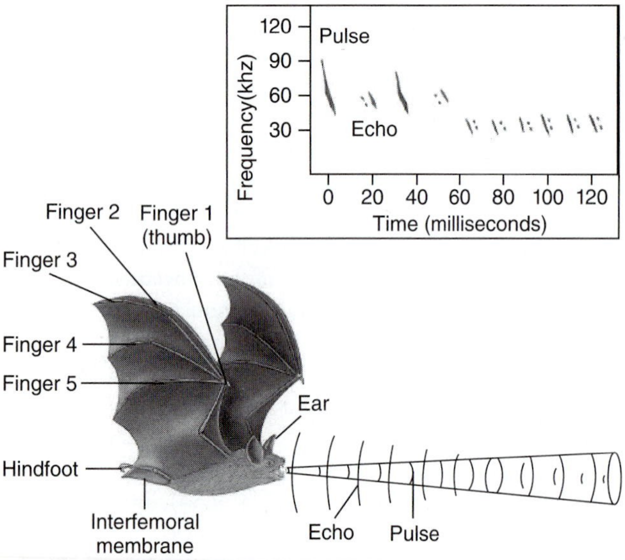

Echolocation by the little brown bat (*Myotis lucifugus*). Frequency-modulated pulses are directed in a narrow beam from the bat's mouth. When the pulses meet an object, their echoes are picked up by the bat's sensitive hearing apparatus. As the bat nears its prey, it emits shorter, lower frequency signals at a faster rate.

auditory bullae allow the ear to act somewhat independently of skull vibrations. The pulses used by most bats during echolocation are produced by the larynx and range in frequency from 30 to 120 kHz, which is above the range of human hearing. When resting, a bat may emit 5 to 10 ultrasonic "cries" per second. When flying and approaching an object, these may increase to more than 100 per second. A discussion of echolocation is presented by Marler and Hamilton (1966).

In bats that use echolocation, the cerebellum is large, and the ears are highly specialized with large pinnae and large cochlea. Within each pinna is a **tragus,** a fleshy projection rising from the inner base of the ear. It has not been shown to function in hearing, and despite much speculation and research, its significance remains unknown.

Toothed whales and dolphins use both audible and ultrasonic sounds in order to communicate, avoid obstacles, and locate food. The larynx of toothed whales (suborder Odontoceti) lacks vocal cords. Most researchers feel that sounds produced by toothed whales originate in the nasal region just below the blowhole in the area of the nasal plugs (Fig. 9.55). The sound then is reflected from the bony surface of the skull, which acts as a parabolic reflector, producing a focused beam. The forehead (melon) probably acts as an acoustic lens, concentrating the sound energy in a narrow beam.

BIO-NOTE 9.16

Sensing Barometric Pressure

Some bats, including eastern pipistrelles (*Pipistrellus subflavus*), apparently can sense changes in barometric (air) pressure and use these changes to govern their foraging strategies during the spring and fall when they are roosting deep inside caves. Low air pressure means light, rising air and warm, cloudy weather—conditions suitable for insect activity. High pressure during spring and fall indicates a cold, clear night and few insects.

Most pipistrelles leave their roosts on hunting forays only when barometric pressure falls below 73.69 cm (29.1 in.) of mercury. They can detect pressure changes smaller than one-tenth of an inch by using a middle ear receptor known as the Vitalli organ. This is the same organ that migrating birds use to detect air pressure changes to avoid bad weather. The Vitalli organ connects to the hypothalamus of the brain, which controls metabolism. Air pressure changes also may be used to control the bat's metabolic activities, which are known to slow down greatly during roosting, with some bats even entering into a hibernation-like torpor on high-pressure days. This ability to maintain a low metabolism while roosting serves as an adaptation for conserving vital energy.

Paige, 1996

Sound reception in delphinids has been reviewed by Mead (1972), Evans and Maderson (1973), and Popper (1980). Some, such as bottlenosed dolphins (*Tursiops truncatus*), produce a narrowband pure-tone "whistle," whereas other species use pulsed sounds. Those species using whistles are typically highly social and often form large herds, whereas those without tend to be solitary or are found most commonly in small groups (Herman and Tavolga, 1980). It is theorized that sound enters the fat-filled lower jaw and is conducted to the auditory bullae (ear bones), apparently bypassing the tiny external auditory meatus (ear canal) (Reynolds and Odell, 1991).

Vocalizations of mysticete whales are very different from those of odontocetes and are more difficult to classify. Generally, they are at lower frequency, of narrower bandwidth, and of longer duration than odontocete sounds (Norris, 1969). Social functions proposed for the sounds of mysticete whales include long-range contact, assembly calls, sexual advertisement, greeting, spacing, threat, and individual identification (Herman and Tavolga, 1980).

BIO-NOTE 9.17

Protecting Porpoises from Fishing Nets

One of the most serious dangers to dolphins and porpoises around the world is the threat from various forms of gill-net fishing (see Fig. 15.16). Acoustic alarms have proven effective in reducing the number of deaths of harbor porpoises (*Phocoena phocoena*) in sink gill-nets. Active alarms were placed 92 m apart and emitted a broadband signal with a frequency of 10 kHz and a loudness level of 132 dB. The signal lasted for approximately 300 milliseconds and was repeated every 4 seconds.

Kraus et al., 1997

FIGURE 9.55

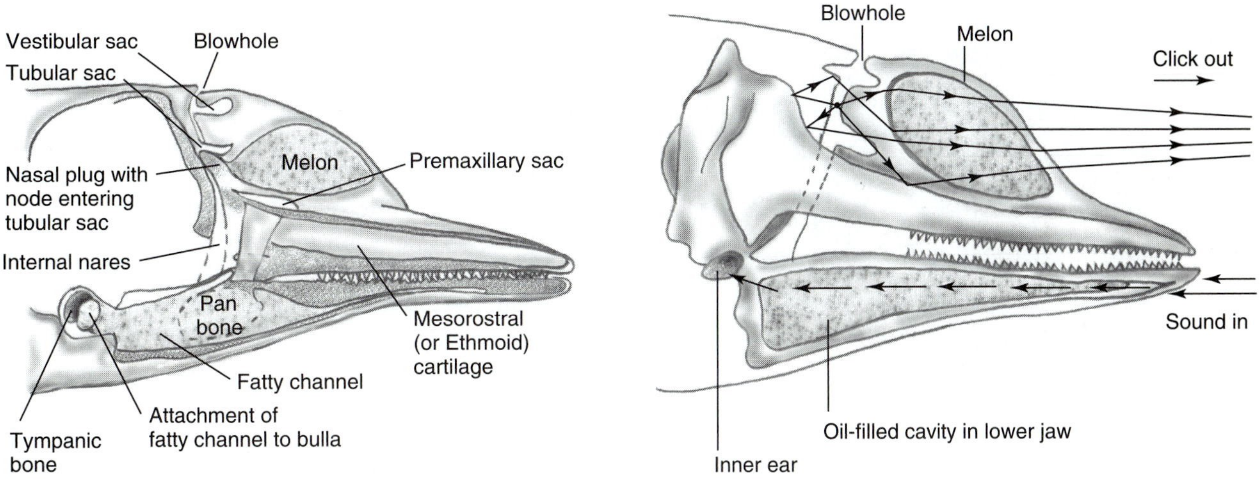

Sound propagation and reception in a dolphin (family Delphinidae). The melon acts as an acoustic lens, focusing the clicks, while the oil-filled cavity of the lower jaw acts as a wave guide to conduct the sound received to the ear. The oil-filled channel passes posteriorly to contact the tympanic bone, which is connected to the malleus.

Manatee sounds described as chirps, whistles, or squeaks have peak energles in the 3 to 5 kHz range and are probably produced in the larynx. It has been suggested that the most sensitive location on the manatee's head for sound reception is an area near the cheek bones, which are large and in direct contact with the ear bones.

Infrasound is utilized by some mammals. Elephants, for example, make calls at frequencies of 14 to 35 Hz that can be heard by other elephants at distances up to 6 km (Poole et al., 1988). These low-frequency calls probably are used to communicate between individuals in a herd of these highly social animals. Anatomical studies suggest that manatees also are capable of hearing infrasound. Infrasound by estrous females may attract males.

Eye

The eyes of mammals are similar to those of most amniotes (Fig. 9.56). Rods (black-and-white vision) are sensitive to low levels of illumination, whereas cones are sensitive to colors in bright light. As in birds, the lens is elastic, and its shape is altered for near and far viewing by action of the ciliary muscle on the suspensory ligament. The eyes of a cat and many other mammals have about 100 million receptor cells in the retina of each eye. Oil droplets, either colorless or colored, are found in the cone photoreceptors of some mammals (Robinson, 1994). Colored droplets absorb particular wavelengths of light. Colorless droplets are found in marsupials and monotremes (Bowmaker, 1986). Many nocturnal species, such as lemurs, galagos, cats, and ungulates, possess a **tapetum lucidum,** a reflective structure in the choroid layer that improves night vision by reflecting light back onto the retina.

FIGURE 9.56

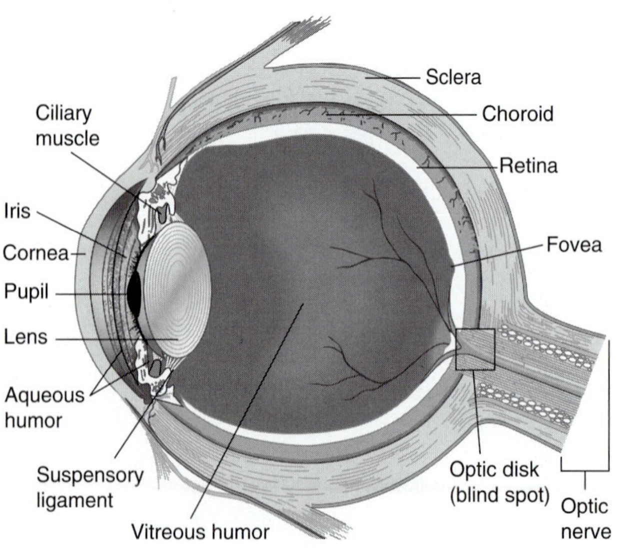

The mammalian eye. The retina contains the photosensitive rods and cones. Rods permit vision under dimly lit situations, whereas cones provide sharpness of vision under brightly lit situations while also providing color vision in some mammals.

Many mammals are color-blind. As they evolved, they seem to have lost the ability to see colors (Stewart, 1995). Because early mammals were active at night, they apparently came to rely more on scent and hearing than vision. **Dichromatic** (two-color) vision is present in dogs (*Canis familiaris*) and probably a number of other mammals (Neitz et al., 1989). Humans and many other primates are diurnal and have three types of cone cells (S, M, and L cones), each containing a different photopigment. Thus, primates have trichromatic vision and can detect blue, green, and red colors. Roorda and Williams (1999) recorded the first images of the arrangement of S, M, and L cones in the living human eye.

Some mammals, such as moles and mole rats, have vestigial eyes. Mole rats (*Spalax*) carry eye reduction to its extreme among mammals. Spending most of their lives in total darkness in underground runways, mole rats have eyes that are reduced to less than 1 mm in diameter, 100 times as small as above-ground rodents of the same body size (Diamond, 1996). Each eye is covered permanently with thick skin and fur, lies embedded in a large gland, and lacks a pupil, functional lens, and eyeball muscles.

Tosini and Menaker (1996) have shown that the retinas of hamsters produce the hormone melatonin in quantities that ebb and flow daily. This retinal biological clock appears to control such daily cycles as the regeneration of light receptors and increased sensitivity to light, rhythms that are maintained on a roughly 24-hour cycle. Williams (cited in Holden, 1996a) has found the first clues to what may be a second circadian clock in the mammalian eye—not in the retina, but in the iris of a rat. How these clocks are coordinated with the biological clock in the brain (pineal gland) and whether such clocks exist in other mammals are still undetermined.

Nose

Olfactory mechanisms are highly developed in most mammals. Nasal passages are large and are elongated by the presence of both a hard palate and a soft palate, so that the internal nares open into the nasopharynx and are lined with ciliated mucous epithelium. The olfactory epithelium occupies the dorsal wall of the nasal cavity. Olfactory cells serve as chemoreceptors and respond to chemicals dissolved in the film of mucus covering the olfactory epithelium. When stimulated, chemoreceptors cause nerve impulses to be transmitted along olfactory nerves and through the cribriform plate to the olfactory bulb near the anterior end of the brain. The impulses then travel along olfactory tracts to portions of the cerebrum where specific odors are interpreted.

Vomeronasal (Jacobson's) organs are well developed in monotremes, marsupials, and in some insectivores. Cats, some rodents, and elephants (Shoshani, 1997) also retain these structures, but in most mammals they are vestigial. Compared with many other mammals, humans have a poor sense of smell. Even so, the human nose can distinguish up to 10,000 different odors (Dajer, 1992).

The trunk of an elephant is a conjoined upper lip and nose, with the nostrils running like two parallel hoses from the trunk's base down to its tip. Most breathing is done through the nostrils. Elephants, which have a keen sense of smell, possess seven turbinal bones within their nasal cavity (Shoshani, 1997).

BIO-NOTE 9.18

The Nose of the Star-Nosed Mole

The nose of the star-nosed mole (*Condylura cristata*) consists of 22 fleshy processes that surround the nostrils (Fig. 9.57). Besides being used for the sense of smell, the mole's unique nose, which is covered with tens of thousands of complex touch sensors known as Eimer's organs, is used to explore the environment by touch. It contains more than five times as many nerve fibers as innervate the mechanoreceptors in the entire human hand and may be one of the most sensitive and highly developed touch organs among mammals. The nose appendages differ in size and shape and also seem to be used differently when food is encountered. The longest appendages usually make the initial contact with the prey, whereas others are used for detailed tactile exploration once the prey has been located.

Catania and Kaas, 1996
Gould et al., 1993

Scent and sound are particularly important to nocturnal animals. Most primitive primates (galagos, lorises, lemurs, and tarsiers) feed and mate in the dark and rely extensively on their senses of smell and hearing (Finkbeiner, 1993). One such primate, the aye-aye, is found only on the island of Madagascar and is considered the world's most endangered primate because it is the only extant member of its family. Adults spend more than 80 percent of their time alone, but recent studies have shown that females have definite home ranges and never stray into another female's range. The invisible boundaries are formed by scent marks left by streaking urine along branches. Chemicals that elicit a response from other members of the same species are known as pheromones and are discussed in greater detail in Chapter 16.

FIGURE 9.57

The nose of the star-nosed mole (*Condylura cristata*). The 22 fleshy processes that surround the nostrils are covered with tens of thousands of touch receptors (Eimer's organs), which are used to explore the environment by touch.

Taste

Taste buds are abundant in mammals (see Table 8.5) and are found mainly on the tongue, posterior palate, and pharynx. Those on the tongue are located on the sides of papillae. Each taste bud consists of a cluster of gustatory cells that are the primary chemoreceptive cells. Each gustatory cell possesses a sensory hairlike process that projects through the epithelium via a taste pore. These chemoreceptors are stimulated by substances dissolved in the saliva. The sense of taste recognizes only four basic qualities: salty, sweet, sour, and bitter.

Electroreceptors

The duck-billed platypus seeks food in muddy streams. When it dives, it usually has its eyes, ears, and nose shut. It long has been known that the bill contains densely packed receptor organs, and it has been assumed that these are extremely sensitive mechanoreceptors. This is undoubtedly true, but recently the presence of electrosensitivity has been demonstrated in the platypus. It can home in on weak electrical fields, whether emitted by prey animals or by an artificial source (Scheich et al., 1986).

Endocrine System

In most cases, the endocrine organs of mammals are homologous to those of the other vertebrates. Some, however, have unique functions in mammals. For example, the hormone prolactin, which is secreted by the anterior lobe of the pituitary (adenohypophysis), stimulates milk production by mammary glands. Oxytocin, produced by the hypothalamus and secreted by the posterior lobe of the pituitary (neurohypophysis), causes uterine contractions and also contractions of muscles in the mammary glands. In some species, oxytocin and antidiuretic hormone play a role in pair bonding and stimulation of a mother's interest in her newborn (Insel and Carter, 1995).

The thyroid gland secretes thyroxin, triiodothyronine, and calcitonin. Thyroxin and triiodothyronine regulate overall metabolism, whereas calcitonin reduces calcium levels in the blood and prevents bone resorption. Parathyroid hormone increases the levels of calcium and phosphate in the blood by increasing the rate of calcium and phosphate absorption from the gastrointestinal tract. It also stimulates the breakdown of bone tissue in order to release additional calcium and phosphate into the blood.

The adrenal medulla produces epinephrine and norepinephrine, vasodilators of the circulatory system and stimulators of increased production of adrenocortical hormones in times of extreme and/or prolonged stress. Epinephrine and norepinephrine also accelerate the heart rate, increase blood pressure, and increase the amount of blood sugar in times of sudden metabolic need. Additionally, epinephrine increases blood flow in the heart muscle, skeletal muscle, and lungs, and it decreases blood flow in the smooth muscle of the digestive tract and skin. Epinephrine is the primary hormone involved in the "fight-or-flight" response to fear, pain, and aggression.

Studies of the "stress hypothesis" (having higher levels of adrenocortical hormones when under stress) have been done on many species with varying conclusions. Healthy free-ranging Chinese water deer (*Hydropotes inermis*), shot as part of a management program, had lower cortisol levels than either diseased deer or free-ranging deer that were either netted and manually restrained or anesthetized by dart (Hastings et al., 1992). Adrenal glands of deer with various, mostly chronic, disease conditions weighed significantly more than those of healthy deer. Koala (*Phascolarctos cinereus*) adrenal glands increased in size in response to the stress of disease (Booth et al., 1991). The increase in size associated with disease varied with the type of disease.

The little brown bat (*Myotis lucifugus*), which has exceptionally high levels of the glucocorticoids cortisol and corticosterone (Gustafson and Belt, 1981; Widmaier and Kunz, 1993), has a pronounced diurnal rhythm in glucocorticoid levels with peak levels just prior to the onset of the active (feeding) phase of the animal's diurnal cycle, a pattern observed in most mammals (Gibson and Krieger, 1981; Widmaier et al., 1994). In addition, *M. lucifugus* has seasonal cycles of steroid hormone levels that are correlated with hibernation periods (Gustafson and Belt, 1981). Certain megachiropteran bats (*Pteropus pumilis* and *P. hypomelanus*) have been shown to respond to stress with increases in glucocorticoids and glucose (Widmaier et al., 1994). Stress probably activates the sympathetic branch of the autonomic nervous system in bats, as it does in other mammals, and is a significant contributing factor in the development of hyperglycemia.

The effects of stress on the adrenal gland and on population size were dramatically illustrated in the 1950s and early 1960s in Maryland. In 1916, four or five Sika deer (*Cervus nippon*) were released on the 115-hectare (ha) James Island near Cambridge, Maryland. By 1955, the population had increased to 300 healthy animals. In 1958, about half of them died, although the food supply was adequate. The population continued to decrease to 80 animals during succeeding years. Animals examined during the years of the decline showed medullary enlargement and histological changes in the adrenal glands that indicated that the stress caused by the overpopulation may have contributed to the decline (Christian 1959, 1963). Other factors and mechanisms undoubtedly also were involved in this complex process.

Density-dependent stress has been studied in tree shrews (*Tupaia glis* and *T. belangeri*) (Autrum and Holst, 1968; Holst, 1969). Stress causes a delay in the development of the young and changes in the behavior and physiology of the adult. Females under stress produce less milk or none at all. The sternal gland, located on the chest, ceases secretion, and females thus cannot mark their young with its secretion. Without this protection, the young get eaten by their cage mates or even by their mother. Under strong stress, females do not reproduce, and the testes of males recede into the body cavity.

Marked adrenocortical and adrenomedullary response to the stress of capture have been reported in the duck-billed platypus (*Ornithorhynchus anatinus*) (McDonald et al., 1992). Seasonal data revealed a distinct influence of time of year on the adrenocortical response to capture. That of males was greatest just prior to mating, whereas that of females was greatest during pregnancy and early lactation.

The pineal gland is a dorsal evagination of the midbrain. Due to its 24-hour cycle, the pineal gland and the hormone melatonin may play a role in the synchronization of other 24-hour circadian cycles such as sleeping, eating, and adrenocortical function. Diurnal rhythms of the melatonin level in the retina have been reported (reviewed by Zachmann et al., 1992). Melatonin may also play a role in regulating the estrous cycles in mammals. Many species inhabiting Arctic or Antarctic regions have been reported to have extremely large pineal glands (reviewed by Miche et al., 1991). Reproductive synchronization is essential in these areas where annual variations in daylength and climatic conditions are drastic. Melatonin also may play an important role in temperature regulation of hibernating mammals and has been shown to be involved in the reproductive process (reviewed by Rismiller and Heldmaier, 1987).

The pineal gland also is known to inhibit gonad function and development in rats, humans, and some other mammals. Sperm production is suppressed significantly by melatonin in adult male deer mice (*Peromyscus maniculatus*) housed under long photoperiod but had no additional suppressive effects on mice housed under a short-day regimen (Blank and Freeman, 1991). Berria et al. (1990) reported that melatonin plays a role in regulating the seasonal testicular cycle of the spotted skunk (*Spilogale putorius*). Melatonin induced fall molt, the growth of white pelage, and testicular regression in male short-tailed weasels (*Mustela erminea*) (Rust and Meyer, 1969). The endogenous circannual reproductive seasonality rhythm in male red deer (*Cervus elaphus*) was affected significantly by melatonin implants; treatment with melatonin implants in November or December advanced reproductive development, whereas treatment from June to August delayed development (Webster et al., 1991).

The cane mouse (*Zygodontomys brevicauda*) is a year-round breeder in Venezuela and is not reproductively responsive to either variation in photoperiod or to continuous exposure to melatonin (Bronson and Heideman, 1992). Thus far, the cane mouse is the only mammal known in which the reproductive system shows no photoresponsiveness. Because Venezuela receives nearly 12 hours of daylight year-round, photoperiod cannot serve as an environmental cue.

Urogenital System

Mammalian embryos pass through both the pronephric and mesonephric stages of kidney development. The mesonephros serves as the functional embryonic kidney; in mammals, it reaches its peak of development earlier than in birds—approximately one-fourth of the way through the gestation period. The number of functional tubules varies among species. Before the last mesonephric tubules have formed near the caudal end of the nephrogenic mesoderm, the earliest ones formed at the anterior end have already been resorbed.

Although basically an embryonic kidney, the mesonephros functions for a short time following birth in monotremes and marsupials. When the metanephros begins to function, the mesonephros degenerates. Mesonephric ducts remain as sperm ducts in male mammals, but they become nonfunctional in females and remain only as short, blind Gartner's ducts embedded in the mesentery of the oviducts.

The metanephros continues to develop from the caudal end of the nephrogenic mesoderm but is displaced anteriorly and laterally. Although some embryonic kidneys are lobulated, most adult kidneys are smooth and more or less bean-shaped. The renal artery, renal vein, nerves, and ureter enter and/or leave at the median notch, or **hilum.**

Nephrons are the functional units of the kidney (see Fig. 1.19). Blood enters each kidney via renal arteries, whose branches (arterioles) each end in a network of specialized capillaries known as a **glomerulus.** Each glomerulus is partially enclosed by one end of a nephron, known as the **renal (Bowman's) capsule.** The functional unit formed by a glomerulus and renal capsule is a **renal corpuscle.**

The process of **filtration** occurs as blood passes through the glomerulus. Waste products such as urea, uric acid, and creatinine, as well as water, glucose, and many other substances, are filtered from the blood and enter the **proximal convoluted tubule.** After passing through this structure, the filtrate flows through the **loop of Henle** and **distal convoluted tubule** before entering a **collecting tubule,** which drains into the **renal pelvis** of the kidney. The renal pelvis is the expanded opening into the ureter. Each kidney is drained by a ureter. During its passage through the nephron, many substances such as water and glucose are reabsorbed by a process known as **tubular reabsorption,** while in some cases, substances that were not filtered out in the glomerulus, such as ammonia or hydrogen ions, are moved from capillaries into the filtrate by a process known as **tubular secretion.**

Ureters empty into the cloaca in monotremes (Fig. 9.58). In placental mammals, however, ureters empty into the urinary bladder, which is an evagination of the ventral wall of the cloaca. The bladder is drained by the urethra. Urine excreted by mammals contains mostly metabolic byproducts (nitrogenous wastes) that collect within the body and must be voided in order to avoid possible toxic effects. Urea is the primary waste product and is excreted in a relatively concentrated urine in order to conserve water. Urine excreted by desert mammals is very concentrated.

Reproduction in most mammals involves numerous adaptations for internal fertilization and viviparity. Unlike birds, both ovaries in most female mammals are functional in the production of ova, or eggs, although only one may function during each breeding cycle. Only the left ovary is functional in the duck-billed platypus, and eggs develop only in the left uterus (Anderson, 1967). Monotremes possess an ovary with fluid-filled cavities (lacunate), whereas other mammals possess a compact ovary with no chambers or lacunae. Growth of the ovary in mammals is controlled by hormones from the pituitary and pineal glands, and the ovary itself produces estrogen and progesterone. The size and shape of the ovary varies with age, reproductive stage, and species.

In female mammals, a Müllerian duct arises parallel and next to the embryonic mesonephric duct before the mesonephric duct regresses. The Müllerian duct, rather than the Wolffian duct, forms the oviducts, uterus, and vagina.

Female mammals possess **oviducts** (also known as uterine tubes or fallopian tubes), a **uterus,** and a **vagina** (Fig. 9.59). The convoluted and ciliated oviducts are relatively short and small

FIGURE 9.58

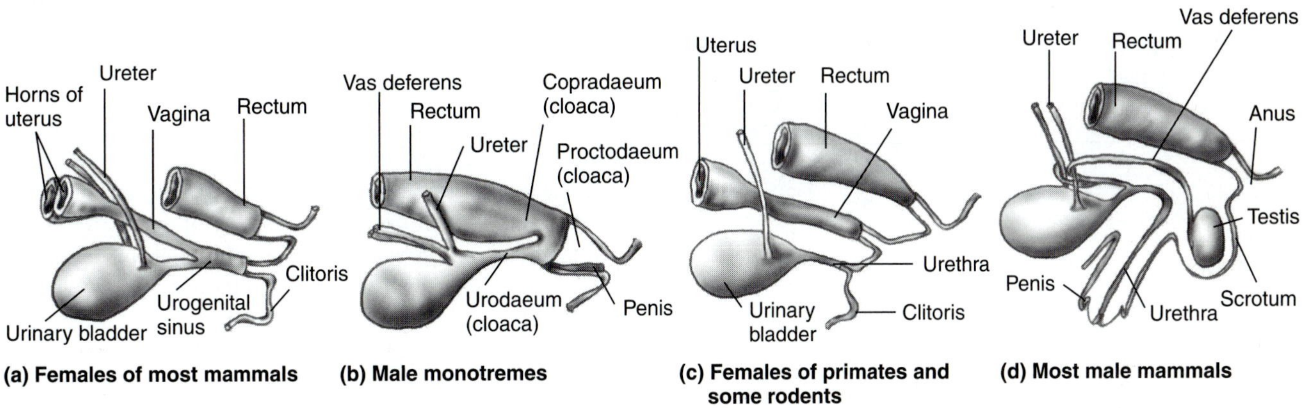

Representative divisions of the cloaca and their relations with urogenital ducts and the urinary bladder in male and female mammals: (*a*) females of most mammals; (*b*) male monotremes; (*c*) females of primates and some rodents; (*d*) most male mammals.
From Hildebrand, Analysis of Vertebrate Structure, *4th edition. Copyright ©1995 John Wiley & Sons, Inc. Reprinted by permission of John Wiley & Sons, Inc.*

FIGURE 9.59

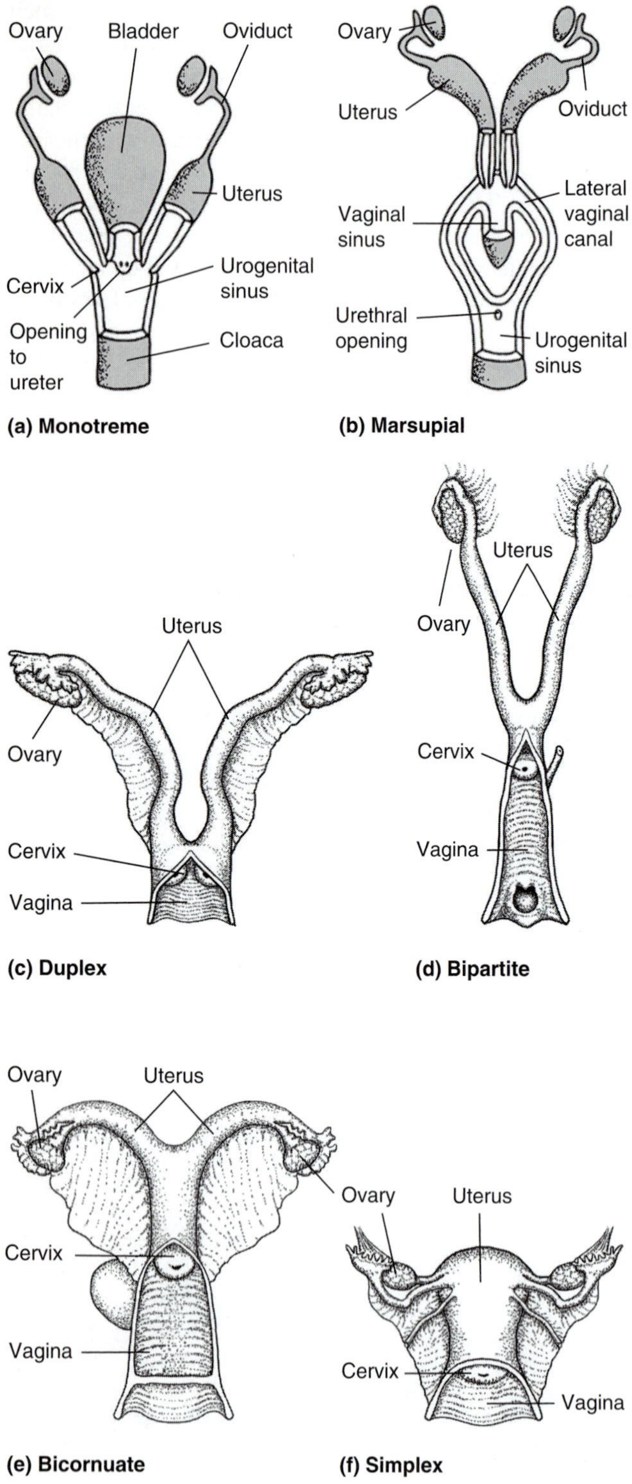

Types of female reproductive tracts: (*a*) monotremes; (*b*) marsupials. (c–f) Placental mammals: (*c*) duplex; (*d*) bipartite; (*e*) bicornuate; (*f*) simplex.

in diameter. In monotremes, there is no fusion of the Müllerian ducts, and the genital tract is double all the way to the urogenital sinus. The posterior ends of the tracts are modified to form shell glands in these oviparous mammals (Fig. 9.59a).

In marsupials, the two Müllerian ducts meet to form a median vagina just beyond the paired uteri (Fig. 9.59b). The median vagina may or may not be paired internally. Beyond the median vagina, the two ducts continue as paired (lateral) vaginae. The median vagina is separated from the urogenital sinus by a septum. At birth, the fetus usually is forced through the septum directly into the urogenital sinus, although in some marsupials both the median and lateral routes may be used as birth canals. The new passageway may remain open throughout life and is known as a pseudovagina. It closes in opossums.

In placental mammals, there are varying degrees of fusion that result in two uterine horns, a single uterine body, and a single vagina. If the uteri are completely separate, each with a separate **cervix** opening from the vagina, but are joined to a single vagina, the uterus is said to be **duplex** (Fig. 9.59c). This arrangement occurs in rodents and rabbits. In **a bipartite** uterus, there is a single cervix and the lower portions of the two uteri are fused; but in the major part of their length, the uteri are still separate (Fig. 9.59d). Implantation of embryos usually occurs in the unfused portions of the uteri. This type of uterus occurs in most carnivores and in some ruminants. If the lower two-thirds of the two uteri are fused into a single structure but the upper third of each uterus remains separate, it is a **bicornuate** uterus (Fig. 9.59e). Implantation usually occurs in the lower portion. This condition exists in many ungulates. In a **simplex** uterus, the uteri are fused into a single structure, there are no uterine horns, and the oviducts empty directly into the body of the uterus (Fig. 9.59f). A simplex uterus is found in some bats, armadillos, and primates.

The inner layer of the uterus, or endometrium, becomes highly vascularized prior to implantation of the blastocyst. It contributes to the formation of the placenta and serves as a favorable environment for the developing young. The thickest layer of the wall (myometrium) of the body of the uterus consists of smooth muscle, which at the end of the gestation period contracts and assists in expelling the fetus.

In all mammals except monotremes, the embryonic cloaca becomes subdivided into a dorsal rectum and a ventral urogenital sinus. The rectum opens to the exterior via the **anus.** In the developing males, the urogenital sinus extends into the penis as a continuation of the urethra running from the bladder. In the developing female of some mammals (some rodents, most primates), the urogenital sinus expands to form a shallow **vestibule** into which the reproductive system (vagina) and excretory system (urethra) open separately.

In mammals, the body of the uterus narrows to form a cervix, which projects into the vagina. Formed by the fusion of the terminal portion of the Müllerian ducts, the vagina usually opens into the urogenital sinus. In many rodents and primates, however, the vagina opens directly to the exterior of the body.

A **clitoris,** composed of columns of spongy erectile tissue, is present in females. Upon sexual stimulation, a series of reflex actions causes the spongy tissue to fill with blood and the clitoris to become erect. In otters, several rodents, rabbits, and a number of other mammals, a bone, the **os clitoris,** develops in the clitoris. An os clitoris is homologous to the baculum and develops in females of those species in which males have a baculum.

In males, the caudal pole of each embryonic testis is connected by a ligament, the **gubernaculum,** to the labioscrotal pouch. As development proceeds, these pouches increase in size and become the **scrotal sacs** (Fig. 9.60). Partly as a result of shortening of the gubernaculum and partly because growth of the gubernaculum does not keep pace with elongation of the trunk, the testes of many species are displaced caudad toward the scrotal sacs.

The testes remain retroperitoneal (behind the peritoneal membrane lining the peritoneal or abdominal cavity) and descend permanently into the scrotal sacs in most mammals. In some, such as bats, rabbits, and rodents, the testes move back and forth seasonally between the abdominal cavity and the scrotal sacs. The passage between the two chambers is the **inguinal canal.** In species that retract their testes, the canal remains broadly open. In others, as in humans, the canal is only wide enough to accommodate the spermatic cord consisting of the vas deferens, arteries, veins, lymphatics, and nerves. These structures descend into the scrotal sacs along with the testes.

Depending on the species, the temperature in scrotal sacs is 1 to 6°C cooler than in the abdominal cavity, thus allowing spermatogenesis to take place and viable sperm to be produced. Each testis consists of interstitial cells and tightly packed seminiferous tubules. Interstitial cells secrete the male sex hormone testosterone, which causes undifferentiated cells in the seminiferous tubules to undergo cell division (meiosis) to form spermatozoa (spermatogenesis).

In some mammals, the testes remain permanently in the abdomen, and scrotal sacs do not develop. Such is the case with monotremes, some insectivores, the xenarthrans (edentates), sirenia, cetaceans, elephants, hyraxes, rhinoceroses and most seals. In others, such as chipmunks, squirrels, many mice, some bats, and some primates, the testes descend temporarily into the scrotum during the breeding season. Why some mammals have evolved a scrotum and others have not is still not understood.

FIGURE 9.60

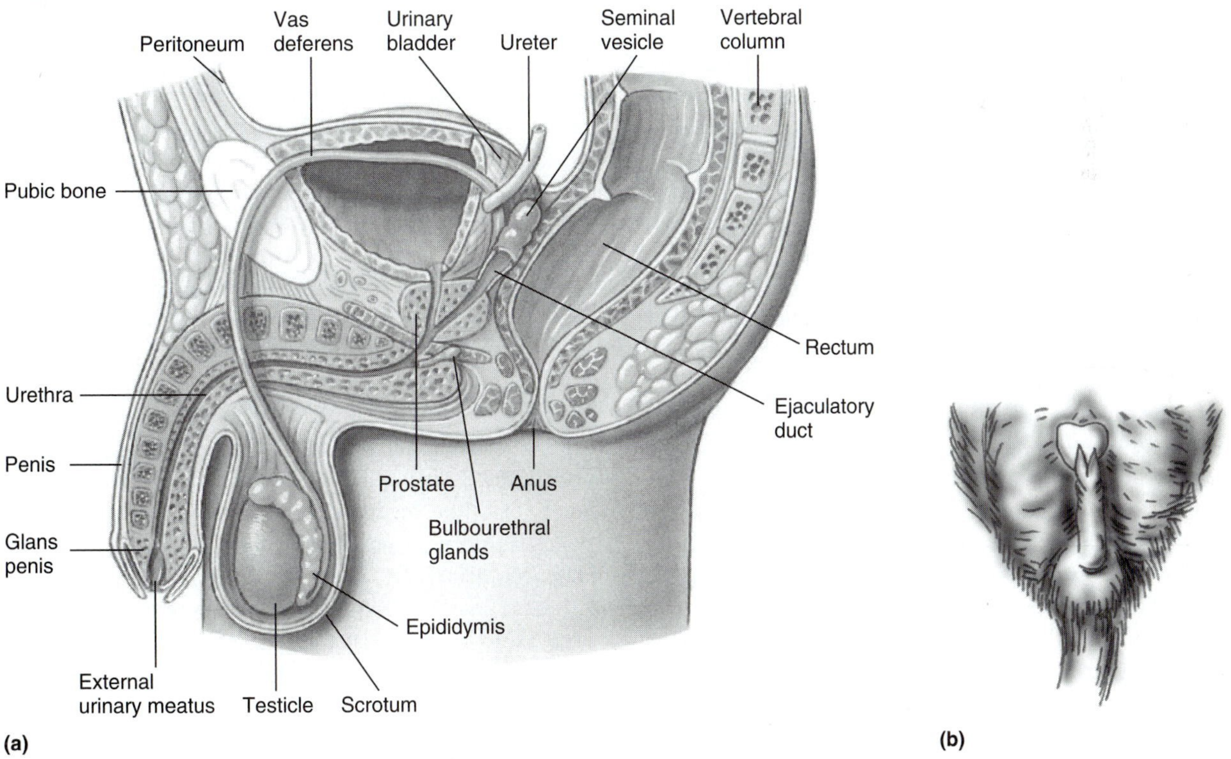

(*a*) Human male reproductive system. (*b*) Ventral view of the male external genitalia of the opossum (*Didelphis virginiana*). The penis is caudal to the scrotum, and its gland is bifurcate. The ureter opens between, not at the tips, of the bifid prong. A groove to direct sperm into each vagina extends along the medial side of each part of the glans penis for more than half its length.

BIO-NOTE 9.19

Spermatogenesis via Transplantation

Mouse donor spermatogonial stem cells from a fertile testis transplanted to an infertile recipient mouse testis were successful in establishing spermatogenesis in the seminiferous tubules of the host, and normal spermatozoa were produced. Testis cells from rats also have been transplanted to testes of immunodeficient mice, with rat spermatogenesis occurring in all recipient mice. The inducing of rat spermatogenesis in mouse testes opens the possibility that spermatogonial stem cells of other species could be transplanted. The mouse could potentially become an *in vivo* host for spermatogenesis of many mammalian species.

Brinster and Avarbock, 1994
Brinster and Zimmermann, 1994
Clouthier et al., 1996

Spermatic ducts (vasa deferentia) empty into the urethra (Fig. 9.60a). Due to descent of the testes, spermatic ducts must loop over the ureters en route to the urethra. One or more accessory sex glands (a single prostate, paired seminal vesicles, and paired bulbourethral or Cowper's glands) are located near the junction of the spermatic ducts and the urethra. Most male mammals have an unpaired erectile penis. Erectile tissue in the penis consists of the single medial **corpus spongiosum** and two lateral masses, the **corpora cavernosa,** that surround the urethra. In monotremes, the penis is reptilelike and nonprotrusible. In more derived mammals, the penis becomes external. The tip of the penis, called the **glans penis,** glandular and richly supplied with sensory nerve endings, is covered with loose skin, known as **prepuce** or **foreskin.**

As an adaptation to the dual vaginae of the female, the penis of male marsupials is forked at the tip (Fig. 9.60b). One tip enters each lateral vagina during copulation. The ureter opens between the prongs, and not at the tips, of the penis. A groove on the medial side of each glans penis directs semen into each vagina. The penis of the platypus (*Ornithorhynchus*) is also bifid as an adaptation to the unpaired Müllerian ducts.

A heterotropic bone, the **os penis or baculum,** is found in the penis of some mammals, including marsupials, insectivores, bats, rodents, carnivores, bovines, and lower primates. The structure and size of the baculum have been used as taxonomic characters and as an indicator of age. The baculum appears to be the ossification of a corpus cavernosum (Gunderson, 1976).

Thus, the trend in reproductive systems in vertebrates has been toward a reduction in number of zygotes that must be produced. This has been accomplished principally through (1) internal as opposed to external fertilization; (2) viviparity as opposed to oviparity; and (3) parental care of the young.

REPRODUCTION

Most mammals are viviparous with the young (embryo and fetus) being retained within the female's uterus. The embryo/fetus is nourished by means of a placenta, to which it is attached by an umbilical cord. Gestation ranges from approximately 12 1/2 days in the opossum (*Didelphis virginiana*) up to 24 months in elephants. The only exceptions are the oviparous monotremes, which incubate their eggs after depositing them in a nest (duck-billed platypus) or in a temporary pouch on their abdomen (echidna).

Many mammals breed at specific times of the year and are known as seasonal breeders. Others, such as elephants, rhinoceroses, giraffes, and humans, have no specific breeding season. Seasonal breeding and reproductive rhythm are the result of not just a single external stimulus, but rather a combination of nutritional, visual, auditory, tactile, environmental, and social factors. Most smaller mammals, which have shorter gestation periods, begin breeding as photoperiods increase. The quantity and quality of food may affect such things as the length of the breeding season and litter size, whereas pheromones can affect the reproductive behavior and actions of individuals by stimulating their central nervous system. Temperature has only an indirect effect on mammalian reproduction.

Naked mole rats (*Heterocephalus glaber*) are unique among mammals in that a single dominant individual (queen) can control the reproduction of all males and females in the colony (Faulkes and Abbott, 1991). The queen selects her breeding partners and initiates mating behavior. She suppresses reproduction in nonbreeding females and may control the reproductive endocrinology of breeding males because testosterone secretion in males becomes synchronized with the ovarian cycle of the queen.

Climatic disruptions thousands of kilometers away can affect the breeding of some species such as Weddell seals (*Leptonychotes weddelli*) in Antarctica. The climatic phenomenon known as the El Niño/Southern Oscillation (ENSO) occurs when a tropical pool of warm water shifts eastward from the western Pacific and alters weather in much of the tropical and temperate latitudes. It apparently causes a drop in the birth rate of Weddell seals living some 6,000 km away (Monastersky, 1992). The lowered birth rate represents the most southern biological effect ever recorded for an ENSO. The decline may result from changes in the fish population caused possibly by shifts in ocean currents and water temperatures.

Seasonal breeding in deer mice is regulated by photoperiod and by food availability (Nelson et al., 1992). Short photoperiods and restricted food intake cause a reduction in gonad size and sperm production. Many small mammals, such as shrews, mice, and voles in North America, may have several litters of young during the warmer months of the year. A female of a given species in northern North America may produce 2 or 3 litters annually, whereas a conspecific female in a southern state may be able to breed throughout the year, thereby producing more litters and young. Females are usually receptive to males immediately following the birth of a litter. There are records of a single meadow vole (*Microtus pennsylvanicus*) having 17 litters in a 12-month period (Bailey, 1924), and a golden mouse (*Ochrotomys nuttalli*) having 17 litters within 18 months (Linzey and Packard, 1977). However, larger mammals tend to have fewer litters per year. The number of young may vary from 1 in many species to approximately 18 in the American opossum. Bats, although small, breed only once a year in temperate regions and normally have either 1 or 2 young, usually in the spring. Many tropical species of bats breed more than once annually. Female canids, felids, and other related groups normally come into breeding condition twice each year, whereas larger mammals (deer, elk, buffalo, sheep) produce young only once a year. Some of the largest mammals, such as elephants, may breed only every 2 years due to the extreme length of their gestation period.

Some mammals that breed only once a year, such as bears, seals, weasels, badgers, and many deer, mate immediately after giving birth to their young in the spring. Although fertilization occurs, development ceases after a short time and the embryo (blastocyst) floats freely in the uterus for several months. Not until implantation occurs does the embryo resume development. Thus, the total gestation period for mammals with such **delayed implantation** may be as long as 350 to 360 days and allows the young to be born at the time of year most favorable for their survival (Table 9.1). A somewhat similar embryonic diapause occurs in many kangaroos (*Macropus*) (Fig. 9.61). As long as a joey is in the pouch, development of the next generation's blastocyst is suppressed. Thus, female kangaroos may have an embryo in diapause, a joey in the pouch, and a young kangaroo that is out of the pouch but still nursing daily. The mammary glands are capable of providing two kinds of milk simultaneously—high-protein, low-fat milk (from one teat) to the joey in the pouch, and low-protein, high-fat milk (from the other teat) to the older offspring.

The length of time required to reach sexual maturity is more variable than that of any other vertebrate group. Female meadow voles are known to breed when 6 weeks of age. At the other extreme, some whales do not mature for 7 to 8 years, and elephants not for 15 years.

Mammals may use most of the sensory processes—smell, vision, hearing, and touch—when seeking a mate. Scent glands that produce pheromones are better developed in mammals than in any other group of vertebrates. Odor is especially advantageous because many mammals are nocturnal, and most individuals are fairly widely separated from each other. A variety of sounds including howling, bellowing, barking, roaring, and squeaking may be used to attract members of the opposite sex. Some, such as drills and mandrills, develop brightly colored buttock pads. Humans adorn their bodies with alluring clothes, jewelry, and perfumes in an attempt to make themselves more attractive to members of the opposite sex. Male members of the family Cervidae (deer, elk, caribou, moose, and others) annually grow sets of antlers on their heads (Figs. 9.16 and 9.62). Antlers are used for defense, but this sexual characteristic also may be an important visual factor in sex recognition and determining male "quality" for the female. Courting may involve grooming, nuzzling, and other forms of bodily contact. Some females, including humans, are presented with "gifts" by courting males. Male cottontop tamarins of South America court females by carrying young cottontops around as a display of paternal devotion.

Mammals may be monogamous, polygamous, or promiscuous. Earlier estimates that 2 to 4 percent of mammalian species were monogamous now seem high: recent studies show that many of these "monogamous" species engage in extra-pair copulations (EPCs) (Stone, 1991). The existence of EPCs does not mean there is no primary bonding relationship between a male and female. We know that over 90 percent of mammalian species are polygynous, with

TABLE 9.1

Periods During Which Blastocysts Remain Dormant in Some Mammals with Delayed Implantation

Species	Dormancy of Blastocyst (in months)
Order Chiroptera	
Equatorial fruit bat (*Eiolon helvum*)	3+
Jamaican fruit bat (*Artibeus jamaicensis*)	2 1/2
Order Edentata	
Nine-banded armadillo (*Dasypus novemcinctus*)	3 1/2–4 1/2
Order Carnivora	
Black bear (*Ursus americanus*)	5–6
Grizzly bear (*Ursus arctos*)	6+
Polar bear (*Ursus maritimus*)	8
Marten (*Martes americana*)	8
Fisher (*Martes pennanti*)	10–11
Badger (*Taxidea taxus*)	6
River otter (*Lutra canadensis*)	9–11
Mink (*Mustela vison*)	1/2–1 1/2
Long-tailed weasel (*Mustela frenata*)	7
Order Pinnipedia	
Alaskan fur seal (*Callorhinus ursinus*)	3 1/2–4
Harbor seal (*Phoca vitulina*)	2–3
Gray seal (*Halichoerus grypus*)	5–6
Walrus (*Odobenus rosmarus*)	3–4
Order Artiodactyla	
Roe deer (*Capreolus capreolus*)	4–5

Table from Mammalogy *by Terry A. Vaughan, Copyright © 1972 by Saunders College Publishing, reproduced by permission of the publisher.*

FIGURE 9.61

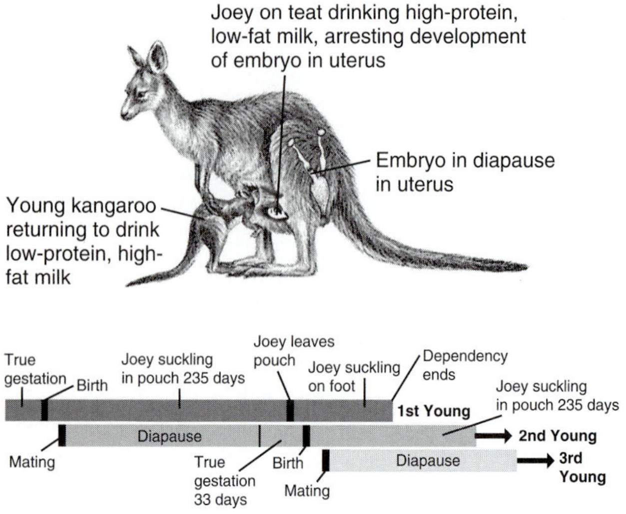

Kangaroos have a complicated reproductive pattern in which the mother may have three young in different stages of development at any one time. As long as a joey is in the pouch, development of the next generation's blastocyst is suppressed.

FIGURE 9.62

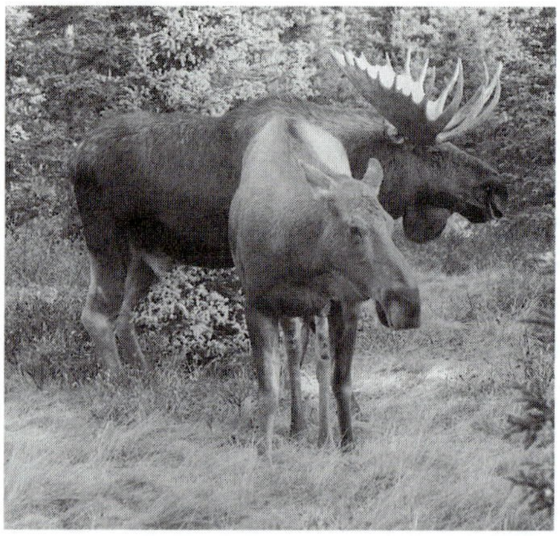

Male members of the family Cervidae, such as moose (*Alces alces*), annually grow sets of antlers on their heads. Antlers are used for defense, but this sexual characteristic may also be an important visual factor in sex recognition by the female.

monogamy restricted to the few species in which paternal care is necessary or in which females are widely dispersed (Amos et al., 1994).

Data obtained by DNA analyses reveals that female chimpanzees (*Pan troglodytes verus*) from the Ivory Coast may seek to mate outside their own social group (Gagneux et al., 1997). High-ranking female chimpanzees were shown to have significantly higher infant survival, faster maturing daughters, and more rapid production of young (Pusey et al., 1997; Wrangham, 1997). High rank probably influences reproductive success by helping females establish and maintain access to good foraging areas.

Female elk (*Cervus canadensis*) will mate with several males a day throughout the breeding season. In contrast, the California mouse (*Peromyscus californianus*) mates for life and is monogamous (Ribble and Salvioni, 1990). Prairie voles (*Microtus ochrogaster*) are also exceptionally monogamous (Insel and Carter, 1995). Many gray seals (*Halichoerus grypus*) mate with previous partners, even though the species has been described as polygynous.

During the 4- to 6-week-long mating season, male red deer (*Cervus elaphus*) roar repeatedly. Besides serving to warn other males, the calls apparently induce females to ovulate sooner (McComb, 1987). This is reproductively advantageous to the male in that it gives him a better chance to mate before possibly being replaced by a stronger male, and he can father more offspring than a noncalling male. The female's advantage of early mating lies in giving birth earlier in the spring so her calf has a better chance of surviving. In addition, the female has more time to recover so that she can give birth the next season.

All mammals use internal fertilization. Storage of viable sperm in the female urogenital tract for variable periods of time prior to fertilization of ova, called **delayed fertilization,** is known in every vertebrate group except the jawless fishes (Howarth, 1974). In some species of hibernating bats, sperm are stored for as long as 6 months prior to ovulation (Wimsatt, 1966, 1969). Copulation in temperate bats usually occurs in September, but fertilization does not occur until emergence from hibernation in April or May. Delayed fertilization means that females do not have to expend energy on mating immediately after emerging from hibernation when their energy stores are low. It also probably allows for embryonic development to begin earlier in the spring. Spermatozoa within the female reproductive tract obtain nourishment from the uterine and oviductal epithelium. The presence of zinc in the tails of spermatozoans, in the cauda epididymes, and in the seminal fluid may suppress sperm metabolism and reduce their need for sustenance during the period of storage (see review by Crichton et al., 1982). The prostate gland has been found to be the major source of zinc in both *Myotis lucifugus* and *M. velifer.*

Sperm competition leads to selection for increased sperm numbers and motility (Parker, 1982; Smith, 1984). When ejaculates from different males compete within the reproductive tract of a single female, males with a numerical sperm advantage statistically would be most successful at fertilizing that female's egg(s) (Parker, 1982). Among those primates and rodents in which females are promiscuous, sperm are longer in total length than in monogamous species. Sperm length is positively correlated with maximum sperm velocity (Gomendio and Roldan, 1991). If longer sperm can swim faster, they can reach ova sooner, thus hypothetically outcompeting rival sperm.

Some sperm produced by a male are better swimmers than others and can reach and fertilize eggs far more easily than others. However, males of all species produce a high percentage of "abnormal" sperm—two heads, two tails, etc. In some species, such as mice, bats, marsupials, and monkeys, a vaginal plug is left behind by the male after copulation. These plugs are thought to prevent leakage and to make more difficult the deposition of sperm by another male. Plugs themselves are composed of abnormal, misshapened sperm meshed together like a spider's web (Baker and Bellis, 1988, 1989a). Thus, although abnormal cells cannot fertilize an egg, they are important in forming a barrier that keeps out sperm from other males. Sperm not capable of fertilization might also be able to deactivate and/or kill foreign sperm with enzymes.

Males can vary their sperm counts quickly when mating situations change. Studies with rats (*Rattus norvegicus*) show that males that have no familiarity with their new mate ("unguarded") ejaculate an average of 51.6 million sperm as opposed to only 29.7 million for males that have been housed in the same divided cage with their mate ("guarded") (Bellis et al., 1990). In humans, sperm counts for ejaculations during intercourse decreased the more time couples spent together and increased after long periods away from their mates. (When faced with females who are out of sight and possibly mating with unknown partners, human males subconsciously perceive sperm competition and release more gametes, but when a female partner is close at hand, males reserve their sperm supplies [Baker and Bellis, 1989b].) No conscious or unconscious physiological mechanism has been suggested to account for this ability of males to alter their sperm counts in response to specific social situations.

Human females have the ability to influence sperm competition by controlling or manipulating the sequence and frequency with which they mate with different males, the time interval between in-pair and extra-pair copulations, and the ejection of sperm following copulation (Baker and Bellis, 1993). The occurrence and timing of female orgasm in relation to copulation and male ejaculation also influences the number of sperm retained.

Ralt et al. (1991) provide evidence of a chemotactic factor in the human female reproductive tract that attracts spermatozoa to the egg. Although the specific chemotactic factors are unknown, they are thought to be contained in the follicular fluid that bathes the oocyte; this fluid is released at the time of ovulation. Until this discovery, it had been thought that sperm swam "upstream against the current" until they reached the Fallopian tubes and a chance collision with an egg.

Relatives of some species that act altruistically toward each other improve the odds that their family gene pool will survive, even though every individual may not breed. This "kinship theory" is supported by studies of birds, wolves, and lions (e.g., Flam, 1991). In African lions (Packer et al., 1991; Packer and Pusey, 1997), brothers who banded together in large prides and participated in group defense as nonreproductive "helpers" often forfeited their own chances for fathering cubs so that their brothers could succeed. Thus, brothers who stick together in large prides stand a better chance of passing on some of their own genes than do others who go their separate ways, even though some brothers don't always breed themselves. Because brothers share about half their genes, the altruistic siblings hypothetically gain an evolutionary advantage from their reproductive sacrifice. Paternity analyses reported by Packer et al. (1991) revealed that resident males fathered all cubs conceived during their tenure in the pride.

■ GROWTH AND DEVELOPMENT

Prenatal Development

Oviparous

The only oviparous mammals are the duck-billed platypus (*Ornithorhynchus anatinus*), which occurs in Australia and Tasmania; the short-beaked echidna (*Tachyglossus aculeatus*), and the long-beaked echidna (*Zaglossus bruijni*), the latter two of which are found in parts of New Guinea, Australia, and Tasmania (see Fig. 9.6). The nest of the platypus is located in a tunnel excavated by the female in the bank of a stream or pond with burrows extending as far as 18 m horizontally and opening above water level. Normally, two spherical eggs about the size of those of a house sparrow are laid 2 weeks after breeding. Eggs are incubated for approximately 10 days by the female. Female echidnas develop a temporary breeding pouch on their abdomens during the breeding season. The one to three eggs are transferred directly from the cloaca to the pouch. The young remain in the pouch for approximately 8 weeks following hatching.

Viviparous

All other mammals are viviparous. Developing young are attached to a placenta by means of an umbilical cord (Fig. 9.63). Gas exchange, nutrient supply, and the removal of wastes all occur through the placenta.

The appearance and reproductive behavior of a female rodent can be affected by the sex of her immediate neighbors in utero (Clark and Galef, 1998). In house mice (*Mus musculus*) and gerbils (*Gerbillus* spp.), male fetuses secrete testosterone during the last week of gestation. The testosterone enters the amniotic fluid and is absorbed by the fetuses on either side of the male. Exposure to the hormone exerts masculinizing effects on both males and females, causing, among other things, both sexes to mature later. Twenty-seven of 28 female gerbils that matured late had been either next to a male or between two males; in contrast, 21 of 22 early-maturing females had developed without an immediate male neighbor. Additional preliminary studies on gerbils show that eggs produced by the right ovary are more likely to develop into males than are eggs from a left ovary (Clark and Galef, 1998).

Duration of Embryonic/Fetal Development

Among mammals, the shortest gestation period of 12 1/2 days occurs in the American opossum (*Didelphis virginiana*),

FIGURE 9.63

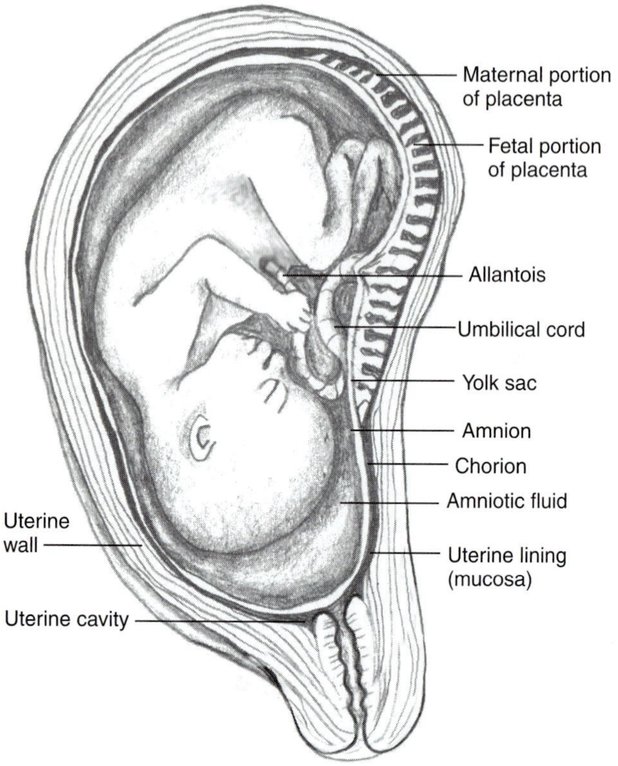

The young of viviparous mammals develop within the uterus. They are attached to a placenta by means of an umbilical cord. Gas exchange, nutrient supply, and waste removal all occur through the placenta.

FIGURE 9.64

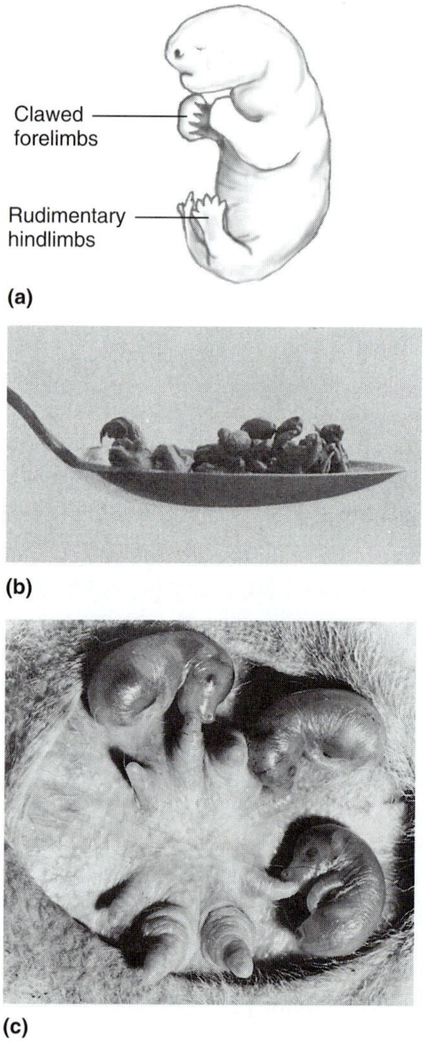

(a) Detail of a newborn opossum (*Didelphis virginiana*). (b) Eighteen newborn opossums in a teaspoon. Actual size. (c) Long-nosed bandicoot (*Perameles nasuta*), a marsupial. Litter of three young at about 4 days after birth.

a marsupial, whereas the longest gestation period of 22–24 months belongs to the elephants. The duck-billed platypus, a monotreme, incubates her eggs for approximately 14 days.

Young marsupials are born in a very undeveloped condition, so much so that newborn opossums have been referred to as "living embryos" (Fig. 9.64). They immediately crawl into the marsupium, attach to a nipple, and undergo the rest of their development within the pouch. All marsupials have greatly delayed tooth development because they spend a considerable part of their early life attached to a nipple. As a result, modern marsupials largely bypass having deciduous (baby) teeth in favor of adult teeth, which erupt after the period of intense suckling. The only deciduous teeth to be replaced are one pair of premolars in each half of the upper and lower jaws.

High-resolution x-ray computer tomography of a fossil baby *Alphadon*, a 70-million-year-old mouse-size mammal, has revealed a pattern of tooth replacement exactly like that of modern marsupials. Discovery of this same tooth pattern in an ancient mammal from the time of the dinosaurs suggests that the marsupial reproductive system already had evolved as far back as the Cretaceous period (Cifelli, 1996).

Female armadillos normally have an 8- to 9-month gestation period, after which identical quadruplets are born. However, females apparently can delay delivering young for up to 24 months in some cases (Storrs et al., 1988). Either the fertilized embryos must remain free-floating in the uterus most of this time, or the female can store viable sperm for extended periods.

Hatching and Birth

Most mammals are enclosed only in the amniotic sac, which ruptures either before or during parturition (Fig. 9.65). Females usually aid newborns in freeing themselves from the ruptured membranes. In monotremes, pressure exerted by the front paws as well as an egg tooth on the top of the bill enable them to break out of their shells.

Parental Care

Parental care is universal among mammals. Altricial species require proportionately longer to develop and become independent than do precocial species. The duration of

FIGURE 9.65

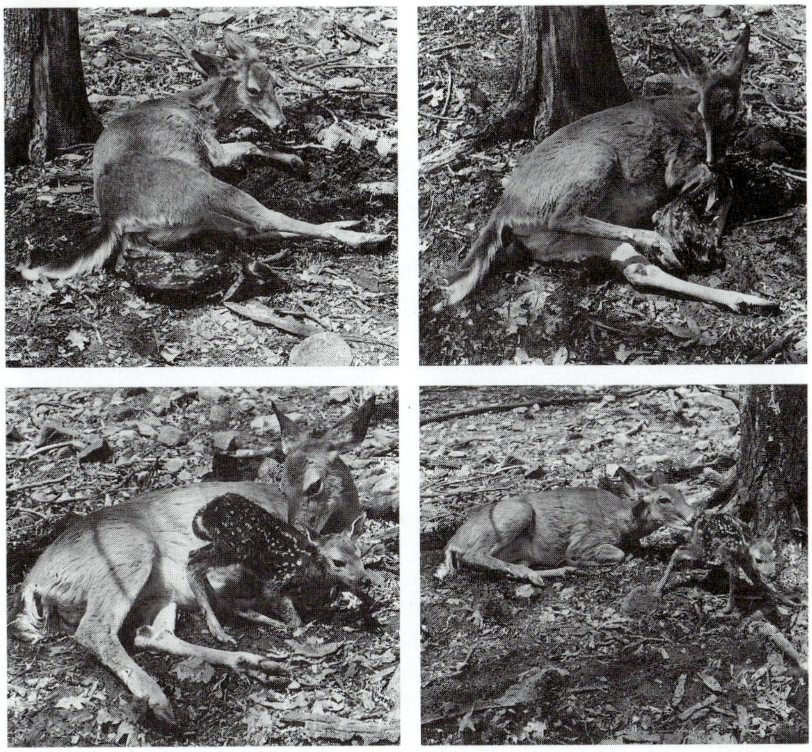

Birth of a white-tailed deer (*Odocoileus virginianus*). The fawn is born with its eyes and ears open and has a full coat of hair. It is able to stand, nurse, and walk shortly after birth.

parental care ranges from several weeks in small rodents to many years in some higher primates.

For their first month of life, Malaysian tree shrews (*Tupaia tana*) reside in a nursery nest in a tree cavity apart from the mother (Emmons and Biun, 1991). The mother visits the nest only once every other day to nurse them. The mean time spent inside the nest cavity is 2.74 minutes per visit. The total time that the mother spends with her young during their first month of life is less than 50 minutes. The short and infrequent visits of *Tupaia* mothers is thought to be an adaptation to prevent the production of heavy scent trails, since the nests of these animals are close to or in the ground and prone to predation.

Communal nursing has been reported in many mammals ranging from bats (Wilkinson, 1992) and prairie dogs (Hoogland et al., 1989) to Hawaiian monk seals (Boness, 1990) and fallow deer (Birgersson et al., 1991). For example, communal nursing in evening bats (*Nycticeius humeralis*) rarely occurred prior to 2 weeks before weaning, but after that time, over 18 percent of nursing bouts involved nondescendant offspring (Wilkinson, 1992). It occurred most frequently when pups began hunting on their own and when lactating females attained their lowest average pre-fed body weight. Wilkinson (1992) hypothesized that if a female with extra milk reduced her weight by dumping milk prior to her next foraging trip, she could obtain an immediate energetic benefit and maintain maximum milk production. Milk production, which is under the control of the hormone prolactin, continues to remain high as long as milk is removed from the mammary glands.

Growth

Oviparous

Female platypuses normally produce two blind, naked young annually in a nesting chamber within a burrow (see Fig. 9.6b). Nipples and teats are absent in monotremes, and breasts do not form. Milk is released from ducts onto the flattened milk patch, or areola, on the surface of the skin, where vigorous suckling permits its ingestion by the young. Their eyes open at about 6 weeks of age, at which time the fur is about 6 to 8 mm in length. The young emerge from the burrow when about 4 months of age (Orr, 1982; Nowak, 1991).

Female echidnas normally produce a single egg, which is transferred from the cloaca into a temporary pouch on the female's abdomen. The young echidna is nourished by thick, yellowish milk produced by the mammary glands that open into the pouch, and it remains in the pouch for 6–8 weeks (Nowak, 1991).

Viviparous

The young of some viviparous species are well developed and alert at birth and are capable of taking care of themselves to some extent (precocial). Fishes, anurans, lizards, snakes, crocodilians, and some mammals such as hares, cetaceans, and ungulates are precocial. Parents often provide protection for these young, and in the case of mammals, the young are nourished by nursing from the mother. Young pronghorns (*Antilocapra americana*) are particularly precocious. Within 4 days following their birth (although still somewhat unsteady on their feet), pronghorn young can outrun a human, and at 1 week of age they can outdistance the average dog (Einarsen, 1948). Wildebeests are able to run with the herd within hours of their birth.

Other newborn mammals, however, require extensive parental care for their survival. Their eyes and ears are sealed at birth; they have little or no hair on their bodies and are usually unable to thermoregulate; and the distal portions of their limbs may not be fully formed. These altricial species include shrews, rabbits, mice, squirrels, dogs, and cats.

Differential growth rates between spring-born and fall-born litters have been recorded for a number of species of small mammals (Meyer and Meyer, 1944; Dunaway, 1959;

TABLE 9.2

Longevity of Some Species of Mammals

Species	Maximum Age (years)
Ornithorhynchidae	
Duck-billed platypus (*Ornithorhynchus anatinus*)	17
Tachyglossidae	
Short-nosed echidna (*Tachyglossus aculetata*)	50
Didelphidae	
American opossum (*Didelphis marsupialis*)	7
Macropodidae	
Red kangaroo (*Macropus rufus*)	16.3 (captive) 22 (wild; estimated)
Soricidae	
Big short-tailed shrew (*Blarina brevicauda*)	2.8
Vespertilionidae	
Big brown bat (*Eptesicus fuscus*)	19
Little brown bat (*Myotis lucifugus*)	34
Pongidae	
Chimpanzee (*Pan troglodytes*)	53
Orangutan (*Pongo pygmaeus*)	59
Gorilla (*Gorilla gorilla*)	54
Bradypodidae	
Two-toed sloth (*Choloepus didactylus*)	27.8+
Leporidae	
European rabbit (*Oryctolagus cuniculus*)	13
Eastern cottontail (*Sylvilagus floridanus*)	9+
Sciuridae	
Fox squirrel (*Sciurus niger*)	13
Gray squirrel (*Sciurus carolinensis*)	23.5
Chipmunk (*Tamias striatus*)	8
Woodchuck (*Marmota monax*)	13–15
Castoridae	
Beaver (*Castor canadensis*)	19
Muridae	
Deer mouse (*Peromyscus maniculatus*)	8.3
Golden mouse (*Ochrotomys nuttalli*)	8.5
House mouse (*Mus musculus*)	6
Norway rat (*Rattus norvegicus*)	4.2
Delphinidae	
Common dolphin (*Delphinus delphis*)	20+
Bottlenosed dolphin (*Tursiops truncatus*)	25+
Canidae	
Gray wolf (*Canis lupus*)	16
Coyote (*Canis latrans*)	14.5
Dog (*Canis familiaris*)	34
Red fox (*Vulpes vulpes*)	12

Species	Maximum Age (years)
Ursidae	
Black bear (*Ursus americanus*)	26
Grizzly bear (*Ursus arctos*)	34
Polar bear (*Ursus maritimus*)	38+
Procyonidae	
Raccoon (*Procyon lotor*)	20.5+
Mustelidae	
Otter (*Lutra canadensis*)	23
Badger (*Taxidea taxus*)	26
Striped skunk (*Mephitis mephitis*)	12.9+
Felidae	
Lion (*Panthera leo*)	30
Mountain lion (*Puma concolor*)	20+
Bobcat (*Felis rufus*)	32.3
House cat (*Felis catus*)	27
Elephantidae	
Indian elephant (*Elephas maximus*)	69
African elephant (*Loxodonta africanus*)	50–70
Equidae	
Horse (*Equus caballus*)	62
Suidae	
Swine (*Sus scrofa*)	27
Hippopotamidae	
Nile hippopotamus (*Hippopotamus amphibius*)	54.3
Camelidae	
Llama (*Lama glama*)	21
Camel (*Camelus bactrianus*)	50
Cervidae	
Wapiti (*Cervus canadensis*)	26.7
Antilocapridae	
Pronghorn (*Antilocapra americana*)	11.9
Bovidae	
Domestic cow (*Bos taurus*)	39
Bison (*Bison bison*)	40
Goat (*Capra hircus*)	18

Sources: Data from E. L. Cockrum, Introduction to Mammalogy, *1962, The Ronald Press; L. S. Crandall,* The Management of Wild Animals in Captivity, *1964, University of Chicago Press; J. Gurnell,* The Natural History of Squirrels, *1987, Facts on File Publication; and R. M. Nowak,* Walker's Mammals of the World, *5th ed., 1991, The Johns Hopkins University Press.*

Davis and Golley, 1963; Linzey, 1967; Martinet and Spitz, 1971). Variation in body growth seems to be primarily dependent on photoperiod and food quality.

Spring-born cotton rats (*Sigmodon hispidus*) were consistently heavier than fall-born rats throughout a 220-day study period (Dunaway, 1959). Spring-born golden mice (*Ochrotomys nuttalli*) were heavier at birth than fall-born mice, but were shorter in tail length and total length (Linzey, 1967). The greater weight of spring-born individuals may be correlated with the smaller litter size at this time, since it might be expected that females would produce either large numbers of lighter young or smaller litters with heavier individuals.

As juvenile mammals grow, they undergo one or more pelage changes. These pelage changes occur in a specific pattern in each species and result in a **juvenal** pelage that may be considerably different in coloration from the adult pelage. For example, young deer mice (*Peromyscus* spp.) are grayish and do not attain their typical tan adult pelage for several months.

Attainment of Sexual Maturity

The time required to reach sexual maturity in mammals ranges from several weeks in some rodents to as long as 10 to 15 years in such species as whales, elephants, gorillas, and humans. The shortest known time required to reach sexual maturity occurs in female meadow voles (*Microtus pennsylvanicus*), which are ready to breed when 3 weeks of age (Hamilton, 1943). The quantity and quality of the food and water supply may affect the age of attaining sexual maturity in some mammals.

Sexual maturation of juvenile female California mice, *Peromyscus californicus,* is delayed if they remain in physical contact with their mother (Gubernick and Nordby, 1992). In contrast, puberty was unaffected by exposure to the father or a strange adult male. Females exposed to their mother, but prevented physical contact by a double wire-mesh barrier, showed an intermediate delay in sexual maturation. Actual physical contact with the mother, and not solely a urinary chemosignal, is necessary to delay sexual maturation in this species. The delay in sexual maturation of juvenile females may be a means of avoiding competition with their mothers until the young females disperse. It may also serve as an incest avoidance mechanism since fathers do not mate with their daughters.

Longevity

The life span of an elephant may be 60 or 70 years (Table 9.2). More and more humans are living past the century mark as medical science continues to find cures for disorders and diseases. Banded little brown bats (*Myotis lucifugus*) have been recorded as living for 32 years. Among small mammals, a golden mouse (*Ochrotomys nuttalli*) lived for 8 years and 5 months in captivity (Linzey and Packard, 1977). This represents the longest life span of any North American cricetine rodent.

Review Questions

1. List at least five major functions of mammalian skin.
2. List five major functions of hair. Give an example of a specific mammal illustrating each function.
3. Differentiate between horns and antlers. Give one example for each.
4. Differentiate between sudoriferous and sebaceous glands. Discuss the functions of each.
5. How does the middle ear of mammals differ from that of all other vertebrates?
6. Compare and contrast the vertebral columns of vertebrates (kinds of vertebrae, types of articulating surfaces, etc.).
7. Distinguish between plantigrade, digitigrade, and unguligrade locomotion. Give examples of mammals that use each type.
8. Differentiate between artiodactyls and perissodactyls. Give examples of each.
9. Describe the anatomy and function of the respiratory system of a terrestrial mammal.
10. Trace a bolus of food in a carnivore from the oral cavity until it is absorbed from the small intestine. List all of the structures and organs involved and their functions.
11. Differentiate between ultrasonic communication and infrasound. Give an example of a mammal that uses each method.
12. Describe a nephron. Trace the filtrate from the glomerulus to the urinary bladder.
13. Differentiate between duplex, bipartite, bicornuate, and simplex uteri. Give an example of each.
14. Differentiate between an altricial species and a precocial species. Give several mammalian examples of each.
15. Define delayed fertilization.
16. Define delayed implantation. What are some of its adaptive advantages? Give several examples of mammals in which delayed implantation occurs.
17. Distinguish between monogamy and polygamy.
18. Which mammalian-like characteristics were possessed by early synapsids?
19. Most mammals are viviparous; a few are oviparous. List several advantages and disadvantages of each of these methods of reproduction.

Supplemental Reading

Alterman, L., G. A. Doyle, and M. K. Izard (eds.). 1995. *Creatures of the Dark: The Nocturnal Prosimians.* New York: Plenum Press.

Altringham, J. D. 1996. *Bats: Biology and Behaviour.* New York: Oxford University Press.

Anton, M., and A. Turner. 1997. *The Big Cats and Their Fossil Relatives: An Illustrated Guide to their Evolution and Natural History.* New York: Columbia University Press.

Augee, M. L. (ed.). 1992. *Platypus and Echidnas.* Mosman, New South Wales: The Royal Zoological Society of New South Wales.

Barbour, R. W., and W. H. Davis. 1969. *Bats of America.* Lexington: The University Press of Kentucky.

Bauer, E. A. 1997. *Bears: Behavior, Ecology, Conservation.* New York: Voyageur Press.

Birkhead, T. R., and A. P. Moller. 1998. *Sperm Competition and Sexual Selection.* New York: Academic Press.

Bronson, F. H. 1989. *Mammalian Reproductive Biology.* Chicago: The University of Chicago Press.

Brown, R. E., and D. W. Macdonald (eds.). 1985. *Social Odours in Mammals.* Oxford, England: Clarendon Press.

Byers, J. 1997. *American Pronghorn.* Chicago: University of Chicago Press.

Catton, C. 1990. *Pandas.* New York: Facts On File.

deWaal, F. 1997. *Bonobo: The Forgotten Ape.* Berkeley: University of California Press.

Dunstone, N. and M. L. Gorman. 1998. *Behavior and Ecology of Riparian Mammals.* New York: Cambridge University Press.

Eisenberg, J. F. 1989. *Mammals of the Neotropics, Volume I.* Chicago: The University of Chicago Press.

Eliot, J. L. 1998. Polar bears: Stalkers of the high Arctic. *National Geographic* 193(1):52–71.

Flannery, T., R. Martin, and A. Szalay. 1996. *Tree Kangaroos.* Australia: Reed Books.

Fogle, B. 1997. *The Encyclopedia of the Cat.* New York: DK Publishing.

Foote, M., J. P. Hunter, C. M. Janis, and J. J. Sepkoski, Jr. 1999. Evolutionary and preservational constraints on origins of biologic groups: Divergence times of eutherian mammals. *Science* 283:1310–1314.

Franzmann, A. W., and C. C. Schwartz. 1998. *Ecology and Management of the North American Moose.* Washington, D.C.: Wildlife Management Institute and Smithsonian Institution Press.

Gittleman, J. L. (ed.). 1989 (Volume 1). 1996 (Volume 2). *Carnivore Behavior, Ecology, and Evolution.* Ithaca, New York: Comstock Publishing Associates.

Gubernick, D. J., and P. H. Klopfer (eds.). 1981. *Parental Care in Mammals.* New York: Plenum Press.

Hairston, N. G., Sr. 1994. *Vertebrate Zoology: An Experimental Field Approach.* New York: Cambridge University Press.

Hall, E. R. 1981. *The Mammals of North America.* 2 volumes. New York: John Wiley and Sons.

Hayssen, V., A. van Tienhoven, and A. van Tienhoven. 1993. *Asdell's Patterns of Mammalian Reproduction.* Ithaca, New York: Cornell University Press.

Hotton, N., III, P. D. MacLean, J. J. Roth, and E. C. Roth (eds.). 1987. *The Ecology and Biology of Mammal-like Reptiles.* Washington, D.C.: Smithsonian Institution Press.

Janis, C. M., K. M. Scott, and L. L. Jacobs. 1998. *Evolution of Tertiary Mammals of North America.* Volume 1: Terrestrial Carnivores, Ungulates, and Ungulatelike Mammals. New York: Cambridge University Press.

Kleiman, D. G., M. E. Allen, K. V. Thompson, S. Lumpkin, and H. Harris. 1997. *Wild Mammals in Captivity.* Chicago: The University of Chicago Press.

McKenna, M. C., and S. K. Bell. 1997. *Classification of Mammals Above the Species Level.* New York: Columbia University Press.

Linzey, D. W. 1995. *Mammals of Great Smoky Mountains National Park.* Blacksburg, Virginia: The McDonald & Woodward Publishing Company.

Linzey, D. W. 1998. *The Mammals of Virginia.* Blacksburg, Virginia: The McDonald & Woodward Publishing Company.

Newman, C. 1997. Cats: Nature's masterwork. *National Geographic* 191(6):54–76.

Nixon, S. W., and C. A Oviatt. 1973. Ecology of a New England salt marsh. *Ecological Monographs* 43(4):463–498.

Nowak, R. M. 1999. *Walker's Mammals of the World.* Baltimore, Maryland: The Johns Hopkins University Press.

Redford, K. H., and J. F. Eisenberg. 1992. *Mammals of the Neotropics, Volume 2.* Chicago: The University of Chicago Press.

Reynolds, J. E., III, and S. A. Romme (eds.) 1999. *Biology of Marine Mammals.* Washington, D. C.: Smithsonian Institution Press.

Ridgway, S. H., and S. R. Harrison (eds.). 1981–1994. *Handbook of Marine Mammals.* 5 volumes. New York: Academic Press.

Schaller, G. B. 1998. *Wildlife of the Tibetan Steppe.* Chicago: University of Chicago Press.

Shoshani, J., and P. Tassy (eds.). 1996. *The Proboscidea: Evolution and Palaeoecology of Elephants and Their Relatives.* New York: Oxford University Press.

Solomon, N. G., and J. A. French. 1996. *Cooperative Breeding in Mammals.* New York: Cambridge University Press.

Szalay, F. S., M. J. Novacek, and M. C. McKenna (eds.). 1993. *Mammal Phylogeny.* 2 volumes. New York: Springer-Verlag.

Thewissen, J. G. M. (ed.). 1998. *The Emergence of Whales: Evolutionary Patterns in the Origin of Cetacea.* New York: Plenum.

Vaughan, T. A. 1986. *Mammalogy.* Third edition. Philadelphia: W. B. Saunders Company.

Ward, G. 1997. Making room for wild tigers. *National Geographic* 192(6):2–35.

Whitaker, J. O., Jr., and W. J. Hamilton, Jr. 1998. *Mammals of the Eastern United States.* Ithaca, New York: Comstock Press.

Wilson, D. E. 1997. *Bats in Question: The Smithsonian Answer Book.* Washington, D.C.: Smithsonian Institution Press.

Wilson, D. E., and D. M. Reeder (eds.). 1993. *Mammal Species of the World.* Washington, D.C.: Smithsonian Institution Press.

SELECTED JOURNALS

Acta Theriologica. Published by the Polish Academy of Sciences. This is an international journal that publishes original research and review papers on all aspects of mammalian biology.

International Journal of Primatology. Published by Plenum Press, New York. This is the official journal of the International Primatological Society and includes articles on laboratory and field studies in basic primatology.

Journal of Mammalogy. Published by the American Society of Mammalogists. Publishes scientific and nontechnical materials. All responsible types of research on mammals are encouraged.

Mammalia. Publishes papers on all aspects of mammalian research.

Mammalian Species. Published by the American Society of Mammalogists. Each issue of this unique work is devoted to a particular species and is authored by experts on that species. Issued irregularly.

Marine Mammal Science. Published by the Society for Marine Mammalogy. Publishes primarily current research in all areas of marine mammal studies.

Vertebrate Internet Sites

Visit the zoology web site at http://www.mhhe.com to find live Internet links for each of the references listed below.

1. **Subphylum Vertebrata, Class Mammalia, from the University of Minnesota.**
2. **Animal Diversity Web, University of Michigan.**
   Mammalia. Links to all orders of mammals, and then more links to specific families and species, complete with information on morphology, geographic distribution, behavior, habitat, reproduction, references, and much more.
3. **American Society of Mammalogists.**
   Information on the organization, on mammals, and sites of interest.
4. **Mammal Species of the World Home Page.**
   Complete taxonomic information on all mammals currently identified.
5. **The Bear Den.**
   Information on ecology of bears, and links to related sites.
6. **Bat Conservation International Home Page.**
   Information on ecology and conservation of bats.
7. **WhaleNet Website.**
   Sponsored by Wheelock College in Boston, this site contains much information on marine mammals, updates on strandings and entanglements, and many links.
8. **Whale Museum.**
   Information from the museum in Friday Harbor, San Juan Island, Washington.
9. **Cetacean Research Unit Site.**
   Information on species, conservation, adoption programs, and other information.
10. **Society for Marine Mammalogy.**
    Information on the society, and many links to more information.
11. **Oceania Project.**
    Information about whales and dolphins and those who are involved in their care, protection, and conservation.
12. **Gorillas Online.**
    Links to pictures, information, FAQs, and more links about gorillas.
13. **Primate Behavior and Ecology.**
    Information on current research.
14. **Predatory Behavior and Ecology of Wild Chimpanzees.**
    A research article on this interesting behavior or chimpanzees. http://www-rcf.usc.edu/~stanford/chimphunt.html
15. **Orangutan Foundation International.**
    This foundation organizes study tours of orangutans.
16. **Fetal Pig Dissection on the WWW, done by instructors at Lakeview High School.**
    Contains 37 photographs in order of the steps of the dissection. Excellent site.

CHAPTER 10

Population Dynamics

■ INTRODUCTION

Animal populations, which are dynamic and constantly changing, depend on successful reproduction to maintain their existence. Other important factors in maintaining viable populations include an adequate food supply, sufficient home sites, and the effects of dispersal, immigration, emigration, climate, predation, disease, and parasites. The impact of some of these factors is **density-dependent**—that is, the effect varies according to the population density; for others, the impact is **density-independent**—that is, unrelated to population size.

■ POPULATION DENSITY

Population density is an important variable that can influence the level of competition for scarce resources. Every habitat has a theoretical maximum number of individuals of a given species that it can support for an extended period of time. This level is known as the **carrying capacity** (Fig. 10.1a) and is determined by environmental resistance factors acting on the reproductive (biotic) potential of a population. It is primarily determined by the availability of food and shelter.

Vertebrates exhibit three basic types of population growth. Once many species reach the environmental carrying capacity of their range, they maintain relatively stable populations (Fig. 10.1b). This is especially true of species inhabiting some tropical regions where temperature and rainfall show little variability. Some species that normally maintain relatively stable populations experience sharp population increases at irregular intervals. Such **irruptions**, which cause the population to exceed its carrying capacity, may be the result of such transient factors as a reduction in predators, an increase in food, a favorable change in the weather, or any combination of these. Still other species experience sharp increases in their population sizes at regular intervals, followed by crashes. Species exhibiting regular cyclic population increases usually do so either every 3 to 4 years or approximately every 10 years.

Reproductive (Biotic) Potential

The maximum number of young that a population can produce under ideal conditions during a particular time period is referred to as the **reproductive (biotic) potential** of that population. In a healthy, natural population, the birth rate will equal, if not exceed, the death rate, but due to environmentally limiting factors, the reproductive potential is rarely, if ever, reached. Dispersal, immigration, and emigration may affect the reproductive potential to a limited degree.

Most populations will level off after the population reaches a certain size (the carrying capacity). The point at which population growth levels off varies with the species, the habitat, and the climate. A natural population will continue to show fluctuations (seasonal, annual), but they will generally not be far removed from the average carrying capacity (Fig. 10.1c).

Each individual can affect the reproductive potential of its species in one or more of the following ways:

1. By producing more offspring at a time.
2. By having a longer reproductive life, so that it reproduces more often during its life span.
3. By reproducing earlier in life. The shorter the generation time of a species (that is, the younger its members when they first reproduce), the higher its reproductive potential (Fig. 10.2).

Reproductive rates vary widely among the vertebrates. Some fishes such as sturgeon and cod may produce several million eggs annually, whereas many mammals normally give birth to only a single young. Factors such as climate and predation of eggs and/or young have undoubtedly been factors in the evolution of egg production. Numerous hypotheses have been proposed to explain clutch size in birds. These were summarized by Lack (1954), who presented arguments for and against each hypothesis. Among the principal hypotheses are the following:

1. Females produce as many eggs as they are physiologically capable of producing.

FIGURE 10.1

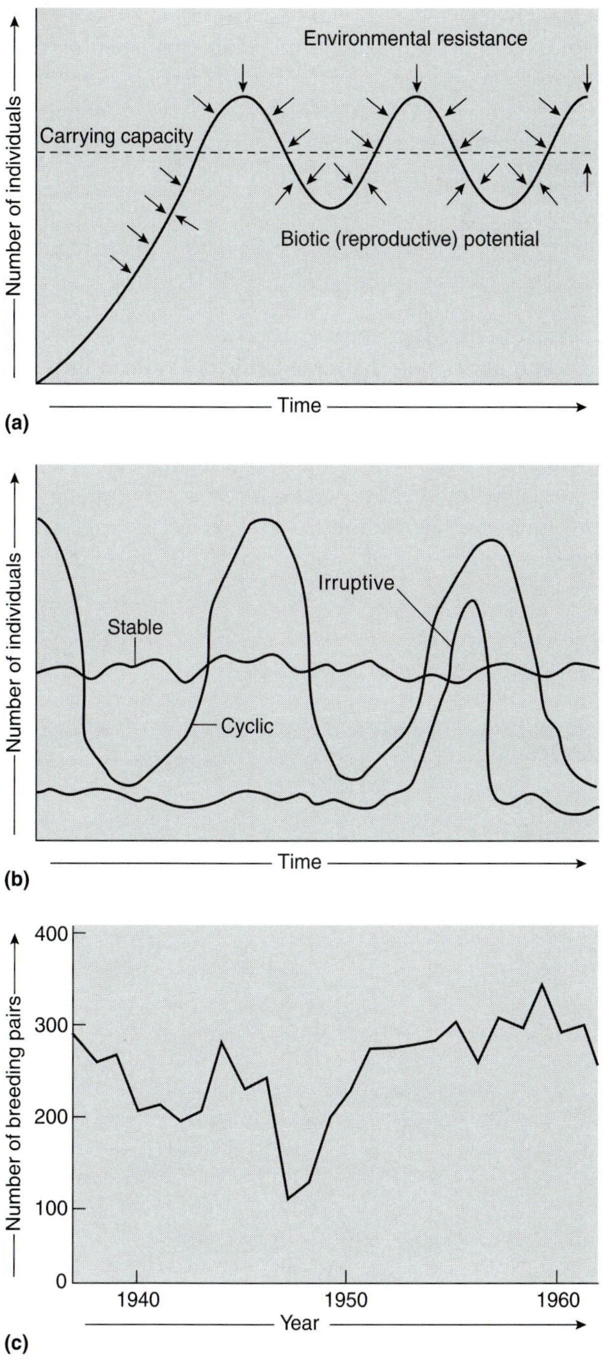

(*a*) The theoretical number of individuals of a given species that can be supported for an extended period of time in a habitat is known as the carrying capacity. It is determined by environmental resistance factors (primarily food and shelter) acting on the reproductive (biotic) potential of the population. (*b*) Some populations remain relatively stable after reaching the carrying capacity of their range; some experience regular cyclic population increases; and others experience sharp population increases (irruptions) at irregular intervals. (*c*) A stable population is illustrated by the number of breeding pairs of gray herons (*Ardea cinerea*) in northwestern England. After recovering from the severe winter of 1947, this population showed little fluctuation over a 15-year period.

Source: (c) *Data from D. Lack,* Population Studies in Birds, *1966, Clarendon Press, New York.*

FIGURE 10.2

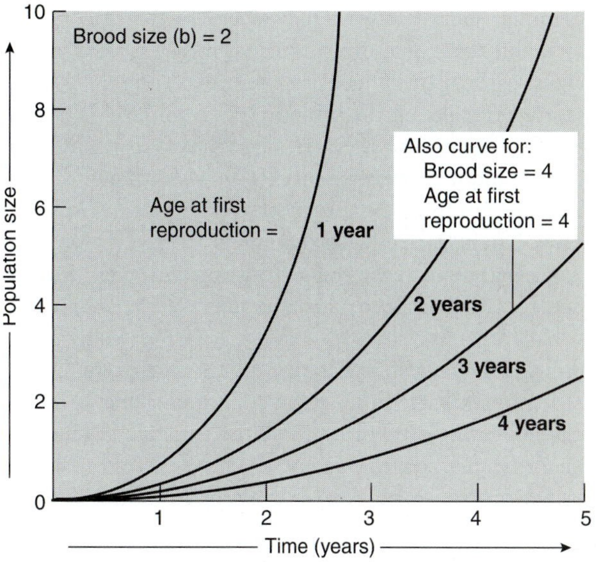

Population growth is dramatically affected by the age at which females first reproduce. In each of these examples, females produce two offspring per year, but the age at which females first reproduce differs for each curve (first reproduction at 1, 2, 3, or 4 years of age). Changing the age of first reproduction from 4 to 3 years has the same effect as doubling the brood size from two to four.

Source: Data from Cole, Quarterly Review of Biology, *29:103, 1954.*

2. Females produce as many eggs as they can successfully incubate.
3. Females produce approximately the number of eggs and young that the parent(s) can satisfactorily feed and care for.

Whereas each of these hypotheses holds true for many species, many exceptions exist. For example, many birds will lay additional eggs in their nests if one or more of the original eggs is removed. This fact has been of extreme importance in the attempt to increase the population of endangered whooping cranes (*Grus americana*). Females normally produce two eggs. When biologists remove one egg for artifical incubation in order to increase the size of captive flocks, the female usually will produce and incubate a third egg.

Whereas many birds apparently produce as many eggs as they can satisfactorily incubate, there are other species that seemingly could incubate more than the number of eggs they produce in the average clutch. Critics of the third hypothesis point out that precocial birds do not need to expend time and energy feeding their offspring.

In studies where clutch size was adjusted experimentally during incubation, larger clutches were associated with significantly lower percentage hatching success in 11 of 19 studies; longer incubation periods in 8 of 10 studies; greater loss of adult body condition in 2 of 5 studies; and higher adult energy expenditures in 8 of 9 studies (Thomson et al., 1998). Since incubation does involve metabolic costs and since the demands of incubation increase sufficiently with clutch size to affect breeding performance, Thomson et al. (1998) proposed that optimal clutch size in birds may in part be shaped by the number of eggs that the parents can afford to incubate.

Among mammals, small prey species such as mice, voles, rabbits, and ground squirrels usually produce several litters annually, each of which consists of several young. Many of these species make up the primary consumer level in a food web or food pyramid; they are subsequently consumed by secondary and tertiary consumers (Fig. 10.3). Bailey (1924) recorded a meadow vole that produced 17 litters within 12 months. Larger species, such as most ungulates, breed only once a year and produce a single offspring. Although smaller species generally produce greater numbers of young annually than larger species, longevity is also a factor. Many small mammals have a life expectancy of approximately 1 year. However, most bats, although small, are long-lived—up to at least 34 years in *Myotis lucifugus* (Keen and Hitchcock, 1980; Tuttle, pers. comm., 1992). With the exception of lasiurine bats, most North American species produce a single young annually. Many predators, such as mustelids, canids, and felids, produce only one or two litters annually.

The ratio of adults to juveniles varies during the year. Juveniles form a larger proportion of the population during and immediately after the breeding season in temperate regions. By fall and early winter, many juveniles have either matured and become part of the adult population or have been lost. Most temperate populations, therefore, reach their largest size in late fall and early winter. Loss of individuals during the harsher conditions of winter usually leads to low population levels in late winter, just prior to the breeding season of many species. In years when the climate is favorable and food is plentiful, many species may breed well into the fall, resulting in larger populations the next year.

Environmental Resistance

Although populations have the biotic potential to increase, a variety of factors act to limit the number of young actually produced or that survive. These factors represent the **environmental resistance.** Climate (including rainfall, flooding, drought, and temperature) is a primary controlling factor. Other controls are exerted by intraspecific aggression, inadequate supply of den sites, predation, disease, and par-

FIGURE 10.3

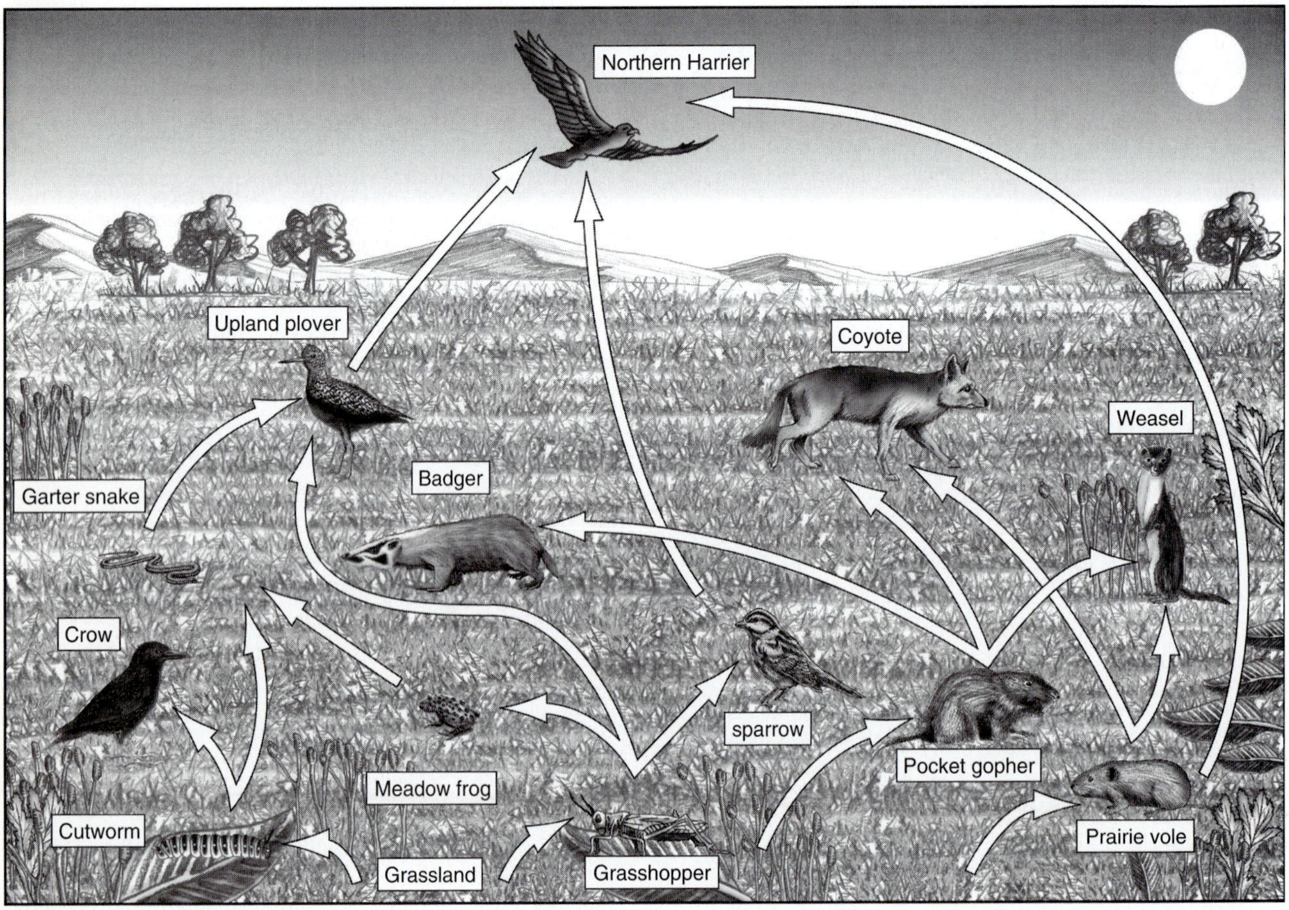

A food web for a prairie grassland in the midwestern United States. Arrows flow from the grassland (producer) to various levels of consumers.

asites. Resistance factors can be grouped into two categories: density-dependent and density-independent.

Density-Dependent Factors

Density-dependent factors are those whose effects vary directly with the density of the population. For example, as population density increases, suitable home sites and food may become scarcer per individual. As the rate of individual contacts increases, intraspecific aggression may increase; females may stop breeding; the rate at which nestling young may be killed and/or cannibalized by their parents may increase; and the rate at which juveniles are forced to disperse may be greater than what occurred at lower densities. Parasites can increase in response to population size of the host, and diseases can spread much more rapidly. For example, when waterfowl congregate in dense flocks, the incidence of infection and the chances of an epizootic are increased (Fig. 10.4).

Approximately 600 Mediterranean monk seals (*Monachus monachus*) remain in the wild, mostly in groups of about 20. From May to August, 1997, a catastrophic epizootic struck the largest social group (Osterhaus, 1997; Harwood, 1998). Of 270 seals living in a pair of caves on West Africa's Mauritanian coast, only about 70 survived the disease. Osterhaus (1997) reported that most of the seals examined harbored a dolphin morbillivirus, a virus similar to the one that causes distemper in dogs. Hernandez et al. (1998), however, carried out histopathological examination of lung and other tissues from 14 fresh carcasses and found no indication of typical morbillivirus lesions. There was no evidence of primary viral damage or secondary opportunistic infections in lung tissue, which are hallmarks of morbillivirus infections in other aquatic mammal species. The terminally ill seals exhibited clinical signs of lethargy, motor incoordination, and paralysis in the water—symptoms consistent with drowning caused by paralysis due to poisoning. Hernandez et al. (1998), who identified three species of toxic dinoflagellates in eight water samples collected from near the colony during the mortality event, suggested that poisoning by paralytic algal toxins may have been the cause of death.

FIGURE 10.4

A concentration of snow and blue geese. When waterfowl congregate in dense flocks, the incidence of infection and the chances of an epizootic such as fowl cholera are increased.

Optimal densities may vary seasonally in temperate areas. For example, the lower flow of many streams during summer determines the annual carrying capacity for species like trout (*Salmo*). The winter food supply may determine the annual carrying capacity for many species, even though more animals can be supported during the summer months. In ungulates, food limitation as a cause of density-dependent population regulation has been shown for roe deer (*Capreolus*), wild reindeer (*Rangifer*), kangaroos (*Macropus*), wildebeests (*Connochaetes*), and white-eared kobs (*Kobus*) (reviewed by Skogland, 1990).

One effect of high population density in herbivores is overgrazing. Skogland (1990) reported increased tooth wear and lowered body size and fat reserves in wild reindeer (*Rangifer tarandus*). Skogland (1990) stated:

> *During late winter foraging, lichen mats in the Loiseleria-Arctostaphylion plant alliance become the only available vegetation type due to snow cover [Skogland, 1978]. As the unrooted lichens are grazed off, the animals substitute easily digestible lichens in their diet by the dead parts of grasses, dwarf shrubs, and also mosses, with insufficient nutrient content [Skogland, 1984a]. Increased use of crustaceous lichens with encrusted small rock particles as well as soil particles and detritus in the ingested diet accelerated molar wear. This lowers chewing efficiency and increases the passage time of larger plant particles into the digestive system whose ability to process energy is slowed down [Skogland, 1988].*

Although adult female survival rate was not affected, a significant negative correlation existed between population density and juvenile winter survival rate. Calves normally were not able to compete successfully with conspecifics of higher rank. Neonatal survival was directly related to maternal condition during the last part of gestation and the calving season (Skogland, 1984b).

Male common toads (*Bufo bufo*) tend to call at low densities, but are more likely to remain silent at high densities (Hoglund and Robertson, 1988). Male wood frog (*Rana sylvatica*) density has a significant effect on the behavior of searching male wood frogs at breeding ponds (Woolbright et al., 1990). When the male population density is low, males are more likely to be stationary. As male density increases, more males actively search for females.

High density reduces growth rates in amphibians, thus lengthening exposure to predators and possible unfavorable environmental conditions (e.g., Petranka and Sih, 1986; Wilbur, 1987, 1988). Van Buskirk and Smith (1991) recorded significantly reduced survival and growth rates and an increase in the skewness of the size distributions of individuals with increasing density in blue-spotted salamanders (*Ambystoma laterale*) in Michigan. Individuals in high-density populations showed an increased skewness in body size, with only a few salamanders becoming large and most remaining small. This tendency did not occur in populations with lower densities.

Some species seem to have an inherent self-regulating population control mechanism. This is especially true in territorial species in which individuals space themselves so that they have an adequate supply of food and shelter. In other species, when the population increases to a certain size, food and home sites become scarcer, intraspecific aggression increases, and breeding decreases or ceases. Many, but not all, members of the population may emigrate from the area. The best known example is the lemming (*Lemmus*) of Norway. As many members leave the area, more home sites and food become available for those left behind. Intraspecific aggression falls, individuals become better nourished, breeding resumes, and the population begins growing again.

Some species exceed the carrying capacity of their range. Because of all the environmental factors acting to control population increase, this is rare for a natural population. It is most common in those populations that are managed by humans, such as herds of deer and elk that are confined to military reservations, parks, and refuges. In addition, herds of elephants whose ancient migration routes and feeding areas are being encroached on by an expanding human population are, through no fault of their own, exceeding the carrying capacity of their dwindling range. In their attempts to locate food, they frequently trample crops and break down fences.

In some areas, predator control measures are undertaken in efforts to increase the numbers of another species, usually a game species. Predators often cull sick, lame, injured, and old individuals from a population. When the predator control measure is implemented, the protected species often increases and may exceed the carrying capacity of its range.

To prevent overpopulation, the levels of many game species are controlled by federal and state agencies. These agencies set "limits" on the number of individuals of each sex of a given species that can be killed during certain seasons of the year. Formerly, many people were subsistance hunters and utilized most parts of an animal that they killed. Today, some hunters still fall into this category, but most hunters are looking for trophy animals (biggest rack of antlers, etc.). They attempt to kill the largest, healthiest male specimens in order to mount their heads. Only in recent years have regulatory agencies promoted efforts to cull females from the population in order to balance the sex ratio and manage reproductive rates. In deer and many other species, one male will breed with multiple females; thus, populations can be more efficiently managed by culling some females rather than by focusing exclusively on males.

A coyote-proof enclosure was erected encompassing 391 hectares (ha) of pasture on the Welder Wildlife Foundation Refuge in Texas in 1972 (Teer et al., 1991). The immediate response was an increased size of the deer (*Odocoileus virginianus texanus*) herd (Fig. 10.5). Fawn survival was 30 percent

FIGURE 10.5

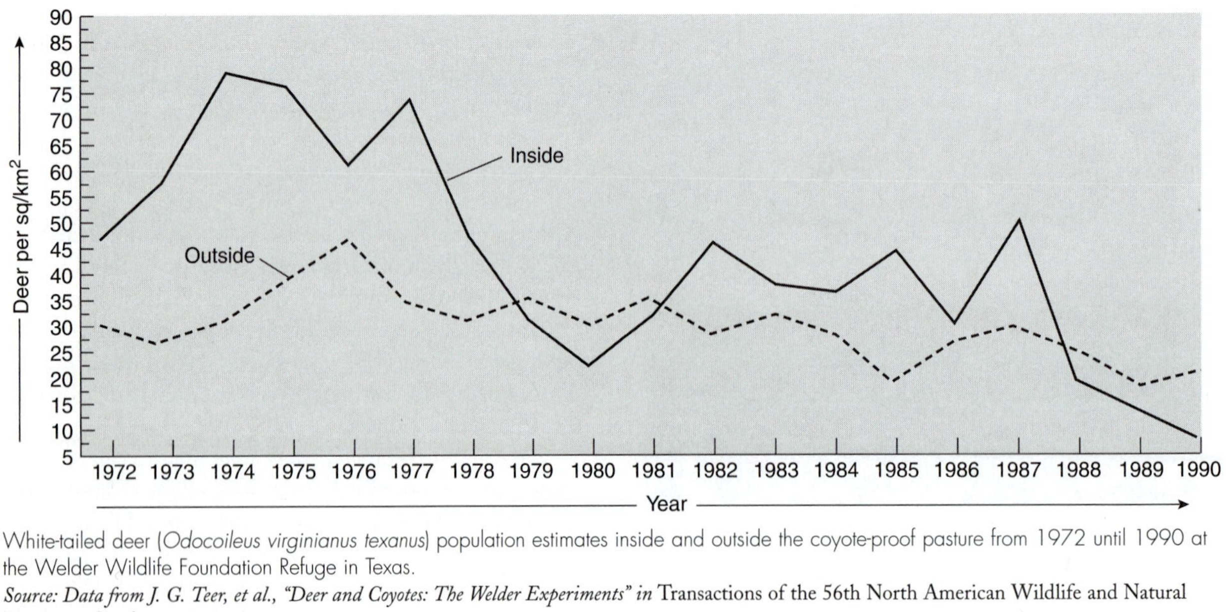

White-tailed deer (*Odocoileus virginianus texanus*) population estimates inside and outside the coyote-proof pasture from 1972 until 1990 at the Welder Wildlife Foundation Refuge in Texas.

Source: Data from J. G. Teer, et al., "Deer and Coyotes: The Welder Experiments" in Transactions of the 56th North American Wildlife and Natural Resources Conference *550–560, 1991.*

higher inside the enclosure than outside, where coyotes (*Canis latrans*) were uncontrolled. After several years, the high number of deer caused forage to become scarce, and deer began to die. Studies showed that mortality was caused primarily by lack of adequate food. Parasite loads also had increased. The deer herd reached a low point in 1980, after which it began to increase as food supplies returned to normal. Beginning in 1982, coyotes were once again present within the enclosure, and predation prevented the herd from increasing as it had during the years of predator control.

A classic example of a species exceeding the carrying capacity of its range involved the Kaibab mule deer (*Odocoileus hemionus*) in northern Arizona. In 1906, President Theodore Roosevelt set aside approximately 750,000 acres as a game refuge. At that time, an estimated 4,000 deer inhabited the area. Not only was hunting prohibited, but a predator control program was begun. Within the next 10 years, 600 mountain lions were killed. The small wolf population living on the area was almost exterminated by 1926, and it was completely eliminated by 1939. By 1939, more than 7,000 coyotes had been killed (Table 10.1). The winter food supply was the limiting factor that determined the annual carrying capacity, which was estimated to be about 30,000 animals. By 1920, however, an estimated 100,000 deer were present on the refuge. During the next two winters, 60 percent of the population died of starvation. An estimated 75 percent of the fawn crop was lost during the winter of 1924–25.

The numbers continued to decrease due to the depleted range. By 1939, the population had declined to 10,000 animals. However, publicly funded coyote control (trapping, shooting, and poisoning) continued in the area from 1940 until May 1963, resulting in phenomenally high deer densities in the early 1950s (McCulloch, 1986). Presumably due to the high density and inadequate food supply, deer were in poor physical condition and reproduction was low. McCulloch (1986) noted that absolute comparisons of herd size estimates could not be made for the different eras such as 1906–1940 vs. 1950–1961 vs. 1972–1979 because the deer inventory methods varied and were not compatible. Private sport hunting and fur trapping continued after 1963. Mountain lions were designated as game animals in 1971, and now can be taken normally only during designated hunting seasons by sport hunters. As of 1977, an estimated 40 adult mountain lions inhabited the Kaibab (McCulloch, 1986). During the period 1972–1979, the deer population experienced a decline of 9 percent per year (Barlow and McCulloch, 1984). Barlow and McCulloch (1984) stated: "Reasons for the decrease in deer abundance from 1972 to 1979 are not yet known. Climatic factors and increased natural predation are both suspected. We know, however, that the decline has not continued. Pellet counts indicate that the number of deer in Kaibab has increased dramatically since 1979, and now may have exceeded the 1972 population size." Thus, high levels of reproduction and deer in good physical condition have been achieved without predator control programs. As of 1996, an estimated 30,000 deer were living on the refuge.

In many areas of the world, increasing wildlife populations are creating problems. The spread of communicable diseases, such as rabies, has been slowed by reducing the populations of striped skunks, raccoons, and foxes (Bickle et al., 1991). Various control methods involving shooting, trapping, and poisoning have been used. More recently, fertility-inhibiting implants and contraceptive vaccines are being tested for birth control purposes (Moore et al., 1997). Norplant implants containing levonorgestrel, a synthetic progestin, have proved to be effective fertility inhibitors in several species, including Norway rats (*Rattus norvegicus*), rabbits (*Oryctolagus cuniculus*), striped skunks (*Mephitis*), and humans (*Homo sapiens*) (Phillips et al., 1987; Brache et al., 1990; Bickle et al., 1991). A single administration of immunocontraceptive vaccine was effective for more than 3 years in gray seals (*Halichoerus grypus*) (Brown et al., 1996). A vaccine made from pig ovaries has been used successfully as a contraceptive to control wild horse populations on Assateague Island in Maryland and white-tailed deer on Long Island (Kemp, 1988; Daley, 1997). The vaccine is made of minced pig ovaries that are distilled until only the membrane of the eggs (zona pellucida) is left. This is then mixed with a substance that helps stimulate the immune system. When the mixture is injected, it causes the horse's body to form antibodies that bind to the outside of the egg when the female ovulates, blocking the sperm receptor

TABLE 10.1

Number of Predators Removed from Kaibab Deer Habitat

Dates	Mountain Lions	Coyotes	Wolves	Bobcats
1906–1923. U.S. Bur. of Biol. Survey	674	3,000	11	120
1929–1939. Private hunters and trappers for fur, sport and lion bounty	142	4,488	20	743
1940–1947. USFWS	36	1,401	0	396
1948–1963. USFWS	18	733	0	282
1964–1976. Private hunters and trappers for fur and sport. Lion bounty ended 1970	13	No records	0	No records

From C. Y. McCulloch, in Southwestern Naturalist, *31(2):217, 1986. Copyright © The Southwestern Naturalist. Reprinted by permission.*

sites there and preventing fertilization (Daley, 1997). In laboratory tests, the vaccinations have proven to be 90 percent effective and reversible. Fertility returns within a year.

BIO-NOTE 10.1

Elephant Birth Control Programs

Female African elephants usually are in heat just 2 days of every 17 weeks. In an attempt to control the expanding elephant population in South Africa's Kruger National Park, testing began on two forms of contraception in 1996. The first method involved injecting specially-designed estrogen implants into 31 non-pregnant females. The implant is designed to slowly release hormones into the bloodstream in much the same manner as the contraceptive pills used by women. In 6 months, no cow became pregnant. However, the contraceptive caused the females to be permanently in heat, which in turn caused the bulls to be in a perpetual state of sexual excitement. These unintended side effects resulted in a breakdown of the close-knit elephant societies and social responsibilities, including the loss of several baby elephants because their mothers were permanently distracted by as many as eight sexually excited males at one time. Although unwanted pregnancies were prevented, the social cost was too high, and this population control program was discontinued in April 1997.

The second method is based on creating an immunological response: a vaccine made from the outer coating of egg cells taken from pigs produces antigens that prevent elephant eggs from recognizing elephant sperm. Initial data from 21 elephant cows has shown the vaccine to be only about 60 percent effective, a problem the researchers predict they can overcome by giving booster inoculations every 10 months.

Daley, 1997

Old World rabbits (*Oryctolagus cuniculus*) were introduced successfully to Australia by British settlers in 1859, and to New Zealand a few years later (Grzimek, 1990). The rabbits reproduced until they numbered in the hundreds of millions, causing an ecological disaster in the southern half of Australia (see Fig. 3.34). Unchecked, the burgeoning rabbit population creates deserts by devouring plants, shrubs, and seedlings. The widespread destruction of vegetation seriously harmed the sheep-raising industry. A number of native Australian marsupial species have been endangered or totally eliminated through competition with, or by having their habitats destroyed by, the Old World rabbit. Competition with rabbits for burrows has caused the extinction of one species of bilbie, or rabbit-eared bandicoot (*Macrotis leucura*), and has caused a second species (*Macrotis lagotis*) to retreat to northern Australia, where it is listed as endangered. Other marsupials adversely affected by rabbits include mulgaras (*Dasycercus cristicauda*), hairy-nosed wombats (*Lasiorhinus latifrons*), long-nosed potaroos (*Potorous tridactylus*), and banded hare wallabys (*Lagostrophus fasciatus*). Livestock, including introduced sheep and cattle, struggle to compete with the rabbits for pasture.

In the early 1950s, Australian government scientists released myxomatosis, a rabbit-killing virus (Kaiser, 1995; Adler, 1996a; Drollette, 1996; Seife, 1996). Although quite successful at first, myxomatosis gradually became less effective, particularly in Australia's dry rangelands. In 1991, researchers began testing a calicivirus known as rabbit hemorrhagic disease (RHD) virus. It kills quickly and fairly painlessly by causing blood clots in the lungs, heart, and kidneys. In March 1995, following laboratory testing, it was injected into rabbits quarantined on Wardang Island in Spencer Gulf, South Australia. By late September, however, the virus had evaded containment (possibly by flying insects) and spread to the mainland, killing rabbits hundreds of kilometers inland. It appears to kill 80 to 95 percent of the adult rabbits it encounters. In September 1996, the Australian government announced a nationwide campaign to reduce the annual $472-million damage that rabbits cause to agriculture. The lethal rabbit virus was to be released at 280 sites. The expectation is that, after the calicivirus kills most of the rabbits, it will remain in the reduced population and act as a long-term regulator of the rabbit population.

The virus appears to be working exactly as animal control and health officials had hoped (Drollette, 1997). The wild rabbit population has dropped by 95 percent in some regions, and native fauna and flora are already staging a comeback.

Opponents fear that the virus could jump the species barrier (Anonymous, 1996). For this reason, the New Zealand Department of Agriculture decided not to introduce the virus pending further study (Duston, 1997). However, in August 1997, officials confirmed that several dead rabbits near Cromwell in New Zealand tested positive for the rabbit calicivirus (Pennisi, 1997d). It is suspected that the virus may have been released intentionally. The virus quickly spread across hundreds of miles, making containment and eradication impossible.

Density-Independent Factors

Climatic factors such as rainfall, flooding, drought, and temperature often play a major role in limiting population growth. Fires and volcanic eruptions also affect populations without regard to their density.

Most species in temperate areas are seasonal breeders, with temperature being a major factor affecting reproduction. They produce their young during the time of year that is most favorable for their survival. Most fishes, amphibians, and reptiles breed in late winter or spring. Birds breed and raise their young during the warmer months of the year. Most mammals produce their young during the same optimum period. Most bats breed in the fall, but because of delayed fertilization (see Chapter 9), the ova are not fertilized until late winter or early spring, and young are born shortly thereafter.

Temperature controls the food supply for many species. A late spring freeze that kills flying insects or forces them to become dormant can have disastrous effects on insectivorous birds such as swallows and purple martins (*Progne subis*), as well as on bats. A freeze that kills the buds of oak, hickory, and other mast-bearing trees can create hardship for many animals in late summer, fall, and winter. For example, turkeys, squirrels, deer, bears, and others depend on acorns, hickory nuts, and other mast for their late summer food supply. Mass emigrations of some forms such as gray squirrels (*Sciurus carolinensis*) have been reported during years of poor food supply (Seton, 1920; Flyger, 1969; Gurnell,1987). During such mass movements, more individuals are susceptible to predation, and many more than normal are struck and killed by vehicles; natural mortality probably also increases. Some species, such as black bears, often leave the protective confines of parks and refuges in search of food. Many are shot as nuisance bears when they wander into civilization; others become victims of hunters or motor vehicles.

Members of a species living in an optimal habitat generally produce more young than members of the same species living in a poor habitat. A study of mule deer (*Odocoileus hemionus*) in California revealed that does in good shrubland habitat produced an average of 1.65 fawns annually, whereas does in poor chaparral habitat averaged 0.77 fawns each (Taber, 1956) (Fig. 10.6). At this rate over a 4-year period, shrubland does will produce an average of 6.48 fawns, whereas does in chaparral habitat will produce only an average of 3.08 fawns.

FIGURE 10.6

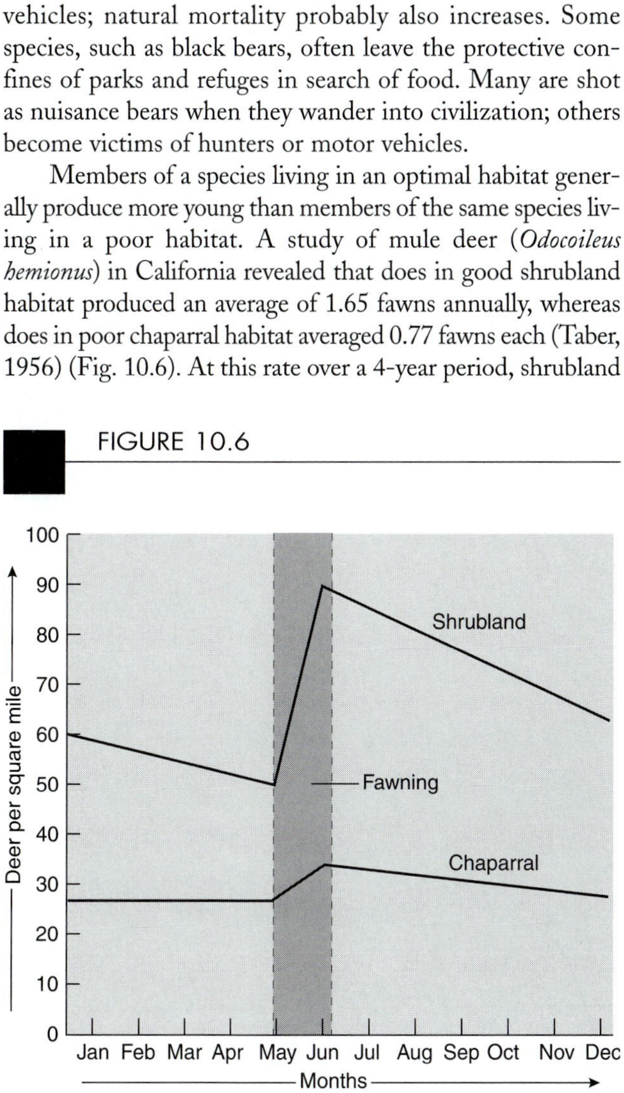

Comparison of mule deer (*Odocoileus hemionus*) population density through the year on poor range (chaparral) and good range (shrubland) in California. Females living in shrubland produced an average of 1.65 fawns annually, whereas does in chaparral habitat averaged 0.77 fawns annually. Over a 4-year period, shrubland does will produce an average of 6.48 fawns, whereas does in chaparral habitat will produce an average of only 3.08 fawns.

Source: Data from R. D. Taber, Transactions of the 21st North American Wildlife Conference, *1956.*

The breeding season following a poor food year also usually results in fewer young being born. Litter and clutch sizes will be smaller in many species. Depending on the severity of the food shortage, female white-tailed deer (*Odocoileus virginianus*), for example, may resorb a developing fetus or give birth to no more than one young. Herd sizes obviously will decrease as the average production per female decreases.

Rainfall, or the lack thereof, can drastically affect the breeding of certain groups, especially amphibians and waterfowl. If breeding ponds and pools dry up before the larvae and tadpoles can successfully metamorphose, annual recruitment may approach zero. Many nesting waterfowl are much more susceptible to predators during periods of drought. Extensive periods of rainfall and flooding also can be disastrous for many species.

The deaths of 158 manatees along Florida's Gulf Coast between Naples and Fort Myers during a 3-month period in the spring of 1996 was caused by red tide algae. A red tide is a natural algal (*Gymnodinium breve*) bloom that sporadically occurs along the coast and produces brevitoxin, a powerful neurotoxin. Unseasonably cold weather farther north brought a large concentration of manatees to Florida's Gulf Coast, and a strong northwest wind blew a potent strain of the red tide algae deep into manatee feeding areas (Fig. 10.7). Manatees swam in contaminated water, drank it, and ate sea grass infected with it. When the toxin level got high enough, it attacked the manatees' nervous system. One of the first nerve centers to be incapacitated was the one that regulates

FIGURE 10.7

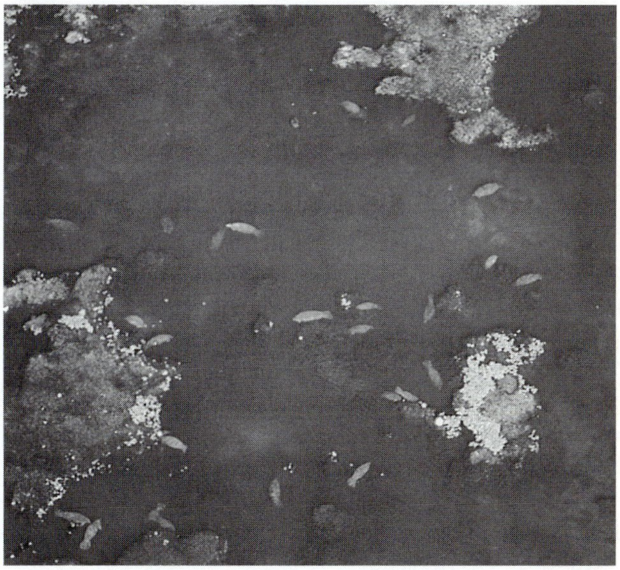

During the winter months, manatees (*Trichechus manatus*) congregate in the Crystal River in Florida, a sanctuary of warm water with an abundance of water hyacinth.

the diaphragm—the major muscle used by mammals for breathing. Many manatees suffocated. Levels of brevitoxin 50 to 100 times normal were found in tissues from the lungs, stomachs, kidneys, and livers (Holden, 1996b). The result was the greatest number of manatee deaths from a single event since record keeping began in 1974. This deadly red tide, along with deaths from other natural causes, cold weather stress, boats on Florida's waterways, and other undetermined factors caused 415 manatee deaths in 1996, more than twice as many as the previous record of 206 deaths in 1990 (Anonymous, 1997d). The total Florida manatee population in 1996 was 2,639.

Cycles

Populations of some species such as lemmings and voles show rhythmic fluctuations (Fig. 10.8). Their populations increase for several years and then fall dramatically. This cycle is repeated with some regularity. Three- or 4-year cycles are characteristic of certain species inhabiting tundra and northern boreal forests, such as lemmings (*Lemmus* and *Dicrostonyx*), voles (*Microtus*), ptarmigan (*Lagopus*), and spruce grouse (*Dendragopus*), as well as some of the birds and mammals that prey on these species. Some species inhabiting the northern coniferous forests, such as lynx (*Lynx canadensis*), hares (*Lepus americanus*), and ruffed grouse (*Bonasa umbellus*), have a longer 10-year cycle.

Due to the intricacies of most food webs, anything affecting one species also will affect one or more additional species. When a prey species is abundant, its numbers will be reflected in increasing numbers of the predatory species (Fig. 10.9). Better-nourished females will be able to produce and successfully care for a larger number of offspring than if they were malnourished and/or emaciated (Madsen and Shine, 1992). In addition, many predators will turn their efforts to a secondary prey if their primary prey becomes scarce. Erlinge

BIO-NOTE 10.2

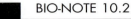

Invasion of the Brown Tree Snakes

The U.S. territory of Guam is being overrun by brown tree snakes (*Boiga irregularis*), a nocturnal, tree-climbing, bird-eating, egg-gobbling, mildly poisonous reptile that can reach 3 m in length. Brown tree snakes, which originally found their way to Guam some 50 years ago, encountered no natural predators and an abundant food supply. The population of these snakes has soared to an estimated 2,000,000 or more—about 10,000 per 1.6 km^2. The snakes hang like vines from trees, fences, and power poles. Power outages caused by electricity arcing across snakes spanning power lines have become a frequent problem. These snakes have eliminated Guam's native lizards and 9 of 18 species of Guam's native forest birds; 6 of the remaining species are endangered, and the other 3 are rare. Research is under way to control the snake population by using a strain of virus that will kill the snakes without affecting other animal life. Extensive efforts are being taken to prevent this snake from invading Hawaii, which is home to 40 percent of the nation's endangered birds (many of which are already threatened by introduced wildlife). Snake-sniffing beagles and their handlers closely inspect every commercial and military flight from Guam.

Douglas, 1997
Allen, 1998
Fritts and Rodda, 1999

FIGURE 10.8

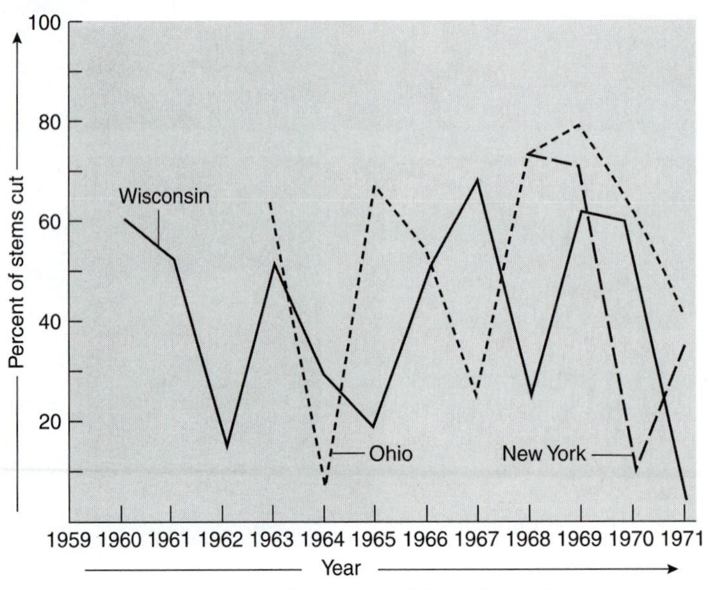

Comparison of cyclic population fluctuations of the meadow vole (*Microtus pennsylvanicus*) in Wisconsin, Ohio, and New York.
Source: Data from U.S. Fish and Wildlife Service, 1971.

FIGURE 10.9

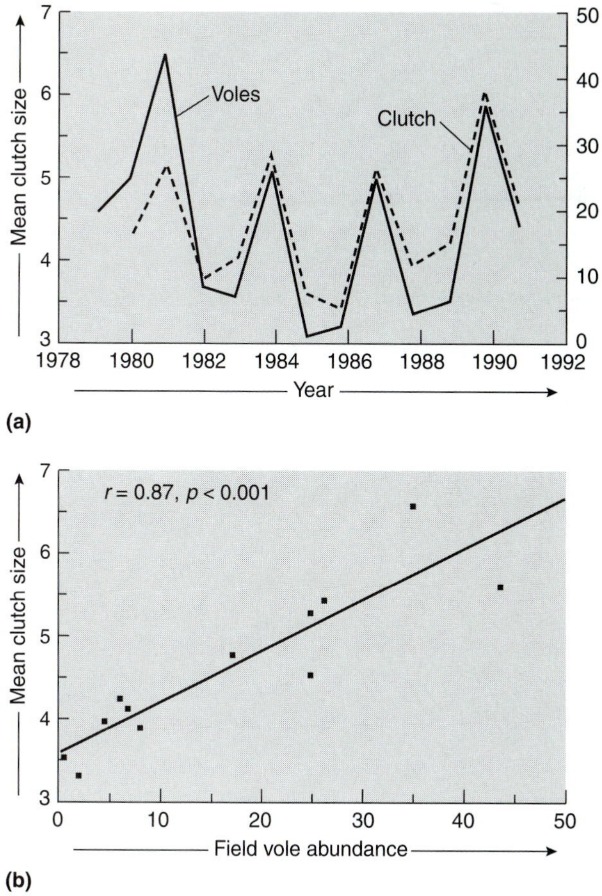

(a) Average annual clutch sizes of barn owls (*Tyto alba*) show a cyclic pattern clearly in synchrony with the field vole (*Microtus agrestis*) cycle near Esk, Scotland. (b) Average barn owl clutch sizes in the Esk study area were closely correlated with spring field vole abundance. *Data from Taylor,* Barn Owls, *1994, Cambridge University Press.*

et al. (1991) suggested that predation has a significant influence on the pattern of change in a population. In ecosystems dominated by predators specializing on a single species, a cyclic pattern is promoted, whereas in ecosystems dominated by switching "generalist" predators, cyclicity is limited.

Numerous studies of snowshoe (varying) hares (*Lepus americanus*) and a variety of predators have shown significant predator responses to hare cycles (Brand et al., 1976; Brand and Keith, 1979; Powell, 1980; Todd et al., 1981; Thompson and Colgan, 1987) (see Chapter 13). For example, snowshoe hares are the primary prey of many fisher (*Martes americana*) populations. Bulmer (1974, 1975) examined fur sale records in Canada and concluded that population fluctuations of fishers were linked to hare cycles. However, a study of fishers in Minnesota during eight winters when the snowshoe hare population declined revealed that fishers consumed less hare as the hare population declined (33% of the diet during 1977–79, but only 3% in 1984). Consumption of small mammals (deer mice, *Peromyscus*; voles, *Microtus*, *Clethrionomys*; lemmings, *Synaptomys*; shrews, *Blarina, Sorex*; and moles, *Condylura*), however, increased from 4 to 5 percent during 1977–79, to 19 percent of the weight of the stomach contents in 1984. Fat deposits and reproduction (proportion of pregnant females, mean number of corpora lutea, and proportion of juveniles in the fisher harvest) by the fishers did not decrease during the period of the study (Kuehn, 1989).

MacLulich (1937) presented the original data on cyclic fluctuations of snowshoe hare and Canadian lynx (*Lynx canadensis*) populations obtained from records of pelts received by the Hudson Bay Company and covering the period from 1845 to 1935 (Fig. 10.10). These data show that these cycles have been going on for as long as records have been kept in North America. It now serves as a classic study of how the cyclic fluctuations of one species (prey) apparently affect another species (predator). More recent studies have shown, however, that lynx are not the primary cause of periodic drops in hare populations, although they may be a contributing factor in the decline. Furthermore, Stenseth et al. (1999) found that the dynamics of lynx populations could be grouped according to three geographical regions of Canada that differed in climate and proposed that external factors such as weather influence lynx population density.

In reference to snowshoe hares, Lack (1954) stated: "It is suggested that the basic cause of the cycles is the dominant rodent [snowshoe hare] interacting with its vegetable food to produce a predator-prey oscillation. When the primary consumers decline in numbers, their bird and mammal predators become short of food, prey upon and cause the decrease of the gallinaceous birds of the same region, and themselves die of starvation and/or emigrate." Keith (1974) and Keith and Windberg (1978) proposed an essentially identical theory to explain the 10-year snowshoe hare and grouse cycles. A similar theory was also proposed to explain the 3- to 4-year vole–predator–small game cycle in Sweden (Hornfeldt, 1978).

Hares normally feed on the bark and twigs of birch, poplar, alder, and black spruce (Fig. 10.11). As hare populations increase, food becomes scarcer, and the hares are forced to feed on the young shoots of these plants, which contain large amounts of toxins (see Chapter 13). The plant toxins act as antifeedants, resulting in a loss of weight and a decline in health in the hares, which causes them to be more susceptible to predation (Joggia et al., 1989; Reichardt et al., 1990a, b). Thus, it appears that the chemical defenses of certain plants serve as a density-dependent means of regulating hare populations, at least indirectly. While hare populations are low, the vegetation recovers, stimulating a resurgence of hare populations and initiating another cycle. It may well be a combination of limited food resources, climatic conditions, and predation—rather than any single phenomenon alone—that explains cycles in hare populations.

Some researchers feel that some cycles can be explained by another type of nutrient recovery, namely, seed production (Pitelka, 1964). Many northern plants have seed cycles of approximately 3 1/2 years. These plants require this time to build up sufficient nutrient material to produce seeds.

FIGURE 10.10

(a)

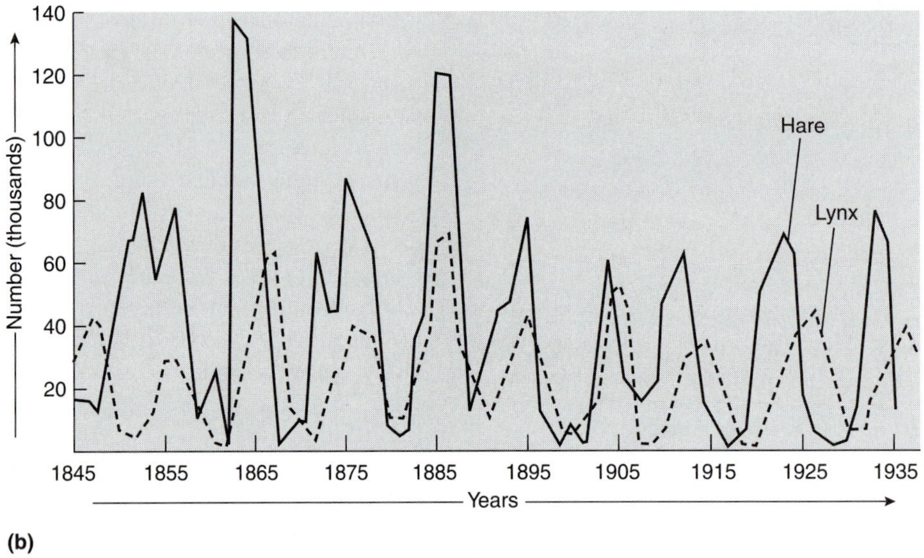

(b)

Population cycles for the snowshoe hare (*Lepus americanus*) and its major predator, the Canadian lynx (*Lynx canadensis*). The 9- to 10-year cycles are based on records of the Hudson Bay Company. Increases and decreases in the hare population are followed by corresponding increases and decreases in the lynx population. *Source: Data based on the number of lynx and snowshoe hare pelts received by the Hudson Bay Company in the years indicated.*

Such a cycle corresponds closely to the population cycles of many small mammals.

In northern Scandinavia, microtine rodent populations are cyclic. Interestingly, the production of willow ptarmigan (*Lagopus lagopus*) chicks is usually low in microtine rodent crash years. Although some plants in this region produce compounds (estrogens) that have negative effects on bird and mammal reproduction, it could not be found that intense herbivory from the microtine rodents induced production of plant estrogens in the spring food plants of the ptarmigan (Hanssen et al., 1991).

Gliwicz (1990) proposed that an important intrinsic factor regulating population diversity is a regular dispersal of first-born young of the year from their natal habitats. Population cycles in microtine rodents are an ecological consequence of this dispersal behavior, which normally results in annual cycles. Multiannual cycles occur only under certain sets of extrinsic factors. Low abundance of predators or sufficient snow cover to reduce predation is required for cycles to occur.

Cyclicity only occurs in communities with few predators (Hansson and Henttonen, 1988). High predation pressure normally prevents small rodents from population cycling by

FIGURE 10.11

Snowshoe hare (*Lepus americanus*) browsing intensively on an early successional shrub. As the density of hare populations increases, trees such as alder, poplar, black spruce, and birch become overbrowsed. The new shoots that emerge contain high concentrations of toxins, which result in a loss of weight and decline in health of the hares, causing them to be more susceptible to predation. As the hare populations decline, the vegetation recovers.

keeping their densities low, especially during winter and early spring (Hansson, 1979; Erlinge et al., 1983; Erlinge, 1987).

Although many possible explanations, including extraterrestrial events such as sunspot cycles, have been proposed to explain cycles (reviewed by Keith, 1963, and Roseberry and Klimstra, 1984), their most immediate and probable cause is the interaction of the population and its environment. In addition, some species may possess an inherent self-regulation (biological clock) that triggers the cyclic events. Cyclic trends in local populations are poorly documented primarily because intensive, long-term data for such populations are generally lacking.

Irruptions

Enormous numbers of animals sometimes occur in a given area for brief periods of time because of certain environmental and climatic conditions. Abnormally high numbers of a given species are known as irruptions and are of limited duration (Figs. 10.1b and 10.12). Conditions leading to irruptions may involve a combination of mild winters, an abundant food supply, and the removal of predators (as with Kaibab deer).

The greatest irruption ever recorded in North America involved the montane vole (*Microtus montanus*). This spectacular irruption occurred in 1906–1908 in Nevada and California (Piper, 1909). In some areas, estimated population density exceeded 25,000 voles/ha. Approximately 10,000 ha of alfalfa—stems, leaves, and roots—were destroyed in Humboldt County, Nevada. In 1957–58, a smaller, but more extensive, outbreak of montane voles occurred in California, Oregon, Washington, Idaho, Nevada, Utah, southwestern Montana, and western Wyoming (Spencer, 1959; White, 1965). Although most densities per hectare were in the hundreds, maximum population density in some areas was estimated to be between 5,000 and 7,500 individuals/ha. Predator populations (owls, hawks, gulls) increased (by immigration) and helped to lower vole densities. A high incidence of tularemia, caused by the bacterium *Pasturella tularensis*, was found among these animals during the subsequent decline and may have been a (density-dependent) factor in the population decline. Another factor in the decline of high populations may be the exhaustion of the adrenal gland (see the Endocrine System section in Chapter 9) brought on by stress caused by the high level of intraspecific interaction (Christian, 1950, 1959, 1963; Autrum and Holst, 1968; Holst, 1969). Stress stimulates the adrenal medulla to secrete epinephrine (adrenalin) and norepinephrine (noradrenalin). These adrenal secretions assist in preparing an animal's body for stressful situations by altering blood flow, adjusting heart and breathing rates, decreasing action of the digestive system, and so forth. Under conditions of continuous stress, this chemical control system may cease functioning.

FIGURE 10.12

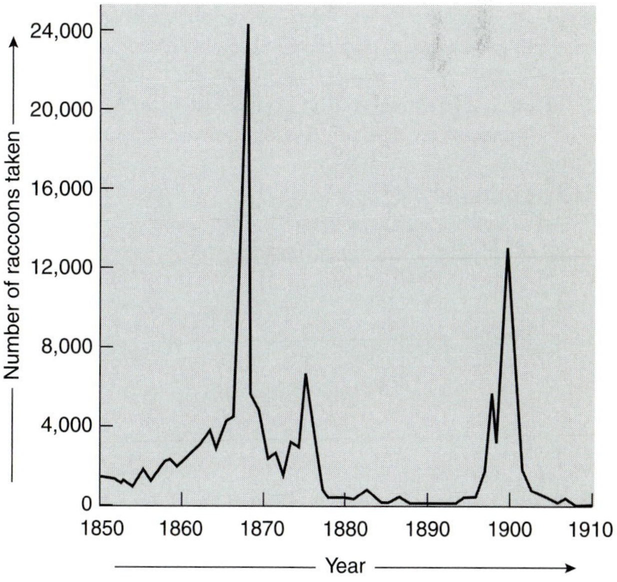

A 60-year record of raccoon pelts taken in the same general area and purchased by the Hudson Bay Company illustrates the irruptive changes in the population. These irruptive changes were caused by occasional improvements in the weather followed by returns to more normal conditions.

Source: Data from Miller, Resource Conservation and Management, *1990, Wadsworth, Inc., Belmont, CA.*

Review Questions

1. What defines the carrying capacity for a particular environment? How are carrying capacities for game species regulated by state wildlife agencies?
2. List several factors that affect carrying capacity.
3. How does carrying capacity relate to humans? List several ways in which humans have increased the carrying capacities for certain regions.
4. Differentiate between density-dependent and density-independent factors. Give examples of limiting factors that come into play when a population of mammals reaches very high density.
5. List some modern techniques that are being used to control exploding vertebrate populations.
6. The release of the rabbit hemorrhagic disease virus in Australia has been very controversial. If the virus should jump the species barrier, what steps would you undertake to keep it under control?
7. List several theories that have been proposed to explain cyclic fluctuations in mammals.
8. How do irruptions differ from cyclic fluctuations?

Supplemental Reading

Chitty, D. 1996. *Do Lemmings Commit Suicide?...* New York: Oxford University Press.

Gunderson, H. L. 1978. A mid-continental irruption of Canada lynx, 1962–1963. *Prairie Naturalist* 10:71–80.

Morris, D. W., Z. Abramsky, B. J. Fox, and M. R. Willig. 1989. *Patterns in the Structure of Mammalian Communities.* Lubbock: Texas Tech University Press.

Slobodkin, L. B. 1980. *Growth and Regulation of Animal Populations.* New York: Dover Publications.

Thompson, H. V., and C. M. King (eds.). 1994. *The European Rabbit: History and Biology of a Successful Colonizer.* New York: Oxford University Press.

Wildlife Conservation. May–June 1996 issue. Entire issue devoted to cats—bobcats, lynx, ocelots, jaguars, cougars, tigers, lions, cheetahs, etc.

Vertebrate Internet Sites

Visit the zoology website at http://www.mhhe.com to find live Internet links for each of the references listed below.

1. **Terrestrial Mammals of the Arctic.**
   Information from a text on Arctic mammals; data on 48 species from the book's appendix.
2. **The Lynx and the Hare.**
   An analysis of the interrelationship between lynx and hares in the Northwest Territory of Canada.
3. **Population Ecology.**
   This site provides on-line data, information from lecture courses, and the names of organizations, people, and journals involved in population ecology.
4. **Predator–Prey Dynamics.**
   Links to information from a class in population dynamics.
5. **Beklemishev.**
   Information on this famous Russian population biologist and his theories.

CHAPTER 11

Movements

INTRODUCTION

Vertebrates are mobile animals that move about to secure food, to locate suitable homes and nesting sites, to avoid unfavorable periods of the year, and to find mates. Some species move very little during their lifetimes, whereas others such as golden plovers (*Pluvialis dominica*) and elephant seals (*Mirounga angustirostris*) may cover over 20,000 km annually. Some movements are seasonal, or annual, whereas other movements occur only once in a lifetime. Orientation consists of two different phenomena: the control of an animal's position and stability in space, and the control of an animal's path through space (Wiltschko and Wiltschko, 1994). Movements undertaken by vertebrates can be categorized on the basis of where and when they occur—home range movements, dispersal, invasions, migration, homing, and emigration. Alternatively, movements can be classified by the mechanisms by which the movement is achieved—vision, hearing, olfaction, navigation, or compass orientation. Our understanding of the way in which animals know how, when, and where to orient and navigate around their environment has grown considerably over the last few decades.

HOME RANGE

Home range is highly variable and is often difficult to define. It is the area around the home of an individual that is covered by the animal in its normal activities of gathering food, mating, and caring for its young. Home ranges may be linear, two-dimensional, or three-dimensional.

Home range generally is correlated with the size of the animal. Small forms usually have relatively small home ranges, whereas larger species normally have larger home ranges. Among mammals of the same size, carnivorous species such as cougars (*Felis concolor*) generally have larger home ranges than herbivorous forms such as white-tailed deer (*Odocoileus virginianus*). A carnivore must expend considerably more energy and cover a much greater area in order to secure sufficient food. However, some small aerial species including bats, hummingbirds, and warblers cover great distances during their daily activities.

Other factors affecting home range size include habitat, population density, sex, age, body size, and season of the year. In polygynous and some monogamous species, males generally have larger home ranges than females; in polyandrous birds, however, the female's home range is larger (Blair, 1940d; Adams, 1959; Linzey, 1968). Very young and very old individuals of many species usually have the smallest home ranges. Animals living in marginal habitats generally need larger ranges than members of the same species living in better habitats. For example, Layne (1954) found that red squirrels (*Tamiasciurus hudsonicus*) living on the maintained portion of the Cornell University campus in central New York had an average home range of 2.0 to 2.5 hectare (ha), while red squirrels living in the more diverse and natural habitats of the nearby gorges had average home ranges of 0.12 to 0.16 ha.

Population density also may play a significant role in determining the home range, with the average size of the home range generally decreasing as population density increases. Linzey (1968) recorded an average home range of 0.26 ha for male golden mice (*Ochrotomys nuttalli*) and 0.24 ha for females over a 3-year period in the Great Smoky Mountains National Park. During a portion of this study, the population decreased drastically in size. During this period, the male home range more than doubled (0.63 ha), but the female home range, possibly because of nesting responsibilities and caring for young, remained approximately constant (0.21 ha). The density of large trees and possibly population density were factors that affected koala (*Phascolarctos cinereus*) home ranges in Australia (males, 1.0 ha; females, 1.18 ha) (Mitchell, 1991b). Some animals that live in northern regions, such as white-tailed deer (*Odocoileus virginianus*), have a larger home range during the warmer months of the year but live in small restricted areas, termed yards, during the winter months.

Few long-term home range studies exist. One such study of three-toed box turtles (*Terrapene carolina triunguis*) covered a period of 25 years. It revealed permanent home ranges varying from 2.2 to 10.6 ha in size for turtles known to have inhabited the study area for all 25 years (Schwartz and Schwartz,1991).

Home range figures are subject to a great deal of variation; therefore, these figures must be used with a great deal of caution. Many methods can be used to calculate the home range of a species; thus, results are somewhat subjective. Figure 11.1 illustrates three methods of calculating home range using the same capture sites. The minimum area method, calculated by computing the area within the actual capture sites, results in the smallest measured range. The boundary strip methods utilize a boundary strip that extends half the distance to each of the nearest traps around it. This method recognizes that even though an animal entered a particular trap, it probably also utilized some of the adjacent areas. The inclusive boundary strip method connects the outer points of the boundary strips, includes the greatest amount of area, and results in the largest home range estimate. The exclusive boundary strip method allows the investigator to utilize his or her judgment about unsuitable areas of habitat when drawing the perimeters of the home range. This home range value will be between the minimum area estimate and the inclusive boundary strip estimate. Though it is possible to gain an approximate idea of the size of the home range of a species, such statistics should not be accepted as absolute. Table 11.1 lists typical home ranges for selected vertebrates.

FIGURE 11.1

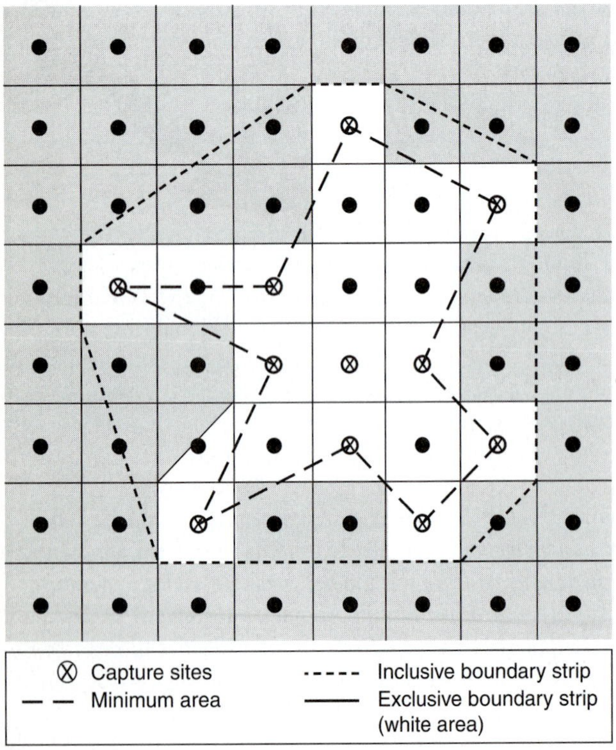

Three standard methods of calculating home range.

Under normal conditions, many animals have permanent ranges and spend their entire lifetimes within these areas. Most frogs, salamanders, lizards, turtles, snakes, moles, shrews, woodchucks, chipmunks, deer mice, and many others establish permanent home ranges. For example, after dispersing from their parental (natal) area, many lizards will remain in the same area throughout their lives. The home range generally will center around a favorable basking site or perch (Fig. 11.2). Migratory species such as sea turtles, many birds, elk, and caribou have seasonal home ranges. Their summer home ranges usually include the locations where they reproduce and care for their young, and their winter range is in a different area in order to allow them to survive adverse seasonal or climatic conditions.

Most home ranges are usually amorphous or amoeboid in shape. Some may be bounded by natural landmarks such as a river, whereas others are bounded by human-made structures such as roads or railroad tracks. Home ranges and even "core areas" (areas of high-intensity use) of several members of the same species often overlap. For example, giant pandas have ranges between 3.9 and 6.2 km^2 that may overlap extensively (Catton, 1990). Most pandas, especially females, tend to concentrate their activity within core areas of 0.3 to 0.4 km^2. Overlapping areas usually are not used at the same time; this helps to avoid conflict. However, in western North Carolina, neighboring black bears often use areas of overlap for the same activities (e.g., feeding, denning) and at the same time (Horner and Powell, 1990). In Alabama, adult home ranges of long-nosed (nine-banded) armadillos (*Dasypus novemcinctus*) overlapped extensively, and there was no indication of territorial or aggressive interactions (Breece and Dusi, 1985). Adults often were seen feeding within 3 m of each other, and on one occasion, three adults were seen leaving one den.

Home ranges often are marked by means of glandular secretions (pheromones), urine, or excrement. Ungulates, such as deer, use secretions from tarsal and metatarsal glands on their lower legs and orbital glands on their head to mark their home ranges. Tenrecs (*Echinops telfairi*) put saliva on the object to be marked and transfer their body odor by alternately scratching themselves with a foot and then rubbing the foot in the saliva. Galagos (*Galago* sp.) urinate on the palms of their hands and rub the urine into the soles of their feet. When climbing about, they leave obvious scent marks that also are visible as dark spots. Some mammals in which the anal glands are well developed, such as martens (*Martes*) and hyenas, use pheromones from anal glands to mark their home range. Gray squirrels (*Sciurus carolinensis*), fox squirrels (*S. niger*), and red squirrels (*Tamiasciurus hudsonicus*) use cheek-rubbing to deposit scent from glands in the oral–labial region (Benson, 1980; Koprowski, 1993). Rabbits use their pheromone-containing chin glands, urine, and feces for marking. Small mounds of fecal pellets indicate that an area is occupied.

TABLE 11.1

Home Ranges of Selected Vertebrates

Species	Home Range	Locality	Reference
Red salamander	6.1–41.2 m^2	California	Stebbins, 1954
Spadefoot toad	30.5 m^2	East North America	Pearson, 1955
Eastern box turtle	101–113 m^2	Maryland	Stickel, 1950
	2.2–10.6 ha^2	Missouri	Schwartz and Schwartz, 1991
Desert tortoise	4–41 ha^2	SW U.S.	Woodbury and Hardy, 1948
White-throated monitor lizard	5–16 km^2	S. Africa	Phillips, 1995
Red kangaroo	4.3–6.3 km^2	Australia	Dawson, 1995
Short-tailed shrew	0.41 ha	Michigan	Blair, 1940a
Meadow vole	0.08–0.21 ha	Michigan	Blair, 1940b
Deermouse	0.21 ha	Michigan	Blair, 1940c
Meadow jumping mouse	0.37 ha	Michigan	Blair, 1940d
Varying hare	7.5–10.1 ha	Montana	Adams, 1959
	5.9 ha	Alaska	O'Farrell, 1965
Red squirrel	1.3–1.5 ha	Saskatchewan	Davis, 1969
Armadillo	7.6–10.8 ha	Florida	Galbreath, 1983
	3.5 ha	Texas	Clark, 1951
	3.5 ha	Alabama	Breece and Dusi, 1985
Fisher	16.3–30.9 km^2	Maine	Arthur et al., 1989
	1,500–1,971 ha	New Hampshire	Kelly, 1977
Least weasel	1–15 ha	England	King, 1975
Cheetah	24–483 km^2	Tanzania	Caro, 1996
Asian lion	77–129 km^2	India	Chellam, 1996
Jaguar	3.2–39 km^2	Central America	Rabinowitz, 1996
Mountain lion	196–453 km^2	Idaho	Seidensticker et al., 1973

The "**home**" is within the home range and serves as a refuge from enemies and competitors. It may be in the form of an underground burrow, a cave, a tree cavity, a rotting log, an arboreal nest, or a brush pile. It may be the nest of a bird, the temporary "form" (nest) of a rabbit, or the more permanent burrow of a gopher tortoise (*Gopherus*) or woodchuck (*Marmota*). It may serve a single animal (cougar, *Felis concolor*), a pair of adults and their offspring (beaver, *Castor canadensis*), or a colony of animals (flying squirrels, *Glaucomys*; golden mice, *Ochrotomys*). Some species such as harvest mice (*Reithrodontomys*) have been shown to have a metabolic rate ranging from 7 percent to as much as 24 percent lower when in their nest than when they are active (Kaye, 1960).

Radio transmitters attached to subterranean naked mole rats (*Heterocephalus glaber*) revealed that the network of tunnels constructed by a colony currently comprising 87 animals was more than 3.0 km long and occupied an area greater than 100,000 m^2—about the size of 20 football fields (Sherman et al., 1992) (Fig. 11.3). Much of the tunneling to dig their vast network of tunnels is a cooperative effort to find food. One animal gnaws at soil, while others, in turn, transport it to a surface opening, where it is ejected by a larger colony mate.

Some vertebrates actively defend a portion of their home range. The defended area is known as the **territory** and contains the home or nest site. In general, an individual or a group of animals is considered to be territorial when it has exclusive use of an area or resource with respect to other members of its species and defends it in some way (either actively through aggression or passively through advertisement). Habitat quality, particularly the availability of food, can influence territorial behavior and territory size. Thus,

FIGURE 11.2

A chuckwalla (*Sauromalus obesus*) basking in the warm Arizona desert sun.

FIGURE 11.3

Naked mole rats (*Heterocephalus glaber*) live in a cooperative eusocial society. These subterranean mammals dig vast networks of tunnels—in some instances, more than 3.0 km long—to locate food.

FIGURE 11.4

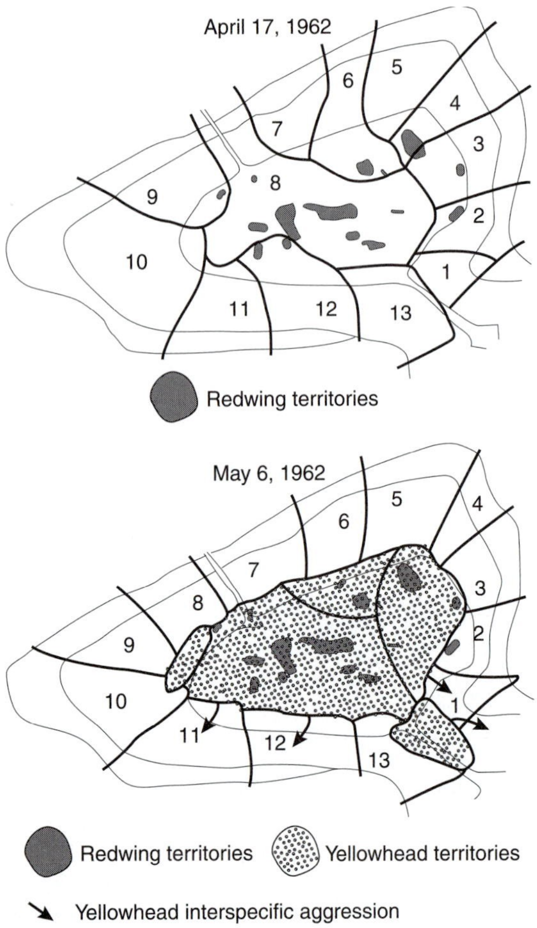

Interspecific territoriality between red-winged blackbirds (*Agelaius phoeniceus*) and yellow-headed blackbirds (*Xanthocephalus xanthocephalus*). Redwings that have established territories in the center of the marsh are evicted by the later-arriving yellowheads. Arrows indicate areas with intensive interspecific aggression.

optimal size may vary from year to year and from locality to locality (Smith, 1990).

The territory may be fixed in space, or it may be mobile as in bison (*Bison bison*), barren ground caribou (*Rangifer tarandus*), and swamp rabbits (*Sylvilagus aquaticus*), where a male may defend an area around an estrous female. Some male cichlid fishes occupy the same territory for as long as 18 months (Hert, 1992). Drifting territoriality has been reported in a red fox (*Vulpes vulpes*) population in England (Doncaster and Macdonald, 1991). Troops of howler monkeys (*Alouatta* spp.) have little or no area of exclusive use, but they do defend the place where they happen to be at a given time. During the breeding season, male northern fur seals (*Callorhinus ursinus*) come onto land, choose and defend a breeding area against other bulls, and then collect a harem within this area.

Territoriality is one of the most important behavioral traits affecting the spatial organization of animal populations and population dynamics. As a result of territorial behavior, some individuals are forced into suboptimal habitat, which reduces the relative fitness of these individuals (Smith, 1990) (Fig. 11.4). Territorial behavior may prevent overpopulation and overexploitation of the available habitat by ensuring a certain amount of living space or hiding places for an individual or a group of animals (Alcock, 1975; Smith, 1990). Territories may be defended by a single individual (Fig. 11.5), by a pair of adults, or by larger groups such as a flock of birds, a pack of wolves, or a troop of baboons or gorillas (Smith, 1990). Although defense is usually by the male, both male and female may share in defending the territory. In some cases, such as the American alligator, the female is the sole defender.

The defended territory is usually much smaller than the home range, although in a few species the territory and the home range may be equivalent. As the size of the territory increases, the cost of defending the territory increases (Smith, 1990). Many fishes, lizards, crocodilians, birds, and mammals, as well as some salamanders, will actively defend an area immediately around their nests and/or homes, particularly during the breeding season and, if they provide parental care, during the time they are caring for their young. Many colonial birds nest just out of range of pecking distance of their neighbors (Fig. 11.6). Both male and female red-backed salamanders (*Plethodon cinereus*) mark their substrates and fecal pellets with pheromones (Jaeger and Gergits, 1979; Jaeger et al., 1986; Horne and Jaeger, 1988) and defend these feeding territories (Jaeger et al., 1982; Horne, 1988; Mathis, 1989, 1990a). Territoriality may affect the mating success of males, because territorial quality has been found to be positively correlated with body size in *Plethodon cinereus* (Mathis, 1990b, 1991a, b).

FIGURE 11.5

The striking wing pattern of the willet (*Catoptrophorus semipalmatus*) is important in advertising its territory and in defense.

FIGURE 11.6

Gannet (*Morus bassanus*) nesting colony. Note the precise spacing of nests so that each bird is just beyond the pecking distance of its neighbors.

Some anurans defend their territories, which may include feeding sites, calling sites, shelter, and oviposition sites. During the breeding season, for example, male bullfrogs (*Rana catesbeiana*) defend an area surrounding their calling site from other males. A resident frog floats high in the water with its head raised to display its yellow throat, and it calls frequently. Initial defensive behavior consists of a vocal challenge followed by an advance toward the intruder. This is followed by another vocal challenge and an advance of a few feet, and so on until the intruder leaves. If the intruder does not leave, the two frogs push and wrestle each other and grasp each other's pectoral regions, each attempting to throw the other on its back. As soon as one frog is forced onto its back, contact is broken and the winner begins calling again. After remaining submerged for several seconds, the loser usually swims away some distance under water before surfacing.

Little owls (*Athene noctua*) of Germany are a non-migrating, all-year territorial species (Finck, 1990); however, distinct seasonal changes in territory size and in intraspecific aggressiveness of males have been observed. Territories were largest during the courtship season (March and April) and averaged 28.1 ha. They reached their smallest size (average 1.6 ha) during July and August, when the fledglings were still being fed in the parents' territory. As the young began to disperse in September, territories again began to increase in size.

Little, if any, evidence of territoriality has been reported among turtles and snakes. A study of male snapping turtles in Ontario revealed they do not occupy a fixed, exclusive, defended area (Galbraith et al., 1987). They do, however, occupy relatively stable home ranges that overlap and whose spacing may in part be determined by aggressive interactions. Even in burrow-dwelling species such as desert tortoises (*Gopherus agassizi*) that rarely share summer holes, there is no evidence for the existence of defended territories.

In most species, territorial boundaries are marked in the same manner as the boundaries of the home range. For example, some salamanders, such as the red-backed salamander (*Plethodon cinereus*), produce fecal pellets that serve as pheromonal territorial markers (Jaeger and Gergis, 1979; Jaeger et al., 1986; Horne and Jaeger, 1988). Birds commonly use song and characteristic display behavior, whereas mammals use scents, urine, and excrement to mark the boundaries of their territories (Smith, 1990).

DISPERSAL/INVASION

Dispersal refers to the movement an animal makes from its point of origin (birthplace) to the place where it reproduces. This type of movement generally occurs just prior to sexual maturity and takes place in all vertebrate groups. Dispersal is significant for a number of reasons. It tends to promote outbreeding in the population; it permits range extension; it may contribute to the reinvasion of formerly occupied areas; and it tends to reduce intraspecific competition. Many of the gradual invasions made by vertebrate species into newly developed or previously occupied territories are the result of dispersal of the young and their selection of breeding territories for the first time.

In many species of vertebrates, dispersal is density-dependent. There is a tendency to move only if the population in a given area is high or if aggression is shown by the parents. This is the case with many amphibians, reptiles, and birds and is also true of mammals such as beavers (*Castor canadensis*), bears (*Ursus*), and many species of mice. Other species, such as spruce grouse, deer mice, voles, and chipmunks, tend to have an innate predisposition to travel away from their place of birth regardless of the density of the population. After reaching a certain age, members of these species tend to wander away in search of unoccupied areas (Fig. 11.7).

Five juvenile emperor penguins (*Aptenodytes forsteri*) were fitted with satellite transmitters and tracked for several weeks after leaving their place of birth at Cape Washington in Antarctica (Fig. 11.8) (Kooyman et al., 1996). The juveniles traveled beyond the Ross Sea, with one individual being recorded 2,845 km from Cape Washington when last located. The fact that juveniles engage in such extensive travels suggests that adequate protection against human disturbance is not being provided during all phases of the life cycle of this species. Of most concern is the impact of commercial fishing around the Antarctic continent.

Among mammals that live in groups, males usually disperse about the time they reach breeding age. Sometimes it is voluntary, but other times they are pushed out of the group by dominant, older males who prevent adolescents from mating with the group's available females. In other groups, both males and females leave their birthplace. In a few species, such as the African hunting dog (*Lycaon pictus*) and chimpanzees (*Pan troglodytes*), only the females leave the security of their home group and disperse. The dynamics of groups favoring female dispersal may be driven, in part, by the relative ages of dominant fathers and maturing daughters. In these groups, females that reach maturity while older-generation males still are breeding run a high risk of mating with their fathers or other close relatives. In these cases, it is genetically advantageous for them to leave in order to avoid inbreeding among closely related individuals.

FIGURE 11.8

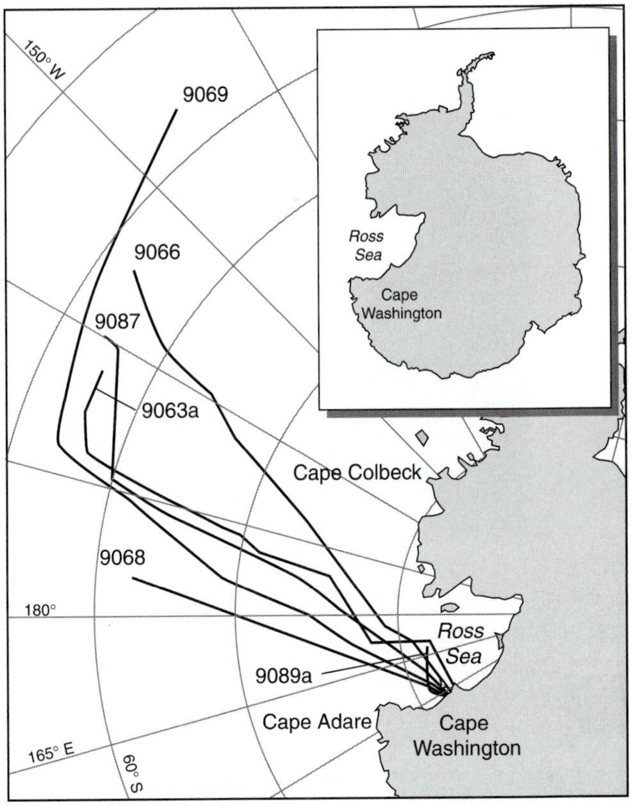

Routes of emperor penguin juveniles (*Aptenodytes fosteri*) obtained from satellite transmitters. From December 15–19, 1994 and 1995, the birds were captured and released near the ice edge of Cape Washington. Within a few hours of release, the birds entered the water. Positions were monitored from January 4, 1995, to March 6, 1996. During this time, all birds had reached positions far enough north to be in the Westwind Drift. Although researchers had expected signals to continue during June, the lack of signal suggests that the birds remained in water north of the pack ice.

FIGURE 11.7

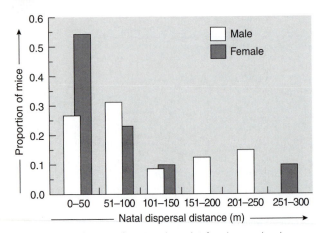

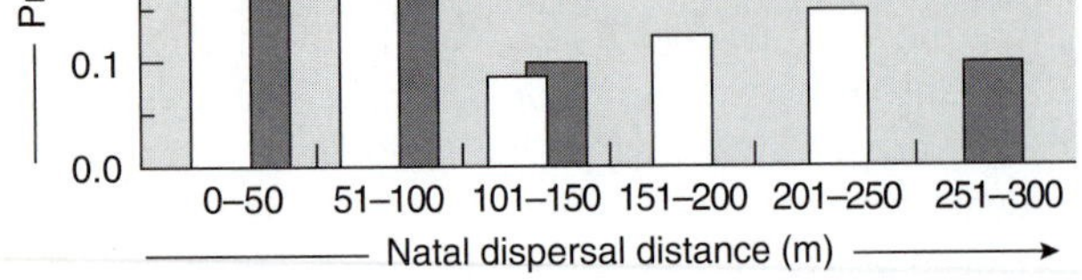

Natal dispersal distances for 22 male and 9 female juvenile white-footed mice (*Peromyscus leucopus*). This species has an innate predisposition to disperse regardless of the density of the population.
Source: Data from Keane, "Dispersal in White-footed Mice," Association for Study of Animal Behavior, 1990.

The **invasion** of the Great Lakes by the sea lamprey (*Petromyzon marinus*) was made possible by the completion, in 1829, of the Welland Canal, which bypassed Niagara Falls. Niagara Falls had served as a natural barrier to aquatic dispersal prior to this time. The lampreys reached Lake Huron in the 1930s and Lake Superior by the mid-1950s. This invasion of lampreys drastically reduced populations of lake trout, lake whitefish, and burbot in most of the Great Lakes. Control measures, including the release of sterile males and the use of a lampricide specific for ammocoete larvae, have allowed the prey species to partially recover and reach an equilibrium with the lampreys.

The rapid spread of the English sparrow (*Passer domesticus*) (Johnston and Selander, 1964) and starling (*Sturnus vulgaris*) (see Fig. 3.35) serve as excellent examples of dispersal/invasion, as does the northward and eastward expansion of the coyote (*Canis latrans*) (see Fig. 3.37). The gradual northward expansion of the range of the gray fox (*Urocyon cinereoargenteus*), opossum (*Didelphis virginiana*), and armadillo (*Dasypus novemcinctus*) are less dramatic examples. All of these movements have resulted in the expansion of the range of the individual species.

The cattle egret (*Bubulcus ibis*) (Fig. 11.9) is a native of Africa. It crossed the South Atlantic under its own power and was first recorded in Dutch Guiana (now Suriname) in 1877 (Line, 1995). By the late nineteenth century, it had become established on the northeastern coast of South America and, since that time, has dispersed rapidly to become one of the most abundant herons in the Americas. The distance from the bulge of West Africa to the northeastern coast of South America is approximately 2,870 km. Taking into consideration the prevailing trade winds, it is estimated that the trip would have required about 40 hours.

FIGURE 11.9

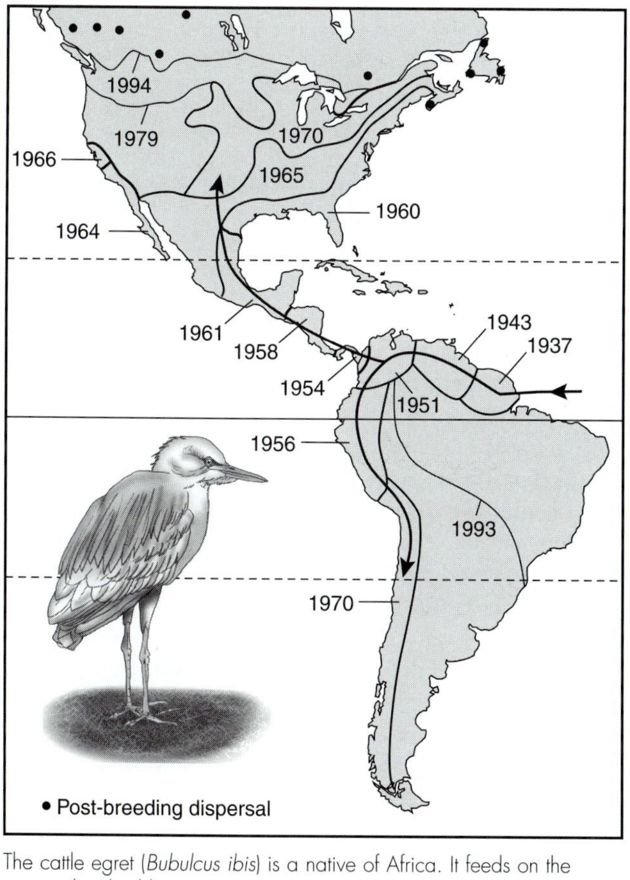

The cattle egret (*Bubulcus ibis*) is a native of Africa. It feeds on the insects disturbed by grazing ungulates. This species apparently crossed the South Atlantic Ocean from Africa under its own power and became established in northeastern South America by the late nineteenth century. It dispersed rapidly and is now one of the most widespread and abundant herons in the New World.

Even today, cattle egrets are routinely sighted at sea between Africa and South America.

MIGRATION

The periodic movement of a population or a part of a population of animals away from a region and their subsequent return to that *same* region is termed **migration.** Migration is a transfer of the home range to a distant region. Many animals travel either at regular times during the year or at a particular time during their lives. Some travel to avoid cold or hot weather, some to find a steady food supply, and others to move to breeding sites or to special places to produce their young. The length of the trip varies from species to species, with many traveling in large groups, whereas others travel alone. Migratory movements—which may be daily, seasonal, or irregular in occurrence—may cover short distances or many thousands of kilometers. They may occur annually, as is the case in many birds and mammals, or they may require a lifetime to complete, as is true of some salmon and freshwater eels. Daily movements commonly occur among fishes that move upward and downward in the water column. Such movements are generally in response to similar movements of zooplankton, although some upward and downward movements are associated with predator avoidance. Hammerhead sharks (*Sphyrna* spp.) in the Gulf of California engage in nightly round trips to feeding sites using magnetic undersea peaks as navigational centers (Klimley, 1995). Crows (*Corvus brachyrhynchos*) and starlings (*Sturnus vulgaris*) move from roosts to feeding areas and back each day.

Some vertebrates inhabit areas that have suitable living conditions during only part of the year. During the colder winter months, these species must either hibernate or migrate. Thus, migration permits a species to leave an area with unfavorable conditions during a period of the year, for one with more favorable conditions, even though this move is only temporary.

Microgeographic (Short-Distance) Migration

Some species migrate only short distances. This local, or **microgeographic**, migration is typical of some ambystomatid salamanders that migrate from their subterranean hibernacula to their breeding pond (Fig. 11.10). They remain active and above ground for several weeks before returning to their underground existence. Many anurans move to breeding ponds in the spring. Norway rats (*Rattus norvegicus*), house mice (*Mus musculus*), and some snakes may move from fields into barns during the winter and then return to the fields in the spring. Mule deer (*Odocoileus hemionus*) in the western mountains move from their summer ranges on north-facing slopes to wintering grounds on south-facing slopes (Taber and Dasmann, 1958).

FIGURE 11.10

Some ambystomid salamanders, such as (clockwise from top) the marbled salamander (*Ambystoma opacum*), tiger salamander (*A. tigrium*), Jefferson's salamander (*A. jeffersonianum*), and spotted salamander (*A. maculatum*).

FIGURE 11.11

Elk (*Cervus elaphus*) spend the summer months in high mountain meadows and descend into lower valleys during the winter months. This altitudinal migration may lower insect harassment, reduce the risk of predation, and enable the elk to take advantage of a more nutritious food supply.

Altitudinal Migration

Some species that live in mountainous regions move between higher and lower elevations in a kind of **altitudinal migration**. For example, many elk (*Cervus elaphus*) in the western United States spend the summer months in high mountain meadows and descend into lower valleys during the winter months (Fig. 11.11). These movements may be to reduce insect harassment (e.g., black flies, mosquitos), to seek more abundant or nutritious forage, and/or to lower the risk of predation. Carolina dark-eyed juncos (*Junco hyemalis carolinensis*) and the Carolina chickadee (*Parus carolinensis*) in the southern Appalachians of North Carolina, Tennessee, and Virginia migrate several thousand meters in elevation, whereas closely related subspecies and species such as the boreal slate-colored junco (*Junco hyemalis hyemalis*) and the black-capped chickadee (*Parus atricapillus*) migrate hundreds, sometimes thousands, of kilometers twice a year between their breeding and wintering areas. Altitudinal migrators face considerably fewer hazards and expend much less energy than long-distance migrators; thus, the survival value of this behavior is great.

Macrogeographic (Long-Distance) Migration

The best known migrators are the **macrogeographic**, or long-distance, migrators such as ducks, geese, swans, cranes, vireos, warblers, flycatchers, swallows, and thrushes. These species feed primarily on aquatic vegetation and/or flying insects, neither of which is available during the winter months in northern regions. Three hundred and thirty-two of the 650 (51%) North American migratory bird species spend from 6 to 9 months of the year in the tropics of the Americas, where they live under environmental conditions very different from those of their breeding grounds. Many of these migratory birds, especially waterfowl and shorebirds, use four major flyways in North America. From east to west, these are the Atlantic flyway, the Mississippi flyway, the Central flyway, and the Pacific flyway (Fig. 11.12).

These four migration flyways were originally proposed by U.S. Fish and Wildlife Service biologist Frederick Lincoln (1935), and many federal and state wildlife refuges have been established along the four routes. Although the concept of flyways is useful, especially for waterfowl and shorebird movements during fall migrations, it is an overly simple depiction of migration patterns of most other birds, particularly passerines. Studies by Bellrose (1968) and Richardson (1974, 1976) suggest that most species migrate over broad geographic fronts, particularly in spring, and do not follow narrow migratory corridors.

Furthermore, not all birds migrate north and south; some fly east and west. For example, Pacific populations of harlequin ducks (*Histrionicus histrionicus*) overwinter in the western coastal waters from northern California to Alaska (Turbak, 1997). In the spring, they fly eastward to nest along mountain streams in Alaska, Washington, Oregon, Montana, Idaho, Wyoming, Alberta, British Columbia, the Yukon, and the Northwest Territories. A few even cross the Continental Divide to nest. Harlequin society is matriarchal, with adult females returning salmonlike to their natal streams to reproduce. The Pacific population of harlequins is the only duck population in the world that divides its time between sea and mountains. A small eastern population breeds in maritime Canada and winters on the New England coast.

The migratory journeys of some species are astounding because of their length and/or duration. Adult Pacific salmon

(*Oncorhynchus*) of the North Pacific breed in freshwater streams or lakes, and the young migrate to the sea within the first 2 years of their lives. After 2 to 4 years at sea (during which time the salmon mature), they then travel back to the river system in which they were born. They swim upstream to the headwaters of rivers such as the Columbia and Yukon, where they will spawn and die. Some of these fish will have covered several thousand kilometers during their migratory travels.

One population of the Atlantic green turtle (*Chelonia mydas*) (Fig. 11.13) nests on Ascension Island in the South Atlantic Ocean (Bowen et al., 1989). After depositing their eggs, females return to the warm shallow waters off the coast of Brazil, a distance of over 1,600 km. After feeding on marine vegetation for several years, they return to the same beach to lay another clutch of eggs.

Recent studies analyzing mitochondrial DNA (mtDNA) from eggs and hatchlings at four green turtle breeding sites in the Atlantic and Caribbean—Florida, Costa Rica, Venezuela, and Ascension Island—have revealed slight differences in their genetic sequences; this may complicate efforts to preserve this endangered species, because each subgroup could be unique and irreplaceable (Bowen et al., 1989; Meylan et al., 1990). This finding lends credence to the natal homing theory, proposed in the 1960s, which holds that, while turtles hatched in different regions may share common feeding grounds away from home, the animals part company at breeding time, each swimming hundreds or thousands of kilometers to breed and nest at their own (natal) birthplace. Female leatherback turtles (*Dermochelys coriacea*) appear to travel along migration corridors leading southwest from their nesting sites in Costa Rica (Morreale et al., 1996) (Fig. 11.14). Travel distances up to 2,700 km have been recorded.

FIGURE 11.12

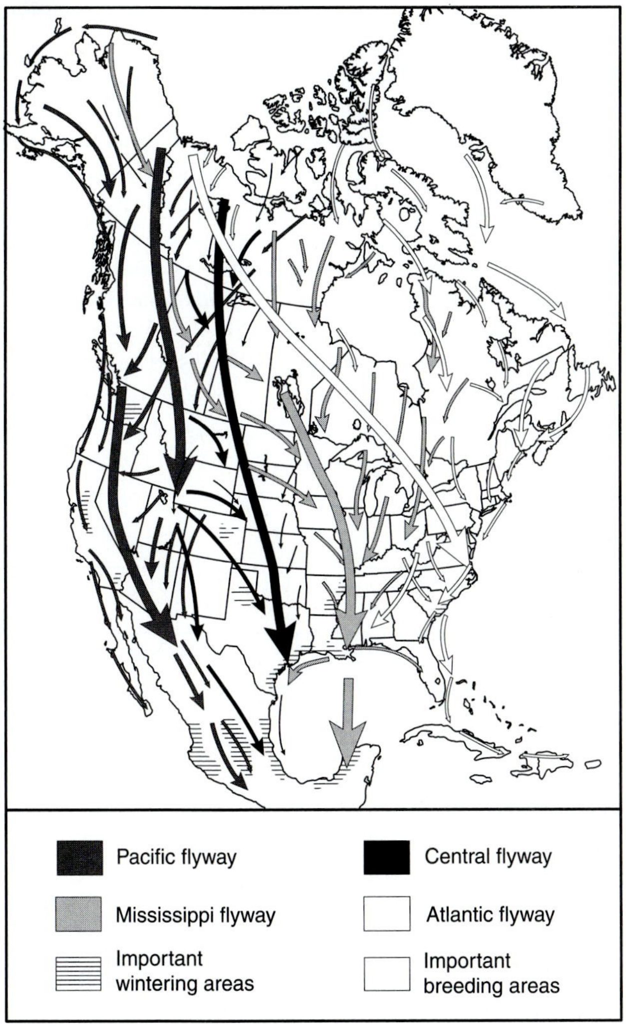

The four major flyways used by migratory birds in North America: the Atlantic, Mississippi, Central, and Pacific flyways.
Source: Data from Miller, Resource Conservation and Management, *1990, Wadsworth Publishing.*

FIGURE 11.13

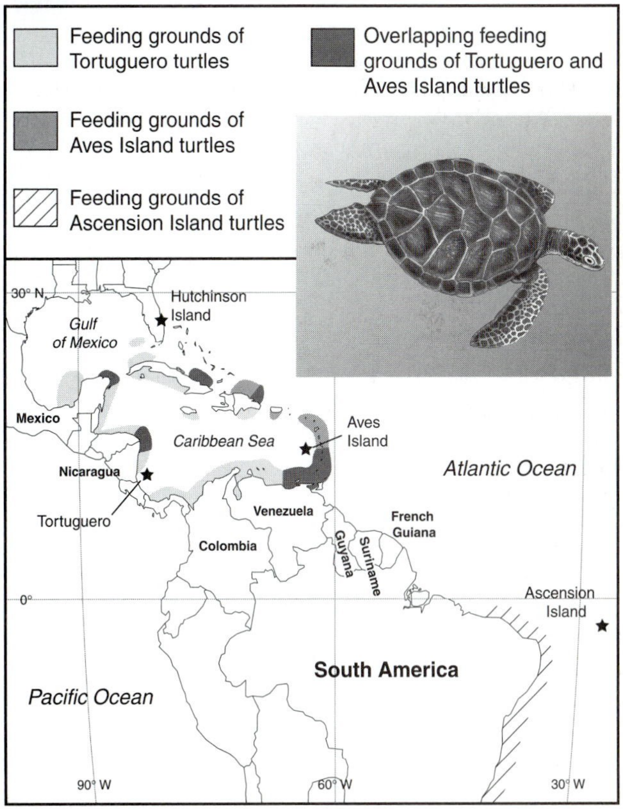

The Atlantic green turtle (*Chelonia mydas*) (inset) nests on Ascension Island in the South Atlantic Ocean but lives most of its life in the warm shallow waters off the coast of Brazil. The foraging grounds in the Caribbean and West Central Atlantic Ocean are used by green turtles that nest at three of the four surveyed rookeries (the foraging grounds of the Florida colony are unknown).
Source: Data from A. B. Meylan, et al., "A Genetic Test of the Natal Homing Versus Social Facilitation Models for Green Turtle Migration" in Science, *248(4956):724–727, May 11, 1990.*

Between 1978 and 1988, scientists collected more than 22,000 eggs of Kemp's ridley sea turtles (*Lepidochelys kempii*) from the species' only known nesting colony at Rancho Nuevo, a Mexican beach 160 km south of Brownsville, Texas. The young were released at Padre Island, Texas, in hopes that the turtles would imprint on the Texas site and return when they reached maturity in 10 to 15 years (Kaiser, 1996). Two turtles returned and nested in 1996. In addition, in May, 1996, a Kemp's ridley turtle, originally tagged in the Chesapeake Bay near the mouth of the Potomac River in 1989, was found on the beach at Rancho Nuevo. This is the first known Kemp's ridley from the Atlantic Ocean to return to the turtles' ancestral nesting ground.

BIO-NOTE 11.1

A Lengthy Turtle Trek

An estimated 10,000 juvenile loggerhead turtles (*Caretta caretta*) feed and develop off the coast of Baja California annually. The nearest known nesting sites, however, lie in Japan and Australia, some 10,000 km away. Mitochondrial DNA samples from Baja turtles, and from another group caught by North Pacific fishermen, revealed that 95 percent (of both groups) carried the same distinctive genetic sequences as the baby turtles in Japan, while the remainder matched the DNA markers of the Australian turtles. If additional data support these findings, the 10,000-km trek to Baja—a distance spanning more than one-third of the globe—would rank among the longest documented marine vertebrate migrations.

Bowen, 1995

The golden plover (*Pluvialis dominica*) breeds in the Arctic and winters in southeastern South America. It is estimated that these birds cover a distance of 25,000 to 29,000 km annually. The Alaskan population of the wheatear (*Oenanthe oenanthe*), which winters in southeastern Africa, can make annual journeys of about 30,000 km (Kiepenheuer, 1984). The champion migrator, however, is the Arctic tern (*Sterna paradisaea*), whose annual round-trip journey from its Arctic breeding grounds near the North Pole to its winter quarters in Antarctica may cover up to 50,000 km per year (Berthold, 1998) (Fig. 11.15).

Only three Southern Hemisphere birds—Wilson's petrel, the sooty shearwater, and the great shearwater—migrate north in large numbers to spend their winters in the Northern Hemisphere, in contrast to the hundreds that go south during our winter. Wilson's petrel (*Oceanites oceanicus*), for example, breeds in the Antarctic and may be found as far north as Labrador, a distance of approximately 11,250 km.

Many neotropical migrants, such as warblers, thrushes, bobolinks, tanagers, orioles, and hummingbirds, fly nonstop some 1,000 km over the Gulf of Mexico from the Gulf Coast of North America to Central America, a journey requiring about 20 hours. Blackpoll warblers (*Dendroica striata*) use a trans-Atlantic route in the fall, but an overland route in the spring (Fig. 11.16). On their southward journey, some use the islands of Bermuda as a resting stop, whereas others fly nonstop from New England to South America, a journey requiring approximately 100 hours of continuous flight time. Many migrants carry at least a 40 percent fat load, which serves as their source of energy for this strenuous journey (Alerstam, 1990). The ruby-throated hummingbird (*Archilochus colubris*) breeds from the Gulf of St. Lawrence and Saskatchewan to the Gulf of Mexico. It normally weighs no more than 2.5 g, but increases its weight with at least 2 g of fat before migrating over the sea (Alerstam, 1990). The rufous hummingbird (*Selasphorus rufus*) of western North America breeds from northern Cal-

FIGURE 11.14

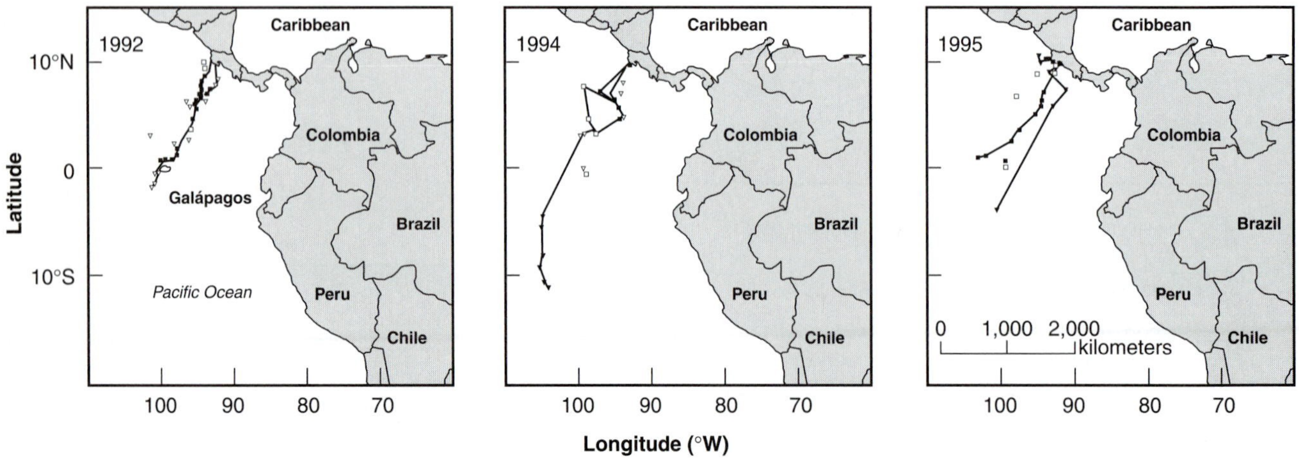

Migratory movements of eight leatherback turtles (*Dermochelys coriacea*) monitored by satellite transmitter after nesting near Playa Grande, Costa Rica. The Cocos Ridge runs beneath the first 1,500 km of the migration corridor, extending out to the Galapagos Islands. Four turtles were tracked as they passed the Galapagos and continued beyond the ridge into deeper Pacific waters.

FIGURE 11.15

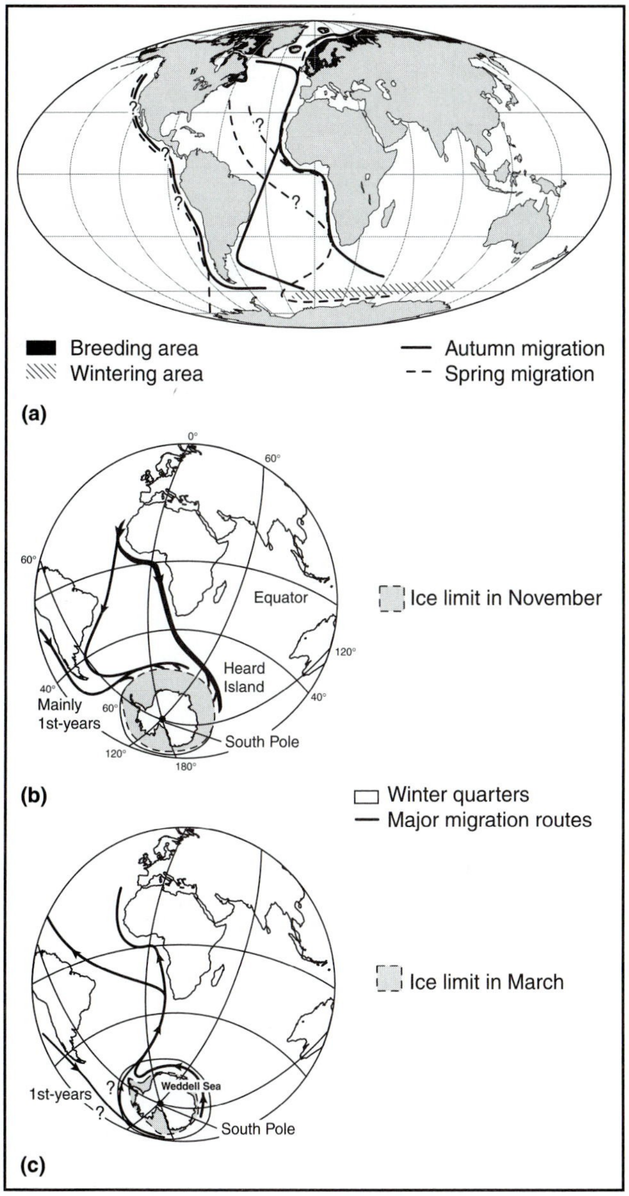

Distribution and migration routes of the Arctic tern (*Sterna paradisaea*). (*a*) Overall routes, with question marks indicating portions of migration pathways that have not been proved. (*b*) The migration routes to wintering areas along the edge of the pack ice in the Antarctic Ocean. Most terns arrive via Africa's western coast, but a significant number migrate along the coasts of South America. Most terns arrive in Antarctica during November and December. (*c*) The migration routes from the wintering areas in March. Adults and intermediate-age terns migrate mainly westward off the Antarctic coast before continuing north in the Atlantic. There are indications that first-years in particular migrate around the South Pole to summer on the Humboldt Current.

ifornia, Oregon, Idaho, and Washington northwestward through British Columbia to the southeastern Alaskan coast, and it winters in Mexico. Thus, this 9-cm-long bird may migrate a total of 3,000 km each way between its summering and wintering grounds (Diamond, 1990b).

FIGURE 11.16

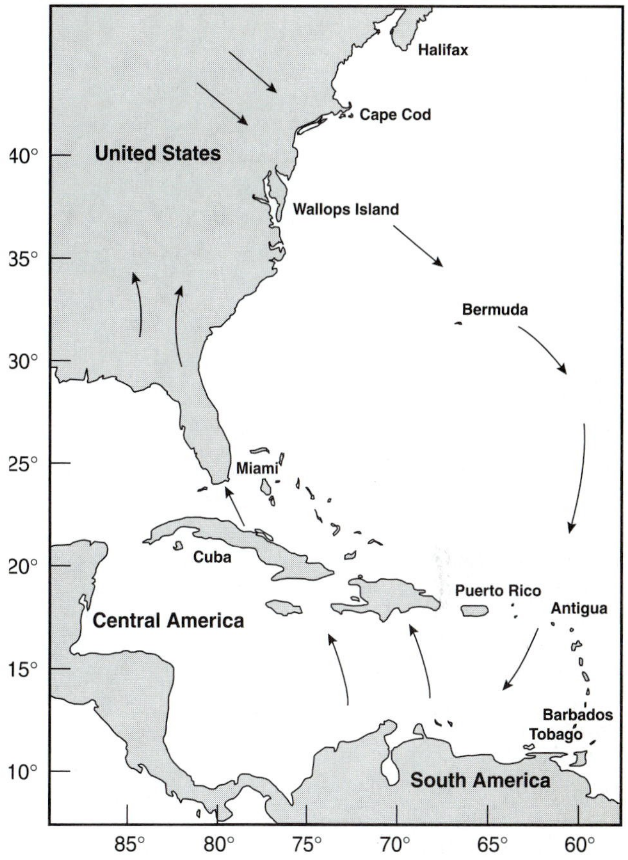

Migratory routes used by the blackpoll warbler (*Dendroica striata*). This species uses a trans-Atlantic route in the fall, but an overland route in the spring.

In spite of these very long migratory distances, birds can perform accurate and direct migratory movements and are capable of regularly commuting between small breeding sites and wintering places or of returning to remote tiny islands even after several years. Banding recoveries have revealed that some species migrate along narrow migration corridors and give evidence of remarkable breeding site, stopover area, and wintering site fidelity (Zink, 1973–1985). Satellite tracking has allowed us to establish the directedness of migratory journeys in actual migrating individuals (Nowak and Berthold, 1991).

In recent years, some populations of migratory species such as Canada geese appear to have lost their migratory instincts. Geese that should be nesting in the subarctic expanses of northern Quebec are laying eggs in New York, New Jersey, Pennsylvania, Virginia, and adjacent states. Flocks that should be overwintering in the Carolinas are staying in the colder Northeast. Some geese are living their entire lives within a few miles of where they hatched. In Virginia, for example, the resident Canada goose population is estimated at 150,000 to 200,000 birds and is growing at a rate of 15 percent annually (Cochran, 1996). Virginia agricultural

BIO-NOTE 11.2

How Hummingbirds Store the Energy to Migrate

Hummingbirds, which are among the smallest endothermic vertebrates, have a high metabolic rate. Flying hummingbirds must fuel their high rate of aerobic metabolism, and migratory hummingbirds have the additional problem of building up large fat deposits to meet the energy demand for their long migratory flights. They rely on different stores for the two activities. Fat stores for migration are built up by synthesizing fatty acids from sugar ingested in nectar. Burning the sugar during foraging spares fat stores and avoids the inefficiency of synthesizing fatty acid from glucose and then burning the fat. Migrating hummingbirds must burn fat because of its higher calorie yield per gram and because of the need to save weight in flight. To support such record-high metabolic rates, hummingbirds far surpass all other animals in their rates of intestinal glucose absorption, hepatic fatty acid biosynthetic capacity (over ten times that of mammals), and muscle levels of the enzymes hexokinase and carnitine palmitoyltransferase.

Suarez et al., 1990

BIO-NOTE 11.3

The Migratory Diet

Many species of migratory songbirds switch from an insect-rich diet to one based on fruit during their autumn migration. In some, such as hermit thrushes (*Hylocichla guttata*), red-eyed vireos (*Vireo olivaceus*), and yellow-rumped warblers (*Dendroica coronata*), fruit makes up 80 percent of their autumn diet. A few, such as American redstarts (*Setophaga ruticilla*), continue to feed exclusively on insects, even though insect eaters put on less weight than fruit eaters. The bigger the proportion of fruit in the diet, the greater the weight gain.

Parrish (in press)

officials estimated goose damage to farm crops at $236,000 for 1996. To reduce the numbers of resident birds, state hunting seasons must be set for times when migratory birds are still in their northern breeding areas and have not yet begun to migrate.

Part of the change in migratory habits is attributable to the establishment of wildlife refuges and preserves providing winter food and protection. Large concentrations of birds increase the risk of human confrontations as well as increasing the potential risk of density-dependent factors such as the rapid spread of avian parasites and disease organisms.

Migratory populations of geese that nest in Canada and fly southward to spend the winter are in an alarming decline. All goose hunting seasons along the Atlantic coast—including Canada—were closed in 1996. A 6-year study of the Canada goose population in eastern North America is being coordinated by the U.S. Fish and Wildlife Service.

Among mammals, elephant seals and whales migrate the longest distances annually. Northern elephant seals (*Mirounga angustirostris*) make two migrations each year between the Channel Islands off the coast of Southern California and Mexico's Baja California northward to the Gulf of Alaska and the waters near the Aleutian Islands (Stewart and DeLong, 1995) (See Fig. 9.34d). The first migration starts at the end of the breeding season in early to mid-February. Females travel to the North Pacific near the Aleutian Islands and back, a round trip of about 6,440 km lasting an average of 73 days. Males leave the California–Mexico area in late February to early March and swim to the Gulf of Alaska, approximately 2,730 km north of the females. The male migration lasts about 120 days and covers approximately 12,000 km. The second female migration begins in May and lasts for approximately 234 days. During this time the females remain at sea and cover a total distance of at least 12,200 km, before returning to their home off the California coast to give birth and mate again. The second male migration begins in late August or early September, lasts for about 126 days, and covers between 9,650 and 11,250 km. Thus, the 19,000 to 22,000 km annual journeys place elephant seals near the top of the list of long-distance mammalian travelers. No other vertebrate is known to undertake such a double migration.

Gray whales (*Eschrichtius robustus*) migrate some 9,600 km from the Bering and Chukchi seas to Baja California. From May to October, the whales inhabit the waters in the Bering Sea region. They begin swimming south in October, arriving off the California coast in mid-December, where they swim slowly and stay near the shore. The whale's migration, which is viewed by millions of people each year, has been called one of the world's outstanding wildlife spectacles. The northward migration begins in mid-February. The population was estimated to have dipped below 2,000 just before the turn of the century, primarily because of excessive whaling. The current population size, which is estimated to be about 21,000 animals, allowed this species to be removed from the endangered species list in June 1994.

Blue, fin, and humpback whales may travel more than 8,000 km during their annual migratory journeys. Five feeding aggregations of humpback whales (*Megaptera novaeangliae*) exist in the North Atlantic Ocean from Maine to Iceland and Denmark. In the fall, whales from all feeding aggregations migrate to breed within the nearshore areas and banks of the West Indies near the Dominican Republic, Puerto Rico, and the Virgin Islands. More than 1,000 photographically verified resightings of individually known humpbacks demonstrate that individual whales return in the spring to their particular feeding region. Acoustic monitoring of humpback whale songs has provided useful information concerning their migratory routes (Clapham and Mattila, 1990).

Stone et al. (1990) reported a humpback whale that migrated from the Antarctic peninsula region to Colombia in South America, a minimum of 8,334 km. It was the first time that a humpback whale had been known to cross the Equator, and the first time an Atlantic humpback had been documented in South American waters.

BIO-NOTE 11.4

A Migration Surprise

Until recently, it was thought that most of the humpback whales along Australia's eastern coast migrated each year from their feeding grounds in Antarctica to breed in tropical waters. When the migratory groups were surveyed by Peter Corkeron of James Cook University in Queensland and Miranda Brown of the University of Cape Town in South Africa, the sex ratio turned out to be 2.4 males : 1 female. Apparently, young females may spend all year in the Antarctic and migrate north only when they are sexually mature. The authors suggest that it may be energetically less taxing for a young whale to remain in cold waters year-round than to migrate to warmer waters. The results of this survey indicate that population estimates need to be increased, because only 60 percent of the population is being counted.

Brown et al., 1995

Northern fur seals (*Callorhinus ursinus*), Steller sea lions (*Eumetopias jubata*), and several other pinnipeds migrate several thousand kilometers annually. Some bats hibernate in northern areas, but others are well-known migrators. The red bat (*Lasiurus borealis*), hoary bat (*Lasiurus cinereus*), and the silver-haired bat (*Lasionycteris noctivagans*) regularly migrate 1,300 to 1,600 km. The two lasiurine bats (red and hoary) have been taken in Bermuda, a destination that required a nonstop flight of at least 950 km (Orr, 1982). A Brazilian free-tailed bat (*Tadarida brasiliensis*) banded at Carlsbad, New Mexico, and recovered in Jalisco, Mexico, had a minimum recorded movement of at least 1,340 km in 69 days (Villa and Cockrum, 1962).

Among North American ungulates, caribou (*Rangifer tarandus*) make the longest migratory journeys, traveling as much as 300 to 650 km annually. They spend the summer on the tundra north of the timberline; however, beginning in July, they move south to the taiga and the edge of the tundra. Vast herds of African gazelles, wildebeests, gnus, and many others travel many kilometers annually in search of food and water.

Navigational Cues Used in Migration

Mechanisms possessed by species such as salmon and birds that enable them to return to precise locations from distances of hundreds and even thousands of kilometers have been debated for centuries. Olfaction, celestial cues, sound, vision, and detection of the earth's magnetic field appear to be involved in different species. Like other animals, birds are equipped with two types of biological or internal clocks, which play an important part in the spatiotemporal and physiological organization of migration as well as in other processes: circadian, that is, endogenous diurnal rhythms, and circannual, that is, endogenous annual rhythms or so-called internal calendars (Gwinner, 1986, 1996; Berthold, 1996). Circadian rhythms are involved in such functions as the seasonal expression of migratory activity and fat deposition during migration, in mechanisms measuring daylength, and in synchronizing processes to the changing ecological requirements of the calendar year. Circannual rhythms seem to fix not only the dates of onset and termination of the first migratory period in naive first-time migrants, but also the course of migration and the approximate distance to be covered.

BIO-NOTE 11.5

A Wandering Manatee

A male Florida manatee (*Trichechus manatus*), captured in Chesapeake Bay in the fall of 1994, was fitted with a radio transmitter and airlifted back to Florida to save it from cold weather. During the summer of 1995, "Chessie" became the first known Florida manatee to swim to New England and back to Florida. He got as far as Point Judith, Rhode Island (a distance of approximately 3,220 km north of Fort Lauderdale) on August 16 before turning around. Satellite readings from his radio transmitter in July 1996 indicated that he had crossed the North Carolina–South Carolina border and was again headed north at an estimated speed of approximately 40 km a day.

Anonymous, 1996e

Some young birds migrate south several weeks after adults have left breeding areas. It is now known that these species have an internal annual rhythm, preprogrammed for the migratory journey with regard to both distance and direction (Able and Able, 1995; Weindler et al., 1996). The migratory bird follows this inborn instinct, which serves as an automatic-pilot system.

Other migratory bird populations have preprogrammed migratory directions (Berthold et al., 1990b., 1992; Helbig, 1991; Helbig et al., 1994). Thus, the primary directions in migratory birds such as blackcaps (*Sylvia atricapilla*) are genetically determined.

The possibility that animals could sense the Earth's magnetic field has been under investigation for over 100 years (Viguier, 1882; review by Gould, 1980; Able and Able, 1995). Studies during the 1970s suggested that birds possibly could sense low-intensity alternating-current electromagnetic fields and that they also could sense natural *fluctuations* in the Earth's magnetic field. Permanently magnetic iron oxide (probably magnetite, Fe_3O_4) has been found concentrated in the head and neck muscles of both migratory and nonmigratory birds, including western grebes (*Aechmophorus occidentalis*), pintails (*Anas acutus*), pigeons (*Columba livia*), white-crowned sparrows (*Zonotrichia leucophrys*), Savannah sparrows (*Passerculus sandwichensis*), northern bobwhites (*Colinus virginianus*), chimney swifts (*Chaetura pelagica*), cliff swallows (*Hirundo pyrrhonota*), tree swallows (*Tachycineta bicolor*), Carolina wrens (*Thryothorus ludovicianus*), European starlings (*Sturnus vulgaris*), red-winged blackbirds (*Agelaius phoniceus*), brown-headed cowbirds (*Molothrus ater*), house sparrows (*Passer domesticus*), and European robins (*Erithacus rubecula*) (Edwards et al., 1992). This magnetic material is thought to be coupled to a magnetic-sensitive muscle receptor such as a muscle spindle.

Edmonds (1992) stated: "There are at least three independent magnetic field detectors in the bird's head, one associated with the pineal gland which helps control melatonin secretions, one in the eye which requires light to function, and a third of high sensitivity for which the transducer is unknown but a magnetite-based detector is suspected." A pigeon's optic nerves may respond to differences in the angles of magnetic fields, and magnetic crystals bound to cranial nerves may detect field strength. Pigeons flying under an overcast sky with magnets attached to their bodies become disoriented, whereas control birds affixed with brass bars of the same weight show no disorientation. Thus, the ability to perceive magnetic fields could provide birds with both a map and a compass sense. So far, no differences in the sensitivity to magnetism have been found between migratory and nonmigratory birds.

Researchers have long thought birds navigate by the stars and use the Earth's magnetic field as a backup on cloudy nights. However, Weindler et al. (1996) showed that young garden warblers (*Sylvia borin*) from Germany need information from both the magnetic field and the stars to choose the right heading on their first migration. Their migratory route is southwest for several weeks, then southeast. This route avoids the Alps, the Mediterranean, and most of the Sahara Desert. Birds raised under an "artificial" sky with only celestial cues flew due south.

Polarized light can be used by some day-migrating species, such as yellow-faced honeyeaters (*Lichenostomus chrysops*), when no other known cues are available (Munro and Wiltschko, 1995). However, in the hierarchy of cues of this species, the polarization pattern clearly ranks lower than information from the geomagnetic field.

Magnetic material also has been found in the brain of loggerhead turtles (*Caretta caretta*) and in the blubber, bone, muscle, and brain of several cetaceans including the Atlantic bottlenosed dolphin (*Tursiops truncatus*), goosebeak whale (*Ziphius cavirostris*), Dall porpoise (*Phocoenoides dalli*), and humpback whale (*Megaptera novaeangliae*) (Klinowska, 1986). Magnetic crystals also have been found in the dura mater of the common dolphin (*Delphinus delphis*) (Zoeger et al., 1981).

Olfaction plays a major role in some species. Salmon can identify their home streams by chemical cues that are present in the stream water (Hasler and Wisby, 1951; Hasler and Scholz, 1983; Brannon and Quinn, 1990) or by pheromones released into the water by resident juveniles (Nordeng, 1971, 1977). Juvenile salmonids are thought to imprint on particular odors specific to home stream waters (Harden-Jones, 1968), which they use as cues when returning as adults. Both olfactory and gustatory receptors are extremely sensitive in fishes, and both may be employed by anadromous species (Keefe and Winn, 1991). In contrast, white bass (*Roccus chrysops*) migrate by means of sun-compass orientation (Hasler et al., 1958).

The newt *Taricha* also uses olfactory mechanisms, as apparently do many other amphibians. Vocalizations by male frogs and toads are especially important in attracting females to breeding ponds. Amphibians also use visual cues and can orient by using solar, lunar, and stellar cues in conjunction with an internal clock mechanism (Ferguson and Landreth, 1966; Ferguson, 1967).

Leatherback (*Dermochelys coriacea*) (Fig. 11.17) and green (*Chelonia mydas*) sea turtle hatchlings emerging from underground nests on oceanic beaches immediately are confronted by two separate problems in orientation. First, hatchlings use their vision to seek out bright, open horizons to find the sea (Lohmann et al., 1990). Both species orient toward violet and blue-green wavelengths (360–500 nm) as opposed to yellow-orange and red light (600–700 nm). Some level of spectral quality assessment must exist for both species, with possibly heightened sensitivity in the short wavelengths being an adaptation for vision in sea water (Witherington and Bjorndal, 1991). Once they reach water, they must orient *into* the waves,

FIGURE 11.17

A leatherback sea turtle (*Dermochelys coriacea*) instinctively makes its way to the sea upon hatching. It seeks out bright, open horizons and orients toward violet and blue-green wavelengths.

swimming toward approaching waves and oceanic swells. Thus, these turtles employ two separate orientation systems, each based on different cues. The sea-finding orientation is not a prerequisite for wave orientation. Although both land and sea orientation systems can function independently, they may both be operational and interact under natural conditions (Lohmann et al., 1990).

Hatchling loggerhead turtles (*Caretta caretta*) use wave direction as their primary cue during the early phases of offshore migration (Salmon and Lohmann, 1989). In addition, particles of the magnetic mineral magnetite have been found in their brains. This allows these turtles to carry a remarkably sophisticated magnetic compass in their heads, enabling them to sense the magnetic field of the Earth and, further, to distinguish between different field intensities found along their migratory route, and orient themselves to them (Lohmann, 1991, 1992; Lohmann and Lohmann 1992, 1993, 1994a, b, 1996a, b). The compass, which also enables them to sense how far north or south they have traveled, gives them the minimal sensory abilities necessary to approximate global position (latitude and longitude) using a bicoordinate magnetic map. Turtles use this magnetic sense along with two other directional cues—light reflecting off the ocean surface and wave motion—to orient themselves in the world. Some researchers also feel that olfaction is another major factor in turtle orientation (Seachrist, 1994).

The annual return migration between summer feeding ranges and winter denning sites is a common behavior in many populations of northern temperate zone snakes. Plains garter snakes (*Thamnophis radix*) use solar cues as an orientation guide (Lawson and Secoy, 1991). The use of solar and other celestial cues also has been demonstrated in rattlesnakes (*Crotalus atrox*), water snakes (*Nerodia sipedon* and *Regina septemvittata*), alligators (*Alligator mississippiensis*), and some turtles (common box turtle, *Terrapene carolina*; spiny softshell turtle, *Apalone* [*Trionyx*] *spinifera*; and the painted turtle, *Chrysemys picta*) (Landreth, 1973; Newcomer et al., 1974; Murphy, 1981; deRosa and Taylor, 1982).

Birds may use visual and memory cues of familiar landmarks. In some species, the young may learn migratory routes from older birds. In others, however, the young leave nesting areas after the adults. In still others, the birds are either solitary migrants, or they migrate at night when terrestrial landmarks would be of little use except for celestial cues. It is now known that migratory bird populations have preprogrammed migratory directions. Thus, the primary directions in migratory birds such as blackcaps (*Sylvia atricapilla*) are genetically determined (Berthold et al., 1990, 1992; Helbig, 1991; Helbig et al., 1994). Migrating birds also rely on interacting compass senses—magnetic, the position of celestial bodies (sun, moon, stars), and polarized light—as major aids in navigation (Wiltschko et al., 1996).

Starlings and white-throated sparrows apparently use the star pattern in the sky (Kramer, 1957, 1959; Sauer, 1958) (Fig. 11.18). Golden-crowned (*Zonotrichia coronata*) and white-crowned sparrows (*Z. leucophrys*) as well as white-throated sparrows caged outdoors show a strong orientation toward the north in the spring and toward the south in the fall, indicating orientation by either the moon or stars. In addition, perception of ultraviolet light and linearly polarized light also may serve as navigational aids.

Infrasound may be another important factor enabling birds to navigate and orient themselves (Wiltschko and Wiltschko, 1994). Homing pigeons have been shown to detect frequencies as low as 0.05 Hz, well below the lower limit of human hearing, which is approximately 10 Hz. See further discussion of infrasound as a means of communication in Chapter 12.

Much less is known about navigation during migration in mammals than in birds. Olfaction is probably a major factor in some mammals, such as large herbivores, because the presence of many scent-producing glands suggests that olfaction is well developed in these species. Such glands may be present on the face (preorbital), rump, and lower limbs (tarsal, metatarsal, interdigital). Pheromones from these glands are left as scents on the ground and on vegetation along the

FIGURE 11.18

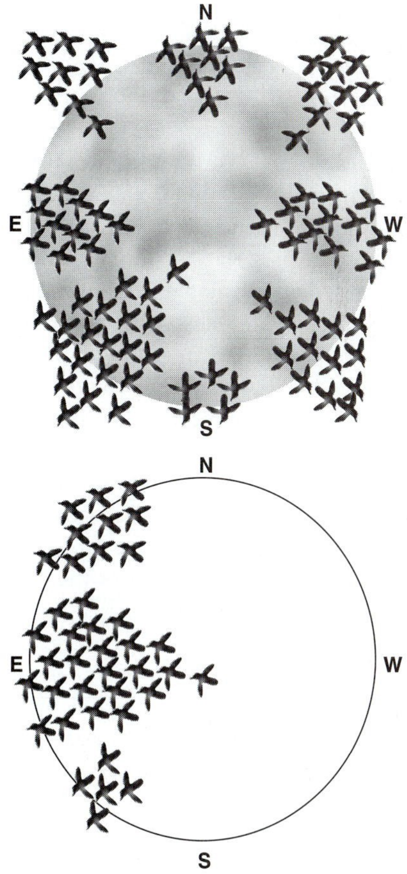

Influence of clouds on the orientation ability of captive starlings (Sturnus vulgaris). Top: Dense clouds; here, the starling wanders at random. Bottom: Clearing; in this situation, the starling immediately gets its bearings in the direction in which it is flying.
Source: Data after Kramer, in V. Dorst, The Migration of Birds, *1962 Houghton Mifflin Co., Boston.*

animal's trail. These serve as a guide to other members of the same species traveling along that route. The recognition of familiar landmarks through memory and vision is undoubtedly important in many mammalian species. Marine species may be influenced by ocean currents, olfactory stimuli, ocean floor topography, and the Earth's magnetic field.

The study of migration continues to intrigue scientists. The mechanisms and processes employed by these long-distance migrators remain a fruitful field for investigation. Humanity's effects on the climate (global warming, acid rain, etc.) and on the world's ecosystems (habitat destruction, pesticide use, etc.) may cause drastic changes in the populations of migratory species during our lifetime. For example, populations of many songbirds in the Americas have been declining because of fragmentation of their breeding habitat in North America, reduction of wintering habitat caused by destruction of tropical habitat, and loss of resting and feeding areas by urban development and/or pollution.

HOMING

Many vertebrates have an innate ability to return to their home after foraging, displacement, or migration. Birds return to their nests after foraging, to a specific place to rest at night, to winter ranges, and to breeding ranges during the appropriate season, sometimes after migratory journeys of thousands of kilometers. Homing pigeons return to their lofts year-round. This ability to return is known as **homing.** Displacement may occur through natural causes (flooding, drought, storm, fire), food-gathering activities (chasing prey into unfamiliar territory), or through human activities (trapping and relocating nuisance animals, restocking, racing pigeons, etc.). Homing success decreases as the distance the individual is displaced increases (Fig. 11.19). Homing ability within some species is also correlated with home range size. For example, Anderson et al. (1977) found that house mice (*Mus musculus*) with larger home ranges were able to home from greater distances than house mice with smaller home ranges. The ability to return home from any location requires information on the position of the present site in relation to the home site. Homing is a component of migration, but it can also be observed in nonmigrating animals.

Homing ability has been demonstrated in all of the vertebrate groups. A number of tidepool fishes can find their way back to their home pools by using olfactory and visual cues after being displaced several hundred meters away (Gibson, 1969,1982; Dooley and Collura, 1988; Horn and Gibson, 1988). Some male rock-dwelling cichlids (*Pseudotropheus aurora*) may home over distances of up to 2,500 m (Hert, 1992). Salmon can find their way back to their natal streams several years later after traveling thousands of kilometers in the open ocean (Braithwaite, 1998).

Homing has been demonstrated in several species of salamanders. An average of 45 percent of mountain dusky salamanders (*Desmognathus ochropheus*) displaced 30 m upstream and downstream from their capture sites in Ohio were able to return, some in less than 24 hours (Holomuzki, 1982). Eight percent of newts (*Taricha*) displaced 0.8 km upstream or downstream returned to their homes within 12 months (Twitty, 1959; Twitty et al., 1967). Within 2 years, 15 percent had returned and by the end of 3 years, 29 percent had returned. A total of 2 percent (18 newts) displaced 4.8 km away from their points of capture in a stream in a deep canyon on the other side of a 305-m mountain ridge found their way back to their homes. Red-cheeked salamanders (*Plethodon jordani*) have returned from distances up to 150 m (Madison, 1969). Forty-three of 83 Pacific tree frogs (*Hyla regilla*) (52%) returned from 275 m to their breeding sites in 1 month (Jameson, 1957). Toads (*Bufo bufo*) have returned 3 km to their home sites (Heusser, 1969). Leopard frogs (*Rana pipiens*) displaced up to 1 km from their capture site generally oriented correctly in their homeward direction and returned to their home range (Dole, 1968).

Adelie penguins (*Pygoscelis adeliae*) leave their young for up to 2 weeks to forage for food at sea and then return. The Manx shearwater (*Puffinus puffinus*) is a European bird that has a north–south migration route. A breeding adult was transported by airplane 5,150 km westward to Boston, Massachusetts, and released; 12 days later, it was recaptured in its nesting burrow in Skokholm, an island off the southwest coast of Wales (Mazzeo, 1953). The bird had returned from outside the range of the species, along an east–west route, and at an average speed of over 400 km per day.

A total of 412 golden-crowned (*Zonotrichia coronata*) and white-crowned (*Z. leucophrys*) sparrows were transported

FIGURE 11.19

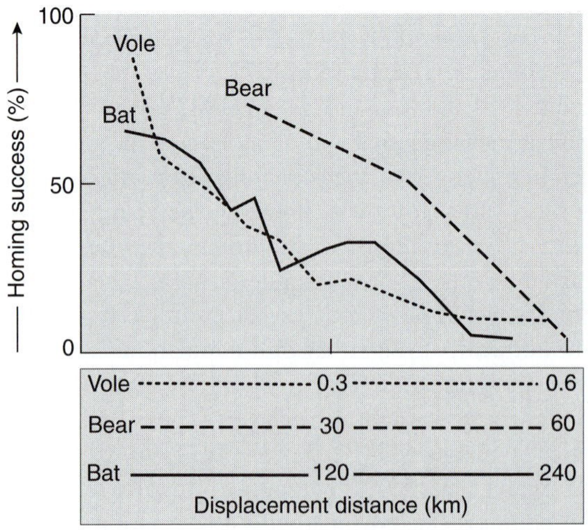

Homing success as a function of displacement distance in three species of mammals: Indiana bat (*Myotis sodalis*), n = 700; meadow vole (*Microtus pennsylvanicus*), n = 460; and black bear (*Ursus americanus*), n = 112.

Source: Data from Robinson and Falls, 1965; MacArthur, 1981; Hassel, 1960; in J. Bovet "Mammals" in Animal Homing, *F. Papi (editor), 1992, Chapman and Hall, New York.*

from their San Jose, California, wintering grounds to Baton Rouge, Louisiana, a distance of 2,900 km, and released (Mewaldt, 1964). The following winter, 26 of these birds were recaptured in San Jose. Mewaldt then transported 660 birds (including 22 that had returned from Baton Rouge) to Laurel, Maryland, a distance of 3,860 km, and released them. The next winter, 15 (including 6 of the 22) had returned to San Jose.

Eleven nonmigratory dark-eyed juncos (*Junco hyemalis*) were transported distances ranging from 55 to 563 km from their capture site at Mountain Lake, Virginia (Nolan et al., 1986). Four of six homed from 328 km or less, and three of five did so from 563 km. Two of the latter individuals returned in 28 and 35 days, respectively. Estimated mean distances traveled per day ranged from 13.0 to 25.2 km.

Among mammals, Hitchcock and Reynolds (1942) showed that half of the little brown bats (*Myotis lucifugus*) returned to the home roost after being released over 113 km away. Schramm (1957) recovered 2 out of 34 little brown bats back at the home roost 17 and 22 days, respectively, after they had been released 435 km away (Fig.11.20).

In a study of the big brown bat (*Eptesicus fuscus*), 155 individuals were displaced 725 km north of their point of capture in Ohio (Smith and Goodpaster, 1958). They were released on July 21. Three were found at the home roost on August 24, and 4 more were found on October 26. Some (2.5%) released 365 km away returned within 1 to 2 years. On August 2, 18 bats were taken to western Tennessee, 547 km southwest of Cincinnati, their point of capture. Fifteen days later, 2 of these, including a juvenile, had returned home.

Cope et al. (1961) released big brown bats (*E. fuscus*) at various distances from the home roost. Nearly all of those released 32 km away returned the same night. Most bats released 64 km away did not return until the second night, and most of those taken 161 km away arrived on the third night. Of bats released 403 km from home, some returned during the fourth night, and nearly all had returned by the end of the fifth night. Eastern gray squirrels (*Sciurus carolinensis*) possess a strong homing tendency and may return from as far as 4.5 km (Hungerford and Wilder, 1941).

FIGURE 11.20

Little brown bats (*Myotis lucifugus*) have returned to their home roost after being released 435 km away.

How do homing animals return? In some cases, the animal can rely on information obtained en route during the outward journey. In others, the animal finds itself in territory that is somewhat familiar. Periodic exploratory movements out of its normal home range may have given the animal some familiarity with the surroundings, so that by using olfaction and/or vision, it may be able to return. One common mechanism is the use of landmarks surrounding the goal to which the animal is returning. This is also called piloting or mnemotaxis. Most species use visual landmarks. Landmarks used by fish may serve as beacons marking a goal, or the geometrical relationships between several landmarks can be used to calculate a trajectory to a destination. Some animals simply may begin wandering from their release site and, through random wandering, eventually return to their homes. Still others may engage in a directed search pattern following their release. Pigeons and sea gulls usually circle to get their correct orientation before heading homeward.

Vertebrates can orient themselves in their environment via several methods. Some use their senses of taste and smell, whereas others, such as bats, elk, and many birds, probably orient by using their sense of sight to recognize familiar landmarks (Fig. 11.21). Other species use either celestial bodies (sun, stars, or moon) or the earth's magnetic field as cues to enable them to successfully return to their home range. The celestial compasses are based on visual information: compensation for the sun's apparent movement is done with the help of an internal clock (Schmidt-Koenig, 1960); the constant spatial relationship between stars is used for star compass orientation (Emlen, 1967).

Kramer (1959) first demonstrated that homing pigeons primarily rely on the sun to orient or determine directions in space (compass), while some other mechanism was responsible for allowing pigeons to determine their position in space relative to home (navigational map). When displaced to some distant, unfamiliar release site, a homing pigeon will first determine its location relative to home by relying on its navigational map. For example, if the navigational map indicated that the pigeon was north of its home, it would rely on the sun as a compass to determine south and fly toward home. Thus, navigation was based on a two-step map and compass process.

A magnetic compass was first described for European robins (*Erithacus rubecula*), a night-migrating passerine (Wiltschko, 1968). During the migration season, captive birds are restless. Their activity was concentrated in the part of their cage that pointed to the migratory direction of their free-flying conspecifics. When the magnetic field was experimentally altered around the test cage, the birds reacted to a shift in magnetic north with a corresponding deflection of their directional tendencies, thus indicating that they used magnetic information in direction finding.

Fishes are known to home by four methods: orientation to gradients of temperature, salinity, and chemicals; sun orientation; orientation to polarized light; and orientation to geomagnetic and geoelectric fields. Salmon hatch in

FIGURE 11.21

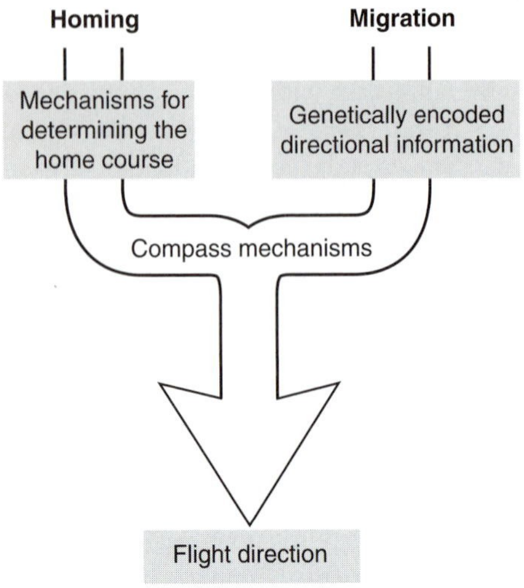

The current view of the avian orientation system. The nature of the first step, the determination of a direction as a compass course, varies according to the behavioral context of the orientation task. In migration, the course depends on the geographical relationship of the breeding ground and the wintering area. Birds possess innate information on this course, which is genetically transmitted from one generation to the next. The birds face the task of transforming this genetically encoded directional information into an actual migratory course. In homing, however, the correct course varies according to the bird's current position relative to its home. Here the bird must be able to interpret environmental information in order to determine the course, which is then located by compass mechanisms. *Source: Data from Wiltschko and Wiltschko, in Davies and Green (editors),* Perception and Motor Control in Birds. *Berlin: Springer-Verlag 1994. Pages 95–119.*

freshwater streams but swim hundreds of kilometers out to sea, where they feed and grow for a year or more. When the fish mature, they return to breed in the stream where they hatched. Through their olfactory apparatus, they can recognize the odor of their home stream, which was imprinted in their brain before they left. By searching out this olfactory stimulus, each fish can return to the precise stream in which it was hatched (Hasler and Wisby, 1951). In addition, there is evidence that salmon may be able to detect pheromones given off by other members of their species and are able to discriminate one population from another (Groot et al., 1986).

Amphibians are known to home by using chemical, visual, and celestial cues, as well as the earth's geomagnetic field. Some salamanders, such as the eastern red-spotted newt (*Notophthalmus viridescens*) possess magnetic-compass orientation that is directly affected by the wavelength of light (Phillips and Borland, 1992). This wavelength dependence is due to a direct effect of light on the underlying magnetoreception mechanism. Celestial cues may be perceived by photoreceptors located outside the retinas of the eyes, either in the pineal body and/or frontal organ (Adler, 1971; Taylor and Auburn, 1978) or possibly in the upper part of the brain (Landreth and Ferguson, 1967a, b; Taylor and Ferguson, 1970; Taylor, 1972; Adler, 1976; Demian and Taylor, 1977). Poison-dart frogs, such as *Dendrobates pumilio,* use both visual and chemical cues (Forester and Wisnieski, 1991).

The moon and stars are of primary importance to night-migrating birds and bats, whereas the sun is used by many diurnal migrators. Beginning in late summer, the night skies of the Northern Hemisphere are filled with birds of all sizes flying south to spend the winter (Fig. 11.22). Nocturnal migration allows diurnal birds to forage during the day. Birds that flap vigorously during their migration generate a great deal of heat, causing their resting body temperatures of about 38°C to increase during strenuous flight to between 41 and 43°C (Kerlinger, 1995). To avoid further overheating, many birds migrate during the night, when temperatures are lower. Some of the insect eaters (such as kingbirds and swallows) that feed on the wing migrate during the day. True goal-oriented homing requires an accurate internal clock in order to compensate for the ever-changing positions of the celestial bodies in the sky.

It is now known that homing pigeons may use visual, olfactory, and magnetic cues, as well as the sun, to get specific information concerning their position with respect to their home (Alerstam, 1990). Olfactory stimuli within approximately 100 km of their home may be of considerable importance. Some may sample olfactory cues, changes in the Earth's magnetic field, and geographic landmarks en route to the release site. These mechanisms interact in a complex way, supplementing as well as replacing each other. With several means of orientation available to it, a pigeon can use the one that is most effective at any given time.

Young, inexperienced homing pigeons and adult, experienced pigeons differ in the weighting of different orientational mechanisms (Alerstam, 1990). The navigational system undergoes a change during the first few months of a pigeon's life.

FIGURE 11.22

Canada geese (*Branta canadensis*) silhouetted against the moon as they migrate south to spend the winter.

Initially, birds must rely on mechanisms such as visual cues to determine their home direction. Later, as their flying experience increases, they develop a navigational map that employs several cue systems such as odor, magnetic cues, and infrasound (Wiltschko and Wiltschko, 1994). During the development of the navigational system, the magnetic compass seems to control the sun compass, whereas later the sun compass dominates. The magnetic compass, however, continues to serve as a backup system for overcast days and in case the sun compass is impaired (Wiltschko and Wiltschko, 1994).

EMIGRATION

Some vertebrates engage in a general movement away from an occupied area. Such a one-way trip is known as an **emigration.** Emigrations are sometimes unidirectional, with all individuals moving in the same direction. In years when the mast crop is sparse, animals such as gray squirrels and black bears may be observed emigrating to other areas in search of food (Flyger, 1969). Mast is divided into two groups. Soft mast consists of fruits such as blackberries, blueberries, and cherries, whereas hard mast consists of acorns, hickory nuts, beech nuts, and formerly, chestnuts. In the southern Appalachians, the mast crop failed in 1992. More than 900 black bears (*Ursus americanus*) were reported killed by hunters and vehicles in the four-state southern Appalachian region (Anonymous, 1996b). Some biologists feel that 40 percent or more of the black bear population was lost.

Countless tales have been told about lemming emigrations. These stories, however, are a myth for most species. One species, *Lemmus sibiricus*, which ranges from the Bering Strait in Alaska to Baffin Island and south in Canada to British Columbia and Alberta, experiences regular fluctuations in population with peak numbers being reached every 2 to 5 years. As the excess population overutilizes its food supply, many animals begin a large-scale movement away from the area. At such times, many animals are seen swimming across rivers and lakes and often appear in human settlements. Many die because they cannot find suitable habitat. The Norwegian lemming (*Lemmus lemmus*) in parts of Fennoscandia, Norway, is best known for its spectacular emigrations. At roughly 30-year intervals, millions of Norwegian lemmings move at speeds of 5 km per hour, covering distances up to 200 km (Stenseth and Ims, 1994).

Orientation and navigation involve a multitude of integrated cues (innate, celestial, magnetic, visual, auditory, and olfactory), with the significance of some factors being adjusted with respect to that of others to bring the entire system into harmony. The different factors must be seen as integrated components of a complex system, which is partially redundant and which tries to use all suitable factors provided by the environment.

Review Questions

1. Define home range. Discuss three methods of calculating home range. Which do you feel gives a more accurate representation?
2. Differentiate between territory and home range. Which is usually larger? Which is usually protected?
3. What are some adaptive advantages of territoriality?
4. Discuss several methods used by vertebrates to return to their home range (homing) if displaced.
5. List several adaptive advantages of dispersal.
6. Differentiate between migration and emigration.
7. Why must some species migrate?
8. List several adaptive advantages of altitudinal migration.
9. List the four major flyways used by waterfowl in North America. Why are they significant?
10. Discuss the two orientation systems used by hatchling leatherback sea turtles immediately after emerging from their underground nests.

Supplemental Reading

Alerstam, T. 1991. *Bird Migration.* New York: Cambridge University Press.

Berthold, P. B. 1994. *Bird Migration.* New York: Oxford University Press.

Brown, M., and P. Corkeron. 1995. Pod characteristics of migrating humpback whales (*Megaptera novaeangliae*) off the East Australian coast. *Behaviour* 132(3–4):163–179.

Dingle, H. 1996. *Migration: The Biology of Life on the Move.* New York: Oxford University Press.

Hasler, A., and A. Scholz. 1983. *Olfactory Imprinting and Homing in Salmon.* New York: Springer-Verlag.

Healy, S. (ed.). 1998. *Spatial Representation in Animals.* New York: Oxford University Press.

Lincoln, F. C. 1950. *Migration of Birds.* Circular 16. Washington, D.C.: U.S. Fish and Wildlife Service.

Madden, J. R. 1974. Female territoriality in a Suffolk County, Long Island, population of *Glaucomys volans. Journal of Mammalogy* 55:647–652.

Stone, G. S., S. K. Katona, and E. B. Tucker. 1987. History, migration and present status of humpback whales *Megaptera novaeangliae* at Bermuda. *Biological Conservation* 42:133–145.

Weidensaul, S. 1999. *Living on the Wind: Across the Hemisphere with Migratory Birds.* New York: North Point Press.

Vertebrate Internet Sites

Visit the zoology website at http://www.mhhe.com to find live internet links for each of the references listed below.

1. **FishFAQ2.**
   This site discusses a variety of facts regarding several species of salmon, including migration, longevity, and body weights.
2. **Migratory Bird Treaty Act.**
   This site contains the most recent listing of bird species in the United States that are protected by the Migratory Bird Treaty Act.
3. **Bird Migration Ecology at the University of Lund, Sweden.**
   Information about patterns of migration, tracking mechanisms, and links.
4. **Movement of Atlantic Sharks.**
   An article on movements of sharks from the NFMS.

CHAPTER 12

Intraspecific Behavior and Ecology

■ INTRODUCTION

Very few animals are not, at one time or another, "social." While the social nature of schools of fish, flocks of migrating geese, and herds of African big game animals is obvious, one might hesitate to use the word "social" to describe the intricate interaction between the members of a breeding pair or between parents and offspring. Likewise, the fighting between rival males in the spring might at first glance seem to deserve the epithet "antisocial" rather than "social." The complex interactions of individuals with kin groups such as Florida scrub jays (*Aphelocoma coerulescens*) is much different from the way individuals of non-kin groups, such as a flock of gulls, interact. Yet all of these interactions have a great deal in common; all contribute to the success of the species and all depend on communication—albeit through many different methods—between individuals. In short, social behavior—the joint activities that make an animal community function—depends on various types of interactions among individuals, each playing its part in communication with others.

The terms for groups of vertebrates are listed in Appendix II. Many have their origins quite far back in history; some descend from the hunting royalty of England, France, and Germany.

■ SOCIAL INTERACTIONS

Social animals do much more than merely stay together. They *do* things together; the activities of all members are jointly timed and oriented, and they do this, too, by influencing each other. A family of ducklings, for example, goes through a common diurnal rhythm. Part of the day they feed, keeping close together wherever they go. On other occasions, they bathe together and swim to the shore together, where they may spend half an hour or so preening, standing next to each other. Then they fall asleep, side by side.

Even while sleeping, ducks and many other birds continue to interact. Half-brain sleep—one cerebral hemisphere alert and the other asleep—has been documented in a wide range of birds and is thought to have evolved as a form of predator detection. Rattenborg et al. (1999) filmed rows of napping mallard ducks (*Anas platyrhynchos*). The end birds tended to keep open the eye on the side away from the other birds. Researchers found outer birds resorting to single-hemisphere sleep rather than total relaxation during 32 percent of napping time versus 12 percent for birds in internal spots, an increase of more than 150 percent. Furthermore, birds at the edge position oriented the open eye away from the group's center 86 percent of the time, whereas birds in the central position showed no preference for gaze direction. This study is believed to be the first evidence for an animal behaviorally controlling sleep and wakefulness simultaneously in different regions of the brain.

On many occasions, there is a division of labor among members of a group. Members of a flock of Canada geese take turns leading the V-shaped formation when migrating. Old, experienced chimpanzees (*Pan*) lead the group and keep a sharp lookout at all times. Perhaps the most extreme social hierarchy known among mammals occurs in naked mole rats.

There is also division of labor in more solitary animals, particularly between male and female. This applies both to different roles in mating and to different parental activities. Numerous examples of such division of labor in all vertebrate groups have been discussed in Chapters 5 through 9.

Social interactions may be beneficial in many ways. It has been estimated that 25 percent of all fishes school throughout their lives, and about half of all fishes spend at least part of their lives in schools (Moyle and Cech, 1996). Schooling can reduce the risk of predation, increase reproductive success, and in some cases, increase the efficiency of finding food for fishes and many marine animals. For example, groups of dolphins and porpoises aid wounded members of their own species, raising them to the surface so they can breathe. They also circle a female giving birth in order to protect the mother and newborn against sharks.

Mobbing behavior, in which one to a few individuals approach and often chase and/or attack a potential predator, is common in birds. The primary purpose of mobbing is to force the predator to move on (Curio, 1978; Curio et al., 1978a, b).

BIO-NOTE 12.1

Mole Rat Societies

Naked mole rats (*Heterocephalus glaber*), which exhibit **eusociality** or "true sociality," usually live in colonies of 75 to 80 animals, although colonies of more than 250 animals have been recorded. Most colonies contain only a single reproductive female (see Fig. 11.3). Chores are performed by both males and females, but not by all individuals equally. For example, the primary role of the breeding female is to produce young, nourish the pups, and keep them clean. Nonbreeders help to clean and carry pups and also to maintain and defend the colony's tunnel system. Labor is divided according to size. Large nonbreeders defend their colony against mole rats from other colonies and also against predators.

Dominance hierarchies exist within colonies: The queen and breeding males dominate the nonbreeders; larger workers dominate smaller ones, regardless of sex.

Chemical, tactile, and acoustic forms of communication are used. At least 17 distinct categories of vocalizations have been recorded, with the vocal repertoire being the most extensive known among rodents.

Naked mole rats, which are ectothermic, are the only known mammals whose body temperature fluctuates with the ambient temperature. The temperature within their tunnels remains near 30°C most of the year. If the animals get colder, they regulate their temperature by huddling with colony mates (social endothermy, like bees).

Inbreeding is a constant problem in such highly organized societies. Recently, a dispersal phenotype was discovered that may occasionally promote outbreeding. These dispersers are morphologically, physiologically, and behaviorally different from other colony members. These rare morphs are fatter than average, have higher than normal levels of luteinizing hormone, have a strong urge to disperse, and will mate only with noncolony members. Although rare, they are essential in producing the gene flow that maintains the heterogenicity required for reproductive compatibility between isolated populations.

Sherman et al., 1992
O'Riain et al., 1996

Clearly, no sexually reproducing species could exist without intricate cooperation between male and female for the purpose of mating. This period of interaction may last only long enough for fertilization to occur, or it may result in a lifetime bond. Many marine fishes simply discharge their gametes into the surrounding water. Most do this in response to an environmental stimulus that induces the synchronized release of gametes by both sexes. This simple mode of reproduction ensures fertilization, genetic recombination in offspring, and hence, variation in the population.

Species in which young receive parental care need close cooperation between parents and young. Mated pairs are usually more successful at raising offspring than a single animal working alone. Each member of a pair can share in food gathering, defending the territory, and protecting its mate and young from predators. Protection is even more effective when a group faces a potential predator. For example, gulls in a breeding colony attack predators in force. This concerted defense, quickly mounted as the birds alert each other by alarm calls, is much more successful than individual attacks. This response is elicited not just because the gulls nest close together, but also because they nest synchronously and will benefit almost equally. Likewise, many mammals, such as musk-ox and elephants, band together to protect their young from potential predators.

Social hierarchies occur in many groups of animals. In some, the female is dominant—a **matriarchal** hierarchy; in others, the male is dominant—a **patriarchal** hierarchy. The dominant individual is usually an older member of the group and controls activities until challenged and deposed by a younger rival. Classic studies of peck-orders in chickens have clearly demonstrated the nature of the dominant–subordinate behavior. Similar studies have been carried out on a variety of other vertebrates. Within a clan of spotted hyenas (*Crocuta crocuta*), for example, the highest-ranking female and her descendants are dominant over all other animals (Nowak, 1991). Although all resident males court females, only the highest-ranking male was observed mating in a study by Frank (1986b). Dominant individuals in non-kin groups, such as flocks of sparrows, have been shown to gain access to better food sources and suffer lower risks of predation than do subordinate individuals. Thus, the value of social behavior accrues to a greater extent among dominant individuals than it does among subordinate individuals.

Some species of birds, such as white-fronted bee-eaters (*Merops bullockoides*) of Africa, are cooperative breeders (Emlen and Wrege, 1992). They live in colonies averaging 200 individuals making up several clans. Young females remain in their parental group (clan) for 1 or 2 years until they begin to breed, at which time they leave their parents and join the clan of their mates. Males, however, do not leave their clans. Each clan establishes its own feeding territory, but all individuals of each clan roost and nest at the colony site.

Not all intraspecific interactions are peaceful. Competition in many birds, for example, begins in the nest as individuals compete for food and space. Intraspecific competition, whether for a mate, food, or territory, however, rarely results in injury to the participants. Most species have ritualized aggressive behaviors that are used in these situations. Many fishes engage in tail-beating, mouth-pulling, or mouth-pushing activities. Red-backed salamanders (*Plethodon cinereus*) raise their trunks off the substratum and look toward their opponent (Fig. 12.1a). A biting lunge directed toward the opponent's tail or nasolabial groove area may follow. Frogs attempt to topple intruders that come into their territory (Fig. 12.1b). Rattlesnakes wrap their bodies around each other and butt each other with their heads. Some lizards whip each other

FIGURE 12.1

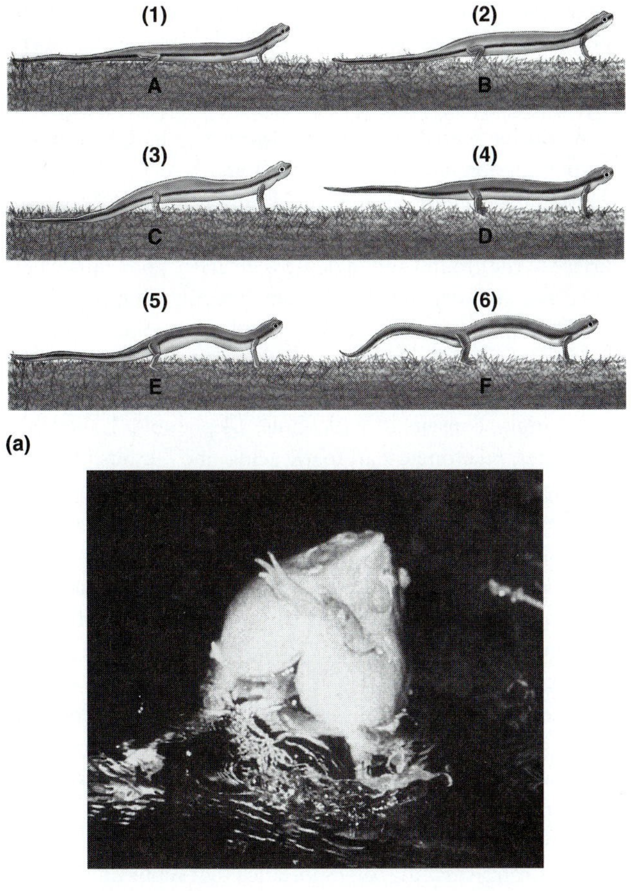

(*a*) Red-backed salamander (*Plethodon cinereus*) escalating the intensity (A–F) of its threat display toward an intruder. (*b*) Male bullfrogs (*Rana catesbeiana*) aggressively defend territories used as egg deposition sites; fights are typically wrestling matches in which the larger male prevails. *Source:* (a) *After Jaeger and Schwartz, 1991,* Journal of Herpetology *in Stebbins and Cohen,* A Natural History of Amphibians, *1995, Princeton University Press.*

with their tails. Turkeys drive off their rivals by means of threatening calls and/or by jumping at them. Giraffes, deer, elk, and bighorn sheep butt each other with their heads (Figs. 12.2 and 12.3). Brown bears may charge, growl, and push one another with their forelegs. Oryx antelope possess sharp-pointed horns with which they stab potential predators such as lions, but when faced with a conspecific adversary, they merely butt heads and do not attempt to stab each other.

In spotted hyenas (*Crocuta crocuta*), sibling rivalry is carried to a deadly extreme. Females generally give birth to twins in underground dens. Sibling fighting begins at the earliest possible moment, sometimes while the second pup is still in the amniotic sac. This instant antagonism lets the pups establish a ranking order that determines which one gets the most of a limited food supply: their mother's milk. The dominant animal generally grows larger and has a better chance of succeeding in the dangerous adult world. The loser often dies. Female twins fight the hardest and longest—probably

FIGURE 12.2

Male Masai giraffes (*Giraffa camelopardalis*) sparring for social dominance. Such bouts are primarily symbolic and rarely result in injury.

FIGURE 12.3

Butting bouts among desert bighorn sheep (*Ovis canadensis*) appear to be contests of skill and stamina with little real antagonism involved. It has nothing to do with the pre-mating collection and maintenance of a "harem," nor does it seem to result in the elimination of one ram from participation in mating activity with a certain ewe. It appears to have no objective whatever except the satisfaction of some deep-seated urge aroused by the mating instinct and demanding and receiving an outlet for its own sake. When males are 12 feet apart, with every muscle bulging for a final effort, and with amazing timing and accuracy, they lunge forward like football tackles. The remarkable synchronization of movement pictured here is the rule, not the exception. Every effort seems to be made to ensure a perfect head-on and balanced contact. Note that both heads are tilted to the same side. Occasionally, one slips or miscalculates and a severe neck-twisting or nose-smashing results. The combined speed at impact has been estimated at 50 to 70 miles per hour and to be the equivalent of a 2,400-pound blow. More than 40 blows between two rams have been counted in one afternoon.

because large size is favored if a female is to give birth to healthy pups of her own. Battles between male–female twins usually are not as intense (Frank et al., 1991).

Animals show submission in various ways. Some fishes collapse their fins and change coloration. Bullfrogs (*Rana catesbeiana*) that maintain a low position in the water are not challenged or attacked. Iguanas flatten themselves to appear as small as possible. Many canids flatten their bodies and bring their ears to lie flat against their heads. The tails often will be tucked between their legs.

BIO-NOTE 12.2

Intraspecific Parasitism

Although parasitism usually is considered an interspecies interaction, intraspecific brood parasitism occurs in a large number of bird species in which females lay eggs in the nests of conspecifics, who then provide parental care. Females without nests, as well as those with viable nests, engage in brood parasitism. In several species, parasitic eggs have been found to be less successful than nonparasitic eggs. Many parasitic females are young birds of poor competitive ability. Some lay eggs in the nests of other females before laying eggs in their own nests.

The addition of parasitic eggs to those already in a nest may result in more young than the host parents can rear successfully. This may lead to reduced incubation efficiency and overcrowding. Antiparasite behaviors include nest guarding, aggression, ejection of eggs, and nest desertion.

Petrie and Moller, 1991

SENSORY RECEPTION AND COMMUNICATION

For effective organization to exist within a population that maintains a social structure such as a family group, school, flock, or herd, some form of unambiguous communication must exist among the members of that population. This exchange of information influences the behavior of both the sender and the recipient. In general, those forms that live in social groups have the more highly developed sets of communication signals. However, even in solitary or unsocial animals, elaborate signals may be required to establish and maintain the species' dispersed spatial patterns (Bradbury and Vehrencamp, 1998).

Sensory reception and communication among vertebrates are accomplished in a variety of ways. They may use pheromones, sound, vision, tactile stimulation, electrical signals, signal patches, or a particular behavior such as the slapping of the tail (beaver) on the surface of the water or foot-drumming (kangaroo rats).

Olfaction

Olfactory communication is widespread among vertebrates and may be the primary mode of communication for many species. Chemical signals exchanged between members of the same species are known as **pheromones** (Greek *pherein,* to carry, and *hormon,* to excite). They *control* a wide variety of behaviors and physiological states and may be detected from considerable distances during both day and night. Normal, or nonpheromonal, chemoreception *influences* behavior. Both pheromonal and nonpheromonal chemoreception are important means of communication. Olfactory communication is effective beneath the surface of the ground and in dense vegetation, both areas where visual and auditory signals would be difficult to detect.

Pheromones may contain steroid or steroidlike organic compounds, which may be part of a mixture of compounds. Castoreum from the castor sacs of beaver (*Castor canadensis*), for example, consists of 6 alcohols, 14 phenols, 1 aldehyde, 15 amines, 6 ketones, 9 aromatic acids, and 5 esters (Müller-Schwarze and Houlihan, 1991). A total of 37 compounds have been identified from the temporal gland secretion of the Asian elephant (*Elephas maximus*) (Rasmussen et al., 1990). This gland, located in the mid-cheek region, is a modified apocrine sweat gland and has been implicated in chemical communication of African (*Loxodonta africana*) as well as Asian elephants. Secretions occur only during the physiological state of *musth,* the strange emotional state that periodically afflicts all male and some female elephants. *Musth* (a state of increased serum testosterone) occurs after elephants reach maturity and is accompanied by great activity of the temporal glands. The temples become puffy, and the glands exude a dark, strong-smelling, oily substance that stains much of the lower part of the face (Fig. 12.4). Elephants in *musth* either become highly excitable or dull and morose.

FIGURE 12.4

Male African elephant (*Loxodonta africana*) showing the temporal gland and its secretion. The glands exude a dark, strong-smelling, oily substance that stains much of the lower part of the face.

Biological activity of several compounds of a mixture may interact in synergistic, redundant, or addictive fashion. In some cases, individual components of a mixture are inactive, but when combined or dissolved in a fluid such as urine, they become effective olfactory signals.

Pheromones may represent a primitive communication technique. They may serve to attract members of the same species, including a mate; they may elicit courtship behavior; they may stimulate ovulation; they may serve as a warning when used to mark the boundaries of a territory; they may be used for defensive purposes; or, in some cases, they may indicate danger.

Among fishes, pheromones are important in species, like catfishes, that lack keen eyesight. By means of pheromones, migrating salmon may be able to discriminate members of their own population from individuals of other populations, thus permitting increased precision in their homing.

Some salamanders can distinguish between odors produced by conspecifics and heterospecifics and distinguish between odors of familiar and nonfamiliar conspecifics (Mathis, 1990). Pheromones, which may also convey information about gender, are used by many salamanders to mark their territories. The nasolabial grooves of plethodontid salamanders serve as specialized channels to transmit chemicals, such as pheromones, to the vomeronasal organs (see Fig. 6.27).

During the breeding season, the glands of some turtles enlarge and are thought to secrete pheromones. Many lizards and snakes use pheromones for species and sex recognition as well as the recognition of eggs. Some, such as male broad-headed skinks (*Eumeces laticeps*), have been shown to follow female conspecific odor trails (Cooper and Vitt, 1986). Chemical trailing of conspecifics occurs widely in snakes.

Skin lipids extracted from female red-sided garter snakes (*Thamnophis sirtalis parietalis*) are attractive to sexually active courting males (see Fig. 8.29). The lipids contained a female sex attractiveness pheromone consisting of a series of nonvolatile long-chain methyl ketones (Mason et al., 1989). When researchers added extracts of male lipids to female extracts, male courtship stopped, suggesting that males emit specific chemical cues that identify them as males. One chemical in the male lipid—squalene—caused a significant drop in courting and is an important part of the male sex recognition pheromone. Preliminary studies of related groups of snakes suggest that some of the same methyl ketones are found in females of several species.

Pheromones are well developed in mammals, especially those with the keenest senses of smell. Scent marking is a well-recognized and important aspect of mammalian communication and has been observed in a variety of mammals (Fig. 12.5). Glandular secretions and urine are used as the principal means of chemical communication. Estrous female mole rats (*Spalax ehrenbergi*) are known to be attracted to substances in adult male urine. Menzies et al. (1992) reported the extraction of sexual pheromones from lipids and other fractions of the urine. Male meadow voles (*Microtus pennsylvanicus*) emit odors that are attractive to females at the beginning, but not at the end, of the breeding season (Ferkin et al., 1992). Some mammals can differentiate between individuals on the basis of odor. Female house mice (*Mus musculus*), for example, use smell to recognize related females (Manning et al., 1992). The similarity in smell results from related females sharing genes of the major histocompatability complex (MHC), which is involved in fighting disease. In addition, if recently mated female mice are exposed to the urine or pheromones of strange males before implantation, pregnancy block occurs and pregnancy fails (Brennan et al., 1990).

In black-tailed deer (*Odocoileus hemionus*), secretions from four glands are considered important in social communication (Müller-Schwarze, 1971) (Fig. 12.6). The scent of the tarsal

FIGURE 12.5

This male cheetah (*Acinonyx*) is spraying a pheromone onto a tree in order to mark his territory. Scent marking is a well-recognized and important aspect of mammalian communication.

FIGURE 12.6

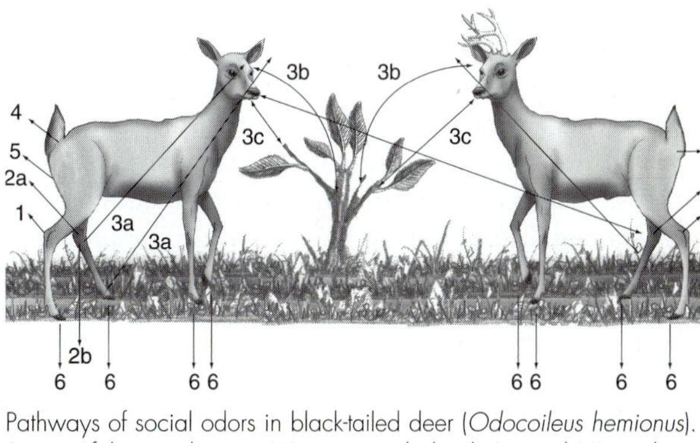

Pathways of social odors in black-tailed deer (*Odocoileus hemionus*). Scents of the tarsal organ (1), metatarsal gland (2a), tail (4), and urine (5) are transmitted through air. When the deer is reclining, the metatarsal gland touches the ground (2b). The deer rubs its hind leg over its forehead (3a). Marked twigs are sniffed and licked (3c). Interdigital glands leave scent on the ground (6).

gland identifies the age and sex of an individual at close range. The scent from the metatarsal gland acts like an alarm pheromone over moderate distances. The scent of the forehead glands is left on branches when a deer rubs its head and serves to mark the home range. Scent from the interdigital glands, which also is used in marking the home range, is left on the ground.

Flehmen is a reaction of some mammals to direct physical contact with a scent mark and its incorporated pheromones (Pough et al., 1996). After sniffing the scent mark, the animal licks it and takes it into its mouth. The upper lip curls, the jaws open, and the head is raised and turned from side to side or is nodded up and down. The animal inhales deeply to move the scent into the vomeronasal organ. Flehmen occurs during the breeding season and is characteristic of many ungulates, especially members of the deer family (Cervidae). It is also known to occur in some cats (Felidae).

Glandular secretions may be deposited on the substrate or on objects in the environment; they may be applied to the animal's own body or to the bodies of other members of the social group; or they may be released into the air. Feces and/or urine often contain pheromonal secretions.

Koalas (*Phascolarctos cinereus*) and other marsupials use sternal glands, paracloacal glands, and urine for marking. Trees are marked by koalas as they climb, by rubbing their sternum on the tree. Mitchell (1991a) noted: "Although koalas produce scent and inspect their environment for scent, there was no direct evidence that they used scent to define space, recognize individuals or recognize physiological states." Whole-body and pouch gland odors are important chemical signals between young Virginia opossums (*Didelphis virginiana*) and their mothers just prior to weaning (Holmes, 1992).

Some pheromones signal the presence of danger. Some wounded fishes release a substance from special cells in the epidermis, which induces other members of the school to flee for shelter. Similar effects have been recorded in amphibian tadpoles (Eibl-Eibesfeldt, 1949; Kulzer, 1954) and in mice (Heintz, 1954). Chemical signals also have been shown to facilitate schooling of young fish (Kuhme, 1964).

Some pheromones are very similar in structure to sex steroid hormones that are used to attract the opposite sex. Humans secrete pheromones, but most humans continually remove the real musks by bathing and then apply scented animal musks (perfumes) as a substitute. The symbolic message is still communicated, and the opposite sex still responds.

The morphology and chemistry of scent glands and the role of pheromones in mammalian social communication have been discussed in Brown and Macdonald (1985) and Gorman and Trowbridge (1989). The influence of selective factors such as substrate, persistence, intensity, and localizability on the signal structure in mammalian chemical communication systems has been reviewed by Alberts (1992).

Sound

The production and reception of sound is most highly developed in anurans, birds, bats, primates, and cetaceans. Many fishes, including grunters and croakers, produce sounds by contracting muscles attached to their swim bladders. Other fishes produce sounds by grinding their teeth or rubbing the base of a fin spine against the socket.

Sound production is limited in salamanders and caecilians, but auditory commmunication is highly developed in male anurans, particularly during the breeding season. Many males possess vocal sacs that serve as resonating chambers. The purpose of most anuran calls is to advertise for mating or to maintain territories or interindividual distances. Male gray tree frogs (*Hyla versicolor*) with long calls—known to be favored by females—sire higher quality young than those with short calls (Welch et al., 1998). For two years, researchers compared how the offspring fared as tadpoles and after they metamorphosed into frogs, measuring their growth rates under regimes of scarce and plentiful food. Offspring of males with long calls always performed significantly better than or not significantly differently from offspring of males with short calls. Because female *H. versicolor* do not gain direct benefits from their choice of mate, the indirect genetic benefits suggest good-genes selection as a probable explanation for the evolution and maintenance of the female preference in this species. Among reptiles, vocal cords are present only in a few lizards, such as geckos (Hildebrand, 1995).

Males of many species of birds have highly characteristic territorial songs announcing that the resident is a sexually mature male attempting to attract a suitable mate and defend an area against other males of the same species. Birds possess a unique modification of the lower trachea, the syrinx. Contraction of muscles attached to membranes within the syrinx produces the characteristic songs and calls of each species, which usually are heard most often during the breeding season. Individuality is common. Extensive studies on a variety of species show that songs differ among individuals in pitch, speed, and details of phrasing. In addition to their voices, some birds, such as ruffed grouse (*Bonasa umbellus*), also communicate by vigorously moving their wings back and forth, creating a drumming sound.

Young birds are predisposed to learn a specific kind of vocal information. Their learning pathways are highly selective and very sensitive to the "right" information (Adler, 1996b). For example, young male white-crowned sparrows (*Zonotrichia leucophrys*) and white-throated sparrows (*Z. albicollis*) possess a neural template that allows them to repeat the songs from males of their species. If the young bird does not receive this information during a critical song-learning period, it will not develop a typical full song 5 to 6 months later (Fig. 12.7). This song learning period extends from the 10th to 50th day of its life. (Some other species do not show this critical learning period.) In addition, juvenile males must be able to hear themselves sing; otherwise, they will develop aberrant songs. While

FIGURE 12.7

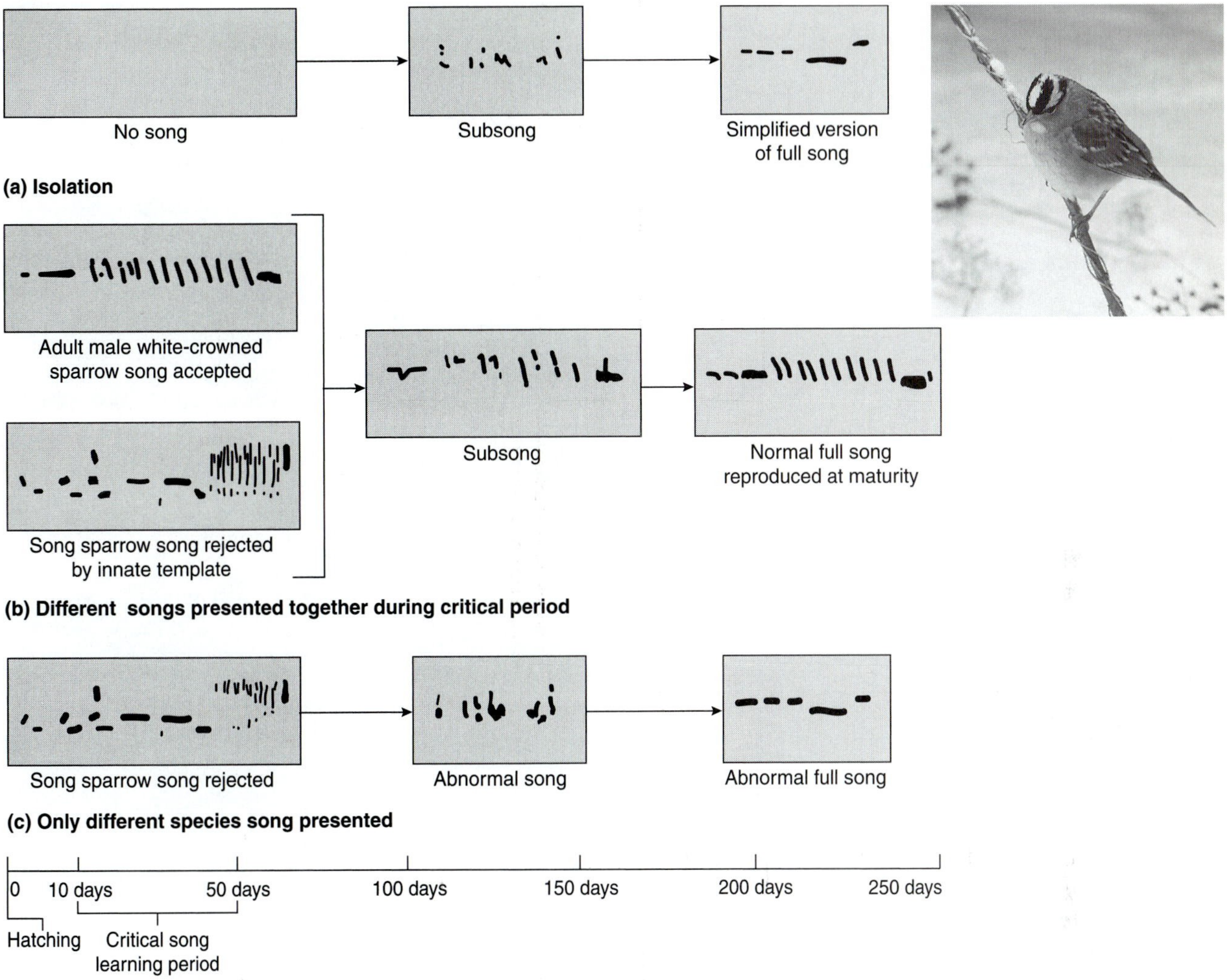

(*a*) Exposed to no song at all, male white-crowned sparrows (*Zonotrichia leucophrys*) produce subsong, but develop only a rudimentary version of their species' normal song. (*b*) Exposed to tapes of both their own species' song and that of the related song sparrow, they produce more complex subsong and a fully developed song characteristic of their own species. (*c*) Exposed only to the other species' song, they fail to learn.

the songs of male white-crowned sparrows within a population are strikingly consistent from year to year, males of other distinct populations have easily recognizable dialects (Marler and Tamura, 1962) (Fig. 12.8a–c).

Sound production and reception is very efficient in mammals. Vocal cords for producing sound are well developed, and the middle ear contains three bones (malleus, incus, and stapes) for receiving sound. The pinnae of many mammals (e.g., deer) are mobile, and each can be controlled independently of the other to enhance hearing. Mammals may emit many sounds. They may squeak, bark, bugle, howl, bellow, roar, neigh, moo, oink, cry, laugh, and speak. They may engage in tooth chattering, tail rattling, and drumming on the ground with their hind feet.

Foot-drumming in kangaroo rats (*Dipodomys*) is individually distinct (Randall, 1989). Individual rates are higher in males than in females. Rates are also higher in young adults than in juveniles and older adults; thus, foot-drumming rates may be used to communicate age, sex, or vitality. Foot-drumming may also be important in territorial defense.

East African vervet monkeys (*Cercopithecus pygerythrus*) give different alarm calls in response to three major predators: leopards, eagles, and snakes (Seyfarth and Cheney, 1992) (Fig. 12.9). Each call elicits a distinct escape response from nearby vervets. Alarm calls about leopards cause vervets to run into trees. Eagle alarms cause them to look upward or run into the bushes, whereas snake alarms cause them to stand on their hind legs and look into the grass.

Prairie dogs (*Cynomys* spp.) have a "vocabulary" of 10 different calls ranging from a commonly used warning bark to a chuckle, a "fear" scream, and a fighting snarl (Waring, 1970; Smith et al., 1977). Each call results in a specific action

FIGURE 12.8

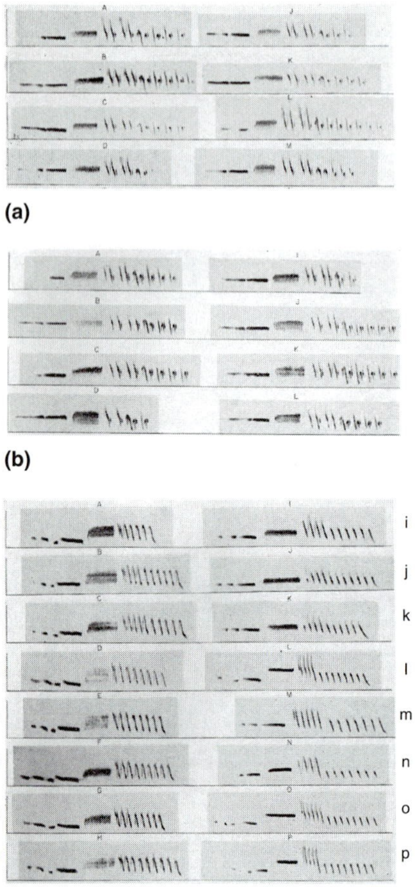

(a)

(b)

(c)

(*a*) Songs of eight male white-crowned sparrows (*Zonotrichia leucophrys*) recorded at Sunset Beach, Santa Cruz County, California in April 1959. The horizontal time scale is marked at 1-second intervals. The vertical frequency scale ranges from 2 to 7 kHz. (*b*) Songs of eight white-crowned sparrows recorded at Sunset Beach in May 1960. Note the consistency of the song when compared with the songs of the same population of males in 1959. (*c*) A–H, songs of eight white-crowned sparrows recorded at Inspiration Point, Contra Costa County, California, in May 1960. I–P, songs of eight birds recorded in Berkeley, Alameda County, in April 1959 and May 1960. Note the consistent difference in dialects in these birds from Contra Costa and Alameda counties from those in Santa Cruz County.

by nearby individuals. Howler monkeys (*Alouatta* sp.) of Panama have a vocabulary of 15 to 20 calls (Sekulic, 1982). Their calls have been heard by people 3 km away through the jungle and 5 km away across lakes (Nowak, 1991). Koalas bellow, squeak, groan, and moan (Mitchell, 1991a). Twelve different social and communicative calls are given by white-tailed deer, including snorts, bawls, grunts, mews, bleats, and whines (Atkeson et al., 1988).

Sherman (1977) found that female Belding's ground squirrels (*Spermophilus beldingi*) (Fig. 12.10a) gave alarm calls when a predator was in the vicinity more often than expected by chance, whereas the converse was true for males (Fig. 12.10b).

FIGURE 12.9

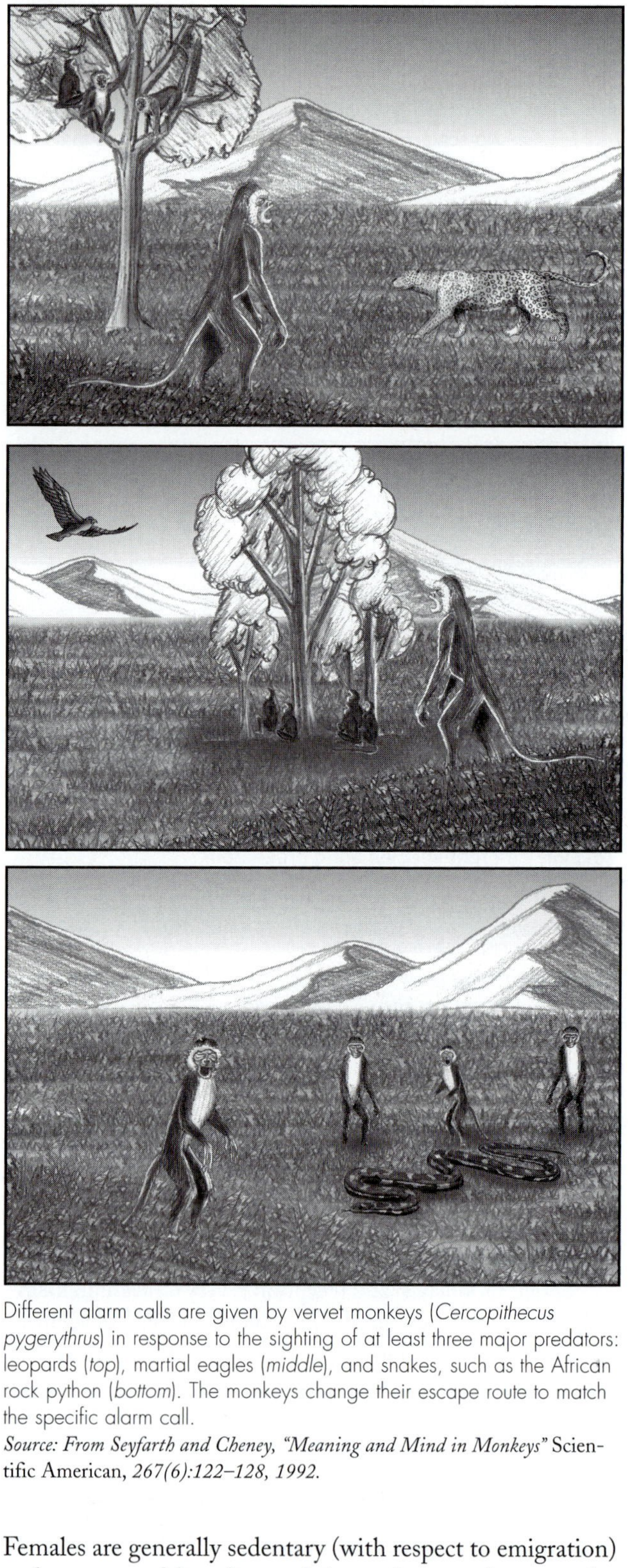

Different alarm calls are given by vervet monkeys (*Cercopithecus pygerythrus*) in response to the sighting of at least three major predators: leopards (*top*), martial eagles (*middle*), and snakes, such as the African rock python (*bottom*). The monkeys change their escape route to match the specific alarm call.

Source: From Seyfarth and Cheney, "Meaning and Mind in Monkeys" Scientific American, *267(6):122–128, 1992.*

Females are generally sedentary (with respect to emigration) and mature and breed near their natal sites, whereas males always emigrate from their birthplace and do not aggregate with siblings after emigration. As such, females were warning close kin (often offspring) by giving such alarm calls, whereas no such benefit accrued to males for warning others about the presence of a potential predator. Further support for the kinship hypothesis includes evidence that "invading" (nonnative) females gave alarm calls less frequently than native females.

The young of some bats and rodents, such as house mice (*Mus musculus*), emit both audible and ultrasonic sounds. These calls elicit search behavior in the female for her young; they also reduce maternal aggression (Ehret, 1983). Many pinnipeds produce a variety of underwater and airborne sounds that appear to be related to breeding activities and social interaction (Riedman, 1990).

Cetaceans produce a variety of pulsed calls and sounds. The eerie and plaintive songs of the humpback whale are repeated according to identifiable patterns. These sounds usually range between 40 Hz and 5 kHz in frequency and can be detected over 30 km away (Winn and Winn, 1978). They may last from 6 to 35 minutes before being repeated. One whale was recorded singing nonstop for at least 22 hours (Winn and Winn, 1978). Singing may take place during migration, as well as during courting. The singers are normally solitary males found in shallow coastal areas of 20 to 40 m in depth (Evans, 1987). One function of the humpback's song is thought to serve "as a spacing mechanism for courting males advertising their sexual availability to females" (Tyack, 1981). Identification is an important function of the sounds made by many baleen and toothed whales. The sounds may give the location of the whale, its sex, status, emotional or activity state, and possibly even its individual identity (Evans, 1987).

The vocal repertoire of many toothed whale cetaceans consists of ultrasonic clicks. Most cetacean strandings, particularly those involving pilot whales, occur on gently sloping beaches. Some biologists believe that the gradual slope of the beaches may not reflect the whales' ultrasonic signals effectively. If the whales do not hear an echo, they may receive a false impression of deep open water ahead and continue swimming toward shore until it is too late.

Bats (order Chiroptera) are the only mammals known to use echolocation as a principal means of locating prey.

FIGURE 12.10

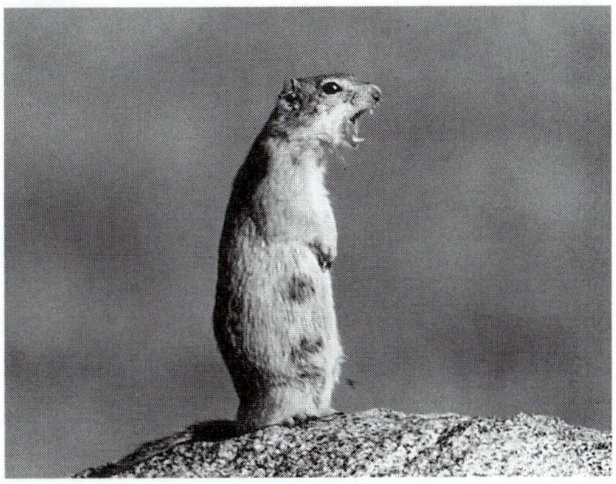

(a)

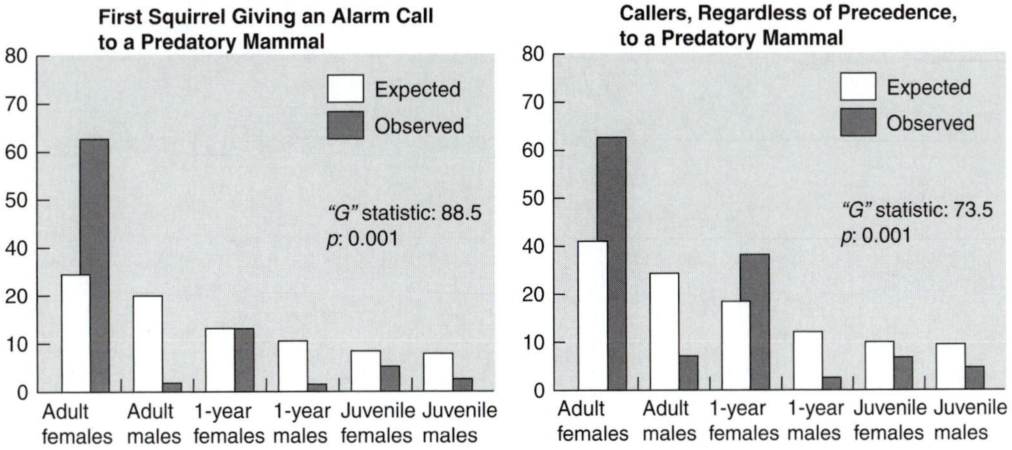

(b)

(*a*) A female Belding's ground squirrel (*Spermophilus beldingi*) emitting a predator alarm call. (*b*) Expected and observed frequencies of alarm calling in Belding's ground squirrel. The overall significance of both comparisons is due to females calling more than expected and males calling less. Data based on 102 observations.
Source: (b) *Data from Sherman, in* Science, *197:1246–1253, 1977.*

FIGURE 12.11

Elephants in the breeding herd at Circus World, Haines City, Florida. The arrow indicates the region of the elephant's forehead where fluttering can be observed during the production of infrasonic calls.

However, baleen whales (order Cetacea) and pinnipeds (order Pinnipedia) may use echolocation to a limited degree in intraspecific interactions. Some terrestrial species, such as shrews, voles, tenrecs, oilbirds, and the cave swiftlet, appear to use echolocation in certain instances.

Calls at frequencies below the level of human hearing—**infrasound**—may provide a significant means of communication in some social species such as elephants (*Loxodonta africana*), hippopotamuses (*Hippopotamus amphibius*), and alligators (*Alligator mississippiensis*) (Payne et al., 1986; Langbauer et al., 1991a, b; Montgomery, 1992). The long wavelengths of low frequency sounds are not reflected or absorbed by vegetation or blocked by obstacles the way higher frequency sounds are. The frequency of most elephant calls ranges from 14 to 24 Hz, with durations of 10 to 15 seconds. Fluttering (Fig. 12.11) in a particular area of the elephant's forehead can be observed during infrasonic calling. Infrasonic calls may be important in coordinating the behavior and activity of animals in thick vegetation or in communicating among separated groups of elephants. Fin whales were the first marine mammals known to produce infrasound; elephants were the first terrestrial mammals known to produce such sounds.

Hippopotamuses can produce infrasonic vocalizations both above and below the surface of the water (Montgomery, 1992). Above-water sounds are transmitted through the animal's nostrils, whereas the underwater signal is delivered close-mouthed and is transmitted through the tissue of the head and neck.

Vision

Visual communication occurs in all vertebrate groups, with the eye being a highly specialized special sensory organ in most species. In most fishes, vision is an important sense for

BIO-NOTE 12.3

The Bark of the Dog

The wolf (*Canis lupus*) is considered to be the ancestor of the modern domestic dog. By comparing mitochondrial DNA from wolves and dogs in different parts of the world today, researchers have found that about 100,000 years ago there was a genetic fork in the road of canine evolution—biologically separating wolf and dog. Previous estimates, based on archaeological findings of bones in Germany and Israel, placed canine domestication back about 13,000 years ago—older than cats, cows, and horses, but not by much. Cattle were domesticated only about 8,000 to 9,000 years ago; horses, 6,000 to 7,000 years; cats, 5,000 to 6,000 years; and chickens, 4,000 years. Over time, dogs have become progressively less wolflike, evolving smaller teeth, a more delicate body, and puppylike juvenile characteristics—traits more appealing to human beings.

Barking is the hallmark of the domestic dog (*Canis familiaris*). Coyotes and wolves, on the other hand, bark only rarely. In one study, only 2.5 percent of 3,256 vocalizations by captive wolves were barks. And when wild canids do bark, their barks tend to be brief and isolated, as opposed to the long, rhythmic barking of the domestic dog.

Repetitive barking in wolf pups is significantly more frequent than it is in adults. As the wild animal matures and develops normal adult behavior, it gradually loses its puppylike characteristics.

Regulatory genes control an organism's overall pattern and growth and the rate at which its individual parts grow. Any change in the timing of these regulatory genes is referred to as **heterochrony** (Greek *hetero-*, "different," and *chronos*, "time"). Heterochronic evolutionary mechanisms can speed up or slow down the rate at which an animal grows from a newborn into an adult. It may slow the rate so much that the animal may not attain its "normal" full adult form. Some biologists believe the dog "is an adolescent in a state of change"—reproductively capable but not yet endowed with the full physical and psychological maturity of a "real" adult. Heterochronic change is believed to have frozen *Canis familiaris* in mid-metamorphosis. It remains a "metamorphic" adolescent for life. Its bark is thought to be a juvenile characteristic serving no real function, but probably is motivated by indecision. Some dogs, however, learn to use barking as a means of communication, adapting this initially functionless behavior to serve specific functions such as indicating when they want to be let in or out of the house, or when they want food or attention.

Coppinger and Feinstein, 1991
Vila et al., 1997
Morell, 1997d

finding food and for communicating with other fishes. In many fishes, bright colors are arranged in a wide variety of elaborate patterns that are easily observed. Only in a few vertebrates have eyes degenerated due to a particular lifestyle (certain cave-dwelling and subterranean species). In some terrestrial vertebrates, vision is associated with special sensory patches of skin that can be displayed during courtship, territorial defense, or aggression, such as the dewlap in anoles (*Anolis*) (see Fig. 8.35), the bright gular pouch in male frigate birds (*Fregata*), and the brightly colored buttock pads in drills and mandrills (*Mandrillus*). In birds, vision is associated with the evolution of special plumages (nuptial plumage) used for display during the breeding season.

In mammals, vision is important for recognizing the movements of facial and ear muscles, which can signal intensities of threat, submission, or greeting; and for recognizing special patches of hair that can be displayed at appropriate times, such as the white rump patch and underside of the tail in white-tailed deer (*Odocoileus*) (see Fig. 9.13) and the white rump patch of pronghorn antelopes (*Antilocapra*) and elk (*Cervus*). Dugatkin (1997) listed six theories concerning tail flagging: (1) to warn conspecifics of potential dangers; (2) to close ranks and tighten group cohesion, perhaps to ensure group-related foraging and antipredator benefits in the future; (3) to announce to a predator that it has been sighted and should therefore abandon any attack; (4) to entice the predator to attack from a distance that is likely to result in an aborted attempt; (5) to cause other group members to respond, thereby confusing the predator and making the flagger itself less likely to be the victim of an attack; and (6) to serve as a sign for appeasing dominants, playing only a secondary role in antipredator behavior. Dugatkin (1997) points out that since data on the success rates of predators on flaggers and nonflaggers do not exist, it is difficult to evaluate which, if any, mechanisms of cooperation might account for signaling to a predator. Vision is also important in recognizing body posture, such as the way a tail is carried in carnivores, and, in all vertebrate classes, interpreting the behavior and actions of other members (see Figs. 8.64 and 12.9).

Tactile Signals

Tactile signals are particularly important in the reproductive behaviors of some vertebrates. The tremble-thrusts of male sticklebacks were described in Chapter 5. The long nails on the front limbs of male sliders (*Chrysemys picta*) are used to stimulate the female during copulation.

Grooming, nuzzling, and licking are means of communication involving touch. These actions may be particularly important to altricial young before their eyes open. Touch receptors—**Meissner's corpuscles**—and pressure receptors—**Pacinian corpuscles**—are especially well developed in mammals. Mutual grooming in primates not only serves to eliminate ectoparasites, it is extremely important in establishing and maintaining social bonds.

Electrical Signals

Some groups of fishes, especially the elephantfishes (Mormyriformes), knifefishes (Notopteridae), and electric eels (Gymnotiformes), use electrical discharges for intraspecific signaling. The discharges can be modified in a variety of ways and can be used for the recognition of individuals, courtship, and agonistic behavior. (See discussion of electric organs in Chapter 5.)

■ FEEDING BEHAVIOR

While some fishes, amphibians, and reptiles protect their eggs and young, they do not procure food for the young. Among most birds and mammals, however, one or both parents either secure food and bring it to young in the nest or teach the young how to secure their own food. Most adult birds and mammals secure their own food independently, but some, such as eagles, buzzards, vultures, hyenas, wolves, and lions, will feed on the same food source with other members of their species.

Female mammals are equipped with mammary glands, which provide nourishment to their young. As the young grow, most parents are instrumental in showing their offspring how to secure proper food on their own. In some cases, this is the sole responsibility of the female. In others, both parents participate.

Feeding is a communal activity in those mammals that have developed highly social societies, such as wolves (*Canis lupus*), which live and hunt cooperatively in packs. Rewards of the hunt are shared among all members of the pack. Lions (*Panthera leo*) are matriarchal, with lion prides consisting of several related adult females, their cubs, and a few males. The kill success rates of solitary females appear to be inversely correlated with prey size (Schaller, 1972; Scheel and Packer, 1991; Packer and Pusey, 1997). Although solitary hunters may have the highest kill success rates when hunting small prey such as warthogs (*Phacochoerus aethiopicus*), female lions usually cooperate and hunt larger prey in groups. Hyenas (*Crocula, Hyaena*) and hunting dogs of Africa (*Lycaon pictus*) are other social carnivores. Some primates, such as chimpanzees (*Pan*), gorillas (*Gorilla*), orangutans (*Pongo pygmaeus*), and baboons (*Papio*), live in troops or family units and share the food source.

Predation usually is considered an interspecific action and is discussed in Chapter 13. However, cannibalism is a form of intraspecific predation and may play an important role in shaping the reproductive ecology of many species. It is surprisingly common in many species of fishes and is found in many of the vertebrate groups. The main ecological factor favoring cannibalism is a low availability of alternative food (Elgar and Crespi, 1992). This, in turn, may be a consequence of the density of the population. Competition for reproductive dominance also may be an important factor in some species. The role of cannibalism in the reproductive ecology of the three-spine stickleback has been discussed by FitzGerald (1991, 1992).

To minimize cannibalism, many species of fish produce their young in areas away from adult feeding areas. Adult bullfrogs (*Rana catesbeiana*) are known to ingest many different animals, including tadpoles and newly metamorphosed members of their own species. Male alligators (*Alligator mississippiensis*) will devour newly hatched alligators. Some female mammals under stress will partially or completely consume some or all members of their litter. Cannibalism is used by some males to eliminate unrelated young in order to increase the male's chances of mating with females who have previously bred (see Labov et al., 1985, for review). Infant rodents produce fewer ultrasonic vocalizations in the presence of odors from males than from females, but it is uncertain whether they can discriminate between infanticidal and non-infanticidal males on the basis of odor (Elwood, 1992).

TORPOR (DORMANCY)

Torpor occurs in amphibians, reptiles, birds, and mammals. It may last for only hours (e.g., as in some hummingbirds and many bats that undergo an almost daily cycle of torpor during which they reduce both their body temperature and metabolic rate in order to conserve energy), or it may last for as long as weeks or months (amphibians, reptiles, many mammals). Certain periods of the year may be environmentally unfavorable for certain species of vertebrates, and they become inactive. If this unfavorable period of the year occurs during the colder winter months, some mammals (including many bats) greatly reduce their body temperatures and metabolic rates in order to survive for a prolonged period without food intake (Schmidt-Nielsen, 1990). Such dormancy during the winter is known as **hibernation**. Some vertebrates become dormant for varying periods of time during the hotter summer months. This dormancy is known as **estivation**.

Although torpor is a physiological state, some vertebrates gain advantage during hibernation by clustering together with other members of their own species. In particular, many species of bats hibernate in clusters in caves, presumably as a means of reducing heat and water loss (see Fig. 12.16). Black bears give birth and nurse their young during dormancy. Flying squirrels (*Glaucomys* spp.), golden mice (*Ochrotomys nuttalli*), and others, although not hibernators, have been found in large aggregations during the winter months in some areas. Thus, torpor/dormancy is included in this chapter.

Winter Survival and Hibernation

Hibernation is a well-regulated physiological state of torpor characteristic of temperate zone animals (Schmidt-Nielsen, 1990). In the case of ectothermic groups, such as amphibians and reptiles, body temperature and metabolic activities decrease as the ambient temperature drops. Thus, they are forced to seek a safe refuge (hibernaculum) either on land or in water in order to survive the colder winter months. Selective pressures over time have resulted in most of these groups seeking protected areas where the temperature will remain

FIGURE 12.12

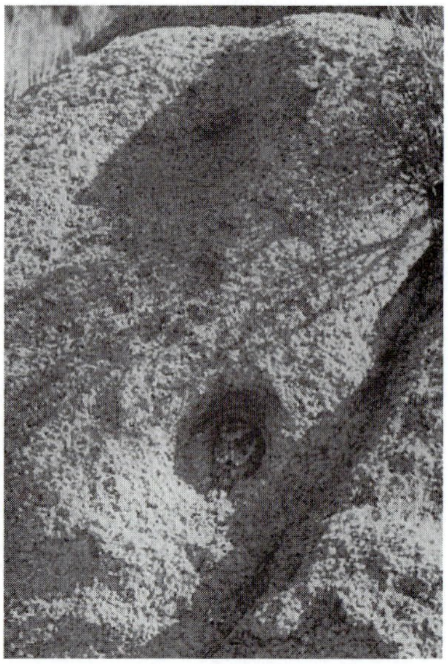

Common poorwill (*Phalaenoptilus nuttalli*) in a rock crevice in California.

FIGURE 12.13

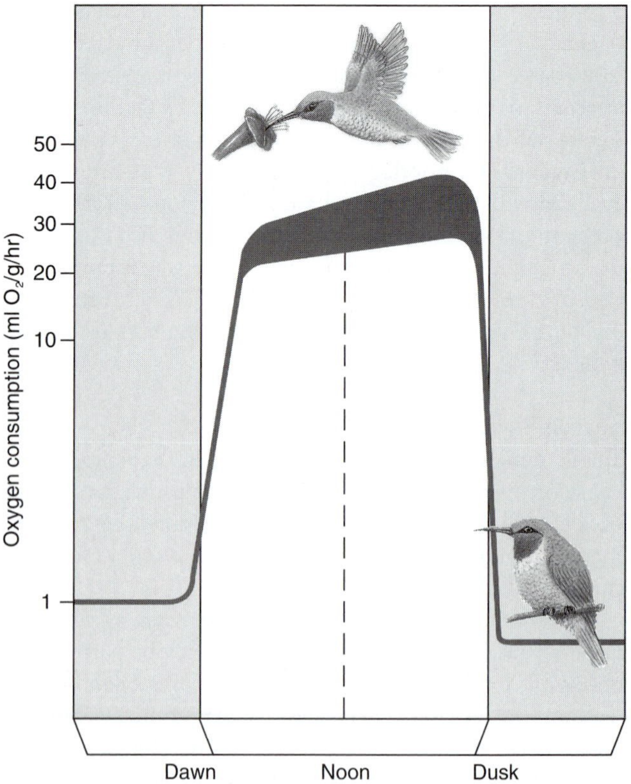

Torpor in hummingbirds. Body temperature and oxygen consumption are high when hummingbirds are active during the day, but may drop to 1/20 of these levels during periods of food shortage at night. Torpor vastly lowers demands on the bird's limited energy reserves.

above freezing and where relative humidity remains fairly constant. Examples of hibernacula include caves, rock crevices, underground burrows, muskrat houses, decaying logs and stumps, sawdust piles, and beneath mud and debris on the bottoms of bodies of water. Several species of anurans including the wood frog *(Rana sylvatica)* and the gray treefrog *(Hyla versicolor)* are freeze-tolerant and can survive temperatures below 0°C (Layne, 1999).

Some closely related species select different types of hibernacula under natural conditions. For example, leopard frogs (*Rana pipiens*) are aquatic hibernators, whereas wood frogs (*Rana sylvatica*) are terrestrial hibernators (Licht, 1991).

Reptiles seek out the same types of protected areas as amphibians in order to survive the winter. Terrestrial turtles may use existing burrows, or they may simply burrow several inches into loose soil or leaf litter. Aquatic turtles may hibernate in beaver lodges, muskrat houses, or in the mud on the bottoms of ponds and lakes. Lizards and snakes may overwinter in caves, hollow trees, sawdust piles, ant hills, houses, and barns.

Hibernating adult Great Basin rattlesnakes (*Crotalus viridis lutosus*) showed differences as great as 4°C in body temperature when measured at the same time in a single den. The lowest body temperature experienced by any Great Basin rattlesnake was 4.2°C (Peterson and Cobb, 1991).

Hibernation is rare among birds. However, the poorwill (*Phalaenoptilus nuttalli*) of western North America is known to hibernate in the wild (Fig. 12.12). During this time, its body temperature may drop to as low as 7°C. Laboratory studies have demonstrated that the trilling nighthawk (*Chordeiles acutipennis*) also is capable of hibernation.

FIGURE 12.14

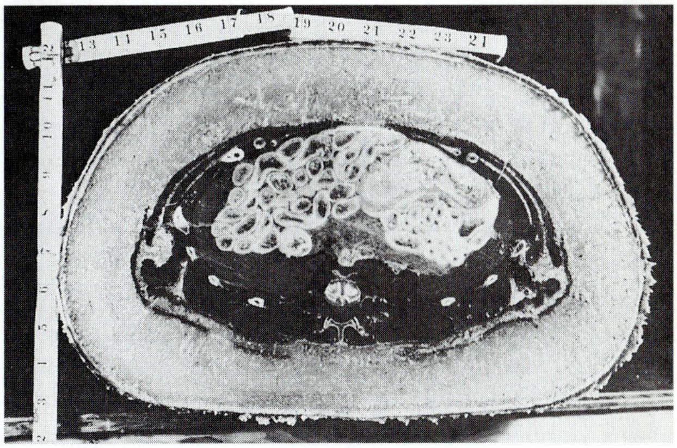

Seal blubber. This cross section of a frozen seal reveals the thick layer of blubber. Of the total area in the photo, 58 percent is blubber and the remaining 42 percent is muscle, bone, and visceral organs. The scale of the ruler is in inches.

Nectar is the main source of energy for hummingbirds, but they also prey on small insects. The amount of food stored during the day determines whether a hummingbird undergoes a period of torpor at night as a means of conserving energy. During torpor, hummingbirds reduce both their metabolic rate and body temperature (Fig. 12.13). During this time their metabolic rate may be reduced up to tenfold (see review by Powers, 1991). Torpor has both advantages and disadvantages. Energy savings is a definite advantage,

FIGURE 12.15

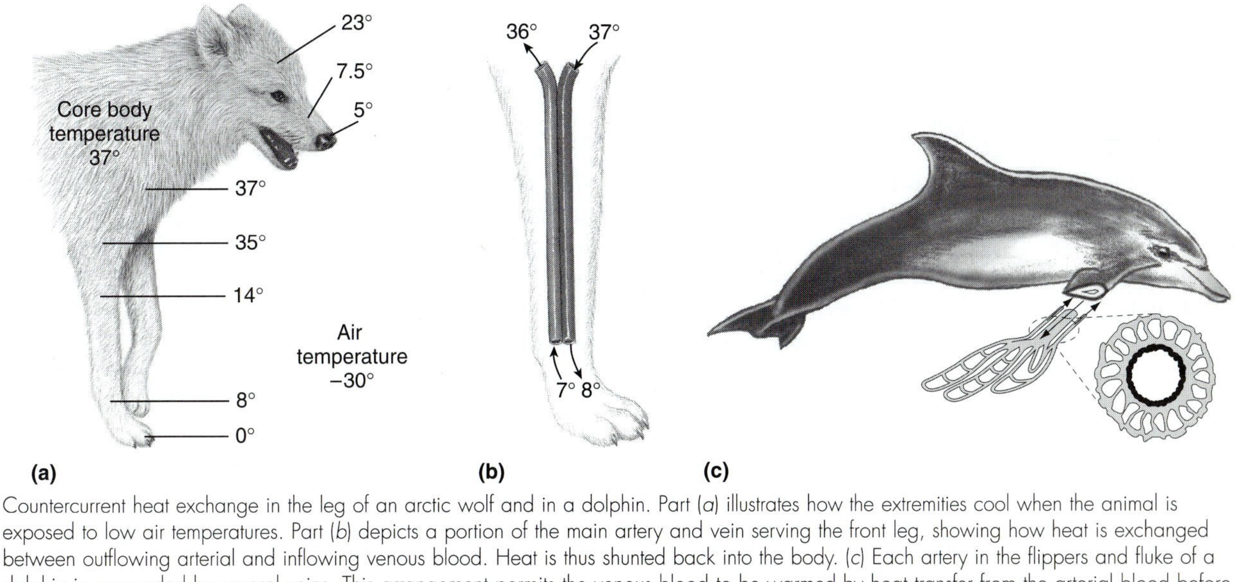

Countercurrent heat exchange in the leg of an arctic wolf and in a dolphin. Part (*a*) illustrates how the extremities cool when the animal is exposed to low air temperatures. Part (*b*) depicts a portion of the main artery and vein serving the front leg, showing how heat is exchanged between outflowing arterial and inflowing venous blood. Heat is thus shunted back into the body. (*c*) Each artery in the flippers and fluke of a dolphin is surrounded by several veins. This arrangement permits the venous blood to be warmed by heat transfer from the arterial blood before it reenters the body.

Source: (c) *Schmidt-Nielson,* Animal Physiology, *4th edition, 1990, Cambridge University Press.*

whereas vulnerability to predation is a major disadvantage in terms of species survival.

Mammals living in cold environments are adapted in several ways to withstand the rigorous winter period. Small mammals use runways beneath the snow as a means of avoiding both cold temperatures and predators. Shivering (muscular thermogenesis) produces some heat through muscle contraction. Some mammals (such as bats) have brown adipose tissue that generates heat rather than ATP when it is metabolized, a process known as **nonshivering thermogenesis**. Fur becomes thicker, and often white, in winter to provide insulation and camouflage. Underfur plays a key role in trapping a layer of air next to the skin, which provides insulation. The limbs and other extremities, however, remain poorly insulated. Seals, sea lions, walruses, narwhals, and whales develop thick insulating layers of blubber around most parts of their bodies (Fig. 12.14). The flippers and flukes, however, lack blubber and are poorly insulated.

Northern species of birds and mammals possess peripheral countercurrent heat exchange mechanisms in their extremities (nose, ears, legs, flippers, and tail) to minimize heat loss (Fig. 12.15) (Schmidt-Nielsen, 1990; Hickman, 1996). These parts of the body are maintained at much cooler temperatures than the rest of the body, sometimes almost approaching the freezing point. The major artery and vein supplying an extremity lie side by side. As warm arterial blood flows into the extremity, heat diffuses into the cooler adjacent vein and is carried back to the core of the body. Thus, while the core body temperature remains relatively high, the temperature in the extremities may actually be below 0°C. Fats in the extremities have much lower freezing points than elsewhere in the body and serve to keep the limbs flexible and supple at low temperatures.

Minor fluctuations in body temperature may occur in northern mammals that remain active through the winter. In northern Minnesota and Michigan, adult beavers had a mean daily body temperature of 36.3°C with a mean daily fluctuation of 1.4°C from late October through early November (Smith et al., 1991). From fall to winter, the body temperature of adult beavers declined at an average rate of 0.01°C/day and the average daily winter body temperature was 35.3°C. In early March, average daily adult body temperatures began increasing at a rate of 0.03°C/day. Juvenile (kit) beavers did not undergo a significant body temperature decline during the winter. Lowering energy demands by reducing body temperature during the winter may be an adaptation by adult beavers that aids in their survival during times of resource scarcity.

Woodchucks (*Marmota*), western ground squirrels (*Spermophilus*), jumping mice (*Zapus, Napaeozapus*), pygmy possums (Burramyidae), and some bats living in northern areas are among the true hibernators. They have a lower metabolic rate, a body temperature near the ambient temperature, much lower heart and breathing rates, lowered blood pressure, and inactive endocrine glands. In certain species, this state of inactivity may last for as long as 6 months in some regions.

FIGURE 12.16

A group of Townsend's big-eared bats (*Plecotus townsendii*) in Jewel Cave, South Dakota, where the thermometer indicates a temperature of 36°F (4°C). Clustering reduces exposed surface areas, thereby reducing evaporative water loss and heat loss.

Hibernating bats and ground squirrels arouse periodically during winter; these arousals are thought to be energetically very costly and account for 75–85 percent of winter fat depletion. It is thought that arousals may be provoked by the depletion of certain metabolic substances, the accumulation of wastes to unacceptable levels, or evaporative water loss (EWL) (Thomas, 1990). Because more than 99 percent of EWL occurs cutaneously, bats can significantly reduce EWL by selecting hibernacula with high humidities and by clustering to reduce exposed surface areas (Thomas and Cloutier, 1992) (Fig. 12.16).

Energy cost and number of arousals are key factors in determining the energy expenditures of hibernators during winter. For a 6.58-g little brown bat (*Myotis lucifugus*), Thomas et al. (1990) calculated that warming from 5° to 37°C required the metabolism of 14.5 mg of fat. During the warming phase, heat is generated by brown adipose tissue, the liver, and muscles. During a hibernation period of 193 days in the Quebec–Ontario region of Canada, little brown bats will arouse about 15 times and require 1,618.5 mg of fat to cover arousal costs. This fat reserve represents 29.3 percent of the bat's mass at the start of hibernation. These data correlate with an earlier study by Fenton (1970), who found that little brown bats in Ontario lost 25 percent of their mass during hibernation.

Golden-mantled ground squirrels (*Spermophilus saturatus*) hibernate for about 64 percent of each year but use only about 17 percent of their annual energy expenditure during that time (Kenagy et al., 1989) (Fig. 12.17). A

FIGURE 12.17

Photograph of a hibernating golden-mantled ground squirrel (*Spermophilus saturatus*). The head is tucked under the body, and the tail is curled over the head.

ground squirrel's core body temperature drops from 36 to 2°C, and its heart slows from 350 beats per minute to as few as 2 per minute.

Semidormant hibernators include such mammals as black bears (*Ursus americanus*), eastern chipmunks (*Tamias striatus*), skunks (*Mephitis mephitis*), raccoons (*Procyon lotor*), badgers (*Taxidea taxus*), and opossums (*Didelphis virginiana*). These species do not experience the drastic changes in their bodily systems that occur in true hibernators. During winter dormancy their body temperatures are not much lower than their normal body temperatures. Although a black bear's heart rate may fall from a normal sleeping rate of 40 beats per minute to 8, its body temperature does not fall more than about 6°C from a normal reading of about 38°C. Whereas small hibernators are slow to rouse, a torpid black bear can become fully alert in moments. They often may be seen moving about outside their dens during the warmer periods of winter.

Skeletal muscle biopsies of bears prior to denning, during denning, and following spring arousal revealed no significant differences in glycogen, triglyceride, and protein concentrations (Koebel et al., 1991). In addition, the activity of citrate synthase, a mitochondrial oxidative enzyme, was not significantly different during these three time periods. A slight degree of muscle atrophy may occur during the denning period, however.

Deer mice (*Peromyscus maniculatus*) undergo daily torpor in some areas. During torpor, they undergo significant intracellular and extracellular acidosis (Nestler, 1991, 1992). Carbohydrate levels are significantly lower, and fatty acid and ketone levels are significantly higher during torpor.

Southern flying squirrels (*Glaucomys volans*) do not hibernate and enter torpor only during extended periods of food shortage or low temperature. To reduce their exposure to cold temperatures and conserve energy, squirrels form small aggregations in nest-lined tree cavities. Research shows that huddling in groups of three and six reduced energy expenditure by 27 and 36 percent, respectively, at 9°C (Stapp et al., 1991), whereas nest insulation decreased heat loss by 37 percent for single squirrels. Aggregating reduced winter daily energy expenditure by 26 to 33 percent.

Factors that trigger an animal to enter hibernation vary. Climatic factors such as colder temperatures, cold rains, and snow are ultimately responsible. Change in daylength has been suggested as a possible triggering mechanism in some species as well as a biological clock that recognizes seasonal rhythms. Recent research efforts have been aimed at isolating the hibernation induction trigger, a chemical compound that appears to be related chemically to opiates such as morphine, which depress certain operational aspects of the nervous system.

Some individuals hibernate singly, whereas others hibernate in groups. Where favorable hibernacula are scarce, snakes of several species may hibernate in large aggregations, or **dens**. One such den in Manitoba was in an active ant hill and contained 257 snakes: 148 green snakes (*Opheodrys vernalis*), 101 red-bellied snakes (*Storeria occipitomaculata*), and 8 garter snakes (*Thamnophis radix*) (Criddle, 1937). An ant hill excavated in Michigan yielded 62 snakes of seven species and 15 amphibians belonging to three species (Carpenter, 1953).

Many mammals hibernate in a characteristic C-shape configuration in order to reduce heat loss and water loss (Orr, 1982) (Fig. 12.17). Many bats hibernate individually, but some hibernate in clusters (Fig. 12.16). It has been suggested that this may be the result of selective pressure to conserve heat. Each individual is kept partially warm by the bodies of surrounding bats.

For a period of weeks and even months prior to hibernation, all endotherms and some ectotherms store fat. Just before entering hibernation, many bats appear to fast. Bears purge their digestive tracts and form an anal plug just prior to entering their hibernaculum. Even though they may become active for short periods during the winter, they will not ingest any food until spring. Those mammals that enter daily torpor have only hours to prepare for significant reductions in metabolism and body temperature.

A major difference between hibernation in endotherms and ectotherms is that endotherms actively terminate torpor by internal heat production. While they all actively raise their body temperature, the rewarming rate is inversely related to their body mass, with most small mammals rewarming faster than larger mammals (Geiser and Baudinette, 1990; Stone and Purvis, 1992). Even though marsupials lack brown adipose tissue, there is no significant difference in warm-up rates between marsupials and eutherian mammals.

Many ectotherms, primarily invertebrates, experience winter temperatures below the freezing point of their body fluids. Those overwintering above the frostline must survive either by extensive supercooling or by tolerating the formation of ice within body tissues. Similar adaptations have been identified in certain vertebrates.

Some fishes avoid freezing solid at temperatures below the freezing point in one of three ways: they produce a solute that lowers the freezing point of body fluids, much as salt lowers the freezing point of slush; they produce an antifreeze protein that binds to ice crystals and inhibits the crystals' growth; or they supercool themselves by lowering the freezing point of their serum and defy the normal freezing temperature (Moyle and Cech, 1996).

Some fish species, such as the naked dragon fish (*Gymnodraco acuticeps*), live in polar oceans with water temperatures ranging between -2 and 0°C. These species manage to remain fluid and flexible in supercooled waters by carrying antifreeze molecules in their blood. A variety of proteins or protein–sugar compounds (antifreeze glycoproteins, or AFGPs) adsorb to the surfaces of forming ice crystals and inhibit crystal growth (Raymond et al., 1989; Rubinsky et al., 1992).

Chen et al. (1997a) report that Arctic cod and Antarctic notothenoid fishes have very similar AFGPs, but that the genes for these antifreeze proteins evolved through completely different pathways. Through convergent evolution, both groups of fishes achieved the same result. Recent studies have shown that fish AFGPs also can protect mammalian cells and organs from damage caused by hypothermic exposure (Rubinsky et al., 1991; Lee et al., 1992). The hypothermic preservation of organs to be used in transplant operations for longer periods of time without damage may significantly affect human medicine.

FIGURE 12.18

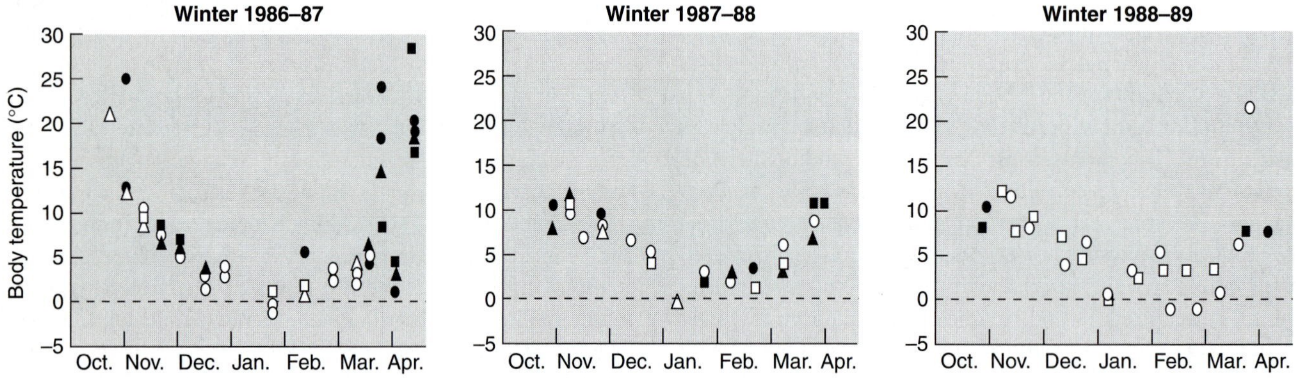

The body temperatures of eastern box turtles (*Terrapene carolina*) in Ohio during three consecutive winters. Five turtles were monitored in 1986–87, three in 1987–88, and two in 1988–89. Filled symbols depict animals on the surface.

FIGURE 12.19

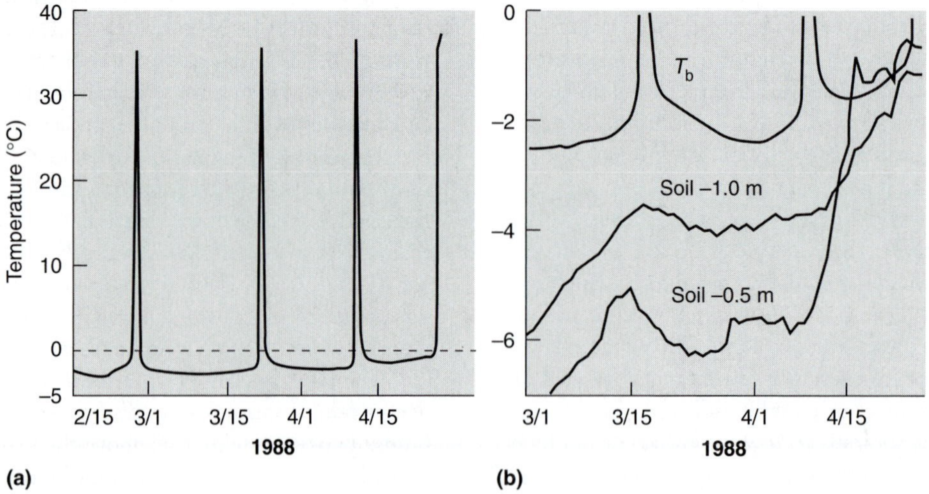

(*a*) Body temperature of a hibernating female arctic ground squirrel (*Spermophilus parryii*) as recorded by an abdominal temperature-sensitive radiotransmitter. (*b*) With an expanded scale, abdominal temperature during the last three arousals from torpor and concurrent adjacent soil temperature.

Source: Brian M. Barnes, "Freeze Avoidance in a Mammal" in Science, *244:1593–1595. Copyright © 1989 American Association for the Advancement of Science.*

FIGURE 12.20

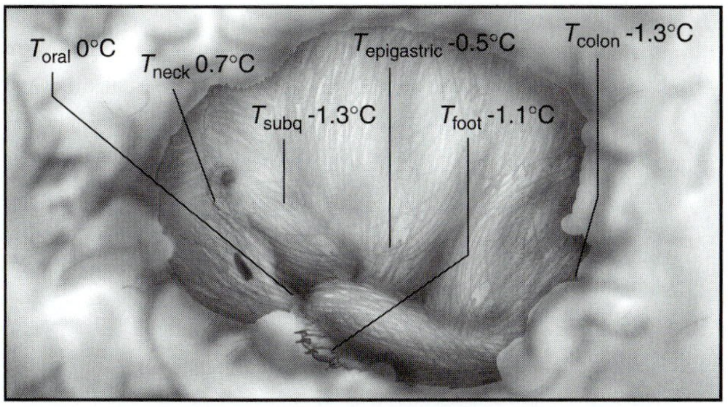

Regional body temperatures of a hibernating arctic ground squirrel housed at an ambient temperature of –4.3°C. Average temperatures (± standard error) and depth of temperature probe for 6 to 11 animals also at –4.3°C were colon: 0.62°C ± 0.11°C, 6 cm; foot: –0.65°C ± 0.15°C, 3 mm; abdominal: –0.59°C ± 0.13°C, 2 cm; thoracic: 0.49°C ± 0.12°C, 1 cm; oral: –0.16°C ± 0.16°C.
From Brian M. Barnes, "Freeze Avoidance in a Mammal" in Science, *244:1593–1595. Copyright © 1989 American Association for the Advancement of Science.*

A few species of terrestrial vertebrates can withstand freezing of part of their body fluids (Claussen et al., 1990). Freeze tolerance has been demonstrated in five frogs (*Rana sylvatica, Pseudacris triseriata, Hyla* [*Pseudacris*] *crucifer, H. versicolor, H. chrysoscelis*), one snake (*Thamnophis sirtalis*), and two turtles (*Chrysemys picta* and *Terrapene carolina*) (Costanzo, Wright, and Lee, 1992).

Eastern box turtles are the largest vertebrates known to withstand freezing. The depth beneath the soil of hibernacula of box turtles in Ohio averaged 4–5 cm and never exceeded 14 cm. Leaf litter averaged 8 cm in depth. Body temperatures of these turtles approximated, but were sometimes lower than, adjacent soil temperatures. Some box turtles experienced body temperatures of -0.3°C or below (Claussen et al., 1991) (Fig. 12.18).

Several anurans (*Rana sylvatica, Pseudacris triseriata, Hyla* [*Pseudacris*] *crucifer,* and *H. versicolor*) have evolved a tolerance to slow freezing by generation of increased blood glucose levels as a cryoprotectant (Licht, 1991; Costanzo et al., 1992). In addition, striated muscle function of *R. sylvatica* remains intact at below-freezing temperatures (Miller and Dehlinger, 1969); its cardiac function remains nearly unchanged under low temperatures (Lotshaw, 1977); and its organs undergo dehydration, presumably to prevent mechanical injury during freezing (Costanzo et al.,1992). Rapid cooling is lethal, because this species inhibits the production and distribution of cryoprotectant and organ dehydration during freezing (Costanzo et al., 1991a, b, 1992).

The hibernating Arctic ground squirrel (*Spermophilus parryiii*) is the only known endothermic animal that can survive without freezing at a body temperature below the freezing point of water (Figs. 12.19 and 12.20). Some individuals have survived after their body temperature dropped as low as -2.9°C (Barnes, 1989). Because the freezing and melting points of plasma from animals held at temperatures below freezing were identical, the presence of antifreeze molecules was ruled out as a freezing survival mechanism. Because no solute was found, it is thought that the animals must have the ability to supercool themselves.

Estivation

Dormancy during periods of great environmental stress such as extreme heat, drought, or lack of food is known as estivation (Latin *aestas*, summer). In many forms, it merely consists of long periods of little or no activity, with no decrease in the metabolic rate below the normal resting rate. In others, it is similar to hibernation in that most of the metabolic activities slow down, but because ambient temperature is not low, the metabolic rate is not as reduced as it is during the winter. Estivation is thought to be triggered by either a reduction in the food supply or by the reduced water content in the food.

South American (*Lepidosiren*) and African (*Protopterus*) lungfishes live in areas subject to flooding and extensive droughts. Both genera can estivate by digging vertical burrows into the mud that end in enlarged chambers. Burrows extend less than 1 m beneath the surface. Heavy mucous secretions cover the lungfish's body, encasing it in a cocoon. Atmospheric air is breathed through a small vent that extends upward through the hardened, sun-baked mud of the dried-up lake or river. Metabolism continues at a very low rate, using muscle proteins as an energy source. Most lungfishes spend less than 6 months in estivation, but some have been revived after 4 years of enforced estivation (Pough et al., 1996). When the rains return and begin filling the sites with water, lungfishes break free from their entombment and resume swimming. Until the next drought, they will inspire air into their lungs and eliminate most of the carbon dioxide through their vestigial gills (Moyle and Cech, 1996).

Some salamanders, frogs, and toads estivate, with some Australian desert frogs remaining in estivation for several years. Box turtles submerge most of their body in a stream and become torpid for periods of up to several days at a time. Several species of snakes have been found estivating. A number of mammals including jerboas (Dipodidae) and pocket mice (*Perognathus*) are known to estivate during the summer months. These mammals usually seek out an underground burrow where the ambient temperature is considerably cooler than it is on the surface.

Review Questions

1. Discuss the characteristics of eusociality in a naked mole rat colony.
2. What are pheromones? What is their significance?
3. Why do male frogs call from their breeding ponds?
4. Why does each species of male frog and toad have a distinctive, species-specific call?
5. Define infrasound. How is it used by some vertebrates?
6. List several reasons that cannibalism may occur.
7. Develop a hypothesis for this observation: Male lions kill the offspring of females they acquire. They do so after they chase away the males that had been the mates of these females.
8. Distinguish between hibernation and estivation.
9. Differentiate between semidormant and true hibernation. Give several examples of each.
10. List several methods used by fish to prevent freezing solid at temperatures below the freezing point of water.
11. Many mammals hibernate either in a characteristic C-shape position or in clusters. What are some adaptive advantages of these behaviors?

Supplemental Reading

Andersen, H. T. (ed.). 1969. *The Biology of Marine Mammals*. New York: Academic Press.

Bagemihl, B. 1998. *Biological Exuberance: Animal Homosexuality and Natural Diversity*. New York: St. Martin's Press.

Bennett, N. C. and C. G. Faulkes. 2000. *African mole-rats*. New York: Cambridge University Press.

Bradbury, J. W., and S. L. Vehrencamp. 1998. *Principles of Animal Communication*. Sunderland, Massachusetts: Sinauer Associates, Inc., Publishers.

Brown, R. E., and D. W. Macdonald (eds.). 1985. *Social Odours in Mammals*. Oxford, England: Clarendon Press.

Busnell, R. G., and J. F. Fish (eds.). 1980. *Animal Sonar Systems*. New York: Plenum Press.

diPrisco, G., B. Maresca, and B. Tota. 1991. *Biology of Antarctic Fish*. Berlin: Springer-Verlag.

Douglas, R. H., and M. B. A. Djamgoz. 1990. *The Visual System of Fish*. London: Chapman and Hall.

Gadagkar, R. 1997. *Survival Strategies: Cooperation and Conflict in Animal Societies*. Boston, Massachusetts: Harvard University Press.

Hara, T. J. 1992. *Fish Chemoreception*. London: Chapman and Hall.

Hoogland, J. L. 1995. *The Black-Tailed Prairie Dog*. Chicago: The University of Chicago Press.

Houck, L. D., and L. C. Drickamer. 1996. *Foundations of Animal Behavior*. Chicago: The University of Chicago Press.

Jellis, R. 1984. *Bird Sounds and Their Meaning*. Ithaca, New York: Cornell University Press.

Lynch, G. R., F. D. Vogt, and H. R. Smith. 1978. Seasonal study of spontaneous daily torpor in the white-footed mouse, *Peromyscus leucopus*. *Physiological Zoology* 51:289–299.

Manning, A., and M. S. Dawkins. 1992. *An Introduction to Animal Behaviour*. New York: Cambridge University Press.

Narins, P. M. 1995. Frog communication. *Scientific American* 273(2):78–83.

Owings, D. H., and E. S. Morton. 1998. *Animal Vocal Communication*. New York: Cambridge University Press.

Payne, K. 1998. *Silent Thunder: In the Presence of Elephants*. New York: Simon and Schuster.

Ruby, N. F., J. Dark, H. C. Heller, and I. Zucker. 1996. Ablation of suprachiasmatic nucleus alters timing of hibernation in ground squirels. *Proceedings of the National Academy of Sciences* 93:9864–9868.

Seigel, R. A., and J. T. Collins. 1993. *Snakes: Ecology and Behavior*. New York: McGraw-Hill, Inc.

Stokes, D. W., and L. Q. Stokes. 1979, 1983, 1989. *A Guide to Bird Behavior*. 3 volumes. Boston, Massachusetts: Little Brown and Company.

Tavolga, W. N., A. N. Popper, and R. R. Fay. 1981. *Hearing and Sound Communication in Fishes*. New York: Springer-Verlag.

Vertebrate Internet Sites

Visit the zoology website at http//www.mhhe.com to find live Internet links for each of the references listed below.

1. **Desert Animal Survival.**
   Excellent site concerning vertebrate adaptations to desert survival, including estivation.
2. **Anabat File Analysis.**
   Call recognition of echolocating bats using the ANABAT bat detection system.
3. **Bioacoustics and Sonar Research Group, University of Loughborough, UK.**
   This site contains information on cetacean echolocation.
4. **International Hibernation Society Home Page.**
   Limited information on hibernation, and updates on upcoming symposia.
5. **Whale-Watching Web/Bioacoustics.**
   Many links to sites with whale songs, and information on bioacoustics of cetaceans.

CHAPTER 13

Interspecific Interactions

■ INTRODUCTION

Because organisms depend on each other for food and other biotic factors, they inevitably interact with each other. Although the most intense relationships exist between members of the same species, individuals do not live apart from members of other species. Living in close association, different species may compete for a shared resource such as food, space, or moisture. These interactions can be classified into several categories: competition, symbiosis (commensalism, mutualism, parasitism), predation, and human interactions. In competition, both species are affected adversely; in commensalism, one species benefits and the other is unaffected; in mutualism, species benefit each other; in parasitism and predation, one species benefits and the other is harmed. Human interactions may benefit both species, only one species, or possibly, either species.

■ COMPETITION

The concept of interspecific competition is one of the cornerstones of evolutionary ecology. Darwin based his idea of natural selection on competition, the struggle to survive.

Whenever different species occupy the same place at the same time, there will likely be competition for common resources such as food, water, or space that are in limited supply. Such interspecific competition consumes both time and energy. Stress caused by such competition may decrease growth and birth rates and/or increase the death rate; if intense, competition can slow or even halt population growth and cause the population to decline. If ecological requirements of two species are similar but not identical, selection pressure will tend to cause the species to diverge from each other through morphological, physiological, and/or behavioral specializations. However, if two species have identical ecological requirements, they will not be able to coexist because of competition for limited resources. Competition is difficult to study and demonstrate in nature because it is such an ephemeral phenomenon.

The fundamental role of an organism in the community is its **niche** (Elton, 1927). The niche is the occupational status of the species in the community—what it does and its relation to its food, its competitors, and its enemies. It is an abstract concept that has not yet been defined and fully measured. A niche should not be confused with a habitat, the physical place where an organism lives. The niche is partially defined by characteristics of the habitat, but also by what the organism eats; how, when, and where it finds and captures its food; the time of the year and time of the day when it is most active; the optimal and extreme climatic factors (heat and cold, sun and shade, wet and dry) it can withstand; its parasites and predators; where, how, and when it reproduces; and so forth. Every aspect of an organism's existence helps define that organism's niche. Interspecific competition may play an important role in shaping a species' niche.

Niches of different species may overlap either temporally or spatially. Niche overlap may promote interspecific competition, but the special adaptations that each species has for its own specific niche should protect it from extinction. For example, downy woodpeckers (*Picoides pubescens*) and hairy woodpeckers (*P. villosus*) are found in similar habitats from Newfoundland to the Gulf of Mexico. Although they often feed on the same tree at the same time, downies feed among the upper and smaller branches, while hairies locate their food on the trunk and larger branches. Niche overlap may occur on the medium-sized branches, but competition is minimized because the *primary* foraging microhabitat is slightly different.

If there is complete niche overlap between two or more species, intense competition for the niche will occur, and one species will outcompete the others. The unsuccessful species will either be excluded from the habitat or forced to shift its niche—usually to a suboptimal habitat. This concept is often known as **Gause's Rule** after the Russian G. F. Gause, who published a study in 1934 showing that when cultured together, one species of *Paramecium* drove a second, competing species to extinction. Hardin (1960) proposed the name **competitive exclusion principle** for this phenomenon. Although competitive exclusion has been demonstrated clearly in the laboratory, it probably is rare in nature because different species seldom compete for precisely the same niche in the same habitat (Mares, 1993).

Fishes may adapt to different foods and to water of different depths, temperatures, salinities, and oxygen contents. Terrestrial species use elevational, macrohabitat, and microhabitat differences as well as food-size partitioning. Studies of plethodontid salamanders in the Appalachian Mountains, for example, have revealed habitat partitioning based on elevation and moisture gradients (Hairston, 1949, 1980, 1983; Dumas, 1956; R. Jaeger, 1971). Daily (diel) and seasonal activity cycles, differences in microhabitats (terrestrial, arboreal), and characteristics of the environment such as temperature or moisture are used by anurans to partition the environment. Within a breeding pond, the difference in time of breeding, egg development, and larval development of different species avoids direct competition with other species. Natural selection favors those individuals that breed at such a time as to avoid competition. Adaptive modifications of salamander larvae and tadpoles, including distinct interspecific differences in mouth structure and whether they inhabit still or flowing water, permit them to gather food in different macro- and microhabitats (see Fig. 6.30).

Anoles of the Greater Antilles (Cuba, Hispaniola, Jamaica, and Puerto Rico) illustrate a classic case of adaptive radiation (Losos and de Queiroz, 1998). Each of six species is adapted to its own ecological niche, in particular to the substrate on which it lives and moves. Anoles that live in the grass have slender bodies and very long tails, whereas a closely related species that must maintain its balance on narrow twigs has evolved a short body and stubby legs. A shorter-limbed species with large toe pads inhabits the upper trunk and canopy of trees, whereas a large species with large toe pads lives high in the crowns of trees. Some prefer shade; others seek out sunny basking sites. Although some overlap occurs, each species consumes differing food items.

Garter snakes and ribbon snakes (both members of the genus *Thamnophis*) often inhabit the same general area (sympatric). Competition between them, however, is reduced by differences in food requirements. Garter snakes feed extensively on earthworms; ribbon snakes usually shun earthworms, but are fond of salamanders, frogs, and small fish.

Interspecific competition is evident among introduced starlings (*Sturnus vulgaris*) and house sparrows (*Passer domesticus*) and native species in North America. Both European species compete with native American hole-nesters such as eastern bluebirds (*Sialia sialis*) and purple martins (*Progne subis*) for suitable nest sites. Due to their aggressiveness, both introduced species often usurp or evict the native species from their nest sites.

BIO-NOTE 13.1

A Competitive Interaction

An interesting example of competitive interaction was reported between bluebirds (*Sialia*) and chickadees (*Parus*). A pair of chickadees began building a nest of green moss in a bluebird house. The next day, the bluebirds entered the house and carried some of the moss away. These actions were repeated. Then the chickadees deposited a single egg on the nearly bare floor. The male bluebird went inside the box and came out with the egg in his beak. He flew to a large tree limb, where the egg balanced momentarily, before falling to the ground below. This action was repeated a second time, after which the chickadees left and did not return.

Reed, 1989

MacArthur's (1967) study of five species of warblers revealed they all fed on the same species of caterpillar prey, but they partitioned spruce trees into preferred foraging regions (Fig. 13.1). Although some overlap occurred, competition was minimal, and all five species were able to coexist during the breeding season.

Among mammals, interspecific competition occurs in some areas between black bears (*Ursus americanus*) and grizzly bears (*U. horribilus*), between red squirrels (*Tamiasciurus*) and gray squirrels (*Sciurus*), and between southern flying squirrels (*Glaucomys volans*) and northern flying squirrels (*G. sabrinus*) (Weigl, 1978; Flyger and Gates, 1983). This competition may be a function of territorial behavior (Layne, 1954; Ackerman and Weigl, 1970; Flyger and Gates, 1983). In the southern Appalachians, four closely related species of mice (*Peromyscus leucopus, P. maniculatus, P. gossypinus,* and *Ochrotomys nuttalli*) inhabited a 6-ha study area (Linzey, 1968). The mice partitioned the habitat by means of spatial orientation (terrestrial vs. arboreal) and by food preference.

Both native and domestic mammals may be affected by indirect competition, a more subtle type of interspecific interaction. Small mammals, such as mice, rabbits, prairie dogs, ground squirrels, gophers, and others, affect the growth of forage plants without competing directly with livestock or game animals for some of the aerial parts of plants. These small mammals consume the roots and early growth of grasses and forbs, and their presence can result in lower forage yields above ground. Long-term investigations of the interactions among rodents, birds, and plants in the Chihuahuan desert of southeastern Arizona have shown a persistent and steady competition among species despite the importance of climatic effects on the numbers of individuals (Brown et al., 1986). Brown et al. (1986) stated: "Our experiments suggest a view of community organization in which virtually all species affect each other through a complex web of direct and indirect interactions. These relationships are

FIGURE 13.1

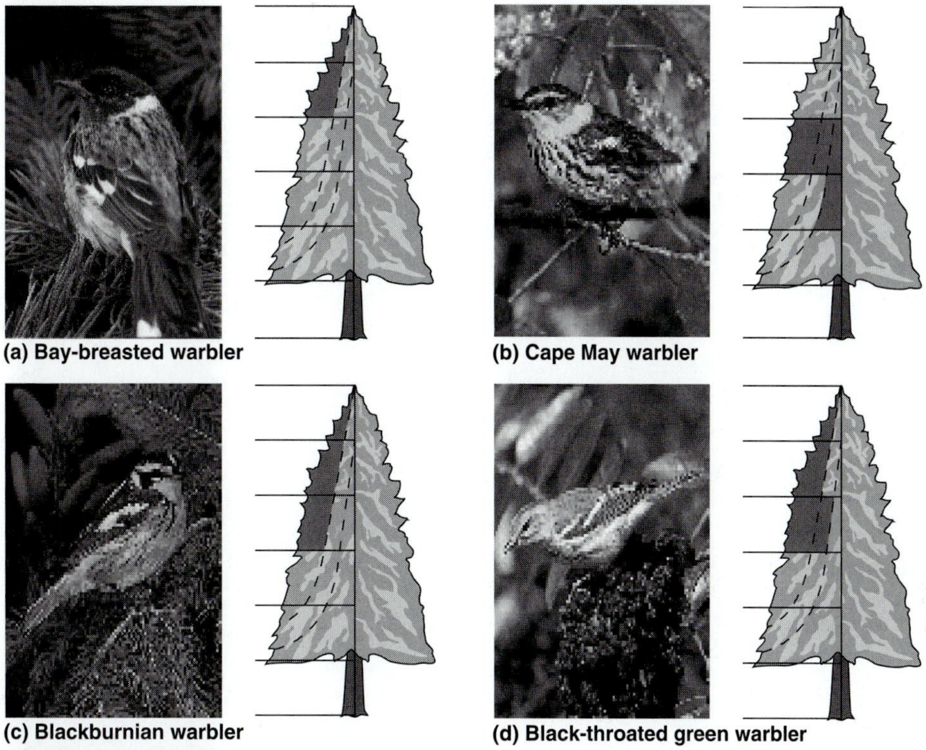

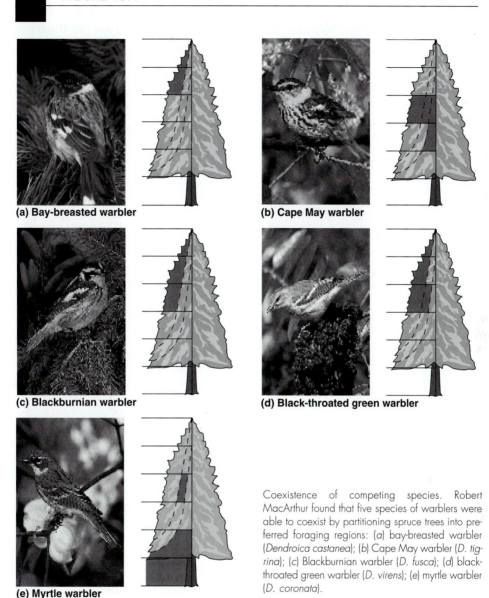

Coexistence of competing species. Robert MacArthur found that five species of warblers were able to coexist by partitioning spruce trees into preferred foraging regions: (*a*) bay-breasted warbler (*Dendroica castanea*); (*b*) Cape May warbler (*D. tigrina*); (*c*) Blackburnian warbler (*D. fusca*); (*d*) black-throated green warbler (*D. virens*); (*e*) myrtle warbler (*D. coronata*).

highly asymmetrical, nonlinear, and influenced importantly by the physical environment as well as by other species."

■ SYMBIOSIS

Symbiosis (*sym*, "together," and *bios*, "life") is the term applied to an intimate relationship between members of different species. Such interactions may be beneficial to one or more members (commensalism, mutualism); other interactions may be detrimental (parasitism). Participants in symbiotic associations often have coevolved with one another and continue to do so.

Commensalism

A **commensal** relationship exists when one member of the association benefits while the other is neither helped nor harmed. Here are some examples: Some fishes, such as jackfish (*Caranx*) and pilot fish (*Naucrates*), seek protection in the vicinity of larger fishes such as barracudas, sharks, and rays (Moyle and Cech, 1996). In most cases, the larger fishes derive no advantage from their companions. A commensal relationship exists between gopher tortoises and many amphibians, reptiles, and mammals that inhabit their burrows (Lips, 1991) (Table 13.1) (Fig. 13.2). Woodchuck (*Marmota monax*) burrows also are used by a wide variety of vertebrates and invertebrates. Turtles often deposit their

TABLE 13.1

Summary of Captures of Amphibians, Reptiles, and Mammals in Gopher Tortoise Burrows in Four Habitats in South-Central Florida

	HABITATS				
			Scrubby Flatwoods		
Species	Turkey Oak	Sand Pine Scrub	Burned	Unburned	Total
Amphibians					
Southern toad (*Bufo terrestris*)	0	1	0	0	1
Greenhouse frog (*Eleutherodactylus planirostris*)	101	70	12	27	210
Narrow-mouthed toad (*Gastrophryne carolinensis*)	0	1	1	0	2
Gopher frog (*Rana areolata*)	9	5	1	0	15
Subtotal	110	77	14	27	228
Reptiles					
Green anole (*Anolis carolinensis*)	1	0	6	1	8
Six-lined racerunner (*Cnemidophorus sexlineatus*)	1	5	4	0	10
Black racer (*Coluber constrictor*)	3	2	2	1	8
Eastern indigo snake (*Drymarchon corais*)	0	2	0	0	2
Southeastern five-lined skink (*Eumeces inexpectatus*)	18	3	0	3	24
Eastern coachwhip snake (*Masticophis flagellum*)	1	1	2	0	4
Eastern coral snake (*Micrurus fulvius*)	0	0	1	0	1
Pine snake (*Pituophis melanoleucus*)	0	0	1	0	1
Fence lizard (*Sceloporus woodi*)	3	6	8	5	22
Subtotal	27	19	24	10	80
Mammals					
Florida mouse (*Podomys floridanus*)	1	1	7	0	9
Cotton rat (*Sigmodon hispidus*)	1	0	0	0	1
Spotted skunk (*Spilogale putorius*)	0	0	1	0	1
Subtotal	2	1	8	0	11
Total	139	97	46	37	319

From K.R. Lips, Journal of Herpetology, *25(4):477–481, 1991. Copyright © Society for the Study of Amphibians and Reptiles, Oxford, OH. Reprinted by permission.*

eggs in alligator nests. The eggs presumably benefit from the alligator's defense of the nest from predators. Aquatic turtles may hibernate inside a beaver lodge. Small birds sometimes construct their nests among the branches and twigs of an eagle's nest.

A unique commensal relationship exists between a bird and a lizard in New Zealand (Carr, 1970). Sooty shearwaters (*Puffinus griseus*) often share their burrows with tuataras (*Sphenodon*) (Fig. 13.3). The diurnal shearwaters occupy their burrows at night while the nocturnal tuatara is out foraging. The tuatara occupies the burrow during the day while the shearwater is fishing. When the bird migrates, the tuatara hibernates in the burrow.

An unusual commensal relationship exists between fossorial blind snakes (*Leptotyphlops dulcis*), which feed on insect larvae in screech owl (*Otus asio*) nests, thus potentially reducing larval parasitism on nestling owls (Gehlbach and Baldridge, 1987). Nestling owls in such nests had a higher survival rate, grew 19 percent faster, and fledged earlier than those in nests without snakes (Table 13.2). Owls transport live snakes to their nests and gain a benefit. There is no evidence, however, that the snakes gain any benefit by being in the nests rather than in the soil.

Squirrel monkeys (*Saimiri sciureus*) find it advantageous to associate with capuchin monkeys (*Cebus*) because the latter provide a better predator warning system than squirrel monkeys possess (Terborgh, 1985). The reciprocal benefit for the capuchins is minimal or nonexistent. Rats (*Rattus rattus* and *R. norvegicus*) and house mice (*Mus musculus*) have benefited by using structures built by

FIGURE 13.2

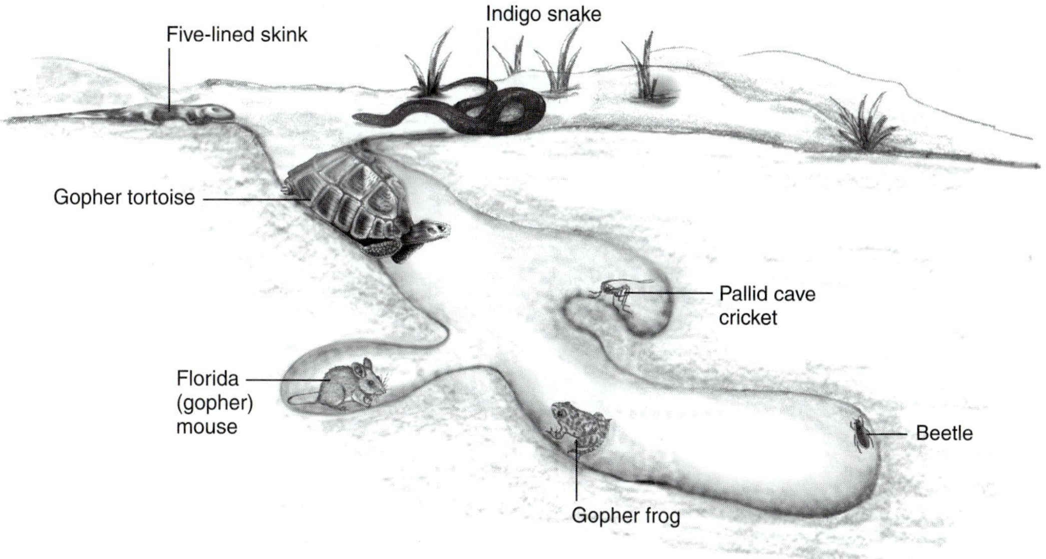

The long, cool burrow of the gopher tortoise (*Gopherus polyphemus*) in Florida provides refuge for a variety of vertebrate and invertebrate species. The tortoise derives neither benefit nor harm from these commensal relationships. Slender beetles feed on tortoise dung; cave crickets eat beetle dung as well as fungus. Gopher frogs (*Rana capito*) eat insects that wander or fall into the burrow. The nocturnal gopher mouse (*Peromyscus floridanus*) may excavate a side burrow in which it constructs its nest. Even the sandy dump pile may provide refuge for a five-lined skink (*Eumeces inexpectatus*). Sometimes, the gopher tortoise defends its burrow against a predatory snake by blocking the entrance with its shell.
Source: Carr, The Reptiles, *Life Nature Library.*

FIGURE 13.3

Sooty shearwaters (*Puffinus griseus*) often share their burrows with tuataras (*Sphenodon*). This unique commensal relationship allows the diurnal shearwaters to occupy their burrows at night while the nocturnal tuatara is out foraging. The tuatara occupies the burrow during the day while the shearwater is fishing. When the bird migrates, the tuatara hibernates in the burrow.

humans and reach their highest densities in agricultural and urban areas. Large grazing animals such as zebras, cattle, buffalo, and horses stir up insects as they feed. Birds, such as cattle egrets (*Caserodius albus*) (see Fig. 11.9) and cowbirds (*Molothrus ater*), live among these mammals and feed on the grasshoppers, leafhoppers, and other insects disturbed by the grazing animals.

Mutualism

In the type of symbiotic relationship known as **mutualism**, both partners benefit from the association. Clownfishes live among anemones. They are not affected by the anemone's sting, which serves to provide them a protected habitat. In turn, the clownfishes defend their homesite from other species that feed on anemones, and also provide anemones with scraps of food. Many other species of fishes allow themselves to be cleaned by cleaner fishes (Fig. 13.4). Some even change color, a procedure that indicates a safe time to be cleaned and also makes parasites more easily visible against a contrasting background. Some pilot fish clean the mouths of manta rays. Reef fishes have been recorded cleaning algae or ectoparasites from sea turtles, and blacknose dace (*Rhinichthys atratulus*) have been observed apparently cleaning wood turtles (*Clemmys insculpta*) (Kaufmann, 1991). Kuhlmann (1966) observed a toothed carp (*Gambusia*) cleaning the mouth of a crocodile (*Crocodylus acutus*). Cleaning symbioses were reviewed and discussed by Feder (1966).

The small ground finch (*Geospiza fuliginosa*) of the Galapagos Islands searches for ticks on marine iguanas. Oxpeckers (*Buphagus africanus*) remove ticks, botfly larvae, and other parasites from zebras, rhinoceroses, and other large mammals (Fig. 13.5).

The intestines of most vertebrates, including humans, provide a suitable environment for beneficial bacteria that aid in food digestion and synthesize certain vitamins. Herbivores, such as cattle, sheep, and deer, depend on bacteria and

TABLE 13.2

Nestling Growth Dynamics in Eastern Screech Owl Nests With One, Undisturbed (by Us), Live Blind Snake at Fledging Time Versus Same-Season Nests Without Blind Snakes but Same-Size Broods. Mean + *SD* and *F* values are from two-way ANOVAs, *N* = 6 each group[a]

	Snake Present	Snake Absent		*F*	*p*
Nestling growth rate (g/day)	4.52 ± 0.54	3.79 ± 0.61	Between groups	49.8	<0.001
			Among broods	11.5	<0.001
Fledging weight (g)	121.3 ± 7.19	123.3 ± 7.73	Between groups	1.1	NS
			Among broods	4.5	<0.01

From F. R. Gehlbach & R. S. Baldridge, "Live Blind Snakes (Leptotyphlops dulcis) *in Eastern Screech Owl* (Otus asio) *Nests." in* Oecologia, *71:560–563.*

[a]N=10 per group in a comparison without brood-size equality; however, mean number of nestlings is no different (3.3±0.7 vs. 2.8±1.0, F=1.6, NS) and results are the same; F=54.8, 7.2 (p <0.001) for growth rate and F=0.8 (NS), 2.5 (p <0.02) for fledging weight, between groups vs. among broods, respectively.

protozoans to help them digest the tough cellulose cell walls of the plant material on which they feed.

Dickman (1992) noted: "Commensal and mutualistic associations among terrestrial vertebrates are clearly dynamic, and form and dissolve under different conditions of predator risk, resource levels, competition, and many other factors. An important assumption is that these associations are favoured only when the benefits to individuals exceed the costs."

Parasitism

Parasitism is a vital interspecific interaction in which one member—the **parasite**—benefits while the other member—the **host**—is harmed in some way. Lampreys parasitize fish by sucking out their blood and body fluids (see Fig. 4.12). Cowbirds of the New World and cuckoos of the Old World are social parasites (Milius, 1998a) (see Fig. 8.72). They both lay their eggs in nests of other bird species, often removing one egg from the host's nest prior to laying their own. Female cuckoos usually lay an average of eight eggs a year. Eggs are laid on alternate days and usually in two batches, separated by several days rest (Davies and Brooke, 1991). Their decep-

FIGURE 13.4

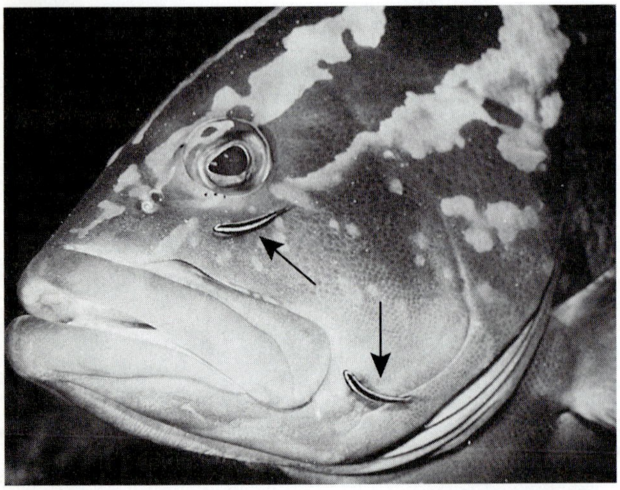

This Nassau grouper (*Epinephelus striatus*) is being cleaned by two gobies. Cleaning symbiosis is a common mutualistic relationship between marine animals.

FIGURE 13.5

Yellow-billed and red-billed oxpeckers (*Buphagus* sp.) perch on the rump of a plains zebra (*Equus* sp.) in Masai Mara National Reserve. The birds help zebras by plucking off ticks and other pests.

tion involves surveillance, stealth, surprise, and speed. In less than 10 seconds, the female cuckoo alights on a nest, lays her own egg, removes one host egg, and is gone. The eggs often mimic the appearance of the host's eggs (Fig. 13.6). Cuckoo eggs have been found in nests of at least 125 bird species in Europe (Wyllie, 1981).

Many researchers have thought that cuckoos imprint on their foster parents and, when adult, choose to parasitize the same host species. However, studies in which newly hatched cuckoos were transferred into nests of other species failed to demonstrate host imprinting (Brooke and Davies, 1991).

Besides having a shorter incubation period than their host species, cowbirds (*Molothrus ater*) hatch before many hosts by disrupting incubation of smaller eggs and, possibly, hatching in response to stimuli from host eggs (McMaster and Sealy, 1998). In addition, young cowbirds and cuckoos are usually larger than the natural young in the parasitized nest, and they either take the lion's share of the food or eject the host young from the nest (see Fig. 8.72). Friedmann and Kiff (1985) recorded 220 species as having been parasitized by brown-headed cowbirds, with 144 species actually rearing young cowbirds. This difference in the number of species parasitized versus those actually rearing cowbirds is due to host recognition and counter-strategies: deserting the nest, rejecting the cowbird egg, or depressing the egg into the bottom of the nest.

In Virginia, 39 percent of dark-eyed junco (*Junco hyemalis*) nests contained at least one cowbird egg (Wolf, 1987). Cowbirds laid an average of 1.7 eggs per nest and removed an average of 1.2 junco eggs per nest. Smaller species such as cedar waxwings (*Bombycilla cedrorum*), Baltimore orioles (*Icterus galbula*), and warbling vireos (*Vireo gilvus*) remove the cowbird egg by puncture-ejection (entire cowbird egg removed or pieces of shell removed after egg contents are consumed) (Sealy, 1996). Larger species generally remove cowbird eggs by grasp-ejection.

In the early 1980s, half of all nests of the least Bell's vireo (*Vireo bellii pusillus*) on the Camp Pendleton military base in southern California were parasitized by cowbirds, and the vireo population was near extinction (Holmes, 1993). When a cowbird trapping program reduced parasitism to near zero, the vireo population increased tenfold. Cowbird populations are being controlled by trapping at Camp Pendleton as well as in the breeding grounds of several endangered songbirds, including the Kirtland's warbler (*Dendroica kirtlandii*) in northern Michigan and the black-capped vireo (*Vireo atricapillus*) in central Texas (Holmes, 1993).

American goldfinches (*Spinus tristis*) are regularly parasitized by cowbirds, but cowbirds do not survive the nestling period because the granivorous diet of the goldfinch provides inadequate protein (Middleton, 1991). It is rare among birds for the diet of nestlings to be composed mainly of seeds, because seeds are relatively low in protein.

The only known parasitic mammals are vampire bats, which feed on fresh blood of sleeping or resting birds and mammals, including humans (Fig. 13.7). Their teeth are sharp, so that the incision is virtually painless and the victim is not awakened or disturbed as blood is "lapped up" rather than "sucked out." The saliva may contain an anticoagulant, so that blood may continue to flow from the wound for several hours. The bitten animal may contract virally caused diseases such as rabies, or secondary infections may develop at the site of the wound.

Male Asian elephants with longer tusks have been found to have fewer internal parasites (Bagla, 1997). Males carrying genes for resistance to parasites are healthier and in a better

FIGURE 13.6

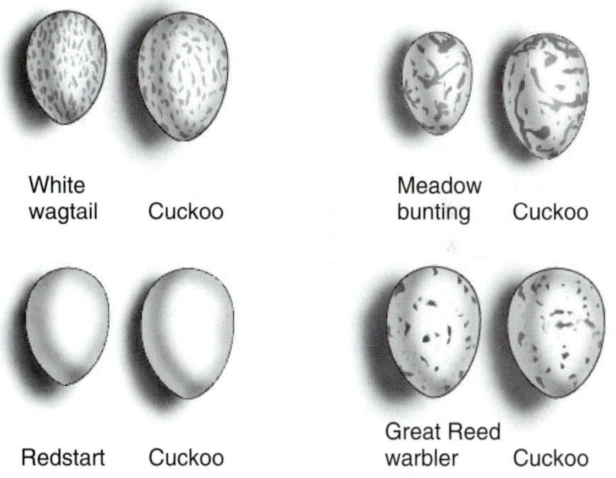

Cuckoo eggs often mimic the appearance of the host's eggs.
Source: Faaborg Ornithology, *1988, Prentice-Hall, Inc.*

FIGURE 13.7

Vampire bat (*Desmodus rotundus*) feeding on the foot of a cow. Observations of vampires indicate that this stance and area of assault illustrate the most frequent method of attack. The stance shows the quadrupedal relationship of bats.

condition to develop secondary sexual characteristics. These better-fit males are then more likely to be chosen by females as mates. Since ivory hunters are likely to poach the best males because of their larger tusks, it is feared that poaching may weaken the elephant gene pool by removing the most fit males and their parasite-resistant genes from the population.

Parasitism of reintroduced and captive endangered species by external (ticks, mites, lice, fleas) and internal (nematodes, trematodes, cestodes) parasites may be a critical danger to their well-being and reestablishment (Phillips and Scheck, 1991). Potential parasites need to be considered when designing and implementing restoration projects.

PREDATION

Predation is an interaction in which one species—the **predator**—benefits from killing and eating a second species—the **prey**. Predators and prey coevolve, with predators becoming specialized to capture their prey, and prey species becoming adapted to evade their predators.

Each vertebrate class has a large number of predaceous species. Sharks feed on other fishes and marine mammals. Largemouth bass (*Micropterus salmoides*) feed primarily on smaller fishes.

Amphibians and reptiles feed on a wide variety of invertebrates and vertebrates, with the choice of food being primarily determined by the size of the mouth-opening—amphibians and reptiles are gape-limited predators. Amphibians feed primarily on invertebrates, although some larger forms such as bullfrogs (*Rana catesbeiana*) will eat almost any suitably sized animal that moves within striking range. Many aquatic turtles feed on invertebrates, as well as on small fish and amphibians. Snapping turtles (Chelydridae) are known to consume amphibians, snakes, small turtles, birds, and small mammals. Most marine turtles are omnivorous. Juvenile green turtles (*Chelonia mydas*), however, are more carnivorous than adults, which subsist mainly on plants. Most lizards feed on invertebrates, although some, such as the Gila monster (*Heloderma suspectum*) and the Komodo dragon (*Varanus komodoensis*), include mammals in their diet. Crocodilians prey on fishes, reptiles, and birds and on mammals as large as antelopes. All snakes are predaceous, feeding on prey ranging from larger invertebrates, birds, and bird eggs (Fig. 13.8) to mammals.

Carnivores, in general, have a more difficult time obtaining food than herbivores. Once a carnivore captures its prey, however, the meal is far higher in nutrition because of its protein and fat content. Thus, meat eaters spend considerably less time eating than plant eaters (Fig. 13.9). In addition, the larger the herbivore, the more time it needs each day to obtain sufficient food.

Many birds and mammals are insectivorous; others feed on a wide variety of fishes, amphibians, reptiles, birds, and mammals. Hawks, eagles, ospreys, and owls feed on fishes, lizards, snakes, other birds, and mammals up to the size of skunks, monkeys, and sloths. Most predatory birds consume their smaller prey whole, later regurgitating the indigestible hair, bones, feathers, scales, or insect parts as pellets (Fig. 13.10). Predatory mammals include such groups as bears, raccoons, cats, wolves, foxes, and weasels. Black bears (*Ursus americanus*) and raccoons (*Procyon lotor*), for example, are major predators on American alligator (*Alligator mississippiensis*) eggs and young (Hunt and Ogden, 1991).

Predators in certain regions have preferred prey (cougar–deer; wolf–moose; fox–rabbit), but most are opportunistic and will kill a variety of prey. They frequently capture older, weaker, debilitated animals, thus acting as a selective agent promoting the genes of those prey animals able to evade capture.

Predators may differentially consume individuals based on age or sex within populations of prey species and thus may have subtle effects on prey-population dynamics. On an island off the coast of western Australia, adult house mice (*Mus musculus*) foraged primarily in dense cover; juveniles, especially females, used areas of open vegetation more than adults and were potentially most at risk of predation (Fig. 13.11a) (Dickman et al., 1991). Barn owls (*Tyto alba*) took a greater number of young female house mice than any other size or sex class. Correlations between the hourly number of hunting owls and the overall hourly capture rates of mice were significant for juvenile female ($r = +0.84$, $p < 0.001$), almost significant for juvenile males ($r = +0.57$, $p \sim 0.05$), and not significant for adults (males: $r = -0.10$; females: $r = +0.51$) (Fig. 13.11b). These data strongly support the hypothesis that juvenile mice, especially females, that use

FIGURE 13.8

A black rat snake (*Elaphe obsoleta*) homes in on a clutch of northern cardinal eggs. Snakes strongly prefer to forage along forest edges. Eggs and nestlings are normally only a minor part of the snake's diet.

FIGURE 13.9

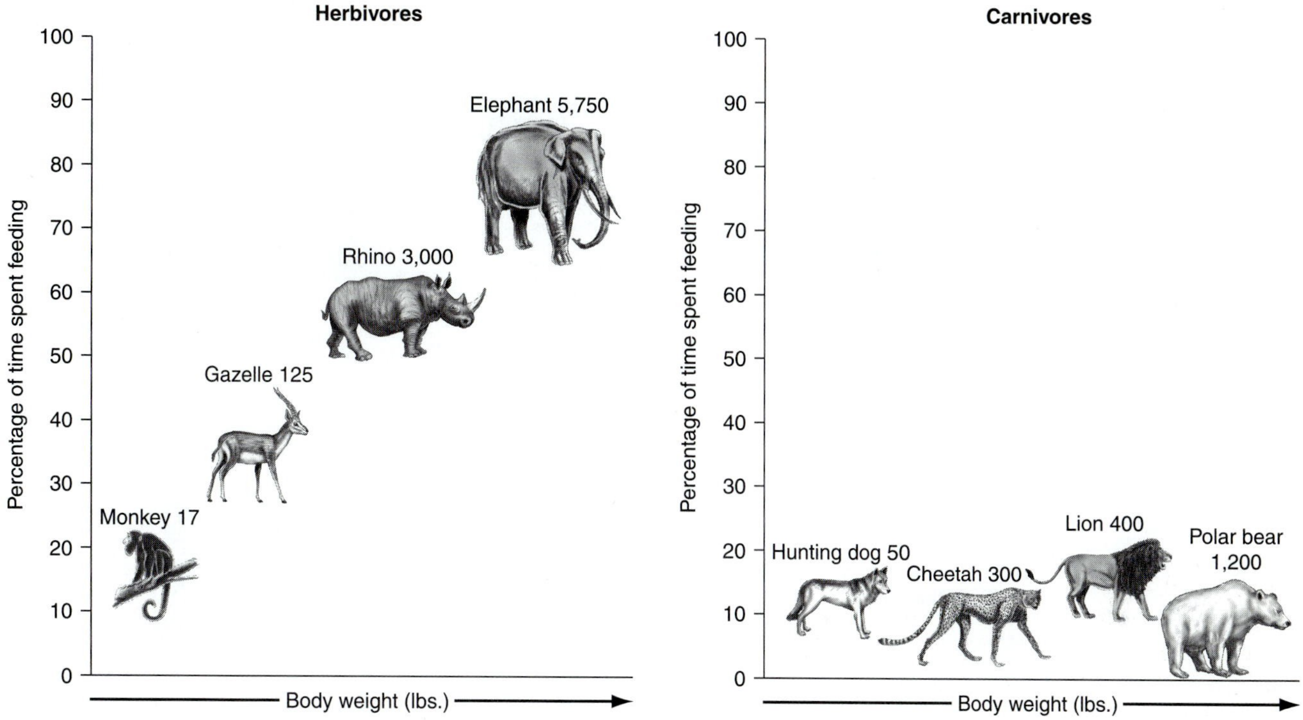

Although carnivores must work harder than herbivores to find a meal, a carnivore's meal is higher in nutrition than an herbivore's. Thus, carnivores spend much less of their time eating than herbivores. In addition, the larger the herbivore, the more time it needs each day simply to stay fed.
Source: Data from Shipman, "What Does it Take to be a Meateater?" in Discover Magazine, *September, 1988.*

FIGURE 13.10

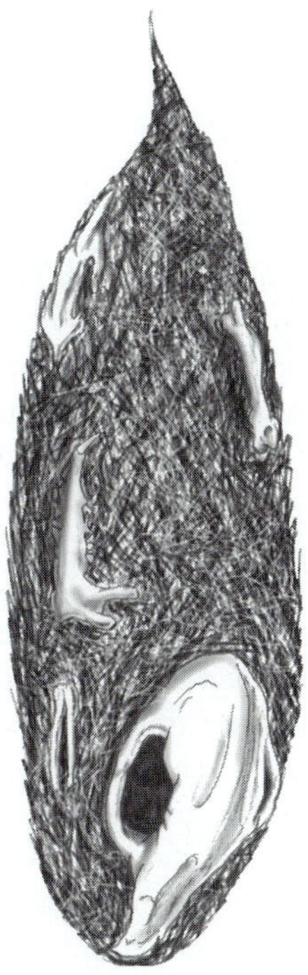

A life-size regurgitated pellet from a short-eared owl (*Asio flammeus*). The pellet contains the remains of a small rodent.

more open vegetation than adults face a higher risk of predation from hunting owls.

Some predators and their prey have developed complex interrelationships. For example, moose (*Alces alces*) colonized Isle Royale in Lake Superior, probably swimming from nearby Ontario in the early part of the 20th century (Mech, 1966) (Fig. 13.12). With no effective predators and an abundant food supply, the population grew to very high levels by the late 1920s. Murie (1934) estimated 1,000–3,000 moose present in 1929 and 1930. Significant mortalities from malnutrition apparently reduced the population to several hundred animals by the mid-1930s (Hickie, 1936). The population again increased, until direct mortality from malnutrition was observed in the late 1940s. In 1947, a population of 600 moose was estimated by aerial strip count (Krefting, 1951). Mech (1966) estimated the 1960 population at 600 animals. The moose population apparently increased during the 1960s (1,300 to 1,600 from 1968–1970) and leveled off, or perhaps even declined, from 1970 to 1974. Mid-winter aerial censuses in 1972 and 1974 produced estimates of 818 ± SE 234 and 875 ± SE 260 moose, respectively (Peterson, 1977).

During the winter of 1948–49, timber wolves (*Canis lupus*) managed to cross the ice from the mainland of Ontario and became established on the island. Their population increased and fluctuated between 20 and 50 animals during

FIGURE 13.11

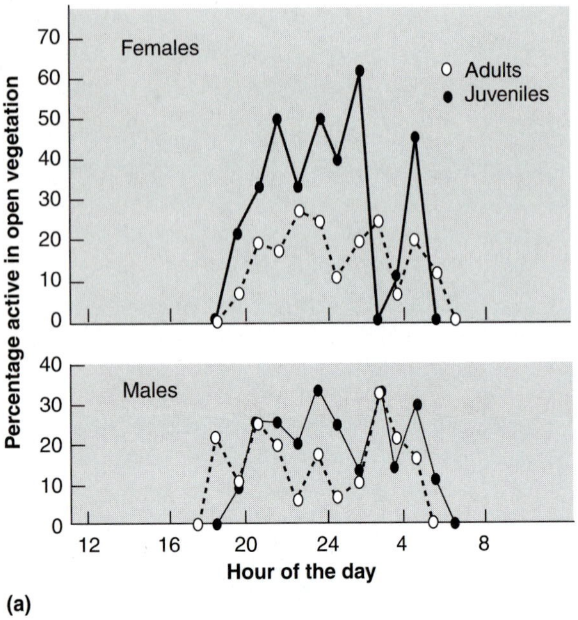

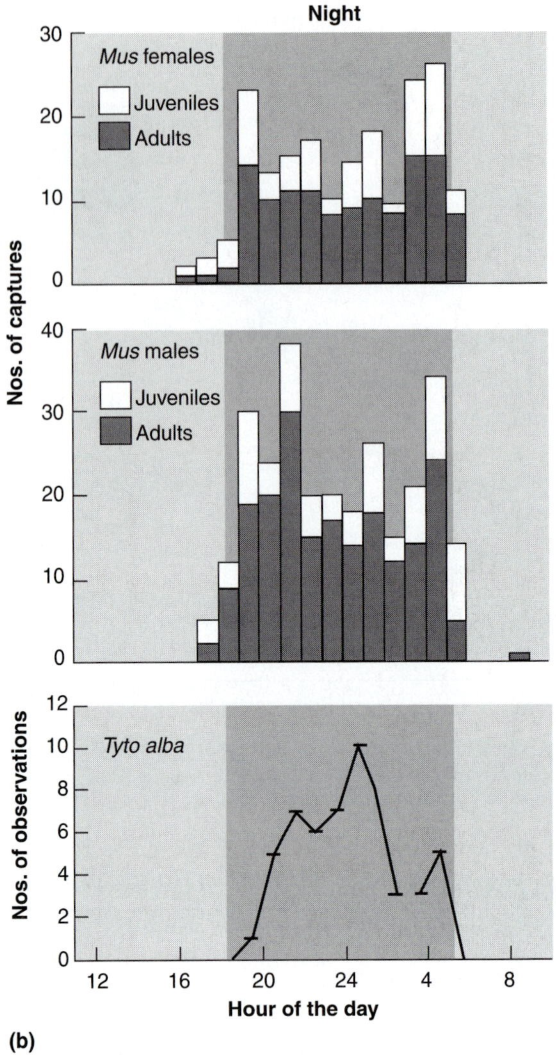

(*a*) Hourly number of captures of *Mus musculus* in open vegetation, expressed as percentages of the total numbers of captures in different sex and size categories. (top) Females. (bottom) Males. Results for three periods of 24 hours are combined. (*b*) Comparison of hourly numbers of captures of *Mus musculus* by sex and size, and numbers of observations of *Tyto alba*. Results for three periods of 24 hours are combined.

the period 1960–1980. Thus, predator and prey had reached a dynamic equilibrium—a stabilization of numbers such that each species could survive without having a detrimental impact on the other. Sufficient resources were available to support the moose population, which was maintained at healthy levels by selective culling of old and weak individuals by the wolves. In 1958, wildlife biologist Durward Allen began tracking the changing population numbers in what has become the longest-studied system of natural predator–prey dynamics in existence.

After the wolf population on Isle Royale reached a peak of 50 animals in 1980, it experienced a decline in the early 1980s, from which it still has not recovered (Fig. 13.12). Only four pups were born, to the same female in one wolf pack, between 1991 and 1993 (Mlot, 1993). As of August 1993, the other two packs were down to just a pair of wolves each. The moose population, which has steadily increased, reached a record high of about 1,900 animals in 1993.

The decline of the wolves was probably the result of two factors: an encounter with canine parvovirus in 1981, and low genetic variability (Mlot, 1993). Because the start of the wolf's decline coincided with a 1981 parvovirus outbreak in nearby Houghton, Michigan, it is thought that the virus could have been carried to Isle Royale on the hiking boots of visitors to the U.S. national park on the island. Restriction enzyme analysis of the wolves' mitochondrial DNA revealed that they were all descended from a single female and had only about half the genetic variability of mainland wolves (Mlot, 1993). Whenever a small number of individuals manage to cross an already existing barrier and found a new geographically isolated colony, they generally carry with them in their own genotypes only a small percentage of the total genetic

FIGURE 13.12

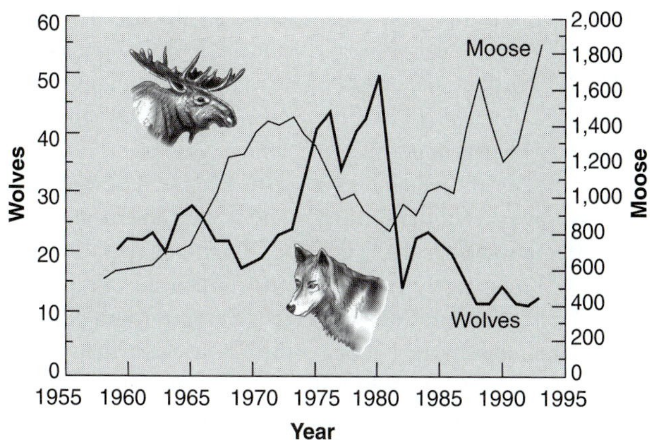

Comparison of wolf (*Canis lupus*) and moose (*Alces alces*) populations on Isle Royale, 1957 to 1993.
Source: Data from C. Mlot, "Predators, Prey, and Natural Disasters Attract Ecologists" in Science, *261:1115, 1993.*

variation present in the gene pool of the parental populations. Thus, the new colony likely will have allelic frequencies very different from those of the parental population. This is a special form of genetic drift, called the **founder effect**. For these reasons, some biologists believe the wolf population is on its way to being extirpated, although others feel it will make a comeback.

Probably the best known predator–prey oscillation is that between lynx (*Lynx canadensis*) and snowshoe hares (*Lepus americanus*) (see Fig. 10.10). Hare populations have a cyclic fluctuation of 8 to 11 years. As the hare population increases in size, so does the lynx population. When the hare population suffers a dramatic decrease, it is followed shortly thereafter by the lynx population. The cycle of the lynx, as well as other predators such as coyotes, great horned owls, and goshawks, lags a year or more behind that of the hare. Various theories have been proposed over the years to explain this predator–prey interaction. Recent evidence has suggested that the hare population size may be limited by their plant food and *not* by a predator (see Cycles, Chapter 10).

Some predators, such as bears (Fig. 13.13), lynx, cougars, owls, and snakes, hunt singly. Others, such as wolves and lions, usually hunt in groups. Female lions normally make the kill, after which food is shared with the rest of the pride, consisting of several males and the young.

Grasshopper mice (*Onychomys* spp.), whose primary diet consists of invertebrates and small mammals, are the only predominately carnivorous rodents in North America. They are extremely aggressive toward other species of small mammals and often include them in their diet. They have a rich repertoire of vocalizations and are reputed to "bay" in pursuit of their prey. When giving this call, the animal often stands on its hind legs with its nose pointed upward. The shrill call is often repeated several times and has been compared to a miniature wolf howl in its qualities of smoothness and duration, and in the posture of the animal when it is calling (Nowak, 1991).

Prey species have developed a variety of antipredator defenses. The schooling of fishes, flocking of birds, and herding of mammals are all adaptations that protect individuals from their predators. Schooling behavior confers considerable antipredator benefit to individual group members through a dilution effect (Morgan and Godin, 1985). The risk of predation is greater for those individuals that stray from a school. Grazing herbivores are almost always alert for predators. If one senses danger, others are alerted by vocalizations or by the display of special patches of hair such as the white rump patch and underside of the tail in white-tailed deer (*Odocoileus virginianus*) and the white rump patch of elk (*Cervus elaphus*) and pronghorn antelope (*Antilocapra americana*).

Crucian carp (*Carassius carassius*) exposed to pike (*Esox lucius*) predation develop deeper bodies—as measured from backbone to belly—than carp that live without pike (Bronmark and Miner, 1992). Over a period of several weeks, carp develop "potbellies," making them too large to fit into the predators' mouths, thus reducing their susceptibility to predation. Although some fish such as the pufferfish (Tetraodontidae), many toads, lizards such as the chuckwalla, and snakes can inflate their lungs and thereby enlarge their body to intimidate a potential predator, this discovery marks the first known instance in which natural selection brings

FIGURE 13.13

The solitary grizzly bear (*Ursus arctos*) is classified as an omnivore, but in coastal regions of the bears' range, salmon become an important component of the diet.

about a change in body dimensions through growth over time in order to avoid predators.

The protective coloration of many species renders them inconspicuous in their normal habitat, whereas some, such as the striped skunk (*Mephitis mephitis*), which have a powerful defensive weapon in its spray, are boldly marked—called **aposematic coloration**—as a warning to potential predators. Some poisonous species also are brightly colored, which serves as a warning to those that might prey on them. The coral snake (*Micruris fulvius*) of the southern United States is one example. The poison-dart frogs of South America are brilliantly patterned with reds, blues, yellows, and blacks. The fire-bellied toad (*Bombinator igneus*) assumes a unique "warning" position when threatened (Fig. 13.14). It exposes its bright red or yellow belly coloration in an attempt to discourage a potential predator.

The frilled lizard of Australia can open up folds of skin around its neck, a behavior which makes it appear much larger and serves as a unique predator-avoidance strategy. In contrast, newly metamorphosed green frogs (*Rana clamitans*) and leopard frogs (*R. pipiens*) crouch and cease to move in the presence of active snakes (Heinen and Hammond, 1997). Green herons (*Butorides striatus*) attempt to conceal themselves by looking like straight tree branches. American bitterns (*Botaurus lentiginosus*) stick their beaks in the air and sway from side to side in an attempt to appear like just another reed in the marsh (Fig. 13.15). Musk-oxen form a circular defensive ring when threatened (Fig. 13.16).

Certain plants have evolved effective antiherbivore chemical defenses. During the winter, snowshoe hares (*Lepus americanus*) are highly selective in their feeding habits in response to such chemical defenses. They often ignore large areas of potential food, even though snow cover puts extreme limitations on the browse available (Sinclair and Smith, 1984). One shunned food source is the Alaskan balsam poplar (*Populus balsamifera*). Buds are ignored, and twigs of juvenile poplar are fed upon only to a slight extent. These poplars produce specific plant metabolites that act as antifeedants. Buds are defended from hares by cineol, benzyl alcohol, and bisabdol, whereas internodes are defended by 6-hydroxycylohexenone and salicaldehyde (Jogia et al., 1989; Reichardt et al., 1990b). The major component of the oil of Labrador tea (*Ledum groenlandicum*), a slow-growing evergreen, consists of germacrone, which also has been shown to be a potent antifeedant to snowshoe hares. Labrador tea has very low palatability for hares, and so this chemical serves as an effective antiherbivore defense (Reichardt et al., 1990a).

Food preferences of wild voles are influenced by the presence of phenolics in certain plants (Bergeron and Jodoin, 1987). Phenolics in the bark of coniferous trees such as white pine (*Pinus strobus*) and white spruce (*Picea glauca*) have been shown to act as deterrents against debarking by meadow voles (*Microtus pennsylvanicus*) (Roy and Bergeron, 1990b).

Voles often cut the branches of young trees and leave them lying in the snow for 2 days or longer before eating them completely. This behavioral manipulation of a potential food source results in a decrease in the concentration of secondary metabolites to a level similar to that found in preferred summer food sources (Roy and Bergeron, 1990a).

Australian koalas (*Phascolarctos cinereus*) feed almost exclusively on the leaves of about a dozen of the 650 native varieties of eucalyptus trees (Stix, 1995) (Fig. 13.17). The leaves are hard to digest and provide one of the least nutritious diets of any mammal. Eucalyptus trees growing in poor soil produce more toxins than trees in good soil. Thus, the trees have evolved leaves that are foul-tasting and toxic, but koalas have evolved complex digestive systems to deal with the poisons.

FIGURE 13.14

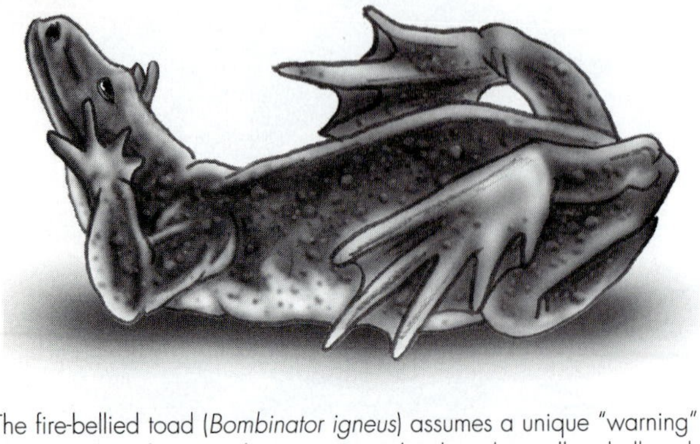

The fire-bellied toad (*Bombinator igneus*) assumes a unique "warning" position when threatened. It exposes its bright red or yellow belly coloration in an attempt to discourage a potential predator.
Source: H. Gadow, Amphibian and Reptiles, *1901, Macmillan and Co., Ltd., England*

FIGURE 13.15

Camouflage behavior of the American bittern (*Botaurus lentiginosus*). The bill is pointed upward in an attempt to be just another "reed" in the marsh.

FIGURE 13.16

Musk oxen (*Ovibos moschatus*) forming a defensive circle with the females and young in the center. Predation pressure can favor the evolution of social life when members of the group are safer than solitary animals.

FIGURE 13.17

Australian koalas (*Phascolarctos cinereus*) feed almost exclusively on the leaves of eucalyptus trees. However, these trees, which provide homes for koalas, are being increasingly destroyed to provide space for urban development.

At times, even biological researchers unwittingly are responsible for the predation of the species upon which they are working. Moen (1973) noted that intrusions by biologists in the process of marking animals, attaching radios, checking nests, and counting eggs may result in increased predation. The effect of human activity depends on the type and amount of cover used by prey species and on the abilities of predators to capitalize on human research activities to locate prey. Moen (1973) noted that crows (*Corvus brachyrhynchos*) have been known to locate nests by watching biologists working at nest sites, and that merely flushing a bird from a nest leaves the eggs exposed to crow predation. Significant losses of ruffed grouse (*Bonasa umbellus*) may have been caused by tags attached to the backs of the grouse, which made them easier to see under field conditions.

HUMAN INTERACTIONS

Humans have had a major impact on many vertebrate species; such effects include the alteration of habitat, the introduction of competitor species into new areas, and the extirpation and extinction of many groups (see Chapters 15 and 16). Possibly the greatest impact of humans on other species will be from global warming. The year 1997 was the hottest year on record (Kerr, 1998). It barely edged out 1990 and 1995, by 0.08°C. All of the 10 warmest years on record have occurred in the last 15 years. In the last 10,000 years through the 19th century, the earth's average temperature appears not to have varied by more than 1.8°F. However, in just the last 100 years, the average surface temperature has risen by about 0.5°F to 1.1°F.

As a result of human impact, populations of some vertebrates have had to be managed to prevent overexploitation and possible extinction. Such is the case with bear, elk, antelope, deer, moose, squirrels, rabbits, muskrats, waterfowl, turkeys, quail, and many other species. Federal and state agencies have been established to propose and enforce legislation defining certain species as "game" animals and establishing open and closed hunting and fishing seasons. Size limits as well as the numbers and kinds (e.g., bucks, does) that can be harvested legally are written into law. Age limits are set for hunters and fishermen, and licenses (in-state, out-of-state) are required.

As a result of these measures, fish hatcheries and game farms have been established to propagate certain species for stocking in order to increase populations of prey for sports hunting and fishing. In some areas, the protective measures begun in the early part of the 20th century have yielded populations of certain species approaching and/or exceeding the carrying capacities of their natural ranges, resulting in damage to crops and ornamental plantings. The white-tailed deer (*Odocoileus virginianus*) in parts of the eastern United States is an excellent example of a species whose population has grown beyond the carrying capacity of its range.

BIO-NOTE 13.2

Seal Die-Offs and Global Warming

Human predation has all but eliminated several regional populations of seals, including harbor seals (*Phoca vitulina*), Mediterranean monk seals (*Monachus monachus*), and Caribbean monk seals (*Monachus tropicalis*). Thousands more seals have succumbed to viral plagues such as phocine distemper virus (PDV), which is caused by a morbillivirus (family Paramyxoviridae). Chemical pollutants known to disarm seal immune systems have been suggested as a possible contributing factor.

Lavigne and Schmitz (1990) proposed two other factors that could explain the 20th century's most conspicuous seal die-offs: unusually warm weather and overcrowding within the seal herd. They note that "four of the six documented mass mortalities in seal populations have occurred in the past 12 years, a period that includes some of the warmest years in the 20th century." They also point out that recent die-offs of dolphins and whales followed unseasonably warm temperatures. Bacteria, algae ("red tide"), and other organisms harmful to this species often flourish as temperatures rise (see Chapter 16).

The apparent global warming resulting from human activities will have the most far-reaching effects of any of humanity's interactions with other species who share the planet with us. Predictions of global temperatures rising as much as 3°C within the next century carry profound implications for the future. Lavigne stated: "Our data show that a 1°C to 3°C increase in average temperature can trigger very significant ecological events." Noting a recent rash of die-offs among seabirds, fish, coral reefs, and sea turtles, Lavigne added: "If the record of the past 12 years is anything to go by, we probably have much more to worry about than seal deaths."

Lavigne and Schmitz, 1990
Osterhaus, 1997

Review Questions

1. Differentiate between the habitat and the niche of a species.
2. Discuss the competitive exclusion principle. Why does it rarely occur in nature?
3. Differentiate between commensalism and mutualism. Give examples.
4. Discuss factors involving the eggs and nestling young of cowbirds that allow them to outcompete the natural young of their host parents.
5. Differentiate between parasitism and predation. Why is it advantageous for parasites not to kill their host?
6. What is the founder effect? How would this affect a population of frogs introduced onto an island devoid of their species?
7. Discuss several antipredator defenses developed by prey species.
8. List several specific pieces of evidence in support of the occurrence of global warming. What effects is global warming having on vertebrate populations?

Supplemental Reading

Ausubel, J. H. 1991. A second look at the impacts of climate change. *American Scientist* 79:210–220.

Byers, J. A. 1997. *American Pronghorn: Social Adaptations and the Ghosts of Predators Past.* Chicago: The University of Chicago Press.

Dietz, R., M.-P. Heide-Jorgensen, and T. Harkonen. 1989. Mass deaths of harbor seals (*Phoca vitulina*), in Europe. *Ambio* 18:258–264.

Furness, R. W., and J. J. D. Greenwood (eds.). 1993. *Birds as Monitors of Environmental Change.* London: Chapman and Hall.

Geraci, J. 1989. Clinical investigation of the 1987/88 mass mortality of bottlenose dolphins along the U. S. central and south Atlantic coast. *Final report to National Marine Fisheries Service and U. S. Navy, Office of Naval Research and Marine Mammal Commission.* April 1989.

Keddy, P. A. 1989. *Competition.* London: Chapman and Hall.

Kerfoot, W. C., and A. Sih (eds.). 1987. *Predation: Direct and Indirect Impacts on Aquatic Communities.* Hanover, New Hampshire: University Press of New England.

Kotler, B. P. 1984. Risk of predation and the structure of desert rodent communities. *Ecology* 65:789–802.

Langland, W. S., and M. V. Price. 1991. Direct observation of owls and heteromyid rodents: Can predation risk explain microhabitat use? *Ecology* 72:2261–2273.

McGowan, C. 1997. *The Raptor and the Lamb: Predators and Prey in the Living World.* New York: Henry Holt.

Schaller, G. B. 1972. *The Serengeti Lion.* Chicago: The University of Chicago Press.

Schneider, S. H. 1989. The changing climate. *Scientific American* 261(3):70–79.

White, R. M. 1990. The great climate debate. *Scientific American* 263(1):36–43.

Vertebrate Internet Sites

Visit the zoology website at http://www.mhhe.com to find live Internet links for each of the references listed below.

1. **Interspecies Communication.**
   Interspecies communication between humans and particularly cetaceans, such as orca and sperm whales; sounds of orcas; information about *The Interspecies Newsletter*, links to other sites.
2. **Predation and Competition.**
   An on-line course with lessons and information on competition, predation, and coevolution from Northern Arizona University.
3. **Skies Above Foundation–Home.**
   Check out the "Related Links" that lead you to much information on global warming and climate change.
4. **Competition.**
   A description of competition and competitive exclusion models from a course at the University of Minnesota.
5. **The Struggle for Existence—A Classic of Mathematical Biology.**
   Gause's classic writings.

CHAPTER 14

Techniques for Ecological and Behavioral Studies

■ INTRODUCTION

Discovery and problem solving in all fields of science employ techniques of gathering, coordinating, organizing, and evaluating information as it relates to a specific subject. In a scientific approach to any problem, the researcher must first ask a question or identify a problem based on observations of objects or events. Then, a hypothesis or potential answer to the question being asked is proposed, and the investigator predicts what the consequences might be if the hypothesis is valid. The investigator then devises ways to test the hypothesis by making observations, developing models, or performing experiments. Hypotheses must be testable; those that are not testable are inadmissible in science. Observations and/or tests should be repeated as often as necessary to determine whether results will be consistent and as predicted. Hypotheses that are found upon testing to be contradicted by the evidence must be modified or abandoned. The investigator must then report objectively on the results and on conclusions drawn from them, presenting both the data and the investigator's intepretation of the information as it relates to the hypothesis. This mode of action is known as the **scientific method**. Critical evaluation of the techniques—or methodology—used in any scientific investigation is extremely important. Thus, scientists constantly must be concerned with the selection and application of the best techniques for use in each of these steps. A deficiency in any of these steps will hinder the interpretation of the results.

Limitations inherent in field and laboratory investigations of animal populations make it even more critical that researchers choose most carefully the techniques to be used. The mobility, secretiveness, and constant fluctuation in numbers of practically all wild animals make precise data difficult to secure. For these reasons, census work often requires a major portion of the time in many field investigations. The secretive nature of most wild animals makes the determination of the influence of pathology, disease, and related factors especially difficult to examine for any species of animal in nature.

Because the objective of any investigation is to gather and evaluate accurate data, the investigator must always bear in mind that the techniques used should yield data that are objective and reliable. If the investigator's approach to the problem is not scientifically sound—if the techniques are inadequate or flawed—the results will be of little value. Critical appraisal of the investigational techniques should be made at the beginning, not at the termination, of the research project.

The difficulties mentioned above highlight the need for careful planning and equally careful collection of data on the part of fish and wildlife investigators. Inexperienced investigators often propose to collect data that, due to field conditions or the characteristics of the animal being investigated, are impossible to secure. Thus, detailed planning, including critical scrutiny of all the proposed techniques to be employed, is necessary to ensure that insurmountable problems are not encountered in the proposed study. Long-term studies are needed to learn these limitations and to collect the necessary information; yet this is rarely accomplished, since most studies are short-term (and are *financed* as short-term projects).

Many studies do not require the capture of individual animals. Techniques employing simple observation, aerial photography, aerial censusing, transect counts, actual counting of individuals such as fishes and hawks that pass a certain point on their migratory journeys, and identification of signs such as tracks and scats can provide valuable data. For some purposes, animals such as amphibians, reptiles, and mammals killed by vehicles (DOR = dead on road) can yield useful data. Many behavioral studies can be done in the animal's natural habitat.

Other studies, however, do require direct contact between animal and investigator. Such studies include those dealing with the collection of anatomical data such as weight, length, condition of molt, or the like; or those dealing with age determination, sex ratios, genetic analyses, home ranges, and parasites.

■ CAPTURE TECHNIQUES

A wide variety of techniques are used for capturing vertebrates. Humane capture techniques should always be employed. They should not injure or increase the mortality of the animals, and they should not cause more than minimal disruption to the animals' normal behavior patterns.

Dip-netting, seining, the use of trap- and gill-nets, and the use of immobilizing chemicals and electroshocking are among the capture methods utilized by fish biologists. Most amphibians can be captured either by hand or with the use of a net. Some terrestrial reptiles also can be taken by hand, although nets, nooses, and tongs frequently are used for some species. Aquatic turtles may be secured through the use of turtle traps, commercial fish trap-nets, and trawls.

Nestling birds can be removed from the nest by hand for weighing, sexing, and tagging. Fine-mesh mist nets are often used to capture small flying birds, which become entangled in the mesh and can be removed uninjured for study. Live traps placed on the ground and baited with seeds are used successfully for some granivorous species. The projection, or cannon, net trap is widely used for turkeys and waterfowl (Fig. 14.1). It consists of a large, light net that is carried over the baited birds by mortar projectiles or rockets. Funnel-entrance traps are used commonly for waterfowl. Hawks can be trapped by using traps baited with live prey. Some carrion-eating species have been immobilized by consuming drug-laden meat.

Many small-mammal researchers employ traps. These should be live traps suitable for the species, although snap traps were extensively used in the past. Traps may be placed on the surface of the ground or in tunnels, or they may be affixed to the branches of trees.

Shrews are taken most effectively in pitfall traps in which a series of containers (cans, plastic cups, etc.) are buried with their tops flush with the ground and loosely covered by a piece of wood or some other object. Due to the shrews' high metabolism, this method of collection will yield live shrews only if the cans are checked several times each day. If the cans cannot be checked frequently, they can be partially filled with a preservative liquid/fixative to kill and preserve animals for future study. All cans should be removed and the holes filled at the conclusion of the study. Small-mammal distribution studies can be augmented by examining discarded bottles along roadsides. Shrews are the most abundant small mammals found in bottles (Morris and Harper, 1965; Glegg, 1966; Pagels and French, 1987). Examination of owl pellets also can yield valuable distributional and population data on shrews, voles, and other small mammals.

Bats usually are captured with mist nets positioned at cave entrances or along watercourse flyways. Nets must be monitored continually, and bats removed as soon as possible in order to prevent injury.

Although mammals as large as bears can be trapped successfully with snares and culvert traps, most large mammals are shot with a tranquilizer-containing dart. Fairly accurate estimation of weights of animals in the field must be made for proper dosages to be administered.

Drift fences and traps are used for studies on a wide range of terrestrial vertebrates, including amphibians, reptiles, and mammals. This method requires the erection of one or more fences with openings at periodic intervals. The object is to direct the movement of an animal into a trap at one of the openings.

FIGURE 14.1

(a)

(b)

(*a*) The cannon net is an effective way of taking gamebirds, unharmed, for scientific purposes. The birds are prebaited at the site; the net is then carefully folded and camouflaged in front of the "cannons." (*b*) When properly deployed, the net is highly effective. The birds are snow geese (*Chen hyperborea*).

IDENTIFICATION TECHNIQUES

To study the movements and behavior of animals in the wild, there must be a means of identifying specific individuals. In some cases, this can be accomplished by noting unique individual characteristics such as distinctive coloration, scars, deformities, injuries, or some aberrant behavior. For example, Schaller (1963) found that the noses of mountain gorillas appeared distinctive and served as the best single character for recognizing individuals.

In most cases, however, it is not possible to distinguish individuals visually. Therefore, some appropriate method of marking or tagging each individual animal must be devised. In deciding on a particular technique, consideration must be given as to whether the study in question is short-term or long-term and how many animals will be involved. The method selected should not injure the animal, alter the animal's behavior or locomotion, or cause increased susceptibility to predation. Many identifying techniques have been devised including marking, tagging, photography, use of radio transmitters, and satellite tracking.

Marking

Marking usually refers to changing a part of the animal's body so that it can be discerned readily from all other members of the population. Moyle and Cech (1996) summarized fish marking methods as follows: "Marks may consist of clipped fin rays, liquid nitrogen 'cold brands,' pigmented epidermis from high-pressure spray painting, or fluorescent rings on bones or scales (visible under ultraviolet light) from incorporation of tetracycline or 2, 4-*bis(N, N'*-dicarboxymethyl-aminomethyl) fluorescein (DCAF) in the diet." Juvenile salmonids have been marked chemically by feeding them dissolved strontium, a biologically rare element, which is then incorportated into their scales (Snyder et al., 1992).

Amphibians usually are marked by toe clipping—that is, excising the terminal phalanx of one or more toes in a specific pattern. They also may be marked by branding and by the use of dyes and phosphorescent powders. Larval (tadpole) stages may be semipermanently marked by injecting an acrylic polymer dye into the fin. A detailed discussion of marking and tagging techniques suitable for amphibians may be found in Heyer et al. (1994).

Lizards and snakes may be marked by toe clipping, by excising specific scales in a prearranged manner, or by using a latex-based house paint. Marking methods for reptiles have been reviewed by Dunham et al. (1988). Individual turtles can be identified by having an identifying mark painted on their shells or by notching specific marginal scutes. Birds may be marked by dyeing their feathers.

Mammals may be marked by toe clipping, fur clipping, ear notching, tattooing, branding, dyeing, painting, or bleaching. In the case of toe clipping, smoked paper affixed to plywood or cardboard can be placed throughout the study area so that whenever a marked animal crosses the surface, it will leave its own distinctive identifying imprint.

Commercial dyes have been employed in various ways to identify mammals. In some species, the dye is applied to the captured animal prior to its release. In other studies, a marking device may be placed in the animal's normal habitat (designed in such a way that the animal triggers the device in its typical pattern of activity). Once triggered, the device discharges a quantity of the dye onto the animal's body. The use of a dye in this manner will provide visual identification until the animal undergoes its next molt. Fluorescent powders have also been used successfully.

Tagging

Tagging requires the attachment of a metal, plastic, or cloth device to the body of an animal to allow for future identification (Fig. 14.2). Any tagging device must anticipate the growth of the animal and must not impede its movements or other normal behavior.

Tagging of fishes can be done by externally and internally attached disks, microtags, dart tags, plates, streamers, and small, implantable metal rods detectable in a magnetic field (Moyle and Cech, 1996). Electronic tags that record depth, water temperature, and light intensity weigh as little as 16 g and can store more than 500,000 data samples (Metcalfe and Arnold, 1997). Data on plaice (*Pleuronectes platessa*) have been recorded continuously by electronic tags for over 200 days.

FIGURE 14.2

A biologist tags a hawksbill turtle (*Eretmochelys imbricata*) in an effort to gather more information on the species' movements and habitat needs. The threats facing this species include habitat destruction and commercial demand for stuffed juveniles and products made out of its shell.

Amphibian studies have used tags and radioactive isotopes for identifying individual animals. The use of isotopes allows the continuous monitoring of an individual without recapture. Passive integrated transponder (PIT) tags are small, glass-encapsulated diodes (0.1 g) and do not require batteries (Camper and Dixon, 1988). When activated by a detector, they transmit a unique code back to the receiver. PIT tags must be implanted in the animal (thus, infection is a major consideration), and current transponder systems have a very short range. Ingested radio transmitters also have been successful in yielding short-term data on amphibians such as the common frog (*Rana temporaria*) and the common toad (*Bufo bufo*) (Oldham and Swan, 1992) as well as on snakes.

Several unique methods of tracking turtles have been employed. Stickel's (1950) attachment of a spool of thread to the carapace of a box turtle yielded valuable data on the movements of this species. The attachment of helium-filled weather balloons to marine turtles allowed tracking of their movements for short periods of time.

For years, ornithologists have been studying migration in birds, as well as many other aspects of avian biology, by using aluminum, stainless, or monel alloy leg bands (Figs. 14.3a, b and 14.4). These tags are numbered and contain the address of an agency to which the finder should mail them. In the United States, the agency is the U.S. Fish and Wildlife Service. Bands come in a variety of sizes, and future growth in the diameter of the leg must be carefully anticipated prior to attachment of the band. These bands have provided valuable data on the migratory habits of many species of birds, but a bird must either be recaptured or found dead in order to be

FIGURE 14.3

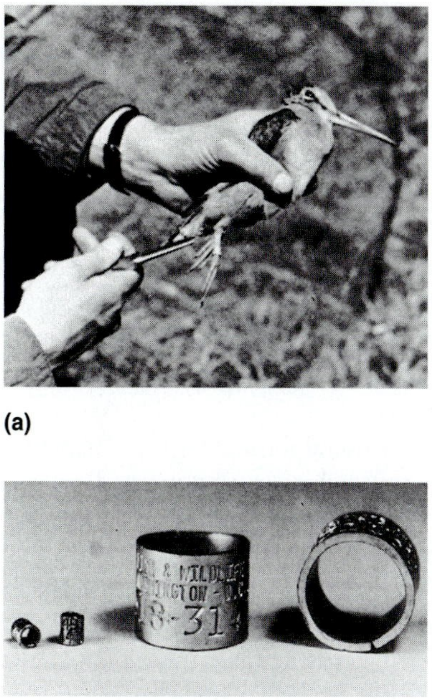

(a)

(b)

(*a*) Banding a woodcock (*Philohela minor*). Future growth of the leg must be anticipated when selecting the proper size band. (*b*) Bands of various sizes, made of soft, lightweight metal, are provided by the U.S. Fish and Wildlife Service for bird banding to determine the migratory movements of various species.

positively identified. A toll-free telephone number (1-800-327-BAND) is now available to report any bird band identified or recovered in North America. This recording service, developed in cooperation with the National Biological Survey, the U.S. Fish and Wildlife Service, and the Canadian Wildlife Service, can be called from anywhere in the United States, Canada, and most parts of the Caribbean.

Colored leg bands, neck bands, or plastic streamers are used in behavioral or home range studies so that individual birds can be identified without recapture (Fig. 14.5). Patagial tags and feather grafts also have been used as field identification tags (Fig. 14.6).

Researchers studying western European populations of the white stork (*Ciconia ciconia*) implanted electronic PIT tags (30 mm long, 3 mm in diameter, 0.8 g mass) beneath the stork's skin (Michard et al., 1995). The tag, which permits automatic individual identification, is long-lasting because it does not require a battery. Body condition also can be assessed, as the birds weigh themselves on scales coupled with tag-identification systems at feeding sites. The tags are read by an antenna-recorder from a distance of approximately 1 m.

Mammals may be tagged in a variety of ways. Studies involving bats utilize lightweight aluminum bands similar to those used for birds. These bands are numbered and are affixed to the forearms of the bats. In some studies, inch-long luminous cyalume rods have been attached to the backs of bats for easier tracking at night. Metal or plastic ear tags

FIGURE 14.4

The neck collars on these parent Canada geese (*Branta canadensisi*) make it possible to keep track of eggs and young up to the migratory stage, yielding information on daily and seasonal habitat preferences.

FIGURE 14.5

Colored patagial tags have been used to study the breeding behavior of mourning doves (*Zenaidura macroura*).

have been used on mammals of all sizes. In some cases, colored plastic streamers have been attached to the tags so that visual identification can be made at a distance. Neck collars are used on larger mammals. Unfortunately, all tags are

FIGURE 14.6

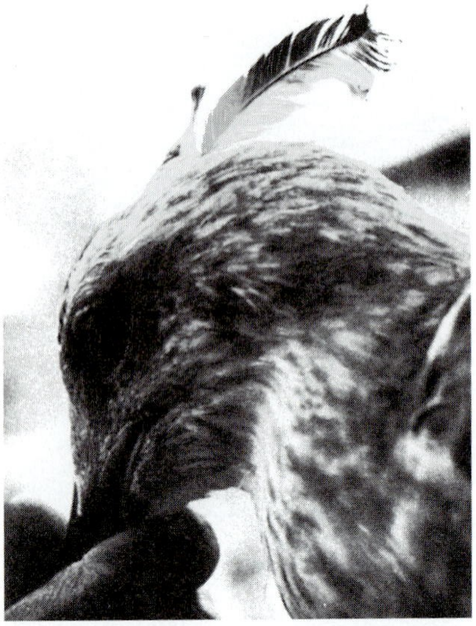

Feather graft of an immature wing feather onto the head of a great black-backed gull (*Larus marinus*). Best results are obtained when the graft is made on immature birds. The grafted feather is permanent and molts with other body feathers, thereby serving as a permanent field identification "tag."

subject to loss; Siniff and Ralls (1991), for example, reported an estimated annual tag loss rate of 26 percent in California sea otters (*Enhydra lutris*).

Spool-and-line tracking has been employed in several mammal studies. This technique utilizes a spool of thread attached to the animal's body. The spool continuously releases thread as the animal moves, thus providing a fairly accurate representation of the animal's travels. For example, Hawkins and Macdonald (1992) used spools attached to webbed collars to investigate the movements of badgers (*Meles meles*). One disadvantage of this method is that it yields only 1 or 2 nights of potentially high-quality data per capture.

Dyes have been incorporated in food in order to stain the feces. In small-mammal studies, dropping boards are placed throughout the study area in order to facilitate the recovery of dyed fecal pellets. This is a temporary technique that depends on the rate of passage of the food material through the animal's alimentary canal.

Radioactive isotopes in the form of wires and pellets have been inserted under the skin of various species of mammals. This method of tagging permits continuous location of the animal with minimal disturbance. Radioactive materials injected into animals will render their feces identifiable.

The use of genetic tagging has revealed individual local and migratory movements and yielded estimations of abundance in humpback whales (*Megaptera novaeangliae*) (Palsboll et al., 1997). Genetic tagging consists of collecting skin samples, removing the DNA, and determining the sex and genotype at six Mendelian-inherited microsatellite loci for each sample. More than 2,300 unique genotypes were identified. Genetic tracking has also allowed the tracking of an individual whale from fishery to market (Cipriano and Palumbi, 1999). This technique, as well as similar genetic tools, will allow new management efforts to focus on the individual, rather than the species, and to distinguish individual "legal" whales (those of a particular sex and size which can be legally harvested) from all others.

Tagging frequently requires specific federal and/or state permits, as well as approval from university and institutional animal care and/or ethics committees in many instances, particularly when dealing with species whose travels cross international boundaries, such as most birds. Researchers must be qualified in identification and handling of particular species, as well as in the tagging/marking techniques to be employed.

Photography

Photography is useful for making a permanent record of the location and/or behavior of a marked or tagged animal. Approximately 80 percent of the manatees (*Trichechus manatus*) in the Homosassa and Crystal rivers in Florida are distinctively scarred, primarily from boat strikes. These scar patterns have been used to identify individual manatees. Photographs were taken at regular intervals (twice a week, weekly, biweekly), both from the water's surface and from beneath the surface, and were incorporated into an identification catalogue (Powell and Rathbun, 1984; Rathbun et al., 1990). Resightings of humpback whales (*Megaptera novaeangliae*) returning to their summer feeding grounds have been verified photographically. Photoidentification of cetaceans is a worldwide ongoing endeavor with regional catalogues and specific repositories of all photographed whales.

Photography also has been used in some studies so that an unsuspecting animal triggers a mechanism and takes its own picture. This not only provides a record of the animal's presence but may also identify food brought to the nest and the frequency and length of absence from the nest. A clock or timing device can be positioned so that it is included in the photograph and records the time the photograph was taken.

Small video systems and data loggers that were mounted on the heads of four adult Weddell seals (*Leptonychotes weddellii*) at McMurdo Sound in Antarctica have revealed some aspects of the secret lives of diving animals (Davis et al., 1999) (Fig. 14.7). The video system recorded images of the seal's head and the environment immediately in front of the animal. Filming was accomplished in near-infrared light emitted from the camera like a flashlight. The light, which was invisible to the seal's eye and its prey, should not have altered either one's behavior. The data logger recorded time, depth, water speed, and compass bearing once per second. Flipper stroke frequency and ambient sound were recorded continuously on the audio channels. Several unknown tactics used by the seals to extract prey from their refuges in the ice were revealed.

Radio Transmitters

One of the earliest reports on the use of radio telemetry to locate free-ranging animals was their use on woodchucks (*Marmota monax*) by LeMunyan (1959). The use of radio transmitters has met with considerable success since that

FIGURE 14.7

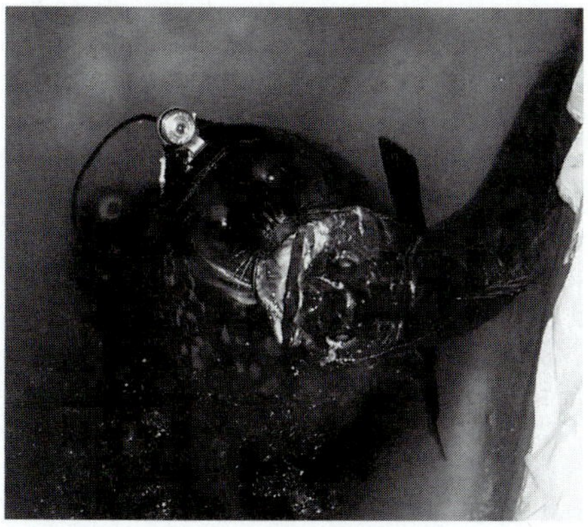

A Weddell seal (*Leptonychotes weddellii*) surfaces in McMurdo Sound, Antarctica, with a 40-pound cod in its mouth and a video camera strapped to its head. Filming was accomplished in near-infrared light emitted from the camera. The video camera and data logger revealed heretofore unknown behavior and tactics used by the seals to secure their food beneath the ice.

FIGURE 14.8

Male elk (*Cervus elaphus*) with a radio transmitter, permitting movement and behavioral studies of animals of known age and sex, even though they may be located several miles away.

time, as transmitters continue to be miniaturized and receiving equipment continues to be improved. In many ways, the use of radio transmitters has revolutionized the study of animal movements. They have been used in studies involving all of the vertebrate groups.

The use of radio telemetry in field studies of vertebrates provides the ability to locate the transmitter regularly, both day and night, to check on the location and condition of the carrier. Radio telemetry is valuable in studying predation, individual behavior patterns, and home ranges. Several telemetry techniques have been designed specifically for detecting mortality in free-ranging animals. For example, transmitters may contain temperature sensors that detect the drop in body temperature upon the death of the animal.

Transmitters may be strapped to the body, attached by means of a collar placed around the neck (Fig. 14.8), wired to the carapace of turtles, or implanted intraperitoneally or subcutaneously (Ralls et al., 1989; Werner, 1991; Rowe and Moll, 1991; and others). For example, surgically implanted temperature-sensitive radio transmitters revealed daily variations in the body temperatures of free-ranging garter snakes (*Thamnophis elegans vagrans*) in eastern Washington (Peterson, 1987) (Fig. 14.9). Collars may be designed to deteriorate after a certain length of time, or in long-term studies, the animal may need to be recaptured and refitted with a new collar. Intraperitoneal radio transmitter implants were found to have no effect on reproductive performance (copulation, embryonic and fetal development, and lactation) in river otters (Reid et al., 1986).

Transmitters have been glued to the bony shells of some species of turtles, but the oily, flexible skin that covers the thin, loosely fused, bony plates in a leatherback sea turtle's carapace resists adhesives. In an experimental technique, the bone is pierced with half-inch-long screws made of a synthetic polymer that slowly dissolves (Raloff, 1998). A nylon suture is threaded through each screw and serves to firmly attach the transmitter. As the screws dissolve, they are replaced by bone that continues to anchor the sutures until they weaken and release the transmitter.

FIGURE 14.9

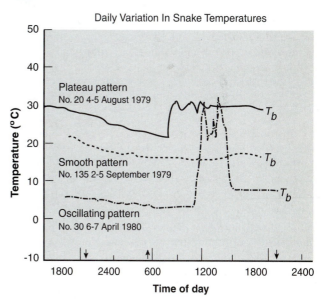

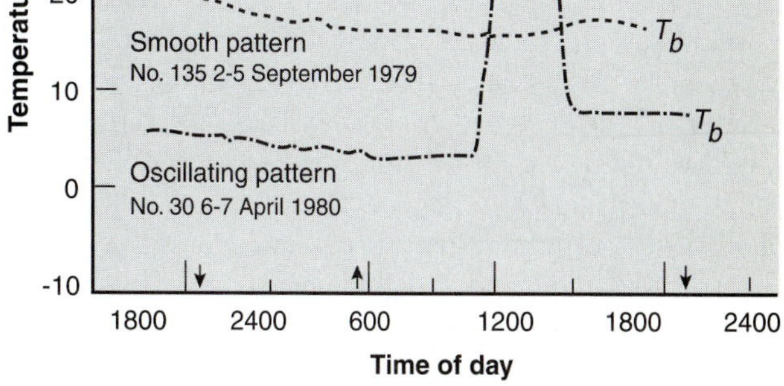

Surgically implanted temperature-sensitive radio transmitters have been used to reveal daily variations in the body temperatures of free-ranging vertebrates including garter snakes (*Thamnophis elegans vagrans*) in eastern Washington. The daily body temperature patterns shown here are classified as (a) plateau pattern, (b) smooth pattern, and (c) oscillating pattern. Sunrise (↑) and sunset (↓) are indicated on the time axis.
Source: Data from Peterson in Ecology, *68(1)1987.*

BIO-NOTE 14.1

Teaching Birds to Migrate

Trumpeter swans (*Cygnus buccinator*) are white with a black beak, weigh up to 13.5 kg, have a 2.5-m wing span, and can stand 1.8 m tall with neck outstretched. They vanished from the Chesapeake Bay nearly 200 years ago. Scientists working with the Defenders of Wildlife, a Washington, D.C.–based conservation organization, and the U.S. Fish and Wildlife Service are now trying to restore America's largest waterfowl to the mid-Atlantic region by reteaching the birds to migrate by using a bright yellow ultralight plane with an overarching white wing. The goal of this project is to reestablish a migration route between upstate New York and Maryland's Eastern Shore. Migration is important because birds that do not fly south for the winter are more likely to exhaust their food supply, become a nuisance to people, get sick, or freeze to death. Trumpeter swans learn to migrate from their parents, but if the older birds in a flock are killed by hunters, the young do not know where to migrate and the knowledge is forever lost.

In December 1997, three trumpeter swans followed the ultralight plane from the Airlie Environmental Center in Warrenton, Virginia, to open tidewater marshes near Cambridge in Dorchester County, Maryland, a distance of 166 km (Fig. 14.10). All three swans returned halfway from their winter site near Cambridge to Airlie in Spring 1998, not deviating by more than 5 miles from the pre-selected route they had been shown in December 1997. One returned to within 10 miles of Airlie; another was injured and was trucked back. The third was also trucked back from Cambridge after backtracking there. None of these three females returned to their winter site near Cambridge in 1998–99. They remained at Airlie.

A second project, begun in December 1998, involved transporting 13 trumpeter swans from where they were trained at a traditional breeding ground at a New York Department of Environmental Conservation site near Buffalo, to the Wildfowl Trust of North America on Chesapeake Bay near Grasonville, Maryland. Instead of flying them the entire 320-mile route, they were trucked between stops with the birds being flown as high as possible at each stop. In Spring 1999, this flock, while showing migration intention behavior, did not explore more than 10 miles from their wintering quarters. Thus, having failed to return to New York on their own, they were trucked back in May to the New York Department of Conservation's Oak Orchard Wildlife Management Area a few miles from where they were trained during Fall 1998. On January 22, 2000, one of the experimental trumpeter swans arrived in Claysburg, Pennsylvania, roughly 210 miles due south from the summer site and approximately 100 miles west of the migration route. Since the male bird arrived shortly after a storm, researchers feel that it may have been blown off course.

In 1993, a motorized ultralight led a Canada goose migration from Ontario, Canada, to the Airlie Center, a trip depicted in the 1996 movie *Fly Away Home*. In October 1995, an ultralight aircraft led eight sandhill cranes on an 11-day, 1,204-km trip from Idaho to the Bosque del Apache National Wildlife Refuge in New Mexico. In October 1997, an ultralight painted to look like a whooping crane guided two of the endangered white birds and six sandhill cranes on a 9-day flight from Idaho to New Mexico.

Lewis, 1996
Rininger, 2000

Radio contact with migrating whooping cranes (*Grus americana*) was maintained by means of leg-band radio transmitters, antennas attached to aircraft struts, and radio receivers carried in the aircraft (Kuyt, 1992). Radio signals, which could be picked up from distances up to 155 km, allowed researchers to follow the cranes. Visual contact was maintained for up to 50 percent of the migration, enabling air crews to obtain data on flight behavior.

In studies of marine species, specific problems arise because of diving and because of the effect of high electrolyte concentrations on the radio signals. A floating transmitter tethered to a swivel strapped to the tail stock was devised and successfully used in studies of manatees (Rathbun et al.,1987). Baits containing acoustic transmitters have been consumed by deepsea fishes. In conjunction with an automatic tracking system and cameras on the sea floor, this technique has allowed the tracking of the speed and direction of travel in deepsea scavenging fishes (Priede et al., 1991).

A summary of standard radio-tracking techniques was presented by Mech (1983). Specific data for amphibians were reviewed in Heyer et al. (1994); data for mammals were reviewed in Wilson et al. (1996).

Satellite Tracking

Satellite tracking is one of the latest tools in the repertoire of wildlife biologists. It allows for monitoring of an individual by providing an update with every pass of the satellite. It has been used successfully on a variety of species, including sea turtles, penguins, whales, elephant seals, elephants, caribou, bears, musk-oxen, and manatees (Mate, 1989; Reynolds, 1989; Rathbun et al., 1990; Holden, 1992; Stewart and DeLong, 1995; Reid, 1997). One of the first successful tracking experiments of a bird using satellite telemetry was reported by Jouventin and Weimerskirch (1990), who showed that wandering albatrosses (*Diomedea exulans*) remain active at night, fly at speeds of up to 80 km/hr, and range over distances of up to 900 km per day. Albatrosses covered from 3,600 to 15,000 km in a single foraging trip during the time their mates had taken over the duties of incubation.

Satellite tracking is best used in situations where conventional tracking techniques are not useful, such as animals that range widely or are in habitats where they cannot be followed. Compared with conventional radio-tracking, satellite tracking is less accurate and more expensive.

FIGURE 14.10

Trumpeter swans (*Cygnus buccinator*) following an ultralight aircraft, a method designed to teach birds the ancient migratory route of their ancestors.

MAPPING TECHNIQUES

Geographic Information Systems (GIS) technology is the computerized recording of data for a region, using geographic coordinates as the primary indexing system. The kinds of data that can be stored include presence or absence of a species, abundance of that species where it is present, ecosystem type, soil type, geology and physiography, land protection status, and many other variables. For example, most home range studies have focused only on the horizontal component of the landscape (planimetric area), where the slope of the terrain is assumed to be zero. Topography adds an important element to landscape because the slope of the terrain often fluctuates throughout the home range, and because changes in topography can increase the surface area. A GIS that incorporates topography can account for topographic changes and yield more accurate estimates of home range size (Stone et al., 1997).

GIS systems are well adapted to using data from remote sensing sources. Detailed data on the actual vegetation of a geographical area are difficult to obtain from traditional vegetation maps, which often show the potential climax vegetation thought to characterize a region rather than the vegetation actually present. With improved satellite imagery and analysis, detailed data on the vegetation that actually exists can be determined on a grid scale and entered into the indexing system. Ideally, a GIS system permits data on a particular feature to be stored for all the geographic units included in the indexing system. With modern GIS systems, it should be possible to develop much more comprehensive databases than previously available for researchers and conservationists.

CENSUSING TECHNIQUES

The word "census" is defined as a count, which usually includes details as to sex and age. A **true census** is a count of all individuals present in a given area. Because such counts of wild animals are rarely possible, estimates usually are made based on some sampling procedure. **Sampling estimates** are derived from counts made on sample plots or a portion of a population. These estimates have variability, but still permit inferences about the population. An **index** is a count of some object that is related in some numerical way to the animal, such as tracks, feces, call counts, or nests. For example, Richard and Karen Barnes developed the first standardized method for gauging elephant populations by counting dung piles along previously identified routes and inserting the results into a mathematical formula that considers rates of defecation and dung decay (Tangley, 1997). Similar methods have been developed for jackrabbit indexing (Blackburn, 1968).

For population estimates to be valid, all members of a population must have an equal probability of being counted, or the relative probabilities of counting different categories of individuals (e.g., sex and age classes) must be known. Animals must not group by sex or age; they must mix randomly; and they must not develop "trap-shyness" or "trap-happiness" if grid live-trapping is being employed (see Chapter 11). In addition, during the period when data are collected, either mortality and recruitment must be negligible, or the estimates must be corrected for these effects.

Data may be gathered by visual observation or by evidences of an animal's presence (tracks, calls, etc.). For instance, haypiles of pikas (*Ochotona*) may be found in late summer and fall and can be used as an index. The average distance between the haypiles of adjacent pikas is approximately 30 m (Smith, 1982). Another method of gathering data is by the use of a transect. **Transects** are predetermined routes that are covered in an effort to estimate a population. All animals that are sighted or heard are recorded. Transect data from different seasons and years provide relative estimates of population size.

Many territorial species can be observed easily within their territories and counted for a specific area. The result usually is expressed as animals per hectare. However, nonterritorial individuals in these species often are hard to count. Animals that congregate in groups or flocks (e.g., coveys of quail, flocks of turkeys and other birds, herds of antelope and bison) are relatively easy to count either on the ground or by means of aerial photographs. In those species of frogs and birds that call or sing, the vocal members of the population can be counted. The National Audubon Society's annual Christmas Bird Count provides an index of species' abundance nationwide. The North American Breeding Bird Survey has provided valuable data on the sizes of breeding bird populations since 1965, especially those of neotropical migrants. Such databases are revealing steady population declines of breeding birds for many species in North America (see Chapters 15 and 16).

Statistical estimates of population size based on sample plots, indices, rates of capture, changes in sex or age ratios, recaptures, or home range data can be calculated by many different methods (Mosby, 1963). The Lincoln Index (also known as the Petersen–Jackson Method because it was first used on wild populations of plaice by Petersen [1896]), for example, is based on the recapture of marked individuals where the population (N) is related to the number marked and released (M) in the same way as the total caught at a subsequent time (n) is related to the number of marked individuals captured (m).

Censusing methods, along with capture and marking techniques, have been discussed for game birds and mammals by Mosby (1963), for terrestrial vertebrates by Davis (1982), for amphibians by Heyer et al. (1994), and for mammals by Wilson et al. (1996).

AGING TECHNIQUES

Many fishes, amphibians, and reptiles grow throughout their lives. This **indeterminate** growth is most rapid in younger individuals, and it may speed up when food and environmental conditions are favorable and slow down when conditions are more stressful, such as during periods of cold, drought, and food shortage.

Birds and mammals generally experience a steady increase in size until they reach maturity, after which growth slows and essentially ceases for the remainder of their lives. This is known as **determinate** growth.

Various methods of determining the age of vertebrates have been developed. Animals that are captured shortly after hatching or birth and that are marked and recaptured at periodic intervals provide the most accurate means of determining age under natural conditions. In some cases, direct observation of an animal's life stage, physical features, and size can give an approximation of its age. Life cycles of most amphibians, for example, involve two, and sometimes three, distinct stages (larval or tadpole, and adult). Few long-term age-determination studies have been reported. In one long-term reptilian study, three-toed box turtles (*Terrapene carolina triunguis*) studied for 25 years had estimated ages ranging from 27 to 59 years of age at the conclusion of the study (Schwartz and Schwartz, 1991).

Most young birds have several distinct juvenile and subadult plumages as they mature (natal down, juvenal plumage, first winter plumage, nuptial plumage; see Chapter 12 for detailed discussion of molts and plumages). Some birds, such as bald eagles, may not attain their full adult plumage until they are 3, 4, or even 5 years old. The pelage of many young mammals also differs from the adult pelage and is known as the **juvenal** pelage. When molting occurs, this pelage usually is replaced by the **postjuvenal** pelage and then by the **adult** pelage.

More precise age-determination techniques vary among the vertebrates and involve features of the integumentary, skeletal, and even the nervous system. Some techniques are useful in field investigations with live animals, whereas others can only be used on dead specimens. For example, temperate zone fishes can be aged by examining the annuli on scales (Fig. 14.11a), bones, and ear-stones (otoliths) (Fig. 14.11b), and in cross sections of fin rays, fin spines, and vertebral centra. Many fish deposit otolith growth increments with a 24-hour periodicity (Pannella, 1971; Prince et al., 1991; Kingsmill, 1993). In some species, such as Atlantic salmon (*Salmo salar*), the scales may contain spawning marks. Examination of such scales can provide information about when the fish first went to sea, its age when it first spawned, how many times it has spawned, and its age at capture. Because growth accelerates in the sea, annuli are more widely spaced.

Annuli also are evident on the scutes of some turtles. Most juvenile turtles add single growth rings each year, whereas rings are added less frequently as adults (Galbraith and Brooks, 1989) (Fig. 14.11c). Moll and Legler (1971) reported that

FIGURE 14.11

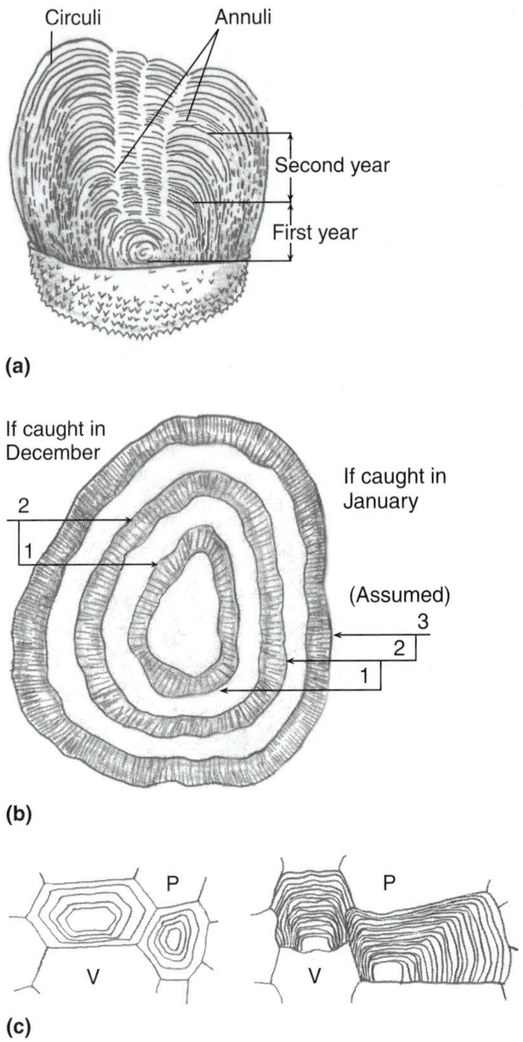

(*a*) A typical ctenoid scale, showing groups of concentric rings that can be classified into annuli and interpreted as seasonal growth marks. Source: Calliet, Love, and Ebeling, *Fishes: A Field and Lab Manual*, 1986, Wadsworth Publishing. (*b*) A fish otolith showing annuli. A year class and/or birth date can be assigned, using the time of year the fish was collected. (*c*) Growth lines (annuli) on the vertebral (V) and the pleural (P) shields of the terrapin (*Malaclemmys*) (left) and the box turtle (*Terrapene*) (right). In *Malaclemys*, embryonic shield areas are near the center of the shields; in *Terrapene*, they are eccentrically located, and growth proceeds primarily anteriorly and laterally.

Source: (a) *Calliet, Love, and Ebeling,* Fishes: A Field and Lab Manual, *1986, Wadsworth Publishing.* (c) *Zangerl, "The Turtle Shell" in C. Gans,* Biology of the Reptiles, *1969.*

FIGURE 14.12

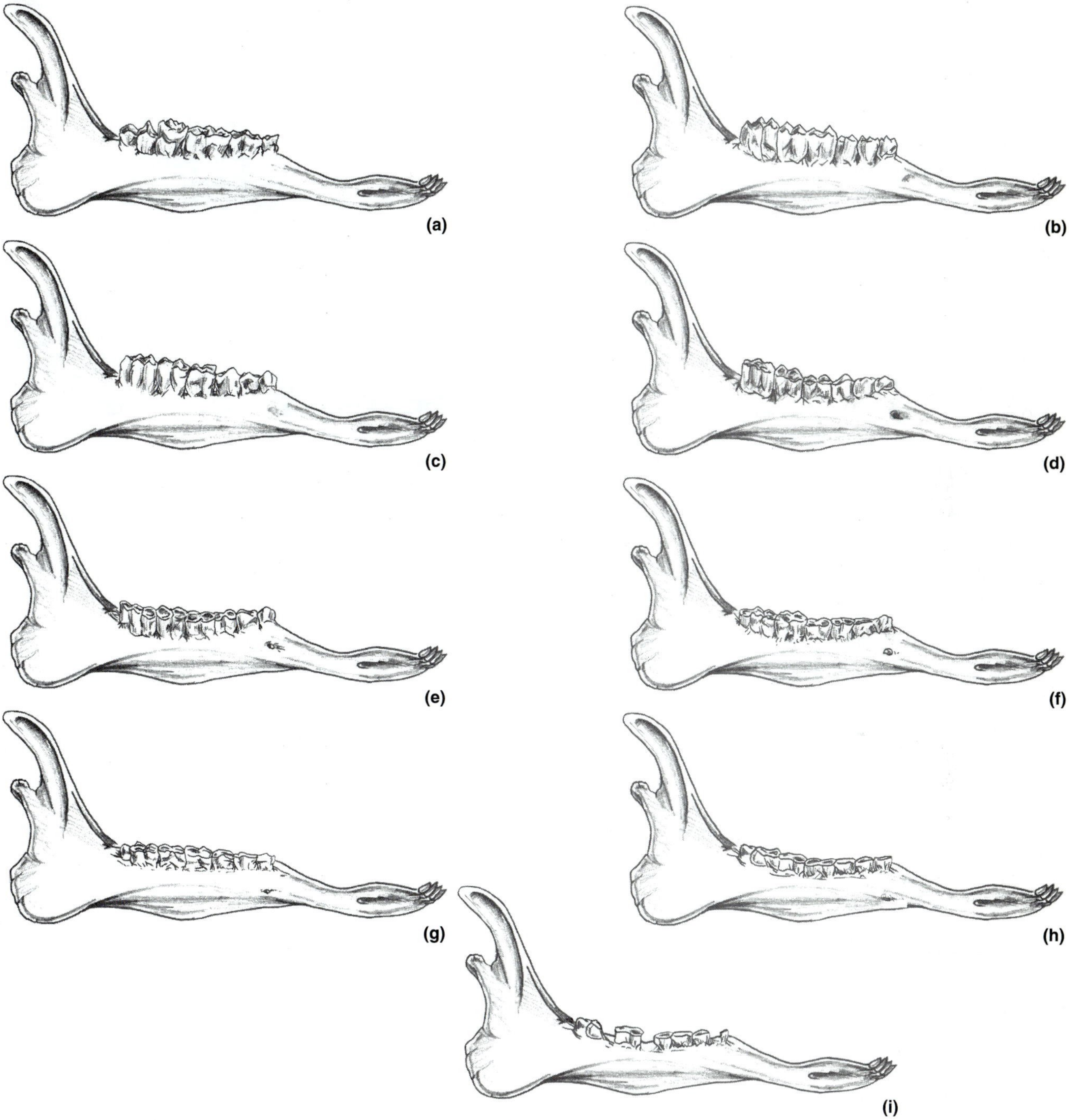

Progressive wear on the molars is used to determine the age of white-tailed deer (*Odocoileus virginianus*): (*a*) 1 year, 7 months; (*b*) 2.5 years; (*c*) 3.5 years; (*d*) 4.5 years; (*e*) 5.5 years; (*f*) 6.5 years; (*g*) 7.5 years; (*h*) 8.5–9.5 years; (*i*) 10.5 or older.
Sources: Halls (ed.) in White-tailed Deer Ecology and Management, *1984, Stackpole Books, and Cockrum,* Mammalogy, *1962, Ronald Press.*

multiple growth lines were added each year in a population of neotropical sliders (*Pseudemys scripta*) in Panama. Annual bone rings in the phalanges and femurs of lizards have been used to age such species as tuataras (*Sphenodon*) (Castenet et al., 1988). Klinger and Musick (1992) injected tetracycline into juvenile loggerhead turtles (*Caretta caretta*) in the Chesapeake Bay area and found annular deposition in bone layers.

The condition of teeth is useful for establishing the age of mammals. Both the deciduous and permanent dentitions usually erupt in a definite sequence and at definite times in different species. Patterns of wear, particularly of the permanent dentition, provide a fairly accurate means of determining age, particularly in large herbivores (Fig. 14.12). In addition, roots of teeth in some mammals form annual growth ridges on their surfaces. The roots of teeth also may be sectioned to reveal the presence of growth rings or annuli. Many studies involving game species, such as bears, use data obtained from an extracted premolar removed at a checking station. In addition to being used as an aging technique, the incremental layering of the dental cementum can be used to reconstruct the reproductive histories of female black bears (Coy and Garshelis, 1992).

Mammals with permanent horns, such as sheep and goats, often possess ridges on their horns (Fig. 14.13). These ridges are the result of periods of good (summer) and poor (winter) food conditions. When forage is good, the horns grow rapidly; when forage is poor, the horns stop growing. The stoppage is marked by a ridge that has proved to be a valuable aging tool in these mammals.

As a mammal grows, its long bones lengthen from the tips. While it is growing, each long bone has a cartilaginous zone, the **epiphyseal plate** near each end covered by a bony cap, the **epiphysis.** Bone is deposited at the inner side of the cartilaginous zone, pushing the cap farther out as the bone grows. When growth is complete, the cartilage ossifies (is replaced by solid bone), so that the cap and the shaft are fused firmly together. The only remaining evidence of the epiphyseal plate is a line known as the epiphyseal line. Thus, examination of the ends of long bones often can provide an indication of the age of a mammal. In many bats, the joints of the bones in the wing remain swollen until ossification is complete.

Other methods of measuring age include the weight of the eye lens, the condition of the baculum (os penis) in those mammals in which it is present, and the development of prominent crests and ridges on the skull, particularly in male mammals.

FIGURE 14.13

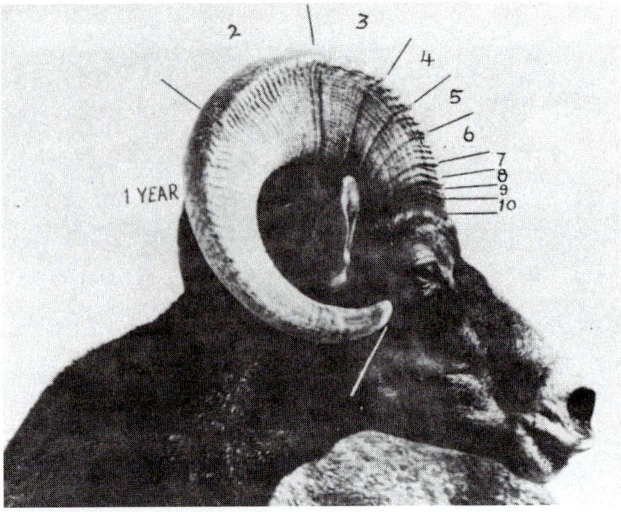

Annual rings in the horn of a 10-year-old desert bighorn sheep (*Ovis canadensis*).

In addition, a number of species-specific aging techniques have been employed. Among these are leg coloration in coots (*Fulica*), the size of spurs on ring-necked pheasants (*Phasianus colchicus*), and the presence of lateral suprasesamoid tubercles on the femurs of adult mink (*Mustela vison*) (Gullion, 1952; Godin, 1960; Lechleitner, 1954).

Review Questions

1. What factors must be taken into consideration when designing a research program for a wide-ranging species such as a cougar? How would this differ from a prairie dog research study?
2. Discuss several techniques for capturing small mammals alive and uninjured.
3. List several techniques for marking reptiles and mammals for short-term studies.
4. List several techniques used to tag birds and mammals so that individuals can be recognized at a distance.
5. List several advantages and several disadvantages of attaching external radio transmitters to vertebrates.
6. What are some advantages and some disadvantages of satellite tracking as compared with the use of radio transmitters?
7. List several ways of censusing vertebrate populations. Which methods would provide the greatest amount of data for the spotted salamander (*Ambystoma maculatum*), which lives underground most of the year but migrates to a breeding pond in late winter?
8. Differentiate between determinate and indeterminate growth. Give several examples.
9. List several techniques used to determine the age of vertebrates.
10. Which methods of age determination would be most appropriate for studying a population of marine toads (*Bufo marinus*) whose life expectancy may be as much as 8 years?

Supplemental Reading

Adler, B., Jr. 1996. *Outwitting Squirrels.* Chicago: Chicago Review Press.

Adler, B., Jr. 1997. *Outwitting Critters.* New York: Lyons and Burford.

Anderson, R. M. 1965. *Methods of Collecting and Preserving Vertebrate Animals.* Fourth edition. Bulletin No. 69. Biological Series No. 18. Ottawa: National Museum of Canada.

Berwick, S. H., and V. B. Saharia (eds.). 1995. *The Development of International Principles and Practices of Wildlife Research and Management.* New York: Oxford University Press.

Bibby, C. J., N. D. Burgess, and D. A. Hill. 1992. *Bird Census Techniques.* London: Academic Press.

Bub, H. 1991. *Bird Trapping and Bird Banding.* Ithaca, New York: Cornell University Press.

Burnham, K. P., D. R. Anderson, and J. L. Laake. 1980. Estimation of density from line transect sampling of biological populations. *Wildlife Monographs* 72: 1–202.

Davis, D. E. (ed.). 1982. *CRC Handbook of Census Methods for Terrestrial Vertebrates.* Boca Raton, Florida: CRC Press.

Emlen, S. T. 1971. Population estimates of birds derived from transect count. *The Auk* 88:323–342.

Heyer, W. R., M. A. Donnelly, R. W. McDiarmid, L. C. Hayek, and M. S. Foster (eds.). 1994. *Measuring and Monitoring Biological Diversity: Standard Methods for Amphibians.* Washington, D.C.: Smithsonian Institution Press.

Kendeigh, S. C. 1944. Measurement of bird populations. *Ecological Monographs* 14:67–106.

Kunz, T. H. (ed.). 1988. *Ecological and Behavioral Methods for the Study of Bats.* Washington, D.C.: Smithsonian Institution Press.

Lebreton, J.-D., and Ph. M. North. 1993. *Marked Individuals in the Study of Bird Populations.* Basel: Birkhauser Verlag.

Marion, W. R., and J. D. Shamis. 1977. An annotated bibliography of bird marking techniques. *Bird-Banding* 48:42–61.

McClure, H. E. 1984. *Bird Banding.* Pacific Groves, California: Boxwood Press.

Mosby, H. S. (ed.). 1963. *Wildlife Investigational Techniques.* Second edition. Washington, D.C.: The Wildlife Society.

Ralph, C. J., and J. M. Scott (eds.). 1981. *Estimating Numbers of Terrestrial Birds.* Studies in Avian Biology No. 6. Columbus, Ohio: Cooper Ornithological Society.

Robbins, C. S. 1970. Recommendations for an international standard for a mapping method in bird census work. *Audubon Field Notes* 24:723–726.

Svenson, S. E. 1979. Census efficiency and number of visits to a study plot when estimating bird densities by the territory mapping method. *Journal of Applied Ecology* 16:61–68.

Verner, J. 1985. Assessment of counting techniques. Pages 247–302. In: Johnston, R. F. (ed.). *Current Ornithology, Volume 2.* New York: Plenum Press.

Wilson, D. E., F. R. Cole, J. D, Nichols, R. Rudran, and M. S. Foster (eds.). 1996. *Measuring and Monitoring Biological Diversity: Standard Methods for Mammals.* Washington, D.C.: Smithsonian Institution Press.

Vertebrate Internet Sites

Visit the zoology website at http://www.mhhe.com to find live Internet links for each of the references listed below.

1. **Scientific Investigations.**
   Lecture notes from a Wildlife Population Ecology course that outline the scientific method, scientific reasoning, and related topics.
2. **United States Fish and Wildlife Service Endangered Species Homepage.**
   Provides program overview, information concerning the Endangered Species Act, a map of regions, contact information, policies, species information, state lists, proposed and candidate species information, questions and answers regarding species recovery, kid's corner, frequently asked questions, and links to other sites.
3. **Geographical Information Systems (GIS).**
   This USGS page on GIS provides information and a useful link to the National Geospatial Data Clearinghouse (NSDI).

CHAPTER 15

Extinction and Extirpation

INTRODUCTION

Extinction is the most obscure and local of all biological processes. We usually do not see the last individual of a species as it dies or is captured by a predator. We hear that a certain animal or plant is imperiled, perhaps already gone. We return to the last known locality to search, and when no individuals are encountered there year after year, we pronounce the species extinct.

Populations decline whenever deaths and emigration exceed births and immigration. The elimination of a species or subspecies from a region, although it continues to exist elsewhere, is known as **extirpation**. The cougar (*Puma concolor*) is thought by many to be extirpated from most of the eastern United States, but cougars remain in Florida and many western states as well as in Canada and Central America. **Extinction** is the total disappearance of a species and has been the fate of most species since the origin of life. Dinosaurs, passenger pigeons, heath hens, dodos (Fig. 15.1), mastodons, and saber-toothed tigers are among the many vertebrates that have become extinct. Disappearance of entire species or even entire families, orders, or classes has occurred at times of extreme environmental change or, more recently, because of human action.

With or without human interference, extinction always has occurred. The last dinosaurs disappeared 65 million years ago, over 60 million years before humans evolved. Judging from the fossil record, Peter Raven, Director of the Missouri Botanical Gardens, calculated the average life span of a species at about 4 million years (Raven, 1995). If there are about 10 million species in the world, Raven calculated the normal rate of extinction at about 4 species a year. Many scientists believe that humans now have increased the pace of extinction far beyond natural levels, so that species are now becoming extinct at rates 1,000 to 10,000 times the natural rate that occurred before our ancestors first appeared on Earth. Raven predicts that animal and plant species will likely become extinct at the rate of 50,000 species a year during the next few decades (Raven, 1995). If Dr. Raven is correct, it will be the greatest mass extinction ever, far surpassing the die-off of the dinosaurs. These extinctions—and the loss of biodiversity—are completely irreversible.

NATURAL EXTINCTION

More than 99 percent of all plant and animal species that ever have lived are extinct (Romer, 1949; Simpson, 1952). Little is known, however, about the immediate causes of extinction, even of species that have become extinct in historic times (Simberloff, 1986).

Natural extinction, a normal ongoing process with a certain number of species steadily disappearing over time, is somewhat balanced with the natural process of speciation. This background extinction usually is localized and may be caused by overspecialization, climatic or other environmental changes, or competition with more adaptable forms. A species must evolve continually to keep pace with a constantly changing environment, simply because other species also are evolving, thus altering the availability of resources and the patterns of biotic interactions. Species that cannot keep pace with this change become extinct.

Mass extinctions, on the other hand, were worldwide events in which a large number of species, and even entire higher taxonomic groups, disappeared within an interval of just a few million years. They have occurred throughout the history of the Earth. Afterward, remaining groups are apt to undergo adaptive radiations as they spread out and fill niches vacated by those that have become extinct. The greatest mass extinctions occurred during Late Ordovician, Late Devonian, Late Permian, Late Triassic, and Late Cretaceous periods (Fig. 15.2a, b). The latter three had significant impacts particularly on terrestrial vertebrates. Whether mass extinctions have followed a periodic pattern over the past 250 million years has not been resolved (Raup and Sepkoski, 1986; Sepkoski and Raup, 1986; Benton, 1995).

A leading extinction theorist, David Jablonski of the University of Chicago, believes that selection pressures are

FIGURE 15.1

The flightless dodo (*Didus ineptus*) is classified in the family Raphidae in the order Columbiformes, which includes the pigeons and doves. It was twice as large as a goose, with thick stubby legs, a poodlelike tail (complete with curls), tight pigeonlike body plumage, a tremendous skull equipped with a stout, heavily plated and deeply hooked bill, a naked face, and a dog-sized mouth. It was perhaps the most unbirdlike bird that ever lived. The true dodo lived on the island of Mauritius in the Indian Ocean, about 500 miles east of Madagascar. It was first discovered by the Portuguese in 1507 and last seen alive in 1681. Dodos lived in deep forest, walked with a ludicrous waddle, and laid one large egg, which both parents incubated. They swallowed gizzard stones as large as chicken eggs. Most people took a skeptical view of the early sketches and accounts that depicted so preposterous a creature—even though a number of live dodos were brought to Europe early in the 17th century. Most of what we know of these birds comes from about 400 remains and from written records, such as this 1848 account: "These birds were of large size and grotesque proportions, the wings too short and feeble for flight, the plumage loose and decomposed, and the general aspect suggestive of gigantic immaturity....So rapid and complete was their extinction that the vague descriptions given of them by early navigators were long regarded as fabulous or exaggerated, and these birds, almost contemporary of our great-grandfathers, became associated in the minds of many persons with the Griffin and Phoenix of mythological antiquity." Dodos were no match for the pigs that were introduced by early settlers. The pigs are thought to have feasted on the eggs and young birds, and the phrase "dead as a dodo" soon became a tragic reality.

changed by mass extinctions (Jablonski, 1986). Often it is the most fortunate, not necessarily the most fit, that survive such an event. Groups that had been healthy may suddenly be at a disadvantage when their environment is disrupted. Other species that had been barely surviving somehow manage to survive the event, find conditions more favorable, and proceed to "inherit the Earth."

Mammals are a good example of the latter scenario. Dinosaurs and mammals originated within 10 million years of each other, about 220 million years ago; however, for 140 million years, dinosaurs were the dominant terrestrial vertebrates, while mammals stayed relatively small and inconspicuous. Most early mammals were shrewlike or squirrel-like, and no larger than woodchucks. Mammals probably began their radiation to fill ecological niches left vacant by the demise of the dinosaurs about 65 million years ago, and within 10 million years, there were mammals of all shapes and lifestyles ranging from moles and bats to elephants and whales (see Fig. 9.1).

In contrast, Hedges et al. (1996) suggest that the continental fragmentation that took place in the Mesozoic may have been a more important mechanism in the diversification of orders of birds and mammals than the Cretaceous/Tertiary (K/T) extinction event of 65 million years ago. The adaptive radiations of birds and mammals occurred rapidly after the K/T extinction event. Nuclear gene comparisons of four bird orders (galliform, anseriform, columbiform, and struthioniform) and three mammal species (human, *Homo sapiens*; house mouse, *Mus musculus*; and cattle, *Bos taurus*) reveal molecular estimates of divergence averaging 50 to 90 percent earlier than fossil-based estimates. The use of molecular time estimation of evolutionary divergence assumes that genera evolve at a relatively constant rate. All molecular estimates of divergence occurred during the Mesozoic rather than the Cenozoic and are considerably older than divergence times suggested by fossil evidence. Hedges et al. (1996) conclude that fragmentation of land areas during the Cretaceous, not the relatively sudden availability of ecological niches following the K/T extinction event, was the mechanism responsible for the diversification of avian and mammalian orders.

Tooth fossils (family Zhelestidae) found in 85-million-year-old sediment in Uzbekistan in Asia bear the marks of animals that grazed, and they could be from the ancestors of modern-day horses, cows, elephants, and other hooved animals (Archibald, 1996). The teeth had flat, squared, grinding surfaces similar to those found in herbivores' teeth. The ancestors of hooved mammals may have evolved *during* the time of the dinosaurs—about 20 million years earlier than previously believed—and the evolution of ungulates probably was well under way before the dinosaurs were gone.

Permian

The Permian extinction was the first to affect terrestrial life significantly and was easily the greatest extinction event of all time. The known genera of tetrapods represented by fossils decreased from 200 in the Late Permian to 50 in the Early Triassic. Between 80 and 95 percent of all marine *species* and about 70 percent of vertebrate *families* on land disappeared (Gore, 1989; Erwin, 1994; Stanley and Yang, 1994; Renne et al., 1995). Among vertebrates, 78 percent of reptile

FIGURE 15.2

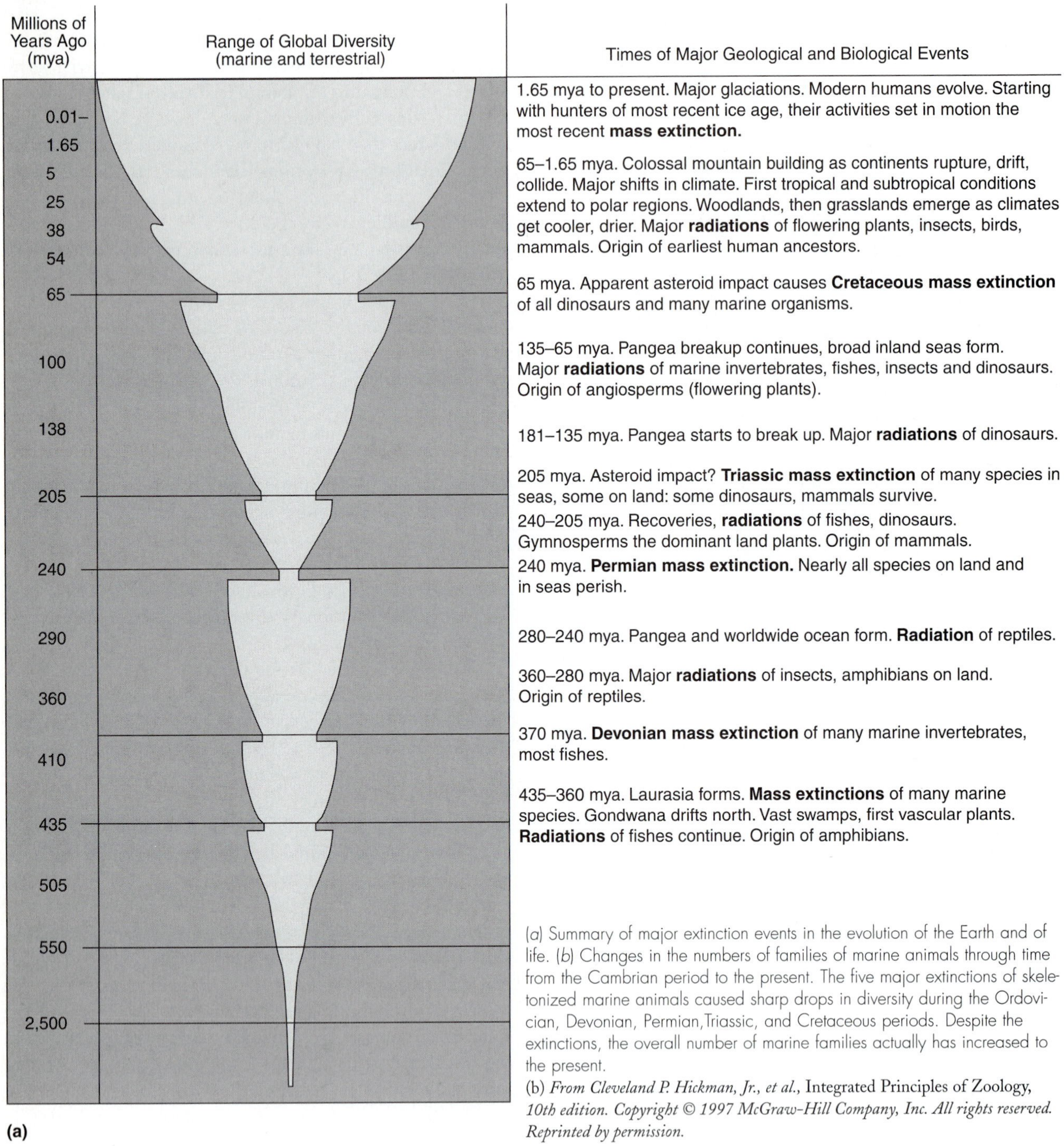
Millions of Years Ago (mya)
Range of Global Diversity (marine and terrestrial)
0.01–
1.65
5
25
38
54
65
100
138
205
240
290
360
410
435
505
550
2,500
(a)
Times of Major Geological and Biological Events
1.65 mya to present. Major glaciations. Modern humans evolve. Starting with hunters of most recent ice age, their activities set in motion the most recent **mass extinction.**
65–1.65 mya. Colossal mountain building as continents rupture, drift, collide. Major shifts in climate. First tropical and subtropical conditions extend to polar regions. Woodlands, then grasslands emerge as climates get cooler, drier. Major **radiations** of flowering plants, insects, birds, mammals. Origin of earliest human ancestors.
65 mya. Apparent asteroid impact causes **Cretaceous mass extinction** of all dinosaurs and many marine organisms.
135–65 mya. Pangea breakup continues, broad inland seas form. Major **radiations** of marine invertebrates, fishes, insects and dinosaurs. Origin of angiosperms (flowering plants).
181–135 mya. Pangea starts to break up. Major **radiations** of dinosaurs.
205 mya. Asteroid impact? **Triassic mass extinction** of many species in seas, some on land: some dinosaurs, mammals survive.
240–205 mya. Recoveries, **radiations** of fishes, dinosaurs. Gymnosperms the dominant land plants. Origin of mammals.
240 mya. **Permian mass extinction.** Nearly all species on land and in seas perish.
280–240 mya. Pangea and worldwide ocean form. **Radiation** of reptiles.
360–280 mya. Major **radiations** of insects, amphibians on land. Origin of reptiles.
370 mya. **Devonian mass extinction** of many marine invertebrates, most fishes.
435–360 mya. Laurasia forms. **Mass extinctions** of many marine species. Gondwana drifts north. Vast swamps, first vascular plants. **Radiations** of fishes continue. Origin of amphibians.

(*a*) Summary of major extinction events in the evolution of the Earth and of life. (*b*) Changes in the numbers of families of marine animals through time from the Cambrian period to the present. The five major extinctions of skeletonized marine animals caused sharp drops in diversity during the Ordovician, Devonian, Permian, Triassic, and Cretaceous periods. Despite the extinctions, the overall number of marine families actually has increased to the present.

(b) *From Cleveland P. Hickman, Jr., et al.,* Integrated Principles of Zoology, *10th edition. Copyright © 1997 McGraw-Hill Company, Inc. All rights reserved. Reprinted by permission.*

and 67 percent of amphibian families disappeared during the Late Permian (Erwin, 1996). Benton (1995) calculated a mean familial extinction rate of 60.9 percent for all life, 62.9 percent for continental organisms, and 48.6 percent for marine life.

Researchers have used isotopic dating to show that extensive volcanic activity in Siberia was contemporary with the Permian extinction. The Siberian traps (after the Swedish word for "stairs," which describes the steplike edges of the deposits) are solidified layers of ancient lava ranging from 400 to 3,700 m in thickness (Erwin, 1996). At least 45 separate flows cover an area of at least 1.5 million km^2. Periodic outpourings of magma occurred for 600,000 to 1,000,000 years (Renne et al., 1995; Erwin, 1996). The world's oceans also became anoxic (depleted of oxygen) in the

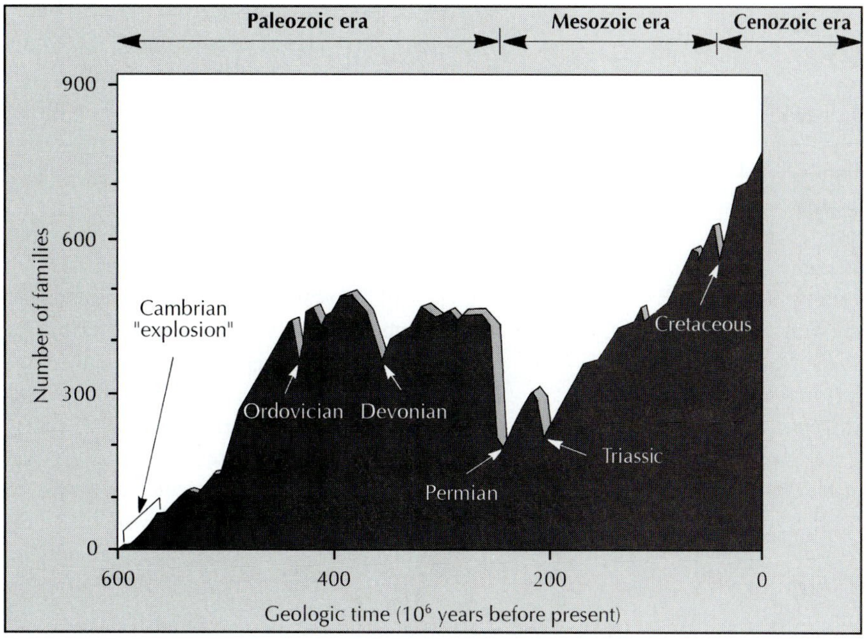

(b)

Late Permian, a condition that could have suffocated some marine life and might have contributed to the extinction of marine organisms (Wignall and Twitchett, 1996). Reductions in oxygen levels occurred throughout a range of depths and extended into shallow waters that serve as critical nurseries for many marine organisms. Wignall and Twitchett (1996) concluded that while oxygen solubility declines in warmer waters, the most probable cause of oxygen-deficient waters was the decline in oceanic circulation as the waters warmed and the equator-to-pole temperature gradient declined.

Triassic

At the family level, the Triassic extinctions were greater than those of the Permian, with an estimated 80 percent of all families becoming extinct at or near the end of the Triassic. Colbert (1986) believed that the Triassic extinctions were caused largely by the loss of long-established taxa, perhaps in part as a result of the appearance and rapid development of new groups better adapted to the warmer environment during the Mesozoic. In addition, some of the Triassic extinctions were the result of the evolution of some lines of therapsid reptiles into early mammals and of some thecodont reptiles into more advanced archosaurian reptiles.

Mollusks, such as the chambered shelled ammonoids, and bivalves such as mussels, clams, scallops, and oysters, were decimated, and conodonts finally disappeared during the Late Triassic extinctions. On land, several families of reptiles disappeared, particularly the last of the basal archosaurs (thecodontians), the group that includes the ancestors of dinosaurs and crocodilians, and some mammal-like reptiles (therapsids), the group that includes the ancestors of the mammals (Benton, 1993).

Cretaceous

An estimated 61 percent of all tetrapod families became extinct at the end of the Cretaceous period (Jablonski and Raup, 1995). This extinction event caused a 70 to 80 percent reduction in marine biodiversity at the species level and a 50 percent reduction at the generic level.

As in previous mass extinctions, some Cretaceous extinctions were the result of the development of better-adapted groups and the evolution of ancestral groups into more derived groups. Others were the result of evolutionary attrition—the disappearance of "experimental" groups such as certain groups of Mesozoic mammals (symmetrodonts, pantotheres, multituberculates) during the early stages of their evolutionary development. However, Cretaceous extinctions were marked largely by the rather sudden disappearance of many members of well-established and seemingly highly successful groups such as microscopic foraminiferans (protozoans), bivalves, gastropods, and cephalopods as well as dinosaurs, pterosaurs, and many marine reptiles.

The extent of terrestrial vertebrate extinctions at the end of the Cretaceous is poorly understood, and estimates have ranged from a mass extinction of many avian and mammalian lineages to limited extinctions of specific groups (Gibbons, 1997a). Colbert (1986), for example, noted that 35 orders of tetrapods lived during Mesozoic times (4 amphibians, 15 reptiles, 7 birds, and 9 mammals):

Amphibia
- *Temnospondyli (labyrinthodont amphibians)
- *Proanura (ancestral to the anurans—frogs and toads)
- Anura (frogs and toads)
- Urodela (salamanders)

Reptilia
- *Cotylosauria (stem reptiles)
- Eosuchia (the first and most primitive diapsids)
- Rhynchocephalia ("beaked reptiles," represented today by the tuatara)
- Chelonia (turtles)
- Squamata (lizards and snakes)
- *Thecodontia (Triassic archosaurs)
- *Pterosauria (archosaurs, flying reptiles)
- Crocodilia (archosaurs, crocodilians)
- *Saurischia (archosaurs, saurischian dinosaurs)
- *Ornithischia (archosaurs, ornithischian dinosaurs)
- *Protorosauria (a "wastebasket group" of Triassic reptiles)
- *Sauroptergia (marine nothosaurs and plesiosaurs)
- *Placodontia (marine, mollusk-eating reptiles of Triassic age)
- *Ichthyosauria (ichthyosaurs, of fishlike form)
- *Therapsida (mammal-like reptiles)

Aves
- *Archaeopterygiformes (*Archaeopteryx*, the first bird, Jurassic)
- *Hesperornithoformes (loonlike toothed birds, Cretaceous)
- *Ichthyornithoformes (ternlike birds, Cretaceous)
- Gaviiformes (the divers, loons and grebes)
- Colymbiformes (doves and pigeons)
- Circoniiformes (waders, storks, and herons)
- Charadriformes (gulls and terns and their relatives)

Mammalia
- *Multituberculata (earliest herbivores, with specialized teeth)
- *Triconodonta (small carnivores with sharp-cusped teeth)
- *Docodonta (Jurassic mammals with expanded tooth crowns)
- *Symmetrodonta (ancient mammals with triangular-shaped cheek teeth)
- *Eupantotheria (possible ancestors of later mammals)
- Marsupialia (pouched mammals)
- Proteutheria (very primitive eutherian mammals)
- Primates (today the lemurs, monkeys, apes, and man)
- *Condylarthra (primitive hooved mammals)

Of the 35 orders, 21 (2 amphibians, 10 reptiles, 3 birds, and 6 mammals) became extinct during the Mesozoic and are designated by an asterisk. During the same period, plants, turtles, crocodiles, fishes, birds, and placental mammals were comparatively unaffected, a fact that has not yet been fully explained (Dodson and Dodson, 1985).

Cooper and Penny (1997) used molecular and paleontological data to show that modern bird orders started diverging in the Early Cretaceous, and that at least 22 avian lineages of modern birds survived the K/T boundary. Using the combined data for other terrestrial vertebrates, Cooper and Penny (1997) estimate that a minimum of 100 terrestrial vertebrate lineages survived the end-Cretaceous extinctions. Incremental changes probably occurred during a Cretaceous diversification of birds and mammals, rather than an explosive radiation in the Early Tertiary.

Various theories have been proposed to explain the K/T event. In a period of time variously estimated from weeks to 50,000 years or more, life on Earth was totally devastated by what probably was the greatest catastrophe in the history of our planet. Theories for the demise of the dinosaurs include racial senescence, bodily disorders, stress, disease, climatic change, an extraterrestrial impact, cosmic radiation, extensive volcanism, major regression of the sea floor from the land, geochemical changes, predation by mammals, and the rise of new flowering plants to which the highly specialized herbivorous dinosaurs could not adapt (Stanley, 1987; Norman, 1991). At one time or another, almost every conceivable catastrophe, terrestrial or extraterrestrial, has been advanced to explain mass extinctions.

Based on paleobotanical evidence (comparisons of modern leaf sizes and shapes with those of fossil leaves), the Cretaceous was a time of global warmth (Herman and Spicer, 1996). The Arctic Ocean was relatively warm, remaining above 0°C even during the winter months. The ocean's warmth implies that there was significant heat transport toward the poles during all seasons of the year. Normal geological events (mountain building, massive volcanic activity, and especially, a major regression of sea level that eliminated the epicontinental seas) also occurred at that time.

The K/T extinction was widespread geographically, but selective in the groups that it affected. Norman (1991) noted the general disappearance of any land-living animal more than 1 m long, and the extinction of nearly all large marine reptiles including marine crocodiles (but excluding marine turtles). All of the ammonites disappeared, as did most brachiopods and clams. All flying reptiles vanished, but birds and freshwater crocodiles survived with few apparent effects. Most bony fishes, sharks, and mammals also seemed to be unaffected. Although some early flowering plants were lost, the majority of plant species seem to have survived. Following the K/T event, however, there seems to have been a brief, extraordinarily diverse flora of ferns.

The Jurassic was characterized by uniform tropical conditions with abundant rainfall and lush vegetation. These conditions continued into the Cretaceous, but beginning about 100 million years ago (Middle Cretaceous), a gradual worldwide cooling trend began. By the early part of the Late Cretaceous, average yearly temperatures were in the range of 18 to 20°C (64–68°F). Norman (1991) noted that periods of prolonged cool temperatures could have caused a fatal drain on the body temperature of large ectotherms, from which they would have had little chance of recovering. Even though small endothermic dinosaurs apparently could generate heat, they may not have been able to control the loss of their body heat during cooler periods because they lacked an insulating layer of fur or feathers.

The end of the Cretaceous was marked by a sudden cooling and drying trend, which resulted from the lowering of sea levels. These climatic changes may have significantly affected the distribution of plant populations that served as food for herbivorous dinosaurs. Some believe the demise of the dinosaurs was an indirect effect of such a climate-induced change in vegetation.

Other researchers (Wignall and Twitchett, 1996; Renne et al., 1995) believe massive volcanism was the cause of the extinction. Ash, sulfur and sulfate aerosols, along with chlorine and other acidic compounds are emitted into the stratosphere by volcanic eruptions. Sulfur dioxide eventually becomes small droplets of sulfuric acid, which condense and remain in the stratosphere as a mist of fine particles that eventually return to Earth in the form of acid precipitation (rain, snow). These particles also reflect the sun's radiation and cool the lower atmosphere while the reflected radiation warms the upper atmosphere. The chlorine interacts chemically with ozone molecules (O_3), breaking them apart and hastening the depletion of the planet's ozone shield. Destruction of the ozone layer has been linked to increased volcanic activity at the end of the Cretaceous. Increased ultraviolet radiation striking the Earth could have destroyed animals living on land and the plankton in the upper layers of the ocean. For example, the eruption of Toba in Sumatra 73,500 years ago was the largest known explosive volcanic event in the Late Quaternary. It could have sent huge quantities of fine ash and sulfur gases to heights of 27 to 37 km, creating dense stratospheric dust and aerosol clouds (Rampino and Self, 1992). The volcanic dust could have caused a "volcanic winter" and several years of decreased surface temperatures.

Significant eruptions are continually occurring. Air waves generated by the eruption of Indonesia's Krakatoa in August 1883 traveled around the world four times (Flannery, 1996). Dust from the explosion circled the globe, producing brilliant-colored sunsets for as long as 2 1/2 years following the eruption (Alvarez et al., 1980). The associated explosion was estimated to be 10,000 times more powerful than the Hiroshima bomb and was heard by people as far away as central Australia, who thought it was a distant cannon. More recently, eruptions of Mount St. Helens (1980) in North America and Mount Pinatubo (1991) in the Philippines spewed great quantities of debris into the stratosphere (Fig. 15.3). Material from these eruptions circled the Earth many times. In the case of Mount Pinatubo, scientists believe that the eruptions are the cause of a temporary cooling of the Earth's climate (at least 1°C over 1 to 2 years) and ozone decreases of 30 percent in certain areas.

Still others, though, believe that a supernova explosion or other extraterrestrial event is the most likely explanation of the extinction, with the ensuing cosmic radiation killing off the large, unprotected dinosaurs. The explosion could have triggered a chain reaction of major changes in the Earth's climate. Many mammals and birds, protected by fur or feathers, survived. Among ectothermic reptiles, only those that could hibernate or seek refuge in riverbanks or under rocks escaped death. To explain the multiple stages of extinctions that occurred near the K/T boundary, some believe Earth was hit not by one great object but by a shower of comets that bombarded the planet over several million years. The movements of the comets or asteroids through the atmosphere could have ionized molecules in the air, which would have fallen to the ground as acid rain. The rain could have made the ocean's surface acidic enough to kill off many tiny marine animals (and, thus, the animals that feed on them) by dissolving their calcium-based shells. This would help explain why species with calcium-based shells suffered at the K/T boundary far more than those with silica-based shells.

Some scientists believe that a meteor impact caused a huge cloud of dust, water droplets, and other debris to ascend into the air (Fig. 15.4). Such a cataclysmic event might have darkened the entire globe for 3 to 9 months and interrupted plant photosynthesis, created acid rain, caused a greenhouse effect that warmed the air and the seas, and burned huge forests, thus causing the extinction of vulnerable species both

FIGURE 15.3

The eruption of Mt. Pinatubo in the Philippines in 1991 spewed hundreds of millions of tons of ash, rocks, and molten lava. An estimated 20 million tons of sulfur from the volcano created an acidic aerosol that circled the Earth for 2 years and cooled the average global temperature by at least 1°C.

on land and in the sea. Researchers who study isotopes from the rocks of the sea floor believe that a period of global warming coincided with the downfall of the dinosaurs. The discovery of an apparent global soot layer at the Cretaceous/Tertiary boundary provides evidence that global wildfires might have been ignited by energy radiated from reentering ejecta from the impact (Wolbach et al., 1985, 1988; Melosh et al., 1990). Solar transmission was reduced to 10–20 percent of normal for a period of 8–13 years (Pope et al., 1994). This reduction also may have caused a cooling of the climate that far exceeded the greenhouse warming caused by the increase in carbon dioxide through the vaporization of carbonates and, therefore, produced a decade or more of freezing and near-freezing temperatures. Several decades of moderate warming followed the decade of severe cooling. The prolonged impact-winter may have been a major cause of the K/T extinctions.

The impact hypothesis originated in 1978, when a team from the University of California and Lawrence Livermore Laboratory discovered a clay layer a few centimeters thick in a 66-million-year-old layer of rock from Italy (Alvarez et al., 1980; Alvarez, 1983). The clay layer contained a relatively high concentration of the element iridium, which is rare on Earth but relatively common in meteors and asteroids. At present, deposits of iridium dating back approximately 65 million years have been found in more than 100 locations around the world. Its distribution over the Earth may have resulted from fallout from a great cloud of dust following the explosive impact of a meteorite. Because the iridium deposits are 65 to 66 million years old, they are coincident with the worldwide die-off of many species at the K/T boundary. Considerable evidence has been cited both for and against this hypothesis (Clemens et al., 1981; VanValen, 1984; Jablonski, 1984, 1986).

The search for a crater big enough (approximately 200 km in diameter) and old enough to explain the demise of the dinosaurs has focused on several sites. Glassy rock from the center of a huge crater with a diameter of at least 100 km in northern Siberia is 66.3 million years old. Because the crater is so large, it could be the point where a meteorite 8 to 16 km in diameter hit the Earth. More recently, however, researchers have focused on a potential site beneath the coast of the Yucatán Peninsula in Mexico (Chicxulub), where an impact crater 180 km in diameter was discovered and dated (Kring and Boynton, 1992; Kerr, 1992; Swisher et al., 1992; Hildebrand et al., 1995; Alvarez et al., 1995). Tiny fragments of glass (shocked quartz) in nearby sediments are thought to be hardened droplets of rocks (melted by the impact and ejected into the atmosphere) that cooled into glass as they rained down. Radioisotopic dating has revealed the crater to have a reported age of 64.98 million years, ± 0.06 million years (Swisher et al., 1992). Impact debris has been dated at 65.06 million years, ± 0.18 million years. Thus, the ages of impact, impact debris, and the heart of the mass extinction are indistinguishable. In addition, a 2.5-mm chip of rock rich in iridium and thought to have been thrown from the crater was found in a sediment core taken from the North Pacific (Kerr, 1996a; Kyte, 1998). The meteor's impact angle was from the southeast to the northwest at a 20° to 30° angle from the horizontal (Schultz and D'Hondt, 1996). Chemical and mineralogical signs in the sediments surrounding the rock chip put it at the base of a 10-cm-thick layer rich in debris particles thrown from the impact crater. Cores of ancient sea-floor sediment, taken off the eastern coast of the United States in early 1997, provide additional evidence that the impact occurred precisely at the time of the extinction of many marine microfossils (Kerr, 1997a).

On the other hand, some scientists point out that the dinosaurs dwindled slowly over a period of many thousands of years, and that the end of the Cretaceous simply marked the end of a long decline (Clemens et al., 1981; Officer and Drake, 1983). An intensive study and analysis of dinosaur bones from the last 2.5 million years of the Cretaceous period in North Dakota and Montana, however, revealed no evidence of a gradual decline (Sheehan et al., 1991). Eight families were represented in lower, middle, and upper portions of the rock formation, and relative strengths of the families remained constant from the earliest portion to the latest. Past studies of pollen fossils also revealed that many species of plants in the same region died out at the end of the Cretaceous.

Most geologists agree that an extraterrestrial body struck the earth at the end of the Cretaceous and that at least some

FIGURE 15.4

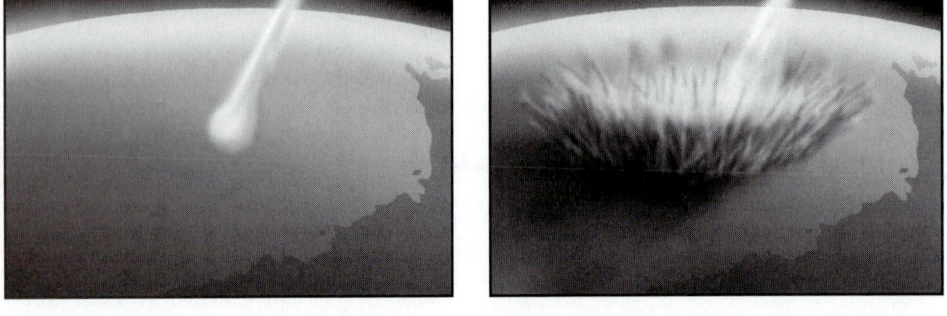

An artist/astronomer's interpretation of what might have happened during the last few minutes of the Cretaceous. Note the angle of impact is 20° to 30°.

major groups of organisms became extinct rather abruptly; but there is still no clear consensus on whether or not an extraterrestrial impact was the principal cause of the entire mass extinction (Futuyma, 1986). While most paleontologists agree that impacts have occurred, many believe that a combination of normal biological, climatic, and geological processes provide the most plausible explanation for the observed faunal changes (Hallam, 1987; Kerr, 1988).

Paleobotanical evidence for a marked temperature increase following the Cretaceous/Tertiary boundary is consistent with inferred greenhouse heating. Wolfe (1990) stated:

> *An oceanic impact site, resulting in the injection of large amounts of water vapour into the stratosphere and the formation of a humid greenhouse is suggested in one model; however, the stratospheric residence time for this water vapour would be of the order of months or years. Because warmth and wetness continued for a far longer time, complex feedback mechanisms in the earth's ocean-atmosphere system may have altered the carbon cycle and may have involved factors such as production of large amounts of carbon dioxide by the bolide impact.*

Bakker (1986), however, believes the extinction was a natural event preceded by the draining of shallow seas, such as the Bering Strait, and the emergence of land bridges as described by Osborn as long ago as 1925. The exchange of species across continents brought new combinations of predators and prey together. In addition, Bakker speculates the exchange also may have transmitted parasites and disease organisms to species that possessed little or no resistance. When two continents mix their faunas, each group is challenged by enemies for which they are unprepared. During the Late Cretaceous, many Asian dinosaurs crossed the Bering land bridge into North America, and many North American species crossed into Asia. Foreign predators might have thrived unchecked until they succumbed to a disease for which they had no immunity. The constantly warm tissue of warm-blooded creatures with high metabolic rates could have provided an ideal habitat for pathogenic organisms. Thus, Bakker speculates that dinosaurs with high metabolic rates would have been at much greater risk of mass extinction during intercontinental exchange than would have been the giant, ectothermic reptiles.

HUMAN IMPACTS ON EXTIRPATION AND EXTINCTION

For decades, there was a consensus that the earliest Americans came from Asia across the Bering Strait "land bridge" (Beringia) near the end of the Ice Age, settling first in the North American high plains, then moving into South America down the Andean chain (Martin, 1973; Patrusky, 1980; Brown and Gibson, 1983) (Fig. 15.5a). Dating of stone tools shows the presence of humans from Montana to Mexico between 11,500 and 11,000 years ago. Fluted points found

FIGURE 15.5

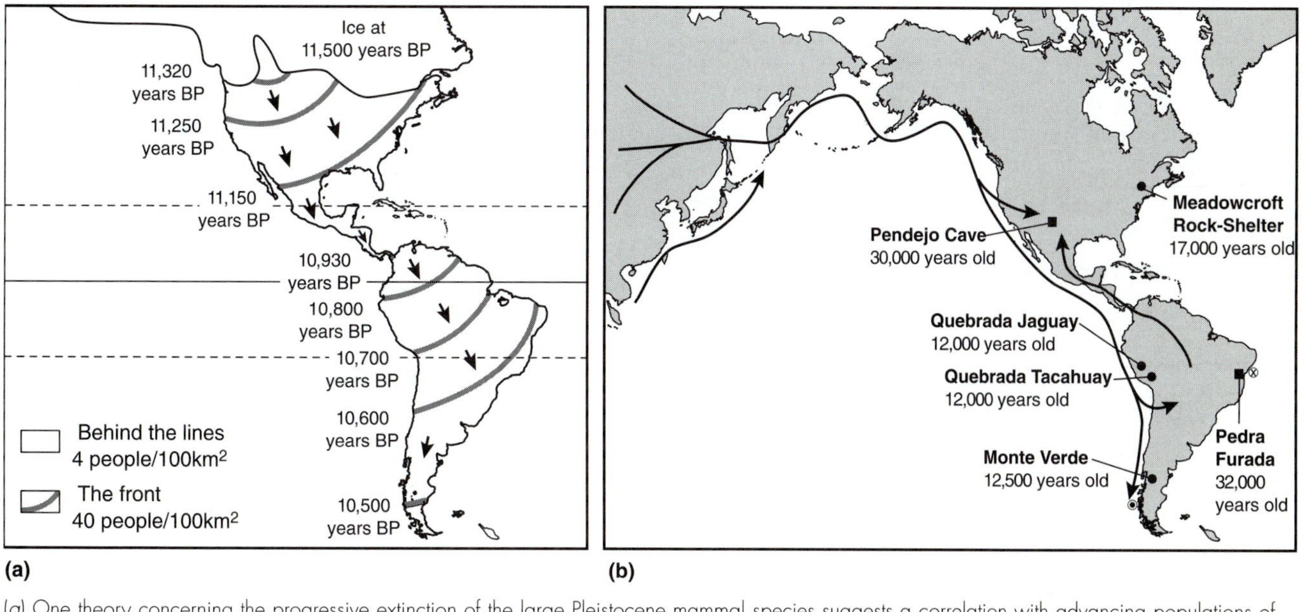

(*a*) One theory concerning the progressive extinction of the large Pleistocene mammal species suggests a correlation with advancing populations of big game hunters who crossed the Bering Strait and moved southward, maintaining a relatively dense front population that subsisted on large mammals. (*b*) Old view of land route into the New World some 11,500 years ago (top). New evidence from various sites (black circle is probable site; black square is possible site) suggests that migrants might have arrived well over 11,500 years ago, perhaps by sea.

Source: (a) *From P. S. Martin, "The Discovery of America" in* Science, *179:969–974. Copyright © 1973 American Association for the Advancement of Science.*
(b) *Wright "First Americans" in* Discover Magazine, *February 1999.*

among the bones of mammoths near the town of Clovis, New Mexico, in 1932 have been dated at 10,900 to 11,200 years old and long have been accepted as the continent's oldest known human artifacts. Other sites in southern Patagonia and in the Brazilian Amazon date back 11,500 years (Roosevelt et al., 1996, 1997; Haynes, 1997; Reanier, 1997; Barse, 1997) (Fig. 15.5b).

Recent evidence, however, indicates that early humans—*Homo erectus*—may have reached Siberia 500,000 years ago (Morell, 1994). Descendants of *Homo erectus* could have pushed from Siberia through Beringia (a continent-sized land mass linking Asia and North America) and into America long before the currently accepted colonization date of around 14,000 years ago. Paleontologists continue to present evidence showing that unspecialized hunters and gatherers may have been present in the New World at least 25,000 years ago and possibly more than 40,000 years ago (Fig. 15.5b) (Patrusky, 1980; Adovasio and Carlisle, 1986; Bower, 1990). Linguists say the diversity of native languages in the Americas—more than 140 language families, each as different as English and Arabic—also attests to a much longer period of occupation, probably at least 30,000 years or more. Evidence has been accumulating that the Clovis people may have shared the Americas with people of a different culture—one based on gathering fruits and nuts, fishing, and hunting small animals rather than felling mammoths (Gibbons, 1996b).

From 1977 to 1985, a site adjacent to a small creek between the Andes and the Pacific Ocean was excavated by an international team led by Dr. Tom D. Dillehay of the University of Kentucky (Dillehay, 1989, 1997). The site, known as Monte Verde, is about 800 km south of Santiago, Chile. As a result of these excavations, Dillehay concluded that tool-using humans lived in southern Chile 12,500 years ago—more than 1,000 years earlier than most scientists had believed possible. In January 1997, a team lead by Alex Barker, Curator of Archaeology at the Dallas Museum of Natural History, worked at Monte Verde and reached a similar conclusion (Anonymous, 1997b). In the same area, Dillehay has also found preliminary evidence—charcoal, stone tools, and clay-lined pits that could be hearths—of an even more ancient settlement in a soil layer more than 30,000 years old (Wright, 1999). Consequently, researchers may have to radically revise their ideas of how and when humans migrated into the New World. The possibility has been raised that some early inhabitants of Chile may have crossed the Pacific from Southeast Asia. The discovery in southeastern Brazil of an 11,500-year-old skull—the oldest in the New World—may help to rewrite the theory of how the Americas were settled.

Humans have greatly increased the rate of extinction through many of their activities. Some investigators believe that humans were at least partially responsible for the extinction of such Late Quaternary species such as mammoths, mastodons, saber-toothed cats, pygmy hippos, dodos, elephant birds, and many others (Martin, 1973; Mosimann and Martin, 1975) (Fig. 15.6). Diamond (1991) noted that Madagascar and several Mediterranean islands are yielding fossil evidence that human arrival on islands always has been accompanied by selective extinction of island megafaunas (large animals), irrespective of whether this arrival was around 1,000 years ago (New Zealand), 1,500 (Madagascar), 3,600 (New Caledonia), 10,000 (Mediterranean islands), or 30,000 years ago (Bismarcks). He suggests that, whenever anatomically and behaviorally modern *Homo sapiens* reached land previously unoccupied by humans—whether it be a continent such as Australia or the Americas, or an island— many of the native large prey have become extinct. Miller et al. (1999) concluded that human impact, not climate, was responsible for the sudden disappearance 50,000 years ago of the large flightless mihirung (*Genyornis newtoni*) in Australia. This was about the same time that humans arrived in Australia.

Steadman (1995) estimated that the prehistoric (2,000-30,000 years before the present) loss of bird life on tropical Pacific islands may have exceeded 2,000 species, many of which were pigeons, doves, parrots, flightless rails, and passerines. If accurate, this represents a 20 percent worldwide reduction in the number of species of birds. Instead of 9,600 + species alive today, there probably would have been about 11,600 species if these extinctions had not occurred. The loss of island birds mainly was due to predation by humans and the nonnative mammals (rats, dogs, pigs) brought with them, removal of native forests and plants, introduction of nonnative plants, and erosion of the soil.

Other factors may have been responsible for the extinctions on Madagascar. Some species may not have been able to adapt to the natural wet-to-dry oscillations of the climate. Clearly, whenever humans invade new territory, many large animals (megafauna) vanish (Diamond, 1991; Steadman, 1995). Direct competition for space and resources could have been responsible for their demise. A most intriguing, but still unsupported, theory is that early humans carried a lethal pathogen to the vulnerable island communities (Culotta, 1995a). A lethal pathogen could have swept rapidly through native animals that had never been exposed to the disease. Because illness usually affects young animals hardest, and because larger species have fewer offspring, the megafauna could have been pushed to extinction. Those species that survived the pandemic would be resistant to future outbreaks. Culotta's theory explains why first contact with humans seems to be the deadliest. It also might be applicable to the North and South American extinctions that occurred 10,000 to 12,000 years ago. During this time, North America lost 73 percent and South America 80 percent of their genera of big mammals (Diamond, 1987). North American losses included 3 genera of elephants, 6 of giant edentates, 15 of ungulates, and various giant rodents and carnivores (Martin, 1967). Culotta (1995a) points out that more than 70 species of large mammals became extinct; since that time, in contrast, no large mammals have been lost.

Diamond (1987) believed many species were quickly exterminated—possibly within just 10 years at any given site—by paleo-Americans arriving in North America from Asia toward the end of the last Ice Age. Evidence of excessive human predation of mammals (overkill) has been shown by computer simulation to be possible. Mosimann and Martin (1975) hypothesized that, perhaps, a new wave of humans immigrated from Asia some 13,000 years ago (Fig. 15.7). Many paleontologists, however, blame the extinctions of America's megafauna on drastic changes in climate and habitat at the end of the Ice Age rather than on human predation. With deglaciation, deserts expanded northward, wiping out huge areas of grassland once used for foraging (Patrusky, 1980). Diamond points out, however, that ice-free habitats for mammals expanded rather than contracted as glaciers yielded to grass and forest; in addition, big American mammals already had survived the ends of many glaciations without such an extinction event; and there were far fewer extinctions in Europe and Asia when the glaciers of those continents melted at around the same time.

Since 1500, more than 200 extinctions have been documented among vertebrates, mostly birds and mammals. Approximately 90 of these have been mammals (Fig. 15.8) (MacPhee and Flemming, 1997); undoubtedly, more have disappeared without a recorded history. In some cases, overhunting resulted in the extirpation of some species from former areas (bison) or in total extinction (Steller's sea cow, passenger pigeon, great auk, dodo; (see Fig. 15.1). Passenger pigeons (Fig. 15.9) formerly traveled in dense flocks numbering in the millions. Ornithologists have estimated that passenger pigeons in precolonial America numbered 2 to 3 billion, making them perhaps the most abundant bird species on the Earth at the time. By 1890, they virtually had disappeared due to overhunting for food and feathers. The last passenger pigeon died in the Cincinnati Zoo in 1914. Overhunting also has greatly reduced populations of alligators, sea turtles, and whales. The American alligator (*Alligator mississippiensis*) benefited from its protection under the Endangered Species Act and has recovered to the point where its status has been changed from endangered to threatened. In some areas, it is being legally harvested for its meat and skin. Many predators have been extirpated from large parts of their former ranges (gray wolf, *Canis lupus*; red wolf, *Canis rufus*; cougar, *Felis concolor*; and grizzly bear, *Ursus arctos*).

The world continues to face a biodiversity crisis, with the sources of current extinction patterns all around us. The most important and undoubtedly the number one modern-day cause of species population declines is habitat alteration and habitat destruction.

Clearing of forest areas for agriculture, subdivisions, shopping centers, and roads destroys the habitat of many species. Such practices have caused the destruction of vast areas of tropical rain forests, as well as temperate forests, worldwide (Fig. 15.10). The forested habitat of gorillas (*Gorilla gorilla*), orangutans (*Pongo pygmaeus*), and other primates, as well as giant pandas (*Ailuropoda melanoleuca*) and many tropical birds is decreasing at an alarming rate. In temperate

FIGURE 15.6

A Pleistocene scene. The extinction of large herbivores (such as mastodons, mammoths, giant sloths, peccaries, beavers, bears, deer, and antelope) by humans who were hunting their way south from Alaska suggests that the two events may be related. If true, this last round of extinction may be attributable to our own species.

FIGURE 15.7

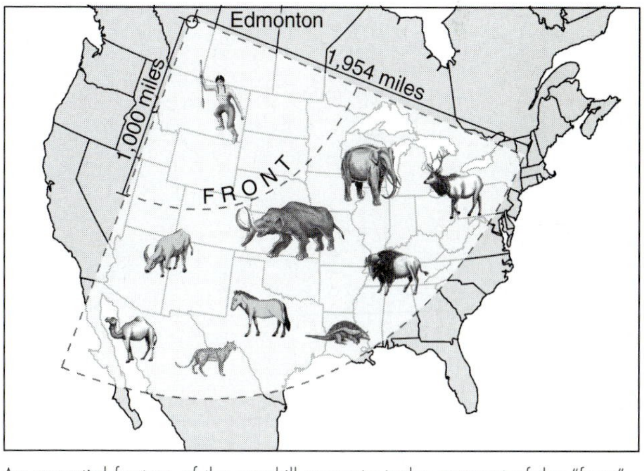

An essential feature of the overkill scenario is the concept of the "front". Upon reaching a critical density, the population of hunters, newly arrived in the New World, expands southward in a quarter circle. As long as some prey remains in the area of human occupation, the front advances smoothly. When the local herds are exhausted, it advances in a jump. The range available to the hunted is steadily reduced. The width of the front prevents survivors from "leaking" back into unoccupied areas behind the front. By the time the front has reached the gulfs of Mexico and of California, the herds of North America have been hunted to extinction.

FIGURE 15.8

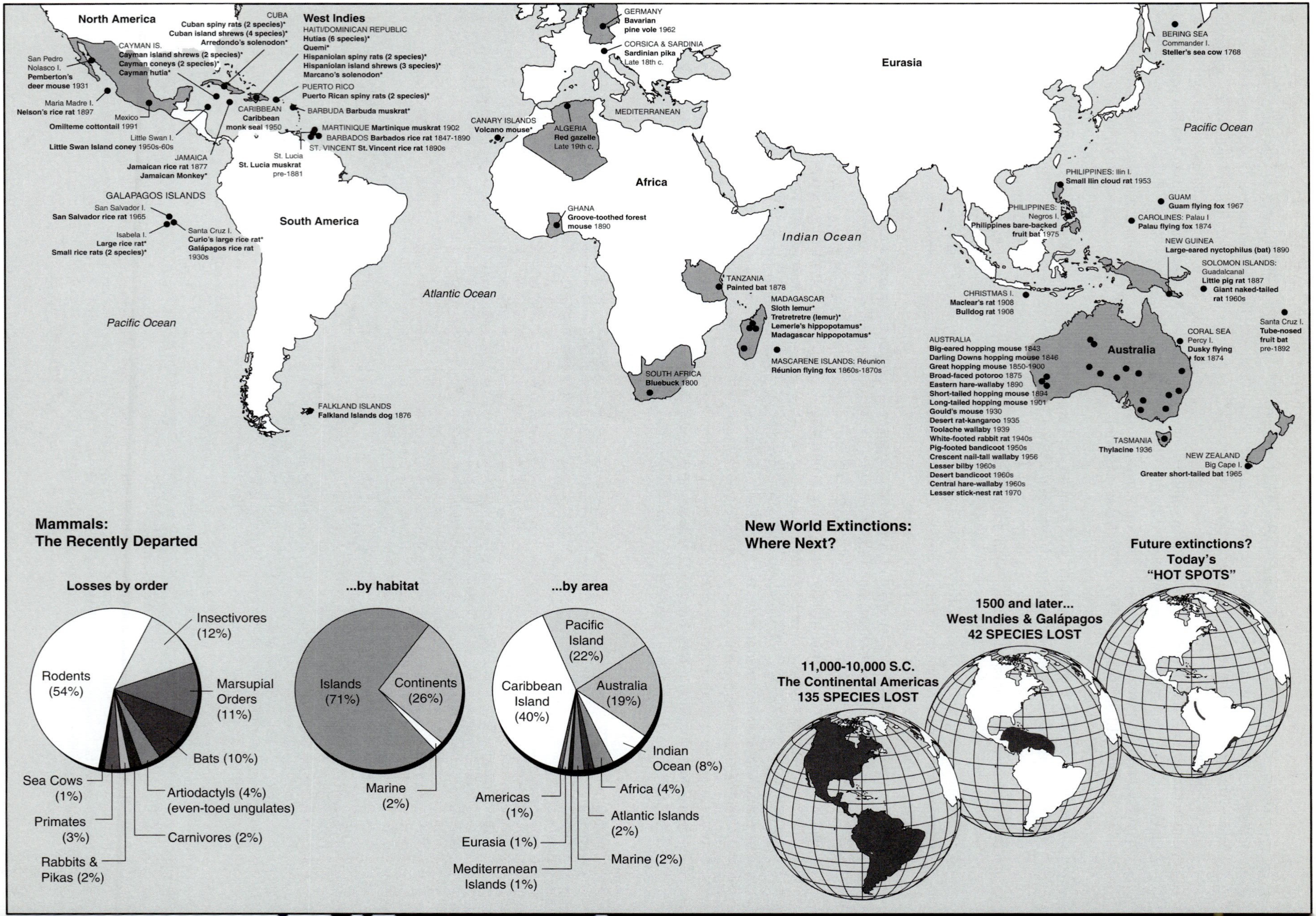
North America
San Pedro Nolasco I.
Pemberton's deer mouse 1931
Maria Madre I.
Nelson's rice rat 1897
Mexico
Omilteme cottontail 1991
Little Swan I.
Little Swan Island coney 1950s-60s
CAYMAN IS.
Cayman island shrews (2 species)*
Cayman coneys (2 species)*
Cayman hutia*
CUBA
Cuban spiny rats (2 species)*
Cuban island shrews (4 species)*
Arredondo's solenodon*
JAMAICA
Jamaican rice rat 1877
Jamaican Monkey*
CARIBBEAN
Caribbean monk seal 1950
West Indies
HAITI/DOMINICAN REPUBLIC
Hutias (6 species)*
Quemi*
Hispaniolan spiny rats (2 species)*
Hispaniolan island shrews (3 species)*
Marcano's solenodon*
PUERTO RICO
Puerto Rican spiny rats (2 species)*
BARBUDA Barbuda muskrat*
MARTINIQUE Martinique muskrat 1902
BARBADOS Barbados rice rat 1847-1890
ST. VINCENT St. Vincent rice rat 1890s
St. Lucia
St. Lucia muskrat pre-1881
GALAPAGOS ISLANDS
San Salvador I.
San Salvador rice rat 1965
Isabela I.
Large rice rat*
Small rice rats (2 species)*
Santa Cruz I.
Curio's large rice rat*
Galápagos rice rat 1930s
South America
Pacific Ocean
Atlantic Ocean
FALKLAND ISLANDS
Falkland Islands dog 1876
CANARY ISLANDS
Volcano mouse*
GERMANY
Bavarian pine vole 1962
CORSICA & SARDINIA
Sardinian pika Late 18th c.
MEDITERRANEAN
ALGERIA
Red gazelle Late 19th c.
Africa
GHANA
Groove-toothed forest mouse 1890
TANZANIA
Painted bat 1878
MADAGASCAR
Sloth lemur*
Tretretretre (lemur)*
Lemerle's hippopotamus*
Madagascar hippopotamus*
MASCARENE ISLANDS: Réunion
Réunion flying fox 1860s-1870s
SOUTH AFRICA
Bluebuck 1800
Indian Ocean
Eurasia
PHILIPPINES: Ilin I.
Small Ilin cloud rat 1953
PHILIPPINES: Negros I.
Philippines bare-backed fruit bat 1975
CHRISTMAS I.
Maclear's rat 1908
Bulldog rat 1908
BERING SEA
Commander I.
Steller's sea cow 1768
Pacific Ocean
GUAM
Guam flying fox 1967
CAROLINES: Palau I
Palau flying fox 1874
NEW GUINEA
Large-eared nyctophilus (bat) 1890
SOLOMON ISLANDS: Guadalcanal
Little pig rat 1887
Giant naked-tailed rat 1960s
Santa Cruz I.
Tube-nosed fruit bat pre-1892
CORAL SEA
Percy I.
Dusky flying fox 1874
Australia
AUSTRALIA
Big-eared hopping mouse 1843
Darling Downs hopping mouse 1846
Great hopping mouse 1850-1900
Broad-faced potoroo 1875
Eastern hare-wallaby 1890
Short-tailed hopping mouse 1894
Long-tailed hopping mouse 1901
Gould's mouse 1930
Desert rat-kangaroo 1935
Toolache wallaby 1939
White-footed rabbit rat 1940s
Pig-footed bandicoot 1950s
Crescent nail-tail wallaby 1956
Lesser bilby 1960s
Desert bandicoot 1960s
Central hare-wallaby 1960s
Lesser stick-nest rat 1970
TASMANIA
Thylacine 1936
NEW ZEALAND
Big Cape I.
Greater short-tailed bat 1965
Mammals:
The Recently Departed
Losses by order
Rodents (54%)
Insectivores (12%)
Marsupial Orders (11%)
Bats (10%)
Artiodactyls (4%) (even-toed ungulates)
Carnivores (2%)
Rabbits & Pikas (2%)
Primates (3%)
Sea Cows (1%)
...by habitat
Islands (71%)
Continents (26%)
Marine (2%)
...by area
Caribbean Island (40%)
Pacific Island (22%)
Australia (19%)
Indian Ocean (8%)
Africa (4%)
Atlantic Islands (2%)
Marine (2%)
Mediterranean Islands (1%)
Eurasia (1%)
Americas (1%)
New World Extinctions:
Where Next?
11,000-10,000 S.C.
The Continental Americas
135 SPECIES LOST
1500 and later...
West Indies & Galápagos
42 SPECIES LOST
Future extinctions?
Today's
"HOT SPOTS"

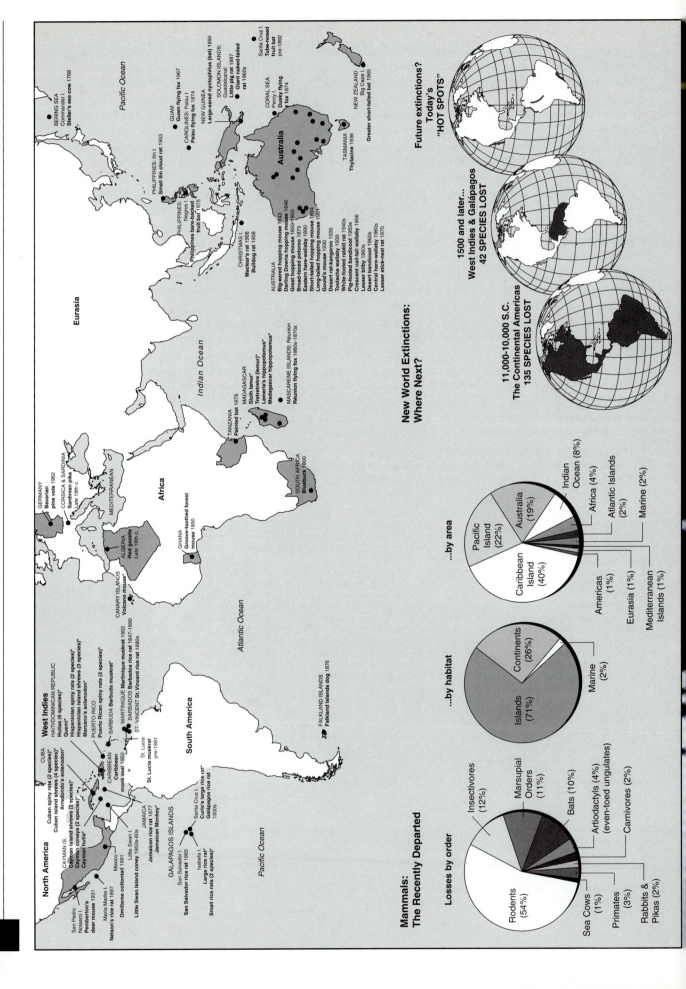

BIO-NOTE 15.1

Birds at Risk

A major study by The World Conservation Union (IUCN) and BirdLife International indicates that 1,107 bird species, approximately 11 percent of the world's total, are at risk of dying out. They list 168 bird species that are "critically endangered"—meaning they face an extremely high risk of extinction in the wild in the immediate future. Like other animal groups, birds are most threatened in island habitats. Of the 104 bird species that became extinct in the past 400 years, approximately 90 percent lived on islands. Island species often are found nowhere else on Earth and therefore cannot be replenished from outside. They have few defenses against introduced predators, and they are vulnerable to introduced diseases. Hawaii, at 33 percent, currently has the highest proportion of threatened bird species. Significant threats to bird life also exist in continental areas due to such factors as population pressure, exploitation of tropical forests, and habitat destruction in nesting areas. Continental species with small ranges are most vulnerable.

Doyle, 1997
Manne et al., 1999

regions, ivory-billed woodpeckers (*Campephilus principalis*), red-cockaded woodpeckers (*Picoides borealis*), Kirtland's warblers (*Dendroica kirtlandii*), Bachman's warblers (*Vermivora bachmanii*), and spotted owls (*Strix occidentalis*) are examples of species whose habitats have almost disappeared. The construction of dams and the straightening and channelization of rivers destroy habitats for many species. In addition, civil unrest in parts of Africa and the resettling of hundreds of thousands of refugees have put the mountain gorilla and the chimpanzee in great peril, according to the World Wildlife Fund.

The populations of long-distance, migrating songbirds have declined in recent years. The excessive fragmentation of forests that has occurred throughout the United States and elsewhere results in the loss of nesting sites and food and also increases the exposure of bird nests to predators such as raccoons, cats, opossums, blue jays, and parasitic birds such as the brown-headed cowbird (Terborgh, 1992).

Deforestation of tropical wintering habitats is another factor causing severe losses in the breeding birds of North America. Robbins et al. (1989) analyzed data from the North American Breeding Bird Survey (BBS), which was begun in 1965. Although populations of neotropical migrants throughout eastern North America were stable or increasing before 1978, there has been a general decline since then, which has not been paralleled in residents or short-distance migrants. Chiefly, those species that winter in tropical forests have been affected. In addition, many migratory species winter in countries that still use DDT and other biologically harmful pesticides.

FIGURE 15.9

The extinct passenger pigeon (*Ectopistes migratorius*). In 1605 the French explorer Samuel de Champlain observed passenger pigeons in "infinite numbers" in Maine. These birds perched on trees in such dense groups that they often broke the branches by their sheer weight. Migratory flocks, miles long, darkened the sun and took hours to pass a given point. Alexander Wilson, the "Father of American ornithology," estimated that one such throng observed by him near Frankfort, Kentucky, about the year 1808, contained over 2 billion passenger pigeons. Humans mass-slaughtered passenger pigeons for food for themselves and for their pigs. It was fantastically easy to shoot the birds, but it was cheaper and quicker to blind them with torches and club them to death, or suffocate them by burning sulfur under their roosts. The last known passenger pigeon died in the Cincinnati Zoo in 1914.

FIGURE 15.10

Ground-level view of tropical rain forest destruction in the Amazon basin in Brazil.

Coffee beans were traditionally grown in shade under a canopy of trees that protected the plants and offered sanctuary for more than 150 species of migratory birds. For many years, however, agriculture experts encouraged large coffee plantations in Latin America to grow more beans at a faster pace by cutting down trees shading the coffee trees and growing high-yield, sun-tolerant hybrid trees that need high doses of pesticides and chemical fertilizers. As the canopy trees disappeared, so did the birds—in some cases by as much as 97 percent. In 1996, the Smithsonian Institution, National Audubon Society, Rainforest Alliance, and other conservation groups began encouraging farmers to cultivate shade-grown coffee, which they hope will save the tall trees where United States and Canadian migratory birds seek refuge from the cold. Shade-grown trees yield benefits not only for migratory birds but also for the farmer and the consumer. For example, the fruit from shade trees such as walnuts, palm fruits, oranges, bananas, lemons, and avocados, together with other products grown beneath the trees, can provide farmers with additional food and supplemental income. The trees also provide firewood. For the consumer, a slower ripening red coffee fruit, known as a cherry, retains more natural sugars and results in better tasting coffee. Since 1996, several U.S. companies have developed coffee brands that are certified as shade-grown by independent monitors.

The Switzerland-based World Wide Fund for Nature announced in 1996 that tigers (*Panthera tigris*) that once roamed much of Asia are dying at a rate of perhaps one a day because of deforestation and poaching. The number of Asian tigers remaining in the wild could be as low as 4,600, down from 5,000 in 1994. A major threat to Asian tigers is an increased illegal trade in tiger bones, skin, and other parts for traditional Chinese medicines. Illegal trading is also a threat to many other vertebrates, especially reptiles, birds, and primates.

In January, 2000, Conservation International and the IUCN released a report stating that after surviving a century with no extinctions, 25 species of apes, monkeys, lemurs, and other primates now risk disappearing forever. In some cases, only a few hundred individuals survive. The main causes are tropical forest habitat destruction and local bushmeat hunting, according to the report. Live capture for the pet trade and export for biomedical research also threaten some species.

Pesticides and heavy metals are present in the blubber, muscle, liver, and brains of many vertebrate species. Beattie and Tyler-Jones (1992) reported that increasing aluminum concentrations and low pH reduced the fertilization success of European common frogs (*Rana temporaria*), increased embryonic mortality, and decreased larval body length. The effects of environmental contamination on the development of snapping turtle eggs was reported by Bishop et al. (1991). Evans (1987) reviewed contaminant levels and their biological effects in cetaceans. Concentrations of DDT up to 2,700 ppm and up to 850 ppm of PCBs have been reported (Gaskin, 1982).

Some species literally have almost been poisoned out of existence, such as California condors (*Gymnogyps californianus*), Utah prairie dogs (*Cynomys parvidens*), and black-footed ferrets (*Mustela nigripes*) (Fig. 15.11), while others have ingested and accumulated such high quantities of chlorinated hydrocarbon compounds like DDT and polychlorinated biphenyls (PCBs) that they can no longer reproduce and maintain viable populations in certain areas (brown pelicans, *Pelecanus occidentalis*; bald eagle, *Haliaeetus leucocephalus*; peregrine falcon, *Falco peregrinus*) (Office of Endangered Species and International Activities of the U.S. Fish and Wildlife Service, 1973). Organochlorine pesticides such as DDT, which are insoluble in water but soluble in fat, accumulate in food chains and food webs, a process called **biological amplification** (Fig. 15.12). Ingested compounds caused thinning of egg shells, which frequently cracked prior to hatching (Fig.15.13). Although the use of DDT and other organochlorines in the United States has been legally banned since 1970, the United States is still the largest producer of DDT, which is exported to developing countries.

FIGURE 15.11

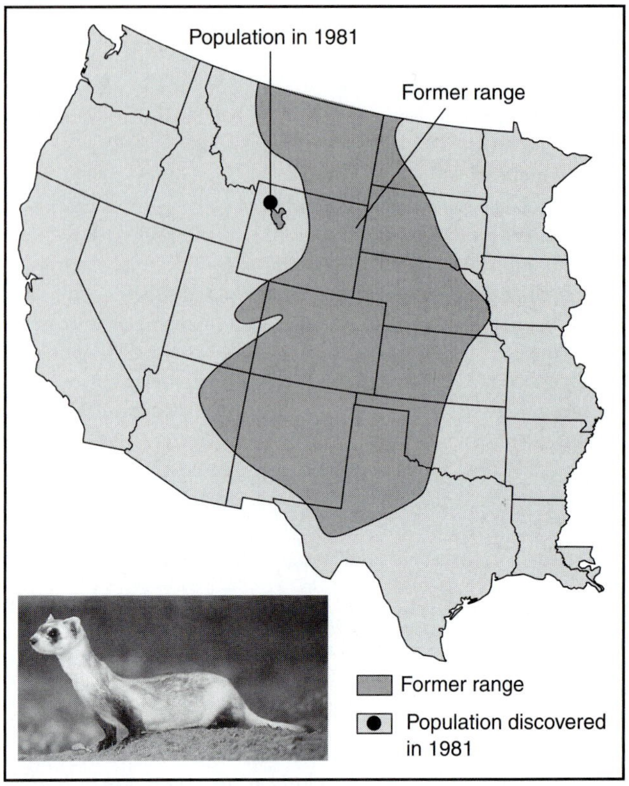

The black-footed ferret (*Mustela nigripes*) (inset) is one of the most endangered mammals in North America. As of March 1997, the wild and captive population numbered approximately 500 animals. Although the former range of the black-footed ferret has been drastically reduced, ferrets have been reintroduced into Montana (1994), South Dakota (1994), and Arizona (1996).
Line Art Source: Data from Defenders of Wildlife.

Green (1998) documented widespread declines in egg shell thickness beginning in 1850 for several species of thrush (*Turdus* sp.) in Britain. These declines, well before the introduction of DDT in 1947, may have been an early consequence of industrialization. Acid deposition, caused by anthropogenic emissions of sulfur dioxide, nitrogen oxides, and ammonia, may have altered soil or water chemistry, affecting the availability of those invertebrates such as snails (or their shells) that are an important source of calcium during egg-shell formation. Decreased availability of dietary cal-

FIGURE 15.12

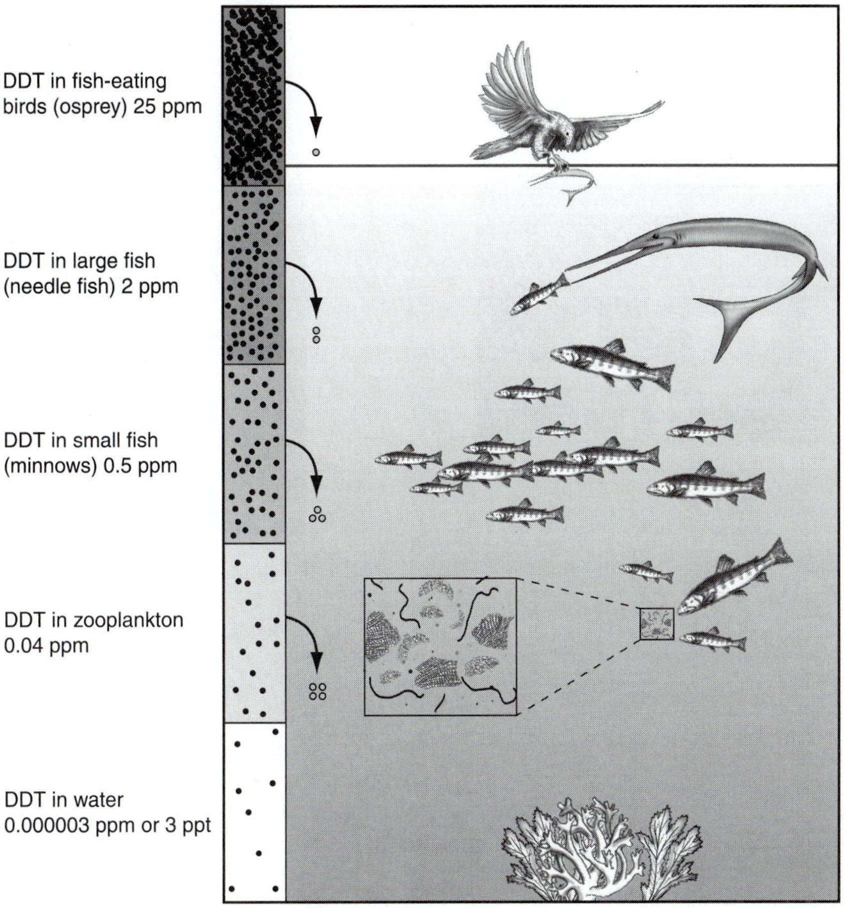

Bioaccumulation. The concentration of DDT in the fatty tissues of organisms was biologically amplified about 10 million times in this food chain of an estuary adjacent to Long Island Sound, near New York City. Dots represent DDT, and arrows show small losses of DDT through respiration and excretion. *Source: Miller,* Living in the Environment, *2000 Brooks-Cole, Inc.*

cium is known to adversely affect egg laying and egg-shell integrity in birds as well as the growth of hatchling birds and neonatal mammals.

Scheuhammer (1991) reviewed the effects of acidification of the environment on the availability of toxic metals and calcium to wild birds and mammals. The dusky seaside sparrow (*Ammospiza maritima nigrescens*) (Fig. 15.14), which formerly resided on Merritt Island and on the adjacent mainland along the St. John's River in east-central Florida, became extinct in the wild in the late 1970s after the habitat was altered for mosquito and flood control. Habitat destruction combined with an extremely low reproductive rate (one egg every other year) were key factors in the almost complete loss of California condors (*Gymnogyps californianus*).

In recent years, many amphibian populations throughout the world have been declining (Doyle, 1998). Frogs living in highland and northern areas seem to be the most drastically affected species. Wake (1991) summarized some of the most dramatic declines. Frogs of the genus *Rana* have almost disappeared from southern California. By the late 1980s, *Rana muscosa,* a montane frog, had disappeared from 98 percent of the ponds in which it had been studied in the mid-1970s in Sequoia–Kings Canyon National Park. Populations of *Rana cascadae* monitored since the mid-1970s have suffered about an 80 percent disappearance in Oregon. The golden toad (*Bufo periglenes*), endemic to the Monteverde Cloud Forest Preserve in Costa Rica, has not been seen since 1989. During this same period, many other species of frogs in the same region have experienced declines and possible extirpation. The gastric-brooding frog (*Rheobatrachus silus*) (see Fig. 6.35), which inhabited rivers in relatively undisturbed regions of Queensland, Australia, has not been seen since 1979, and several other sympatric species of frogs now are thought to be extinct. Sometime after 1981, 8 of 13 species of frogs that had been present in Reserva Atlantica, Brazil, disappeared. Do these declines represent normal population fluctuations, or could they be the result of habitat destruction and/or climatic fluctuations? Many researchers feel that amphibians could be serving as "indicator" species, warning of serious environmental changes brought about by pesticide pollution, increased ultraviolet radiation, acid precipitation, and/or global warming.

FIGURE 15.13

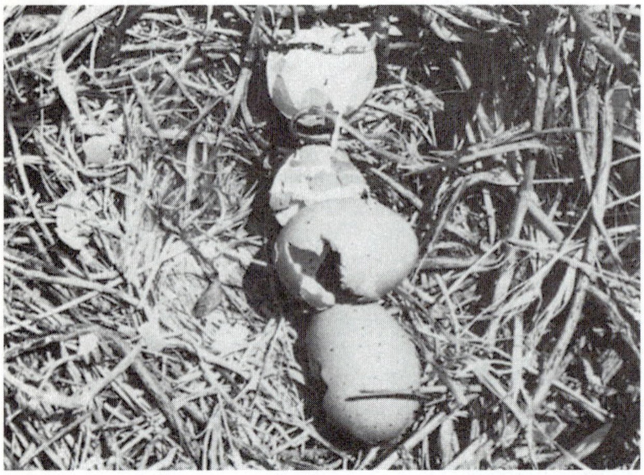

(a)

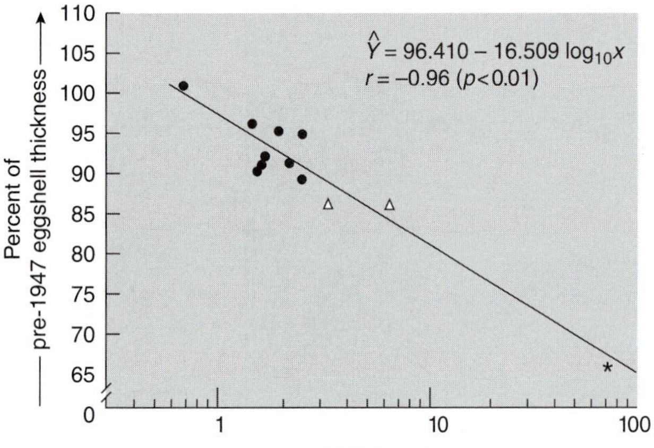

Association of DDE residues in brown pelican eggs from nine colonies in Florida (•), two colonies in South Carolina [△], and one colony in California [*]

(b)

(*a*) Abandoned ibis nest with broken thin-shelled eggs. (*b*) DDE content and egg-shell thinning of brown pelican eggs from Florida, South Carolina, and California are strongly correlated.
Source: Data from the U.S. Fish and Wildlife Service, 1971.

While concern about newly evolving diseases has focused on humans, scientists warn that wildlife faces the same threats as animals are forced into more crowded areas and moved to new locations. Globalization of the planet leads to the spread of diseases as animals are taken abroad from one country to another. Emerging infectious diseases (EIDs) can cause the extirpation of local populations of wild animals and even lead to extinction of a species (Daszak et al., 2000). Such diseases not only threaten wildlife, but also may form a reservoir of germs that could harm humans and domestic animals. This kind of transmission may have been a factor in the emergence of the virus that causes AIDS from chimpanzees and mangabeys to humans. Researchers have determined that a type of herpes virus that is harmless to African elephants can be fatal to Indian elephants. Measles contracted from humans threatens wild mountain gorillas habituated to tourists, and poliovirus has killed chimpanzees in Gombe National Park in Tanzania.

FIGURE 15.14

The dusky seaside sparrow (*Ammospiza maritima nigrescens*) was a resident on Merritt Island and on the opposite mainland along the St. Johns River in east-central Florida. It is now extinct because the habitat has been altered for mosquito and flood control.

Cape buffalo in Kruger National Park in South Africa contracted tuberculosis by mingling with infected cattle herds in the 1950s near southern Kruger, before the park's boundary was fenced off. The spread of the disease is slow among buffalo, with many of those affected being able to live out a normal lifespan. However, African lions feeding on infected buffalo, their natural prey, get a massive dose of TB. Researchers say that more than 90 percent of the lions tested during 1998 in the southern part of the park are infected with the incurable disease which causes the lions to develop intestinal lesions before the disease spreads to lymph nodes and other organs. A lion that contracts bovine tuberculosis can take several years to die from the disease itself, but its weakened condition makes it an easier target for rival lions. Scientists believe that most of the infected lions will die, and that the entire lion population is at risk.

A number of species of desert pupfish (*Cyprinodon* sp.) have been exterminated due to the introduction of predatory fishes, the construction of bathhouses over the sources of thermal springs, and the filling or draining of springs for agricultural purposes (Brown, 1971). Pumping of water from natural underground reservoirs lowers the water level in the springs and is threatening the extinction of seven additional species (U.S. Fish and Wildlife Service, 1990).

Pollution of water by oil or other toxic chemicals is detrimental to many marine species as well as to shorebirds (e.g., Exxon Valdez spill) (Fig. 15.16a). Mead (1997) reviewed European and American studies on the fate of seabirds cleaned and released after being caught in oil slicks. The best efforts to clean up oiled birds are not very successful. Fishes

BIO-NOTE 15.2

Restoring Endangered Species

The last wild California condor (*Gymnogyps californianus*) was trapped by biologists in 1987 (Fig. 15.15). At that time, only 27 birds remained as genetic "founders" for a breeding program. By December 1991, the program had produced 25 additional birds, including the first 2 freed in January 1992. As many populations decline, they often experience a significant loss in genetic variation that may decrease fitness, limit the long-term capacity of the population to respond to environmental challenges, and allow chance environmental and demographic events to pose a more immediate threat. Using DNA fingerprinting, researchers found that the 52 living condors comprised three distinct ancestral groups (Hedrick, 1992). The existing condor gene pool is more diverse than some had believed, and this could prove invaluable in developing successful reintroduction plans for the condor into its native habitat.

By November 1996, three captive breeding flocks contained 104 California condors. These flocks are maintained for propagation at the Los Angeles Zoo, San Diego Wild Animal Park, and World Center for Birds of Prey in Boise, Idaho. In addition, another 17 condors were flying free in California.

On October 29, 1996, six captive-reared condor chicks were transferred to a release site located at Vermilion Cliffs, approximately 48 km north of Grand Canyon National Park in northern Arizona. These birds were released in December 1996, and they became the first free-flying California condors in Arizona since 1924. This was the eighth release of captive-bred condors since 1992, and the first release outside California. Plans call for a series of releases of captive-reared condors at the Vermilion Cliffs each fall to establish a population in the region of 150 condors. Released birds will be approximately 6 months old and will be monitored through the use of radio transmitters and wing markers. If this reintroduction project is successful, it will achieve one of the primary goals of the California Condor Recovery Plan: to establish a second self-sustaining population in the wild. The population of wild and captive individuals was 162 as of September 1999. This includes 49 condors in the wild—29 in California and 20 in northern Arizona. A total of 19 chicks were hatched from the captive breeding flocks during 1998.

Only a few endangered species have prospered enough to be removed from the endangered species list (e.g., American alligators, bald eagles, peregrine falcons). Peregrine falcon (*Falco peregrinus*) populations, for example, have rebounded from a low of 324 nesting pairs in 1975 to at least 1,593 breeding pairs in the United States and Canada in 1998, well above the overall recovery goal of 631 pairs (Hoffman, 1998; Reichhardt, 1998). The species has recolonized most of its former range and has even expanded into some new areas. Other species have also begun to make notable comebacks. A total of 158 whooping cranes (*Grus americana*) overwintered on the Aransas National Wildlife Refuge on the Texas coast during the winter of 1995–96 (up from 15–16 in 1941; Collar et al., 1993). An attempt to establish a separate nonmigrating flock of whooping cranes in Florida in January 1993 failed when the 14 birds were killed by bobcats and other predators. However, as of April 1997, a nonmigratory population was in the process of being established in central Florida's Kissimmee Prairie and currently consists of approximately 60 cranes. Older birds have begun showing territorial behavior characteristic of sexual maturity. During the spring of 1996, a pair constructed a nest and were observed copulating and defending a several-hundred-acre marsh. It was not until April 1999, however, that the first eggs were observed.

One of the last known populations of black-footed ferrets was extirpated in South Dakota during the early 1970s. The species was feared to be extinct until a small population was found in northwestern Wyoming, in 1981. Several animals from the Wyoming population were caught for a captive breeding program before the wild population disappeared. Successes in the captive breeding program allowed reintroductions at a south-central Wyoming site from 1991 until 1995, but disease and other factors caused that program to be suspended. Beginning in 1994, reintroduction efforts were begun in Montana and South Dakota. In March 1996, Arizona became the recovery program's fourth reintroduction site when 44 captive-bred ferrets were sent to Aubrey Valley for acclimation. They were released to the wild between September 5 and November 19, 1996. Each animal had a surgically implanted passive integrated transponder tag; 9 adults and 7 kits also had radio collars affixed to track their movements (Reading et al., 1996). As of March 1997, approximately 90 ferrets, including reintroduced captive-born animals and their offspring born in the wild, inhabited these sites. The total number of ferrets alive (captive and wild) in March 1997 was approximately 500.

In January 1995, a restoration project was begun to reestablish gray wolves (*Canis lupus*) in the ecosystem of Yellowstone National Park. As of fall 1997, a total of at least 94 wolves inhabited the Greater Yellowstone area; 47 of the 94 were pups born during the 1997 breeding season. As one biologist stated, "Wolf restoration is evidence of a new way of thinking that will lead to protection of ecosystems, biological diversity, and humankind".

During 1998, Mexican gray wolves (*Canis lupus baileyi*) were reintroduced into Arizona in an attempt to reestablish a population that was hunted to near extinction about 50 years ago. The goal of the recovery program is to build a population of about 100 wild wolves.

Mesta, 1996
Gober and Lockhart, 1996
Anonymous, 1996d
Wheelwright, 1997
Reading et al., 1996
Guernsey, 1997
Chadwick, 1998

FIGURE 15.15

California condor (*Gymnogyps californianus*) breeding pair in captivity at the San Diego Zoo. When biologists trapped the last remaining California condor in 1987, only 27 birds remained as genetic "founders" for a breeding program.

living in polluted water often develop lesions due to bacterial or viral infection (Fig. 15.15b). Viruses have been identified as the pathogen in mass mortalities of seals in Antarctica in 1955, in northwestern Europe in 1988, in Lake Baikal in 1989, and in West Africa in 1997; porpoises in Ireland in 1988; and dolphins in the western Mediterranean in 1990 (Harvell et al, 1999). Increased industrialization in many areas poses a severe threat to cetacean populations, especially the smaller species that inhabit semienclosed waters.

Penguins and alcids are more vulnerable to oil spills than are flying birds, because they must swim through any oil that occurs between their breeding colonies and feeding grounds. Resting Adelie penguins (*Pygoscelis adeliae*) contaminated with oil have reductions in heart rate, body temperature, and energy expenditure. In the water, the swimming speed of oiled penguins is lower, their heart rates are increased, and their metabolic rates are 50 percent higher while swimming than in control penguins (Culik et al., 1991).

The most serious danger to dolphins, porpoises, sea turtles, and other marine species around the world is the threat from various methods of gill-net fishing. More than 80,000 small cetaceans are killed annually in coastal waters around the world. Active acoustic alarms spaced 92 m apart and emitting a broadband signal with a frequency of 10 kHz and a source level of 132 dB have proven effective in reducing the number of deaths of harbor porpoises (*Phocoena phocoena*) in gill-nets along the coasts of New Hampshire and Maine (Kraus et al., 1997). The signal lasted for approximately 300 milliseconds and was repeated every 4 seconds.

Sea otters (*Enhydra lutris*) living off the Alaskan coast play a major role in marine ecosystems by limiting the distribution and abundance of sea urchins, thereby promoting kelp forest development. By feeding on sea urchins, otters help preserve kelp forests that provide food for a wide range of species, from barnacles to bald eagles. Since 1990, however, Estes et

FIGURE 15.16

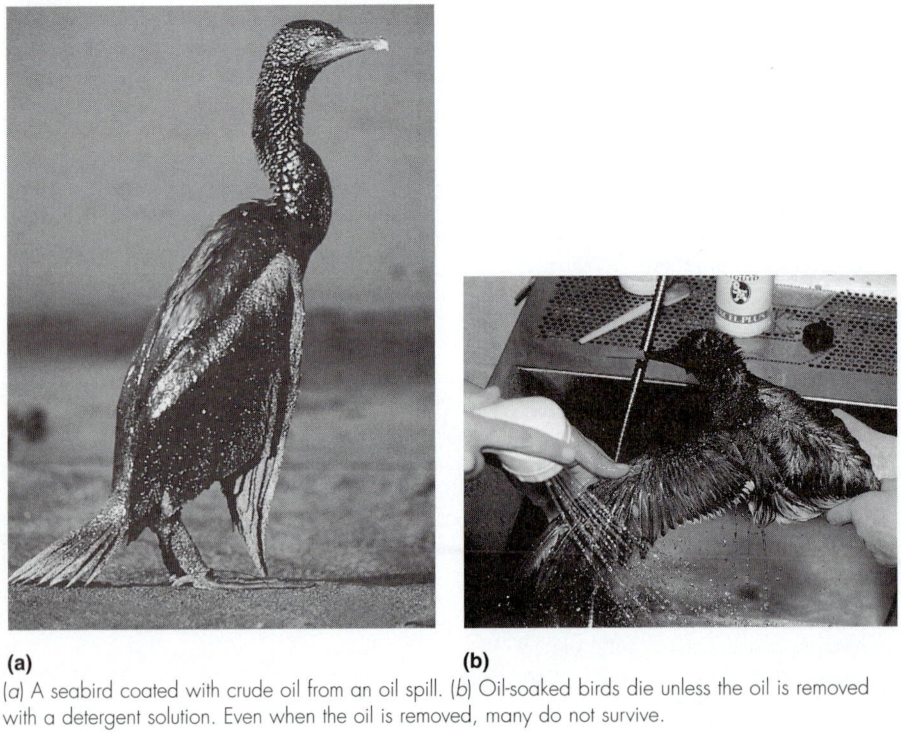

(a) (b)

(*a*) A seabird coated with crude oil from an oil spill. (*b*) Oil-soaked birds die unless the oil is removed with a detergent solution. Even when the oil is removed, many do not survive.

al. (1998) have documented a 90 percent decline in sea otter populations in western Alaska's Aleutian Islands. Evidence points to increased predation by killer whales (*Orcinus orca*). Killer whales, whose usual food sources are seals and sea lions, have begun feeding on sea otters, whose numbers have plummeted from about 53,000 in the 1970s to 6,000 in 1998 along an 800-km span of coastline. As sea otter numbers decline, sea urchin density and kelp deforestation increase. Estes et al. (1998) believe that whales are feeding on otters because of a decline in the numbers of sea lions and seals. The reason for the seal and sea lion decline is controversial. It may be due to intensified trawler fishing in the Bering Sea, which has sharply curtailed or altered the food supply for the whales, or changes in fish populations could also be related to warmer ocean temperatures and/or the local extirpation of baleen whales, which has allowed pollock, a fish low in fat, to flourish.

Ingestion of human-related debris also causes the deaths of many aquatic animals each year. Monofilament fishing line, plastic bags, string, rope, fishhooks, wire, paper, cellophane, synthetic sponges, rubber bands, and stockings have been recovered from intestinal tracts of endangered Florida manatees (Beck and Barros, 1991). Additional deaths have resulted from entanglement in lines and nets (Fig. 15.17).

FIGURE 15.17

This sea lion, as well as seals, die by the thousands each year after becoming entangled in plastic debris, especially broken and discarded fishing nets.

Paul Erlich (1986) noted that the earth's biota appears to be entering an era of extinction that may rival or greatly surpass in scale that which occurred at the end of the Cretaceous. As he stated:

> *As far as is known, for the first time in geologic history, a major extinction episode will be entrained by a global overshoot of carrying capacity by a single species—* Homo sapiens. *The episode, if it culminates as projected, will produce a crash in the population size of the species that caused it. Unfortunately, few laypeople are aware of the utter dependence of our species on the free services provided by natural ecosystems—and thus on other organisms that are key components of those systems. Ironically, for the first time, a species engendering its own collapse has the knowledge necessary to avoid its fate but may not be able to disseminate that knowledge and act on it in time.*

Review Questions

1. Differentiate between extirpation and extinction. Give two examples each of extirpated and extinct species.
2. Discuss several theories for the Cretaceous extinctions.
3. Give several pieces of evidence in support of the meteor impact theory.
4. List several species of vertebrates whose extinction is believed to have been at least partially caused by humans.
5. What do you think would have happened if extinction had never occurred during evolutionary history? Do you think there would be a greater or a lesser number of species on Earth than are currently present? Why?
6. Define biological amplification. Discuss its significance and give several examples.
7. What are some of the theories concerning the cause of declining amphibian populations?
8. Prepare a brief research proposal outlining the methods you would employ in an attempt to determine the cause(s) of a regional amphibian decline.
9. List several ways in which human activities affect aquatic species.
10. Why is it important to vertebrate populations to limit forest destruction?
11. List several factors contributing to the decline of long-distance migratory songbirds.

Supplemental Reading

Alvarez, W. 1997. *T.* rex *and the Crater of Doom.* Princeton, New Jersey: Princeton University Press.

Archibald, J. D. 1996. *Dinosaur Extinction and the End of an Era: What the Fossils Say.* New York: Columbia University Press.

Burton, J. F. 1995. *Birds and Climate Change.* London: Christopher Helm Limited.

Chadwick, D. H. 1998. Return of the gray wolf. *National Geographic* 193(5):72–99.

Collar, N. J., and S. N. Stuart. 1985. *Threatened Birds of Africa and Related Islands.* The ICBP/IUCN Red Data Book. Washington, D.C.: Smithsonian Institution Press.

Collar, N. J., L. P. Gonzaga, N. Krabbe, A. Madrono Nieto, L. G. Naranjo, T. A. Parker, III, and D. C. Wege. 1993. *Threatened Birds of the Americas.* The ICBP/IUCN Red Data Book, Part 2, Third edition. Washington, D.C.: Smithsonian Institution Press.

Courtillot, V. 1999. *Evolutionary Catastrophes: The Science of Mass Extinctions.* New York: Cambridge University Press.

Cunningham, C., and J. Berger. 1997. *Horn of Darkness: Rhinos on the Edge.* New York: Oxford University Press.

DeGraaf, R. M., and J. H. Rappole. 1995. *Neotropical Migratory Birds.* Ithaca, New York: Comstock Publishing Associates.

Frankel, C. 1999. *The End of Dinosaurs: Chicxulub Crater and Mass Extinctions.* London: Cambridge University Press.

Fuller, E. 1987. *Extinct Birds.* New York: Facts-on-File Publications.

Heaney, L. R., and J. C. Regalado, Jr. 1998. *Vanishing Treasures of the Philippine Rain Forest.* Chicago: University of Chicago Press.

Keast, A. (ed.). 1990. *Biogeography and Ecology of Forest Bird Communities.* The Hague, Netherlands: SPB Academic Publishing.

Lawton, J. H., and R. M. May. 1995. *Extinction Rates.* New York: Oxford University Press.

Long, M. E. 1998. The vanishing prairie dog. *National Geographic* 193(4):116–131.

MacLeod, N., and G. Keller. 1996. *Cretaceous-Tertiary Mass Extinctions: Biotic and Environmental Changes.* New York: Norton.

Miller, B., R. P. Reading, and S. Forrest. 1996. *Prairie Night: Black-Footed Ferrets and the Recovery of Endangered Species.* Washington, D.C.: Smithsonian Institution Press.

Mulvaney, J. and J. Kamminga. 1999. *Prehistory of Australia.* Washington, D.C.: Smithsonian Institution Press.

Nettleship, D. N., J. Burger, and M. Gochfeld (eds.). 1994. *Seabirds on Islands.* Washington, D.C.: Smithsonian Institution Press.

Officer, C., and J. Page. 1996. *The Great Dinosaur Extinction Controversy.* New York: Addison-Wesley.

Quammen, D. 1996. *The Song of the Dodo: Island Biogeography in an Age of Extinctions.* New York: Scribner.

Quay, T. L., J. B. Funderburg, Jr., D. S. Lee, E. F. Potter, and C. S. Robbins. 1983. *The Seaside Sparrow, Its Biology and Management.* Proceedings of a Symposium, Raleigh, North Carolina. October 1–2, 1981. Occasional Paper of the North Carolina Biological Survey 1983–5.

Radziewicz, J. (ed.) 1999. *Swift as a Shadow: Extinct and Endangered Animals.* Boston: Houghton Mifflin.

Safina, C. 1995. The world's imperiled fish. *Scientific American* 273(5):46–53.

Stuart, C., and T. Stuart. 1996. *Africa's Vanishing Wildlife.* Washington, D.C.: Smithsonian Institution Press.

Wheelwright, J. 1997. Condors: Back from the brink. *Smithsonian* 28(2):48–55.

Wilson, E. O. (ed.). 1988. *Biodiversity.* Washington, D.C.: National Academy Press.

Vertebrate Internet Sites

Visit the zoology website at http://www.mhhe.com to find live Internet links for each of the references listed below.

1. **Patuxent Wildlife Research Center. Laurel, Maryland.**
   This site provides access to hundreds of ongoing studies including amphibian surveys based on vocalizations, breeding bird censusing, and endangered species; includes biological monitoring information bases as well as information concerning many other baseline studies; images; links.
2. **EE-Link Endangered Species.**
   Extensive information including "what's new," endangered and extinct species lists that may be sorted by region or taxonomic group, both in the U.S. and internationally, information on particular endangered species, and much more.
3. **Endangered Species.**
   The U.S. Fish and Wildlife Service maintains information on endangered species. This site provides answers to FAQs, information on the Endangered Species Act, and species lists.
4. **Sierra Club–Habitat Report–How the ESA Works.**
   A plain-language explanation of the Endangered Species Report.
5. **1994 IUCN Red List of Threatened Animals.**
   This site lists animals from around the world whose numbers are going down. To use this list, it's best to know the scientific name of the animal you're researching, but you can also look up endangered animals by country.
6. **Duke University Primate Center.**
   This site provides information on primate research, tools for researchers, captive breeding, biodiversity, and endangered species.
7. **WWF Global Network.**
   The World Wildlife Federation home page. Information on conserving endangered species. For example, did you know that by eating Chilean Sea Bass, you are actually eating the soon-to-be-endangered Patagonian toothfish, much of which is caught illegally? At this site, you can sign up for periodic e-mail updates on endangerment.

CHAPTER 16

Conservation and Management

INTRODUCTION

Vertebrates, which have been a part of the Earth's fauna for more than 500 million years, have evolved into many different groups that have successfully adapted to virtually every habitat. They can swim, crawl, walk, run, climb, glide, and fly. Insects (phylum Arthropoda: class Insecta) are the only other group in the Animal kingdom whose evolution surpasses that of the vertebrates in terms of global distribution and adaptation to such a wide variety of habitats.

Humans are vertebrates and represent one of the most recent products in the evolution of placental mammals. Although the first modern humans—*Homo sapiens sapiens* (as "Cro-Magnon" man)—appeared only about 165,000 years ago in Africa, the impact that our species has made has been greater than that of any other species in the history of the Earth. Not only have humans had a direct impact on many other species (extirpation, extinction, propagation, dispersal), but indirectly actions by humans may ultimately threaten the continued existence of vertebrates and even life as we know it. Activities that threaten biodiversity, such as destruction of the rain forests, damaging of the ozone layer, global warming, production of acid rain, and pollution of waterways, are major global concerns. Humanity ultimately may determine whether *Homo sapiens* and all other organisms on Earth will continue to survive. The fate of 500 million years of vertebrate evolution seemingly rests in the hands of one species.

REGULATORY LEGISLATION AFFECTING VERTEBRATES IN THE UNITED STATES

Conservation of plants and animals is a global concern. Although the conservation movement did not start in the United States, this country has been the leader in the realm of conservation both within North America and throughout the world.

Efforts to regulate hunting in the United States can be traced back to the 17th century. Virginia, for example, was one of the earliest colonies to offer a bounty for wolf control. This law was adopted by the Grand Assembly at Jamestown on September 4, 1632 (Green, 1940). The first major federal legislation in this area, however, came with the passage of the Lacey Act of 1900. This law, among other things, prohibited the interstate transportation of "any wild animals or birds" killed in violation of state laws. It also authorized the Secretary of Agriculture to adopt all measures necessary for the "preservation, distribution, introductions, and restoration of game birds and other wild birds" subject to the laws of the various states. In 1916, a Convention of Migratory Birds produced a treaty adopting a uniform system of protection for certain species of birds that migrate between the United States and Canada. The Migratory Bird Treaty Act of 1918 implemented the treaty of 1916, providing for regulations to control the taking, selling, transporting, and importing of migratory birds for their feathers. The Act played an important role in protecting many species such as the snowy egret (*Egretta thula*) (Fig. 16.1).

The Migratory Bird Conservation Act of 1929 provided for the acquisition and development of land for migratory bird refuges and also authorized investigations and publications on North American birds. The Act, however, provided no funds for these purposes. Funding was not provided until 1934, when the Migratory Bird Hunting and Conservation Stamp Act (commonly known as the Duck Stamp Act) was passed. This legislation requires all waterfowl hunters 16 years of age or older to possess a valid federal hunting stamp. Receipts from the sales of this stamp are set aside in a special account known as the Migratory Bird Conservation Fund, from which funds are appropriated for the acquisition and management of migratory bird refuges and waterfowl production areas. These two acts have played an important role in protecting such birds as the trumpeter swan (*Olor buccinator*) and the whooping crane (*Grus americana*).

The Fish and Wildlife Coordination Act of 1934 was the first major federal statute to employ the strategy of requiring consideration of humanity's impact on wildlife. This forward-looking legislation authorized federal water resource agencies

FIGURE 16.1

Snowy egret (*Egretta thula*) in breeding plumage. The Migratory Bird Treaty Act of 1918 played an important role in protecting many species, including the snowy egret.

to acquire lands or interests in connection with water-use projects specifically for protection and enhancement of fish and wildlife.

In 1936, a Convention between the United States and Mexico for the Protection of Migratory Birds and Game Mammals was held. This meeting was similar to the Convention with Canada in 1916, and it was similarly implemented under the Migratory Bird Treaty Act of 1918. This Convention was amended in 1972, to add 32 additional families of birds, including eagles, hawks, owls, and members of the family Corvidae (jays, magpies, and crows).

The Federal Aid in Wildlife Restoration Act (commonly referred to as the Pittman–Robertson Act) was passed in 1937. This Act provides federal aid to states for wildlife restoration work with funds being raised by an excise tax on sporting arms and ammunition. Funds are apportioned to the states on a 75%–25% matching basis and can be used for approved land acquisition, wildlife research, and development and management projects. Amendments in 1970 and 1972 added excise taxes on pistols, revolvers, bows, arrows, parts, and accessories used in wildlife projects or hunter safety programs. The majority of all wildlife refuges have been purchased through funding provided by the Pittman–Robertson Act.

The Bald Eagle Act of 1940 was designed to provide for the protection of bald and golden eagles. The Convention on Nature Protection and Wildlife Preservation in the Western Hemisphere produced a 1940 treaty that stated that the government of the United States and 11 other American republics wished to "protect and preserve in their natural habitat representatives of all species and genera of their native flora and fauna, including migratory birds." This treaty covers the wintering grounds of many birds that nest in the United States.

The federal Aid in Fish Restoration Act (commonly referred to as the Dingell–Johnson Act) was passed in 1950. It provides federal aid to the states on a 75%–25% matching basis for approved land acquisition, research, and development and management projects involving fish (Fig. 16.2). Funds are obtained by means of an excise tax on certain items of sport fishing tackle. Most fish hatcheries, including those now used to raise endangered species, were built using funds from the Dingell–Johnson Act. Hunters and fisherman that buy Duck Stamps, hunting licenses, and hunting and fishing

FIGURE 16.2

The Aid in Fish Restoration Act (commonly known as the Dingell–Johnson Act) passed in 1950 and provides funds for management projects involving fishes—for example, fish ladders that provide access around a dam for fish migrating upstream. Fish ladders are often constructed with a 10-percent-graded flume interrupted with vertical, slotted partitions. The maximum 1-foot drop in water level at each partition produces a flow that fishes instinctively pursue. Setting the slots at an angle directs the flow exclusively into the pools behind the partitions, so that the dropping water never has more energy than the fish can resist. Changes in water level do not disrupt the ladder. Higher water increases the flow through the slots as well as the amount of energy-absorbing water in the pools.

equipment have been indirectly responsible for most of the species management and protection efforts in this country.

The Fish and Wildlife Act of 1956 established a comprehensive national fish and wildlife policy. It directs a program of continuing research, extension, and information services on fish and wildlife matters of national and international importance. This Act was responsible for the establishment of Cooperative Fisheries and Wildlife Units at many of the nation's universities. It designated a Fish and Wildlife Service (USFWS) made up of the Bureau of Sport Fisheries and Wildlife (BSFW) and the Bureau of Commercial Fisheries. This Act was amended in 1970, to transfer the Bureau of Commercial Fisheries to the National Oceanic and Atmospheric Administration. A 1974 amendment redesignated the BSFW as the USFWS under the Assistant Secretary of Interior for Fish and Wildlife and Parks.

In an effort to accelerate the acquisition of migratory waterfowl habitat, Congress passed the Wetlands Loan Act in 1961, which authorized $100 million to be added to the Migratory Bird Conservation Fund. Advances are to be repaid to the Treasury using Duck Stamp receipts.

The Wilderness Act of 1964 provided for the formal preservation of wilderness areas. Areas within the National Wildlife Refuge System and areas within the National Parks and National Forests were to be reviewed for wilderness designation and recommendations submitted to Congress.

Additional environmental protection under such laws as the National Wild and Scenic Rivers Act (1968), the National Environmental Policy Act (1969), the Marine Mammal Protection Act (1972), the Endangered Species Act (1973), the Fishery Conservation and Management Acts (1976, 1978, 1982), the Whale Conservation and Protection Study Act (1976), the Fish and Wildlife Improvement Act (1978), and the Fish and Wildlife Conservation Act (Nongame Act) (1980) has helped preserve habitats of endangered species as well as other wildlife. This protection, however, has been insufficient for many species.

ENDANGERED SPECIES IN THE UNITED STATES

In January 1964, the Bureau of Sport Fisheries and Wildlife circulated a tentative list of rare and endangered fish and wildlife among some 300 knowledgeable persons and organizations. Comments and suggestions were solicited. A revised list based on these suggestions was reviewed further and the additional comments incorporated into the first edition of the "Red Book," as the Federal List of Rare and Endangered Fish and Wildlife of the United States was popularly known. This was issued in July 1966, and revised in 1968. Species were classified as **endangered, rare, peripheral,** or **status undetermined**.

A second revision of the Red Book in March 1973, combined endangered and rare species into a single category termed **threatened**. This change was made primarily to indicate that the Red Book did not comprise the official list of endangered species. The official list is found in the U.S. Department of the Interior's list of endangered native fish and wildlife, published annually in the *Federal Register*.

In 1966, the Bureau of Sport Fisheries and Wildlife began a special research program for endangered species. This program was centered at the Patuxent Wildlife Research Center in Laurel, Maryland, and had two primary objectives: (1) to learn how to propagate certain species in captivity; and (2) to seek, through field studies, key factors that threatened the existence of certain species. With enactment of the Endangered Species Preservation Act of 1966, which authorized use of land and water conservation funds for the acquisition of endangered species habitat, refuge lands began to be purchased specifically for endangered species.

In 1969, Congress passed the Endangered Species Conservation Act (Public Law 91-135). This Act provided broad authority to the federal government to establish a comprehensive program for the conservation, restoration, and propagation of selected fish and wildlife in the United States that are threatened with extinction. The Act also provided assistance on an international level for the preservation of wild animals in other nations.

The Endangered Species Act of 1973 (Public Law 93-205; 87 Stat. 884) became effective on December 28, 1973, and thereby supplanted the Endangered Species Conservation Act of 1969. The new law sought "to provide a means whereby the ecosystems upon which endangered species and threatened species depend may be conserved, to provide a program for the conservation of such endangered species and threatened species, and to take such steps as may be appropriate to achieve the purposes of the treaties and conventions" in which the United States has pledged its support for the conservation of wild flora and fauna worldwide. This law encompasses all species of the animal and plant kingdoms, with the term "species" including any species, any subspecies, and any smaller taxonomic unit of plant or animal, and also any viable population segment thereof. Furthermore, the law established two categories of endangerment:

1. *Endangered Species*—those species in danger of extinction throughout all or a significant portion of their range
2. *Threatened Species*—those species that are likely to become endangered within the foreseeable future throughout all or a significant portion of their range

This law also emphasized the need to preserve critical habitats on which endangered species depend for their continued existence. Individual states were encouraged to establish guidelines to complement the goals outlined in the 1973 Act.

Also in 1973, the United States was one of 44 nations attending the Convention on International Trade in Endangered Species of Wild Fauna and Flora. The treaty and species lists negotiated at the Convention were implemented by the United States on February 22, 1977. The scientific authority for the United States is an autonomous committee of representatives of six federal agencies. Known as the Endangered Species Scientific Authority (ESSA), this committee's primary responsibility is to establish biological criteria on which to base findings for individual species protected by the Convention so that it may advise the management authority (Federal Wildlife Permit Office of the U.S. Fish and Wildlife Service) on the issuance of appropriate U.S. export and import permits.

In mid-1978, the U.S. Supreme Court handed down a decision upholding the applicability of the Endangered Species Act of 1973 in a case involving the Tennessee Valley Authority. Following this decision, Congress subsequently amended the Act (the Endangered Species Act Amendments of 1978) reauthorizing administration of the Endangered Species Act of 1973 and, among other things, providing for a review board and a cabinet-level committee to act as the final decision-making authorities in those cases where a seemingly irresolvable impasse has been reached regarding approval of a project that might destroy the habitat and last remaining members of an endangered species. Whether this amendment permitting exemptions from the Act's stringent requirements will seriously weaken the Endangered Species Act remains to be seen. The Endangered Species Act was reauthorized in 1988 and 1995.

Currently, 1,186 native species and/or subspecies of plants and animals are listed as endangered and/or threatened in the United States and/or Trust Territories. These include 69 mammals, 90 birds, 35 reptiles, 17 amphibians, and 110 fishes (U.S. Fish and Wildlife Service, 1999). Nearly 4,000 other dwindling species, most of them little known plants and invertebrates, are awaiting classification. Appendix II contains a complete listing of the vertebrates currently classified as endangered by the U.S. Fish and Wildlife Service. Each listed species has a Recovery Plan—a species management plan developed by a Recovery Team that is designed to restore habitats or do whatever else is necessary to enhance the survival of the species.

In 1998, the U.S. Department of the Interior announced that 29 species of fish, reptiles, birds, and mammals had recovered enough to be seriously considered for removal from the endangered list over a 2-year period. Some of the species will be downgraded to threatened and others removed from the law's protection altogether, although states may still regulate them. In 1999, the American peregrine falcon (*Falco peregrinus anatum*) and the American bald eagle (*Haliaeetus leucocephalus*) were proposed for removal. Since the American alligator was the first to be removed from the endangered species list in the late 1970s, only 6 other species have recovered enough to be taken off the list entirely. Another 14 species were removed after they either disappeared or new information was uncovered indicating they never should have been included in the list in the first place.

In addition, most states have held one or more endangered species symposia. Statewide symposia allow researchers to pool the most accurate and up-to-date information about the endangered and threatened species that reside within their borders.

■ SANCTUARIES AND REFUGES

The establishment of wildlife refuges, parks, and sanctuaries has been a major component in the survival of many species. Protected areas allow species to breed, rest during migration, or winter with minimum disturbance. Some refuges have been established specifically to provide critical habitat for endangered species.

The first U.S. federal wildlife refuge was established at Pelican Island in Florida in 1903. Its purpose was to protect a large heronry from plume hunters. Since 1903, the National Wildlife Refuge System has grown until it now consists of 508 units (Miller, 1999) (Fig. 16.3). Approximately 85 percent of the land area of these refuges is located in Alaska. Many refuges permit sport and commercial fishing, hunting, trapping, mining, oil and gas development, timber harvesting, farming, and livestock grazing. Regulations usually prohibit such activities on refuges during periods of the year when protected species are present. Approximately 25 percent of the land within national wildlife refuges is protected as wilderness, a designation that prohibits all of the above activities.

Yellowstone National Park, established in 1872, was the first public national park in the world. Currently, the National Park System consists of 55 major parks and 323 national recreation areas, monuments, battlefields, historic sites, parkways, trails, rivers, seashores, and lakeshores (Miller, 1999). Although open to the public for certain types of recreational activities, these areas, to some extent, preserve and protect wildlife and its habitat. Sport fishing is allowed in all units, but hunting is prohibited except in national recreation areas. Approximately 50 percent of the land is protected as wilderness.

The U.S. Forest Service oversees 156 national forests and 20 national grasslands (Miller, 1999). These areas are managed under the concepts of sustained yield and multiple use. Uses include timber harvesting, fishing, hunting, grazing, mining, and oil and gas development. Only about 15 percent of national forest lands are protected as wilderness.

The National Wilderness Preservation System is the most protective and restrictive of all federal lands. Wilderness areas are located within national wildlife refuges, parks, and forests. They are managed by the agency that oversees the area in which they are located. At present, 630 wilderness areas have been designated (Miller, 1999). These areas are roadless and, generally, have little or no evidence of human intrusion or presence.

The National Wild and Scenic Rivers Act was passed by Congress in 1968. It prohibits development along portions of rivers that have unique and outstanding wildlife, geological, scenic, historical, or cultural values. At present, approximately 150 rivers and portions of rivers are protected,

FIGURE 16.3

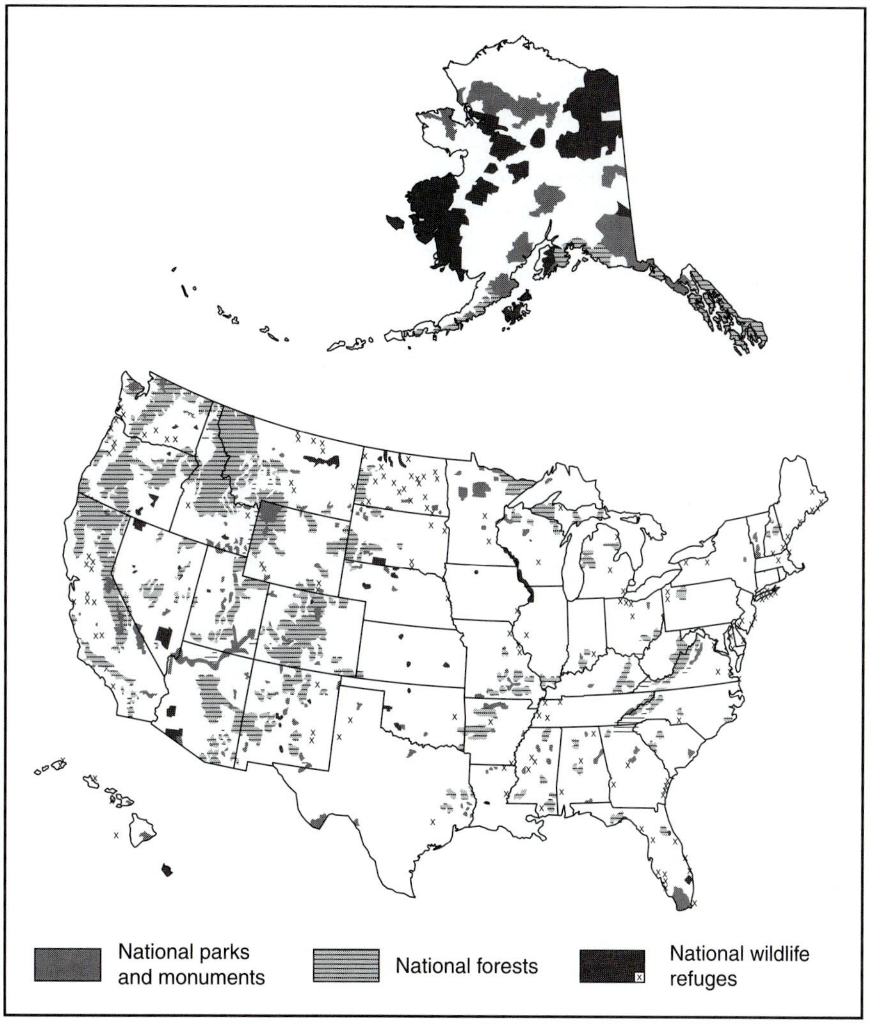

As of January 2000, the National Wildlife Refuge system consists of 508 units. The National Park Service consists of 55 major parks and 322 national recreation areas, monuments, battlefields, historic sites, parkways, trails, rivers, seashores, and lakeshores. The U.S. Forest Service oversees 156 national forests and 20 national grasslands (January 2000).

encompassing only 0.2 percent of the nation's 6 million km of rivers. In contrast, 17 percent of the length of wild rivers in the United States have been modified by dams (Miller, 1999).

Although not managed primarily for wildlife, many federal military installations provide habitat for numerous species. In addition to federal lands, there are numerous state wildlife management areas, parks, and forests that provide protection for many species. These are generally smaller than their federal counterparts and place more emphasis on managed human recreational activities such as hunting, fishing, timber management, and camping.

The Nature Conservancy, a private nonprofit environmental organization, has done more than any other private organization to preserve unique and fragile habitats. The Conservancy acquires land either through donation or purchase and sets it aside to protect endangered wildlife and plants and to provide future generations with opportunities to enjoy the out-of-doors. Since 1951, the Conservancy has acquired over 1 million hectares of land. Some tracts are maintained and managed by the Conservancy; others are donated to appropriate government agencies or universities.

Two private organizations, Ducks Unlimited and Trout Unlimited, have played extremely important roles in purchasing critical wetland habitats and protecting them from development. Both organizations also sponsor research on species management.

The National Audubon Society also maintains a network of refuges. Many colleges and universities maintain research stations and protected areas for faculty and student research. The American Museum of Natural History maintains two research stations and sanctuaries in Florida and Arizona. The Smithsonian Institution has a primary research station in Panama and others in various locations around the world.

On a global basis, many countries are seeing the value of setting aside areas as sanctuaries and refuges. A unique program that began in 1987 was designed to help less-developed countries preserve vital habitat. Known as the debt-for-nature swap, international nongovernmental organizations such as the Nature Conservancy, World Wildlife Fund, and Conservation International purchase part of the debt of a foreign country from the bank to whom the debt is owed. Many banks are willing to sell these debts for between 5 cents and 60 cents on each dollar owed. In exchange for not having to repay a portion of their debt, the country and the conservation organization enter into an agreement whereby the country agrees to protect a portion of its tropical forest and its inhabitants.

Bolivia was the first country to participate in a debt-for-nature swap when the government agreed to protect 1.5 million hectares of tropical forest surrounding its Beni Biosphere Reserve. Conservation International purchased $650,000 of debt at an 85 percent discount. Implementation of the plan was delayed for several years, however, due to internal opposition and Bolivia's tight financial resources. By 1992, ten other countries had participated in debt-for-nature swaps, with Costa Rica, Ecuador, Madagascar, and the Philippines having accounted for 95 percent of the funds expended (Miller, 1994). Some of the countries have been much more successful than others. Costa Rica, for example, has made great strides in protecting its remaining tropical forests and restoring ecologically damaged areas. Guanacaste National Park in Costa Rica is designated an international biosphere reserve. Debt-for-nature swaps represent just one way of protecting biodiversity.

Although problems remain, such as human encroachment, overpopulation, and poaching, many countries are recognizing economic benefits by attracting wildlife tours and expeditions. This is **ecotourism**, defined by conservationists as travel that promotes conservation of natural resources. With the promotion and regulation of ecotourism, people gain from the creation of jobs, countries see increases in their economies, and wildlife benefits because of the preservation of habitat and the improvement of human awareness. Bolivia is among the countries with the most biological diversity in the world, with over 40 percent of the bird species of South America being found within its boundaries. In late 1994, the Bolivian government teamed with private investors to build a program of "ecological tourism" it hopes will bring in $1 billion a year. An important part of the plan involves Noel Kempff National Park, a huge wilderness park the size of West Virginia, carved out of a remote area on the border with Brazil. The park contains more than 500 species of birds, as many as all of North America. The project is financed by the Nature Conservancy of Arlington, Virginia, and the Parks in Peril program of the U.S. Agency for International Development. The government of Bolivia will provide the basic infrastructure needed for tourism, while the private sector will supply the basic services.

Tourism has become the salvation for mountain gorillas in Rwanda and Congo (former Zaire), and ecotourism has funded the Monteverde Cloud Forest Preserve in Costa Rica. (However, recent internal strife in Rwanda, including a civil war, has created a potentially disastrous problem for continued survival of the gorillas.) Some countries, however, have been lax about monitoring the impact of tourism, and in some cases, unregulated ecotourism has led to habitat destruction. Ecuador, for instance, has tripled the number of tourists permitted to visit the Galápagos Islands, and in Nepal, forests have been devastated to provide firewood and lodging for trekkers.

VALUE OF MUSEUM COLLECTIONS

Vertebrate collections range in size from those maintained by such institutions as the American Museum of Natural History and the Smithsonian Institution to collections of only a few specimens maintained by single individuals. The curatorial care provided ranges from excellent to nonexistent. Professional societies, such as the American Society of Mammalogists and the American Society for Systematics Collections have attempted to develop standards for curatorial care. Properly maintained collections are inspected at intervals and are certified by the appropriate society. In cases where a collection cannot be maintained, the institution is encouraged to donate the collection to an organization that can provide the necessary curatorial care to prevent the possibly irreplaceable loss of data.

Vertebrate collections are used in many ways. They serve as a permanent repository for voucher specimens—that is, type specimens, locality records, and so forth (Fig. 16.4). They provide a reference collection for researchers needing to examine a large series of specimens from a given area or areas. In the case of extinct species, such as passenger pigeons (*Ectopistes migratorius*), museums often retain the only evidence of their existence. Natural history collections are absolutely essential to biodiversity research. Creating a complete inventory of life on Earth is impossible without reference to museum specimens, and the associated taxonomy and systematics is needed to make rational decisions about conservation.

During the 1960s, populations of several species of fish-eating birds (bald eagle, *Haliaeetus leucocephalus*; brown pelican, *Pelecanus occidentalis*; osprey, *Pandion haliaetus*), and others became seriously threatened because of the birds' inability to fledge young. Although the birds were breeding and females were laying eggs, a high percentage of egg shells broke, resulting in the loss of young prior to hatching. By measuring the thickness of eggshells collected prior to 1940 in museum collections and comparing them with recent shells

FIGURE 16.4

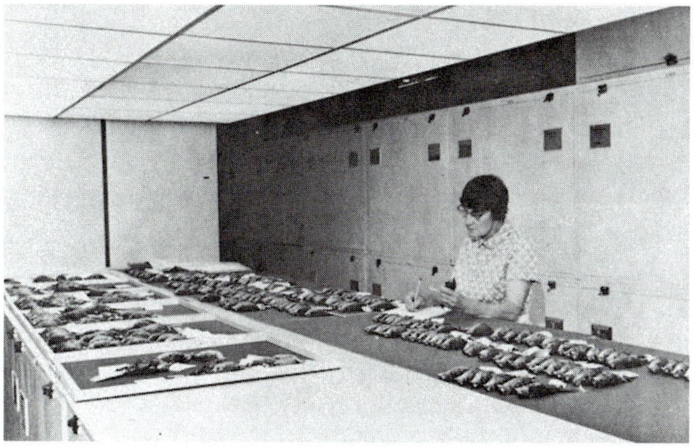

Museum collections serve as permanent repositories for voucher specimens and provide reference collections for researchers who need to examine a large series of specimens from a given area. The collection shown here is part of the mammal collection in the Museum of Natural History at the Smithsonian Institution in Washington, D.C. These specimens are available for a variety of studies.

from broken eggs in nests, researchers showed that egg shells had become much thinner and, thus, more susceptible to breakage. Further investigation linked the manufacture and use of the chlorinated hydrocarbon DDT (dichlorodiphenyltrichloroethane) to the egg shell thinning. In 1939, entomologist Paul Mueller had discovered that DDT was a potent insecticide, and it soon became the world's most-used pesticide. Birds at the top of the food chain were subject to the cumulative effects of all the chlorinated hydrocarbons previously ingested by their prey. This discovery was critical in the successful effort to ban DDT and either restrict or ban other chlorinated hydrocarbons such as aldrin, dieldrin, chlordane, mirex, and kepone in the United States. These broad-spectrum insecticides kill insects by causing convulsions, paralysis, and death, and they persist in the environment for up to 15 years. The discovery of the cause of egg-shell thinning undoubtedly saved a number of species from extirpation, if not extinction. Unfortunately, chlorinated hydrocarbons are still widely used in many other parts of the world, especially in less-developed countries.

By means of the polymerase chain reaction (PCR), molecular biologists are now able to extract DNA from many parts of museum specimens, including dried skin, bone, hair, feathers, and egg shells. DNA can be extracted, amplified, and sequenced from archaeological and museum specimens in order to study the relationships of extinct and extant forms. In addition, researchers now can compare modern and historical specimens of the *same* extant species in terms of individual and geographical variation in mitochondrial DNA (Thomas et al., 1990). Thus, historical trends of genotype frequencies can be directly measured.

Concerning the value of museum collections, Diamond (1990a) stated:

> *Old specimens constitute a vast, irreplaceable source of material for directly determining historical changes in gene frequencies, which are among the most important data in evolutionary biology. Until PCR became available, those data were lost forever as soon as the gene-bearing individual died. Now, however, museums with large, well-run collections of specimen series large enough for statistical analysis will be at the forefront of research in molecular evolution.*

WILDLIFE CONSERVATION IN A MODERN WORLD

Earth's capacity to support humans is determined both by natural constraints and by human choices concerning economics, environment, and culture. Human choice is not governed by ecological factors such as those that affect the carrying capacity of nonhuman populations; therefore, human carrying capacity is more dynamic and uncertain. The expanding human population requires increasing amounts of space, food, and resources—all to the detriment of other vertebrate species. Thus, emphasis must be placed on controlling human population growth. According to the United Nations, the world's population in 1999 just exceeded 6 billion, with 4.7 billion in developing (Third World) countries and 1.3 billion in industrialized nations (Fig. 16.5). The population of

FIGURE 16.5

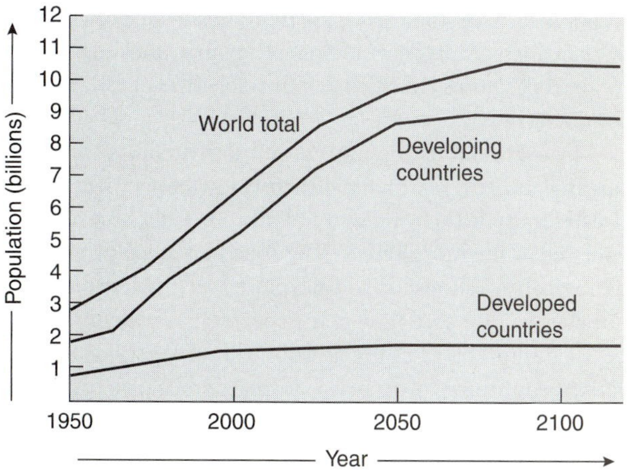

The United Nations broadly classifies the world's countries as "developing" or "developed." Developed countries are highly industrialized and usually have a high gross national product (GNP) per capita, whereas developing countries have low to moderate industrialization and GNPs per capita. Past and projected population sizes for developed countries, developing countries, and the world for the period 1950–2120 are shown.
Source: United Nations.

less-developed regions is currently growing at 1.9 percent per year, whereas that of more-developed regions is growing at 0.3 to 0.4 percent per year (United Nations, 1994). The present growth is nearly 90 million people a year, 7.5 million a month, 1.7 million a week, 245,000 a day, 10,000 an hour, or 170 a minute—with 164 of the 170 being born in poor countries. If human population growth can be brought under control, many of the problems facing other species of vertebrates would resolve themselves. Unilateral efforts by only a few countries, no matter how well intentioned, will not be sufficient to preserve Earth's vital habitats and species. The coordinated efforts of most, if not all, countries will be required to sustain life as we know it.

Worldwatch Institute's annual *State of the World* (Flavin, 1997) reported that 5 years after the Earth Summit in Rio de Janeiro in 1992, millions of acres of tropical and deciduous forest still disappear every year, carbon dioxide emissions are at record highs, and population growth is outpacing food production. The report noted that governments lag badly in meeting goals set at the Rio summit: "Unfortunately, few governments have even begun the policy changes that will be needed to put the world on an environmentally sustainable path," the report declared. On the positive side, the report found hope in increasing numbers of grassroots groups, particularly in Bangladesh and India, and in the fact that more than 1,500 cities in 51 countries have adopted local plans and rules, often more stringent than their national governments proposed at Rio de Janeiro.

The National Heritage Network of the Nature Conservancy has developed a consistent method for evaluating the health and condition of species and ecological communities. Using this method, it is possible to rank their conservation status. Conservation status ranks are based on a scale from one to five, ranging from critically imperiled (G1) to demonstrably secure (G5) (see Table 16.1 for more details). Species known to be extinct, or missing and possibly extinct, are also recorded. In general, species classified as vulnerable (G3 or rarer) may be considered to be at risk.

Emphasis must be placed on biodiversity and on preserving and/or restoring entire ecosystems (Fig. 16.6). **Biodiversity** is the total of all plants, animals, and microorganisms in the biosphere or in a specified area. The planet's biological storehouse is so unexplored that researchers cannot even say for sure how many species exist; the total could be 10 million or as many as 100 million. Yet global biodiversity continues to decrease, primarily because of the loss of habitat. Of the 2,536 vertebrates tracked by the National Heritage Network, 563 (22%) are deemed rare to critically imperiled (Stein and Flack, 1997a, b) (Fig. 16.7). Forty-two percent of turtles worldwide are threatened with extinction, and 6,000 of the approximately 9,000 living species of birds are declining in numbers (Hanks, 1996). In recent decades, populations of migratory songbirds in the mid-Atlantic states decreased by 50 percent. Frogs and other amphibians are declining throughout the world (Blaustein and Wake, 1995).

TABLE 16.1

Definition of Conservation Status Ranks

Acronym	Explanation
GX	Presumed extinct (not located despite intensive searches)
GH	Possibly extinct (of historical occurrence; still some hope of rediscovery)
G1	Critically imperiled (typically 5 or fewer occurrences or 1,000 or fewer individuals)
G2	Imperiled (typically 6 to 20 occurrences or 1,000 to 3,000 individuals)
G3	Vulnerable (rare; typically 21 to 100 occurrences or 3,000 to 10,000 individuals)
G4	Apparently secure (uncommon, but not rare; some cause for long-term concern; usually more than 100 occurrences and 10,000 individuals)
G5	Secure (common; widespread and abundant)

From Stein and Flack, (1997b).

Note: "G" refers to the global or rangewide status of a species. Both national (N) and state (S) status ranks are also assessed.

FIGURE 16.6

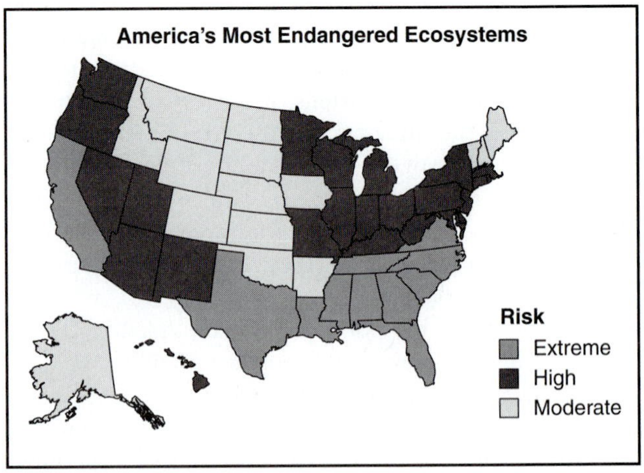

The Defenders of Wildlife, a national conservation organization, assessed the health of ecosystems state by state in 1995. They factored in the number of endangered ecosystems, numbers of endangered species, and the rates at which both are losing ground. Florida was the hands-down leader. Despite conservation efforts, advances are being overwhelmed by rapid population growth and development. The Southeast, from its longleaf pine forests to its coastal wetlands, is in particular jeopardy.

World Wildlife Fund scientists have identified approximately 200 key "ecoregions" that are all important for their rich diversity of species and other unique biological features (Fig. 16.8a, b). The Global 200 is a centerpiece in the World Wildlife Fund's campaign to stimulate public action aimed at protecting the Earth's most valuable harbors of biological diversity. A critical feature is the emphasis on conserving a full representation of the world's diverse ecosystems, including not only tropical rain forests and coral reefs, but also tundra, tropical lakes, and temperate broadleaf forests. **Ecoregions** define geographically distinct assemblages of natural communities that share a large majority of their species, ecological processes, and environmental conditions. Ecoregions function efficiently as conservation units at regional scales, because they encompass similar biological communities and roughly coincide with the regions over which key ecological processes most strongly interact. They can be defined for marine, freshwater, and terrestrial ecosystems. The Global 200 will hopefully serve as a powerful tool for conservation planning and priority setting.

Studies of the geographic distribution of endangered species in the United States were undertaken by Dobson et al. (1997) to determine whether significant correlations exist in the geographic distributions of different groups of endangered species and whether such correlations, if present, could be used as indicators for identifying potential protected areas for other poorly known taxa. These studies revealed "hot spots" of threatened diversity, where the ranges of many endemic species overlap with intensive urbanization and agriculture, but no consistent correlations existed in the distributions of endangered species from different taxa. However, the presence of endangered birds, amphibians, and reptiles provided a more sensitive indication of overall endangered biodiversity within any region. Conserving plant species maximizes the incidental protection of all other species groups. The greatest number of endangered species occur in Hawaii, southern California, the southeastern coastal states, and southern Appalachia (Fig. 16.9). Two counties were hot spots for three groups: San Diego, California (fish, mammals, and plants) and Santa Cruz, California (arthropods, amphibians, reptiles, and plants). Nine counties were found to be hot spots for two groups: Hawaii, Honolulu, Kauai, and Maui, Hawaii; Los Angeles, California; San Francisco, California; Highlands, Florida; Monroe, Florida; and Whitfield, Georgia. Except for these locations, the key areas for most groups overlap only weakly, which suggests that the endangered species hot spots of one group do not necessarily correspond with those for other groups.

Czech and Krausman (1997) compiled a database of the 877 American threatened and endangered species listed by the U.S. Fish and Wildlife Service up to 1995 and the causes of their endangerment that have been operational since passage of the Endangered Species Act. Eighteen causes of endangerment were identified and ranked (Table 16.2, p. 432). Most of these result from habitat destruction caused by human activities.

For a species to survive, its habitat must provide all of its needs. When a habitat is divided or decreased in size, a

FIGURE 16.7

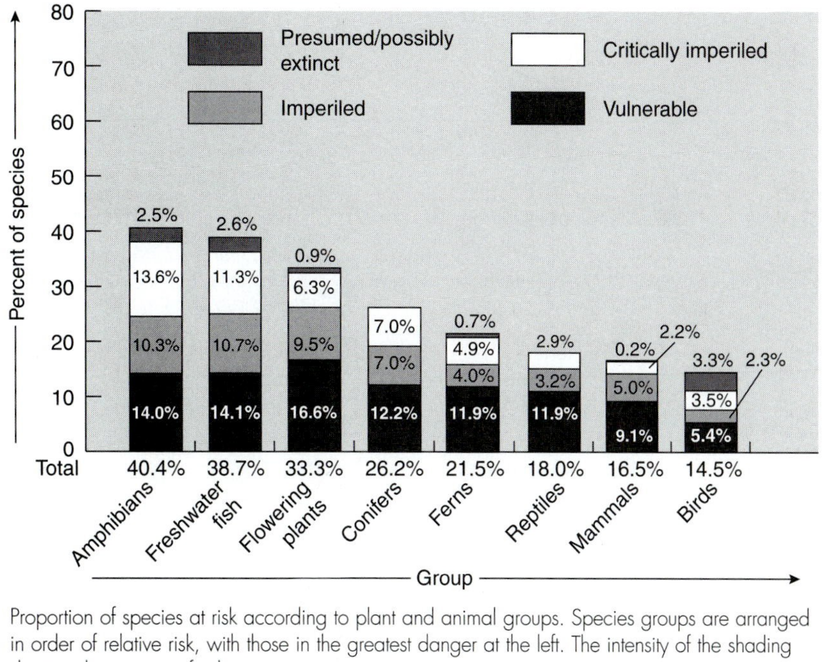

Proportion of species at risk according to plant and animal groups. Species groups are arranged in order of relative risk, with those in the greatest danger at the left. The intensity of the shading denotes the severity of risk.

Source: Data from B. A. Stein and S. R. Flack. 1997 Species Report Card: The State of U.S. Plants and Animals, *The Nature Conservancy, 1997, Arlington, VA.*

FIGURE 16.8

(a) Terrestrial ecoregions (approximate original extent)

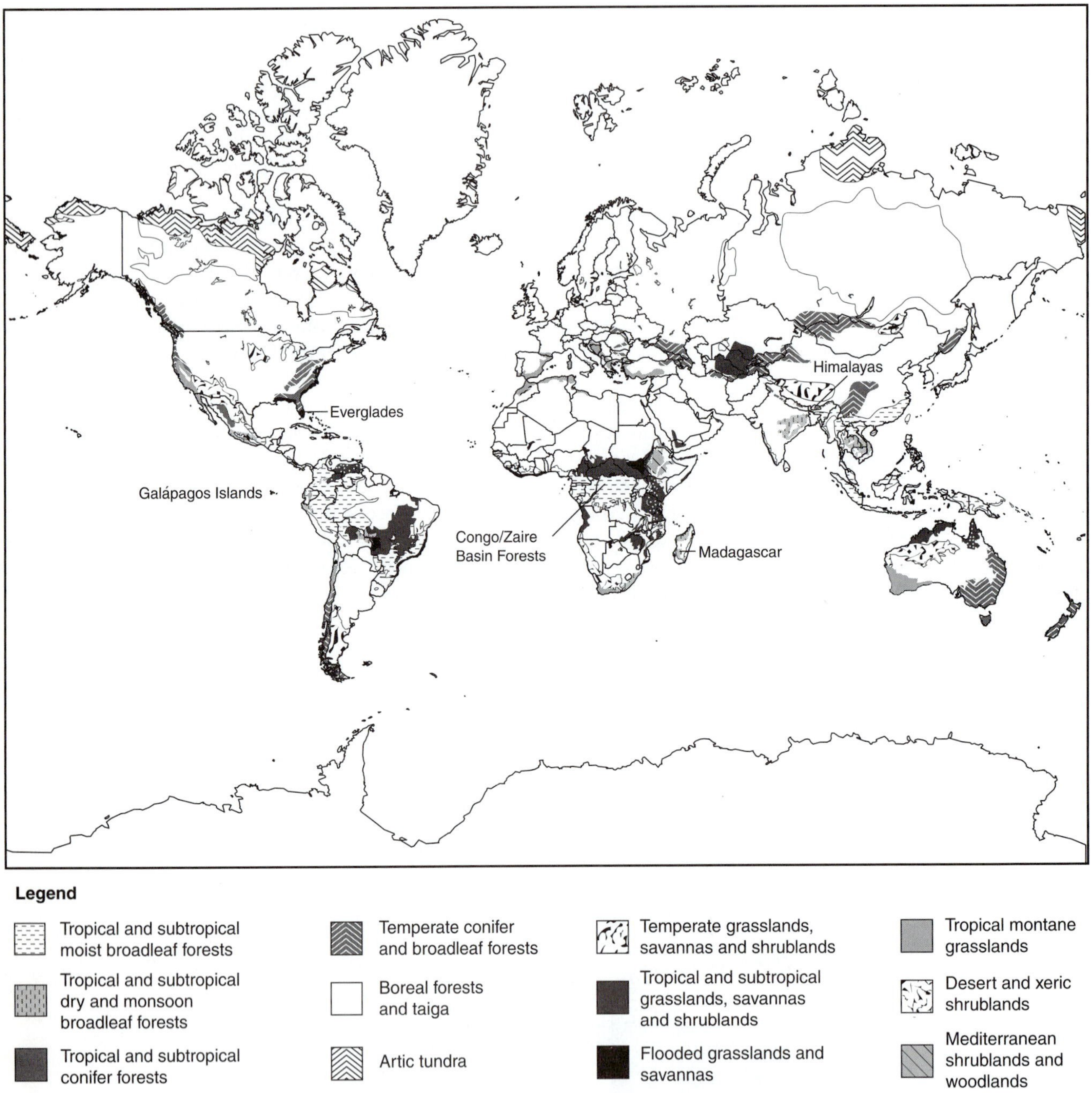

(a) Terrestrial ecoregions identified by World Wildlife Fund scientists. Ecoregions were selected on the basis of their rich diversity of species and other unique biological features.
Source: Focus World Wildlife Fund, *March–April 1997, Vol. 19(2): 4–5.*

process known as **fragmentation**, it may no longer be large enough to meet the needs of all the species that formerly occupied it. Even moderate habitat destruction can cause time-delayed, but predictable, extirpation of the dominant species in the remaining patches (Tilman et al., 1994). As habitat continues to be destroyed, additional species may be extirpated. Because some extirpations may occur generations after fragmentation, they represent a debt—a future ecological cost of current habitat destruction.

Hanks (1996) noted that biodiversity is related to the size of an area. For example, an area of 10 square hectares generally contains twice as many species as an area of 1 square

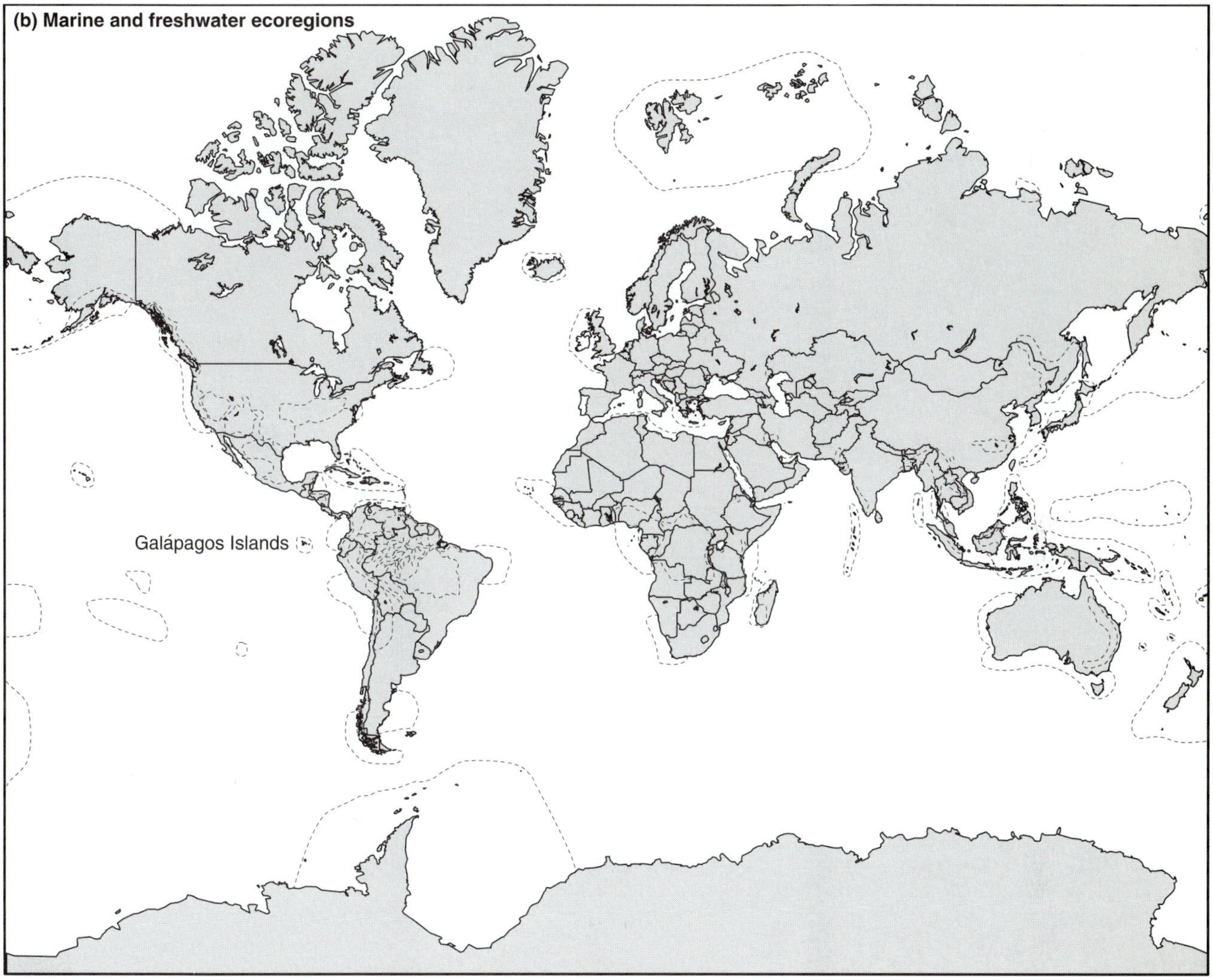

(*b*) Marine and freshwater ecoregions identified by World Wildlife Fund scientists.

hectare. If an area is reduced to one-tenth of its original size, half of the species in that area will no longer be able to live there. Loss of biodiversity has insidious consequences. Ecosystems with fewer species tend to be less stable in the long term, and in the event of climatic changes, diversity will help determine which ecosystems collapse and which ones flourish. In a long-term study of grasslands, Tilman and Downing (1994) showed that primary productivity in more diverse plant communities was more resistant to, and recovered more fully from, a major drought. Thus, preservation of biodiversity is essential for the maintenance of stable productivity in ecosystems (Naeem et al., 1994). Fossil evidence suggests that ponderosa pine, now prevalent throughout America's Rocky Mountain forests, was a marginal species at the end of the last Ice Age (Wilson, 1992). Thus, some of today's rare and apparently "insignificant" species may be the ones best able to cope with the climate of the next century.

Tropical forests, which grow near the equator in Latin America, Africa, and Asia, cover about 6 percent of the land surface of the earth. Three countries—Brazil, Zaire, and Indonesia—contain more than half of the world's total. Some forests receive almost daily rainfall and are known as **tropical rain forests**; others have one or two dry seasons each year and are known as **tropical seasonal deciduous forests**. Tropical rain forests have the highest species diversity of any biome. Each hectare may contain from 100 to 300 different tree species, as opposed to generally fewer than 10 species in a hectare of temperate forest.

At least half of the world's species—and perhaps up to 90 percent—inhabit tropical rain forests, which are being destroyed at the rate of approximately 15 to 20 million hectares a year (Wilson, 1992; Chiras, 1994). Other time frames equate the destruction to 37 city blocks per minute or to the area of two football fields each second (Miller, 1994).

FIGURE 16.9

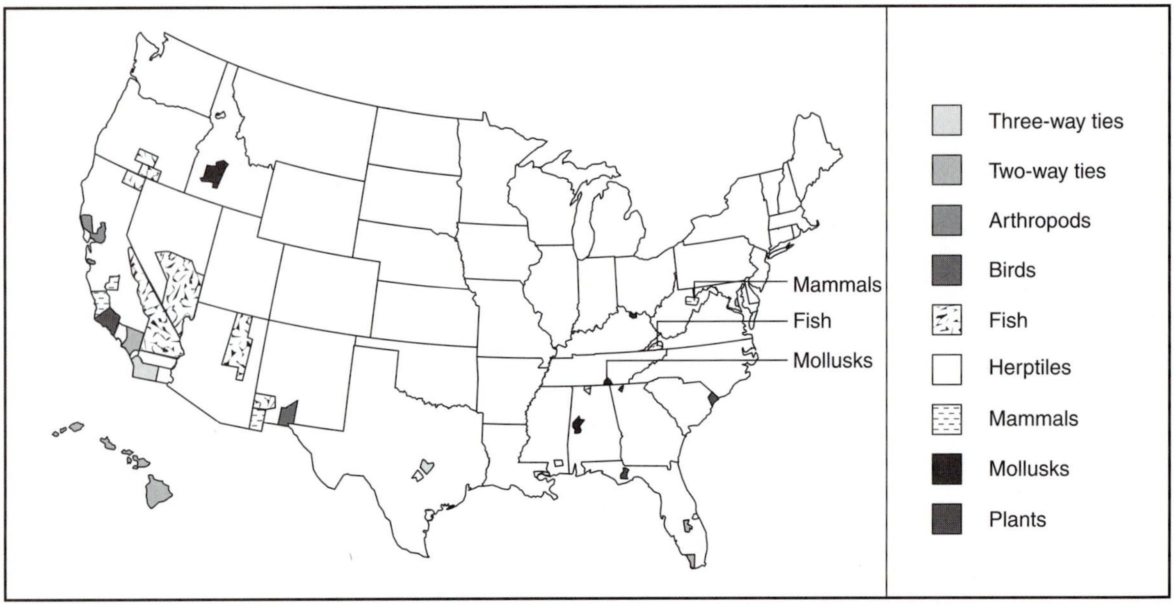

"Hot spots" of threatened diversity, as determined by Dobson et al. (1997). These are areas where the ranges of many endemic species overlap with intensive urbanization and agriculture. The complementary set of counties shown contains 50 percent of the listed species for each taxonomic group. The analysis identified two counties that contain large numbers of endangered species from three groups and nine counties that contain large numbers of species from two groups. (Hawaii not to scale.)
From A. P. Dobson, et al., "Geographic Distribution of Endangered Species in the U.S." in Science *275:550–553, January 24, 1997. Copyright © 1997 American Association for the Advancement of Science.*

TABLE 16.2

Causes of Endangerment for Species Classified as Threatened or Endangered by the U.S. Fish and Wildlife Service

	NUMBER OF SPECIES ENDANGERED (WITH FREQUENCY RANK)	
Cause	Including Hawaiian and Puerto Rican Species	Not Including Hawaiian and Puerto Rican Species
Interactions with nonnative species	305 (1)	115 (8)
Urbanization	275 (2)	247 (1)
Agriculture	224 (3)	205 (2)
Outdoor recreation and tourism development	186 (4)	148 (4)
Domestic livestock and ranching activities	182 (5)	136 (6)
Reservoirs and other running water diversions	161 (6)	160 (3)
Modified fire regimes and silviculture	144 (7)	83 (10)
Pollution of water, air, or soil	144 (8)	143 (5)
Mineral, gas, oil, and geothermal extraction or exploration	140 (9)	134 (7)
Industrial, institutional, and military activities	131 (10)	81 (12)
Harvest, intentional and incidental	120 (11)	101 (9)
Logging	109 (12)	79 (13)
Road presence, construction/maintenance	94 (13)	83 (11)
Loss of genetic variability, inbreeding depression, or hybridization	92 (14)	33 (16)
Aquifer depletion, wetland draining or filling	77 (15)	73 (15)
Native species interactions, plant succession	77 (16)	74 (14)
Disease	19 (17)	7 (18)
Vandalism (destruction without harvest)	12 (18)	11 (17)

From Czech and Krausman (1997).

BIO-NOTE 16.1

Biosphere Reserves

The United Nations Educational, Scientific and Cultural Organization (UNESCO) organized the first intergovernmental conference on "the rational use and conservation of the resources of the biosphere" in 1968. Out of this pioneering gathering emerged the Man and the Biosphere Programme (MAB), an ongoing international research and training effort. The MAB, which is supervised by an international coordinating council, was designed to address three major objectives. The first was to reinforce the extent and relevance of the conservation of biological diversity, including genetic resources, through a world system of protected areas. The second objective was to ensure the harmonious coexistence of rural populations and the ecosystems from which they derive their subsistence and income. The third was to provide basic and applied researchers with a number of permanent field sites that could be used as a network of information exchange.

These basic objectives corresponded with the three basic functions of what came to be known as a **biosphere reserve.** Today, 337 biosphere reserves cover a total of more than 200 million hectares in 85 countries (Fig. 16.10) (Batisse, 1997). Each reserve consists of three zones (Fig. 16.11). A **core area** is devoted to long-term protection. One or more **buffer zones** surround the core area. Only activities compatible with the conservation objectives (research, education, nondestructive recreation, tourism, resource use) may take place in the buffer zone. A **transition area** is a flexible outer region where sustainable resource management practices are promoted and developed, and where local communities cooperate in managing the biosphere reserve, possibly deriving some benefits from it. These three areas vary considerably depending on geographic conditions and local constraints.

Although the biosphere reserve concept has been successful, continual monitoring is necessary to ensure that they fulfill their functions properly, and that each country continues to afford them legal protection. In cooperation with local communities, biosphere reserves should be key components in maintaining biodiversity.

Landsat satellite imagery covering the entire forested portion of the Brazilian Amazon basin showed an increase in annual deforestation from 78,000 km^2 in 1978 to 230,000 km^2 in 1988 (Skole and Tucker, 1993). In a study released in 1982, the United Nations Food and Agriculture Organization (FAO) estimated that 11.3 million hectares of tropical forest had been lost each year during the 1970s (Aldhous, 1993). A 1993 assessment indicated that the destruction jumped by nearly 40 percent in the 1980s, reaching an average of 15.4 million hectares per year (Fig. 16.12). South America accounted for the largest absolute losses—6.2 million hectares a year—which translates to an annual deforestation rate of 0.6 percent, but continental southeast Asia is losing a staggering 1.6 percent and Central America 1.5 percent a year. A Brazilian–American study is currently under way to determine minimum effective reserve areas for wildlife species (Fig. 16.13).

As population demands have increased in areas containing tropical rain forests, trees have been harvested for lumber and the land used for agriculture. However, converting tropical rain forests to agricultural land is not sustainable because the soils are nutrient-poor, erode easily, and are compacted by heavy machinery. Approximately 95 percent of the nutrients in tropical forests are in the biomass and only 5 percent are in the soil, just the opposite of a temperate forest. Thus, these soils can produce food for only 3 or 4 years before becoming unusable. Clear-cutting of tropical rain forests exposes the soil to intense rains that wash it away, filling nearby streams and rivers with sediment.

The Brazilian government plans to repair and pave 2,170 miles of roads as part of an economic development program known as "Advance Brazil." These projects could destroy up to 72,000 sqaure miles of rain forest over the next 25 to 30 years.

In addition to providing a unique habitat and playing a significant role in maintaining biodiversity, tropical forests are a critical component in the global recycling of oxygen and carbon dioxide. An estimated 13 percent of the world's annual increase in carbon dioxide is brought about by global deforestation (Miller, 1997). Although efforts are being made to protect the remaining tropical forests, their destruction continues to proceed at a rate much faster than their protection.

Temperate forests also suffer from overharvesting and deforestation (Fig. 16.14). During the 19th century in the United States, over 80 percent of hardwood forests were destroyed. Many pine forests also were destroyed, along with most of the prairie ecosystems (Smith, 1996). Ninety percent of the old-growth forests in the northwestern United States already have been logged (Hanks, 1996). This massive deforestation is thought to have played a role in the extirpation and/or extinction of some wildlife species, including passenger pigeons (*Ectopistes migratorius*) and ivory-billed woodpeckers (*Campephilus principalis*).

Every spring, more than half of the 650 bird species that breed in the United States return to North America from their wintering grounds in Central and South America. These neotropical migrants include such familiar species as the ruby-throated hummingbird (*Archilochus colubris*), gray catbird (*Dumetella carolinensis*), purple martin (*Progne subis*), chimney swift (*Choetura pelagica*), red-eyed vireo (*Vireo olivaceus*), scarlet tanager (*Piranga olivacea*), and wood thrush (*Hylocichla mustelina*). Many are strictly forest-nesters, whereas others—such as the Tennessee warbler (*Vermivora peregrina*), eastern meadowlark (*Sturnella magna*), and dickcissel (*Spiza americana*)—prefer scrub or grasslands. They are some of our best insect controllers, eating tons of immatures and adults annually.

FIGURE 16.10

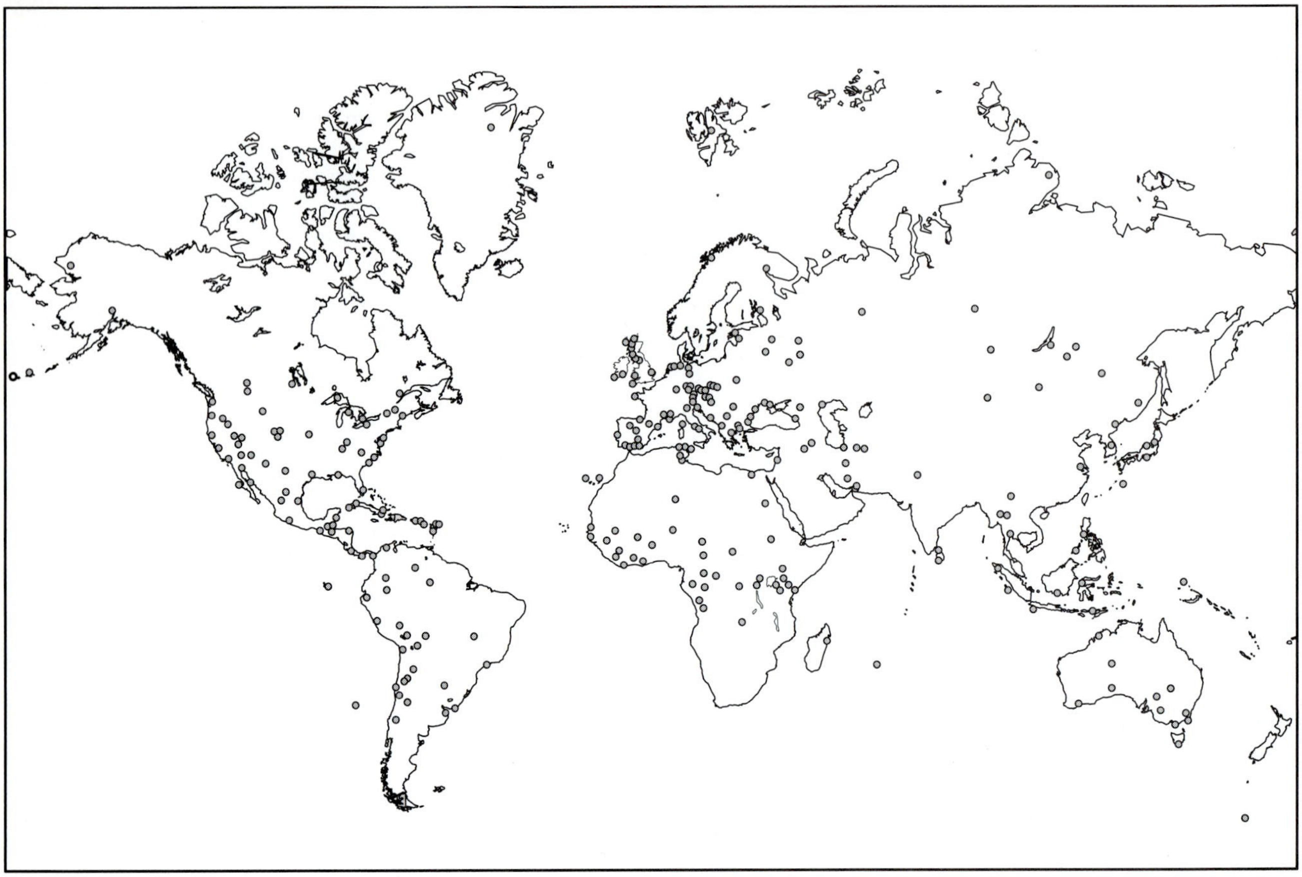

World network of biosphere reserves. The dots mark specific biosphere reserves.
Source: Batisse, "Biosphere Reserves" in Environment, *39(5):11.*

Over recent years, scientists have noticed an alarming drop in the numbers of neotropical migrants—particularly in the populations of forest-nesting migrants (Terborgh, 1992; Askins, 1995; Robinson et al., 1995). Although threats to populations are many and complex, habitat loss throughout North, Central, and South America has had the greatest impact on their decline (Temple, 1998). The excessive fragmentation of forests that has occurred throughout the United States increases the exposure of bird nests to predators such as raccoons (*Procyon lotor*), opossums (*Didelphis virginiana*), cats, blue jays (*Cyanocitta cristata*), and crows (*Corvus brachyrhynchus*), as well as to parasitic birds such as the brown-headed cowbird (*Molothrus ater*). Fewer chicks survive to adulthood to replace the adults killed by natural causes.

Overwintering habitats for migratory birds in Central and South America also are being altered and are disappearing, in some cases faster than their breeding habitats. Robbins et al. (1989) analyzed data from the North American Breeding Bird Survey (BBS), which was begun in 1965. Although populations of neotropical migrants throughout North America were stable or increasing before 1978, they have declined since. In contrast, residents or short-distance migrants have maintained stable populations. Species wintering in tropical forests have been affected chiefly, whereas those overwintering in scrub remain common. Some of the species most affected in the northeastern United States include the American redstart (*Setophaga ruticilla*), black and white warbler (*Mniotilta varia*), black-throated blue warbler (*Dendroica caerulescens*), Canada warbler (*Wilsonia canadensis*), magnolia warbler (*Dendroica magnolia*), Kentucky warbler (*Oporornis formosus*), crested flycatcher (*Myiarchus crinitus*), northern waterthrush (*Seiurus noveboracensis*), ovenbird (*Seiurus aurocapillus*), Swainson's thrush (*Catharus ustulatus*), wood thrush (*Hylocichla mustelina*), veery (*Hylocichla fuscescens*), scarlet tanager (*Piranga olivacea*), summer tanager (*Piranga rubra*), and the yellow-billed cuckoo (*Coccyzus americanus*).

Many biologists who study migratory songbirds have concentrated their efforts on northern breeding grounds rather than southern wintering areas. In a study of American redstarts (*Setophaga ruticilla*), however, Marra et al. (1998) showed for the first time that the quality of a migratory songbird's tropical breeding grounds can affect its survival and breeding success when it arrives in the north. For many migratory species, males arrive at breeding habitats before females, and

FIGURE 16.11

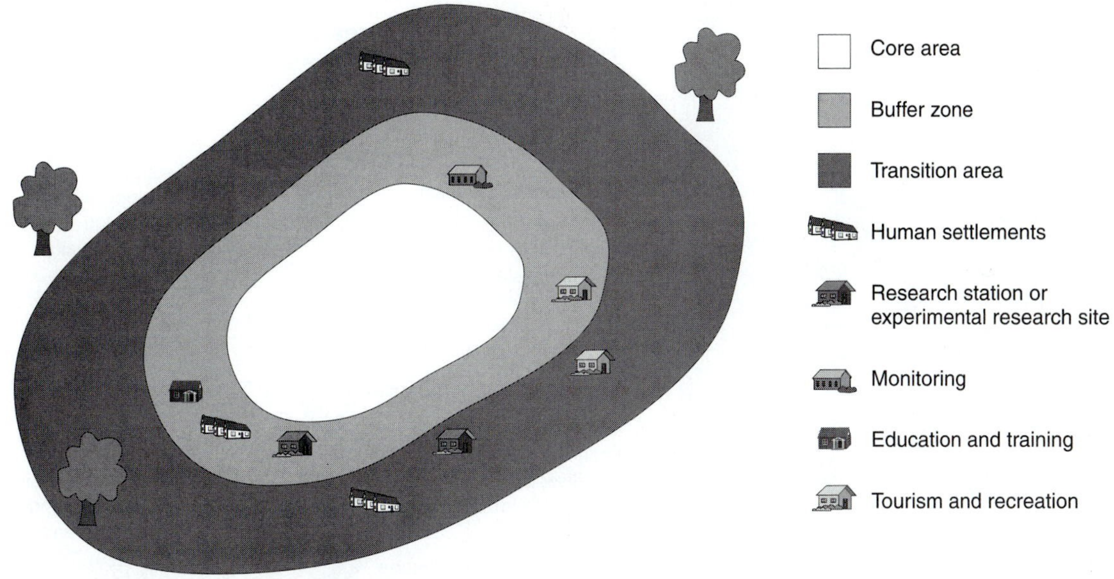

Pattern of zoning for a biosphere reserve.
Source: Data from United Nations Educational, Scientific and Cultural Organization.

FIGURE 16.12

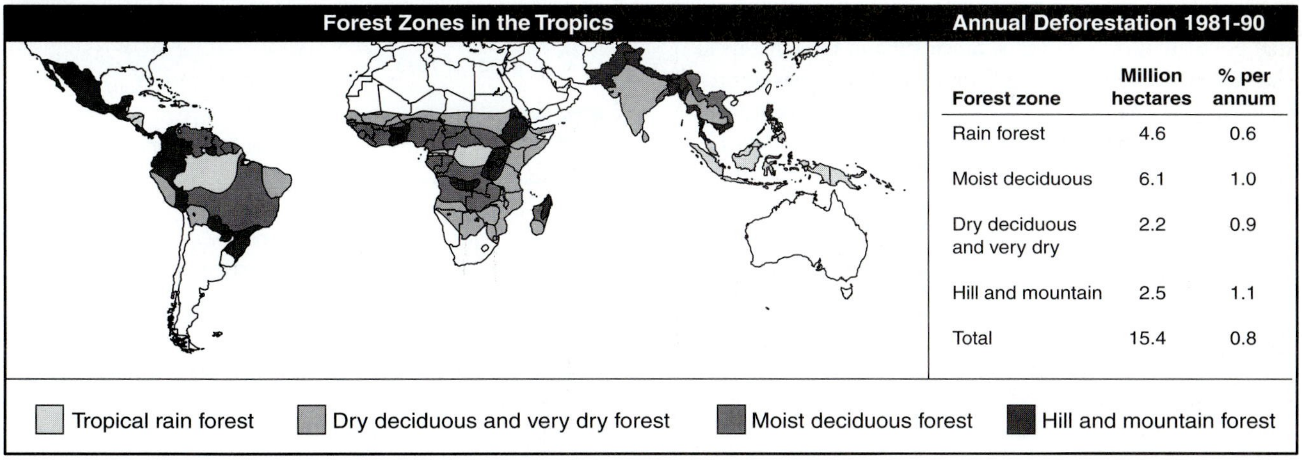

Annual Deforestation 1981-90

Forest zone	Million hectares	% per annum
Rain forest	4.6	0.6
Moist deciduous	6.1	1.0
Dry deciduous and very dry	2.2	0.9
Hill and mountain	2.5	1.1
Total	15.4	0.8

Annual deforestation in tropical forest zones, 1981–1990. Deforestation reached an average of 15.4 million hectares per year by 1990.
From P. Aldhouse, "Tropical Deforestation: Not Just a Problem in Amazonia" in Science, *259: 1390, March 3, 1993. Copyright © 1993 American Asociation for the Advancement of Science.*

breeding success and physical condition decline with arrival date. Thus, early arrival and physical condition are important determinants of reproductive success and fitness. Through the use of a carbon isotope marker (birds incorporate habitat-specific carbon isotopes into their tissues from the plant-eating insects they consume) in the bird's blood that is keyed to winter habitat type, Marra and his team showed that wet forest birds—which were 65 percent male—maintained or gained weight during the winter months, whereas the scrub-dwellers—which were 70 percent female—had lost up to 11 percent of their body mass and had elevated levels of the stress hormone corticosterone. Wet forest birds—mostly males—were the earliest arrivals at the breeding grounds. Dry-scrub females and younger males arrived later. The wet forest habitat in this study consisted of mangroves, which shelter large numbers of migratory species, but which are declining worldwide. Thus, as prime habitat grows scarce, more and more birds will be forced into the scrub or other suboptimal habitat. Marra's study has provided a crucial piece of information for conservation.

FIGURE 16.13

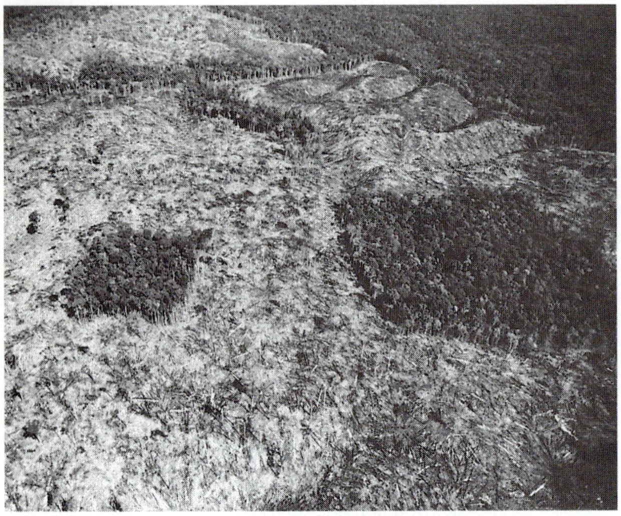

A joint Brazilian–American study sponsored by the World Wildlife Federation is attempting to determine the minimum effective reserve sizes for both plants and animals. These 1-ha and 10-ha plots in Amazonia are surrounded by clear-cut areas. Some plots are linked to virgin forest by corridors.

FIGURE 16.14

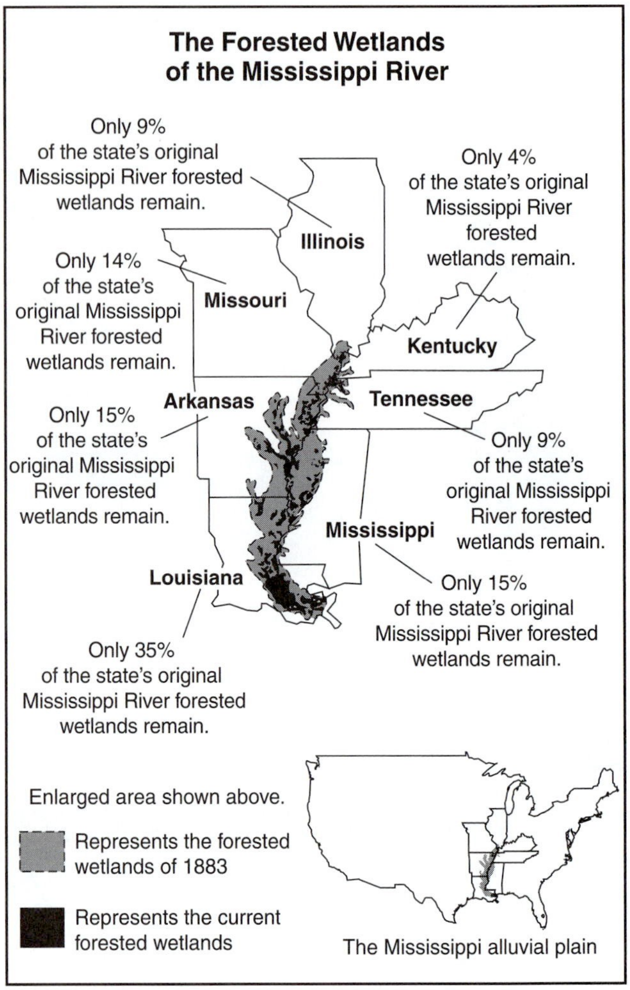

Temperate forests, such as those along the Mississippi River, also suffer from overharvesting and deforestation. During the 19th century in the United States, over 80 percent of hardwood forests were destroyed. *Source: Data from The Nature Conservancy.*

Though some of the annual decreases in bird populations may seem small, a steady trend of population decrease over several years can be devastating. For instance, a bird species declining 2 percent per year translates into a 50 percent decline in that population over 35 years. Since 1965 in the Chesapeake Bay watershed, the Breeding Bird Survey (BBS) has recorded yellow-billed cuckoos, gray catbirds, scarlet tanagers, least flycatchers (*Empidonax minimus*), and barn swallows (*Hirundo rustica*) declining at less than 1 percent per year. Wood thrushes, whip-poor-wills (*Caprimulgus vociferus*), wood pewees (*Contopus virens*), chipping sparrows (*Spizella passerina*), and Baltimore orioles (*Icterus galbula*) have been declining at a rate of 1 to 3 percent annually. Warblers, however, have been affected most drastically. Populations of prothonotary (*Protonotaria citrea*) and black-throated blue warblers are declining by more than 3 percent; the Cerulean warbler (*Dendroica cerulea*) is declining by 4 percent; the black-and-white warbler by more than 5 percent; and the golden-winged warbler (*Vermivora chrysoptera*) by an average of 7 percent per year.

Sydney Gauthreaux, a Clemson University biologist, has studied bird migrations using radar for nearly 30 years. He and his students used the U.S. Navy's weather radar on Dauphin Island off the coast of Alabama to monitor neotropical migrants flying over the Gulf of Mexico (Watson, 1992). By comparing radar pictures from a 3-year period in the 1960s with images taken over a 3-year span in the 1980s, Watson concluded that migrants from the tropics appeared to have declined by 40 to 50 percent.

The European Union's Habitats Directive, enacted in 1992, is a bold conservation law containing a master list of threatened European habitat types that member nations are supposed to identify and protect. The law has foundered, however, because a lack of scientific data about local habitats and species distributions has prevented Spain and some southern European countries from determining which habitats to protect.

In terms of water quality, forests are, hectare for hectare where they naturally occur, the most beneficial land use because they help regulate stream flow, control runoff, filter nutrients, and create the stream environment needed by many fish and other aquatic species. In addition, forests improve air quality. One hectare of trees can remove over 95 tons of carbon dioxide annually, while producing enough oxygen to sustain more than 1,000 people during the year (Anonymous, 1996e). Woodlands also provide much of the watershed's habitat for animal and plant life.

In addition to population control, preservation of biodiversity, and the conservation of tropical and temperate forests, air and water pollution laws must be significantly strengthened and enforced worldwide. Increasing quantities of diverse litter are being discarded in the world's freshwater and marine environments. Nonbiodegradable plastics, in particular, have caused mortality in sea turtles, waterfowl, and various aquatic mammals when they become entangled in plastic products or ingest them (Wallace, 1985; Walker and Coe, 1990) (see Fig. 14.12).

Global warming is thought to be caused primarily by an increase in the amounts of carbon dioxide, methane, and some industrial gases such as chlorofluorocarbons (CFCs) spewed into the atmosphere through human activities, although scientists say it will be a decade before computer models can confidently link the warming to human activities (Kerr, 1997b). Carbon dioxide is increasing in the atmosphere as a result of the burning of fossil fuels, the destruction of forests, and some agricultural activities. These **greenhouse gases** warm the planet by trapping heat in the atmosphere, thus causing the so-called greenhouse effect. Most experts agree that the accumulation of greenhouse gases in the atmosphere is warming the Earth's climate at an unprecedented rate (Ramanathan, 1988; Schneider, 1989). Just exactly how fast is still controversial. Some climate models are predicting that within 100 years the Earth will not only be warmer than it has been during the past million years, but the change will have occurred more rapidly than any on record.

In addition, since 1978 an upward trend in total solar irradiance of 0.036 percent per decade has been verified (Willson, 1997). If sustained, this upward trend could also result in rising global temperatures.

Analysis of global weather data by National Oceanographic and Atmospheric Administration's (NOAA) National Climatic Data Center revealed that 1990 and 1995 were the hottest years since at least 1400 A.D. (Kerr, 1995a, b; Hansen et al., 1996; Mann et al., 1998) (Fig. 16.15). However, 1997 came in 0.08°C warmer (Kerr, 1998; Monastersky, 1998a; Mann et al., 1998) and, according to NOAA, 1998 was the hottest year both globally and nationally. 1999 was the fifth hottest year ever. Globally, the 1990s are the hottest decade ever recorded with the five hottest years being 1998, 1997, 1995, 1990, and 1999, in that order (Mann and Bradley, 1999). Eleven of the top twelve warmest years on record have occurred since 1987. Each year of the 1990s ranks among the 15 hottest years since 1880, when records began to be kept. As the world warms, ocean levels and temperatures increase. The predicted changes could drastically affect ecosystems and the distribution and existence of many vertebrate species (Poiani and Johnson, 1991). For example, a comparison of breeding distributions of British birds from 1968 to 1991 revealed that the northern margins of many species moved farther north by an average of 18.9 km (Thomas and Lennon, 1999). Hotter, drier tropics may cause some rain forests to become grasslands, whereas cold tundra areas may warm enough to grow forests. Most deserts are likely to become hotter and drier (although the Sahara is expected to get more rain), and agricultural areas will shift. One-third to one-half of the mountain glaciers and polar ice caps will disappear, in some cases changing river flows. Rising sea levels due to melting of sea ice and polar ice caps and expansion of water will flood some coastal areas where most of the world's population resides. Higher temperatures may cause not only growth in the number of violent storms, such as hurricanes and typhoons, but also an increase in their strengths.

Researchers are suggesting that the 4 to 5°C mid-winter warming of the western Antarctic Peninsula climate observed over the last five decades may be a factor in the significant decrease in the Adelie penguin population (Kaiser, 1997). Over the past 11 years, the active growing season of plants has advanced by approximately 8 days in northern latitudes (Crick et al., 1997). In addition, Crick et al. (1997) presented data from 65 species of birds in the United Kingdom for a 25-year period from 1971 to 1995 showing a significant trend toward earlier laying dates for 20 species (31%), with only 1 species laying significantly later. The shift toward earlier laying for the 20 species averaged 8.8 days.

Polar bears along the Hudson River are threatened with starvation because the pack ice season, the time when they feed on seals, is dwindling. The pack ice season has been reduced by three weeks over the past two decades, according to a study by the Canadian Wildlife Service. Polar bears hibernate or fast for up to eight months a year and depend on hunting on the pack ice to sustain themselves and bring up their weight. The reduced ice season has resulted in polar bears returning to the mainland in worse shape with females giving birth less often. Starving bears will also become progressively more likely to enter areas of human habitation in search of food, threatening human life and property. Canada's indigenous people, the Inuit, could find their hunting prey out of reach, their water supplies contaminated, and their coastal communities subjected to erosion from seas no longer covered by ice. Eventually, forests could spread much farther north, taking range land from some of the world's biggest caribou herds.

Based on an analysis of temperature measurements taken daily since 1916 along the Pacific coast, researchers have

FIGURE 16.15

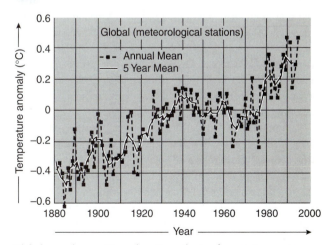

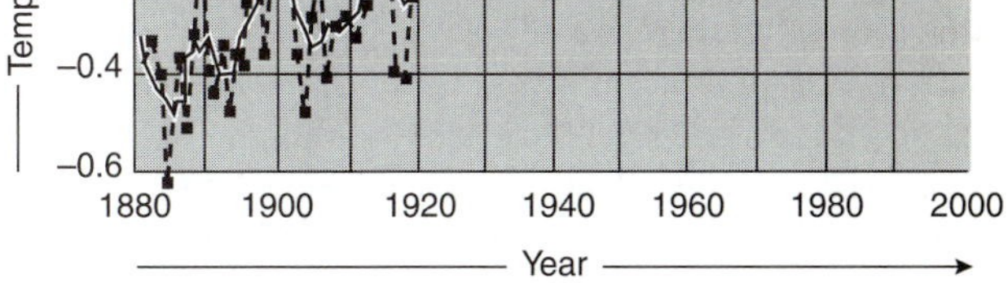

Global annual mean (December–November) surface air temperature change based on NOAA's meteorological station network, 1880 to 1995. These data from NOAA's National Climatic Data Center show that 1995 was the hottest year in recorded history. It has since been surpassed by 1997 and 1998.

recorded a trend of warmer water that may signal a climate change deeper than just a temporary El Niño effect (McGowan et al., 1998). Within just 10 months in 1977, sea surface temperatures suddenly jumped upward, and the average has remained about 2°C warmer than previously. The whole temperature range shifted upward, so that the lows are not as low as they used to be and the highs are higher. Rising sea surface temperatures, which are affecting the entire eastern half of the northern Pacific and the Gulf of Alaska, are causing severe declines in fish, birds, seaweed, and mammals. Biological changes linked to the warmer water since 1977 include a 70 percent decline in zooplankton; a 90 percent decline in sooty shearwater (*Puffinus griseus*) populations; warm-water fishes and other animals have migrated northward and are now common in places they once shunned; nearshore species such as abalone, sea urchins, and kelp plants have been severely affected; warm surface waters are blocking the upwelling of nutrient-rich cold waters, resulting in the warmer surface waters lacking some of the chemicals that support plankton, which is at the base of the food chain; and fish populations have declined about 5 percent per year since 1986. Although the rise in sea surface temperature is consistent with global warming, additional data are necessary to confirm the cause.

The Kyoto Protocol, announced in December 1997, is the latest attempt by governments to set legal, binding limits on greenhouse gas emissions. However, the binding commitments for nations to reduce carbon dioxide and other greenhouse gases—6 to 8 percent below 1990 levels—applies only to industrial countries, including the United States, Japan, and the European Union. Although it is understood that at some point developing nations will also reduce emissions, they argued that they should not be required to take any action now because most of the gases in the atmosphere were put there by Europe and the United States. Much opposition from groups representing agriculture, manufacturing, transportation, and the coal mining industries has developed against the treaty. The U.S. Senate approved a resolution to not accept a global warming treaty that does not also require developing countries (and economic competitors) such as China and India to make a binding commitment to curtail greenhouse gases. Although the United States attempted to add limits on developing countries to the treaty when it came up for further review in November 1998, at the worldwide environmental conference in Brazil, the treaty has still not been approved by the U.S. Senate.

Ozone (O_3) is a gas that filters out harmful ultraviolet radiation and keeps it from reaching the Earth's surface. However, increasing amounts of human-made chlorofluorocarbons and other chemicals in the atmosphere have caused the destruction of ozone molecules, resulting in ozone holes over certain regions of the planet. When a CFC molecule absorbs ultraviolet light, it gives up a chlorine atom. The chlorine can react with ozone to form an oxygen molecule and a chlorine monoxide molecule. When the chlorine monoxide reacts with a free oxygen atom, another chlorine atom is released that can attack another ozone molecule. Each chlorine atom released in the reaction can convert as many as 10,000 molecules of ozone to oxygen.

Ozone holes are areas of severe ozone loss, but losses are occurring throughout the stratosphere (Rex et al., 1997). Between 1985 and 1995, the ozone holes over the Arctic and Antarctic regions increased in size, and concentrations of ozone fell to levels 10 to 20 percent lower than previously recorded over much of central Asia, western Europe, and the western United States. Overall, 1992 ozone levels north of the equator were 2 to 3 percent lower than ever before recorded. During late 1994, ozone levels were as much as 30 percent lower than normal over Finland and Siberia. According to measurements by a NASA satellite, the ozone hole over Antarctica grew to a record 27.3 million square kilometers on September 19, 1998, larger than the North American continent (Monastersky, 1998). The biggest previous ozone hole reached 26 million km^2 in 1996. Although ground-level measurements on three continents and on two Pacific islands have indicated a reduction in the concentration of the group of industrial chemicals that erode the ozone layer (Montzka et al., 1996), it will require many years to reduce the effect of the chemical molecules in the stratosphere.

Reactive chlorine and bromine are expected to reach a maximum in the stratosphere between 1997 and 1999, and they will decline thereafter if all countries comply with their agreements in the revised Montreal Protocol, which sets limits on the manufacture and use of ozone-destroying chemicals. This reduction, which is the first step toward the goal of eliminating ozone loss, means that the holes could start closing within 10 years. In the meantime, increasing amounts of biologically harmful ultraviolet-B (UV-B) radiation now reaching the Earth's surface will disrupt plant growth and damage the health of animals by producing more severe sunburns, causing an increase in skin cancers and cataracts, as well as damaging the DNA within cells. Of special concern are those species of fishes and amphibians whose embryos develop in eggs directly exposed to sunlight.

Certain synthetic compounds, such as the estrogen diethylstilbestrol (DES), widely used before it was banned in 1971, can mimic the activity of naturally occurring hormones and act as endocrine and reproductive disruptors in various species (Smolen and Colborn, 1997). At least 51 chemicals disrupt hormones in one way or another (Raloff, 1995; Colborn et al., 1996). Some mimic estrogen, as DES does, and others interfere with different parts of the endocrine system, such as thyroid and testosterone metabolism. These substances in fishes, reptiles, and birds potentially could have drastic consequences including sterility, sperm abnormalities, a reduction in sperm count, and the alteration of male genitalia (Colborn et al., 1995; Sharpe, 1995; Dold, 1996). They also cause the development of oviducts in male carp during sexual differentiation (Gimeno et al., 1996).

Hormone-disrupting chemicals may have synergistic effects, and seemingly insignificant quantities of individual chemicals can have a major cumulative effect. At present, 100,000 synthetic chemicals are on the market, with approximately 1,000 new substances being introduced each year. Many of these are ingredients in pesticides that are applied worldwide not only to agricultural fields, but in parks, schools, homes, gardens, restaurants, and supermarkets. Besides pesticides, they include certain detergent breakdown products and various plastics additives and stabilizers. The hormonal activity of these chemicals usually bears little relationship to their intended function. There is no known way of predicting—based on structure or function—which compounds will mimic naturally produced hormones.

■ WILDLIFE MANAGEMENT IN A MODERN WORLD

The manipulation and/or restoration of habitats and populations have been discussed in many chapters. Most endangered species require some form of habitat restoration for their survival. Wildlife management is a separate professional field of study with its own professional organization (the Wildlife Society) and a rigid certification requirement.

Considerable effort has been devoted to restoring and/or maintaining wildlife populations. In February 1989, a domestic cat gave birth to an endangered Indian desert cat, *Felis sylvestris*, derived from an artifically fertilized embryo at the Cincinnati Zoo (Fig. 16.16). This was the first interspecies birth of a cat, and the first birth of an exotic cat from a test-tube-fertilized egg. Additional breeding experiments have been undertaken as well as research into freezing embryos and sperm. On November 24, 1999 the first successful transplant of a frozen embryo from one species to another was recorded when a domestic cat delivered a prefrozen Indian desert cat at the Audubon Center for Research of Endangered Species in New Orleans (Holden, 1999). As one Cincinnati Zoo researcher stated: "Once they're frozen, we can keep a species from going extinct" (Anonymous, 1989).

In February 1997, researchers in Scotland reported the development of a "viable offspring" named Dolly, the first to be derived from the fusion of an adult sheep mammary gland epithelial cell and a fertilized egg cell from which researchers had removed the nucleus—a technique the researchers call nuclear transfer (Wilmut et al., 1997) (Fig. 16.17). Although the word "clone" is never used in the journal article, these researchers had cloned a lamb from the cells of an adult sheep. Some researchers feel that cloning may be a useful technique for saving rare species (Cohen, 1997). Cells of rare species such as Przewalski's horse (*Equus caballus przewalskii*) and the Sumatran rhinoceros (*Dicerorhinus sumatrensis*) are banked at the San Diego Zoo's Center for Reproduction of Endangered Species (CRES) and at other sites and could be used to preserve and increase the genetic diversity of rare animals.

Other lambs cloned from fetal, rather than adult, cells have now been born (Pennisi, 1997d). In some of these lambs, the fetal-cell procedure has been combined with genetic engineering so that they carry extra genes, with a few even having a human gene. This achievement could aid efforts to develop livestock that produce human proteins, such as blood-clotting factors, for therapeutic use.

Researchers in the United States have successfully produced clones of cloned mice. In January 2000, Japanese scientists produced the clone of a cloned bull, the first time a large cloned animal has itself been cloned.

FIGURE 16.16

Desi, the black domestic cat, was a surrogate mom to Noah, a rare Indian desert cat. This was the first interspecies birth of a cat, and the first birth of an exotic cat from a test-tube-fertilized egg.

FIGURE 16.17

Dolly, a lamb cloned from the cells of an adult sheep. The technique, known as nuclear transfer, was reported in February 1997, from Scotland. Since then, other sheep, as well as cows and mice, have been cloned using similar techniques.

Captive breeding programs involving whooping cranes (*Grus americana*) (see Chapter 11), California condors (*Gymnogyps californianus*) (see Chapter 15), black-footed ferrets (*Mustela nigripes*) (see Chapter 15), gray wolves (*Canis lupus*) (see Chapter 15), red wolves (*Canis rufus*), and others have shown short-term success. But what about the long-term prospects for such species? What good does it do to produce condors in captivity if sufficient suitable habitat no longer exists into which they can be released? How much habitat currently exists for the red wolf? How much habitat remains for the desert cat? If the last remaining members of a species cannot be reestablished successfully in their native habitats, what then? Will an entire species be relegated to being maintained solely in captivity? We temporarily reached this level with three species—the California condor, the black-footed ferret, and the red wolf—although the ultimate goal was, and still is, the reestablishment of the species in its native range. Or will the species exist only in the form of frozen gametes? If either of these situations come to be, it will be a sad day in the history of humankind.

Homo sapiens has emerged as the species that ultimately can control the fate of the earth and all of its inhabitants. As Walt Kelly's cartoon character Pogo once said: "We have met the enemy and it are us." Only time will tell if we make the right decisions.

The frog does not drink up the pond in which he lives—
Indian Proverb

Review Questions

1. Do you feel that it is advantageous or disadvantageous for one species to control the destiny of all plants and animals on the Earth? How might this affect future efforts to establish colonies on other planets?
2. In your opinion, has the Endangered Species Act been beneficial or detrimental to (a) wildlife species and (b) landowners? Explain your answer.
3. Do you feel that the provision for a review board and a cabinet-level committee to act as the final decision-making authority for certain endangered species issues will strengthen or weaken the Endangered Species Act? Why?
4. Define ecotourism. How can it benefit the wildlife, the people, and the country?
5. List several ways in which vertebrate museum collections are valuable to researchers.
6. Will a baby born in a developed country have more or less impact on the world's resources than a baby born in a developing, or Third World, country? Why?
7. Give several reasons that human population growth is causing many of the problems that are facing other vertebrate species.
8. How does the ecoregion approach differ from previous efforts to preserve a particular species?
9. Define fragmentation in ecological terms. What are its effects?
10. Define biodiversity. Why is it important to maintain biodiversity?
11. Give several reasons that tropical forests are vital to the world's ecosystem.
12. Give several reasons for the alarming decline in the numbers of many species of neotropical migrating birds.
13. What difference could the loss of essentially all the remaining old-growth tropical forests and old-growth forests in North America have on your life and the lives of your descendants?
14. Discuss the major features of a biosphere reserve. What is the significance of each?
15. Is a protected area "useless" because it does not add significantly to economic growth? Explain. Describe the usefulness of an area in ecological terms.
16. What are some of the changes that are predicted to occur as global warming increases?
17. Do you feel that global warming is occurring? Why? Support your reasoning.
18. Do you believe that global warming from an enhanced natural greenhouse effect caused at least partially by human activities is a serious problem or one that has been greatly exaggerated? Explain.
19. What do you think would have happened to biodiversity if extinction rates occurred throughout evolutionary history at the same rate they have been occurring during the past 50 years?
20. List ten activities you engage in each day that harm the environment. List ten ways in which you might reduce the harmful effects of these activities.

Supplemental Reading

Anderson, S.H. 1990. *Managing Our Wildlife Resources.* New York: Prentice-Hall.

Aplin, G., P. Mitchell, H. Cleugh, A. Pitman, and D. Rich. 1995. *Global Environmental Crises: An Australian Perspective.* New York: Oxford University Press.

Bean, M. J., and M. J. Rowland (eds.). 1997. *The Evolution of National Wildlife Law.* Third edition. Westport, Connecticut: Greenwood Publishing.

Beissinger, S. R., and N. F. R. Snyder. 1991. *New World Parrots in Crisis.* Washington, D.C.: Smithsonian Institution Press.

Bender, D., and B. Leone (eds.). 1996. *Endangered Species: Opposing Viewpoints.* San Diego: Greenhaven Press.

Bjorndal, K. A. (ed.). 1995. *Biology and Conservation of Sea Turtles.* Washington, D.C.: Smithsonian Institution Press.

Bolen, E. G., and W. L. Robinson. 1995. *Wildlife Ecology and Management.* Third edition. New York: Macmillan.

Bowles, M. L. and C. J. Whelan (editors). 1995. *Restoration of Endangered Species.* New York: Cambridge University Press.

Bruenig, E. F. 1996. *Conservation and Management of Tropical Rainforests.* New York: Oxford University Press.

Carson, R. 1962. *Silent Spring.* Boston: Houghton Mifflin Co.

Clemmons, J. R. and R. Buchholz. 1997. *Behavioral Approaches to Conservation in the Wild.* New York: Cambridge University Press.

Colborn, T., D. Dumanoski, and J. P. Myers. 1996. *Our Stolen Future: Are We Threatening our Fertility, Intelligence, and Survival?* New York: Dutton.

DeGraaf, R. M., and J. H. Rappole. 1995. *Neotropical Migratory Birds: Natural History, Distribution, and Population Change.* Ithaca, New York: Comstock Publishing Associates.

Di Castri, F., and T. Younos (eds.). 1996. *Biodiversity: Science and Development.* New York: Oxford University Press.

Dobkin, D. S. 1994. *Conservation and Management of Neotropical Migrant Landbirds in the Northern Rockies and Great Plains.* Moscow: University of Idaho Press.

Dobson, A., and R. Carper. 1992. Global warming and potential changes in host-parasite and disease-vector relationships. Pages 201–217. In: Peters, R. L., and T. E. Lovejoy (eds.). *Global Warming and Biodiversity.* New Haven, Connecticut: Yale University Press.

Eldredge, N. 1998. *Life in the Balance: Humanity and the Biodiversity Crisis.* Princeton, New Jersey: Princeton University Press.

Emmons, L. H. 1997. *Neotropical Rainforest Mammals.* Chicago: The University of Chicago Press.

Grifo, F., and J. Rosenthal (eds.). 1997. *Biodiversity and Human Health.* Washington, D.C.: Island Press.

Hagan, J. M., III, and D. M. Johnston. 1992. *Ecology and Conservation of Neotropical Migrant Landbirds.* Washington, D.C.: Smithsonian Institution Press.

Hornocker, M. 1997. Siberian tigers. *National Geographic* 191(2):100–109.

Kirkpatrick, J. 1995. *A Continent Transformed: Human Impact on the Natural Vegetation of Australia.* New York: Oxford University Press.

Kramer, R., C. van Schaik, and J. Johnson (eds.). 1997. *Last Stand: Protected Areas and the Defense of Tropical Biodiversity.* New York: Oxford University Press.

Laurance, W. F., and R. O. Bierregaard, Jr. (eds.). 1997. *Tropical Forest Remnants.* Chicago: The University of Chicago Press.

Long, M. E. 1998. The vanishing prairie dog. *National Geographic* 193(4):116–130.

Martin, T. E., and D. M. Finch (eds.). 1995. *Ecology and Management of Neotropical Migratory Birds.* New York: Oxford University Press.

Maxwell, J. 1995. Swimming with salmon. *Natural History* 104(9):26–39.

May, R. M. 1992. How many species inhabit the earth? *Scientific American* 267(4): 42–48.

Myers, N., R. A. Mittermeier, C.G. Mittermeier, G. A. B. da Fonseca, and J. Kent. 2000. Biodiversity hotspots for conservation priorities. *Nature* 403: 853–858.

Ormund, R. F. G., J. D. Gage, and M. V. Angel (eds.). 1997. *Marine Biodiversity: Patterns and Processes.* New York: Cambridge University Press.

Philander, S. G. 1998. *Is the Temperature Rising?* Princeton, New Jersey: Princeton University Press.

Phillips, K. 1994. *Tracking the Vanishing Frogs.* New York: St. Martin's Press.

Pimm, S. L., G. J. Russell, J. L. Gittleman, and T. M. Brooks. 1995. The future of biodiversity. *Science* 269:347–350.

Primack, R. B. 1998. *Essentials of Conservation Biology.* Second edition. Sunderland, Massachusetts: Sinauer Associates.

Pulliam, H. R., and B. Babbitt. 1997. Science and the protection of endangered species. *Science* 275:499–500.

Rappole, J. 1995. *The Ecology of Migrant Birds.* Washington, D.C.: Smithsonian Institution Press.

Raustiala, K., and D. G. Victor. 1996. The future of the Convention on Biological Diversity. *Environment* 38(4):17–20, 37–45.

Reid, J. W., and I. A. Bowles. 1997. Reducing the impacts of roads on tropical forests. *Environment* 39(8):10–13, 32–35.

Reid, W. V. 1997. Strategies for conserving biodiversity. *Environment* 39(7):16–20, 39–43.

Rosenzweig, C., and D. Hillel. 1998. *Climate Change and the Global Harvest: Potential Impacts of the Greenhouse Effect on Agriculture.* New York: Oxford University Press.

Sellars, R. W. 1997. *Preserving Nature in the National Parks: A History.* New Haven, Connecticut: Yale University Press.

Sharp, I. 1994. *Green Indonesia: Tropical Forest Encounters.* New York: Oxford University Press.

Simon, N. 1995. *Nature in Danger: Threatened Habitats and Species.* New York: Oxford University Press.

Suplee, C. 1998. Unlocking the climate puzzle. *National Geographic* 193(5):38–71.

Szaro, R. C., and D. W. Johnston. 1996. *Biodiversity in Managed Landscapes.* New York: Oxford University Press.

Terborgh, J. 1989. *Where Have All the Birds Gone?* Princeton, New Jersey: Princeton University Press.

Tuxill, J. 1998. *Losing Strands in the Web of Life: Vertebrate Declines and the Conservation of Biological Diversity.* Washington, D.C.: Worldwatch Paper no.141.

United States Geological Survey. 1999. *Status and Trends of the Nation's Biological Resources.* 2 volumes. Washington, D.C.: U.S. Government Printing Office.

Vitousek, P. M., H. A. Mooney, J. Lubchenco, and J. M. Melillo. 1997. Human domination of earth's ecosystems. *Science* 277: 494–499.

Watson, J. (ed.). 1996. *Endangered Species Update.* Entire issue devoted to conservation and management of the southern sea otter.

World Commission on Forests and Sustainable Development. 1999. *Our Forest, Our Future.* New York: Cambridge University Press.

JOURNALS AND MAGAZINES

Animal Conservation. Published for the Zoological Society of London. New research into the factors which influence the conservation of animal species and their habitats.

Natural History. Published by the American Museum of Natural History. Wide-ranging general articles on all aspects of natural history.

International Wildlife. Published by the National Wildlife Federation. Articles concerning legislative issues, wildlife, and other natural resources throughout the world.

National Wildlife. Published by the National Wildlife Federation. Excellent photographs and articles concerning North American species.

Wildlife Conservation. Published by the Wildlife Conservation Society. Articles concerning conservation of species throughout the world.

Vertebrate Internet Sites

Visit the zoology website at http://www.mhhe.com to find live Internet links for each of the references listed below.

1. **Wired for Conservation.**
   This Nature Conservancy ecology site serves as a valuable introduction to ecosystems and endangered species.
2. **Effects of Air Pollutants on Wildlife.**
   National Park Service site; includes discussion of declining amphibians.
3. **National Wildlife Federation Home Page.**
   Informative site for organizing college students, faculty, and staff for such activities as recycling, composting, landscaping, Earth Day; many campus ecology links.
4. **IGC: Forest Resources.**
   Links to the latest in forest actions and activism, with special focus on the Pacific Northwest.
5. **Natural Resources Defense Council Homepage.**
   Links to Congress and congressional actions; annual summary of laws that affect ancient forests, endangered species, national parks and refuges, etc.
6. **Canadian Biodiversity Information Network (CBIN).**
   Databases, biodiversity strategy and research, resource network, museums, and publications.
7. **The Student Conservation Association, Inc.**
   Provides high-quality conservation job listings for students; publishes *Guide to Graduate Environmental Programs*; publishes *Earth Work* magazine.
8. **Biodiversity Flashpoints in the Americas.**
   This Natural Resources Defense Council site identifies endangered Western Hemisphere ecosystems especially rich in species, such as Alaska's Tongass National Forest.
9. **National Gap Analysis Program.**
   Program attempting to map areas in each state to identify existing "gaps" in biodiversity protection.

Appendix I

Classification of Living Vertebrates

Phylum Chordata
Subphylum Vertebrata

■ FISHES

The classification of fishes is based primarily on the system presented by Nelson (1994), but with modifications in the classification of the Chondrichthyes based on Compagno (1991).

- **Superclass Agnatha** (Jawless Fishes)
  - **Class Myxini**
    - Order Myxiniformes—hagfishes (60 species)
  - **Class Cephalaspidomorphi**
    - Order Petromyzontiformes—lampreys (40 species)
- **Superclass Gnathostomata** (Jawed Fishes)
  - **Class Chondrichthyes** (Cartilaginous Fishes)—800+ species
    - Subclass Holocephali
      - Order Chimaeriformes—chimaeras
    - Subclass Elasmobranchii
      - Order Hexanchiformes—frill and cow sharks
      - Order Squaliformes—dogfish sharks
      - Order Pristiophoriformes—saw sharks
      - Order Squatiniformes—angel sharks
      - Order Pristiformes—sawfishes
      - Order Rhinobatiformes—guitarfishes
      - Order Torpediniformes—electric rays
      - Order Myliobatiformes—sting, eagle, manta, and devil rays
      - Order Heterodontiformes—horn or bullhead sharks
      - Order Orectolobiformes—whale, nurse, and carpet sharks
      - Order Lamniformes—mackerel, thresher, and basking sharks
      - Order Carchiniformes—requiem, great white, and hammerhead sharks
  - **Class Sarcopterygii** (Lobe-Finned Fishes)—6 species
    - Subclass Coelacanthimorpha
      - Order Coelacanthiformes—coelacanth
    - Subclass Porolepimorpha and Dipnoi
      - Order Ceratodontiformes—Australian lungfish
      - Order Lepidosireniformes—South American and African lungfishes

Class Actinopterygii (Ray-Finned fishes)
- Subclass Chondrostei
  - Order Polypteriformes—bichirs
  - Order Acipenseriformes—sturgeons and paddlefishes
- Subclass Neopterygii
  - Order Semionotiformes—gars
  - Order Amiiformes—bowfin
  - Division Teleostei
    - Subdivision Osteoglossomorpha
      - Order Osteoglossiformes—bonytongues
    - Subdivision Elopomorpha
      - Order Elopiformes—tarpons
      - Order Albuliformes—bonefishes
      - Order Anguilliformes—eels
      - Order Saccopharyngiformes—gulpers
    - Subdivision Clupeomorpha
      - Order Clupeiformes—herrings and anchovies
    - Subdivision Euteleostei
      - Superorder Ostariophysi
        - Order Gonorhynchiformes—milkfish
        - Order Cypriniformes—carps and minnows
        - Order Characiformes—characins
        - Order Siluriformes—catfishes
        - Order Gymnotiformes—knifefishes
      - Superorder Protacanthopterygii
        - Order Esociformes—pikes
        - Order Osmeriformes—smelts
        - Order Salmoniformes—salmon, trout, and whitefish
      - Superorder Stenopterygii
        - Order Stomiiformes—bristlemouths and dragonfishes
        - Order Ateleopodiformes—jellynose fishes
      - Superorder Cyclosquamata
        - Order Aulopiformes—aulopiforms
      - Superorder Scopelomorpha
        - Order Myctophiformes—lanternfishes
      - Superorder Lampridiomorpha
        - Order Lampridiformes—opah and oarfishes
      - Superorder Polymixiomorpha
        - Order Polymixiiformes—beardfishes
      - Superorder Paracanthopterygii
        - Order Percopsiformes—troutperches and cavefishes
        - Order Ophidiiformes—cuskeels and brotulas
        - Order Gadiformes—cods, hakes, and grenadiers
        - Order Batrachoidiformes—toadfishes
        - Order Lophiiformes—anglerfishes
      - Superorder Acanthopterygii
        - Order Mugiliformes—mullets
        - Order Atherinomorpha—silversides and rainbow fishes
        - Order Beloniformes—flying fishes
        - Order Cyprinodontiformes—killifishes, four-eyed fishes, and rivulines
        - Order Stephanoberyciformes—gibberfishes and whalefishes
        - Order Beryciformes—berycoids
        - Order Zeiformes—dories
        - Order Gasterosteiformes—sticklebacks, seahorses, and pipefishes
        - Order Synbranchiformes—swamp eels and spiny eels
        - Order Scorpaeniformes—scorpionfishes

Order Perciformes—perches, basses, mackerels, tunas, and barracudas
Order Pleuronectiformes—flounders and soles
Order Tetraodontiformes—trunkfishes, porcupinefishes, and puffers

■ AMPHIBIANS

Many classification schemes have been presented to portray the relationships of amphibians. The following classification system follows Zug (1993).

Class Amphibia
Subclass Lissamphibia
Order Gymnophiona—caecilians—162 living species
Family Caeciliidae—common caecilians
Family Ichthyophiidae—Asian aquatic caecilians
Family Rhinatrematidae—Beaked caecilians
Family Scolecomorphidae — African caecilians
Family Typhlonectidae — aquatic caecilians
Family Uraeotyphlidae — Indian caecilians
Order Caudata (Urodela) —salamanders—390 species
Family Cryptobranchidae—giant salamanders
Family Hynobiidae—Asian salamanders
Family Ambystomatidae—mole salamanders
Family Amphiumidae—amphiumas
Family Dicamptodontidae—American giant salamanders
Family Plethodontidae—lungless salamanders
Family Proteidae—mudpuppies and olms
Family Rhyacotritonidae—Torrent salamanders
Family Salamandridae—newts
Family Sirenidae—sirens
Order Anura—frogs and toads—3,800+ living species
Family Ascaphidae—tailed frog
Family Discoglossidae—painted frogs
Family Bombinatoridae—fire-bellied toads
Family Leiopelmatidae—New Zealand frogs
Family Pipidae—clawed frogs
Family Rhinophrynidae—burrowing toads
Family Pelobatidae—spadefoot toads
Family Megaphryidae—Asian toads
Family Pelodytidae—parsley frogs
Family Brachycephalidae—saddle-backed toads
Family Bufonidae—toads
Family Centrolenidae—glass frogs
Family Dendrobatidae—dart-poison frogs
Family Heleophrynidae—ghost frogs
Family Hylidae—treefrogs
Family Leptodactylidae—Neotropical frogs
Family Allophrynidae—turkeit hill treefrog
Family Myobatrachidae—myobatrachid frogs
Family Pseudidae—harlequin frogs
Family Rhinodermatidae—mouth-breeding frogs
Family Sooglossidae—seychelles frogs
Family Microhylidae—narrowmouthed frogs
Family Hyperoliidae—reed frogs
Family Ranidae—true frogs
Family Rhacophoridae—Asian treefrogs
Family Arthroleptidae—screeching frogs

REPTILES

The system of classification for turtles follows Ernst and Barbour (1989). The remaining classification is based on Carroll (1988) and Zug (1993).

Class Reptilia
- Subclass Anapsida
  - Order Testudines—turtles and tortoises—230 species
    - Family Pelomedusidae—side-necked turtles
    - Family Chelidae—snake-necked turtles
    - Family Kinosternidae—mud and musk turtles
    - Family Dermatemydidae—Central American river turtle
    - Family Carettochelyidae—pig-nosed turtle
    - Family Trionychidae—soft-shelled turtles
    - Family Dermochelyidae—leatherback sea turtle
    - Family Cheloniidae—marine turtles
    - Family Chelydridae—snapping turtles
    - Family Platysternidae—big-headed turtle
    - Family Emydidae—pond, river, and box turtles
    - Family Testudinidae—tortoises
- Subclass Diapsida
  - Order Sphenodonta—tuatara—2 species
    - Family Sphenodontidae—tuatara
  - Order Squamata—lizards and snakes—6,300 species
    - Suborder Lacertilia—lizards—3,900 species
      - Family Iguanidae—iguanids
      - Family Agamidae—agamids
      - Family Chamaeleonidae—true chameleons
      - Family Eublepharidae—eublepharid geckos
      - Family Gekkonidae—geckos
      - Family Pygopodidae—snake lizards
      - Family Xantusiidae—night lizards
      - Family Teiidae—teiids
      - Family Scincidae—skinks
      - Family Lacertidae—lacertids
      - Family Cordylidae—girdle-tailed lizards
      - Family Dibamidae—dibamids
      - Family Gymnophthalmidae—gymnophthalmids
      - Family Amphisbaenidae—worm lizards
      - Family Rhineuridae—worm lizards
      - Family Bipedidae—worm lizards
      - Family Trogonophidae—worm lizards
      - Family Anguidae—anguids
      - Family Xenosauridae—xenosaurids
      - Family Helodermatidae—Gila monsters
      - Family Varanidae—monitor lizards
    - Suborder Serpentes—snakes—2,400 species
      - Family Typhlopidae—blind snakes
      - Family Anomalepididae—American blind snakes
      - Family Leptotyphlopidae—slender blind snakes
      - Family Aniliidae—cylinder snakes
      - Family Uropeltidae—shieldtail snakes
      - Family Xenopeltidae—sunbeam snake
      - Family Boidae—boas and pythons
      - Family Acrochordidae—wart snakes
      - Family Colubridae—colubrids

Family Elapidae—cobras, mambas, kraits, and coral snakes
Family Hydrophiidae—sea snakes
Family Viperidae—vipers
Order Crocodylia—alligators, crocodiles, and gavials—24 species
Family Alligatoridae—alligators and caimans
Family Crocodylidae—crocodiles
Family Gavialidae—gavials

■ BIRDS

Many ornithologists in North America prefer to use the "Check-list of North American Birds" prepared by the American Ornithologists' Union (1998). This well-respected checklist includes "the species of birds of North America from the Arctic through Panama, including the West Indies and Hawaiian Islands." For the purposes of this text, however, it is necessary to use a worldwide checklist.

The following taxonomy is based on the DNA hybridization studies of Sibley and Ahlquist (1990) and the resulting classification (Sibley and Monroe, 1990). Sibley and Monroe (1990) noted: "This system is the first to be based on a single objective criterion, namely, the degrees of similarity between the DNAs of species representing the major groups of living birds.... Like all others, this system is imperfect, but we believe that its merits outweigh its deficiencies and that it probably represents a closer approach to the 'true phylogeny' than any other system that has been proposed." The number of genera and species as given by Sibley and Monroe (1990) follows the name of each infraclass and order.

Infraclass Eoaves—terrestrial, mostly flightless birds—57 species
Order Struthioniformes—ostriches, rheas, emus, and kiwis
Family Struthionidae—ostriches
Family Rheidae—rheas
Family Casuariidae—cassowaries and emus
Family Apterygidae—kiwis
Order Tinamiformes—tinamous
Family Tinamidae—tinamous
Infraclass Neoaves—9,615 species
Order Craciformes—chachalacas, curassows, and megapodes—69 species
Family Cracidae—guans, chachalacas, and curassows
Family Megapodiidae—megapodes and scrubfowl
Order Galliformes—pheasants, grouse, chickens, turkeys, guineafowl, and peacocks—214 species
Family Phasianidae—pheasants, turkeys, and grouse
Family Numididae—guineafowl
Family Odontophoridae—New World quails
Order Anseriformes—ducks, geese, swans, and screamers—161 species
Family Anhimidae—screamers
Family Anseranatidae—magpie goose
Family Dendrocygnidae—whistling ducks
Family Anatidae—ducks, geese, and swans
Order Turniciformes—buttonquails or hemipodes—17 species
Order Piciformes—honeyguides, woodpeckers, barbets, and toucans—355 species

DNA data suggest that Piciformes are the descendants of what may be one of the oldest lineages in the avian tree (Sibley and Ahlquist, 1990). Buckholz (1986) described a fossil cavity and entrance hole preserved in a piece of petrified wood from the Eocene of Arizona that "even in details...is very close to recent cavities produced by representatives of the genus *Picoides*. It is...concluded that woodpeckers...must be at least of Eocene age." Sibley and Ahlquist (1990) noted: "This places woodpeckers in the record ca. 40–50 MYA [million years ago], although fossil bones of woodpeckers are known only back to the Miocene, less than 25 MYA. Of course, there is nothing in this fossil cavity that proves it was made by an ancestor of the Picidae; these may have been another lineage, now extinct, that made similar nesting cavities."

Family Indicatoridae—honeyguides
Family Picidae—woodpeckers, wrynecks, and piculets
Family Megalaimidae—Asian barbets

Family Lybiidae—African barbets
Family Ramphastidae—New World barbets and toucans
Order Galbuliformes—jacamars and puffbirds—51 species
Family Galbulidae—jacamars
Family Bucconidae—puffbirds
Order Bucerotiformes—hornbills—56 species
Family Bucerotidae—typical hornbills
Family Bucorvidae—ground hornbills
Order Upupiformes—hoopoes, woodhoopoes, and scimitar bills—10 species
Family Upupidae—hoopoes
Family Phoeniculidae—woodhoopoes
Family Rhinopomastidae—scimitar bills
Order Trogoniformes—trogons—39 species
Family Trogonidae—trogons
Order Coraciiformes—rollers, motmots, todies, kingfishers, and bee-eaters—152 species
Family Coraciidae—typical rollers
Family Brachypteraciidae—ground-rollers
Family Leptosomidae—courol or cuckoo-roller
Family Momotidae—motmots
Family Todidae—todies
Family Alcedinidae—alcedinid kingfishers
Family Dacelonidae—dacelonid kingfishers
Family Cerylidae—cerylid kingfishers
Family Meropidae—bee-eaters
Order Coliiformes—mousebirds—6 species
Family Coliida—mousebirds or colies
Order Cuculiformes—cuckoos, anis, hoatzins, roadrunners, and ground cuckoos—133 species

Sibley and Ahlquist (1990) concluded that the hoatzin is a highly modified cuckoo. After comparing more than 1,850 base pairs of DNA—including three different genes, two of them mitochondrial—Hedges et al. (1995) presented statistically significant evidence supporting Sibley and Ahlquist's conclusion that the hoatzin is more closely related to the cuckoo than to the gallinaceous birds (order Galliformes) with whom it had been classified for over 200 years.

Family Cuculidae—Old World cuckoos
Family Centropodidae—coucals
Family Coccyzidae—American cuckoos
Family Opisthocomidae—hoatzin
Family Crotophagidae—anis and guira cuckoo
Family Neomorphidae—roadrunners and ground cuckoos
Order Psittaciformes—Parrots, macaws, cockatoos, and parakeets—358 species
Family Psittacidae—parrots, macaws, and allies
Order Apodiformes—swifts—103 species
Family Apodidae—typical swifts
Family Hemiprocnidae—treeswifts or crested swifts
Order Trochiliformes—hummingbirds—319 species
Family Trochilidae—hummingbirds
Order Musophagiformes—touracos—23 species
Family Musophagidae—touracos and plantain-eaters
Order Strigiformes—owls, owlets, frogmouths, oilbirds, potoos, nightjars, nighthawks, and whip-poor-wills—291 species
Family Tytonidae—barn owls and grass owls
Family Strigidae—typical owls
Family Aegothelidae—owlet-nightjars
Family Podargidae—Australian frogmouths
Family Batrachostomidae—Asian frogmouths
Family Steatornithidae—oilbird
Family Nyctibiidae—potoos

Family Eurostopodidae—eared-nightjars
Family Caprimulgidae—nighthawks, nightjars, whip-poor-wills, and poorwills
Order Columbiformes—pigeons, doves, dodos, and solitaires—313 species
Family Raphidae—dodos and solitaires
Family Columbidae—pigeons and doves
Order Gruiformes—sunbitterns, bustards, cranes, limpkins, trumpeters, seriemas, kagus, rails, gallinules, coots, and mesites—196 species
Family Eurypygidae—sunbitterns
Family Otididae—bustards
Family Gruidae—cranes
Family Heliornithidae—limpkin and New World sungrebe
Family Psophiidae—trumpeters
Family Cariamidae—seriemas
Family Rhynochetidae—kagus
Family Rallidae—rails, gallinules, and coots
Family Mesitornithidae—mesites, monias, and roatelos
Order Ciconiiformes
Suborder Charadrii—oystercatchers, plovers, snipes, sandpipers, phalaropes, gulls, terns, skimmers, and auks—366 species
Family Pteroclidae—sandgrouse
Family Thinocoridae—seedsnipes
Family Pedionomidae—plains-wanderers
Family Scolopacidae—woodcock, snipe, sandpipers, and curlews
Family Rostratulidae—painted snipes
Family Jacanidae—jacanas, lily-trotters
Family Chionididae—sheathbills
Family Burhinidae—thick-knees
Family Charadriidae—oystercatchers, avocets, stilts, plovers, and lapwings
Family Glareolidae—crab plover, coursers, and pratincoles
Family Laridae—jaegers, skuas, skimmers, gulls, terns, auks, murres, puffins, and guillemots
Suborder Ciconii—hawks, eagles, vultures, kites, falcons, caracaras, grebes, tropicbirds, boobies, anhingas, cormorants, herons, egrets, bitterns, flamingos, ibises, spoonbills, pelicans, storks, frigatebirds, penguins, loons, petrels, and albatrosses—661 species
Family Accipitridae—ospreys, eagles, Old World vultures, hawks, kites, and harriers
Family Sagittariidae—secretary-birds
Family Falconidae—falcons and caracaras
Family Podicipedidae—grebes
Family Phaethontidae—tropicbirds
Family Sulidae—boobies and gannets
Family Anhingidae—anhingas and darters
Family Phalacrocoracidae—cormorants and shags
Family Ardeidae—herons, egrets, and bitterns
Family Scopidae—hamerkops or hammerheads
Family Phoenicopteridae—flamingos
Family Threskiornithidae—ibises and spoonbills
Family Pelecanidae—shoebill and pelicans
Family Ciconiidae—New World vultures, condors, storks, openbills, adjutants, and jabiru
Family Fregatidae—frigatebirds
Family Spheniscidae—penguins
Family Gaviidae—loons or divers
Family Procellariidae—storm-petrels, shearwaters, petrels, diving-petrels, and albatrosses
Order Passeriformes—5,712 species. This order contains 59 percent of the 9,672 species of birds recognized by Sibley and Monroe (1990).
Family Acanthisittidae—New Zealand wrens
Family Pittidae—pittas
Family Eurylaimidae—broadbills

Family Philepittidae—asities
Family Tyrannidae—mionectine flycatchers, tyrant flycatchers, tityras, becards, cotingas, plantcutters, sharpbills, and manakins
Family Thamnophilidae—typical antbirds
Family Furnariidae—horneros, ovenbirds, spinetails, woodcreepers, and scythebills
Family Formicariidae—ground antbirds
Family Conopophagidae—gnateaters
Family Rhinocryptida—tapaculos
Family Climacteridae—Australo-Papuan treecreepers
Family Menuridae—lyrebirds and scrub-birds
Family Ptilonorhynchidae—bowerbirds
Family Maluridae—fairywrens, emuwrens, and grasswrens
Family Meliphagidae—honeycreepers, *Ephthianura,* and *Ashbyia*
Family Pardalotidae—pardalotes, bristlebirds, scrub wrens, mouse-warblers, thornbills, whitefaces, and gerygones
Family Eopsaltriidae—Australo-Papuan robins and *Drymodes*
Family Irenidae—fairy-bluebirds and leafbirds
Family Orthonychidae—logrunner and chowchilla
Family Pomatostomidae—Australo-Papuan babblers
Family Laniidae—true shrikes
Family Vireonidae—vireos, greenlets, peppershrikes, and shrike-vireos
Family Corvidae—quail-thrushes, whipbirds, wedgebills, white-winged chough, apostlebird, sittellas, whiteheads, yellowheads, pipipi, shrike-tits, whistlers, shrike-thrushes, pitohuis, crows, ravens, magpies, jays, nutcrackers, choughs, birds-of-paradise, currawongs, wood-swallows, orioles, cuckoo-shrikes, fan tails, drongos, monarchs, magpie-larks, ioras, bushshrikes, boubous, tchagras, gonoleks, helmetshrikes, and vangas
Family Callaeatidae—New Zealand wattlebirds
Family Picathartidae—*Picathartes* and *Chaetops*
Family Bombycillidae—palmchat, silky-flycatchers, and waxwings
Family Cinclidae—dippers
Family Muscicapidae—thrushes, Old World flycatchers, and chats
Family Sturnidae—starlings, mynas, oxpeckers, mockingbirds, thrashers, and American catbirds
Family Sittidae—nuthatches and wallcreepers
Family Certhiidae—tree-creepers, spotted creeper, wrens, gnatcatchers, verdin, and gnatwrens
Family Paridae—penduline-tits, titmice, and chickadees
Family Aegithalidae—long-tailed tits, bushtits, and pygmy tits
Family Hirundinidae—river-martins, swallows, and martins
Family Regulidae—kinglets, firecrest, and goldcrests
Family Pycnonotidae—bulbuls, greenbuls
Family Hypocoliidae—gray hypocolius
Family Cisticolidae—African warblers
Family Zosteropidae—white-eyes
Family Sylviidae—leaf-warblers, reed-warblers, grassbirds, songlarks, fernbirds, laughingthrushes, liocichlas, babblers, minlas, fulvettas, yuhinas, parrotbills, wrentits, and sylvine warblers
Family Alaudidae—larks
Family Nectariniidae—sugarbirds, flowerpeckers, mistletoebirds, sunbirds, and spider-hunters
Family Melanocharitidae—berrypeckers and longbills
Family Paramythiidae—tit berrypickers (*Paramythia*) and crested berrypicker (*Oreocharis*)
Family Passeridae—sparrows, rock sparrows, snowfinches, wagtails, pipits, longclaws, accentors, weavers, malimbes, fodys, bishops, widowbirds, waxbills, firefinches, parrotfinches, munias, indigobirds, and whydahs
Family Fringillidae—olive warbler, chaffinches, bramblings, goldfinches, crossbills, linnets, bullfinches, Hawaiian honeycreepers, buntings, longspurs, towhees, American sparrows, wood warblers including *Zeledonia,* tanagers, swallow-tanagers, neotropical honeycreepers, plushcaps, tanager-finches, cardinals, grosbeaks, troupials, meadowlarks, New World blackbirds, cowbirds, oropendolas, caciques, and bobolinks

■ MAMMALS

The following mammalian taxonomy is based on Wilson and Reeder (1993).

Class Mammalia (Synapsida)

Order Monotremata—3 species
- Family Tachyglossidae—echidnas, spiny anteaters
- Family Ornithorhynchidae—duck-billed platypus

Order Didelphimorphia—63 species
- Family Didelphidae—American opossums

Order Paucituberculata—5 species
- Family Caenolestidae—"shrew" opossums

Order Microbiotheria—1 species
- Family Microbiotheriidae—monitos del Monte

Order Dasyuromorphia—63 species
- Family Thylacinidae—thylacines and Tasmanian wolves
- Family Myrmecobiidae—numbats and banded anteaters
- Family Dasyuridae—marsupial "mice" and "cats," and Tasmanian devils

Order Peramelemorphia—21 species
- Family Peramelidae—bandicoots
- Family Peroryctidae—bandicoots

Order Notoryctemorphia—2 species
- Family Notoryctidae—marsupial "moles"

Order Diprotodontia—117 species
- Family Phascolarctidae—koalas
- Family Vombatidae—wombats
- Family Phalangeridae—possums and cuscuses
- Family Potoroidae—rat kangaroos
- Family Macropodidae—wallabies and kangaroos
- Family Burramyidae—pygmy possums
- Family Pseudocheiridae—ring-tailed possums
- Family Petauridae—gliding and striped possums
- Family Tarsipedidae—honey possums
- Family Acrobatidae—pygmy flying possum

Order Xenarthra—29 species
- Family Bradypodidae—three-toed tree sloths
- Family Megalonychidae—West Indian and two-toed tree sloths
- Family Dasypodidae—armadillos
- Family Myrmecophagidae—anteaters

Order Insectivora—428 species
- Family Solenodontidae—solenodons
- Family Nesophontidae—West Indian shrews (extinct?)
- Family Tenrecidae—tenrecs and Madagascar hedgehogs
- Family Chrysochloridae—golden moles
- Family Erinaceidae—gymnures and hedgehogs
- Family Soricidae—shrews
- Family Talpidae—moles

Order Scandentia—19 species
- Family Tupaiidae—tree shrews

Order Dermoptera—2 species
- Family Cynocephalidae—flying lemurs and colugos

Order Chiroptera—925 species
- Family Pteropodidae—Old World fruit bats
- Family Rhinopomatidae—mouse-tailed bats and long-tailed bats
- Family Craseonycteridae—bumble-bee bats
- Family Emballonuridae—sac-winged bats, sheath-tailed bats, and ghost bats
- Family Nycteridae—slit-faced bats and hollow-faced bats

Family Megadermatidae—false vampire bats
Family Rhinolophidae—horseshoe bats
Family Noctilionidae—bulldog bats and fisherman bats
Family Mormoopidae—spectacled bats
Family Phyllostomidae—American leaf-nosed bats
Family Natalidae—funnel-eared bats
Family Furipteridae—smoky, thumbless bats
Family Thyropteridae—disc-winged bats and New World sucker-footed bats
Family Myzopodidae—Old World sucker-footed bats
Family Vespertilionidae—vespertilionid bats and mouse-eared bats
Family Mystacinidae—New Zealand short-tailed bats
Family Molossidae—free-tailed bats and mastiff bats

Order Primates—233 species
Family Cheirogaleidae—dwarf lemurs and mouse lemurs
Family Lemuridae—lemurs
Family Megaladapidae—sportive lemurs and weasel lemurs
Family Indridae—avahi, sifakas, and indris
Family Daubentoniidae—aye-ayes
Family Loridae—loris and pottos
Family Galagonidae—galagos
Family Tarsiidae—tarsiers
Family Callitrichidae—marmosets and tamarins
Family Cebidae—New World monkeys
Family Cercopithecidae—Old World monkeys
Family Hylobatidae—gibbons and lesser apes
Family Hominidae—bipedal primates and humans

Order Carnivora—271 species
Family Canidae—dogs, wolves, coyotes, jackals, and foxes
Family Felidae—cats
Family Herpestidae—mongooses
Family Hyaenidae—aardwolves and hyenas
Family Mustelidae—weasels, badgers, and otters
Family Mephitidae—skunks and stink-badgers
Family Odobenidae—walruses
Family Otariidae—eared seals, fur seals, and sea lions
Family Phocidae—true, earless, or hair seals
Family Procyonidae—raccoons, ringtails, coatis, and kinkajous
Family Ursidae—bears and pandas
Family Viverridae—civets, genets, linsangs, and fossas

Order Cetacea—78 species
Family Balaenidae—right whales
Family Balaenopteridae—rorquals
Family Eschrictiidae—gray whales
Family Neobalaenidae—pygmy right whales
Family Delphinidae—dolphins and porpoises
Family Monodontidae—belugas and narwhals
Family Phocoenidae—porpoises
Family Physeteridae—sperm whales
Family Platanistidae—freshwater dolphins
Family Ziphiidae—beaked whales

Order Sirenia—5 species
Family Dugongidae—dugongs and sea cows
Family Trichechidae—manatees

Order Proboscidea—2 species
Family Elephantidae—elephants

Order Perissodactyla—18 species
- Family Equidae—horses, zebras, and asses
- Family Tapiridae—tapirs
- Family Rhinocerotidae—rhinoceroses

Order Hyracoidea—6 species
- Family Procaviidae—hyraxes and dassies

Order Tubulidentata—1 species
- Family Orycteropodidae—aardvarks and ant bears

Order Artiodactyla—220 species
- Family Suidae—pigs and hogs
- Family Tayassuidae—peccaries
- Family Hippopotamidae—hippopotamuses
- Family Camelidae—camels, guanacos, llamas, alpacas, and vicuñas
- Family Tragulidae—chevrotains and mouse deer
- Family Giraffidae—okapis and giraffes
- Family Moschidae—musk deer
- Family Cervidae—deer
- Family Antilocapridae—pronghorns
- Family Bovidae—antelopes, cattle, bison, buffalo, goats, and sheep

Order Pholidota—7 species
- Family Manidae—pangolins and scaly anteaters

Order Rodentia—2,021 species
- Suborder Sciurognathi
  - Family Aplodontidae—sewellels and mountain beavers
  - Family Sciuridae—squirrels, chipmunks, marmots, and prairie dogs
  - Family Castoridae—beavers
  - Family Geomyidae—pocket gophers
  - Family Heteromyidae—pocket mice, and kangaroo rats and mice
  - Family Dipodidae—jerboas, birch mice, and jumping mice
  - Family Muridae—rats, mice, hamsters, voles, lemmings, and gerbils
  - Family Anomaluridae—scaly-tailed squirrels
  - Family Pedetidae—springhares and springhaas
  - Family Ctenodactylidae—gundis
  - Family Myoxidae—dormice
- Suborder Hystricognathi
  - Family Bathyergidae—African mole rats and blesmois
  - Family Hystricidae—Old World porcupines
  - Family Petromuridae—dassie rats
  - Family Thryonomyidae—cane rats
  - Family Erethizontidae—New World porcupines
  - Family Chinchillidae—viscachas and chinchillas
  - Family Dinomyidae—pacaranas
  - Family Caviidae—cavies and Patagonian "hares"
  - Family Hydrochaeridae—capybaras
  - Family Dasyproctidae—agoutis
  - Family Agoutidae—pacas
  - Family Ctenomyidae—tuco-tucos
  - Family Octodontidae—octodonts
  - Family Abrocomidae—chinchilla rats and chinchillones
  - Family Echimyidae—spiny rats
  - Family Capromyidae—hutias
  - Family Myocastoridae—nutrias and coypus

Order Lagomorpha—80 species
- Family Ochotonidae—pikas, mouse hares, and conies
- Family Leporidae—hares and rabbits

Order Macroscelidea—15 species
- Family Macroscelididae—elephant shrews

Appendix II

Endangered Vertebrate Species in the United States and/or Trust Territories

The data in Appendix II was originally compiled from data in Report to Congress, U.S. Fish and Wildlife Service, 1994 and from the List of Endangered and Threatened Wildlife, Special Reprint 50 CFR 17.11 and 17.12, 1996. It was updated on February 2, 2000 by using the U.S. Fish and Wildlife Service World Wide Web site (http://www.fws.gov/~r9endspp/endspp.html).

MAMMALS

Bat, gray—*Myotis grisescens*
Bat, Hawaiian hoary—*Lasiurus cinereus semotus*
Bat, Indiana—*Myotis sodalis*
Bat, lesser (= Sanborn's) long-nosed—*Leptonycteris curasoae yerbabuenae*
Bat, little Mariana fruit—*Pteropus tokudae*
Bat, Marianas fruit—*Pteropus mariannus mariannus*
Bat, Mexican long-nosed—*Leptonycteris nivalis*
Bat, Ozark big-eared—*Plecotus townsendii ingens*
Bat, Virginia big-eared—*Plecotus townsendii virginianus*
Caribou, woodland—*Rangifer tarandus caribou*
Deer, Columbian white-tailed—*Odocoileus virginianus leucurus*
Deer, key—*Odocoileus virginianus clavium*
Ferret, black-footed—*Mustela nigripes*
Fox, San Joaquin kit—*Vulpes macrotis mutica*
Jaguar—*Panthera onca*
Jaguarundi (2 subspecies)—*Felis yagouaroundi cacomitli; F. y. tolteca*
Kangaroo rat, Fresno—*Dipodomys nitratoides exilis*
Kangaroo rat, giant—*Dipodomys ingens*
Kangaroo rat, Morro Bay—*Dipodomys heermanni morroensis*
Kangaroo rat, Stephen's—*Dipodomys stephensi*
Kangaroo rat, Tipton—*Dipodomys nitratoides nitratoides*
Kangaroo rat, San Bernadino Merriam's—*Dipodomys merriami parvus*
Manatee, West Indian (Florida)—*Trichechus manatus*
Mountain beaver, Point Arena—*Aplodontia rufa nigra*
Mouse, Alabama beach—*Peromyscus polionotus ammobates*
Mouse, Anastasia Island beach—*Peromyscus polionotus phasma*
Mouse, Choctawhatchee beach—*Peromyscus polionotus allophrys*
Mouse, Key Largo cotton—*Peromyscus gossypinus allopaticola*
Mouse, Pacific pocket—*Perognathus longimembris pacificus*
Mouse, Perdido Key beach—*Peromyscus polionotus trissyllepsis*
Mouse, salt marsh harvest—*Reithrodontomys raviventris*
Mouse, St. Andrew beach—*Peromyscus polionotus peninsularis*
Ocelot—*Leopardus(= Felis) pardalis*
Panther, Florida—*Puma (= Felis) concolor coryi*
Pronghorn, Sonoran—*Antilocapra americana sonoriensis*
Puma (= cougar), eastern—*Puma (= Felis) concolor cougar*
Rabbit, Lower Keys—*Sylvilagus palustris hefneri*
Rice rat, silver—*Oryzomys palustris natator*
Sea-lion, Steller (= northern), western population—*Eumetopias jubatus*
Seal, Caribbean monk—*Monachus tropicalis*
Seal, Hawaiian monk—*Monachus schauinslandi*
Sheep, bighorn (California Peninsular Range and Sierra Nevada populations)—*Ovis canadensis*
Squirrel, Carolina northern flying—*Glaucomys sabrinus coloratus*
Squirrel, Delmarva Peninsula fox—*Sciurus niger cinereus*
Squirrel, Mount Graham red—*Tamiasciurus hudsonicus grahamensis*
Squirrel, Virginia northern flying—*Glaucomys sabrinus fuscus*
Vole, Amargosa—*Microtus californicaus scirpensis*
Vole, Florida salt marsh—*Microtus pennsylvanicus dukecampbelli*
Vole, Hualapai Mexican—*Microtus mexicanus hualapaiensis*
Whale, blue—*Balaenoptera musculus*
Whale, bowhead—*Balaena mysticetus*
Whale, finback—*Balaenoptera physalus*
Whale, humpback—*Megaptera novaeangliae*
Whale, right—*Balaena glacialis*
Whale, Sei—*Balaenoptera borealis*
Whale, sperm—*Physeter macrocephalus (= catodon)*
Wolf, gray (eastern timber)—*Canis lupus lycaon*
Wolf, gray (northern Rocky Mountain)—*Canis lupus irremotus*
Wolf, gray (Mexican)—*Canis lupus baileyi*
Wolf, red—*Canis rufus*
Woodrat, Key Largo—*Neotoma floridana smalli*

BIRDS

'Akepa, Hawaii (honeycreeper)—*Loxops coccineus coccineus*
'Akepa, Maui (honeycreeper)—*Loxops coccineus ochraceus*
'Akialoa, Kauai (honeycreeper)—*Hemignathus procerus*
'Akiapola'au (honeycreeper)—*Hemignathus munroi*
Blackbird, yellow-shouldered—*Agelaius xanthomus*
Bobwhite, masked (quail)—*Colinus virginianus ridgwayi*
Broadbill, Guam—*Myiagra freycineti*
Condor, California—*Gymnogyps californianus*
Coot, Hawaiian—*Fulica americana alai*
Crane, Mississippi sandhill—*Grus canadensis pulla*
Crane, whooping—*Grus americana*
Creeper, Hawaii—*Oreomystis mana*
Creeper, Molokai—*Oreomystis flammea*
Creeper, Oahu—*Paroreomyza maculata*
Crow, Hawaiian—*Corvus hawaiiensis*
Crow, Mariana—*Corvus kubaryi*
Curlew, Eskimo—*Numenius borealis*
Duck, Hawaiian—*Anas wyvilliana*
Duck, Laysan—*Anas laysanensis*
Falcon, northern Aplomado—*Falco femoralis septentrionalis*
Finch, Laysan (honeycreeper)—*Telespyza cantans*
Finch, Nihoa (honeycreeper)—*Telespyza ultima*
Flycatcher, Southwestern willow—*Empidonax traillii extimus*
Goose, Hawaiian (nene)—*Nesochen sandvicensis*
Hawk, Hawaiian (lo)—*Buteo solitarius*
Hawk, Puerto Rican broad-winged—*Buteo platypterus brunnescens*
Hawk, Puerto Rican sharp-shinned—*Accipiter striatus venator*
Honeycreeper, crested—*Palmeria dolei*
Kingfisher, Guam Micronesian—*Halcyon cinnamomina cinnamomina*
Kite, Everglade snail—*Rostrhamus sociabilis plumbeus*
Mallard, Mariana—*Anas oustaleti*
Megapode, Micronesian—*Megapodius laperouse*
Millerbird, Nihoa—*Acrocephalus familiaris kingi*
Moorhen (= gallinule), Hawaiian common—*Gallinula chloropus sandvicensis*
Moorhen (= gallinule), Mariana common—*Gallinula chloropus guami*
Nightjar, Puerto Rican (whip-poor-will)—*Caprimulgus noctitherus*
Nukupu'u (honeycreeper)—*Hemignathus lucidus*
'O'o, Kauai (honeyeater) —*Moho braccatus*
'O'u (honeycreeper)—*Psittirostra psittacea*
Palila (honeycreeper)—*Loxioides bailleui*
Parrot, Puerto Rican—*Amasona vittata*
Parrottbill, Maui (honeycreeper)—*Pseudonestor xanthophrys*
Pelican, brown—*Pelecanus occidentalis*
Petrel, Hawaiian dark-rumped—*Pterodroma phaeopygia sandwichensis*
Pigeon, Puerto Rican plain—*Columba inornata wetmorei*
Plover, piping—*Charadrius melodus*
Po'ouli (honeycreeper)—*Melamprosops phaeosoma*
Prairie-chicken, Attwater's greater—*Tympanuchus cupido attwateri*
Pygmy-owl, cactus ferruginous—*Glaucidium brasilianum cactorum*
Rail, California clapper—*Rallus longirostris obsoletus*
Rail, Guam—*Rallus owstoni*
Rail, light-footed clapper—*Rallus longirostris levipes*
Rail, Yuma clapper—*Rallus longirostris yumanensis*
Shrike, San Clemente loggerhead—*Lanius ludovicianus mearnsi*
Sparrow, Cape Sable seaside—*Ammodramus maritimus mirabilis*
Sparrow, Florida grasshopper—*Ammodramus savannarum floridanus*
Stilt, Hawaiian—*Himantopus himantopus knudseni*
Stork, wood—*Mycteria americana*
Swiftlet, Mariana gray—*Aerodramus vanikorensis bartschi*
Tern, California least—*Sterna antillarum browni*
Tern, least (interior population)—*Sterna antillarum*
Tern, roseate—*Sterna dougalli dougalli*
Thrush, large Kauai—*Myadestes myadestina*
Thrush, Molokai—*Myadestes lanaiensis rutha*
Thrush, small Kauai—*Myadestes palmeri*
Vireo, black-capped—*Vireo atricapillus*
Vireo, least Bell's—*Vireo belli pusillus*
Warbler, Bachman's—*Vermivora bachmanii*
Warbler, golden-cheeked—*Dendroica chrysoparia*
Warbler, Kirtland's—*Dendroica kirtlandii*
Warbler, nightingale reed—*Acrocephalus luscinia*
White-eye, bridled—*Zosterops conspicillatus conspicillatus*
Woodpecker, ivory-billed—*Campephilus principalis*
Woodpecker, red-cockaded—*Picoides borealis*

REPTILES

Anole, Culebra Island giant—*Anolis roosevelti*
Boa, Puerto Rican—*Epicrates inornatus*
Boa, Virgin Islands tree—*Epicrates monensis granti*
Crocodile, American—*Crocodylus acutus*
Gecko, Monito—*Sphaerodactylus micropithecus*
Lizard, blunt-nosed leopard—*Gambelia silus*
Lizard, St. Croix ground—*Ameiva polops*
Sea turtle, green (2 populations)—*Chelonia mydas*
Sea turtle, hawksbill—*Eretmochelys imbricata*
Sea turtle, Kemp's (= Atlantic) ridley—*Lepidochelys kempii*
Sea turtle, leatherback—*Dermochelys coriacea*
Snake, San Francisco garter—*Thamnophis sirtalis tetrataenia*
Turtle, Alabama redbelly—*Pseudemys alabamensis*
Turtle, Plymouth redbelly (= red-bellied)—*Pseudemys rubriventris bangsii*

AMPHIBIANS

Salamander, Barton Springs—*Eurycea sosorum*
Salamander, desert slender—*Batrachoseps aridus*
Salamander, Santa Cruz long-toed—*Ambystoma macrodactylum croceum*
Salamander, Shenandoah—*Plethodon shenandoah*
Salamander, Sonoran tiger—*Ambystoma tigrinum stebbinsi*
Salamander, Texas blind—*Typhlomolge rathbuni*
Toad, arroyo—*Bufo microscaphus californicus*
Toad, Houston—*Bufo houstonensis*
Toad, Wyoming—*Bufo hemiophrys baxteri*

FISHES

Cavefish, Alabama—*Speoplatyrhinus poulsoni*
Chub, bonytail—*Gila elegans*
Chub, Borax Lake—*Gila boraxobius*

Chub, humpback—*Gila cypha*
Chub, Mohave tui—*Gila bicolor mohavensis*
Chub, Oregon—*Oregonichthys (= Hybopsis) crameri*
Chub, Owens tui—*Gila bicolor snyderi*
Chub, Pahranagat roundtail—*Gila robusta jordani*
Chub, Virgin River—*Gila robusta semidnuda*
Chub, Yaqui—*Gila purpurea*
Cui-ui—*Chasmistes cujus*
Dace, Ash Meadows speckled—*Rhinichthys osculus nevadensis*
Dace, Clover Valley speckled—*Rhinichthys osculus oligoporus*
Dace, Independence Valley speckled—*Rhinichthys osculus lethoporus*
Dace, Kendall Warm Springs—*Rhinichthys osculus thermalis*
Dace, Moapa—*Moapa coriacea*
Darter, amber—*Percina antesella*
Darter, bluemask (= jewel)—*Etheostoma* (*Doration*) sp.
Darter, boulder (Elk River)—*Etheostoma wapiti*
Darter, duskytail—*Etheostoma* (*Catonotus*) sp.
Darter, Etowah—*Etheostoma etowahae*
Darter, fountain—*Etheostoma fonticola*
Darter, Maryland—*Etheostoma sellare*
Darter, Okaloosa—*Etheostoma okaloosae*
Darter, relict —*Etheostoma* (*Catonotus*) *chienense*
Darter, watercress—*Etheostoma nuchale*
Gambusia, Big Bend—*Gambusia gaigei*
Gambusia, Clear Creek—*Gambusia heterochir*
Gambusia, Pecos—*Gambusia nobilis*
Gambusia, San Marcos—*Gambusia georgei*
Goby, tidewater—*Eucyclogobius newberryi*
Logperch, Conasauga—*Percina jenkinsi*
Logperch, Roanoke—*Percina rex*
Madtom, pygmy—*Noturus stanauli*
Madtom, Scioto—*Noturus trautmani*
Madtom, Smoky—*Noturus baileyi*
Minnow, Rio Grande silvery—*Hybognathus amarus*
Pikeminnow (= squawfish) —*Ptychocheilus lucius*
Poolfish (= killifish), Pahrump—*Empetrichthys latos*
Pupfish, Ash Meadows Amargosa—*Cyprinodon nevadensis mionectes*
Pupfish, Comanche Springs—*Cyprinodon elegans*
Pupfish, desert—*Cyprinodon macularius*
Pupfish, devils hole—*Cyprinodon diabolis*
Pupfish, Leon Springs—*Cyprinodon bovinus*
Pupfish, Owens—*Cyprinodon radiosus*
Pupfish, Warm Springs—*Cyprinodon nevadensis pectoralis*
Salmon, chinook (Sacramento River)—*Onchorhynchus tshawytscha*
Salmon, sockeye (= red, = blueback)—*Oncorhynchus nerka*
Shiner, Cahaba—*Notropis cahabae*
Shiner, Cape Fear—*Notropis mekistocholas*
Shiner, Palezone—*Notropis* sp.
Shiner, Topeka—*Notropis topeka* (= *tristis*)
Spinedace, White River—*Lepidomeda albivallis*
Springfish, Hiko White River—*Crenichthys baileyi grandis*
Springfish, White River—*Crenichthys baileyi baileyi*
Stickleback, unarmored threespine—*Gasterosteus aculeatus williamsonii*
Sturgeon, pallid—*Scaphirhynchus albus*
Sturgeon, shortnose—*Acipenser brevirostrum*
Sturgeon, white (Kootenai River pop.)—*Acipenser transmontanus*
Sucker, June—*Chasmistes liorus*
Sucker, Lost River—*Deltistes luxatus*
Sucker, Modoc—*Catostomus microps*
Sucker, razorback—*Xyrauchen texanus*
Sucker, shortnose—*Chasmistes brevirostris*
Topminnow, Gila—*Poeciliopsis occidentalis*
Trout, Gila—*Oncorhynchus (= Salmo) gilae*
Trout, Umpqua River cutthroat—*Oncorhynchus* (= *Salmo*) *clarki clarki*
Woundfin—*Plagopterus argentissimus*

Appendix III

Common Names for the Sexes, Young, and Groups of Animals

(Excerpted from Lipton, 1968; Burns, 1972; and Mark and Menning, 1984)

Animal	Male	Female	Young	Group
Antelope	buck	doe	kid	herd
Bear	boar	sow	cub	sloth
Beaver	boar	sow	pup	colony
Bison	bull	cow	calf	herd
Bobcat	tom	lioness/queen	kitten	litter
Cat	tomcat (tom)	pussy/queen	kitten	clowder
Cattle	bull	cow/heifer	calf	herd/drove
Chicken	rooster	hen/pullet	chick	flock/peep
Deer	buck/stag	doe	fawn	herd
Dog	dog	bitch	pup	kennel
Duck	drake	duck	duckling	flock
Eagle	cock	hen	eaglet	convocation
Elephant	bull	cow	calf	herd
Fishes	cock	hen	fry	school
Fox	reynard	vixen	cub/pup	earth/skulk
Giraffe	bull	cow	calf	herd
Goat	billy/buck	nanny	kid	trip
Goose	gander	goose/dame	gosling	gaggle
Hog	boar	sow/gilt	piglet/shoat	herd/drove
Horse	stallion	mare	foal, colt (male), filly (female)	stable/herd
Kangaroo	buck	doe	joey	troop/herd
Leopard	leopard	leopardess	cub	leap
Lion	lion	lioness	cub	pride
Otter	dog	bitch	cub	—
Owl	owl	jenny	owlet	parliament
Peafowl	peacock	peahen	—	flock
Pheasant	cock	hen	—	nye/bouquet
Pigeon	cock	hen	squab	flock
Polecat	hob	jill	kitten	—
Rabbit	buck	doe	kitten	colony
Rat	buck	doe	—	colony
Seal	bull	cow	pup	herd/harem/rookery
Sheep	ram	ewe	lamb	flock/hurtle
Swan	cob	pen	cygnet	flock
Tiger	tiger	tigress	cub	—
Turkey	tom	hen	poult	flock/rafter
Walrus	bull	cow	cub	herd
Whale	bull	cow	calf	herd/pod
Wolf	he-wolf	she-wolf	pup	pack/rout
Zebra	stallion	mare	colt	herd

Additional Group Terms:

Animal	Group	Animal	Group
Apes	shrewdness	Moles	labor
Birds (small)	dissimulation	Monkeys	troupe
Camels	flock/herd	Mules	barren
Coots	covert	Nightingales	watch
Crows	murder	Oxen	team/yoke/drove/herd
Doves	flight	Parrots	flock
Ducks in flight	team	Porpoises	school/gam/herd
Ducks on water	paddling	Quails	bevy
Ducks (two)	brace	Ravens	unkindness
Eels	swarm	Redwings	crowd
Elk	gang	Rhinoceroses	crash
Ferrets	business	Snakes (young)	bed
Finches	charm	Sparrows	host/tribe
Frogs	army	Squirrels	dray
Geese (in flight)	flock	Starlings	chattering/murmuration
Geese (on water)	gaggle	Storks	mustering
Grouse	covey	Swallows	flight
Herons	siege	Toads	knot
Jays	band	Turtledoves	pitying
Larks	exaltation	Turtles	bale
Magpies	tidings	Woodpeckers	descent
Mallards	flush/puddling	Wrens	herd

Glossary

A

Abomasum. Fourth and last chamber of the ruminant stomach.

Abyssopelagic zone. Region of the oceanic zone from approximately 4,000 to 6,000 m; almost constant physical environment; continually dark, cold (4°C), and virtually unchanging in chemical composition.

Accidental parthenogenesis. Development of a new individual from an unfertilized egg due to the physical or chemical stimulation of the egg.

Acid rain. Precipitation with a pH value less than 5.6.

Acinar. Cells in the pancreas that produce and secrete digestive enzymes.

Acoelous. Describes a vertebra having the anterior and posterior articular surfaces of the centrum (body) flattened.

Acrodont. Type of tooth attachment in which there are no sockets; teeth are attached to the summit of the jaw.

Adpressed limbs. In salamanders, the situation in which the forelimbs are pushed backward and the hindlimbs are pushed forward along the sides of the body.

Adipose fin. Fleshy fin, without rays, located behind the dorsal fin.

Adrenal. Endocrine gland near or on the kidney; secretes epinephrine, norepinephrine, aldosterone, cortisone, and gonadocorticoids.

Adrenocorticotropic hormone (ACTH). Hormone produced by the anterior pituitary gland that stimulates the secretion of hormones by the adrenal cortex.

Adult. Mature individual; capable of producing sex cells (eggs, sperm).

Agnatha. Taxon of vertebrates, comprising those without jaws.

Aldosterone. Mineralocorticoid hormone produced by the adrenal cortex that induces sodium and water reabsorption and potassium excretion.

Allantois. Extraembryonic saclike extension of the hindgut of amniotes, aiding in excretion and respiration.

Allen's Rule. Ecological principle describing a general trend among homeotherms for limbs to become longer and extremities (such as ears) to become less compact in warmer climates than in colder ones; best applied within those species having wide north–south geographic ranges.

Allopatric. Occupying different geographic regions.

Altitudinal migration. Vertical migration; generally seasonal, as is seen in elk and some birds.

Altricial. Young bird or mammal hatched or born in a helpless condition, and requiring extensive parental care in order to survive.

Alveolus. Respiratory pocket in the lungs; site of gas exchange (diffusion of oxygen and carbon dioxide across an alveolar –capillary membrane).

Ammocoetes. Larval form of the lamprey; occurs in streams with sandy bottoms.

Amnion. Fluid-filled innermost extraembryonic sac surrounding the embryo of reptiles, birds, and mammals.

Amniote. Vertebrate whose embryo possesses an amnion, chorion, and allantois (reptiles, birds, mammals).

Amphicoelous. Describes a vertebra having concave anterior and posterior articular surfaces of the centrum (body).

Amphistylic. Type of jaw suspension found in some sharks where the jaws and hyoid arch are braced directly against cranium.

Amplexus. Sexual embrace of frogs in which the male mounts the female so that he is dorsal to her; male's forelimbs may grasp female around waist (inguinal) or around pectoral region (axillary).

Ampulla. Dilation at end of each semicircular canal containing sensory epithelium; low-frequency electroreceptor in certain fishes (e.g., ampulla of Lorenzini).

Anadromous. Fish that typically inhabit seas or lakes but ascend freshwater streams to spawn (e.g., salmon).

Anal fin. Median unpaired fin situated posterior to the anus (vent) and in front of the caudal peduncle.

Analogy. Features of two or more organisms that perform a similar function; similarity of function but not of embryonic (evolutionary) origin (e.g., a butterfly wing and a bat wing).

Anamniote. Vertebrate that lacks an amnion, allantois, and chorion during development; agnathans, fishes, and amphibians.

Anapsid. Amniotes in which the skull lacks temporal openings; turtles are the only living representatives.

Angular. Dermal bone that ensheathes part of Meckel's cartilage.

Antler. Deciduous, usually branched, bony outgrowth on the head of various members of the deer family (Cervidae).

Anuran. Tailless amphibian from the order Anura; a frog or toad.

Aphotic. Without light; that portion of a body of water lying at a depth beyond the penetration of sunlight.

Aplacental viviparity. See *ovoviviparous*.

Apodan. Legless amphibian; a member of the tropical order Apoda.

Aposematic coloration. Adaptation of some species' bright colors that serve as a warning to potential predators (e.g., skunks, poison-dart frogs, coral snakes).

Appendix. Saclike structure attached to the caecum in mammals; may contain bacteria in some herbivores; vestigial in some species including humans.

Apterium (pl. **apteria**). Area of skin in birds devoid of feathers located between the pterylae.

Arachnoid layer. Middle of the three meninges surrounding the brain and spinal cord in birds and mammals.

Arboreal. Living in trees.

Archeopteryx. Earliest known birdlike vertebrate; from the Jurassic.

Archeornithes. Subclass containing the oldest known fossil birds.

Archinephros (holonephros). Ancestral vertebrate kidney, existing today only in hagfish embryos; extends the length of the coelomic cavity; composed of segmentally arranged tubules, each opening into the coelom via a nephrostome.

Arciferous. Nonrigid type of pectoral girdle in anurans in which the two epicoracoids overlap; as opposed to firmisternal.

Argenine vasotocin. See *oxytocin*.

Arrector pili (pl. **arrectores pilorum**). Smooth muscle attached to a hair follicle; contraction pulls hair into a more vertical position; cause of "goose bumps" in humans.

Arrector plumari (pl. **arrectores plumarum**). Smooth muscle attached to every feather follicle; permits "fluffing" of feathers.

Artery. Blood vessel conducting blood away from the heart.

Articular. Ossified posterior tip of Meckel's cartilage; becomes the malleus in the middle ear of mammals.

Artiodactyla. Order of the ungulate or hooved mammals having an even number of toes (either two or four).

Atlas. First cervical vertebra; articulates with the skull.

Auricle. Earlike lobe of an atrium of the heart; also, the external ear (pinna).

Autostylic. Type of jaw suspension in which the jaws articulate directly with the cranium.

Autotomy. The breaking off of a part of the body as a defensive escape maneuver by the organism itself (e.g., the tail of glass lizards).

B

Baculum. Penis bone (os penis) present in some mammals.

Baleen. Horny plates of epidermal origin in the upper jaws of certain whales; serve to filter plankton from sea water.

Barb. Branch from the shaft of a feather, which with other barbs form the vane.

Barbule. Projection that fringes the barbs of a typical feather.

Barrier. Impediment restricting the distribution of one or more species; may be physical (land, water, elevation, topography), climatic (temperature, humidity, rainfall, sunlight), or biological (lack of food, presence of predators or effective competitors).

Bathypelagic zone. Region of the oceanic zone from 1,000 to 4,000 m; cold, quiet water; characterized by permanent darkness and great pressure.

Benthic. Pertaining to the bottoms of oceans, seas, and lakes.

Bergman's Rule. Ecological principle stating that populations of homeotherms living in cooler climates tend to have a larger body size and a smaller surface area–volume ratio than conspecific populations living in warmer climates.

Bicornuate. Describes a uterus having two horns or extensions; the lower two-thirds of uterus is fused while the upper third remains separate; found in many ungulates.

Biodiversity. All living organisms (microorganisms, fungi, plants, and animals) in the biosphere or in a specified area.

Biogeographic region. One of six worldwide areas proposed by Wallace in 1876, in an attempt to divide the land masses into a classification reflecting the affinities of the terrestrial flora and fauna; the six are called the Palearctic, Nearctic, Oriental, Neotropical, Ethiopian, and Australian regions.

Biological amplification. Process by which pesticides, toxic metals, and other substances become more concentrated in each successive trophic level of a food web.

Biome. Major regional ecological community of plants and animals.

Biotic potential. See *reproductive potential.*

Biotic province. Geographic area used in classifying North American plant and animal communities and their distribution; proposed by Dice (1943).

Bipartite. Describes a uterus in which paired uteri are separate for most of their length, but join to form a single cervix; found in most carnivores and some ruminants.

Boreal forest. Needle-leaved evergreen or coniferous forest bordering subpolar regions; also called *taiga.*

Bowman's capsule (= glomerular capsule). Spherical structure of each nephron of the kidney enclosing the glomerulus.

Brachiation. Arboreal form of locomotion with grasping hands and arm swings; body suspended below tree branches.

Brackish. Water that has a salt concentration greater than fresh water and less than sea water.

Branchial. Pertaining to gills.

Branchiomeric. Muscles modified by pharyngeal system anterior (superior) to pectoral girdle.

Bridge (= lateral bridge). In turtles, the narrow connection between the plastron and carapace on each side of the body.

Bristle. Modified, usually vaneless, feather consisting of only a shaft.

Buccopharyngeal mucosa. Highly vascularized epithelium in the pharynx of some amphibians; utilized in respiration.

Bulb. Enlarged, layered base of a hair follicle.

C

Caecum (pl. **caeca**). Blind sac arising from the digestive tract; in fishes, the pyloric caeca are slender fingerlike structures arising from the junction between the stomach and intestines; in tetrapods, there are usually one or two colic caeca at the junction of the small and large intestines.

Calamus. Hollow, cylindrical basal portion of a feather shaft.

Calcitonin. Hormone secreted by the thyroid gland; lowers calcium and phosphate levels in the blood by inhibiting bone breakdown and accelerating calcium absorption by the bones.

Canine. Member of the dog family (Canidae); also the long, stout, cone-shaped pointed tooth just behind the incisors in mammals.

Carapace. Dorsal shell of a turtle.

Carina. Longitudinal ridge or plate of bone on the ventral side of the sternum in birds; also called the *keel.*

Carinate. Term referring to all birds that possess a keeled sternum; all birds exclusive of the ratites.

Carnassial. Modified premolar or molar tooth in the jaw of carnivores; possesses a sharp, bladelike cutting edge; used for cracking bones and shearing tendons.

Carnivore. Animal that feeds on animal tissue; taxonomically, a member of the order Carnivora (Mammalia).

Carotenoid. Group of fat-soluble pigments (yellows, browns, reds, oranges).

Carpometacarpus. Bone formed by the fusion of carpal and metacarpal bones in a bird.

Carpus. The wrist.

Carrying capacity (K). Maximum number of individual organisms that the resources of a given area can support for an extended period of time; the most unfavorable period of the year is the most critical to the reproductive success and survival of a species.

Catadromous. Fish that live in fresh water but migrate down a river or stream to the ocean to spawn.

Caudal. Pertaining to the tail or rear.

Caudal fin. Most posterior unpaired fin of fishes and some amphibians.

Caudal peduncle. Slender portion of a fish behind the anal fin and bearing the caudal fin.

Centrum. (pl. **centra**) Thick, disk-shaped ventral (anterior) portion of a vertebra that is the weight-bearing part; also known as the body.

Ceratotrichia. Fin rays in cartiliaginous fishes.

Cerebellum. Portion of the brain lying posterior to the medulla oblongata and pons; concerned with muscular control and coordination of movements.

Cerebral hemisphere. One of a pair of dorsal portions of the forebrain.

Cerebrospinal fluid. Fluid secreted in the ventricles of brain; serves as a transport medium; circulates within the ventricles and spinal cord, also around the brain and spinal cord.

Cerebrum. Part of the forebrain; consists of two hemispheres; composed of areas that receive sensory impulses (sensory areas), areas that control muscular movement (motor areas), and areas that deal with complex integrative functions such as memory, emotions, reasoning, and intelligence (association areas).

Cervical. Pertaining to the neck.

Cervix. Lower portion of the mammalian uterus that projects into the vagina.

Chaparral. Dense, shrubby, fire-resistant scrubland in regions of world where most of the rain falls in the cool winter, and summers are hot and dry; shrubs have small, thick, evergreen leaves and thick underground stems; found in California and in parts of Africa, Australia, Chile, and Europe.

Chief cell. Cell in the stomach that secretes pepsinogen, a precursor of pepsin.

Choana (pl. **choanae**). Internal naris; the opening of the nasal passage into the pharynx.

Chondrocranium. Cartilaginous region of skull surrounding the brain and special sense organs in Chondrichthyes.

Chordata. Phylum of animals with a notochord, dorsal hollow nerve cord, and pharyngeal slits at some time during their development.

Chorion. Outer of the double membrane that surrounds the embryo of reptiles, birds, and mammals; contributes to the placenta in mammals.

Chromatophore. Pigment-containing cell.

Clade. Taxon or other group consisting of an ancestral species and all of its descendants, forming a distinct branch on a phylogenetic tree.

Cladogram. Branching diagram representing the hypothesized relationships of a group of taxa; developed through cladistic analysis.

Claspers. Paired intromittent organs in elasmobranchs and chimaeras; assist in the transfer of sperm into the female reproductive tract.

Claw. Sharp, curved, laterally compressed nail at the end of a digit; present as a talon in some birds.

Cleithrum. Bone of the pectoral girdle.

Climax. Stable end of succession; a community that is capable of self-perpetuation under prevailing environmental conditions.

Cline. Gradual change in a biological character along a geographic gradient.

Clitoris. Female homologue of the male penis in mammals.

Cloaca. Common chamber that receives the products of the digestive and urogenital ducts in monotremes, birds, reptiles, amphibians, and some fishes.

Cochlea (= spiral organ). Tubular auditory organ in the inner ear of crocodiles, birds, and mammals; spirally coiled in mammals.

Coevolution. Joint evolution of two or more species that have a close ecological relationship; the evolution of one species in the relationship is partially directed or controlled by the evolution of the other.

Colon. Portion of the large intestine; in mammals, divided into ascending, transverse, descending, and sigmoid portions.

Columella. Slender bone connecting the tympanum with the internal ear in amphibians, reptiles, and birds; homologous with the hyomandibular bone of fishes and the stapes of mammals.

Commensalism. Symbiotic relationship between species that is beneficial to one species but is neither beneficial nor harmful to the other.

Comparative zoogeography. Study of the distribution of related groups of animals according to their external features.

Competition. Interaction among individuals that are competing for the same space or resources.

Competitive exclusion principle. Ecological rule stating that when there is competition for a niche between two or more species, only one species will be successful; to avoid competition, character displacement may occur among one or both species, or one species will be excluded from the habitat; see *Gause's Rule.*

Concertina. Type of locomotion in snakes; consists of alternate curving and straightening of the body; adaptive to living in burrows and tunnels.

Concha (pl. **conchae**). Bone shaped like a scroll; found in the nasal cavity.

Cone. Photoreceptor cell in the retina specialized for sharpness of vision and color.

Contour feathers. Outermost body and flight feathers that form the contour or outline of a bird.

Convergent evolution. Independent development of similar characteristics in unrelated species due to similar selective pressures caused by living under similar environmental conditions.

Coprodeum. Most anterior region of former cloaca; receives the large intestine.

Cornea. Nonvascular, transparent fibrous coat over the anterior portion of the eye through which the iris can be seen; continuous with the sclera.

Corpora cavernosa. Paired columns of erectile tissue in the penis.

Corpus callosum. Broad transverse sheet of nerve fibers connecting the cerebral hemispheres.

Corpus spongiosum. Single column of erectile tissue in the penis.

Cortex. Outer layer of an organ (e.g., adrenal cortex); also, the convoluted layer of gray matter covering each cerebral hemisphere.

Corticosterone. See *glucocorticoids.*

Cortisol. See *glucocorticoids.*

Cortisone. See *glucocorticoids.*

Cosmoid scale. Small, thick fish scale composed of cosmine and covered by a thin layer of enamel; found today only on the coelacanth (*Latimeria*).

Costal fold. Area between two costal grooves.

Costal groove. Vertical grooves in the sides of salamanders.

Countercurrent exchange (= counterflow). Exchange of heat and/or oxygen between two fluids moving past each other in opposite directions.

Crista (pl. **cristae**). Patches of sensory cells in the ampulla of a semicircular canal; functions in dynamic equilibrium.

Crop. Membranous sac in the lower portion of the esophagus in some birds; used for the temporary storage of food.

Crossopterygian. Primitive lobe-finned bony fishes (order Sarcopterygii) ancestral to the amphibians; only living representative is the coelacanth (*Latimeria*).

Ctenoid scale. Thin, overlapping dermal scale of fish; posterior margin with fine, toothlike spines.

Cycloid scale. Thin, overlapping dermal scale of fish; posterior margin smooth.

Cutaneous. Pertaining to the skin.

Cuticle. Outermost layer of a hair.

Cytology. Study of cells and their internal structure and physiology.

D

Deciduous. Shed during life.

Delayed fertilization. Fertilization following an extended storage of sperm within the female's body.

Delayed implantation. Following fertilization, the mammalian embryo (blastocyst) ceases development and floats freely in the uterus for several months; occurs in some bats, armadillos, carnivores, pinnipeds, and artiodactyls.

Deme. Local population of closely related animals.

Dens (= odontoid process). Process on the anterior (superior) end of the second cervical vertebra (axis); articulates with the first cervical vertebra (atlas).

Density-dependent. Mortality that varies directly with population density.

Density-independent. Mortality that is unaffected by population density.

Dentary. One of a pair of dermal bones making up part of the lower jaw; in mammals, the only bones making up the lower jaw.

Derived. Referring to a trait that evolved later than an ancestral trait.

Dermatocranium. Collectively the superficial bones of the skull that develop in the dermis without cartilaginous precursors.

Dermis. Layer of skin beneath the epidermis.

Determinate growth. Steady increase in size until maturity, after which growth slows and essentially ceases for the remainder of life; characteristic of birds and mammals.

Dewlap. Extensible reddish throat fan in some male lizards.

Diadromous. Migrating either from fresh water to sea water to spawn (catadromous) or from sea water to fresh water to spawn (anadromous).

Diaphragm. Muscular partition between the abdominal and thoracic cavities in mammals.

Diapsid. Amniote in which the skull has two pairs of temporal fossae; extant reptiles (except turtles) and birds.

Diastema. Space separating the premolars from the incisor teeth in mammals that lack canines, such as rabbits and rodents.

Dichromatism. Having two or more color phases.

Diencephalon. Part of the brain consisting primarily of the thalamus and the hypothalamus; posterior region of the prosencephalon.

Digit. A finger or toe.

Digitigrade. Condition in which the animal walks on the ends of its metacarpals and metatarsals; only the toes contact the ground in walking.

Dimorphism. A species having two different structural forms or two color phases in a population.

Dioecious. Having male and female sex organs in separate individuals.

Diphycercal. Tail that tapers to a point as in lungfishes; vertebral column extends to its tip without upturning.

Diphyletic. A group whose members are derived from more than one ancestor; not of monophyletic origin.

Direct development. In some anurans, the elimination of a free-living feeding tadpole stage; all development occurs inside the egg; hatchlings are fully formed, four-legged froglets.

Disjunct. Species consisting of two or more isolated populations.

Dispersal. Generalized movement of individuals within a population away from their original home range; nondirected movement in general.

Dormancy. State of inactivity; torpidity; see *hibernation* and *estivation*.

Dorsal fin. Median unpaired fin on the back; may be supported by spines and/or rays.

Down feather. Small, fluffy feather lying beneath and between the contour feathers; principal function is insulation.

Duplex. Describes a uterus in which the uteri are completely separate, but joined to a single vagina; found in rabbits and rodents.

E

Ecdysis. Shedding of the skin, as in a snake.

Echolocation. Radarlike system used by some birds and some mammals (especially bats and cetaceans) for maneuvering and locating food.

Ecological zoogeography. Study of the analogies between animal communities occupying similar habitats.

Ecology. Science of the relationships between organisms and their environments.

Ecotone. Transition zone between two adjacent ecosystems.

Ecotourism. Travel that aims to increase the understanding of ecological (or natural) systems; ideally results in creation of jobs, increased economies, and preservation of habitat.

Embryo. Developing organism, especially in the early stages; generally still contained within the egg or uterus.

Emigration. Movement of an individual or part of a population permanently out of an area; a one-way movement.

Endangered species. Species in imminent danger of extinction throughout all or a significant portion of its range.

Endotherm. Vertebrate that maintains a relatively high body temperature primarily by internal heat production.

Environmental resistance. Environmental factors that limit the number of young produced by a population; includes climatic factors such as rainfall, flooding, drought, and temperature; intraspecific aggression; available den sites; predation; disease; and parasites; divided into two categories—density-dependent and density-independent.

Epaxial. Muscle mass dorsal to the horizontal skeletogenous septum in fishes and amphibians.

Epibranchial. Upper gill cartilage located between the pharyngobranchial and ceratobranchial cartilages; also, muscles above the gills in fishes.

Epidermis. Outer epithelial portion of the skin.

Epididymis. Part of the sperm duct that is coiled and lying adjacent to the testis; serves as a storage area for sperm.

Epiglottis. Large leaf-shaped cartilage lying on top of the larynx in mammals; covers the glottis during swallowing.

Epinephrine. Hormone secreted by the adrenal medulla that prepares the body for stressful or emergency situations; also called adrenaline.

Epipelagic zone. Region of the oceanic zone that receives abundant sunlight; phytoplankton and zooplankton are abundant.

Epipubic bone. One of a pair of small bones in marsupials that articulate with the pubic bone and extend forward in the abdominal wall to provide additional support for the abdominal pouch.

Equilibrium. State of balance.

Estivation. State of dormancy due to external stress resulting from long-term periods of heat.

Estuary. Partially enclosed embayment where fresh water and sea water meet and mix.

Erythrophore. Cell containing red pigment.

Ethology. Scientific study of animal behavior.

Eustachian tube (auditory tube). Passageway from the middle ear to the pharynx; serves to equalize pressure on both sides of the tympanum.

Extinction. Total disappearance of a species or a higher taxon from the face of the Earth.

Extirpation. Total disappearance of a taxon from a geographic area but not from its entire range.

Extrinsic. In anatomy, originating on the appendages (e.g., extrinsic musculature).

F

Facultative parthenogenesis. Development of a new individual from an unfertilized egg (only when this mode of reproduction is necessary as a last resort to produce offspring), as an adaptation to changing conditions.

Faunal zoogeography. Faunal lists of animal populations for specific areas.

Feather. Epidermal derivative; a modified reptilian scale; the most distinctive characteristic of a bird.

Femoral pores. Integumental glands that appear as openings in the scales on the undersurface of the thigh of most lizards.

Femur. Proximal bone of the pelvic appendage; articulates with the pelvic girdle.

Fenestra (pl. **fenestrae**). An opening within the bony braincase.

Fibula. Lateral and smaller of the two distal bones of the lower hind leg.

Filoplume. Very specialized, hairlike or bristlelike feather.

Fin rays. Slender, flexible rods that stiffen the fins of fishes distal to the skeletal components; may be composed of cartilage, keratin, elastoidin fibers, or bone.

Firmisternal. Type of pectoral girdle in amphibians that lacks epicoracoid horns; sternum fused to pectoral arch; found in ranids, microhylids, and dendrobatids.

Fluke. The dorsoventrally compressed tail of a cetacean.

Follicle. Small secretory sac or cavity; also invaginated portion of epidermis that gives rise to feathers and hairs.

Follicle-stimulating hormone (FSH). Hormone secreted by the anterior pituitary gland that initiates development of ova and stimulates the ovaries to secrete estrogens; also initiates sperm production in males.

Food chain. Sequence of organisms through which energy and nutrients move from one trophic (feeding) level of organisms to another in a series that normally begins with plants and ends with carnivores, detritus feeders, and decomposers.

Food web. Interlocking pattern formed by a series of interconnecting food chains.

Fossorial. Adapted for digging or burrowing through the soil.

Founder effect (= event; principle). Principle that populations on oceanic islands and other isolated places may be established by a very small sample from a continent or another island. Such a sample may include only limited variability. Due to its small size, the genetic structure of the new population may be dramatically different from that of its ancestral population. Phenotypic characteristics that were stable in the ancestral population often reveal wide variation in the new population. As natural selection acts on the newly expressed variation, large changes in phenotype and reproductive properties occur, hastening the evolution of reproductive barriers between the ancestral and newly founded populations.

Fragmentation. Divided habitat or one that is decreased in size; may no longer be large enough to meet the needs of all the species that formerly occupied it.

Furculum. Fused clavicles, or wishbone, of a bird.

G

Ganoid scale. Type of fish scale covered with an enamel-like substance known as ganoin; seldom overlapping; occurs in gars (Lepisosteidae).

Gartner's duct. In birds, a short, blind vessel embedded in the mesentery of the oviducts; vestige of mesonephric duct.

Gastric juice. Digestive juice consisting of the combined secretions of chief cells, mucous cells, and parietal cells of the stomach.

Gastrosteges. Ventral scales anterior to the anal plate; found in snakes.

Gause's Rule. Ecological principle stating that when there is competition for a niche between two or more species, only one species will be successful, and the others will be excluded from the habitat; see *competitive exclusion principle*.

Genus (pl. **genera**). Taxon in which all species sharing certain characteristics are grouped; taxonomic level above species and below family and subfamily.

Geographic race. Same as *subspecies*.

Geographic range. Specific land or water area where a species normally occurs.

Geologic range. Past and present distribution of a taxon over time.

Gestation. Period during which an embryo is developing in the reproductive tract of the mother.

Gill. Specialized structure covered by a thin, vascular epithelium that functions in the aquatic exchange of respiratory gases; may be internal or external; mainly confined to the pharyngeal region.

Gill raker. Projection from the anterior surface of a gill arch; aids in filtering food from water.

Gill slit. Paired opening from the pharynx through the body wall to the exterior.

Girdle. Skeletal elements joining limbs to the body; pectoral girdle is associated with the forelimbs, pelvic girdle with the hindlimbs.

Gizzard. Muscular portion of the stomach in birds; serves as a grinding chamber in crocodilians, some dinosaurs, and some birds.

Glacial lake. Steep-sided lake gouged out of previously existing valley by advancing glacier; e.g., Finger Lakes in central New York.

Glans penis. Slightly enlarged region at the distal end of the penis.

Glenoid fossa. Depression serving as a point of articulation for the pectoral appendage with the scapula.

Glomerulus (pl. **glomeruli**). Tuft of capillaries at the beginning of each nephron; enclosed by glomerular (Bowman's) capsule.

Glottis. Anterior (superior) opening between the pharynx and the trachea.

Glucagon. Hormone produced by the pancreas; increases blood sugar level.

Glucocorticoids. Hormones secreted by the adrenal cortex, especially corticosterone, cortisol, and cortisone, that influence glucose metabolism.

Gonad. Gland that produces gametes and hormones; the ovary in the female and the testis in the male.

Gonopodium. Intromittent organ found in some teleost fishes.

Greenhouse gases. Certain gases (including carbon dioxide, methane, nitrous oxide, halons, and chlorofluorocarbons) produced on Earth by burning of fossil fuels and other processes; allow rays of sun to pass through but absorb and reradiate heat back to Earth, causing the Earth to warm.

Growth hormone (GH). Hormone produced by the anterior pituitary gland that causes body cells to grow; also stimulates protein synthesis and inhibits protein breakdown; promotes tissue repair; stimulates the breakdown of triglycerides into fatty acids and glycerol and the elevation of blood glucose concentration.

Guard hair. Outer, coarser, and usually longer hairs making up the pelage of a mammal.

Gubernaculum. Ligament in male mammals connecting the caudal pole of each embryonic testis to the labioscrotal pouch; assists in descent of testes into scrotum in some mammals.

Gustatory. Related to the sense of taste.

H

Habenula. Structure in diencephalon; assists in coordinating the olfactory reflexes.

Habitat. Place where an animal normally lives or where individuals of a population live.

Hadopelagic zone. Region of the oceanic zone below 6,000 m; areas of ocean trenches.

Hallux. First digit of the posterior limb; usually directed backward in birds.

Hamulate. Having a small hook (hamulus).

Hamulus (pl. **hamuli**). Hooked barbule of a feather.

Hemal arch. Arch formed by paired transverse projections ventral to the centra of the caudal vertebrae and enclosing the caudal blood vessels.

Hemal spine. Ventral projection from the ventral bony arch (hemal arch) of a caudal vertebra.

Hemipenis. Male copulatory organ in lizards and snakes.

Hepatic. Relating to the liver.

Herbivore. Organism that feeds on plant tissue.

Hermaphrodite. Organism with both male and female functional reproductive organs.

Heterocercal. Type of tail characteristic of sharks in which the dorsal lobe is larger than the ventral lobe, and has caudal vertebrae extending into it.

Heterocoelous. Describes a vertebra having the anterior and posterior articulating surfaces of the centrum (body) shaped like a saddle.

Heterodont. Having teeth differentiated for various functions; tooth types include incisors, canines, premolars, and molars.

Hibernation. Winter dormancy; condition of passing the winter in a torpid state during which the body temperature drops near freezing (or below) and metabolic activities are drastically reduced.

Historical zoogeography. Study of historical animal distribution; establishment of present-day distributions over geologic time.

Holonephros. See *archinephros.*

Home. Place of refuge within a home range; nest site.

Homeostasis. Condition of physiological equilibrium with regard to temperature, fluid content, pressure, etc.

Homeotherm. Endothermic animal with a fairly constant body temperature.

Home range. Area over which an animal moves during its normal daily activities.

Homing. Ability to find the way home; returning home.

Homocercal. Type of tail characteristic of most modern bony fishes in which the upper and lower lobes are about the same size.

Homodont. Having teeth similar in form to one another; teeth may differ in size.

Homology. Organs or structural features in different species of animals with common embryonic and evolutionary origins, but perhaps with different functions (e.g., a bat's wing and a human arm).

Hoof. Horny sheath encasing the ends of the digit or foot in ungulate mammals.

Horizontal undulatory. Type of locomotion used by snakes; body glides along in a series of waves, with each part of the body passing along the same track; serpentine.

Hormone. Chemical substance formed in one organ or body part and carried in the blood to another organ or body part, which it stimulates to functional activity.

Host. Animal that harbors another as a parasite.

Humerus. Proximal bone of the pectoral appendage; articulates with the glenoid fossa of scapula.

Hyoid. Second visceral arch; serves to support the tongue.

Hyomandibula. Uppermost segment of the hyoid arch.

Hyostylic. Type of jaw suspension wherein the hyomandibula is inserted between the jaws and cranium.

Hypaxial. Muscle mass ventral to the horizontal skeletogenous septum in fishes and amphibians.

Hypobranchial. Describes muscles below the gill region.

Hypothalamus. Portion of the diencephalon lying beneath the thalamus and forming the floor and part of the wall of the third ventricle of the brain.

I

Imprinting. Period of rapid and usually stable learning during a critical period of early development of a member of a social species, involving recognition of its own species; may involve attraction to the first moving object seen.

Incisor. Chisel-like cutting tooth at front of jaw.

Incus. Middle ear bone of mammals, derived evolutionarily from the quadrate.

Indeterminate growth. Pattern of growth that continues indefinitely in an animal or a structure; common in many fishes, amphibians, and reptiles.

Index. Count of some object (tracks, leaf nests, etc.) that is related the numerical population size of the animal being studied; can be used to estimate the population.

Infrasound. Sound below the range of human hearing; below 20 Hz.

Inguinal canal. Passageway in the abdominal wall for the spermatic cord in male mammals and for the round ligament in females.

Innominate. Bone forming one-half of the mammalian pelvic girdle.

Insectivorous. Insect-eating.

Insulin. Hormone secreted by the pancreas that lowers blood glucose level.

Integument. External covering or enveloping layer of the body; the skin.

Intercalary. Cartilage between the ultimate and penultimate phalanges in hylid anurans.

Intercostal. Between the ribs (e.g., intercostal muscles).

Interfemoral membrane. Flight membrane of bats located between the hindlimbs and often involving the tail.

Interstitial cell-stimulating hormone. See *luteinizing hormone.*

Intrinsic. In anatomy, originating on the body (e.g., intrinsic musculature).

Iridophore (guanophore). Cell containing crystals that reflect and disperse light.

Isolating mechanism. Structural, behavioral, or physiological mechanism that blocks or inhibits gene exchange between two populations.

J

Jugular. Pertaining to the throat; in fishes, situated in front of the pectoral fins.

Juvenal. Plumage or pelage acquired following the postnatal molt.

Juvenile. Immature stage of development.

K

Keel. See *carina.*

Keratin. Relatively insoluble protein found in hair, feathers, nails, and other keratinized tissues of the epidermis; replaces cytoplasm in epidermal cells as they become cornified.

Keratinocyte. Most numerous of the specialized epidermal cells; produces keratin.

L

Labyrinthodont. A primitive amphibian; one of the first land vertebrates; with complex, "folde" teeth.

Lagena. Auditory receptor of fishes, amphibians, and reptiles; homologous with cochlea of birds and mammals.

Lamella (pl.**lamellae**). Thin, layered structure.

Larva (pl. **larvae**). Immature stage of life between the embryo and the adult; often sharply different in form from the adult.

Larynx. Modified upper portion of respiratory tract of air-breathing vertebrates; bounded by the glottis above and the trachea below; voice box.

Lateral bridge. See *bridge.*

Lateral line. Part of the sensory system of fishes and amphibians; series of tubes in the lateral scales (variously developed or absent in fishes; present in larval amphibians) that sense water movements.

Lentic system. Freshwater system consisting of flowing water.

Lepidotrichia. Dermal fin rays of bony fishes.

Life zone. Region of the ecological classification scheme proposed by C. Hart Merriam in 1890; zone boundaries are determined by mean annual temperatures.

Limnetic zone. Upper portion of the deeper open water of a lake where sunlight is sufficient to support photosynthesis.

Lipophore. Pigment-bearing cell containing carotenoids responsible for yellow, orange, and red colors; pigment granules are soluble in lipids.

Littoral. That portion of the sea floor between high and low tides—intertidal; in lakes, the shallow part from the shore to the lakeward limit of rooted aquatic plants.

Longitudinal fissure. Deep median groove separating the cerebrum into two cerebral hemispheres.

Lotic system. Freshwater system consisting of still water.

Luteinizing hormone (LH). Hormone secreted by the anterior pituitary gland that stimulates ovulation and progesterone secretion by the corpus luteum; stimulates production of "pigeon milk" in some birds; prepares mammary glands for milk secretion in female mammals; stimulates testosterone secretion by the testes in males; also called *interstitial cell-stimulating hormone* in males.

Lymph. Fluid circulating in the lymphatic system.

M

Macrogeographic migration. Long-distance migration (e.g., ducks, geese, many passerines, humpback whales).

Macula (pl. **maculae**). Patches of sensory cells in the utricle and saccule; functions in the maintenance of static equilibrium.

Malleus. One of the three middle ear bones in mammals; evolutionarily derived from the articular.

Mammillary body. Small rounded body in the diencephalon posterior to the hypophysis; involved in reflexes related to the sense of smell.

Mandible. Lower jaw.

Mandibular arch. First pair of visceral arches in Chondrichthyes.

Manus. Hand.

Marginal. One of the lateral rows of scales in the carapace of turtles.

Marsh. Shallow wetland dominated by grassy vegetation such as cattails and sedges.

Marsupium. Pouch of female marsupials.

Masseter. Mammalian jaw muscle that moves the lower jaw forward and upward.

Mass extinction. Catastrophic, worldwide event in which a large number of taxa disappear within an interval of just a few million years.

Maxilla. Bone of the upper jaw lying above or behind and parallel to the premaxilla; the upper bill of birds.

Meckel's cartilage. Ventral portion of the mandibular arch in Chondrichthyes; becomes encased by dermal bones in higher vertebrates.

Medulla. Inner layer of either an organ (e.g., adrenal medulla) or a structure such as a hair.

Melanin. Brown-black pigment produced by skin melanocytes.

Melanocyte-stimulating hormone (MSH). Hormone secreted by the anterior pituitary gland that stimulates the dispersion of melanin granules in melanocytes in amphibians, increasing skin pigmentation.

Melanophore. Chromatophore containing the pigment melanin.

Melatonin. Hormone produced by the pineal gland; thought to control adrenocortical function and cyclic activities such as sleeping and eating.

Meninges. Protective membranes enclosing the central nervous system; in mammals, these are the dura mater, arachnoid, and pia mater.

Mental. Pertaining to the chin.

Meristic. Divided into segments or serial parts; see *metamerism*.

Mesaxonic. Type of foot in which the axis passes through the middle digit, which is larger than the others; characteristic of perissodactyls (horses, zebras, asses, tapirs, and rhinoceroses).

Mesencephalon. Middle of the three embryonic divisions of the brain.

Mesonephros. Kidney formed of nephric tubules arising in the middle of the nephric ridge; usually a transient embryonic stage that replaces the pronephros, and is itself replaced by the metanephros.

Mesopelagic zone. Region of the oceanic zone between 200 and 1,000 m; semidark; also known as the twilight zone.

Metacarpal. Referring to that region of the hand or forelimb lying between the digits and the wrist; also, a bone of this region.

Metamerism. Serial segmentation; made up of serially repeated parts.

Metamorphosis. Transformation of an immature animal into an adult (e.g., tadpole to frog); change in the body form and way of life that ends the larval stage; also called transformation.

Metanephros. Embryonic renal organs arising posterior to the mesonephros; the functional kidney of reptiles, birds, and mammals; drained by a ureter.

Metatarsal. Referring to that region of the foot or hindlimb lying between the digits and the ankle; also, a bone of this region.

Metatarsal tubercle. Small protuberance on the sole of the hind foot in amphibians; typically there are two—a large inner tubercle and a smaller outer tubercle; sometimes modified for digging.

Microgeographic migration. Short-distance migration (e.g., salamanders, anurans, and others).

Migration. Intentional, directional, usually seasonal movement of animals between two regions or habitats; a round-trip directional movement.

Molar. Grinding tooth in the posterior region of the jaw.

Monophyodont. Having a single set of teeth without replacement during the animal's lifetime.

Monotypic. Describes species without subspecies.

Morphology. Study of the form and structure of living organisms.

Mucus. Clear, viscid secretion of certain types of cells.

Mutualism. Symbiotic relationship between two species in which both benefit from the association.

Myelin. Fatty sheath insulating a nerve fiber.

Myomere. Muscle segment of the successive segmental trunk musculature.

Myosepta (pl. **myoseptae**). Membrane separating adjacent myomeres.

N

Naris (pl. **nares**). Opening of the nasal cavity; may be external or internal.

Nasolabial groove. Groove from the nostril to the upper lip in plethodontid salamanders.

Natural extinction. Normal disappearance of species over time.

Nearctic. North America, Greenland, and Iceland.

Nekton. Aquatic organisms that move actively through the water, rather than drifting or floating passively like plankton.

Neornithes. Subclass of extinct and living birds with well-developed sternae and reduced tails, and with metacarpals and some carpals fused together.

Neoteny. Retention of one or more juvenile characteristics into adulthood.

Neotropical. Pertaining to Central and South America.

Nephron. Basic functional unit of the vertebrate kidney.

Nephrostome. Ciliated, funnel-shaped opening of primitive glomeruli.

Neritic. Portion of the sea overlying the continental shelf, specifically from the subtidal zone to a depth of 200 m.

Neural spine. Dorsal projection from the dorsal bony arch of a vertebra.

Neurocranium. Part of the skull enclosing the brain.

Neuromast. Mechanoreceptor cells on or near the surface of a fish or amphibian that are sensitive to vibrations in the water.

Niche. Functional role and position of an organism in the ecosystem; how it lives and its relation to other species in a food chain or web.

Nictitating membrane. More or less transparent eyelid located medial to the eye; cleanses and lubricates cornea; "third" eyelid.

Norepinephrine (noradrenalin). Hormone secreted by the adrenal medulla that prepares the body for stressful or emergency situations.

Notochord. Longitudinal dorsal rod of tissue that gives support to the bodies of lampreys, some adult fishes, and the embryos of chordates.

Nuchal. Pertaining to the back of the neck; in turtles, the median anterior shield of the carapace; in lizards, the enlarged scales immediately posterior to the head.

Nuptial plumage. Breeding plumage.

Nuptial tubercle. Hardened process on the skin of the head of a fish, usually a breeding male; used to assist the male in maintaining contact with the female during breeding and to stimulate the female.

O

Oblique septum. Tendinous, transverse partition separating the pleural cavities from the coelom in crocodilians, and some snakes and lizards.

Oceanic zone. Portion of the sea beyond the continental shelf.

Odontoid process. See *dens*.

Olfactory. Relating to the sense of smell.

Omasum. Third chamber of the ruminant stomach.

Omnivore. Animal that feeds on both plant and animal matter.

Oogenesis. Egg cell production in females.

Operculum. Covering flap, as the gill cover of fishes and larval amphibians.

Opisthocoelous. Type of vertebra with a convex anterior (superior) end of the centra and a concave posterior (inferior) end.

Opisthoglyph. Rear-fanged snake.

Opisthonephros. Adult kidney formed from the mesonephros and additional tubules from the posterior region of the nephric ridge; found in most adult fishes and amphibians.

Organ of Corti. See *spiral organ*.

Origin. In fish, the most anterior end of the dorsal fin or anal fin base; in muscle terminology, the site of attachment that usually remains fixed when the muscle contracts.

Otolith. Calcified body in the sacculus of the inner ear.

Outgroup. In cladistics, a species or group of species closely related to, but not included within, a taxon whose phylogeny is being studied; used as a reference for determining whether characters in other groups are ancestral or derived.

Ovary. Female reproductive glands producing the ova and associated hormones.

Oviducal glands (= shell glands). Glands that secrete albumin and shell materials around an egg.

Oviduct. Tube serving to transport the eggs from the ovary to the cloaca or uterus and/or house the eggs and embryos.

Oviparous. Egg-laying; fertilized eggs develop outside the mother.

Ovipositor. Papilla-like terminal portion of oviduct in some teleost fishes.

Ovisac. Enlarged caudal portion of the oviduct in some amphibians; provides for the temporary storage of eggs prior to oviposition.

Ovoviviparous. Eggs retained and develop within the body of the female, but without placental attachment; also known as *aplacental viviparity*.

Oxytocin. Hormone secreted by the posterior pituitary gland; stimulates smooth muscle contraction.

P

Paedogenesis. See *paedomorphosis*.

Paedomorphosis. Condition whereby a larva becomes sexually mature without transforming into the adult body form.

Palatoquadrate (pterygoquadrate). Cartilage forming the upper jaw of primitive fishes and Chondrichthyes; portions ossify and contribute to the palate, jaw articulation, and middle ear of other vertebrates.

Palearctic. Asia and Europe north of the Sahara Desert.

Pancreas. Abdominal digestive (secretes pancreatic juice) and endocrine (secretes insulin and glucagon) gland.

Pancreatic islets. Cells in pancreas that produce the hormones insulin and glucagon.

Pangaea. Supercontinent comprising entire land mass of Earth approximately 250 million years ago.

Panniculus carnosus. Integumentary muscle derived from the hypaxial musculature.

Papilla (pl. **papillae**). Small, pimplelike protuberance.

Parasitism. Symbiotic relationship between two species in which one (the parasite) benefits while the other (the host) is harmed.

Parathyroid. Endocrine gland adjacent to, or embedded in, the thyroid; secretes parathyroid hormone (parathormone) to raise the calcium level in blood.

Paraxonic. Type of foot in which the axis passes between the third and fourth digits, which are almost equally developed.

Parotoid gland. Glandular swelling behind the eye of some anurans and salamanders; exudes a poisonous secretion.

Parthenogenesis. Development of ovum without fertilization.

Patagium. Thin, often furry flight membrane found in gliding and flying mammals.

Patella. Kneecap.

Pectoral. Pertaining to the chest; in turtles, one of a pair of shields of the plastron.

Pectoral fin. Usually the most anterior of the paired fins in fishes.

Pectoral girdle. Bones that attach the forelimb to the axial skeleton.

Pelagic. Pertaining to the open ocean waters; oceanic.

Pelvic fin. One of a pair of fins in a ventral position well posterior to the pectoral fins (abdominal position) or ventral to the pectoral fins (thoracic position).

Pelvis. Ring formed by the pelvic girdle around the caudal ends of the digestive and urogenital systems.

Penis. Male organ used for intromission and sperm transfer in turtles, crocodilians, some birds, and mammals.

Pentadactyl. Five-toed.

Penultimate. Next to the last.

Pericardium. Area around the heart; membrane around heart.

Peripheral. Term used for endangered species or subspecies at the edge of its range (threatened with extinction at the edge of its range, although not in its range as a whole).

Peritoneum. Epithelial lining of the body cavity.

Permafrost. Permanently frozen soil.

Perissodactyla. Order of odd-toed ungulates; includes horses, zebras, asses, tapirs, and rhinoceroses.

Phalanx (pl. **phalanges**). One of the bones of a digit.

Pharyngeal. Pertaining to the pharynx; a toothed bone of the throat region of fishes.

Phenogram. Diagram used by pheneticists for grouping species on the basis of overall similarity; dendrogram.

Pheromone. Chemical substance released by an animal that controls or alters the behavior of others of the same species.

Photic. Relating to light; underwater region penetrated by sunlight.

Photophore. Light-emitting organ in some fishes.

Physoclistic. In bony fishes, not having a connection between the pharynx and the swim bladder.

Physiology. Science that deals with the functions of an organism and its organs, tissues, and cells.

Physostomous. In bony fishes, having a connection between the pharynx and the swim bladder.

Phytoplankton. Microscopic photosynthetic life in aquatic ecosystems; may be on or below the surface of the water.

Pia mater. Most interior of the meninges surrounding the brain and spinal cord.

Pineal eye. Median, light-sensitive structure (retina, lens, and cornea may be developed) found in agnathans, primitive fishes, and amphibians; an analogous organ, the parapineal or parietal eye, is found in *Sphenodon* and many lizards; in lampreys, both parapineal and pineal organs form eyelike structures.

Pineal gland. Endocrine gland that produces the hormone melatonin.

Pinna. External ear exclusive of the ear canal.

Pit organ (= loreal pit). Specialized heat receptor between the eye and the nostril in crotalid snakes (Viperidae), and boas and their relatives.

Pituitary (hypophysis). Endocrine gland at the base of the brain; composed of two lobes: anterior (adenohypophysis) produces and secretes growth hormone, thyroid-stimulating hormone, adrenocorticotropic hormone, follicle-stimulating hormone, luteinizing hormone, prolactin, and melanocyte-stimulating hormone; posterior lobe (neurohypophysis) secretes antidiuretic hormone and oxytocin.

Placenta. Structure composed of tissues through which an embryo receives nourishment and respiratory gases from the mother and has its wastes removed; typical of viviparous vertebrates.

Placoid scale. Type of scale found in cartilaginous fishes; consists of basal plate of dentin embedded in the skin and a posteriorly pointing spine tipped with enamel.

Plankton. Microscopic or weakly swimming plants and animals in aquatic ecosystems.

Plantigrade. Type of locomotion in which the entire sole of the foot contacts the ground.

Plastron. Ventral shell of a turtle.

Plate tectonics. Arrangement and movements of rigid, slablike plates making up the Earth's crust.

Pleural. Pertaining to the lung.

Pleurodont. Teeth attached to the side of the jaw.

Pleuroperitoneal cavity. Body cavity of lower vertebrates containing the lungs and viscera.

Plicae vocales. Vocal organ of Pacific giant salamander (*Dicamptodon*).

Poikilotherm. Organism that does not maintain a constant body temperature; an organism whose body temperature fluctuates with ambient thermal conditions.

Pollex. Thumb or inner digit of the hand.

Polyandry. Mating of one female with several males within a breeding season.

Polygyny. Mating of one male with several females within a breeding season.

Polyphyletic. Taxon having a number of evolutionary origins; members of a taxon that do not share a common ancestor.

Polyphyodont. Ability to continually replace teeth throughout the animal's lifetime; examples are found in fishes, amphibians, and reptiles.

Polytypic. Describes a species with two or more subspecies.

Postjuvenal molt. Partial molt in birds in which the juvenal plumage, except for the flight feathers, is replaced by the first winter plumage.

Postnatal molt. First molt in a bird or mammal; replaces the natal down or fur with the juvenal plumage or pelage.

Powder down. Modified down feathers that grow continuously, disintegrating at the tips; keratin is given off as a fine powder of minute scalelike particles; used in preening plumage; may protect feathers from moisture; may affect the color of the bird.

Prairie. A level or rolling tract of treeless land covered with coarse grass and rich soil.

Precocial. Pertaining to birds and mammals born with their eyes and ears open, covered by down or fur, and able to run about shortly after hatching or birth.

Predation. Act of one living organism killing and consuming another living organism.

Predator. Animal that kills and eats other animals.

Prehallux. Small bone on the medial side of the hind foot of some anurans; may have a sharp-edged tubercle for digging.

Prehensile. Capable of grasping, as with the prehensile tail of an opossum or a New World monkey.

Premaxilla (pl. premaxillae). Most anterior bone of the upper jaw; paired dermal bones.

Premolar. Grinding tooth anterior to the molars; may be modified into a cutting tooth (carnassial) in carnivores.

Prepuce. Loose-fitting skin covering the glans of the penis and the clitoris of mammals.

Prey. Animal consumed by another for food.

Primary feather. One of the flight feathers attached to the hand (manus).

Procoelous. Pertaining to the centrum (body) of a vertebra with a concave anterior (articular) surface and a convex posterior surface.

Proctodeum. Terminal portion of the rectum formed in the embryo by an ectodermal invagination.

Profundal. Deep zone in aquatic ecosystems below the limnetic zone; in deep lakes, the region below the depth of light penetration.

Prolactin. Hormone from the anterior pituitary gland; regulates a wide range of parental behavior patterns including nest building, the incubation of eggs, and the protection of young; promotes the secretion of "pigeon milk" in certain birds and milk in mammals.

Pronephros. Most anterior portion of holonephros; functional only in adult hagfishes and larval fishes and amphibians; vestigial in amniote embryos.

Prosencephalon. Most anterior of the three embryonic divisions of the brain.

Proteroglyph. Venomous snake with rigid fangs; includes coral and sea snakes (Elapidae).

Proventriculus. Glandular portion of the stomach of a bird; between the esophagus and the muscular portion (gizzard) of the stomach.

Pseudobranch. Vestigial gill in the spiracle of elasmobranchs.

Pterygiophore. Segment of cartilaginous skeleton supporting paired fins in cartilaginous fishes.

Pterygoquadrate (= palatoquadrate). Dorsal portion of the mandibular arch in Chondrichthyes; cartilaginous.

Pteryla (pl. **pterylae**). Area of skin in birds from which a group of feathers grows; a feather tract.

Puboischiac plate. Ventral portion of the pelvic girdle of salamanders.

Purine. Crystalline substance in chromatophores that reflects light.

Pygostyle. Fused terminal caudal vertebrae of birds; supports the tail feathers.

Pylorus. Posterior portion of the stomach that leads into the small intestine.

Q

Quadrate. Skull bone that articulates with the lower jaw in bony fishes, amphibians, reptiles, and birds; in mammals, it has become an ear ossicle, the incus.

R

Rachis. Vane-bearing shaft of a feather.

Radius. Forearm bone on the lateral (thumb) side of the arm.

Raptor. Bird of prey; includes hawks, owls, eagles, and condors.

Rare. Of infrequent occurrence; formerly a category of classification along with endangered and status undetermined; later combined with endangered.

Ratite. Flightless bird with a flat (unkeeled) sternum; includes the ostrich, rhea, and kiwi.

Rectilinear. Type of locomotion used by snakes; axis of body is essentially straight and movement is effected by alternate movements of the ventral scutes and the body itself; snake moves along a straight path without any lateral motion.

Rectrix (pl. **rectrices**). Tail feather.

Renal. Pertaining to the kidney.

Reproductive potential. Maximum number of individuals a population can produce; also called *biotic potential.*

Rete mirabile. Mass of intertwined capillaries specialized for heat and/or gas exchange between blood flowing in opposite directions.

Reticulum. Second in the series of four chambers of the ruminant stomach.

Retina. Light-sensitive layer at the back of the eye.

Rhombencephalon. Posterior of the three embryonic brain divisions.

Rift lake. Long, narrow lake formed by a fissure in the Earth's crust or by the sinking of a narrow strip of land.

Rod. Photoreceptor cell in the retina specialized for light reception in dim light.

Root. The basal portion of a hair.

Rostrum. Preorbital part of the skull.

Rugae. Internal folds of the stomach.

Rumen. First and largest chamber of the four-chambered ruminant stomach; food here is subjected to bacterial action.

S

Sacculus. Smaller of the two sacs (sacculus, utriculus) in the inner ear; contains maculae; functions in static equilibrium.

Sacral. Pertaining to vertebrae modified for articulation with the pelvic girdle.

Sacrum. Structure formed by the fusion of the sacral vertebrae; articulates with the pelvic girdle.

Saltatorial. Adapted for jumping.

Sampling estimate. Estimate derived from counts made on sample plots; has variability but permits a statistical measurement of the total population.

Savanna. Large area of tropical or subtropical grassland, covered in part with trees and spiny shrubs.

Scapula. Bone of the pectoral girdle.

Scent gland. Modified sudoriferous or sebaceous gland; used to mark an individual's territory, to attract members of the opposite sex, or to serve in defense.

Sclera. Outer hardened layer of the eyeball.

Sclerotic ring. Series of 10 to 18 overlapping platelike bones found in the lateral (or anterior) part of the sclera of the eyeball in birds; ringlike in most birds, but conelike in a few (hawks, owls).

Scutes. Scales, especially the broad, belly scales of snakes; also, scales on turtles.

Sebaceous gland. Epidermal exocrine gland located in the dermis of mammals; almost always associated with a hair follicle; secretes sebum; also called an oil gland.

Sebum. Secretion of sebaceous (oil) gland.

Secondary feather. Flight feather attached to the ulna.

Sectorial. Modified teeth (canine and premolar) in some primates; cutting edges present on upper canine and lower premolar in each half of jaw.

Semidormant hibernator. Species that enters a sleeplike state during cold weather but does not experience the drastic physiological changes that occur in a true hibernator.

Semiplume. Loose-webbed contour feather.

Septum. Wall between two cavities; also, a sheet of tissue dividing groups of muscles (e.g., horizontal skeletogenous septum).

Shaft. Long, tapering central portion of a feather that consists of a hollow basal portion (calamus) and a solid, angular portion (rachis); the superficial portion of a hair that projects from the surface of the skin; also, the diaphysis portion of a bone.

Sidewinder progression. Type of locomotion used by certain desert snakes; series of lateral, looping movements in which only a vertical force is applied and no more than two parts of the body contact the ground at any one time; resulting tracks are a series of parallel, diagonal, J-shaped marks.

Simplex. Type of uterus in which the uterine horns are fused into a single structure; oviducts empty directly into the body of the uterus; found in some bats, the armadillo, and primates.

Sinus. Cavity or space in tissues or in bone.

Sister group. Relationship between two taxa that are each other's closest phylogenetic relatives.

Solenoglyph. Venomous snake with hinged fangs; includes vipers and pit vipers.

Speciation. Evolution of populations of a species into reproductively isolated groups and, ultimately, new species.

Spectacle. Transparent, permanently fused upper and lower eyelids in snakes, some lizards, and a few turtles.

Speculum. Distinctively colored area on the secondary feathers of ducks.

Spermatheca. Storage receptacle for sperm in the roof of the cloaca of some salamanders.

Spermatogenesis. Formation of spermatozoa.

Spermatophore. Packet enclosing spermatozoa; found in certain salamanders.

Spermatozoa. Male sex cells; sperm.

Sphincter. Constrictor muscle that serves to close an opening.

Spiracle. Modified first gill opening of a shark; also, excurrent channel for a tadpole's gills.

Spiral organ (= organ of Corti). Sensory epithelium within the cochlea.

Spiral valve. Helical membrane in the intestine of sharks and primitive fishes that increases the absorptive surface area; also, a membrane that separates oxygenated from deoxygenated blood in the conus arteriosus of an amphibian heart.

Splanchnocranium. Cartilage that forms jaws and visceral arches in fishes and gill-breathing amphibians.

Spleen. Large abdominal gland belonging to the circulatory and lymphatic systems; serves as a blood reservoir and the site for the formation of some white blood cells.

Squamosal. Dermal bone forming part of the posterior skull wall; in mammals, the site of articulation with the lower jaw.

Stapes. Smallest of the three middle ear bones in mammals; evolutionarily derived from the columella (hyomandibula).

Status undetermined. Category containing species that may be threatened or endangered with extinction, but about which there is not enough information to determine their status.

Sternebra (pl. **sternebrae**). Bony segment of a mammalian sternum.

Sternum. Breastbone.

Subspecies. Genetically distinct geographic subunit of a species.

Sudoriferous gland. Epidermal exocrine gland in the dermis or subcutaneous layer of a mammal that produces perspiration; also called a sweat gland.

Swim bladder. Membranous gas-filled sac present in the dorsal portion of the abdominal cavity of some fishes; assists in regulating buoyancy.

Symbiosis. Intimate relationship between members of different species; includes commensalism, mutualism, and parasitism.

Sympatric. Describes two or more species living in the same general area.

Symplesiomorphy. Sharing of ancestral characteristics among species.

Synapomorphy. In cladistics, a homologous trait that is assumed to be uniquely derived because it occurs in two or more groups being classified, but not in the outgroups.

Synapsid. Amniote in which the skull has a single pair of temporal openings bordered above by the postorbital and squamosal bones in mammal-like reptiles; variously modified in mammals.

Synsacrum. Unique structure in birds in which the posterior thoracic vertebrae together with the lumbar, sacral, and anterior caudal vertebrae fuse with the pelvic girdle.

Syrinx. Vocal organ of birds; located near the junction of the trachea and bronchi.

Systematics. Science of classification and reconstruction of phylogeny.

T

Taiga. Coniferous forest bordering the northern subpolar regions; also called *boreal forest*.

Tapetum lucidum. Light-reflecting layer in the eyes of animals that returns light to the photoreceptor cells; best developed in nocturnal forms.

Tarsometatarsus. Bone formed by the fusion of the distal tarsal elements with the metatarsals; found in birds and some dinosaurs.

Tarsus. Ankle.

Taxon (pl. **taxa**). Category such as phylum, order, etc., in which organisms are placed according to shared similarities and homologies.

Taxonomy. Study of the principles of classification of organisms.

Tectonic lake. Lake created by movement of the Earth's crust.

Telencephalon. Anterior division of the forebrain.

Terrestrial. Inhabiting land; opposite to aquatic.

Territory. Defended area within an animal's home range.

Testis (pl. **testes**). Male reproductive gland that produces the male sex hormone testosterone and sperm.

Tetrapod. Vertebrate with four legs.

Thalamus. Large oval structure located superior to the midbrain; part of the diencephalon; principal relay station for sensory impulses that reach the cerebral cortex from the spinal cord.

Thecodont. Teeth set in bony sockets in the jaw.

Thoracic. Relating to the thorax or chest.

Threatened species. Species that are likely to become endangered within the foreseeable future throughout all or a significant portion of their range.

Thymosin. Hormone produced by the thymus gland; stimulates the lymph glands to produce lymphocytes.

Thymus. Endocrine gland in the neck that secretes thymosin.

Thyroid. Endocrine gland in the neck region that secretes thyroxine, triiodothyronine, and calcitonin.

Thyroid-stimulating hormone (TSH). Hormone secreted by the anterior pituitary gland that stimulates the synthesis and secretion of thyroxine and triiodothyronine by the thyroid gland.

Thyroxine. Hormone secreted by the thyroid gland; assists in regulating metabolic activities.

Tibia. Bone on the medial (big toe) side of the distal portion of the hindlimb.

Tibiofibula. Bone in the hindlimb of anurans formed by the fusion of the tibia and fibula.

Tibiotarsus. Bone formed by the fusion of the tibia and the proximal tarsal elements in birds and some dinosaurs.

Trachea. Air tube extending from the pharynx to the bronchi of the lungs.

Tragus. Fleshy projection inside the pinna of bats; arises from the inner base of the ear; function is unknown.

Triiodothyronine. Hormone secreted by the thyroid gland; assists in regulating metabolic activities.

Troglodyte. Organism that lives in caves.

Tropical rain forest. Tropical forest that receives almost daily rainfall; contains the highest species diversity of any biome.

Tropical seasonal deciduous forest. Tropical forest with one or two dry seasons each year, during which time the leaves are dropped.

Tropic hormone. Hormone whose target is another endocrine gland.

True census. Count of all individuals present in a given area.

Tundra. Area in arctic and alpine regions characterized by bare ground and absence of trees; dominated by mosses, lichens, sedges, forbs, and low shrubs; permafrost.

Tympanum (= tympanic membrane). Eardrum.

U

Ulna. The bone on the medial (little-finger) side of the forearm.

Ultimobranchial bodies. Endocrine glands that develop from the last pair of pharyngeal pouches; may produce the hormone calcitonin.

Uncinate process. A posteriorly projecting process of the vertebral ribs of birds and certain reptiles.

Underfur. Inner, finer, and usually shorter hairs making up the pelage of a mammal.

Ungulate. Collective term used to designate hooved mammals.

Unguligrade. Type of locomotion in which only the tips of the digits contact the ground.

Ureter. Tube through which urine is conducted from the kidney to the cloaca or urinary bladder.

Urethra. Tube through which urine is voided from the body of a mammal.

Urodeum. Ventral portion of the cloaca in some vertebrates; receives the urogenital duct.

Urogenital. Pertaining to the organs, ducts, and structures of the urinary and reproductive systems.

Uropygial gland. Oil gland on the dorsal surface of the body at the base of the tail of a bird.

Urosteges. Ventral scales posterior to the anal plate; found in snakes.

Urostyle. Rodlike bone, representing a number of fused vertebrae, making up the posterior part of the vertebral column in anurans.

Uterus. Hollow, muscular organ in which the fertilized ovum develops.

Utriculus. Larger of the two sacs (sacculus, utriculus) in the inner ear; contains maculae; functions in static equilibrium.

V

Vagina. Muscular, tubular organ that leads from the uterus to the vestibule of female mammals; situated between the urethra and the rectum.

Vasa efferentia. Modified mesonephric tubules in some male fishes that carry sperm from the testis to the mesonephric duct.

Vas deferens (= ductus deferens). Sperm duct; tube through which sperm are ejaculated.

Vasopressin. Hormone produced by the hypothalamus that stimulates water reabsorption from kidney cells into the blood as well as vasoconstriction of arterioles; also called antidiuretic hormone (ADH); released from the posterior pituitary gland.

Vein. Blood vessel conducting blood toward the heart.

Velvet. Vascularized skin covering the antlers during their development.

Vent. External opening of the cloaca; used especially in reference to amphibians and reptiles.

Ventricle. Cavity in the brain; also, a chamber of the heart.

Ventricular trabecula (pl. **ventricular trabeculae**). Ridge in the ventricular wall of an amphibian heart that separates oxygenated from deoxygenated blood.

Vermiform. Having a wormlike shape.

Vertebra (pl. **vertebrae**). Segment of the vertebral, or spinal, column.

Vestibule. Shallow space into which the vagina and urethra open in some female rodents and primates; also part of the inner ear.

Viviparous. Giving birth to nonshelled young.

Vocal sacs. Paired or unpaired resonating chambers in most male anurans.

Volcanic lake. Circular lake formed in the crater of an extinct volcano.

W

Weberian apparatus. Modified anterior vertebrae joining the ear with the swim bladder in suckers, minnows, catfishes, and characins.

X

Xanthophore. Cell containing yellow pigments.

Xeric. Characterized by dry environmental conditions.

Y

Yolk sac. Membrane enclosing the yolk in developing vertebrates.

Ypsiloid cartilage. Cartilage extending forward from the pelvic girdle in the ventral body wall of certain salamanders.

Z

Zoogeography. Study of the geographic distribution of animals.

Zooplankton. Microscopic animals in aquatic ecosystems.

Zygodactyl. Condition in some birds in which two of the toes are oriented forward and two are oriented backward; found in woodpeckers, cuckoos, and some other birds.

Zygomatic. Relating to the cheekbone.

Zygote. Fertilized ovum; first cell of the next generation.

Bibliography

A

Abbott, L. A., F. A. Bisby, and D. J. Rogers. 1985. Taxonomic Analysis in Biology. New York: Columbia University Press.

Able, K. P., and M. A. Able. 1995. Interactions in the flexible orientation system of migratory bird. *Nature* 375:230–231.

Abler, W. L. 1999. The teeth of the tyrannosaurs. *Scientific American* 281(3):50–51.

Ackerman, R., and P. D. Weigl. 1970. Dominance relations of red and grey squirrels. *Ecology* 51(2):332–334.

Adams, L. 1959. An analysis of a population of snowshoe hares in northwestern Montana. *Ecological Monographs* 29(2): 141–170.

Adler, K. 1970. The role of extraoptic photoreceptors in amphibian rhythms and orientation: A review. *Journal of Herpetology* 4:99–112

Adler, K. 1971. Pineal end organ: Role in extraoptic entrainment of circadian locomotor rhythm in frogs. In: Menaken, M. (ed.). *Biochronometry: Proceedings of a symposium at Friday Harbor, September, 1969.* Washington, D. C.: U.S. National Academy of Science.

Adler, K. 1976. Extraocular photoreception in amphibians. *Photochemistry and Photobiology* 23:275–298.

Adler, K., and D. H. Taylor. 1973. Extraocular perception of polarized light by orienting salamanders. *Journal of Comparative Physiology* 87:203–212.

Adler, T. 1996a. Hippity hop goes the virus. *Science News* 149(13):206–207.

Adler, T. 1996b. How songbirds get their tunes. *Science News* 149(18):280–281.

Adovasio, J. M., and R. C. Carlisle. 1986. The first Americans, Pennsylvania pioneers. *Natural History* 95(12):20–27.

Ahlberg, P. E. 1991. Tetrapod or near-tetrapod fossils from the Upper Devonian of Scotland. *Nature* 354:298–301.

Ahlberg, P. E. 1995. *Elginerpeton pancheni* and the earliest tetrapod clade. *Nature* 373:420–425.

Ahlberg, P. E., and Z. Johanson. 1998. Osteolepiforms and the ancestry of tetrapods. *Nature* 395:792–794.

Aimi, M., and H. Inagaki. 1988. Grooved lower incisors in flying lemurs. *Journal of Mammalogy* 69:138–140.

Alberts, A. C. 1992. Constraints on the design of chemical communication systems in terrestrial vertebrates. *American Naturalist* 139:S62–S69.

Alcock, J. 1975. *Animal Behavior: An Evolutionary Approach.* Sunderland, Massachusetts: Sinauer Associates.

Alcorn, G. D. 1991. *Birds and Their Young.* Harrisburg, Pennsylvania: Stackpole Books.

Aldhous, P. 1993. Tropical deforestation: Not just a problem in Amazonia. *Science* 259:1390.

Alerstam, T. 1990. *Bird Navigation.* New York: Cambridge University Press.

Alexander, R. M., G. M. O. Maloiy, R. F. Ker, A. S. Jayes, and C. N. Warui. 1982. The role of tendon elasticity in the locomotion of the camel (*Camelus dromedarius*). *Journal of Zoology* 198:293–313.

Allee, W. C., and K. P. Schmidt. 1951. *Ecological Animal Geography.* Second Edition. New York: John Wiley and Sons Inc.

Allen, J. A. 1917. The skeletal characters of *Scutisorex* Thomas. *Bulletin of the American Museum of Natural History* 378:769–784.

Allen, L. 1998. Alien predator poised to invade Hawaii. *Nature Conservancy* 48(4):7.

Alvarez, L. W. 1983. Experimental evidence that an asteroid impact led to the extinction of many species 65 million years ago. *Proceedings of the National Academy of Science* 80:627–642.

Alvarez, L. W., W. Alvarez, F. Asaro, and H. V. Michel. 1980. Extraterrestrial cause for the Cretaceous-Tertiary extinction. *Science* 208(4448):1095–1108.

Alvarez, W., P. Claeys, and S. W. Kieffer. 1995. Emplacement of Cretaceous-Tertiary boundary shocked quartz from Chicxulub crater. *Science* 269:930–935.

Alvarez-Buylla, A., J. R. Kirn, and F. Nottebohm. 1990. Birth of projection neurons in adult avian brain may be related to perceptual or motor learning. *Science* 249:1444–1446.

Amos, B., S. Twiss, P. Pomeroy, and S. Anderson. 1994. Evidence for mate fidelity in the gray seal. *Science* 268:1897–1899.

Ancel, A., G. L. Kooyman, P. J. Ponganis, J. -P. Gendner, J. Ligon, X. Mestre, N. Huin, P. H. Thorson, P. Robisson, and Y. LeMaho. 1992. Foraging behavior of emperor penguins as a resource detector in winter and summer. *Nature* 360:336–339.

Andersen, K. K., and D. T. Bernstein. 1975. Some chemical constituents of the scent of the striped skunk (*Mephitis mephitis*). *Journal of Chemical Ecology* 1:493–499.

Anderson, A. 1991. Early bird threatens *Archeopteryx's* perch. *Science* 253(5017):35.

Andersson, S., J. Ornborg, and M. Andersson. 1997. Ultraviolet sexual dimorphism and assortative mating in blue tits. *Proceedings of the Royal Society, London B* 265:445–450.

Andrews, R. 1932. *The New Conquest of Asia.* New York: American Museum of Natural History.

Anonymous. 1989. Domestic cat bears exotic kitten. *Science News* 135(11):172.

Anonymous. 1991. Living thin and long. *Science* 254:373.

Anonymous. 1992. Shocking behavior. *Discover* 13(3):13.

Anonymous. 1996a. Large and lungless. *Discover* 17(4):23.

Anonymous. 1996b. Mast crop can be life or death for Park bears. *Smokies Guide*, Autumn 1996:1.

Anonymous. 1996c. Rabbit virus set for release. *Nature* 383:210.

Anonymous. 1996d. Regional news and recovery updates. *Endangered Species Bulletin* 21(3):31.

Anonymous. 1996e. Chessie heading our way. *Bay Journal* 6(5):9.

Anonymous. 1997a. Small monkey is a big discovery. *Roanoke Times*, August 19, 1997:A7.

Anonymous. 1997b. 415 Florida manatees died in 1996, twice as many as previous record. *Bay Journal* 7(1):8.

Anonymous. 1997c. Dinos for dinner. *Discover* 18(7):21.

Anonymous. 1997d. A bridge to Madagascar. *Discover* 18(12):26–27.

Appenzeller, T. 1994. Argentine dinos vie for heavyweight titles. *Science* 266:1805.

Archibald, J. D. 1996. Fossil evidence for a Late Cretaceous origin of "hoofed" mammals. *Science* 272:1150–1153.

Arnold, E. N. 1988. Caudal autotomy as a defense. Pages 235-273. In: Gans, C. and R. B. Huey (eds.). *Biology of the Reptilia,16 (Ecology B. Defense and Life History)*. New York: Alan R. Liss.

Arnold, S. J. 1976. Sexual behavior, sexual interference and sexual defense in the salamanders *Ambystoma maculatum, Ambystoma tigrinum* and *Plethodon jordani. Zeitschrift Tierpsychologie* 42:247–300.

Arnold, S. J. 1977. The evolution of courtship behavior in New World salamanders with some comments on Old World salamandrids. Pages 141–183. In: Taylor, D. H. and S. I. Guttman (eds.), *The Reproductive Biology of Amphibians*. New York: Plenum Press.

Arthur, S. M., W. B. Krohn, and J. R. Gilbert. 1989. Home range characteristics of adult fishers. *Journal of Wildlife Management* 53(3):674–679.

Askins, R. A. 1995. Hostile landscapes and the decline of migratory songbirds. *Science* 267:1956–1957.

Astheimer, L. B., W. A. Buttemer, and J. C. Wingfield. 1995. Seasonal and acute changes in adrenocortical responsiveness in an Arctic-breeding bird. *Hormones and Behavior* 29:442–457.

Astier, H., F. Halberg, and I. Assenmacher. 1970. Rhythmes circanniens de l'activitie thyroidienne chez le canard Pekin. *Journal of Physiology* (Paris) 62:219.

Atkeson, T. D., R. L. Marchinton, and K. V. Miller. 1988. Vocalizations of white-tailed deer. *American Midland Naturalist* 120(1):194–200.

Auburn, J. S., and D. H. Adler. 1976. Orientation by means of polarized light in bullfrog tadpoles, *Rana catesbeiana*. *Herpetological Review* 7:74.

Autrum, H., and D. von Holst. 1968. Sozialer "stress" bei Tupajas (*Tupaja glis*) und seine Wirkung auf Wachstum, Korpergewicht und Fortpflanzung. *Zeitshrift vergleich Physiologie* 58:347–355.

B

Bagla, P. 1997. Longer tusks are healthy signs. *Science* 276:1972.

Bagnara, J. T. 1960. Pineal regulation of the body lightening reaction in amphibian larvae. *Science* 132:1481–1483.

Bailey, V. 1924. Breeding, feeding and other life habits of meadow mice (*Microtus*). *Journal of Agricultural Research* 27(8): 523–536.

Baker, P. C., W. B. Quay, and J. Axelrod. 1965. Development of hydroxyindole-O-methyl transferase activity in the eye and brain of the amphibian *Xenopus laevis*. *Life Sciences* 4:1981–1987.

Baker, R. H. 1968. Habitats and distribution. Pages 98-126. In: King, J.A. (ed.), *Biology of* Peromyscus. American Society of Mammalogists Special Publication No. 2.

Baker, R. R., and M. A. Bellis. 1988. "Kamikaze" sperm in mammals? *Animal Behaviour* 36(3):936–939.

Baker, R. R., and M. A. Bellis. 1989a. Elaboration of the kamikaze sperm hypothesis: A reply to Harcourt. *Animal Behaviour* 37(5):865–866.

Baker, R. R., and M. A. Bellis. 1989b. Number of sperm in human ejaculates varies in accordance with sperm competition theory. *Animal Behaviour* 37(5):867–869.

Baker, R. R., and M. A. Bellis. 1993. Human sperm competition: Ejaculate manipulation by females and a function for the female orgasm. *Animal Behavior* 46(5):887–909.

Bakker, R. T. 1986. *The Dinosaur Heresies*. New York: William Morrow and Company.

Bakker, R. T. 1992. Inside the head of a tiny *T. rex*. *Discover* 13(3):58–62, 67, 69.

Bakst, M. R. 1987. Anatomical basis of sperm-storage in the avian oviduct. *Scanning Microscopy* 1:1257–1266.

Balfour, F. M. 1876. The development of elasmobranch fishes. *Journal of Anatomy and Physiology* (London) 11:128–172.

Ballinger, R. E. 1977. Reproductive strategies: Food availability as a source of proximal variation in a lizard. *Ecology* 58:628–635.

Ballinger, R. E., L. Holy, J. W. Rowe, F. Karst, C. L. Ogg, and D. W. Stanley-Samuelson. 1992. Seasonal change in lipid composition during the reproductive cycle of the red-chinned lizard, *Sceloporus undulatus erythrocheilus*. *Comparative Biochemistry and Physiology. Part B. Comparative Biochemistry* 103B(3):527–553.

Balter, M. 1997. Morphologists learn to live with molecular upstarts. *Science* 276:1032–1034.

Bannon, A. W., M. W. Decker, M. W. Holladay, P. Curzon, D. Donnelly-Roberts, P. S. Puttfarcken, R. S. Bitner, A. Diaz, A. H. Dickenson, R. D. Porsolt, M. Williams, and S. P. Arneric. 1998. Broad-spectrum, non-opioid analgesic activity by selective modulation of neuronal nicotinic acetylcholine receptors. *Science* 279:77–81.

Bardack, D. 1991. First fossil hagfish (Myxinoidea): A record from the Pennsylvanian of Illinois. *Science* 254:701–703.

Bardack, D., and R. Zangerl. 1968. First fossil lamprey: A record from the Pennsylvanian of Illinois. *Science* 162:1265–1267.

Barinaga, M. 1999. Salmon follow watery odors home. *Science* 286:705–706.

Barlow, J., and C. Y. McCulloch. 1984. Recent dynamics and mortality rates of the Kaibab deer herd. *Canadian Journal of Zoology* 62:1805–1812.

Barnea, A., Y. Yom-Tov, and J. Friedman. 1991. Does ingestion by birds affect seed germination? *Functional Ecology* 5:394–402.

Barnes, B. M. 1989. Freeze avoidance in a mammal: Body temperatures below 0°C in an Arctic hibernator. *Science* 244: 1593–1595.

Barreto, C., R. M. Albrecht, D. E. Bjorling, J. R. Horner, and N. J. Wilsman. 1993. Evidence of the growth plate and the growth of long bones in juvenile dinosaurs. *Science* 262:2020–2023.

Barrick, R. E., and W. J. Showers. 1994. Thermophysiology of *Tyrannosaurus rex*: Evidence from oxygen isotopes. *Science* 265:222–224.

Barse, W. P. 1997. Dating a paleoindian site in the Amazon in comparison with Clovis culture. *Science* 275:1949–1950.

Bartlett, H. H. 1940. The concept of the genus. *Bulletin of the Torrey Botanical Club* 67(5):349–389.

Batisse, M. 1997. Biosphere reserves. A challenge for biodiversity conservation and regional development. *Environment* 39(5):6–15, 31–33.

Baylis, H. A. 1939. Delayed reproduction in the spotted salamander. *Proceedings of the Zoological Society* (London) 109A:243–246.

Beachy, C. K., and R. C. Bruce. 1992. Lunglessness in plethodontid salamanders is consistent with the hypothesis of a mountain stream origin: A response to Ruben and Boucot. *American Naturalist* 139(4):839–847.

Beard, K. C., L. Krishtalka, and R. K. Stucky. 1991. First skulls of the Early Eocene primate *Shoshonius cooperi* and the anthropoid-tarsier dichotomy. *Nature* 349:64–67.

Beattie, R. C., and R. Tyler-Jones. 1992. The effects of low pH and aluminum on breeding success in the frog *Rana temporaria*. *Journal of Herpetology* 26(4):353–360.

Beck, C. A., and N. B. Barros. 1991. The impact of debris on the Florida manatee. *Marine Pollution Bulletin* 22(10):508–510.

Bellairs, A. 1970. *The Life of Reptiles*. 2 volumes. New York: Universe Books.

Bellis, M. A., R. R. Baker, and M. J. G. Gage. 1990. Variation in rat ejaculates consistent with the kamikaze-sperm hypothesis. *Journal of Mammalogy* 71(3):479–480.

Bellrose, F. C. 1968. Waterfowl migration corridors east of the Rocky Mountains in the United States. *Biological Notes 61*. Urbana: Illinois Natural History Survey.

Benkman, C. W. 1987. Crossbill foraging behavior, bill structure, and patterns of food profitability. *Wilson Bulletin* 99(3):351–368.

Benkman, C. W. 1988a. On the advantages of crossed mandibles: An experimental approach. *Ibis* 130:288–293.

Benkman, C. W. 1988b. Seed handling ability, bill structure, and the cost of specialization for crossbills. *The Auk* 105:715–719.

Benkman, C. W. 1993. Adaptation to single resources and the evolution of crossbill (*Loxia*) diversity. *Ecological Monographs* 63(3):305–325.

Bennett, A. T. D., and I. C. Cuthill. 1994. Ultraviolet vision in birds: What is its function? *Vision Research* 34:1471–1478.

Bennett, A. T. D., I. C. Cuthill, J. C. Partridge, and E. J. Maier. 1996. Ultraviolet vision and mate choice in zebra finches. *Nature* 380:433–435.

Bennett, A. T. D., I. C. Cuthill, J. C. Partridge, and K. Lunau. 1997. Ultraviolet plumage colors predict mate preferences in starlings. *Proceedings of the National Academy of Sciences, USA* 94:8618–8621.

Benson, B. N. 1980. Dominance relationships, mating behaviour and scent marking in fox squirrels (*Sciurus niger*). Mammalia 44:143–160.

Benson, S. B. 1933. Concealing coloration among some desert rodents of the southwestern United States. *University of California Publications in Zoology* 40:1–70.

Benton, M. J. 1990. Phylogeny of the major tetrapod groups: Morphological data and divergence dates. *Journal of Molecular Evolution* 30:409–424.

Benton, M. J. 1993. Late Triassic extinctions and the origin of the dinosaurs. *Science* 260:769–770.

Benton, M. J. 1995. Diversification and extinction in the history of life. *Science* 268:52–58.

Berg, L. S. 1940. (Reprint 1947). Classification of fishes both recent and fossil. *Trav. Inst. Zool. Acad. Sci. URSS* 5:87–517. Reprint 1947, J. W. Edwards, Ann Arbor, Michigan.

Berger, M., J. S. Hart, and O. Z. Roy. 1970. Respiration, oxygen consumption, and heart rate in some birds during rest and flight. *Zeitschrift vergleich Physiologie* 66:201.

Bergeron, J. M. and L. Jodoin. 1987. Defining "high quality" food resources of herbivores: The case for meadow voles (*Microtus pennsylvanicus*). *Oecologia* 71:510–517.

Berkson, H. 1967. Physiological adjustments to deep diving in the Pacific green turtle (*Chelonia mydas*). *Comparative Biochemical Physiology* 21:507–524.

Berria, M., M. DeSantis, and R. A. Mead. 1990. Testicular response to melatonin or suprachiasmatic nuclei ablation in the spotted skunk. *Journal of Experimental Zoology* 255:72–79.

Berthold, J. 1996. *Control of Bird Migration*. London: Chapman and Hall.

Berthold, P. 1998. Spatiotemporal aspects of avian long-distance migration. Pages 103–118. In: Healy, S. (ed.). *Spatial Representation in Animals*. New York: Oxford University Press.

Berthold, P., W. Wiltschko, H. Miltenberger, and U. Querner. 1990. Genetic transmission of migratory behavior into a nonmigratory bird population. *Experientia* 46:107–108.

Berthold, P., A. J. Helbig, G. Mohr, and U. Querner. 1992. Rapid microevolution of migratory behaviour in a wild bird species. *Nature* 360:668–669.

Bickle, C. A., J. F. Kirkpatrick, and J. W. Turner, Jr. 1991. Contraception in striped skunks with Norplant implants. *Wildlife Society Bulletin* 19:334–338.

Billo, R., and M. Wake. 1987. Tentacle development in *Dermophis mexicanus* (Amphibia, Gymnophiona) with an hypothesis of tentacle origin. *Journal of Morphology* 192:101–111.

Birgersson, B., K. Ekvall, and H. Temrin. 1991. Allosuckling in fallow deer, *Dama dama*. *Animal Behaviour* 42:326–327.

Birkhead, T. R. 1988. Behavioural aspects of sperm competition in birds. *Advances in the Study of Behaviour* 18:35–72.

Birkhead, T. R., and A. P. Moller. 1992. Numbers and size of sperm storage tubules and the duration of sperm storage in birds: A comparative study. *Biological Journal of the Linnean Society* 45:363–372.

Birkhead, T. R., L. Atkin, and A. P. Moller. 1987. Copulation behaviour of birds. *Behaviour* 101:101–138.

Bisazza, A., C. Cantalupo, A. Robins, L. J. Rogers, and G. Vallortigara. 1996. Right-pawedness in toads. *Nature* 373:408.

Bishop, C. A., R. J. Brooks, J. H. Carey, P. Ng, R. J. Norstrom, and D. R. S. Lean. 1991. The case for a cause-effect linkage between environmental contamination and development in eggs of the common snapping turtle (*Chelydra s. serpentina*) from Ontario, Canada. *Journal of Toxicology and Environmental Health* 33:521–547.

Bishop, M. J., and A. E. Friday. 1988. Estimating the interrelationships of tetrapod groups on the basis of molecular sequence data. Pages 33-58. In: Benton, M. J. (ed.). *The Phylogeny and Classification of the Tetrapods. Volume 1. Amphibians, Reptiles, Birds*. Oxford, England: Clarendon Press.

Blackburn, D. G. 1982. Evolutionary origins of viviparity in the Reptilia. I. Sauria. *Amphibia-Reptilia* 3:185–205.

Blackburn, D. G. 1985. Evolutionary origins of viviparity in the Reptilia. II. Serpentes, Amphisbaenia, and Ichthyosauria. *Amphibia-Reptilia* 6:259–291.

Blackburn, D. G. 1992. Convergent evolution of viviparity, matrotrophy, and specializations for fetal nutrition in reptiles and other vertebrates. *American Zoologist* 32:313–321.

Blackburn, D. R. 1968. Behavioral ecology of black-tailed and white-tailed jackrabbits of eastern Oregon. MS thesis. University of Idaho, Moscow.

Blair, W. F. 1940a. Home ranges and populations of the jumping mouse. *American Midland Naturalist* 23:244–250.

Blair, W. F. 1940b. A study of prairie deer mouse populations in southern Michigan. *American Midland Naturalist* 24:273–305.

Blair, W. F. 1940c. Notes on home ranges and populations of the short-tailed shrew. *Ecology* 21:285–288.

Blair, W. F. 1940d. Home ranges and populations of the meadow vole in southern Michigan. *Journal of Wildlife Management* 4:149–161.

Blair, W. F. 1976. Amphibians, their evolutionary history, taxonomy, and ecological adaptations. Pages 1–28. In Fite, K. V., (ed.). *The Amphibian Visual System*. New York: Academic Press.

Blanchard, F. N. 1933. Eggs and young of the smooth green snake, *Liopeltis vernalis* (Harlan). *Paper of the Michigan*

Academy of Science, Arts and Letters 17:493–514.

Block, B. A., J. R. Finnerty, A. F. R. Stewart, and J. Kidd. 1993. Evolution of endothermy in fish: Mapping physiological traits on a molecular phylogeny. *Science* 260:210–214.

Boatright-Horowitz, S. S., and A. M. Simmons. 1997. Transient "deafnesss" accompanies auditory development during metamorphosis from tadpole to frog. *Proceedings of the National Academy of Sciences USA* 94:14877–14882.

Boggs, D. F., J. J. Seveyka, D. L. Kilgore, Jr., and K. P. Dial. 1997. Coordination of respiratory cycles with wingbeat cycles in the black-billed magpie (*Pica pica*). *Journal of Experimental Biology* 200:1413–1420.

Bonaparte, J. F., and M. Vince. 1974. El Hallazgo del primer nido de dinosaurios Triasicos (Saurischia, Prosauropoda), Triasico Superior de Patagonia, Argentina. *Ameghiniana* 16:173–182.

Bond, C. E. 1996. *Biology of Fishes*. Second edition. Philadelphia: W. B. Saunders.

Boness, D. J. 1990. Fostering behavior in Hawaiian monk seals: Is there a reproductive cost? *Behavioral Ecology and Sociobiology* 27:113–122.

Booth, R. J., F. N. Carrick, and P. A. Addison. 1991. The structure of the koala adrenal gland and the morphological changes associated with the stress of disease. Pages 281–288. In: Lee, A. K., K. A. Handasyde, and G. D. Sanson (eds.). *Biology of the Koala*. Chipping Norton, New South Wales: Surrey Beatty and Sons.

Bourne, D., and M. Coe. 1978. The size, structure and distribution of the giant tortoise population of Aldabra. *Philosophical Transactions of the Royal Society of London, Series B* 282:139–175.

Bowen, B. W., F. A. Abreu-Grobois, G. H. Balazs, N. Kamezaki, C. J. Limpus, and R. J. Ferl. 1995. Trans-Pacific migrations of the loggerhead turtle (*Caretta caretta*) demonstrated with mitochondrial DNA markers. *Proceedings of the National Academy of Sciences, USA* 92:3731–3734.

Bowen, B. W., A. B. Meylan, and J. C. Avise. 1989. An odyssey of the green sea turtle: Ascension Island revisited. *Proceedings of the National Academy of Sciences, USA* 86:573–576.

Bowen, B. W., A. B. Meylan, and J. C. Avise. 1991. Evolutionary distinctiveness of the endangered Kemp's ridley sea turtle. *Nature* 352:709–711.

Bower, B. 1990. America's talk: The great divide. Do Indian languages hold clues to the peopling of the New World? *Science News* 137(23):360–362.

Bower, B. 1995. Egyptian fossils illuminate primate roots. *Science News* 148:6.

Bowers, M. A. 1909. Histogenesis and histolysis of the intestinal epithelium of *Bufo lentiginosus*. *American Journal of Anatomy* 9(2):263–279.

Bowmaker, J., L. Heath, S. Wilkie, and D. Hunt. 1997. Visual pigments and oil droplets from six classes of photoreceptors in the retinas of birds. *Vision Research* 37:2183–2194.

Bowmaker, J. K. 1986. Avian colour vision and the environment. Pages 1284–1294. In: Ouellet, H. (ed.). *Proceedings of the 19th International Ornithological Congress*.

Bowmaker, J. K., and G. R. Martin. 1985. Visual pigments and oil droplets in the penguin, *Spheniscus humboldti*. *Journal of Comparative Physiology* A 156:71–77.

Boxer, S. 1987. Prehistoric pelves: Bones of contention. *Discover* 8(7):6–7.

Boyd, C. E. 1962. Waif dispersal in toads. *Herpetologica* 18:269.

Bracke, V., F. Alvarez-Sanchez, A. Faundes, A. S. Tejada, and L. Cochon. 1990. Ovarian endocrine function through five years of continuous treatment with Norplant subdermal contraceptive implants. *Contraception* 41:169–177.

Bradley, D. 1993. Frog venom cocktail yields a one-handed painkiller. *Science* 261:1117.

Braithwaite, V. A. 1998. Spatial memory, landmark use and orientation in fish. Pages 86-102. In: Healy, S. (ed.). *Spatial Representation in Animals*. New York: Oxford University Press.

Branch, W. R. 1989. Alternative life-history styles in reptiles. Pages 127-151. In: Bruton, M. N., (ed.). *Alternative Life-History Styles of Animals*. Dordrecht, Netherlands: Kluwer Academic Publishers.

Brand, C. J., and L. B. Keith. 1979. Lynx demography during a snowshoe hare decline in Alberta. *Journal of Wildlife Management* 43:827–849.

Brand, C. J., L. B. Keith, and C. A. Fischer. 1976. Lynx responses to changing snowshoe hare densities in central Alberta. *Journal of Wildlife Management* 40: 416–428.

Brandon, R. A., and J. E. Huheey. 1981. Toxicity in the plethodontid salamanders *Pseudotriton ruber* and *Pseudotriton montanus* (Amphibia, Caudata). *Toxicon* 19:25–31.

Brannon, E. L., and T. P. Quinn. 1990. Field test of the pheromone hypothesis for homing by Pacific salmon. *Journal of Chemical Ecology* 16:603–609.

Breece, G. A., and J. L. Dusi. 1985. Food habits and home ranges of the common long-nosed armadillo *Dasypus novemcinctus* in Alabama. Pages 419–427. In: Montgomery, G. G. (ed.). *The Evolution and Ecology of Armadillos, Sloths, and Vermilinguas*. Washington, D.C.: Smithsonian Institution Press.

Brewer, R. 1988. *The Science of Ecology*. Philadelphia: Saunders College Publishing.

Brinster, R. L., and M. R. Avarbock. 1994. Germline transmission of donor haplotype following spermatogonial transplantation. *Proceedings of the National Academy of Science, USA*. 91:11303–11307.

Brinster, R. L., and J. W. Zimmermann. 1994. Spermatogenesis following male germ-cell transplantation. *Proceedings of the National Academy of Sciences, USA*. 91:11298–11302.

Briskie, J. V. 1992. Copulation patterns and sperm competition in the polygynandrous Smith's longspur. *The Auk* 109(3):563–575.

Briskie, J. V., and R. Montgomerie. 1992. Sperm size and sperm competition in birds. *Proceedings of the Royal Society of London B* 247:89–95.

Britt, B. B., P. J. Makovicky, J. Gauthier, and N. Bonde. 1998. Postcranial pneumatization in *Archaeopteryx*. *Nature* 395:374–376.

Brodie, E. D., Jr., J. L. Hensel, Jr., and J. A. Johnson. 1974. Toxicity of the urodele amphibians *Taricha, Notophthalmus, Cynops,* and *Paramesotriton* (Salamandridae). *Copeia* 1974:506–511.

Brodie, E. D., Jr., P. K. Ducey, and E. A. Baness. 1991. Antipredator skin secretions of some tropical salamanders (*Bolitoglossa*) are toxic to snake predators. *Biotropica* 23(1):58–62.

Brodie, E. D., III. 1989. Genetic correlations between morphology and antipredator behaviour in natural populations of the garter snake *Thamnophis ordinoides*. *Nature* 342:542–543.

Brodkorb, P. 1951. The number of feathers in some birds. *Quarterly Journal of the Florida Academy of Sciences* 12 for 1949 (1951):241–245.

Bronmark, C., and J. G. Miner. 1992. Predator-induced phenotypical change in body morphology in crucian carp. *Science* 258:1348–1350.

Bronson, F. H., and P. D. Heideman. 1992. Lack of reproductive photoresponsiveness and correlative failure to respond to melatonin in a typical rodent, the cane mouse. *Biology of Reproduction* 46:246–250.

Brooke, M. d., and N. B. Davies. 1991. A failure to demonstrate host imprinting in the cuckoo (*Cucculus canorus*) and alternative hypotheses for the maintenance of egg mimicry. *Ethology* 89:154–166.

Brooks, R. J., M. L. Bobyn, D. A. Galbraith, J. A. Layfield, and E. G. Nancekivell. 1991. Maternal and environmental influences on growth and survival of embryonic and hatchling snapping turtles (*Chelydra serpentina*). *Canadian Journal of Zoology* 69:2667–2676.

Broom, R. 1910. A comparison of the Permian reptiles of North America with those of South Africa. *Bulletin of the American Museum of Natural History* 28:197–234.

Brosseau, L. 1998. Regional news and recovery updates. *Endangered Species Bulletin* 23(5):29.

Brown, B., and E. Schlaikjer. 1940. The structure and relationships of *Protoceratops*. *Annals of the New York Academy of Sciences* 40:133–266.

Brown, C. W. 1968. Additional observations on the function of the nasolabial grooves of plethodontid salamanders. *Copeia* 1968:728–731.

Brown, H. A. 1977. Oxygen consumption of a large, cold-adapted frog egg (*Ascaphus truei* [Amphibia: Ascaphidae]). *Canadian Journal of Zoology* 55:343–348.

Brown, J. H. 1971. The desert pupfish. *Scientific American* 225(5):104–110.

Brown, J. H., and A. C. Gibson. 1983. *Biogeography*. St. Louis, Missouri: C. V. Mosby.

Brown, J. H., D. W. Davidson, J.C. Munger, and R. S. Inouye. 1986. Experimental community ecology: The desert granivore system. Pages 41–61. In: Diamond, J. and T. J. Case (eds.). *Community Ecology*. New York: Harper and Row.

Brown, M. R., P. J. Corkeron, P. T. Hale, K. W. Schultz, and M. M. Bryden. 1995. Evidence for a sex-segregated migration in the humpback whale (*Megaptera novaeangliae*). *Proceedings of the Royal Society of London B* 259:229–234.

Brown, R. E., and D. W. Macdonald (eds.). 1985. *Social Odours in Mammals. Volume 1*. Oxford, England: Clarendon Press.

Brown, R. G., W. C. Kimmins, M. Menzel, J. Parsons, B. Pohajdak, and W. D. Bowen. 1996. Birth control for grey seals. *Nature* 379:30–31.

Brown, W. S. 1987. Hidden life of the timber rattler. *National Geographic* 172:(1)128–138.

Browne, M. W. 1992. A dinosaur named Sue divides fossil hunters. *New York Times*, July 21,1992:C1,8.

Buckholz, H. 1986. Die Hôhle eines Spechtvôgels aus dem Eozän von Arizonas USA (Aves, Piciformes). *Verh. Naturwiss*. Ver. Hamburg N. F. 28:5–25.

Buffetaut, E., and J. LeLoeuff. 1989. La première découverte d'oeufs de dinosaures. *Pour la Science* 143:22.

Bugden, S. C., and R. M. Evans. 1991. Vocal responsiveness to chilling in embryonic and neonatal American coots. *Wilson Bulletin* 103(4):712–717.

Bull, J. J. 1980. Sex determination in reptiles. *Quarterly Review of Biology* 55:3–21.

Bull, J. J. 1983. *Evolution of Sex-Determining Mechanisms*. Menlo Park, California: Benjamin/Cummings.

Bullock, T. H., and F. P. J. Diecke. 1956. Properties of an infra-red receptor. *Journal of Physiology* 134:47–87.

Bulmer, M. G. 1974. A statistical analysis of the 10-year cycle in Canada. *Journal of Animal Ecology* 43:701–718.

Bulmer, M. G. 1975. Phase relations of the ten-year cycle. *Journal of Animal Ecology* 44:609–621.

Bundle, M. W., H. Hoppeler, R. Vock, J. M. Tester, and P. G. Weyand. 1999. High metabolic rates in running birds. *Nature* 397:31–32.

Burger, J. 1989. Incubation temperature has long-term effects on behaviour of young pine snakes (*Pituophis melanoleucus*). *Behavioral Ecology and Sociobiology* 24:201–207.

Burgers, P., and L. M. Chiappe. 1999. The wing of *Archaeopteryx* as a primary thrust generator. *Nature* 399:60–62.

Burghardt, G. M. 1970. Chemical perception in reptiles. Pages 241–308. In: Johnston, J. W., D. G. Moulton, and A.Turk (eds.). *Advances in Chemoreception. Volume 1. Communication by Chemical Signals*. New York: Appleton-Century-Crofts.

Burghardt, G. M. 1980. Behavioral and stimulus correlates of vomeronasal functioning in reptiles; feeding, grouping, sex, and tongue use. Pages 275-301. In: Muller-Schwarze, D., and R. M. Silverstein (eds.). *Chemical Signals: Vertebrates and Aquatic Invertebrates*. New York: Plenum Press.

Burke, A. C., and A. Feduccia. 1997. Developmental patterns and the identification of homologies in the avian hand. *Science* 278:666–668.

Burns, W. A. 1972. For your nature scrapbook. *Environment Southwest*, Spring, 1972:8–9. San Diego: San Diego Museum of Natural History.

Burroughs, D. 1999. New life for a vanished zebra? *International Wildlife* 29(2):46–51.

C

Caccone, A., and J. R. Powell. 1989. DNA divergence among hominoids. *Evolution* 43(5):925–942.

Cailliet, G. M., M. S. Love, and A. W. Ebeling. 1986. *Fishes: A Field and Laboratory Manual on their Structure, Identification, and Natural History*. Belmont, California: Wadsworth Publishing Company.

Caldwell, J. P. 1997. Pair bonding in spotted poison frogs. *Nature* 385:211.

Caldwell, M. W., and M. S. Y. Lee. 1997. A snake with legs from the marine Cretaceous of the Middle East. *Nature* 386:705–709.

Cameron, J. N. 1975. Morphometric and flow indicator studies of the teleost heart. *Canadian Journal of Zoology* 53:691–698.

Campbell, K. E., and C. D. Frailey. 1991. Uncovering the mysteries of the Amazon. *Terra* 29(2–3):36–49.

Camper, J. D., and J. R. Dixon. 1988. Evaluation of a microchip marking system for amphibians and reptiles. *Research Publication 7100–159*. Texas Parks and Wildlife Department, Austin, Texas.

Capranica, R. R. 1976. Morphology and physiology of the auditory system. Pages 551–575. In: Llinas, R., and W. Precht (eds.). *Frog Neurobiology*. New York: Springer-Verlag.

Caro, T. 1996. An elegant enigma. *Wildlife Conservation* 99(3):44–47.

Carpenter, C. C. 1953. A study of hibernacula and hibernating associations of snakes and amphibians in Michigan. *Ecology* 34:74–80.

Carpenter, C. C. 1978. Tongue display by the common bluetongue (*Tiliqua scincoides*: Reptilia, Lacertilia, Scincidae). *Journal of Herpetology* 12:428–429.

Carpenter, K., and K. Alf. 1994. Global distribution of dinosaur eggs, nests, and babies. Pages 15–30. In: Carpenter, K., K. F. Hirsch, and J. R. Horner (eds.). *Dinosaur Eggs and Babies*. New York: Cambridge University Press.

Carpenter, K., K. F. Hirsch, and J. R. Horner (eds.). 1994. *Dinosaur Eggs and Babies*. New York: Cambridge University Press.

Carr, A. 1970. *The Reptiles*. Alexandria, Virginia: Time Incorporated.

Carroll, R. L. 1969. Origin of reptiles. Pages 1–44. In: Gans, C., A. d. Bellairs, and T. S. Parsons (eds.). *Biology of the Reptilia. Volume 1. Morphology*. New York: Academic Press.

Carroll, R. L. 1988. *Vertebrate Paleontology and Evolution*. New York: W. H. Freeman.

Carter, G. S. 1967. *Structure and Habit in Vertebrate Evolution*. Seattle: University of Washington Press.

Cartmill, M. 1985. Climbing. Pages 73-88. In: Hildebrand, M., D. M. Bramble, K. F. Liem, and D. B. Wake (eds.). *Functional Vertebrate Morphology*. Cambridge, Massachusetts: Belknap Press.

Case, T. J. 1997. Natural selection out on a limb. *Nature* 387:15–16.

Castanet, J., D. G. Newman, and H. Saint Girons. 1988. Skeletochronological data of the growth, age, and population structure of the tuatara, *Sphenodon punctatus*, on the Stephens and Lady Alice Islands, New Zealand. *Herpetologica* 44(1):25–37.

Catania, K. C., and J. H. Kaas. 1996. The unusual nose and brain of the star-nosed mole. *BioScience* 46(8):578–586.

Catton, C. 1990. *Pandas*. New York: Facts on File.

Catzeflis, F. M., F. H. Sheldon, J. E. Ahlquist, and C. G. Sibley. 1987. DNA-DNA hybridization evidence of the rapid rate of muroid rodent DNA evolution. *Molecular Biology and Evolution* 4(3):242–253.

Censky, E. J., K. Hodge, and J. Dudley. 1998. Over-water dispersal of lizards due to hurricanes. *Nature* 395:556.

Chadwick, D. H. 1998. Return of the gray wolf. *National Geographic* 193(5):72-99.

Chan, S. 1994. New tree kangaroo species discovered. *Animal Keepers' Forum* 21(10):359–360.

Charland, M. B., and P. T. Gregory. 1990. The influence of female reproductive status on thermoregulation in a viviparous snake, *Crotalus viridis*. *Copeia* 1990(4):1089–1098.

Chatterjee, S. 1991. Cranial anatomy and relationships of a new Triassic bird from Texas. Royal Society of London. *Philosophical Transactions, Series B. Biological Sciences* B332:277–342.

Chellam, R. 1996. Lions of the Gir Forest. *Wildlife Conservation* 99(3):40–43.

Chen, D.-M., J. S. Collins, and T. H. Goldsmith. 1984. The ultraviolet receptor of bird retinas. *Science* 225:337–340.

Chen, J. Y., J. Dzik, G. D. Edgecombe, L. Ramsköld, and G. Q. Zhou. 1995. A possible Early Cambrian chordate. *Nature* 377:720–722.

Chen, J.-Y., D.-Y. Huang, and C.-W. Li. 1999. An early Cambrian craniate-like chordate. *Nature* 402:518–522.

Chen, L., A. L. DeVries, and C. C. Cheng. 1997a. Evolution of antifreeze glycoprotein gene from a trypsinogen gene in Antarctic notothenioid fish. *Proceedings of the National Academy of Science, USA.* 94:3811–3816.

Chen, L., A. L. DeVries, and C. C. Cheng. 1997b. Convergent evolution of antifreeze glycoproteins in Antarctic notothenioid fish and Arctic cod. *Proceedings of the National Academy of Science, USA.* 94:3817–3822.

Chen, P.-j., Z. Dong, and S. Zhen. 1998. An exceptionally well-preserved theropod dinosaur from the Yixian formation of China. *Nature* 391:147–152.

Cheng, M.-F. 1992. For whom does the female dove coo? A case for the role of vocal self-stimulation. *Animal Behaviour* 43:1035–1044.

Chiappe, L., M. Norell, and J. Clark. 1997. *Mononykus* and birds: Methods and evidence. *The Auk* 114(2):300–302.

Chiappe, L. M., R. A. Coria, L. Dingus, F. Jackson, A. Chinsamy, and M. Fox. 1998. Sauropod dinosaur embryos from the Late Cretaceous of Patagonia. *Nature* 396:258–261.

Chin, K., T. T. Tokaryk, G. M. Erickson, and L. C. Calk. 1998. A king-sized theropod coprolite. *Nature* 393:680–682.

Chinsamy, A. 1995. Within the bone. *Natural History* 104(6):62–63.

Chinsamy, A., L. M. Chiappe, and P. Dodson. 1994. Growth rings in Mesozoic birds. *Nature* 368:196–197.

Chiszar, D., V. Lipetz, K. Scudder, and E. Pasanello. 1980. Rate of tongue flicking by bull snakes and pine snakes (*Pituophis melanoleucus*) during exposure to food and non-food odors. *Herpetologica* 36(3):225–231.

Christian, J. J. 1950. The adreno-pituitary system and population cycles in mammals. *Journal of Mammalogy* 31(3):247–259.

Christian, J. J. 1959. The roles of endocrine and behavioral factors in the growth of mammalian populations. Pages 71–97. In: Gorbman, A. (ed.). *Comparative Endocrinology*. New York: John Wiley.

Christian, J. J. 1963. Endocrine adaptive mechanisms and the physiologic regulation of population growth. Pages 189–353. In: Meyer, M. V., and R. van-Gelder (eds.). *Physiological Mammalogy. 1.* New York: Academic Press.

Cifelli, R. L., T. M. Rowe, W. P. Luckett, J. Banta, R. Reyes, and R. I. Howes. 1996. Fossil evidence for the origin of the marsupial pattern of tooth replacement. *Nature* 379:715–718.

Cipriano, F. and S. R. Palumbi. 1999. Genetic tracking of a protected whale. *Nature* 397:307–308.

Clack, J. A. 1998. A new Early Carboniferous tetrapod with a melange of crown-group characters. *Nature* 394:66–69.

Clapham, P. J., and D. K. Mattila. 1990. Humpback whale songs as indicators of migration routes. *Marine Mammal Science* 6(2):155–160.

Clark, J. M., J. A. Hopson, R. Hernandez R., D. E. Fastovsky, and M. Montellano. 1998. Foot posture in a primitive pterosaur. *Nature* 391:886–889.

Clark, M. M., and B. G. Galef, Jr. 1998. Where the males are. *Natural History* 107(9):22–24.

Clark, W. K. 1951. Ecological life history of the armadillo in eastern Edwards Plateau region. *American Midland Naturalist* 46:337–358.

Claussen, D. L., M. D. Townsley, and R. G. Bausch. 1990. Supercooling and freeze-tolerance in the European wall lizard, *Podarcis muralis*, with a revisional history of the discovery of freeze-tolerance in vertebrates. *Journal of Comparative Physiology B* 160:137–143.

Claussen, D. L., P. M. Daniel, S. Jiang, and N. A. Adams. 1991. Hibernation in the eastern box turtle, *Terrapene c. carolina*. *Journal of Herpetology* 25(3):334–341.

Clemens, W. A., J. D. Archibald, and L. J. Hickey. 1981. Out with a whimper not a bang. *Paleobiology* 7(3):293–298.

Clements, F. E., and V. E. Shelford. 1939. *Bio-ecology*. New York: John Wiley and Sons.

Clouthier, D. E., M. R. Avarbock, S. D. Maika, R. E. Hammer, and R. L. Brinster. 1996. Rat spermatogenesis in mouse testis. *Nature* 381:418–421.

Coates, M. I. 1996. The Devonian tetrapod *Acanthostega gunnari* Jarvik: Postcranial skeleton, basal tetrapod relationships and patterns of skeletal evolution. *Transactions of the Royal Society of Edinburgh. Earth Science.* 87:363–421.

Coates, M. I., and J. A. Clack. 1990. Polydactyly in the earliest known tetrapod limbs. *Nature* 347:66–69.

Coates, M. I., and J. A. Clack. 1991. Fish-like gills and breathing in the earliest known tetrapod. *Nature* 352:234–236.

Cochran, B. 1996. Virginia's geese not cooked yet. *Roanoke Times*, July 21, 1996:C12.

Cockrum, E. L. 1962. *Introduction to Mammalogy*. New York: The Ronald Press Company.

Cohen, J. 1997. Can cloning help save beleaguered species? *Science* 276:1329–1330.

Colbert, E. H. 1962. The weights of dinosaurs. *American Museum Novitates* 2076:1–16.

Colbert, E. H. 1986. Mesozoic tetrapod extinctions: a review. Pages 49-62. In: Elliott, D. K. (ed.). *Dynamics of Extinction*. New York: John Wiley and Sons.

Colborn, T., D. Dumanoski, and J. P. Meyers. 1996. *Our Stolen Future: How We Are Threatening Our Fertility, Intelligence, and Survival—a Scientific Detective Story*. New York: Penguin Books.

Collar, N. J., L. P. Gonzaga, N. Krabbe, A. Madrono Nieto, L. G. Naranjo, T. A. Parker, III, and D. C. Wege. 1993. *Threatened Birds of the Americas.* The ICBP/IUCN Red Data Book. Washington, D.C.: Smithsonian Institution Press.

Compagno, L. J. V. 1973. Interrelationships of living elasmobranchs. Pages 15-62. In: Greenwood, P. H., R. S. Miles, and C. Patterson (eds.). *Interrelationships of Fishes.* New York: Academic Press.

Compagno, L. J. V. 1977. Phyletic relationships of living sharks and rays. *American Zoologist* 17(2):303–322.

Conover, D. O., and B. E. Kynard. 1981. Environmental sex determination: Interaction of temperature and genotype in a fish. *Science* 213:577–579.

Cooke, J. 1991. *The Restless Kingdom: An Exploration of Animal Movement.* New York: Facts on File.

Cooper, A., and D. Penny. 1997. Mass survival of birds across the Cretaceous-Tertiary boundary: Molecular evidence. *Science* 275:1109–1113.

Cooper, H. S. F., Jr. 1996. Peter Moller and his talking fish. *Natural History* 105(1):62, 64–65.

Cooper, W. E., Jr., and L. J. Vitt. 1986. Tracking of female conspecific odor trails by male broad-headed skinks (*Eumeces laticeps*). *Ethology* 71:242–248.

Cope, J. B., K. Koontz, and E. Churchwell. 1961. Notes on homing of two species of bats, *Myotis lucifugus* and *Eptesicus fuscus. Proceedings of the Indiana Academy of Sciences* 70:270–274.

Coppinger, R., and M. Feinstein. 1991. Hark! Hark! The dogs do bark...and bark and bark. *Smithsonian* 21:119–129.

Coria, R. A., and L. Salgado. 1995. A new giant carnivorous dinosaur from the Cretaceous of Patagonia. *Nature* 377:224–226.

Cortelyou, J. R., and D. J. McWhinnie. 1967. Parathyroid glands of amphibians: 1. Parathyroid structure and function in the amphibian, with emphasis on regulation of mineral ions in body fluids. *American Zoologist* 7:843–855.

Cortelyou, J. R., A. Hibner-Owerko, and J. Mulloy. 1960. Blood and urine calcium changes in totally parathyroidectomized *Rana pipiens. Endocrinology* 66:441–450.

Cossins, A. R., and N. Roberts. 1996. The gut in feast and famine. *Nature* 379:23.

Costanzo, J. P., R. E. Lee, Jr., and M. F. Wright. 1991a. Glucose loading prevents freezing injury in rapidly cooled wood frogs. *American Journal of Physiology* 261: R1549–R1553.

Costanzo, J. P., R. E. Lee, Jr., and M. F. Wright. 1991b. Effect of cooling rate on the survival of frozen wood frogs, *Rana sylvatica. Journal of Comparative Physiology* 161:225–229.

Costanzo, J. P., R. E. Lee, Jr., and M. F. Wright. 1992a. Cooling rate influences cryoprotectant distribution and organ dehydration in freezing wood frogs. *Journal of Experimental Zoology* 261:373 –378.

Costanzo, J. P., M. F. Wright, and R. E. Lee, Jr. 1992b. Freeze tolerance as an overwintering adaptation in Cope's grey treefrog (*Hyla chrysoscelis*). *Copeia* 1992(2):565–569.

Costanzo, J. P., P. A. Callahan, R. E. Lee, Jr., and M. F. Wright. 1997. Frogs absorb glucose from urinary bladder. *Nature* 389:343–344.

Coy, P. L., and D. L. Garshelis. 1992. Reconstructing reproductive histories of black bears from the incremental layering in dental cementum. *Canadian Journal of Zoology* 70:2150–2160.

Cracraft, J. 1971. Continental drift and Australian avian biogeography. *Emu* 72(4):171–174.

Cracraft, J. 1973. Continental drift, paleoclimatology, and the evolution and biogeography of birds. *Journal of Zoology* (London) 169(4):455–545.

Cracraft, J. 1974. Phylogeny and evolution of the ratite birds. *Ibis* 116:494–521.

Cracraft, J. 1983. The significance of phylogenetic classifications for systematic and evolutionary biology. Pages 1-17. In: Felsenstein, J. (ed.). *Numerical Taxonomy.* Berlin: Springer-Verlag.

Cracraft, J. 1986. The origin and early diversification of birds. *Paleobiology* 12(4):383–389.

Crandall, L. S. 1964. *The Management of Wild Mammals in Captivity.* Chicago: University of Chicago Press.

Crews, D., and L. J. Young. 1991. Pseudocopulation in nature in a unisexual whiptail lizard. *Animal Behaviour* 42:512–514.

Crichton, E. G., P. H. Krutzsch, and M. Chvapil. 1982. Studies on prolonged spermatozoa survival in Chiroptera—II. The role of zinc in the spermatozoa storage phenomenon. *Comparative Biochemistry and Physiology. Part A. Comparative Physiology* 71A:71–77.

Crick, H. Q. P., C. Dudley, D. E. Glue, and D. L. Thomson. 1997. UK birds are laying eggs earlier. *Nature* 388:526.

Criddle, S. 1937. Snakes from an ant hill. *Copeia* 1937(2):142.

Crowe, M. 1995. Upwardly mobile. *Natural History* 104(8):86–87. (Oct. 1995)

Crompton, A. W., and F. A. Jenkins, Jr. 1968. Molar occlusion in Late Triassic mammals. *Biological Review* 43:427–458.

Crompton, A. W., and F. A. Jenkins, Jr. 1979. Origin of mammals. Pages 59–73. In: Lillegraven, J. A., Z. Kielan-Jaworowski, and W. A. Clemens (eds.). *Mesozoic Mammals: The First Two-Thirds of Mammalian History.* Berkeley: University of California Press.

Culik, B. M., R. P. Wilson, A. T. Woakes, and F. W. Sanudo. 1991. Oil pollution of Antarctic penguins: Effects on energy metabolism and physiology. *Marine Pollution Bulletin* 22(8):388–391.

Culotta, E. 1994. Ninety ways to be a mammal. *Science* 266:1161.

Culotta, E. 1995a. Many suspects to blame in Madagascar extinctions. *Science* 268:1568–1569.

Culotta, E. 1995b. New finds rekindle debate over anthropoid origins. *Science* 268:1851.

Culver, M. 1999. Molecular genetic variation, population structure, and natural history of free-ranging pumas *(Puma concolor).* Ph.D. dissertation. University of Maryland, College Park.

Curio, E. 1978. The adaptive significance of avian mobbing. I. Teleonomic hypotheses and predictions. *Zeitschrift fur Tierpsychologie* 48:175–183.

Curio, E., U. Ernest, and W. Vieth. 1978a. The adaptive significance of avian mobbing. II. Cultural transmission and enemy recognition in blackbirds: effectiveness and some constraints. *Zeitschrift fur Tierpsychologie* 48:185–202.

Curio, E., U. Ernest, and W. Vieth. 1978b. Cultural transmission of enemy recognition: One function of mobbing. *Science* 202:899–901.

Czech, B., and P. R. Krausman. 1997. Distribution and causation of species endangerment in the United States. *Science* 277:1116.

D

Daeschler, E. B., N. H. Shubin, K. S. Thomson, and W. W. Amaral. 1994. A Devonian tetrapod from North America. *Science* 265:639–642.

Dajer, T. 1992. How the nose knows. *Discover* 13(1):67.

Daley, S. 1997. Park uses contraception, not killing, to keep elephants in check. *New York Times*, July 22, 1997:C3.

Dal Sasso, C., and M. Signore. 1998. Exceptional soft-tissue preservation in a theropod dinosaur from Italy. *Nature* 392:383–387.

Daltry, J. C., W. Wuster, and R. S. Thorpe. 1996. Diet and snake venom evolution. *Nature* 379:537–540.

Daly, J. W., G. B. Brown, M. Mensah-Dwumah, and C. W. Myers. 1978. Classification of skin alkaloids from neotropical poison-dart frogs (Dendrobatidae). *Toxicon* 16:163–188.

Daly, J. W., S. I. Secunda, H. M. Garraffo, T. F. Spande, A. Wisnieski, C. Nishihira, and J. F. Cover, Jr. 1992. Variability in alkaloid profiles in neotropical poison frogs (Dendrobatidae): Genetic versus environmental determinants. *Toxicon* 30(8):887–898.

Daly, J. W., H. M. Garraffo, T. F. Spande, C. Jaramillo, and A. S. Rand. 1994a. Dietary source for skin alkaloids of poison frogs (Dendrobatidae)? *Journal of Chemical Ecology* 20(4):943–955.

Daly, J. W., S. I. Secunda, H. M. Garaffo, T. F. Spande, A. Wisnieski, and J. F. Cover, Jr. 1994b. An uptake system for dietary alkaloids in poison frogs (Dendrobatidae). *Toxicon* 32(6):657–663.

Dashzeveg, D., M. J. Novacek, M. A. Norell, J. M. Clark, L. M. Chiappe, A. Davidson, M. C. McKenna, L. Dingus, C. Swisher, and P. Altangerel. 1995. Extraordinary preservation in a new vertebrate assemblage from the Late Cretaceous of Mongolia. *Nature* 374:446–449.

Daszak, P., A. A. Cunningham, and A. D. Hyatt. 2000. Emerging infectious diseases of wildlife—threats to biodiversity and human health. *Science* 287:443–449.

Davies, N. B., and M. Brooke. 1991. Coevolution of the cuckoo and its hosts. *Scientific American* 264(1):92–98.

Davies, N. B., R. M. Kilner, and D. G. Noble. 1998. Nestling cuckoos, *Cuculus canorus*, exploit hosts with begging calls that mimic a brood. *Proceedings of the Royal Society of London B* 265:673–678.

Davis, D. E. (ed.). 1982. *CRC Handbook of Census Methods for Terrestrial Vertebrates.* Boca Raton, Florida: CRC Press.

Davis, D. E., and F. B. Golley. 1963. *Principles in Mammalogy.* New York: Reinhold Publishing.

Davis, D. W. 1969. The behavior and population dynamics of the red squirrel, *Tamiasciurus hudsonicus*, in Saskatchewan. Ph.D. dissertation. University of Arkansas, Fayetteville.

Davis, R. W., L. A. Fuiman, T. M. Williams, S. O. Collier, W. P. Hagey, S. B. Kanatous, S. Kohin, and M. Horning. 1999. Hunting behavior of a marine mammal beneath the Antarctic fast ice. *Science* 283:993–996.

Dawley, E. M. 1984. Recognition of individual, sex, and species odours by salamanders of the *Plethodon glutinosus–P. jordani* complex. *Animal Behaviour* 32:353–361.

Dawley, E. M. 1987. Salamander vomeronasal systems: Why plethodontids smell well. *American Zoologist* 27:166A (Abstract).

Dawley, E. M. 1992. Sexual dimorphism in a chemosensory system: The role of the vomeronasal organ in salamander reproductive behavior. *Copeia* 1992(1):113–120.

Dawley, E. M., and A. H. Bass. 1989. Chemical access to the vomeronasal organs of a plethodontid salamander. *Journal of Morphology* 200:163–174.

Dawson, T. 1995. Kangaroos, the kings of cool. *Natural History* 104(4):38–45.

Dawson, W. R., and J. W. Hudson. 1970. Birds. Page 223. In: Whittow, G. C. (ed.). *Comparative Physiology of Thermoregulation.* New York: Academic Press.

Deban, S. M., and K. C. Nishikawa. 1992. The kinematics of prey capture and the mechanism of tongue protraction in the green tree frog *Hyla cinerea. Journal of Experimental Biology* 170:235–256.

Deban, S. M., and T. Theimer. 1991. The mechanism of defensive inflation in the chuckwalla. *American Zoologist* 31(5):18A.

Deban, S. M., D. B. Wake, and G. Roth. 1997. Salamander with a ballistic tongue. *Nature* 389:27–28.

Dearing, D. 1993. The manipulation of secondary compounds by the North American pika. *Bulletin of the Ecological Society of America* 74 (Supplement 2):210 (Abstract).

Dehnhardt, G., B. Mauck, and H. Bleckmann. 1998. Seal whiskers detect water movements. *Nature* 394:235–236.

Delius, J. D., and J. Emmerton. 1979. Visual performance in pigeons. Pages 51–70. In: Granda, A. M., and J. H. Maxwell (eds.). *Neural Mechanisms of Behaviour in the Pigeon.* New York: Plenum Press.

del Pino, E. M., M. L. Galarza, C. M. de Abuja, and A. A. Humphries, Jr. 1975. The maternal pouch and development in the marsupial frog *Gastrotheca riobambae* (Fowler). *Biological Bulletin* 149:480–491.

Demas, S., M. Duronslet, S. Wachtel, C. Caillouet, and D. Nakamura. 1990. Sex-specific DNA in reptiles with temperature sex determination. *Journal of Experimental Zoology* 253:319–324.

Demian, J. J., and D. H. Taylor. 1977. Photoreception and locomotor rhythm entrainment by the pineal body of the newt, *Notophthalmus viridescens* (Amphibia, Urodela, Salamanderidae). *Journal of Herpetology* 11:131–139.

Demski, L. S., and R. G. Northcutt. 1983. The terminal nerve: A new chemosensory system in vertebrates? *Science* 220:435–437.

Derickson, W. K. 1974. Lipid deposition and utilization in the sagebrush lizard *Sceloporus graciosus*: Its significance for reproduction and maintenance. *Comparative Biochemistry and Physiology* 49A:267–272.

Derickson, W. K. 1976. Lipid storage and utilization in reptiles. *American Zoologist* 16:711–723.

DeRosa, C. T., and D. H. Taylor. 1982. A comparison of compass orientation mechanisms in three turtles (*Trionyx spinifer, Chrysemys picta* and *Terrapene carolina*). *Copeia* 1982:394–399.

deQueiroz, K., 1988. Systematics and the Darwinian revolution. *Philosophy of Science* 55:238–259.

deQueiroz, K., and J. Gauthier. 1992. Phylogenetic taxonomy. *Annual Review of Ecology and Systematics* 23:449–480.

Dial, B. E., and L. C. Fitzpatrick. 1983. Lizard tail autotomy: Function and energetics of postautotomy tail movement in *Scincella lateralis. Science* 219:391–393.

Diamond, J. 1987. The American blitzkrieg: A mammoth undertaking. *Discover* 8(6):82-88.

Diamond, J. 1988. Double trouble. *Discover* 9:64–70.

Diamond, J. 1994. Dining with the snakes. *Discover* 15(4):48–49, 52–59.

Diamond, J. M. 1988. Relationships of humans to chimps and gorillas. *Nature* 334:656.

Diamond, J. M. 1990a. Old dead rats are valuable. *Nature* 347:334–335.

Diamond, J. M. 1990b. How to fuel a hummingbird. *Nature* 348:392.

Diamond, J. M. 1991. Twilight of Hawaiian birds. *Nature* 353:505–506.

Diamond, J. M. 1996. Competition for brain space. *Nature* 382:756–757.

Dice, L. R. 1943. *The Biotic Provinces of North America.* Ann Arbor: University of Michigan Press.

Dickman, C. R. 1992. Commensal and mutualistic interactions among terrestrial vertebrates. *Trends in Ecology and Evolution* 7(6):194–197.

Dickman, C. R., M. Predavec, and A. J. Lynam. 1991. Differential predation of size and sex classes of mice by the barn owl, *Tyto alba. Oikos* 62(1):67–76.

Dillehay, T. D. 1989. *Monte Verde: A Late Pleistocene Settlement in Chile. Volume 1. Paleoenvironment and Site Context.* Washington, D.C.: Smithsonian Institution Press.

Dillehay, T. D. 1997. *Monte Verde: A Late Pleistocene Settlement in Chile. Volume 2. The Archaeological Context and Interpretation.* Washington, D.C.: Smithsonian Institution Press.

Dixon, A., D. Ross, S. L. C. O'Malley, and T. Burke. 1994. Paternal investment inversely related to degree of extra-pair paternity in the reed bunting. *Nature* 371:698–700.

Dobson, A. P., J. P. Rodriguez, W. M. Roberts, and D. S. Wilcove. 1997. Geographic distribution of endangered species in the United States. *Science* 275:550–553.

Dodson, E. O. and P. Dodson. 1985. *Evolution: Process and Product.* Third edition. Boston: Prindle, Weber, and Schmidt.

Dolk, H. E., and N. Postma. 1927. Ueber die Hautund die Lungenatmung von *Rana temporaria. Zeitschrift Vergleich Physiologie* 5:417–444.

Dodt, E., and E. Heerd. 1962. Mode of action of pineal nerve fibers in frogs. *Journal of Neurophysiology* 25:405–429.

Dodt, E., and M. Jacobson. 1963. Photosensitivity of a localized region of the frog diencephalon. *Journal of Neurophysiology* 26:752–758.

Dold, C. 1996. Hormone hell. *Discover* 17(9):52–59.

Dole, J. W. 1967. The role of substrate moisture and dew in the water economy of leopard frogs, *Rana pipiens. Copeia* 1967:141–149.

Dole, J. W. 1968. Homing in leopard frogs, *Rana pipiens. Ecology* 49(1):386–399.

Doncaster, C. P., and D. W. Macdonald. 1991. Drifting territoriality in the red fox *Vulpes vulpes. Journal of Animal Ecology* 60:423–439.

Dooley, J. K., and J. Collura. 1988. Homing behavior in tidepool fishes. *Underwater Naturalist* 17:3–6.

Douglas, E. L., K. S. Peterson, J. R. Gysi, and D. J. Chapman. 1985. Myoglobin in the heart tissue of fishes lacking hemoglobin. *Comparative Biochemistry and Physiology.* Part A. 81:855–888.

Douglas, J. S. 1997. Handling the snake problem in Guam. *Virginia Tech Magazine* 19(2):27–28.

Doyle, R. 1997. Threatened birds. *Scientific American* 277(3):28.

Doyle, R. 1998. Amphibians at risk. *Scientific American* 279(2):27.

Dragoo, J. W., and R. L. Honeycutt. 1997. Systematics of mustelid-like carnivores. *Journal of Mammalogy* 78(2):426–443.

Drent, R. H. 1970. Functional aspects of incubation in the Herring Gull. *Behaviour Supplement* 17:1–132.

Drollette, D. 1996. Australia fends off critic of plan to eradicate rabbits. *Science* 272:191–192.

Drollette, D. 1997. Wide use of rabbit virus is good news for native species. *Science* 275:154.

Drummond, F. H. 1946. Pharyngeoesophageal respiration in the lizard, *Trachysaurus rugosus. Proceedings of the Zoological Society of London* 116:225–228.

Dubach, J., and A. Sajewicz. 1997. Parthenogenesis in the Arafuran filesnake (*Acrochordus arafurae*). *Herpetological Natural History* 5(1):11–18.

Duellman, W. E. 1989. Alternative life-history styles in anuran amphibians: evolutionary and ecological implications. Pages 101–126. In: Bruton, M. N. (ed.). *Alternative Life-History Styles of Animals.* Dordrecht, Netherlands: Kluwer Academic Publishers.

Duellman, W. E. 1992. Reproductive strategies of frogs. *Scientific American* 267(1):80–87.

Duellman, W. E., and L. Trueb. 1986. *Biology of Amphibians.* New York: McGraw-Hill.

Dugatkin, L. A. 1997. *Cooperation Among Animals: An Evolutionary Perspective.* New York: Oxford University Press.

Dumas, P. C. 1956. The ecological relations of sympatry in *Plethodon dunni* and *Plethodon vehiculum. Ecology* 37:484–495.

Dumbacher, J., B. M. Beehler, T. F. Spande, H. M. Garraffo, and J. W. Daly. 1992. Hemobatrachotoxin in the genus *Pitohui*: Chemical defense in birds? *Science* 258:799–801.

Dunaway, P. B. 1959. Growth of cotton rats. Pages 33-37. In: Ecological Research, Health Physics Division. Annual Progress Report, Oak Ridge National Laboratory for period ending July 31, 1959. Oak Ridge, Tennessee.

Dunham, A. E., P. J. Morin, and H. M. Wilbur. 1988. Methods for the study of reptile populations. Pages 331-386. In: Gans, C., and R. B. Huey (eds.). *Biology of the Reptilia. Volume 16, Ecology B, Defense and Life History.* New York: Alan R. Liss.

Dunning, J. S. 1987. *South American Birds.* Newton Square, Pennsylvania: Harrowood Books.

Dunson, W. A., and R. D. Weymouth. 1965. Active uptake of sodium by softshell turtles (*Trionyx spinifer*). *Science* 149:67–69.

Duston, T. H. 1997. Rabbit control in New Zealand. *Science* 276:20.

Dymond, J. R. 1922. A provisional list of the fishes of Lake Erie. *University of Toronto Studies in Biology Series* 20:57–73.

E

Eastman, J. T. 1993. *Antarctic Fish Biology.* San Diego: Academic Press.

Eckert, S. A., D. W. Nellis, K. L. Eckert, and G. L. Kooyman. 1986. Diving patterns of two leatherback sea turtles (*Dermochelys coriacea*) during interesting intervals at Sandy Point, St. Croix, U.S. Virgin Islands. *Herpetologica* 42:381–388.

Edmonds, D. T. 1992. A magnetite null detector as the migrating bird's compass. *Proceedings of the Royal Society of London B* 249:27–31.

Edmund, A. G. 1969. Dentition. Pages 117-200. In: Gans, C., A. d. Bellairs, and T. S. Parsons (eds.). *Biology of the Reptilia. Volume I.* New York: Academic Press.

Edwards, H. H., G. D. Schnell, R. L. DuBois, and V. H. Hutchison. 1992. Natural and induced remanent magnetism in birds. *The Auk* 109(1):43–56.

Edwards, J. 1989. Two perspectives on the evolution of the tetrapod limb. *The American Zoologist* 29:235–254.

Ehret, G. 1983. Auditory processing and perception of ultrasounds in house mice. Pages 911–917. In: Ewert, J.-P., R. R. Capranica, and D. J. Ingle (eds.). *Advances in Vertebrate Neuroethology.* New York: Plenum Press.

Ehrhart, L. M. 1981. A review of sea turtle reproduction. Pages 117–125. In: Bjorndal, K. A. (ed.). *Biology and Conservation of Sea Turtles.* Washington, D.C.: Smithsonian Institution Press.

Ehrlich, P. H. 1986. Extinction: What is happening now and what needs to be done. Pages 157–164. In: Elliott, D. K. (ed.). *Dynamics of Extinction.* New York: John Wiley and Sons.

Eibl-Eibesfeldt, I. 1949. Über das Vorkommen von Shreckstoffen bei Erdkrotenquappen. *Experientia* 5:236.

Einarsen, A. S. 1948. *The Pronghorn Antelope.* Washington, D.C.: Wildlife Management Institute.

Ekman, S. 1941. Ein laterales Flossensaumrudiment bein Haiembryonen. *Nova Acta R. Soc. Scient. Upsal.* IV. 12(7):5–44.

Eldredge, N., and J. Cracraft. 1980. *Phylogenetic Patterns and the Evolutionary Process.* New York: Columbia University Press.

Elgar, M. A. 1990. Evolutionary compromise between a few large and many small eggs: Comparative evidence in teleost fish. *Oikos* 59(2):283–287.

Elgar, M. A., and B. J. Crespi. 1992. Ecology and evolution of cannibalism. Pages 1-12. In: Elgar, M. A., and B. J. Crespi (eds.). *Cannibalism.* New York: Oxford University Press.

Elton, C. S. 1927. *Animal Ecology.* London: Sidgwick and Jackson.

Elwood, R. 1992. Pup-cannibalism in rodents: Causes and consequences. Pages 299-322. In: Elgar, M. A., and B. J. Crespi (eds.). *Cannibalism.* New York: Oxford University Press.

Elzanowski, A., and P. Wellnhofer. 1992. A new link between theropods and birds from the Cretaceous of Mongolia. *Nature* 359:821–823.

Emerson, S. B., and D. Diehl. 1980. Toe pad morphology and mechanisms of sticking in frogs. *Biological Journal of the Linnean Society* 13:199–216.

Emlen, S. T. 1967. Migratory orientation in the indigo bunting, *Passerina cyanea.* Part II. Mechanism of celestial orientation. *The Auk* 84:463–489.

Emlen, S. T., and P. H. Wrege. 1992. Parent -offspring conflict and the recruitment of helpers among bee-eaters. *Nature* 356:331–333.

Emmons, L. H., and A. Biun. 1991. Malaysian treeshrews. *National Geographic Research and Exploration* 7(1):70–81.

Erdmann, M. V., R. L. Caldwell, and M. Kasim Moosa. 1998. Indonesian "king of the sea" discovered. *Nature* 395:335.

Erlinge, S. 1987. Predation and non-cyclicity in a microtine population in southern Sweden. *Oikos* 50:347–352.

Erlinge, S., J. Agrell, J. Nelson, and M. Sandell. 1991. Why are some microtine populations cyclic while others are not? *Acta Theriologica* 36(1–2):63–71.

Erlinge, S., G. Göransson, L. Hansson, G. Högstedt, O. Liberg, I. Nilsson, T. Nilsson, T. von Schantz, and M. Sylvén. 1983. Predation as a regulating factor in small rodents in southern Sweden. *Oikos* 40:36–52.

Ernst, V. V. 1973. The digital pads of the treefrog, *Hyla cinerea.* I. The epidermis. *Tissue and Cell* 5:83–96.

Ernst, V. V., and R. Ruibal. 1966. The structure of the digital lamellae of lizards. *Journal of Morphology* 120:233–265.

Erwin, D. H. 1994. The Permo-Triassic extinction. *Nature* 367:231–236.

Erwin, D. H. 1996. The mother of mass extinctions. *Scientific American* 275(1): 72–78.

Estes, J. A., M. T. Tinker, T. M. Williams, and D. F. Doak. 1998. Killer whale predation on sea otters linking oceanic and nearshore ecosystems. *Science* 282: 473–476.

Estes, R. 1981. Gymnophiona, Caudata. *Handbuch de Paläherpetologie* 2:1–115.

Estes, R., and O. A. Reig. 1973. The early fossil record of frogs: A review of the evidence. Pages 11–63. In: Vial, J. L. (ed.). *Evolutionary Biology of the Anurans: Contemporary Research on Major Problems.* Columbia: University of Missouri Press.

Estes, R., and M. H. Wake. 1972. The first fossil record of caecilian amphibians. *Nature* 239:228–231.

Estrada, A. R., and S. B. Hedges. 1996. At the lower size limit in tetrapods: A new diminutive frog from Cuba (Leptodactylidae: Eleutherodactylus). *Copeia* 1996(4):852–859.

Etchberger, C. R., M. A. Ewert, B. A. Raper, and C. E. Nelson. 1992. Do low incubation temperatures yield females in painted turtles? *Canadian Journal of Zoology* 70(2):391–394.

Evans, C. P. 1938. Observations on hibernating bats with especial reference to reproduction and splenic adaptation. *American Naturalist* 72:480–484.

Evans, P. G. H. 1987. *The Natural History of Whales and Dolphins.* New York: Facts on File.

Evans, R. M. 1988. Embryonic vocalizations as care-soliciting signals, with particular reference to the American white pelican. *Proceedings of the 19th International Ornithological Congress*:1467–1475.

Evans, R. M. 1990a. Terminal-egg chilling and hatching intervals in the American white pelican. *The Auk* 107(2):431–434.

Evans, R. M. 1990b. Terminal-egg neglect in the American white pelican. *Wilson Bulletin* 102(4):684–692.

Evans, R. M. 1990c. Embryonic fine tuning of pipped egg temperature in the American white pelican. *Animal Behaviour* 40:963–968.

Evans, R. M. 1990d. Vocal regulation of temperature by avian embryos: A laboratory study with pipped eggs of the American white pelican. *Animal Behaviour* 40:969–979.

Evans, R. M. 1992. Embryonic and neonatal vocal elicitation of parental brooding and feeding responses in American white pelicans. *Animal Behaviour* 44:667–675.

Evans, W. E., and P. F. Maderson. 1973. Mechanisms of sound production in delphinid cetaceans: A review and some anatomical considerations. *American Zoologist* 13:1205–1213.

Ewert, M. A. 1985. Embryology of turtles. Pages 75–267. In: Gans, C. (ed.). *Biology of the Reptilia. Volume 14. Development A.* New York: John Wiley and Sons.

Ewert, M. A., and C. E. Nelson. 1991. Sex determination in turtles: Diverse patterns and some possible adaptive values. *Copeia* 1991:50–59.

Ezzell, C. 2000. Jumbo trouble. *Scientific American* 282(1):41–42.

F

Faaborg, J. 1988. *Ornithology: An Ecological Approach.* Englewood Cliffs, New Jersey: Prentice Hall.

Fackelmann, K. A. 1991. Randy reptiles. *Science News* 139(19):300–301.

Fackelmann, K. A. 1992. Bluebird fathers favor pink over blue. *Science News* 141(1):7.

Falcon, J., J. F. Guerlotte, P. Voisin, and J.-P. H. Collin. 1987. Rhythmic melatonin biosynthesis in a photoreceptive pineal organ: A study in the pike. *Neuroendocrinology* 45:479–486.

Fänge, R. 1976. Gas exchange in the swimbladder. Pages 189-211. In: Hughes, G. M. (ed.). *Respiration of Amphibious Vertebrates.* New York: Academic Press.

Farlow, J. O. 1975. The behavioral significance of frill and horn morphology in ceratopsian dinosaurs. *Evolution* 29:353–361.

Faulkes, C. G., and D. H. Abbott. 1991. Social control of reproduction in breeding and non-breeding male naked mole-rats (*Heterocephalus glaber*). *Journal of Reproduction and Fertility* 93:427–435.

Fay, R. R. 1988. *Hearing in Vertebrates: A Psychophysics Databook.* Winnetka, Illinois: Hill-Fay.

Feder, H. M. 1966. Cleaning symbiosis in the marine environment. Pages 327–380. In: Henry, S. M. (ed.). *Symbiosis I.* New York: Academic Press.

Feder, M. E., and W. W. Burggren. 1992. *Environmental Physiology of Amphibians.* Chicago: University of Chicago Press.

Feduccia, A. 1980. *The Age of Birds.* Cambridge, Massachusetts: Harvard University Press.

Feduccia, A. 1993. Evidence from claw geometry indicating arboreal habits of *Archaeopteryx. Science* 259:790–793.

Feduccia, A. 1995. Explosive evolution in Tertiary birds and mammals. *Science* 267:637–638.

Feduccia, A. 1996. *The Origin and Evolution of Birds.* New Haven, Connecticut: Yale University Press.

Feduccia, A., and E. McCrady. 1991. *Torrey's Morphogenesis of the Vertebrates.* Fifth edition. New York: John Wiley and Sons.

Fenton, M. B. 1970. Population studies of *Myotis lucifugus* (Chiroptera: Vespertilionidae) in Ontario. *Royal Ontario Museum of Life Sciences Contribution* 77:1–34.

Ferguson, D. E. 1967. Sun-compass orientation in anurans. Pages 21–32. In: Storm, R. M. (ed.). *Animal Orientation and Navigation.* Corvallis: Oregon State University Press.

Ferguson, D. E., and H. F. Landreth. 1966. Celestial orientation of Fowler's toad, *Bufo fowleri. Behaviour* 26:105–123.

Ferguson, M. W. J., and T. Joanen. 1982. Temperature of egg incubation determines sex in *Alligator mississippiensis. Nature* 296(5860):850–853.

Ferkin, M. H., M. R. Gorman, and I. Zucker. 1992. Influence of gonadal hormones on odours emitted by male meadow voles (*Microtus pennsylvanicus*). *Journal of Reproduction and Fertility* 95:729–736.

Finck, P. 1990. Seasonal variation of territory size with the little owl (*Athene noctua*). *Oecologia* 83:68–75.

Finkbeiner, A. 1993. Nocturnal researchers tune their ears to our ancestors. *Science* 261:30–31.

Fischman, J. 1993a. Little bird, big find. *Discover* 14(1):66–67.

Fischman, J. 1993b. A closer look at the dinosaur-bird link. *Science* 262:1975.

Fischman, J. 1995a. Were dinos cold-blooded after all? The nose knows. *Science* 270:735–736.

Fischman, J. 1995b. Why mammal ears went on the move. *Science* 270:1436.

Fite, K. V. (ed.). 1976. *The Amphibian Visual System.* New York: Academic Press.

FitzGerald, G. J. 1991. The role of cannibalism in the reproductive ecology of the threespine stickleback. *Ethology* 89:177–194.

FitzGerald, G. J. 1992. Filial cannibalism in fishes: Why do parents eat their offspring? *Trends in Ecology and Evolution* 7(1):7–10.

Flam, F. 1991. Brotherhood of lions. *Science* 253:28.

Flannery, T. 1996. Up from the ashes. *Natural History* 105(3):6–7.

Flannery, T. F., Boeadi, and A. L. Szalay. 1995. A new tree-kangaroo (*Dendrolagus*: Marsupialia) from Irian Jaya, Indonesia, with notes on ethnography and the evolution of tree-kangaroos. *Mammalia* 59(1):65–84.

Flavin, C. 1997. The legacy of Rio. Pages 1–22. In: Brown, L. R., C. Flavin, H. F. French, and L. Starke (eds.). *State of the World, 1997.* New York: W. W. Norton.

Fleskes, J. P. 1991. Two incubating mallards move eggs to drier nest sites. *Prairie Naturalist* 23(1):49–50.

Flood, P. R. 1966. A peculiar mode of muscular innervation in Amphioxus: Light and electron microscopic studies of the so-called ventral roots. *Journal of Comparative Neurology* 125:181–217.

Flyger, V. 1969. The 1968 squirrel "migration" in the eastern United States. *Transactions of the Northeast Section of the Wildlife Society* 26:69–79.

Flyger, V., and J. E. Gates. 1983. Pine squirrels. Pages 230–238. In: Chapman, J. A., and G. A. Feldhamer (eds.). *Wild Mammals of North America.* Baltimore, Maryland: The Johns Hopkins University Press.

Flynn, J.J., J.M. Parrish, B. Rakotosamimanana, W.F. Simpson, R.L. Whatley, and A.R. Wyss. 1999. A Triassic fauna from Madagascar, including early dinosaurs. *Science* 286:763–765.

Foran, D. 1991. Evidence of luminous bacterial symbionts in the light organs of myctophid and stomiiform fishes. *Journal of Experimental Zoology* 259:1–8.

Forester, D. C., and A. Wisnieski. 1991. The significance of airborne olfactory cues to the recognition of home area by the dart-poison frog *Dendrobates pumilio. Journal of Herpetology* 25(4):502–504.

Forey, P. L. 1984. Yet more reflections on agnathan-gnathostome relationships. *Journal of Vertebrate Paleontology* 4(3): 330–343.

Forey, P. L. 1986. Relationships of lungfishes. *Journal of Morphology Supplement* 1:75–91.

Forey, P. 1991. Blood lines of the coelacanth. *Nature* 351:347–348.

Forster, C. A., S. D. Sampson, L. M. Chiappe, and D. W. Krause. 1998. The theropod ancestry of birds: New evidence from the Late Cretaceous of Madagascar. *Science* 279:1915–1919.

Fox, R. C., G. P. Youzwyshyn, and D. W. Krause. 1992. Post-Jurassic mammal-like reptile from the Paleocene. *Nature* 358:233–235.

Frank, L. G. 1986. Social organization of the spotted hyaena *Crocuta crocuta.* II. Dominance and reproduction. *Animal Behaviour* 34:1510–1527.

Frank, L. G., S. E. Glickman, and P. Licht. 1991. Fatal sibling aggression, precocial development, and androgens in neonatal spotted hyenas. *Science* 252:702–704.

Fraser, N. C. 1996. Are dinosaurs really extinct? *Virginia Explorer* 12(4):9–13.

Frey, E., H-D. Sues, and W. Munk. 1997. Gliding mechanism in the Late Permian reptile *Coelurosauravus. Science* 275: 1450–1452.

Fricke, H., K. Hissmann, J. Schauer, and R. Plante. 1995. Yet more danger for coelacanths. *Nature* 374:314.

Friedmann, H., and L. F. Kiff. 1985. The parasitic cowbirds and their hosts. *Proceedings of the Western Foundation of Vertebrate Zoology* 2:225–304.

Fritts, T. H., and G. H. Rodda. 1998. Alien snake threatens Pacific islands. *Endangered Species Bulletin* 23(6):10–11.

Futuyma, D. J. 1986. *Evolutionary Biology.* Second edition. Sunderland, Massachusetts: Sinaur Associates.

G

Gabbotts, S. E., R. J. Aldridge, and J. N. Theron. 1995. A giant conodont with preserved muscle tissue from the Upper Ordovician of South Africa. *Nature* 374:800–803.

Gaeth, A. P., R.V. Short, and M. B. Renfree. 1999. The developing renal, reproductive, and respiratory systems of the African elephant suggest aquatic ancestry. *Proceedings of the National Academy of Sciences,* 96(10):5555–5558.

Gagneux, P., D. S. Woodruff, and C. Boesch. 1997. Furtive mating in female chimpanzees. *Nature* 387:358–359.

Galbraith, D. A., and R. J. Brooks. 1989. Age estimates for snapping turtles. *Journal of Wildlife Management* 53(2):502–508.

Galbraith, D. A., M. W. Chandler, and R. J. Brooks. 1987. The fine structure of home ranges of male *Chelydra serpentina*: Are snapping turtles territorial? *Canadian Journal of Zoology* 65:2623–2629.

Galbreath, G. J. 1983. Armadillo. Pages 71–79. In: Chapman, J. A., and G. A. Feldhamer (eds.). *Wild Mammals of North America.* Baltimore, Maryland: The Johns Hopkins University Press.

Gans, A., and T. S. Parsons. 1970. *Biology of the Reptilia. Volume II. Morphology B.* New York: Academic Press.

Gans, A., d'A. Bellairs, and T. S. Parsons (eds.). 1969. *Biology of the Reptilia. Volume I. Morphology.* New York: Academic Press.

Gans, C., and G. C. Gorniak. 1982. How does the toad flip its tongue? Test of two hypotheses. *Science* 216:1335–1337.

Gans, C., and G. M. Hughes. 1967. The mechanism of lung ventilation in the tortoise *Testudo graeca* Linné. *Journal of Experimental Biology* 47:1–20.

Garcia-Fernandez, J., and P. W. H. Holland. 1994. Archetypal organization of the amphioxus Hox gene cluster. *Nature* 370:563–565.

Gardiner, B. G. 1982. Tetrapod classification. *Zoological Journal of the Linnean Society* 74:207–232.

Garstang, W. 1928. The morphology of Tunicata. *Quarterly Journal of the Microscopical Society* 72:51–87.

Gasc, J.-P. 1967. Introduction a l'etude de la musculature axiale des squamates serpentiformes. *Museum National d'Histoire Naturelle, Paris. Memoires (N.S.), A* 48(2):69–125.

Gaskin, D. E. 1982. *The Ecology of Whales and Dolphins*. London: Heinemann.

Gatten, R. E., Jr., K. Miller, and R. J. Full. 1992. Energetics at rest and during locomotion. Pages 314–377. In: Feder, M. E., and W. W. Burggren (eds.). *Environmental Physiology of the Amphibians*. Chicago: University of Chicago Press.

Gaudin, T. J., and A. A. Biewener. 1992. The functional morphology of xenarthrous vertebrae in the armadillo *Dasypus novemcinctus* (Mammalia, Xenarthra). *Journal of Morphology* 214:63–81.

Gauthier, J. A., A. G. Kluge, and T. Rowe. 1988a. The early evolution of the Amniota. Pages 103–155. In: Benton, M. J. (ed.). *The Phylogeny and Classification of the Tetrapods. Volume I: Amphibians, Reptiles, Birds*. Oxford, England: Clarendon Press. Systematics Association Special Volume 35A.

Gauthier, J. A., A. G. Kluge, and T. Rowe. 1988b. Amniote phylogeny and the importance of fossils. *Cladistics* 4:104–209.

Gebo, D. L., M. Dagosto, K. C. Beard, and T. Qi. 2000b. The smallest primates. *Journal of Human Evolution* 38:585–594.

Gebo, D. L., M. Dagosto, K. C. Beard, T. Qi, and J. Wang. 2000a. The oldest known anthropoid postcranial fossils and the early evolution of higher primates. *Nature* 404:276–278.

Gee, H. 1994. Return of the amphioxus. *Nature* 370:504–505.

Gegenbaur, C. 1872. Über das Archipterygium. *Jena Z. Naturw.* 7:131–141.

Gegenbaur, C. 1876. Zur Morphologie der Gliedmaassen der Wirbelthiere. *Morphologische Jahrbuch* 2:396–420.

Gehlbach, F. R., and R. S. Baldridge. 1987. Live blind snakes (*Leptotyphlops dulcis*) in eastern screech owl (*Otus asio*) nests: A novel commensalism. *Oecologia* 71:560–563.

Geiser, F., and R. V. Baudinette. 1990. The relationship between body mass and rate of rewarming from hibernation and daily torpor in mammals. *Journal of Experimental Biology* 151:349–359.

Geist, N. R., and T. D. Jones. 1996. Juvenile skeletal structure and the reproductive habits of dinosaurs. *Science* 272:712–714.

Gibbons, A. 1994. Dino embryo recasts parents' image. *Science* 266:731.

Gibbons, A. 1996a. The species problem. *Science* 273:1501.

Gibbons, A. 1996b. First Americans: Not mammoth hunters, but forest dwellers? *Science* 272:346–347.

Gibbons, A. 1996c. New feathered fossil brings dinosaurs and birds closer. *Science* 274:720–721.

Gibbons, A. 1996d. Early birds rise from China fossil beds. *Science* 274:1083.

Gibbons, A. 1997a. Did birds sail through the K-T extinction with flying colors? *Science* 275:1068.

Gibbons, A. 1997b. Feathered dino wins a few friends. *Science* 275:1731.

Gibbons, A. 1997c. Lung fossils suggest dinos breathed in cold blood. *Science* 278:1229.

Gibson, M. J., and D. T. Krieger. 1981. Circadian corticosterone rhythm and stress response in rats with adrenal autotransplants. *American Journal of Physiology* 249:E363–E366.

Gibson, R. N. 1969. The biology and behavior of littoral fish. *Oceanographic and Marine Biology Annual Review* 7:367–410.

Gibson, R. N. 1982. Recent studies on the biology of intertidal fishes. *Oceanographic and Marine Biology Annual Review* 20:363–414.

Giffin, E. B. 1990. Gross spinal anatomy and limb use in living and fossil reptiles. *Paleobiology* 16(4):448–458.

Giffin, E. B. 1991. Endosacral enlargements in dinosaurs. *Modern Geology* 16:101–112.

Gilbert, D. A., N. Lehman, S. J. O'Brien, and R. K. Wayne. 1990. Genetic fingerprinting reflects population differentiation in the California Channel Island fox. *Nature* 344:764–767.

Gillette, D. D. 1995. True grit. *Natural History* 104(6):41–43.

Gimeno, S., A. Gerritsen, T. Bowmer, and H. Komen. 1996. Feminization of male carp. *Nature* 384:221–222.

Gingerich, P. D. 1993. Rates of evolution in Plio-Pleistocene mammals: Six case studies. Pages 84–106. In: Martin, R. A., and A. D. Barnosky (eds.). *Morphological Changes in Quaternary Mammals of North America*. New York: Cambridge University Press.

Gingerich, P. D., B. H. Smith, and E. L. Simons. 1990. Hind limbs of Eocene *Basilosaurus*: Evidence of feet in whales. *Science* 249:154–157.

Girgis, S. 1961. Aquatic respiration in the common Nile turtle, *Trionyx triunguis* (Forskal). *Comparative Biochemistry and Physiology* 3:206–217.

Glausiusz, J. 1998. The frog solution. *Discover* 19(11):88–90, 92, 94.

Glegg, T. M. 1966. The abundance of shrews as indicated by trapping and remains in discarded bottles. *American Naturalist* 899:122.

Gliwicz, J. 1990. The first born, their dispersal, and vole cycles. *Oecologia* 83:519–522.

Gnam, R. S. 1991. Underground parrots: Nesting habits of the Bahama parrot. *American Zoologist* 31(5):11A (Abstract).

Gober, P., and M. Lockhart. 1996. As goes the prairie dog...so goes the ferret. *Endangered Species Bulletin* 21(6):4–5.

Godin, A. J. 1960. A compilation of diagnostic characteristics used in aging and sexing game birds and mammals. Master's thesis. University of Massachusetts, Amherst.

Goin, C. J., O. B. Goin, and G. R. Zug. 1978. *Introduction to Herpetology*. Third edition. New York: W. H. Freeman.

Goldsmith, T. H. 1980. Hummingbirds see near ultraviolet light. *Science* 207:786–788.

Goldsmith, T. H., J. S. Collins, and S. Licht. 1984. The cone oil droplets of avian retinas. *Vision Research* 24:1661–1671.

Gomendio, M., and E. R. S. Roldan. 1991. Sperm competition influences sperm size in mammals. *Proceedings of the Royal Society of London B* 243:181–185.

Gondo, M., and H. Ando. 1995. Comparative histophysiological study of oil droplets in the avian retina. *Japanese Journal of Ornithology* 44:81–91.

Goodman, B. 1991. Holy phylogeny! Did bats evolve twice? *Science* 253:36.

Goodman, M., J. Czelusniak, and J. E. Beeber. 1985. Phylogeny of primates and other eutherian orders: A cladistic analysis using amino acid and nucleotide sequence data. *Cladistics* 1:171–185.

Goodrich, E. S. 1930. *Studies on the Structure and Development of Vertebrates*. London: Macmillan.

Gore, R. 1989. The march toward extinction. *National Geographic* 175(6):662–699.

Gorman, M. L., and B. J. Trowbridge. 1989. The role of odor in the social lives of carnivores. Pages 57–88. In: Gittleman, J. L. (ed.). *Carnivore Behavior, Ecology, and Evolution*. Ithaca, New York: Comstock Publishing.

Gorr, R., T. Kleinschmidt, and H. Fricke. 1991. Close tetrapod relationships of the coelacanth *Latimeria* indicated by haemoglobin sequences. *Nature* 351:394–397.

Goss, R. J. 1983. *Deer Antlers: Regeneration, Function, and Evolution*. New York: Academic Press.

Gould, E., W. McShea, and T. Grand. 1993. Function of the star in the star-nosed mole, *Condylura cristata*. *Journal of Mammalogy* 74:108–116.

Gould, J. L. 1980. The case for magnetic sensitivity in birds and bees (such as it is). *American Scientist* 68:256–267.

Gould, S. J. 1989. *Wonderful Life: The Burgess Shale and the Nature of History*. New York: W. W. Norton.

Gowaty, P. A. 1997. Birds face sexual discrimination. *Nature* 385:486–487.

Grafe, T. U., and K. E. Linsenmair. 1989. Protogynous sex change in the reed frog *Hyperolius viridiflavus*. *Copeia* 1989(4):1024–1029.

Graham, J. B. 1974. Aquatic respiration in the sea snake, *Pelamis platurus*. *Respiration Physiology* 21:1–7.

Graham, R. R. 1934. The silent flight of owls. *Journal of the Royal Aeronautical Society* 38:837–843.

Grajal, A., S. D. Strahl, R. Parra, M. G. Dominguez, and A. Neher. 1989. Foregut fermentation in the hoatzin, a neotropical leaf-eating bird. *Science* 245:1236–1238.

Graur, D., W. A. Hide, and W-H. Li. 1991. Is the guinea-pig a rodent? *Nature* 351:649–652.

Green, D. D. 1940. Controlling predatory animals. *American Wildlife* 29(1):35–40.

Green, D. M. 1979. Treefrog toe-pads: Comparative surface morphology using scanning electron microscopy. *Canadian Journal of Zoology* 57:2033–2046.

Green, D. M. 1981. Adhesion and the toe-pads of treefrogs. *Copeia* 1981:790–796.

Green, D. M., and J. Carson. 1988. The adhesion of treefrog toe-pads to glass: Cryogenic examination of a capillary adhesion system. *Journal of Natural History* 22:131–135.

Green, R. E. 1998. Long-term decline in the thickness of eggshells of thrushes, *Turdus* spp., in Britain. *Proceedings of the Royal Society of London B* 265:679–684.

Greenwalt, C. H. 1962. Dimensional relationships for flying animals. *Smithsonian Miscellaneous Collection* 144:1–46.

Greer, A. E., and F. Parker. 1979. On the identity of the New Guinea scincid lizard *Lygosoma fragile* Maclean 1877, with notes on its natural history. *Journal of Herpetology* 13(3):221–225.

Griffin, D. R. 1953. Acoustic orientation in the oil bird, *Steatornis*. *Proceedings of the National Academy of Sciences USA*. 39:884–893.

Griffin, D. R. 1954. Bird sonar. *Scientific American* 190(3):78–83.

Griffiths, M. 1978. *The Biology of the Monotremes*. New York: Academic Press.

Grillner, S. 1996. Neural networks for vertebrate locomotion. *Scientific American* 274(1):64, 66–69.

Griscom, L. 1937. A monographic study of the red crossbill. *Proceedings of the Boston Society of Natural History* 41:77–210.

Groot, C., T. P. Quinn, and H. J. Hara. 1986. Responses of migrating adult sockeye salmon (*Onchorhynchus nerka*) to population-specific odours. *Canadian Journal of Zoology* 64:926–932.

Grubb, P. 1971. The growth, ecology and population structure of giant tortoises on Aldabra. *Philosophical Transactions of the Royal Society of London, Series B* 260:327–372.

Gruber, S. H., D. I. Hamasaki, and B. L. Davis. 1975. Window to the epiphysis in sharks. *Copeia* 1975(2):378–380.

Grzimek, B. (ed.). 1990. *Grzimek's Encyclopedia of Mammals*. 5 volumes. New York: McGraw-Hill.

Gubernick, D. J., and J. C. Nordby. 1992. Parental influences on female puberty in the monogamous California mouse, *Peromyscus californicus*. *Animal Behaviour* 44:259–267.

Guernsey, D. 1997. Wolf update. *Yellowstone Today*. Autumn 1997:7.

Guillette, L. J., Jr. 1991. The evolution of viviparity in amniote vertebrates: New insights, new questions. *Journal of Zoology* (London) 223:521–526.

Gullion, G. W. 1952. Sex and age determination in the American coot. *Journal of Wildlife Management* 16:191–197.

Gunderson, H. L. 1976. *Mammalogy*. New York: McGraw-Hill.

Gurnell, J. 1987. *The Natural History of Squirrels*. New York: Facts On File.

Gustafson, A. W., and W. D. Belt. 1981. The adrenal cortex during activity and hibernation in the male little brown bat, *Myotis lucifugus lucifugus*: Annual rhythm of plasma cortisol levels. *General and Comparative Endocrinology* 44:269–278.

Gutzke, W. H. N., and D. Crews. 1988. Embryonic temperature determines adult sexuality in a reptile. *Nature* 332:832–834.

Gwinner, E. 1986. *Circannual Rhythms*. Berlin: Springer-Verlag.

Gwinner, E. 1996. Circadian and circannual programmes in avian migration. *Journal of Experimental Biology* 199:39–48.

H

Hagelberg, E., M. G. Thomas, C. E. Cook, Jr., A. V. Sher, G. F. Baryshnikov, and A. M. Lister. 1994. DNA from ancient mammoth bones. *Nature* 370:333–334.

Hailman, J. P. 1976. Oil droplets in the eyes of adult anuran amphibians: A comparative survey. *Journal of Morphology* 148:453–468.

Hairston, N. G. 1949. The local distribution and ecology of the plethodontid salamanders of the southern Appalachians. *Ecological Monographs* 19:47–73.

Hairston, N. G. 1980. Species packing in the salamander genus *Desmognathus*: What are the interspecific interactions involved? *American Naturalist* 115:354–366.

Hairston, N. G. 1983. Growth, survival and reproduction of *Plethodon jordani*: Trade-offs between selective pressures. *Copeia* 1983:1024–1035.

Hall, R. J., P. F. P. Henry, and C. M. Bunck. 1999. Fifty-year trends in a box turtle population in Maryland. *Biological Conservation* 88(1999):165–172.

Hallam, A. 1987. End-Cretaceous mass extinction event: Argument for terrestrial causation. *Science* 238:1237–1242.

Halliday, T. R., and K. Adler. (eds.). 1986. *The Encyclopedia of Reptiles and Amphibians*. New York: Facts on File.

Halpern, M. 1983. Nasal chemical senses in snakes. Pages 141–176. In: Ewert, J-P., R. R. Capranica, and D. J. Ingle (eds.). *Advances in Vertebrate Neuroethology*. New York: Plenum Press.

Halpern, M. 1992. Nasal chemical senses in reptiles: Structure and function. Pages 423–523. In: Gans, C., and D. Crews (eds.). *Biology of the Reptilia. Volume 18, Physiology. E. Hormones, Brain, and Behavior*. Chicago: University of Chicago Press.

Halstead, L. B. 1982. Evolutionary trends and the phylogeny of the Agnatha. Pages 159–196. In: Joysey, K. A., and A. E. Friday (eds.). *Problems of Phylogenetic Reconstruction*. New York: Academic Press.

Hamilton, W. J., Jr. 1943. *The Mammals of Eastern United States*. Ithaca, New York: Comstock Publishing.

Hammer, W. R., and W. J. Hickerson. 1994. A crested theropod dinosaur from Antarctica. *Science* 264:828–830.

Handrich, Y., R. M. Bevan, J. B. Charrassin, P. J. Butler, K. Putz, A. J. Woakes, J. Lage, and Y. LeMaho. 1997. Hypothermia in foraging king penguins. *Nature* 388:64–67.

Hanks, S. L. 1996. *Ecology and the Biosphere*. Delray Beach, Florida: St. Lucie Press.

Hansen, J., R. Ruedy, M. Sato, and R. Reynolds. 1996. Global surface air temperature in 1995: Return to pre-Pinatubo level. *Geophysical Research Letters* 23(13):1665–1668.

Hanssen, I., H. C. Pedersen, and T. Lundh. 1991. Does intense herbivory from microtine rodents induce production of plant estrogens in the spring food plants of willow ptarmigan *Lagopus l. lagopus*? *Oikos* 62:77–79.

Hansson, L. 1979. Condition and diet in relation to habitat in bank voles *Clethrionomys glareolus*: Population or community approach? *Oikos* 33:55–63.

Hansson, L., and H. Henttonen. 1988. Rodent dynamics as community processes. *Trends in Ecology and Evolution* 3(8):195–200.

Harden-Jones, F. R. 1968. *Fish Migration.* London: Edward Arnold.

Hardin, G. 1960. The competitive exclusion principle. *Science* 131:1292–1297.

Harding, C. 1993. Going to extremes. *National Wildlife* 31(5):38–45.

Hardisty, M. W., 1982. Lampreys and hagfishes: Analysis of cyclostome relationships. Pages 165–260. In: Hardisty, M. W., and I. C. Potter (eds.). *The Biology of Lampreys. Volume 4B.* New York: Academic Press.

Hardisty, M. W., and I. C. Potter. 1971. The behaviour, ecology and growth of larval lampreys. Pages 85–125. In: Hardisty, M. W., and I. C. Potter (eds.). *The Biology of Lampreys.* San Diego: Academic Press.

Hardy, J. D., Jr. 1982. Biogeography of Tobago, West Indies, with special reference to amphibians and reptiles: A review. *Bulletin of the Maryland Herpetological Society* 18(2):37–142.

Harrison, P., G. Zummo, F. Farina, B. Tota, and I. A. Johnston. 1991. Gross anatomy, myoarchitecture, and ultrastructure of the heart ventricle in the haemoglobinless icefish *Chaenocephalus aceratus. Canadian Journal of Zoology* 69:1339–1347.

Hartline, P. H. 1971. Physiological basis for detection of sound and vibration in snakes. *Journal of Experimental Biology* 54:349–371.

Harvell, C. D., K. Kim, J. M. Burkholder, R. R. Colwell, P. R. Epstein, D. J. Grimes, E. E. Hofmann, E.K. Lipp, A. D. M. E. Osterhaus, R. M. Overstreet, J. W. Porter, G. W. Smith, and G. R. Vasta. 1999. Emerging marine diseases—climate links and anthropogenic factors. *Science* 285:1505–1510.

Harwood, J. 1998. What killed the monk seals? *Nature* 393:17–18.

Hasler, A. D., and A. T. Scholz. 1983. *Olfactory Imprinting and Homing in Salmon.* New York: Springer-Verlag.

Hasler, A. D., and W. J. Wisby. 1951. Discrimination of stream odors by fishes and relation to parent stream behavior. *American Naturalist* 85:223–238.

Hasler, A. D., R. M. Horrall, W. J. Wisby, and W. Braemer. 1958. Sun-orientation and homing in fishes. *Limnology and Oceanography* 3:353–361.

Hasler, A. D., A. T. Scholz, and R. M. Horrall. 1978. Olfactory imprinting and homing in salmon. *American Scientist* 66:347–355.

Hasselquist, D., S. Bensch, and T. von Schantz. 1996. Correlation between male song repertoire, extra-pair paternity and offspring survival in the great reed warbler. *Nature* 381:229–232.

Hastings, B. E., D. E. Abbott, L. M. George, and S. G. Stadler. 1992. Stress factors influencing plasma cortisol levels and adrenal weights in Chinese water deer (*Hydropotes inermis*). *Research in Veterinary Science* 53:375–380.

Hawkins, C. E., and D. W. Macdonald. 1992. A spool-and-line method for investigating the movements of badgers, *Meles meles. Mammalia* 56(2):322–325.

Hayes, J. M. 1996. The earliest memories of life on Earth. *Nature* 384:21–22.

Haynes, C. V., Jr. 1997. Dating a paleoindian site in the Amazon in comparison with Clovis culture. *Science* 275:1948.

Haynes, G. 1991. *Mammoths, Mastodonts, and Elephants.* New York: Cambridge University Press.

Haywood, S., and C. M. Perrins. 1992. Is clutch size in birds affected by environmental conditions during growth? *Proceedings of the Royal Society of London B* 249:195–197.

Hazelhoff, E. H., and H. H. Evenhuis. 1952. Importance of the "counter-current principle" for the oxygen uptake in fishes. *Nature* 169:77.

Hedges, S. B. 1994. Molecular evidence for the origin of birds. *Proceedings of the National Academy of Sciences,* USA 91:2621–2624.

Hedges, S. B., and L. L. Poling. 1999. A molecular phylogeny of reptiles. *Science* 283:998–1001.

Hedges, S. B., M. D. Simmons, M. A. M. vanDijk, G-J. Caspers, W. W. deJong, and C. G. Sibley. 1995. Phylogenetic relationships of the hoatzin, an enigmatic South American bird. *Proceedings of the National Academy of Sciences,* USA 92:11662–11665.

Hedges, S. B., P. H. Parker, C. G. Sibley, and S. Kumar. 1996. Continental breakup and the ordinal diversification of birds and mammals. *Nature* 381:226–229.

Hedrick, P. 1992. Shooting the RAPDs. *Nature* 355:679–680.

Heiligenberg, W. 1977. *Principles of Electrolocation and Jamming Avoidance in Electric Fish.* Berlin: Springer-Verlag.

Heinen, J. T., and G. Hammond. 1997. Antipredator behaviors of newly metamorphosed green frogs (*Rana clamitans*) and leopard frogs (*Rana pipiens*) in encounters with eastern garter snakes (*Thamnophis s. sirtalis*). *American Midland Naturalist* 137:136–144.

Heintz, E. 1954. Actions repulsives exercees sur divers animaux par des substances contenues dans la peau ou le corps d'animaux de meme espece. *Comptes Rendus des Séances de la Société de Biologie et de ses filiales, Paris* 148:585–588, 717–719.

Helbig, A. J. 1991. Inheritance of migratory direction in a bird species: A cross-breeding experiment with SE- and SW-migrating blackcaps (*Sylvia atricapilla*). *Behavioral Ecology and Sociobiology* 28:9–12.

Helbig, A. J., P. Berthold, G. Mohr, and U. Querner. 1994. Inheritance of a novel migratory direction in central European blackcaps. *Naturwissenschaften* 81:184–186.

Heldmaier, G. 1991. Pineal involvement in avian thermoregulation. *Acta XX Congressus Internationalis Ornithologici,* Christchurch, New Zealand. December 1990.

Hemelaar, A. S. M. 1986. Demographic study on *Bufo bufo* L. (Anura, Amphibia) from different climates, by means of skeletochronology. Ph.D. dissertation. Katholie Universiteit te Nijmegen, Nijmegen, Netherlands.

Herald, E. S. 1967. *Living Fishes of the World.* Garden City, New York: Doubleday and Company.

Herbert, C. V., and D. C. Jackson. 1985. Temperature effects on the responses to prolonged submergence in the turtle *Chrysemys picta bellii*—II. Metabolic rate, blood acid-base and ionic changes, and cardiovascular function in aerated and anoxic water. *Physiological Zoology* 58:670–681.

Herman, A. B., and R. A. Spicer. 1996. Palaeobotanical evidence for a warm Cretaceous Arctic Ocean. *Nature* 380:330–333.

Herman, L. M., and W. N. Tavolga. 1980. The communication systems of cetaceans. Pages 149–209. In: Herman, L. M. (ed.). *Cetacean Behavior: Mechanisms and Functions.* New York: John Wiley and Sons.

Hernandez, M., I. Robinson, A. Aguilar, L. M. Gonzalez, L. F. Lopez-Jurado, M. I. Reyero, E. Cacho, J. Franco, V. Lopez-Rodas, and E. Costas. 1998. Did algal toxins cause monk seal mortality? *Nature* 393:28–29.

Herreid, C. F., W. L. Bretz, and K. Schmidt-Nielsen. 1968. Cutaneous gas exchange in bats. *American Journal of Physiology* 215:506–508.

Hert, E. 1992. Homing and home-site fidelity in rock-dwelling cichlids (Pisces: Teleostei) of Lake Malawi, Africa. *Environmental Biology of Fishes* 33:229–237.

Heulin, B., K. Osenegg, and M. Lebouvier. 1991. Timing of embryonic development and birth dates in oviparous and viviparous strains of *Lacerta vivipara*: Testing the predictions of an evolutionary hypothesis. *Acta Oecologica* 12(4):517–528.

Heusser, H. 1969. Die Lebensweise der Erdkrote, *Bufo bufo* (L.); Wanderunger und Sommerquartiere. *Revue Suisse de Zoologie* 75:927–982.

Hevesy, G. D., D. Lockner, and K. Sletten. 1964. Iron metabolism and erythrocyte formation in fish. *Acta Physiologica Scandinavica* 60:256–266.

Heyer, W. R., M. A. Donnelly, R. W. McDiarmid, L. C. Hayek, and M. S. Foster. 1994. *Measuring and Monitoring Biological Diversity: Standard Methods for Amphibians.* Washington, D.C.: Smithsonian Institution Press.

Heyning, J. E., and J. G. Mead. 1997. Thermoregulation in the mouths of feeding gray whales. *Science* 278:1138–1139.

Hickie, P. F. 1936. Isle Royale moose studies. *Transactions of the North American Wildlife Conference* 1:396–399.

Highton, R. 1991. Molecular phylogeny of plethodonine salamanders and hylid frogs: Statistical analysis of protein comparisons. *Molecular Biology and Evolution* 8(6):796–818.

Highton, R., and A. Larson. 1979. The genetic relationships of the salamanders of the genus *Plethodon. Systematic Zoology* 28:579–599.

Highton, R., and T. P. Webster. 1976. Geographic protein variation and divergence in populations of the salamander *Plethodon cinereus. Evolution* 30:33–45.

Hildebrand, A. R., M. Pilkington, M. Connors, G. Ortiz-Alaman, and R. E. Chavez. 1995. Size and structure of the Chicxulub crater revealed by horizontal gravity gradients and cenotes. *Nature* 376:415–417.

Hildebrand, S. F. 1929. Review of experiments on artificial culture of diamondback terrapin. *Bulletin of the United States Bureau of Fisheries* 45:25–70.

Hill, G. E. 1990. Female house finches prefer colourful males: Sexual selection for a condition-dependent trait. *Animal Behaviour* 40:563–572.

Hillenius, W. J. 1994. Turbinates in therapsids: Evidence for late Permian origins of mammalian endothermy. *Evolution* 48(2):207–229.

Hirayama, R. 1998. Oldest known sea turtle. *Nature* 392:705–708.

Hitchcock, H. B., and K. Reynolds. 1942. Homing experiments with the little brown bat, *Myotis lucifugus lucifugus* (LeConte). *Journal of Mammalogy* 23:258–267.

Hoar, W. S. 1969. Reproduction. Pages 1–72. In: Hoar, S., and D. J. Randall, (eds.). *Fish Physiology. Volume 3.* New York: Academic Press.

Hockey, P. 1997. Eyes in the sky. Understanding avian vision. *Africa-Birds and Birding* 2(6):50–53.

Hoffman, C. 1998. Peregrine to soar off endangered list. *Endangered Species Bulletin* 23(5):20–21.

Hoglund, J., and J. G. M. Robertson. 1988. Chorusing behaviour, a density-dependent alternative mating strategy in male common toads (*Bufo bufo*). *Ethology* 79:324–332.

Holden, C. 1992. Tracking elephants by satellite. *Science* 257:1342.

Holden, C. 1996a. Tales from a dead rat's eye. *Science* 272:655.

Holden, C. 1996b. Manatee killer revealed. *Science* 273:191.

Holden, C. 1997. Oldest dino embryo unearthed. *Science* 277:41.

Holden, C. 1999. Wild thing. *Science* 286:2447.

Holland, P. 1992. Homeobox genes in vertebrate evolution. *BioEssays* 14(4):267–273.

Holmes, B. 1993. An avian arch-villain gets off easy. *Science* 262:1514–1515.

Holmes, D. J. 1992. Odors as cues for orientation to mothers by weanling Virginia opossums. *Journal of Chemical Ecology* 18(12):2251–2259.

Holmgren, N. 1933. On the origin of the tetrapod limb. *Acta Zoologica* (Stockholm) 14:185–295.

Holmgren, N. 1939. Contributions to the question of the origin of the tetrapod limb. *Acta Zoologica* (Stockholm) 20:89–124.

Holmgren, N. 1949. Contributions to the question of the origin of tetrapods. *Acta Zoologica* (Stockholm) 30:459–484.

Holmgren, N. 1952. An embryological analysis of the mammalian carpus and its bearing upon the question of the origin of the tetrapod limb. *Acta Zoologica* (Stockholm) 33:1–115.

Holomuzki, J. R. 1982. Homing behavior of *Desmognathus ochrophaeus* along a stream. *Journal of Herpetology* 16(3):307–309.

Hoogland, J. L., R. H. Tamarin, and C. K. Levy. 1989. Communal nursing in prairie dogs. *Behavioral Ecology and Sociobiology* 24:91–95.

Hopson, J. A. 1970. The classification of non-therian mammals. *Journal of Mammalogy* 51:1–9.

Hopson, J. A., and H. R. Barghusen. 1986. An analysis of therapsid relationships. Pages 83–106. In: Hotten, N., III, P. D. MacLean, J. J. Roth, and E. C. Roth (eds.). *The Ecology and Biology of Mammal-like Reptiles.* Washington, D.C.: Smithsonian Institution Press.

Hopson, J. A., and W. W. Crompton. 1969. Origin of mammals. Pages 15–72. In: Dobzhansky, T., M. K. Hecht, and W. C. Steere (eds.). *Evolutionary Biology. III.* New York: Appleton-Century-Crofts.

Hori, M. 1993. Frequency-dependent natural selection in the handedness of scale-eating cichlid fish. *Science* 260:216–219.

Horn, M. H., and R. N. Gibson. 1988. Intertidal fishes. *Scientific American* 258 (1):64–70.

Horne, E. A. 1988. Aggressive behavior of female red-backed salamanders. *Herpetologica* 44:203–209.

Horne, E. A., and R. G. Jaeger. 1988. Territorial pheromones of female red-backed salamanders. *Ethology* 78:143–152.

Horner, J. 1984. The nesting behavior of dinosaurs. *Scientific American* 250(4): 130–137.

Horner, J. R. 1998. Dinosaur behavior. *The Phi Kappa Phi Journal* 78(3):12–15.

Horner, J. R. 1999. Egg clutches and embryos of two hadrosaurian dinosaurs. *Journal of Vertebrate Paleontology* 19(4):607–611.

Horner, J. R., and J. Gorman. 1988. *Digging Dinosaurs.* New York: Workman.

Horner, M. A., and R. A. Powell. 1990. Internal structure of home ranges of black bears and analyses of home-range overlap. *Journal of Mammalogy* 71(3):402–410.

Hornfeldt, B. 1978. Synchronous population fluctuations in voles, small game, owls, and tularemia in northern Sweden. *Oecologia* 32:141–152.

Hou, L., L. D. Martin, Z. Zhou, and A. Feduccia. 1996. Early adaptive radiation of birds: evidence from fossils from northeastern China. *Science* 274:1164–1167.

Hou, L., Z. Zhou, L. D. Martin, and A. Feduccia. 1995. A beaked bird from the Jurassic of China. *Nature* 377:616–618.

Hou, L., L. D. Martin, Z. Zhou, A. Feduccia, and F. Zhang. 1999. A diapsid skull in a new species of the primitive bird *Confuciusornis. Nature* 399:679–682.

Houck, L. D., and S. K. Woodley. 1995. Field studies of steroid hormones and male reproductive behavior in amphibians. Pages 677-703. In: Heatwole, H. (ed.). *Amphibian Biology.* Chipping Norton, New South Wales: Surrey Beatty and Sons.

Howarth, B. 1974. Sperm storage as a function of the female reproductive tract.

Pages 237–270. In: Johnson, A. D., and C. E. Foley (eds.). *The Oviduct and its Functions.* New York: Academic Press.

Hull, D. L. 1988. *Science as a Process.* Chicago: University of Chicago Press.

Hungerford, K. E., and N. G. Wilder. 1941. Observations on the homing behavior of the gray squirrel (*Sciurus carolinensis*). *Journal of Wildlife Management* 5:458–460.

Hunt, R. H., and J. J. Ogden. 1991. Selected aspects of the nesting ecology of American alligators in the Okefenokee Swamp. *Journal of Herpetology* 25(4):448–453.

Hunt, S., I. C. Cuthill, J. P. Swaddle, and A. T. D. Bennett. 1997. Ultraviolet vision and band-colour preferences in female zebra finches, *Taeniopygia guttata. Animal Behaviour* 54(6):1383–1392.

Hunt, S., A. T. D. Bennett, I. C. Cuthill, and R. Griffiths. 1998. Blue tits are ultraviolet tits. *Proceedings of the Royal Society of London B* 265:451–455.

Hyvärinen, H. 1989. Diving in darkness: Whiskers as sense organs of the ringed seal (*Phoca hispida saimensis*). *Journal of Zoology* (London) 218:663–678.

I

Insel, T. R., and C. S. Carter. 1995. The monogamous brain. *Natural History* 104(8):12, 14.

Irwin, D. M., T. D. Kocher, and A. C. Wilson. 1991. Evolution of the cytochrome *b* gene of mammals. *Journal of Molecular Evolution* 32:128–144.

Izhaki, I., and A. Maitav. 1998. Blackcaps *Sylvia atricapilla* stopping over at the desert edge; physiological state and flight-range estimates. *Ibis* 140(2):223–233.

J

Jablonski, D. 1984. Keeping time with mass extinctions. *Paleobiology* 10:139–145.

Jablonski, D. 1986. Causes and consequences of mass extinctions: A comparative approach. Pages 183–229. In: Elliott, D. K. (ed.). *Dynamics of Extinction.* New York: John Wiley and Sons.

Jablonski, D., and D. M. Raup. 1995. Selectivity of end-Cretaceous marine bivalve extinctions. *Science* 268:389–391.

Jacobs, L. F., S. J. C. Gaulin, D. F. Sherry, and G. E. Hoffman. 1990. Evolution of spatial cognition: Sex-specific patterns of spatial behaviour predict hippocampus size. *Proceedings of the National Academy of Science USA* 87(16):6349–6352.

Jaeger, C. B., and D. E. Hillman. 1976. Morphology of gustatory organs. Pages 588–606. In: Llinás, R., and W. Precht (eds.). *Frog Neurobiology.* New York: Springer-Verlag.

Jaeger, R. G. 1971. Moisture as a factor influencing the distributions of two species of terrestrial salamanders. *Oecologia* 6:191–207.

Jaeger, R. G., and W. F. Gergits. 1979. Intra- and inter-specific communication in salamanders through chemical signals on the substrate. *Animal Behaviour* 27(1):150–156.

Jaeger, R. G., and S. E. Wise. 1991. A reexamination of the male salamander "sexy faeces hypothesis." *Journal of Herpetology* 25(3):370–373.

Jaeger, R. G., D. Kalvarsky, and N. Shimizu. 1982. Territorial behaviour of the red-backed salamander: Expulsion of intruders. *Animal Behaviour* 30:490–496.

Jaeger, R. G., J. M. Goy, M. Tarven, and C. Marquez. 1986. Salamander territoriality: Pheromonal markers as advertisement by males. *Animal Behaviour* 34(3):860–864.

Jameson, D. L. 1957. Population structure and homing responses in the Pacific tree frog. *Copeia* 1957:221–228.

Janke, A., N. J. Gemmell, G. Feldmaier-Fuchs, A. von Haeseler, and S. Pääbo. 1996. The mitochondrial genome of a monotreme—the platypus (*Ornithorhynchus anatinus*). *Journal of Molecular Evolution* 42:153–159.

Janvier, P. 1981. The phylogeny of the Craniata, with particular reference to the significance of fossil "agnathans." *Journal of Vertebrate Paleontology* 1(2):121–159.

Janvier, P. 1995. Conodonts join the club. *Nature* 374:761–762.

Janvier, P. 1996. Fishy fragments tip the scales. *Nature* 383:757–758.

Janiver, P. 1999. Catching the first fish. *Nature* 402:21–22.

Janzen, D. H. 1967. Why mountain passes are higher in the tropics. *American Naturalist* 101(919):233–249.

Janzen, F. J., and G. L. Paukstis. 1991a. Environmental sex determination in reptiles: Ecology, evolution, and experimental design. *Quarterly Review of Biology* 66(2):149–179.

Janzen, F. J., and G. L. Paukstis. 1991b. A preliminary test of the adaptive significance of environmental sex determination in reptiles. *Evolution* 45(2):435–440.

Jarvik, E. 1980. *Basic Structure and Evolution of Vertebrates.* 2 volumes. London: Academic Press.

Jarvik, E. 1986. The origin of the Amphibia. Pages 1–24. In: Rocek, Z. (ed.). *Studies in Herpetology.* Prague: Charles University.

Jarvik, E. 1996. The Devonian tetrapod *Ichthyostega. Fossils and Strata* no. 40:1–213.

Jaslow, A. P., T. E. Hetherington, and R. E. Lombard. 1988. Structure and function of the amphibian middle ear. Pages 69–91. In: Fritzsch, B., M. J. Ryan, W. Wilczynski, T. E. Hetherington, and W. Walkowiak (eds.). *The Evolution of the Amphibian Auditory System.* New York: John Wiley and Sons.

Jefferies, R. P. S. 1986. *The Ancestry of Vertebrates.* Dorchester, United Kingdom: British Museum of Natural History.

Jermann, T., and D. G. Senn. 1992. Amphibious vision in *Coryphoblennius galerita* L. (Perciformes). *Experientia* 48:217–218.

Ji, Q., P. J. Currie, M. A. Norell, and J. Shu-An. 1998. Two feathered dinosaurs from northeastern China. *Nature* 393:753–761.

Ji, Q., Z. Luo, and S. Ji. 1999. A Chinese triconodont mammal and mosaic evolution of the mammalian skeleton. *Nature* 398:326–330.

Jogia, M. K., A. R. E. Sinclair, and R. J. Andersen. 1989. An antifeedant in balsam poplar inhibits browsing by snowshoe hares. *Oecologia* 79:189–192.

Johansen, K. 1960. Circulation in the hagfish, *Myxine glutinosa* L. *Biological Bulletin* 118:289–295.

Johnston, R. F., and R. K. Selander. 1964. House sparrows: Rapid evolution of races in North America. *Science* 144:548–550.

Johnston, R. F., and R. K. Selander. 1971. Evolution in the house sparrow. II. Adaptive differentiation in North American populations. *Evolution* 25 (1):1–28.

Jordan, J. 1976. The influence of body weight on gas exchange in the air-breathing fish *Clarias batrachus. Comparative Biochemistry and Physiology* 53A:305–310.

Jorgensen, C. B. 1992. Growth and reproduction. Pages 439–466. In: Feder, M. E., and W. W. Burggren (eds.). *Environmental Physiology of the Amphibians.* Chicago: The University of Chicago Press.

Jorgensen, C. B. 1998. Role of urinary and cloacal bladders in chelonian water economy: Historical and comparative perspectives. *Biological Review* 73:347–366.

Jorgensen, C. B., and L. O. Larsen. 1961. Molting and its hormonal control in toads. *General and Comparative Endocrinology* 1:145–153.

Jouventin, P., and H. Weimerskirch. 1990. Satellite tracking of wandering albatrosses. *Nature* 343:746–748.

K

Kaiser, J. 1995. Aussie rabbit virus causes ruckus. *Science* 270:583.

Kaiser, J. 1996. Turtle project scores a success. *Science* 273:435.

Kaiser, J. 1997. Is warming trend harming penguins? *Science* 276:1790.

Kapoor, B. G., H. Smit, and I. A. Verighina. 1975. The alimentary canal and digestion in teleosts. *Advances in Marine Biology* 13:109–239.

Kardong, K. V. 1998. *Vertebrates: Comparative Anatomy, Function, Evolution.* Boston: McGraw-Hill.

Kare, M. R., and J. G. Rogers, Jr. 1976. Sense organs. Pages 29–52. In: Sturkie, P. D. (ed.). *Avian Physiology.* New York: Springer-Verlag.

Katzir, G. 1993. Visual mechanisms of prey capture in water birds. Pages 301–315. In: Zeigler, H. P., and H.-J. Bischof (eds.). *Vision, Brain, and Behavior in Birds.* Cambridge, Massachusetts: MIT Press.

Katzir, G. 1994. Tuning of visuomotor coordination during prey capture in water birds. Pages 315–338. In: Davies, M. N. O., and P. R. Green (eds.): *Perception and Motor Control in Birds. An Ecological Approach.* Berlin: Springer-Verlag.

Kaufmann, J. H. 1991. *Clemmys insculpta* (wood turtle). Cleaning symbiosis. *Herpetological Review* 22(3):98.

Kaye, S. V. 1960. Gold-198 wires used to study movements of small mammals. *Science* 131:824.

Keefe, M., and H. E. Winn. 1991. Chemosensory attraction to home stream water and conspecifics by native brook trout, *Salvelinus fontinalis*, from two southern New England streams. *Canadian Journal of Fisheries and Aquatic Sciences* 48:938–944.

Keen, R., and H. B. Hitchcock. 1980. Survival and longevity of the little brown bat (*Myotis lucifugus*) in southeastern Ontario. *Journal of Mammalogy* 61(1):1–7.

Keith, L. B. 1963. *Wildlife's Ten-Year Cycle.* Madison: University of Wisconsin Press.

Keith, L. B. 1974. Some features of population dynamics in mammals. *Proceedings of the International Congress of Game Biologists* 11:17–58.

Keith, L. B., and L. A. Windberg. 1978. A demographic analysis of the snowshoe hare cycle. *Wildlife Monographs* 58:1–70.

Kelly, G. M. 1977. Fisher (*Martes pennanti*) biology in the White Mountain National Forest and adjacent areas. Ph.D. dissertation. University of Massachusetts, Amherst.

Kemp, M. 1988. Wild-animal contraception. *Discover* 9:28–29.

Kemp, T. S. 1982. *Mammal-like Reptiles and the Origin of Mammals.* New York: Academic Press.

Kemp, T. S. 1988. Haemothermia or Archosauria? The interrelationships of mammals, birds and crocodiles. *Zoological Journal of the Linnean Society* 92:67–104.

Kenagy, G. J., S. M. Sharbaugh, and K. A. Nagy. 1989. Annual cycle of energy and time expenditure in a golden-mantled ground squirrel population. *Oecologia* 78:269–282.

Kenrick, P., and P. R. Crane. 1997. The origin and early evolution of plants on land. *Nature* 389:33–39.

Kerlinger, P. 1995. Night flight. *Natural History* 104(9):66–69.

Kermack, K. A. 1967. The interrelations of early mammals. *Zoological Journal of the Linnean Society* 47:241–249.

Kermack, K. A., F. Mussett, and H. W. Rigney. 1981. The skull of *Morganucodon. Zoological Journal of the Linnean Society* 71:1–158.

Kerr, R. A. 1988. Was there a prelude to the dinosaurs' demise? *Science* 239:729–730.

Kerr, R. A. 1992. Isotopic thermometer hints at warm-blooded dinosaurs. *Science* 257:486–487.

Kerr, R. A. 1994. How high was Ice Age ice? A rebounding earth may tell. *Science* 265:189.

Kerr, R. A. 1995a. Studies say—tentatively—that greenhouse warming is here. *Science* 268:1567–1568.

Kerr, R. A. 1995b. It's official: First glimmer of greenhouse warming seen. *Science* 270:1565–1567.

Kerr, R. A. 1996a. A piece of the dinosaur killer found? *Science* 271:1806.

Kerr, R. A. 1996b. 1995 the warmest year? Yes and no. *Science* 271:137–138.

Kerr, R. A. 1997a. Cores document ancient catastrophe. *Science* 275:1265.

Kerr, R. A. 1997b. Greenhouse forecasting still cloudy. *Science* 276:1040–1042.

Kerr, R. A. 1998. The hottest year, by a hair. *Science* 279:315–316.

Kezuka, H., K. Furukawa, K. Aida, and I. Hanyu. 1988. Daily cycles in plasma melatonin levels under long and short photoperiod in the common carp, *Cyprinus carpio. General and Comparative Endocrinology* 72:296–302.

Kiepenheuer, J. 1984. The magnetic compass mechanism of birds and its possible association with the shifting course directions of migrants. *Behavioral Ecology and Sociobiology* 14:81–99.

Kikuyama, S., F. Toyoda, Y. Ohmiya, K. Matsuda, S. Tanaka, and H. Hayashi. 1995. Sodefrin: A female-attracting peptide pheromone in newt cloacal glands. *Science* 267:1643–1645.

King, C. 1989. *The Natural History of Weasels and Stoats.* Ithaca, New York: Comstock Publishing.

King, C. M. 1975. The home range of the weasel (*Mustela nivalis*) in an English woodland. *Journal of Animal Ecology* 44:639–668.

King, W. 1962. The occurrence of rafts for dispersal of land animals into the West Indies. *Quarterly Journal of the Florida Academy of Sciences* 25:45–52.

Kingsmill, S. 1993. Ear stones speak volumes to fish researchers. *Science* 260:1233–1234.

Kley, N. J., and E. L. Brainerd. 1999. Feeding by mandibular raking in a snake. *Nature* 402:369–370.

Klicka, J., and R. M. Zink. 1997. The importance of recent Ice Ages in speciation: A failed paradigm. *Science* 277:1666–1669.

Klimkiewicz, M. K., and A. G. Futcher. 1989. Longevity records of North American birds. Supplement 1. *Journal of Field Ornithology* 60(4):469–494.

Klimley, A. P. 1995. Hammerhead city. *Natural History* 104(10):32–39.

Klinger, R. C., and J. A. Musick. 1992. Annular growth layers in juvenile loggerhead turtles (*Caretta caretta*). *Bulletin of Marine Science* 51(2):224–230.

Klinowska, M. 1986. The cetacean magnetic sense—evidence from strandings. Pages 401–432. In: Bryden, M. M., and R. Harrison (eds.). *Research on Dolphins.* Oxford, England: Clarendon Press.

Kluger, J. 1991. Pretty poison. *Discover* 12:68–71.

Kobayashi, Y., J.-C. Lu, Z.-M. Dong, R. Barsbold, Y. Azuma, and Y. Tomida. 1999. Herbivorous diet in an ornithomimid dinosaur. *Nature* 402:480–481.

Koebel, D. A., P. G. Miers, R. A. Nelson, and J. M. Steffen. 1991. Biochemical changes in skeletal muscles of denning bears (*Ursus americanus*). *Comparative Biochemistry and Physiology* 100B(2):377–380.

Komdeur, J., S. Daan, J. Tinbergen, and C. Mateman. 1997. Extreme adaptive modification in sex ratio of the Seychelles warbler's eggs. *Nature* 385:522–525.

Konishi, M. 1973. How the owl tracks its prey. *American Scientist* 61:414–424.

Kooyman, G. L., and T. G. Kooyman. 1995. Diving behavior of emperor penguins nurturing chicks at Coulman Island, Antarctica. *The Condor* 97:536–549.

Kooyman, G. L., and P. J. Ponganis. 1990. Behavior and physiology of diving in emperor and king penguins. Pages 229–242. In: Davis, L. S., and J. T. Darby (eds.). *Penguin Biology*. New York: Academic Press.

Kooyman, G. L., Y. Cherel, Y. LeMaho, J. P. Croxall, P. H. Thorson, V. Ridoux, and C. A. Kooyman. 1992. Diving behavior and energetics during foraging cycles in king penguins. *Ecological Monographs* 62(1):143–163.

Kooyman, G. L., T. G. Kooyman, M. Horning, and C. A. Kooyman. 1996. Penguin dispersal after fledging. *Nature* 383:397.

Koprowski, J. L. 1993. Sex and species biases in scent marking by fox squirrels and eastern gray squirrels. *Journal of Zoology* (London) 230:319–323.

Kraak, S. B., and E. P. Van Den Berghe. 1992. Do female fish assess paternal quality by means of test eggs? *Animal Behaviour* 43:865–867.

Kramer, G. 1957. Experiments on bird orientation and their interpretation. *Ibis* 99:196–227.

Kramer, G. 1959. Recent experiments in bird orientation. *Ibis* 101:399–416.

Kraus, S. D., A. J. Read, A. Solow, K. Baldwin, T. Spradlin, E. Anderson, and J. Williamson. 1997. Acoustic alarms reduce porpoise mortality. *Nature* 388:525.

Krefting, L. W. 1951. What is the future of the Isle Royale moose herd? *Transactions of the North American Wildlife Conference* 1:461–470.

Kreithen, M. L., and D. B. Quine. 1979. Infrasound detection by the homing pigeon: A behavioral audiogram. *Journal of Comparative Physiology* 129:1–4.

Kring, D. A., and W. V. Boynton. 1992. Petrogenesis of an augite-bearing melt rock in the Chicxulub structure and its relationship to K-T impact spherules in Haiti. *Nature* 358:141–144.

Kruuk, H. 1992. *Wild Otters*. Oxford, England: Oxford University Press.

Kuehn, D. W. 1989. Winter foods of fishers during a snowshoe hare decline. *Journal of Wildlife Management* 53(3):688–692.

Külmann, D. H. H. 1966. Putzerfische säubern Krokodile. *Zeitschrift Tierpsychologie* 23:853–854.

Kühme, W. 1964. Eine chemisch ausgelöste Schwarmreaktion bei jungen Cichliden (Pisces). *Naturwissenschaften* 51:120–121.

Kulzer, E. 1954. Untersuchungen uber die Schreckreaktion der Erdkrotenkaulquappen (*Bufo bufo* L.). *Zeitschrift vergleiche Physiologie* 36:443–463.

Kumar, S., and S. B. Hedges. 1998. A molecular timescale for vertebrate evolution. *Nature* 392:917–920.

Kuyt, E. 1992. Aerial radio-tracking of whooping cranes migrating between Wood Buffalo National Park and Aransas National Wildlife Refuge, 1981–84. *Occasional Paper No. 74. Canadian Wildlife Service*, Ottawa.

Kyte, F. T. 1998. A meteorite from the Cretaceous/Tertiary boundary. *Nature* 396:237–239.

L

Labov, J. B., U. W. Huck, R. W. Elwood, and R. J. Brooks. 1985. Current problems in the study of infanticidal behaviour of rodents. *Quarterly Review of Biology* 60:1–20.

Lack, D. 1954. *The Natural Regulation of Animal Numbers*. Oxford, England: Clarendon Press.

Lam, T. L. 1983. Environmental influences on gonadal activity in fish. Pages 65–116. In: Hoar, W. S., D. J. Randall, and E. M. Donaldson (eds.). *Fish Physiology, Volume 9B*. New York: Academic Press.

Land, M. F. 1995. Fast-focus telephoto eye. *Nature* 373:658–659.

Landreth, H. F. 1973. Orientation and behavior of the rattlesnake, *Crotalus atrox*. *Copeia* 1973:26–31.

Landreth, H. F., and D. E. Ferguson. 1967a. Newt orientation by sun-compass. *Nature* 215:516–518.

Landreth, H. F., and D. E. Ferguson. 1967b. Sun-compass orientation. *Science* 158:1459–1461.

Lang, J. W. 1987. Crocodilian thermal selection. Pages 301–317. In: Grigg, G., R. Shine, and E. Ehmann (eds.). *Wildlife Management Crocodiles and Alligators*. Chipping Norton, N. S. W., Australia: Surrey Beatty and Sons.

Langbauer, W. R., Jr., R. A. Charif, K. B. Payne, and R. B. Martin. 1991a. Movements of African elephants as observed by radio telemetry. *American Zoologist* 31(5):98A.

Langbauer, W. R., Jr., R. A. Charif, K. B. Payne, and R. B. Martin. 1991b. Vocalizations of African elephants recorded by radio-telemetry. *American Zoologist* 31(5):98A.

Larsen, L. O. 1992. Feeding and digestion. Pages 378–394. In: Feder, M. E., and W. W. Burggren (eds.). *Environmental Physiology of the Amphibians*. Chicago: The University of Chicago Press.

Larson, A. 1991. A molecular perspective on the evolutionary relationships of the salamander families. *Evolutionary Biology* 25:211–277.

Larson, A., D. B. Wake, L. R. Maxson, and R. Highton. 1981. A molecular phylogenetic perspective on the origins of morphological novelties in the salamanders of the tribe Plethodontini (Amphibia, Plethodontidae). *Evolution* 35(3):405–422.

Lasiewski, R. C. 1963. Oxygen consumption of torpid, resting, active, and flying hummingbirds. *Physiological Zoology* 36:122.

Lavigne, D. M., and O. J. Schmitz. 1990. Global warming and increasing population densities: A prescription for seal plagues. *Marine Pollution Bulletin* 21(6):280–284.

Lawson, P. A., and D. M. Secoy. 1991. The use of solar cues as migratory orientation guides by the plains garter snake, *Thamnophis radix*. *Canadian Journal of Zoology* 69:2700–2702.

Layne, J. N. 1954. The biology of the red squirrel, *Tamiasciurus hudsonicus loquax* (Bangs), in central New York. *Ecological Monographs* 24:227–267.

Layne, J. R., Jr. 1999. Freeze tolerance and cryoprotectant mobilization in the gray treefrog (*Hyla versicolor*). *Journal of Experimental Zoology* 283:221–225.

Lechleitner, R. R. 1954. Age criteria in Mink, *Mustela vison*. *Journal of Mammalogy* 35:496–503.

Lee, C. Y., B. Rubinsky, and G. L. Fletcher. 1992. Hypothermic preservation of whole mammalian organs with "antifreeze" proteins. *Cryo-Letters* 13:59–66.

Lee, M. S. Y. 1993. The origin of the turtle body plan: Bridging a famous morphological gap. *Science* 261:1716–1720.

Lee, M. S. Y. 1997. Reptile relationships turn turtle… *Nature* 389:245–246.

LeMunyan, C. D., W. White, E. Nyberg, and J. J. Christian. 1959. Design of a miniature radio transmitter for use in animal studies. *Journal of Wildlife Management* 23(1):107–110.

Lewis, J. C. 1996. Teaching cranes to migrate. *Endangered Species Bulletin* 21(5):10–11.

Licht, L. E. 1991. Habitat selection of *Rana pipiens* and *Rana sylvatica* during exposure to warm and cold temperatures. *American Midland Naturalist* 125:259–268.

Lighthill, M. J. 1969. Hydromechanics of aquatic animal propulsion. *Annual Review of Fluid Mechanics* 1:413–446.

Lighthill, M. J. 1971. Large-amplitude elongated-body theory of fish locomotion. *Proceedings of the Royal Society of London B* 179:125–138.

Lincoln, F. C. 1935. The waterfowl flyways of North America. *United States Department of Agriculture Circular 342.* Washington, D.C.: U.S. Fish and Wildlife Service.

Lindemann, B., and C. Voute. 1976. Structure and function of the epidermis. Pages 169–210. In: Llinas, R., and W. Precht (eds.). *Frog Neurobiology.* New York: Springer-Verlag.

Lindstedt, S. L., J. F. Hokanson, D. J. Wells, S. D. Swain, H. Hoppeler, and V. Navarro. 1991. Running energetics in the pronghorn antelope. *Nature* 353:748–750.

Line, L. 1995. African egrets? Holy cow! *International Wildlife* 25(6):44–54.

Linzey, D. W. 1968. An ecological study of the golden mouse, *Ochrotomys nuttalli*, in the Great Smoky Mountains National Park. *American Midland Naturalist* 79(2):320–345.

Linzey, D. W., and A. V. Linzey. 1967. Growth and development of the golden mouse, *Ochrotomys nuttalli nuttalli. Journal of Mammalogy* 48(3):445–458.

Linzey, D. W., and R. L. Packard. 1977. *Ochrotomys nuttalli. Mammalian Species* no. 75:1–6.

Lips, K. R. 1991. Vertebrates associated with tortoise (*Gopherus polyphemus*) burrows in four habitats in south-central Florida. *Journal of Herpetology* 25(4):477–481.

Lipton, J. 1968. *An Exaltation of Larks.* New York: Grossman Publishers.

Lohmann, K. J. 1991. Magnetic orientation by hatchling loggerhead sea turtles (*Caretta caretta*). *Journal of Experimental Biology* 155:37–49.

Lohmann, K. J. 1992. How sea turtles navigate. *Scientific American* 266(1):100–106.

Lohmann, K. J., and C. M. F. Lohmann. 1993. A light-independent magnetic compass in the leatherback sea turtle. *Biological Bulletin* 185:149–151.

Lohmann, K. J., and C. M. F. Lohmann. 1994a. Acquisition of magnetic directional preference in hatchling loggerhead sea turtles. *Journal of Experimental Biology* 190:1–8.

Lohmann, K. J., and C. M. F. Lohmann. 1994b. Detection of magnetic inclination angle by sea turtles: A possible mechanism for determining latitude. *Journal of Experimental Biology* 194:23–32.

Lohmann, K. J., and C. M. F. Lohmann. 1996a. Orientation and open-sea navigation in sea turtles. *Journal of Experimental Biology* 199:73–81.

Lohmann, K. J., and C. M. F. Lohmann. 1996b. Detection of magnetic field intensity by sea turtles. *Nature* 380:59–61.

Lohmann, K. J., and M. F. Lohmann. 1992. Orientation to oceanic waves by green turtle hatchlings. *Journal of Experimental Biology* 171:1–13.

Lohmann, K. J., M. Salmon, and J. Wyneken. 1990. Functional autonomy of land and sea orientation systems in sea turtle hatchlings. *Biological Bulletin* 179:214–218.

Long, J. A. 1995. *The Rise of Fishes.* Baltimore, Maryland: The Johns Hopkins University Press.

Longmire, J. L., G. F. Gee, C. L. Hardekopf, and G. A. Mark. 1992. Establishing paternity in whooping cranes (*Grus americana*) by DNA analysis. *The Auk* 109(3):522–529.

Losos, J. B. and K. de Queiroz. 1998. Darwin's lizards. *Natural History* 106(11): 34–39.

Losos, J. B., K. I. Warheit, and T. W. Schoener. 1997. Adaptive differentiation following experimental island colonization in *Anolis* lizards. *Nature* 387:70–73.

Lotshaw, D. P. 1977. Temperature adaptation and effects of thermal acclimation in *Rana sylvatica* and *Rana catesbeiana. Comparative Biochemistry and Physiology B* 56:287–294.

Love, J. A. 1992. *Sea Otters.* Golden, Colorado: Fulcrum Publishing.

Lovtrup, S. 1985. On the classification of the taxon Tetrapoda. *Systematic Zoology* 34:463–470.

Lowe, P. R. 1944. An analysis of the characters of *Archeopteryx* and *Archeornis.* Were they reptiles or birds? *Ibis* 86:517–543.

Lowenstein, J. M. 1991. The remaking of the president. *Discover* 12(8):18–20.

Luer, C. A., and P. W. Gilbert. 1991. Elasmobranch fish. Oviparous, viviparous, and ovoviviparous. *Oceanics Magazine* 34(3):47–53.

Luo, Z. 2000. In search of the whales' sisters. *Nature* 404:235–237.

Lutcavage, M. E., P. G. Bushnell, and D. R. Jones. 1990. Oxygen transport in the leatherback sea turtle, *Dermochelys coriacea. Physiological Zoology* 63:1012–1024.

Lutcavage, M. E., P. G. Bushnell, and D. R. Jones. 1992. Oxygen stores and aerobic metabolism in the leatherback sea turtle. *Canadian Journal of Zoology* 70:348–351.

Lutz, G. J., and L. C. Rome. 1994. Built for jumping: The design of the frog muscular system. *Science* 263:370–372.

M

MacArthur, R. H. 1958. Population ecology of some warblers of northeastern coniferous forests. *Ecology* 39:599–619.

MacLulich, D. A. 1937. Fluctuations in the numbers of the varying hare (*Lepus americanus*). *University of Toronto Studies, Biological Series* no. 43:1–136.

MacPhee, R., and C. Flemming. 1997. Brown-eyed, milk-giving...and extinct. *Natural History* 106 (3):84–88.

Maddock, M., and G. S. Baxter. 1991. Breeding success of egrets related to rainfall: A six-year Australian study. *Colonial Waterbirds* 14(2):133–139.

Maderson, P. F. A. 1985. Some developmental problems of the reptilian integument. Pages 523–598. In: Gans, C., F. Billett, and P. F. A. Maderson (eds.). *Biology of the Reptilia. Volume 14. Development A.* New York: John Wiley and Sons.

Madison, D. M. 1969. Homing behaviour of the red-cheeked salamander, *Plethodon jordani. Animal Behaviour* 17:25–39.

Madsen, T., and R. Shine. 1992. Determinants of reproductive success in female adders, *Vipera berus. Oecologia* 92:40–47.

Madsen, T., R. Shine, J. Loman, and T. Hakansson. 1992. Why do female adders copulate so frequently? *Nature* 355:440–441.

Maisey, J. G. 1986. Heads and tails: A chordate phylogeny. *Cladistics* 2:201–256.

Mallatt, J. 1996. Ventilation and the origin of jawed vertebrates: A new mouth. *Zoological Journal of the Linnaean Society* 117:329–404.

Mann, D. A., Z. Lu, and A. N. Popper. 1997. A clupeid fish can detect ultrasound. *Nature* 389:341.

Mann, M. E., and R. S. Bradley. 1999. Northern hemisphere temperatures during the past millennium: Inferences, uncertainties, and limitations. *Geophysical Research Letters* 26(6):759–762.

Mann, M. E., R. S. Bradley, and M. K. Hughes. 1998. Global-scale temperature patterns and climate forcing over the past six centuries. *Nature* 392:779–787.

Manne, L. L., T. M. Brooks, and S. L. Pimm. 1999. Relative risk of extinction of passerine birds on continents and islands. *Nature* 399:258–261.

Manning, C. J., E. K. Wakeland, and W. K. Potts. 1992. Communal nesting patterns in mice implicate MHC genes in kin recognition. *Nature* 360:581–583.

Mares, M. A. 1993. Desert rodents, seed consumption, and convergence. *BioScience* 43(6):372–379.

Mark, L. D., and E. L. Menning. 1984. Common names for animals. Pages 598–599. In: *1984 Yearbook of Agriculture: Animal Health.* Washington, D.C.: U.S. Department of Agriculture.

Marks, J. 1988. Relationships of humans to chimps and gorillas. *Nature* 334:656.

Marks, J. 1992. Genetic relationships among the apes and humans. *Current Opinions on Genetic Developments* 2(6):883–889.

Marler, P., and W. J. Hamilton, III. 1966. *Mechanisms of Animal Behavior.* New York: John Wiley and Sons.

Marler, P., and M. Tamura. 1962. Song "dialects" in three populations of white-crowned sparrows. *Condor* 64:368–377.

Marra, P. P., K. A. Hobson, and R. T. Holmes. 1998. Linking winter and summer events in a migratory bird by using stable-carbon isotopes. *Science* 282:1884–1886.

Marshall, L. G. 1979. Evolution of metatherian and eutherian (mammalian) characters: A review based on cladistic methodology. *Zoological Journal of the Linnean Society* 66:369–410.

Marshall, P. T., and G. M. Hughes. 1980. *Physiology of Mammals and Other Vertebrates.* Second edition. New York: Cambridge University Press.

Martin, G. R. 1993. Producing the image. Pages 5–24. In: Zeigler, H. P., and H-J. Bischof (eds.). *Vision, Brain, and Behavior in Birds.* Cambridge, Massachusetts: MIT Press.

Martin, G. R., and G. Katzir. 1995. Visual fields in ostriches. *Nature* 374:19–20.

Martin, P. S. 1967. Prehistoric overkill. Pages 75–120. In: Martin, P. S., and H. E. Wright, Jr. (eds.). *Pleistocene Extinctions: The Search for a Cause.* New Haven, Connecticut: Yale University Press.

Martin, P. S. 1973. The discovery of America. *Science* 179:960–974.

Martinet, L., and F. Spitz. 1971. Variations saisonnieres de la croissance et de la mortalite du campagnol des champs, *Microtus arvalis.* Role du photoperiodisme et de la vegetation sur ces variations. *Mammalia* 35:38–84.

Martini, F. H. 1998. Secrets of the slime hag. *Scientific American* 279(4):70–75.

Mason, R. T., H. M. Fales, T. H. Jones, L. K. Pannell, J. W. Chinn, and D. Crews. 1989. Sex pheromones in snakes. *Science* 245:290–292.

Mate, B. 1989. Satellite-monitored radio tracking as a method for studying cetacean movements and behaviour. *Report of the International Whaling Commission* 39:389–391.

Mathis, A. 1989. Do seasonal spatial distributions in a terrestrial salamander reflect reproductive behavior or territoriality? *Copeia* 1989:788–791.

Mathis, A. 1990a. Territoriality in a terrestrial salamander: The influence of body size and resource quality. *Behaviour* 112:162–175.

Mathis, A. 1990b. Territorial salamanders assess sexual and competitive information using chemical signals. *Animal Behaviour* 40:953–962.

Mathis, A. 1991a. Territories of male and female terrestrial salamanders: Costs, benefits, and intersexual spatial associations. *Oecologia* 86:433–440.

Mathis, A. 1991b. Large male advantage for access to females: Evidence of male-male competition and female discrimination in a territorial salamander. *Behavioral Ecology and Sociobiology* 29:133–138.

Maxson, L. R., and W. R. Heyer. 1988. Molecular systematics of the frog genus *Leptodactylus* (Amphibia: Leptodactylidae). *Fieldiana—Zoology New Series* no. 41:1–13.

Mayr, E. 1942. *Systematics and the Origin of Species.* New York: Columbia University Press.

Mayr, E. 1969. *Principles of Systematic Zoology.* New York: McGraw-Hill.

Mayr, E., E. G. Linsley, and R. L. Usinger. 1953. *Methods and Principles of Systematic Zoology.* New York: McGraw-Hill.

Mazzeo, R. 1953. Homing of the Manx shearwater. *The Auk* 70:200–201.

McClanahan, L. L., and R. Baldwin. 1969. Rate of water uptake through the integument of the desert toad, *Bufo punctatus. Comparative Biochemistry and Physiology* 28:381–389.

McComb, K. 1987. Roaring in red deer stags advances the oestrus in hinds. *Nature* 330:648–649.

McCulloch, C. Y. 1986. A history of predator control and deer productivity in northern Arizona. *Southwestern Naturalist* 31(2):215–220.

McDonald, I. R., K. A. Handasyde, and B. K. Evans. 1992. Adrenal function in the platypus. Pages 127–133. In: Augee, M. L. (ed.). *Platypus and Echidnas.* Sydney, Australia: The Royal Zoological Society of New South Wales.

McGowan, G., and S. E. Evans. 1995. Albanerpetontid amphibians from the Cretaceous of Spain. *Nature* 373:143–145.

McGowan, J. A., D. R. Cayan, and L. M. Dorman. 1998. Climate-ocean variability and ecosystem response in the northeast Pacific. *Science* 281:210–217.

McKnight, M. L., C. K. Dodd, Jr., and C. M. Spolsky. 1991. Protein and mitochondrial DNA variation in the salamander *Phaeognathus hubrichti. Herpetologica* 47(4):440–447.

McMaster, D. G., and S. G. Sealy. 1998. Short incubation periods of brown-headed cowbirds: How do cowbird eggs hatch before yellow warbler eggs? *The Condor* 100:102–111.

Mead, C. 1997. Poor prospects for oiled birds. *Nature* 390:449–450.

Mead, J. G. 1972. On the anatomy of the external nasal passages and facial complex in the family Delphinidae of the order Cetacea. Ph.D. dissertation. The University of Chicago, Chicago, Illinois.

Mech, L. D. 1966. *The Wolves of Isle Royale.* Washington D.C.: United States National Park Service Fauna Series, Number 7.

Mech, L. D. 1983. *Handbook of Animal Radio-Tracking.* Minneapolis: University of Minnesota Press.

Melosh, H. J., N. M. Schneider, K. J. Zahnie, and D. Latham. 1990. Ignition of global wildfires at the Cretaceous/Tertiary boundary. *Nature* 343:251–254.

Menzies, R. A., G. Heth, R. Ikan, V. Weinstein, and E. Nevo. 1992. Sexual pheromones in lipids and other fractions from urine of the male mole rat, *Spalax ehrenbergi. Physiology and Behavior* 52:741–747.

Merriam, C. H. 1890. The geographic distribution of life in North America. *Proceedings of the Biological Society of Washington* 7:1–64.

Merriam, C. H. 1898. Life zones and crop zones of the United States. *Bulletin of the United States Biological Survey* 10:1–79.

Mertens, R. 1955. Die Amphibien und Reptilien Südwestafrikas. *Abh. Senckenb. Naturforsch. Ges.* 490:1–172.

Mesta, R. 1996. Condors return to Arizona. *Endangered Species Bulletin* 21(6):16–17.

Metcalfe, J. D., and G. P. Arnold. 1997. Tracking fish with electronic tags. *Nature* 387:665–666.

Mewaldt, L. R. 1964. California sparrows return from displacement to Maryland. *Science* 146:941–942.

Meyer, A., and A. C. Wilson. 1991. Coelacanth relationships. *Nature* 353:218–219.

Meyer, A., T. D. Kocher, P. Basasibwaki, and A. C. Wilson. 1990. Monophyletic origin of Lake Victoria cichlid fishes suggested by mitochondrial DNA sequences. *Nature* 347:550–553.

Meyer, B. J., and R. K. Meyer. 1944. Growth and reproduction of the cotton rat, *Sigmodon hispidus hispidus,* under laboratory conditions. *Journal of Mammalogy* 25(2):107–129.

Meylan, A. B., B. W. Bowen, and J. C. Avise. 1990. A genetic test of the natal homing versus social facilitation models for green turtle migration. *Science* 248(4956):724–727.

Michard, D., A. Ancel, J.-P. Gendner, J. Lage, Y. LeMaho, T. Zorn, L. Gangloff, A Schierrer, K. Struyf, and G. Wey. 1995. Non-invasive bird tagging. *Nature* 376:649–650.

Miché, F., B. Vivien-Roels, P. Pévet, C. Spehner, J. P. Robin, and Y. LeMaho.

1991. Daily pattern of melatonin secretion in an Antarctic bird, the emperor penguin, *Aptenodytes forsteri*: Seasonal variations, effect of constant illumination and of administration of isoproterenol or propranolol. *General and Comparative Endocrinology* 84:249–263.

Middleton, A. L. A. 1991. Failure of brown-headed cowbird parasitism in nests of the American goldfinch. *Journal of Field Ornithology* 62(2):200–203.

Milinkovitch, M. C., and J. G. M. Thewissen. 1997. Even-toed fingerprints on whale ancestry. *Nature* 388:622–624.

Milinski, M., and T. C. M. Bakker. 1990. Female sticklebacks use male coloration in mate choice and hence avoid parasitized males. *Nature* 344:330–333.

Milius, S. 1998a. Stealth, lies, and cowbirds. *Science News* 153:345–347.

Milius, S. 1998b. New bird species found in surprising place. *Science News* 153:372.

Milius, S. 1998c. Second group of living fossil reported. *Science News* 154:196.

Millard, A. R. 1995. The body temperature of *Tyrannosaurus rex*. *Science* 267:1666.

Miller, G. H., J. W. Magee, B. J. Johnson, M. L. Fogel, N. A. Spooner, M. T. McCulloch, and L. K. Ayliffe. 1999. Pleistocene extinction of *Genyornis newtoni*: Human impact on Australian megafauna. *Science* 283:205–208.

Miller, G. T., Jr. 1997. *Living in the Environment*. Tenth edition. Belmont, California: Wadsworth Publishing Company.

Miller, G. T., Jr. 1999. Environmental Science. Seventh edition. Belmont, California: Wadsworth Publishing Company.

Miller, L. K., and J. P. Dehlinger. 1969. Neuromuscular function and low temperatures in frogs from cold and warm climates. *Comparative Biochemistry and Physiology B* 28:915–921.

Mitchell, P. 1991a. Social behaviour and communication of koalas. Pages 151–170. In: Lee, A. K., K. A. Handasyde, and G. D. Sanson (eds.). *Biology of the Koala*. Chipping Norton, N. S. W., Australia: Surrey Beatty and Sons.

Mitchell, P. 1991b. The home ranges and social activity of koalas—a quantitative analysis. Pages 171–187. In: Lee, A. K., K. A. Handasyde, and G. D. Sanson (eds.). *Biology of the Koala*. Chipping Norton, N. S. W., Australia: Surrey Beatty and Sons.

Mlot, C. 1993. Predators, prey, and natural disasters attract ecologists. *Science* 261:1115.

Moen, A. N. 1973. *Wildlife Ecology*. San Francisco: W. H. Freeman.

Moffat, A. S. 1997. Teeth and bones tell their stories at Chicago meeting. *Science* 278:801–802.

Mojzsis, S. J., G. Arrhenius, K. D. McKeegan, T. M. Harrison, A. P. Nutman, and C. R. L. Friend. 1996. Evidence for life on Earth before 3,800 million years ago. *Nature* 384:55–59.

Moll, E. O., and J. M. Legler. 1971. The life history of a neotropical slider turtle *Pseudemys scripta* (Schoepff) in Panama. *Bulletin of the Los Angeles County Museum of Natural History and Science* 11:1–102.

Moller, A. P. 1992. Female swallow preference for symmetrical male sexual ornaments. *Nature* 357:238–240.

Mommsen, T. P., and E. M. Plisetskaya. 1991. Insulin in fishes and agnathans: History, structure, and metabolic regulation. *Reviews in Aquatic Sciences* 4(2–3):225–259.

Monastersky, R. 1989a. Huge dinosaur bones discovered hollow. *Science News* 135(17):261.

Monastersky, R. 1989b. Dinosaurs used their heads to beat the heat. *Science News* 136(20):309.

Monastersky, R. 1990a. Chinese bird fossil: Mix of old and new. *Science News* 138(16): 246–247.

Monastersky, R. 1990b. Dinosaur digestive aids. *Science News* 138(16):255.

Monastersky, R. 1990c. Rare fossils of enigmatic amphibian. *Science News* 138(17):270.

Monastersky, R. 1992. Do Antarctic seals feel El Niño? *Science News* 142(22):382.

Monastersky, R. 1996a. The first shark: To bite or not to bite? *Science News* 149(7):101.

Monastersky, R. 1996b. Hints of a downy dinosaur in China. *Science News* 150(17):260.

Monastersky, R. 1996c. Early kin of vertebrates found in China. *Science News* 150(20):311.

Monastersky, R. 1996d. The lost tribe of the mammals. *Science News* 150(24):378–379.

Monastersky, R. 1997a. Paleontologists deplume feathery dinosaur. *Science News* 151(18):271.

Monastersky, R. 1997b. A fowl flight. Fossil finds recharge debate about birds and dinosaurs. *Science News* 152(8):120–121.

Monastersky, R. 1997c. *T. rex* bested by Argentinian beast. *Science News* 151(21):317.

Monastersky, R. 1998a. Planet posts temperature record for 1997. *Science News* 153(3):38.

Monastersky, R. 1998b. Antarctic ozone hole reaches record size. *Science News* 154 (1):246.

Monastersky, R. 1999. Out of the swamps: How early vertebrates established a foothold—with all 10 toes—on land. *Science News* 155(21):328–329.

Monastersky, R. 2000. All mixed up over birds and dinosaurs. *Science News* 157(3):38.

Montgomery, S. 1992. Infrasound: Catching the vibes. *Animals* 125(2):24–27.

Montzka, S. A., J. H. Butler, R. C. Myers, T. M. Thompson, T. H. Swanson, A. D. Clarke, L., T. Lock, and J. W. Elkins. 1996. Decline in the tropospheric abundance of halogen from halocarbons: Implications for stratospheric ozone depletion. *Science* 272:1318–1322.

Moore, H. D. M., N. M. Jenkins, and C. Wong. 1997. Immunocontraception in rodents: A review of the development of a sperm-based immunocontraceptive vaccine for the grey squirrel (*Sciurus carolinensis*). *Reproduction, Fertility, and Development* 9:125–129.

Morgan, M. J., and J. J. Godin. 1985. Antipredator benefits of schooling behaviour in a cyprinodontid fish, the banded killifish (*Fundulus diaphanus*). *Zeitschrift Tierpsychologie* 70:236–246.

Morell, V. 1993a. Dino DNA: The hunt and the hype. *Science* 261:160–162.

Morell, V. 1993b. *Archaeopteryx*: Early bird catches a can of worms. *Science* 259: 764–765.

Morell, V. 1994. Did early humans reach Siberia 500,000 years ago? *Science* 263: 611–612.

Morell, V. 1996. New mammals discovered by biology's new explorers. *Science* 273:1491.

Morell, V. 1997a. Predator-free guppies take an evolutionary leap forward. *Science* 275:1880.

Morell, V. 1997b. Catching lizards in the act of adapting. *Science* 276:682–683.

Morell, V. 1997c. Fossilized hatchling heats up the bird-dinosaur debate. *Science* 276:1501.

Morell, V. 1997d. The origin of dogs: Running with the wolves. *Science* 276: 1647–1648.

Morell, V. 1997e. The origin of birds: The dinosaur debate. *Audubon* 99(2):36–45.

Morgan, M. J., and J.-G. J. Godin. 1985. Antipredator benefits of schooling behaviour in a cyprinodontid fish, the banded killifish (*Fundulus diaphanus*). *Zeitschrift Tierpsychologie* 70:236–246.

Morreale, S. J., E. A. Standora, J. R. Spotila, and F. V. Paladino. 1996. Migration corridor for sea turtles. *Nature* 384:319–320.

Morris, P. A., and J. F. Harper. 1965. The occurrence of small mammals in discarded bottles. *Proceedings of the Zoological Society of London* 145:148–153.

Mortola, J. P., P. B. Frappell, and P. A. Woolley. 1999. Breathing through skin in a newborn mammal. *Nature* 397:660.

Mosauer, W. 1935. The myology of the trunk region of snakes and its significance for ophidian taxonomy and phylogeny. *University of California at Los Angeles Publications in Biological Science* 1:81–120.

Mosby, H. S. (ed.). 1963. *Wildlife Investigational Techniques*. Washington, D.C.: The Wildlife Society.

Mosimann, J. E., and P. S. Martin. 1975. Simulating overkill by paleoindians. *American Scientist* 63:304–313.

Moss, M. L. 1969. Evolution of mammalian dental anatomy. *American Museum Novitates* 2360:1–39.

Motani, R., N. Minoura, and T. Ando. 1998. Ichthyosaurian relationships illuminated by new primitive skeletons from Japan. *Nature* 393:255–257.

Moyle, P. B., and J. J. Cech, Jr. 1996. *Fishes: An Introduction to Ichthyology*. Third edition. Upper Saddle River, New Jersey: Prentice Hall.

Müller-Schwarze, D. 1971. Pheromones in black-tailed deer (*Odocoileus hemionus columbianus*). *Animal Behaviour* 27(3): 422–427.

Müller-Schwarze, D., and P. W. Houlihan. 1991. Pheromonal activity of single castoreum constituents in beaver, *Castor canadensis*. *Journal of Chemical Ecology* 17(4):715–734.

Munro, U., and R. Wiltschko. 1995. The role of skylight polarization in the orientation of a day-migrating bird species. *Journal of Comparative Physiology A Sensory, Neural, and Behavioral Physiology* 177(3):357–362.

Münz, H., B. Claas, and B. Fritzsch. 1984. Electroreceptive and mechanoreceptive units in the lateral line of the axolotl *Ambystoma mexicanum*. *Journal of Comparative Physiology A* 154:33–44.

Murie, A. 1934. The moose of Isle Royale. *University of Michigan Museum of Zoology Miscellaneous Publications* no. 25:1–44.

Murphy, J. B., and R. D. Nance. 1992. Mountain belts and the supercontinent cycle. *Scientific American* 266(4):84–91.

Murphy, P. 1981. Celestial compass orientation in juvenile American alligators. *Copeia* 1981:638–645.

Murrish, D. E., and K. Schmidt-Nielsen. 1970. Exhaled air temperature and water conservation in lizards. *Respiration Physiology* 10:151–158.

Myers, C. W., and J. W. Daly. 1983. Dart-poison frogs. *Scientific American* 248(2):120–133.

Myers, C. W., and J. W. Daly. 1993. Tropical poison frogs. *Science* 262:1193.

Mylvaganam, S. E., C. Bonaventura, J. Bonaventura, and E. D. Getzoff. 1996. Structural basis for the Root effect in haemoglobin. *Nature Structural Biology* 3(3):275–283.

N

Naeem, S., L. J. Thompson, S. P. Lawler, J. H. Lawton, and R. M. Woodfin. 1994. Declining biodiversity can alter the performance of ecosystems. *Nature* 368:734–737.

Naitoh, T., and R. Wassersug. 1996. Why are toads right-handed? *Nature* 380:30–31.

Naples, V. L. 1999. Morphology, evolution and function of feeding in the giant anteater (*Myrmecophaga tridactyla*). *Journal of Zoology* 249(1):19–41.

Neitz, J., T. Geist, and G. H. Jacobs. 1989. Color vision in the dog. *Visual Neuroscience* 3:119–125.

Nelson, G., and N. Platnick. 1981. *Systematics and Biogeography: Cladistics and Vicariance*. New York: Columbia University Press.

Nelson, J. S. 1994. *Fishes of the World*. Third edition. New York: John Wiley and Sons.

Nelson, R. J., M. Kita, J. M. C. Blom, and J. Rhyne-Grey. 1992. Photoperiod influences the critical caloric intake necessary to maintain reproduction among male deer mice (*Peromyscus maniculatus*). *Biology of Reproduction* 46:226–232.

Nero, R. W. 1992. New great horned owl longevity record. *Blue Jay* 50(2):91–92.

Nestler, J. R. 1991. Metabolic substrate change during daily torpor in deer mice. *Canadian Journal of Zoology* 69:322–327.

Nestler, J. R. 1992. Tissue-specific metabolism during normothermy and daily torpor in deer mice (*Peromyscus maniculatus*). *Journal of Experimental Zoology* 261:406–413.

Nevitt, G. A., R. R. Veit, and P. Kareiva. 1995. Dimethyl sulphide as a foraging cue for Antarctic procellariiform seabirds. *Nature* 376:680–682.

Newcomer, R. T., D. H. Taylor, and S. I. Guttman. 1974. Celestial orientation in two species of water snakes (*Natrix sipedon* and *Regina septemvittatus*). *Herpetologica* 30:194–200.

Nicholson, R. E. 1996. Just like pheromones. *Environment* 38(3):22.

Nielsen, J. G. 1977. The deepest living fish *Abyssobrotula galatheae*. *Galathea Report* 14:41–48.

Niemitz, C., A. Nietsch, S. Warter, and Y. Rumpler. 1991. *Tarsius dianae*: A new primate species from central Sulawesi (Indonesia). *Folia Primatologica* 56:105–116.

Nilsson, G. E. 1999–2000. The cost of a brain. *Natural History* 108(10):66–73.

Nishikawa, K. C., and D. C. Cannatella. 1991. Kinematics of prey capture in the tailed frog *Ascaphus truei* (Anura: Ascaphidae). *Zoological Journal of the Linnean Society* 103:289–307.

Nishikawa, K. C., and G. Roth. 1991. The mechanism of tongue protraction during prey capture in the frog *Discoglossus pectus*. *Journal of Experimental Biology* 159:217–234.

Noble, G. K. 1931. *The Biology of the Amphibia*. New York: McGraw-Hill.

Noble, G. K., and P. G. Putnam. 1931. Observations on the life history of *Ascaphus truei* Stejneger. *Copeia* 1931:97–101.

Nogales, M., J. D. Delgado, and F. M. Medina. 1998. Shrikes, lizards and *Lycium intricatum* (Solanaceae) fruits: A case of indirect seed dispersal on an oceanic island (Alegranza, Canary Islands). *Journal of Ecology* 86:866–871.

Nolan, V., Jr., E. D. Ketterson, and L. Wolf. 1986. Long-distance homing by nonmigratory dark-eyed juncos. *The Condor* 88:539–542.

Norberg, R. A. 1995. Feather asymmetry in *Archaeopteryx*. *Nature* 374:221.

Nordeng, H. 1971. Is the local orientation of anadromous fishes determined by pheromones? *Nature* 233:411–413.

Nordeng, H. 1977. A pheromone hypothesis for homeward migration in anadromous salmonids. *Oikos* 28:155–159.

Norell, M. A., J. M. Clark, D. Demberelyin, B. Rhinchen, L. M. Chiappe, A. R. Davidson, M. C. McKenna, P. Altangerel, and M. J. Novacek. 1994. A theropod dinosaur embryo and the affinities of the Flaming Cliffs dinosaur eggs. *Science* 266:779–782.

Norell, M. A., P. Makovicky, and J. M. Clark. 1997. A *Velociraptor* wishbone. *Nature* 389:447.

Norman, D. 1991. *Dinosaur!* New York: Prentice-Hall.

Normile, D. 1998. New views of the origins of mammals. *Science* 281:774–775.

Norris, K. S. 1969. The echolocation of marine mammals. Pages 391–423. In: Anderson, H. T. (ed). *The Biology of Marine Mammals*. New York: Academic Press.

Norris, K. S., and G. W. Harvey. 1972. A theory of the function of the spermaceti organ of the sperm whale (*Physeter catodon* L.). Pages 397-417. In: Galler, S. R., K. Schmidt-Koenig, G. J. Jacobs, and R. E. Belleville (eds.). *Animal Orientation and Navigation*. Washington, D.C.: National Aeronautics and Space Administration (NASA).

Novacek, M. J. 1992. Mammalian phylogeny: Shaking the tree. *Nature* 356:121–125.

Novacek, M. J., G. W. Rougler, J. R. Wible, M. C. McKenna, D. Dashzeveg, and I. Horovitz. 1997. Epipubic bones in eutherian mammals from the Late Cretaceous of Mongolia. *Nature* 389:483–486.

Novas, F. E., and P. F. Puerta. 1997. New evidence concerning avian origins from the Late Cretaceous of Patagonia. *Nature* 387:390–392.

Nowak, E., and P. Berthold. 1991. Satellite tracking, a new method in orientation research. Pages 307–321. In: Berthold, P. (ed.). *Orientation in Birds*. Basel: Birkhauser Verlag.

Nowak, R. M. 1991. *Walker's Mammals of the World*. Fifth edition. 2 volumes. Baltimore, Maryland: The Johns Hopkins University Press.

Nowak, R. M., and J. L. Paradiso. 1983. *Walker's Mammals of the World*. Fourth edition. Baltimore, Maryland: Johns Hopkins University Press.

Nussbaum, R. A. 1969. Nest and eggs of the Pacific giant salamander, *Dicamptodon ensatus* (Eschscholtz). *Herpetologica* 25:257–261.

O

O'Brien, C. 1995. Dinosaur embryos spark excitement, concern. *Science* 267:1760.

Obst, B. S. 1991. The avian feeding system: Intestinal nutrient absorption in birds. *Acta XX Congressus Internationalis Ornithologici*, Christchurch, New Zealand. December 2–9, 1990:920–926.

O'Farrell, T. P. 1965. Home range and ecology of snowshoe hares in interior Alaska. *Journal of Mammalogy* 46(3):406–418.

Office of Endangered Species and International Activities of the U.S. Fish and Wildlife Service. 1973. *Threatened Wildlife of the United States*. Washington, D.C.: Bureau of Sport Fisheries and Wildlife.

Officer, C. B., and C. L. Drake. 1983. The Cretaceous-Tertiary transition. *Science* 219:1383–1390.

Oldham, R. S., and M. J. S. Swan. 1992. Effects of ingested radio transmitters on *Bufo bufo* and *Rana temporaria*. *Herpetological Journal* 2:82–85.

O'Leary, M. A. and J. H. Geisler. 1999. The position of Cetacea within Mammalia: phylogenetic analysis of morphological data from extinct and extant taxa. *Systematic Biology* 48(3):455–490.

Oliver, J. A. 1955. *The Natural History of North American Amphibians and Reptiles*. Princeton, New Jersey: Van Nostrand.

Olsson, M., R. Shine, T. Madsen, A. Gullberg, and H. Tegelstrom. 1996. Sperm selection by females. *Nature* 383:585.

O'Reilly, J. C., R. A. Nussbaum, and D. Boone. 1996. Vertebrate with protrusible eyes. *Nature* 382:33.

O'Reilly, J. C., D. A. Ritter, and D. R. Carrier. 1997. Hydrostatic locomotion in a limbless tetrapod. *Nature* 386:269–272.

O'Riain, M. J., J. U. M. Jarvis, and C. G. Faulkes. 1996. A dispersive morph in the naked mole-rat. *Nature* 380:619–621.

Oring, L. W. 1995. The early bird gives the sperm. *Natural History* 104(8):58–61.

Oring, L. W., R. C. Fleischer, J. M. Reed, and K. E. Marsden. 1992. Cuckoldry through stored sperm in the sequentially polyandrous spotted sandpiper. *Nature* 359:631–633.

Orr, R. T. 1982. *Vertebrate Biology*. Fifth edition. Philadelphia: W. B. Saunders.

Osterhaus, A., J. Groen, H. Niesters, M. van de Bildt, B. Martina, L. Vedder, J. Vos, H. van Egmond, B. A. Sidi, and M. E. O. Barham. 1997. Morbillivirus in monk seal mass mortality. *Nature* 388:838–839.

Ostrom, J. H. 1972. Were some dinosaurs gregarious? *Palaeogeography, Palaeoclimatology, Palaeoecology* 11:287–301.

Ostrom, J. H. 1974. *Archeopteryx* and the origin of flight. *The Quarterly Review of Biology* 49:27–47.

Ostrom, J. H. 1985. The meaning of *Archeopteryx*. Pages 161–176. In: Hecht, M. K., J. H. Ostrom, G. Viohl, and P. Wellnhofer (eds.). *The Beginnings of Birds*. Eichstätt, Germany: Freunde des Jura-Museums.

Ostrom, J. H. 1991. The bird in the bush. *Nature* 353:212.

Ostrom, J. H. 1994. On the origin of birds and of avian flight. Pages 160–189. In: Spencer, R. S. (ed.). *Major Features of Vertebrate Evolution*. 17th Annual Short Course of the Paleontological Society, October 23, 1994, Seattle, Washington. Knoxville: University of Tennessee at Knoxville Publication E01-1040-001-95.

Ott, M., and F. Schaeffel. 1995. A negatively powered lens in the chameleon. *Nature* 373:692–694.

Oxford, A. E. 1958. Rumen microorganisms and other products. *New Zealand Science Review* 16:38–44.

P

Pacheco, J. F., B. M. Whitney, and L. P. Gonzaga. 1996. A new genus and species of furnariid (Aves: Furnariidae) from the cocoa-growing region of southeastern Bahia, Brazil. *Wilson Bulletin* 108(3):397–433.

Packard, G. C., M. J. Packard, and L. Benigan. 1991. Sexual differentiation, growth, and hatchling success by embryonic painted turtles in wet and dry environments at fluctuating temperatures. *Herpetologica* 47:125–132.

Packer, C., and A. E. Pusey. 1997. Divided we fall: Cooperation among lions. *Scientific American* 276(5):52–59.

Packer, C., D. A. Gilbert, A. E. Pusey, and S. J. O'Brien. 1991. A molecular genetic analysis of kinship and cooperation in African lions. *Nature* 351:562–565.

Packer, W. C. 1963. Dehydration, hydration and burrowing behaviour in *Heleioporus eyrei* (Gray) (Leptodactylidae). *Ecology* 44:643–651.

Padian, K. 1998. When is a bird not a bird? *Nature* 393:729–730.

Padian, K., and L. M. Chiappe. 1998. The origin of birds and their flight. *Scientific American* 278(2):38–47.

Page, J., and E. S. Morton. 1989. *Lords of the Air*. Washington, D.C.: Smithsonian Books.

Pagels, J. F., and T. W. French. 1987. Discarded bottles as a source of small mammal distribution data. *American Midland Naturalist* 118(1):217–219.

Paige, K. 1996. Pressure in the bat cave. *Discover* 17(3):13–14.

Palmer, B. D., and L. J. Guillette. 1992. Alligators provide evidence for the evolution of an archosaurian mode of oviparity. *Biology of Reproduction* 46:39–47.

Palsboll, P. J., J. Allen, M. Berube, P. J. Clapham, T. P. Feddersen, P. S. Hammond, R. R. Hudson, H. Jorgensen, S. Katona, A. H. Larsen, F. Larsen, J. Lien, D. K. Mattila, J. Sigurjonsson, R. Sears, T. Smith, R. Sponer, P. Stevick, and N. Oien. 1997. Genetic tagging of humpback whales. *Nature* 388:767–769.

Panchen, A. L., and T. R. Smithson. 1987. Character diagnosis, fossils and the origin of tetrapods. *Biological Review* 62:341–438.

Panchen, A. L., and T. R. Smithson. 1988. The relationships of the earliest tetrapods. Pages 1–32. In: Benton, M. J. (ed.). *The Phylogeny and Classification of the Tetrapods. Volume I. Amphibians, Reptiles, Birds*. Oxford, England: Clarendon Press.

Panella, G. 1971. Fish otoliths: Daily growth layers and periodical patterns. *Science* 173:1124–1127.

Pang, S. F., S. Y. W. Shui, and S. F. Tse. 1985. Effect of photic manipulation on the level of melatonin in the retinas of frogs (*Rana tigrina regulosa*). *General and Comparative Endocrinology* 58:464–470.

Parker, G. A. 1982. Why so many sperm? The maintenance of two sexes with internal fertilization. *Journal of Theoretical Biology* 96:281–294.

Parrington, F. R. 1967. The origins of mammals. *Advancement of Science* (London) 24:165–173.

Parrington, F. R. 1971. On the Upper Triassic mammals. *Philosophical Transactions of the Royal Society* (London) B 261:231–272.

Parrington, F. R. 1973. The dentitions of the earliest mammals. *Zoological Journal of the Linnean Society* (London) 52:85–95.

Parsons, R. H., and F. Mobin. 1989. Regional osmotic water flow in Amphibia: Effect of dehydration. *American Zoologist* 29:103A.

Paton, R. L., T. R. Smithson, and J. A. Clack. 1999. An amniote-like skeleton from the Early Carboniferous of Scotland. *Nature* 398:508–513.

Patrusky, B. 1980. Pre-Clovis man: Sampling the evidence. *Mosaic* 11(5):2–10.

Payne, K. B., W. R. Langbauer, Jr., and E. M. Thomas. 1986. Infrasonic calls of the Asian elephant (*Elephas maximus*). *Behavioural Ecology and Sociobiology* 18:297–301.

Payne, R. S. 1971. Acoustic location of prey by barn owls. *Journal of Experimental Biology* 54:535–573.

Pearson, P. G. 1955. Population ecology of the spadefoot toad, *Scaphiopus h. holbrooki* (Harlan). *Ecological Monographs* 25:233–267.

Peltier, R. 1994. Ice Age paleotopography. *Science* 265:195–201.

Pennisi, E. 1996. Superhero shrew. *Science* 271:149.

Pennisi, E. 1997a. How male animals gain an edge in the mating game. *Science* 277:317–318.

Pennisi, E. 1997b. Transgenic lambs from cloning lab. *Science* 277:631.

Pennisi, E. 1997c. Brazil's new monkey. *Science* 277:1207.

Pennisi, E. 1997d. Rampaging rabbit virus—again. *Science* 277:1441.

Pennisi, E. 2000. Zoology naming rules eased. *Science* 287:26.

Perle, A. M., A. Norell, L. M. Chiappe, and J. M. Clark. 1993. Flightless bird from the Cretaceous of Mongolia. *Nature* 362:623–626.

Peters, D. 1995. Wing shape in pterosaurs. *Nature* 374:315–316.

Petersen, C. G. J. 1896. The yearly immigration of young plaice into the Limfjord from the German Sea. *Report of the Danish Biological Station for 1895*:1–77.

Peterson, A. T. 1992. Phylogeny and rates of molecular evolution in the *Aphelocoma* jays (Corvidae). *The Auk* 109(1):133–147.

Peterson, C. R. 1987. Daily variation in the body temperatures of free-ranging garter snakes. *Ecology* 68(1):160–169.

Peterson, C. R., and V. A. Cobb. 1991. Thermal ecology of hibernation in Great Basin rattlesnakes. *American Zoologist* 31(5):42A.

Peterson, R. O. 1977. Wolf ecology and prey relationships on Isle Royale. *National Park Service Scientific Monograph Series* 11:1–210.

Petranka, J. W., and A. Sih. 1986. Environmental stability, competition, and density-dependent growth and survivorship of a stream-dwelling salamander. *Ecology* 67:729–736.

Petrie, M., and A. P. Moller. 1991. Laying eggs in others' nests: Intraspecific brood parasitism in birds. *Trends in Ecology and Evolution* 6(10):315–320.

Phillips, A. D., W. Hahn, S. Klimek, and J. L. McGuire. 1987. A comparison of potencies and activities of progestins used in contraceptives. *Contraception* 6:181–192.

Phillips, J. A. 1995. Rhythms of a desert lizard. *Natural History* 104(10):50–55.

Phillips, J. B., and S. C. Borland. 1992. Behavioural evidence for use of a light-dependent magnetoreception mechanism by a vertebrate. *Nature* 359:142–144.

Phillips, M. K., and J. Scheck. 1991. Parasitism in captive and reintroduced red wolves. *Journal of Wildlife Diseases* 27(3):498–501.

Piersma, T., and R. E. Gill, Jr. 1998. Guts don't fly: Small digestive organs in obese bar-tailed godwits. *The Auk* 115(1):196–203.

Pine, R. H. 1994. New mammals not so seldom. *Nature* 368:593.

Piper, S. E. 1909. The Nevada mouse plague of 1907–8. *Farmers Bulletin* 352:1–23.

Pitelka, F. A. 1964. The nutrient-recovery hypothesis for Arctic microtine cycles. In: Crisp,. D. J. (ed). *Grazing in Terrestrial and Marine Environments*. Oxford, England: Blackwell Scientific Publications.

Pitman, C. R. S. 1974. *A Guide to the Snakes of Uganda*. Revised edition. Codicote, United Kingdom: Wheldon and Wesley.

Platz, J.E. 1993. *Rana subaquabocalis*, a remarkable new species of leopard frog (*Rana pipiens* complex) from southeastern Arizona that calls under water. *Journal of Herpetology* 27(2):154–162.

Platz, J. E., and J. M. Conlon. 1997. Reptile relationships turn turtle...and turn back again. *Nature* 389:246.

Poiani, K. A., and W. C. Johnson. 1991. Global warming and prairie wetlands. *BioScience* 41(9):611–618.

Poinar, H. N., M. Hofreiter, W. G. Spaulding, P. S. Martin, B. A. Stankiewicz, H. Bland, R. P. Evershed, G. Possnert, and S. Paabo. 1998. Molecular coproscopy: Dung and diet of the extinct ground sloth *Nothrotheriops shastensis*. *Science* 281:402–406.

Poole, J. H., K. Payne, W. R. Langbauer, and C. J. Moss. 1988. The social contexts of some very low frequency calls of African elephants. *Behavioural Ecology and Sociobiology* 22:385–392.

Pope, K. O., K. H. Baines, A. C. Ocampo, and B. A. Ivanov. 1994. Impact winter and the Cretaceous/Tertiary extinctions: Results of a Chicxulub asteroid impact model. *Earth and Planetary Science Letters* 128:719–725.

Popper, A. N. 1980. Sound emission and detection by delphinids. Pages 1–52. In: Herman, L. M. (ed.). *Cetacean Behavior: Mechanisms and Functions*. New York: John Wiley and Sons.

Porter, K. R. 1972. *Herpetology*. Philadelphia: W. B. Saunders.

Pouyard, L., S. Wirjoatmodjo, I. Rachmatika, A. Tjakrawidjaja, R. Hadiaty, and H. Wartono. 1999. Une nouvelle espece de coelacanthe. Preuves genetiques et morphologiques. *Comptes Rendus de l'Academie des Sciences* 322(4):261–267.

Powell, J. A., and G. B. Rathbun. 1984. Distribution and abundance of manatees along the northern Gulf of Mexico. *Northeast Gulf Science* 7:1–28.

Powell, R. A. 1980. Stability in a one-predator-three-prey community. *American Naturalist* 115:567–579.

Powers, D. R. 1991. Diurnal variation in mass, metabolic rate, and respiratory quotient in Anna's and Costa's hummingbirds. *Physiological Zoology* 64(3):850–870.

Presley, R. 1997. Pelvic problems for mammals. *Nature* 389:440–441.

Priede, I. G., P. M. Bagley, J. D. Armstrong, K. L. Smith, Jr., and N. R. Merrett. 1991. Direct measurement of active dispersal of food-falls by deep-sea demersal fishes. *Nature* 351:647–649.

Prince, E. D., D. W. Lee, J. R. Zweifel, and E. B. Brothers. 1991. Estimating age and growth of young Atlantic blue marlin *Makaira nigricans* from otolith microstructure. *Fishery Bulletin* 89(3):441–459.

Prinzinger, R., and Ch. Hinninger. 1992. (Endogenous?) diurnal rhythm in the energy metabolism of pigeon embryos. *Naturwissenschaften* 79:278–279.

Proctor, N. S., and P. J. Lynch. 1993. *Manual of Ornithology: Avian Structure and Function*. New Haven, Connecticut: Yale University Press.

Purgue, A. P. 1997. Tympanic sound radiation in the bullfrog *Rana catesbeiana*. *Journal of Comparative Physiology A* 181:438–445.

Purnell, M. A. 1995. Microwear on conodont elements and macrophagy in the first vertebrates. *Nature* 374:798–800.

Pusey, A., J. Williams, and J. Goodall. 1997. The influence of dominance rank on the reproductive success of female chimpanzees. *Science* 277:828–831.

Q

Quinn, J. S., G. E. Woolfenden, J. W. Fitzpatrick, and B. N. White. 1999. Multilocus DNA fingerprinting supports genetic monogamy in Florida scrub-jays. *Behavioral Ecology and Sociobiology* 45(1):1–10.

R

Rabinowitz, A. 1996. Spirit of the jaguar. *Wildlife Conservation* 99(3):32–35.

Ralls, K., D. B. Siniff, T. D. Williams, and V. B. Kuechle. 1989. An intraperitoneal radio transmitter for sea otters. *Marine Mammal Science* 5:376–381.

Raloff, J. 1995. Beyond estrogens. *Science News* 148(3):44–46.

Raloff, J. 1998. New tags may help diagnose turtle losses. *Science News* 153(26):406.

Ralt, D., M. Goldenberg, P. Fetterolf, D. Thompson, J. Dor, S. Mashiach, D. L. Garbers, and M. Eisenbach. 1991. Sperm attraction to a follicular factor(s) correlates with human egg fertilizability. *Proceedings of the National Academy of Sciences, USA,* 88:2840–2844.

Ramachandran, V. S., C. W. Tyler, R. L. Gregory, D. Rogers-Ramachandran, S. Duensing, C. Pillsbury, and C. Ramachandran. 1996. Rapid adaptive camouflage in tropical flounders. *Nature* 379:815–818.

Ramanathan, V. 1988. The greenhouse theory of climate change: A test by an inadvertent global experiment. *Science* 240:293–299.

Rampino, M. R., and S. Self. 1992. Volcanic winter and accelerated glaciation following the Toba super-eruption. *Nature* 359:50–52.

Randall, D. J. 1970a. The circulatory system. Pages 133–172. In: Hoar, W. S., and D. J. Randall (eds.). *Fish Physiology. Volume 4.* New York: Academic Press.

Randall, D. J. 1970b. Gas exchange in fish. Pages 253–292. In: Hoar, W. S., and D. J. Randall (eds.). *Fish Physiology. Volume 4.* New York: Academic Press.

Randall, J. A. 1989. Individual footdrumming signatures in banner-tailed kangaroo rats *Dipodomys spectabilis*. *Animal Behaviour* 38:620–630.

Rasmussen, L. E. L., D. L. Hess, and J. D. Haight. 1990. Chemical analysis of temporal gland secretions collected from an Asian bull elephant during a four-month musth episode. *Journal of Chemical Ecology* 16(7):2167–2181.

Rathbun, G. B., J. P. Reid, and J. Bourassa. 1987. Design and construction of a tethered, floating radio-tag assembly for manatees. Report prepared by the United States Fish and Wildlife Service, Denver Wildlife Research Center, Gainesville Field Station. *NTIS Document no. PB87-161345.*

Rathbun, G. B., J. P. Reid, and G. Carowan. 1990. Distribution and movement patterns of manatees (*Trichechus manatus*) in northwestern peninsular Florida. *Florida Marine Resources Publication* no. 48.

Rattenborg, N. C., S. L. Lima, and C. J. Amlaner. 1999. Half-awake to the risk of predation. *Nature* 397:397–398.

Raup, D. M., and J. J. Sepkoski, Jr. 1986. Periodic extinction of families and genera. *Science* 231:833–836.

Raven, P. 1995. A time of catastrophic extinction. *The Futurist* 29:38–41.

Raymond, J. A., P. Wilson, and A. L. DeVries. 1989. Inhibition of growth of nonbasal planes in ice by fish antifreezes. *Proceedings of the National Academy of Sciences*, USA. 86(3):881–885.

Reading, R. P., T. W. Clark, A. Vargas, L. R. Hanebury, B. J. Miller, and D. Biggins. 1996. Recent directions in black-footed ferret recovery. *Endangered Species Update* 13(10-11):1–6.

Reanier, R. E. 1997. Dating a paleoindian site in the Amazon in comparison with Clovis culture. *Science* 275:1948–1949.

Reed, M. A. 1989. Experiences along a Texas bluebird trail. *Quarterly Journal of the North American Bluebird Society* 11(1):1–2.

Reichardt, P. B., J. P. Bryant, B. J. Anderson, D. Phillips, T. P. Clausen, M. Meyer, and K. Frisby. 1990a. Germacrone defends Labrador tea from browsing by snowshoe hares. *Journal of Chemical Ecology* 16(6):1961–1970.

Reichardt, P. B., J. P. Bryant, B. R. Mattes, T. P. Clausen, F. S. Chapin, III, and M. Meyer. 1990b. Winter chemical defense of Alaskan balsam poplar against snowshoe hares. *Journal of Chemical Ecology* 16(6):1941–1959.

Reichardt, T. 1998. Peregrine leads flight from endangered list. *Nature* 395:3.

Reid, D. G., W. E. Melquist, J. D. Woolington, and J. M. Noll. 1986. Reproductive effects of intraperitoneal transmitter implants in river otters. *Journal of Wildlife Management* 50(1):92–94.

Reid, J. 1997. Navy tracks manatees with satellites. *Endangered Species Bulletin* 21(5):22–23.

Reisz, R. 1972. Pelycosaurian reptiles from the Middle Pennsylvanian of North America. *Bulletin of the Harvard Museum of Comparative Zoology* 144:27–62.

Reisz, R. R. 1981. A diapsid reptile from the Pennsylvanian of Kansas. *Special Publication of the University of Kansas Museum of Natural History* no. 7:1–74.

Reisz, R. R., and M. Laurin. 1991. *Owenetta* and the origin of turtles. *Nature* 349: 324–326.

Renne, P. R., Z. Zhang, M. A. Richards, M. T. Black, and A. R. Basu. 1995. Synchrony and causal relations between Permian-Triassic boundary crises and Siberian flood volcanism. *Science* 269:1413–1416.

Rennie, J. 1991. Are species specious? *Scientific American* 265(5):26.

Rennie, J. 1992. Life in the fast lane. *Scientific American* 266(2):31.

Repasky, R. R., R. J. Blue, and P. D. Doerr. 1991. Laying red-cockaded woodpeckers cache bone fragments. *The Condor* 93:458–461.

Repetski, J. E. 1978. A fish from the upper Cambrian of North America. *Science* 200:529–531.

Rex, M., N. R. P. Harris, P. von der Gathen, R. Lehmann, G. O. Braathen, E. Reimer, A. Beck, M. P. Chipperfield, R. Alfier, M. Allaart, F. O'Connor, H. Dier, V. Dorokhov, H. Fast, M. Gil, E. Kyro, Z. Litynska, I. S. Mikkelsen, M. G. Molyneux, H. Nakane, J. Notholt, M. Rummukainen, P. Viatte, and J. Wenger. 1997. Prolonged stratospheric ozone loss in the 1995–96 Arctic winter. *Nature* 389:835–838.

Reynolds, J., III, and D. Odell. 1991. *Manatees and Dugongs*. New York: Facts on File.

Reynolds, P. E. 1989. An experimental satellite collar for muskoxen. *Canadian Journal of Zoology* 67:1122–1124.

Reznick, D. N., F. H. Shaw, F. H. Rodd, and R. G. Shaw. 1997. Evaluation of the rate of evolution in natural populations of guppies (*Poecilia reticulata*). *Science* 275:1934–1937.

Ribble, D. O., and M. Salvioni. 1990. Social organization and nest co-occupancy in *Peromyscus californicus*, a monogamous rodent. *Behavioural Ecology and Sociobiology* 26(1):9–15.

Rice, D. W. 1967. Cetaceans. Pages 291–324. In: Anderson, S., and J. K. Jones, Jr. (eds.). *Recent Mammals of the World.* New York: Ronald Press Company.

Richardson, W. J. 1974. Spring migration over Puerto Rico and the western Atlantic: A radar study. *Ibis* 116:172–193.

Richardson, W. J. 1976. Autumn migration over Puerto Rico and the western Atlantic: A radar study. *Ibis* 118:309–332.

Riedman, M. 1990. *The Pinnipeds: Seals, Sea Lions and Walruses.* Berkeley: University of California Press.

Rieppel, O. 1999. Turtle origins. *Science* 283:945–946.

Rieppel, O., and M. deBraga. 1996. Turtles as diapsid reptiles. *Nature* 384:453–455.

Rininger, D. 2000. Experimental ultraswan arrives in Pennsylvania. Environmental studies at Airlie Press Release, January 27, 2000.

Ringer, R. K. 1976. Thyroids. Pages 348–358. In: Sturkie, P. D. (ed.). *Avian Physiology,* Third edition. New York: Springer-Verlag.

Rismiller, P. D., and G. Heldmaier. 1987. Melatonin and photoperiod affect body temperature selection in the lizard *Lacerta viridis. Journal of Thermal Biology* 12(2):131–134.

Robbins, C. A. 1932. The advantage of crossed mandibles: A note on the American Red Crossbill. *The Auk* 49:159–165.

Robbins, C. S., J. R. Sauer, R. S. Greenberg, and S. Droege. 1989. Population declines in North American birds that migrate to the neotropics. *Proceedings of the National Academy of Sciences,* USA 86:7658–7662.

Robinson, S. K., F. R. Thompson, III, T. M. Donovan, D. R. Whitehead, and J. Faaborg. 1995. Regional forest fragmentation and the nesting success of migratory birds. *Science* 267:1987–1990.

Robinson, S. R. 1994. Early vertebrate colour vision. *Nature* 367:121.

Rollmann, S. M., L. D. Houck, and R. C. Feldhoff. 1999. Proteinaceous pheromone affecting female receptivity in a terrestrial salamander. *Science* 285:1907–1909.

Rome, L. C., D. Swank, and D. Corda. 1993. How fish power swimming. *Science* 261:340–343.

Rome, L. C., D. A. Syme, S. Hollingworth, S. L. Lindstedt, and S. M. Baylor. 1996. The whistle and the rattle: The design of sound producing muscles. *Proceedings of the National Academy of Sciences,* USA. 93:8095–8100.

Romer, A. S. 1949. Time series and trends in animal evolution. Pages 103–120. In: Jepsen, G. L., E. Mayr, and G. G. Simpson (eds.). *Genetics, Paleontology, and Evolution.* Princeton, New Jersey: Princeton University Press.

Romer, A. S. 1957. Origin of the amniote egg. *The Scientific Monthly* 85:57–63.

Romer, A. S. 1966. *Vertebrate Paleontology.* Chicago: University of Chicago Press.

Romer, A. S., and L. I. Price. 1939. The oldest vertebrate egg. *American Journal of Science* 237:826–829.

Romer, A. S., and L. I. Price. 1940. Review of the Pelycosauria. *Special Paper, Geological Society of America* no. 28:1–538.

Romm, J. 1994. A new forerunner for continental drift. *Nature* 367:407–408.

Roorda, A., and D. R. Williams. 1999. The arrangement of the three cone classes in the living human eye. *Nature* 397:520–522.

Roosenburg, W. M. 1996. Maternal condition and nest site choice: An alternative for the maintenance of environmental sex determination. *American Zoologist* 36:157–168.

Roosevelt, A. C., M. Lima da Costa, C. Lopes Machado, M. Michab, N. Mercier, H. Valladas, J. Feathers, W. Barnett, M. Imazio da Silveira, A. Henderson, J. Sliva, B. Chernoff, D. S. Reese, J. A. Holman, N. Toth, and K. Schick. 1996. Paleoindian cave dwellers in the Amazon: The peopling of the Americas. *Science* 272:373–384.

Roosevelt, A. C., M. L. deCosta, L. J. Brown, J. E. Douglas, M. O'Donnell, E. Quinn, J. Kemp, C. L. Machado, M. I. daSilveira, J. Feathers, and A. Henderson. 1997. Dating a paleoindian site in the Amazon in comparison with Clovis culture. *Science* 275:1950–1952.

Roseberry, J. L., and W. D. Klimstra. 1984. *Population Ecology of the Bobwhite.* Carbondale: Southern Illinois University Press.

Rosen, D. E., P. L. Forey, B. G. Gardiner, and C. Patterson. 1981. Lungfishes, tetrapods, paleontology, and plesiomorphy. *Bulletin of the American Museum of Natural History* 167:159–276.

Rosenberg, L. L., H. Astier, G. LaRoche, J. D. Bayle, A. Tixier-Vidal, and I. Assenmacher. 1967. The thyroid function of the drake after hypophysectomy or hypopthalmic pituitary disconnection. *Neuroendocrinology* 2:113.

Rougier, G. W., M. S. de la Fuente, and A. B. Arcucci. 1995. Late Triassic turtles from South America. *Science* 268:855–858.

Rowe, J. W., and E. O. Moll. 1991. A radiotelemetric study of activity and movements of the Blanding's turtle (*Emydoidea blandingi*) in northeastern Illinois. *Journal of Herpetology* 25(2):178–185.

Rowe, T. 1996. Coevolution of the mammalian middle ear and neocortex. *Science* 273:651–654.

Rowe, T. 1999. At the roots of the mammalian family tree. *Nature* 398:283–284.

Roy, J., and J.-M. Bergeron. 1990a. Branch-cutting behavior by the vole (*Microtus pennsylvanicus*). *Journal of Chemical Ecology* 16(3):735–741.

Roy, J., and J.-M. Bergeron. 1990b. Role of phenolics of coniferous trees as deterrents against debarking behavior of meadow voles (*Microtus pennsylvanicus*). *Journal of Chemical Ecology* 16(3):801–808.

Ruben, J. A., and A. J. Boucot. 1989. The origin of lungless salamanders (Amphibia: Plethodontidae). *American Naturalist* 134:161–169.

Ruben, J. A., T. D. Jones, N. R. Geist, and W. J. Hillenius. 1997. Lung structure and ventilation in theropod dinosaurs and early birds. *Science* 278:1267–1270.

Ruben, J. A., C. Dalsasso, N. R. Geist, W. J. Hillenius, T. D. Jones, and M. Signore. 1999. Pulmonary function and metabolic physiology of theropod dinosaurs. *Science* 283:514–516.

Rubinsky, B., A. Arav, and G. L. Fletcher. 1991. Hypothermic protection—a fundamental property of "antifreeze" proteins. *Biochemical and Biophysical Research Communications* 180(2):566–571.

Rubinsky, B., A. Arav, and A. L. DeVries. 1992. The cryoprotective effect of antifreeze glycopeptides from Antarctic fishes. *Cryobiology* 29:69–79.

Ruibal, R., and V. V. Ernst. 1965. The structure of the digital setae of lizards. *Journal of Morphology* 117:271–276.

Ruibal, R., L. Tevis, Jr., and V. Roig. 1969. The terrestrial ecology of the spadefoot toad *Scaphiopus hammondii. Copeia* 1969:571–584.

Russell, I. J. 1976. Amphibian lateral line receptors. Pages 513–550. In: Llinas, R., and W. Precht (eds.). *Frog Neurobiology.* New York: Springer-Verlag.

Rust, C. C., and R. K. Meyer. 1969. Hair color, molt, and testis size in male, short-tailed weasels treated with melatonin. *Science* 165(3896):921–922.

Ryan, M. J. 1991. Sexual selection and communication in frogs. *Trends in Ecology and Evolution* 6(11):351–354.

Ryder, M. L. 1962. Structure of rhinoceros horn. *Nature* 193:1199–1201.

S

Salmon, M., and K. J. Lohmann. 1989. Orientation cues used by hatchling loggerhead sea turtles (*Caretta caretta* L.)

during their offshore migration. *Ethology* 83:215–228.

Sansom, I. J., M. P. Smith, and M. M. Smith. 1994. Dentine in conodonts. *Nature* 368:591.

Sanz, J. L., J. J. Moratalla, M. Díaz-Molina, N. López-Martínez, O. Kälin, and M. Vianey-Liaud. 1995. Dinosaur nests at the sea shore. *Nature* 376:731–732.

Sanz, J. L., L. M. Chiappe, B. P. Perez-Moreno, A. D. Buscalioni, J. J. Moratalla, F. Ortega, and F. J. Poyato-Ariza. 1996. An early Cretaceous bird from Spain and its implications for the evolution of avian flight. *Nature* 382:442–445.

Sanz, J. L., L. M. Chiappe, B. P. Perez-Moreno, J. J. Moratalla, F. Hernandez-Carrasquilla, A. D. Buscalioni, F. Ortega, F. J. Poyato-Ariza, D. Rasskin-Gutman, and X. Martinez-Delcios. 1997. A nestling bird from the Lower Cretaceous of Spain: Implications for avian skull and neck evolution. *Science* 276:1543–1546.

Sato, M. 1976. Physiology of the gustatory system. Pages 576–587. In: Llinás, R., and W. Precht (eds.). *Frog Neurobiology*. New York: Springer-Verlag.

Sauer, E. G. F. 1958. Celestial navigation in birds. *Scientific American* 199(2):42–47.

Sawyer, J. R. 1991. Highly conserved segments in mammalian chromosomes. *Journal of Heredity* 82:128–133.

Schaller, G. B. 1963. *The Mountain Gorilla*. Chicago: The University of Chicago Press.

Schaller, G. B. 1972. *The Serengeti Lion: A Study of Predator Prey Relations*. Chicago: The University of Chicago Press.

Scheel, D., and C. Packer. 1991. Group hunting behaviour of lions: A search for cooperation. *Animal Behaviour* 41:697–709.

Scheich, H., G. Langner, C. Tidemann, R. B. Coles, and A. Guppy. 1986. Electroreception and electrolocation in platypus. *Nature* 319:401–402.

Scheuhammer, A. M. 1991. Effects of acidification on the availability of toxic metals and calcium to wild birds and mammals. *Environmental Pollution* 71:329–375.

Schmidt, A., and M. Wake. 1990. Olfactory and vomeronasal systems of caecilians (Amphibia: Gymnophiona). *Journal of Morphology* 205:255–268.

Schmidt-Koenig, K. 1960. Internal clocks and homing. *Cold Spring Harbor Symposium on Quantitative Biology* 25:389–393.

Schmidt-Nielsen, K. 1990. *Animal Physiology: Adaptation and Environment*. Cambridge, England: Cambridge University Press.

Schmidt-Nielsen, K., W. L. Bretz, and C. R. Taylor. 1970. Panting in dogs: Unidirectional air flow over evaporative surfaces. *Science* 169:1102–1104.

Schneider, D. 1996. Rain forest crunch. *Scientific American* 274(3):19.

Schneider, S. H. 1989. The greenhouse effect: Science and policy. *Science* 243:771–781.

Schramm, P. 1957. A new homing record for the little brown bat. *Journal of Mammalogy* 38:514–515.

Schuett, G. W., P. J. Fernandez, W. F. Gergits, N. J. Casna, D. Chiszar, H. M. Smith, J. B. Mitton, S. P. Mackessy, R. A. Odum, and M. J. Demlong. 1997. Production of offspring in the absence of males: Evidence for facultative parthenogenesis in bisexual snakes. *Herpetological Natural History* 5(1):1–10.

Schultz, P. H., and S. D'Hondt. 1996. Cretaceous-Tertiary (Chicxulub) impact angle and its consequences. *Geology* 24(11):963–967.

Schwartz, E. R., and C. W. Schwartz. 1991. A quarter-century study of survivorship in a population of three-toed box turtles in Missouri. *Copeia* 1991(4):1120–1123.

Schwenk, K. 1994. Why snakes have forked tongues. *Science* 263:1573–1576.

Schwenk, K. 1995. The serpent's tongue. *Natural History* 104(4):48–55.

Seachrist, L. 1994. Sea turtles master migration with magnetic memories. *Science* 264:661–662.

Sealy, S. G. 1996. Evolution of host defenses against brood parasitism: Implications of puncture-ejection by a small passerine. *The Auk* 113(2):346–355.

Secor, S. M., and J. Diamond. 1995. Adaptive responses to feeding in Burmese pythons: Pay before pumping. *Journal of Experimental Biology* 198:1313–1325.

Secunda, R. C., and T. W. Sherry. 1991. Polyterritorial polygyny in the American redstart. *Wilson Bulletin* 103(2):190–203.

Seidensticker, J. C., IV, M. G. Hornocker, W. V. Wiles, and J. P. Messick. 1973. Mountain lion social organization in the Idaho Primitive Area. *Wildlife Monographs* 35:1–60.

Seife, C. 1996. A harebrained scheme. *Scientific American* 274(2):24, 26.

Sekulic, R. 1982. The function of howling in red howler monkeys (*Alouatta seniculus*). *Behaviour* 81:38–54.

Sepkoski, J. J., Jr., and D. M. Raup. 1986. Periodicity in marine extinction events. Pages 3–36. In: Elliott, D. K. (ed.). *Dynamics of Extinction*. New York: John Wiley and Sons.

Sereno, P. C. 1991. Basal archosaurs: phylogenetic relationships and functional implications. *Journal of Vertebrate Paleontology* 11(4) (Supplement):1–53.

Sereno, P. C. 1999. The evolution of dinosaurs. *Science* 284:2137.

Sereno, P. C., and F. E. Novas. 1992. The complete skull and skeleton of an early dinosaur. *Science* 258:1137–1140.

Sereno, P. C., C. A. Forster, R. R. Rogers, and A. M. Monetta. 1993. Primitive dinosaur skeleton from Argentina and the early evolution of Dinosauria. *Nature* 361(6407):64–66.

Sereno, P. C., J. A. Wilson, H. C. E. Larsson, D. B. Dutheil, and H.-D. Sues. 1994. Early Cretaceous dinosaurs from the Sahara. *Science* 266:267–271.

Seton, E. T. 1920. Migrations of the gray squirrel (*Sciurus carolinensis*). *Journal of Mammalogy* 1(2):53–58.

Sever, D. M. 1991a. Comparative anatomy and phylogeny of the cloacae of salamanders (Amphibia: Caudata). I. Evolution at the family level. *Herpetologica* 47(2):165–193.

Sever, D. M. 1991b. Comparative anatomy and phylogeny of the cloacae of salamanders (Amphibia: Caudata). II. Cryptobranchidae, Hynobiidae, and Sirenidae. *Journal of Morphology* 207:283–301.

Sever, D. M. 1992. Comparative anatomy and phylogeny of the cloacae of salamanders (Amphibia: Caudata). III. Amphiumidae. *Journal of Morphology* 211:63–72.

Seyfarth, R. M., and D. L. Cheney. 1992. Meaning and mind in monkeys. *Scientific American* 267(6):122–128.

Seymour, R. S., and D. F. Bradford. 1992. Temperature regulation in the incubation mounds of the Australian brush-turkey. *Condor* 94:134–150.

Shaffer, H. B., J. M. Clark, and F. Kraus. 1991. When molecules and morphology clash: A phylogenetic analysis of the North American ambystomatid salamanders (Caudata: Ambystomatidae). *Systematic Zoology* 40(3):284–303.

Sharpe, R. M. 1995. Another DDT connection. *Nature* 375:538–539.

Sheehan, P. M., D. E. Fastovsky, R. G. Hoffmann, C. B. Berghaus, and D. L. Gabriel. 1991. Sudden extinction of the dinosaurs: Latest Cretaceous, upper Great Plains, U. S. A. *Science* 254:835–839.

Shepherd, V. 1989. The loggerhead turtle. *Virginia Wildlife* 50(11):34–35.

Sherman, P. W. 1977. Nepotism and the evolution of alarm calls. *Science* 197: 1246–1253.

Sherman, P. W., J. U. M. Jarvis, and S. H. Braude. 1992. Naked mole rats. *Scientific American* 267(2):72–78.

Shimamuro, M., H. Yasue, K. Ohshima, H. Abe, H. Kato, T. Kishiro, M. Goto, I. Munechika, and N. Okada. 1997. Molecular evidence from retroposons that whales form a clade within even-toed ungulates. *Nature* 388:666–670.

Shine, R. 1983. Reptilian reproductive modes: The oviparity-viviparity continuum. *Herpetologica* 39(1):1–8.

Shine, R. 1988. Parental care in reptiles. Pages 275–329. In: Gans, C., and R. B. Huey (eds.). *Biology of the Reptilia. Volume 16, Ecology B. Defense and Life History*. New York: Alan R. Liss.

Shipman, P. 1989. Sixth find is a feathered friend. *Discover* 10(1):63.

Shoemaker, V. H., S. S. Hillman, S. D. Hilyard, D. C. Jackson, L. L. McClanahan, P. C. Withers, and M. L. Wygota. 1992. Exchange of water, ions, and respiratory gases in terrestrial amphibians. Pages 125–150. In: Feder, M. E., and W. W. Burggren (eds.). *Environmental Physiology of Amphibians*. Chicago: The University of Chicago Press.

Shoshani, J. 1997. It's a nose! It's a hand! It's an elephant's trunk! *Natural History* 106(10):36–44.

Shu, D.-G., S. C. Morris, and X.-L. Zhang. 1996a. Reinterpretation of *Yunnanozoon* as the earliest known hemichordate. *Nature* 380:428–430.

Shu, D.-G., S. C. Morris, and X.-L. Zhang. 1996b. A *Pikaia*-like chordate from the Lower Cambrian of China. *Nature* 384:157–158.

Shu, D.-G., H.-L. Luo, S. C. Morris, X.-L. Zhang, S.-X. Hu, L. Chen, J. Han, M. Zhu, Y. Li, and L.-Z. Chen. 1999. Lower Cambrian vertebrates from south China. *Nature* 402:42–46.

Shubin, N. H., and F. A. Jenkins, Jr. 1995. An Early Jurassic jumping frog. *Nature* 377:49–52.

Shubin, N. H., A. W. Crompton, H.-D. Sues, and P. E. Olsen. 1991. New fossil evidence on the sister-group of mammals and early Mesozoic faunal distributions. *Science* 251:1063–1065.

Shubin, N., C. Tabin, and S. Carroll. 1997. Fossils, genes and the evolution of animal limbs. *Nature* 388:639–648.

Sibley, C. G., and J. E. Ahlquist. 1984. The phylogeny of the hominoid primates, as indicated by DNA-DNA hybridization. *Journal of Molecular Evolution* 20:2–15.

Sibley, C. G., and J. E. Ahlquist. 1987. DNA hybridization evidence of hominoid phylogeny: Results from an expanded data set. *Journal of Molecular Evolution* 26:99–121.

Sibley, C. G., and J. E. Ahlquist. 1990. *Phylogeny and Classification of the Birds: A Study in Molecular Evolution*. New Haven, Connecticut: Yale University Press.

Sibley, C. G., and B. L. Monroe, Jr. 1990. *Distribution and Taxonomy of Birds of the World*. New Haven, Connecticut: Yale University Press.

Sick, H. 1993. *Birds in Brazil*. Princeton, New Jersey: Princeton University Press.

Silverstone, P. A. 1973. Observations on the behavior and ecology of a Colombian poison-arrow frog, the kokoe-pa (*Dendrobates histrionicus* Berthold). *Herpetologica* 29(4):295–301.

Simberloff, D. 1986. The proximate causes of extinction. Pages 259–276. In: Raup, D. M., and D. Jablonski (eds.). *Patterns and Processes in the History of Life*. Berlin: Springer-Verlag.

Simons, E. L. 1995. Skulls and anterior teeth of *Catopithecus* (Primates: Anthropoidea) from the Eocene and anthropoid origins. *Science* 268:1885–1888.

Simpson, G. G. 1945. The principles of classification and a classification of mammals. *Bulletin of the American Museum of Natural History* 59:1–350.

Simpson, G. G. 1952. How many species? *Evolution* 6:342.

Simpson, G. G. 1961. *Principles of Animal Taxonomy*. New York: Columbia University Press.

Sinclair, A. R. E., and J. N. M. Smith. 1984. Do plant secondary compounds determine feeding preferences of snowshoe hares? *Oecologia* 61:403–410.

Sinha, G. 1996. A new king and his tiny minion. *Scientific American* 275(2):24.

Siniff, D. B., and K. Ralls. 1991. Reproduction, survival and tag loss in California sea otters. *Marine Mammal Science* 7(3):211–229.

Sissom, D. E. F. 1991. Purring in cats. *Journal of Zoology* 223:67–78.

Sivak, J. G. 1986. Super accommodation in diving birds. *Proceedings of the 19th International Ornithological Congress*: 1276–1283.

Skogland, T. 1978. Characteristics of the snow cover and its relationship to wild mountain reindeer feeding strategies. *Arctic Alpine Research* 10:569–580.

Skogland, T. 1984a. Wild reindeer foraging niche organization. *Holarctic Ecology* 7:347–354.

Skogland, T. 1984b. The effects of food and maternal condition on fetal growth and size in wild reindeer. *Rangifer* 4:39–46.

Skogland, T. 1988. Tooth wear by food limitation and its life history consequences in wild reindeer. *Oikos* 51:238–242.

Skogland, T. 1990. Density dependence in a fluctuating wild reindeer herd:Maternal vs. offspring effects. *Oecologia* 84:442–450.

Skole, D., and C. Tucker. 1993. Tropical deforestation and habitat fragmentation in the Amazon: Satellite data from 1978 to 1988. *Science* 260:1905–1910.

Skutch, A. F. 1991. *Life of the Pigeon*. Ithaca, New York: Cornell University Press.

Sloan, C. P. 1999. Feathers for *T. rex*? National Geographic 196(5):98–107.

Smith, A. T. 1982. Pika (*Ochotona*). Pages 131–133. In: Davis, D. E. (ed.). *CRC Handbook of Census Methods for Terrestrial Vertebrates*. Boca Raton, Florida: CRC Press.

Smith, D. W., R. O. Peterson, T. D. Drummer, and D. S. Sheputis. 1991. Overwinter activity and body temperature patterns in northern beavers. *Canadian Journal of Zoology* 69:2178–2182.

Smith, E., and W. Goodpaster. 1958. Homing in nonmigratory bats. *Science* 127:644.

Smith, H. G., and R. Montgomerie. 1992. Male incubation in barn swallows: The influence of nest temperature and sexual selection. *Condor* 94:750–759.

Smith, R. L. (ed.). 1984. *Sperm Competition and the Evolution of Animal Mating Systems*. London: Academic Press.

Smith, R. L., Jr. 1996. *Ecology and Field Biology*. Fifth edition. New York: Harper Collins.

Smith, W. J., S. L. Smith, E. C. Oppenheimer, and J. G. Devilla. 1977. Vocalizations of the black-tailed prairie dog, *Cynomys ludovicianus*. *Animal Behaviour* 25:152–164.

Smithson, T. R. 1989. The earliest known reptile. *Nature* 342:676–678.

Smolen, M., and T. Colborn. 1997. Endocrine disruption: Hidden threats to wildlife. *Endangered Species Update* 14 (9–10):6–10.

Sneath, P. H., and R. R. Sokal. 1973. *Numerical Taxonomy*. San Francisco: W. H. Freeman.

Snyder, R. J., B. A. McKeown, K. Colbow, and R. Brown. 1992. Use of dissolved strontium in scale marking of juvenile salmonids: Effects of concentration and exposure time. *Canadian Journal of Fisheries and Aquatic Sciences* 49:780–782.

Sokal, R. R., and P. H. A. Sneath. 1963. *Principles of Numerical Taxonomy*. San Francisco: W. H. Freeman.

Speakman, J. R., and S. C. Thomson. 1994. Flight capabilities of *Archeopteryx*. *Nature* 370:514.

Spencer, D. A. 1959. Biological and control aspects. In the Oregon meadow mouse irruption of 1957–1958. *Federal Cooper-*

ative Extension Service. Oregon State College, Corvallis.

Spotila, J. R. 1972. Role of temperature and water in the ecology of lungless salamanders. *Ecological Monographs* 42:95–125.

Springer, M. S., G. C. Cleven, O. Madsen, W. W. deJong, V. G. Waddell, H. M. Amrine, and M. J. Stanhope. 1997. Endemic African mammals shake the phylogenetic tree. *Nature* 388:61–64.

Stanley, S. 1987. *Extinction*. New York: Scientific American Books.

Stanley, S. M., and X. Yang. 1994. A double mass extinction at the end of the Paleozoic Era. *Science* 266:1340–1344.

Stapp, P., P. J. Pekins, and W. W. Mautz. 1991. Winter energy expenditure and the distribution of southern flying squirrels. *Canadian Journal of Zoology* 69: 2548–2555.

Steadman, D. W. 1995. Prehistoric extinctions of Pacific island birds: Biodiversity meets zooarchaeology. *Science* 267: 1123–1131.

Stebbins, R. C. 1954. Natural history of the salamanders of the plethodontid genus *Ensatina*. *University of California Publications in Zoology* 54:47–124.

Stefano, F. J. E., and A. O. Donoso. 1964. Hypophyso-adrenal regulation of moulting in the toad. *General and Comparative Endocrinology* 4:473–480.

Stein, B. A., and S. R. Flack. 1997a. *Species Report Card: The State of U. S. Plants and Animals*. Arlington, Virginia: The Nature Conservancy.

Stein, B. A., and S. R. Flack. 1997b. Conservation priorities. The state of U. S. plants and animals. *Environment 39*(4):6–11.

Stenseth, N. C, and R. A. Ims (eds.). 1994. *The Biology of Lemmings*. New York: Academic Press.

Stenseth, N. C., K.-S. Chan, H. Tong, R. Boonstra, S. Boutin, C. J. Krebs, E. Post, M. O'Donoghue, N. G. Yoccoz, M. C. Forchhammer, and J. W. Hurrell. 1999. Common dynamic structure of Canada lynx populations within three climatic regions. *Science* 285:1071–1073.

Stevens, K. A., and J. M. Parrish. 1999. Neck posture and feeding habits of two Jurassic sauropod dinosaurs. *Science* 284:798–800.

Stewart, B. S. 1996. Uncommon commuters. *Natural History* 105(2):58–63.

Stewart, B. S., and R. L. DeLong. 1995. Double migrations of the northern elephant seal, *Mirounga angustirostris*. *Journal of Mammalogy* 76(1):196–205.

Stewart, D. 1995. Do lions purr? And why are there no green mammals? *National Wildlife* 33(5):30–39.

Stiassny, M. L., and A. Meyer. 1999. Cichlids of the rift lakes. *Scientific American* 280(2):64–69.

Stickel, L. F. 1950. Population and home range relationships of the box turtle, *Terrapene c. carolina* (Linnaeus). *Ecological Monographs* 20:351–378.

Stix, G. 1995. Broken dreamtime. *Scientific American* 272(2):14–15.

Stock, D. W., and G. S. Whitt. 1992. Evidence from 18S ribosomal RNA sequences that lampreys and hagfishes form a natural group. *Science* 257:787–789.

Stokstad, E. 1998. Young dinos grew up fast. *Science* 282:603–604.

Stone, G. N., and A. Purvis. 1992. Warm-up rates during arousal from torpor in heterothermic mammals: physiological correlates and a comparison with heterothermic insects. *Journal of Comparative Physiology B* 162:284–295.

Stone, G. S., L. Florez-Gonzalez, and S. Katona. 1990. Whale migration record. *Nature* 346:705.

Stone, J. 1991. Wild—very wild—kingdom. *Discover* 12(5):34–36.

Stone, K. D., G. A. Heidt, P. T. Caster, and M. L. Kennedy. 1997. Using geographic information systems to determine home range of the southern flying squirrel (*Glaucomys volans*). *American Midland Naturalist* 137:106–111.

Stone, P. A., J. L. Dobie, and R. P. Henry. 1992. Cutaneous surface area and bimodal respiration in soft-shelled (*Trionyx spiniferus*), stinkpot (*Sternotherus odoratus*), and mud turtles (*Kinosternon subrubrum*). *Physiological Zoology* 65(2):311–330.

Storrs, E. E., H. P. Burchfield, and R. J. W. Rees. 1988. Superdelayed parturition in armadillos: A new mammalian survival strategy. *Leprosy Review* 59:11–15.

Strauss, E. 1998. New nonopioid painkiller shows promise in animal tests. *Science* 279:32–33.

Sturkie, P. D. 1976. Heart and circulation: Anatomy, hemodynamics, blood pressure, blood flow, and body fluids. Pages 76–101. In: Sturkie, P. D. (ed.). *Avian Physiology*. New York: Springer-Verlag.

Suarez, R. K. 1992. Hummingbird flight: Sustaining the highest mass-specific metabolic rates among vertebrates. *Experientia* 48:565–570.

Suarez, R. K., J. R. B. Lighton, C. D. Moyes, G. S. Brown, C. L. Gass, and P. W. Hochachka. 1990. Fuel selection in rufous hummingbirds: Ecological implications of metabolic biochemistry. *Proceedings of the National Academy of Sciences*, USA 87:9207–9210.

Sues, H.-D. 1992. No Palaeocene "mammal-like reptile." *Nature* 359:278.

Summers, K. 1989. Sexual selection and intra-female competition in the green dart-poison frog, *Dendrobates auratus*. *Animal Behaviour* 37:797–805.

Svitil, K. A. 1996. We are all Panamanians. *Discover* 17(4):30–31.

Swisher, C. C., III, J. M. Grajales-Nishimura, A. Montanari, S. V. Margolis, P. Claeys, W. Alvarez, P. Renne, E. Cedillo-Pardo, F. J.-M. R. Maurrasse, G. H. Curtis, J. Smit, and M. O. McWilliams. 1992. Coeval $^{40}Ar/^{39}Ar$ ages of 65.0 million years ago from Chicxulub crater melt rock and Cretaceous-Tertiary boundary tektites. *Science* 257:954–958.

Swisher, C. C., III, Y. Wang, X. Wang, X. Xu, and Y. Wang. 1999. Cretaceous age for the feathered dinosaurs of Liaoning, China. *Nature* 58–61.

Szalay, F.S., M.J. Novacek, and M.C. McKenna (eds.). 1993. *Mammal Phylogeny*. (2 vols.). New York: Springer-Verlag.

T

Taber, R. D. 1956. Deer nutrition and population dynamics in the North Coast Range of California. *Transactions of the North American Wildlife Conference* 21:159–172.

Taber, R. D., and R. F. Dasmann. 1958. The black-tailed deer of the chaparral. California Department of Fish and Game. *Game Bulletin* no. 8.

Tablin, F., A. E. Oliver, N. J. Walker, L. M. Crowe, and J. H. Crowe. 1996. Membrane phase transition of intact human platelets: Correlation with cold-induced activation. *Journal of Cellular Physiology* 168:305–313.

Tangley, L. 1997. In search of Africa's forgotten forest elephant. *Science* 275:1417–1419.

Tarduno, J. A., D. B. Brinkman, P. R. Renne, R. D. Cottrell, H. Scher, and P. Castillo. 1998. Evidence for extreme climatic warmth from Late Cretaceous arctic vertebrates. *Science* 282:2241–2243.

Taylor, D. H. 1972. Extraoptic photoreception and compass orientation in larval and adult salamanders (*Ambystoma tigrinum*). *Animal Behaviour* 20:233–236.

Taylor, D. H., and K. Adler. 1973. Spatial orientation by salamanders using plane-polarized light. *Science* 181:285–287.

Taylor, D. H., and J. S. Auburn. 1978. Orientation of amphibians by linearly polarized light. Pages 334–346. In: Schmidt-Koenig, K., and W. T. Keeton (eds.). *Animal Migration, Navigation, and Homing (Symposium)*. New York: Springer-Verlag.

Taylor, D. H., and D. E. Ferguson. 1970. Extraoptic celestial orientation in the southern cricket frog *Acris gryllus*. *Science* 168:390–392.

Taylor, S., and D. Ewer. 1956. Moulting in the Anura: The normal moulting cycle of *Bufo regularis* Reuss. *Proceedings of the Royal Society of London* 127:435–440.

Teer, J. G., D. L. Drowe, T. L. Blankenship, W. F. Andelt, R. S. Cook, J. G. Kie, F. F. Knowlton, and M. White. 1991. Deer and coyotes: The Welder experiments. *Transactions of the 56th North American Wildlife and Natural Resources Conference*:550–560.

Telfair, R. C., II. 1994. Cattle egret. *The Birds of North America*, no. 113:1–32.

Temple, S. A. 1998. Easing the travails of migratory birds. *Environment* 40(1):6–9, 28–32.

Templeton, J. R. 1966. Responses of the lizard nasal salt gland to chronic hypersalemia. *Comparative Biochemistry and Physiology* 18:563–572.

Terborgh, J. 1985. *Five New World Primates*. Princeton, New Jersey: Princeton University Press.

Terborgh, J. 1992. Why American songbirds are vanishing. *Scientific American* 266(5):98–104.

Thacher, J. K. 1876. Medial and paired fins, a contribution to the history of vertebrate limbs. *Transactions of the Connecticut Academy of Arts and Sciences* 3:281–310.

Thewissen, J. G. M. (ed.). 1998. *The Emergence of Whales: Evolutionary Patterns in the Origin of Cetacea*. New York: Plenum.

Thewissen, J. G. M., and S. A. Etnier. 1995. Adhesive devices on the thumb of vespertilionid bats (Chiroptera). *Journal of Mammalogy* 76(3):925–936.

Thewissen, J. G. M., S. I. Madar, and S. T. Hussain. 1998. Whale ankles and evolutionary relationships. *Nature* 395:452.

Thomas, C. D., and J. J. Lennon. 1999. Birds extend their ranges northwards. *Nature* 399:213.

Thomas, D. 1990. Is evaporative water loss the underlying cause of winter arousals in bats and other hibernating mammals? *Bat Research News* 31(4):95–96.

Thomas, D. W., and D. Cloutier. 1992. Evaporative water loss by hibernating little brown bats, *Myotis lucifugus*. *Physiological Zoology* 65(2):443–456.

Thomas, D. W., M. Dorais, and J.-M. Bergeron. 1990. Winter energy budgets and cost of arousals for hibernating little brown bats, *Myotis lucifugus*. *Journal of Mammalogy* 71:475–479.

Thomas, W. K., S. Pääbo, F. X. Villablanca, and A. C. Wilson. 1990. Spatial and temporal continuity of kangaroo rat populations shown by sequencing mitochondrial DNA from museum specimens. *Journal of Molecular Evolution* 31:101–112.

Thompson, I. D., and P. W. Colgan. 1987. Numerical responses of martens to a food shortage in northcentral Ontario. *Journal of Wildlife Management* 51:824–835.

Thomson, D. L., P. Monaghan, and J. R. W. Furness. 1998. The demands of incubation and avian clutch size. *Biological Review* 73:293–304.

Thomson, K. S. 1991. *Living fossil: The Story of the Coelacanth*. New York: W. W. Norton.

Thorpe, W. H., and D. R. Griffin. 1962. The lack of ultrasonic components in the flight noise of owls compared with other birds. *Ibis* 104:256–257.

Tihen, J. A. 1960. Comments on the origin of the amniote egg. *Evolution* 14:528–531.

Tilman, D., and J. A. Downing. 1994. Biodiversity and stability in grasslands. *Nature* 367:363–365.

Tilman, D., R. M. May, C. L. Lehman, and M. A. Nowak. 1994. Habitat destruction and the extinction debt. *Nature* 371:65–66.

Todd, A. W., L. B. Keith, and C. A. Fischer. 1981. Population ecology of coyotes during fluctuation of snowshoe hares. *Journal of Wildlife Management* 45:629–640.

Tosini, G., and M. Menaker. 1996. Circadian rhythms in cultured mammalian retina. *Science* 272:419–421.

Townsend, D. S., M. M. Stewart, F. H. Pough, and P. F. Brussard. 1981. Internal fertilization in an oviparous frog. *Science* 212:469–471.

Trainer, J. E. 1947. The pterylography of the ruffed grouse. Pages 741–748. In: Bump, G., R. W. Darrow, F. C. Edminster, and W. F. Crisse (eds.). *The Ruffed Grouse*. Albany: New York State Conservation Department.

Tucker, V. A. 1966. Oxygen consumption of a flying bird. *Science* 154:150.

Tucker, V. A. 1968. Respiratory exchange and evaporative water loss in the flying budgerigar. *Journal of Experimental Biology* 48:67.

Tucker, V. A. 1972. Metabolism during flight in the laughing gull, *Larus atricilla*. *American Journal of Physiology* 222:237.

Turbak, G. 1997. The bizarre life of the harlequin duck. *National Wildlife* 35(1):34–39.

Twitty, V. C. 1959. Migration and speciation in newts. *Science* 130:1735–1743.

Twitty, V. C., D. Grant, and O. Anderson. 1967. Long distance homing in the newt *Taricha rivularis*. *Proceedings of the National Academy of Sciences, USA* 50:51–58.

Tyack, P. 1981. Interactions between singing Hawaiian humpback whales and conspecifics nearby. *Behavioural Ecology and Sociobiology* 8:105–116.

U

United Nations. 1994. World population 1994. New York: United Nations.

United States Fish and Wildlife Service. 1990. *Report to Congress: Endangered and Threatened Species Recovery Program*. Washington, D.C.: U.S. Department of the Interior, Fish and Wildlife Service.

United States Fish and Wildlife Service. 1996. *Report to Congress: Endangered and Threatened Species Recovery Program*. Washington, D.C.: U.S. Department of the Interior, Fish and Wildlife Service.

United States Fish and Wildlife Service. 1996. *List of Endangered and Threatened Wildlife*. Special Reprint of 50 CFR 17.11 and 17.12. October 31, 1996. Washington, D.C.: U.S. Government Printing Office.

United States Fish and Wildlife Service. 1999. Box score. *Endangered Species Bulletin* 24(3):32.

Unwin, D. M. 1987. Joggers or waddlers? *Nature* 327:13–14.

Unwin, D. M., and N. N. Bakhurina. 1994. *Sordes pilosus* and the nature of the pterosaur flight apparatus. *Nature* 371:62–64.

V

Van Buskirk, J., and D. C. Smith. 1991. Density-dependent population regulation in a salamander. Ecology 72(5):1747–1756.

Van Damme, R., D. Bauwens, F. Brana, and R. F. Verheyen. 1992. Incubation temperature differentially affects hatching time, egg survival, and hatchling performance in the lizard *Podarcis muralis*. *Herpetologica* 48(2):220–228.

van Roosmalen, M. G. M., T. van Roosmalen, R. A. Mittermeir, and G. A. B. da Fonseca. 1998. A new and distinctive species of marmoset (Callitrichidae, Primates) from the lower Rio Aripuana, State of Amazonas, Central Brazilian Amazonia. *Goeldianna Zoologia* 22:1–27.

Van Tyne, J., and A. J. Berger. 1961. *Fundamentals of Ornithology.* New York: John Wiley and Sons.

Van Valen, L. M. 1984. Catastrophes, expectations, and the evidence. *Paleobiology* 10:121–137.

Varricchio, D. J., F. Jackson, J. J. Borkowski, and J. R. Horner. 1997. Nest and egg clutches of the dinosaur *Troodon formosus* and the evolution of avian reproductive traits. *Nature* 385:247–250.

Vaughn, T. A. 1972. *Mammalogy.* Philadelphia: W. B. Saunders.

Vaughn, T. A. 1986. *Mammalogy.* Third Edition. Philadelphia: W. B. Saunders.

Vazquez, R. J. 1992. Functional osteology of the avian wrist and the evolution of flapping flight. *Journal of Morphology* 211:259–268.

Vergano, D. 1996. Smallest frog leaps into the limelight. *Science News* 150:357.

Viguier, C. 1882. Le sens d'orientation et ses organes chez les animaux et chez l'homme. *Philosophical Review* 14:1–36.

Viitala, J., E. Korpimaki, P. Palokangas, and M. Kolvula. 1995. Attraction of kestrels to vole scent marks visible in ultraviolet light. *Nature* 373:425–427.

Vila, C., P. Savolainen, J. E. Maldonado, I. R. Amorim, J. E. Rice, R. L. Honeycutt, K. A. Crandall, J. Lundeberg, and R. K. Wayne. 1997. Multiple and ancient origins of the domestic dog. *Science* 276:1687–1689.

Villa-R., B., and E. L. Cockrum. 1962. Migration in the guano bat *Tadarida brasiliensis mexicana* (Saussure). *Journal of Mammalogy* 43(1):43–64.

Vitt, L. J., and W. E. Cooper, Jr. 1989. Maternal care in skinks (*Eumeces*). *Journal of Herpetology* 23:29–34.

Vogel, G. 1998. Doubled genes may explain fish diversity. *Science* 281:1119-1120.

von der Emde, G., S. Schwarz, L. Gomez, R. Budelli, and K. Grant. 1998. Electric fish measure distance in the dark. *Nature* 395:890–894.

von Holst, D. 1969. Sozialer Stress bei Tupajas (*Tupaia belangeri*). Die Aktivierung des sympathischen Nervensystems und ihre Beziehung zu hormonal ausgelosten ethologischen und physiologischen Veranderungen. *Zeitschrift Vergleiche Physiologie* 63:1–58.

W

Wabnitz, P. A., J. H. Bowie, M. J. Tyler, J. C. Wallace, and B. P. Smith. 1999. Aquatic sex pheromone from a male tree frog. *Nature* 401:444–445.

Wagner, W. E., Jr. 1989a. Fighting, assessment, and frequency alteration in Blanchard's cricket frog. *Behavioural Ecology and Sociobiology* 25:429–436.

Wagner, W. E., Jr. 1989b. Graded aggressive signals in Blanchard's cricket frog: Vocal responses to opponent proximity and size. *Animal Behaviour* 38:1025–1038.

Wake, D. B. 1991. Declining amphibian populations. *Science* 253:860.

Wake, D. B., and I. G. Dresner. 1967. Functional morphology and evolution of tail autotomy in salamanders. *Journal of Morphology* 122:265–306.

Wake, M. H. 1977. The reproductive biology of caecilians: An evolutionary perspective. Pages 73–101. In: Taylor, D. H., and S. I. Guttman (eds.). *The Reproductive Biology of Amphibians.* New York: Plenum Press.

Wake, M. H. 1980. The reproductive biology of *Nectophrynoides malcolmi* (Amphibia: Bufonidae), with comments on the evolution of reproductive modes in the genus *Nectophrynoides. Copeia* 1980(2):193–209.

Walker, C. A. 1981. New subclass of birds from the Cretaceous of South America. *Nature* 292:51–53.

Walker, M. M., C. E. Diebel, C. V. Haugh, P. M. Pankhurst, J. C. Montgomery, and C. R. Green. 1997. Structure and function of the vertebrate magnetic sense. *Nature* 390:371–376.

Walker, W. A. and J. M. Coe. 1990. Survey of marine debris ingestion by odontocete cetaceans. Pages 747–774. In: Shomura, R. S. and M. L. Godfrey (eds.). *Proceedings of the Second International Conference on Marine Debris,* April 2–7, 1989, Honolulu, Hawaii. United States Department of Commerce, National Oceanic and Atmospheric Administration. Technical Memorandum NMFS, NOAA-TM-NMFS-SWFSC-154.

Wallace, A. R. 1876. *The Geographical Distribution of Animals with a Study of the Relations of Living and Extinct Faunas as Elucidating the Past Changes of the Earth's Surface.* 2 volumes. London: Macmillan.

Wallace, G. J. 1963. *An Introduction to Ornithology.* Second edition. New York: Macmillan.

Wallace, N. 1985. Debris entanglement in the marine environment: A review. Pages 259–277. In: Shomura, R. S., and H. O. Yoshida (eds.). *Proceedings of the Workshop on the Fate and Impact of Marine Debris.* November 26–29, 1984, Honolulu, Hawaii. United States Department of Commerce, National Oceanic and Atmospheric Administration Technical Memorandum NMFS, NOAA-TM-NMFS-SWFC-54.

Walls, S. C., A. Mathis, R. G. Jaeger, and W. F. Gergits. 1989. Male salamanders with high-quality diets have faeces attractive to females. *Animal Behaviour* 38(3):546–548.

Walsberg, G. E., and C. A. Schmidt. 1992. Effects of variable humidity on embryonic development and hatching success of mourning doves. *The Auk* 109(2): 309–314.

Ward, G. 1997. Making Room for Wild Tigers. *National Geographic* 192(6):2–35.

Waring, G. H. 1970. Sound communications of black-tailed, white-tailed, and Gunnison's prairie dogs. *American Midland Naturalist* 83:167–185.

Watson, J. 1992. Fading images. *National Wildlife* 30(5):8.

Weathers, W. W. 1997. Energetics and thermoregulation by small passerines of the humid, lowland tropics. *The Auk* 114(3):341–353.

Webb, G. J. W., and H. Cooper-Preston. 1989. Effects of incubation temperature on crocodiles and the evolution of reptilian oviparity. *American Zoologist* 29:953–971.

Webb, J. K., and R. Shine. 1992. To find an ant: Trail-following in Australian blindsnakes (Typhlopidae). *Animal Behaviour* 43:941–948.

Webster, J. R., J. M. Suttie, and I. D. Corson. 1991. Effects of melatonin implants on reproductive seasonality of male red deer (*Cervus elaphus*). *Journal of Reproduction and Fertilization* 92:1–11.

Weibel, E. R. 1984. *The Pathway for Oxygen.* Cambridge, Massachusetts: Harvard University Press.

Weichert, C. K. 1965. *Anatomy of the Chordates.* New York: McGraw-Hill.

Weigl, P. D. 1978. Resource overlap, interspecific interactions and the distribution of the flying squirrels, *Glaucomys volans* and *G. sabrinus. American Midland Naturalist* 100(1):83–96.

Weindler, P., R. Wiltschko, and W. Wiltschko. 1996. Magnetic information affects the stellar orientation of young bird migrants. *Nature* 383:158–160.

Weiner, G. S., C. R. Shreck, and H. W. Li. 1986. Effects of low pH on reproduction in rainbow trout. *Transactions of the American Fisheries Society* 115:75–82.

Weiss, R. 1990. Frog finds empty bandwidth, then croaks. *Science News* 138(6):93.

Welch, A. M., R. D. Semlitsch, and H. C. Gerhardt. 1998. Call duration as an indicator of genetic quality in male gray tree frogs. *Science* 280:1928–1930.

Wellnhofer, P. 1990. *Archeopteryx. Scientific American* 262:70–77.

Wells, K. D. 1977. The social behaviour of anuran amphibians. *Animal Behaviour* 25:666–693.

Welty, J. C. 1963. *The Life of Birds.* New York: Alfred A. Knopf.

Werner, D. I. 1982. Social organization and ecology of land iguanas, *Conolophus subcristatus*, on Isla Fernandina, Galapagos. Pages 342–365. In: Burghardt, G. M., and A. S. Rand (eds.). *Iguanas of the World.* Park Ridge, New Jersey: Noyes Publications.

Werner, D. I. 1983. Reproduction in the iguana *Conolophus subcristatus* on Fernandina Island, Galapagos: Clutch size and migration costs. *American Naturalist* 121(6):757–775.

Werner, J. K. 1991. A radiotelemetry implant technique for use with *Bufo americanus. Herpetological Review* 22(3):94–95.

Westneat, D. F. 1987. Extra-pair fertilizations in a predominately monogamous bird: Genetic evidence. *Animal Behaviour* 35:877–886.

Wetmore, A. 1936. The number of contour feathers in passeriform and related birds. *The Auk* 53:159–169.

Wetmore, A. 1967. Re-creating Madagascar's giant extinct bird. *National Geographic* 132(4):488–493.

Wever, E. G., and J. A. Vernon. 1956a. The sensitivity of the turtle's ear as shown by the cochlear potentials. *Proceedings of the National Academy of Sciences,* USA 42:213–220.

Wever, E. G., and J. A. Vernon. 1956b. Auditory responses in the common box turtle. *Proceedings of the National Academy of Sciences,* USA 42:962–965.

Wever, E. G. and J. A. Vernon. 1960. The problem of hearing in snakes. *Journal of Auditory Research* 1:77–83.

Wheeler, A. 1985. *The World Encyclopedia of Fishes.* London: Macdonald and Company.

Wheelwright, J. 1997. Condors: Back from the brink. *Smithsonian* 28(2):48–55.

Whitaker, J. O., Jr. 1970. The biological subspecies: An adjunct of the biological species. *The Biologist* 52:12–15.

Whitaker, J. O., Jr. and W. J. Hamilton, Jr. 1998. *Mammals of the Eastern United States.* Ithaca, NY: Comstock.

White, L. 1965. Biological and ecological considerations in meadow mouse population management. *California Department of Agriculture Bulletin* 54(3):161–167.

White, T. D. 1989. An analysis of epipubic bone function in mammals using scaling theory. *Journal of Theoretical Biology* 139:343–357.

Whitfield, J. 1999. Heard but not seen. *Nature* 399:24.

Whitford, W. G., and V. H. Hutchison. 1966. Cutaneous and pulmonary gas exchange in ambystomatid salamanders. *Copeia* 1966(3): 573–577.

Wibbels, T., R. E. Martin, D. W. Owens, and M. S. Amoss, Jr. 1991. Female-biased sex ratio of immature loggerhead sea turtles inhabiting the Atlantic coastal waters of Florida. *Canadian Journal of Zoology* 69:2973–2977.

Widmaier, E. P., and T. H. Kunz. 1993. Basal, diurnal, and stress-induced levels of glucose and glucocorticoids in captive bats. *Journal of Experimental Zoology* 265:533–540.

Widmaier, E. P., T. L. Harmer, A. M. Sulak, and T. H. Kunz. 1994. Further characterization of the pituitary-adrenocortical responses to stress in Chiroptera. *Journal of Experimental Zoology* 269:442–449.

Wignall, P. B., and R. J. Twitchett. 1996. Oceanic anoxia and the End Permian mass extinction. *Science* 272:1155–1158.

Wilbur, H. M. 1987. Regulation of structure in complex systems: Experimental temporary pond communities. *Ecology* 68:1437–1452.

Wilbur, H. M. 1988. Interactions between growing predators and growing prey. Pages 157–172. In: Ebenman, B., and L. Person (eds.). *Size-Structured Populations.* Berlin: Springer-Verlag.

Wilder, I. W., and E. R. Dunn. 1920. The correlation of lunglessness in salamanders with a mountain brook habitat. *Copeia* 1920:63–68.

Wiley, E. O. 1981. *Phylogenetics: The Theory and Practice of Phylogenetic Systematics.* New York: John Wiley and Sons.

Wilkinson, G. S. 1992. Communal nursing in the evening bat, *Nycticeius humeralis. Behavioural Ecology and Sociobiology* 31:225–235.

Wilkinson, M., J. Thorley, and M. J. Benton. 1997. Uncertain turtle relationships. *Nature* 387:466.

Williams, R. W., C. Cavada, and F. Reinoso-Suarez. 1993. Rapid evolution of the visual system: A cellular assay of the retina and dorsal lateral geniculate nucleus of the Spanish wildcat and the domestic cat. *Journal of Neuroscience* 13(1):208–228.

Williams, T. M., R. W. Davis, L. A. Fuiman, J. Francis, B. J. LeBoeuf, M. Horning, J. Calambokidis, and D. A. Croll. 2000. Sink or swim: strategies for cost-efficient diving by marine mammals. *Science* 288:133–136.

Willson, R. C. 1997. Total solar irradiance trend during solar cycles 21 and 22. *Science* 277:1963–1965.

Wilmut, I., A. C. Schnieke, J. McWhir, A. J. Kind, and K. Campbell. 1997. Viable offspring derived from fetal and adult mammalian cells. *Nature* 385:810–813.

Wilson, D. E., 1997. *Bats in Question: The Smithsonian Answer Book.* Washington, D.C.: Smithsonian Institution Press.

Wilson, D. E., and D. M. Reeder (eds.). 1993. *Mammal Species of the World.* Washington, D.C.: Smithsonian Institution Press.

Wilson, D. E., and D. M. Reeder (eds.). 1993. *Mammal Species of the World: A Taxonomic and Geographic Reference.* Washington, D.C.: Smithsonian Institution Press.

Wilson, D. E., F. R. Cole, J. D. Nichols, R. Rudran, and M. S. Foster (eds.). 1996. *Measuring and Monitoring Biological Diversity: Standard Methods for Mammals.* Washington, D.C.: Smithsonian Institution Press.

Wilson, D. E., and S. Ruff (eds.). 1999. *The Smithsonian Book of North American Mammals.* Washington, D.C.: Smithsonian Institution Press.

Wilson, E. O. 1992. *The Diversity of Life.* Cambridge, Massachusetts: Harvard University Press.

Wiltschko, R., and W. W. Wiltschko. 1994. Avian orientation: Multiple sensory cues and the advantage of redundancy. Pages 95–119. In: Davies, M. N. O., and P. R. Green (eds.). *Perception and Motor Control in Birds.* Berlin: Springer-Verlag.

Wiltschko, W. 1968. Uber den Einfluss Statischer Magnetfelder auf die Zugorientierung der Rotkehlchen (*Erithacus rubecula*). *Zeitschrift Tierpsychologie* 25:537–558.

Wimsatt, W. A. 1966. Some problems of reproduction in relation to hibernation in bats. *Bulletin of the Harvard Museum of Comparative Zoology* 124:249–267.

Wimsatt, W. A. 1969. Some interrelations of reproduction and hibernation in mammals. *Symposium of the Society of Experimental Biology* 23:511–549.

Wingfield, J. C. 1984. Influence of weather on reproduction. *Journal of Experimental Zoology* 232:589–594.

Wingfield, J. C. 1985a. Influence of weather on reproduction of male song sparrows, *Melospiza melodia. Journal of Zoology, London A* 205:525–544.

Wingfield, J. C. 1985b. Influence of weather on reproduction of female song sparrows, *Melospiza melodia. Journal of Zoology, London A* 205:545–558.

Wingfield, J. C., and D. S. Farner. 1993. Endocrinology of reproduction in wild species. Pages 163–327. In: Farner, D. S., J. R. King, and K. C. Parkes (eds.).

Avian Biology, Volume IX. New York: Academic Press.

Wingfield, J. C., M. C. Moore, and D. S. Farner. 1983. Endocrine responses to inclement weather in naturally breeding populations of white-crowned sparrows (*Zonotrichia leucophrys pugetensis*). *The Auk* 100:56–62.

Wingfield, J. C., H. Schwabl, and P. W. Mattocks, Jr. 1990. Endocrine mechanisms of migration. Pages 232–256. In: Gwinner, E. (ed.). *Bird Migration*. Berlin: Springer-Verlag.

Wingfield, J. C., T. P. Hahn, M. Wada, L. B. Astheimer, and S. Schoech. 1996. Interrelationship of day length and temperature on the control of gonadal development, body mass, and fat score in white-crowned sparrows, *Zonotrichia leucophrys gambelii*. *General and Comparative Endocrinology* 101:242–255.

Wingfield, J. C., T. P. Hahn, M. Wada, and S. J. Schoech. 1997. Effects of day length and temperature on gonadal development, body mass, and fat depots in white-crowned sparrows, *Zonotrichia leucophrys pugetensis*. *General and Comparative Endocrinology* 107:44–62.

Winn, H. E., and L. K. Winn. 1978. The song of the humpback whale *Megaptera novaengliae* in the West Indies. *Marine Biology* 47:97–114.

Witherington, B. E., and K. A. Bjorndal. 1991. Influences of wavelength and intensity on hatchling sea turtle phototaxis: Implications for sea-finding behavior. *Copeia* 1991(4):1060–1069.

Woese, C. R. 1981. Archaebacteria. *Scientific American* 244(6):106–122.

Wolbach, W. S., R. S. Lewis, and E. Anders. 1985. Cretaceous extinctions: Evidence for wildfires and search for meteoritic material. *Science* 230:167–170.

Wolbach, W. S., I. Gilmour, E. Anders, C. J. Orth, and R. R. Brooks. 1988. Global fire at the Cretaceous-Tertiary boundary. *Nature* 334:665–669.

Wolf, L. 1987. Host-parasite interactions of brown-headed cowbirds and dark-eyed juncos in Virginia. *Wilson Bulletin* 99(3):338–350.

Wolfe, J. A. 1990. Paleobotanical evidence for a marked temperature increase following the Cretaceous-Tertiary boundary. *Nature* 343:153–156.

Wood, W. F. 1990. New components in defensive secretion of the striped skunk, *Mephitis mephitis*. *Journal of Chemical Ecology* 16:2057–2065.

Wood, W. F., C. G. Morgan, and A. Miller. 1991. Volatile components in defensive spray of the spotted skunk, *Spilogale putorius*. *Journal of Chemical Ecology* 17(7):1415–1420.

Woodburne, M. O., and W. J. Zinsmeister. 1982. Fossil land mammal from Antarctica. *Science* 218:284–286.

Woodburne, M. O., and W. J. Zinsmeister. 1984. The first land mammal from Antarctica and its biogeographic implications. *Journal of Paleontology* 58:913–948.

Woodbury, A. M., and R. Hardy. 1948. Studies of the desert tortoise, *Gopherus agassizii*. *Ecological Monographs* 18:145–200.

Woolbright, L. L., E. J. Greene, and G. C. Rapp. 1990. Density-dependent mate searching strategies of male woodfrogs. *Animal Behaviour* 40:135–142.

Worden, A. N. 1964. Alimentary system. Pages 45–47. In: Thomson, A. L. (ed.). *A New Dictionary of Birds*. New York: McGraw-Hill.

Wrangham, R. W. 1997. Subtle, secret female chimpanzees. *Science* 277:774–775.

Wright, K. 1999. First Americans. *Discover* 20(2):52–58, 60, 62–63.

Wu, X., H.-D. Sues, and A. Sun. 1995. A plant-eating crocodyliform reptile from the Cretaceous of China. *Nature* 376:678–680.

Wurtman, R. J., J. Axelrod, and D. E. Kelly. 1968. *The Pineal*. New York: Academic Press.

Wygoda, M. L., and A. A. Williams. 1991. Body temperature in free-ranging green tree frogs (*Hyla cinerea*): A comparison with "typical" frogs. *Herpetologica* 47(3):328–335.

Wyllie, I. 1981. *The Cuckoo*. London: Batsford.

X

Xing, X. 2000. Feathers for *T. Rex*? *National Geographic* 197(3):Forum 2–3.

Xu, X., X.-L. Wang, and X.-C. Wu. 1999. A dromaeosaurid dinosaur with a filamentous integument from the Yixian Formation of China. *Nature* 401:262–266.

Y

Yager, D. D. 1992a. A unique sound production mechanism in the pipid anuran *Xenopus borealis*. *Journal of the Linnean Society* 104:351–375.

Yager, D. D. 1992b. Underwater acoustic communication in the African pipid frog *Xenopus borealis*. *International Journal of Animal Sound and Its Recording* 4:1–24.

Young, G. C., V. N. Karatajute-Talimaa, and M. M. Smith. 1996. A possible Late Cambrian vertebrate from Australia. *Nature* 383:810–812.

Young, J. Z. 1962. *The Life of Vertebrates*. New York: Oxford University Press.

Young, S. R., and G. R. Martin. 1984. Optics of retinal oil droplets: A model of light collection and polarization detection in the avian retina. *Vision Research* 24:129–137.

Z

Zachmann, A., S. C. M. Knijff, M. A. Ali, and M. Anctil. 1992. Effects of photoperiod and different intensities of light exposure on melatonin levels in the blood, pineal organ, and retina of the brook trout (*Salvelinus fontinalis* Mitchell). *Canadian Journal of Zoology* 70:25–29.

Zardoya, R., and A. Meyer. 1997a. The complete DNA sequence of the mitochondrial genome of a "living fossil," the coelacanth (*Latimeria chalumnae*). *Genetics* 146:995–1010.

Zardoya, R., and A. Meyer. 1997b. Molecular phylogenetic information on the identity of the closest living relative(s) of land vertebrates. *Naturwissenschaften* 84:389–397.

Zeigler, H. P., and H.-J. Bischof. 1993. *Vision, Brain, and Behavior in Birds*. Cambridge, Massachusetts: MIT Press.

Zhou, Z. 1995. Is *Mononykus* a bird? *The Auk* 112(4):958–963.

Zhou, Z., F. Jin, and J. Zhang. 1992. Preliminary report on a Mesozoic bird from Liaoning, China. *Chinese Science Bulletin* 37(16):1365–1368.

Zhu, M., X. Yu, and P. Janvier. 1999. A primitive fossil fish sheds light on the origin of bony fishes. *Nature* 397:607–610.

Zimmer, C. 1992. Ruffled feathers. *Discover* 13(5):44–54.

Zimmer, C. 1997. Dinosaurs in motion. *Discover* 18(11):96–104, 109.

Zimmer, C. 1999. Fossil offers a glimpse into mammals' past. *Science* 283:198–199.

Zink, G. 1973–1985. *Der Zug europaisher Singvogel*. Moggingen: Vogelzug-Verlag.

Zischler, H., M. Höss, O. Handt, A. von Haeseler, A. C. van der Kuyl, J. Goudsmit, and S. Pääbo. 1995. Detecting dinosaur DNA. *Science* 268:1192–1193.

Zoeger, J., J. R. Dunn, and M. Fuller. 1981. Magnetic material in the head of the common dolphin *Delphinus delphis*. *Science* 213:892–894.

Zug, G. R. 1993. *Herpetology*. New York: Academic Press.

Credits

PHOTOS

CHAPTER 1

Fig. 1.7: © S. Bohleber; **Fig. 1.20**: © Edward Degginger / Bruce Coleman, Inc.; **Fig. 1.22**: Bat Conservation International; **Fig. 1.23**: © George Grill / National Aquarium in Baltimore; **Fig. 1.25a**: AP / Wide World Photos; **Fig. 1.25b**: Tim Flannery / Nature Focus.

CHAPTER 2

Fig. 2.1: © John Shaw; **Fig. 2.12a**: Melanie L.J. Stiassny; **Fig. 2.12b1,b2**: Prof. Dr. Axel Meyer; **Fig. 2.14**: from *Vertebrate Biology*, Fifth Edition by Robert T. Orr, copyright © 1982 by Harcourt, Inc., reproduced by permission of the publisher; **Fig. 2.16**: © The Zoological Society of London.

CHAPTER 3

Fig. 3.3: from *Vertebrate Biology*, Fifth Edition by Robert T. Orr, copyright © 1982 by Harcourt, Inc., reproduced by permission of the publisher.; **Fig. 3.15**: © Roger Cole / Visuals Unlimited; **Fig. 3.16a**: © Max & Bea Hunn / Visuals Unlimited; **Fig. 3.16b**: George Silk / Life Magazine © Time, Inc.; **Fig. 3.21**: © Dan Guravich / Photo Researchers, Inc.; **Fig. 3.22**: © Leonard Lee Rue III / Visuals Unlimited; **Fig. 3.23**: Christina Bird-Holenda; **Fig. 3.24a**: Corel Scenics #199092; **Fig. 3.25**: © S. Bohleber; **Fig. 3.28**: © Geral & Buff Corsi / Visuals Unlimited; **Fig. 3.29**: © James Hanken / Bruce Coleman Inc.; **Fig. 3.30**: Digital Stock Trees #013; **Fig. 3.31**: © Adrienne T. Gibson / Animals, Animals; **Fig. 3.32a**: from *Vertebrate Biology*, Fifth Edition by Robert T. Orr, copyright © 1982 by Harcourt, Inc., reproduced by permission of the publisher; **Fig. 3.32b**: © Jack Wilburn / Animals, Animals / Earth Scenes; **Fig. 3.33a**: © Max & Bea Hunn / Visuals Unlimited; **Fig. 3.33b**: © Charlie Ott / Photo Researchers, Inc.; **Fig. 3.34:** Australian Information Service; **Fig. 3.36a**: photo by Ernest P. Walker-Kuntz; **Fig. 3.36b**: © Leonard Lee Rue III / Visuals Unlimited; **Fig. 3.37**: © Charles G. Summers, Jr., / Visuals Unlimited.

CHAPTER 4

Fig. 4.3: Dr. D. Collins; **Fig. 4.5b**: Dr. Mark A. Purnell / Nature; **Fig. 4.8a**: Alex Ritchie / The Australian Museum; **Fig. 4.13**: U.S. Fish & Wildlife Service.

CHAPTER 5

Fig. 5.28: © Tom McHugh / Photo Researchers, Inc.; **Fig. 5.37**: © Rudie Kuiter / Innerspace Visions.

CHAPTER 6

Fig. 6.3: photo F.L. Jaques, Courtesy Dept. of Library Services American Museum of Natural History Neg. No. 322871; **Fig. 6.7c**: D.W. Linzey; **Fig. 6.7d**: Clint Farlinger / Visuals Unlimited; **Fig. 6.8a**: Alvin E. Staffan / Photo Researchers, Inc.; **Fig. 6.9**: Michael Fogden / Bruce Coleman, Inc.; **Fig. 6.10a–c**: S. Emerson - Univ. of Utah / Academic Press; **Fig. 6.18b**: © Karl H. Maslowski / Visuals Unlimited; **Fig. 6.18c**: Photo Researchers, Inc.; **Fig. 6.24a,b**: Wayne Van Devender; **Fig. 6.25**: © Dwight R. Kuhn; **Fig. 6.26**: © Paul Freed / Animals Animals; **Fig. 6.32a**: © Dwight R. Kuhn; **Fig. 6.32b**: © Herbert Schwind / Okapia / Photo Researchers, Inc.; **Fig. 6.33**: Courtesy Dr. Edward J. Zalisko; **Fig. 6.34b**: Dr. Charles Stine, Jr.; **Fig. 6.34c**: © Harry Rogers / Photo Researchers, Inc.; **Fig.6.34d**: © Stephen Dalton / Photo Researchers, Inc.; **Fig. 6.35**: © Michael Tyler.

CHAPTER 7

Fig. 7.7c: Courtesy of the Peabody Museum of Natural History, Yale University; **Fig. 7.8**: © Staatliches Museum fur Naturkunde Stuttgart, Neg #8266 SMNS 6293; **Fig. 7.10**: Courtesy Dept. of Library Services American Museum of Natural History Neg. #34809; **Fig. 7.12a**: Courtesy Dept. of Library Services, American Museum of Natural History Neg. #324393; **Fig. 7.17**: Prof. Anusuya Chinsamy-Turan; **Fig. 7.25**: Courtesy Dept. of Library Services, American Museum of Natural History Neg. #325288; **Fig. 7.27b**: Luis Chiappe. Natural History Museum of Los Angeles County; **Fig. 7.27c**: U.S. National Museum / A. Feduccia; **Fig. 7.29**: Chatterjee; **Fig. 7.30**: © Russll Mittermeyer / Bruce Coleman, Inc.

CHAPTER 8

Fig. 8.3: Chris Mattison; **Fig. 8.4a**: © Dwight R. Kuhn; **Fig. 8.5**: Dr. E.R. Degginger; **Fig. 8.8a**: John Crawley; **Fig. 8.8b**: Chris Mattison; **Fig. 8.14b**: © John Visser / Bruce Coleman, Inc.; **Fig. 8.15**: Gunter Ziesler / Bruce Coleman Collection; **Fig. 8.18**: © Stephen Dalton / Animals Animals / Earth Scenes; **Fig. 8.21b**: © David D. Gillette; **Fig. 8.24**: from Bellairs 1970 *The Life of Reptiles* fig 55; **Fig. 8.28**: © James H. Robinson / Animals, Animals; **Fig. 8.29**: Bianca Lavies / National Geographic Society Image Collection; **Fig. 8.30**: © Moira & Rod Borland / Bruce Coleman Inc.; **Fig. 8.31**: © Edward Degginger / Bruce Coleman, Inc.; **Fig. 8.32**: © NYZA / The Wildlife Conservation Society; **Fig. 8.37**: Chris Jeffree / Edinburgh University Division of Biological Sciences EM Facility; **Fig. 8.39**: © Stuart D. Strahl; **Fig. 8.57a**: Thomas McAvoy / LIFE magazine © 1955 Time, Inc.; **Fig. 8.57b**: Dr. Frederic Schutz; **Fig. 8.58**: © Silvestris / Gross / Peter Arnold, Inc.; **Fig. 8.62**: © D.B. Carter /A.N.T. Photo Library; **Fig. 8.63a**: © Gordon & Cathy Ill / Animals Animals; **Fig. 8.63b**: © Leonard Lee Rue III / Visuals Unlimited; **Fig. 8.64**: © Patti Murray / Animals Animals; **Fig. 8.66a**: Kristi Taylor; **Fig. 8.66b**: © Stephen Dalton / Photo Researchers, Inc.; **Fig. 8.67**: © Wendell Metzen / Bruce Coleman, Inc.; **Fig. 8.68**: © Don W. Fawcett / Visuals Unlimited; **Fig. 8.69**: G.R. Roberts; **Fig. 8.72**: David Hosking / Frank Lane Picture Agency; **Fig. 8.73a,b**: Jeffrey W. Lang.

CHAPTER 9

Fig. 9.5: © 1999 Mark A. Klingler / Zhexi Luo, Carnegie Museum of Natural History; **Fig. 9.6a**: © J. Alcock / Visuals Unlimited; **Fig. 9.6b**: Jean Philippe Varin © Jacan / Photo Researchers, Inc.; **Fig. 9.7**: All Rights Reserved, Photo Archives, Denver Museum of Natural History; **Fig. 9.9a**: © Steve Maslowski / Visuals Unlimited; **Fig. 9.9b**: Len Rue, Jr. / Leonard Rue Enterprises; **Fig. 9.11**: © Steve Maslowski / Visuals Unlimited; **Fig. 9.12**: Photograph by Jon E. Cates; **Fig. 9.13b**: © Jeff Foott / Tom Stack & Associates; **Fig. 9.19a**: from *Vertebrate Biology*, Fifth Edition by Robert T. Orr, copyright © 1982 by Harcourt, Inc., reproduced by permission of the publisher; **Fig. 9.19b**: © S. Bohleber; **Fig. 9.19c1**: from *Vertebrate Biology*, Fifth Edition by Robert T. Orr, copyright © 1982 by Harcourt, Inc., reproduced by permission of the publisher; **Fig. 9.19c2**: Marineland of Florida; **Fig. 9.23**: Dr. Dennis M. Cullinane; **Fig. 9.31b**: © Leonard Lee Rue III / Visuals Unlimited; **Fig. 9.39**: © David M. Phillips / Visuals Unlimited; **Fig. 9.4**: © Ken Lucas / Visuals Unlimited; **Fig. 9.42a–b**: Prof. Ewald R. Weibel / Universitat Bern; **Fig. 9.45**: Leonard Lee Rue III; **Fig. 9.50**: © Cleveland P. Hickman, Jr. / Visuals Unlimited; **Fig. 9.57**: © Dwight R. Kuhn; **Fig. 9.62**: © Cleveland P. Hickman, Jr. / Visuals Unlimited; **Fig. 9.64b**: Hartman, Possums Univ. of Texas Press 1952 page 90; **Fig. 9.64c**: A. Gordon Lyne through Australian Commonwealth Scientific and Industrial Research Organization; **Fig. 9.65a–d**: Leonard Lee Rue, III.

CHAPTER 10

Fig. 10.4: Glen Sherwood; **Fig. 10.7**: © Robert Rattner; **Fig. 10.10a**: © Alan Carey / Photo Researchers, Inc.; **Fig. 10.11**: © D. Cavagnaro / Visuals Unlimited.

CHAPTER 11

Fig. 11.2: D.W. Linzey; **Fig. 11.3**: © Gregory G. Dimijian / Photo Researchers, Inc.; **Fig. 11.5**: © Rob & Ann Simpson; **Fig. 11.6**: © Leonard Lee Rue III / Visuals Unlimited; **Fig. 11.10**: Dr. Charles Stine, Jr.; **Fig. 11.11**: Idaho Fish and Game Department; **Fig. 11.17**: © Gary Braasch; **Fig. 11.20**: Dr. E.R. Degginger; **Fig. 11.22**: © Scott Nielsen / Bruce Coleman, Inc.

CHAPTER 12

Fig. 12.1b: Tim Clutton-Brock; **Fig. 12.2**: Leonard Lee Rue III; **Fig. 12.3**: Irene Vandermolen © Leonard Rue Enterprises; **Fig. 12.4**: © Leonard Lee Rue III / Visuals Unlimited; **Fig. 12.5**: © Joe McDonald / Bruce Coleman, Inc.; **Fig. 12.7b**: Maslowski; **Fig. 12.8a–c**: Marler & Tamura *Condor* Sept. 1962 figs 2,3,4 page 371-374; **Fig. 12.10a**: © G.D. Lepp / www.leppstock.com; **Fig. 12.11**: © Leonard Lee Rue III / Visuals Unlimited; **Fig. 12.12**: Kenneth Middleham; **Fig. 12.14**: Courtesy of P.F. Scholander, University of California, San Diego; **Fig. 12.16**: Charles E. Mohr; **Fig.12.17**: Leonard Lee Rue III.

CHAPTER 13

Fig. 13.1a: © Forest W. Buchanan / Visuals Unlimited; **Fig. 13.1b**: © Wayne Lankinen / Bruce Coleman, Inc.; **Fig. 13.1c**: © Steve Maslowski / Visuals Unlimited; **Fig. 13.1d**: © Rob Simpson / Visuals Unlimited; **Fig. 13.1e**: © Edgar Jones / Bruce Coleman, Inc.; **Fig. 13.4**: © Marty Snyderman / Visuals Unlimited; **Fig. 13.5**: Jonathan Scott / Planet Earth Pictures; **Fig. 13.7**: F. Gomez-M. / USFWS; **Fig. 13.8**: Dave Maslowski; **Fig. 13.12**: Leonard Lee Rue III / Visuals Unlimited; **Fig. 13.13**: © Len Rue, Jr. / Visuals Unlimited; **Fig. 13.15**: © Leonard Lee Rue III / Visuals Unlimited; **Fig. 13.16**: Corbis #ANI0029 Animals; **Fig. 13.17**: © Ken Lucas / Visuals Unlimited.

CHAPTER 14

Fig. 14.1a,b: Science Visuals Unlimited; **Fig. 14.2**: U.S. Fish and Wildlife Service; **Fig. 14.3a**: Photo by Massachusetts Cooperative Wildlife Research Unit / U.S. Fish and Wildlife Service; **Fig. 14.3b**: U.S. Fish and Wildlife Service; **Fig. 14.4**: © Dave Spier / Visuals Unlimited; **Fig. 14.5**: © David H. Ellis / Visuals Unlimited; **Fig. 14.6**: Photo by B.C. Wentworth / U.S. Fish and Wildlife Service; **Fig. 14.7**: Dr. Randall Davis Prof. Dept. of Marine Biology / Texas A & M University at Galveston; **Fig. 14.8**: Photo by John J. Craighead / U.S. Fish and Wildlife Service; **Fig. 14.10**: Perry Thorsvik, Baltimore Sun; **Fig. 14.13**: National Park Service.

CHAPTER 15

Fig. 15.3: Courtesy K. Jackson , USAF; **Fig. 15.9**: Photo. J.G. Hubbard. Courtesy Dept. of Library Services American Museum of Natural History Neg. No. 326030; **Fig. 15.10**: © Erwin and Peggy Bauer / Bruce Coleman, Inc.; **Fig. 15.12a**: Photo by Kirke A. King / U.S. Fish and Wildlife Service; **Fig. 15.13**: U.S. Fish and Wildlife Service / P.W. Sykes; **Fig. 15.14**: © Kenneth Fink / Bruce Coleman, Inc.; **Fig. 15.15a**: © Gamma Images / Photo Researchers, Inc.; **Fig. 15.15b**: © Still Pictures / Peter Arnold, Inc.; **Fig. 15.16**: Thomas Kitchin / Tom Stack & Associates; **Fig. 15.17**: © G. Prance / Visuals Unlimited.

CHAPTER 16

Fig. 16.1: © Leonard Lee Rue III / Visuals Unlimited; **Fig. 16.2**: © Eric Carle / Bruce Coleman, Inc.; **Fig. 16.4**: Photo by Luther Goldman / U.S. Fish and Wildlife Service; **Fig. 16.13**: Photo by Rob Bierregaard; **Fig. 16.16**: Ron Austin / Cincinnati Zoo; **Fig. 16.17**: AP / Wide World Photos; **Fig. 16.18**: © Joe McDonald / Visuals Unlimited.

FIGURES

CHAPTER 1

Fig. 1.1, 1.2, 1.3a,c: From Cleveland P. Hickman, Jr., et al., *Integrated Principles of Zoology*, 10th edition. Copyright © 1997 McGraw-Hill Company, Inc. All Rights Reserved. Reprinted by permission.; **Fig. 1.14**: From *Biology of Fishes* by Carl E. Bond, copyright © 1979 by Saunders College Publishing, adapted by permission of the publisher.; **Fig. 1.16**: After Tuchmann-Duplessis, et al., 1974 in Kenneth V. Kardong, *Vertebrates: Comparative Anatomy, Function, Evolution*, 2nd edition. Copyright © 1998 McGraw-Hill Company, Inc. All Rights Reserved. Reprinted by permission.; **Fig. 1.17**: (c,d,e) After Walls, (f) After H.M. Smith, from Kenneth V. Kardong, *Vertebrates: Comparative Anatomy, Function, Evolution*, 2nd edition. Copyright © 1998 McGraw-Hill Company, Inc. All Rights Reserved. Reprinted by permission.; **Fig. 1.21**: From Cleveland P. Hickman, Jr., et al., *Integrated Principles of Zoology*, 10th edition. Copyright © 1997 McGraw-Hill Company, Inc. All Rights Reserved. Reprinted by permission.

CHAPTER 2

Fig. 2.3 and 2.4: From Stephen A. Miller and John P. Harley, *Zoology*, 3rd edition. Copyright © 1996 McGraw-Hill Company, Inc. All Rights Reserved. Reprinted by permission.; **Fig. 2.13**: From Stiassny & Meyer, "Cichlids of the Rift Lakes" in

Scientific American Feb. 1999, pp. 64-69. Reprinted by permission of the illustrator, Roberto Osti.; **Fig. 2.15**: From Sylvia S. Mader, *Biology*, 5th edition. Copyright © 1996 McGraw-Hill Company, Inc. All Rights Reserved. Reprinted by permission.; **Fig. 2.17a,b**: From *Biology: The Network of Life* by Michael C. Mix, et al. Copyright © by Michael C. Mix, Paul Farber and Keith I. King. Reprinted by permission of Addison-Wesley Educational Publishers, Inc.; **Fig. 2.19**: From Ricki Lewis, *Life*, 2nd edition. Copyright © 1995 McGraw-Hill Company, Inc. All Rights Reserved. Reprinted by permission.

CHAPTER 3

Fig. 3.4: From *Concepts in Zoology*, 2nd edition by C. Leon Harris. Copyright © 1996 by HarperCollins College Publishers, Inc. Reprinted by permission of Addison-Wesley Educational Publishers, Inc.; **Fig. 3.6**: From Cleveland P. Hickman, Jr., et al., *Integrated Principles of Zoology*, 10th edition. Copyright © 1997 McGraw-Hill Company, Inc. All Rights Reserved. Reprinted by permission.; **Fig. 3.18**: From Starr, *Biology Concepts & Applications*, 1991. Copyright © Wadsworth, Inc., Belmont, CA. Reprinted by permission.

CHAPTER 4

Fig. 4.1: From Miller, *Living in the Environment*. Copyright © 2000 Brooks-Cole Publishing.; **Fig. 4.2, 4.6, 4.7**: From Cleveland P. Hickman, Jr., et al., *Integrated Principles of Zoology*, 10th edition. Copyright © 1997 McGraw-Hill Company, Inc. All Rights Reserved. Reprinted by permission.; **Fig. 4.9a,b**: From Kenneth V. Kardong, *Vertebrates: Comparative Anatomy, Function, Evolution*, 2nd edition. Copyright © 1998 McGraw-Hill Company, Inc. All Rights Reserved. Reprinted by permission.; **Fig. 4.14 a,b**: (a) After A.S. Romer & T.S. Parsons, 1985 in Kenneth V. Kardong, *Vertebrates: Comparative Anatomy, Function, Evolution*, 2nd edition. Copyright © 1998 McGraw-Hill Company, Inc. All Rights Reserved. Reprinted by permission.; **Fig. 4.15**: Data from J. Gauthier, 1986, *Memoirs Cal. Acad. Sci.* No. 18, 1986 and J.M.V. Rayner, *Evolution and the Fossil Record*, from Cleveland P. Hickman, Jr., et al., *Integrated Principles of Zoology*, 10th edition. Copyright © 1997 McGraw-Hill Company, Inc. All Rights Reserved. Reprinted by permission.

CHAPTER 5

Fig. 5.1a: From Lawrence Ashley & Robert B. Chiasson, *Lab Anatomy of the Shark*, 5th edition. Copyright © 1988 McGraw-Hill Co. All Rights Reserved. Reprinted by permission.; **Fig. 5.1b**: From Samuel L. Eddy and James C. Underhill, *How to Know the Freshwater Fishes*, 3rd edition. Copyright © 1978 McGraw-Hill Co. All Rights Reserved. Reprinted by permission.; **Fig. 5.5**: After J.Z. Young in Kenneth V. Kardong, *Vertebrates: Comparative Anatomy, Function, Evolution*, 2nd edition. Copyright © 1998 McGraw-Hill Company, Inc. All Rights Reserved. Reprinted by permission.; **Fig. 5.10**: From Kenneth V. Kardong, *Vertebrates: Comparative Anatomy, Function, Evolution*, 2nd edition. Copyright © 1998 McGraw-Hill Company, Inc. All Rights Reserved. Reprinted by permission.; **Fig. 5.11**: From Lawrence Ashley & Robert B. Chiasson, *Lab Anatomy of the Shark*, 5th edition. Copyright © 1988 McGraw-Hill Co. All Rights Reserved. Reprinted by permission.; **Fig. 5.15 and 5.16**: Figure from *Biology of Fishes*, Second edition, by Carl E. Bond, copyright ©1996 by Saunders College Publishing, reproduced by permission of the publisher.; **Fig. 5.18b**: From Cleveland P. Hickman, Jr., et al., *Integrated Principles of Zoology,* 10th edition. Copyright © 1997 McGraw-Hill Company, Inc. All Rights Reserved. Reprinted by permission.; **Fig. 5.19a**: After Goodrich in Kenneth V. Kardong, *Vertebrates: Comparative Anatomy, Function, Evolution*, 2nd edition. Copyright © 1998 McGraw-Hill Company, Inc. All Rights Reserved. Reprinted by permission.; **Fig. 5.20**: From Kenneth V. Kardong, *Vertebrates: Comparative Anatomy, Function, Evolution*, 2nd edition. Copyright © 1998 McGraw-Hill Company, Inc. All Rights Reserved. Reprinted by permission.; **Fig. 5.23**: (a) After Youson (b-f) After Dean in Kenneth V. Kardong, *Vertebrates: Comparative Anatomy, Function, Evolution*, 2nd edition. Copyright © 1998 McGraw-Hill Company, Inc. All Rights Reserved. Reprinted by permission.; **Fig. 5.25**: From Cleveland P. Hickman, Jr., et al., *Integrated Principles of Zoology*, 10th edition. Copyright © 1997 McGraw-Hill Company, Inc. All Rights Reserved. Reprinted by permission.; **Fig. 5.26a**: After Q. Bone and N.B. Marshall in Kenneth V. Kardong, *Vertebrates: Comparative Anatomy, Function, Evolution*, 2nd edition. Copyright © 1998 McGraw-Hill Company, Inc. All Rights Reserved. Reprinted by permission.; **Fig. 5.29**: From Cleveland P. Hickman, Jr., et al., *Integrated Principles of Zoology,* 9th edition. Copyright © 1993 McGraw-Hill Company, Inc. All Rights Reserved. Reprinted by permission.; **Fig. 5.30**: From Kenneth V. Kardong, *Vertebrates: Comparative Anatomy, Function, Evolution*, 2nd edition. Copyright © 1998 McGraw-Hill Company, Inc. All Rights Reserved. Reprinted by permission.; **Fig. 5.31**: After Romer and Parsons in Kenneth V. Kardong, *Vertebrates: Comparative Anatomy, Function, Evolution,* 2nd edition. Copyright © 1998 McGraw-Hill Company, Inc. All Rights Reserved. Reprinted by permission.

CHAPTER 6

Fig. 6.1: From Cleveland P. Hickman, Jr., et al., *Integrated Principles of Zoology*, 10th edition. Copyright © 1997 McGraw-Hill Company, Inc. All Rights Reserved. Reprinted by permission.; **Fig. 6.2**: After E.S. Gaffney, 1979 in Cleveland P. Hickman, Jr., et al., *Integrated Principles of Zoology*, 10th edition. Copyright © 1997 McGraw-Hill Company, Inc. All Rights Reserved. Reprinted by permission.; **Fig. 6.5a**: From John W. Kimball, *Biology*, 6th edition. Copyright © 1994 McGraw-Hill Company, Inc. All Rights Reserved. Reprinted by permission.; **Fig. 6.5b**: From Stephen A. Miller and John P. Harley, *Zoology*, 3rd edition. Copyright © 1996 McGraw-Hill Company, Inc. All Rights Reserved. Reprinted by permission.; **Fig. 6.7a,b**: From Kenneth V. Kardong, *Vertebrates: Comparative Anatomy, Function, Evolution,* 2nd edition. Copyright © 1998 McGraw-Hill Company, Inc. All Rights Reserved. Reprinted by permission.; **Fig. 6.12c**: After M.S. Gordon, et al., *Animal Function: Principles and Adaptations*, 1968 Macmillan, NY as appeared in Cleveland P. Hickman, Jr., et al., *Integrated Principles of Zoology*, 10th edition. Copyright © 1997 McGraw-Hill Company, Inc. All Rights Reserved. Reprinted by permission.; **Fig. 6.16**: From Cleveland P. Hickman, Jr., et al., *Integrated Principles of Zoology,* 10th edition. Copyright © 1997 McGraw-Hill Company, Inc. All Rights Reserved. Reprinted by permission.; **Fig. 6.17**: From Rogers, *Looking at Vertebrates*. Copyright © 1986 Halsted Press. Reprinted by permission of John Wiley & Sons, Inc.; **Fig. 6.19**: From

Cleveland P. Hickman, Jr., et al., *Integrated Principles of Zoology,*10th edition. Copyright © 1997 McGraw-Hill Company, Inc. All Rights Reserved. Reprinted by permission.; **Fig. 6.22**: From Eingert, Bruce D. *Frog Dissection Manual*, pp. 46, fig. 9.1. Copyright © 1988 John Hopkins University Press.; **Fig. 6.23a,b**: From Kenneth V. Kardong, *Vertebrates: Comparative Anatomy, Function, Evolution*, 2nd edition. Copyright © 1998 McGraw-Hill Company, Inc. All Rights Reserved. Reprinted by permission.; **Fig. 6.23c**: After A.S. Romer and T.S. Parsons, 1986 in Kenneth V. Kardong, *Vertebrates: Comparative Anatomy, Function, Evolution*, 2nd edition. Copyright © 1998 McGraw-Hill Company, Inc. All Rights Reserved. Reprinted by permission.; **Fig. 6.29a,b**: From Kenneth V. Kardong, *Vertebrates: Comparative Anatomy, Function, Evolution*, 2nd edition. Copyright © 1998 McGraw-Hill Company, Inc. All Rights Reserved. Reprinted by permission.; **Fig. 6.29c,d**: After Romer and Parsons in Kenneth V. Kardong, *Vertebrates: Comparative Anatomy, Function, Evolution*, 2nd edition. Copyright © 1998 McGraw-Hill Company, Inc. All Rights Reserved. Reprinted by permission.; **Fig. 6.36**: From Moore, *Ecology*, 20 (4): 462-463, 1939. Copyright © Ecological Society of America. Reprinted by permission.

CHAPTER 7

Fig. 7.1: From Cleveland P. Hickman, Jr., et al., *Integrated Principles of Zoology,* 10th edition. Copyright © 1997 McGraw-Hill Company, Inc. All Rights Reserved. Reprinted by permission.; **Fig. 7.2**: Data from F.H. Pough, et al., *Vertebrate Life,* 3rd edition, 1989, Macmillan, New York, from Cleveland P. Hickman, Jr., et al., *Integrated Principles of Zoology*, 10th edition. Copyright © 1997 McGraw-Hill Company, Inc. All Rights Reserved. Reprinted by permission.; **Fig. 7.12**: From Kenneth V. Kardong, *Vertebrates: Comparative Anatomy, Function, Evolution*, 2nd edition. Copyright © 1998 McGraw-Hill Company, Inc. All Rights Reserved. Reprinted by permission.; **Fig. 7.15**: After D. Lambert, 1983 in Kenneth V. Kardong, *Vertebrates: Comparative Anatomy, Function, Evolution*, 2nd edition. Copyright © 1998 McGraw-Hill Company, Inc. All Rights Reserved. Reprinted by permission.; **Fig. 7.23 and 7.24**: From Cleveland P. Hickman, Jr., et al., Integrated *Principles of Zoology,* 10th edition. Copyright © 1997 McGraw-Hill Company, Inc. All Rights Reserved. Reprinted by permission.

CHAPTER 8

Fig. 8.1: From Ernst, *Turtles of the World*, 1989. Copyright © 1989 Smithsonian Institution Press. Reprinted by permission.; **Fig. 8.6**: From Kenneth V. Kardong, *Vertebrates: Comparative Anatomy, Function, Evolution*, 2nd edition. Copyright © 1998 McGraw-Hill Company, Inc. All Rights Reserved. Reprinted by permission.; **Fig. 8.9a,b,c**: From Cleveland P. Hickman, Jr., et al., *Integrated Principles of Zoology,* 10th edition. Copyright © 1997 McGraw-Hill Company, Inc. All Rights Reserved. Reprinted by permission.; **Fig. 8.11**: From Kenneth V. Kardong, *Vertebrates: Comparative Anatomy, Function, Evolution,* 2nd edition. Copyright © 1998 McGraw-Hill Company, Inc. All Rights Reserved. Reprinted by permission.; **Fig. 8.17**: After Kubie, et al.; Halpern and Kubie, from Kenneth V. Kardong, *Vertebrates: Comparative Anatomy, Function, Evolution,* 2nd edition. Copyright © 1998 McGraw-Hill Company, Inc. All Rights Reserved. Reprinted by permission.; **Fig. 8.20a–d**: (a,c) After Romer and Parsons; (b) after Bellairs, from Kenneth V. Kardong, *Vertebrates: Comparative Anatomy, Function, Evolution*, 2nd edition. Copyright © 1998 McGraw-Hill Company, Inc. All Rights Reserved. Reprinted by permission.; **Fig. 8.25**: From Kenneth V. Kardong, *Vertebrates: Comparative Anatomy, Function, Evolution*, 2nd edition. Copyright © 1998 McGraw-Hill Company, Inc. All Rights Reserved. Reprinted by permission.; **Fig. 8.27**: (a,c–e) After A.S. King, (b) After van Tienhoven; from Kenneth V. Kardong, *Vertebrates: Comparative Anatomy, Function, Evolution,* 2nd edition. Copyright © 1998 McGraw-Hill Company, Inc. All Rights Reserved. Reprinted by permission.; **Fig. 8.33**: From Proctor and Lynch, *Manual of Ornithology*. Copyright © 1993 Yale University Press. Reprinted by permission.; **Fig. 8.35**: From Proctor and Lynch, *Manual of Ornithology*. Copyright © 1993 Yale University Press. Reprinted by permission.; **Fig. 8.36**: From Cleveland P. Hickman, Jr., et al., *Integrated Principles of Zoology*, 10th edition. Copyright © 1997 McGraw-Hill Company, Inc. All Rights Reserved. Reprinted by permission.; **Fig. 8.43**: From Barbara J. Stahl, *Vertebrate History*, 1985, Dover Publications. Reprinted by permission of the author.; **Fig. 8.46**: After Evans, from Kenneth V. Kardong, *Vertebrates: Comparative Anatomy, Function, Evolution*, 2nd edition. Copyright © 1998 McGraw-Hill Company, Inc. All Rights Reserved. Reprinted by permission.; **Fig. 8.48**: From Kenneth V. Kardong, *Vertebrates: Comparative Anatomy, Function, Evolution*, 2nd edition. Copyright © 1998 McGraw-Hill Company, Inc. All Rights Reserved. Reprinted by permission.; **Fig. 8.50**: After Pooley and Gans from Kenneth V. Kardong, *Vertebrates: Comparative Anatomy, Function, Evolution,* 2nd edition. Copyright © 1998 McGraw-Hill Company, Inc. All Rights Reserved. Reprinted by permission.; **Fig. 8.52b**: Figure from *Vertebrate Biology* by Robert T. Orr, copyright © 1976 by Saunders College Publishing, reprinted by permission of the publisher; adapted from *General Zoology*, 5th edition by C.A. Villee, W. Walker, and R.D. Barnes. Copyright © 1978 W.B. Saunders, Philadelphia, PA.; **Fig. 8.53**: Reprinted by permission from *Nature* from Matthew Bundle, et al., "High Metabolic Rates in Running Birds" in *Nature*, 397:31, January 1999. Copyright © 1999 Macmillan Magazines Ltd.; **Fig. 8.56c,d**: After Romer and Parsons, from Kenneth V. Kardong, *Vertebrates: Comparative Anatomy, Function, Evolution*, 2nd edition. Copyright © 1998 McGraw-Hill Company, Inc. All Rights Reserved. Reprinted by permission.; **Fig. 8.60**: b,c From Kenneth V. Kardong, *Vertebrates: Comparative Anatomy, Function, Evolution*, 2nd edition. Copyright © 1998 McGraw-Hill Company, Inc. All Rights Reserved. Reprinted by permission.; **Fig. 8.60**: (inset) From "The Hearing of the Barn Owl" edited by E.I. Knudson, drawn by Tom Prentiss in *Scientific American*, December 1981. Reprinted by permission of the estate of Tom Prentiss; **Fig. 8.59**: After H. Evans, from Kenneth V. Kardong, *Vertebrates: Comparative Anatomy, Function, Evolution*, 2nd edition. Copyright © 1998 McGraw-Hill Company, Inc. All Rights Reserved. Reprinted by permission.

CHAPTER 9

Fig. 9.1: Data from R.L. Carroll, *Vertebrate Paleotology and Evolution*, 1988, from Cleveland P. Hickman, Jr., et al., *Integrated Principles of Zoology*, 10th edition.

Copyright © 1997 McGraw-Hill Company, Inc. All Rights Reserved. Reprinted by permission.; **Fig. 9.2**: Data from T.S. Kemp, 1982; Z. Kielan-Jaworowska, 1987; J. Gauthier, 1988; R.L. Carroll, 1988; and F.H. Pough, 1989, from Cleveland P. Hickman, Jr., et al., *Integrated Principles of Zoology*, 10th edition. Copyright © 1997 McGraw-Hill Company, Inc. All Rights Reserved. Reprinted by permission.; **Fig. 9.8**: From Kenneth V. Kardong, *Vertebrates: Comparative Anatomy, Function, Evolution*, 2nd edition. Copyright © 1998 McGraw-Hill Company, Inc. All Rights Reserved. Reprinted by permission.; **Fig. 9.15**: After Modell, from Kenneth V. Kardong, *Vertebrates: Comparative Anatomy, Function, Evolution*, 2nd edition. Copyright © 1998 McGraw-Hill Company, Inc. All Rights Reserved. Reprinted by permission.; **Fig. 9.21**: From George A. Feldhamer, et al., *Mammalogy*. Copyright © 1991 McGraw-Hill Company, Inc. All Rights Reserved. Reprinted by permission.; **Fig. 9.32**: After Hildebrand in Kenneth V. Kardong, *Vertebrates: Comparative Anatomy, Function, Evolution*, 2nd edition. Copyright © 1998 McGraw-Hill Company, Inc. All Rights Reserved. Reprinted by permission.; **Fig. 9.33**: From George A. Feldhammer, et al., *Mammalogy*. Copyright © 1999 McGraw-Hill Company, Inc. All Rights Reserved. Reprinted by permission; **Fig. 9.44**: From Kenneth V. Kardong, *Vertebrates: Comparative Anatomy, Function, Evolution*, 2nd edition. Copyright © 1998 McGraw-Hill Company, Inc. All Rights Reserved. Reprinted by permission.; **Fig. 9.46a,b,c**: After Smith in Kenneth V. Kardong, *Vertebrates: Comparative Anatomy, Function, Evolution*, 2nd edition. Copyright © 1998 McGraw-Hill Company, Inc. All Rights Reserved. Reprinted by permission.; **Fig. 9.46d,e,f**: From Kenneth V. Kardong, *Vertebrates: Comparative Anatomy, Function, Evolution*, 2nd edition. Copyright © 1998 McGraw-Hill Company, Inc. All Rights Reserved. Reprinted by permission.; **Fig. 9.48**: After M.E. Miller, et al., in Kenneth V. Kardong, *Vertebrates: Comparative Anatomy, Function, Evolution*, 2nd edition. Copyright © 1998 McGraw-Hill Company, Inc. All Rights Reserved. Reprinted by permission.; **Fig. 9.49**: After E. Rogers, *Looking at Vertebrates*, 1986, from Cleveland P. Hickman, Jr., et al., *Integrated Principles of Zoology*, 10th edition. Copyright © 1997 McGraw-Hill Company, Inc. All Rights Reserved. Reprinted by permission.; **Fig. 9.51a,b**: From Kenneth V. Kardong, *Vertebrates: Comparative Anatomy, Function, Evolution*, 2nd edition. Copyright © 1998 McGraw-Hill Company, Inc. All Rights Reserved. Reprinted by permission.; **Fig. 9.52a,b**: After A.S. Romer and T.S. Parsons, 1985 in Kenneth V. Kardong, *Vertebrates: Comparative Anatomy, Function, Evolution*, 2nd edition. Copyright © 1998 McGraw-Hill Company, Inc. All Rights Reserved. Reprinted by permission.; **Fig. 9.54**: After N. Suga, *Scientific American*, 1990, from Cleveland P. Hickman, Jr., et al., *Integrated Principles of Zoology*, 10th edition. Copyright © 1997 McGraw-Hill Company, Inc. All Rights Reserved. Reprinted by permission.; **Fig. 9.59**: From George A. Feldhamer, et al., *Mammalogy*. Copyright © 1991 McGraw-Hill Company, Inc. All Rights Reserved. Reprinted by permission.; **Fig. 9.60a**: From Cleveland P. Hickman, Jr., et al., *Integrated Principles of Zoology*, 10th edition. Copyright © 1997 McGraw-Hill Company, Inc. All Rights Reserved. Reprinted by permission.; **Fig. 9.61**: After C.R. Austin and R.V. Short, eds., *Reproduction in Mammals* vol. 4, *Reproductive Patterns,* 1972 in Cleveland P. Hickman, Jr., et al., *Integrated Principles of Zoology*, 10th edition.Copyright © McGraw-Hill Company, Inc. All Rights Reserved. Reprinted by permission.

CHAPTER 11

Fig. 11.4: From G.H. Orians and M.F. Wilson, "Interspecific Territories of Birds" in *Ecology*, 1964, 45:736-745. Copyright © Ecological Society of American. Reprinted by permission.; **Fig. 11.8**: Reprinted by permission of *Nature* from Kooyman, et al., "Penguin Dispersal" in *Nature,* 13:83, 1996. Copyright © Macmillan Magazine, Ltd.; **Fig. 11.9**: From *Ecology and Field Biology* 2nd ed. By Robert Leo Smith. Copyright 1986, 1974 by Robert Leo Smith. Reprinted by permission of Addison-Wesley Longman, Inc.; **Fig. 11.14**: Reprinted by permission of *Nature* from S.J. Morreale, et al., "Migration Corridor for Sea Turtles" in *Nature,* 384:319-320, November 28, 1996. Copyright © 1996 Macmillan Magazine, Ltd.; **Fig. 11.15**: From Thomas Alerstam, *Bird Migration*. Copyright © 1994 Cambridge University Press. Reprinted by permission of Cambridge University Press.; **Fig. 11.16**: From Faaborg, *Orinthology*. Copyright © 1988 Prentice-Hall, Inc., Upper Saddle River, NJ.

CHAPTER 12

Fig. 12.6: From Muller-Schwarze, "Pheromones in Black-tailed Deer" in *Animal Behavior* 27(3): 422. Copyright © Academic Press Ltd.; **Fig. 12.7**: From Levine, Joseph and Kenneth Miller, *Biology: Discovering Life*. Copyright © 1991 by D.C. Heath Company. Used by permission of Houghton Mifflin Co.; **Fig. 12.13 and 12.15**: After R.C. Lasiewski, *Physiol. Zool.*, 1963, from Cleveland P. Hickman, Jr., et al., *Integrated Principles of Zoology,* 10th edition. Copyright © 1997 McGraw-Hill Company, Inc. All Rights Reserved. Reprinted by permission.; **Fig. 12.18**: From D.L. Claussen, et al., *Journal of Herpetology*, 25(3):334-341, 1991. Reprinted by permission.

CHAPTER 13

Fig. 13.1: From Stephen A. Miller and John P. Harley, *Zoology,* 3rd edition. Copyright © 1996 McGraw-Hill Company, Inc. All Rights Reserved. Reprinted by permission.; **Fig. 13.10**: From Proctor and Lynch, *Manual of Ornithology*. Copyright © 1993 Yale University Press. Reprinted by permission.; **Fig. 13.11a,b**: From Dickman, et al., in *Oikos* 62(1), 1991. Reprinted by permission.

CHAPTER 14

Fig. 14.11a,b: From Calliet, Love, and Ebeling, *Fishes: A Field and Lab Manual*, 1986. Copyright © Wadsworth Publishing.

CHAPTER 15

Fig. 15.1: From Ricki Lewis, *Life*, 3rd edition. Copyright © 1998 McGraw-Hill Company, Inc. All Rights Reserved. Reprinted by permission.; **Fig. 15.2a**: From Starr and Taggart, *Biology*, 8th edition, 1995. Copyright © Wadsworth Publishing.; **Fig. 15.7**: From Mosimann in *American Scientist*, 63:304-31. Copyright © 1975 Sigma XI, The Scientific Research Society.; **Fig. 15.8**: From MacPhee and Flemming in *Natural History*, April 1997, Vol. 106(3): 86–87. Courtesy of Ross MacPhee and Clare Flemming.

Index

A

Abdominal position (fins), **100**, 102*f*
Abomasum, **302**
Abyssobrotula galatheae, 57*f*
Abyssopelagic zone, 56*f*, **57**
Acanthodians (spiny sharks), 92
Accessory nerve (XI), **151**
Acinar cells, **117**
Acoustic alarms, 307
Acrodont dentition
  fish, **108**
  reptilian, **208**
Actinopterygii (ray-finned fishes), 4, 94, 95
Adenohypophysis (anterior pituitary gland), 155, **276**
ADH (antidiuretic hormone), **116**
Adipose fin, **100**
Adrenal glands, **116**, 156
Adrenocorticotropic hormone (ACTH), **116**, 155
Adult pelage, **398**
Adult stage, jawed fishes, **127**
Aerial locomotion, 4–6
Aging techniques for wildlife, 398–401
Agnathans. *See* Jawless fishes (agnathans)
Aid in Fish Restoration Act (Dingell-Johnson Act) of 1950, 422
Air bladders, 110
Albanerpetontids, 135
Aldosterone, **117**
Alimentary canal, reptilian, 212*f*
Allantois, reptilian egg, **173**, **223**
Allen's rule, 33, **38**
Alligator
  alimentary canal of, 212*f*
  brain of, 245*f*
  longevity of, 261*t*
  respiratory system of, 239*f*
  skull of, 231*f*
Allopatric populations, **34**
Allopatric speciation, **36**
Allopatry, contiguous and disjunct, 36
Alroy, John, 270
Altitudinal migration, **344**
Altricial species, **18**, **260**
  birds, comparison of, 260*t*
  mammals, 318–19
  tactile communications in, 367
Altruism, mammalian, 317
Alveoli
  amphibian, **149**
  mammalian, **294**, 295*f*
American Museum of Natural History, 425, 426, 427*f*
Amino-acid sequences, evolutionary studies comparing, 38, 40, 41*f*
Amnion, **17**
  reptilian egg, **173**, **222**
Amniotes, **17**, **223**. *See also* Mammals (Mammalia); Reptiles (Reptilia)
  circulatory system of embryonic, 12*f*
  cladogram of living, 171*f*
  double circulatory system of, 106*f*
  egg of, 172, 173*f*
  evolutionary origin of, 169, 170*f*1
  evolutionary *versus* cladistic systematics of, 29*f*
Amphibian(s), 129–68
  commensalism of, 377, 378*t*
  evolution of, 129–36
  eye of, 14*f*
  growth and development in, 161–67
  homing ability in, 352, 354
  as "indicator species" of environmental threats, 415
  introduction to, 129
  Labyrinthodontia subclass, 129, 130, 132*f*, 133–35
  Lepospondyli subclass, 135
  limbs of, compared to sarcopterygian and reptilian, 133*f*
  Lissamphibia subclass, 135–36
  migrations in, 343, 344*f*
  morphology of, 135, 136–58
  niches of, 376
  reproduction in, 135, 158–61
  reptiles compared to, 172–73
  tetrapods and rise of, 130*f*, 131*f*
Amphicoelous condition, **100**
Amphioxus, 2, 3*f*
Amphistylic suspension, **99**
Amphiumas, 149
Amplexus, **159**, 160
Ampulla (neuromast), **86**, **114**
Ampullae, mammalian, **305**
Anadromous fish, **60**, **119**
Anal fin, **100**, 101
Analogous structures, **8**, 10*f*
Anamniotes, **17**, **223**. *See also* Amphibian(s); Fishes
Anapsid (Anapsida), 174, **201**. *See also* Turtles
Ancestral characters, **28**
Androgen hormones, 156, 251
Anguilliform body, 4, 5*f*
Angulars (bones), **99**
Animal kingdom (animals), 1
Ankylosaurs, 184, 185*f*
Anomodonts, 264, 266
Antarctica, dinosaurs in, 183
Anthracosauria (anthracosaurs), 134–35, 169, 172
Antibiotic peptides in amphibian skin, 137
Antifreeze, natural, in fishes, 119, 371–72
Antlers, **276**, 316*f*
Anurans. *See also* Frogs; Toad(s)
  aortic arch of, 206*f*
  hibernation and winter survival in, 369, 373
  integument of, 136–40
  larval development in, 161, 162*f*, 163, 166*f*
  locomotion in, 144–45
  muscular system of, 145
  projectile tongue of, 150
  reproduction in, 158–61
  respiratory system of, 147, 148, 149*f*
  sense organs of, 152–55
  sexual maturity in, 166–67
  skeletal system of, 141, 143–44
  territory of, 341
Anus, mammalian, **312**
Aortic arches
  amphibian, 146, 147*f*
  fish, 106, 146
  reptilian, 206*f*

Aphotic zone, marine, **55**, 56*f*
Aposematic (warning) coloration, **138**, **386**
Appendages, vertebrate, 7–8. *See also* Limb(s)
analogous, to other organisms, 8*f*
homologous bones in, 8*f*, 10*f*
locomotion and, 7, 9*f*
tetrapod, 130, 133*f*
Appendicular skeleton, mammalian, 283–88
Apteria, **227**
Arachnoid layer, **245**
Arandaspids (*Arandaspis*), 78, 81*f*
Archaeopteryx lithographica, 190, 191–93
Archinephros, hagfish, 86, 87*f*
Archosauromorpha, 169, 174, 185–91. *See also* Birds (Aves); Crocodilians
cladogram of, 188*f*
extinct, 185–86
Arciferous type pectoral girdle, amphibian, **143**, 144*f*
Arctic-Alpine zone, North American, 66, 67*f*
Arctic tern (*Sterna paradisaea*), migrations of, 346, 347*f*
Argenine vasotocin (oxytocin), **116**
Aristotelean essentialism, classification as, 26–27
Aristotle, 26
Armadillos (*Dasypus novemcinctus*), 6*f*, 318, 338
Arrectores plumarum muscle, **228**
Arrector pili muscle, 271*f*, **272**, 290
Arthropod, 133
Articular bone, **98**
mammalian, **281**
Artiodactyls, 270, **288**, 289*f*
foot structure in, 288, 289*f*
Ascending colon, **302**
Atlantic green turtle (*Chelonia mydas*), migrations of, **345**
Atlas cervical vertebra, **143**, 281, 282*f*
Atrium
jawed fishes, **104**, 105*f*
jawless fishes, **84**
Auditory bulla, mammalian, **305**
Auditory lobes, 214
Auditory ossicles, 268, 280–81, 305, 306*f*
Auricles (heart), **292**
Australian region, 48*f*, 49
Austral region, North American, 66*f*, 67–69
Austroiparian life zone, North American, 68, 69*f*
Autostylic suspension
amphibian, **142**
jawed fishes, **99**
Aves, 169, 186–91, 194*f*. *See also* Birds (Aves)
Axial skeleton, mammalian, 279–83

B

Baculum, mammalian, **314**
Bald Eagle Act of 1940, 422
Baleen, whale, **274**, 275*f*, **298**
Baltica, ancient continent of, **49**
Banding birds, 393, 394*f*
Baphetids, 169
Barbs, feather, **226**
Barbules, feather, **226**
Barriers
to dispersal of species, 45–46
to speciation, 34–35
Basking poikilothermy, **212**, 214*f*
Bass, largemouth (*Micropterus salmoides*)
anatomy of, 90*f*
musculature of, 10*f*
Bathypelagic zone, 56*f*, **57**
Bats
echolocation by, 305, 306*f*
hibernation in, 368, 370*f*
homing ability in, 353
pollen transfer by long-nosed, 19*f*
sensing of barometric pressure by eastern pipistrelle, 307
vampire, as parasites, 381
Bears, 338, 355, 385*f*
semidormant hiberation in black, 368, 370–71
Benthic region, marine, 56*f*, **57**
Benthic zone, lake, 59*f*, **60**
Bergman's rule, **37**–38
Bering land bridge, 52, 54*f*
Bichir fish
gas bladder-gut relationship in, 110*f*
oviduct of, 120*f*
Bicornate uterus, mammalian, **312**
Binomial nomenclature, 23–25
rules of, 26
Biodiversity, **428**
Biogeographic regions, **47**–49
Biological amplification as threat to vertebrates, **414**, 415*f*
Biological barriers to dispersal of species, **46**
Biological species concept, **25**, 34
Bioluminescent light organs, 97
Biome(s), 60–65
deserts, 64, 65*f*
of Earth, 62*f*
effect of latitude and altitude on, 63*f*
grasslands, 63–64
shifts in, caused by climate changes, 54–55
taiga, 61–62
temperate deciduous forests, 62–63
transition between (ecotone), 60
tropical forests, 64–65
tundra, 60–61, 63*f*
Biosphere reserves, 426, **433**
world network of, 434*f*
zoning of, 433, 435*f*
Biota, defined, 50
Biotic districts, **69**
Biotic provinces, North American, **69**–72
Bipartite uterus, mammalian, **312**
Bipedal locomotion, 4
Birds (Aves), 4, 169
ancestral, 191–95
aortic arch of, 206*f*
bills (beaks) of, 232–33, 234*f*, 235
cardiovascular system of, 238–39
classification of, 28, 29*f*, 194*f*
clutch size of, 258, 324–25
commensalism and, 378, 379*f*, 380*t*
digestive system of, 212*f*, 242–44
earliest (*Archaeopteryx*), 190, 191–93
endocrine system of, 250–51
evolution of, 178, 186–91
external anatomy of, 226*f*
extinction of, 403*f*, 411, 413–15
eye of, 14*f*
feathers of, 226–30, 260
feet of, 229, 230*f*
flight in, 195–96, 237*f*, 238
gas bladder-gut relationship in, 110*f*
genetic divergence of North American songbirds, 35
giant elephant *Aepyornis*, 195
growth and development in, 255–61
half-brain sleep in, 357
homing ability in, 352, 353, 354, 355
imprinting in, 245, 246*f*
inner ear of, 15*f*
integumentary system of, 226–30
lekking in, 255
longevity of, 261*t*
migrating, 242, 244, 247, 344–45, 346–48, 350, 351–52, 396, 397*f*
modern, 195
muscular system of, 236–38
nervous system of, 244–45, 246*f*
niches of, 376, 377*f*
as parasites, 258, 259, 380–81
poisonous, 228

ratite, living families of, 53*f*
relationship of, to dinosaurs, 186–91
relationship of, to mammals, 191
reproduction in, 17, 18*f*, 252–55
respiratory system of, 239–42
sense organs and sensing of, 245–50
skeletal system of, 232–36
skull of, 189, 232, 233*f*
social behavior in, 357, 358, 359
songbird species (*see* Songbird species)
songs of, 362, 363*f*, 364*f*
temporal fenestrae of, 175*f*
territoriality among, 340*f*, 341*f*, 363
torpor in, 368*f*, 369
urogenital system of, 251–52
Birth
amphibian, 164
mammalian, 318, 319*f*
reptilian, 223–24
Blackbirds, interspecific territoriality between species of, 340*f*
Blackpool warbler (*Dendroica striata*), migrations of, 346, 347*f*
Bladder
amphibian, 151, 156
fish, **119**
mammalian, 311*f*
reptilian, 217
Blood, 9–10
amphibian, 146–47
effect of hemolytic poisons on, 209–10
hemoglobin lacking in icefish, 107
jawed fishes, 104–6
mammalian, 292, 293
myoglobin in blood of diving animals, 297–98
as oxygen pumps (Root effect), 111
reptilian, 206
Blood pressure, fish, 106
Body armor, 6
Body of vertebrates, 4–18
appendages of, 7–8
body form and locomotion in, 4–6
circulation in, 9–10, 11–12*f* (*see also* Cardiovascular system)
control and coordination in, 11–13, 14–15*f*
digestion in, 10–11 (*see also* Digestive system)
growth and development in, 18 (*see also* Development; Growth)
heart size and, 294
integument of (*see* Integumentary system)
kidney excretion in, 13–15 (*see also* Kidney(s))
musculature of (*see* Muscular system)
nervous system of (*see* Nervous system)
reproduction in, 15–18 (*see also* Reproduction)
respiratory system of (*see* Respiratory system)
skeleton of (*see* Skeletal system)
urogenital system of (*see* Urogenital system)
weights of brain and, in select vertebrates, 213*t*
Body temperature, 4
Bolivia, debt-for-nature swaps in, 426
Bones
amphibian, 144
fish, 98, 99, 100
fossil, 78
homologous, 8*f*, 10*f*
mammalian, 279–86
reptilian/bird, 182–83, 198*f*, 201–4, 232–33
Bony fishes. *See* Osteichthyes (bony fishes)
Booted sheath on bird legs, 229, 230*f*
Boreal region, North American, 66–67
Boundary strip methods of calculating home range, 338
Bowfin
gas bladder-gut relationship in, 110*f*
oviduct of, 120*f*
Brain, 11–12
amphibian, 151
jawed fishes, 113–14
lamprey, 86*f*
mammalian, 302–5
olfaction and, 13*f*
reptilian/bird, 212, 214, 244, 245*f*
weights of body and, in select vertebrates, 213*t*
Brain stem, mammalian, **303**, 304*f*
Branchiometric muscles, 290
Breathing, 10
frog, 149*f*
mammalian, 295, 296*f*
reptilian/bird, 239–42
Breeding, 358. *See also* Reproduction
captive programs for, 440
cooperative, 358
food supply and success of, 331
mammalian, 314–16
reptilian/bird, 252–54
strategies for, 254, 315–16
Bristles, feather, **227**
Bronchioles, mammalian, **294**, 295*f*
Buccal funnel, **84**–85
Buccopharyngeal mucosa, **150**
Buffer zones of biosphere reserves, **433**, 434*f*, 435*f*
Bulb, hair, **272**
Bulbus arteriosus, **106**
Buoyancy, fish, 110–12

C

Caecae, pyloric, 11
Caecilians (Microsauria), 4, 135–36
chemoreception and tactile function in, 155
forward motion in, 145
skull of, 141, 142*f*
vertebra of, 143
Calamus, feather, **226**
Calcichordates, **76**
Calcitonin, **116**, 117, 155, 309
Calcium, egg shells and, 255
Canadian zone, North American, 66*f*, 67
Canals (neuromast), **114**
Canines (teeth), **7**, 299
Cannibalism, 367
Cannon net, 391*f*
Cape buffalo, 416
Captorhinids (Captorhinomorpha), 172
Capture techniques, 390–391
Carapace, turtle, 199*f*, **200**
Cardinal heart, **84**
Cardiovascular system, 9–10
amphibian, 145–47
jawed fishes, 104–6
jawless fishes, 84
mammalian, 292–94
reptilian/bird, 205–6, 238–39
Caribou, 349
Carina, **233**
Carinate birds, **233**
Carnassial teeth, mammalian, **300**
Carnivores, nutritional value of food for, 382, 383*f*
Carolinian life zone, North American, 68*f*
Carpal bones, amphibian, 141*f*, **144**
Carpometacarpus, **233**
Carrying capacity, **324**, 325*f*
as factor in population density, 328, 329
Cartilaginous fishes. *See* Chondrichthyes (cartilaginous fishes)
Caruncle, reptilian, **223**
Cat (*Felis*)
brain of, 305
Indian desert (*Felis sylvestris*), 439*f*

Cat (*Felis*)—*Cont.*
vertebrae of, 282*f*
vocalizations of, 294
Catadromous fishes, **60**, **119**
Cathaymyrus diadexus, 74–75, 77*f*
Cauda equina, mammalian, **305**
Caudal artery, **106**
Caudal autonomy, reptilian, **202**
Caudal fin, **100**–101, 102*f*
Caudal heart, **84**
Caudal peduncle, **101**
Caudal-sacral region, amphibian, **143**
Caudal (tail) vertebrae
amphibian, 141*f*, **143**
bird, 233
fish, **100**
mammalian, 282
Cecum, **302**
Celestial bodies (sun, moon, stars)
animal homing ability and, 353, 354
animal navigation and, 350, 351
Censusing techniques for wildlife, 397–98
Cephalochordates (Cephalochordata), 1, **2**, 3*f*, 74
Ceratopsians, 184,
Cerebellum, 12
mammalian, **303**, 304*f*, 305
Cerebrospinal fluid, mammalian, **304**
Cerebrum, mammalian, **303**, 304*f*, 305
Cervical vertebrae, amphibian, 141*f*, **143**
Cervix, mammalian, **312**
Cetaceans, 270–71, 284, 289, 418
vocalizations of, 365–66
Chameleon, 201, 210*f*, 216
Channichthyidae family, icefishes of, 107
Chaparral, **64**
Cheetah (*Acinonyx*), 361*f*
Chemical defenses, amphibian, 19*f*, 137–38
Chemoreception, 116. *See also* Olfaction; Taste
Chickens, peck-orders in, 358
Chief cells, **301**
Chimaeras (ratfishes), 93, 94*f*
digestive tract of, 109*f*
Chimpanzees, 357
breeding strategies in, 316
cytochrome *c* and DNA studies comparing humans and, 40, 41*f*, 42*f*, 43
China, ancient continent of, 49
Chondrichthyes (cartilaginous fishes), 93, 94*f*. *See also* Rays; Sharks
circulatory system of, 106
skull of, 98
Chondrocranium, **7**, **97**
Chondrostei (primitive ray-finned fishes), 95
Chordata, **1**. *See also* Chordates
Chordates, 1–4
early, 74–78
protochordates, 1–2, 74
subphylum Vertebrata, 2–4 (*see also* Vertebrate(s))
Chorion, reptilian egg, **173**, **222**
Christmas Bird Count, National Audubon Society, 398
Chromatophores
in amphibian dermis, 140
in fish dermis, **96**
in reptilian dermis, **200**–201
Cichlids
left- and right-mouthed scale-eating, 109
rapid evolution in, 36, 37*f*
Circadian cycles, **117**
migrations and, 349
Circulation, vertebrate, 9–10, 11–12*f*. *See also* Cardiovascular system
Clades, **28**
Cladistic (phylogenetic) classification, 27–32
of amphibians, 129, 131*f*
evolutionary systematics *versus*, 28, 29*f*
of reptiles, 169, 171*f*
Cladogram, **28**
of Archosauria, 188*f*
construction of, 31–32
evolutionary relationship of tuna, lungfish, and pig shown in, 31*f*
of fishes, 78, 80*f*
for four vertebrates (fish, amphibian, two mammals), 31*f*
group divergences indicated by, 29*f*
hypothetical, 28*f*
of living amniotes, 171*f*
of synapsids with mammalian emphasis, 266*f*
of Tetrapoda with amphibian emphasis, 131*f*
Claspers, **101**, 121, 123, 125*f*
Class, **25**
of C. Linnaeus, 24
Classification of vertebrates, 4, 25–26
binomial nomenclature, taxonomy, and, 23–25
evolution and, 32–43 (*see also* Evolution)
methods of, 26–32
systematics as study of, 23
Clavicle, mammalian, 283
Claws
bird, 230
mammalian, **274**
reptilian, **199**
Climate, as factor in population density, 326, 331
Climatic barriers to dispersal of species, **46**
Climatic changes, 53–55
biome shifts caused by, 54–55
effects of, on animal migrations, 352
effects of, on mammalian reproduction, 314
global warming and (*see* Global warming)
Ice Age, and North-South American isthmus, 52–53
Cline, **33**, 34*f*
Clitoris
mammalian, **313**
reptilian, **218**
Cloaca
amphibian, 151
mammalian, 311*f*
reptilian, 217, 244
Cloacal gas exchange, reptilian, **207**
Cloning animals, 439, 440*f*
Cloven hoof, **288**
Clutch size, bird, 258, 325–26
Coccyx, **282**
Cochlea
mammalian, **305**, 306*f*
reptilian, **215**, 249
Code of Zoological Nomenclature, 26
Coelacanths (*Latimeria* spp.), 94, 119
evolutionary relationship of, to tetrapods, 130, 132
habitat of, 95*f*
Coelurosauravus jaekeli, 185, 186*f*
Coffee cultivation, 414
Collecting tubule, mammalian kidney, **311**
Colon, **302**
Color
ability to change, in amphibians, 140
aposematic, 138, 386
in fish integument, 93, 96, 97
green, in frogs, 140
in reptilian integument, 200–201
Colored oil droplets in reptilian eye, 216, 247–48
Columella, **142**, 152*f*, 215, 281
Commensalism, **377**–79, 380*t*
Communication among vertebrates, 360–67
olfaction as, 360–62
sound as, 362–66

tactile signals as, 367
vision as, 366–67
Comparative zoogeography, **45**
Competition
interspecific, 375–77
intraspecific, 358–59
Competitive exclusion principle, **376**
Compressiform body, 5*f*
Concertina, reptilian locomotion, **203**, 204*f*
Condors, California, 40, 245, 260
species restoration of, 417, 418*f*
Cones (eye), **115**
bird, 247
Confuciusornis sanctus, 192
Conodonts, **76**–77
restoration of living, 77*f*
Conservation and management, 421–42
endangered species in United States and, 423–24, 428, 428*f*, 429, 429*f*
issues related to modern wildlife conservation, 427–39
issues related to modern wildlife management, 439–40
sanctuaries and refuges for, 424–26
U.S. regulatory legislation related to, 421–23
value of vertebrate museum collections for, 426–27
Conservation International, 426
Conservation Status Ranks, 428*t*
Contiguous allopatry, **36**
Continental drift, 49–53
Contour feathers, **227**
Contraceptive programs for wildlife, 329–30
Control and coordination, vertebrate, 11–13, 14–15*f*
Conus arteriosus
jawed fishes, **106**
jawless fishes, **84**
Convention of Migratory Birds (1916), 421
Convention on International Trade in Endangered Species of Wild Fauna and Flora, 423–24
Convergence, **8**
Convergent evolution, **8**
Coprodeum, **217**, **251**
Coprolites, 180
DNA from, 303
Core area of biosphere reserve, **433**, 435*f*
Cornified (keratinized) epidermal tissue, 6
Corpus callosum, mammalian, **304**
Corpus cavernosa, mammalian, **314**
Corpus spongiosum, mammalian brain, **314**
Cortex, hair shaft, 271*f*, **272**
Corticoids, **117**
Corticosterone, **117**, 155, 251
Cortisol, **117**
Cortisone, **117**
Cosmoid scales, **95**
Costal cartilage, **282**
Costa Rica, conservation in, 426
Cotylosaurs, 200, 201
Countercurrent flow, **108**
Countercurrent heat exchange, **239**, 369*f*, 370
Courtship behaviors
amphibian, 159, 160
fish, 112, 122, 123*f*
mammalian, 315
reptilian/bird, 219–21, 222, 253–54
Coverts, feather, 226*f*, **227**
Cowbird, brown-headed (*Molothrus ater*), as parasite, 258, 259, 380, 381
Coyote (*Canis latrans*), 72*f*
dispersal of, 343
as predator, 328–29
skull and mandible of, 280*f*
Cranes, whooping, 40, 396
species restoration of, 417
Cranial nerves
amphibian, 151
jawed fishes, 113*f*, 114
reptilian/bird, 245
Craniata, 2, 81
Cranium, 2. *See also* Skull
hagfish, 81
Cretaceous/Tertiary (K/T) extinctions, 403, 405–9
Cristae
amphibian, **153**
fish, **86**
mammalian, **305**
Crocodilians, 4, 169, 226–61
cardiovascular system of, 238
digestive system of, 242
endocrine system of, 250
evolution of, 186
growth and development in, 255, 258, 259, 260, 261*t*
integumentary system of, 226
muscular system of, 236
nervous system and brain of, 244, 245*f*
reproduction in, 252
respiratory system of, 239, 240*f*
sense organs of, 248, 249
skeletal system of, 230–31
skull of, 231*f*
urogenital system of, 251
Crop, bird, **243**–44
Crossbill (*Loxia*), beak of, 234*f*, 235
Ctenoid scales, **96**
Cuckoo (*Cuculus canorus*) as parasite, 259*f*, 381*f*
Cutaneous gas exchange, reptilian, **207**
Cutaneous receptors, reptilian, 214–15
Cuticle, 82
Cuticle, hair shaft, 271*f*, **272**
Cuticular scales, **272**
Cycloid scales, **96**
Cynodonts, 266, 267*f*, 268
Cynognathus, 266, 267*f*
Cystovarian ovaries, **120**
Cytochrome *c*, molecular evolution and studies of, 38, 40, 41*f*
Cytological barriers to speciation, **35**

D

Darwin, Charles, on evolution through natural selection, 32
DDT (pesticide), threat of, to vertebrates, 413, 414, 415*f*, 426–27
Debt-for-nature swaps, 426
Deer, black-tailed (*Odocoileus hemionus*), social orders in, 361*f*
Deer, Chinese water, and musk, 277*f*
Deer, Kaibab mule (*Odocoileus hemionus*)
food supply and population of, 331*f*
predator control program and population of, 329*t*
Deer, white-tailed (*Odocoileus virginianus*), 72, 387
birth in, 319*f*
coloration on tail of, 274*f*, 367
determining age of, using teeth, 399*f*
musculature of, 291*f*
predator control programs and increased population of, 328*f*, 329
skeletal structure of, 279*f*
Defenders of Wildlife, 428*f*
Delayed fertilization, **16**
mammalian, **316**
Delayed implantation, **315**
Deme, **32**–33
Dens (hibernation), **371**
Dens (vertebra)
mammalian, **281**
reptilian, **201**
Density-dependent factors in populations, **324**, 327–30, 341
Density-independent factors in populations, **324**, 330–32
Dentaries, **99**

Dentin, 78
Depressiform body, 5*f*
Derived characters, **28**
Dermal scales (plates), **271**
Dermatocranium, **98**
Dermis, **6**
  amphibian, 136, 137*f*, 139–40
  jawed fishes, 95–96
  mammalian, 279
  reptilian, 200–201
Descending colon, **302**
Deserts, 64, 65*f*
Determinate growth, **398**
Development, 18. *See also* Growth
  amphibian, 161–65
  jawed fishes, 125–26
  jawless fishes, 88
  mammalian, 317–19
  reptilian, 222–25, 255–59
Dewlap, reptilian, **220**
Diadromous fishes, **119**
Diaphragm, mammalian, **295**, 296*f*
Diapsids (Diapsida), 4, **201**. *See also* Birds (Aves); Crocodilians; Dinosaurs; Lizards; Snakes; Tuataras
Diastema, **299**
Dicencephalon, mammalian, **303**, 304*f*
Dichromatic vision, mammalian, **308**
Dichromatism, **273**
Diet. *See* Food
Digestion, vertebrate, 10–11
Digestive system
  amphibian, 150–51
  jawed fishes, 108–13
  jawless fishes, 84–85
  mammalian, 298–302
  reptilian/bird, 207–12, 242–44
Digestive tube, 10
Digitigrade locomotion, **287**, 289*f*
Dihydrotestosterone (DHT), 156
Dinocephalians, 264
Dinosauria, 184, 189
Dinosaurs, 173, 176
  in Antarctica, 183
  bones of, 182*f*, 183
  classification of, 176, 178–85
  coprolites of, 180
  eggs and nests of, 174
  extinction of, 408–9
  feathered, 189–90
  fossilized, 179
  gastroliths (stomach stones) in, 211, 213*f*
  giant, 178, 180*f*, 181, 183*f*
  hip bones (pelvic girdle) of, 181*f*
  mammal-eating, 190
  ornithischian, 176, 184–85
  protodinosaurs, 178, 181*f*
  relationship of, to birds, 186–91
  sauischian, 176, 178–83
  size comparison of mammals, reptiles, and, 180*f*
  skull of, 181–82
  theropods, 179, 181*f*, 182, 190
  tracks of, 179*f*
  Utahraptor, as deadly, 179
Dioecious organisms, **16**
Diphycercal caudal fins, **100**, 102*f*
Diphyodont condition, mammalian, 268, **298**
Direct development, larval, **135**, **163**
Disease
  affecting seals, 388
  animal dislocations and threat of, to humans, 416
  control of wildlife by using introduced, 330
  as factor in population density, 327
  wildlife, as threat to humans, 329
Disjunct allopatry, **36**
Disjunct distribution of species, **54**, 55*f*
Dispersal as animal movement, **341**–42
Distal convoluted tubule, mammalian kidney, **311**
Distribution of species, 47–55
  climatic changes and, 53–55
  ecological, 45, 55–72
  geographic, 47–49
  geologic, 49–53
  human impact on, 70–72
Diving
  birds (penguin), 242
  mammals, 292, 297
DNA fingerprinting, **40**
DNA hybridization, **42**–43
DNA studies
  animal migration and, 345
  molecular evolution and, 38–40, 42–43, 169, 270
  of vertebrate museum collections, 426–27
Dodo, flightless (*Didus ineptus*), 403*f*
Dogs (*Canis familiaris*)
  salivary glands of, 300*f*
  vocalizations of, 366
Dolphins, 327, 357
  countercurrent heat exchange in, 369*f*
  protection of, from fishing nets, 307
  sound propagation and reception in, 307*f*
Dormancy. *See* Torpor (dormancy)
Dorsal aorta, **106**
Dorsal fins, **100**
Dorsalis trunci muscle, amphibian, **145**
Down feathers, **227**
Drugs
  antibiotic, 137
  from tropical frogs, 138
Ducks, harlequin (*Histrionicus histrionicus*), migration of, 344
Duck Stamp Act of 1934, 421
Ducks Unlimited, 425
Ductus deferens
  amphibian, 157*f*, **158**
  reptilian, **218**
Duodenum, **302**
Duplex, mammalian, **312**
Dynamic equilibrium, **115**

E

Ear
  amphibian, 142, 147, 152–53
  inner, 15*f*
  mammalian, 280–81, 305–8
  reptilian/bird, 215, 248–49
Ear ossicles, mammalian, 268, 280–81, 305, 306*f*
Earth, animal navigation and magnetic field of, 349–50
Ecdysis, reptilian, **199**, 216
Echolocation, **248**, **305**–6, 366
Ecological and behavioral studies, 390–401
  aging techniques, 398–400
  capture techniques, 390
  censusing techniques, 397–98
  identification techniques, 391–96
  mapping techniques, 397
  scientific method and introduction to, 390
Ecological barriers to speciation, **35**
Ecological distribution of species, **45**, 55–72
  in fresh water environments, 58–60
  in marine environments, 55–58
  in terrestrial environments, 60–72
Ecological zoogeography, **47**
Ecology of vertebrates
  competition, interspecific, 375–77
  competition, intraspecific, 358–59
  conservation and management, 421–42
  dispersals and invasions of vertebrates, 341–43
  extinction and extirpation, 402–20
  feeding behaviors, 150, 151, 201, 208*f*, 209*f*, 302, 367–68
  home ranges, 337–41
  homing behaviors, 352–55
  human interactions and, 387–88

migration, 343–52
niches, 375–76, 377*f*
population dynamics (*see* Population(s); Population density)
predation, interspecific, 382–87 (*see also* Predation)
reproduction and (*see* Breeding; Reproduction)
social behavior, intraspecific, 357–67
studies of (*see* Ecological and behaviorial studies)
symbiosis, interspecific, 377–82
torpor (dormancy), 368–73
Ecoregions, **429**
key, 430–31*f*
Ecosystem(s)
America's most endangered, 428*f*
freshwater, 58–60
marine, 55–58
ranking health of species and communities in, 428*t*
role of vertebrates in, 18–19
terrestrial, 60–72
Ecotone, **60**
Ecotourism, **19**, **426**
Ectothermy, **4**
hibernation and, 371
Eel, American (*Anguilla rostrata*), as catadromous, 119
Egg(s)
amniotic, 172, 173*f*
dinosaur, 174
hatching of (*see* Hatching)
meroblastic and holoblastic cleavage of, 88
reptilian/bird, 173, 218, 222–23, 255, 256*f*, 381*f*
shells, 120, 426–27
threat of pesticides to bird, 414, 415*f*, 416*f*, 426–27
Egg deposition
amphibian, 161*f*
fish, 122, 124*f*
reptilian/bird, 256–58
Egret
cattle (Bubulcus ibis), dispersal of, 343*f*
snowy (*Egretta thula*), 421, 422*f*
Elasmobrachs
gill slits of, 107*f*, 108
inner ear of, 15*f*
male urogenital system of, 121*f*
ovaries of, 120
Elasmobranchii (sharks, skates, rays), 93, 94*f*
Electrical signals as communication, 367
Electric organs in fish derived from muscle fibers, 104
Electrocytes, 104
Electroreceptors
amphibian, 152
fish neuromast systems as, **114**
mammalian, 309
Elephants, 381
birth control programs for, 330
molecular studies of species of, 39
must state and pheromone secretions of, 360
sound production by, 308, 366
trunks of, 308–9
tusks of, 299
Elk, 276*f*, 277*f*, 316
migrating, 344
radio transmitters on, 395*f*
Embryo
amphibian, 164–66
fish, 126
mammalian, 315, 317
Embryonic development, duration of
amphibian, 163–64
mammalian, 317–18
reptilian/bird, 223, 258
Embryonic stage, jawed fishes, **126**
Emerging infectious diseases (EIDs), 416
Emigration of animals, **355**
Enantiornithine birds, 193
Endangered species, 20, 411–19
causes for endangerment of, 429–31, 432*f*, 433–39
classification of, **423**
restoration of, 417
in United States, 423–24, 429, 432*f*
Endangered Species Act of 1973, 423, 424
Endangered Species Conservation Act of 1969, 423
Endangered Species Preservation Act of 1966, 423
Endangered Species Scientific Authority (ESSA), 424
Endocrine system, 11, 13, 86
amphibian, 155–56
fish, 116–17
glands and hormones secreted by, 116–17
mammalian, 309–10
reptilian, 216–17, 250–51
Endolymph, **114**
Endothermy, **4**, 6
hibernation and, 371
Envenomation, **210**
Environmental problems
animal extinctions and extirpations, 409–19
global warming, 55*f*, 387, 388, 437–38
role of humans in creating, 18–19
Environmental resistance, population density and, **326**–35
Epaxial (dorsal) muscles, **8**, 10*f*, 204, 288
Epicercal caudal fins, **100**, 102*f*
Epicontinental seas, **52**
Epidermis, **6**
amphibian, 136, 137*f*
jawed fishes, 97
mammalian, 271–79
reptilian, 198, 199*f*
Epididymus
fish, **119**
reptilian, **218**
Epiglottis, mammalian, **294**
Epinephrine (adrenaline), **116**, 309–10
Epipelagic zone, 56*f*, **57**
Epiphyseal plate, **400**
Epiphysis, **400**
Epipubic bones, **285**, 288*f*
Equilibrium
inner ear and, 86
jawed fishes, 114–15
reptilian/bird, 248*f*
static *versus* dynamic, 114–15
Erlich, Paul, 419
Esophogus, 109
mammalian, 301
reptilian/bird, 211, 243–44
Estivation, **119**, **368**, 373
Estrogen, **117**, 217
Estuaries, 57, 58*f*
productivity of, 57*f*
Ethiopian region, 48*f*, 49
Ethological barriers to speciation, **35**
Eusociality, **358**
Eustachian (auditory) tube
amphibian, **153**
mammalian, **305**, 306*f*
Eusthenopteron foordi, 129, 132*f*
Evolution, **32**–43, 74–78
amphibian, 129–36
cases of very rapid, 33, 36–37
early chordate, 74–78
geographic variation and, 37–48
homeobox (*hox*) genes and, 78
jawed fishes, 90–95
jawless fishes, 78–82
of jaws, 90–91, 99*f*
mammalian, 264–71
molecular, 38–43
of paired fins, 91

Evolution—*Cont.*
of primates, 270
reptilian, 169–96
species and speciation in, 32–37
tetrapod, 129, 130*f*, 131*f*
of vertebrate skull, 280*f*
Evolutionary (classical/traditional) classification, 27, 30
cladistic classification *versus*, 28, 29*f*
Exclusive boundary method of calculating home range, 338
Exocrine gland, **117**
External auditory meatus, mammalian, **305**
External fertilization, **16**
amphibian, 159
External nares, **107**, 239
Extinction, **402**–20
human impacts on, 409–19
natural, 402–9
Extirpation, **402**
human impacts on, 409–19
Extrinsic muscles, **8**
in birds, **237**
Eye
amphibian, 153–54
jawed fishes, 115
mammalian, 247, 308*f*
reptilian/bird, 215–16, 245–48
structure of, 14*f*
Eyelids, amphibian, 153, 154*f*

F

Facial expressions, muscles and, 292
Facultative parthenogenesis, **222**
Family, **25**
Fangs, reptilian, 209, 210*f*
Faunal zoogeography, **45**
Faveoli, reptilian, **206**
Feather(s), 6, **226**–30
in *Archaeopteryx*, 192
development of, 228, 229*f*
dinosaur, 189–90
molting of, 228, 229, 260
muscles attached to, 228
types of, 226, 227
in young birds, 260
Feather follicles, **227**, **228**
Federal Aid in Wildlife Restoration Act of 1937, 422
Feeding behaviors, 367–68
amphibian, 150, 151
mammalian, 302, 367
reptilian, 201, 208*f*, 209*f*
Feet
bird, 229, 230*f*
mammalian, 286, 287*f*, 288, 289*f*
Female(s)
amphibian reproductive system, 157*f*
bird reproductive system, 251, 252*f*
fish oviducts, 120*f*
mammalian reproductive system, 311–13
Femur bone, amphibian, 141*f*, **144**
Fenestrae, **201**. *See also* Temporal fenestrae
Ferret, black-footed, as endangered species, 414*f*, 417
Fertilization
amphibian, 159–61
jawed fishes, 123
reptilian/bird, 218, 219*f*, 220*f*, 221, 254
Fibula bone, amphibian, 141*f*, **144**
Filiform body, 5*f*
Filoplumes, **228**
Filtration, mammalian kidney, **311**
Fin(s), 7
caudal, 100–101, 102*f*
convergent evolution in tetrapods and, 8
evolution of paired, 91
pectoral, 100, 101*f*
pelvic, 100, 101, 102*f*
skeleton of, 100–101, 102*f*
Fin Fold Theory, **91**
Fin Spine Theory, **91**
Firmisternal type pectoral girdle, amphibian, **144**
Fish and Wildlife Act of 1956, 423
Fish and Wildlife Coordination Act of 1934, 421–22
Fish and Wildlife Service (USFWS), 423
Fishes, 4, 6–7, 8–9, 11*f*, 13–14, 16–17
antifreeze glycoproteins in, 119, 372–73
body forms of, 4, 5*f*
bony (*see* Osteichthyes (bony fishes))
cartilaginous (*see* Chondrichthyes (cartilaginous fishes))
catadromous, and anadromous, 60
cladogram of, 80*f*
courtship behaviors in, 112
deepest-living, 57*f*
determining age of, 398, 400*f*
evolution of, 78, 79*f*, 80*f*, 90–95
extinct, 129–32
family tree of, 79*f*
heart of, 84, 104–6, 146, 147*f*
homing ability in, 352, 353–54
jawed (*see* Jawed fishes (gnathostomes))
jawless (*see* Jawless fishes (agnathans))
migrating, 118–19, 344–45, 350
niches of, 376
pheromones in, 361
schooling in, 357
Fishlike fossils, early Cambrian, 77
Flehmen, **362**
Flicking, reptilian tongue, **210**
Flight, 4–6
bird feathers and, 229, 230*f*
bird muscles and metabolism and, 237–38
in hummingbirds, 238
origin of, 195–96
Florida, vertebrates in, 71
Flounders (order Pleuronectiformes), color change in, 97
Flyways, bird migration, 344, 345*f*
Follicle-stimulating hormone (FSH), **116**, 156
Food
human population growth and production of, 428
migration and, 348
nutritional value of herbivore *versus* carnivore, 383*f*
population density and supply of, 331, 333–34
Food web, 326*f*
Foramen magnum, **114**
Forebrain (telencephalon, diencephalon), 11–12
Foreskin, mammalian penis, **314**
Fossil(s)
earliest chordate, 74–76, 77*f*
earliest mammalian, 268, 269*f*
early Cambrian fishlike, 77
missing links in, to modern amphibians, 129–33
Saltville, Virginia, deposits of, 269
Founder effect, **36**, **385**
Fox, California Channel Island (*Urocyon littoralis*), 34, 40
Fragmentation, habitat, 429, **430**
Fresh water environments, 58–60
Frogs, 4, 327
antibiotic substances in skin of, 137
arboreal adaptations in, 139, 140*f*
blue dart poison (*Dendrobates azureus*), 19*f*
brain of, 151*f*
breathing in, 149*f*
competition among, 358, 359*f*
development rates in wood and leopard, 163*f*
gastric brooding in *Rheobatrachus silus*, 163*f*

green color in, 140
hairy (*Astylosternus robustus*), 138, 139*f*
homing ability in, 352
integumentary system of, 136–40
intromittent organ in *Ascaphus truei*, 158*f*
locomotion in, 5*f*, 144–45
nictitating membrane in, 154*f*
poisonous, 19*f*, 136–38
reproduction in, 157*f*, 158*f*, 159, 160
reproduction modes in, 162*t*
sample reproduction date for, 165*t*
sexual maturity in, 166–67
skeleton of, 141*f*
skull of, 142*f*
tailed (*Ascaphus truei*), cold-water habitat of, 60
territories of, 341
toe pads and webbing of, 139, 140*f*
vertebral column of, 143
vocalization in, 143, 160
water storage in urinary bladder of, 156
Fungi kingdom (fungi), 1
Fur (pelage), 273, 321, 370, 398
Furcula, **233**
Fusiform body, 4, 5*f*

G

Ganoid scales, **96**
Gars, gas bladder-gut relationship in, 110*f*
Garstang, thesis on tunicates and vertebrate evolution by, 74, 76*f*
Gartner's ducts, **251**
Gas bladders, relationship of, to gut, 110*f*
Gas exchange, 8–9, 11–12*f*
amphibian, 138, 139*f*, 148–49
fish, 107–8
mammalian, 294–95
reptilian/bird, 206*f*, 207, 240–41
Gastric brooders, 163
Gastric juice, **302**
Gastroliths, 211, 213*f*
Gause, G. F., 375
Gause's Rule, **375**–76
Geese, 327*f*, 357
migration of Canada, 347, 348, 354*f*, 396
Gene(s)
homeobox (*Hox*), 78
total number of, in vertebrates *versus* invertebrates, 92
Gene flow, **34**
Genus, **24**, **25**
Geographic information systems (GIS), 397
Geographic isolation, **34**
Geographic range of species, **45**
Geographic variation in species, rules governing, 37–38
Geologic distribution of species, 49–53
Geologic range of species, **45**
Gill(s), 9, 11*f*
jawed fishes, 107–8
larval amphibian, 142–43, 148, 149*f*
Gill Arch Theory, **91**
Gill net fishing, threat of, 418
Gill rakers, **107**
Giraffes
competition among, 359*f*
horns of, 277*f*
Gizzard, **244**
Glacial lakes, **58**
Glaciations, 53–54, 55*f*
Glans penis, mammalian, **314**
Global warming, 55*f*
effects of, on interspecies interactions, 387
effects of, on species and ecosystems, 437–38
seal die-offs and, 388
Globiform body, 5*f*
Gloger's rule, **38**
Glomerulus, mammalian kidney, **311**
Glottis, amphibian, **147**, 148*f*
Glucagon, **117**
Glycoproteins as natural antifreeze in body fluids of Antarctic fishes, 119, 371–72
Gnathostome, **91**. *See also* Jawed fishes (gnathostomes); Tetrapod(s) (Tetrapoda)
Gonadotropic hormones, **116**, 156
Gonadotropin-releasing hormone (GnRH), 156
Gonads as endocrine organs, 117. *See also* Ovaries; Testes
Gondwana, ancient continent of, **49**, 52
Gonopodium, **101**, 123
Gorgonopsians, 264
Granular glands, **136**
Grasslands, 63–64
Great Lakes, sea lamprey in, 71, 85, 342
Greenhouse gases, **437**, 438
Growth, 18. *See also* Development
amphibian, 166–67
jawed fishes, 125–27
jawless fishes, 88
mammalian, 319–21
reptilian, 225–26, 260–61
Growth hormone (GH), **116**
Guard hairs, **272**
Gubernaculum, mammalian, **313**
Guppies (*Poecilia reticulata*), high-speed evolution in, 33
Gustatory cells (taste buds)
amphibian, 155
fish, **116**
mammalian, 309
reptilian, 216, 250*t*
Gymnovarian ovaries, **120**
Gyri ridges, mammalian brain, **304**

H

Habenula, mammalian brain, **304**
Habitat
carrying capacity of, 324–25
fragmentation of, 429–**30**
loss of, and species extinction, 72, 411–14, 429–38
restoration of, 439–40
Hadopelagic zone, 56*f*, **57**
Hagfish (class Myxini), 4, 80–82
growth and development in, 88
lampreys compared to, 82*t*
male urogenital system of, 121*f*
morphology of, 82–87
reproduction in, 88
Haikouella lanceolata, 75–76
Hair, 6, 271*f*, 272–74
Hallux, **286**, 287*f*
Hamuli (hooklets), feather, **226**
Hands, mammalian, 284, 285*f*
Harderian glands, 153
Hare, snowshoe (*Lepus americanus*), 273*f*
food supply of, 335*f*
predation of, by lynx, 333, 334*f*, 385
Hatching
amphibian, 164
fish, 126
mammalian, 318
reptilian/bird, 223–24, 258–59
Hawaii, threatened birds and bird extinctions in, 413
Head, 4
Hearing, 13
amphibian, 142, 152*f*
inner ear structure and, 15*f*, 86
jawed fishes, 114–15
mammalian, 305–8
range of, in humans and birds, 250*t*
reptilian/bird, 215, 248–49
swim bladder in fishes and, 112
Heart
amphibian, 145, 146*f*

Heart—*Cont.*
evolution of vertebrate, 146*f*
hagfish and lamprey, 84
jawed fishes, 104–6
mammalian, 292
reptilian, 205–6, 238–39
size of, and body size, 294
size of, and body weight of birds, 238, 239*t*
Heat exchange, 214*f*, 239, 293
countercurrent, 239, 369*f*, 370
Heat exchanger, **239**
Hematopoiesis, **147**
Hemipenes, reptilian, **218**, 219*f*
Hemispheres, mammalian brain, **304**
Hemoglobin
lacking in icefish, 107
molecular evolution and studies of, 38
as oxygen pumps (Root effect), 111
Hemolytic poisons, reptilian, **209**
Hennig, Willi, 27
Herbivores
nutritional value of meals of, 383*f*
Pleistocene extinction of, 411*f*
Hermaphroditic organisms, **16**
fishes as, 120, **123**
sequential, 123, 125
synchronous, 123
Heterocercal caudal fins, **100**, 102*f*
Heterochrony, **366**
Heterocoelous vertebrae, **233**
Heterodont dentition, **7**, 208
mammalian, **298**
Hibernation, **368**–73
Hilum, mammalian, **311**
Hindbrain (rhombencephalon), 12
Hindgut fermenters, 302, 303*f*
Hippocampus, mammalian brain, **305**
Hippopotamuses, 366
Historical zoogeography, **47**
Holarctic region, 47, 48*f*, 49
Holoblastic egg cleavage, **88**
Holocephali (ratfishes), 93, 94*f*
Homeothermy, **4**, 10
in dinosaurs, 182–83
Home range, 337–41
calculation of, 338
factors affecting, 337–38
marking of, 338
of select vertebrates, 339*t*
territory within, 339–41
"Home" within home range, **339**
Homing, **352**–55
Homocercal caudal fins, **101**, 102*f*
Homodont dentition, **7**, **150**, 208, 298
Homologous structures, **8**
Hooves, mammalian, **274**
Horizontal undulations, reptilian locomotion, **203**, 204*f*
Hormone(s)
amphibian metamorphosis caused by, 155
disruption of vertebrate system by synthetic, 438–39
endocrine system secretion of, 116–17
stress and secretion of, 251
Horns
determining animal age using, 400*f*
lizard, 199–200
mammalian, 274, 275*f*, 276
Horse
brain of, 304*f*
digestive system of, 302, 303*f*
Host (parasite), **380**
Hudsonian zone, North American, 66*f*, 67
Human(s), 421
cytochrome *c* and DNA studies comparing chimpanzees and, 40, 41*f*, 42*f*, 43
effects of, on animal extinctions and extirpations, 409–19
effects of, on animal interspecific interactions, 387–88
fetal development of, 318*f*
hearing range of, 250*t*
impact of, on vertebrate distribution, 70–72
male reproductive system of, 313*f*
pheromones in, 362
population growth among, 427–28
pulmonary ventilation in, 296*f*
respiratory system of, 295*f*
role of, in ecosystems, 18–19, 421, 440
Humerus bone, amphibian, 141*f*, **144**
Hummingbirds, 238
energy for migrations of, 348
torpor in, 368*f*, 369
Hunting
legislation regulating, 421
overhunting, 411
prehistoric human, and animal extinctions, 409–11
Hydrostylic suspension, **100**
Hyenas, 358
Hyoid apparatus, **281**
Hyoid arch, **98**
Hyomandibular cartilage, **98**, 142
Hypaxial (ventral) muscles, **8**, 10*f*, 204, 288
Hypocercal caudal fins, **100**, 102*f*
Hypoglossal nerve (XII), **151**
Hypophysis (pituitary gland), mammalian, **304**
Hypothalamic-pituitary-gonadal axis, 156*f*
Hypothalamus, 155, 156
mammalian, **304**

I

Ice Age, 52–54, 55*f*
Icefishes, 107
Ichthyornis, 193*f*
Ichthyosaurs (Ichthysauria), 175–76, 177*f*
front limb of, 202, 203*f*
Ichthyostegalia, 133–35
Identification techniques, 391–96
marking, 392
photography, 394
radio transmitters, 394–96
satellite tracking, 396
tagging, 392–94
Ileum, **302**
Imprinting
bird, **245**, 246*f*
fish, 113
Incisors (teeth), **7**, 298, 299*f*
Inclusive boundary strip method of calculating home range, 338
Incus, **280**, **305**, 306*f*
Indeterminate growth, 158, **398**
Index (censusing technique), **397**
Infrasound, **13**, 308, **366**
navigation and, 351
Inguinal canal, **313**
Inner ear, 86
amphibian, 152*f*, 153
structure of, 15*f*
Innominate (coxal, hip) bones, **284**
Insulin, **117**
molecular evolution and comparison studies of amino acid sequence in, 40–42
Integumentary system, 6
amphibian, 135, 136–40
jawed fishes, 95–97
jawless fishes, 82–83
mammalian, 271–79
reptilian/bird, 198–201, 226–30
Intercalary bone, **139**, 140*f*
Intercostal muscles, reptilian, **205**
Intermolt, **136**, 155
Internal fertilization, **16**
amphibian, 160, 161
mammalian, 316–17
Internal nares (choanae), **107**, 239
Interspecific interactions, 375–89

- competition, 375–77
- human interactions affecting, 387–88
- introduction to, 375
- predation, 382–87 (*see also* Predation)
- symbiosis, 377–82

Interstitial cell stimulating hormone, **116**
Intertidal (littoral) zone, 56f, **57**
Intestine, 10–11
Intraspecific interactions
- feeding behavior, 367–68
- sensory reception and communication, 360–67
- social interactions, 357–60

Intrinsic muscles, **8**
- in birds, **237**

Intromittent organ
- frog, 158f
- reptilian, 218, 219f, 220f

Invasion as animal movement, **342**–43
Invertebrates, number of genes in, *versus* in vertebrates, 92
Iridophores, **96**
Irruptions, population, **324**, 335
Islands as interbreeding barriers, 34

J

Jablonski, David, 402–3
Jaw(s)
- amphibian, 141–42
- evolution of, 90–91, 99f
- fishes (*see* Jawed fishes (gnathostomes))
- mammalian, 280
- reptilian, 207, 208
- suspension (*see* Jaw suspension)
- synapsid and mammalian, 264, 267f

Jawed fishes (gnathostomes), 90–128
- cichlids, rapid evolution of, 36, 37f
- evolution of, 90–95
- growth and development in, 125–27
- morphology of, 95–121
- reproduction in, 121–25

Jawless fishes (agnathans), 77, 78–88. *See also* Hagfish (class Myxini); Lamprey(s) (class Cephalaspidomorphi)
- evolution of, 78–82
- growth and development in, 88
- morphology of, 82–87
- reproduction of, 88

Jaw suspension
- amphibian, 142
- fish, 99–100

Jeholodens jenkinsi, 268, 269f
Jejunum, **302**
Jugular position, **100**, 102f
Juvenile pelage, **321**, **398**
Juvenile plumage, bird, **260**
Juvenile stage, jawed fishes, 127

K

Kangaroo (*Macropus*)
- locomotion in, 5f
- reproduction in, 315, 316f
- tree (*Dendrolagus mbaiso*), 20f

Kangaroo rats (*Dipodomys*), 363
Karyotypes, **40**
Kazakhstania, ancient continent of, **49**
Keel, sternum, 193, 195, **233**
Kemp's ridley sea turtles (*Lepidochelys kempii*), migrations of, 346
Keratin, **271**
Keratinocytes, **271**
Kestrels, use of ultraviolet light in predation by, 247
Kidney(s), 2, 13–15
- amphibian, 156–57
- jawed fishes, 119
- jawless fishes, 86–87
- mammalian, 310–11
- reptilian, 217

Kidney excretion, vertebrate, 13–16
Kingdom, **1**, **25**
Koalas, 362
- plant predation by, 386, 387f

K-strategists, 15, 17t
Kyoto Protocol of 1997, 438

L

Labor, animal division of, 357
Labyrinthodontia (Labyrinthodonts), 129, 130, 132f, 133–35, 172
- comparison of salamander to, 134f
- teeth of, 130, 132f

Lacey Act of 1900, 421
Lacrimal glands
- amphibian, 153
- reptilian, **215**

Lagena, **249**
Lakes and reservoirs, 58–60
- effect of, on fauna, 71
- formation of, 58
- life zones in, 58, 59f, 60

Lamellae, **107**
Lamprey(s) (class Cephalaspidomorphi), 2, 4, 81–82
- digestive tract of, 84–85
- eye of, 14f
- gills of, 11f
- growth and development in, 88
- hagfishes compared to, 82t
- inner ear of, 15f
- larval, 3f
- morphology of, 82–87
- reproduction in, 88
- sea, in Great Lakes, 71, 85, 342

Lancelet, 2, 74
Land (filter) bridges, species distribution and, 46–47, 52, 53, 54f
Lantern-eye fish (*Anomalops katoptron*), 97f
Large intestine, mammalian, 302
Larva, amphibian
- direct development of, 135
- gills in, 142–43, 148–49
- growth and metamorphosis of, 166

Larva, lamprey, 3, 88
Larval period, jawed fishes, **126**–27
Larynx
- amphibian, **147**, 148f
- mammalian, **281**, 294, 306
- reptilian/bird, 239, 241f

Late Quaternary, volcanism and species extinction in, 407, 410
Lateral bridges, turtle, **200**
Lateral line system, 86, 152
Laurasia, ancient supercontinent of, **50**
Laurentia, ancient continent of, **49**
Leatherback turtles (*Dermochelys coriacea*), migrations of, 345, 346f, 350–51
Legislation, U.S. wildlife conservation and management, 421–23
Lekking, 255
Lemmings, 328
- emigrations of, 355
- population cycles in, 332

Lentic system, **58**
Lepidosauromorpha, 169, 174, 175–76. *See also* Lizards; Snakes; Tuataras
Lespospondyli, 135
Life belts, **69**
Life zones, North American, C. Hart Merriam on, **65**–69
Light-emitting organs, jawed fishes, 97
Limb(s). *See also* Appendages, vertebrate
- bird, 229, 230f, 234, 235f
- comparison of sarcopterygian, amphibian, and reptilian, 133f
- dominance of right forelimb in toads, 151
- evolution of, 202f
- mammalian forelimbs, 283f, 284f, 285f
- mammalian hindlimbs, 285–86, 290f

Limb(s)—*Cont.*
reptilian, 202–3
of therapsids, 266
Limnetic zone of lakes, **58**, 59*f*
Lincoln Index, 397
Linnaeus, Carolus, 23–24
Lions (*Panthera leo*), 40, 416
Lipophores, **96**
Lissamphibia, 135–36
Litter size, factors affecting mammalian, 314–15, 326
Littoral zone of lakes, **58**, 59*f*
Lizards, 4, 169
alimentary canal of, 212*f*
cardiovascular system of, 205–6
classification of, 176
digestive system of, 207–12
endocrine system of, 216–17
eye of, 14*f*, 216*f*
gas exchange in, 206*f*
growth and development in, 222–26
high-speed evolution in brown anole (*Anolis sagres*), 33
inner ear of, 15*f*
integumentary system of, 198–201
muscular system of, 204–5
nervous sytem of, 212–14
reproduction in, 218–22
respiratory system of, 206–7
sense organs of, 214–16
sex changes in whiptail (*Cnemidophorus uniparens*), 222
skeletal system of, 201–4
stratum corneum modifications in, 199, 200*f*
temporal fenestrae of, 175*f*
urogenital system of, 217–18, 219*f*
venomous, 209
Lobe-finned fish (Sarcopterygii), 4, 94
Locomotion, 4, 5*f*
amphibian, 144, 145
appendages and, 7, 9*f*
evolution of amphibians and problem of, 135
fish, 101
flight, 4–6, 195–96
mammalian, 10*f*, 286–87, 289*f*
musculature for, 8, 10*f*
in snakes and lizards, 203, 204*f*
swimming in birds, 234, 236
Loggerhead turtles (*Caretta caretta*), migrations of, 346, 350, 351
Longevity
fish, 127
mammalian, 320*t*, 321
reptilian/bird, 225*t*, 226, 261*t*
Longitudinal fissure, mammalian brain, **304**
Loop of Henle, mammalian kidney, **311**
Loreal pits, reptilian, **215**
Lorenz, Konrad, 245, 246*f*
Lotic system, **58**
Lower Austral zone, North America, 66*f*, 68–69
Luciferase, 97
Lumbar vertebrae, reptilian, **202**
Lung(s), 9, 10, 12*f*
amphibian, 148–50, 152
mammalian, 294–97
reptilian/bird, 206, 207, 239–42
Lungfishes, 94, 119
digestive tract of, 109*f*
estivation in, 373
evolutionary relationship of amphibians to, 132
evolutionary relationship of tuna, pig, and, 31*f*
gas bladder-gut relationship in, 110*f*
gills of, 11*f*
male urogenital system of, 121*f*
Luteinizing hormone (LH), **116**, 156
Lynx (*Lynx canadensis*), predator-prey relationship of, with snowshoe hare, 333, 334*f*, 385

M

MacArthur, Robert, 376, 377*f*
Macrogeographic (long-distance) migration, **344**–49
Maculae
amphibian, **153**
mammalian, **305**
Madagascar, animal extinctions on, 410
Magainin, 137
Magnetic field, Earth's
homing ability and, 353
migrations and, 349–50
Magnetoreceptor cells, 116
Male(s)
amphibian reproductive system, 157*f*, 158
fish urogenital system, 120, 121*f*
mammalian reproductive organs, 313–14
reptilian/bird reproductive system, 252
Malleus, **281**, 305, 306*f*
Mammals (Mammalia), 4, 264–323
aortic arch of, 206*f*
cardiovascular system of, 292–94
commensalism and, 377, 378*t*, 379
competition and niches of, 376–77
digestive system of, 298–302, 303*f*
endocrine system of, 309–10
equilibrium in diversity of, in North America, 270
evolution of, 202*f*, 264–71
extinction and threatened extinction of, 412*f*, 414, 416, 417, 418–19
gas bladder-gut relationship in, 110*f*
growth and development in, 317–21
hibernation and torpor in, 368, 370, 371
homing ability in, 352*f*, 353
inner ear of, 15*f*
integumentary system of, 271–79
introduction to, 264
litter size of, 314–15, 326
locomotion in, 10*f*
migrations of, 348–49, 351
morphological convergence of tropical-forest, 64, 66*f*
muscular system of, 288–92
nervous system of, 302–5
new species of, 20*f*, 21
as parasites, 381
relationship of, to birds, 191
reproduction in, 17–18, 314–17
reproductive organs of, 311–14
respiratory system of, 294–98
sense organs of, 305–9
size comparison of reptiles, dinosaurs, and, 180*f*
skeletal system of, 279–88
temporal fenestrae of, 175*f*
urogenital system of, 310–14
Mammary glands, **277**, 278*f*, 315, 367
Mammillary bodies, mammalian brain, **304**
Man and Biosphere Programme (MAB), 433
Manatees
deaths of, caused by red tide algae, 331–32
migration of, 349
sound production by, 308
Mandibular arch, fish, **98**
Mandibular salivary gland, **300**
Mapping of wildlife, 397
Marine environments, distribution of species in, 55–58
Marine vertebrates, extinctions and changes in diversity of, 405*f*
Marking techniques for wildlife identification, 392
Marmoset, sagui dwarf, 20*f*

Marsupials
distribution of, 46*f*, 47
embryonic/fetal development in, 318
reproductive organs of, 312, 314
Mass extinctions, **402**–3, 404*f*, 405*f*
Matriarchal hierarchy, **358**
Maxillae bones, fish, **98**, 99*f*
Mayr, E., biological species concept of, 25, 34
Mechanoreceptors, **272**
Meckel's cartilage, **98**, 99, 141
Median fins, **100**
Medulla, hair shaft, 271*f*, **272**
Medulla oblongata, mammalian brain, **304**
Meissner's corpuscles, **367**
Melanin, 96
Melanocytes, **271**
Melanocyte-stimulating hormone (MSH), **116**
Melanophores, **96**
Melatonin, **117**, 140, 156, 217, 272, 304, 308, 310
Meninges, **12**
amphibian, 151
mammalian, 305
reptilian, **214**
Meroblastic egg cleavage, **88**
Merriam, C. Hart, on North American life zones, 65–69
Mesaxonic foot, **288**
Mesonephros
lamprey, fish, and amphibian, 87*f*
mammalian, 311
Mesopelagic zone, 56*f*, **57**
Mesozoic extinctions, 405–6
Metacarpal bones, amphibian, 141*f*, **144**
Metamorphosis
amphibian, 155, **166**
jawed fishes, 126–27
lamprey, 88
Metanephros, 13
reptilian, bird, and mammalian, 87*f*
reptilian, **217**
Metatarsals, amphibian, 144
Meteorite and greenhouse effect as cause of mass extinctions, 407–8
Micrographic migration, **343**, 344*f*
Microvilli, small intestine, **302**
Midbrain (mesencephalon), 12
Migration, **343**–52
altitudinal, 344
amphibian, 343, 344*f*
bird, 242, 244, 247, 344–45, 346–48, 350, 351–52
fish, 118–19, 344–45, 350
macrographic (long-distance), 344–49
mammalian, 348–49, 351
micrographic (short-distance), 343, 344*f*
navigational cues used in, 349–52
reptilian, 345, 346, 350–51
teaching birds to migrate, 396, 397*f*
Migratory Bird Conservation Fund, 421, 423
Migratory Bird Hunting and Conservation Stamp Act (Duck Stamp Act) of 1934, 421
Migratory Bird Treaty Act of 1918, 421, 422
Milk (deciduous) teeth, mammalian, **298**
Minimum area method of calculating home range, 338
Mississippi River forested wetlands, 436*f*
Mitochondrial DNA, 38
Mobbing behavior, 357
Molars (teeth), 7, 299, 399–400*f*
Mole, star-nosed (*Condylura cristata*), sense organs in nose of, 309
Molecular evolution studies, 38–43, 169, 270–71
Molecular taxonomy, 27
Mole rats, naked (*Heterocephalus glaber*)
eusociality, breeding, and vocalizations of, 358
home range of, 339, 340*f*
reproduction of, 314
vertebral column of queen, 282
Molting
amphibian, 136, 155
bird, **228**, 229, 260
mammalian, 272, 273*f*
Monera kingdom (bacteria), 1
Mongoose (*Herpestes* spp.), 70, 71*f*
Monkeys
commensalism among capuchin and squirrel, 378
vervet (*Cercopithecus pygerythrus*), alarm calls of, 363, 364*f*
Monophyletic taxon, **26**, 30
Monophyodont dentition, reptilian, **209**
Monotremes, 268
oviparity in, 269, 314
reproductive organs of, 312
Moose (*Alces alces*), 316*f*
predator-prey relationship of, with wolf, 383–85
Morphological barriers to speciation, **35**
Mouse
dispersal of white-footed, 342*f*
golden (*Ochrotomys nuttalli*), 273*f*
predation by grasshopper (*Onychomys* spp.), 385
as prey, 382–83
Mouthbrooding in fish, 123
Movements of vertebrates, 337–55
dispersals as, 341–42
emigration as, 355
home range and, 337–41
homing as, 352–55
as invasions, 342–43
migrations as, 343–52
Mucous cells, **301**
Mueller, Paul, 427
Multituberculates (Multituberculata), 268
Mummichog (*Fundulus heteroclitus*), embryo of, 125*f*
Muscle(s), 8, 10*f*
erector, 228, 272, 290
flight and, 237–38
mammalian, 288–92
myomeres (*see* Myomeres)
myospta, 8, 83, 84*f*
white *versus* red, in birds, 236–37
white *versus* red, in jawed fishes, 101–3
Muscular system, 8, 10*f*
amphibian, 145
jawed fishes, 101–4
jawless fishes, 83–84
mammalian, 288–92
reptilian/bird, 204–5, 236–38
Museum collections, vertebrate, 426–27
Musk oxen (*Ovibos moschatus*), 387*f*
Mutualism, **379**–80
Myoglobin, 297–98
Myomeres, **2**, 8, 10*f*
jawed fishes, **101**, 103*f*
jawless fishes, **83**
Myosepta, **8**, **83**, 84*f*

N

Nail, mammalian, **274**
Narrowly endemic species, **45**
Nasal conchae, **294**
Nasohypophyseal sac, **84**
Nasolabial grooves, amphibian, 154*f*, **155**
Nasopharynx, mammalian, **294**
National Audubon Society, 425
National Forests, U.S., 423, 424, 425*f*
National Oceanic and Atmospheric Administration (NOAA), 423, 437

National Park System, U.S., 423, 424, 425*f*
National Wild and Scenic Rivers Act of 1968, 424–25
National Wilderness Preservation System, 424
National Wildlife Refuge System, U.S., 423, 424, 425, 425*f*
Natural extinction, **402**–9
Natural selection, evolution and, **32**–43
Nature Conservancy, 425, 426, 428
Nearctic region, 47, 48*f*, 49
Nekton, **57**
Neopterygii (advanced ray-finned fishes), 95
Neoteny, **149**, **166**
Neotropical region, 48*f*, 49
Nephrons, 2
  mammalian, 16*f*, **311**
Neritic zone, 56*f*, **57**
Nervous system, 11–12, 13*f*
  amphibian, 151
  jawed fishes, 113–16
  jawless fishes, 86
  mammalian, 302–5
  reptilian/bird, 212–14, 244–45
Nests
  bird, 256, 257*f*, 258
  crocodile, 252*f*
  dinosaur, 174
Neuromast(s), **13**
  fish, **86**
Neuromast organs, amphibian, **152**
Neuromast system, jawed fishes, **114**
Newt, 151, 159*f*
Niche, **375**–76, 377*f*
Nictitating membrane
  amphibian, **153**, 154*f*
  reptilian, **215**
Nipple, mammalian, 277, 278*f*
Nitrogenous wastes, removal of, 13–16
Nonshivering thermogenesis, **370**
Norepinephrine (noradrenaline), **116**, 309–10
North America
  biotic provinces of, 69
  bird migration flyways of, 344, 345*f*
  genetic divergence of songbird species in, 35
  glaciation in, 55*f*
  Great American Interchange between South America and, 52*f*
  introduction and distribution of European starling in, 70*f*
  life zones of, 65–69
  prehistoric human hunting and extinction of megafauna in, 409–11
North American Breeding Survey, 398
Nose. *See also* Olfaction
  amphibian, 154–55
  mammalian, 308–9
  reptilian/bird, 216, 249–50
Notochord, 2
Notothenioidei, suborder, 107
Nuptial (breeding) plumage, bird, **260**
Nursing, mammalian communal, 319
Nutria (*Myocastor coypu*), 71

O

Oblique septum, reptilian, **207**
Ocean
  epicontinental seas, 52
  pollution in, as threat to wildlife, 416, 418–19
  rising temperature of, 438
  seafloor spreading of, 49, 50*f*
  zones of, 55, 56*f*, 57, 58
Oceanic zone, 56*f*, **57**
Odontoid process, reptilian vertebra, **201**
Oil spills, threat of, to wildlife, 416, 418*f*
Olfaction, 12
  amphibian, 154–55
  brain and, 13*f*
  as communication, 360–62
  jawed fishes, 107, 116
  jawless fishes, 86
  mammalian, 308–9
  migration and, 350, 351
  pheromones and (*see* Pheromones)
  reptilian, 216
Olfactory lobes, mammalian brain, **305**
Omasum, **302**
Operculum
  amphibian, **153**
  fish, **99**
Opisthocoelous vertebrae, amphibian, **143**
Opisthoglyphs, reptilian, **209**
Opisthonephric (mesonephric) duct, fish, **119**
Opisthonephros, 13
  fish, **119**
  reptilian, 217
Opossum (*Didelphis*)
  embryonic/fetal development in, 318*f*
  male reproductive organs of, 313*f*
  skeleton of, 288*f*
Optic chiasma, mammalian, **304**
Optic lobes, 214, 244
Oral cavity
  mammalian, 293, 298
  reptilian, 208–11
Orbital glands, amphibian, **153**
Order, **25**
Organ of Corti, mammalian, **305**
Oriental region, 48*f*, 49
Ornithischia, 176, 181*f*, 184–85. *See also* Dinosaurs
Ornithopods, 184
Oropharynx, mammalian, **294**
Ortelius, Abraham, 49
Os clitoris, mammalian, **313**
Osmoregulation
  amphibian, 136
  marine and freshwater fish, 117–19
Os penis, mammalian, **314**
Osteichthyes (bony fishes), 90, 92, 93–95, 129. *See also* Teleosts
  gills of, 11*f*, 107*f*, 108*f*
  skull of, 98–99
  subgroups of (lungfishes, coelacanths), 94–95
  swim bladders of, 111
Osteolepiforms (*Osteolepis*), 129–30, 132
Ostracoderms, 78, 80, 81*f*
Ostrich, 234
Otoliths, fish, **114**
Otters, sea (*Enhydra lutris*), threats to, 418–19
Outgroup, **28**, **31**
Ovaries, 87, 119–20. *See also* Reproductive organs
  hormones secreted by, 117
  reptilian/bird, 218, 251, 252*f*
Overgrazing as factor in population density, 327
Overhunting as cause of species extinction, 411
Oviducal glands, fish, **120**
Oviducts
  mammalian, **311**, 312*f*
  reptilian, 218
Oviparous development, **16**, 17
  amphibian, 161–63
  jawed fishes, 125
  mammalian (monotreme), 269, 314, 317, 319
  reptilian/bird, 221, 252–55
Ovisacs, amphibian, **157**
Ovotestes, 120
Owls
  clutch size of barn, 333*f*
  commensalism between snakes and, 378, 380*t*
  feathers of, 229, 230*f*
  hearing in, 248, 249*f*

as predators, 382–83
territories of, 341
Oxytocin, 309
Ozone, atmospheric, 407, 438–39

P

Pachycephalosaurs, 184, 185*f*
Pacinian corpuscles, **367**
Paedogenesis, **166**
Paedomorphosis, **74**, **158**
Paired fins, evolution of, 91
Palate, mammalian, 281, 294
Palatoquadrate (pterygoquadrate) cartilage, fish, **98**
Palearctic region, 47, 48*f*, 49
Pallium (hippocampus), mammalian brain, **305**
Pancreas, 117
Pancreatic islets, fish, **117**
Pandas, giant, 338
pseudothumb of, 284, 286*f*
Pangaea, ancient supercontinent of, **50**, 51*f*
Pangolins (*Manis*), 6
Panniculus carnosus, **290**
Pantotheres (Pantotheria), 268
Papilla, feather, **228**
Parakeet, musculature of, 236*f*
Paraphyletic taxon, **26**
Parapineal organ
fish, **115**
reptilian, 214
Parasite, **380**
Parasitism, **380**–82
birds engaging in, 258, 259, 360
intraspecific, 360
Parathormone, 216
Parathyroid gland, 155, 216, 309
Paraxonic foot, **288**
Parental care, 18
amphibian, 165
jawed fishes, 126
mammalian, 310, 318–19
reptilian/bird, 225, 259, 260*f*
Parietal cells, **302**
Parietal organ
amphibian, 154
reptilian, **216**
Parotid salivary gland, **300**
Parthenogenesis, reptilian, **222**
Parthenogenetic species, **16**, 222
Passive integrated transponder (PIT) tags, 392
Patella, reptilian, **203**
Patriarchal hierarchy, **358**
Pectoral girdle
amphibian, 143, 144*f*
fish, **100**
mammalian, 283–84
Pelagic region, marine, 56*f*, **57**
Pelvic fins, 100, 101, 102*f*
Pelvic girdle, 100
bird, 233–34
dinosaur, 178, 181*f*
mammalian, 284–86
reptilian, 203
Pelvis, mammalian, **285**
Pelycosaurs (Pelycosauria), 172
Penguins
deep diving by, 242
dispersal of emperor, 342*f*
homing ability in, 352
swimming by, 234, 236
threats to, 418
Penis
mammalian, 313*f*, 314
reptilian, **218**, 220*f*
Pentadactyl, **284**
Perch
digestive tract of, 109*f*
gas bladder-gut relationship in climbing, 110*f*
musculature of yellow (*Perca flavescens*), 103*f*
Pericardial cavity, fish, **84**
Peripheral species, classification of, **423**
Perissodactyls, **288**, 289*f*
Permafrost, **60**
Permanent (adult) dentition, mammalian, **298**
Permian extinctions, 403–5
Pesticides, threat of, to vertebrates, 413, 414, 415*f*, 426–27
pH, freshwater environments and, 58
Phalanges, amphibian, 141*f*, **144**
Pharyngeal (buccopharyngeal) gas exchange, reptilian, **207**
Pharyngeal pouches
amphibian, 152
reptilian, 207
Phenetic (numerical) classification, 27
Phenogram, **27**
Phenon, **27**
Pheromones, **360**
amphibian, 138–39, 151, **155**, 361
animal communication using, 360–62
fish, 361
as home range markers, 338
mammalian, **279**, 315, 361–62
as migratory navigation cues, 351–52
reptilian, 210–11, 219, 220–21, 361
Photic zone, marine, **55**, 56*f*
Photography, wildlife identification through, 394
Photophores, **97**
Photoreceptor
lamprey, 86
pineal organ as, 113, 117
Phylum, **1**, **25**
Physical barriers to dispersal of species, **46**
Physiological barriers to speciation, **35**
Physoclists, **111**
Physostomes, **111**
Phytoplankton, **57**
Phytosauria, 184–85
Pia mater, **245**
Pigeon
homing, 352, 354–55
passenger (*Ectopistes migratorius*), 413*f*
Pigment cells in fish integument, 93, 96–97
Pikas (*Ochotona* sp.), food storage by, 302
Pinatubo, Mount, eruption of, 407*f*
Pineal body (gland), mammalian, **304**, 310
Pineal gland, 117, 217
Pineal organ (epiphysis)
amphibian, 154
fish, **115**
as photoreceptor, 113, 117
reptilian, 244
Pinna, mammalian, **305**, 306*f*
Pits (neuromast system), **114**
Pits, reptilian, 215
Pituitary gland
amphibian, 156
fish, **116**
mammalian, 304
Placental mammals, reproductive organs of, 312
Placoderms, 92, 93*f*
Placoid scales, **95**, 96*f*
Plankton, **57**
Plantigrade locomotion, **286**, 289*f*
Plant kingdom (plants), 1
defenses of, against predators, 386
Plastron, turtle, 198*f*, **200**
Plate tectonics, **49**, 50*f*
Pleistocene, animal extinctions in, 409–11
Plesiomorphic character, **31**
Plesiomorphies, **28**
Plesiosaurs (Plesiosauria), 176, 178*f*, 202
Plethodontidae, 135
Pleural cavities
bird, **240**

Pleural cavities—*Cont.*
reptilian, **207**
Pleurodont dentition
amphibian, **150**
reptilian, **208**
Pleuroperitoneal cavity, reptilian, **206**
Plicae vocales, amphibian, **148**
Pliny, 27
Plover, American golden (*Pluvialis dominica*), 47*f*
Poikilothermy, **4**
basking, 212, 214*f*
in dinosaurs, 182
Poisonous amphibians, 136–38
Poisonous birds, 228
Poisonous reptiles, 209–10
Polyandry, **254**
Polygyny, **254**
Polyphyletic taxon, **26**
Polyphyodont condition, **298**
Polyphyodont dentition, **150**
fish, **108**
reptilian, **209**
Population(s), **23**
allopatric, 34–36
carrying capacity of habitat for, 324, 325*f*
clines in, 33, 34*f*
crashes of, 324
demes in, 32–33
dynamics of (*see* Population density)
sympatric, 36–37
Population cycles, 332–35
Population density, 324–35
cycles and, 332–35
density-dependent factors and, 324, 327–30
density-independent factors and, 324, 330–32
environmental resistance as factor in, 326–35
home range and, 337
irruptions of, 324, 335
reproductive (biotic) potential and, 324–26
Population growth, 324, 325*f*, 326
Porolepiformes, 132
Portal heart, jawless fishes, **84**
Postcaudal vertebrae, reptilian, **202**
Postjuvenal molt, **260**
Postjuvenal pelage, **398**
Post-mating (postzygotic) barriers to speciation, **35**
Postnatal molt, **260**
Postnuptial molt, **228**
Postzygotic (post-mating) barriers to speciation, **35**
Prairie dogs (*Cynomys* spp.), vocalizations of, 363–64
Prairies, **63**–64
Precaudal vertebrae, reptilian, **202**
Precipitation in terrestrial environments, 60, 61*f*
Precocial species, **18**, **260**
birds, characteristics of, 260*t*
Predation, 382–87
aposematic coloration against, 138, 386
as cause of high-speed evolution, 33
chemical defense against, 19*f*, 137–38
as factor in population size, 332, 333, 334*f*
kestral use of ultraviolet light in, 247
Predator(s), **382**–87
defenses against, 138, 385, 386
Predator control programs, 328
Predatory fishes, light organs in, 97
Prehallux bone, amphibian, 141*f*, **144**
Pre-mating (prezygotic) barriers to speciation, **35**
Premaxillae bones, **98**, 99*f*
Premolars (teeth), **7**, 299
Prenatal development
amphibian, 161–64
jawed fishes, 125–26
mammalian, 317, 318*f*
reptilian, 222–23
Prenuptial molt, **228**
Prepuce, mammalian, **314**
Prey, **382**–87
Prezygotic (pre-mating) barriers to speciation, **35**
Primaries, feather, 226*f*, **227**
Primary isolating barrier to speciation, **34**
Primary productivity of estuaries, mangroves, tidal marshes, and coral reefs, 57*f*
Primate(s). *See also* Human(s)
classification of, based on cladistics, **30**
evolution of, 270
facial expressions of, 292
feet modifications in, 286, 287*f*
hand modifications in, for grasping, 285*f*
molecular comparison of genomes of, 39*f*
threat of extinction to, 414
Proanua, 136
Probainognathus, 268
Procoelous vertebrae, amphibian, **143**
Procolophonids, 175, 176*f*
Proctodeum, reptilian, **217**
Profundal zone of lakes, 59*f*, **60**
Progesterone, 217
Prolactin (PRL), **116**, 309
Pronephros, 13, 217
hagfish, 87*f*
Pronghorn antelopes, 367
oxygen metabolism and running ability in, 292
Protandry, **125**
Protein, evolutionary studies using amino-acid sequences of, 38, 40, 41*f*
Proteroglyphs, reptilian, **209**, 210*f*
Protista kingdom (single-cell organisms), 1
Protoavis texensis, 194*f*, 195
Protochordates, 1–2, 74, 76*f*
Protogyny, **125**
Protonephridia, 2
Prototheria, 267
Proventriculus, **244**
Proximal convoluted tubule, mammalian kidney, **311**
Pterosaurs (Pterosauria), 185–86
Pterylae, **227**, 228*f*
Puboischiac plate, **144**
Puma, molecular studies of species of, 39
Puma concolor, 24–25
Purines, **96**
Pygostyle, **233**
Pyloric caeca, **109**

Q

Quadrate bones, fish, **98**, 99*f*
Quagga (*Equus quagga*), classification of, 40

R

Rabbit
brush (*Sylvilagus bachmani*), Allen's and Gloger's rule applying to, 38
population of European, in Australia, 70*f*, 330
Raccoons, population irruptions among, 335*f*
Races (subspecies), **34**–36
Rachis, feather, **226**
Radio transmitters, wildlife identification using, 394–96
Radio-ulna bone, amphibian, 141*f*, **144**
Radius bone, amphibian, 141*f*, **144**
Rafting, species dispersal by, 47, 48*f*
Rare species, classification of, **423**
Ratfishes. *See* Chimaeras (ratfishes)
Ratite birds, **233**
living families of, 53*f*

Rattles, snake, 199, 200*f*
Ray-finned fishes (Actinopterygii), 4, 94, 95
Rays, 4, 93, 94*f*
electric (*Torpedo*), 104
uterine milk secreted by, 126
Rectilinear locomotion, reptilian, **203**, 204*f*
Refuges, wildlife, 424–26
Renal (Bowman's) capsule, mammalian kidney, **311**
Renal corpuscle, mammalian kidney, **311**
Renal pelvis, mammalian kidney, **311**
Reproduction, 15–18
amphibian, 135, 157, 158–61
dioecious, hermaphroditic, and parthenogenetic forms of, 16
fertilization methods and, 16–17
jawed fishes, 121–25
jawless fishes, 88
mammalian, 314–17
oviparous *versus* viviparous modes of, 16 (*see also* Oviparous development; Viviparous development)
reptilian/bird, 218–22, 252–55
r strategists and *K*-strategists and, 15, 17*t*
Reproductive organs
amphibian, 157–58
mammalian, 311–14
reptilian/bird, 218, 219*f*, 220*f*, 251, 252*f*
Reproductive (biotic) potential, **324**–26
Reproductive strategies, 254, 315–17
Reptiles (Reptilia)
amphibians compared to, 172–73
aortic arch in, 206*f*
classification of, 28, 29*f*, 174
commensalism and, 377, 378*t*, 379, 380*t*
crocodilians and birds (Archosaurs), 226–61
dinosaurs (*see* Dinosaurs)
evolution of (*see* Reptiles, evolution of)
gas bladder-gut relationship in, 110*f*
growth and development in, 222–26, 255–61
limbs of, compared to amphibian and sarcopterygian, 133*f*
mammal-like, 264–69
migrating, 345, 346, 350–51
morphology of, 198–218, 226–52
niches of, 376
phylogenetic systematics of, 169, 171*f*, 174–91
reproduction in, 17–18, 218–22, 252–55
turtles, tuataras, lizards, and snakes (Testudomorpha, Lepidosauromorpha), 198–226
Reptiles, evolution of, 169–96
ancestral birds and, 191–95
ancestral reptiles and, 169–73
ancient and living reptiles and, 173–91
introduction to, 169
modern birds and, 195
origin of amniotes and, 170*f*
origin of flight and, 195–96
winter survival and hibernation in, 369
Research on vertebrates, future, 19–20
Respiration, vertebrate, 8–9, 11–12*f*
Respiratory system
amphibian, 147–50
jawed fishes, 107–8
jawless fishes, 84
mammalian, 294–98
reptilian/bird, 206–7, 239–42
Rete mirabile, 111
Retia mirabilia, **239**, 292, 293
Reticulate sheath on bird legs, 230*f*
Reticulum, **302**
Retina, **115**
mammalian, 247
Retrices, feather, 226*f*, **227**
Rhinoceros horns, **276**
Rhipidistian fishes, 129–32
Ribs
amphibian, 189
mammalian, 282
reptilian, 202, 205
Rift lakes, **58**
Rivers and streams, 58, 59*f*
Rodents, adaptation of, to desert biome, 64, 65*f*
Rods, eye, **115**
Root, hair, **272**
Root effect, 111
r-strategists, 15, 17*t*
Rugae, **301**
Rumen, **302**
Ruminants
bird, 243*f*
mammalian, digestive system of, 302, 303*f*

S

Saccule, mammalian, **305**
Sacral vertebra, amphibian, 141*f*, **143**
Sacrum, reptilian, **202**
Saddle joint, **284**
Sagittiform body, 5*f*
Salamanders, 4
arboreal adaptations in, 140*f*
comparison of labyrinthodont to, 134*f*
evolution of limb structure in, 202*f*
gills of, 11*f*, 146, 148, 149*f*
homing ability in, 352
inner ear and hearing in, 152*f*
intraspecific competition among, 358, 359*f*
larval, 146, 148–49
locomotion in, 144
migrating, 343, 344*f*
nasolabial grooves and vomeronasal organs in plethodontid, 154*f*, 155
pheromone production by, 138–39
poison-secreting, 136
projectile tongue in, 150
reproduction in, 17, 18*f*, 135, 157*f*, 159, 160
sample reproduction data for, 164*t*
skeleton of, 141*f*
skull of, 141
territory of, 340
vertebra of, 143
Salivary glands, 300
Salmon
as anadromous, 119
imprinting of juvenile, 113
migration of, 344–45, 350, 361
musculature of chinook, 104*f*
Saltatorial (jumping) locomotion, 4, 5*f*
Salt gland, reptilian, **207**
Salt marshes, 57, 58*f*
productivity of, 57*f*
Saltville, Virginia, fossils, 269
Sampling estimates, **397**
Sanctuaries, wildlife, 424–26
San Diego Zoo Center for Reproduction of Endangered Species, 439
Sarcopterygii (lobe-finned fishes), 4, 94
limbs of, compared to amphibian and reptilian, 133*f*
Satellite tracking of wildlife, 396
Saurischians (Saurischia), 176, **178**–83. *See also* Dinosaurs
Sauropodomorphs, **183**
Sauropsida, 169. *See also* Reptiles (Reptilia)
Savannas, **64**
Scales, 6
on bird feet, 229, 230*f*
cichlid fishes as eaters of, 109
fish, 95, 96*f*, 398*f*
reptilian, 198
Scent glands, **279**

Scientific method, **390**
Sclerotic ring, **232**, 233*f*
Scrotal sac, mammalian, **313**
Scutellate sheath on bird legs, 229, 230*f*
Sea horses, 126
Seals, 293, 327
blubber of, 369*f*
global warming and die-offs of, 388
hindlimbs of, 290*f*
migration of, 348, 349
threats to, 388, 418
vibrissae in ringed (*Phoca hispida saimensis*), 274
video camera attached to, 394*f*
Sea squirts. *See* Tunicate(s)
Sebaceous (oil) glands, **279**
Sebum, **279**
Secondaries, feather, 226*f*, **227**
Secondary isolating barriers to speciation, **34**, 35*f*
Sectorial teeth, mammalian, **300**
Seed
animal population cycles and production of, 333–34
bird's digestive system and germination of, 244
Semicircular canals, 86
mammalian, **305**
Semidormant hibernators, **370**–71
Semiplumes, **227**
Senses and sense organs, 12–13
amphibian, 152–55
jawed fishes, 114–16
jawless fishes, 86
mammalian, 273, 305–9
reptilian, 214–16, 245–50
Sensory reception and communication, 360–67
Sequential hermaphrodites, **123**, 125
Sex changes
fish and amphibian, 160
reptilian, 222
Sex determination, reptilian, 224
Sex roles, bird, 254–55
Sexual dimorphism
amphibian, 153*f*, 154*f*, 159*f*
bird, 253
jawed fishes, **122**
Sexual maturity
amphibian, 166–67
jawed fishes, 127
mammalian, 315, 321
reptilian/bird, 225, 260
Sexual recognition, jawed fishes, 122
Shaft
feather, **226**
hair, 271*f*, **272**
Sharks, 4, 93, 94*f*
anatomy of dogfish (*Squalus*), 90*f*
brain, cranial nerves, and spinal chord of, 113*f*
circulatory system of, 12*f*, 105*f*
digestive tract of, 109*f*
eye of, 14*f*
gill coverings of, 107*f*
gills of, 11*f*
male clasper in horn of (*Heterodontus francisci*), 125*f*
musculature of dogfish (*Squalus*), 103*f*
oviducts of, 120*f*
skeleton of dogfish (*Squalus*), 98*f*
viviparity in, 126
Sheep, desert bighorn (*Ovis canadensis*), butting behavior among, 359*f*
Shrew
armored (*Scutisorex somereni*), vertebral column of, 281*f*
Etruscan (*Suncus etruscus*), respiratory system of, 296
Siberia, ancient continent of, **49**
Sidewinding locomotion, reptilian, 203, 204*f*
Simplex uterus, mammalian, **312**
Sinus impar, **112**
Sinus venosus
jawed fishes, **104**, 105*f*
jawless fishes, **84**
Sirens, 149
Sister group, **28**
Skates (Chondrichthyes), 4
Skeletal system, 6–7
amphibian, 140–45
jawed fishes, 97–101
jawless fishes, 83, 84*f*
mammalian, 279–88
reptilian/bird, 201–4, 230–36
Skin. *See* Integumentary system
Skull, 7
amphibian, 141, 142*f*
bird, 189, 232, 233*f*
evolution of vertebrate, 280*f*
jawed fishes, 97–100
jawless fishes, 83
mammalian, 280–81, 295*f*
reptilian, 181–82, 201, 208*f*, 231*f*
Skunk, 279
Sleep, half-brain, in birds, 357
Small intestine
mammalian, 302
reptilian/bird, 244
Smell. *See* Olfaction
Smithsonian Institution, 426
Snakes, 4, 169
alimentary canal of, 212*f*
cardiovascular system of, 205–6
classification of, 176
digestive system of, 207–12
endocrine system of, 216–17
eye of, 216*f*
growth and development in, 222–26
integumentary system of, 198–201
locomotion in, 203, 204*f*
migrating, 351
muscular system of, 204–5
nervous system of, 212–14
predation by, 332
radio transmitter tracking of, 394–396
reproduction in, 218–22
respiratory system of, 206–7
sense organs of, 214–16
skeletal system of, 201–4
skull of, 208*f*
territories of, 341
tongue flicking in, 210*f*
urogenital system of, 217–18, 219*f*
venomous, 209–10
Social hierarchies, 358
Social interactions, 357–60
dog barking, 366
intraspecific parasitism, 360
naked mole rat society, 358
sensory reception and communications, 360–67
Sodefin, 151
Solenoglyphs, reptilian, **209**, 210*f*
Sonar system. *See* Echolocation
Songbird species
divergence of North American, 35
migrations of, 346
threat of extinction to migratory, 413–14, 428, 433–37
Sonoran life zone, North American, 68*f*, 69*f*
Sound production. *See also* Vocalization
amphibian, 143, 147–48
as communication, 362–66
fish, 113
mammalian, 305–8, 363–66
South America, Great American Interchange between North America and, 52*f*
Sparrow
dusky seaside (*Ammospiza martima nigrescens*), extinction of, 416*f*
English (*Passer domesticus*), dispersal of, 342–43
homing abililty in, 352–53
house (*Passer domesticus*), 33, 70

song (*Melospiza melodia*), 34*f*, 38
white-crowned (*Zonotrichia leucophrys*), 362, 363*f*, 364*f*
white-throated (*Zonotrichia albicollis*), 351, 362
Speciation, 32–37
Species, **23**
development of (speciation), 32–37
discovery of new, 20*f*, 21
distribution of (*see* Distribution of species)
endangered and threatened, 20, 411–19, 423–24
geographic variation in, 37–38
group of (genus), 25
molecular evolution of, 38–43
populations of (*see* Population(s); Population density)
Spectacle, reptilian eye, **215**, 216*f*
Sperm
jawed fishes, 120, 121
mammalian, 310, 316–17
Spermaceti organ, **297**
Spermatheca
amphibian, **157**, 160
reptilian, **218**
Spermatogenesis, 156
via transplantation, 314
Spermatophore, amphibian, 135, **160**
Sphenodontidae, 176. *See also* Tuataras
Spinal chord
amphibian, 151
jawed fishes, 113*f*, 114
mammalian, 305
reptilian/bird, 214, 245
Spinal nerves, 2
amphibian, 151
fish, 86
Spiny sharks (acanthodians), 92
Spiracle, amphibian, **148**
Spiral valve, **109**
Splanchnocranium, **97**
Splendipherin, 155
Squamates (Squamata), 176, 198. *See also* Lizards; Snakes
Squirrels, 355
Belding's ground (*Spermophilus beldingi*), 364, 365*f*
hibernation in ground, 370, 371, 372*f*, 373*f*
northern flying (*Glaucomys sabrinus*), 54, 55*f*, 274*f*
Stanley, Steven, 52–53
Stapes, **305**, 306*f*
Starling, European (*Sturnus vulgaris*)
introduction of, into North America, 70*f*
migration of, 351
Static equilibrium, **114**–15
Status undetermined species, **423**
Stegosaurs, 184, 185*f*
Sterebrae, **282**
Sternum
amphibian, **143**
keeled, in birds, 193, 195, 233
mammalian, 282–83
Sticklebacks, three-spined (*Gasterosteus aculeatus*), courtship behavior in, 122, 123*f*
Stomach, 10, 109
mammalian, 301–2, 303*f*
reptilian/bird, 211, 212*f*, 242, 244
Stomach stones, 211, 213*f*
Stratum basale (germinativum), **271**
Stratum corneum, **6**
amphibian, **135**
bird, 229*f*
mammalian, **271**
reptilian, 198–99, 200*f*
Stratum germinativum, **6**
Stratum granulosum, **271**
Stratum lucidum, **271**
Stratum spinosum, **271**
Stress, hormones and
bird, 251
mammalian, 310
Sturgeon
digestive tract of, 109*f*
gas bladder-gut relationship in, 110*f*
oviduct of, 120*f*
Subabdominal position (fins), **100**, 102*f*
Sublingua salivary gland, **300**
Submissive behavior, 360
Subphyla, **1**
Subspecies (races), **34**–36
Sudoriferous (sweat) glands, 271*f*, **277**
Sulci grooves, mammalian brain, **304**
Supernova explosion as cause of mass extinctions, 407
Swamp, freshwater, 59*f*
Swim bladders, 9, 110–12
courtship and, 112
Root effect and, 111
sound transfer and, 112*f*
in teleosts, 111*f*
Symbiosis, **377**–82
commensalism as, 377–80
mutualism as, 379–80
parasitism as, 380–82
Symmetrodonts (Symmetrodonta), 268
Sympatric populations, **36**–37
Symplesiomorphies, **28**
Synapomorphies, **28**
Synapsids, 174, **201**
as ancestors of mammals, 264–69
cladogram of, 266*f*
evolution of major groups of, 265*f*
Synchronous hermaphrodites, **123**
Synsacrum, **233**
Syrinx, **239**, 241*f*
Systematics, **23**–32

T

Tactile sense, 309
communication through, 367
Taeniform body, 5*f*
Tagging techniques for wildlife identification, 392–94
Taiga, 61–62
Tail, locomotion and, 4, 5*f*
Tapetum lucidum, mammalian eye, **308**
Tarsals, amphibian, **144**
Tarsometatarsus, **234**
Taste
amphibian, 155
fish, **116**
mammalian, 309
number of taste buds in select animals, 250*t*
reptilian/bird, 216, 250
Taxon (taxa), **26**
three classes of, 26*f*
Taxonomy, **23**
dental formulas used for, 300
of fishes, 78–79
hierarchical order of, 25–26
Taylor, F. B., 49
Teat, mammalian, 277, 278*f*
Tectonic lake, **58**
Teeth, 7
amphibian, 150–51
determining animal age using, 399, 400*f*
jawed fishes, 108
labyrinthodont amphibian, 130, 132*f*
mammalian, 298, 299*f*, 300
reptilian, 176, 208
Teleostei (advanced ray-finned fishes), 95, 99*f*. *See also* Teleosts
Teleosts, 95, 99
amphibious eye of, 115
circulatory system of, 105*f*
eye of, 14*f*
gill coverings of, 107*f*
gill morphology in, 108*f*
inner ear of, 15*f*
male urogenital system of, 121*f*
oviduct of, 120*f*
swim bladder of, 111*f*

Temnospondyli (temnospondyls), 134–35, 169
Temperate deciduous forests, 62–63, 433, 436*f*
Temporal fenestrae
  in reptiles and descendants, **174**, 175*f*
  in reptilian skull, **201**
Terrestrial environments, 60–72
  adaptation to, 172
  biomes, 60–65
  biotic provinces, 69
  human impact on, 70–72
  life zones, 65–69
Terrestrial vertebrates, water conservation and excretion in, 15
Territory, **339**–41
  bird songs as announcement of, 362
  defense of, 340, 341*f*
  interspecific, between blackbird species, 340*f*
  marking of, 279, 340
Testes, 87, 119–20. *See also* Reproductive organs
  hormones secreted by, 117
  mammalian, 313
Testosterone, **117**, 156, 217, 251, 276
Testudomorpha, 169, 174. *See also* Turtles
Tetrapod(s) (Tetrapoda)
  in Antarctica, 183
  aortic arches in, 206*f*
  appendages (limbs) of, 7–8, 130, 133*f*
  embryo, 3*f*
  evolution of, 129–33
  primitive, 169–72
  vertebral column of, 7
Thalamus, mammalian brain, **304**
Thecodont dentition, **108**, **230**
Thecodonts (Thecodontia), 176–78, 191
  Saltoposuchus, 178*f*
Therapsids (Therapsida), 264–68
Therians, 268
Thermoregulation
  amphibian, 138
  mammalian, 293
  reptilian/bird, 212, 214*f*, 242
Therocephalians, 266
Theropods, **178**, 181*f*
  feathered, 190
  thoracic cavities of, 182
Thoracic position (fins), **100**, 102*f*
Thoracic vertebra, reptilian, **202**
Thorax, reptilian, **205**
Threatened species, **423**
Thymosin, **117**
Thymus gland, **117**
Thyroid gland, **116**, 155, 216
  bird, 250–51
  mammalian, 309
Thyroid-stimulating hormone (TSH), **116**
Thyroxin, **116**, 119, 155, 309
Tibia bone, amphibian, 141*f*, **144**
Tibiofibula bone, amphibian, 141*f*, **144**
Tibiotarsus bone, bird, **234**
Tigers (*Panthera tigris*), threat of extinction to, 414
T lymphocytes (T cells), **117**
Toad(s), 327
  defenses of, against predation, 386*f*
  projectile tongue in, 150*f*
  respiratory structures of, 148*f*
  right forelimb dominance in, 151
  sexual maturity in *Bufo bufo*, 167*t*
Toe pads, **139**
Tongue
  bird, 242, 243*f*
  mammalian, 29, 294, 300
  projectile, in amphibians, 150
  reptilian, 210–11
  taste and, in amphibians, 155
Torpor (dormancy), 368–73
  estivation, 119, 368, 373
  winter survival and hibernation, 368–73
Toxic substances
  amphibian skin secretions of, 136–38
  biological amplification of, and threats to vertebrates, 414, 415*f*
Trachea
  amphibian, **147**, 148*f*
  mammalian, 294
  reptilian/bird, 240
Tragus, **306**
Transects, **398**
Transition area of biosphere reserve, **433**, 435*f*
Transition zone of Austral region, North American, 67, 68*f*
Transverse colon, **302**
Tree frogs, 139, 140*f*
Triassic extinctions, 405
Triconodonts, 268
Triiodothyronine, **116**, 155, 309
Tritheledontids, 268
Tritylodontids, 268
Trivial name, **24**
Tropical rain forests, 64–65, **431**
  deforestation of, and species extinction, 411–14, 431, 433–37, 435*f*
  morphological convergence of mammals living in, 64, 66*f*
Tropical region, North American, 66*f*, 69
Tropical seasonal deciduous forests, **431**
Trout Unlimited, 425
True census, **397**
True horns, mammalian, **274**, 275*f*
Trumpeter swans (*Cygnus buccinator*), teaching migration routes to, 396, 397*f*
Trunk vertebrae
  amphibian, 141*f*, **143**
  fish, **100**
Tuataras, 4, 169, 178*f*
  commensalism between birds and, 378, 379*f*
  reproduction in, 218–22
  reptilian morphology in, 198–218
Tubal bladders, **119**
Tubular reabsorption, mammalian kidney, **311**
Tubular secretion, mammalian kidney, **311**
Tuna
  evolutionary relationship of pig, lungfish, and, 31*f*
  growth and development in, 222–26
  musculature of skipjack, 104*f*
Tundra, 60–61, 63*f*
Tunicate(s), 1–2
  Garstang thesis on larval, 74, 76*f*
  larval, 3*f*
  metamorphosis of, **2**
  structure of, 1*f*
Turtles, 4, 6, 169
  carapace of, 198*f*, 200
  cardiovascular system of, 205–6
  classification of, 174–75
  digestive system of, 207, 208
  endocrine system of, 216–17
  fossil of largest, 177*f*
  gas exchange in, 206*f*, 207
  growth and development in, 222–26
  integumentary system of, 198*f*, 200
  migrating, 345, 346, 350, 351
  muscular system of, 204–5
  nervous system of, 212–14
  reproduction in, 218–22
  respiratory system of, 206–7
  sea, 202, 203*f*, 207, 222*f*, 224
  sense organs of, 214–16
  skeletal system of, 201–3, 284*f*
  skull of, 177*f*
  temporal fenestrae of, 175*f*
  territories of, 341
  threat of extinction to, 428
  torpor and freeze resistance in, 372*f*, 373
  urogenital system of, 217–18, 220*f*

Tympanic cavity, amphibian, **153**
Tympanic membrane (tympanum)
amphibian, **152**
reptilian, 215

U

Ulna bone, amphibian, 141*f*, **144**
Ultimobranchial bodies, **117**, **251**
Ultrasound, **13**
Ultraviolet light
courtship behavior and, 253
ozone depletion and, 438
predation using, 247
Uncinate processes
amphibian, 189
reptilian, **202**, **231**
Underfur, **272**, 370
Ungulates
artiodactyl and perissodactyl types of, 270, 288, 289*f*
locomotion in, 287, 289*f*
whales related to, 289
Unguligrade locomotion, **287**, 289*f*
United Nations Educational, Scientific, and Cultural Organization (UNESCO), 433
United States
endangered species in, 423–24, 429, 429*f*
food web for prairie in, 326*f*
hot spots of threatened biodiversity in, 429, 432*f*
issues related to modern wildlife conservation and management in, 427–40
regulatory legislation affecting vertebrates in, 421–23
sanctuaries and refuges in, 424–26, 425*f*
vertebrate museum collections in, 426–27
Upper Austral zone, North American, 66*f*, 68
Urinary bladder. *See* Bladder
Urochordata, **1**–2. *See also* Tunicate(s)
Urodeum, **217**, **251**
Urogenital system, 13–15
amphibian, 156–58
jawed fishes, 117–21
jawless fishes, 86–87
mammalian, 310–14
reptilian/bird, 217–18, 219*f*, 220*f*, 251–52
Uropygial gland, **230**
Urostyle, 141*f*, **143**
Uterine milk, **126**
Uterus, mammalian, **311**, 312*f*
Utricle, mammalian, **305**

V

Vagina, mammalian, **311**, 312*f*, 313
Vane, feather, **226**
Vasa efferentia
amphibian, 157*f*, **158**
fish, **120**
Vas deferens
amphibian, 157*f*, **158**
mammalian, 314
Vasopressin, **116**
"Velvet," antler, **276**, 277*f*
Vent, **110**
Ventral aorta, **106**
Ventricle
jawed fishes, **104**, 105*f*
jawless fishes, **84**
Ventricles, mammalian brain, **303**, 304*f*
Ventricular trabeculae, amphibian, **145**
Vermiform appendix, **302**
Vertebra, 6*f*
amphibian, 143
bird, 233
fish, 100
mammalian, 281, 282*f*
reptilian, 201–2, 211*f*
Vertebral column, 2, 6–7
amphibian, 143
early bird (*Archeopteryx*), 191
jawed fishes, 98*f*, 99*f*, 100
mammalian, 281–82
reptilian, 201–2, 231
Vertebrata, **2**. *See also* Vertebrate(s)
Vertebrate(s), 1–22
conservation and management of (*see* Conservation and management)
discovering new species of, 21
ecological role of, 18–19
ecology of (*see* Ecology of vertebrates)
evolution of (*see* Evolution)
features of, 4–18
flying, 185–91 (*see also* Birds (Aves))
four-legged (*see* Tetrapod(s) (Tetrapoda))
future research on, 19–20
geographic distribution of (*see* Zoogeography)
introduction to, 2–4
locomotion in (*see* Locomotion)
major groups of, 4
movements of (*see* Movements of vertebrates)
number of genes in, *versus* in invertebrates, 92
representative, 3*f*
social behavior of (*see* Social interactions)
studying (*see* Ecological and behavioral studies)
systematics and classification of, 4, 23–32
Vestibule, mammalian, **305**, **312**
Vibration receptors, 13
Villi, small intestine, **302**
Vision, 13
amphibian, 153–54
animal communication through, 366–67
eye structure and, 14*f*
fish, 115, 366–67
mammalian, 308, 367
reptilian/bird, 215–16, 245–48, 367
Viviparous development, **16**, 17
amphibian, 163
jawed fishes, 125–26
mammalian, 269, 314, 317, 318*f*, 319–21
reptilian, 221, 223
Vocalization
amphibian, 143, 147, 148*f*, 159, 160, 362
mammalian, 294, 305–8, 358, 363–66
reptilian/bird, 239, 241*f*, 258, 362, 363*f*
Vocal sacs, amphibian, **147**, 148*f*
Volcanic lakes, **58**
Volcanism as cause of vertebrate extinctions, 404–5, 407
Voles
cyclic population fluctuations of meadow, 332*f*
plant predation by, 386
population irruptions among, 335
predation of, 247
Vomeronasal (Jacobson's) organs
amphibian, 154–55
mammalian, 308, 362
reptilian, 210, 216, 217*f*

W

Walking catfish (*Clarias batrachus*), 108
Wallace, A. R., on biogeographic regions, 47, 48*f*, 49
Wallace's line, 48*f*, 49

Warblers, competition and niches of, 376, 377*f*
Water conservation, 15
  reptilian, 207
Water resources, 436
Water storage in frog urinary bladder, 156
Webbing
  bird, 234
  frog, 129, 140*f*
Weberian ossicles, **112**, 115
Wegener, Alfred, 49
Wetlands Loan Act of 1961, 423
Whales, 293, 419
  baleen, 274, 275*f*, 298
  brain of, 303
  migration of, 348, 349
  origin and evolution of, 289, 291*f*
  skeletons of, 286*f*, 291*f*
  spermaceti organ in, 297
  vocalization by, 306, 307, 365–66
Whitaker, J. O., Jr., 34, 35
Wilderness Act of 1964, 423
Wildlife conservation and management. *See* Conservation and management
Winter survival, 368–73
Wisconsin glaciation, **53**, 55*f*
Wolf (*Canis lupus*)
  countercurrent heat exchange in Arctic, 369*f*
  heterodont dentition of, 7*f*
  predator-prey relationship of, with moose, 383–85
  species restoration of, 417
  vocalizations of, 366
Woodpecker, hyoid apparatus and tongue of, 243*f*
Worldwatch Institute, 428
World Wildlife Fund, 426

Y

Yolk sac, reptilian egg, **173, 223**
Ypsiloid (prepublic) cartilage, **144**

Z

Zoogeography, **45**–73
  branches of, 45
  distribution of species, 47–55
  freshwater environments, 58–60
  introduction to, 45–46
  marine environments, 55–58
  terrestrial environments, 60–72
Zooplankton, **57**
Zygapophyses, reptilian, **202**
Zygomatic arch, 281